Exploring Quantum Mechanics

Exploring Quantum Mechanics

A Collection of 700+ Solved Problems
for
Students, Lecturers, and Researchers

Victor Galitski, Boris Karnakov, Vladimir Kogan,
and Victor Galitski, Jr.

OXFORD
UNIVERSITY PRESS

Great Clarendon Street, Oxford, OX2 6DP,
United Kingdom

Oxford University Press is a department of the University of Oxford.
If furthers the University's objective of excellence in research, scholarship,
and education by publishing worldwide. Oxford is a registered trade mark of
Oxford University Press in the UK and in certain other countries

© V. Galitski, B. Karnakov, V. Kogan, and V. Galitski, Jr. 2013

The moral rights of the authors have been asserted

First Edition published in 2013

Impression: 1

All rights reserved. No part of this publication may be reproduced, stored in
a retrieval system, or transmitted, in any form or by any means, without the
prior permission in writing of Oxford University Press, or as expressly permitted
by law, by licence or under terms agreed with the appropriate reprographics
rights organization. Enquiries concerning reproduction outside the scope of the
above should be sent to the Rights Department, Oxford University Press, at the
address above

You must not circulate this work in any other form
and you must impose this same condition on any acquirer

British Library Cataloguing in Publication Data

Data available

ISBN 978–0–19–923271–0 (Hbk.)
ISBN 978–0–19–923272–7 (Pbk.)

Printed and bound in Great Britain by
CPI Group (UK) Ltd, Croydon, CR0 4YY

Preface to the first English edition

The book you are now reading contains arguably the world's largest collection of problems in quantum physics, which span just about the full scope of subjects in quantum theory ranging from elementary single-particle quantum mechanics in one dimension to relativistic field theory and advanced aspects of nuclear physics. There are more than 700 problems of various difficulty, accompanied by detailed solutions. While some problems are elementary and may be accessible to advanced undergraduate students, some are at the research level that would be of challenge and interest to a practicing theoretical physicist. Consequently, the book will be useful to a wide range of readers including both students studying quantum mechanics and professors teaching it. In fact, this problem book has a proven record of being a scientific bestseller in other countries, notably in Russia, where its earlier versions have been used for over thirty years as a standard text to learn and teach quantum physics at all leading universities.

Even though this is the first English edition of the book, several editions in the Russian, French, and Japanese languages have been available since 1981 (a much smaller collection of elementary problems by Victor Galitski, Sr. and Vladimir Kogan was first published in 1956 and is available in English too). Following the fine tradition of Russian theoretical physics, this book of problems has usually been used in conjunction with the legendary Landau and Lifshitz theoretical physics course. However, the latter is certainly not the only option for a basic supporting text, and many other excellent textbooks are now available. Therefore, I do not reiterate here the strong recommendation of the first Russian edition, and suggest instead that a choice of the supporting quantum textbook be left to the reader or professor teaching a course. Furthermore, each chapter of this book begins with a concise summary of the underlying physics and main equations, which makes the book almost self-contained. Therefore, for a student or researcher, who has had even minimal exposure to quantum mechanics, it may become a useful and rich independent resource to master the field and develop actual problem-solving skills.

Let me also emphasize two points about the text. The first is that the level of mathematical abstraction here is generally pretty low (with the exception of the first chapter perhaps) not due to a lack of expertise by the original authors, but because their target audience has been practicing physicists interested in understanding and solving real-life problems. Consequently, if your motivation for doing or studying quantum physics is to understand actual phenomena and explain experiments, this book will certainly be useful. For example, the chapters on nuclear physics contain a large number of problems that are directed to quantitative explanation of specific experimental data. The second warning is that the terminology may be somewhat outdated at times, and no references to modern literature are provided. I intentionally decided to preserve the integrity of the book (a number of new problems that are included here compared to the earlier editions have been incorporated into the old structure of chapters). Because I believe that the "old-school style" of the book is not at all a downside.

Quite on the contrary, we can certainly learn a lot from people who worked in the good old days of theoretical physics. These were people living and working in completely different times, and they were quite different from us, today's scientists: with their attention spans undiminished by constant exposure to email, internet, and television, and their minds free of petty worries about citation counts, indices, and rankings, they were able to devote 100% of their attention to science and take the time to focus on difficult problems that really mattered, making a difference on an absolute scale. We are now standing on their shoulders, and in fact most quantum phenomena that we are fascinated with today go back to their accomplishments in decades past.

With this in mind, I would like to say a few words about the rather fascinating history of this book and the original authors, particularly about the lead author – Victor Galitski, Sr. – who was the main driving force behind the book and who conceived most of the problems you will find here. Even though I admittedly cannot be completely unbiased in talking about him (he was my late grandfather), I can say with confidence that apart from being a first-rate teacher and educator, he was also a giant of theoretical physics and an amazingly creative scientist. His contributions to quantum physics are manifold, and include the first use of Feynman diagrams in solid-state physics, the first microscopic derivation of Landau's Fermi liquid theory, various fundamental results in nuclear physics (*e.g.*, Galitskii–Feynman equations), the first theory of unconventional superconductivity and many other famous results. He co-authored papers with such luminaries as Lev Landau and Arkady Migdal, and trained a generation of theoretical physicists (my own late adviser, the great Anatoly Larkin, among them).

The sheer amount of work by Galitski, Sr. is very impressive in and by itself, but it is even more so considering his very short and difficult life and his short career as a physicist. I hope the reader will excuse my mentioning it briefly in the following couple of pages. Victor Galitski, Sr. was born in 1924 in Moscow. In 1942, shortly after graduating from high school, he enlisted in the army, having married my grandmother, Tatiana, just a few days before his leaving. (They did not know if they would see each other again. Fortunately they did, and she was his devoted wife for almost forty years.) He was sent to the front shortly after, and eventually became a commander of an artillery unit. In 1943 he was severely wounded during the Battle of Kursk. He hardly survived, and spent over a year in a hospital. After being honorably discharged from the army, he entered the Moscow Mechanical Institute of War Ammunition (which later was renamed the Moscow Engineering Physics Institute, MEPhI, and where years later – in 1960 – he became Head of the Theoretical Nuclear Physics Department). While studying at MEPhI he discovered his gift and interest in theoretical physics, and after graduating in 1949 he joined the theory group led by Arkady Migdal at the Institute of Atomic Energy in Moscow.

This is where his creative research finally took off, and where he, in collaboration with Migdal, developed, among other things, the first microscopic quantum theory of solids based on methods of quantum-field theory that brought him world-wide recognition. I will not talk about the details of his research here, but will just mention that this was also the time when he started putting together the first problems that eventually led to this book. I will also mention that these times after the war were not at all easy for him and his family. My grandmother's brother, Vsevolod Leont'ev, was arrested by the Stalin regime (his only "crime" being that he was being wounded and captured during the war), and he first received the death sentence, which was later changed to 25 years in a work camp (due to a temporary and brief moratorium on capital punishment in the Soviet Union, which saved his life. He was released and acquitted after Stalin's death, and lived a long life well into his nineties, having been among the most optimistic and fun people I knew). After this arrest, my grandparents were given the ultimatum to either make a public statement and denounce the "enemy of the people", or face very serious consequences. They refused to follow orders, and my grandmother was immediately fired from Moscow State University, where she had been on a fast-track to becoming a young Professor of Economics (she never returned to work after that). Galitski, Sr. was also facing similar prospects.

However, help came from Igor Kurchatov (the Russian counterpart of Robert Oppenheimer, leading the Soviet effort to develop an atomic bomb), who had a very high opinion of Galitski, Sr. Apparently, he was also involved in classified research on the nuclear bomb, about which little is known apart from that (according to my grandmother) while working on the project, my grandfather was exposed to a significant radiation amount that resulted in health problems and might have contributed to his developing cancer at an early age.

After being saved from the Gulag by Kurchatov, Galitski, Sr. entered the most productive and successful part of his research career, which eventually propelled him to the position of Head of the nuclear theory division at the Kurchatov Institute of Atomic Energy. Let me now skip twenty or so very interesting years to 1975, when the idea of writing this book was first conceived. Surprisingly, the venue where it happened was very unlikely, and deserves some explanation. Physics was certainly my grandfather's main passion in life, but he also had many other interests beyond science. In particular, he was a big sports fan, with ice hockey being one of his other passions. In fact, he was a friend with many leading Soviet hockey players. Many members of the national hockey team were frequent guests at his house. (I still remember that as a child I was completely unimpressed with my grandfather's physics library, which included books signed by the authors, such as Bohr, Dirac, and Landau, but was fascinated by his collection of hockey clubs autographed by famous hockey players, such as his good friend Vyacheslav Starshinov – a top forward in the Soviet national hockey team who later became an athletic director in the Moscow Engineering Physics Institute. When I was a student there in the 1990s, Starshinov, for old time's sake perhaps, cut me some slack, letting me take a tennis class instead of fulfilling some stricter requirements and going through less desirable courses. I certainly appreciated that!)

Returning to the book, in the fall of 1975, Galitski, Sr. was hosting a reception for the hockey team "Spartak", which was also attended by a young physics researcher and big hockey fan, Boris Karnakov. Apparently, it turned out during a discussion between Galitski, Sr. and Karnakov at this reception that they both shared not only a passion for hockey but also for creating and collecting exciting problems in quantum physics. By that time, they both had accumulated large collections of unpublished problems, and so they decided to combine them into a large book (on the basis of the Kogan–Galitski short book published 20 years prior) to cover all key aspects of quantum physics.

It took Galitski, Sr. and Karnakov five years to put together this book. Unfortunately, this difficult but I am sure rewarding work was dampened by the fact that my grandfather was diagnosed with a serious cancer, and had to have a lung removed. The cancer returned in an incurable form shortly after the first complete manuscript of the book was finalized – in the summer of 1980. He was battling against it for six long months, and died in January of 1981, at the age of only 56. He never saw the book published, as its first Russian edition appeared in print in March 1981. The current English translation is close to that first Russian edition, but contains some additional material added by Boris Karnakov. I have also added new problems, but have resisted the temptation to make significant additions and structural changes at this stage. (My own collection of problems on non-equilibrium quantum mechanics, single-particle quantum mechanics in the presence of a random potential, coherent states, and path integrals, might be published in a follow-up edition of this book or separately, if there is interest.)

It was actually surprising to me at first that an English edition of this book had not appeared earlier. However, I was surprised only until I signed up with Oxford University Press to prepare its first expanded English edition and actually began work on editing and translating the material, the enormity of which became frightening to me. For as much as I wanted to do it, partly in my grandfather's memory, I quickly realized that the patience needed for this type of work is not among my virtues. (Apparently, this gene that Galitski, Sr. had I was not fortunate enough to inherit.) It would have taken me much longer to complete this work without the help of students and postdocs in my group: Dr. Brandon Anderson, Dr. Greg Boyd, Mr. Meng Cheng, Mr. Joe Mitchell, Dr. Andrew Robertson, Dr. So Takei, Ms. Alena Vishina, and Mr. Justin Wilson. I would like to specifically emphasize the significant contributions of Alena and Joe, who have reviewed a considerable part of the text, and Justin, who helped put together a small new section on coherent-state spin path-integrals. I am grateful to Professor Khmelnitskii of Cambridge University for his constant encouragement during this long project, and to Ms. Anna Bogatin for putting together a very nice and fun image for the book cover. I would also like to thank the U.S. Army Research Office and the National Science Foundation, whose CAREER award and support through the Physics Frontier Center allowed me to effectively share time between my research activities and the work on the book, thereby contributing to the success of this project. Finally, I am grateful to Ms. Catherine Cragg, Production Editor at Oxford University Press (OUP), for her help with finalizing the book for publication and Dr. Sönke Adlung, Commissioning Editor at OUP, for being so patient with me over the course of several years while this English edition of the book was in preparation.

Washington DC   Professor Victor Galitski, Jr.
September, 2011

Contents

Symbols used in the book	xiv
Notations often used in the book	xv
Universal constants	xvi

1 Operators in quantum mechanics   1
  1.1 Basic concepts of the theory of linear operators   2
  1.2 Eigenfunctions, eigenvalues, mean values   8
  1.3 The projection operators   20
  1.4 Quantum-mechanical representations of operators and wave-functions; Unitary operators   22

2 One-dimensional motion   32
  2.1 Stationary states in discrete spectrum   33
  2.2 The Schrödinger equation in momentum space; The Green function and integral form of the Schrödinger equation   46
  2.3 The continuous spectrum; Reflection from and transmission through potential barriers   56
  2.4 Systems with several degrees of freedom; Particle in a periodic potential   74

3 Orbital angular momentum   84
  3.1 General properties of angular momentum   86
  3.2 Angular momentum, $l=1$ 94
  3.3 Addition of angular momenta   99
  3.4 Tensor formalism in angular momentum theory   109

4 Motion in a spherically-symmetric potential   116
  4.1 Discrete spectrum states in central fields   117
  4.2 Low-energy states   141
  4.3 Symmetries of the Coulomb problem   153
  4.4 Systems with axial symmetry   158

5 Spin   165
  5.1 Spin $s=1/2$ 166
  5.2 Spin-orbital states with spin $s=1/2$; Higher spins   180
  5.3 Spin density matrix; Angular distributions in decays   191
  5.4 Bound states of spin-orbit-coupled particles   198
  5.5 Coherent-state spin path-integral   202

6 Time-dependent quantum mechanics — 213
6.1 The Schrödinger representation; The motion of wave packets — 214
6.2 Time-dependent observables; Constants of motion — 225
6.3 Time-dependent unitary transformations; The Heisenberg picture of motion — 232
6.4 The time-dependent Green function — 247
6.5 Quasistationary and quasi-energy states; Berry phase — 252

7 Motion in a magnetic field — 270
7.1 Stationary states in a magnetic field — 271
7.2 Time-dependent quantum mechanics in a magnetic field — 286
7.3 Magnetic field of the orbital currents and spin magnetic moment — 291

8 Perturbation theory; Variational method; Sudden and adiabatic theory — 296
8.1 Stationary perturbation theory (discrete spectrum) — 298
8.2 Variational method — 316
8.3 Stationary perturbation theory (continuous spectrum) — 325
8.4 Non-stationary perturbation theory; Transitions in continuous spectrum — 336
8.5 Sudden perturbations — 353
8.6 Adiabatic approximation — 358

9 Quasi-classical approximation; $1/N$-expansion in quantum mechanics — 374
9.1 Quasi-classical energy quantization — 381
9.2 Quasi-classical wavefunctions, probabilities, and mean values — 408
9.3 Penetration through potential barriers — 420
9.4 $1/N$-expansion in quantum mechanics — 436

10 Identical particles; Second quantization — 447
10.1 Quantum statistics; Symmetry of wavefunctions — 448
10.2 Elements of the second quantization formalism (the occupation-number representation) — 456
10.3 The simplest systems with a large number of particles ($N \gg 1$) — 472

11 Atoms and molecules — 485
11.1 Stationary states of one-electron and two-electron atoms — 486
11.2 Many-electron atoms; Statistical atomic model — 509
11.3 Principles of two-atom-molecule theory — 525
11.4 Atoms and molecules in external fields; Interaction of atomic systems — 536
11.5 Non-stationary phenomena in atomic systems — 565

12	**Atomic nucleus**	598
	12.1 Nuclear forces—the fundamentals; The deuteron	600
	12.2 The shell model	612
	12.3 Isotopic invariance	627
13	**Particle collisions**	637
	13.1 Born approximation	642
	13.2 Scattering theory: partial-wave analysis	663
	13.3 Low-energy scattering; Resonant scattering	675
	13.4 Scattering of fast particles; Eikonal approximation	713
	13.5 Scattering of particles with spin	726
	13.6 Analytic properties of the scattering amplitude	737
	13.7 Scattering of composite quantum particles; Inelastic collisions	747
14	**Quantum radiation theory**	776
	14.1 Photon emission	779
	14.2 Photon scattering; Photon emission in collisions	790
15	**Relativistic wave equations**	810
	15.1 The Klein–Gordon equation	812
	15.2 The Dirac equation	838
16	**Appendix**	866
	16.1 App.1. Integrals and integral relations	866
	16.2 App.2. Cylinder functions	867
Index		871

Symbols used in the book

$\propto$	–	proportionality sign
$\sim$	–	order of magnitude sign
$\langle m\|\hat{f}\|n\rangle \equiv f_{mn} \equiv f_n^m$	–	matrix element of the operator $\hat{f}$
$\overline{f}$, $\langle f \rangle$	–	mean value of $\hat{f}$
$\left[\hat{f},\ \hat{g}\right] \equiv \hat{f}\hat{g} - \hat{g}\hat{f}$	–	commutator of the operators $\hat{f}$ and $\hat{g}$

A summation over repeated vector or spinor indices is assumed unless noted otherwise.

Notations often used in the book

The following notations are often used throughout the book without additional clarification and definitions.

$\psi_f(q)$	–	wave-function in q-representation labelled by quantum number, f
ψ_n^{osc}	–	eigenfunction of linear oscillator, see Eq. (II.2)
e	–	particle charge
c	–	speed of light
$\hat{H}$	–	Hamiltonian
E, ε	–	energy
$\mathcal{E}$ and $\mathcal{H}$	–	electric and magnetic fields
$\mathbf{A}$	–	vector potential for electromagnetic field
U	–	potential energy (interaction potential)
$\hat{V}$	–	perturbation operator
$\mathbf{d}$, d	–	dipole moment
φ, A_0	–	scalar potential for electromagnetic field
a_B	–	Bohr radius
δ_l	–	phase shift
$\hat{\boldsymbol{\sigma}} = (\hat{\sigma}_x, \hat{\sigma}_y, \hat{\sigma}_z)$	–	Vector of Pauli matrices
w, W	–	transition probability, transition probability per unit of time
Z, Ze	–	nucleus charge
R	–	radius of an interaction or scattering potential
m, M	–	mass or magnetic quantum number
μ	–	reduced mass, or magnetic moment
$\mathbf{p}$, $\mathbf{P}$	–	momentum
$\mathbf{k}$	–	wave vector
A	–	nucleus mass number
ω	–	frequency
l, L, j, J	–	angular moment (orbital and total)
s, S	–	spin
$J_\nu(z)$	–	Bessel function
$H_n(x)$	–	Hermite polynomial
$Y_{lm}(\theta, \varphi)$	–	spherical function
$\Gamma(z)$	–	Γ-function
$\delta(x)$, $\delta(\mathbf{r})$	–	One-dimensional and three-dimensional Dirac delta-function
δ_{ik}	–	unit tensor, Kronecker δ
ε_{ikl}	–	Levi-Civita symbol, totally antisymmetric pseudotensor, $\varepsilon_{123} = 1$, $\varepsilon_{213} = -1$, *etc.*

Universal constants

A list of some fundamental physical constants.

Planck constant	–	$\hbar = 1.055 \cdot 10^{-27}$ erg·s
Elementary charge	–	$e = 4.80 \cdot 10^{-10}$ CGS unit
Electron mass	–	$m_e = 9.11 \cdot 10^{-28}$ g
Speed of light	–	$c = 3.00 \cdot 10^{10}$ cm/s
Bohr radius	–	$a_0 = 0.529 \cdot 10^{-8}$ cm
Atomic unit of energy	–	$\frac{m_e e^4}{\hbar^2} = 4.36 \cdot 10^{-11}$ erg $= 27.21$ eV
Atomic unit of frequency	–	$\frac{m_e e^4}{\hbar^3} = 4.13 \cdot 10^{16}$ s^{-1}
Atomic unit of electric field strength	–	$\frac{e}{a_0^2} = 5.14 \cdot 10^9$ V/cm
Fine structure constant	–	$\alpha = \frac{e^2}{\hbar c} = \frac{1}{137}$
Proton mass	–	$m_p = 1836 m_e = 1.673 \cdot 10^{-24}$ g
Electron rest energy	–	$m_e c^2 = 0.511$ MeV

$1 \; eV = 1.602 \cdot 10^{-12}$ erg

1
Operators in quantum mechanics

Mathematical formalism of quantum mechanics is closely connected with the theory of linear operators. A key quantum-mechanical principle is that physical quantities (observables) are represented by Hermitian linear operators that act on state vectors that belong to a Hilbert vector space. A state vector (or equivalently, a wavefunction, $|\psi\rangle$) completely describes a state of the physical system.

For every linear operator, $\hat{L}$, one can define its Hermitian-adjoint operator, $\hat{L}^\dagger$, as follows:

$$\langle \psi_2 | \hat{L} \psi_1 \rangle \equiv \int \psi_2^*(q) \hat{L} \psi_1(q) d\tau_q = \int \left[\hat{L}^\dagger \psi_2(q) \right]^* \psi_1(q) d\tau_q \equiv \langle \hat{L}^\dagger \psi_2 | \psi_1 \rangle. \tag{I.1}$$

Here and below, the variable q corresponds to a complete set of parameters characterizing the system, which define a specific representation of the wavefunctions, $\psi(q) = \langle q | \psi \rangle$, with the states, $|q\rangle$, satisfying the following resolution of identity, $\int |q\rangle\langle q| d\tau_q = \hat{1}$, and $\hat{1}$ being the identity operator. If $\hat{L}^\dagger = \hat{L}$, then the operator is called Hermitian or self-adjoint.[1]

A physical quantity, f, associated with the quantum mechanical operator, $\hat{f}$, can only take on values that belong to the spectrum of $\hat{f}$ – i.e., a measurement of f can only result in one of the eigenvalues, f_n, determined by the eigenvalue problem:

$$\hat{f} \psi_{f_n} = f_n \psi_{f_n} \tag{I.2}$$

The corresponding eigenfunction, ψ_{f_n}, describes a physical state that has a certain value of the physical quantity, $f = f_n$ (while in a generic state, the physical quantity has no definite value). For an Hermitian operator, $\hat{f}$, its eigenvalues are all real. Also, for a generic quantum-mechanical Hermitian operator, its linearly-independent eigenfunctions form a complete set of states and can be chosen mutually-orthogonal. The latter property allows to expand the wavefunction of an arbitrary state, $\psi(q)$, in a series of eigenfunctions:[2]

$$\psi(q) = \sum_n c(f_n) \psi_{f_n}(q), \tag{I.3}$$

---

[1] There exists a subtle distinction between the notion of Hermitian and self-adjoint operators, which is discussed in Problems 1.28 and 1.29.

[2] For the sake of brevity, the expansion is written as a sum (which assumes a discrete spectrum). In general, this expression may include a sum over discrete states and/or integral over continuous spectrum.

where

$$c(f_n) = \langle \psi_{f_n} | \psi \rangle \equiv \int \psi_{f_n}^*(q)\psi(q) d\tau_q. \quad (I.4)$$

We assume here and below, unless noted otherwise, that eigenfunctions ψ_{f_n} are chosen to be orthonormal and normalized to unity for the discrete part of the spectrum, $\langle \psi_f | \psi_{f'} \rangle = \delta_{f,f'}$ and to the δ-function $\langle \psi_f | \psi_{f'} \rangle = \delta(f - f')$ for the continuous spectrum. In the former case, the coefficients $c(f_n)$ directly determine the probability, $w(f_n) = |c(f_n)|^2$, that $f = f_n$ is measured in the state, $\psi_{f_n}(q)$. In the latter case – continuous spectrum – the coefficients determine the probability density, $dw/df = |c(f)|^2$. The mean value of f, $\bar{f} = \sum_n f_n w(f_n)$, follows from the formula

$$\bar{f} = \langle \psi^* | \hat{f} | \psi \rangle \equiv \int \psi^*(q) \hat{f} \psi(q) d\tau_q \quad (I.5)$$

and can be used without reference to the individual probabilities.

If the Hermitian operator, $\hat{f}(\lambda)$, depends on some real parameter λ, then the derivative of the eigenvalues $f_n(\lambda)$ with respect to this parameter satisfies the relation (for the discrete part of the spectrum)

$$\frac{\partial f_n(\lambda)}{\partial \lambda} = \left\langle \psi_{f_n(\lambda)} \left| \frac{\partial \hat{f}(\lambda)}{\partial \lambda} \right| \psi_{f_n(\lambda)} \right\rangle, \quad (I.6)$$

which has a variety of useful applications.

1.1 Basic concepts of the theory of linear operators

Problem 1.1

Consider the following operators (below: $-\infty < x < \infty$):

1. The translation operator, $\hat{T}_a$: $\hat{T}_a \psi(x) \equiv \psi(x + a)$.
2. The inversion operator, $\hat{I}$: $\hat{I}\psi(x) \equiv \psi(-x)$.
3. The scaling operator, $\hat{M}_c$: $\hat{M}_c \psi(x) \equiv \sqrt{c}\psi(cx)$, $c > 0$.
4. The complex-conjugation operator, $\hat{K}$: $\hat{K}\psi(x) \equiv \psi^*(x)$.
5. The permutation operator of identical particles $\hat{P}_{12}$: $\hat{P}_{12}\psi(x_1, x_2) \equiv \psi(x_2, x_1)$.

a) Are these operators linear?

For the operators, 1–5, above, find:

b) The Hermitian conjugate operators.
c) The inverse operators.

Solution
a) All the operators except $\hat{K}$ are linear.
b) The operator $\hat{T}_a^\dagger$ can be found from the following sequence of equations

$$\int \psi^*(x)\hat{T}_a\psi(x)dx \equiv \int \psi^*(x)\psi(x+a)dx$$
$$= \int \psi^*(x-a)\psi(x)dx \equiv \int \left[\hat{T}_a^\dagger \psi(x)\right]^* \psi(x)dx$$

(integration is performed in the infinite limits). Therefore, $\hat{T}_a^\dagger \psi(x) \equiv \psi(x-a) \equiv \hat{T}_{-a}\psi(x)$ and $\hat{T}_a^\dagger = \hat{T}_{-a}$.
Similarly, we find:
$$\hat{I}^\dagger = \hat{I}, \quad \hat{M}_c^\dagger = \hat{M}_{1/c}, \quad \hat{P}_{12}^\dagger = \hat{P}_{12},$$

The operator $\hat{K}$ is non-linear, so $\hat{K}^\dagger$ does not exist.

c) All the operators have the inverse:
$$\hat{I}^{-1} = \hat{I}, \quad \hat{T}_a^{-1} = \hat{T}_{-a}, \quad \hat{M}_c^{-1} = \hat{M}_{1/c}, \quad \hat{K}^{-1} = \hat{K}, \quad \hat{P}_{12}^{-1} = \hat{P}_{12}.$$

Problem 1.2

$\hat{A}$ and $\hat{B}$ are Hermitian operators and $\hat{L}$ is an arbitrary linear operator.[3] Prove that the following operators are Hermitian:

1) $\hat{L}^\dagger \hat{L}$ and $\hat{L}\hat{L}^\dagger$; 2) $\hat{L} + \hat{L}^\dagger$; 3) $i(\hat{L} - \hat{L}^\dagger)$; 4) $\hat{L}\hat{A}\hat{L}^\dagger$; 5) $\hat{A}\hat{B} + \hat{B}\hat{A}$; 6) $i(\hat{A}\hat{B} - \hat{B}\hat{A})$.

Hint
Use the following relations $(\hat{L}^\dagger)^\dagger = \hat{L}$ and $(\hat{F}\hat{L})^\dagger = \hat{L}^\dagger \hat{F}^\dagger$.

Problem 1.3

Prove that an arbitrary linear operator $\hat{L}$ can be expressed in the form $\hat{L} = \hat{A} + i\hat{B}$ where $\hat{A}$ and $\hat{B}$ are Hermitian operators.

Result
$\hat{A} = \frac{1}{2}(\hat{L} + \hat{L}^\dagger)$, $\hat{B} = \frac{1}{2i}(\hat{L} - \hat{L}^\dagger)$.

Problem 1.4

Express the commutators $[\hat{A}, \hat{B}\hat{C}]$ and $[\hat{A}\hat{B}, \hat{C}]$ in terms of the commutators $[\hat{A}, \hat{B}]$, $[\hat{A}, \hat{C}]$, and $[\hat{B}, \hat{C}]$.

Solution
$[\hat{A}, \hat{B}\hat{C}] = \hat{A}\hat{B}\hat{C} - \hat{B}\hat{C}\hat{A} = \hat{A}\hat{B}\hat{C} - \hat{B}\hat{A}\hat{C} + \hat{B}\hat{A}\hat{C} - \hat{B}\hat{C}\hat{A} = [\hat{A}, \hat{B}]\hat{C} + \hat{B}[\hat{A}, \hat{C}]$. In the same way, $[\hat{A}\hat{B}, \hat{C}] = \hat{A}[\hat{B}, \hat{C}] + [\hat{A}, \hat{C}]\hat{B}$.

---

[3] Hereafter all operators are assumed to be linear. For brevity's sake, the term "linear" is omitted throughout this chapter.

Problem 1.5

Is it possible for two $N \times N$ matrices $\hat{P}$ and $\hat{Q}$ (with a finite, N) to satisfy the canonical commutation relation, $[\hat{P}, \hat{Q}] = -i\hbar \hat{1}$?

Solution

No, it is not possible. If we calculate the trace of both sides of the equation, $\hat{P}\hat{Q} - \hat{Q}\hat{P} = -i\hbar\hat{1}$, and use the relations, $\text{Tr}\,(\hat{P}\hat{Q}) = \text{Tr}\,(\hat{Q}\hat{P})$, and $\text{Tr}\,\hat{1} = N$, we come to the contradiction.[4]

Problem 1.6

Assume that λ is a small parameter. Derive an expansion of the operator $(\hat{A} - \lambda\hat{B})^{-1}$ in the powers of λ.

Solution

If we multiply both sides of the equation, $(\hat{A} - \lambda\hat{B})^{-1} = \sum\limits_{n=0}^{\infty} \lambda^n \hat{C}_n$, by $(\hat{A} - \lambda\hat{B})$, and equate the coefficients that contain the same power of λ, we find

$$\hat{A}\hat{C}_{n+1} = \hat{B}\hat{C}_n, \quad \hat{C}_{n+1} = \hat{A}^{-1}\hat{B}\hat{C}_n, \quad \hat{C}_0 = \hat{A}^{-1}.$$

Therefore, the expansion has the form

$$(\hat{A} - \lambda\hat{B})^{-1} = \hat{A}^{-1} + \lambda\hat{A}^{-1}\hat{B}\hat{A}^{-1} + \cdots = \hat{A}^{-1}\sum_n \lambda^n \hat{B}^n \hat{A}^{-n}.$$

Problem 1.7

If a function, $F(z)$ has a Taylor expansion, $F(z) = \sum\limits_n c_n z^n$, then knowing an operator $\hat{f}$, the operator $\hat{F} \equiv F(\hat{f})$ is defined as follows: $\hat{F} = \sum\limits_n c_n \hat{f}^n$. Use this rule to determine the explicit form of the following operators:

1) $\exp(ia\hat{I})$;
2) $\hat{T}_a \equiv \exp\left(a\frac{d}{dx}\right)$;
3) $\hat{L}_a \equiv \exp\left(ax\frac{d}{dx}\right)$.

Here a is a real parameter and $\hat{I}$ is the inversion operator. See also Problems 1.8, 1.22, and 1.24.

Solution

1) Taylor-expanding the exponential and using the fact that $\hat{I}^2 = 1$, we find that $\exp(ia\hat{I}) = \cos a + i(\sin a)\hat{I}$.

---

[4] In the case of $N = \infty$ there is no contradiction, since, $\text{Tr}\,(\hat{P}\hat{Q}) = \infty$.

2) Expanding the operator, $\hat{T}_a$, we obtain

$$e^{a\frac{d}{dx}}\psi(x) = \sum_{n=0}^{\infty}\frac{a^n}{n!}\left(\frac{d}{dx}\right)^n\psi(x) \equiv \sum_{n=0}^{\infty}\frac{a^n}{n!}\psi^{(n)}(x) = \psi(x+a).$$

The operator $\exp\left(a\frac{d}{dx}\right)$ is a translation operator.

3) Apply the operator

$$\hat{L}_a = \sum_{n=0}^{\infty}\frac{1}{n!}\left(ax\frac{d}{dx}\right)^n$$

to the term, x^k (to appear in the Taylor expansion of a wavefunction, $\psi(x) = \sum_{k=0}^{\infty}\frac{c_k}{k!}x^k$ on which the operator acts). Since $(xd/dx)x^k = kx^k$ and $(axd/dx)^n x^k = (ak)^n x^k$, we have $\hat{L}_a x^k = (e^a x)^k$. Using a Taylor expansion for $\psi(x)$, we obtain

$$\hat{L}_a\psi(x) = \hat{L}_a\sum_{k=0}^{\infty}\frac{c_k}{k!}x^k = \sum_{k=0}^{\infty}\frac{c_k}{k!}(e^a x)^k = \psi(e^a x).$$

This operator corresponds to the operator, $\hat{M}_c$, from Problem 1.1. with $c = e^a$ (up to a numerical factor, $\sqrt{c}$).

Problem 1.8

Determine the explicit form of the operator $\hat{T}[g(x)] \equiv \exp\left[g(x)\frac{d}{dx}\right]$ where $g(x)$ is a monotonic function of x. Consider two special cases: 1) $g = ax$, and 2) $g = a^3/3x^2$.

Solution

Define the new variable, $y = y(x)$, via the following relation: $y = \int\limits_{b}^{x}\frac{dx}{g(x)}$. Then, $\hat{T} = \exp\left[\frac{d}{dy}\right]$. Therefore, the operator considered is the translation operator along the "y-axis" over the "distance", $\Delta y = 1$ (see Problem 1.7) and

$$\hat{T}[g(x)]\psi(x) = \exp\left[\frac{d}{dy}\right]\psi[x(y)] = \psi[x(y+1)].$$

Here, $x = x(y)$ is an inverse function for $y = y(x)$.

For the special cases we have:

1) $y = \frac{1}{a}\ln|x|$, so $|x| = e^{ay}$ and $x(y+1) = e^a x$. Therefore, $\hat{T}(ax)\psi(x) = \psi(e^a x)$ (cf. Problem 1.7).
2) $x = ay^{1/3}$ and $\hat{T}(a^3/3x^2)\psi(x) = \psi\left[(x^3 + a^3)^{1/3}\right]$.

Problem 1.9

Prove the relation

$$\text{Tr}\left\{\frac{\partial}{\partial \lambda}\exp(\lambda\hat{A}+\hat{B})\right\} = \text{Tr}\left\{\hat{A}\exp(\lambda\hat{A}+\hat{B})\right\},$$

where $\hat{A}$ and $\hat{B}$ are arbitrary matrices (of an equal rank). Is taking the trace crucial in the above relation or does the latter hold more generally as an operator identity?

Hint

To prove the relation, Taylor-expand the exponential, and after differentiating it and applying the trace, equate the coefficients containing the same powers of λ. The expression above does not generalize to the operator identity, if the operators $\hat{A}$ and $\hat{B}$ do not commute.

Problem 1.10

Operators $\hat{A}$ and $\hat{B}$ commute to a "c-number", which is a corollary of the following commutation relation: $[\hat{A}, \hat{B}] = ic\hat{1}$, with $\hat{1}$ being the identity operator (which is often omitted in the physics literature and below). Prove the equation

$$\exp(\hat{A}+\hat{B}) = \exp(\hat{A})\exp(\hat{B})\exp\left(-i\frac{c}{2}\right).$$

Solution

Let us introduce the following operator $e^{\lambda(\hat{A}+\hat{B})}$ and present it as

$$e^{\lambda(\hat{A}+\hat{B})} = e^{\lambda\hat{A}}e^{\lambda\hat{B}}\exp\left(-\frac{i}{2}\lambda^2 c\right)\hat{G}(\lambda), \quad (1)$$

where the operator function $\hat{G}$ is to be determined. Differentiating both parts of Eq. (1) with respect to λ and using the notation

$$e^{\lambda\hat{A}}\hat{B} = (\hat{B}+i\lambda c)e^{\lambda\hat{A}},$$

(which is easy to establish by expanding the term, $e^{\lambda\hat{A}}$, and using $[\hat{A}, \hat{B}] = ic$), we find that $\frac{d}{d\lambda}\hat{G}(\lambda) = 0$: *i.e.* the operator $\hat{G}$ does not depend on λ. Substituting $\lambda = 0$ into Eq. (1), we find $\hat{G} = 1$. This proves the relation.

Problem 1.11

Action of a linear operator, $\hat{L}$, in a Hilbert space spanned by functions, $\psi(\xi)$, can be represented as action of an integral operator as

$$\varphi(\xi) = \hat{L}\psi(\xi) \equiv \int L(\xi,\xi')\psi(\xi')d\xi',$$

where $L(\xi,\xi')$ is the *kernel* of the operator, $\hat{L}$ (ξ represents a set of variables of a particular representation used). For such a generic linear operator (not necessarily Hermitian), $\hat{L}$, find the relation between its kernel and that of its Hermitian-conjugate, $L^\dagger(\xi,\xi')$. Find the kernels of the following operators $\hat{I}$, $\hat{T}_a$, $\hat{M}_c$, $\hat{x} \equiv x$, $\hat{p} \equiv -i\hbar\, d/dx$ (for a description of the operators $\hat{I}$, $\hat{T}_a$, and $\hat{M}_c$, see Problem 1.1).

Solution

1) $L^\dagger(\xi,\xi') = L^*(\xi',\xi)$.
2) By writing

$$\hat{M}_c\psi(x) = \sqrt{c}\psi(cx) = \sqrt{c}\int_{-\infty}^{\infty} \delta(cx-x')\psi(x')dx',$$

we find that the kernel of $\hat{M}_c$ has the form $M_c(x,x') = \sqrt{c}\delta(cx-x')$. Similarly, the other kernels are as follows:

$$I(x,x') = \delta(x+x'),$$
$$T_a(x,x') = \delta(x-x'+a),$$
$$X(x,x') = x\delta(x-x'),$$
$$P(x,x') = -i\hbar\frac{\partial}{\partial x}\delta(x-x').$$

Problem 1.12

An operator, $\hat{L}$, acts in the Hilbert space of wavefunctions, $\psi(x)$, associated with single-particle quantum mechanics in one dimension. Assuming that $\hat{L}$ commutes with both the coordinate-operator, $\hat{x}$, and the momentum-operator, $\hat{p}$, find its kernel $L(x,x')$. Prove that the operator $\hat{L}$ that commutes with both $\hat{x}$ and $\hat{p}$ is proportional to the identity operator, i.e. $\hat{L} \equiv L_0 = $ const $(\propto \hat{1})$.

Solution

a) Taking into account that the kernel of an operator product, $\hat{C} = \hat{A}\hat{B}$, is a convolution of the corresponding kernels

$$C(x,x') = \int_{-\infty}^{\infty} A(x,x'')B(x'',x')dx'',$$

and that the kernel of $\hat{x}$ is $X(x,x') = x\delta(x-x')$, and using the relation $\hat{L}\hat{x} - \hat{x}\hat{L} = 0$, we find

8 Exploring Quantum Mechanics

$$(x' - x)L(x, x') = 0 \text{ and therefore } L(x, x') = f(x)\delta(x - x'). \tag{1}$$

Here $f(x)$ is arbitrary function at this stage.

b) Similarly using $\hat{L}\hat{p} - \hat{p}\hat{L} = 0$ and $P(x, x') = -i\hbar\frac{\partial}{\partial x}\delta(x - x')$, we find

$$\left(\frac{\partial}{\partial x} + \frac{\partial}{\partial x'}\right)L(x, x') = 0, \text{ so that } L(x, x') = g(x - x'), \tag{2}$$

where $g(x - x')$ is an arbitrary function.

c) Equations (1) and (2) are simultaneously valid if and only if, $f(x) \equiv L_0 = \text{const}$, so that $L(x, x') = L_0\delta(x - x')$ describes a multiplication by a constant, which commutes with all other operators acting in the corresponding Hilbert space. The operator with such a kernel is $\hat{L} = L_0\hat{1}$.

1.2 Eigenfunctions, eigenvalues, mean values

Problem 1.13

The state of a particle is described by the following wavefunction:

$$\psi(x) = C \exp\left[i\frac{p_0 x}{\hbar} - \frac{(x - x_0)^2}{2a^2}\right], \tag{1}$$

where p_0, x_0, a are real parameters. Find the probability distribution of the particle coordinates, mean values, and the standard deviations of the coordinate and momentum.

Solution

The normalization condition yields $C^2 = 1/\sqrt{\pi a^2}$. The corresponding probability density is $dw = |\psi(x)|^2 dx$. From Eq. (I.5) we find the mean values:

$$\bar{x} = x_0, \quad \overline{x^2} = x_0^2 + \frac{a^2}{2}, \quad \overline{(\Delta x)^2} = \frac{a^2}{2},$$

$$\bar{p} = p_0, \quad \overline{p^2} = p_0^2 + \frac{\hbar^2}{2a^2}, \quad \overline{(\Delta p)^2} = \frac{\hbar^2}{2a^2}.$$

As we see, $\sqrt{\overline{(\Delta p)^2} \cdot \overline{(\Delta x)^2}} = \hbar/2$, so the wavefunction (1) minimizes the Heisenberg uncertainty relation.

Problem 1.14

Find the relation between the mean values of coordinate and momentum in the quantum states $\psi_1(x)$ and $\psi_2(x)$, related by

a) $\psi_2(x) = \psi_1(x+a)$, b) $\psi_2(x) = \exp\left(i\,\dfrac{p_0 x}{\hbar}\right)\psi_1(x)$.

Solution

a) $\overline{x}_2 = \overline{x}_1 - a$, $\overline{p}_2 = \overline{p}_1$, b) $\overline{x}_2 = \overline{x}_1$, $\overline{p}_2 = \overline{p}_1 + p_0$, where the indices 1 and 2 determine the mean values corresponding to the wavefunctions $\psi_{1,2}(x)$.

Problem 1.15

Prove that the mean values of Hermitian operators $\hat{L}\hat{L}^\dagger$ and $\hat{L}^\dagger\hat{L}$ in an arbitrary state are non-negative.

Solution

$$\overline{\hat{L}\hat{L}^\dagger} = \int \psi^* \hat{L}\hat{L}^\dagger \psi \, d\tau = \int (\hat{L}^\dagger \psi)^* (\hat{L}^\dagger \psi) d\tau \geq 0.$$

Problem 1.16

Prove that the mean value of the dipole moment of a system containing charged particles vanishes in a state with a certain-parity: $\psi(\{\mathbf{r}_a\}) = I\psi(\{-\mathbf{r}_a\})$, with $I = \pm 1$.

Solution

The mean value of the dipole moment is given by

$$\overline{\mathbf{d}} = \int_{-\infty}^{\infty} \psi^*(\mathbf{r}_1, \ldots, \mathbf{r}_n) \sum_a e_a \mathbf{r}_a \psi(\mathbf{r}_1, \ldots, \mathbf{r}_n) \prod_b d^3 \mathbf{r}_b. \tag{1}$$

If we make the substitution of variable $\mathbf{r}'_a = -\mathbf{r}_a$, we obtain

$$\overline{\mathbf{d}} = -\int_{-\infty}^{\infty} \psi^*(-\mathbf{r}'_1, \ldots, -\mathbf{r}'_n) \sum_a e_a \mathbf{r}'_a \psi(-\mathbf{r}'_1, \ldots, -\mathbf{r}'_n) \prod_b d^3 \mathbf{r}'_b. \tag{2}$$

Since $\psi(-\mathbf{r}'_1, \ldots, -\mathbf{r}'_n) = I\psi(\mathbf{r}_1, \ldots, \mathbf{r}_n)$ with the parity $I = \pm 1$, from Eqs. (1) and (2), we obtain $\overline{\mathbf{d}} = -\overline{\mathbf{d}} = 0$.

Problem 1.17

An Hermitian operator $\hat{f}$ satisfies the operator identity:

a) $\hat{f}^2 = p^2$; b) $\hat{f}^2 = p\hat{f}$; c) $\hat{f}^3 = p^2 \hat{f}$,

where p is a real parameter. Find the eigenvalues of such an operator.

Solution

The operator relation $A(\hat{f}) = B(\hat{f})$ with some arbitrary functions $A(z)$ and $B(z)$ leads to the similar one $A(f_1) = B(f_1)$ for its eigenvalues. Therefore, the operator(s) $\hat{f}$ could have only eigenvalues given by

a) $f_{1,2} = \pm p$,
b) $f_1 = 0, \; f_2 = p$,

and

c) $f_1 = 0, \; f_{2,3} = \pm p$. No other eigenvalues are possible.

Problem 1.18

Find the eigenfunctions and eigenvalues of the following operator: $\hat{f} = \alpha \hat{p}_x + \beta \hat{x}$, where $\hat{x}$ and $\hat{p}_x$ are the canonically-conjugate coordinate and momentum and α and β are real parameters. Prove that the eigenfunctions of this operator are orthogonal to each other, and normalize them properly.

Solution

The equation for the eigenfunctions and eigenvalues of the operator, $\hat{f}$, and its solution are given by

$$-i\tilde{\alpha}\frac{d}{dx}\psi_f(x) + \beta x \psi_f(x) = f\psi_f(x), \quad (1)$$

$$\psi_f(x) = C_f \exp\left\{-i\frac{(\beta x - f)^2}{2\tilde{\alpha}\beta}\right\}, \quad (2)$$

where $\tilde{\alpha} = \hbar\alpha$. From (2), it follows that the eigenvalues, f, can take arbitrary real values (if the values of f are complex, the wavefunction (2) increases at large distances and is not normalizable). Also, the parameters α and β above are real, since $\hat{f}$ is Hermitian. The spectrum of $\hat{f}$ is continuous, and the eigenvalues are non-degenerate. The proper normalization condition is as follows

$$\int \psi_{f'}^*(x)\psi_f(x)dx = \delta(f - f'),$$

which gives $C = (2\pi\tilde{\alpha})^{-1/2}$. The proof of completeness of the system of eigenfunctions (2) is left to the reader.

Problem 1.19

Solve the previous problem for the Hermitian operator,[5] $\hat{F}$, with a kernel of the following form: $F(x, x') = f(x) \cdot f^*(x')$ (see Problem 1.11). Determine the degeneracy of its eigenvalues.

---

[5] Such operators appear in atomic and nuclear physics in the context of particles in so-called separable potentials (see Problems 2.19, 2.34, 4.12). Also note that the operator considered could be written as $\hat{F} = |f\rangle\langle f|$.

Solution
The eigenvalue problem for $\hat{F}$ reads

$$\hat{F}\psi_n(x) \equiv f(x)\int f^*(x')\psi_n(x')dx' = f_n\psi_n(x).$$

This equation has the following solutions: 1) One of the eigenfunctions $\psi_0 = Cf(x)$ corresponds to the non-degenerate eigenvalue $f_0 = \int |f(x)|^2 dx > 0$. 2) The second eigenvalue $f_1 = 0$ is infinitely-degenerate. The corresponding eigenfunctions, $\psi_i(x)$, have the following properties $\int f^*(x)\psi_i(x)dx = 0$ (*i.e.*, these functions are orthogonal to the eigenfunction, $\psi_0(x)$, corresponding to the different eigenvalue, as it ought to be). There are no other eigenvalues.

Problem 1.20

Find the eigenfunctions and eigenvalues of the complex-conjugation operator, $\hat{K}$ (see Problem 1.1.)

Result
The eigenfunctions of the operator, $\hat{K}$, are given by $\psi_\alpha(x) = e^{i\alpha}g(x)$, where $g(x)$ is an arbitrary real function and α is a real number. The corresponding eigenvalues are $k_\alpha = e^{-2i\alpha}$.

Problem 1.21

An Hermitian operator (matrix) $\hat{f}$ has only N eigenvalues. Prove that the operator $\hat{f}^N$ can be expressed as a linear combination of the operators $\hat{1}, \hat{f}, \ldots, \hat{f}^{N-1}$. To illustrate the result, consider the inversion operator, $\hat{I}$.

Solution
Applying the operator $\hat{G} \equiv (\hat{f} - f_1)(\hat{f} - f_2)\ldots(\hat{f} - f_N)$ to arbitrary state-vector $|\psi\rangle$ we have $\hat{G}|\psi\rangle = 0$. This state, $|\psi\rangle$, can be decomposed into a linear combination of the eigenfunctions of $\hat{F}$, $|\psi_{f_k}\rangle$ that form a complete set: $|\psi\rangle = \sum_k c_k |\psi_{f_k}\rangle$, with $(\hat{f} - f_k)|\psi_{f_k}\rangle = 0$.

Therefore, $\hat{G} \equiv 0$, and consequently we prove the desired statement:

$$\hat{f}^N - \sum_{i=1}^N f_i \hat{f}^{N-1} + \frac{1}{2}\sum_{i,k;i\neq k} f_i f_k \hat{f}^{N-2} + \cdots + (-1)^N \prod_{i=1}^N f_i = 0. \quad (1)$$

In the case of $N = 2$, Eq. (1) gives $\hat{f}^2 = (f_1 + f_2)\hat{f} - f_1 f_2$. Hence, for the inversion operator with the eigenvalues ± 1, we have $\hat{I}^2 = 1$, as expected.

Problem 1.22

The operator $\hat{f}$ has only N different eigenvalues, $f_1, f_2, \ldots, f_N$. Find the operator $\hat{F} = F(\hat{f})$, where $F(z)$ is an arbitrary function. To illustrate the general result, consider the cases of $N = 2$ and $N = 3$ (in the latter case, assume that the spectrum consists of 0 and $\pm f_0$).

Solution

Using the result of the previous problem we have

$$\hat{F} = F(\hat{f}) = \sum_{n=0}^{N-1} c_n \hat{f}^n; \quad F(f_i) = \sum_{n=0}^{N-1} c_n f_i^n, \quad i = 1, 2, \ldots, N \tag{1}$$

(compare this with Problem 1.17). The second relation (1) above gives way to a linear system of equations that determines the coefficients, c_n.

If $N = 2$, Eq. (1) yields

$$\hat{F} = \frac{f_2 F(f_1) - f_1 F(f_2)}{f_2 - f_1} + \frac{F(f_1) - F(f_2)}{f_1 - f_2} \hat{f}. \tag{2}$$

Similarly, for $N = 3$, we obtain

$$\hat{F} = F(0) + \frac{F(f_0) - F(-f_0)}{2f_0} \hat{f} + \frac{F(f_0) + F(-f_0) - 2F(0)}{2f_0^2} \hat{f}^2. \tag{3}$$

Problem 1.23

Prove the relation (I.6) from the introductory part of this chapter.

Solution

Differentiating both sides of Eq. (1.6) for eigenfunctions and eigenvalues with respect to λ: $\hat{f}(\lambda)\psi_n(q, \lambda) = f_n(\lambda)\psi_n(q, \lambda)$, we have

$$\left(\frac{\partial \hat{f}}{\partial \lambda}\right)\psi_n(\lambda) + \hat{f}\frac{\partial}{\partial \lambda}\psi_n(\lambda) = \left(\frac{\partial f_n}{\partial \lambda}\right)\psi_n(\lambda) + f_n \frac{\partial}{\partial \lambda}\psi_n(\lambda). \tag{1}$$

Multiply both sides of Eq. (1) by ψ_n^* from the left, and integrate over q. Using the relation

$$\int \psi_n^* \hat{f} \frac{\partial}{\partial \lambda} \psi_n d\tau_q = \int (\hat{f}\psi_n)^* \frac{\partial}{\partial \lambda} \psi_n d\tau_q = f_n \int \psi_n^* \frac{\partial}{\partial \lambda} \psi_n d\tau_q,$$

which follows from the Hermiticity of $\hat{f}$, we prove Eq. (I.6).

Problem 1.24

How would you define the operator, $\hat{F} = F(\hat{f})$, where $\hat{f}$ is a Hermitian operator and $F(z)$ is an arbitrary function of z, which does not necessarily have a regular Taylor expansion in the entire parameter range of interest? How important is the assumption of Hermiticity of $\hat{f}$ in this construction? Consider specifically the operator $\sqrt{-\Delta}$, where Δ is the Laplacian in the three-dimensional Euclidean space.

Solution

A way to define the operator $\hat{F} = F(\hat{f})$ is to demand that its spectrum and eigenfunctions coincide with those of the operator, $\hat{f}$, and the corresponding eigenvalues are given by $F_n = F(f_n)$. Since the system of eigenfunctions ψ_f is complete (the Hermiticity of the operator $\hat{f}$ is important here), we can write

$$\hat{F}\psi = \hat{F}\sum_n c(f_n)\psi_{f_n}(q) \equiv \sum_n c(f_n)F(f_n)\psi_{f_n}(q). \tag{1}$$

Using Eq. (I.4) for $c(f)$ we can find that $\hat{F}$ is an integral operator with the kernel[6]

$$F(q,q') = \sum_n F(f_n)\psi_{f_n}(q)\psi_{f_n}^*(q'). \tag{2}$$

Since $(-\Delta)^{-1/2} = \hbar \left(\hat{\mathbf{p}}^2\right)^{-1/2} \equiv \hbar/|\hat{\mathbf{p}}|$, we use Eq. (2) to obtain the kernel

$$\hbar P^{-1}(\mathbf{r},\mathbf{r}') = \frac{\hbar}{(2\pi\hbar)^3} \int \frac{1}{p} e^{i\mathbf{p}\cdot(\mathbf{r}-\mathbf{r}')/\hbar} d^3p = \frac{1}{2\pi^2(\mathbf{r}-\mathbf{r}')^2}.$$

To calculate the integral it is convenient to use spherical coordinates and choose the polar axis along the vector $\mathbf{r} - \mathbf{r}'$.

Problem 1.25

Hermitian operators $\hat{A}, \hat{B}$, and $\hat{L}$ satisfy the following commutation relations: $[\hat{A}, \hat{L}] = 0$, $[\hat{B}, \hat{L}] = 0$, but $[\hat{A}, \hat{B}] \neq 0$. Prove that the spectrum of the operator $\hat{L}$ has degenerate eigenvalues. Give examples.

Solution

Applying the equation $\hat{A}\hat{L} - \hat{L}\hat{A} = 0$ to the eigenfunction ψ_L of the operator $\hat{L}$ (L_i are its eigenvalues), we find that the function $\hat{A}\psi_L$ is also the eigenfunction of $\hat{L}$ with the same eigenvalue L_i (or $\hat{A}\psi_{L_i} = 0$). If the eigenvalue L_i is non-degenerate, then $\hat{A}\psi_{L_i} = A_i\psi_{L_i}$, i.e., ψ_{L_i} is the eigenfunction of the operator $\hat{A}$ too. In a similar way, it is also an eigenfunction of $\hat{B}$, i.e., $\hat{B}\psi_{L_i} = B_i\psi_{L_i}$. If all the eigenvalues L_i were non-degenerate, then the relation $(\hat{A}\hat{B} - \hat{B}\hat{A})\psi_{L_i} = (A_iB_i - B_iA_i)\psi_{L_i} = 0$ would be valid for all states. But if such a relation were valid for all eigenfunctions that form a complete set, then it would result in $\hat{A}\hat{B} - \hat{B}\hat{A} = 0$, which contradicts the initial data.

---

[6] The system of eigenfunctions $\psi_{f_n}(q)$ is considered to be orthonormal.

To illustrate this result, consider a free quantum particle in one dimension. Its Hamiltonian $\hat{H} = \hat{p}^2/2m$ commutes with both the momentum $\hat{p}$ and inversion $\hat{I}$ operators. However, these two operators do not commute with each other, $[\hat{p}, \hat{I}] \neq 0$, and this fact is related to the two-fold degeneracy of the free-particle spectrum, $E(p) = E(-p)$.

Problem 1.26

Give examples of a quantum state, where:

a) Two physical quantities, whose operators do not commute, simultaneously have a definite values;
b) Only one of two physical quantities, whose operators commute, has a definite value.

Solution

a) Different components of the angular momentum do not commute with each other, but in the state with the angular momentum, $L = 0$, they do have the same definite value $L_{x,y,z} = 0$. For another example, see Problem 1.27.
b) The momentum and kinetic-energy operators commute with each other. However, the function $\psi = C \sin(\mathbf{p} \cdot \mathbf{r}/\hbar)$ is an eigenfunction for the kinetic-energy operator, but not for the momentum operator.

These examples do not contradict the general quantum mechanical statements about the simultaneous measurability of two physical quantities, nor the *uncertainty relation* (see Problem 1.30).

Problem 1.27

Consider a quantum state, ψ_{ab}, where the physical quantities A and B have certain values. What can you say about the eigenvalues a and b of the corresponding operators, $\hat{A}$ and $\hat{B}$, if they anticommute with each other. To illustrate the general result, consider the operators $\hat{x}$ (coordinate) and $\hat{I}$ (inversion).

Solution

We have the relation $(\hat{A}\hat{B} + \hat{B}\hat{A})\psi_{ab} = (ab + ba)\psi_{ab} = 2ab\psi_{ab} = 0$. Therefore, either a or b vanishes. For example, $\hat{I}\hat{x} + \hat{x}\hat{I} = 0$; but there is only one wavefunction $\psi_0(x) = C\delta(x)$ that is the eigenfunction of both $\hat{x}$ and $\hat{I}$. Eigenvalues of the coordinate operator are $x_0 = 0$. Note that anticommuting operators can not have a common eigenfunction (e.g., such is the case for the anticommuting Pauli matrices, see Chapter 7).

Problem 1.28

Find an operator corresponding to the radial component of momentum $\hat{p}_r$ (using spherical coordinates). Prove the Hermiticity of this operator. Find its eigenfunctions and eigenvalues. Are these eigenvalues real? Are the eigenfunctions orthogonal? Explain the results obtained. See also Problem 1.29.

Solution

In classical mechanics we have $p_r = m\dot{r} = \mathbf{p} \cdot \mathbf{n}$, where $\mathbf{n} = \frac{\mathbf{r}}{r}$. The Hermitian operator

$$\hat{p}_r \equiv \frac{1}{2}(\hat{\mathbf{p}} \cdot \hat{\mathbf{n}} + \hat{\mathbf{n}} \cdot \hat{\mathbf{p}}) = \mathbf{n} \cdot \hat{\mathbf{p}} + \frac{\hbar}{2i}\text{div } \mathbf{n} = \frac{\hbar}{i}\frac{1}{r}\frac{\partial}{\partial r}r \qquad (1)$$

is the quantum-mechanical analog of this relation.

The solution of the eigenvalue problem for this operator is given by $\psi(\mathbf{r}) = \frac{C(\vartheta,\varphi)}{r}e^{\frac{i}{\hbar}p_r r}$, where $C(\vartheta,\varphi)$ is an arbitrary function of the angles. Technically, the eigenvalues p_r could take on complex values $p_r = p_1 + ip_2$, with $p_2 \geq 0$, and one can show that the corresponding eigenfunctions are not orthogonal.

These conclusions about the eigenfunctions and eigenvalues of the operator $\hat{p}_r$ rule out the possibility of its direct physical interpretation, and demonstrate subtleties behind the quantum-mechanical statement about the relation between physical quantities (or observables, according to Dirac) and Hermitian (or self-adjoint) operators. Physically, it shows that not every physical quantity from classical mechanics has a well-defined quantum-mechanical equivalent (and vice versa, not every quantum operator, e.g. the parity, has a classical equivalent). From the mathematical point of view, this result demonstrates that there is a difference between the notion of an Hermitian operator and self-adjoint operator. In this interpretation, the operator $\hat{p}_r$ is Hermitian but not self-adjoint (see also the following problem).

Problem 1.29

Use the operator $-i\hbar\frac{d}{dx}$ that acts in the space of the functions defined on

1) the entire real axis, $-\infty < x < \infty$;
2) the finite interval, $a \leq x \leq b$;
3) the half-axis, $0 \leq x < \infty$;

to demonstrate the difference between the Hermitian and self-adjoint operators. Discuss the eigenvalues and eigenfunctions of such operators.

Discussion

The Hermitian and self-adjoint operators, $\hat{f} = \hat{f}^\dagger$, are defined via the same relation:

$$\int \psi_2^* \hat{f}\psi_1 d\tau = \int (\hat{f}^\dagger \psi_2)^* \psi_1 d\tau = \int (\hat{f}\psi_2)^* \psi_1 d\tau. \qquad (1)$$

However, the difference between them can be formulated in terms of different restrictions on the class of functions, ψ_1 and ψ_2, where the relation (1) is required to hold.

If relation (1) is satisfied for some class of functions, $\tilde{D}_f$ (the same for both ψ_1 and ψ_2), then the operator $\hat{f}$ is called Hermitian in this particular class of functions. We call the operator, $\hat{f}$, self-adjoint if it is Hermitian in the class of functions, D_f, subject to the constraints below. (Here we are not making an attempt to provide a full mathematical theory, but simply outlining key concepts relevant to the quantum-mechanical problem at hand).

D_f includes all functions $\psi_{\{f\}}(q)$ that satisfy

$$\int |\psi_{\{f\}}(q)|^2 d\tau_q < \infty \qquad (2')$$

and

$$\int |\hat{f}\psi_{\{f\}}(q)|^2 d\tau_q < \infty, \quad \int \psi^*(q)\hat{f}\psi_{\{f\}}(q) d\tau < \infty, \qquad (2'')$$

where $\psi(q)$ is an arbitrary function from the Hilbert space. Importantly, the eigenvalues of a self-adjoint operator are real, and its eigenfunctions are mutually orthogonal and form a complete system.[7]

Furthermore, Hermitian but not self-adjoint operators may be separated into the following two groups: i) essentially self-adjoint operators that allow a self-adjoint extension, and ii) Hermitian in some domain, but not essentially self-adjoint operators that do not allow such an extension. More precisely, if a realization of an operator $\hat{f}$ as Hermitian in $\tilde{D}_f$ has the property that the relation (1) is fulfilled by $\forall \psi_{1,2} \in D_f$, but is not fulfilled if at least one of the functions – ψ_1 or ψ_2 – does not belong to $\tilde{D}_f$, then the operator is said to allow a self-adjoint extension and is called essentially self-adjoint (the extension can usually be obtained by putting additional constraints on the set of functions, e.g. boundary conditions). The properties of eigenvalues and eigenfunctions of essentially self-adjoint operators are the same as those of self-adjoint operators. Operators that are not essentially self-adjoint do not necessarily share these nice properties (positive-definite eigenvalues and orthonormal eigenfunctions forming a complete set of states). These subtleties are illustrated in the examples below.

Solution

1) Returning to the example in the problem, if the operator $\hat{p} = -i\hbar d/dx$ acts in the space of functions defined for $x \in (-\infty, +\infty)$, we have

$$\int_{-\infty}^{\infty} \psi_2^* \left(-i\hbar \frac{d}{dx}\right) \psi_1 dx = \int_{-\infty}^{\infty} \left(-i\hbar \frac{d}{dx} \psi_2\right)^* \psi_1 dx - i\hbar \psi_2^* \psi_1 \big|_{-\infty}^{\infty}. \qquad (3)$$

For the functions $\psi_{1,2}$ that belong to the definition domain of the operator, $D_p(-\infty, +\infty)$, the last term in (3) vanishes per Eq. (2'). Therefore, the operator, $\hat{p}_x$, is self-adjoint. Its eigenvalues p are real, and eigenfunctions $\psi_p(x)$ are orthogonal and form a complete set.

---

[7] However, the eigenfunctions of a self-adjoint operator are not necessarily normalized to unity, i.e., the integrals (2) for them may be divergent, and the orthogonality condition is defined with the help of the δ-function.

2) For the operator $\hat{p} = -i\hbar \cdot d/dx$ that acts in the space of the functions defined on a finite interval, $x \in (a, b)$, we have

$$\int_a^b \psi_2^* \left(-i\hbar \frac{d}{dx}\right) \psi_1 dx = \int_a^b \left(-i\hbar \frac{d}{dx}\psi_2\right)^* \psi_1 dx - i\hbar\psi_2(x)^*\psi_1(x)\big|_a^b. \quad (4)$$

Here the last term is not necessarily zero, and therefore the operator is not self-adjoint. However, it has (multiple) Hermitian realizations and allows a self-adjoint extension.

For example, the operator is Hermitian on the class of functions that obey the boundary conditions of the form $\psi(a) = \psi(b) = 0$. However, such conditions do not give a self-adjoint extension. Indeed, to make the integrated term equal to zero it would be enough for only one of the functions $\psi_1(x)$ and $\psi_2(x)$ to obey this condition. With such a choice of boundary conditions, the operator $\hat{p} = -i\hbar d/dx$ does not have any eigenfunctions.

Another Hermitian realization of this operator is to impose the following boundary condition

$$\frac{\psi_1(b)}{\psi_1(a)} = \left[\frac{\psi_2(a)}{\psi_2(b)}\right]^* = e^{i\beta} = \text{const}, \quad (5)$$

where β is a real value. The choice of such a boundary condition gives the self-adjoint extension of the operator $\hat{p} = -i\hbar \cdot d/dx$ on the finite interval. The corresponding eigenvalues and eigenfunctions are given by

$$\lambda_n = \frac{\hbar}{b-a}(\beta + 2\pi n), \quad \psi_n(x) = \frac{1}{\sqrt{b-a}} \exp\left\{\frac{ip_n x}{\hbar}\right\}, \quad n = 0, \pm 1, \pm 2, \ldots.$$

Eigenfunctions are mutual orthogonal and form a complete system. Notably, the Hermitian operator $\hat{l}_z = -i\partial/\partial\varphi$ – the projection of angular momentum onto the z-axis – with $a = 0$, $b = 2\pi$, $\beta = 0$ is an essentially self-adjoint operator of this type.

3) Finally, the operator $\hat{p}_r = -i\hbar \cdot d/dr$ that acts in the space of functions defined on the half-axis, $r \in (0, +\infty)$, has the following property:

$$\int_0^\infty \psi_2^*(r)[-i\hbar\psi_1'(r)]dr = \int_0^\infty [-i\hbar\psi_2'(r)]^*\psi_1(r)dr + i\hbar\psi_2(0)^*\psi_1(0). \quad (6)$$

Its unique Hermitian realization is obtained by enforcing the boundary condition $\psi(0) = 0$. However, it is sufficient that only one function in (6) obeys the condition, so this does not give a self-adjoint extension. This operator does not have any eigenfunctions that obey the boundary condition $\psi(0) = 0$. If we ignore the boundary condition, we can find non-trivial solutions to the eigenvalue problem. However, the eigenvalues are complex and the eigenfunctions are not orthogonal, which provides an explanation of the puzzling result for the spectrum of the operator naïvely associated with the radial component of momentum, as discussed in the previous problem.

Final comments An Hermitian operator $\hat{f}$ could be classified by the deficiency index (N_+, N_-), where $N_\pm$ is the number of linearly-independent eigenfunctions normalized to 1 such that $\hat{f}\psi = \pm i f_0 \psi$, where f_0 is real. If $N_+ = N_- = 0$, the operator is self-adjoint. If $N_+ = N_- = N \neq 0$, then it allows a self-adjoint extension made by imposing N additional conditions. If $N_+ \neq N_-$, the operator is not essentially self-adjoint.

Note that in quantum-mechanical problems we often have to deal with a self-adjoint extension of Hermitian operators. In such cases the additional constraints must be chosen from physical considerations. Apart from the example of the operator $\hat{l}_z$ mentioned above, another example is the operator $\hat{p}^2/2m$ on a finite interval with the boundary conditions $\psi(0) = \psi(a) = 0$ (these conditions appear in the canonical problem of a particle in an infinite potential well). Furthermore, restrictions on the wavefunction in the origin (*i.e.*, with $r = 0$) often appear in the problems dealing with bound states in and scattering off central potentials. Note that even in the case of "good" potentials $U(r)$, these constraints in effect realize self-adjoint extensions of the Hamiltonian operator. In this case, a general condition of a self-adjoint extension is given by

$$\frac{[r\psi(r)]'}{r\psi(r)} \to \alpha \quad \text{with} \quad r \to 0,$$

and physically corresponds to inclusion of an additional interaction in the form of a potential or the zero radius (see Problem 4.10). In the case of *singular* attractive potentials such that allow "falling into the origin" – these boundary conditions do not realize a self-adjoint extension and must be modified (see Problem 9.14).

Problem 1.30

A commutator of operators $\hat{A}$ and $\hat{B}$ of two physical quantities has the form $[\hat{A}, \hat{B}] = i\hat{C}$, where $\hat{C}$ is an Hermitian operator. Prove (with certain restrictions on the wavefunctions) the *uncertainty relation*

$$\overline{(\hat{A} - \bar{A})^2} \cdot \overline{(\hat{B} - \bar{B})^2} \geq \frac{1}{4}(\bar{C})^2,$$

where all mean values refer to the same state of the system.

Consider specifically the operators $\hat{x}$ and $\hat{p}_x$ to find an explicit form of the wavefunctions that minimize the uncertainty relation.

Consider also the operators $\hat{l}_z$ and $\hat{\varphi}$.

Solution

Let us consider the integral $J(\alpha) = \int |(\alpha \hat{A}_1 - i\hat{B}_1)\psi|^2 d\tau \geq 0$, where $\hat{A}_1 = \hat{A} - a$, $\hat{B}_1 = \hat{B} - b$; with α, a, and b are some real parameters. Using the Hermiticity of the operators $\hat{A}_1$ and $\hat{B}_1$, the relation $[\hat{A}_1, \hat{B}_1] = i\hat{C}$, and considering a normalized wavefunction, ψ, we can rewrite the integral in the form

$$J = \int ((\alpha \hat{A}_1 - i\hat{B}_1)\psi)^*(\alpha \hat{A}_1 - i\hat{B}_1)\psi d\tau = \int \psi^*(\alpha^2 \hat{A}_1^2 - i\alpha[\hat{A}_1, \hat{B}_1] + \hat{B}_1^2)\psi d\tau$$

$$= \alpha^2 \overline{(\hat{A} - a)^2} + \alpha \bar{C} + \overline{(\hat{B} - b)^2} \geq 0. \qquad (1)$$

Set $a = \overline{A}$ and $b = \overline{B}$. The condition of non-negativeness of the trinomial, quadratic in α, gives the uncertainty relation as stated in the problem:

$$\overline{(\hat{A} - \overline{A})^2} \cdot \overline{(\hat{B} - \overline{B})^2} \geq \frac{1}{4} \left(\overline{C}\right)^2. \tag{2}$$

The relation (2) is realized only when $(\alpha\hat{A}_1 - i\hat{B}_1)\psi = 0$. In particular, for the operators $\hat{A} = \hat{x} = x$, $\hat{B} = \hat{p} = -i\hbar d/dx$, $\hat{C} = \hbar$, this condition takes the form $\psi' + [(x - x_0)/d^2 - ip_0/\hbar]\,\psi = 0$ (where instead of $\alpha < 0$, a and b we have more convenient real combinations involving x_0, p_0, and d). So, we have

$$\psi = \frac{1}{(\pi d^2)^{1/4}} \exp\left\{i\frac{p_0 x}{\hbar} - \frac{(x - x_0)^2}{2d^2}\right\},$$

which gives the explicit form of the wavefunction that minimizes the uncertainty relation for the coordinate and momentum (see also Problem 1.13).

One should exercise care in using Eq. (2): e.g., in the case of operators $\hat{A} = \hat{l}_z = -id/d\varphi$ and $\hat{B} = \hat{\varphi} = \varphi$ that do satisfy the canonical commutation relations, a blind application of Eq. (2) yields $\overline{(\Delta l_z)^2} \cdot \overline{(\Delta \varphi)^2} \geq 1/4$. This result is physically meaningless, because the uncertainty in the angle can not possibly exceed $\pi^2 - \overline{(\Delta\varphi)^2} < \pi^2$, and there exist physical states with a well-defined projection of the angular momentum, i.e., $\overline{(\Delta l_z)^2}$ can be zero.

The paradox is resolved by noting that while deriving Eq. (1) we used the relations

$$\int (\hat{A}\psi)^*(\hat{A}\psi)d\tau = \int \psi^* \hat{A}^2 \psi d\tau, \quad \int (\hat{A}\psi)^*(\hat{B}\psi)d\tau = \int \psi^* \hat{A}\hat{B}\psi d\tau$$

which were based on the Hermiticity of all operators involved. But if we take into account the result of the previous problem, we see that this assumption is applicable only in the case of self-adjoint operators. For physical operators that represent a self-adjoint extension of an Hermitian operator ($\hat{l}_z$ is an example of such operator), a more strict condition is needed: it is necessary that not only the wavefunction ψ but also $\hat{B}\psi$ belongs to the appropriate domain, where $\hat{A}$ is Hermitian (and similarly, $\hat{A}\psi$ must remain in the domain of wavefunctions where $\hat{B}$ is Hermitian). If these conditions are fulfilled, the relation (2) is valid, otherwise it is not necessarily so. Particularly, in the case of operators $\hat{l}_z$ and $\hat{\varphi}$, we must require that the states involved in the uncertainty relation satisfy the condition $\psi(0) = \psi(2\pi) = 0$ (which ensures that the function $\tilde{\psi}(\varphi) = \varphi\psi(\varphi)$ belongs to the domain of the Hermiticity of $\hat{l}_z$). For such states the inequality $\overline{(\Delta l_z)^2} \cdot \overline{(\Delta \varphi)^2} \geq 1/4$ is indeed valid. Otherwise, the uncertainty relation has to be generalized as follows:

$$\overline{(\Delta l_z)^2} \cdot \overline{(\Delta \varphi)^2} \geq \frac{1}{4}\left(1 - 2\pi|\psi(0)|^2\right)^2.$$

1.3 The projection operators

Problem 1.31

An Hermitian operator $\hat{P}$ is called a *projection operator or projector* if it satisfies the relation[8] $\hat{P}^2 = \hat{P}$. Consider the operator, $\hat{P}(f_i)$, acting on the eigenfunctions of another Hermitian operator, $\hat{f}$, as follows:[9]

$$\hat{P}(f_i)\psi_{f_k} = \delta_{f_i,f_k}\psi_{f_i} = \begin{cases} \psi_{f_i}, & f_i = f_k, \\ 0, & f_i \neq f_k. \end{cases}$$

a) Prove that this operator is a projector (since the system of eigenfunctions ψ_{f_i} is complete, the relations above determine action of $\hat{P}(f_i)$ on an arbitrary wavefunction, ψ).
b) On which states does this operator project? What physical meaning does the mean value $\overline{\hat{P}(f_i)}$ have for an arbitrary physical state, ψ?
c) Using the projectors $\hat{P}(f_i)$, construct the projection operator, $\hat{P}(\{f\})$, that projects on the states, where the physical quantity f takes some value from an array of eigenvalues $\{f\} = \{f_{i_1}, f_{i_2}, \ldots, f_{i_N}\}$? Show that $\hat{P}^2(\{f\}) = \hat{P}(\{f\})$.
d) Find explicitly the projector, $\hat{P}(f_i, g_k, \ldots, h_l)$, that projects on a state with definite values of $f_i, g_k, \ldots, h_l$ of physical quantities that form a complete set of operators (*i.e.*, express it in terms of the operators, $\hat{P}(f_i), \hat{P}(g_k), \ldots$).

Solution

a) Consider two arbitrary wavefunctions ψ and ϕ and express them as follows: $\psi = \sum_k c_k \psi_{f_k}$ and $\phi = \sum_k b_k \phi_{f_k}$ (for the sake of simplicity, we assume that the eigenvalues of $\hat{f}$ are non-degenerate). First, verify that the operator $\hat{P}(f_i)$ is indeed Hermitian:

$$\int \phi^* \hat{P}(f_i)\psi d\tau = c_i \int \phi * \psi_{f_i} d\tau = c_i b_i^* = \int \left[\hat{P}(f_i)\phi\right]^* \psi d\tau \equiv \int \left[\hat{P}^\dagger(f_i)\phi\right]^* \psi d\tau.$$

---

[8] Eigenvalues of such an operator are 0 and 1. Using this operator we can "divide" the Hilbert space into two mutually orthogonal subspaces: $\hat{P}|\psi>$ and $(1-\hat{P})|\psi>$. The operator $\hat{P}' = 1 - \hat{P}$ is also a projection operator, and it projects on the second of two subspaces.

[9] The relation above belongs to the discrete part of eigenvalues spectrum. A generalization for the continuous part of spectrum is given by the projection on some finite interval $(f, f + \Delta f)$ of eigenvalues according to

$$\hat{P}(f, \Delta f)\psi_{f'} = \begin{cases} \psi_{f'}, & f < f' < f + \Delta f, \\ 0, & f' < f, \ f' > f + \Delta f. \end{cases}$$

And $\overline{\hat{P}(f, \Delta f)}$ also gives the probability of the value f being included in the interval (see 1.32.).

Here we have used the orthogonality of the eigenfunctions of the operator $\hat{f}$. From $\hat{P}^2(f_i)\psi = \hat{P}(f_i)(c_i\psi(f_i)) = c_i\psi_{f_i} = \hat{P}(f_i)\psi$, it follows that $\hat{P}^2(f_i) = \hat{P}(f_i)$. Hence, $\hat{P}(f_i)$ is a projection operator.

b) $\hat{P}(f_i)$ projects onto the state with the definite value f_i of physical quantity f. Further, we find

$$\overline{\hat{P}(f_i)} = \int \psi^* \hat{P}(f_i)\psi d\tau = |c_i|^2 \equiv |c(f_i)|^2$$

Here we have assumed that the wavefunctions ψ are normalized to 1: i.e. the mean value $\overline{\hat{P}(f_i)}$ gives the probability to measure $f = f_i$ in the given state.

c) According to the interpretation above, $\hat{P}(\{f\}) = \sum_a \hat{P}(f_{i_a})$. Since $\hat{P}(f_i)\hat{P}(f_k) = \delta_{ik}\hat{P}(f_i)$, then $\hat{P}(\{f\})^2 = \hat{P}(\{f\})$, as it has to be.

d) The operators from the complete set commute with each other. Therefore,

$$\hat{P}(f_i, g_k, \ldots, t_l) = \hat{P}(f_i) \cdot \hat{P}(g_k) \ldots \hat{P}(t_l).$$

Problem 1.32

Find the projection operator that projects onto states with the values of a particle's coordinate such that $x_0 \geq a$.

Solution

From the definition of the projection operator, $\hat{P}(x_0 \geq a)$, we have $\hat{P}\psi(x) = \psi(x)$, if $x \geq a$, and $\hat{P}\psi(x) = 0$, if $x < a$. Hence, $\hat{P}(x_0 \geq a) = \eta(x - a)$, where $\eta(x)$ is the Heaviside step function, which is equal to 1 for $x > 0$ and to 0 for $x < 0$. It is evident that $\hat{P}(x_0 \geq a)$ is an Hermitian operator and $\hat{P}^2(x_0 \geq a) = \hat{P}(x_0 \geq a)$.

Problem 1.33

Find the projection operators $\hat{P}_\pm$ that project onto states with a definite even/odd parity with respect to coordinate inversion, $\mathbf{r} \to -\mathbf{r}$. Express the projectors in terms of the inversion operator, $\hat{I}$.

Solution

An arbitrary function can be written as superposition of odd and even components:

$$\psi(\mathbf{r}) = \frac{1}{2}[\psi(\mathbf{r}) + \psi(-\mathbf{r})] + \frac{1}{2}[\psi(\mathbf{r}) - \psi(-\mathbf{r})].$$

Per definition of $\hat{P}_\pm$, it must be that $\hat{P}_\pm\psi = \frac{1}{2}[\psi(\mathbf{r}) \pm \psi(-\mathbf{r})]$. Hence, we find $\hat{P}_\pm = \frac{1}{2}(1 \pm \hat{I})$. We also have $\hat{P}_\pm^2 = \hat{P}_\pm$ and $\hat{P}_+ + \hat{P}_- = 1$.

Problem 1.34

Prove that the Hermitian operator, $\hat{F}$, from Problem 1.19, is proportional to a projection operator: i.e., $\hat{P} = c\hat{F}$, where c is a constant. On which state does $c\hat{F}$ project?

Solution

An operator $\hat{P}$ with the kernel $P(x, x') = c \cdot f(x)f^*(x')$ where $c^{-1} = \int |f(x)|^2 dx$ is a projection operator. It projects onto the state described by the wavefunction $\psi_0(x) \equiv f(x)$.

Problem 1.35

An Hermitian operator $\hat{f}$ has only N different eigenvalues. Find the explicit form of the projection operator $\hat{P}(f_i)$ for the states with a given value f_i of physical quantity f.

Solution

Let $N = 2$. From the condition $\hat{P}(f_1)\psi_{f_2} = 0$, it follows that $\hat{P}(f_1) = a(\hat{f} - f_2)$ and the relation $\hat{P}(f_1)\psi_{f_1} = \psi_{f_1}$ gives $a = (f_1 - f_2)^{-1}$. The result generalizes to an arbitrary N, as follows:

$$\hat{P}(f_i) = \prod_{k=1}^{N}{}' \frac{1}{f_i - f_k}(\hat{f} - f_k),$$

where the prime implies that the factor with $k = i$ is absent in the product.

1.4 Quantum-mechanical representations of operators and wave-functions; Unitary operators

Problem 1.36

Find the properly normalized eigenfunctions of the three-dimensional coordinate, $\psi_{\mathbf{r}_0}$, and the three-dimensional momentum, $\psi_{\mathbf{p}_0}$, in the coordinate and momentum representations.

Result:

$$\psi_{\mathbf{r}_0}(\mathbf{r}) = \delta(\mathbf{r} - \mathbf{r}_0), \quad \psi_{\mathbf{p}_0}(\mathbf{r}) = \frac{1}{(2\pi\hbar)^{3/2}} e^{\frac{i}{\hbar}\mathbf{p}_0 \cdot \mathbf{r}},$$

$$\phi_{\mathbf{r}_0}(\mathbf{p}) = \frac{1}{(2\pi\hbar)^{3/2}} e^{-\frac{i}{\hbar}\mathbf{r}_0 \cdot \mathbf{p}}, \quad \phi_{\mathbf{p}_0}(\mathbf{p}) = \delta(\mathbf{p} - \mathbf{p}_0).$$

Problem 1.37

Find the momentum-representation wavefunction of a particle in the state defined in Problem 1.13.

Result:

$$\phi(p) = \sqrt{\frac{a^2}{\hbar}} C \exp\left\{-\frac{i}{\hbar}(p-p_0)x_0 - \frac{a^2}{2\hbar^2}(p-p_0)^2\right\}.$$

Problem 1.38

Given a wavefunction, $\psi(x,y,z)$, find the probability of finding a particle with the z-coordinate in the interval, $z_1 < z < z_2$, and the y-component of momentum in the interval, $p_1 < p_y < p_2$.

Result

The sought probability is given by

$$w = \int_{z_1}^{z_2}\int_{p_1}^{p_2}\int_{-\infty}^{\infty} |F(x,p_y,z)|^2 dx\, dp_y\, dz,$$

where

$$F(x,p_y,z) = \frac{1}{(2\pi\hbar)^{1/2}} \int_{-\infty}^{\infty} \psi(x,y,z) e^{-\frac{i}{\hbar}p_y y} dy.$$

The wavefunction $\psi(\mathbf{r})$ is assumed to be normalized to unity.

Problem 1.39

Express the operators from Problem 1.1 in the momentum representation.

Solution

In the coordinate representation we have $\psi_2(x) = \hat{I}\psi_1(x) \equiv \psi_1(-x)$. Multiply these relations by $\psi_p^*(x) = \frac{1}{(2\pi\hbar)^{1/2}} \exp\left\{-\frac{i}{\hbar}px\right\}$, and integrate with respect to x. The resulting Fourier transform yields

$$\phi_2(p) = \hat{I}\phi_1(p) = \frac{1}{(2\pi\hbar)^{1/2}} \int \exp\left\{-\frac{i}{\hbar}px\right\}\psi_1(-x)dx, \tag{1}$$

where $\phi_{1,2}(p) = \int \psi_p^*(x)\psi_{1,2}(x)dx$ are wavefunctions in the momentum representation. Since the integral in (1) is equal to $\phi_1(-p)$, we have $\hat{I}\phi_1(p) \equiv \phi_1(-p)$: i.e., the operator $\hat{I}$ in the momentum representation remains the inversion operator.

Similarly, we find for the other operators:

$$\hat{T}_a \phi(p) = \exp\left\{\frac{i}{\hbar}pa\right\}\phi(p),$$

$$\hat{M}_c \phi(p) = \frac{1}{\sqrt{c}}\phi\left(\phi(p/c)\right),$$

$$\hat{K}\phi(p) = \phi^*(-p),$$

$$\hat{P}_{12}\phi(p_1, p_2) = \phi(p_2, p_1).$$

Problem 1.40

Find explicitly the inverse-momentum operator, $\hat{p}^{-1}$, in coordinate representation, and the inverse-coordinate operator, $\hat{x}^{-1}$, in momentum representation (in one-dimensional quantum mechanics).

Solution

Since $\hat{p}\hat{p}^{-1} = 1$, we have the equation $\frac{d}{dx}\left(\hat{p}^{-1}\psi(x)\right) = (i/\hbar)\psi(x)$. By integrating this over x from $-\infty$ to x we find the explicit form of the operator $\hat{p}^{-1}$ in the coordinate representation:

$$(\hat{p})^{-1}\psi(x) = \frac{i}{\hbar}\int_{-\infty}^{x}\psi(x')dx'. \tag{1}$$

On the other hand, by integrating from x to ∞ we obtain a slightly different expression:

$$(\hat{p})^{-1}\psi(x) = \frac{i}{\hbar}\int_{x}^{\infty}\psi(x')dx'. \tag{2}$$

The contradiction is resolved by noting that for the functions contained in the definition domain of the operator $\hat{p}^{-1}$, the two results coincide. To belong to this domain, the functions must satisfy the equation $\int_{-\infty}^{\infty}\psi(x)dx = 0$, which ensures that the function $\hat{p}^{-1}\psi(x)$ vanishes in the limits $x \to \pm\infty$, as required by the condition, $\int |\hat{p}^{-1}\psi(x)|^2 dx < \infty$.[10] (See Problem 1.29.)

It should be noted that the eigenfunctions of the inverse-momentum operator $\hat{p}^{-1}$ are also the eigenfunctions of the momentum operator, as we might expect.

The problem for the inverse-coordinate operator, $\hat{x}^{-1}$, is dual to the one we just solved, and we find

---

[10] In momentum representation we have $\hat{p}^{-1} \equiv \widehat{\frac{1}{p}}$, and this condition takes the form $\int_{-\infty}^{\infty} p^{-1}|\phi(p)|^2 dp < \infty$, which gives $\phi(0) = 0$, or equivalently, $\int_{-\infty}^{\infty}\psi(x)dx = 0$.

$$(\hat{x})^{-1}\phi(p) = -\frac{i}{\hbar}\int_{-\infty}^{p}\phi(p')dp' = \frac{i}{\hbar}\int_{p}^{\infty}\phi(p')dp', \quad \int_{-\infty}^{\infty}\phi(p)dp = 0.$$

(See also Problem 4.15, where the result above is used to solve the Schrödinger equation for a particle in the Coulomb potential in momentum representation.)

Problem 1.41

Find the relation between the kernels $L(\mathbf{r}, \mathbf{r}')$ and $L(\mathbf{p}, \mathbf{p}')$ of the same linear operator $\hat{L}$ in the **r**- and **p**-representations. (See also Problem 1.11.)

Result

$$L(\mathbf{p}, \mathbf{p}') = \frac{1}{(2\pi\hbar)^3}\int\int e^{i(\mathbf{p}'\cdot\mathbf{r}' - \mathbf{p}\cdot\mathbf{r})/\hbar} L(\mathbf{r}, \mathbf{r}') dV dV'$$

$$L(\mathbf{r}, \mathbf{r}') = \frac{1}{(2\pi\hbar)^3}\int\int e^{i(\mathbf{p}\cdot\mathbf{r} - \mathbf{p}'\cdot\mathbf{r}')/\hbar} L(\mathbf{p}, \mathbf{p}') d^3p\, d^3p'$$

Problem 1.42

Determine the form of the operators $\hat{r}^{-1}$ and $\hat{r}^{-2}$ in momentum representation.

Solution

In the coordinate representation the operator $\hat{G}_1 \equiv \hat{r}^{-1}$ has the following kernel: $G_1(\mathbf{r}, \mathbf{r}') = \frac{1}{r}\delta(\mathbf{r} - \mathbf{r}')$. Using the result of Problem 1.41, we find that the momentum-representation kernel is given by

$$G_1(\mathbf{p}, \mathbf{p}') = \frac{1}{(2\pi\hbar)^3}\int\int \frac{1}{r}e^{i(\mathbf{p}' - \mathbf{p})\cdot\mathbf{r}/\hbar}dV = \frac{1}{2\pi^2\hbar(\mathbf{p} - \mathbf{p}')^2}.$$

Similarly, we find for the operator $\hat{r}^{-2}$:

$$G_2(\mathbf{r}, \mathbf{r}') = \frac{1}{r^2}\delta(\mathbf{r} - \mathbf{r}'), \text{ and } G_2(\mathbf{p}, \mathbf{p}') = \frac{1}{4\pi\hbar^2|\mathbf{p} - \mathbf{p}'|}.$$

A useful exercise is to use the momentum-representation results above to prove that $\hat{G}_2 = \hat{G}_1\hat{G}_1$.

Problem 1.43

Given two Hermitian operators $\hat{A}$ and $\hat{B}$, find a relation between the eigenfunctions of the operator $\hat{A}$ in the B-representation and the eigenfunctions of the operator $\hat{B}$ in the A-representation. Provide examples to illustrate the result.

Solution

Let us denote by $\psi_{A_n}(q)$ and $\psi_{B_n}(q)$ the eigenfunctions of operators, $\hat{A}$ and $\hat{B}$, in the q-representation, and by $\psi(q)$ the wavefunction of an arbitrary state. The wavefunctions of this state in the A- and B-representations, denoted as $a(A_n)$ and $b(B_n)$ correspondingly, can be determined from

$$\psi(q) = \sum_n a(A_n)\psi_{A_n}(q), \quad a(A_n) = \int \psi^*_{A_n}\psi d\tau,$$
$$\psi(q) = \sum_m b(B_m)\psi_{B_m}(q), \quad b(B_m) = \int \psi^*_{B_m}\psi d\tau. \tag{1}$$

For simplicity, we only consider here the case where the spectra of the operators $\hat{A}$ and $\hat{B}$ are discrete and non-degenerate.

Setting $\psi = \psi_{B_k}$ in Eq. (1) above, we find its form in the A-representation as follows:

$$a_{B_k}(A_n) = \int \psi^*_{A_n}(q)\psi_{B_k}(q) d\tau_q. \tag{2}$$

The eigenfunction ψ_{A_n} in the B-representation is obtained from Eq. (2) by permuting the A and B indices:

$$b_{A_n}(B_k) = \int \psi^*_{B_k}(q)\psi_{A_n}(q) d\tau_q. \tag{3}$$

From relations (2) and (3) we have $a_{B_k}(A_n) = b^*_{A_n}(B_k)$. One possible example of this relation is given in Problem 1.36. As a corollary to this result we obtain the following relation between the probabilities $w_{B_k}(A_n) = |a_{B_k}(A_n)|^2 = |b_{A_n}(B_k)|^2 = w_{A_n}(B_k)$. (This result is used in Problems 3.14 and 3.33.)

Problem 1.44

Which of the operators considered in 1.1 are unitary operators?

Result

The operators $\hat{I}$, $\hat{T}_a$, $\hat{M}_c$, and $\hat{P}_{12}$ are unitary.

Problem 1.45

Consider a unitary operator that satisfies the equation $\hat{U}^2 = \hat{U}$. Find its explicit form.

Solution

From the equations $\hat{U}^2 = \hat{U}$ and $\hat{U}\hat{U}^\dagger = \hat{U}^\dagger\hat{U} = 1$ it follows that $\hat{U} = 1$.

Problem 1.46

An operator $\hat{U}$ is unitary. Can the operator $\hat{U}' = c\hat{U}$ (c is some complex constant) be unitary as well; and if so, what are the conditions on the constant, c?

Result

$|c| = 1$: *i.e.*, $c = e^{i\alpha}$, where α is a real parameter (c an overall $U(1)$ phase factor).

Problem 1.47

Prove that the product $\hat{U}_1 \hat{U}_2$ of two unitary operators is also a unitary operator.

Solution

From the equation $\hat{U} = \hat{U}_1 \hat{U}_2$, it follows that $\hat{U}^\dagger = \hat{U}_2^\dagger \hat{U}_1^\dagger$ and $\hat{U}\hat{U}^\dagger = \hat{U}^\dagger \hat{U} = 1$ (we took into account the unitarity of the operators $\hat{U}_{1,2}$).

Problem 1.48

Is it possible that some unitary operator (matrix) is at the same time an Hermitian one? Provide examples.

Solution

From the conditions of the unitarity $\hat{U}\hat{U}^\dagger = 1$ and of the Hermiticity $\hat{U}^\dagger = \hat{U}$ of the operator, it follows that $\hat{U}^2 = 1$. An operator with only two eigenvalues ± 1 has such a properties (compare with Problem 1.17). Examples: the inversion operator $\hat{I}$ and the particle exchange operator $\hat{P}_{12}$, from Problem 1.1; Pauli matrices (see Chapter 5).

Problem 1.49

Prove that the Hermitian and anti-Hermitian parts of any unitary operator commute with each other (consequently, the unitary operator can always be diagonalized). What are the properties of its eigenvalues (compare with Problem 1.50)?

Solution

Let us write $\hat{U} = \frac{1}{2}(\hat{U} + \hat{U}^\dagger) + \frac{1}{2}(\hat{U} - \hat{U}^\dagger)$. Since $\hat{U}\hat{U}^\dagger = \hat{U}^\dagger \hat{U} = 1$ we have $[(\hat{U} + \hat{U}^\dagger), (\hat{U} - \hat{U}^\dagger)] = 0$. Therefore, the Hermitian operators $(\hat{U} + \hat{U}^\dagger)$ and $(\hat{U} - \hat{U}^\dagger)/i$ and the operator $\hat{U}$ can be diagonalized simultaneously. The eigenvalues u_k of an operator $\hat{U}$ satisfy the equations $|u_k| = 1$.

Problem 1.50

Show that an operator of the form $\hat{U} = \exp(i\hat{F})$ is unitary, if $\hat{F}$ is an Hermitian operator. Express the unitary operators $\hat{I}$, $\hat{T}_a$, $\hat{M}_c$ from Problem 1.1 in such a form.

Solution

Since $\hat{U}^\dagger = \exp(-i\hat{F}^\dagger) = \exp(-i\hat{F})$, then $\hat{U}\hat{U}^\dagger = \hat{U}^\dagger\hat{U} = 1$. The eigenvalues u_k of an operator $\hat{U}$ are related to the eigenvalues, f_k, of the operator $\hat{F}$ as follows: $u_k = e^{if_k}$.

Therefore, we have:

$$\hat{I} = \exp\left\{i\frac{\pi}{2}(\hat{I} - 1)\right\},$$

$$\hat{T}_a = \exp\left\{i\frac{a}{\hbar}\hat{p}\right\},$$

$$\hat{M}_c = \exp\left\{i\frac{\ln c}{2\hbar}(\hat{x}\hat{p} + \hat{p}\hat{x})\right\}.$$

These relations follow from Problem 1.7 (and see also Problem 1.57).

Problem 1.51

Square matrices $\hat{A}$ and $\hat{A}'$ are of the same rank and are related by a unitary transformation as follows: $\hat{A}' = \hat{U}\hat{A}\hat{U}^\dagger$. Prove that the traces and determinants of these matrices are the same.

Solution

From the equations $\hat{U}^\dagger\hat{U} = 1$ and $\hat{A}' = \hat{U}\hat{A}\hat{U}^\dagger$, it follows that

$$\text{Tr }\hat{A}' = \text{Tr}(\hat{U}\hat{A}\hat{U}^\dagger) = \text{Tr}(\hat{A}\hat{U}^\dagger\hat{U}) = \text{Tr }\hat{A}.$$

Analogously,

$$\det \hat{A}' = \det(\hat{U}\hat{A}\hat{U}^\dagger) = \det(\hat{A}\hat{U}^\dagger\hat{U}) = \det \hat{A}.$$

Problem 1.52

Prove the relation

$$\det\|\exp \hat{A}\| = \exp(\text{Tr }\hat{A})$$

where $\hat{A}$ is an Hermitian matrix.

Solution

For any Hermitian matrix there exists a unitary transform that brings it to a diagonal form. In the new diagonal basis, where $\left(\exp \hat{A}\right)_{nm} = (\exp A_n)\,\delta_{nm}$, the above identity

Problem 1.53

becomes obvious. Combining this with the result of the previous problem, we prove the relation in an arbitrary representation.

What is the determinant of a unitary matrix, $\hat{U}$? Calculate it explicitly for the matrix $\hat{U} = \exp(i\hat{F})$, where $\hat{F}$ is an Hermitian matrix with a known spectrum. Prove that there always exists a transform $\hat{U}' = c\hat{U}$ that gives rise to a unimodular matrix with $\det \hat{U}' = 1$.

Solution

On the one hand we have $\det(\hat{U}\hat{U}^\dagger) = \det \hat{1} = 1$. But at the same time, $\det(\hat{U}\hat{U}^\dagger) = \det \hat{U} \cdot \det \hat{U}^\dagger$ and $\det \hat{U}^\dagger = (\det \hat{U})^*$. Hence, $|\det \hat{U}|^2 = 1$ or equivalently $\det \hat{U} = e^{i\alpha}$, where α is a real number. (The same result follows from the property of eigenvalues, u_k, of the unitary matrix; see Problem 1.50.)

For the operator $\hat{U} = e^{i\hat{F}}$ (according to Problem 1.52) we have $\det \hat{U} = \exp(i\mathrm{Tr}\hat{F})$.

If we consider the matrix $\hat{U}' = e^{-i\alpha/N}\hat{U}$, where N is its rank, then we indeed have $\det \hat{U}' = 1$.

Problem 1.54

Find the number of independent square matrices of rank N that are: a) Hermitian, b) unitary. How many unitary unimodular matrices (that is, matrices with the determinant equal to $+1$) of rank N exist?

Solution

There are N^2 linearly-independent matrices of rank N. There is the same number of independent Hermitian matrices of rank N. The number of independent unitary matrices is also N^2, since there is a relation between them and Hermitian matrices $\hat{U} = e^{i\hat{F}}$ (see Problem 1.50). For a unitary matrix to be unimodular, the relation, $\mathrm{Tr}\hat{F} = 0$, holds (see Problem 1.53). Therefore, the number of unimodular matrices, as well as the number of the Hermitian matrices $\hat{F}' = \hat{F} - N^{-1}\left(\mathrm{Tr}\,\hat{F}\right)\hat{1}$, with the trace equal to zero, is $N^2 - 1$.

Problem 1.55

Prove that the algebraic relations between operators of the form

$$F(\hat{A}_i) \equiv c_0 + \sum_i c_i \hat{A}_i + \sum_{i,k} c_{ik} \hat{A}_i \hat{A}_k + \cdots = 0$$

are invariant under an arbitrary unitary transformation of the operators $\hat{A}' = \hat{U}\hat{A}\hat{U}^\dagger$: i.e., prove that $F(\hat{A}'_i) = 0$.

Solution

$$\hat{F}' \equiv \hat{U}\hat{F}\hat{U}^\dagger = \hat{U}\left[c_0 + \sum_i c_i \hat{A}_i + \sum_{ik} c_{ik}\hat{A}_i\hat{A}_k + \ldots\right]\hat{U}^\dagger$$

$$= c_0 + \sum_i c_i \hat{U}\hat{A}_i\hat{U}^\dagger + \sum_{ik} c_{ik} \hat{U}\hat{A}_i\hat{A}_k\hat{U}^\dagger + \cdots = 0. \quad (1)$$

Taking into account that $\hat{U}^\dagger\hat{U} = 1$, we can write an arbitrary term in the sum (1) as follows:

$$c_{ik\ldots n}\hat{U}\hat{A}_i\hat{A}_k\ldots\hat{A}_n\hat{U}^\dagger = c_{ik\ldots n}\hat{U}\hat{A}_i\hat{U}^\dagger\hat{U}\hat{A}_k\hat{U}^\dagger\ldots\hat{U}\hat{A}_n\hat{U}^\dagger = c_{ik\ldots n}\hat{A}'_i\hat{A}'_k\ldots\hat{A}'_n.$$

Therefore, Eq. (1) takes the form

$$c_0 + \sum_i c_i\hat{A}'_i + \sum_{ik} c_{ik}\hat{A}'_i\hat{A}'_k + \cdots \equiv F(\hat{A}'_i) = 0,$$

which is identical to the initial expression.

Problem 1.56

Determine the transformation law for the operators $\hat{x}$ and $\hat{p}$ under the following unitary transforms: a) inversion operator $\hat{I}$, b) translation operator $\hat{T}_a$, c) scaling operator $\hat{M}_c$. The form of the operators is given in Problem 1.1.

Solution

The operators $\hat{x}' = \hat{U}\hat{x}\hat{U}^\dagger$ and $\hat{p}' = \hat{U}\hat{p}\hat{U}^\dagger$ are given by:

a) $\hat{x}' = -\hat{x}$, $\hat{p}' = -\hat{p}$;
b) $\hat{x}' = \hat{x} + a$, $\hat{p}' = \hat{p}$;
c) $\hat{x}' = c\hat{x}$, $\hat{p}' = \frac{1}{c}\hat{p}$.

These relations can be proven in the coordinate representation. For $\hat{U} = \hat{T}_a$, we have $\hat{U}^\dagger = \hat{T}^\dagger_a = \hat{T}_{-a}$ (see Problem 1.1), and

$$\hat{x}'\psi(x) = \hat{U}\hat{x}\hat{U}^\dagger\psi(x) = \hat{T}_a\hat{x}\hat{T}_{-a}\psi(x) = \hat{T}_a[x\psi(x-a)] = (x+a)\psi(x) = (\hat{x}+a)\psi(x).$$

Hence we have $\hat{x}' = \hat{x} + a$. Furthermore,

$$\hat{p}'\psi(x) = \hat{U}\left(-i\hbar\frac{\partial}{\partial x}\right)\hat{U}^\dagger\psi(x) = -i\hbar\hat{T}_a\frac{\partial}{\partial x}\hat{T}_{-a}\psi(x) = -i\hbar\hat{T}_a\frac{\partial}{\partial x}\psi(x-a) = -i\hbar\frac{\partial}{\partial x}\psi(x),$$

so $\hat{p}' = \hat{p}$. The other relations are obtained in a similar way.

Problem 1.57

A family of unitary operators, $\hat{U}(a)$, parameterized by a real continuous parameter, a, has the following properties $\hat{U}(0) = \hat{1}$ and $\hat{U}(a_3) = \hat{U}(a_1)\hat{U}(a_2)$, if $a_3 = a_1 + a_2$. Assume that for an infinitesimally small $\delta a \to 0$, there holds $\hat{U}(\delta a) \approx 1 + i\hat{F}\,\delta a$, and prove that $\hat{U}(a) = \exp(ia\hat{F})$. As an illustration, consider the operators $\hat{T}_a$ and $\hat{M}_c$ (see Problem 1.1), and find the corresponding generators of infinitesimally small transformations.

Solution

Set $a_1 = a$ and $a_2 = da \to 0$ in the relation $\hat{U}(a_1 + a_2) = \hat{U}(a_2)\hat{U}(a_1)$. Taking into account $\hat{U}(da) \approx 1 + ida\hat{F}$, we obtain

$$d\hat{U} = \hat{U}(a + da) - \hat{U}(a) \approx i\hat{F}\hat{U}(a)da. \tag{1}$$

Therefore, $\hat{U}(a)$ satisfies the differential equation, $d\hat{U}/da = i\hat{F}\hat{U}$, with the initial condition, $\hat{U}(0) = \hat{1}$. The exponential $\hat{U}(a) = e^{ia\hat{F}}$ solves this equation.

For an infinitesimal translation, we have

$$\hat{T}_{da}\psi(x) = \psi(x + da) \approx \left(1 + da\frac{\partial}{\partial x}\right)\psi(x).$$

Therefore, the corresponding generator is proportional to the momentum operator $\hat{F}_{T_a} \equiv \hat{p}_x/\hbar = -i\frac{\partial}{\partial x}$, and the operator of finite translations is $\hat{T}_a = e^{a\partial/\partial x}$.

In the case of the operator $\hat{M}_c$ let us first introduce $c = e^a$ and write $\hat{M}_c \equiv \hat{M}(a)$. The dependence of $\hat{M}(a)$ on a satisfies the conditions of the problem. So, we have

$$\hat{M}(da)\psi(x) \equiv e^{da/2}\psi(e^{da}x) \approx \left[1 + da\left(\frac{1}{2} + x\frac{\partial}{\partial x}\right)\right]\psi(x),$$

and we find the generator $i\hat{F} = \left(\frac{1}{2} + x\frac{\partial}{\partial x}\right)$ and the exponential expression for the scaling operator as follows, $\hat{M}_c = \exp\left\{\frac{-i}{2}\ln c \cdot \left(ix\frac{\partial}{\partial x} + i\frac{\partial}{\partial x}x\right)\right\}$ (see Problem 1.7).

2

One-dimensional motion

The time-independent Schrödinger equation

$$\hat{H}\psi_E \equiv \left[-\frac{\hbar^2}{2m}\frac{d^2}{dx^2} + U(x)\right]\psi_E(x) = E\psi_E(x) \tag{II.1}$$

supplemented by the appropriate boundary conditions determines the energy spectrum and wavefunctions of the stationary states of a particle moving in the potential field, $U(x)$.

The energy spectrum, E_n, in the domain $\min U(x) < E_n < U(\pm\infty)$ is always discrete, which corresponds to the finiteness of classical motion.[11] These energy levels, E_n, are non-degenerate, and the corresponding eigenfunctions, $\psi_n(x)$, are square-integrable (they describe localized states of the particle and correspond to finite motion in classical theory).

For a linear oscillator, $U(x) = kx^2/2$, $\omega = \sqrt{k/m}$, the spectrum is given by $E_n = \hbar\omega\left(n + \frac{1}{2}\right)$, and the corresponding eigenfunctions are as follows:

- $$\psi_n^{(\text{osc})}(x) = \left(\frac{1}{\pi a^2}\right)^{1/4} \frac{1}{\sqrt{2^n n!}} \exp\left(-\frac{x^2}{2a^2}\right) H_n\left(\frac{x}{a}\right), \tag{II.2}$$

where $a = \sqrt{\hbar/m\omega}$ and $H_n(z)$ are the Hermite polynomials; for example, $H_0(z) = 1$, $H_1(z) = 2z$, $H_2(z) = 4z^2 - 2$, etc. The non-zero matrix elements of the coordinate operator are

$$x_{n,n+1} = x_{n+1,n} = \sqrt{(n+1)/2}\,a, \tag{II.3}$$

and the non-zero matrix elements of the momentum operator are given by $p_{nk} = im\omega_{nk}x_{nk}$, where $\omega_{nk} = \pm\omega$ for $n = k \pm 1$.

For an arbitrary potential, $U(x)$, the spectrum with the energies $E > \min U(\pm\infty)$ is continuous. Such continuum states with energy $E > \max U(\pm\infty)$ are doubly degenerate, which corresponds to classical motion extending to infinity in both directions: $x \to -\infty$ and $x \to +\infty$. In this case, the choice of independent solutions of the

---

[11] We label discrete energy levels as E_n and the corresponding eigenfunctions as ψ_n (the ground state is labelled by $n = 0$). Note that in this enumeration convention, n also gives the number of zeros of the eigenfunction, $\psi_n(x)$ (except, possibly, zeros at $x = \pm\infty$). For a particle in a spherically-symmetric potential, the label n_r is used to label the states.

Schrödinger equation (II.1) is usually made on physical grounds to represent particles reflected by and transmitted through a potential, and these wavefunctions are uniquely defined by their asymptotic form at $x \to \pm\infty$. For example, reflection and transmission of quantum particles propagating from the left towards a potential can be described by the following wavefunction:

$$\psi^+_{k_1}(x) \approx \begin{cases} e^{ik_1 x} + A(E) e^{-ik_1 x}, & x \to -\infty, \\ B(E) e^{ik_2 x}, & x \to +\infty, \end{cases} \quad (\text{II.4})$$

where $k_{1,2} = \sqrt{2m[E - U(\mp\infty)]}/\hbar^2$. The amplitudes $A(E)$ and $B(E)$ determine the transmission coefficient, $D(E) = \frac{k_2}{k_1}|B|^2$, and the reflection coefficient, $R(E) = |A|^2$, of the particles. These coefficients have the following properties:

$$D(E) + R(E) = 1; \quad D_+(E) = D_-(E);$$
$$D(E) \to 1 \text{ at } E \to \infty; \quad (\text{II.5})$$
$$D(E) \to 0 \text{ at } E \to \max U(\mp\infty).$$

The second symmetry relation, $D_+(E) = D_-(E)$, implies independence of the transmission coefficient with a certain energy E of the initial direction of particle propagation. The latter property is also discussed in Problems 2.37 and 2.39.

2.1 Stationary states in discrete spectrum

Problem 2.1

Find energy levels and normalized wavefunctions of stationary states of a particle in the infinitely deep potential well of the width, a (i.e., $U(x) = 0$, if $0 < x < a$, and $U(x) = \infty$, if $x < 0$ or $x > a$). Find also the mean value and the standard deviation of the coordinate and momentum of the particle.

In the state descried by the wavefunction, $\psi = Ax(x - a)$, for $0 < x < a$, find the probability distribution function of the particle energy and its mean value.

Solution

1) The energy levels and the eigenfunctions are as follows:

$$E_n = \frac{\pi^2 \hbar^2}{2ma^2}(n+1)^2, \quad \psi_n(x) = \sqrt{\frac{2}{a}} \sin\frac{\pi(n+1)x}{a}, \quad 0 < x < a,$$

where $n = 0, 1, \ldots$ ($\psi \equiv 0$ for $x < 0$ and for $x > a$).

The average coordinate and momentum in the nth state are given by

$$\bar{x} = \frac{a}{2}, \quad \overline{(\Delta x)^2} = a^2 \left[\frac{1}{12} - \frac{1}{2\pi^2(n+1)^2}\right],$$

$$\bar{p} = 0, \quad \overline{(\Delta p)^2} = \frac{\hbar^2 \pi^2 (n+1)^2}{a^2}.$$

2) Normalization condition for the given wavefunction yields $A = \sqrt{30/a^5}$. From Eq. (I.4) we find coefficients c_n in its expansion into a series in terms of the eigenfunctions, ψ_n, as follows:

$$c_n = \frac{\sqrt{60}}{a^3} \int_0^a x(x-a) \sin \frac{\pi(n+1)x}{a} dx = -\frac{\sqrt{240}}{\pi^3} \frac{1 + (-1)^n}{(n+1)^3}.$$

These coefficients determine the probability of finding the particle in the nth quantum state and the probability of the corresponding energy values, E_n: $w(E_n) = |c_n|^2$; specifically, $w_0 \approx 0.999$. Finally, from relation (I.5) we have $\overline{E} = 5\hbar^2/ma^2 \approx 1.013 E_0$.

See also Problem 8.23.

Problem 2.2

Find the change in energy levels and wavefunctions of a charged linear oscillator in a uniform electric field applied along the oscillation axis. Find the *polarizability* of the oscillator in these eigenstates.[12]

Solution

The potential energy of the charged oscillator in the uniform electric field, $\mathcal{E}$, reads

$$U = kx^2/2 - e\mathcal{E}x.$$

Substitution $z = x - e\epsilon_0/k$ leads to the Schrödinger equation, which is identical to that for the unperturbed linear oscillator. This determines the energy spectrum and eigenfunctions as follows:

$$E_n = \hbar\omega\left(n + \frac{1}{2}\right) - \frac{1}{2}\frac{e^2}{k}\mathcal{E}^2, \quad n = 0, 1, \ldots$$

$$\psi_n(x) = \psi_n^{(osc)}(z) = \psi_n^{(osc)}\left(x - \frac{e}{k}\mathcal{E}\right), \quad \omega = \sqrt{\frac{k}{m}}$$

(see Eq. (II.2)). This shows that as well as in the classical case, the effect of a uniform field on a harmonic oscillator reduces to a shift of its equilibrium point. Polarizabilities of all stationary states of an oscillator therefore are the same and have the form, $\beta_0 = e^2/m\omega^2$.

---

[12] Recall that the polarizability, β_0, determines the mean dipole moment, $\mathbf{d} \approx \beta \mathcal{E}$, induced by the weak external electric field. It also determines the quadratic term, $\Delta E = -\beta_0 \mathcal{E}^2/2$, of the energy shift in such a field.

Problem 2.3

Calculate the expectation value of the particle energy, $\overline{E}(\alpha)$, in the state, $\psi(x,\alpha) = \sqrt{\alpha} e^{-\alpha|x|}$ (with $\alpha > 0$), to show that there exists at least one bound state with negative energy, $E_0 < 0$, for any one-dimensional potential,[13] $U(x)$, such that $U(x) \to 0$ for $x \to \pm\infty$ and $\int U(x) dx < 0$.

Solution

Let us first prove that $\overline{E}(\alpha) < 0$ for small enough values of α. Since $E_0 \leq \overline{E}$ where E_0 is the ground-state energy, this will automatically prove the existence of a negative-energy state. We easily find that $\overline{T} = \frac{1}{2m}\overline{p^2} = \frac{\hbar^2 \alpha^2}{2m} \propto \alpha^2$, and $\overline{U} \approx \alpha \int U(x) dx \propto \alpha$, while $\alpha \to 0$, so indeed, $\overline{E}(\alpha) \approx \overline{U} < 0$.

Problem 2.4

E_n and $\tilde{E}_n$ are the energies of the nth level in the potentials $U(x)$ and $\tilde{U}(x) = U(x) + \delta U(x)$, correspondingly. Assuming that $\delta U(x) \geq 0$, prove that $\tilde{E}_n \geq E_n$.

Solution

Let $E_n(\lambda)$ and $\psi_n(x,\lambda)$ be the energy levels and the corresponding eigenfunctions of the Hamiltonian, $\hat{H}(\lambda) = \frac{\hat{p}^2}{2m} + U(x) + \lambda \delta U(x)$. From Eq. (I.6) we have

$$\frac{d}{d\lambda} E_n(\lambda) = \int \delta U(x) |\psi_n(x,\lambda)|^2 dx \geq 0.$$

Since $E_n = E_n(\lambda = 0)$ and $\tilde{E}_n = E_n(\lambda = 1)$, the statement of the problem is proven.

Problem 2.5

Consider a symmetric potential, $U(x)$, and the same potential in the half-space, $x > 0$,

$$\tilde{U}(x) = \begin{cases} U(x), & x > 0; \\ \infty, & x < 0, \end{cases}$$

as shown in Fig. 2.1. Find the relation between the energy levels of bound states and the corresponding normalized wavefunctions for a particle in the potentials $U(x)$ and $\tilde{U}(x)$.

Solution

The energy levels, E_n, in the case of the symmetrical potential, $U(x)$, have a definite parity equal to $(-1)^n$. For the odd-parity states with $x \geq 0$, the Schrödinger equation and the boundary conditions $\psi(0) = \psi(\infty) = 0$ are satisfied for the potential, $\tilde{U}(x)$, as

---

[13] Compare with the results of Problems 4.21 and 4.33.

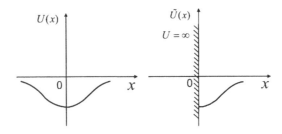

Fig. 2.1

well. Therefore, for odd values of n, the energy spectrum, $\tilde{E}_n$, for the two potentials is identical, and the normalized eigenfunctions differ by a numerical normalization constant only:

$$\tilde{E}_n = E_{2n+1},$$
$$\tilde{\psi}_n(x) = \sqrt{2}\psi_{2n+1}(x),$$
$$x \geq 0, \ n = 0, \ 1, \ldots,$$

where we have taken into account that the even and odd levels alternate and the lowest level is even (see Fig. 2.2).

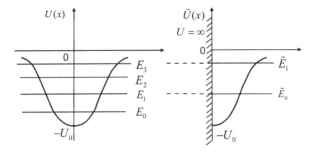

Fig. 2.2

Problem 2.6

A potential has the form $U(x) = \tilde{U}(x) + \alpha\delta(x - x_0)$ where $\delta(x)$ is the Dirac δ-function, while $\tilde{U}(x)$ is some finite function. Find the behavior of the solution $\psi_E(x)$ of the Schrödinger equation and its derivative at the point x_0.

Solution

From the Schrödinger equation,

$$-\frac{\hbar^2}{2m}\psi_E''(x) + [\tilde{U}(x) + \alpha\delta(x - x_0)]\psi_E(x) = E\psi_E(x), \tag{1}$$

it follows that the wavefunction, $\psi_E(x)$, is continuous everywhere, including at x_0, while its derivative has a discontinuity at x_0. The jump of $\psi'_E(x)$ at $x = x_0$ must have such a value that the δ-term in the $\psi''_E(x)$ (the derivative of the discontinuous function is proportional to the δ-function) compensates the term $\alpha\delta(x - x_0)\psi_E(x_0)$ in the left-hand side of Eq. (1). By integrating (1) over a narrow interval $x_0 - \varepsilon \leq x \leq x_0 + \varepsilon$, and taking the limit $\varepsilon \to 0$, we find:

$$\delta\psi'_E(x_0) \equiv \psi'_E(x_0 + 0) - \psi'_E(x_0 - 0) = \frac{2m\alpha}{\hbar^2}\psi_E(x_0), \quad \psi_E(x_0 + 0) = \psi_E(x_0 - 0). \quad (2)$$

Problem 2.7

Find the energy level(s) of the discrete states and the corresponding normalized wavefunction(s) of a particle in the δ-potential well,[14] $U(x) = -\alpha\delta(x)$ (see Fig. 2.3). Find the expectation values of the kinetic and potential energy in these states. Determine the product of the uncertainties of the coordinate and momentum. Find the wavefunction in momentum representation.

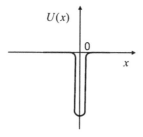

Fig. 2.3

Solution

1) The physically-meaningful solution[15] of the Schrödinger equation with $U(x) = -\alpha\delta(x)$ has the form: $\psi = Ae^{-\kappa x}$, for $x > 0$, and $\psi = Be^{\kappa x}$, for $x < 0$, where $\kappa = (-2mE/\hbar^2)^{1/2} > 0$. Using the relations (2) from the previous problem (with the appropriate change, $\alpha \to -\alpha$) we find that $A = B$ and $\kappa = m\alpha/\hbar^2$. It follows from this equation that for $\alpha < 0$ (δ-barrier), there are no bound states, and for $\alpha > 0$ (δ-well), there is only one bound state with $E_0 = -\frac{m\alpha^2}{2\hbar^2}$. The corresponding normalized wavefunction has the form:

$$\psi_0(x) = \sqrt{\kappa_0}e^{-\kappa_0|x|}, \quad \kappa_0 = \frac{m\alpha}{\hbar^2}.$$

---

[14] An attractive δ-potential in one dimension provides a faithful representation of a shallow potential well, $U(x)$, of an arbitrary shape; that is, the potential, $U(x)$, with $ma^2U_0/\hbar^2 \ll 1$, where U_0 and a are the typical value of the potential and its range, and $\alpha = -\int U(x)dx < 0$. See also Problems 2.17, 2.20, and 2.23.

[15] Solutions of the Schrödinger equation that increase exponentially as $x \to \pm\infty$ have been omitted here.

2) The mean values of the coordinate and momentum are given by:

$$\overline{U} = 2E_0, \quad \overline{T} = -E_0, \quad \overline{x} = 0, \quad \overline{(\Delta x)^2} = \frac{1}{2\kappa_0^2}, \quad \overline{p} = 0, \quad \overline{(\Delta p)^2} = \hbar^2 \kappa_0^2,$$

3) The ground-state wavefunction in the momentum representation is:

$$\phi_0(p) = \frac{1}{\sqrt{2\pi\hbar}} \int_{-\infty}^{\infty} e^{-ipx/\hbar} \psi_0(x) dx = \frac{\sqrt{2\kappa_0^3 \hbar^3}}{\sqrt{\pi}(p^2 + \hbar^2 \kappa_0^2)}. \qquad (3)$$

Compare with Problem 2.17.

Problem 2.8

Find the energy spectrum and wavefunctions of the stationary states of a particle in the potential shown in Fig. 2.4.

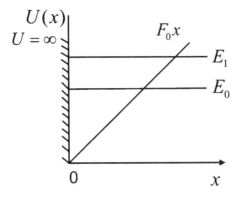

Fig. 2.4

Solution

By making the change of variables, $z = \beta(x - E/F_0)$, with $\beta = (2mF_0/\hbar^2)^{1/3}$, we transform the Schrödinger equation for $x \geq 0$ into the form $\psi''(z) - z\psi(z) = 0$. Its solution that decreases, as z (and x) $\to +\infty$, is given by the Airy function, $\mathrm{Ai}(z)$. Therefore, $\psi(x) = c\mathrm{Ai}[\beta(x - E/F_0)]$, and the boundary condition $\psi(0) = c\mathrm{Ai}(-\beta E/F_0) = 0$ determines the energy spectrum. Denoting by $-\alpha_k$ (with $k = 1, 2, \ldots$), the sequence of zeros for the Airy functions (they can be proven to be all negative) in increasing order, α_k, we find the energy levels as follows:

$$E_n = \left(\frac{\hbar^2 F_0^2}{2m}\right)^{1/3} \alpha_{n+1}, \quad n = 0, 1, \ldots. \qquad (1)$$

In particular, $\alpha_1 \approx 2.338$, and the ground-state energy is $E_0 \approx 1.856 \left(\hbar^2 F_0^2/m\right)^{1/3}$.

Problem 2.9

Find the energy levels of the discrete spectrum and the corresponding wavefunctions of a particle in the potential, $U(x) = U_0(e^{-2x/a} - be^{-x/a})$, where $U_0 > 0$, $a > 0$, and $b > 0$.

Solution

We introduce a new variable, $z = 2\beta e^{-x/a}$, and a function, $w(z)$, defined by $\psi(x) = z^{\kappa a} e^{-z/2} w(z)$ with $\kappa = (-2mE/\hbar^2)^{1/2}$ and $\beta = (2mU_0 a^2/\hbar^2)^{1/2}$. In these variables, the Schrödinger equation takes the form of a hypergeometric equation:

$$zw'' + (1 + 2\kappa a - z)w' + \left(-\kappa a - \frac{1}{2} + \frac{\beta b}{2}\right)w = 0. \tag{1}$$

Since the function $\psi(x) \propto e^{-\kappa x} \propto z^{\kappa a}$ tends to zero, as $x \to +\infty$ (or $z \to 0$), the solution to Eq. (1) must be chosen in the form (the other solution diverges in the limit $z \to 0$):

$$w(z) = cF\left(\kappa a + \frac{1}{2} - \frac{\beta b}{2}, 1 + 2\kappa a, z\right). \tag{2}$$

The condition of decreasing $\psi(x)$ as $x \to -\infty$ (or $z \to +\infty$) demands that the function $F(\alpha, \beta, z)$ in (2) reduces to a polynomial. It gives the spectrum:

$$\alpha \equiv \kappa_n a + \frac{1}{2} - \frac{\beta b}{2} = -n, \; n = 0, 1, \ldots, \left[\frac{b\beta}{2} - \frac{1}{2}\right],$$

or

$$E_n = -\frac{\hbar^2 \kappa_n^2}{2m} = -\frac{\hbar^2}{2ma^2}\left[\left(\frac{mb^2 a^2 U_0}{2\hbar^2}\right)^{1/2} - \left(n + \frac{1}{2}\right)\right]^2.$$

The condition $\sqrt{\frac{mb^2 a^2 U_0}{2\hbar^2}} = (N - \frac{1}{2})$ determines the parameters of the potential that correspond to the appearance of a new Nth discrete level with increasing depth of the potential well.

Problem 2.10

The same as in the previous problem, but for the potential

$$U(x) = \frac{U_1}{(1 + e^{x/a})^2} - \frac{U_2}{1 + e^{x/a}}; \; U_{1,2} > 0, \; a > 0.$$

Solution

A discrete spectrum exists only if $U_1 > U_2/2$ (otherwise the potential has no minimum) and for $E < \min(0, U_1 - U_2)$. To solve the Schrödinger equation, we make the change of variable, $z = -e^{x/a}$ and $\psi = (1-z)^{-\varepsilon} z^{\mu} w(z)$. Then the Schrödinger equation takes the form ($\kappa_{1,2}^2 = 2mU_{1,2}/\hbar^2$, $\kappa = (-2mE/\hbar^2)^{1/2}$):

$$z^2 w''(z) + \left[\frac{2\varepsilon z^2}{1-z} + (2\mu+1)z\right] w'(z) + \left[\frac{\varepsilon(\varepsilon+1)z^2 - (\kappa_1 a)^2}{(1-z)^2} + \right.$$

$$\left. \frac{\varepsilon(2\mu+1)z + (\kappa_2 a)^2}{1-z} - (\kappa a)^2 + \mu^2 \right] w(z) = 0. \tag{1}$$

If we take the parameters ε and μ in the form

$$\varepsilon = -\frac{1}{2} + \left(\frac{1}{4} + \kappa_1^2 a^2\right)^{1/2}, \quad \mu = \sqrt{(\kappa^2 + \kappa_1^2 - \kappa_2^2)a^2}$$

we convert Eq. (1) into the standard hypergeometric equation:

$$z(1-z)w'' + [2\mu + 1 - (2\mu - 2\varepsilon + 1)z]w' - [\mu^2 + \kappa_1^2 a^2 - 2\varepsilon\mu - \varepsilon - \kappa^2 a^2]w = 0 \tag{2}$$

with the parameters

$$\alpha = \mu - \varepsilon + \kappa a, \quad \beta = \mu - \varepsilon - \kappa a, \quad \gamma = 1 + 2\mu.$$

Since $\psi(x)$, as $x \to -\infty$ ($z \to 0$), has the form $\psi \propto e^{\mu x/a} = z^{\mu}$, we should choose the solution of Eq. (2) in the form $w = cF(\alpha, \beta, \gamma, z)$. Then

$$\psi = c \frac{1}{(1-z)^{\varepsilon}} z^{\mu} F(\alpha, \beta, \gamma, z). \tag{3}$$

From this equation, it follows that in the limit z (and x) $\to -\infty$,

$$\psi \approx \bar{c} z^{-\varepsilon+\mu} \left[\frac{\Gamma(\gamma)\Gamma(\beta-\alpha)}{\Gamma(\beta)\Gamma(\gamma-\alpha)} \frac{1}{(-z)^{\alpha}} + \frac{\Gamma(\gamma)\Gamma(\alpha-\beta)}{\Gamma(\alpha)\Gamma(\gamma-\beta)} \frac{1}{(-z)^{\beta}}\right]. \tag{4}$$

Since $z^{-\varepsilon+\mu-\beta} = e^{\kappa x}$ increases as $x \to +\infty$, then it requires: that $\alpha = -n$, with n being an integer that labels the energy spectrum:

$$\sqrt{-E_n} + \sqrt{U_1 - U_2 - E_n} =$$

$$= \sqrt{U_1 + \frac{\hbar^2}{8ma^2}} - \sqrt{\frac{\hbar^2}{2ma^2}} \left(n + \frac{1}{2}\right), \quad n = 0, 1, \ldots. \tag{5}$$

Note also that if $U_1 = U_2 \equiv U_0$, the potential turns into $U = -\frac{U_0}{4\text{ch}^2(x/2a)}$.

Problem 2.11

Find the energy spectrum of a particle in the potential $U(x) = \alpha\delta(x)$, $\alpha > 0$ for $|x| < a$ and $U = \infty$ for $|x| > a$ (see Fig. 2.5). Prove that in the limit, $m\alpha a/\hbar^2 \gg 1$, the low-energy part of the spectrum consists of a set of closely-positioned pairs of energy levels. What is the spectrum of the highly excited states of the particle? What is the structure of the energy levels with $\alpha < 0$?

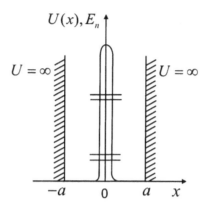

Fig. 2.5

Solution

1) The stationary states in this symmetric potential have a *definite* parity. The wavefunctions of *even* levels for $0 < |x| \leq a$ have the form $\psi^+ = A\sin(k(|x|-a))$ (the boundary condition $\psi(a) = 0$ is taken into account), where $k = \sqrt{2mE/\hbar^2}$. The matching conditions at $x = 0$ (see Eq. (2) in Problem 2.6) give the equation that determines the energy spectrum of the even levels:

$$\tan ka = -\frac{ka}{\xi}, \quad \xi = \frac{m\alpha a}{\hbar^2}. \tag{1}$$

In the limit $\xi \gg 1$, the right-hand side of Eq. (1) is small for the low-lying levels (with $ka \ll \xi$). Therefore, $k_n a = n\pi - \varepsilon$, where $\varepsilon \ll 1$ and $n = 1, 2, \ldots$. From (1), we have $\varepsilon \approx n\pi/\xi$, so

$$E_n^+ \approx \frac{\pi^2\hbar^2 n^2}{2ma^2}\left(1 - \frac{2}{\xi}\right)$$

(the index "+" means that the level is even).

For the *odd* levels, the wavefunctions have the form $\psi^-(x) = B\sin kx$, and the conditions $\psi^-(\pm a) = 0$ give the energy spectrum:

$$E_n^- = \frac{\pi^2\hbar^2 n^2}{2ma^2}, \quad n = 1, 2, \ldots$$

(a particle is not affected by the δ-potential at all in an odd state).

A comparison of E_n^+ and E_n^- confirms the properties of low-lying pairs of levels, as stated in the problem.

2) For the energies $ka \gg \xi$ from (1), and if we put $k_n a = (n - 1/2)\pi + \varepsilon$, $\varepsilon \ll 1$, it is easy to find the spectrum of even levels:

$$E_n^+ \approx \frac{\pi^2 \hbar^2 (2n-1)^2}{8ma^2} + \frac{\alpha}{a}. \tag{2}$$

Here, the first term corresponds to the even energy levels in an infinitely deep potential well of the width, $2a$, while the second term determines their shift under the action of the potential, $U(x) = \alpha \delta(x)$.

3) For the case of $\alpha > 0$ in each pair of close levels, the lower one is even and

$$\delta E_n = E_n^- - E_n^+ \approx \left(\frac{2\hbar^2}{m\alpha a}\right) E_n^\pm > 0. \tag{3}$$

For the case of $\alpha < 0$, we have a different situation – now the higher-energy level is even. But in this case, the low-lying part of the spectrum also contains an additional single even level with the energy $E_0^+ = -\kappa_0^2 \hbar^2/2m$ and wavefunctions $\psi_0^+ \approx \sqrt{\kappa_0} e^{-\kappa_0|x|}$, where $\kappa_0 = m|\alpha|/\hbar^2$. This level corresponds to a particle that is "bound" by the δ-well, $U = -|\alpha|\delta(x)$ (see Problem 2.7).

Problem 2.12

Generalize the results of the previous problem to the case of a δ-barrier separating the square well in a non-symmetrical manner.

Solution

The solution of the Schrödinger equation has the form $\psi(x) = A \sin k(x + a)$ for $-a \leq x < 0$ and $\psi(x) = B \sin k(x - b)$ for $0 < x < b$, where $k = \sqrt{2mE/\hbar^2}$. Here, the boundary conditions $\psi(-a) = \psi(b) = 0$ have been used. The matching conditions at $x = 0$ (see Eqs. (1) and (2) in Problem 2.6) lead to

$$A \sin ka = -B \sin kb, \quad B \cos kb - A \cos ka = \frac{2m\alpha}{k\hbar^2} A \sin ka.$$

This yields an equation that determines the energy spectrum:

$$\sin k(a+b) = -\frac{2m\alpha}{k\hbar^2} \sin ka \cdot \sin kb. \tag{1}$$

Note that if $b = a$, Eq. (1) becomes identical to the corresponding equation in Problem 2.11.

Let us point out some properties of the spectrum.

1) In the energy range where $m\alpha/k\hbar^2 \ll 1$, the right-hand side of equation (1) is small and therefore $k_n(a+b) \approx \pi(n+1)$, as in the case of "free" motion of a particle in the well of width $(a+b)$.

2) In the opposite case, where $m\alpha/k\hbar^2 \gg 1$, the product of the sine-functions in Eq. (1) is small, so $k_{n_1} \approx \pi(n_1+1)/a$ or $k_{n_2} \approx \pi(n_2+1)/b$. In this case, the spectrum represents a superposition of the spectra that correspond to independent motion of a particle in the left and right wells with widths a and b.

Problem 2.13

Investigate the asymptotic behavior at $x \to \pm\infty$ of the zero-energy solution $E = 0$ to the Schrödinger for the potential that $U(x) \to 0$ for $x \to \pm\infty$. Show that the zero-energy state, $\psi_{E=0}(x)$, that has the property of not increasing at $x \to \pm\infty$, exists only for special values of the parameters of the potential (that correspond to the appearance of a new bound state).

What is the number of discrete levels, N_{bound}, of a particle in

a) a rectangular potential well of the depth U_0 and width a,
b) the potential, $U(x) = -\alpha\delta(x) - \alpha\delta(x-a)$,

expressed in terms of the parameters of these potentials?

Solution

If the potential, $U(x)$, decreases sufficiently rapidly,[16] then in the limit $x \to \pm\infty$, the Schrödinger equation and its solution take the form $\psi_{E=0}'' = 0$, $\psi = A_\pm + B_\pm x$, i.e., the solution actually increases. In the case of an arbitrary potential there exists no solution of Schrödinger equation that does not increase at both $x \to +\infty$ and $x \to -\infty$ (also, there is no such solution with $E < 0$ that would decrease at $x \to \pm\infty$). Such solutions do exist, however, for exceptional values of the potential parameters that correspond to the appearance of new discrete states with increasing well depth.

To prove the above statements, let us examine the highest level in the discrete spectrum, E_n. Its wavefunction has the form, $\psi_n \propto e^{-\kappa|x|}$, at $x \to \pm\infty$ with $\kappa = \sqrt{2m|E_n|/\hbar^2}$. As the well depth decreases, all the levels shift up, and for some critical depth, the highest energy level reaches the value $E_n = 0$. Its wavefunction tends to a constant, $\psi_n \to const$, as $x \to \pm\infty$. The number of zeros of the wavefunction is equal to the number of existing discrete spectrum states whose energy is negative, $E_n < 0$. As an illustration, consider a free particle. The Schrödinger equation in this case has a bounded solution $\psi_{E=0}$ =const which has no zeros. In accordance with the discussion above, it follows that an arbitrarily shallow well binds the particle (see Problem 2.3).

a) Let us find the condition for the appearance of a new discrete state. The solution of the Schrödinger equation with $E = 0$ that does not increase at infinity has the form, $\psi = A$ for $x < 0$, $\psi(x) = B\cos(\gamma x + \delta)$ for $0 < x < a$ (the well domain), where

---

[16] It is necessary that the potential decreases faster than $\propto 1/x^2$. In the case of an attractive potential, which decreases faster than a power law, $U(x) \approx -\alpha/x^s$ for $x \to \infty$ with $s \leq 2$, the solution to the Schrödinger equation for $E = 0$ has a completely different asymptotic behavior (see Problems 9.9 and 9.14). If an attractive potential decreases slowly, the number of bound states diverges due to level condensation with $E_n \to -0$.

$\gamma = \sqrt{2mU_0/\hbar^2}$ and $\psi(x) = C$ for $x > a$. The continuity of the wavefunction and its derivative at $x = 0$ and $x = a$ points gives $A = B$, $\delta = 0$, $\gamma a = \pi n$, where n is an integer; $C = (-1)^n B$. This wavefunction has n zeros (as the argument of the cosine-function varies from 0 to πn), and therefore the condition $\gamma a = \pi n$ corresponds to the appearance of the $(n+1)$th level. Hence, it follows that the number of bound states, N_{bound}, inside the well is $\frac{\gamma a}{\pi} < N_{\text{bound}} < \frac{\gamma a}{\pi} + 1$.

b) There is one bound state if $0 < m\alpha a/\hbar^2 < 1$, and two bound states if $m\alpha a/\hbar^2 > 1$.

Problem 2.14

What is the number of bound states of a particle confined in the potential, $U(x)$, of the form a) $U = \infty$, for $x < 0$, $U = -U_0$, for $0 < x < a$, and $U = 0$, for $x > a$ (see Fig. 2.6 a) b) $U = \infty$, for $x < 0$, and $U = -\alpha \delta(x - a)$, for $x > 0$ (see Fig. 2.6 b) Express your answer in terms of the parameters, U_0, a, and α.

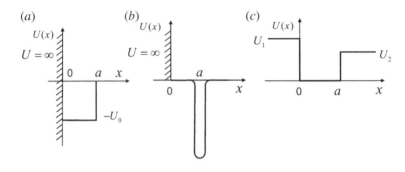

Fig. 2.6

Result

a) $\frac{\sqrt{2mU_0 a^2}}{\pi \hbar} - \frac{1}{2} < N_{\text{bound}} < \frac{\sqrt{2mU_0 a^2}}{\pi \hbar} + \frac{1}{2}$;

b) There is a single bound state that appears if $m\alpha a/\hbar^2 \geq 1/2$, and no bound state otherwise.

Problem 2.15

Find the condition for the existence of a bound state inside the potential well in Fig. 2.6c. Consider also the following limiting cases: a) $U_1 = \infty$ and b) $U_1 = U_2$.

Solution

The bound states correspond to $E \leq U_2$. The signature of the appearance of a new bound state is the existence of a solution to the Schrödinger equation with $E = U_2$ that does not increase at $x \to \pm \infty$ (see Problem 2.13). The corresponding condition has the form:

$$\tan\sqrt{\frac{2mU_2a^2}{\hbar^2}} = \sqrt{\frac{U_1}{U_2} - 1},$$

where the ordinal number, N, of the level is given by

$$\left(N - \frac{3}{2}\right)\pi < \sqrt{\frac{2mU_2a^2}{\hbar^2}} < \left(N - \frac{1}{2}\right)\pi.$$

So, the condition for the appearance of a bound states becomes[17]

$$\sqrt{\frac{2mU_2a^2}{\hbar^2}} \geqq \arctan\sqrt{\frac{U_1 - U_2}{U_2}}.$$

In particular, it requires that $U_2 \geq \pi^2\hbar^2/8ma^2$ for the limiting case, a) $U_1 = \infty$, and it shows that at least one bound state always exists for b) $U_1 = U_2$.

Problem 2.16

A particle moves in a field formed by two identical symmetrical potential wells separated by some distance (see Fig. 2.7). Assume that the wells do not overlap, and that $U(0) = 0$. Show that the average force, which the particle exerts on each of the wells in a stationary bound state, leads to mutual attraction of the wells for the even-parity states and mutual repulsion for the odd-parity states.

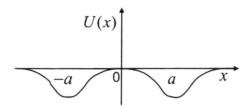

Fig. 2.7

Solution

The average force that the particle exerts on the right well is given by the integral

$$\overline{(F_r)}_{nn} = \int_0^\infty \psi_n^2(x)\frac{dU}{dx}dx.$$

Integrating by parts and using the Schrödinger equation, we obtain

$$\overline{(F_r)}_{nn} = E_n\psi_n^2(0) + \frac{\hbar^2}{2m}[\psi_n'(0)]^2. \tag{1}$$

---

[17] Note that the discrete states $E_n \leq U_2$ shift up both with an increase in U_1 and a decrease in a; see also Problem 2.4.

For an even state we have $\psi'_n(0) = 0$, and since $E_n < 0$ we obtain $\overline{(F_r)}_{nn} < 0$; for odd levels, $\psi_n(0) = 0$, and therefore $\overline{(F_r)}_{nn} > 0$. Note that the force acting on the left well has the opposite sign to that in Eq. (1). This confirms the statement in the problem.

2.2 The Schrödinger equation in momentum space; The Green function and integral form of the Schrödinger equation

Problem 2.17

Find the form of the Schrödinger equation in momentum representation describing a quantum particle influenced by a potential, $U(x)$, that dies off at $x \to \pm\infty$; i.e., $U(\pm\infty) = 0$. Use this equation to investigate bound state(s) in the δ-potential $U = -\alpha\delta(x)$. Compare your results with Problem 2.7.

Solution

The action of the kinetic energy operator, $\hat{T} = \hat{p}^2/2m \equiv p^2/2m$, in momentum space reduces simply to a multiplication, while the potential energy operator, $\hat{U}$, becomes an integral operator with the kernel $U(p, p')$ (see also Problem 1.41):

$$U(p, p') \cong \tilde{U}(p - p'), \quad \tilde{U}(p) = \frac{1}{2\pi\hbar} \int U(x) e^{-ipx/\hbar} dx. \tag{1}$$

Therefore, the Schrödinger equation in momentum space takes the form:

$$\hat{H}\phi(p) \equiv \frac{p^2}{2m}\phi(p) + \int_{-\infty}^{\infty} \tilde{U}(p - p')\phi(p') dp' = E\phi(p). \tag{2}$$

In the specific case of $U = -\alpha\delta(x)$, we have $\tilde{U} = -\alpha/(2\pi\hbar)$, which is momentum-independent. Therefore, Eq. (2) becomes

$$\frac{p^2}{2m}\phi(p) - \frac{\alpha}{2\pi\hbar} C = E\phi(p), \quad C = \int_{-\infty}^{+\infty} \phi(p) dp. \tag{3}$$

Hence (below, $E = -|E| < 0$),

$$\phi(p) = \frac{m\alpha}{\pi\hbar} \frac{C}{p^2 + 2m|E|}. \tag{4}$$

From Eqs. (3) and (4), we have the self-consistency equation

$$1 = \frac{m\alpha}{\pi\hbar} \int_{-\infty}^{\infty} \frac{dp}{p^2 + 2m|E|} = \frac{\alpha}{\hbar}\sqrt{\frac{m}{2|E|}} \tag{5}$$

that determines the discrete spectrum. This equation has only one solution: $E_0 = -m\alpha^2/(2\hbar^2)$ (in the case $\alpha > 0$). The wavefunction (4) corresponds to that level, and with $C = \sqrt{2\pi m\alpha/\hbar}$ it is normalized to unity. The inverse Fourier transform to the coordinate representation recovers the results of Problem 2.7.

Problem 2.18

Use the Schrödinger equation in momentum space to analyze bound states of a particle in the potential, $U(x) = -\alpha[\delta(x-a) + \delta(x+a)]$.

Solution

Here, $\tilde{U}(p) = -\frac{\alpha}{2\pi\hbar}\left(e^{ipa/\hbar} + e^{-ipa/\hbar}\right)$ and the Schrödinger equation takes the form (see the previous problem)

$$\frac{p^2}{2m}\phi(p) - \frac{\alpha}{2\pi\hbar}\left(e^{ipa/\hbar}C_+ + e^{-ipa/\hbar}C_-\right) = E\phi(p), \tag{1}$$

where

$$C_\pm = \int_{-\infty}^{\infty} e^{\mp ipa/\hbar}\phi(p)dp. \tag{2}$$

Hence, denoting $\kappa^2 = -2mE/\hbar^2 > 0$ and $\tilde{\alpha} = m\alpha/\hbar^2$, we find

$$\phi(p) = \frac{\tilde{\alpha}\hbar}{\pi}\left(e^{ipa/\hbar}C_+ + e^{-ipa/\hbar}C_-\right)\frac{1}{p^2 + \hbar^2\kappa^2}, \tag{3}$$

Plugging Eq. (3) into Eq. (2) and calculating the corresponding integrals (see also Eq. (A1.3)), we obtain

$$C_+ = \frac{\tilde{\alpha}}{\kappa}\left(C_+ + e^{-2\kappa a}C_-\right), \quad C_- = \frac{\tilde{\alpha}}{\kappa}\left(e^{-2\kappa a}C_+ + C_-\right). \tag{4}$$

This system of equations has a non-trivial solution if one of the two conditions

$$\kappa = \tilde{\alpha}\left(1 \pm e^{-2\kappa a}\right) \tag{5}$$

is satisfied.

Equation (5) with the "+" sign has only one root, if $\tilde{\alpha} > 0$. Therefore, Eq. (4) yields $C_+ = C_-$, i.e., the corresponding state is even (see Eq. (3)). The energy of this level in the case $\tilde{\alpha}a \ll 1$ is equal to $E_0^+ \approx -2m\alpha^2/\hbar^2$ (two closely-positioned δ-wells effectively act as a single well with a doubled value of α; see Problem 2.7). In the opposite limit, $\tilde{\alpha}a \gg 1$, we have

$$E_0^+ \approx -\frac{m\alpha^2}{2\hbar^2}\left(1 + e^{-2\tilde{\alpha}a}\right). \tag{6}$$

Note that the exponential term in Eq. (5) is small, and by neglecting it we obtain $\kappa_0^+ \approx \tilde{\alpha}$.

Equation (5) also describes the odd levels. An odd level exists only if $\tilde{\alpha}a > 1/2$ (see Problem 2.13). If $0 < \tilde{\alpha}a - 1/2 \ll 1$, and its energy is given by

$$E_1^- \approx -(2\tilde{\alpha}a - 1)^2 \frac{\hbar^2}{2ma^2},$$

and for $\tilde{\alpha}a \gg 1$,

$$E_1^- \approx -\frac{m\alpha^2}{2\hbar^2}\left(1 - e^{-2\tilde{\alpha}a}\right).$$

In the limit, $a \to \infty$, both even and odd levels merge into a single level.

Problem 2.19

Use the Schrödinger equation in momentum space to analyze bound states of a particle in a *separable potential* given by a non-local integral operator, $\hat{U}$, with the kernel, $U(x, x') = -\lambda f(x)f^*(x')$ (it is assumed $|f(x)| \to 0$, for $x \to \pm\infty$).

Solution

The kernel of the operator $\hat{U}$ in momentum space remains separable and reads

$$U(p, p') = -\lambda g(p)g^*(p'), \quad g(p) = \frac{1}{\sqrt{2\pi\hbar}} \int_{-\infty}^{\infty} e^{-ipx/\hbar} f(x) dx, \qquad (1)$$

and the Schrödinger equation (see Problem 2.17) takes the form

$$\frac{p^2}{2m}\phi(p) - \lambda g(p) \int_{-\infty}^{\infty} g^*(p')\phi(p')dp' = E\phi(p). \qquad (2)$$

Hence, it follows that

$$\phi(p) = \frac{2m\lambda C}{p^2 - 2mE} g(p), \quad C = \int_{-\infty}^{\infty} g^*(p)\phi(p)dp. \qquad (3)$$

The self-consistency condition then can be found as follows:

$$2m\lambda \int_{-\infty}^{\infty} \frac{|g(p)|^2}{p^2 - 2mE} dp = 1, \qquad (4)$$

which determines discrete energy levels of the particle in the separable potential.

Let us consider some consequences of this equation:

1) If $E < 0$, the integral in Eq. (4) is a monotonous positive function of $|E|$ that is equal to zero as $|E| \to \infty$. Therefore, if $\lambda < 0$, this equation has no roots, which means bound states are absent. If $\lambda > 0$, there are two possibilities:

1a) If $g(0) \neq 0$ and the integral in Eq. (4) is equal to $+\infty$ as $E \to 0$. In this case there is only one bound state. In the limiting case, $\lambda \to 0$, we also have $E_0 \to 0$; in this case, small values of p play the dominant role in the integral in Eq. (4); therefore,

$$E_0 \approx -2\pi^2 m \lambda^2 |g(0)|^4, \quad \lambda \to 0. \tag{5}$$

In the opposite limiting case, where both $\lambda \to \infty$ and $-E_0 \to \infty$, we have

$$E_0 \approx -\lambda \int_{-\infty}^{\infty} |g(p)|^2 dp. \tag{6}$$

We should note that $|E_0(\lambda)|$ is a monotonically increasing function of the parameter, λ.

1b) If $g(0) = 0$ with $\int_{-\infty}^{\infty} |g|^2 \frac{dp}{p^2} \equiv A$, there also exists just one bound state if $\lambda > (2mA)^{-1}$, and no bound states for $\lambda < (2mA)^{-1}$.

2) When $E > 0$, an unusual situation is possible for separable potentials, which have $g(p_0) = 0$ for some value of $p_0 \neq 0$, so that

$$\int_{-\infty}^{\infty} \frac{|g(p)|^2}{p^2 - p_0^2} dp \equiv B < \infty.$$

In this case, when $\lambda = \lambda_0 = 1/(2mB)$, there appears a bound state with the energy $\tilde{E} = p_0^2/2m > 0$. Interestingly, this discrete level lies within the continuous spectrum.

Problem 2.20

1) Find the Green function, $G_E(x, x')$, of the Schrödinger equation for a free particle with energy, $E < 0$. The Green function obeys the equation

$$(\hat{H} - E)G_E \equiv -\frac{\hbar^2}{2m} \frac{\partial^2}{\partial x^2} G_E - E G_E = \delta(x - x')$$

and is required to vanish as, $|x - x'| \to \infty$. 2) Use the Green function to formulate an integral form of the Schrödinger equation that determines discrete levels in a short-range potential, $U(x)$ [$U(x) \to 0$, as $x \to \pm\infty$]. 3) Use this equation to find bound state(s) of a particle in the δ well, and compare your results with Problem 2.7. 4) Find the form of the Green function in momentum space.

Solution

1) A general solution of the equation for the Green function, G_E, has the form $G_E = A(x')e^{\kappa(x-x')} + B(x')e^{-\kappa(x-x')}$ for $x < x'$. Here, $\kappa = \sqrt{-2mE/\hbar^2} > 0$. Since G_E must decrease as $x \to -\infty$, we conclude that $B(x') = 0$. Similarly, we have $G_E = C(x')e^{-\kappa(x-x')}$ for $x > x'$. G_E is a continuous function at the point $x = x'$, and the derivative G'_E has the "jump" equal to (see Problem 2.6)

$$G'_E(x = x' + 0, x') - G'_E(x = x' - 0, x') = -\frac{2m}{\hbar^2}.$$

Hence,

$$G_E(x, x') = \frac{m}{\kappa \hbar^2} e^{-\kappa|x-x'|}. \tag{1}$$

Using the Green function we can write the general solution of the equation

$$-\frac{\hbar^2}{2m}\psi''(x) - E\psi(x) = f(x) \tag{2}$$

as follows ($E < 0$);

$$\psi(x) = Ae^{-\kappa x} + Be^{\kappa x} + \int_{-\infty}^{\infty} G_E(x, x')f(x')dx'. \tag{3}$$

2) If we now set $f(x) = -U(x)\psi(x)$, in Eq. (2), we obtain the Schrödinger equation and its formal solution (3) in an integral form. Since for physical applications, only solutions that do not increase at $x \to \pm\infty$ matter, and since the integral term in Eq. (3) decreases in this case, we set $A = B = 0$, and the integral form of the Schrödinger equation then reads:

$$\psi_E(x) = -\frac{m}{\kappa\hbar^2} \int_{-\infty}^{\infty} e^{-\kappa|x-x'|} U(x')\psi_E(x')dx'. \tag{4}$$

This formulation of the theory is equivalent to that using the differential Schrödinger equation supplemented by the boundary conditions $\psi(x \to \pm\infty) = 0$ for the values of the energy $E < 0$ that belong to the discrete spectrum.

3) For $U(x) = -\alpha\delta(x)$, Eq. (4) becomes

$$\psi_E(x) = \frac{m\alpha}{\kappa\hbar^2}\psi_E(0)e^{-\kappa|x|},$$

which gives the wavefunction and the energy $E_0 = -m\alpha^2/2\hbar^2$ of the single discrete level in the δ-well.

4) Note that the Green function may be viewed as a linear operator $\hat{G}_E$ defined by its kernel, $G_E(x, x')$, in the coordinate representation. Consequently, from the equation for $G_E(x, x')$, it follows that

$$(\hat{H} - E)\hat{G}_E = \hat{1}, \quad \hat{H} = \frac{\hat{p}^2}{2m}. \tag{5}$$

This operator equation is valid in an *arbitrary representation*. Its formal solution has the form $\hat{G}_E = (\hat{H} - E)^{-1}$. In the momentum representation therefore, the Green function is $G_E = (p^2/2m - E)^{-1}$, which is a multiplication operator. Using the result of Problem 1.41, we can obtain its kernel in the coordinate representation:

$$G_E(x, x') = \int_{-\infty}^{\infty} \frac{e^{ip(x-x')/\hbar} dp}{2\pi\hbar(p^2/2m + |E|)} = \frac{m}{\kappa\hbar^2} e^{-\kappa|x-x'|}, \tag{6}$$

which coincides with Eq. (1).

Problem 2.21

Use the Schrödinger equation in the integral form to investigate bound states in a separable potential (see also Problem 2.19).

Solution

In the case of a separable potential, the integral form of the Schrödinger equation is given by

$$\psi_E(x) = \frac{\lambda m}{\kappa\hbar^2} \int\int e^{-\kappa|x-x'|} f(x') f^*(x'') \psi_E(x'') dx' dx''. \tag{1}$$

Using the notation

$$C = \int_{-\infty}^{\infty} f^*(x) \psi_E(x) dx, \tag{2}$$

and Eq. (1), we obtain the wavefunction

$$\psi_E(x) = \frac{\lambda m C}{\kappa\hbar^2} \int_{-\infty}^{\infty} e^{-\kappa|x-x'|} f(x') dx'. \tag{3}$$

Using this result and Eq. (2), we obtain

$$\kappa = \frac{\lambda m}{\hbar^2} \int_{-\infty}^{\infty}\int_{-\infty}^{\infty} e^{-\kappa|x-x'|} f(x') f^*(x) dx dx', \tag{4}$$

which determines the spectrum. We now consider the limiting cases:

a) If $\lambda \to 0+$, then $\kappa \to 0$ as well. There is a single discrete level present, and its energy is

$$E_0 \approx -\frac{m\lambda^2}{2\hbar^2} \left| \int f(x)dx \right|^4. \tag{5}$$

b) If $\lambda \to \infty$, then $\kappa \to \infty$ as well. For the integral in Eq. (4), we find that only the region $x' \approx x$ is important. Substituting $f(x') \approx f(x)$ and calculating the integral with respect to x', we obtain

$$E_0 \approx -\lambda \int |f(x)|^2 dx. \tag{6}$$

For a more detailed analysis of Eq. (4), it is convenient to transform this equation using formula (A1.3). The relation we obtained reproduces the corresponding result of Problem 2.19.

Problem 2.22

Using the integral form of the Schrödinger equation, prove that the discrete energy levels for a particle in an arbitrary potential $U(x) \leq 0$ [$U(x) \to 0$, as $x \to \pm\infty$] satisfy the following condition:

$$|E_n| \leq \frac{m}{2\hbar^2} \left[\int_{-\infty}^{\infty} U(x)dx \right]^2.$$

Solution

Let us consider the ground-state wavefunction, $\psi_0(x)$, with $E_0 < 0$ ($|E_n| \leq |E_0|$). This function has no zeroes for any finite value of x, and therefore $\psi_0(x) > 0$ (the reality condition could always be satisfied by the appropriate choice of the phase factor). Now, we use the integral Eq. (4) from Problem 2.20. Let us set $x = x_0$, where x_0 corresponds to a maximum of $\psi_0(x)$:

$$\psi_0(x) = \frac{m}{\kappa_0 \hbar^2} \int_{-\infty}^{\infty} e^{-\kappa_0|x-x'|} |U(x')| \psi_0(x') dx'. \tag{1}$$

Function in the integral is non-negative, and the substitution of the factor $e^{-\kappa_0|x_0-x'|}\psi_0(x')$, en lieu $\psi_0(x_0)$ may only increase the right-hand side of the equation. Therefore, we obtain the following inequality:

$$1 \leq \frac{m}{\kappa_0 \hbar^2} \int |U(x)|dx.$$

From this equation it follows that

$$|E_n| \le |E_0| = \frac{\hbar^2 \kappa_0^2}{2m} \le \frac{m}{2\hbar^2}\left[\int U(x)dx\right]^2. \qquad (2)$$

Note that the approximate equality in Eq. (2) is fulfilled for any "shallow" potential well. See also Problem 2.23.

Problem 2.23

There is only one bound state with the approximate energy $E_0 \approx -\frac{m}{2\hbar^2}\left[\int U(x)dx\right]^2$ in a shallow potential well for which $U_0 \ll \hbar^2/ma^2$ (where U_0 and a are the characteristic strength of the potential and its radius). Using the integral form of the Schrödinger equation, find the correction to this relation that is of the order of $ma^2 U_0/\hbar^2$.

Solution

We use the integral form of the Schrödinger equation, as in Eq. (4) in Problem 2.20. Multiply both its sides by $U(x)$, and integrate in the infinite limits. The dominant contribution to the corresponding integrals comes from the region where both x and $x' \sim a$. Since $\kappa a \ll 1$, it is possible to expand the exponential and keep just the first two leading terms. We get

$$\int U(x)\psi(x)dx \approx -\frac{m}{\kappa\hbar^2}\int\int (1-\kappa|x-x'|)U(x)U(x')\psi(x')dxdx'.$$

From this equation, it follows that

$$\kappa \approx -\frac{m}{\hbar^2}\int U(x)dx \left\{1 + \frac{\frac{m}{\hbar^2}\int\int |x-x'|U(x)U(x')\psi(x')dxdx'}{\int U(x)\psi(x)dx}\right\}.$$

The correction – the second one in the brackets in the above equation – contains the wavefunction, whose variation can be neglected in the integration domain, and it can be set to $\psi(0)$ (we restrict ourselves here to the required accuracy, and drop all higher-order terms). Therefore, a correction to the parameter, κ (and thereby to the energy, $E_0 = -\hbar^2\kappa^2/2m$), reads

$$\kappa \approx -\frac{m}{\hbar^2}\int U(x)dx - \left(\frac{m}{\hbar^2}\right)^2 \int\int |x-x'|U(x)U(x')dxdx'. \qquad (1)$$

Note that the correction is negative, as expected from the previous problem.

Problem 2.24

Find the Green function of a free particle moving in half-space bounded by an impenetrable wall; i.e., $U(x) = 0$ for $x > 0$ and $U(x) = \infty$ at $x < 0$ (see also Fig. 2.8a).

The Green function satisfies the boundary condition $G_E(x = 0, x') = 0$ and decreases at $|x - x'| \to \infty$.

Using this Green function, write the integral form of the Schrödinger equation that determines bound states ($E_n < 0$) of a particle in the potential: $U(x) = \tilde{U}(x)$ at $x > 0$ and $U(x) = \infty$ at $x < 0$ (see also Fig. 2.8b)

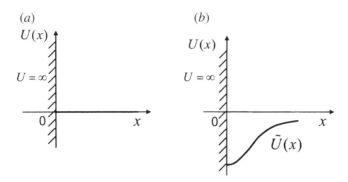

Fig. 2.8

Solution

The Green function could be obtained from the solution in free space (see Problem 2.20), using the method of mirror-image charges (well-known in electrostatics). This method allows us to immediately "guess" the solution that satisfies the appropriate boundary conditions (and per the theorem of existence and uniqueness, it represents the sought-after exact solution), as follows:

$$G_E(x, x') = \frac{m}{\kappa \hbar^2} \left(e^{-\kappa|x-x'|} - e^{-\kappa|x+x'|} \right). \tag{1}$$

The integral form of the Schrödinger equation that automatically takes into account the boundary conditions $\psi(0) = \psi(\infty) = 0$ is given by (compare with Problem 2.20)

$$\psi(x) = -\int_0^\infty G_E(x, x') \tilde{U}(x') \psi(x') dx'. \tag{2}$$

Problem 2.25

Using the integral form of the Schrödinger equation, show that the inequality

$$\int_0^\infty x |\tilde{U}(x)| dx \geq \frac{\hbar^2}{2m}$$

is the *necessary* condition for the existence of a bound state in the potential, $U(x)$, shown in Fig. 2.8b: $U(x) = \infty$ if $x < 0$, $U(x) = \tilde{U}(x)$ for $x > 0$ (assume here that $\tilde{U} \leq 0$ and $\tilde{U}(x) \to 0$ for $x \to \infty$).

Apply the general result specifically to the following potentials: a) $\tilde{U} = -U_0$ for $x < a$, $\tilde{U} = 0$ for $x > a$; b) $\tilde{U} = -\alpha\delta(x-a)$; see Figs. 2.6a, b.

Solution

We can follow here the same logic as in Problem 2.22. First, we estimate the exponential terms in the integral Schrödinger equation (see the previous problem). Since $x, x' \geq 0$ and $|x + x'| - |x - x'| \leq 2x'$, we obtain

$$0 \leq e^{-\kappa|x-x'|} - e^{-\kappa|x+x'|} = e^{-\kappa|x-x'|}[1 - e^{-\kappa|x+x'|+\kappa|x-x'|}] \leq$$
$$e^{-\kappa|x-x'|}[1 - e^{-2\kappa x'}] \leq 2\kappa x'.$$

From here the statement in the problem follows.

a) For the rectangular potential well, the necessary condition of existence of a discrete spectrum becomes $ma^2 U_0/\hbar^2 \geq 1$ (while the exact condition is $ma^2 U_0/\hbar^2 \geq \pi^2/8 \approx 1.24$). b) For the δ-well, the necessary condition has the form $2m\alpha a/\hbar^2 \geq 1$, which coincides with the exact result.

Problem 2.26

Find the Green function, $G_E(x, x')$, for a particle in an infinite potential well of the width, a. Discuss the analytic properties of G_E as a function of the *complex* variable, E. Prove that the Green function has poles in the complex E-plane, and establish a relation between the location of the poles and the energy levels, E_n.

Solution

The equation for the Green function $G_E(x, x')$ and its solution have the form

$$-\frac{\hbar^2}{2m}\frac{d^2}{dx^2}G_E(x, x') - EG_E(x, x') = \delta(x - x')$$

and

$$G_E(x, x') = \begin{cases} A(x') \sin kx, & 0 \leq x < x', \\ B(x') \sin k(x - a), & x' < x \leq a. \end{cases}$$

Here we used the following boundary conditions: $G_E(x = 0) = G_E(x = a) = 0$. Matching conditions for $G_E(x, x')$ at the point $x = x'$ (see also Problem 2.20) allows us to find the coefficients A and B and obtain the following expression for the Green function:

$$G_E(x, x') = -\frac{2m}{k\hbar^2 \sin ka} \sin\left[\frac{k}{2}(x + x' - |x' - x|)\right] \cdot \sin\left[\frac{k}{2}(x + x' + |x' - x| - 2a)\right].$$

From here, it follows that $G_E(x, x')$ is an analytic function of the variable E ($k = \sqrt{2mE/\hbar^2}$) that has the following singular points:

a) $E = \infty$ is an essential singularity.
b) The points $E_n = \hbar^2 k_n^2/2m$, where $k_n a = (n+1)\pi$, $n = 0, 1, \ldots$ are the poles of G_E. The locations of these poles directly correspond to the energy levels of the particle in the well.

Problem 2.27

Consider a class of potentials, $U(x)$, with the following properties:

$$U(x) \leq 0, \quad U(x) \to 0 \text{ at } x \to \pm\infty, \quad \int U(x)dx = \alpha = \text{const}.$$

Find the specific form of the potentials:

a) Where the binding energy of the ground state, $|E_0|$, is maximal;
b) which contains the maximum possible number of discrete levels among all possible potentials within this class.

Solution

a) The solution can be found in Problem 2.22: the deepest-lying level in the δ-well, $U(x) = -\alpha\delta(x - x_0)$, with the energy $E_0 = -m\alpha^2/2\hbar^2$.
b) The maximum number of discrete states is infinite due to their possible condensation when $E \to 0$. This situation is for the potentials that decrease with $x \to \pm\infty$ as $U(x) \approx -\tilde{\alpha}|x|^{-\nu}$ with $\tilde{\alpha} > 0$ and $0 < \nu < 2$. When $1 < \nu < 2$, such potentials satisfy the conditions specified, in the problem.

2.3 The continuous spectrum; Reflection from and transmission through potential barriers

Problem 2.28

Consider a free particle in half-space (*i.e.*, in the presence of the potential, $U(x)$: $U(x) = 0$ for all $x > 0$ and $U = \infty$ for $x < 0$; see Fig. 2.8a). Find the wavefunctions of the stationary states and normalize them to the δ-function of the energy. Prove that these functions form a complete set (for the corresponding Hilbert space in the interval $0 < x < \infty$).

Solution

We have $\psi_E(x) = A(E)\sin(\sqrt{2mE/\hbar^2}x)$, which satisfies the boundary condition $\psi_E(0) = 0$ at the wall. In order to normalize these functions to $\delta(E - E')$ we should choose $A(E) = (2m/\pi^2\hbar^2 E)^{1/4}$. The condition of completeness reads

$$\int_0^\infty \psi_E^*(x)\psi_E(x')dE = \delta(x-x'),$$

and it is indeed satisfied for these functions, which can be verified explicitly using relation (A1.1).

Problem 2.29

Find the reflection coefficient of the potential wall shown in Fig. 2.9. Examine, in particular, the limiting cases $E \to U_0$ and $E \to \infty$.

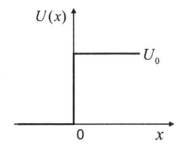

Fig. 2.9

Solution

The solution of the Schrödinger equation that describes the transmission and reflection of particles with $E > U_0$, propagating towards the wall from the left, has the form

$$\psi_k^+(x) = \begin{cases} e^{ikx} + A(k)e^{-ikx}, & x < 0 \ (k = \sqrt{2mE/\hbar^2} > 0), \\ B(k)e^{ik'x}, & x > 0 \ (k' = \sqrt{2m(E-U_0)/\hbar^2} > 0). \end{cases}$$

From continuity of the wavefunctions, ψ_k^+ and $\psi_k^{(+)'}$, in the point $x = 0$, it follows that

$$1 + A = B, \ k(1-A) = k'B; \ A(k) = \frac{k-k'}{k+k'}, \ B(k) = \frac{2k}{k+k'}.$$

Hence, using the relations $R = |A|^2$ and $D = k'|B|^2/k$, we find

$$R(E) = \left(\frac{\sqrt{E} - \sqrt{E-U_0}}{\sqrt{E} + \sqrt{E-U_0}}\right)^2, \quad D(E) = \frac{4\sqrt{E(E-U_0)}}{(\sqrt{E} + \sqrt{E-U_0})^2}, \tag{1}$$

where $R(E) + D(E) = 1$ as expected, and

a) $R(E) \approx U_0^2/(16E^2) \to 0$ as $E \to \infty$,
b) $D(E) \approx 4\sqrt{(E-U_0)/U_0} \propto \sqrt{E-U_0} \to 0$ at $E \to U_0$.

Problem 2.30

Find the reflection and transmission coefficients of the δ-potential barrier, $U(x) = \alpha\delta(x)$.

Discuss the analytic properties of the reflected $A(E)$ and transmitted $B(E)$ amplitudes as functions of the complex variable, E. Pay attention to the special points $E = 0$ and $E = \infty$, which should be properly treated as the branching points of these functions. In the complex E-plane make a branch-cut from the point $E = 0$ along the real semi-axis $E > 0$. Find the singularities of the functions $A(E)$ and $B(E)$ in the first, *physical* sheet as well as in the other sheets of the Riemann surface of these functions. (The physical sheet is defined by the condition that the complex phase of the variable, E, is zero on the upper part of the branch-cut, $E > 0$.) Find a connection between the location of the poles and the physical energy levels.

Solution

1) The wavefunction has the form $\psi_k^+(x) = e^{ikx} + A(k)e^{-ikx}$ for $x < 0$ and $\psi_k^+(x) = B(k)e^{ikx}$ for $x > 0$ (here $k = \sqrt{2mE/\hbar^2} > 0$, which corresponds to the incident particle moving to the right). Matching $\psi_k^+(x)$ and $\left[\psi_k^+(x)\right]'$ in the point $x = 0$ (see also Eq. (2) in Problem 2.6) gives

$$1 + A = B, \quad ik(B - 1 + A) = \frac{2m\alpha}{\hbar^2}B,$$

$$A(k) = \frac{m\alpha}{ik\hbar^2 - m\alpha}, \quad B(k) = \frac{ik\hbar^2}{ik\hbar^2 - m\alpha}. \tag{1}$$

We see that the transmission coefficient, $D(E) = |B|^2$, and the reflection coefficient, $R(E) = |A|^2$, indeed satisfy the constraint $R + D = 1$. Here

a) $R(E) \approx m\alpha^2/2E\hbar^2 \to 0$ as $E \to \infty$; b) $D(E) \approx 2E\hbar^2/m\alpha^2 \propto E \to 0$ as $E \to 0$.

2) Since $k = \sqrt{2mE/\hbar^2}$, it follows from Eq. (1) that $A(E)$ and $B(E)$ are analytic functions of E, which have the following singularity points:

a) $E = 0$ and $E = \infty$, which are their branching points as discussed above;
b) $E = E_0$, where $i\sqrt{2mE_0} = m\alpha/\hbar$, is a pole.

Note that $A(E)$ and $B(E)$ are multivalued functions (in this particular case of the δ-potential, they are double-valued). We now introduce a branch-cut in the complex E-plane along the real semi-axis $E > 0$; see Fig. 2.10a. Since on the physical sheet, $k = \sqrt{2mE/\hbar^2} > 0$, the values of the analytic functions, $A(E)$ and $B(E)$, coincide with the values of the physical amplitudes on the upper part of the branch-cut. The complex phase of E located on the negative semi-axis $E_0 < 0$ of the physical sheet is equal to π, and therefore $\sqrt{E} = i\sqrt{|E|}$.

Therefore, the pole E_0 of the amplitudes with $\alpha < 0$ (δ-well) is located on the physical sheet, and coincides with the energy of the only level present in the well. In the case of a barrier, $\alpha > 0$, the bound states are absent and the pole of amplitudes

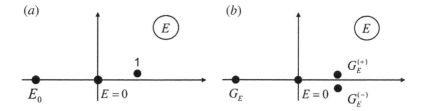

Fig. 2.10

is on the non-physical sheet (the phase of E_0 is equal to 3π). Such poles correspond to so-called *virtual levels* that do not represent proper bound states.

Problem 2.31

Find the transmission coefficient of a particle for the potential barrier shown in Fig. 2.11. How does the transmission coefficient change if the potential barrier ($U_0 > 0$) is "flipped over" and becomes a potential well, ($[U(x) = U_0 < 0$ for $0 < x < a]$?

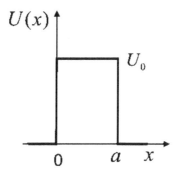

Fig. 2.11

Solution

The transmission coefficient is given by

$$D(E) = \begin{cases} \dfrac{4E(E-U_0)}{4E(E-U_0) + U_0^2 \sin^2 \sqrt{2m(E-U_0)a^2/\hbar^2}}, & E > U_0, \\ \dfrac{4E(U_0-E)}{4E(U_0-E) + U_0^2 \sinh^2 \sqrt{2m(U_0-E)a^2/\hbar^2}}, & E < U_0. \end{cases} \quad (1)$$

Note that the first relation is valid for the potential well if we set $U_0 = -|U_0|$.

Note that $D(E) \to 1$ as $E \to \infty$, which is a natural behavior. On the other hand, $D(E) \propto E \to 0$ as $E \to 0$. This property of $D(E)$ is a rather general quantum-mechanical result (see Problem 2.39). However, in the case of the potential well for the special cases, when

$$\frac{1}{\hbar}\sqrt{2m|U_0|a^2} = \pi n, \ n = 1, \ 2, \ldots,$$

the latter relation manifestly breaks down, and we find $D(E \to 0) \to 1$ instead. These special values of the energy correspond to the emergence of a new level in the discrete spectrum, upon increasing the depth of the well (see Problem 2.13).

Problem 2.32

Determine the values of the particle energy, for which the particles are not reflected from the following potential, $U(x) = \alpha[\delta(x) + \delta(x - a)]$; see Fig. 2.12.

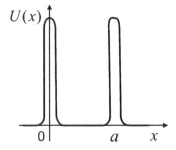

Fig. 2.12

Result
The values of E for which particles do not reflect from the barrier are the roots of the equation

$$\tan ka = -\frac{\hbar^2 k}{m\alpha}, \ k = \frac{1}{\hbar}\sqrt{2mE} > 0.$$

This equation is obtained from the asymptotic form of the solution to the Schrödinger equation (II.4), with the coefficient, A, set to zero: $A = 0$. The matching conditions of Problem 2.6 are used for $x = 0$ and $x = a$.

Problem 2.33

Prove that the reflection coefficient, $R(E)$, is only a function of the particle energy, E, and does not depend on whether the incident particles propagate towards the scattering center from the left or from the right.

Solution

Let us consider the case when $U(x) \to 0$ as $x \to -\infty$ and $U(x) \to U_0$ as $x \to +\infty$. We denote by $\psi_\pm(x)$ the wavefunctions corresponding to the same energy but with the opposite directions of motion of the incident particles (right/left, correspondingly). These functions have the following asymptotic behaviors:

$$\psi_+(x) \approx \begin{cases} e^{ikx} + A(k)e^{-ikx}, & x \to -\infty \ (k = \sqrt{2mE/\hbar^2}), \\ B(k)e^{ik_1 x}, & x \to +\infty \ (k_1 = \sqrt{2m(E-U_0)/\hbar^2}), \end{cases} \tag{1}$$

and

$$\psi_-(x) \approx \begin{cases} \tilde{B}(k)e^{-ikx}, & x \to -\infty, \\ e^{-ik_1 x} + \tilde{A}(k)e^{ik_1 x}, & x \to +\infty, \end{cases}$$

and they obey the Schrödinger equation, $-\frac{\hbar^2}{2m}\psi''_\pm + U(x)\psi_\pm = E\psi_\pm$.

Multiplying the equation for ψ_+ by ψ_- and the equation for ψ_- by ψ_+, and subtracting them from one another, we obtain

$$\psi_-(x)\psi'_+(x) - \psi_+(x)\psi'_-(x) = \text{constant}. \tag{2}$$

Calculating the left-hand side of Eq. (2) at $x \to \pm\infty$ and using the asymptotes (1), we obtain $k\tilde{B} = k_1 B$. Hence, it follows that

$$D_+(E) = \frac{k_1}{k}|B|^2 = \frac{k}{k_1}|\tilde{B}|^2 = D_-(E).$$

Problem 2.34

Find the transmission and reflection coefficients of a particle in a separable potential (see Problem 2.19). Verify that the general properties (II.5) of these coefficients are indeed satisfied in this case.

Solution

We use the integral form of the Schrödinger equation (see Problem 2.42), which, for a separable potential, takes the following form (below, $k = |p|/\hbar$):

$$\psi_p^+(x) = e^{ipx/\hbar} + \frac{i\lambda m}{k\hbar^2} \int\int e^{ik|x-x'|} f(x') f^*(x'') \psi_p^+(x'') dx' dx''.$$

Hence, it follows that

$$\psi_p^+(x) = e^{ipx/\hbar} + \frac{i\lambda m C(p)}{k\hbar^2} \varphi_k(x), \tag{1}$$

where

$$C(p) = \int f^*(x)\psi_p^+(x)dx, \quad \varphi_k(x) = \int e^{ik|x-x'|}f(x')dx'. \qquad (2)$$

Using Eqs. (1) and (2) we obtain

$$C(p) = g^*(p)\left[1 - \frac{i\lambda m}{k\hbar^2}\int\int e^{ik|x-x'|}f^*(x)f(x')dxdx'\right]^{-1}, \qquad (3)$$

where

$$g(p) = \int e^{-ipx/\hbar}f(x)dx.$$

The relations (1), (2), and (3) determine the wavefunction. Calculating its asymptotic behavior at $x \to \pm\infty$, we find the amplitudes of both the transmitted wave, $B(p)$, and the reflected wave, $A(p)$, as follows:

$$B(p) = 1 + \frac{i\lambda m C(p)g(p)}{k\hbar^2}, \quad D(p) = |B(p)|^2 \qquad (4)$$

and

$$A(p) = \frac{i\lambda m C(p)g(-p)}{k\hbar^2}, \quad R(p) = |A(p)|^2.$$

Now, we transform Eq. (4) using Eqs.(A1.3) and (A1.2). We first find

$$\int \frac{F(x)dx}{x - x_0 - i\varepsilon} = \text{V.P.} \int \frac{F(x)dx}{x - x_0} + i\pi F(x_0),$$

where the symbol "V.P. $\int \ldots$" corresponds to the principal value of the integral and $\varepsilon > 0$ is infinitely small. We can rewrite Eq. (3) in the form

$$C(p) = g^*(p)\left[\frac{C_1(p)}{2p\hbar} - i\frac{C_2(p)}{2k\hbar^2}\right]^{-1},$$

where

$$C_1(p) = 2p\hbar\left(1 - \frac{\lambda m}{\pi\hbar}\text{V.P.}\int_{-\infty}^{\infty}\frac{|g(\kappa)|^2 d\kappa}{\kappa^2 - p^2}\right), \quad C_2(p) = \lambda m(|g(p)|^2 + |g(-p)|^2),$$

and after that we have, from Eq. (4),

$$D(p) = \frac{C_1^2(p) + \lambda^2 m^2(|g(p)|^2 - |g(-p)|^2)^2}{C_1^2(p) + C_2^2(p)}, \quad R(p) = \frac{4\lambda^2 m^2|g(p)g(-p)|^2)}{C_1^2(p) + C_2^2(p)}. \qquad (6)$$

Finally, we find the following properties:

1) $D(p) + R(p) = 1$,

2) $D(p) = D(-p)$,
3) $R(E) \approx (\lambda m/\hbar p)^2 \cdot |g(p)g(-p)|^2 \to 0$ at $E \to \infty$,
4) $D(E) \to 0$ at $E \to 0$.
(Compare these results with Problem 2.39.)

Problem 2.35

Find the transmission coefficient, $D(E)$, for the potential barrier shown in Fig. 2.13. Consider various limiting cases, where $D(E)$ can be expressed in terms of elementary functions.

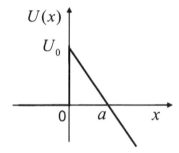

Fig. 2.13

Solution

For $x < 0$, the wavefunction has the form $\psi_k^+(x) = e^{ikx} + A(k)e^{-ikx}$ (we assume that the incident particles are moving from left to right, $k = \sqrt{2mE/\hbar^2} > 0$). For $x > 0$, by virtue of the substitution

$$z = \xi\left(\frac{x}{a} - 1 + \frac{E}{U_0}\right), \text{ where } \xi = \left(\frac{2ma^2 U_0}{\hbar^2}\right)^{1/3},$$

the Schrödinger equation takes the form $[\psi_k^+(z)]'' + z\psi_k^+(z) = 0$. Its solution (that asymptotically takes the form of a plane-wave moving to the right, as $x \to +\infty$) should be chosen as

$$\psi_k^+(x) = C(E)[\text{Bi}(-z) + i\text{Ai}(-z)] \approx z(x) \to \infty \, C(E) \frac{1}{\sqrt{\pi} z^{1/4}} e^{i\frac{2}{3} z^{3/2} + i\frac{\pi}{4}},$$

where $\text{Ai}(z)$ and $\text{Bi}(z)$ are the Airy functions. Since the wavefunction and its derivative with respect to x are continuous at $x = 0$, we can determine the values of A and C. We find

$$C(E) = \frac{2}{\text{Bi}(-z_0) + i\text{Ai}(-z_0) + i\frac{\xi}{ka}\left[\text{Bi}'(-z_0) + i\text{Ai}'(-z_0)\right]}, \quad (1)$$

where $z_0 = \xi(E/U_0 - 1)$.

Calculating the current density, $j = (\hbar/2mi)(\psi^*\psi' - \psi\psi^{*\prime})$, at $x \to \infty$: $j_{tr} = \xi\hbar|C|^2/(\pi m a)$, and taking into account that for the incident particles, $j_{in} = \hbar k/m$, we obtain the transmission coefficient

$$D(E) = \frac{j_{tr}}{j_{in}} = \frac{\xi}{\pi k a}|C(E)|^2. \tag{2}$$

Eqs. (1) and (2) give the solution. Let us now consider some special cases.

1) $E < U_0$, $\xi(1 - E/U_0)| \gg 1$ (and $\xi \gg 1$)

$$D(E) \approx \frac{4\sqrt{E(U_0 - E)}}{U_0} \exp\left\{-\frac{4}{3}\sqrt{\frac{2ma^2(U_0 - E)^3}{\hbar^2 U_0^2}}\right\} \ll 1. \tag{3}$$

2) $E > U_0$, $\xi(E/U_0 - 1)| \gg 1$ (and $ka \gg \xi$)

$$D(E) \approx \frac{4\sqrt{E(E - U_0)}}{(\sqrt{E} + \sqrt{E - U_0})^2}. \tag{4}$$

See Problem 2.29.

3) At $E \to 0$

$$D(E) \approx \frac{4ka}{\pi\xi[(\text{Bi}'(\xi))^2 + (\text{Ai}'(\xi))^2]} \propto \sqrt{E} \to 0.$$

Problem 2.36

The same as in the previous problem, but for the barrier $U = -F_0|x|$, shown in Fig. 2.14.

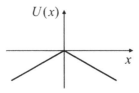

Fig. 2.14

Solution

The wavefunction has the form

$$\psi^+(x) = \begin{cases} [\text{Bi}(z_1) - i\text{Ai}(z_1)] + a(E)[\text{Bi}(z_1) + i\text{Ai}(z_1)], & x < 0, \\ b(E)[\text{Bi}(-z_2) + i\text{Ai}(-z_2)], & x > 0, \end{cases}$$

where $z_{1,2} = \xi(x \mp E/F_0)$, $\xi = (2mF_0/\hbar^2)^{1/3}$, and $a(E)$ and $b(E)$ are the amplitudes of the reflected and transmitted waves respectively, so that $R = |a(E)|^2$ and $D = |b(E)|^2$. The continuity requirement yields

$$b(E) = -\left\{\pi[\text{Bi}(\eta) + i\text{Ai}(\eta)][\text{Bi}'(\eta) + i\text{Ai}'(\eta)]\right\}^{-1},$$

where $\eta = -\xi E/F_0$ (here we used the following result for the Wronskian of the Airy functions: $W\{Ai(z), Bi(z)\} = 1/\pi$).

Using the known asymptotic behavior of the Airy functions, we obtain the following asymptotic expressions for $D = |b(E)|^2$:

1) for $E < 0$ and $\xi|E|/F_0 \gg 1$

$$D(E) \approx \exp\left\{-\frac{8}{3}\sqrt{\frac{2m|E|^3}{\hbar^2 F_0^2}}\right\}, \qquad (1)$$

2) for $E > 0$ and $\xi E/F_0 \gg 1$,

$$D(E) \approx 1 - \frac{\hbar^2 F_0^2}{32mE^3}, \qquad (2)$$

3) $D(E=0) = \frac{3}{4}$, $R(E=0) = \frac{1}{4}$.

Problem 2.37

Consider a potential, $U(x)$, with the following properties: $U(x) \to 0$, as $x \to -\infty$, and $U(x) \to U_0 > 0$, as $x \to +\infty$; see Fig. 2.15. Determine the energy dependence of the transmission coefficient as $E \to U_0+$. Compare with the results of Problem 2.29.

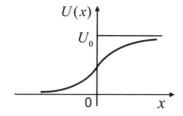

Fig. 2.15

Solution

The wavefunction at $x \to \pm\infty$ is

$$\psi_E^+(x) \approx \begin{cases} e^{ikx} + A(k)e^{-ikx}, & x \to -\infty \ (k = \sqrt{2mE/\hbar^2}), \\ B(k)e^{ik_1 x}, & x \to +\infty \ (k_1 = \sqrt{2m(E-U_0)/\hbar^2}), \end{cases} \qquad (1)$$

and the transmission coefficient is given by $D(E) = (k_1/k)|B(k)|^2$. In the limit, $E \to U_0+$, we have $k_1 \to 0$, $B(k_1) \to B(0) \neq 0$ and therefore $D(E) \propto (E-U_0)^{1/2} \to 0$.

Problem 2.38

Find the reflection and transmission coefficients of slow particles, $ka \ll 1$, in the case of a "weak" potential with $U_0 \ll \hbar^2/ma^2$ (U_0 and a denote the typical strength and radius of the potential). Compare your results with Problem 2.30 (δ-potential).

Solution

In the region where the potential is non-zero, the wavefunction has the form

$$\psi_k^+(x) = \begin{cases} e^{ikx} + A(k)e^{-ikx}, & x < -a, \\ B(k)e^{ikx}, & x > a. \end{cases} \quad (1)$$

If $|x| \leq a$, the Schrödinger equation, $\psi''(x) = [\frac{2m}{\hbar^2}U(x) - k^2]\psi(x)$, yields the approximate relation as follows: $\psi_k^+(x) \approx C_1 + C_2 x$ (since $\psi'' \sim \psi/a^2$ and $ka \ll 1$, the leading approximation of the Schrödinger equation is simply $\psi'' = 0$). Matching this solution with Eq. (1) gives $C_2 \approx 0$ and $C_1 \approx B \approx 1 + A$.

Hence, it follows that Eq. (1), which ensures that the relation, $\psi_k^+(x) \approx$ const, remains approximately valid for all values of x. Taking this into account, we integrate the Schrödinger equation inside the interval $-b < x < b$ where $b > a$. This leads to

$$\int_{-b}^{b} \psi''(x) dx = ikBe^{-ikb} + ikAe^{ikb},$$

$$\int_{-b}^{b} U(x)\psi(x) dx \approx B \int_{-b}^{b} U(x) dx \approx B \int_{-\infty}^{\infty} U(x) dx,$$

$$\int_{-b}^{b} \psi(x) dx \approx B \int_{0}^{b} e^{ikx} dx + \int_{-b}^{0} (e^{ikx} + Ae^{-ikx}) dx,$$

$$= i\frac{1}{k}\{A + B - 1 - (A+B)e^{ikb} + e^{-ikb}\}.$$

It gives the relation

$$ik(A + B - 1) \approx \frac{2m\alpha}{\hbar^2} B, \quad \alpha = \int U(x) dx,$$

and since $1 + A \approx B$, we obtain

$$A \approx -\frac{im\alpha}{\hbar^2 k + im\alpha}, \quad B \approx \frac{\hbar^2 k}{\hbar^2 k + im\alpha}.$$

This result implies that reflection properties of the shallow barrier are identical (in the leading order) to those of the δ-potential, $\tilde{U} = \alpha\delta(x)$, with $\alpha = \int U(x) dx$ (see Problem 2.30).

Problem 2.39

Prove that the transmission coefficient of an arbitrary potential satisfying the condition $U(x) = 0$, for $|x| > a$, generally vanishes linearly as $E \to 0$: $D(E) \approx cE$. Find the special conditions where this relation breaks down.

Express the coefficient c in terms of the parameters that characterize the zero-energy solution to the Schrödinger equation. Use this result for a square potential well and compare it with the exact solution (see Problem 2.31).

Solution

The asymptotic behavior of the wavefunction is $\psi_k^+(x) \approx e^{ikx} + A(k)e^{-ikx}$ (for $x < 0$, $|x| \gg a$) and $\psi_k^+(x) = B(k)e^{ikx}$ (for $x \gg a$). Now consider the limiting case $k \to 0$:

$$\psi^+(x) = \begin{cases} 1 + A(k) + ikx[1 - A(k)], & x < 0, |x| \gg a, \\ (1 + ikx)B(k), & x \gg a, \end{cases} \quad (1)$$

The zero-energy solution of the Schrödinger equation satisfies the boundary condition $\psi_{E=0}(+\infty) = 1$. As $x \to -\infty$, this solution has the form $\psi_{E=0}(x) = bx + d$, where the constants b and d depend on the details of the potential. Comparing this asymptotic behavior of $\psi_{E=0}$ to Eq. (1), we find that $ik(1 - A) \approx bB$ and $1 + A \approx dB$. Hence, it follows $A \approx -1$, $B \approx 2ik/b$, so that[18]

$$D(E) = |B|^2 \approx \frac{8m}{b^2\hbar^2} E \propto E, \quad \text{for } E \to 0. \quad (2)$$

This result breaks down in the case of $b = 0$. In this special case the Schrödinger equation has a zero-energy solution that does not increase as $x \to \pm\infty$. It implies that a new bound state emerges with only an infinitesimal deepening of the potential (see Problem 2.13).

For the potential barrier from Problem 2.31, we have: $\psi_{E=0} = 1$ for $x > a$, $\psi_{E=0} = \cosh[\xi(x-a)]$. for $0 < x < a$ (here $\xi = \sqrt{2ma^2 U_0/\hbar^2}$), $\psi_{E=0} = \cosh \xi a - (\xi \sinh \xi a)x$ for $x < 0$. Hence, it follows $b = -\xi \sinh \xi a$ and $D(E) \approx \frac{4}{U_0 \sinh^2 \xi a} E$ as $E \to 0$, in accordance with the exact result (in the case of a potential well, we should replace $\sinh \xi a$ with $\sin \xi a$ in the above results).

Problem 2.40

Find the transmission coefficient of "slow" particles in the potential, $U(x) = -U_0 a^4 / \left(x^2 + a^2\right)^2$.

---

[18] Equation (2), as well as the asymptotic form of the wavefunction in Eq. (1), is valid for potentials that decrease at large distances, $x \to \pm\infty$, faster than $\propto 1/|x|^3$.

Solution

We introduce the following new variable, $z = \arctan x/a$, and function, $w = (x^2 + a^2)^{-1/2}\psi(x)$. Focusing on the zero-energy solution only with $E = 0$, the Schrödinger equation becomes

$$w''(z) + \xi^2 w(z) = 0, \text{ where } \xi = \sqrt{1 + 2ma^2 U_0/\hbar^2}.$$

The wavefunction, $\psi_{E=0}(x)$, that obeys the boundary condition $\psi_{E=0}(+\infty) = 1$ is

$$\psi_{E=0}(x) = \frac{\sqrt{x^2 + a^2}}{\xi a} \sin\left[\xi\left(\frac{\pi}{2} - \arctan\frac{x}{a}\right)\right]. \quad (1)$$

Since $\psi_{E=0}(x) \approx -x\sin(\pi\xi)/\xi a$ at $x \to -\infty$, then according to the previous problem we find that for slow particles, $D(E) \approx \frac{8m(\xi a)^2}{\hbar^2 \sin^2(\pi\xi)} E$, $E \to 0$. Note that this expression does not apply if $\pi\xi = \pi N$ (N is an integer) or

$$\frac{2ma^2 U_0}{\hbar^2} = N^2 - 1. \quad (2)$$

This Eq. (2) determines the parameters corresponding to the emergence of a new (Nth) discrete state.

Problem 2.41

Use the Schrödinger equation in momentum space to find the wavefunctions of a particle in a uniform field with $U(x) = F_0 x$. Normalize them by the δ-function of the energy and prove the completeness of the obtained set of functions. Use these results to determine the energy spectrum of the potential considered in Problem 2.8.

Solution

1) The Schrödinger equation in the momentum representation and its solution normalized to the δ-function of energy have the form

$$\frac{p^2}{2m}\phi_E(p) + iF_0 \frac{d}{dp}\phi_E(p) = E\phi_E(p),$$

$$\phi_E(p) = \frac{1}{\sqrt{2\pi\hbar F_0}} \exp\{i\frac{p^3}{6m\hbar F_0} - i\frac{Ep}{\hbar F_0}\}.$$

2) The values of E for which the corresponding wavefunction in the coordinate representation obeys the condition $\psi_E(x=0) = 0$, i.e.,

$$\psi_E(x=0) = \frac{1}{\sqrt{2\pi\hbar}} \int_{-\infty}^{\infty} \phi_E(p) dp \equiv C \int_0^{\infty} \cos\left(\frac{Ep}{\hbar F_0} - \frac{p^3}{6m\hbar F_0}\right) dp = 0,$$

determine the discrete energy spectrum of the potential considered in Problem 2.8.

Problem 2.42

Find the Green functions, $G_E^\pm(x, x')$, of a free particle with the energy $E > 0$. Here the indices "$\pm$" refer to the asymptotic behavior of the Green function:

$$G_E^\pm(x, x') \propto \exp\left(\pm i\sqrt{\frac{2mE}{\hbar^2}}\right), \quad \text{as } |x| \to \infty.$$

Use these Green functions to formulate the integral Schrödinger equation, whose solution describe the transmission and reflection processes for particles with momentum p ($-\infty < p < +\infty$). Assume that the potential, $U(x)$, goes to zero as $x \to \pm\infty$. Use the integral Schrödinger equation to describe scattering off of the δ-potential.

Solution

1) Using Eq. (1) of Problem 2.20 (where the Green function, $G_E(x, x')$, for $E < 0$ was found) and the substitution,

$$\kappa = \sqrt{-\frac{2mE}{\hbar^2}} = \mp ik, \text{ where } k = \sqrt{\frac{2mE}{\hbar^2}} > 0,$$

we obtain $G_E^\pm$ for $E > 0$ as follows:

$$G_E^\pm(x, x') = \pm \frac{im}{\hbar^2 k} e^{\pm ik|x-x'|}. \tag{1}$$

We should note that the Green functions, $G_E^\pm$, for $E > 0$ and $E < 0$ could be considered as different boundary values of a single analytic function $\tilde{G}_E$ of the complex variable, E;

$$\tilde{G}_E = i\sqrt{\frac{m}{2\hbar^2 E}} \exp\left\{i\frac{\sqrt{2mE}}{\hbar}|x - x'|\right\}.$$

The point $E = 0$ is its branching point. We introduce a branch-cut in the E-plane along the real semi-axis, $E > 0$, see Fig. 2.10b. Note that the upper part of the branch-cut lies on the physical sheet (see Problem 2.30.) The function, $\tilde{G}_E$, coincides with G_E^- on the lower edge of the branch-cut, while for real $E < 0$, it coincides with the Green function, G_E, as in Problem 2.20.

The Green functions, $G_E^\pm(p, p')$, in the momentum representation have the form

$$G_E^\pm(p, p') = \frac{\delta(p - p')}{p^2/2m - E \mp i\varepsilon}, \tag{2}$$

(see Problem 2.20), here $\varepsilon > 0$ is an infinitesimally small quantity.

2) The Schrödinger equation describing the transmission and reflection processes for the particle with momentum, p, is

$$\psi_p^+(x) = e^{ipx/\hbar} - \int G_E^+(x, x') U(x') \psi_p^+(x') dx'. \tag{3}$$

The first term in the right-hand side of Eq. (3) describes the incident particles, while the second, integral term describes at $x \to \pm\infty$ both the reflected particles and the change of the transmitted wavefunction under the action of the potential.

Specifically for the potential, $U(x) = \alpha\delta(x)$, Eq. (3) takes the form

$$\psi_p^+(x) = e^{ipx/\hbar} - i\frac{m\alpha}{\hbar^2 k}e^{ik|x|}\psi_p^+(0). \tag{4}$$

Therefore,

$$\psi_p^+(0) = \frac{\hbar^2 k}{\hbar^2 k + im\alpha}, \tag{5}$$

and the transmission and reflection coefficient D and R follow from Eqs. (4) and (5), reproducing the results of Problem 2.20.

Problem 2.43

For the δ-barrier, $U(x) = \alpha\delta(x)$ with $\alpha > 0$, prove by a direct calculation the completeness of the set of functions $\psi_p^{(+)}(x)$, that describe the process of reflection and transmission of particles with momentum, p ($-\infty < p < +\infty$).

Solution

Let us consider the integral $I(x, x') = \int\limits_{-\infty}^{\infty} \psi_p^{+*}(x')\psi_p^+(x)dp$, assuming that the wavefunctions, $\psi_p^+(x)$, are normalized to $\delta(p - p')$. (Note that they differ from those in the previous problem – see Eq. (4) – by the factor, $(2\pi\hbar)^{-1/2}$). Therefore, the integral can be rewritten in the form

$$\frac{1}{2\pi\hbar}\int\limits_{-\infty}^{\infty} e^{ik(x-x')/\hbar}dp - \frac{i\tilde{\alpha}}{2\pi}\int\limits_{-\infty}^{\infty} dk \left\{ \frac{\exp[-i(kx' - |kx|)]}{|k| + i\tilde{\alpha}} - \right.$$

$$\left. \frac{\exp[i(kx - |kx'|)]}{|k| - i\tilde{\alpha}} + \frac{\exp[-i(|kx'| - |kx|)]}{2(|k| - i\tilde{\alpha})} - \frac{\exp[-i(|kx'| - |kx|)]}{2(|k| + i\tilde{\alpha})} \right\}, \tag{1}$$

where $\tilde{\alpha} = m\alpha/\hbar^2$ and $p = \hbar k$. The first integral in (1) is equal to $\delta(x - x')$. To analyze the second term in Eq. (1) above, we perform the following transformations. Taking into account the fact that this integral is an even function of the variables, x and x', we can replace them by their absolute values, $|x|$ and $|x'|$, and divide the integration domain into two parts: $(-\infty, 0)$ and $(0, \infty)$. With this, the second integral in Eq. (1) takes on the form

$$-\frac{i\tilde{\alpha}}{2\pi}\int\limits_{-\infty}^{\infty} \frac{\exp[ik(|x| + |x'|)]}{k + i\tilde{\alpha}}dk. \tag{2}$$

Since $\tilde{\alpha} > 0$ (for the δ-barrier), we can close the integration contour in the upper half-plane to find that the corresponding integral is equal to zero. So, $I(x, x') = \delta(x - x')$, which proves that the system of functions $\psi_p^{(+)}(x)$ is indeed complete.

Problem 2.44

Generalize the result of the previous problem for the case of the attractive δ-potential, $U(x) = -\alpha\delta(x)$ with $\alpha > 0$.

Solution

By changing α by $-\alpha$ in the equations of the previous problem, we have

$$\int_{-\infty}^{\infty} \psi_p^{+*}(x')\psi_p^{+}(x)dp = \delta(x - x') - \frac{i\tilde{\alpha}}{2\pi}\int_{-\infty}^{\infty} \frac{\exp[ik(|x| + |x'|)]}{k + i\tilde{\alpha}}dk, \quad (1)$$

($\tilde{\alpha} = m\alpha/\hbar^2 > 0$). Taking into account the value of the integral in the right-hand side[19] and the form of the normalized wavefunction, $\psi_0(x)$, of the single bound state in the δ-well, we see that the second term in the right-hand side of Eq. (1) is equal to

$$-\tilde{\alpha}\exp[-\tilde{\alpha}(|x| + |x'|)] = -\psi_0^*(x')\psi_0(x).$$

Therefore, we obtain the equation

$$\psi_0^*(x')\psi_0(x) + \int_{-\infty}^{\infty} \psi_p^{+*}(x')\psi_p^{+}(x)dp = \delta(x - x'),$$

which is the required completeness condition of the eigenfunctions in the case of the δ-well.

Problem 2.45

For a particle in the repulsive δ-potential, $U(r) = \alpha\delta(x)$ with $\alpha > 0$, find the Green functions, $G_E(x, x')$ for $E < 0$ and $G_E^{(\pm)}(x, x')$ for $E > 0$. Discuss their analytical properties as functions of the complex variable, E. Compare with the free-particle case see Problem 2.42.

Solution

The Green functions of interest obey the following equation;

$$\left[-\frac{\hbar^2}{2m}\frac{d^2}{dx^2} + \alpha\delta(x) - E\right]G_E(x, x') = \delta(x - x'), \quad (1)$$

---

[19] The integral can be calculated using the residue theorem by closing the integration contour in the upper half-plane.

72 *Exploring Quantum Mechanics*

with the corresponding boundary conditions. Using the general method of their construction, and taking into account that there is no discrete spectrum in a purely repulsive potential, we find

$$G_E^{\pm} = \int_{-\infty}^{\infty} \psi_p^{+*}(x')\psi_p^{+}(x)\frac{2m\,dp}{p^2 - 2m(E \pm i\gamma)} \qquad (2)$$

($\gamma > 0$ is an infinitesimally small quantity). Here $\psi_p^{+}(x)$ are the wavefunctions normalized to $\delta(p - p')$ that describe the reflection process. By setting their explicit form (see Problem 2.42) we obtain

$$G_E^{\pm}(x, x') = \frac{m}{\pi\hbar^2}\int_{-\infty}^{\infty}\frac{\exp(ik(x-x'))}{k^2 - (k_0^2 \pm i\gamma)}dk - \frac{im\tilde{\alpha}}{\pi\hbar^2}\int_{-\infty}^{\infty}\frac{dk}{k^2 - (k_0^2 \pm i\gamma)}\left\{\frac{\exp(-i(kx' - |kx|))}{|k| + i\tilde{\alpha}}\right.$$

$$\left. -\frac{\exp(i(kx - |kx'|))}{|k| - i\tilde{\alpha}} + \frac{\exp(-i(|kx'| - |kx|))}{2(|k| - i\tilde{\alpha})} - \frac{\exp(-i(|kx'| - |kx|))}{2(|k| + i\tilde{\alpha})}\right\}, \qquad (3)$$

where $\tilde{\alpha} = m\alpha/\hbar^2 > 0$, $k_0 = \sqrt{2mE/\hbar^2}$. The first integral here is the Green function of a free particle (see A1.3 and Problem 2.42):

$$\tilde{G}_E(x, x') = \pm i\sqrt{\frac{m}{2\hbar^2 E}}\exp\left\{\pm i\frac{\sqrt{2mE}}{\hbar}|x - x'|\right\}. \qquad (4)$$

Note that $\pm\sqrt{E} = \sqrt{E \pm i\gamma}$.

The second integral in Eq. (3) (represented as a sum of four integrals) could be simplified by taking into account that it is an even function of the variables, x and x'. We replace these variables by their absolute values, $|x|$ and $|x'|$, and then we can divide the integration domain in two regions: $(-\infty, 0)$ and $(0, \infty)$. After this, several terms in Eq. (3) cancel each other out and we obtain

$$-\frac{im\tilde{\alpha}}{\pi\hbar^2}\int_{-\infty}^{\infty}\frac{\exp\{ik(|x| + |x'|)\}\,dk}{[k^2 - (k_0^2 \pm i\gamma)](k + i\tilde{\alpha})}. \qquad (5)$$

The remaining integral can be easily calculated using the residue theorem by closing the integration contour in the upper half-plane. In this case, there exists only one pole inside the contour located in the point $k = \pm k_0 + i\gamma$ (if $E < 0$, the pole is in the point $k = i|k_0|$). Therefore, Eq. (5) becomes

$$\pm\frac{m\tilde{\alpha}}{\hbar^2 k_0}\frac{\exp\{\pm ik_0(|x| + |x'|)\}}{\pm k_0 + i\tilde{\alpha}}.$$

As a result, the final expression for the Green function takes the form:

$$G_E^{\pm} = \pm\sqrt{\frac{m}{2\hbar^2 E}} \left\{ i\exp\left\{\pm i\frac{\sqrt{2mE}}{\hbar}|x-x'|\right\} + \frac{m\alpha \exp\left\{\pm i\sqrt{2mE/\hbar^2}(|x|+|x'|)\right\}}{\pm\sqrt{2m\hbar^2 E} + im\alpha} \right\}. \tag{6}$$

Just as in the free-particle case, these Green functions could be considered as different limits of a single analytic function, $\tilde{G}_E$, of the complex variable, E, and could be found from Eq. (6) by removing the "$\pm$" labels (see Problem 2.42). But a difference appears due to the presence of a pole at $\sqrt{E_0} = -i\alpha\sqrt{m/2\hbar^2}$ (or equivalently at $E_0 = -m\alpha/2\hbar^2$). Since $\alpha > 0$ this pole is located in the non-physical sheet and corresponds to a virtual level.

Problem 2.46

The same as in the previous problem, but for the δ-well.

Solution

The equation for the Green function supplemented by the appropriate boundary conditions are valid for any sign of α; that is, for both a barrier and a well. In the case of an attractive potential, the pole in $\tilde{G}_E$ lies within the physical sheet and E_0 coincides with the energy of the discrete level in the δ-well.

We emphasize that to properly modify Eq.(2) from the previous problem, it does not suffice to change $\alpha \to -\alpha$, but we also need to add the term, $\psi_0^*(x')\psi_0(x)/(E-E_0)$, corresponding to the bound state. However, when calculating the integral (5) with $\alpha < 0$ inside the contour, another pole appears in the point $k_1 = i|\tilde{\alpha}|$ and the contribution of this pole compensates the other additional term. This justifies the validity of Eq. (6) for any sign of α.

Problem 2.47

Find the Green function in momentum representation for a particle moving in the δ-potential, $U(x) = \alpha\delta(x)$.

Solution

The equation for Green's function in the momentum representation has the form

$$\left[\frac{p^2}{2m} - (E \pm i\gamma)\right] G_E^{\pm}(p,p') + \frac{\alpha}{2\pi\hbar} \int_{-\infty}^{\infty} G_E^{\pm}(p'',p')dp'' = \delta(p-p'). \tag{1}$$

Here the form of the operator, $\hat{U}$, is taken into account (see Problem 2.17). The "additions" of $\pm i\gamma$ to the energy is required to enforce the appropriate boundary conditions (see Problem 2.45). Using the notation

$$C_E^{\pm}(p') = \frac{\alpha}{2\pi\hbar} \int G_E^{\pm}(p'',p')dp'', \tag{2}$$

we obtain from Eq. (1),

$$G_E^\pm(p,p') = \frac{\delta(p-p') - C_E^\pm(p')}{p^2/2m - (E \pm i\gamma)}. \tag{3}$$

Integrating this with respect to p in the infinite limits and taking into account Eq. (2), we find the explicit form of $C_E^\pm(p')$ and the sought-after Green function

$$G_E^\pm(p,p') = \frac{\delta(p-p')}{p^2/2m - E \mp i\gamma} - \frac{\alpha\sqrt{2m\hbar^2 E}}{2\pi\left(\sqrt{2m\hbar^2 E} \pm im\alpha\right)(p^2/2m - E \mp i\gamma)(p'^2/2m - E \mp i\gamma)}. \tag{4}$$

We should note that $G_E^\pm(p,p')$ could have been found from $G_E^\pm(x,x')$ from Problem 2.45 by changing the representation according to Problem 1.41.

2.4 Systems with several degrees of freedom; Particle in a periodic potential

Problem 2.48

Find the energy levels and the corresponding wavefunctions of a two-dimensional isotropic oscillator. What is the degeneracy of the oscillator levels?

Solution

Since the operators

$$\hat{H}_1 = -\frac{\hbar^2}{2m}\frac{\partial^2}{\partial x^2} + \frac{kx^2}{2} \quad \text{and} \quad \hat{H}_2 = -\frac{\hbar^2}{2m}\frac{\partial^2}{\partial y^2} + \frac{ky^2}{2}$$

commute with each other, the eigenfunctions of the planar oscillator, $\hat{H} = \hat{H}_1 + \hat{H}_2$, may be chosen as the eigenfunctions of both $\hat{H}_1$ and $\hat{H}_2$. Taking this into account and using the Schrödinger equation for a linear oscillator (see (II.2)), we obtain the energy levels and the eigenfunctions of the planar oscillator in the form (see Problem 10.25)

$$\psi_{n_1 n_2}(x,y) = \psi_{n_1}^{(osc)}(x)\psi_{n_2}^{(osc)}(y), \quad E_N = \hbar\omega(N+1), \quad N = 0, 1, \ldots, \tag{2}$$

where

$$\omega = \sqrt{\frac{k}{m}}, N = n_1 + n_2, n_1 = 0, 1, \ldots, n_2 = 0, 1, \ldots.$$

Since there exist $(N+1)$ independent eigenfunctions $\psi_{n_1 n_2}$ with $n_1 = 0, 1, \ldots, N$ (in our case, $n_2 = N - n_1$), the degeneracy of the level, E_N, is equal to $g_N = N + 1$ (the ground state, $N = 0$, is non-degenerate).

Problem 2.49

Find the energy spectrum of a particle in the two-dimensional potential, $U(x,y) = k\left(x^2 + y^2\right)/2 + \alpha xy$, $|\alpha| < k$.

Solution

Let us write the potential energy in the form $U(x,y) = k_1(x+y)^2/4 + k_2(x-y)^2/4$, where $k_{1,2} = k \pm \alpha > 0$. Introducing new variables $x_1 = (x+y)/\sqrt{2}$, $y_1 = (-x+y)/\sqrt{2}$ (this transformation corresponds to a rotation in the xy-plane by the angle, $\pi/4$), we cast the Hamiltonian into the form that involves two independent oscillator Hamiltonians, as in the previous problem:

$$\hat{H} = -\frac{\hbar^2}{2m}\frac{\partial^2}{\partial x_1^2} + \frac{1}{2}k_1 x_1^2 - \frac{\hbar^2}{2m}\frac{\partial^2}{\partial y_1^2} + \frac{1}{2}k_2 y_1^2.$$

Therefore, the energy spectrum has the form

$$E_{n_1 n_2} = \hbar\sqrt{\frac{k+\alpha}{m}}\left(n_1 + \frac{1}{2}\right) + \hbar\sqrt{\frac{k-\alpha}{m}}\left(n_2 + \frac{1}{2}\right), \quad n_{1,2} = 0,\ 1,\ \ldots,$$

and the eigenfunctions that correspond to these levels could be expressed in terms of the eigenfunctions of the linear oscillator.

Problem 2.50

Find the spectrum of the Hamiltonian

$$\hat{H} = \frac{1}{2M}\hat{p}_1^2 + \frac{1}{2m}\hat{p}_2^2 + \frac{1}{2}k\left(x_1^2 + x_2^2\right) + \alpha x_1 x_2, \quad |\alpha| < k.$$

Solution

Introducing the new variables $y_1 = x_1/\gamma$, $y_2 = x_2$ with $\gamma = \sqrt{m/M}$, we have

$$\hat{H} = -\frac{\hbar^2}{2m}\frac{\partial^2}{\partial y_1^2} - \frac{\hbar^2}{2m}\frac{\partial^2}{\partial y_2^2} + \frac{k}{2}(\gamma^2 y_1^2 + y_2^2) + \alpha\gamma y_1 y_2.$$

Rotating the coordinate system in the $y_1 y_2$-plane, the potential can be reduced to the diagonal form: $U = \frac{1}{2}k_1\tilde{y}_1^2 + \frac{1}{2}k_2\tilde{y}_2^2$.

In order to determine $k_{1,2}$, we notice that for a potential of the form $U = k_{ij}y_i y_j/2$, k_{ij} transforms as a tensor under rotation. In the initial coordinate system, $k_{11} = \gamma^2 k$, $k_{22} = k$, $k_{12} = k_{21} = \alpha\gamma$, and in the rotated frame of reference, $k'_{11} = k_1$, $k'_{22} = k_2$, $k'_{12} = k'_{21} = 0$. Recall that the trace and determinant of matrix are invariant with respect to rotation of the matrix components, we have

$$k_{11} = k_1 + k_2 = k(1+\gamma^2),\ \det||k_{1k}|| = k_1 k_2 = (k^2 - \alpha^2)\gamma^2.$$

Hence,
$$k_{1,2} = \frac{(1+\gamma^2)k \pm \sqrt{(1-\gamma^2)^2 k^2 + 4\alpha^2\gamma^2}}{2}.$$

In the new variables, $\tilde{y}_1, \tilde{y}_2$, the Hamiltonian reduces to a sum of two independent linear oscillator Hamiltonians, which immediately determines the energy spectrum,
$$E_{n_1 n_2} = \hbar \sqrt{\frac{k_1}{m}} \left(n_1 + \frac{1}{2}\right) + \hbar \sqrt{\frac{k_2}{m}} \left(n_2 + \frac{1}{2}\right), \quad n_{1,2} = 0, 1, \ldots.$$

Problem 2.51

Two identical particles, placed in the same one-dimensional potential, $U(x_{1,2})$, interact with each other as mutually "impenetrable" points. Find the energy spectrum and the corresponding wavefunctions, assuming that the solution of the single-particle problem for the potential, $U(x)$, is known. To illustrate the general results, consider two such particles in an infinitely deep potential well.

Solution

The Schrödinger equation for $x_1 \leq x_2$ (we assume that the first particle is to the left of the second particle; so, $\psi(x_1, x_2) = 0$ for $x_1 \geq x_2$) has the form
$$[\hat{H}(1) + \hat{H}(2)]\psi = E\psi, \quad \text{where} \quad \hat{H} = \frac{1}{2m}\hat{p}^2 + U(x).$$

Let us now consider the new function, $\tilde{\psi}(x_1, x_2)$, which coincides with $\psi(x_1, x_2)$ for $x_1 \leq x_2$ and is equal to $-\psi(x_2, x_1)$ if $x_1 > x_2$ (that is, $\tilde{\psi}$ is an antisymmetric continuation of ψ into the region, $x_1 > x_2$). Since the resulting function and its derivatives are continuous,[20] we conclude by inspection that $\tilde{\psi}$ indeed satisfies the two-particle Schrödinger equation for any values of x_1 and x_2. The general solution is
$$\tilde{\psi}_{n_1, n_2} = \psi_{n_1}(x_1)\psi_{n_2}(x_2), \quad E_{n_1 n_2} = E_{n_1} + E_{n_2},$$

where E_n and $\psi_n(x)$ are the spectrum and the corresponding eigenfunctions for the single-particle Hamiltonian. The antisymmetric character of the wavefunctions $\tilde{\psi}$ makes it necessary to choose them in form
$$\tilde{\psi}_{n_1, n_2} = \frac{1}{\sqrt{2}} [\psi_{n_1}(x_1)\psi_{n_2}(x_2) - \psi_{n_2}(x_1)\psi_{n_1}(x_2)]$$

and gives restrictions on $n_{1,2}$: $n_1 \neq n_2$. So
$$\psi_{n_1, n_2} = \psi_{n_1}(x_1)\psi_{n_2}(x_2) - \psi_{n_2}(x_1)\psi_{n_1}(x_2), \quad E_{n_1 n_2} = E_{n_1} + E_{n_2}, \quad n_1 < n_2$$

---

[20] The continuity of the derivatives of $\tilde{\psi}$ with respect to $x_{1,2}$ at $x_1 = x_2$ can be explicitly verified by differentiating the equality, $\tilde{\psi}(x, x) = 0$.

($x_1 \leq x_2$) The energy level is two-fold degenerate, with the second independent solution of the Schrödinger equation corresponding to the first particle moving to the right of the second particle.

Problem 2.52

Generalize the result of the previous problem to the case of an N-particle system.

Result

The energy spectrum of the system is given by

$$E_{n_1 \ldots n_N} = \sum_{a=1}^{N} E_{n_a}, \text{ where } n_1 < n_2 < \cdots < n_N.$$

Problem 2.53

For a particle in the periodic potential of the form, $U(x) = \alpha \sum_{n=-\infty}^{\infty} \delta(x - na)$ (this potential can be viewed as a model of an ideal one-dimensional "crystal"; see Fig. 2.16), find a system of independent solutions of the Schrödinger equation for an arbitrary value of E and determine the energy spectrum.

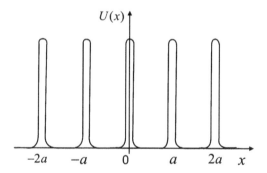

Fig. 2.16

Solution

A general solution of the Schrödinger equation in the region, $n < x/a < (n+1)$, can be written in the form

$$\psi(x) = A_n e^{ik(x-na)} + B_n e^{-ik(x-na)}, \tag{1}$$

where $k = \sqrt{2mE/\hbar^2}$. Let us consider independent solutions that satisfy the relation $\psi(x+a) = \mu\psi(x)$. We have

$$A_{n-1} = \frac{A_n}{\mu}, \quad B_{n-1} = \frac{B_n}{\mu}, \tag{2}$$

Using the matching conditions at $x = na$ (see Problem 2.6), we find

$$A_n + B_n = e^{ika} A_{n-1} + e^{-ika} B_{n-1},$$

$$\left(1 + \frac{2im\alpha}{\hbar^2 k}\right) A_n - \left(1 - \frac{2im\alpha}{\hbar^2 k}\right) B_n = e^{ika} A_{n-1} - e^{-ika} B_{n-1}. \tag{3}$$

Using Eq. (2) we obtain a system of linear equations for the coefficients A_n and B_n. A non-trivial solution exists only if

$$\mu^2 - 2\mu f(E) + 1 = 0, \text{ where } f(E) \equiv \cos ka + \frac{m\alpha}{\hbar^2 k} \sin ka \tag{4}$$

and

$$B_n = \frac{\mu - e^{ika}}{e^{-ika} - \mu} A_n. \tag{5}$$

From that, it follows that

$$\mu_{1,2} = f(E) \pm \sqrt{f^2(E) - 1}. \tag{6}$$

For any fixed value of the energy, E, Eq. (6) determines two values of μ that correspond to two independent solutions of the Schrödinger equation, and their product is $\mu_1 \cdot \mu_2 = 1$. In the case $f^2(E) > 1$, both values, $\mu_{1,2}$, are real. Consequently, the corresponding two solutions of the Schrödinger equation increase at large distances (corresponding to solutions with $\mu_1 > 1$, as $x \to +\infty$, and with $\mu_2 < 1$, as $x \to -\infty$). Therefore, such solutions are unphysical.

The values of E for which $|\mu| = 1$ (see $f^2(E) \leq 1$), however, do describe physical states. This condition gives rise to the following equation:

$$-1 \leq \cos ka + \frac{m\alpha}{\hbar^2 k} \sin ka \leq 1. \tag{7}$$

Hence the allowed values of E arrange themselves into a band structure. If we set[21] $\mu \equiv e^{iqa}$, where $-\pi \leq qa \leq \pi$, and $\hbar q$ is called a *quasi-momentum* (not to be confused with the "real" momentum, $\hbar k$, which is well-defined in the absence of a lattice only), then the equation for $E_n(q)$ takes the form (here, n is a band index, with $n+1$ being the ordinal number of the band; see Fig. 2.17 for $\alpha > 0$):

$$\cos qa = \cos \frac{a}{\hbar} \sqrt{2m E_n(q)} + \frac{\alpha}{\hbar} \sqrt{\frac{m}{2 E_n(q)}} \sin \frac{a}{\hbar} \sqrt{2m E_n(q)}. \tag{8}$$

---

[21] The solutions of the Schrödinger equation that correspond to a definite quasi-momentum are called *Bloch functions*.

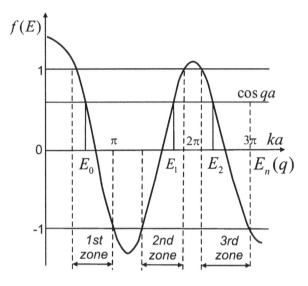

Fig. 2.17

Let us briefly discuss properties of the spectrum following from Eq. (8), (see also Problem 8.32, which focuses on the weak-field limit, $m\alpha a/\hbar^2 \ll 1$):

1) The q-dependence of $E_n(q)$ is even. Therefore, the states that differ only by the sign of quasi-momentum are independent and correspond to the two-fold degeneracy of the spectrum, $E_n(q)$.

2) The energy bands do not overlap. In the case $\alpha > 0$, they all lie in the region $E_n > 0$ and $\pi n < k_n a < \pi(n+1)$, $n = 0, 1, \ldots$. In the case $m\alpha a/[(n+1)\hbar^2] \gg 1$, the bands are narrow, but their width increases with an increase in n. In the case $m\alpha a/[(n+1)\hbar^2] \ll 1$, they fill almost the entire interval mentioned above. With the change of sign of α, the lower band moves down into the negative-energy region, $E < 0$ (in this case, k has a non-zero imaginary part).

3) Near the top or the bottom of a band (e.g., near the points $q_1 = 0$ and $q_2 = \pm \pi/a$), the q-dependence of the spectrum $E_n(q)$ has the parabolic form, i.e., $E_n(q) - E_n(q_{1,2}) \propto (q - q_{1,2})^2$ (see Problem 8.32).

In conclusion we note that eigenfunctions in this problem cannot be normalized to unity, so the localized stationary states of a particle in periodic potential are absent; wavefunctions (1) and (2) correspond to a particle with the quasi-momentum, $\hbar q$, moving "freely" (i.e., without reflection) through the infinite crystal.

Problem 2.54

Find the energy spectrum of a particle in the potential, $U(x) = \alpha \sum_{n=-\infty}^{\infty}{}' \delta(x - na)$, where the prime in the sum sign indicates that the term with $n = 0$ is omitted. This

80 *Exploring Quantum Mechanics*

potential is a model of a one-dimensional crystal with a defect (a *vacancy* at $n = 0$; see Fig. 2.18).

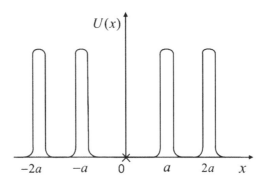

Fig. 2.18

Show that apart from the allowed energy bands in the case of a perfect crystal (see Problem 2.53), new discrete levels appear and correspond to states localized near the lattice defect.

Solution

The allowed energy bands found in the previous problem are relevant here as well. Indeed, an arbitrary solution of the Schrödinger equation with an allowed energy, $E_n(q)$, correspond, for $x > 0$ and $x < 0$, to a superposition of two independent solutions of the perfect-lattice problem above with definite quasi-momenta, $\pm \hbar q$, and such that the resulting function does not increase with $x \to \pm\infty$. However, in contrast to the case of a perfect crystal, the independent solutions in this problem do not have a certain value of the quasi-momentum (the physical reason being that a change in quasi-momentum is now possible due to scattering off of the lattice defect). Note, however, that the two-fold degeneracy remains in the case under consideration.

Furthermore, new energy levels appear that correspond to states localized in the vicinity of the defect. To find these levels, consider a solution of the Schrödinger equation with a certain parity (with respect to inversion $x \to -x$).

In the case of even solutions in the region $|x| < a$, we have $\psi_E^+(x) = C \cos kx$. On the other hand, in the region $x > 0$, a solution of the Schrödinger equation must coincide with that for a periodic potential and satisfy $\psi(x + a) = \mu \psi(x)$ for $\mu < 1$ (another independent solution corresponds to $\mu' = 1/\mu > 1$; such a solution increases as $x \to +\infty$). This solution in the regions $n < x/a < (n + 1)$ has the form (below, $k = \sqrt{2mE/\hbar^2}$)

$$\psi^+(x) = \mu^n [A \cos k(x - na) + B \sin k(x - na)] \tag{1}$$

Since it must coincide with $\psi_E^+(x)$ for $0 \leq x < a$, we find that $A = C$ and $B = 0$. Using the matching conditions at the point, $x = a$, for the wavefunction (1) (see Problem 2.6), we obtain the following solutions:

$$\cos ka = \mu, \quad ka \sin ka = \frac{2m\alpha a}{\hbar^2} \cos ka. \tag{2}$$

The latter of these equations determines the required even levels. We now discuss properties of these levels:

1) The levels are discrete.
2) These levels are positioned between the neighboring bands of the continuous spectrum and in the case $\alpha > 0$ the lower level occurs below the lowest band.
3) As the energy of the level increases – as is seen from Eq. (2) – we notice that $\mu \to 1$. In this case, localization length of the particle in vicinity of the defect expands.
4) Normalizing the wavefunction of the localized level to unity, we have

$$C^2 = \frac{2(1-\mu^2)k}{2ka + \sin\ 2ka}.$$

Note that in the case of $m\alpha a/\hbar^2 \gg 1$, the wavefunctions of the low-lying levels, E_s ($s = 0, 1, \ldots$) with $s \ll m\alpha a/\hbar^2$, are localized in the region $|x| \leq a$ (in this case $\mu \ll 1$) and are close to the wavefunctions of stationary states of a particle in an infinitely deep potential well with the width $2a$. Note also that "new" odd-parity levels do not appear in this problem.

Problem 2.55

Find the energy spectrum and degeneracy of the levels of a particle in the potential of the form

$$U(x) = \begin{cases} \alpha \sum_{n=1}^{\infty} \delta(x - na), & x > 0 \\ U_0 > 0, & x \leq 0 \end{cases}$$

as illustrated in Fig. 2.19. Compare your results with the case of an ideal infinite crystal (see Problem 2.53). Pay special attention to the appearance of states, localized near the boundary of the crystal. These, states you will find, are called *surface* or *Tamm* states, and they play an important role in semiconductor physics. (I. E. Tamm was the first to point out the existence of such states.)

Solution

For $x > 0$, the two independent solutions of the Schrödinger equation for any value of E have the property $\psi_{1,2}(x+a) = \mu_{1,2}\psi_{1,2}(x)$, with $\mu_1 \cdot \mu_2 = 1$. For the values of energy $E_n(q)$ that belong to the allowed energy bands of the infinite crystal (see Problem 2.53), both of these solutions do not increase as $x \to +\infty$. However, for all other values of E there is only one non-increasing solution with $\mu_1 < 1$, which decreases as $x \to \infty$. Relying on these arguments, let us analyze the particle spectrum.

82 *Exploring Quantum Mechanics*

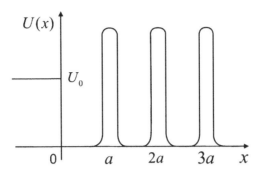

Fig. 2.19

1) For $E > U_0$ the energy spectrum is continuous. The values of energy that belong to the allowed energy bands of the infinite crystal are two-fold degenerate (and correspond to a particle moving "freely" through space, reflecting from the crystal boundaries with a certain probability). The states with any other value of the energy are non-degenerate, and their wavefunction decreases as the particle moves deep into the crystal (a particle with such a "forbidden" energy experiences a total reflection from the crystal).

2) For $E < U_0$, the spectrum has the same band structure as in the case of an infinite crystal. In this case however, the levels are non-degenerate: when $x > 0$, the wavefunction is given by a superposition of states with the quasi-momenta, $\pm \hbar q$ (a particle with such the energy moves inside the crystal reflecting from its boundaries).

3) Furthermore, in the case $E < U_0$, isolated levels might exist that correspond to the particle states localized near the crystal boundaries. To find these levels, consider the solution of the Schrödinger equation that decreases as $x \to \pm \infty$. For $x < 0$, it has the form $\psi = Ce^{\kappa x}$, where $\kappa = \sqrt{2m(U_0 - E)/\hbar^2}$. While in the case $x > 0$ and $n < x/a < (n+1)$ it can be written as follows

$$\psi = A\mu^n \sin[k(x - na) + \delta], \quad k = \sqrt{2mE/\hbar^2}, \quad |\mu| < 1. \tag{1}$$

Matching the solution at $x = 0$ and $x = a$ gives

$$A \sin \delta = 1, \, kA \cos \delta = \kappa, \, \sin(ka + \delta)$$

$$= \mu \sin \delta, \, \mu k \cos \delta - k \cos(ka + \delta) = \frac{2m\alpha}{\hbar^2} \mu \sin \delta$$

(we put $C = 1$). Hence, it follows that

$$ka \cos ka = (\sin ka)\left[\frac{U_0 a}{\alpha} - \frac{\sqrt{2m(U_0 - E)a^2}}{\hbar}\right], \quad \mu = \frac{U_0 a}{ka\alpha} \sin ka. \tag{2}$$

This equation determines the energy spectrum of the states under consideration; the number of the levels depends on the parameters of the potential (there exist parameter regimes, where there are no such levels at all). These levels are situated between the allowed energy bands of the infinite crystal. If we change the potential parameters, the location of the levels also changes. At the same time, new bound states might appear or the existing ones might disappear, as the level moves to the nearest band and delocalizes.

We leave the reader the further analysis of the spectrum that follows from Eq. (2), and illustrate only one special case, where $U_0 \gg \hbar^2/ma^2$, $\alpha < 0$ (this describes a crystal that consists of δ-wells) and $m|\alpha|/\hbar^2 \sim 1$. If $E \ll U_0$, it follows from Eq. (2) that $ka = n\pi + \varepsilon$, where $n = 1, 2, \ldots$ and $|\varepsilon| \ll 1$. So, we obtain $\varepsilon \approx n\pi\alpha/U_0 a$. For such states (positioned between the bands)

$$\mu = \cos ka + \sqrt{\frac{U_0}{E_n} - 1} \sin ka \approx (-1)^n \left(1 + \alpha\sqrt{\frac{2m}{\hbar^2 U_0}}\right), \tag{3}$$

i.e., $|\mu| < 1$ (here $|\mu| \approx 1$), and the delocalization domain extends far inside the crystal. In the case of $\alpha > 0$, there are no bound states in this energy band ($|\mu| > 1$ for the solutions of Eq. (2)), although such states appear with increasing U_0 (such states first appear with the energy, $E = U_0$) and then the level "merges" the band.

3
Orbital angular momentum

The operator for the orbital angular momentum of a particle, $\hbar \hat{\mathbf{l}} = [\hat{\mathbf{r}} \times \hat{\mathbf{p}}]$, obeys the following commutation relations:[22]

$$[\hat{l}_i,\ \hat{l}_k] = i\varepsilon_{ikn}\hat{l}_n,\quad [\hat{\mathbf{l}}^2,\ \hat{l}_i] = 0, \tag{III.1}$$

as well as

$$[\hat{l}_i,\ \hat{x}_k] = i\varepsilon_{ikn}\hat{x}_n,\quad [\hat{l}_i,\ \hat{p}_k] = i\varepsilon_{ikn}\hat{p}_n, \tag{III.2}$$

In spherical coordinates, operators $\hat{l}_i$ depend only on the angular variables ϑ, φ. So the eigenfunctions and eigenvalues of the operator $\hat{l}_z = -i\frac{\partial}{\partial \varphi}$ have the form ($m \equiv l_z$):

$$\phi_m(\varphi) = \frac{e^{im\varphi}}{\sqrt{2\pi}},\quad m = 0, \pm 1, \pm 2, \ldots. \tag{III.3}$$

We can express the operator of the angular momentum squared, $\hat{\mathbf{l}}^2$, in terms of the angular terms of the Laplace operator. Its eigenvalues are $l(l+1)$, $l = 0, 1, 2, \ldots$. Operators $\hat{\mathbf{l}}^2$ and $\hat{l}_z$ form a complete operator set for the angular part of the wavefunctions. *Spherical functions* $Y_{lm}(\vartheta, \varphi)$ are the normalized eigenfunctions of these operators.

$$\hat{\mathbf{l}}^2 Y_{lm} \equiv -\left[\frac{1}{\sin\vartheta}\frac{\partial}{\partial\vartheta}\left(\sin\vartheta\frac{\partial}{\partial\vartheta}\right) + \frac{1}{\sin^2\vartheta}\frac{\partial^2}{\partial\varphi^2}\right]Y_{lm} = l(l+1)Y_{lm},$$

$$\hat{l}_z Y_{lm} \equiv -i\frac{\partial}{\partial\varphi}Y_{lm} = mY_{lm}. \tag{III.4}$$

They have the form ($|m| \leq l$)

$$Y_{lm}(\vartheta, \varphi) = (-1)^{\frac{m+|m|}{2}} i^l \sqrt{\frac{2l+1}{4\pi} \cdot \frac{(l-|m|)!}{(l+|m|)!}} P_l^{|m|}(\cos\vartheta)e^{im\varphi}, \tag{III.5}$$

$$P_l^{|m|}(\cos\vartheta) = \sin^{|m|}\vartheta \frac{d^{|m|}}{d(\cos\vartheta)^{|m|}} P_l(\cos\vartheta).$$

---

[22] $[\hat{l}_x,\ \hat{l}_y] = i\hat{l}_z$, *etc.* In what follows, we assume that the angular momentum is measured in the units of $\hbar$, so the corresponding operators and their eigenvalues are dimensionless. We should stress that relations (III.1) or (III.8)–(III.10) are valid for the angular momentum of any system, irrespective of its nature (orbital, spin, or total).

P_l and $P_l^{|m|}$ are the Legendre polynomials and the associated Legendre polynomials, respectively. Note that $Y_{lm}^* = (-1)^{l-m} Y_{l\,-m}$, $\int Y_{lm}^* Y_{l'm'} d\Omega = \delta_{ll'} \delta_{mm'}$.

The spherical harmonics have a definite parity $I = (-1)^l$. For them, the *addition theorem* holds:

$$\frac{2l+1}{4\pi} P_l(\mathbf{n} \cdot \mathbf{n}') = \sum_{m=-l}^{l} Y_{lm}(\mathbf{n}) Y_{lm}^*(\mathbf{n}'), \qquad \text{(III.6)}$$

where $\mathbf{n}$ and $\mathbf{n}'$ are unit vectors, and in this case

$$Y_{lm}(\mathbf{n}) \equiv Y_{lm}(\vartheta, \varphi), \qquad \mathbf{n} \cdot \mathbf{n}' = \cos\vartheta \cos\vartheta' + \sin\vartheta \sin\vartheta' \cos(\varphi - \varphi').$$

The spherical functions for the angular momenta with $l = 0, 1, 2$ are

$$Y_{00} = \frac{1}{\sqrt{4\pi}} \,;\, Y_{10} = i\sqrt{\frac{3}{4\pi}} \cos\vartheta \,;\, Y_{20} = \sqrt{\frac{5}{16\pi}} (1 - 3\cos^2\vartheta);$$

$$Y_{1\pm 1} = \mp i\sqrt{\frac{3}{8\pi}} \sin\vartheta \, e^{\pm i\varphi}; \; Y_{2\pm 1} = \pm\sqrt{\frac{15}{8\pi}} \sin\vartheta \cos\vartheta \, e^{\pm i\varphi}; \qquad \text{(III.7)}$$

$$Y_{2\pm 2} = -\sqrt{\frac{15}{32\pi}} \sin^2\vartheta \, e^{\pm 2i\varphi}.$$

It may be useful to have the spherical functions in Cartesian coordinates:

$$Y_{10} \propto \cos\vartheta = \frac{z}{r}, \; Y_{1\pm 1} \propto \sin\vartheta \, e^{\pm i\varphi} = \frac{x \pm iy}{r}, \; Y_{20} \sim \frac{x^2 + y^2 - 2z^2}{r^2},$$

etc.

The *raising and lowering operators*, $\hat{l}_\pm = \hat{l}_x \pm i\hat{l}_y$, obey the commutation relations $[\hat{l}_z, \hat{l}_\pm] = \pm\hat{l}_\pm$. It follows that the only matrix elements, $\langle lm'|\hat{l}_\pm|lm\rangle$, that are not equal to zero are[23]

$$(l_+)_{m,m-1} = (l_-)_{m-1,m} = \sqrt{(l+m)(l-m+1)}. \qquad \text{(III.8)}$$

Accordingly, the non-vanishing matrix elements for $\hat{l}_x$ and $\hat{l}_y$ are

$$(l_x)_{m,m-1} = (l_x)_{m-1,m} = \frac{1}{2}\sqrt{(l+m)(l-m+1)},$$

$$(l_y)_{m,m-1} = -(l_y)_{m-1,m} = -\frac{i}{2}\sqrt{(l+m)(l-m+1)}. \qquad \text{(III.9)}$$

These equations determine the form of the operators in the l_z-representation. Finally, we have

$$(l_z)_{mm'} = m\delta_{mm'}. \qquad \text{(III.10)}$$

---

[23] To fully define the value of matrix element, we should also use $\hat{\mathbf{l}}^2 = \hat{l}_-\hat{l}_+ + \hat{l}_z^2 + \hat{l}_z$. A choice of the phase factor in (III.8) fixes the relative phase of the wavefunctions with different m but equal l.

3.1 General properties of angular momentum

Problem 3.1

Show that the relation $\mathbf{L}^2 = l(l+1)$ could be obtained by using elementary equations of probability theory. Assume that the only possible values of angular momentum projection on an arbitrary axis are $m = -l, -l+1, \ldots, l$, and that all of these values have equal probability and all the axes are equivalent.

Solution

Since the probabilities of different L_z are the same, we have[24]

$$\overline{L_z^2} = \frac{1}{2l+1} \sum_{m=-l}^{l} m^2 = \frac{l(l+1)}{3}$$

Due to equivalence of the axes x, y, z, we have

$$\mathbf{L}^2 \equiv \overline{\mathbf{L}^2} = \overline{L_x^2} + \overline{L_y^2} + \overline{L_z^2} = 3\overline{L_z^2} = l(l+1).$$

Note that the change from the discrete probability distribution, $w(m) = \frac{1}{2l+1}$, to the continuous distribution, $dw = dl_z/2l$ with $-l \leq l_z \leq l$, gives the classical result $\mathbf{L}^2 = l^2$.

Problem 3.2

Find the stationary wavefunctions and energy levels of a planar (two-dimensional) rotor[25] with a moment of inertia I. What is the degeneracy multiplicity?

Find the probabilities for different energy and angular momentum projection values, as well as the mean values and fluctuations of these quantities, for the rotor with the wavefunction $\psi = C \cos^2 \varphi$.

Solution

1) The Hamiltonian for the planar rotor is $H = \frac{1}{2I} M_z^2$, where $M_z \equiv p_\varphi$ is the projection of its angular momentum onto the axis z that is perpendicular to the plane of rotation. The Hamiltonian operator has the form $\hat{H} = \frac{\hat{M}_z^2}{2I} \equiv \frac{\hbar^2 \hat{l}_z^2}{2I}$. Since $\hat{H}$

---

[24] This sum can be calculated as follows:

$$\sum_{m=-l}^{l} m^2 = 2 \left[\frac{d^2}{da^2} \sum_{m=0}^{l} e^{am} \right]_{a=0} = 2 \left[\frac{d^2}{da^2} \frac{1 - e^{a(l+1)}}{1 - e^a} \right]_{a=0} = \frac{1}{3} l(l+1)(2l+1).$$

[25] A *rotor* is a system of two rigidly connected particles that rotate around the center of inertia. The moment of inertia for the rotor is equal to $I = \mu a^2$, where μ is the reduced mass of the particles and a is a distance between them.

commutes with $\hat{l}_z$, then the eigenfunctions of $\hat{H}$ could be chosen simultaneously with the eigenfunctions of $\hat{l}_z$. We immediately write the spectrum and eigenfunctions of the Hamiltonian:

$$E_{|m|} = \frac{\hbar^2 m^2}{2I}, \quad \psi_m = \frac{1}{\sqrt{2\pi}} e^{im\varphi}, \quad m = 0, \pm 1, \dots. \tag{1}$$

All levels except for the ground one are two-fold degenerate. See that it is possible to choose $\hat{H}$ eigenfunctions as $\psi^+_{|m|} = \frac{1}{\sqrt{\pi}} \cos m\varphi$, $\psi^-_{|m|} = \frac{1}{\sqrt{\pi}} \sin m\varphi$. In this case, they have a definite parity ($+1$ or -1) with respect to reflection through the x-axis.

2) Since $\cos\varphi = (e^{i\varphi} + e^{-i\varphi})/2$, we have

$$\psi = C\cos^2\varphi = \frac{C}{4}(e^{2i\varphi} + 2 + e^{-2i\varphi}) \equiv \sum_m c_m \frac{1}{\sqrt{2\pi}} e^{im\varphi}.$$

This gives the probability distribution for different values of rotor's angular momentum, $w(m) = |c_m|^2$, and then the probability distribution for energies, $w(E_{|m|}) = w(m) + w(-m)$ (with $m \neq 0$). We have $C^2 = 4/3\pi$ due to normalization.

$$w(0) = 4w(\pm 2), \quad w(E_0) = w(0) = \frac{2}{3}, \quad w(E_2) = 2w(2) = 2w(-2) = \frac{1}{3},$$

The probabilities of other values are equal to zero. Finally, we have

$$\overline{m} = 0, \quad \overline{(\Delta m)^2} = \frac{4}{3}, \quad \overline{E} = \frac{2\hbar^2}{3I}, \quad \overline{(\Delta E)^2} = \frac{8\hbar^4}{9I^2}.$$

Problem 3.3

Find the wavefunctions and energy levels of the stationary states of a spherical (three-dimensional) rotor with a momentum of inertia I. What is the degeneracy multiplicity?

Let the wavefunction be $\psi = C\cos^2\vartheta$. Find the probability distribution for energy, angular momentum, and z-axis angular momentum. Find the mean values and fluctuations of these quantities.

Solution

1) The Hamiltonian operator has the form $\hat{H} = \frac{\hbar^2}{2I}\hat{\mathbf{l}}^2$, and its eigenvalues and eigenfunctions are given by

$$E_l = \frac{\hbar^2 l(l+1)}{2I}, \quad \psi_{lm} = Y_{lm}(\vartheta, \varphi), \tag{1}$$

where $l = 0, 1, \dots$; $m = 0, \pm 1, \dots, \pm l$; Y_{lm} are the spherical harmonics; and ϑ, φ are the polar and azimuth angles of the rotor axis. The energy levels are $(2l+1)$-fold degenerate and have the parity equal to $(-1)^l$.

2) The wavefunction describing the rotor with of $l_z = 0$ has the form

$$\psi(\vartheta) = C\cos^2\vartheta = \frac{\sqrt{4\pi}C}{3}\left[\frac{1}{\sqrt{4\pi}} - \frac{1 - 3\cos^2\vartheta}{\sqrt{4\pi}}\right] \equiv \sum_l c_l Y_{l0}.$$

Taking into account the expressions for Y_{00} and Y_{20} (III.7), we find that the rotor momentum could take only two values: $l = 0$ and $l = 2$, with the probabilities $w(0) = 5/9$ and $w(2) = 4/9$. Finally, $\overline{E} = 4\hbar^2/3I$, $\overline{(\Delta E)^2} = \frac{20\hbar^4}{9I^2}$, and $|C|^2 = 5/4\pi$.

Problem 3.4

Give a simple explanation of

a) the commutativity of different components of the momentum operator;
b) the non-commutativity of the angular momentum components;
c) the commutativity of the momentum projection and angular momentum projection on the same axis, and their non-commutativity for different axes, using kinematic interpretation of these operators in terms of infinitely small translations and rotations.

Solution

As it is known (see Ch. 1, secs. 15, 26), the linear momentum operator $\hat{\mathbf{P}}$ and the angular momentum operator $\hat{\mathbf{L}}$ of the system are connected with the infinitely small translations and rotations:

$$\hat{T}(\delta\mathbf{a}) \approx 1 + \frac{i}{\hbar}\delta\mathbf{a}\cdot\hat{\mathbf{P}} \quad \text{and} \quad \hat{R}(\delta\boldsymbol{\varphi}_0) \approx 1 + i\boldsymbol{\varphi}_0\cdot\hat{\mathbf{L}}.$$

Any translation commutes with any other translation and therefore the operators of different momentum components commute with each other. The same can be said about a translation and rotation over the same axis. On the other hand, two rotations or a translation and rotation about two nonparallel axes do not commute, which implies the non-commutativity of the corresponding operators.

Problem 3.5

Find the following commutators:

a) $[\hat{l}_i, \hat{\mathbf{r}}^2]$, $[\hat{l}_i, \hat{\mathbf{p}}^2]$, $[\hat{l}_i, \hat{\mathbf{p}}\cdot\hat{\mathbf{r}}]$, $[\hat{l}_i, (\hat{\mathbf{p}}\cdot\hat{\mathbf{r}})^2]$;
b) $[\hat{l}_i, (\hat{\mathbf{p}}\cdot\hat{\mathbf{r}})\hat{p}_k]$, $[\hat{l}_i, (\hat{\mathbf{p}}\cdot\hat{\mathbf{r}})\hat{x}_k]$, $[\hat{l}_i, (c_1\hat{x}_k + c_2\hat{p}_k)]$;
c) $[\hat{l}_i, \hat{x}_k\hat{x}_l]$, $[\hat{l}_i, \hat{p}_k\hat{p}_l]$, $[\hat{l}_i, \hat{x}_k\hat{p}_l]$.

Here, c_1, c_2 are some constants.

Solution

For calculation of the operators we can use the results of Problem 1.4 and Eq. (III.2).

a) All the commutators are equal to zero due to the general equation $[\hat{l}_i, \hat{f}] = 0$, where $\hat{f}$ is a scalar quantity operator.

b) These commutators have the form $[\hat{l}_i, \hat{f}_k] = i\varepsilon_{ikl}\hat{f}_l$, where $\hat{f}_k$ is an operator of the kth projection of the corresponding *vector* operator $\hat{\mathbf{f}}$.

c) These commutators have $[\hat{l}_i, \hat{f}_{kl}] = i(\varepsilon_{ikp}\delta_{nl} + \varepsilon_{iln}\delta_{kp})\hat{f}_{pn}$, where $\hat{f}_{ik}$ are the corresponding second-rank tensor components operators.

This universal structure of the commutators of angular momentum components, $\hat{l}_i$, with scalar, vector, and tensor operators is due to the $\hat{\mathbf{l}}$ operator describing transformation of wavefunctions with the coordinate system rotations. Commutators involving tensors of the same rank are transformed in the same way (independently of a specific tensor form).

Problem 3.6

Find the normalized wavefunctions $\psi_{r_0 lm}$ that describe a particle located at a distance r_0 from the origin with angular momentum l and its projection m onto the axis z.

Solution

The desired functions have the form $\psi_{r_0 lm} = C(r_0)\delta(r - r_0)Y_{lm}(\vartheta, \varphi)$. From the normalization condition $\langle r'_0, l', m'|r_0, l, m\rangle = \delta(r_0 - r'_0)\delta_{ll'}\delta_{mm'}$ we get $C(r_0) = r_0^{-1}$.

Problem 3.7

Find general eigenfunctions of particle momentum and angular momentum projections on the z axis.

Solution

$\psi_{p_z m}(\mathbf{r}) = \frac{1}{\sqrt{2\pi\hbar}}e^{ip_z z/\hbar} \cdot \frac{1}{\sqrt{2\pi}}e^{im\varphi} \cdot f(\rho)$, where $f(\rho)$ is an arbitrary function of ρ (distance to the z-axis) in a cylindrical coordinate system.

Problem 3.8

Show that the mean values of the vectors $\bar{\mathbf{L}}$, $\bar{\mathbf{r}}$, $\bar{\mathbf{p}}$ for the particle state with wavefunction $\psi = \exp(i\mathbf{p}_0 \cdot \mathbf{r}/\hbar)\varphi(\mathbf{r})$ are connected by the classical relation $\bar{\mathbf{L}} = \bar{\mathbf{r}} \times \bar{\mathbf{p}}$. Here, $\mathbf{p}_0$ is a real vector and $\varphi(\mathbf{r})$ is a real function.

Solution

Assuming that the wavefunction is normalized to unity, we find $\bar{\mathbf{r}} = \int \mathbf{r}\varphi^2(\mathbf{r})dV$, $\bar{\mathbf{p}} = \mathbf{p}_0$. Since $\hat{L}_i = \varepsilon_{ikl}x_k\hat{p}_l$, we have

$$\bar{L}_i = \varepsilon_{ikl} \int \phi(\mathbf{r}) \{x_k p_{0l} + x_k \hat{p}_l\} \phi(\mathbf{r}) dV. \tag{1}$$

The second term can be transformed:

$$\phi(\mathbf{r}) x_k \hat{p}_l \phi(\mathbf{r}) = -\frac{i}{2}\hbar \frac{\partial}{\partial x_l}(\phi^2 x_k) + \frac{i}{2}\hbar \delta_{kl} \phi^2. \tag{2}$$

We can see that this contribution to $\bar{L}_i$ is equal to zero by using Gauss' law for the first term and the convolution of δ_{kl} and ε_{ikl} for the second. From (1), we get $\bar{L}_i = \varepsilon_{ikl} \bar{x}_k p_{0l}$ or $\bar{\mathbf{L}} = \bar{\mathbf{r}} \times \bar{\mathbf{p}}$.

Problem 3.9

Find the eigenfunctions of the operators $\hat{\mathbf{l}}^2$ and $\hat{l}_z$ in the momentum representation. Show that $\bar{\mathbf{p}} = 0$ for the states with definite values of l and m.

Solution

In the momentum representation, we have $\hat{\mathbf{p}} = \mathbf{p}$ and $\hat{\mathbf{r}} = i\hbar \nabla_\mathbf{p}$ so that $\hbar \hat{\mathbf{l}} = \hat{\mathbf{r}} \times \hat{\mathbf{p}} = -i\hbar \mathbf{p} \times \nabla_\mathbf{p}$, which is the same as the form of vector in the position representation, with only the replacement of $\mathbf{r}$ by $\mathbf{p}$. It allows us to write down the eigenfunctions, $\psi_{lm}(\mathbf{p}) = Y_{lm}(\tilde{\vartheta}, \tilde{\varphi})$, of operators $\hat{\mathbf{l}}^2$ and $\hat{l}_z$, where $\tilde{\vartheta}$ and $\tilde{\varphi}$ are the polar and azimuthal angles of the $\mathbf{p}$-direction in spherical coordinates (in the p-representation just as in the r-representation, the angular momentum operator acts only on the angular values).

Since the spherical harmonics have a definite parity, all matrix elements of the form $\langle lm|\mathbf{p}|lm'\rangle$ are equal to zero (compare 1.16).

Problem 3.10

Prove that the functions produced by the action of operators $\hat{l}_\pm = \hat{l}_x \pm i\hat{l}_y$ on the eigenfunctions ψ_m of $\hat{l}_z$, are also the eigenfunctions of $\hat{l}_z$, corresponding to eigenvalues $m \pm 1$.

Show also that for eigenfunctions of $\hat{l}_z$, we have

a) $\overline{\hat{l}_x} = \overline{\hat{l}_y} = 0$, b) $\overline{\hat{l}_x^2} = \overline{\hat{l}_y^2}$, c) $\overline{\hat{l}_x \hat{l}_y + \hat{l}_y \hat{l}_x} = 0$.

Solution

From the commutation relations of the angular momentum components, it follows that $\hat{l}_z \hat{l}_\pm = \hat{l}_\pm(\hat{l}_z \pm 1)$. If we apply this to ψ_m, we get $\hat{l}_z(\hat{l}_\pm \psi_m) = (m \pm 1)(\hat{l}_\pm \psi_m)$. The functions $\hat{l}_\pm \psi_m$ are also the eigenfunctions of $\hat{l}_z$ corresponding to the eigenvalues $m \pm 1$ (in particular when $m = \pm l$, one of these functions is identically zero).

We can use the orthogonality of eigenfunctions to obtain

$$\langle m|\hat{l}_\pm|m\rangle \propto \langle m|m\pm 1\rangle = 0 \text{ and } \langle m|\hat{l}_\pm^2|m\rangle = 0. \qquad (1)$$

It follows that $\overline{\hat{l}_x \pm i\hat{l}_y} = 0$ or $\overline{\hat{l}_x} = \overline{\hat{l}_y} = 0$. The second equation of (1) is equal to

$$\overline{\hat{l}_x^2 - \hat{l}_y^2} \pm i\overline{(\hat{l}_x\hat{l}_y + \hat{l}_y\hat{l}_x)} = 0$$

so that

$$\overline{\hat{l}_x^2} = \overline{\hat{l}_y^2}, \quad \overline{\hat{l}_x\hat{l}_y + \hat{l}_y\hat{l}_x} = 0. \qquad (2)$$

If we average the commutator $[\hat{l}_x, \hat{l}_y] = i\hat{l}_z$ and use (2), we get $\overline{\hat{l}_x\hat{l}_y} = -\overline{\hat{l}_y\hat{l}_x} = im/2$.

Notice that these properties of the mean values are connected with the fact that states with a definite value of the angular momentum l_z-projection are axially symmetric. All directions in the xy-plane are then equivalent.

Problem 3.11

In the state ψ_{lm} with definite angular momentum l and its z-component m, find the mean values $\overline{\hat{l}_x^2}$, $\overline{\hat{l}_y^2}$ as well as the mean values $\overline{\hat{l}_{\tilde{z}}}$ and $\overline{\hat{l}_{\tilde{z}}^2}$ of the angular momentum projection along the $\tilde{z}$-axis making an angle α with the z-axis.

Solution

Since $\hat{l}_x^2 + \hat{l}_y^2 \equiv \hat{l}^2 - \hat{l}_z^2 = l(l+1) - m^2$, then using the result of the previous problem we have $\overline{\hat{l}_x^2} = \overline{\hat{l}_y^2} = \frac{1}{2}[l(l+1) - m^2]$.

The angular momentum $\tilde{z}$-projection operator has the form

$$\hat{l}_{\tilde{z}} = \cos\alpha \cdot \hat{l}_z + \sin\alpha\cos\beta \cdot \hat{l}_x + \sin\alpha\sin\beta \cdot \hat{l}_y, \qquad (1)$$

where α and β are the polar and azimuthal angles of the $\tilde{z}$-axis. If we average $\hat{l}_{\tilde{z}}$ over the state ψ_{lm}, we find that $\overline{\hat{l}_{\tilde{z}}} = m\cos\alpha$. According to Problem 3.10 we have $\overline{\hat{l}_x} = \overline{\hat{l}_y} = 0$. We should note that the validity of this relation does not require l to be definite. We can use the results of the previous problem while now averaging $\hat{l}_{\tilde{z}}^2$, and we find

$$\overline{\hat{l}_{\tilde{z}}^2} = \frac{1}{2}[l(l+1) - 3m^2]\sin^2\alpha + m^2. \qquad (2)$$

Problem 3.12

Prove the relation

$$\sum_{m=-l}^{l} |Y_{lm}(\vartheta, \varphi)|^2 = \frac{2l+1}{4\pi}.$$

Solution

This equation follows directly from Eq. (III.6) with $\theta' = \theta$, $\varphi' = \varphi$. In that case, $\cos\alpha = 1$ and $P_l(1) = 1$.

Problem 3.13

Find the form of wavefunction $\psi_{l,\tilde{m}=0}(\mathbf{n})$ of a particle with total angular momentum l and projection along the $\tilde{z}$-axis $\tilde{m} = 0$. In this state, determine the probabilities of different values for the z-component of the angular momentum.

Solution

1) The wavefunction of a state with a given angular momentum l and projection $l_z = 0$ has the form $\psi_{l,\,l_z=0}(\mathbf{n}) = ((2l+1)/4\pi)^{1/2} P_l(\cos\vartheta)$. If we note that $\cos\vartheta = \mathbf{k}\cdot\mathbf{n}$, where $\mathbf{k}$ is the unit vector directed along the z-axis, and if we take into account the equivalence of different spatial directions, we obtain $\psi_{l,\,l_{\tilde{z}}=0} = \left(\frac{2l+1}{4\pi}\right)^{1/2} P_l(\tilde{\mathbf{k}}\cdot\mathbf{n})$ where $\tilde{\mathbf{k}}$ is a unit vector along the $\tilde{z}$-axis.

2) We can get the expansion of this wavefunction into a series of the spherical harmonics $Y_{lm}(\mathbf{n})$ immediately from Eq. (III.6). The desired probability for $l_z = m$ is given by $w(m) = (4\pi/(2l+1))|Y_{lm}(\tilde{\mathbf{k}})|^2$. It appears to depend only on the angle α between the axes z and $\tilde{z}$.

Problem 3.14

Let $w_l(m_1; m_2, \alpha)$ be the probability to measure a particle's projection of the angular momentum on the $\tilde{z}$-axis as m_2, if the particle is in the state with a definite angular momentum, m_1, along the z axis, where both states have definite angular momentum l and the angle between the axes is α. Prove that $w_l(m_1; m_2, \alpha) = w_l(m_2; m_1, \alpha)$.

Solution

We use symmetries of the space to prove. Switching axes z and $\tilde{z}$ should not change the probabilities, so $w_l(m_1; m_2, \alpha) = w_l(m_2; m_1, -\alpha)$. There cannot be any probability dependence on the azimuthal angle, so we can rotate one axis around the other to get $w_l(m_2; m_1, -\alpha) = w_l(m_2; m_2, \alpha)$, and we have proven the relation.

Problem 3.15

For an angular momentum L, find the projection operators $\hat{P}_L(M)$ that project onto the state with a definite value M for its z-component. The operators act in the space of the state vectors with a given value L of the angular momentum.

Solution

The form of projection operator is

$$\hat{P}_L(M) = \prod_{m=-L}^{L}{}' \frac{\hat{L}_z - m}{M - m},$$

where the prime means the absence of a multiplier with $m = M$ in the product. This follows from the result of Problem 1.35.

Problem 3.16

Using only the commutation relations for the components of the angular momentum, find $\text{Tr}(\hat{l}_i)$, where $\hat{l}_i$ is a matrix of the ith component of angular momentum l.

Solution

From the relations $\hat{l}_i \hat{l}_k - \hat{l}_k \hat{l}_i = i\varepsilon_{ikl} \hat{l}_l$, and using the equation $\text{Tr}(\hat{A}\hat{B}) = \text{Tr}(\hat{B}\hat{A})$, we have $\text{Tr}(\hat{l}_i) = 0$. Compare with Problem 1.5.

Problem 3.17

Determine the *traces* of the following matrices:

a) $\hat{L}_i$, b) $\hat{L}_i \hat{L}_k$, c) $\hat{L}_i \hat{L}_k \hat{L}_l$, d) $\hat{L}_i \hat{L}_k \hat{L}_l \hat{L}_m$,

where $\hat{L}_i$ is a matrix of the ith component of angular momentum L.

Solution

The matrices $\hat{L}_i$ are vector (more precisely, pseudovector) operators and their products $\hat{L}_i \hat{L}_k \ldots \hat{L}_n$ are tensor operators. After the calculation of trace, such an operator becomes an ordinary numerical tensor which can be expressed in terms of the universal tensors δ_{ik} and ε_{ikl}, since there exist no other vectors or tensors in the conditions of the problem. Thus:

a) $\text{Tr} \, \hat{L}_i = 0$.
b) $\text{Tr}(\hat{L}_i \hat{L}_k) = A\delta_{ik}$.
  We obtain a value of A by making a convolution over the indices i and k:

$$3A = \text{Tr} \, \hat{\mathbf{L}}^2 = L(L+1)\text{Tr} \, \hat{1} = L(L+1)(2L+1).$$

c) $\text{Tr}(\hat{L}_i \hat{L}_k \hat{L}_l) = B\varepsilon_{ikl}$.
  For determining B, we have

$$2B = \text{Tr}(\hat{L}_1 \hat{L}_2 \hat{L}_3) - \text{Tr}(\hat{L}_2 \hat{L}_1 \hat{L}_3) = i\text{Tr}(\hat{L}_3^2) = \frac{i}{3}\text{Tr}(\hat{L}^2) = iL(L+1)(2L+1)/3.$$

We used the relation $\hat{L}_1 \hat{L}_2 - \hat{L}_2 \hat{L}_1 = i\hat{L}_3$.

d) $\text{Tr}(\hat{L}_i\hat{L}_k\hat{L}_l\hat{L}_m) = C_1\delta_{ik}\delta_{lm} + C_2\delta_{il}\delta_{km} + C_3\delta_{im}\delta_{kl}$.
To obtain C_n, we should first perform convolution over i and k and then over l and m, so we obtain

$$9C_1 + 3C_2 + 3C_3 = \text{Tr}(\hat{L}^2\hat{L}^2) = (2L+1)L^2(L+1)^2. \tag{1}$$

And now we perform convolution over i and m and over k and l:

$$3C_1 + 3C_2 + 9C_3 = \text{Tr}(\hat{L}^2\hat{L}^2) = (2L+1)L^2(L+1)^2. \tag{2}$$

At last, we perform convolution[26] over i and l and then over k and m:

$$3C_1 + 9C_2 + 3C_3 = (2L+1)L^2(L+1)^2 - L(L+1)(2L+1). \tag{3}$$

From (1), (2), and (3), it follows that[27]

$$C_1 = C_3 = \frac{2L^2(L+1)^2(2L+1) + L(L+1)(2L+1)}{30}, \tag{4}$$

$$C_2 = \frac{L^2(L+1)^2(2L+1) - 2L(L+1)(2L+1)}{15}. \tag{5}$$

3.2 Angular momentum, $l = 1$

Problem 3.18

For the case of a particle with the angular momentum $l = 1$, find the wavefunction $\psi_{\tilde{m}=0}(\vartheta, \varphi)$ of the state with a definite projection $\tilde{m} = 0$ of the angular momentum on the $\tilde{z}$-axis whose polar and azimuthal angles are α and β.

Solution

The wavefunction of a state with $l = 1$ and $l_z = 0$ is $Y_{10}(\mathbf{n}) \propto \cos\theta \equiv \mathbf{n} \cdot \mathbf{k}$, with $\mathbf{k}$ being the unit vector directed along the z-axis. Due to the equivalence of spatial directions, in order to get $l_{\tilde{z}} = 0$ we should replace $\mathbf{k}$ by $\tilde{\mathbf{k}}$, the unit vector directed along the $\tilde{z}$-axis. Compare with Problem 3.13.

$$\psi_{l=1,\tilde{m}=0} = \sqrt{\frac{3}{4\pi}}\left(\tilde{\mathbf{k}} \cdot \mathbf{n}\right) = \sqrt{\frac{3}{4\pi}}\left[\cos\alpha\cos\theta + \sin\alpha\cos(\phi - \beta)\sin\theta\right].$$

---

[26] In this case we can put $\hat{L}_k\hat{L}_l = \hat{L}_l\hat{L}_k + i\varepsilon_{kls}\hat{L}_s$ and use $\varepsilon_{kls}\varepsilon_{lsk} = 6$.
[27] There is another way of obtaining these relations. Using $\text{Tr}(\hat{L}_i\hat{L}_k\hat{L}_l\hat{L}_m) = \text{Tr}(\hat{L}_k\hat{L}_l\hat{L}_m\hat{L}_i)$ we have $C_1 = C_2$. Multiplying by $\delta_{ik}\delta_{lm}$, we obtain $12C_1 + 3C_2 = (2L+1)L^2(L+1)^2$. Multiplying by $\varepsilon_{ikn}\varepsilon_{lmn}$ gives $6C_1 - 6C_2 = (2L+1)L(L+1)$. So we have (4) and (5).

Problem 3.19

Find the wavefunctions $\psi_{lx}(\vartheta,\varphi)$ and $\psi_{ly}(\vartheta,\varphi)$ of a particle having a given value $l=1$ of angular momentum and a definite value of its projection onto x and y axes. Use the specific form of the spherical harmonics $Y_{1m}(\vartheta,\varphi)$. See Eq. (III.7).

Solution

Using the relations for Y_{1m} and the equivalence of different coordinate system orientations, the wavefunctions can be obtained by a permutation of variables x, y, z. For example:

$$\psi_{l=1,\ l_x=\pm 1} = \mp i \left(\frac{3}{8\pi}\right)^{1/2} \frac{y \pm iz}{r} = \mp i \left(\frac{3}{8\pi}\right)^{1/2} (\sin\phi \sin\theta \pm i\cos\theta).$$

We can find all other wavefunctions in a similar way. See also Problem 3.18.

Problem 3.20

A particle is in a state with angular momentum $l=1$ and z-projection m ($m=0,\pm 1$). For such a state, determine the probabilities, $w(\tilde{m},m)$, of different values $\tilde{m}$ of the angular-momentum projections onto the $\tilde{z}$-axis making an angle α with z-axis.

You can use one of the following two approaches to the problem:

a) by using the result of Problem 3.11;
b) by finding the expansion coefficient $c(\tilde{m},m)$ of the given wavefunction into a series of eigenfunctions of the operator $\hat{l}_{\tilde{z}}$.

Solution

We denote the angular momentum projection probabilities with $\tilde{m} = \pm 1$ by $w(\pm 1)$. From Problem 3.11, we have

$$\overline{l_{\tilde{z}}} = \sum_{\tilde{m}} w(\tilde{m})\tilde{m} = w(1) - w(-1) = m\cos\alpha,$$

$$\overline{l_{\tilde{z}}^2} = \sum_{\tilde{m}} w(\tilde{m})\tilde{m}^2 = w(1) + w(-1) = m^2 + (1 - \frac{3m^2}{2})\sin^2\alpha.$$

The solution follows:

$$w(1,m) \equiv w(1) = \frac{1}{4}[2m^2 + 2m\cos\alpha + (2-3m^2)\sin^2\alpha],$$

$$w(-1,m) \equiv w(-1) = \frac{1}{4}[2m^2 - 2m\cos\alpha + (2-3m^2)\sin^2\alpha],$$

$$w(0,m) = 1 - w(1) - w(-1).$$

Problem 3.21

Show that for a particle with angular momentum $l = 1$, the three functions $\psi_{l_x=0}(\vartheta, \varphi)$, $\psi_{l_y=0}(\vartheta, \varphi)$, and $\psi_{l_z=0}(\vartheta, \varphi)$, that correspond to the states where the projection of the angular momentum onto the x-, y-, and z-axis correspondingly is zero, form a complete set of functions.

What is the meaning of the expansion coefficients of an arbitrary state with $l = 1$ in terms of these functions?

Solution

The wavefunctions considered, $\psi_{l_i=0}(\vartheta, \varphi)$, $(i = 1, 2, 3)$, have the form $(a = i\sqrt{3/4\pi})$ $\psi_{l_z=0} = Y_{10} = az/r = a\cos\vartheta$, $\psi_{l_x=0} = ax/r = a\sin\vartheta\cos\varphi$, $\psi_{l_y=0} = ay/r = a\sin\vartheta\sin\varphi$. Their independence and completeness are obvious. We can see that different wavefunctions $\psi_{l_i=0}$ are orthogonal:

$$\int \psi^*_{l_i=0} \psi_{l_k=0} d\Omega = \delta_{ik}.$$

This is why coefficients C_i in an expansion of an arbitrary wavefunction $\psi_{l=1}$ with these functions determine the probability $w(i) = |C_i|^2$ of the i−projection of angular momentum being equal to zero. We should note that this result does not have a direct analogy to an expansion in terms of operator eigenfunctions.

Problem 3.22

For the angular momentum $l = 1$, write expressions for the operators of angular momentum components, as well as for raising $\hat{l}_+$ and lowering $\hat{l}_-$ operators, in the l_z-representation.

Find the wavefunction of a state with $l_x = 0$ in the l_z-representation from the solution of an eigenfunction equation.

Solution

1) Using Eq. (III.9) for $l = 1$ we obtain:

$$\hat{l}_x = \begin{pmatrix} 0 & \frac{1}{\sqrt{2}} & 0 \\ \frac{1}{\sqrt{2}} & 0 & \frac{1}{\sqrt{2}} \\ 0 & \frac{1}{\sqrt{2}} & 0 \end{pmatrix}, \hat{l}_y = \begin{pmatrix} 0 & -\frac{i}{\sqrt{2}} & 0 \\ \frac{i}{\sqrt{2}} & 0 & -\frac{i}{\sqrt{2}} \\ 0 & \frac{i}{\sqrt{2}} & 0 \end{pmatrix}, \hat{l}_z = \begin{pmatrix} 1 & 0 & 0 \\ 0 & 0 & 0 \\ 0 & 0 & -1 \end{pmatrix},$$

(1)

$$\hat{l}_+ = \begin{pmatrix} 0 & \sqrt{2} & 0 \\ 0 & 0 & \sqrt{2} \\ 0 & 0 & 0 \end{pmatrix}, \hat{l}_- = \begin{pmatrix} 0 & 0 & 0 \\ \sqrt{2} & 0 & 0 \\ 0 & \sqrt{2} & 0 \end{pmatrix}.$$

2) With $\psi_{l_x=0} = \begin{pmatrix} a \\ b \\ c \end{pmatrix}$, we have an eigenfunction equation in the form

$$\hat{l}_x \psi_{l_x=0} \equiv \frac{1}{\sqrt{2}} \begin{pmatrix} 0 & 1 & 0 \\ 1 & 0 & 1 \\ 0 & 1 & 0 \end{pmatrix} \begin{pmatrix} a \\ b \\ c \end{pmatrix} = \frac{1}{\sqrt{2}} \begin{pmatrix} b \\ a+c \\ b \end{pmatrix} = 0.$$

Hence it follows: $b = 0$, $a = -c$; and $|a| = 1/\sqrt{2}$ for a wavefunction normalized to unity.

Problem 3.23

For a state with the value of angular momentum $l = 1$ and its z-projection m, find the mean values $\overline{l_x^n}$ and $\overline{l_y^n}$ (n is integer).

Solution

Since, when $l = 1$, the eigenvalues l_x and l_y are equal to 0, ± 1, it follows that $\hat{l}_x^3 = \hat{l}_x$ and $\hat{l}_y^3 = \hat{l}_y$ (compare with Problem 1.17). In a state with $l = 1$ and $l_z = m$, we have $\overline{l_x} = \overline{l_y} = 0$ and $\overline{l_x^2} = \overline{l_y^2} = 1 - \frac{1}{2}m^2$ (see, for example, Problem 3.11). It follows that $\overline{l_x^n} = \overline{l_y^n} = 0$ for the odd values of n and $\overline{l_x^n} = \overline{l_y^n} = 1 - \frac{1}{2}m^2$ for the even n ($n > 0$).

Problem 3.24

Find an explicit form of the operator $\hat{R}(\boldsymbol{\varphi}_0) = \exp(i\boldsymbol{\varphi}_0 \cdot \hat{\mathbf{l}})$ (a coordinate system rotation over the angle, $\boldsymbol{\phi}_0$) that acts in the space of state vectors with angular momentum $l = 1$. Using this operator, obtain in terms of the spherical function Y_{10}, the wavefunction, $\psi_{\tilde{m}=0}(\vartheta, \varphi)$, of a state with $l = 1$ and $\tilde{m} \equiv l_{\tilde{z}} = 0$, where the $\tilde{z}$-axis is defined by its polar α and azimuth β angles. Compare with Problem 3.18.

Solution

Since the operator $\boldsymbol{\varphi}_0 \cdot \hat{\mathbf{l}}$, acting in the subspace of state vectors with $l = 1$, has only three eigenvalues equal to 0 and $\pm \varphi_0$, then (from Problem 1.22) we have

$$\hat{R} \equiv e^{i\boldsymbol{\varphi}_0 \cdot \hat{\mathbf{l}}} = 1 + i \sin \varphi_0 \cdot (\mathbf{n}_0 \cdot \hat{\mathbf{l}}) - (1 - \cos \varphi_0)(\mathbf{n}_0 \cdot \hat{\mathbf{l}})^2, \tag{1}$$

where $\mathbf{n}_0 = \boldsymbol{\varphi}_0/\varphi_0$. We choose the rotation vector $\boldsymbol{\varphi}_0$ so that after rotation, the z-axis of the initial coordinate system with respect to the axes of the rotated system has the same orientation as does the $\tilde{z}$-axis with respect to the initial coordinate system. Wavefunction $\psi_{\tilde{m}}(\vartheta, \varphi) = \tilde{R} Y_{lm}(\vartheta, \varphi)$ will describe a state with momentum l and $\tilde{z}$-projection m. The operator $\hat{R}$ a rotation. It is easily seen that for this case, we should choose $\boldsymbol{\varphi}_0 = (\alpha \sin \beta, -\alpha \cos \beta, 0)$. So

$$\hat{R} = 1 + i \sin \alpha (\hat{l}_x \sin \beta - \hat{l}_y \cos \beta) - (1 - \cos \alpha)(\hat{l}_x \sin \beta - \hat{l}_y \cos \beta)^2.$$

For the wavefunctions $\psi_{\tilde m=0} = \hat R Y_{10}$, we obtain

$$\psi_{\tilde m=0} = i\left(\frac{3}{4\pi}\right)^{1/2}(\cos\alpha\cos\theta + \sin\alpha\sin\theta\cos(\varphi-\beta)), \qquad (2)$$

in accordance with the result of Problem 3.18.

Problem 3.25

In the space of states with angular momentum $l = 1$, find the projection operators, $\hat P(m)$, to states with a definite z-component of the angular momentum, m.

Generalize the results obtained to the case of an arbitrarily directed $\tilde z$-axis. By using the operator $\hat P_{\tilde m}$, obtain both in the l_z and in the coordinate representations the wavefunction, $\psi_{1,\tilde m=0}$, of a state with angular momentum $l = 1$ and $\tilde z$-projection $\tilde m = 0$. Compare with Problem 3.18 and Problem 3.24.

Solution

For the $\hat P_m$, we have the relations (compare with Problem 3.15):

$$\hat P(0) = 1 - \hat l_z^2, \quad \hat P(\pm 1) = \frac{1}{2}(\hat l_z^2 \pm \hat l_z) \qquad (1)$$

Projection operators $\hat P(\tilde m)$ are obtained from (1) by the substitution of $\hat l_{\tilde z}$ for $\hat l_z$, where $\hat l_{\tilde z}$ is

$$\hat l_{\tilde z} = \tilde{\mathbf k}\cdot\hat{\mathbf l} = \cos\alpha\cdot\hat l_z + \sin\alpha\cos\beta\cdot\hat l_x + \sin\alpha\sin\beta\cdot\hat l_y.$$

$\tilde{\mathbf k}$ is the unit vector directed along the $\tilde z$-axis, and α and β are the polar and azimuthal angles of the $\tilde{\mathbf k}$ direction. In particular, for the $\hat P(\tilde m = 0)$ operator in the $\hat l_z$-representation, we can use Eq. (1) from Problem 3.22 to obtain

$$\hat P(\tilde m = 0) = \begin{pmatrix} \frac{1}{2}\sin^2\alpha & -\frac{1}{2\sqrt{2}}e^{-i\beta}\sin 2\alpha & -\frac{1}{2}e^{-2i\beta}\sin^2\alpha \\ -\frac{1}{2\sqrt{2}}e^{i\beta}\sin 2\alpha & \cos^2\alpha & \frac{1}{2\sqrt{2}}e^{-i\beta}\sin 2\alpha \\ -\frac{1}{2}e^{2i\beta}\sin^2\alpha & \frac{1}{2\sqrt{2}}e^{i\beta}\sin 2\alpha & \frac{1}{2}\sin^2\alpha \end{pmatrix}.$$

Acting with this operator on an arbitrary function, for example the function $\psi = \begin{pmatrix}1\\0\\0\end{pmatrix}$, we find the eigenfunction, $\psi_{\tilde m=0} = C\hat P(\tilde m = 0)\psi$, of the operator $\hat l_{\tilde z}$ that corresponds to the eigenvalue $l_{\tilde z} = 0$:

$$\psi_{\tilde m=0} = \begin{pmatrix} \frac{1}{\sqrt 2}\sin\alpha \\ -e^{i\beta}\cos\alpha \\ -\frac{1}{\sqrt 2}e^{2i\beta}\sin\alpha \end{pmatrix}, \qquad (1)$$

where $C = \sqrt{2}/\sin\alpha$ has been chosen to ensure proper wavefunction normalization. In the case of $\alpha = \pi/2$ and $\beta = 0$, function (1) reproduces the result for $\psi_{l_x} = 0$ from Problem 3.22.

By taking into account the form of spherical function $Y_{1m}(n)$ (see (III.7)), we see that wavefunctions of state (1) in the coordinate representation, $\psi = \sum c_m Y_{1m}$, differs from the ones found earlier in Problem 3.18 and Problem 3.24 only by a phase factor.

3.3 Addition of angular momenta

Problem 3.26

Write down the total angular momentum operator of two particles as a sum of two terms, corresponding to the angular momentum in the center of inertia system (*i.e.*, the angular momentum of relative motion) and the angular momentum in the frame of reference associated with the system's translational motion as a whole.

Solution

An angular momentum operator for a two-particle system has the form

$$\hat{\mathbf{L}} = \hat{\mathbf{l}}_1 + \hat{\mathbf{l}}_2 = -i\mathbf{r}_1 \times \nabla_1 - i\mathbf{r}_2 \times \nabla_2. \tag{1}$$

Now we define new variables $\mathbf{r}$ and $\mathbf{R}$ in terms of $\mathbf{r}_1$ and $\mathbf{r}_2$

$$\mathbf{r} = \mathbf{r}_2 - \mathbf{r}_1, \quad \mathbf{r}_1 = \mathbf{R} - \frac{m_2}{m_1 + m_2}\mathbf{r},$$

$$\mathbf{R} = \frac{1}{m_1 + m_2}(m_1\mathbf{r}_1 + m_2\mathbf{r}_2), \quad \mathbf{r}_2 = \mathbf{R} + \frac{m_1}{m_1 + m_2}\mathbf{r}.$$

Since

$$\nabla_1 = \frac{m_1}{m_1 + m_2}\nabla_\mathbf{R} - \nabla_\mathbf{r}, \quad \nabla_2 = \frac{m_2}{m_1 + m_2}\nabla_\mathbf{R} + \nabla_\mathbf{r},$$

operator (1) could be written in the form

$$\hat{\mathbf{L}} = -i\mathbf{r} \times \nabla_\mathbf{r} - i\mathbf{R} \times \nabla_\mathbf{R},$$

where the first term is the angular momentum operator for the two particles in the center-of-inertia reference frame, while the second one corresponds to the angular momentum operator that is connected with the overall translational motion.

Problem 3.27

Angular momenta l_1 and l_2 of two weakly interacting systems are combined into a resulting angular momentum with the value L. Show that in such states (with a definite L) the products $\hat{\mathbf{l}}_1 \cdot \hat{\mathbf{l}}_2$, $\hat{\mathbf{l}}_1 \cdot \hat{\mathbf{L}}$, $\hat{\mathbf{l}}_2 \cdot \hat{\mathbf{L}}$ have definite values as well.

Solution

From the relation $\hat{\mathbf{L}} = \hat{\mathbf{l}}_1 + \hat{\mathbf{l}}_2$ it follows that

$$\hat{\mathbf{l}}_1 \cdot \hat{\mathbf{l}}_2 = \frac{1}{2}[\hat{\mathbf{L}}^2 - l_1(l_1+1) - l_2(l_2+1)],$$

$$\hat{\mathbf{l}}_1 \cdot \hat{\mathbf{L}} = \frac{1}{2}[\hat{\mathbf{L}}^2 + l_1(l_1+1) - l_2(l_2+1)], \qquad \hat{\mathbf{l}}_2 \cdot \hat{\mathbf{L}} = \frac{1}{2}[\hat{\mathbf{L}}^2 - l_1(l_1+1) + l_2(l_2+1)].$$

We took into account the commutativity of $\hat{\mathbf{L}}$ and $\hat{\mathbf{l}}_{1,2}$. In all states with definite values of $\mathbf{L}^2$, $\mathbf{l}_1^2$, $\mathbf{l}_2^2$, the considered scalar products have definite values also.

Problem 3.28

Find the following commutators:

1) $[\hat{L}_i, (\hat{\mathbf{l}}_1 \cdot \hat{\mathbf{l}}_2)]$, $[\hat{L}_i, (\hat{\mathbf{r}}_1 \cdot \hat{\mathbf{p}}_2)]$, $[\hat{L}_i, (\hat{\mathbf{r}}_1 \cdot \hat{\mathbf{r}}_2)]$;
2) $[\hat{L}_i, \hat{x}_{1k}]$, $[\hat{L}_i, \hat{g}_k]$ where $\hat{\mathbf{g}} = [\hat{\mathbf{l}}_1 \times \hat{\mathbf{l}}_2]$;
3) $[\hat{L}_i, \hat{x}_{1k}\hat{x}_{2l}]$, $[\hat{L}_i, \hat{x}_{1k}\hat{p}_{2l}]$,

$\hat{\mathbf{l}}_1$ and $\hat{\mathbf{l}}_2$ are the angular momentum operators of particles 1 and 2, $\hat{\mathbf{L}} = \hat{\mathbf{l}}_1 + \hat{\mathbf{l}}_2$ is the operator of their total angular momentum. Note that the commutators have a universal structure (inside each group of expressions presented above). Compare with Problem 3.5.

Solution

The commutators considered have the same vector structure as in Problem 3.5.

Problem 3.29

Two weakly interacting systems have states characterized by quantum numbers (l_1, m_1) and (l_2, m_2) of their angular momenta and its z-projection. Give possible values, L, of the total angular momentum of a composite system $(1+2)$ and calculate the mean values $\bar{\mathbf{L}}$ and $\overline{\mathbf{L}^2}$. For the specific case where $m_1 = l_1$, $m_2 = l_2 - 1$ find the probabilities for the different possible values of the total angular momentum.

Solution

1) The possible values of the composite system's total angular momentum are:

$$\max\{|l_1 - l_2|, |m_1 + m_2|\} \leq L \leq l_1 + l_2.$$

Taking into account the mutual commutativity of $\hat{l}_{1i}$ and $\hat{l}_{2k}$, the relation $\hat{\mathbf{L}}^2 = \hat{\mathbf{l}}_1^2 + \hat{\mathbf{l}}_2^2 + 2\hat{\mathbf{l}}_1\hat{\mathbf{l}}_2$, and the mean values $\bar{l}_x = \bar{l}_y = 0$ in the states with a definite value of l_z (see 3.10), we find the mean values $\bar{L}_x = \bar{L}_y = 0$, $\bar{L}_z = m_1 + m_2$, and also

$$\overline{\mathbf{L}^2} = l_1(l_1+1) + l_2(l_2+1) + 2m_1m_2. \tag{1}$$

2) In the case where $m_1 = l_1$, $m_2 = l_2 - 1$, the only possible values of the total angular momentum are $L_1 = l_1 + l_2$ and $L_2 = l_1 + l_2 - 1$. We have $w(L_2) = 1 - w(L_1)$. Using (1), we obtain

$$\overline{\mathbf{L}^2} = \sum_L w(L)L(L+1) = L_1^2 - L_1 + 2L_1 w(L_1)$$

$$= l_1(l_1+1) + l_2(l_2+1) + 2l_1(l_2-1).$$

Thus we have

$$w(L_1) = \frac{l_2}{l_1+l_2}, \; w(L_2) = \frac{l_1}{l_1+l_2}.$$

Problem 3.30

Show that when we add two angular momenta of the same value $(l_1 = l_2 = l)$ to produce a total angular momentum, L, the wavefunction $\psi_L(m_1, m_2)$ in the $l_{1z}l_{2z}$-representation has a symmetry with respect to interchange of m_1 and m_2. Indicate how the nature of the symmetry depends on the value of L.

Solution

First we consider the wavefunction, ψ_{LM}, of a state with $L = 2l$ and $M = 2l$, so that we must have $m_1 = m_2 = l$. It is obviously a symmetric function with respect to the interchange of m_1 and m_2. The wavefunctions with given $L = 2l$ but with other values of M are also symmetric functions. This follows, for example, from the relation $\psi_{L,M=L-n} = C\hat{L}_-^n \psi_{L,\,L}$. We have $\hat{L}_- = (\hat{l}_{1x} + \hat{l}_{2x}) - i(\hat{l}_{1y} + \hat{l}_{2y})$, and in the case $l_1 = l_2$ there is symmetry with respect to permutation of the momenta.

Next we consider the states with $M = 2l - 1$ and write down the most general wavefunction for such states in the form of a sum of symmetric and anti-symmetric terms:

$$\psi_{M=2l-1} = \frac{c_1}{\sqrt{2}}(\delta_{m_1,l}\delta_{m_2,l-1} + \delta_{m_1,l-1}\delta_{m_2,l}) + \frac{c_2}{\sqrt{2}}(\delta_{m_1,l}\delta_{m_2,l-1} - \delta_{m_1,l-1}\delta_{m_2,l}).$$

We see that the symmetric term corresponds to the total angular momentum $L_1 = 2l$, while the anti-symmetric one corresponds to $L_2 = 2l - 1$ (if we had both angular momenta present in the first term it would have contradicted the orthogonality of eigenfunctions corresponding to different eigenvalues). So the wavefunction $\psi_{2l-1,2l-1}$, and any other state corresponding to $L = 2l - 1$ (see above), is anti-symmetric with respect to the mutual interchange of m_1 and m_2.

Similarly, we can consider the states with $M = 2l - 2$. Now the wavefunction $\psi_{M=2l-2}$ includes three independent terms, and two of them (with m_1 and m_2 equal to l and $l - 2$, and also $m_1 = m_2 = l - 1$) are symmetric while one (with $m_{1,2}$ corresponding to l and $l - 2$) is anti-symmetric. The anti-symmetric state corresponds to the angular momentum $L = 2l - 1$, while the two symmetric states correspond to the angular momenta, $L_1 = 2l$ and $L_2 = 2l - 2$.

Therefore, we conclude that the wavefunctions symmetric with respect to the interchange of m_1 and m_2 correspond to the states with $L = 2l, 2l - 2, 2l - 4, \ldots$, while the anti-symmetric wavefunctions correspond to $L = 2l - 1, 2l - 3, \ldots$.

These symmetry considerations apply not only when the values of l are integer but also when they are half-integer. This circumstance is important when considering particle spin. See Chapter 5.

Problem 3.31

A system with the z-projection of the angular momentum M is made up of two particles with the same total angular momentum values, $l_1 = l_2$. Prove that the probabilities of $m_{1(2)} = m$ and $m_{1(2)} = M - m$ are the same.

Solution

The proof follows directly from the two facts: 1) due to the symmetry of wavefunction $\psi_{LM}(m_1, m_2)$ with respect to the interchange of m_1 and m_2 (see Problem 3.30), the probabilities of the same value of m for the both angular momenta are the same, i.e., $w_1(m) = w_2(m) \equiv w(m)$; 2) since $m_1 + m_2 = M$ we have $w_1(m_1) = w_2(M - m_1)$. Hence it follows that $w_{1,2}(m) = w_{1,2}(M - m)$.

Problem 3.32

Two subsystems which have the same values of their angular momenta, $l_1 = l_2 = 1$, are in states with definite values m_1 and m_2 of the angular momentum projections. Determine the probabilities for different values, L, of the total angular momentum in such states. Use the result of Problem 3.29 for the value of $\overline{L}^2$ and take into account the symmetry of the state wavefunction with a definite value of L shown in Problem 3.30. We should note that with arbitrary values of $l_{1,2}$ and $m_{1,2}$, the desired probability is $w(L) = |C_{l_1 m_1 l_2 m_2}^{L, m_1 + m_2}|^2$, where $C_{l_1 m_1 l_2 m_2}^{LM}$ are the Clebsch–Gordan coefficients. See Problem 3.38.

Solution

We will look at different cases for m_1, m_2.

a) When $m_1 = m_2 = \pm 1$, we must have $L = 2$.
b) If $m_1 = \pm 1$, $m_2 = 0$ (or $m_1 = 0$, $m_2 = \pm 1$), then the possible values of angular momentum are $L_1 = 2$ and $L_2 = 1$. Their probabilities, $w(2) = 1/2$ and $w(1) = 1/2$, follow from the results of Problems 3.29 and 3.30.
c) In the case where $m_1 = m_2 = 0$, the total angular momentum may take the values 0 and 2 only; $L = 1$ is excluded due to the wavefunction symmetry with respect to the interchange of m_1 and m_2 (see 3.30). From the condition $\overline{\mathbf{L}^2} = 6w(L = 2) = 4$, it follows that $w(2) = 2/3$ and $w(0) = 1/3$.

d) When $m_1 = -m_2 = \pm 1$, the total angular momentum may take all the three values: 0, 1 and 2. Writing the wavefunctions in the $l_{1x}l_{2x}$-representation for the case of $m_1 = -m_2 = 1$ in the form

$$\psi = \delta_{l_{1z},1}\delta_{l_{2z},-1} = \frac{1}{\sqrt{2}}\left\{\frac{1}{\sqrt{2}}(\delta_{l_{1z},1}\delta_{l_{2z},-1} + \delta_{l_{1z},-1}\delta_{l_{2z},1})\right.$$

$$\left. +\frac{1}{\sqrt{2}}(\delta_{l_{1z},1}\delta_{l_{2z},-1} - \delta_{l_{1z},-1}\delta_{l_{2z},1})\right\}, \tag{1}$$

we see that the probability of the value $L = 1$ (the second, antisymmetric term in (1) corresponds to this value) is equal to $w(L=1) = 1/2$. Then we have $\overline{\mathbf{L}^2} = \sum L(L+1)w(L) = 6w(2) + 1 = 2$. Hence it follows that $w(2) = 1/6$, $w(0) = 1/3$.

Problem 3.33

Illustrate the relation established in Problem 1.43 and its probabilistic interpretation by the example of the addition of the angular momenta l_1 and l_2 for two weakly interacting subsystems with the total angular momentum, L.

Solution

In the statement of Problem 1.43, by $\hat{A}$ we take a set of commuting operators $\hat{l}_{1z}$ and $\hat{l}_{2z}$ with eigenvalues m_1 and m_2, while by $\hat{B}$ we take a set of $\hat{L}^2$ and $\hat{L}_z = \hat{l}_{1z} + \hat{l}_{2z}$. Therefore we have the probability relation $w_{LM}(m_1, m_2) = w_{m_1, m_2}(L, M)$, i.e., the probability of the values m_1 and m_2 in a state with given values L and M (here $M = m_1 + m_2$) is equal to the probability of the values L, M in a state with given values m_1 and m_2. Compare, for example, with the results of Problems 3.32 and 3.35.

Problem 3.34

For a system of two particles with equal angular momenta $l_1 = l_2 = l$, find the wavefunction of a state with $L = 0$ in the $l_{1z}l_{2z}$-representation. Use the operators $\hat{L}_\pm$. Find the wavefunction in the coordinate representation also.

Solution

Let us write down the desired wavefunction of the combined system in the form $\psi_{L=0} = \sum_m C_m \psi_m^{(1)} \psi_{-m}^{(2)}$, where $\psi_m^{(1,2)}$ are the normalized wavefunctions in systems 1 and 2 with angular momentum l and z-projection m. It is obvious that

$$\hat{L}_\pm \psi_{l=0} \equiv (\hat{l}_{1\pm} + \hat{l}_{2\pm})\psi_{L=0} = 0, \tag{1}$$

where $\hat{L}_\pm = \hat{L}_x \pm i\hat{L}_y = \hat{l}_{1\pm} + \hat{l}_{2\pm}$. Now we use the relation (see Eq. (III.8)) $\hat{l}_+\psi_{lm} = \sqrt{(l-m)(l+m+1)}\psi_{l,m+1}$. From (1) we obtain

$$\hat{L}_+\psi_{L=0} = \sum_m \sqrt{(l-m)(l+m+1)}(C_m + C_{m+1})\psi_{m+1}^{(1)}\psi_{-m}^{(2)} = 0.$$

It follows that $C_{m+1} = -C_m$. From this, we get $|C_m| = const = (2l+1)^{-1/2}$ from the normalization of $\psi_{L=0}$ to unity. So, in the state with $L=0$ the probabilities of the different values of the both angular momentum z-projections (and any other axis projection) are the same and are equal to $w = (2l+1)^{-1}$.

The form of the wavefunction, $\psi_{L=0}$, in the $l_{1z}l_{2z}$-representation follows from the fact that in this representation, $\psi_m^{(1,2)} = \delta_{l_{1(2)z},m}$. In the coordinate representation, $\psi_m^{(1,2)} = Y_{lm}(\mathbf{n}_{1,2})$. Using the fact that $C_m = (-1)^{l-m}(2l+1)^{-1/2}$, the relation between the spherical functions $Y_{lm}^*(\mathbf{n}) = (-1)^{l-m}Y_{l,-m}(\mathbf{n})$, and the addition theorem, Eq. (III.6), we find

$$\psi_{L=0} = \sum_m \frac{1}{\sqrt{2l+1}} Y_{l,m}(\mathbf{n}_1) Y_{l,-m}^*(\mathbf{n}_2) = \frac{\sqrt{2l+1}}{4\pi} P_l(\mathbf{n}_1 \cdot \mathbf{n}_2).$$

Let us note that such a view of wavefunctions, $\psi_{L=0}$, could also be seen from the following considerations. Due to the fact that the wavefunction is invariant under rotation ($L=0$), it is a scalar of the form $\psi_{L=0} = f(\mathbf{n}_1 \cdot \mathbf{n}_2)$. The reason $f(x)$ is the Legendre polynomial $P_l(x)$ is that the angular momenta that are being added have the specific value, l. Compare, for example, with Problem 3.13.

Problem 3.35

The angular momenta of two particles are $l_1 = l_2 = 1$. For such a system, find the wavefunctions ψ_{LM} of states with given values L and M of the total angular momentum and its z-projection. Use the results of Problems 3.30 and 3.34.

Solution

In the $l_{1z}l_{2z}$-representation, the expressions for wavefunctions $\psi_{2,\pm 2}$ are as follows

$$\psi_{2,2} = \begin{pmatrix} 1 \\ 0 \\ 0 \end{pmatrix}_1 \begin{pmatrix} 1 \\ 0 \\ 0 \end{pmatrix}_2, \quad \psi_{2,-2} = \begin{pmatrix} 0 \\ 0 \\ 1 \end{pmatrix}_1 \begin{pmatrix} 0 \\ 0 \\ 1 \end{pmatrix}_2. \tag{1}$$

From here on, the columns $\psi_{1(2)} = \begin{pmatrix} c_1 \\ c_0 \\ c_{-1} \end{pmatrix}_{1(2)}$ are the wavefunctions of a particle 1(2) or a subsystem with angular momentum $l=1$ in the l_z-representation. The expressions for wavefunctions ψ_{LM} with quantum numbers $L=1$ or 2, $M=\pm 1$ and also $L=1, M=0$ follow directly from wavefunction symmetry with respect to the interchange of m_1 and m_2, as established in Problem 3.30:

$$\psi_{2\,(1),\,1} = \frac{1}{\sqrt{2}} \left\{ \begin{pmatrix} 1 \\ 0 \\ 0 \end{pmatrix}_1 \begin{pmatrix} 0 \\ 1 \\ 0 \end{pmatrix}_2 \pm \begin{pmatrix} 0 \\ 1 \\ 0 \end{pmatrix}_1 \begin{pmatrix} 1 \\ 0 \\ 0 \end{pmatrix}_2 \right\}, \tag{2}$$

$$\psi_{2\,(1),\,-1} = \frac{1}{\sqrt{2}}\left\{\begin{pmatrix}0\\0\\1\end{pmatrix}_1\begin{pmatrix}0\\1\\0\end{pmatrix}_2 \pm \begin{pmatrix}0\\1\\0\end{pmatrix}_1\begin{pmatrix}0\\0\\1\end{pmatrix}_2\right\}, \qquad (3)$$

$$\psi_{1,\,0} = \frac{1}{\sqrt{2}}\left\{\begin{pmatrix}1\\0\\0\end{pmatrix}_1\begin{pmatrix}0\\0\\1\end{pmatrix}_2 - \begin{pmatrix}0\\0\\1\end{pmatrix}_1\begin{pmatrix}1\\0\\0\end{pmatrix}_2\right\}. \qquad (4)$$

The sign $+$ corresponds to $L = 2$ while the sign $-$ corresponds to $L = 1$.

The form of wavefunction $\psi_{0,0}$ is given by the result of the previous problem:

$$\psi_{0,\,0} = \frac{1}{\sqrt{3}}\left\{\begin{pmatrix}1\\0\\0\end{pmatrix}_1\begin{pmatrix}0\\0\\1\end{pmatrix}_2 - \begin{pmatrix}0\\1\\0\end{pmatrix}_1\begin{pmatrix}0\\1\\0\end{pmatrix}_2 + \begin{pmatrix}0\\0\\1\end{pmatrix}_1\begin{pmatrix}1\\0\\0\end{pmatrix}_2\right\}. \qquad (5)$$

The wavefunction $\psi_{2,0}$ is symmetric in the variables of m_1 and m_2 and may be written in the form

$$\psi_{2,\,0} = C_1\left\{\begin{pmatrix}1\\0\\0\end{pmatrix}_1\begin{pmatrix}0\\0\\1\end{pmatrix}_2 + \begin{pmatrix}0\\0\\1\end{pmatrix}_1\begin{pmatrix}1\\0\\0\end{pmatrix}_2\right\} + C_2\begin{pmatrix}0\\1\\0\end{pmatrix}_1\begin{pmatrix}0\\1\\0\end{pmatrix}_2. \qquad (6)$$

From the condition of its orthogonality to $\psi_{0,0}$, it follows that $C_2 = 2C_1$. We choose $C_1 = 1/\sqrt{6}$, $C_2 = 2/\sqrt{6}$ in (6) and obtain the normalized wavefunction $\psi_{2,0}$. The probabilities of different z-projections of angular momenta which are being added in states ψ_{LM} follow directly from the determined form of wavefunctions (1)–(6).

Problem 3.36

For a system of two angular momenta, $l_1 = l_2 = 1$, find the wavefunction, $\psi_{L=0}$, with $L = 0$ total angular momentum, using the projection operators. Compare with Problem 3.34.

Solution

In the case $l_1 = l_2 = 1$, the operator, $\hat{\mathbf{l}}_1 \cdot \hat{\mathbf{l}}_2$, has the following values in the states with given L: 1 for $L = 2$, -1 for $L = 1$ and -2 for $L = 0$, therefore the projection operator for a state with $L = 0$ has the form $\hat{P}(L = 0) = \frac{1}{3}((\hat{\mathbf{l}}_1 \cdot \hat{\mathbf{l}}_2)^2 - 1)$ (compare with Problem 1.35). Acting on an arbitrary wavefunction Ψ of a state with $l_1 = l_2 = 1$ with this operator, we obtain an (unnormalized) eigenfunction of the operator of a squared total momentum corresponding to $L = 0$, i.e., $\psi_{L=0} = C\hat{P}(L = 0)\Psi$ (C is a normalization coefficient). Writing down

$$\hat{\mathbf{l}}_1 \cdot \hat{\mathbf{l}}_2 = \hat{l}_{1z}\hat{l}_{2z} + \frac{1}{2}(\hat{l}_{1+}\hat{l}_{2-} + \hat{l}_{1-}\hat{l}_{2+})$$

(expressions for $\hat{l}_\pm$ in the case of $l = 1$ are given in Problem 3.22) and taking for convenience the wavefunction Ψ to be equal $\Psi = \begin{pmatrix} 0 \\ 1 \\ 0 \end{pmatrix}_1 \begin{pmatrix} 0 \\ 1 \\ 0 \end{pmatrix}_2$ in the $l_{z1}l_{z2}$-representation, we obtain as a result of straightforward calculation the desired wavefunction:

$$\psi_{L=0} = C\hat{P}(L=0) \begin{pmatrix} 0 \\ 1 \\ 0 \end{pmatrix}_1 \begin{pmatrix} 0 \\ 1 \\ 0 \end{pmatrix}_2$$

$$= \frac{1}{\sqrt{3}} \left\{ \begin{pmatrix} 1 \\ 0 \\ 0 \end{pmatrix}_1 \begin{pmatrix} 0 \\ 0 \\ 1 \end{pmatrix}_2 - \begin{pmatrix} 0 \\ 1 \\ 0 \end{pmatrix}_1 \begin{pmatrix} 0 \\ 1 \\ 0 \end{pmatrix}_2 + \begin{pmatrix} 0 \\ 0 \\ 1 \end{pmatrix}_1 \begin{pmatrix} 1 \\ 0 \\ 0 \end{pmatrix}_2 \right\}.$$

If we choose $C = \sqrt{3}$, we have the normalized wavefunction of a state with $L = 0$, which coincides with the result of Problem 3.34.

Problem 3.37

Classify the independent states of a system which consists of three weakly interacting subsystems whose angular momenta are $l_1 = l_2 = 1$ and $l_3 = l$, by the value of the total angular momentum L.

Solution

There are $3 \cdot 3 \cdot (2l + 1) = 9(2l + 1)$ independent states. Their classification by values of the total angular momentum, L, is listed in the following table

L	$l+2$	$l+1$	l	$l-1$	$l-2$
Numbers of states	$2l+5$	$2 \cdot (2l+3)$	$3 \cdot (2l+1)$	$2 \cdot (2l-1)$	$2l-3$

In order to solve the problem it is convenient to add the angular momenta of the two subsystems that have $l = 1$ into their total angular momentum L_{12} which takes the values equal to 0, 1, 2, and then to add L_{12} and $l_3 = l$ into the total angular momentum L of the entire system.

Problem 3.38

As is known, the problem of the addition of angular momenta of two systems l_1 and l_2 into the total angular momentum L could be solved by the following relation

$$\psi_{LM} = \sum_{m_1 m_2} C^{LM}_{l_1 m_1 l_2 m_2} \psi^{(1)}_{l_1 m_1} \psi^{(2)}_{l_2 m_2}, \quad M = m_1 + m_2,$$

where $C^{LM}_{l_1 m_1 l_2 m_2}$ are the Clebsch–Gordan coefficients. Using the "raising" ("lowering")-operators, $\hat{L}_\pm$, determine these coefficients for the special case of $L = l_1 + l_2$.

Solution

In this problem, by L we shall understand the definite value $L = l_1 + l_2$. The corresponding the wavefunction is $\psi_{LL} = \psi^{(1)}_{l_1 l_1} \psi^{(2)}_{l_2 l_2}$. Using the property of the $\hat{L}_-$-operator,

$$\hat{L}_- \psi_{LM} = \sqrt{(L - M + 1)(L + m)} \psi_{L,\, M-1},$$

we obtain

$$\psi_{LM} = \left[\frac{(L+M)!}{(L-M)!(2L)!} \right]^{1/2} (\hat{L}_-)^{L-M} \psi_{LL}. \tag{1}$$

Since $\hat{L}_- = \hat{l}_{1-} + \hat{l}_{2-}$ and since $\hat{l}_{1-}$ and $\hat{l}_{2-}$ commute with each other, then from (1) it follows that

$$\psi_{LM} = G(L, M) \sum_m C^m_{L-M} (\hat{l}_{1-})^m (\hat{l}_{2-})^{L-M-m} \psi^{(1)}_{l_1 l_1} \psi^{(2)}_{l_2 l_2}$$

$$= G(L, M) \sum_m C^m_{L-M} G^{-1}(l_1, l_1 - m) G^{-1}(l_2, M + m - l_1) \psi^{(1)}_{l_1,\, l_1-m} \psi^{(2)}_{l_2,\, l_2-L+M+m}$$

$$\equiv \sum_{m_1} C^{LM}_{l_1 m_1 l_2 m_2} \psi^{(1)}_{l_1 m_1} \psi^{(2)}_{l_2 m_2}, \tag{2}$$

where the following notations are used:

$$G(L, M) = \left[\frac{(L+M)!}{(L-M)!(2L)!} \right]^{1/2}, \quad C^m_L = \frac{L!}{m!(L-m)!}. \tag{3}$$

From (2), the values of the desired Clebsch–Gordan coefficients follow:

$$C^{LM}_{l_1 m_1 l_2 m_2} = G(LM) G^{-1}(l_1 m_1) G^{-1}(l_2, m_2) C^{l_1 - m_1}_{L-M}.$$

Using (3), we find the final expression:

$$C^{LM}_{l_1 m_1 l_2 m_2} = \left[\frac{(2l_1)!(2l_2)!(L+M)!(L-M)!}{(2L)!(l_1+m_1)!(l_1-m_1)!(l_2+m_2)!(l_2-m_2)!} \right]^{1/2}.$$

Problem 3.39

The same as for the previous problem but for the case where $l_1 = l_2$, $L = 0$.

Solution

The Clebsch–Gordan coefficients for this case are given by the results of Problem 3.34. Putting $C_1 = (2l+1)^{-1/2}$, we find:

$$C^{00}_{l,m,l,-m} = C_m = (-1)^{l-m}\frac{1}{\sqrt{2l+1}}.$$

Problem 3.40

For two weakly interacting systems with j_1 and j_2 for their angular momenta, average the operators

a) $\hat{j}_{1(2)i}$, b) $\hat{j}_{1i}\hat{j}_{2k} - \hat{j}_{1k}\hat{j}_{2i}$; c) $\hat{j}_{1i}\hat{j}_{2k} + \hat{j}_{1k}\hat{j}_{2i}$; d) $\hat{j}_{1i}\hat{j}_{1k} + \hat{j}_{1k}\hat{j}_{1i}$

over the states characterized by a given value, J, of the total angular momentum.

Obtain an explicit expression for the magnetic moment operator of the system, $\hat{\boldsymbol{\mu}} = g_1\hat{\mathbf{j}}_1 + g_2\hat{\mathbf{j}}_2$, in a state with a given total angular momentum, J (here $g_{1,2}$ are the gyro-magnetic ratios for the subsystems considered that relates their magnetic and mechanic angular momenta).

Solution

After averaging, these tensor operators act in the space with an angular momentum J. Each of them is expressed in terms of the vector operator $\hat{J}_i$ and universal tensors δ_{ik} and ε_{ikl}. The condition that the tensor character of the initial and averaged operators are the same puts strict constraints on the form of such expressions.

a) $\overline{\hat{j}_{1(2)i}} = a_{1(2)}\hat{J}_i$ (vectors of the form $\hat{J}_k\hat{J}_i\hat{J}_k$, $\varepsilon_{ikl}\hat{J}_k\hat{J}_l$, etc. all are reduced to $\hat{J}_i$ from the commutation relations for $\hat{J}_i$ components). Multiplying by $\hat{J}_i$, we obtain[28] $a_{1(2)} = \frac{\mathbf{j}_{1(2)}\cdot\mathbf{J}}{J(J+1)}$. From now on, for the sake of brevity, the scalar terms $(\mathbf{j}_{1,2} \cdot \mathbf{J})$ and $(\mathbf{j}_1 \cdot \mathbf{j}_2)$ which have definite values simultaneously with $\mathbf{j}_1^2, \mathbf{j}_2^2, \mathbf{J}^2$ are not written in explicit form. See Problem 3.27.

b) Taking into account the antisymmetric nature of the tensor, we have

$$\overline{\hat{j}_{1i}\hat{j}_{2k} - \hat{j}_{1k}\hat{j}_{2i}} = b\varepsilon_{ikl}\hat{J}_l. \tag{1}$$

We multiply both sides of this equation by $\hat{J}_k$ from the right and by $\hat{J}_i$ from the left, and we find that the left-hand side of the resulting equation is equal to zero while the right-hand side takes the form $b\varepsilon_{ikl}\hat{J}_i\hat{J}_l\hat{J}_k = -iJ(J+1)b$, so $b = 0$.

c) Due to the symmetry property of the tensor we have

$$\overline{\hat{j}_{1i}\hat{j}_{2k} + \hat{j}_{1k}\hat{j}_{2i}} = A_1\delta_{ik} + A_2(\hat{J}_i\hat{J}_k + \hat{J}_k\hat{J}_i). \tag{2}$$

---

[28] Multiplication by $\hat{j}_{1i}$ (or $\hat{j}_{2i}$) is not meaningful, because the operators $\hat{j}_{1,2}$, unlike $\hat{J}_i$ 'mix up' states with different values of J.

First we perform a convolution by i and k, and second we multiply this by $\hat{J}_k$ from the left and by $\hat{J}_i$ from the right. Using $\hat{J}_i\hat{J}_k\hat{J}_i\hat{J}_k = J^2(J+1)^2 - J(J+1)$, we obtain two relations:

$$3A_1 + 2J(J+1)A_2 = 2(\mathbf{j}_1 \cdot \mathbf{j}_2),$$

$$J(J+1)A_1 + J(J+1)(2J^2 + 2J - 1)A_2 = 2(\mathbf{j}_1 \cdot \mathbf{J})(\mathbf{j}_2 \cdot \mathbf{J}).$$

We get

$$A_1 = \frac{(4J^2 + 4J - 2)(\mathbf{j}_1 \cdot \mathbf{j}_2) - 4(\mathbf{j}_1 \cdot \mathbf{J})(\mathbf{j}_2 \cdot \mathbf{J})}{(2J-1)(2J+3)},$$

$$A_2 = \frac{6(\mathbf{j}_1 \cdot \mathbf{J})(\mathbf{j}_2 \cdot \mathbf{J}) - 2J(J+1)(\mathbf{j}_1 \cdot \mathbf{j}_2)}{J(J+1)(2J-1)(2J+3)}. \tag{3}$$

d) In a similar manner, we obtain

$$\overline{\hat{j}_{1i}\hat{j}_{1k} + \hat{j}_{1k}\hat{j}_{1i}} = B_1 \delta_{ik} + B_2(\hat{J}_i\hat{J}_k + \hat{J}_k\hat{J}_i),$$

$$B_1 = \frac{j_1(j_1+1)(4J^2+4J-2) - 4(\mathbf{j}_1 \cdot \mathbf{J})^2 + 2(\mathbf{j}_1 \cdot \mathbf{J})}{(2J-1)(2J+3)}, \tag{4}$$

$$B_2 = \frac{6(\mathbf{j}_1 \cdot \mathbf{J})^2 - 2j_1(j_1+1)J(J+1) - 3(\mathbf{j}_1 \cdot \mathbf{J})}{J(J+1)(2J-1)(2J+3)}.$$

For (4), we used the equation

$$\hat{J}_i\hat{j}_{1(2)k}\hat{j}_{1(2)i}\hat{J}_k = (\hat{\mathbf{j}}_{1(2)} \cdot \hat{\mathbf{J}})^2 - (\hat{\mathbf{j}}_{1(2)} \cdot \hat{\mathbf{J}}).$$

For the magnetic moment operator of the total system, we have

$$\overline{\hat{\boldsymbol{\mu}}(J)} = \overline{g_1 \hat{\mathbf{J}}_1 + g_2 \hat{\mathbf{J}}_2},$$

and using the result from (a) we have

$$\overline{\hat{\boldsymbol{\mu}}(J)} \equiv g(J)\hat{\mathbf{J}} = \frac{(g_1+g_2)J(J+1) + (g_1-g_2)[j_1(j_1+1) - j_2(j_2+1)]}{2J(J+1)} \hat{\mathbf{J}}. \tag{5}$$

3.4 Tensor formalism in angular momentum theory

Problem 3.41

Prove that the function of the form

$$\psi_l(\mathbf{n}) = \varepsilon_{ik...n} n_i n_k \ldots n_n,$$

where $\mathbf{n} = \mathbf{r}/r$ and $\varepsilon_{ik...n}$ is a completely symmetric tensor[29] of the lth rank with a trace $\varepsilon_{iik...n} = 0$, is an eigenfunction of the squared angular momentum operator of a particle whose angular momentum is l.

Further, show that the number of independent tensor components is equal to $2l + 1$, which coincides with the number of the spherical functions, $Y_{lm}(\mathbf{n})$. This shows that the angular dependence given above is the most general such dependence for the particle states with the angular momentum, l.

For the special cases where $l = 1$ and $l = 2$, indicate explicit expressions for the corresponding tensor components $\varepsilon_i(m)$ and tensors $\varepsilon_{ik}(m)$ that make the wavefunction (1) equal to the spherical function Y_{lm}.

Solution

Let us consider wavefunctions of the form

$$\tilde{\psi}_l(\mathbf{n}) = \varepsilon_{ik...n} x_i x_k \ldots x_n \equiv \varepsilon_{ik...n} n_i n_k \ldots n_n r^l.$$

Using the connection of operator $\hat{\mathbf{l}}^2$ with the Laplace operator,

$$\hat{\mathbf{l}}^2 = -r^2 \triangle_{\theta,\ \phi} = r^2(\triangle_r - \triangle),\ \triangle_r = \frac{1}{r^2}\frac{\partial}{\partial r}r^2\frac{\partial}{\partial r}, \tag{1}$$

we obtain

$$r^2 \triangle_r \tilde{\psi}_l = \varepsilon_{ik...n} n_i n_k \ldots n_n \triangle r^l = l(l+1)\tilde{\psi}_l,$$

$$\triangle \tilde{\psi}_l = \frac{\partial}{\partial x_m}\frac{\partial}{\partial x_m} \varepsilon_{ikp...n} x_i x_k x_p \ldots x_n = \tag{2}$$

$$\varepsilon_{ikp...n}(\delta_{im}\delta_{km}x_p \ldots x_n + \ldots) = \varepsilon_{mmp...n} x_p \ldots x_n + \ldots = 0.$$

From (1) and (2) it follows that $\hat{\mathbf{l}}^2 \tilde{\psi}_l = l(l+1)\tilde{\psi}_l$. We see that the function ψ_l given in the problem statement is an eigenfunction of $\hat{\mathbf{l}}^2$.

2) First let us find the number, $\tilde{g}(l)$, of independent components for any symmetric tensor of the lth rank $\tilde{\varepsilon}_{ik...n}$. We make the notations: n_1 is the number of tensor component indices that are equal to 1, n_2 is the number of indices equal to 2, and $n_3 = (l - n_1 - n_2)$ is the number equal to 3. Due to the tensor symmetry, its components with the equal numbers n_1 and n_2 are the same. When n_1 is fixed, n_2 could be equal to $0, 1, \ldots, l - n_1$ so the number of non-equal components with the given n_1 is $(l - n_1 + 1)$. The total number of non-equal components is

$$\tilde{g}(l) = \sum_{n_1=0}^{l}(l - n_1 + 1) = (l+1)^2 - \sum_{n_1=0}^{l} n_1 = \frac{(l+1)(l+2)}{2}.$$

The number of independent components $g(l)$ of a symmetric tensor with rank l and zero trace comes from the relation $\tilde{\varepsilon}_{iik...n} = 0$, and so it is given by a set of $\tilde{g}(l-2)$

---

[29] Do not confuse with the antisymmetric tensor ε_{ikl}!

linear relations between the $\tilde{g}(l)$ independent components of $\tilde{\varepsilon}_{ik...n}$. So $g(l) = \tilde{g}(l) - \tilde{g}(l-2) = 2l+1$.

3) Comparing $\psi_{l=1,\,m} = (\varepsilon(m) \cdot \mathbf{n})$ with Eq. (III.7) we find

$$\varepsilon(0) = \sqrt{3/4\pi}(0,\,0,\,i),\quad \varepsilon(\pm 1) = \sqrt{3/8\pi}(\mp i,\,1,\,0). \tag{3}$$

Similarly for the case $l = 2$, we obtain the tensor components $\varepsilon_{ik}(m)$:

$$\varepsilon_{ik}(2) = -\left(\tfrac{15}{32\pi}\right)^{1/2}\begin{pmatrix} 1 & i & 0 \\ i & -1 & 0 \\ 0 & 0 & 0 \end{pmatrix},\quad \varepsilon_{ik}(-2) = \varepsilon_{ik}^*(2), \tag{4}$$

$$\varepsilon_{ik}(1) = \left(\tfrac{15}{32\pi}\right)^{1/2}\begin{pmatrix} 0 & 0 & 1 \\ 0 & 0 & i \\ 1 & i & 0 \end{pmatrix},\quad \varepsilon_{ik}(-1) = -\varepsilon_{ik}^*(1),$$

$$\varepsilon_{ik}(0) = \left(\tfrac{15}{16\pi}\right)^{1/2}\begin{pmatrix} 1 & 0 & 0 \\ 0 & 1 & 0 \\ 0 & 0 & -2 \end{pmatrix}.$$

Problem 3.42

According to the previous problem, the most general angular dependence of a state with the angular momentum $l = 1$ has the form $\psi_{l=1} = \varepsilon \cdot \mathbf{n}$, where ε is some arbitrary complex vector. Find

a) a condition on the vector ε for the wavefunction to be normalized to unity;
b) mean values of the tensor components $\overline{n_i n_k}$;
c) mean values of the angular momentum vector components $\bar{\mathbf{l}}$;
d) a condition on the vector ε for being able to find such a $\tilde{z}$-axis in space that angular momentum $\tilde{z}$-projection has a definite value $\tilde{m} = 0$ or $\tilde{m} = \pm 1$.

Solution
a) Since

$$|\varepsilon \cdot \mathbf{n}|^2 = \varepsilon_i \varepsilon_k^* n_i n_k \text{ and } \int n_i n_k d\Omega = \frac{4\pi}{3}\delta_{ik},$$

then for the wavefunction normalization to unity, we should choose $\varepsilon = \sqrt{3/4\pi}\mathbf{a}$ with $|\mathbf{a}| = 1$.

b)
$$\overline{n_i n_k} = \varepsilon_l^* \varepsilon_n \int n_l n_i n_k n_n d\Omega = \varepsilon_l^* \varepsilon_n \cdot \frac{4\pi}{15}(\delta_{ik}\delta_{ln} + \delta_{in}\delta_{kl} + \delta_{il}\delta_{kn})$$

$$= \frac{1}{5}(\delta_{ik} + a_i^* a_k + a_k^* a_i).$$

c) Since $\hat{l}_i\psi_{l=1} \equiv -i\varepsilon_{ikn}x_k\partial/\partial x_n(e_m x_m/r) = -i\varepsilon_{ikn}e_n n_k$, it follows that

$$\bar{l}_i = -i\varepsilon_{ikn}e_m^* e_n \int n_m n_k d\Omega = -i\varepsilon_{ikn}a_k^* a_n.$$

In other words, $\bar{\mathbf{l}} = -i[\mathbf{a}^* \times \mathbf{a}]$ or, writing $\mathbf{a} = \mathbf{a}_1 + i\mathbf{a}_2$ (where $\mathbf{a}_{1,2}$ are the real vectors and $\mathbf{a}_1^2 + \mathbf{a}_2^2 = 1$), we obtain $\bar{\mathbf{l}} = 2[\mathbf{a}_1 \times \mathbf{a}_2]$.

d) Remember that $\varepsilon = \sqrt{3/4\pi}(\mathbf{a}_1 + i\mathbf{a}_2)$. It is easily seen that if $\mathbf{a}_1 \parallel \mathbf{a}_2$, the angular momentum projection along a common direction of the vectors $\mathbf{a}_{1,2}$ has a definite value $\tilde{m} = 0$. If $\mathbf{a}_1 \perp \mathbf{a}_2$ and $a_1 = a_2$, then the angular momentum projection along the $\mathbf{a}_1 \times \mathbf{a}_2$ vector has a definite value $\tilde{m} = 1$ (or $\tilde{m} = -1$ for the opposite direction).

Problem 3.43

For the conditions of the previous problem, find the probabilities, $w(\tilde{m})$, of different values of $\tilde{m}$ of the angular-momentum projection on the $\tilde{z}$-axis directed along the unit vector $\tilde{\mathbf{k}}$. Show that for an arbitrary state with the angular momentum $l = 1$, there exists a spatial direction that the probability of angular momentum projection $\tilde{m} = 0$ onto it is equal to zero.

Solution

Let us write the wavefunction in the form $\psi = \sqrt{3/4\pi}(\mathbf{a} \cdot \mathbf{n})$ with $|\mathbf{a}|^2 = 1$. Then we have

$$w(\tilde{m} = 0) = |(\mathbf{a} \cdot \mathbf{n}_0)|^2, \quad w(\tilde{m} = \pm 1) = \frac{1}{2}|(\mathbf{a} \cdot \mathbf{n})_1 \mp i(\mathbf{a} \cdot [\tilde{\mathbf{k}} \times \mathbf{n}_1])|^2. \quad (1)$$

Here $\mathbf{n}_1$ is some real unit vector that is perpendicular to $\tilde{\mathbf{k}}$ (the choice of $\mathbf{n}_1$ is non-unique but relations in (1) do not depend on it). Writing $\mathbf{a} = \mathbf{a}_1 + i\mathbf{a}_2$ shows that the probability of the value $\tilde{m} = 0$ of projection onto the axis directed along vector $[\mathbf{a}_1 \times \mathbf{a}_2]$ is equal to zero. For the case of $\mathbf{a}_1 \parallel \mathbf{a}_2$, projection onto an any axis perpendicular to $\mathbf{a}_1$ can not take the values $\tilde{m} = 0$.

Problem 3.44

According to Problem 3.41, the angular dependence of an arbitrary state with the angular momentum $l = 1$ has the form $\psi_{l=1} = (\mathbf{a} \cdot \mathbf{n})$, i.e., it is completely determined by some complex vector $\mathbf{a}$. Therefore in the case of states with the angular momentum $l = 1$, we can use a representation (let us call it the *vector representation*) in which the wavefunction coincides with components of the vector $\mathbf{a}$, i.e., $\psi(k) \equiv a_k$, ($k = 1, 2, 3$).

Determine an explicit form of the angular momentum component operators in the vector representation.

Solution

Acting by the operator $\hat{l}_i = -i\varepsilon_{ikn}x_k\partial/\partial x_n$ on a wavefunction of the form $\psi = (\mathbf{a} \cdot \mathbf{n})$, we obtain

$$\phi_i = \hat{l}_i\psi = \hat{l}_i a_m n_m = -i\varepsilon_{ikm}a_m n_k \equiv b_{i,k}n_k,$$

which is equivalent to the relation $b_{i,k} = \hat{l}_i a_k \equiv -i\varepsilon_{ikm}a_m$ in the vector representation. If in this representation we write down the wavefunction in the form of the column

$\psi = \begin{pmatrix} a_1 \\ a_2 \\ a_3 \end{pmatrix}$, then the matrices, $\hat{l}_i$, with elements $(\hat{l}_i)_{km} = -i\varepsilon_{ikm}$ are the angular momentum component operators:

$$\hat{l}_x = \begin{pmatrix} 0 & 0 & 0 \\ 0 & 0 & -i \\ 0 & i & 0 \end{pmatrix}, \quad \hat{l}_y = \begin{pmatrix} 0 & 0 & i \\ 0 & 0 & 0 \\ -i & 0 & 0 \end{pmatrix}, \quad \hat{l}_z = \begin{pmatrix} 0 & -i & 0 \\ i & 0 & 0 \\ 0 & 0 & 0 \end{pmatrix}. \quad (1)$$

It is clear that the commutation relations for these matrices have the standard form, i.e., $[\hat{l}_i, \hat{l}_k] = -i\varepsilon_{ikl}\hat{l}_l$ and $\hat{\mathbf{l}}^2 = 2 \cdot \hat{\mathbf{1}}$ ($\hat{\mathbf{1}}$ is the unity matrix).

Problem 3.45

For a system of two particles which have the same angular momenta $l_1 = l_2 = 1$, indicate:

a) the most general angular dependence of the wavefunction;
b) the most general angular dependence of the wavefunction ψ_L that describes the system states with a given value L ($L = 0, 1, 2$) of the total angular momentum;
c) the angular dependence of the wavefunctions ψ_{LM} for the system states with a given value of the total angular momentum L and z-projection M.

Use the results of Problem 3.41.

Solution

a) The most general angular dependence has the form $\psi = a_{ik}n_{1i}n_{2k}$, where $\mathbf{n}_1 = \mathbf{r}_1/r_1, \mathbf{n}_2 = \mathbf{r}_2/r_2$ and a_{ik} is an arbitrary tensor of the second rank, which has nine independent components. This corresponds to the nine independent states of a system consisting of two particles with the angular momenta $l_1 = l_2 = 1$.
b) If we write a_{ik} in the form

$$a_{ik} \equiv \frac{1}{3}a_{nn}\delta_{ik} + \frac{1}{2}(a_{ik} - a_{ki}) + \frac{1}{2}(a_{ik} + a_{ki} - \frac{2}{3}a_{nn}\delta_{ik}), \quad (1)$$

the wavefunction becomes

$$\psi = C(\mathbf{n}_1 \cdot \mathbf{n}_2) + (\boldsymbol{\varepsilon} \cdot [\mathbf{n}_1 \times \mathbf{n}_2]) + \varepsilon_{ik} n_{1i} n_{2k}, \tag{2}$$

$$C = \frac{1}{3} a_{nn}, \quad \varepsilon_{ik} = \frac{1}{2}(a_{ik} + a_{ki} - \frac{2}{3} a_{nn} \delta_{ik}), \quad 2\varepsilon_i = \varepsilon_{ikl} a_{kl}, \quad a_{ik} - a_{ki} = \varepsilon_{ikl} \varepsilon_l$$

ε_{ik} is the symmetric tensor with a trace equal to zero. Taking into account the result of Problem 3.41, it is easy to see that the wavefunction in (2) consists of three terms and each of them corresponds to a definite value of total angular momentum $L = 0, 1, 2$. In particular, the expression for the function $\psi_{L=0} = C(\mathbf{n}_1 \cdot \mathbf{n}_2)$ coincides with the result of Problem 3.34 for the case $l = 1$.

c) For the wavefunctions, ψ_L, in (2) to correspond to states with a definite value, M, of the total angular momentum z-projection, it is necessary that components of the vector $\varepsilon_i(M)$ or tensor $\varepsilon_{ik}(M)$ are in the form presented in Problem 3.41. In particular, for the wavefunction ψ_{22}, this gives the expression

$$\psi_{22} = -\sqrt{\frac{15}{32\pi}} \sin\theta_1 \sin\theta_2 e^{i\varphi_1} e^{i\varphi_2} = \sqrt{\frac{10\pi}{3}} Y_{11}(\theta_1, \varphi_1) Y_{11}(\theta_2, \varphi_2),$$

which is the sought-after (unnormalized) eigenfunction of the operators $\hat{\mathbf{L}}^2$ and L_z. It corresponds to the eigenvalues $L = 2$ and $M = 2$.

Problem 3.46

For a system of two particles, one having the angular momentum, $l_1 = 1$, find the angular dependence of the wavefunction, $\psi_{JJ_z\Lambda}$, describing system states that correspond to definite values of the total angular momentum $J = 0$ and 1, the z-projection J_z, and angular momentum projection Λ onto the direction of the second particle's radius-vector, specifically considering $\Lambda = 0$. What are the parities of these states? What are the possible angular momenta, l_2, of the second particle in such states? Generalize the result to the case of arbitrary values of l_1, J, J_z (but $\Lambda = 0$).

Solution

The conditions $l_1 = 1$ and $\Lambda = 0$ uniquely determine the wavefunction dependence on the angle variables of the first particle in the form $\psi \propto (\mathbf{n}_1 \cdot \mathbf{n}_2)$ where $\mathbf{n}_1 = \mathbf{r}_1/r_1$ and $\mathbf{n}_2 = \mathbf{r}_2/r_2$. (Compare with Problem 3.16. It should be remembered that the total angular momentum projection Λ along the radius-vector $\mathbf{r}_2$ is completely determined by the projection of the first particle, since $(\mathbf{n}_2 \cdot \hat{\mathbf{l}}_2) \equiv 0$). Since $(\mathbf{n}_1 \cdot \mathbf{n}_2)$ is a scalar, as well as the wavefunction for the state with $J = 0$, then $\psi_{000} = const(\mathbf{n}_1 \cdot \mathbf{n}_2)$.

For a state with $J = 1$, the wavefunction is some linear combination of components of a vector which depends only on $\mathbf{n}_1$ and $\mathbf{n}_2$ (compare with Problem 3.41). Since $l_1 = 1$ and $\Lambda = 0$, the only such vector is of the form $\mathbf{v} = A(\mathbf{n}_1 \cdot \mathbf{n}_2)\mathbf{n}_2$. By constructing linear combinations of its components (which correspond to momentum projections J_z), we find that

$$\psi_{1J_z0} = C(\mathbf{n}_1 \cdot \mathbf{n}_2)Y_{1J_z}(\mathbf{n}_2). \tag{1}$$

Wavefunction (1) has a definite parity equal to -1 and describes a state where the angular momentum of the second particle can only be equal to 0 and 2.

The generalization of (1) to arbitrary values of l_1, J, J_z, with $\Lambda = 0$ is

$$\psi_{JJ_z0} = CP_l(\mathbf{n}_1 \cdot \mathbf{n}_2)Y_{JJ_z}(\mathbf{n}_2), \tag{2}$$

where $P_l(z)$ is the Legendre polynomial.

Problem 3.47

For a system consisting of three particles, prove that any state with total angular momentum $L = 0$ (in the center-of-mass system) has a definite, positive parity.

Solution

The wavefunction with $L = 0$ does not change under rotations of the coordinate system, *i.e.*, it is a scalar (or pseudo-scalar, which depends on the parity of the state) function. In the center-of-mass system of the three particles, only the radius-vectors of two particles, $\mathbf{r}_{1,2}$, are independent, while $\mathbf{r}_3 = -(\mathbf{r}_1 + \mathbf{r}_2)$ (we consider all the masses to be the same for the sake of simplicity). From the two vectors $\mathbf{r}_{1,2}$ it is possible to form the following scalar quantities: $\mathbf{r}_1^2, \mathbf{r}_2^2, \mathbf{r}_1 \cdot \mathbf{r}_2$. These are real scalars, not pseudo-scalars. Scalar functions that depend on vectors $\mathbf{r}_{1,2}$ could be functions only of the scalars mentioned above. So the wavefunction is a function of the form $\psi_{L=0} = f(\mathbf{r}_1^2, \mathbf{r}_2^2, \mathbf{r}_1 \cdot \mathbf{r}_2)$. With an inversion of the coordinates, $\mathbf{r}_{1,2} \to -\mathbf{r}_{1,2}$, this function does not change: $\hat{I}\psi_{L=0} = \psi_{L=0}$. The state with $L = 0$ has positive parity.[30]

---

[30] We should note that we mean here the *orbital* parity. Also, if the number of particles in the system is more than three then there is a possibility of making a pseudo-scalar quantity of the form $\mathbf{r}_1 \cdot [\mathbf{r}_2 \times \mathbf{r}_3]$. The states of such systems with $L = 0$ are allowed to have an arbitrary parity.

4

Motion in a spherically-symmetric potential

We are interested in the solutions of the stationary Schrödinger equation for a spherically-symmetric potential:

$$\left[-\frac{\hbar^2}{2m}\triangle + U(r)\right]\psi_E(\mathbf{r}) = E\psi_E(\mathbf{r}). \tag{IV.1}$$

Due to the mutual commutativity of the operators $\hat{H}$, $\hat{\mathbf{l}}^2$, $\hat{l}_z$, we look for a solution in the form[31] $\psi_E(\mathbf{r}) \equiv \psi_{n_r l m}(\mathbf{r}) = R_{n_r l}(r) Y_{lm}(\mathbf{n})$, where Y_{lm} are the spherical functions. Eqn. (IV.1) then reduces to the one-dimensional radial Schrödinger equation:

$$\left[\frac{1}{r}\frac{d^2}{dr^2}r + \frac{2m}{\hbar^2}\left(E_{n_r l} - \frac{\hbar^2 l(l+1)}{2mr^2} - U(r)\right)\right] R_{n_r l}(r) = 0. \tag{IV.2}$$

The boundary condition at $r \to 0$ has the form[32] $R_{n_r,0}(0) = \text{const} < \infty$ for $l = 0$ and $R_{n_r l}(0) = 0$ for $l \neq 0$.

For a particle moving in the attractive Coulomb potential, $U(r) = -\alpha/r$, the energy levels and the radial functions for the states of a discrete spectrum are given by

$$E_n = -\frac{m\alpha^2}{2\hbar^2 n^2} \quad \text{and}$$

$$R_{nl} = -\frac{2}{a^{3/2} n^2} \sqrt{\frac{(n-l-1)!}{[(n+l)!]^3}} \left(\frac{2r}{na}\right)^l e^{-r/na} L_{n+l}^{2l+1}\left(\frac{2r}{na}\right), \tag{IV.3}$$

---

[31] In this chapter we consider only the states of a discrete spectrum and denote the energy levels by $E_{n_r l}$ (where $n_r = 0, 1, \ldots$ is a *radial quantum number*.)

[32] We are discussing the *regular* potentials, for which $r^2 U \to 0$ when $r \to 0$. For these potentials, the two independent solutions for short distances have the form $R_1 \propto r^l$ and $R_2 \propto r^{-l-1}$. We exclude from the consideration the increasing solution for $l \neq 0$ due to its unnormalizability. At $l = 0$ for the increasing solution, $R_2 \propto 1/r$ and we have $\triangle R_2 \propto \delta(\mathbf{r})$, so it do not obey the relation (IV.1) when $r \to 0$. Such a solution that is square integrable for short distances is used while modeling a short-range center by a potential of zero radius. See Problem 4.10. For a *singular* attractive potential, 'falling onto the center' appears, so the choice of boundary condition for $r \to 0$ demands some addition investigation. See Problem 9.14.

where $n = n_r + l + 1$ is the *principle quantum number*, $a = \hbar^2/m\alpha$ (for a hydrogen atom, a defines the Bohr radius) and $L_n^k(z)$ is the associated Laguerre polynomial, which is connected with the hypergeometric function:

$$L_n^k(z) = (-1)^k \frac{(n!)^2}{k!(n-k)!} F(k-n, k+1, z).$$

In particular for a few lowest states, we have

$$R_{10}(r) = \frac{2}{\sqrt{a^3}} e^{-r/a} \qquad \text{(the ground, 1s-state)},$$

$$R_{20}(r) = \frac{1}{\sqrt{(2a^3)}} \left(1 - \frac{r}{2a}\right) e^{-r/2a} \quad \text{(2s-state)}, \tag{IV.4}$$

$$R_{21}(r) = \frac{1}{\sqrt{24a^5}} r e^{-r/2a} \qquad \text{(2p-state)}.$$

To solve Eq. (IV.2), it is often convenient to introduce the new radial functions $\chi_{n_r l} = r R_{n_r l}$, which satisfy the following differential equation

$$\left[-\frac{\hbar^2}{2m} \frac{d^2}{dr^2} + \frac{\hbar^2 l(l+1)}{2mr^2} + U(r) \right] \chi_{n_r l} = E_{n_r l} \chi_{n_r l} \tag{IV.5}$$

with the boundary condition $\chi_{n_r l}(0) = 0$. This equation is similar in form to the ordinary one-dimensional Schrödinger equation.

The following substitution is also often used: $u_{n_r l} = \sqrt{r} R_{n_r l}$. In this case, the equation becomes

$$u''_{n_r l} + \frac{1}{r} u'_{n_r l} - \left[\frac{(l+1/2)^2}{r^2} + \frac{2m}{\hbar^2} \left(U(r) - E_{n_r l} \right) \right] u_{n_r l} = 0 \tag{IV.6}$$

with the boundary condition $u_{n_r l}(0) = 0$.

4.1 Discrete spectrum states in central fields

Problem 4.1

Relate the energy levels, $E_{n_r 0}$, and the normalized wavefunctions, $\psi_{n_r 00}(r)$, of discrete stationary s-states of a particle in a central potential, $U(r)$, to the levels, E_n, and normalized wavefunctions, $\psi_n(x)$, of a particle in the one-dimensional potential, $\tilde{U}(x)$, of the form $\tilde{U}(x) = U(x)$ for $x > 0$ and $\tilde{U}(x) = \infty$ for $x < 0$ (see also Problem 2.5).

By using this relationsip, find:

a) the s-state levels in a spherical infinitely deep potential well, *i.e.*: $U(r) = 0$ for $r < a$ and $U(r) = \infty$ for $r > a$;

b) the condition for existence of a bound state in a potential of the form: $U = -U_0$ for $r < a$ and $U = 0$ for $r > a$.

Solution

The radial Schrödinger equation (IV.5) and its boundary conditions $\chi(0) = \chi(\infty) = 0$ have the same form as for the ordinary Schrödinger equation in a one-dimensional potential, $U(r)$. It follows that $E_{n_r 0} = E_{n_r}$ (the spectra coincide) and $\psi_{n_r 00}(r) = \psi_{n_r}(r)/(\sqrt{4\pi}\, r)$ with $n_r = 0, 1, \ldots$.

These relations immediately allow us to extend some results from one-dimensional quantum mechanics to the case of central potentials. In particular, for the case a), we have $E_{n_r 0} = \pi^2 \hbar^2 (n_r + 1)^2 / 2ma^2$ (compare with Problem 2.1). In the case b), the condition of bound s-states (and hence the bound states in general) existence is given by $U_0 \geq \pi^2 \hbar^2 / 8ma^2$ (compare with Problem 2.14).

Problem 4.2

Describe the character of the change of the energy levels, $E_{n_r l}$, for a particle's discrete spectrum,

a) for a given value of l with increasing of n_r;
b) for a given value of n_r with increasing of l.

Solution

a) Since Eq. (IV.5) is similar in form to the ordinary one-dimensional Schrödinger, we conclude that $E_{n_r l}$ (for a given value of l) increases with the increase in n_r.

b) Considering formally l in the Schrödinger equation (IV.5) as some continuous parameter, from Eq. (I.6) we obtain

$$\frac{\partial}{\partial l} E_{n_r l} = \overline{\frac{\partial}{\partial l} \hat{H}} = \overline{\frac{(2l+1)\hbar^2}{2mr^2}} > 0,$$

which means that $E_{n_r l}$ increases with increasing l.

Problem 4.3

Let N be a number of levels in a central potential in increasing order (for the ground state, $N = 1$). For the Nth level, indicate

a) the maximum value of the angular momentum;
b) the maximum level degeneracy;
c) the maximum level degeneracy if this level has a definite parity.

Solution

a) Taking into account the increase of $E_{n_r l}$ with the increase in l (for a given value of n_r; see Problem 4.2), we see that independently of a specific form of the potential,

$U(r)$ the value of particle's angular momentum in Nth discrete spectrum state cannot exceed the value $l_{max} = N - 1$. For such an angular momentum, the value $n_r = 0$.

b) The maximum degree of level degeneracy appears in the case when it corresponds to the states with $0 \leq l \leq l_{max}$ and is equal to

$$g_{max}(N) = \sum_{l=0}^{N-1}(2l+1) = N^2. \tag{1}$$

Such a situation takes place for the Coulomb potential. For the states with a given l, we have $n_r = N - 1 - l$.

c) Since the parity is $I = (-1)^l$, in (1) we should sum over the values of l with a definite parity (even or odd). Then we have $\tilde{g}_{max}(N) = \frac{1}{2}N(N+1)$ and for the degenerate states with given $l = l_{max}$, $l_{max} - 2$, ..., $1(0)$, we have $n_r = (l_{max} - l)/2$. Such a situation takes place for a spherical oscillator. See Problems 4.4 and 4.5.

Problem 4.4

Find the energy levels and the corresponding normalized wavefunctions for a spherical oscillator, $U(r) = kr^2/2$. Use Cartesian coordinates to solve the Schrodinger equation. Determine the level degeneracy and classify the states with respect to the quantum numbers n_r, l and parity.

Relate the "accidental" level degeneracy and the commutativity of the operators $\hat{T}_{ik} = \frac{1}{m}\hat{p}_i\hat{p}_k + k\hat{x}_i\hat{x}_k$ with the oscillator Hamiltonian.

Solution

Taking into account the consideration used in Problem 2.48 (for a planar oscillator), we obtain the solution in the form

$$\psi_{n_1 n_2 n_3}(\mathbf{r}) = \psi_{n_1}^{(osc)}(x)\psi_{n_2}^{(osc)}(y)\psi_{n_3}^{(osc)}(z); \quad n_1, n_2, n_3 = 0, 1, 2, \ldots, \tag{1}$$

$$E_n = \hbar\omega\left(n + \frac{3}{2}\right), \quad n = n_1 + n_2 + n_3, \quad n = 0, 1, 2, \ldots.$$

The oscillator energy levels have a definite parity that is equal to $I_n = (-1)^n$ and their degeneracy is given by (compare with Problem 3.41)

$$g(n) = \sum_{n_1=0}^{n}(n - n_1 + 1) = \frac{1}{2}(n+1)(n+2).$$

For a given value of n_1 there are $n - n_1 + 1$ degenerate states with $n_2 = 0, 1, \ldots, n - n_1$ and $n_3 = n - n_1 - n_2$.

Since the potential is spherically symmetric, the stationary states can be classified with respect to the values l of the angular momentum. As is seen from (1) and the expressions for the linear oscillator wavefunctions (see Eq. (II.2)), the ground state wavefunction, $n = 0$, is spherically symmetric, $\psi_{000} \propto e^{-r^2/2a^2}$ with $a = \sqrt{\hbar/m\omega}$, and describes the s-state, as expected. For the first excited level, $n = 1$, the wavefunctions (1) have the form $\psi_{n=1} \propto x_i e^{-r^2/2a^2}$ with $i = 1, 2, 3$; they describe the p-level (see Eq. (III.7)).

However, in the case of $n \geq 2$ these wavefunctions do not correspond to a definite value[33] of the angular momentum, l.

This fact reflects the existence of an *accidental degeneracy* of the spherical oscillator energy levels (see Problems 4.3 and 4.5). Such a degeneracy can be understood if we take into account the commutativity of the operators, $\hat{T}_{ik}$, specified in the problem condition with the oscillator Hamiltonian and their non-commutativity with $\hat{l}^2$. See Problem 1.25.

Problem 4.5

Analyze the stationary states of a spherical oscillator (see the previous problem). Use spherical coordinates to solve the Schrödinger equation.

Solution

The Schrödinger equation, (IV.2), for a spherical oscillator, $U(r) = kr^2/2$, by introducing the new variable $x = \frac{m\omega}{\hbar} r^2$ becomes ($\omega = \sqrt{k/m}$)

$$\left\{ x \frac{d^2}{dx^2} + \frac{3}{2}\frac{d}{dx} + \left[\frac{E}{2\hbar\omega} - \frac{l(l+1)}{4x} - \frac{x}{4} \right] \right\} R_{n_r l} = 0. \tag{1}$$

By making the substitution $R_{n_r l} = e^{-x/2} x^{l/2} w(x)$, we convert (1) into the hypergeometric equation

$$xw'' + \left(l + \frac{3}{2} - x \right) w' + \left(\frac{E}{2\hbar\omega} - \frac{l}{2} - \frac{3}{4} \right) w = 0. \tag{2}$$

Since $R \propto r^l \propto x^{1/2}$ when $r \to 0$, the solution of (2) must be of the form

$$w = cF\left(-\frac{E}{2\hbar\omega} + \frac{l}{2} + \frac{3}{4},\ l + \frac{3}{2},\ x \right), \tag{3}$$

where $F(\alpha, \beta, x)$ is the degenerate hypergeometric function. In this case the constraints on the wave-function to decrease for $r \to \infty$ demand that function (3) reduces to a polynomial (in the opposite case $F \propto e^x$ and $R \propto e^{x/2}$ diverge as $x, r \to \infty$). So we have

---

[33] The wavefunctions with a definite value of l are described by some superpositions of the functions (1). For example, in the case of $n = 2$, the s-state wavefunction has the form $\psi_{n=2, l=0} = \frac{1}{\sqrt{3}}(\psi_{200} + \psi_{020} + \psi_{002})$, while the five other independent combinations of (1) which are orthogonal to it correspond to $l = 2$.

$$-\frac{E_n}{2\hbar\omega} + \frac{l}{2} + \frac{3}{4} = -n_r, \quad n_r = 0, 1, 2, \ldots,$$

which determines the energy spectrum:

$$E_{n_r,l} = \hbar\omega\left(2n_r + l + \frac{3}{2}\right) \equiv \hbar\omega\left(n + \frac{3}{2}\right), \quad (4)$$

$$n = 2n_r + l = 0, 1, 2, \ldots.$$

For the level with a given n, we have states with the angular momentum $l = n, n-2, \ldots, 1(0)$, so if the level has a certain parity $I_n = (-1)^n$, then the degeneracy, $g(n) = \sum(2l+1)$, becomes equal to $g(n) = \frac{1}{2}(n+1)(n+2)$, in accordance with the result of the previous problem.

In conclusion, we indicate the value of the coefficient c in (3):

$$c^2 = 2\left(\frac{m\omega}{\hbar}\right)^{3/2} \frac{\Gamma(l + n_r + 3/2)}{n_r! \Gamma^2(l + 3/2)}, \quad (5)$$

which corresponds to the normalization condition $\int_0^\infty R_{n_r,l}^2(r) r^2 dr = 1$.

Problem 4.6

For the ground state of a hydrogen atom, determine:

a) $\overline{r^n}$ for the electron, where n is an integer;
b) the mean kinetic and potential energies of the electron;
c) the momentum distribution for the electron;
d) the effective (average) potential $\varphi(r)$ created by the atom.

Solution

The wavefunction has the form $\psi_0(r) = (\pi a^3)^{-1/2} e^{-r/a}$, where $a = \hbar^2/me^2$. Hence,

a)
$$\overline{r^n} = \int r^n |\psi_0(r)|^2 dV = \frac{(n+2)!}{2}\left(\frac{a}{2}\right)^n. \quad (1)$$

b)
$$\overline{U(r)} = -\int \frac{e^2}{r} |\psi_0(r)|^2 dV = -\frac{e^2}{a}. \quad (2)$$

Since $\overline{T} + \overline{U} = E_0 = -\frac{e^2}{2a}$, we obtain $\overline{T} = \frac{e^2}{2a} = -\frac{1}{2}\overline{U}$.

c) The wavefunction in the momentum representation,

$$\phi_0(p) = \frac{1}{(2\pi\hbar)^{3/2}} \int e^{-i\frac{\mathbf{p}\mathbf{r}}{\hbar}} \psi_0(r) dV = \frac{2\sqrt{2}\hbar^{5/2}}{\pi a^{5/2}}\left(p^2 + \frac{\hbar^2}{a^2}\right)^{-2}, \quad (3)$$

determines the momentum probability distribution for the electron: $dw = |\phi_0(p)|^2 d^3p$.

d) The desired potential, $\varphi(r)$, is an electrostatic potential, characterized by the charge density of the form

$$\rho(r) = e\delta(\mathbf{r}) - e|\psi_0(r)|^2,$$

where the first term describes the point-like nucleus (at the origin of coordinates), while the second describes an electron "cloud". The Poisson equation, $\Delta \varphi = -4\pi\rho$ for $r \neq 0$, takes the form:

$$\frac{d^2\chi(r)}{dr^2} = \frac{4er}{a^3} e^{-2r/a},$$

where $\chi(r) = r\varphi(r)$. By integrating this equation and using the boundary conditions[34] $\chi(\infty) = 0$, $\chi'(\infty) = 0$, we obtain

$$\chi(r) = \frac{4e}{a^3} \int_r^\infty dr' \int_{r'}^\infty r'' e^{-2r''/a} dr''.$$

So it follows that

$$\varphi(r) = \frac{\chi(r)}{r} = e\left(\frac{1}{r} + \frac{1}{a}\right) e^{-2r/a}. \tag{4}$$

In particular, as $r \to 0$ we have $\varphi(r) \approx \frac{e}{r} - \frac{e}{a}$. The first dominant term, $\varphi_p(r) = e/r$, describes the electrostatic potential created by the proton while the second term, $\varphi_{el}(0) \equiv -\frac{e}{a}$, describes the potential of the electron "cloud." We should note that the value of $e\varphi_{el}(0)$ coincides, of course, with $\overline{U}$.

On the other hand, at the large distances, $r \to \infty$, we obtain from Eq. (4) the exponential decrease of the potential, which corresponds to total screening of the proton charge by the spherically symmetric electron "cloud". We should emphasize that this result takes place just for the averaged value of the potential. The "true" value of the field decreases much slower. See the next problem.

Problem 4.7

For the ground state of a hydrogen atom, determine the average electric field $\mathcal{E}$ and its fluctuations (i.e., correlators of the field components) at large distances.

---

[34] The condition $\chi(\infty) = 0$ means that the total charge of the system is equal to zero.

Solution

Utilizing Eq. (4) of the previous problem we see that the average electric field decreases exponentially

$$\overline{\mathcal{E}}(\mathbf{R}) = -\nabla \varphi(R) \underset{R \gg a}{\approx} \frac{2e\mathbf{R}}{a^2 R} e^{-2R/a}. \tag{1}$$

The field $\mathcal{E}(\mathbf{R})$ created by the proton at the origin and the electron at the point $\mathbf{r}$ has the form

$$\mathcal{E}(\mathbf{R}) = \frac{e\mathbf{R}}{R^3} - \frac{e(\mathbf{R}-\mathbf{r})}{|\mathbf{R}-\mathbf{r}|^3} \underset{R \gg r}{\approx} \frac{er(\mathbf{n} - 3\mathbf{N}(\mathbf{n}\cdot\mathbf{N}))}{R^3}.$$

We defined $\mathbf{n} = \mathbf{r}/r$, $\mathbf{N} = \mathbf{R}/R$, and we obtain

$$\overline{\mathcal{E}_i(\mathbf{R})\mathcal{E}_k(\mathbf{R})} =$$

$$\frac{e^2}{R^6} \int\int |\psi_0(r)|^2 r^4 (n_i - 3N_i n_l N_l)(n_k - 3N_k n_m N_m) dr d\Omega_\mathbf{n}$$

$$= e^2 a^2 (\delta_{ik} + 3N_i N_k)\frac{1}{R^6}, \quad R \gg a.$$

Remember that averaging is carried out over the positions of an electron in the hydrogen atom ground state. In particular, $\overline{\mathcal{E}^2(R)} = \frac{6e^2 a^2}{R^6}$.

So the electric field fluctuations decrease according to $\sqrt{\overline{\mathcal{E}^2(R)}} \propto 1/R^3$. This result is reflected in that the interaction between atoms and molecules at the large distances (for example, Van der Waals forces) decreases as a power law, not exponentially.

Problem 4.8

Find the s-levels for the following potentials:

a) $U(r) = -\alpha \delta(r-a)$;
b) $U(r) = -U_0 e^{-r/a}$;
c) $U(r) = -U_0/(e^{r/a} - 1)$ (the *Hulthén potential*).

Solution

For $E < 0$, the spectrum is discrete. Set $\kappa = \sqrt{-2mE_{n_r 0}/\hbar^2}$.

a) If we take into account the boundary conditions at $r = 0$ and $r = \infty$, we obtain the solution of Eq. (IV.5) for $l = 0$ and $U = -\alpha\delta(r-a)$:

$$\chi_{n_r 0} = \begin{cases} A \sinh \kappa r, & r < a, \\ B e^{-\kappa r}, & r > a. \end{cases}$$

We use the matching conditions from Problem 2.6 at the point $r = a$ to obtain the equation

$$\frac{\hbar^2 \kappa}{m\alpha} = (1 - e^{-2\kappa a}),$$

which determines the s-level spectrum.[35] In the case where $\xi \equiv m\alpha a \hbar^{-2} < 1/2$, this equation has no roots, so bound states are absent. At $\xi > 1/2$, there exists one, and only one, discrete s-level. The limiting value of its energy is

$$E_0 \approx \begin{cases} -\left(\frac{\hbar^2}{2ma^2}\right)\left(\frac{2m\alpha a}{\hbar^2} - 1\right)^2, & 0 < \xi - 1/2 \ll 1 \\ -\frac{m\alpha^2}{2\hbar^2}, & \xi \gg 1. \end{cases} \quad (1)$$

Pay attention to the slow quadratic dependence of the depth of the shallow s-level, $E_{n_r 0} \propto -(\xi - \xi_0)^2$, with respect to making the potential well deeper. This is because in the case of $E \to 0$ the s-level wavefunction delocalizes: the particle "moves" outward and has a small probability of being near the origin. Compare with the case $l \geq 1$ considered in the next problem.

b) After performing the substitution $x = \exp(-r/2a)$, Eq. (IV.5) takes the form of the Bessel equation:

$$\left[\frac{d^2}{dx^2} + \frac{1}{x}\frac{d}{dx} + \left(\lambda^2 - \frac{p^2}{x^2}\right)\right]\chi_{n_r 0} = 0, \quad (2)$$

where $p = 2\kappa a$, $\lambda = (8mU_0 a^2/\hbar^2)^{1/2}$. The condition of the wavefunction being zero when $r \to \infty$ (so that $x = 0$) demands choosing the solution of (2) in the form $\chi_{n_r 0} = c J_p(\lambda x)$. Now the condition $\chi(r = 0) = 0$ leads to the equation

$$J_p(\lambda) = 0, \quad \text{or} \quad J_{2\kappa a}(\sqrt{8mU_0 a^2/\hbar^2}) = 0, \quad (3)$$

which determines the s-level spectrum.

When the potential well has grown just enough for a level to appear, the energy of this level is arbitrarily small. So the condition $J_0(\lambda) = 0$ determines the values of the well parameters that correspond to the appearance of new discrete spectrum states with the deepening of the well. For the Nth level to appear we must have a potential factor of at least $U_{0,N} = (\hbar^2 x_N^2/8ma^2)$, where x_N is Nth zero of the function $J_0(x)$. Since $x_1 \approx 2.40$, the condition for a discrete spectrum s-state's existence (and so the existence of a bound state) takes the form $U_0 \geq 0.72\hbar^2/ma^2$.

---

[35] Taking into account the results of Problem 4.1, compare with a spectrum of the odd levels in the Problem 2.18 condition.

The upper s-level is "shallow" when $0 < (\lambda - x_N) \ll 1$. In this case, we use the formulae (J_ν, N_ν are Bessel and Neumann functions)

$$J_0'(x) = -J_1(x) \text{ and } \left(\left(\frac{\partial}{\partial \nu}\right) J_\nu(x)\right)_{\nu=0} = \frac{\pi}{2} N_0(x),$$

and we find the state energy according to (3) ($n_r = N - 1$):

$$E_{n_r 0} \approx -\frac{\hbar^2}{2\pi^2 m a^2} \frac{J_1^2(x_N)}{N_0^2(x_N)} \left(\frac{\sqrt{8mU_0 a^2}}{\hbar} - x_N\right)^2. \tag{4}$$

c) We take Eq. (IV.5) with $l = 0$, and use a change of variable $x = \exp(-r/a)$ (here $U = -U_0 x/(1-x)$) and the substitution $\chi_{n_r 0} = x^\varepsilon y$ where $\varepsilon = \kappa a$. We obtain the equation for the hypergeometric function, $F(\alpha, \beta, \gamma, x)$,

$$(1-x)xy'' + (2\varepsilon + 1)(1-x)y' + \lambda^2 y = 0, \tag{5}$$

with the parameters

$$\alpha = \varepsilon + \sqrt{\varepsilon^2 + \lambda^2}, \ \beta = \varepsilon - \sqrt{\varepsilon^2 + \lambda^2}, \ \gamma = 2\varepsilon + 1; \ \lambda = (2ma^2 U_0/\hbar^2)^{1/2}.$$

The condition of the wavefunction being zero at $r \to \infty$ ($x \to 0$) requires choosing the solution of (5) in the form $y = cF(\alpha, \beta, \gamma, x)$. The condition $\chi(r=0) = \chi(x=1) = 0$ gives

$$F(\alpha, \beta, \gamma, x = 1) \equiv \frac{\Gamma(\gamma)\Gamma(\gamma - \alpha - \beta)}{\Gamma(\gamma - \alpha)\Gamma(\gamma - \beta)} = 0, \tag{6}$$

which determines the s-level spectrum. It follows that $\gamma - \alpha = -n_r$, where $n_r = 0, 1, \ldots$, since $\Gamma(-n_r) = \infty$. Energies for the s-levels take the form

$$E_{n_r 0} = -\frac{\hbar^2}{8ma^2(n_r+1)^2}(\lambda^2 - (n_r+1)^2)^2, \tag{7}$$

where $n_r \leq \lambda - 1$. The condition $\lambda = N$ (here N is integer) determines the potential parameters that correspond to emergence of the Nth level with $l = 0$ with a deepening of the potential well. In the case of $a \to \infty$, $U_0 \to 0$, but $aU_0 = \text{const} \equiv \alpha$, this potential takes the form of the Coulomb potential $U = -\alpha/r$, and (7) gives the known s-levels spectrum. See Eq. (IV.3).

Problem 4.9

Determine the levels with an arbitrary angular momentum l in the potentials:

a) $U(r) = -\alpha \delta(r-a)$;
b) $U(r) = 0$ for $r < a$ and $U(r) = \infty$ for $r > a$.

Solution

a) The solution of Eq. (IV.6) with $\kappa = \sqrt{-2mE_{n_r l}/\hbar^2}$, which satisfies the boundary conditions $u(0) = u(\infty) = 0$ is

$$u_{n_r l} = A I_{l+1/2}(\kappa r) \text{ for } r < a \text{ and } u_{n_r l} = B K_{l+1/2}(\kappa r) \text{ for } r > a,$$

where I_ν and K_ν are the Bessel functions of an imaginary argument (the modified Bessel functions).

Matching the wavefunction at the point $r = a$ is the same as in the case of the one-dimensional δ-potential in Problem 2.6 and gives[36]

$$I_{l+1/2}(\kappa a) K_{l+1/2}(\kappa a) = \frac{\hbar^2}{2m\alpha a} \equiv (2\xi)^{-1}, \tag{1}$$

which determines the particle energy spectrum.

The left-hand side of (1) for $\kappa \to 0$, when the level has an arbitrarily small energy, takes the definite value equal to $1/(2l+1)$. This means (as for the case $l = 0$ seen in Problem 4.8a) that for $\xi \geq \xi_l^{(0)} = (l+1/2)$ there is only one discrete level with a given value of l. Using the asymptotic formulae for $I_\nu(z)$ and $K_\nu(z)$, we obtain the generalization of (1) from Problem 4.8 for the case of states with $l \neq 0$:

$$E_{0l} \approx \begin{cases} -\frac{(2l-1)(2l+3)}{2(2l+1)} \frac{\hbar^2}{ma^2} (\xi - \xi_l^{(0)}), & \xi \to \xi_l^{(0)} \\ -\frac{m\alpha^2}{2\hbar^2} + \frac{\hbar^2 l(l+1)}{2ma^2}, & \xi \to \infty. \end{cases} \tag{2}$$

Note that when $l \geq 1$ the deepening of the shallow level corresponds to the deepening of the potential well, in contrast to the case of $l = 0$ (see the previous problem). This distinction is because of the centrifugal potential, $U_{cf} = \hbar^2 l(l+1)/2mr^2$. A state with $l \geq 1$ remains bound as $E \to 0$. The centrifugal barrier prevents removing a particle to infinity.

b) Eq. (IV.6) in the case considered for $r < a$ reduces to the Bessel equation. Since $u_{n_r l}(0) = 0$, the only solution must be of the form $u_{n_r l} = c J_{l+1/2}(kr)$. The condition $u_{n_r l}(a) = 0$ gives the particle energy levels:

$$E_{n_r l} = \frac{\hbar^2 k^2}{2ma^2} = \frac{\hbar^2 \alpha_{n_r+1,l}^2}{2ma^2}, \tag{3}$$

where α_{nl} is the nth zero (ignoring $x = 0$) of the Bessel function $J_{l+1/2}(x)$.

Problem 4.10

A *zero-range potential* (the three-dimensional analog of a one-dimensional δ-potential; see Problem 2.7) is introduced by imposing on a wavefunction the following boundary

---

[36] We used the Wronskian $W = [I_\nu(z), K_\nu(z)] = I_\nu(z) K_\nu'(z) - I_\nu'(z) K_\nu(z) = -\frac{1}{z}$.

condition[37]:

$$\frac{(r\psi(\mathbf{r}))'}{r\psi(\mathbf{r})} \to -\alpha_0 \quad \text{at } r \to 0$$

i.e.

$$\psi \propto \left(-\frac{1}{\alpha_0 r} + 1 + \ldots\right). \tag{1}$$

Consider the possibility of particle bound states (depending on the sign of α_0) in such a potential. Find the wavefunction of these bound states in the momentum representation. Determine the mean values $\bar{T}$ and $\bar{U}$.

Solution

The solution of the Schrödinger equation for this problem with $E < 0$, has the form

$$\psi_0(r) = A\frac{e^{-\kappa r}}{\sqrt{4\pi}\, r}, \quad \text{where } \kappa = \sqrt{-2mE/\hbar^2} > 0, \tag{2}$$

(it describes a particle with $l = 0$). For $r \to 0$

$$\psi_0(r) = \frac{A}{\sqrt{4\pi}} \left(\frac{1}{r} - \kappa + \ldots\right).$$

By comparing this with the expansion (1), we have $\kappa = \alpha_0$. So for $\alpha_0 < 0$, there are no bound states in a zero-range potential.

In the case of $\alpha_0 > 0$ there is one, and only one, bound state with energy $E_0 = -\hbar^2\alpha_0^2/2m$. To normalize the wavefunction of this state, we should choose $A = \sqrt{2\alpha_0}$. Then the wavefunction in the momentum representation is

$$\phi_0(p) = \frac{\sqrt{\hbar\alpha_0}}{\pi} \frac{1}{p^2 + \hbar^2\alpha_0^2}.$$

It follows[38] that $\bar{T} = \overline{p^2}/2m = \infty$ and $\bar{U} = -\infty$ ($\bar{T} + \bar{U} = E_0$).

We should make several comments:

1) The zero-range potential has an attractive character independently of the sign of α_0. The case of $\alpha_0 < 0$ corresponds to a shallow well that does not bind the particle. However, two or more such wells situated close to each other may lead to the formation of a bound state. See Problem 11.28.
2) The parameter α_0 connected to the *scattering length*, a_0, and for the zero-range potential: $a_0 = 1/\alpha_0$. See Problem 13.20.

---

[37] Such a "potential", which acts only on a particle with the angular momentum $l = 0$, models a potential well $U(r)$ with a finite radius r_S, which has a shallow (possibly virtual) level with the energy $\varepsilon_0 \ll \hbar^2/mr_S^2$. In this case, the properties of these states with the angular momentum $l = 0$ and energy $E \ll \hbar^2/mr_S^2$ depend weakly on the explicit form of the potential $U(r)$. Applications of zero-range potentials in atomic and nuclear physics are considered in Chapters 11 and 13.

[38] The value $\bar{T} = \infty$ also follows from the condition $\Delta\psi_0(r) \sim \Delta r^{-1} = -4\pi\delta(\mathbf{r})$ as $r \to 0$.

3) The limiting case $\alpha_0 \to \pm\infty$ corresponds to "turning off" a zero-range potential.
4) Note the following property of (1) that arises for zero-range potentials: there is no dependence on the particle energy. Compare with Problem 9.14.

Problem 4.11

Determine the energy spectrum of a particle moving in a combined field of an infinitely deep spherical potential well with radius a and a zero-range potential (z.r.p.) at the point $r = 0$. Compare your results with the spectra for the well and with the z.r.p. separately. Note that the former spectrum may be modified under the influence of the z.r.p.

Solution

The energy spectrum of states with $l \neq 0$ is the same as in a single well. See Problem 4.9 b.

For $l = 0$, the solution of the Schrödinger equation satisfying the boundary condition $\psi(a) = 0$ is of the form $\psi(r) = \frac{1}{r} A \sin k(r-a)$. As $r \to 0$, its asymptote is $\psi(r) \approx -A \sin ka \; (r^{-1} - k \cot ka)$. By comparing this with the relation that defines a z.r.p., see Eq. (1) from Problem 1.40, we obtain the equation determining the s-level spectrum:

$$ka \cot ka = \alpha_0 a, \quad k = \sqrt{2mE/\hbar^2}. \tag{1}$$

When $\alpha_0 = \pm\infty$ we have $ka = (n_r + 1)\pi$ which gives the well spectrum (see Problem 4.1). When $a = \infty$, in the case $\alpha_0 > 0$ we have $k_0 = i\alpha_0$, i.e., $E_0 = -\hbar^2 \alpha_0^2/2m$, the level in an isolated z.r.p.

Some conclusions from the equation obtained are:

1) In the case where $|\alpha_0 a| \gg 1$ and $ka \ll |\alpha_0 a|$ (the levels are not highly excited), well levels have only a small shift due to the z.r.p. We can write $ka = (n_r + 1)\pi + \varepsilon$ where $|\varepsilon| \ll 1$, and from (1) we get[39] $E_{n_r,0} \approx E^{(0)}_{n_r,0}(1 + 2/\alpha_0 a)$. If $\alpha_0 > 0$ then the level E_0 that exists in the z.r.p. also "feels" a slight shift that is equal to

$$\Delta E_0 \approx -4e^{-2\alpha_0 a} E_0.$$

2) We have a totally different situation if $|\alpha_0 a| \leq 1$. The energy of a level (real or virtual) existing in z.r.p. is of the order of the lower-energy levels of the well, and as is seen from (1), the particle spectrum for the combined action of the z.r.p. and the well is drastically different from the spectra of either the isolated z.r.p. or the well: a reconstruction of the spectrum takes place. of the spectrum appears. In particular, for $\alpha_0 = 0$ (when there is a level with zero binding energy in z.r.p.)

---

[39] This result corresponds to perturbation theory with respect to the scattering length. Compare with Problem 4.29.

the spectrum has the form $E_{n_r,0} = \frac{\pi^2 \hbar^2 (n_r+1/2)^2}{2ma^2}$. This formula also describes the energy spectrum of the highly excited levels for arbitrary values of α_0.

Problem 4.12

Investigate the bound states of a particle in a *separable* potential which is defined as an integral operator (compare with Problem 2.19) with the kernel $U(r,r') = -\lambda f(r) f^*(r')$, where $f(r) \to 0$ at $r \to \infty$. Consider the case $U(r,r') = -\frac{\lambda}{rr'} e^{-\gamma(r+r')}$ (*Yamaguchi potential*).

Solution

1) It is convenient to solve the Schrödinger equation using the momentum representation. Compare with Problems 2.19 and 1.41.

$$\frac{p^2}{2m}\phi(\mathbf{p}) - \lambda g(p) \int g^*(p')\phi(\mathbf{p'})d^3p' = E\phi(\mathbf{p}), \tag{1}$$

$$g(p) = \frac{1}{(2\pi\hbar)^{3/2}} \int f(r) e^{-i\mathbf{p}\mathbf{r}/\hbar} dV.$$

From this, we see that the potential considered acts only on a particle with $l = 0$ (the wavefunction is spherically symmetric). The Problem then can be solved using the same methods as in Problem 2.19. For example, the energy spectrum of the bound states can be used here using the substitution, $|g(p)|^2 \to 4\pi p^2 |g(p)|^2 \eta(p)$ where $\eta(p)$ is the Heaviside step function. The equation for the energy spectrum takes the form

$$2m\lambda \int_0^\infty \frac{4\pi p^2 |g(p)|^2 dp}{p^2 - 2mE} = 1. \tag{2}$$

2) For the Yamaguchi potential, we have $f = e^{-\gamma r}/r$, $g(p) = \sqrt{2\hbar/\pi}(p^2 + \hbar^2\gamma^2)^{-1}$. We calculate the integral to obtain

$$(\hbar\gamma + \sqrt{2m(-E)})^2 = \frac{4\pi m\lambda}{\gamma}. \tag{3}$$

As is seen, a bound state exists only in the case $\lambda > \lambda_0 \equiv \hbar^2\gamma^3/4\pi m$, and its energy is given by

$$E_0 \equiv -\frac{\hbar^2 \kappa_0^2}{2m} = -\frac{\hbar^2 \gamma^2}{2m}\left[\sqrt{\frac{\lambda}{\lambda_0}} - 1\right]^2. \tag{4}$$

The normalized wavefunctions in the momentum and coordinate representations are

$$\phi_0(p) = \frac{\sqrt{\hbar^2 \kappa_0 \gamma (\gamma + \kappa_0)^3}}{\pi(p^2 + \hbar^2 \kappa_0^2)(p^2 + \hbar^2 \gamma^2)},$$

$$\psi_0(r) = \sqrt{\frac{\kappa_0 \gamma (\gamma + \kappa_0)}{2\pi (\gamma - \kappa_0)^2}} \frac{e^{-\kappa_0 r} - e^{-\gamma r}}{r}.$$

Problem 4.13

Consider the bound s-states of a particle in a δ-potential of the form $U(r) = -\alpha \delta(r - a)$ by solving the Schrödinger equation in the momentum representation.

Solution

The Schrödinger equation in the momentum representation is

$$\frac{\mathbf{p}^2}{2m}\phi(\mathbf{p}) + \int \tilde{U}(\mathbf{p} - \mathbf{p}')\phi(\mathbf{p}')d^3p' = E\phi(\mathbf{p}), \tag{1}$$

$$\tilde{U}(q) = \frac{1}{(2\pi\hbar)^3} \int U(r) e^{-i\mathbf{q}\mathbf{r}/\hbar} dV.$$

Compare with Problem 2.17. For a δ-potential we have

$$\tilde{U}(q) = -\frac{\alpha a}{2\pi^2 \hbar^2 q} \sin\left(\frac{aq}{\hbar}\right).$$

Since for $l = 0$ the wavefunction is angle-independent, (1) becomes

$$\left(\frac{p^2}{2m} - E\right)\phi(p) = \frac{\alpha a}{\pi \hbar^2} \int\int \frac{\sin(a\sqrt{p^2 + \tilde{p}^2 - 2p\tilde{p}\cos\vartheta}/\hbar)}{\sqrt{p^2 + \tilde{p}^2 - 2p\tilde{p}\cos\vartheta}} \phi(\tilde{p})\tilde{p}^2 \sin\vartheta d\vartheta d\tilde{p}. \tag{2}$$

After the integration with respect to ϑ, we obtain

$$\left(\frac{p^2}{2m} - E\right)\phi(p) = \frac{2\alpha}{\pi \hbar p} \sin\left(\frac{pa}{\hbar}\right) \int_0^\infty \sin\left(\frac{\tilde{p}a}{\hbar}\right) \phi(\tilde{p})\tilde{p}d\tilde{p}. \tag{3}$$

It follows that

$$\phi(p) = \frac{4m\alpha C \sin(pa/\hbar)}{\pi \hbar p (p^2 - 2mE)}, \quad C = \int_0^\infty \sin\left(\frac{pa}{\hbar}\right)\phi(p)pdp. \tag{4}$$

The condition for the compatibility of these expressions leads to an equation for the s-level energy spectrum ($E = -\hbar^2 \kappa^2/2m$):

$$\frac{4m\alpha}{\pi\hbar} \int_0^\infty \frac{\sin^2(pa/\hbar)dp}{p^2 - 2mE} = 1, \text{ or } \frac{m\alpha a}{\hbar^2}(1 - e^{-2a\kappa}) = a\kappa. \tag{5}$$

The approximate solution of (5) is given in Problem 4.8 a. See also Problem 2.18. Notice that the bound state only exists when $m\alpha a/\hbar^2 > 1/2$.

Problem 4.14

Find the solution of the Schrödinger equation from the previous problem with the boundary condition $\phi(p) = 0$ for $p \leq p_0$, $p_0 > 0$.

Prove that in this case and in a well of an arbitrary depth, there exists a bound state, so that the particle is localized in a bounded region.[40] Find the bound state energy for a shallow well.

Solution

The solution can be obtained along the same lines as in the previous problem. But we should take into account the following two facts: 1) Due to the condition $\phi(p) = 0$ for $p \leq p_0$ in Eqs. (2)–(5) from the previous problem, the lower limit of the integration with respect to p must be equal to p_0. 2) The bound state of the particle now has the energy $E < E_0 = p_0^2/2m$ instead of $E < 0$, and satisfies the equation

$$\frac{4m\alpha}{\pi\hbar} \int_{p_0}^\infty \frac{\sin^2(pa/\hbar)dp}{p^2 - 2mE} = 1. \tag{1}$$

We assume that $p_0 \neq n\pi\hbar/a$ and $\alpha > 0$. Since the left-hand side of (1) increases monotonically from zero as $E \to -\infty$, to $+\infty$ as $E \to E_0$, then for any well parameters there is only one bound s-state with energy $E < E_0$.

Let us consider two limiting cases. 1) In the case where $m\alpha/\hbar \gg p_0, \hbar/a$ (deep well), from (1) it follows[41] that $E_0 \approx -m\alpha^2/2\hbar^2$ as in the case of a one-dimensional δ-well. See Problems 2.7 and 4.8. 2) In the opposite limiting case, $m\alpha a/\hbar^2 \ll 1$ (shallow well), the level $E \to E_0$; the value of the integral is given by

---

[40] The formation of a bound state in this problem with an arbitrary small attraction is the essence of the Cooper pairing, which is the basis of the microscopic mechanism for superconductivity formation.

[41] In this case, the dominant contribution to the integral in (1) comes from the region $p \sim \sqrt{2mE} \sim m\alpha/\hbar$. For the approximate calculation of the integral, we replace the rapidly oscillating sine square by its mean value equal to $1/2$ and put the lower limit of the integration, p_0, equal to zero; the integral is equal to $\pi/4\sqrt{2mE}$.

$$\int_{p_0}^{\infty} \frac{\sin^2(pa/\hbar)dp}{p^2 - p_0^2 + 2m\varepsilon} \approx \sin^2\left(\frac{p_0 a}{\hbar}\right) \int_{p_0}^{\infty} \frac{dp}{p^2 - p_0^2 + 2m\varepsilon} \approx \frac{\sin^2(p_0 a/\hbar)}{2p_0} \ln\frac{4E_0}{\varepsilon}. \quad (2)$$

Here $\varepsilon = E_0 - E > 0$ is the particle binding energy. From (1) and (2) with $\xi \ll 1$, we have

$$\varepsilon \sim E_0 \exp\left\{-\frac{\pi p_0 a}{2\xi \hbar} \sin^2(p_0 a/\hbar)\right\}. \quad (3)$$

The exact determination of the factor E_0 in (3) requires a more involved calculation of the integral. As $\xi \to 0$, the binding energy goes to zero exponentially $\propto e^{-c/\xi}$.

Problem 4.15

Determine the energy levels and the normalized wavefunctions of the discrete spectrum in a one-dimensional potential of the form: $U(x) = -\alpha/x$ for $x > 0$ and $U(x) = \infty$ for $x < 0$, by solving the Schrödinger equation in the momentum representation.

Using the results obtained, find the normalized wavefunctions of s-states in the momentum representation for the Coulomb potential $U(r) = -\alpha/r$.

Solution

1) We first write the Schrödinger equation for $U(x) = -\alpha/x$ for the semi-axis $x \geq 0$, with the boundary condition $\psi(0) = 0$, in the form of an equation for the entire axis, which for $x \geq 0$ is equivalent to the initial equation and for $x < 0$ automatically enforces the condition $\psi(x) \equiv 0$:

$$\left(\frac{\hat{p}^2}{2m} - \frac{\alpha}{x} - E\right)\psi(x) = -\frac{\hbar^2}{2m}\psi'(0+)\delta(x). \quad (1)$$

This works because $\psi(0-) = \psi(0+) = 0$ and $\psi'(0-) = 0$ (compare with Problem 2.6) and we have $\psi(x) \equiv 0$ for $x \leq 0$. Using the result of problem 1.40, (1) can be written in the momentum representation as

$$\frac{p^2}{2m}\phi(p) - \frac{i\alpha}{\hbar}\int_p^{\infty} \phi(p')dp' - E\phi(p) = -\frac{\hbar^2}{2\sqrt{2\pi\hbar}\,m}\psi'(+0). \quad (2)$$

We obtain an equation with separable variables by differentiating (2) with respect to p. Its solution has the form ($E < 0$):

$$\phi(p) = \frac{C}{p^2 + 2m|E|}\exp\left\{\frac{-2im\alpha}{\sqrt{2m|E|}\,\hbar}\arctan\frac{p}{\sqrt{2m|E|}}\right\}. \quad (3)$$

The condition $\psi(0) = (2\pi\hbar)^{-1/2} \int_{-\infty}^{\infty} \phi(p)dp = 0$ determines the energy spectrum:

$$\sin \frac{\pi m\alpha}{\sqrt{2m|E|}\,\hbar} = 0 \text{ and } E_n = -\frac{m\alpha^2}{2\hbar^2(n+1)^2},\ n = 0, 1, \ldots \quad (4)$$

For the normalization of wavefunction (3) we should choose

$$C = \sqrt{\frac{2}{\pi}} \left(\frac{m\alpha}{\hbar}(n+1)\right)^{3/2}.$$

2) In order to use the one-dimensional results above for the case of the s-states in the Coulomb potential, we use the wavefunctions in coordinate representation:

$$\psi_{n_r 00}(r) = \frac{1}{\sqrt{4\pi}} \frac{\psi_{n_r}(r)}{r}.$$

See Problem 4.1. Taking this to the momentum representation, we obtain

$$\phi_{n_r 00}(p) = \frac{1}{(2\pi\hbar)^{3/2}} \int \psi_{n00}(r) e^{-i\mathbf{p}\cdot\mathbf{r}/\hbar} dV =$$

$$\frac{i}{\sqrt{8\pi^2\hbar}\,p} \int_0^\infty \psi_{n_r}(r) \left(e^{-ipr/\hbar} - e^{ipr/\hbar}\right) dr = i\frac{\phi_{n_r}(p) - \phi_{n_r}(-p)}{\sqrt{4\pi}\,p}. \quad (5)$$

The $\phi_{n_r}(p)$ are given by (3) and (4) with $n \equiv n_r$. Using the equation

$$\exp(i\arctan\varphi) = \frac{1}{\sqrt{1+\varphi^2}} + i\frac{\varphi}{\sqrt{1+\varphi^2}},$$

we rewrite (5) to be (omitting the phase multiplier (-1)):

$$\phi_{n_r 00}(p) = \frac{\sqrt{2}}{\pi p} \frac{p_{n_r}^{3/2}}{p^2 + p_{n_r}^2} \sin\left[2(n_r+1)\arctan\left(\frac{p}{p_{n_r}}\right)\right],\ p_{n_r} = \frac{m\alpha}{\hbar(n_r+1)}. \quad (6)$$

For $n_r = 0$ and $\alpha = e^2$, this gives the result of Problem 4.6 c.

Problem 4.16

Find the behavior of the discrete state wavefunction, $\phi_{n_r lm}$, in the momentum representation as $p \to 0$.

Solution

We have the relation

$$\int e^{-ikr\mathbf{n}\cdot\mathbf{n}'} Y_{lm}(\mathbf{n}')d\Omega_{\mathbf{n}'} = (-i)^l 2\pi \sqrt{\frac{2\pi}{kr}} J_{l+1/2}(kr) Y_{lm}(\mathbf{n}), \tag{1}$$

following from the known expansion of the plane wave as a series of the Legendre polynomials. We use Eq. (III.6) and the asymptotic form of the Bessel function $J_\nu(z)$ for $z \to 0$, to obtain

$$\phi_{n,lm}(\mathbf{p}) = \frac{1}{(2\pi\hbar)^{3/2}} \int e^{-i\mathbf{p}\cdot\mathbf{r}/\hbar} \psi_{n,lm}(\mathbf{r}) dV \underset{p\to 0}{\approx} C_l p^l Y_{lm}(\mathbf{p}/p), \tag{2}$$

$$C_l = \frac{(-i)^l}{2^{l+1/2}\Gamma(l+3/2)\hbar^{l+3/2}} \int_0^\infty r^{l+2} R_{n_r l}(r) dr.$$

Compare the result obtained, $\phi_l \propto p^l$ as $p \to 0$, with the known relation $\psi_l \propto r^l$ as $r \to 0$ in the coordinate representation.

Problem 4.17

Prove that the asymptotic form of the s-state wavefunction in the momentum representation and for $p \to \infty$ has the form

$$\phi_{n_r 00}(p) \approx -2(2\pi\hbar)^{3/2} m \psi_{n_r 00}(0) \frac{\tilde{U}(p)}{p^2}, \tag{1}$$

where ψ is the state's wavefunction in the coordinate representation and

$$\tilde{U}(p) = \frac{1}{(2\pi\hbar)^3} \int e^{-i\mathbf{p}\cdot\mathbf{r}/\hbar} U(r) dV$$

is the Fourier component of the potential. Assume that $\tilde{U}(p)$ decreases at $p \to \infty$ as a power law, $\tilde{U}(p) \propto p^{-n}$ with $n > 1$.

Solution

An analysis of the Schrödinger equation in the momentum representation,

$$\frac{\mathbf{p}^2}{2m}\phi(\mathbf{p}) + \int \tilde{U}(|\mathbf{p}-\mathbf{p}'|)\phi(\mathbf{p}')d^3 p' = E\phi(\mathbf{p}), \tag{2}$$

shows that for $\tilde{U}(p) \propto p^{-n}$ as $p \to \infty$ with $n > 1$, the wavefunction, $\phi(p)$, decreases more rapidly than $\tilde{U}(p)$. Therefore, the main contribution to the integral in (2) is given by the integration region $|\mathbf{p}'| \leq \hbar/a$, where a is the radius of the potential.

Therefore, by factoring $\tilde{U}(\mathbf{p} - \mathbf{p}')$ outside the integral sign with[42] $p' \approx 0$ and using the relation between wavefunctions $\psi(\mathbf{r})$ and $\phi(\mathbf{p})$, we obtain asymptote (1) for the s-states.

We illustrate this result for the Coulomb potenial, $U = -\alpha/r$. According to Eq. (6) from Problem 4.15, we have (in the Coulomb units $m = \hbar = \alpha = 1$)

$$\phi_{n_r 0}(p) \approx (-1)^{n_r} 2^{3/2} \left(\pi(n_r + 1)^{3/2} p^4 \right)^{-1}$$

while $\psi_{n_r s}(0) = (\pi(n_r + 1)^{3/2})^{-1/2}$ and $\tilde{U}(p) = (2\pi^2 p^2)^{-1}$. The relation (1) is fulfilled up to a phase.

Problem 4.18

Show that the previous problem may be extended to describe a state with an arbitrary angular momentum l:

$$\phi_{n_r l m}(\mathbf{p}) \approx -2(2\pi\hbar)^{3/2}(2i\hbar)^l m \tilde{R}_{n_r l}(0) Y_{lm}\left(\frac{\mathbf{p}}{p}\right) p^{l-2} \frac{\partial^l}{\partial (p^2)^l} \tilde{U}(p), \tag{1}$$

where $\tilde{R}_{n_r l}(r)$ is related to the wavefunction in the coordinate representation by $\psi_{n_r l m}(\mathbf{r}) = r^l \tilde{R}_{n_r l}(r) Y_{lm}\left(\frac{\mathbf{r}}{r}\right)$.

Solution

1) We first transform $\tilde{U}(p)$ to the form ($\hbar = 1$)

$$\tilde{U}(p) = \frac{1}{4\pi^2 ip} \int_{-\infty}^{\infty} r e^{ipr} U(|r|) dr. \tag{2}$$

Note that the function $U(|r|)$, which is an even extension of $U(r)$ in the region $r < 0$, is an analytical function of r and has a *singular point* at $r = 0$. The *singularity* of $U(r)$ is subject to the condition $U(r) r^{2-\varepsilon} \to 0$ as $r \to 0$ where $\varepsilon > 0$. For example, for $U = \alpha/r^2$, we have $\tilde{U} = \alpha/4\pi p$. If we write a bound state wavefunction for such a potential in the form

$$\psi_{n_r l m}(\mathbf{r}) = r^l Y_{lm}(\mathbf{r}/r) \tilde{R}_{n_r l}(r),$$

then $\tilde{R}_{n_r l}(0) = \text{const} \neq 0$ and $\tilde{R}_{n_r l}(0) < \infty$.

2) The singularity at the point $r = 0$ manifests itself in the radial wavefunction $\tilde{R}_{n_r l}(r)$. It is important that the singularity in $\tilde{R}_{n_r l}(r)$ is weaker than that in the potential. This comes from the Schrödinger equation. Consider for example the singular part of $U(r)$ in the form $U^{(s)}(r) \approx \alpha r^\nu$, with $\nu > -2$ and is not an even

---

[42] It is important that the asymptote of $\tilde{U}(p)$ does not contain a rapidly oscillating factor of the form $\sin(\alpha p^k)$ with $k \geq 1$. See the following problem.

integer. Then, the singular part of the radial function (see Problem 4.19) has the form

$$\tilde{R}_{n_r l}^{(s)}(r) \approx \frac{2m\alpha}{(\nu+2)(\nu+2l+3)} \tilde{R}_{n_r l}(0) r^{\nu+2}$$

and goes to zero when $r \to 0$ unlike $U(0)$.

3) For further transformations it is convenient to write the spherical function in the form (see Problem 3.41)

$$r^l Y_{lm}(\mathbf{n}) = \varepsilon_{i\ldots n}(m) x_i \ldots x_n,$$

where $\varepsilon_{i\ldots n}(m)$ is a completely symmetric tensor of the lth rank with a trace equal to zero, $\varepsilon_{ii\ldots n}(m) = 0$. To obtain the desired asymptote, we should multiply the both sides of the Schrödinger equation

$$\left(-\frac{\Delta}{2m} - E_{n_r l}\right) \psi_{n_r l m}(\mathbf{r}) = -U(r) \psi_{n_r l m}(\mathbf{r})$$

by $(2\pi)^{-3/2} \exp(-i\mathbf{p} \cdot \mathbf{r})$, and integrate with respect to coordinates. We obtain

$$\left(\frac{p^2}{2m} - E_{n_r l}\right) \phi_{n_r l m}(\mathbf{p}) = -\frac{\varepsilon_{ik\ldots n}(m)}{(2\pi)^{3/2}} \int e^{-i\mathbf{p}\cdot\mathbf{r}} x_i x_k \ldots x_n U(r) \tilde{R}_{n_r l}(r) dV =$$

$$-\frac{i^l}{(2\pi)^{3/2}} \varepsilon_{ik\ldots n}(m) \frac{\partial}{\partial p_i} \ldots \frac{\partial}{\partial p_n} \int e^{-i\mathbf{p}\cdot\mathbf{r}} U(r) \tilde{R}_{n_r l}(r) dV. \quad (3)$$

As $p \to \infty$, the integral in (3) is determined by the singularity in the function $U(r)\tilde{R}_{n_r l}(r)$ at the point $r=0$ (to be more precise, in its even extension, as in (2)). The most singular part of this function is contained in $U(r)$. So we factor $\tilde{R}_{n_r l}$ outside the integral sign in (3) at the point $r=0$, and note that

$$\frac{\partial \tilde{U}(p)}{\partial p_i} = 2p_i \frac{\partial \tilde{U}(p)}{\partial p^2}.$$

This relation, along with the fact that the trace of the tensor $\varepsilon_{1\ldots n}$ is equal to zero, give the wavefunction asymptote as in Eq. (1).

4) Putting everything together, we see that this result could be easily generalized to the case where the singular points of the odd potential extension are $r = \pm a$ on the real axis. This would correspond to different potentials with distinct boundaries or kinks. In this case, we should make a change as follows: $\tilde{R}_{n_r l}(0) \to \tilde{R}_{n_r l}(a)$.

However, despite the similarities of the asymptotes in these cases, there are significant differences. For any critical point of the form $r = \pm a \neq 0$, the Fourier component $\tilde{U}(p)$ contains a rapidly oscillating factor of the form $\sin(\alpha p)$. Its existence leads to fact that all derivatives of $\tilde{U}(p)$ decrease as $\tilde{U}(p)$ does. And so the wavefunctions corresponding to different values of l for $p \to \infty$ all decrease in the same fashion. In contrast with the critical point at $r = 0$, the wave functions for states with larger angular momentum, l, decrease faster.

Problem 4.19

A particle moves in a potential which for $r \to 0$ has the form $U(r) \approx \alpha/r^s$ with $s < 2$. In this case, the radial wavefunction of the state with a given value l has the form $R_{n_r l}(r) \approx C_{n_r l} r^l$. Find a correction to this expression in the next order in $1/r$.

Solution

Omitting the terms with U and E in Eq. (IV.2) we easily obtain the asymptote:

$$R_{n_r l}(r) \approx R^{(0)}_{n_r l}(r) = C_{n_r l} r^l \quad \text{as } r \to 0.$$

To find the correction, $R^{(1)}_{n_r l}(r)$, we use the equation

$$R^{(1)''}_{n_r l} + \frac{2}{r} R^{(1)'}_{n_r l} - \frac{l(l+1)}{r^2} R^{(1)}_{n_r l} - \frac{2m\alpha}{\hbar^2} C_{n_r l} r^{l-s} = 0.$$

Therefore,

$$R^{(1)}_{n_r l} = \frac{2m\alpha}{(2-s)(2l+3-s)\hbar^2} C_{n_r l} r^{l+2-s}. \tag{1}$$

Problem 4.20

Find the Green function, $G_E(\mathbf{r}, \mathbf{r}')$, of a free particle with energy $E < 0$ which vanishes as $r \to \infty$. Using this Green function, write the Schrödinger equation for the discrete spectrum states in a potential $U(r)$ that vanishes as $r \to \infty$ in the form of an integral equation.

Solution

The Green function obeys the equation

$$\hat{H} G_E \equiv \frac{\hbar^2}{2m}(-\Delta + \kappa^2) G_E(\mathbf{r}, \mathbf{r}') = \delta(\mathbf{r} - \mathbf{r}'). \tag{1}$$

We use $\kappa = \sqrt{-2mE/\hbar^2} > 0$. From the considerations of symmetry, it must be a function of the form $G_E = f(|\mathbf{r} - \mathbf{r}'|)$. We see that for $\mathbf{r} \neq \mathbf{r}'$, equation (1) and its solution become

$$\left(\frac{d^2}{dr^2} - \kappa^2\right) r f(r) = 0, \quad f(r) = C \frac{e^{-\kappa r}}{r}. \tag{2}$$

An exponentially increasing term in the expression for $f(r)$ is omitted. The relation $\Delta(1/r) = -4\pi\delta(\mathbf{r})$ allows us to obtain the value of C in (2) and the final expression

138 Exploring Quantum Mechanics

for G_E:

$$G_E(\mathbf{r} - \mathbf{r}') = \frac{m}{2\pi\hbar^2} \frac{e^{-\kappa|\mathbf{r}-\mathbf{r}'|}}{|\mathbf{r} - \mathbf{r}'|}. \qquad (3)$$

We can use this Green function to write the Schrödinger equation for the discrete spectrum states in the form of the integral equation. Compare with Problem 2.20.

$$\psi_E(\mathbf{r}) = -\int G_E(\mathbf{r},\ \mathbf{r})U(r')\psi_E(\mathbf{r}')dV' = -\frac{m}{2\pi\hbar^2}\int \frac{e^{-\kappa|\mathbf{r}-\mathbf{r}'|}}{|\mathbf{r}-\mathbf{r}'|}U(r')\psi_E(\mathbf{r}')dV'. \qquad (4)$$

Problem 4.21

For the three-dimensional case of a particle in the attractive potential $U(r) \leq 0$ ($U(r) \to 0$ as $r \to \infty$), bound states do not always exist. Prove that the inequality

$$\int_0^\infty r|U(r)|dr \geqq \frac{\hbar^2}{2m} \qquad (1)$$

is a necessary condition for their existence. Compare this condition with the exact condition for existence of a discrete spectrum in a rectangular potential well (see Problem 4.1), a δ-potential, and an exponential well (see Problem 4.8). See also Problem 4.32.

Solution

We apply Eqn. (4) from the previous problem to the ground state with $E_0 < 0$ (we assume that a bound state does exist). The corresponding wavefunction, $\psi_0(r)$, is spherically symmetric ($l = 0$) and, since it has no zeros, we may consider $\psi_0(r) \geqq 0$. So in the equation

$$\psi_0(r) = \frac{m}{2\pi\hbar^2}\int \frac{e^{-\kappa_0|\mathbf{r}-\mathbf{r}'|}}{|\mathbf{r}-\mathbf{r}'|}[-U(r')]\psi_0(r')dV', \qquad (2)$$

the integrand is also non-negative.

We choose $|\mathbf{r}| = r_0$, where the function $\psi_0(r)$ takes its maximum value, and make the substitution $\psi_0(r') \to \psi_0(r)$. If we also omit the exponent (this omission cannot decrease the value of the integral), then we obtain

$$\frac{1}{4\pi}\int\frac{|U(r')|}{|\mathbf{r}-\mathbf{r}'|}(r')^2 d\Omega' dr' \geq \frac{\hbar^2}{2m}. \tag{3}$$

the integration over the angles (we choose the polar axis along the vector **r**) gives[43]

$$\int\frac{d\Omega'}{|\mathbf{r}-\mathbf{r}'|} = \min\left\{\frac{4\pi}{r},\frac{4\pi}{r'}\right\} \leq \frac{4\pi}{r'}$$

and we obtain the problem statement. We can see from Problem 4.1 that the result is analogous to the result of Problem 2.25 for a one-dimensional motion.

For the rectangular well, this necessary condition for a bound state existence takes the form $\xi \equiv ma^2 U_0/\hbar^2 \geq 1$ while the exact condition is $\xi \geq \pi^2/8 \approx 1.24$. For the δ-well, this necessary condition coincides with the exact result. For the exponential potential well the necessary condition is $\xi \equiv ma^2 U_0/\hbar^2 \geq 1/2$, while the exact result is $\xi \geq 0.72$.

Problem 4.22

Show that the fulfilment of the following inequality

$$\frac{m}{2\hbar^2}\left\{\int_0^\infty U(r)\left[1-\exp\left(-\frac{2}{\hbar}\sqrt{2m\varepsilon_0}r\right)\right]dr\right\}^2 \geq \varepsilon_0 \tag{1}$$

is a necessary condition for the existence of a particle bound state with the binding energy ε_0 in an attractive central potential $U(r) \leq 0$, where $U(r) \to 0$ as $r \to \infty$. As $\varepsilon_0 \to 0$, this condition corresponds to the result of the previous problem.

Solution

First we perform the integration over the angles of the vector $\mathbf{r}'$ for Eqn. (2) from the previous problem. We choose the direction of $\mathbf{r}$ as the polar axis. We obtain

$$\int\frac{\exp(-\kappa_0\sqrt{r^2+r'^2-2rr'\cos\theta'})}{\sqrt{r^2+r'^2-2rr'\cos\theta'}}d\Omega' = \frac{2\pi}{\kappa_0 rr'}\left[e^{-\kappa_0|r-r'|}-e^{-\kappa_0|r+r'|}\right] =$$

$$\frac{2\pi}{\kappa_0 rr'}e^{-\kappa_0|r-r'|}\left(1-e^{-\kappa_0(|r+r'|-|r-r'|)}\right) \leq \frac{2\pi}{\kappa_0 rr'}(1-e^{-2\kappa_0 r'}).$$

Similarly to the previous problem (for the function $r\psi_0(r)$), we obtain the inequality

$$\int_0^\infty |U(r)|(1-e^{-2\kappa_0 r})dr \geq \frac{\hbar^2 \kappa_0}{m}. \tag{2}$$

---

[43] Since the integral describes an electrostatic potential created by a sphere of radius r' which is charged with the constant surface charge density $\sigma_0 = 1$, we can obtain the value of the integral without calculations.

This is equivalent to Eq. (1) (with $\varepsilon_0 = \hbar^2\kappa_0^2/2m$).

Problem 4.23

Find the Green function, $G_{l,0}(r,r')$, of the radial Schrödinger Eq. (IV.5) for a free particle with $E = 0$ on the interval $[a,b]$. In this case, $0 \leq a < b \leq \infty$.

It corresponds to the equation

$$\hat{H}_l G_{l,E=0} \equiv -\frac{\hbar^2}{2m}\left[\frac{\partial^2}{\partial r^2} - \frac{l(l+1)}{r^2}\right] G_{l,0}(r,r') = \delta(r-r') \tag{1}$$

and boundary conditions $G_{l,0}(a,r') = G_{l,0}(b,r') = 0$.

Solution

The solution of Eq. (1) that satisfies the boundary conditions and is continuous at the point $r = r'$ has the form

$$G_{l,E=0}(r,r') = C \cdot (abrr')^{-l} \cdot \begin{cases} \left(b^{2l+1} - r'^{2l+1}\right)\left(r^{2l+1} - a^{2l+1}\right), & r < r' \\ \left(a^{2l+1} - r'^{2l+1}\right)\left(r^{2l+1} - b^{2l+1}\right), & r > r' \end{cases} \tag{2}$$

The value

$$C = \frac{2m}{(2l+1)\hbar^2} \frac{a^l b^l}{b^{2l+1} - a^{2l+1}} \tag{3}$$

follows from the condition of the jump of the derivative $\partial G_{l,0}/\partial r$ at the point $r = r'$: $\delta G'_{l,0} = -2m/\hbar^2$. Compare with Problem 2.6.

Relations (2) and (3) determine the form of the Green function. For the case $a = 0$ or $b = \infty$, these relations are simplified.

Problem 4.24

Prove that the fulfilment of the inequality

$$\int_0^\infty r|U(r)|dr \geq (2l+1)n_l \frac{\hbar^2}{2m} \tag{1}$$

is a necessary condition for there to exist n_l levels of a particle with angular momentum l in a short-range attractive potential $U(r) \leq 0$, $U(r) \to 0$ as $r \to \infty$.

Solution

Let us consider the emergence of the n_lth bound state with the angular momentum l and denote the state by $\psi^{(0)}_{n_r,lm} = \frac{1}{r}\chi^{(0)}_{n_r,l}Y_{lm}$, with energy $E_{n_r,l} = 0$. Here $n_r = n_l - 1$. The radial function, $\chi^{(0)}_{n_r,l}(r)$, has $(n_l + 1)$ zeroes, including zeroes at $r = 0$ and $r = \infty$, and satisfies Eq. (IV.5). Let a and b be neighboring zeroes of $\chi^{(0)}_{n_r,l}(r)$. Recalling the result of the previous problem, we see that on the interval $[a, b]$, $\chi_{n_r l}^{(0)}$ satisfies the integral equation

$$\chi_{n_r l}^{(0)}(r) = \int_a^b G_{l,0}(r, r') \left[-U(r') \chi_{n_r l}^{(0)}(r') \right] dr'. \tag{2}$$

On this interval, the function $\chi_{n_r l}^{(0)}$ does not change its sign, and we will consider $\chi_{n_r l}^{(0)} \geq 0$. We should note that the Green function from Problem 4.23 is positive and takes its maximum value at $r = r'$. So the function under the integral sign in (2) is non-negative. We choose in (2) the value $r = r_0$ corresponding to the maximum of the function $\chi_{n_r l}^{(0)}$ on the interval $[a, b]$ and replace $\chi_{n_r l}^{(0)}(r')$ in the integrand by $\chi_{n_r l}^{(0)}(r_0)$ (this can only increase the value of integral). We obtain

$$\int_a^b G_{l,0}(r_0, r') |U(r')| dr' \geq 1. \tag{3}$$

Now we replace $G_{l,0}$ by its maximum value, so that $r_0 \to r'$ in this function. Taking into account (2) and (3) from the previous problem, we obtain

$$G_{l,0}(r', r') \geq \frac{2mr'}{(2l+1)\hbar^2}. \tag{4}$$

Putting the result from (4) into (3), we have

$$\int_a^b r(-U(r)) dr \geq \frac{(2l+1)\hbar^2}{2m}, \tag{5}$$

and since there are n_l intervals on the semi-axis $(0, \infty)$, on which $\chi_{n_r l}^{(0)}$ does not change its sign, and on each of them the inequality analogous to (5) is valid. Then by taking the sum over all such integrals we can obtain the statement of the problem.

4.2 Low-energy states

Problem 4.25

Extend the result of Problem 2.13 to the case of particle s-states in a spherically-symmetric field. Find the conditions of the existence and emergence of new discrete s-levels in the following potentials:

a) $U = -\dfrac{\alpha}{r^4}$ for $r > a$ and $U = \infty$ for $r < a$ (see Fig. 4.1);

b) $U = -\dfrac{\alpha}{(r+a)^4}$, $a > 0$;

c) $U = -\dfrac{U_0 a^4}{(r^2+a^2)^2}$;

d) $U = -\dfrac{\alpha}{r^s}$ for $r > a$ and $U = \infty$ for $r < a$, $s > 2$ (see Fig. 4.1);

e) $U = -\dfrac{\alpha}{r^s}$ for $r < a$ and $U = 0$ for $r > a$, $0 < s < 2$ (see Fig. 4.2).

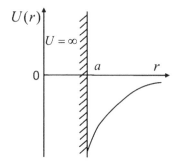

Fig. 4.1

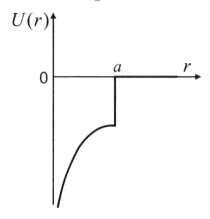

Fig. 4.2

Solution

The condition for the emergence of a new Nth bound state with $l = 0$ with respect to the depth of the potential well corresponds to the existence of a zero-energy solution to the Schrödinger equation. This solution is bounded at finite values of r and has the asymptote $\psi \approx C/r$ as $r \to \infty$ (more generally for a completely arbitrary potential, $\psi \approx A + C/r$ as $r \to \infty$). We should look for solutions that satisfy these conditions.

a) Eq. (IV.2) for $U = -\alpha/r^4$ in the case $l = 0$ and $E = 0$ can be simplified by making the substitution $x = 1/r$:

$$\dfrac{d^2 R}{dx^2} + \tilde{\alpha} R = 0, \quad \tilde{\alpha} = \dfrac{2m\alpha}{\hbar^2}. \tag{1}$$

Its solution, due to the condition $r \to \infty$, should be chosen in the form $R = B \sin\left(\sqrt{\tilde{\alpha}}/r\right)$. The condition $R(a) = 0$ determines the desired values of the potential parameters by $\sqrt{\tilde{\alpha}}/a = \pi N$, or

$$\frac{m\alpha}{\hbar^2 a^2} = \frac{\pi^2}{2} N^2. \tag{2}$$

b) Taking into account Eq. (IV.5) and its boundary conditions, we see that for the s-level, spectra in the potentials $U_1(r) = f(r+a)$ and $U_2(r) = f(r)$ for $r > a > 0$ and $U_2(r) = \infty$ for $r < a$ coincide (note that this is not so for $l \neq 0$). Therefore, the parameters of the potential are given by (2).

c) Equation (IV.5) for this potential in the case $l = 0$ and $E = 0$ can be transformed using the substitutions $w = \chi/\sqrt{r^2 + a^2}$ and $x = \arctan(r/a)$. It takes the form

$$\frac{d^2 w}{dx^2} + \xi^2 w = 0, \quad \xi = \sqrt{1 + 2mU_0 a^2/\hbar^2}.$$

Due to the condition $\chi(r = 0) = w(x = 0) = 0$, the solution must be of the form $w = C \sin \xi x$, or

$$\psi = C\left(1 + \frac{a^2}{r^2}\right)^{1/2} \sin\left(\xi \arctan \frac{r}{a}\right). \tag{3}$$

As $r \to \infty$ we have $\arctan(r/a) \approx \pi/2 - a/r$, so

$$\psi \approx C\left[\sin \frac{\pi \xi}{2} - \frac{\xi a}{r} \cos \frac{\pi \xi}{2}\right], \quad r \to \infty,$$

so the asymptotic condition ($\psi \propto 1/r$) demands $\sin(\pi \xi/2) = 0$ and leads to the relation $\xi/2 = N$. The wavefunction here, (3), has $(N-1)$ zeros for finite r.

For the potentials d and e, the solution of (IV.5) with $l = 0$ and $E = 0$, according to A.II.11, is described in terms of cylindrical functions. Below, we present only the final results. Here $x_{\nu,N}$ is the Nth zero of the Bessel function $J_\nu(x)$, disregarding a zero at $x = 0$.

d)

$$\frac{2}{s-2}\left(\frac{2m\alpha}{\hbar^2 a^{s-2}}\right)^{1/2} = x_{\nu N}, \quad \nu = \frac{1}{s-2} \tag{4}$$

The wavefunction at the threshold of emergence of a new level is $\psi = cJ_\nu(\beta r^{-1/2\nu})/\sqrt{r}$ for $r > a$, with $\beta = \nu\sqrt{8m\alpha/\hbar^2}$.

When $s = 4$ we have $\nu = 1/2$, and (4) coincides with (2).

e)

$$2(\nu + 1)\left(\frac{2m\alpha a^{2-s}}{\hbar^2}\right)^{1/2} = x_{\nu N}, \quad \nu = \frac{1}{2-s} \tag{5}$$

The wavefunction at the threshold of emergence of a new level emergence is $\psi(r) = cJ_{\nu+1}(\beta r^{1/2(\nu+1)})\sqrt{a/r}$ for $r < a$, and $\psi(r) = cJ_{\nu+1}(\beta a^{1/2(\nu+1)})a/r$ for $r < a$.

In the case $s = 1$ (the "truncated" Coulomb potential), from (5) we have $m\alpha a/\hbar^2 = x_{0N}^2/8$. Since the zero of the relevant Bessel function is $x_{01} \approx 2.40$, we find the condition for bound-state existence in such a potential: $m\alpha a/\hbar^2 \geq 0.72$.

Problem 4.26

Discuss the conditions for the existence and emergence of new bound states with non-zero values of angular momentum by increasing the depth of the potential well. Use the Schrödinger equation for $E = 0$. What is the difference between the wavefunctions with $l = 0$ and with $l \neq 0$ at the threshold of emergence? Consider the potentials: a) $U = -\alpha \delta(r - a)$; b) $U = -\frac{\alpha}{r^4}$ for $r > a$ and $U = \infty$ for $r < a$ (see Fig. 4.1).

Solution

A new bound state with an arbitrary angular momentum l emerges if the Schrödinger equation with $E = 0$ has a solution whose radial function has an asymptote of the form $R \approx Cr^{-l-1}$ as $r \to \infty$ (for more general states, the asymptote has the form $R \approx Ar^l + C/r^{l+1}$). Compare with Problem 4.25. Here in the case $l \neq 0$, the wavefunction at the threshold is normalized to unity, *i.e.*, it corresponds to a truly bound state.

a) The radial wavefunction at the threshold of emergence of a new level (*i.e.*, $E = 0$) according to (IV.5) has the form $\chi = Ar^{l+1}$ for $r < a$ and $\chi = C/r^l$ for $r > a$. Matching the solutions at $r = a$ (see 2.6) gives us $C = Aa^{2l+1}$ and $2m\alpha a/\hbar^2 = 2l + 1$, which is the condition of emergence of a unique discrete level with the angular momentum l as the δ-well deepens. We should note that to normalize the wavefunction to unity, it is necessary to choose

$$C^2 = A^2 a^{4l+2} = \frac{(2l-1)(2l+3)}{2(2l+1)} a^{2l-1}.$$

b) After substituting $x = 1/r$, Eq. (IV.2) with $U = -\alpha/r^4$ and $E = 0$ becomes

$$\left(\frac{d^2}{dx^2} + \tilde{\alpha} - \frac{l(l+1)}{x^2}\right) R_l = 0, \quad \tilde{\alpha} = \frac{2m\alpha}{\hbar^2}. \tag{1}$$

Its solution, $R = CJ_{l+1/2}\left(\sqrt{\tilde{\alpha}}/r\right)/\sqrt{r}$ for $r > a$, gives the radial wavefunction at the threshold of emergence. The condition $R(a) = 0$ leads to the relation $\sqrt{\tilde{\alpha}}/a = x_{l+1/2,N}$, which determines the condition of the level's existence. Here $x_{l+1/2,N}$ is the Nth root of the Bessel function $J_{l+1/2}(x)$.

Problem 4.27

The parameters of central potential[44] $U_0(r)$ are chosen such that there exists a discrete spectrum state with angular momentum $l = 0$ and energy $E = 0$. The wavefunction, $\psi_0 = \chi_0(r)/\sqrt{4\pi}r$, of such a state (at the moment of level emergence) is known and normalized, for concreteness, by the condition that $\chi_0(r) \to 1$ as $r \to \infty$. Show that the shift of this level, δE_0, due to a small perturbation, $\delta U(r) \leq 0$, is given by

$$\delta E_0 \approx -\frac{2m}{\hbar^2} \left[\int_0^\infty \delta U(r) \chi_0^2(r) dr \right]^2. \tag{1}$$

Apply this result to the potential $U = -\alpha \delta(r - a)$ and compare it with the exact solution obtained in Problem 4.8 a.

Solution

Compare the Schrödinger Eq. (IV.5) for the potential $U_0(r)$ with $E = 0$ and for the potential $U_0 + \delta U$ with $\delta E_0 = -\hbar^2 \kappa^2/2m$:

$$-\chi_0'' + \tilde{U}_0(r)\chi_0 = 0, \quad -\chi'' + (\tilde{U}_0(r) + \delta \tilde{U}(r) + \kappa^2)\chi = 0. \tag{1}$$

We use $\tilde{U}(r) \equiv 2mU(r)/\hbar^2$. By multiplying the first equation by $\chi(r)$ and the second by $\chi_0(r)$ and then substracting term by term, we obtain

$$\frac{d}{dr}(\chi_0 \chi' - \chi_0' \chi) = (\delta \tilde{U}(r) + \kappa^2)\chi \chi_0. \tag{2}$$

Integrate (2) over r between $r = 0$ and $r = a$, where a is a radius of the potentials $U_0(r)$ and $\delta U(r)$. Take into account that 1) $\chi_0(0) = \chi(0) = 0$; 2) $\chi_0(a) = 1$ and $\chi_0'(a) = 0$; 3) $\chi(r) \approx \chi_0(r)$ for $r \leq a$. This last condition essentially determines the normalization of the wavefunction $\chi(r)$, while the normalization of the function $\chi_0(r)$ is determined by its asymptote, $\chi_0(r) = 1$ for $r > a$. Here $\chi(a) \approx e^{-\kappa a} \approx 1$ and $\chi'(a) \approx -\kappa e^{-\kappa a} \approx -\kappa$, and as the result of integration we obtain

$$-\kappa = \int_0^a \chi(r)\chi_0(r)\delta \tilde{U}(r)dr + \kappa^2 \int_0^a \chi(r)\chi_0(r)dr. \tag{3}$$

For the first of the integrals in (3), we may set $\chi \approx \chi_0$ and then take $a \to \infty$. The second integral is $\propto \kappa^2$ and can be omitted compared to κ. We obtain

$$\kappa \approx \int_0^\infty (-\delta \tilde{U}(r))\chi_0^2(r)dr.$$

---

[44] It is assumed that $U \equiv 0$ for $r > a$, where a is radius of the potential. The problem statement remains for the case of potentials that decrease as $r \to \infty$ faster than $\propto 1/r^2$. See also Problem 13.49.

This is equivalent to the expression for the level shift given in the problem statement.

For the δ-potential, the level with $E = 0$ emerges as $\alpha = \alpha_0$ such that $m\alpha_0 a/\hbar^2 = 1/2$. See, for example, Problem 4.26. Here the wavefunction at the threshold of level emergence has the form $\chi_0 = 1$ for $r > a$ and $\chi_0 = r/a$ for $r < a$, while $\delta U = -(\alpha - \alpha_0)\delta(r - a)$. Hence the level shift with small $\alpha - \alpha_0 > 0$ is equal to

$$\delta E_0 \approx -\frac{2m}{\hbar^2}\left\{\int_0^\infty \delta U(r)\chi_0^2(r)dr\right\} = -\frac{2m}{\hbar^2}(\alpha - \alpha_0)^2.$$

This is in accordance with the exact result. See Problem 4.8 a.

Finally, we give another derivation of the equation for the level shift, this time based on the relation (I.6). We can write the potential in the form $U(r) = U_0(r) + \lambda\delta U(r)$, with $\lambda \geq 0$. Here the level energy $\delta E_0(\lambda) \equiv -\hbar^2\kappa^2/2m$ also depends on λ and $\delta E_0(\lambda = 0) = 0$. From Eq. (I.6) we have

$$\frac{\partial}{\partial\lambda}\delta E_0(\lambda) = \int_0^\infty \delta U(r)\tilde{\chi}_0^2(r, \lambda)dr, \tag{4}$$

where $\tilde{\chi}_0(r; \lambda)$ is a wavefunction normalized to unity. This function is related to $\chi_0(r)$:

$$\tilde{\chi}_0(r; \lambda) \approx C(\kappa)e^{-\kappa r}\chi_0(r).$$

As you see, for $r \leq a$ the function $\tilde{\chi}_0(r;\lambda)$ differs from $\chi_0(r)$ only by a multiplicative factor (we use $e^{-\kappa r} \approx 1$), while for $r > 0$ we have $\tilde{\chi}_0(r;\lambda) = C(\kappa)e^{-\kappa r}$ (here $\chi_0(r) = 1$). To normalize the wavefunction $\tilde{\chi}_0(r;\lambda)$ we should choose $C^2(\kappa) \approx 2\kappa$, since the dominant contribution in the normalization integral comes from the region $r \sim 1/\kappa \gg a$, where $\tilde{\chi}_0(r;\lambda) = C(\kappa)e^{-\kappa r}$. Taking into account everything mentioned above, we transform (4) to the form:

$$\frac{\partial}{\partial\lambda}\delta E_0(\lambda) \equiv -\frac{\hbar^2\kappa}{m}\frac{\partial\kappa}{\partial\lambda} \approx 2\kappa\int_0^\infty \delta U(r)\chi_0^2(r,\lambda)dr.$$

By integrating this relation and using $\kappa(\lambda = 0) = 0$, we find

$$\kappa(\lambda) = -\frac{2m}{\hbar^2}\lambda\int_0^\infty \delta U(r)\chi_0^2(r,\lambda)dr.$$

Inserting $\lambda = 1$ gives the relation for the level shift.

Problem 4.28

Prove that the main result of the previous problem can be generalized to the case with $l \neq 0$ is as follows

$$\delta E_l \approx \int_0^\infty \delta U(r) \left(\chi_l^{(0)}(r)\right)^2 dr,$$

where $\chi_l^{(0)}$ is the wavefunction at the threshold of emergence of a new level $\left(\psi^{(0)} = \chi_l^{(0)} Y_{lm}/r\right)$ and is normalized to unity by $\int_0^\infty \left(\chi_l^{(0)}(r)\right)^2 dr = 1$.

Note that the dependences of the level shift on the perturbation, $\delta E_l \propto \delta U$ for $l \neq 0$ and $\delta E_0 \propto -(\delta U)^2$ for $l = 0$, are different. Use the result obtained for the δ-potential and compare it with the exact solution. See Problem 4.9 a.

Solution

The problem could be solved similarly to the previous one. But now Eqn. (1) of the latter includes terms with centrifugal energy. Relation (2) remains the same for $l \neq 0$. By integrating this relation over r between 0 and ∞, we obtain

$$\int_0^\infty \chi_l(r) \chi_l^{(0)}(r) \delta \tilde{U}(r) dr + \kappa^2 \int_0^\infty \chi_l(r) \chi_l^{(0)}(r) dr = 0. \tag{1}$$

We have $\chi_l \approx \chi_l^{(0)}$ in the region $r \sim a$ which is significant to the integral. This does not work as well for the case $l = 0$ because a divergence appears in the second integral. Here, taking into account the normalization $\int (\chi_l^{(0)})^2 dr = 1$ for the wavefunction $\chi_l^{(0)}$, we immediately find the value κ^2. Putting $\delta \tilde{U}(r) = 2mU(r)/\hbar^2$ reproduces the result for the level shift given in the problem statement.

Note the reasons for the difference in the dependence of level deepening on the perturbation, $\delta E \propto \delta U$ for $l \neq 0$ and $\delta E \propto -(\delta U)^2$ for $l = 0$ are as follows: For $l \neq 0$ the state with $E = 0$ is truly a bound state and corresponds to a wavefunction that is normalized to unity. For $l = 0$, the situation is different: the wavefunction is not normalized. The physical reason for normalization and the lack of it is connected with the existence of a centrifugal barrier that prevents the particle from escaping to infinity.

For the δ-well at the threshold of emergence of a single level (with given l), we have $2m\alpha_0 a/\hbar^2 = (2l + 1)$ (see 4.26), and there is a normalized wavefunction at the threshold. Here $\delta U = -(\alpha - \alpha_0)\delta(r - a)$ and the energy of the level for the small $(\alpha - \alpha_0) > 0$ is equal to

$$\delta E_l \approx \int_0^\infty \delta U \left(\chi_l^{(0)}\right)^2 dr = -\frac{(2l-1)(2l+3)}{2(2l+1)} \frac{(\alpha - \alpha_0)}{a}, \quad l \neq 0,$$

which coincides with the result of the exact solution. See Problem 4.9 a.

Note that for $l \neq 0$, the expression for the level shift has a form like the first term in perturbation theory in the interaction $\delta U(r)$ (see Eq. (VIII.1)). Furthermore, since for $l \neq 0$ the wavefunction $\chi_l^{(0)}(r)$ is normalized to unity, then for the region $r \leq a$ which gives the main contribution to the level value in the (4), we can set $\chi_l(r, \lambda) \approx \chi_l^{(0)}(r)$. Compare this with $l = 0$ when the functions $\chi_0(r; \lambda)$ and $\chi_0(r)$ have different normalization. In the new case, relation (4) from Problem 4.27 immediately gives the expression for the level shift.

Problem 4.29

Find[45] the shift of the energy levels of a particle in a central field $U(r)$ caused by a zero-range potential (z.r.p., see Problem 4.10), assuming these shifts to be small with respect to the unperturbed level spacing. Assume that the spectrum and the eigenfunctions of the Hamiltonian for the potential $U(r)$ are known. Determine the condition of applicability of this result. As an illustration, consider its application to Problem 4.11.

Solution

A zero-range potential causes only s-levels to shift. The normalized radial wavefunction of an unperturbed state for the small r is $R_n^{(0)}(r) \approx R_n^{(0)}(0)$. Let L be the interval where we can consider $R_n^{(0)}(r)$ to be constant. The precise value of L depends on the concrete form of $U(r)$ and the energy $E_n^{(0)}$. In the presence of the z.r.p., according to Problem 4.10, the radial function for small r has the form:[46]

$$R_n(r) \approx R_n^{(0)}(0) \left(-\frac{1}{\alpha_0 r} + 1 + \ldots\right). \tag{1}$$

This differs greatly from $R_n^{(0)}(r)$ for $r \to 0$ due to the term $\propto 1/r$. But if $|\alpha_0 L| \gg 1$ then for $r \sim L$ the function $R_n(r)$ differs only slightly from the $R_n^{(0)}(r)$ (see Fig. 4.3). This means that the level shift caused by the zero-range potential is small,[47] and the wavefunctions R_n and $R_n^{(0)}$ hardly differ from each other for all $r \geq L$. This is so, because after all, this region that provides the main contribution in the normalization integral.

---

[45] For the questions discussed in Problem 4.29–31, see also Problems 11.4 and 9.3.
[46] We assume that $rU(r) \to 0$ as $r \to 0$. For more singular potentials, the asymptote (1) is modified and the boundary condition from Problem 4.10 that determines the z.r.p. cannot be fulfilled. For example, it is impossible to simulate a strong short-range potential by a zero-range potential in the presence of Coulomb interaction at small distances, (although, perturbation theory with respect to the scattering length remains valid).
[47] We should note that Fig. 4.3 corresponds to the value $\alpha_0 > 0$ when in the z.r.p. there exists a discrete at small distances level with the energy $E_s^{(0)} = -\hbar^2 \alpha_0^2 / 2m$, where $|E_s^{(0)}| \gg |\delta E_n^{(0)}| \sim \hbar^2 / mL^2$. If the condition $\alpha_0 L \gg 1$ is not fulfilled, the level energy in the z.r.p. is of the same order of magnitude as the level spacing $\delta E_n^{(0)}$ in the potential $U(r)$. But in this case the levels shifts are as large as the level spacings, so a *reconstruction of the spectrum takes place* (see Problems 4.11 and 9.3). Eqs. (4) in that case cannot be used.

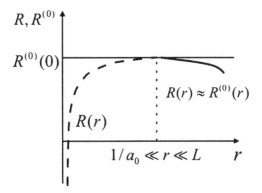

Fig. 4.3

Now, we address the level shifts. Let $\chi = rR$.

$$\frac{\hbar^2}{2m}\chi_n''^{(0)} + (E_n^{(0)} - U(r))\chi_n^{(0)} = 0,$$

$$\frac{\hbar^2}{2m}\chi_n'' + (E_n^{(0)} + \Delta E_n - U(r))\chi_n = 0. \qquad (2)$$

Though these equations have the same form, they differ in the boundary condition for $r \to 0$: without the z.r.p. we have $\chi_n^{(0)} \approx R_n^{(0)} r \propto r$, while in the presence of the z.r.p. we have $\chi_n \approx R_n r \approx \text{const}$, where $R_n(r)$ is determined by (1). We multiply the first of Eqs. (2) by χ_n, the second by $\chi_n^{(0)}$, and then subtract one from the other term by term. If we then perform the integration between $r = 0$ and $r = \infty$, we obtain

$$\frac{\hbar^2}{2m}\left[\chi_n'^{(0)}\chi_n - \chi_n^{(0)}\chi_n'\right]\Big|_0^\infty = \Delta E_n \int_0^\infty \chi_n^{(0)}(r)\chi_n(r)dr. \qquad (3)$$

The left part here is equal to $\hbar^2 R_n^{(0)2}(0)/2m\alpha_0$. In the right part we can replace $\chi_n(r)$ by $\chi_n^{(0)}(r)$ (such a substitution does not work for small r, but this region does not significantly contribute), so the integral value is close to unity. Hence, (3) gives us the desired expression for the level shift:

$$\Delta E_n \approx \frac{\hbar^2 R_n^{(0)2}(0)}{2m\alpha_0} \equiv 2\pi\frac{\hbar^2}{m}\left(\psi_n^{(0)}(0)\right)^2 a_0. \qquad (4)$$

Here $a_0 \equiv 1/\alpha_0$ is the scattering length for z.r.p. See Problem 13.20.

This equation could be applied in case of non-central potentials, $U(r)$. See Problems 4.31 and 8.61.

Consider the application of (4) to Problem 4.11. Here we have

$$R_n^{(0)} = \sqrt{\frac{2}{a}} \frac{\sin \kappa_n^{(0)} r}{r} \underset{r \to 0}{\approx} \sqrt{\frac{2}{a}} \kappa_n^{(0)},$$

where $\kappa_n^{(0)} = \pi(n_r + 1)/a$ (do not confuse the well radius a with the scattering length a_0) and

$$\Delta E_n \approx \frac{\hbar^2 \kappa_n^{(0)2}}{m a \alpha_0} = \frac{2 E_n^{(0)}}{a \alpha_0},$$

which for $\kappa_n^{(0)} \ll |\alpha_0|$ coincides with the result of the exact solution. The last inequality, corresponds to the condition of (4) applicability $|\alpha_0 L| \gg 1$.

Problem 4.30

Show that the generalization of the result of the previous problem to the case of a weakly bound state, $\psi_n^{(0)}(r)$, is given by

$$E_n \approx - \left[\sqrt{-E_n^{(0)}} - \frac{1}{\alpha_0} \sqrt{\frac{\hbar^2}{2m}} \left(\tilde{R}_n^{(0)}(0) \right)^2 \right]^2,$$

where $\tilde{R}_n^{(0)}(r)$ is the radial wavefunction at the threshold of emergence of an s-level and is normalized by the condition $r\tilde{R}_n^{(0)}(r) \to 1$ at $r \to \infty$.

Solution

The fact that we can use Eq. (4) from the previous problem means that the level shift is small in comparison with the particle binding energy. But if the unperturbed state has an anomalously small energy, then we need to make a simple generalization for the case where the shift is comparable to the binding energy. This generalization can be deduced from Eq. (3) of problem 4.29 if we take into account the following circumstances.

First, the wavefunction, $R_n^{(0)}(r)$, of an unperturbed state in the field $U(r)$ with the small binding energy is simply related to the wavefunction, $\tilde{R}_n^{(0)}(r)$, at the threshold of level emergence (compare with Problem 4.27):

$$R_n^{(0)} \approx \sqrt{2\kappa_n^{(0)}} \tilde{R}_n^{(0)}(r) e^{-\kappa_n^{(0)} r}, \quad \kappa_n^{(0)} = \sqrt{-\frac{2m E_n^{(0)}}{\hbar^2}}. \tag{1}$$

Analogously, the wavefunction of the state perturbed by the z.r.p. for $r \geq L$ takes the form

$$R_n \approx \sqrt{2\kappa_n} \tilde{R}_n^{(0)}(r) e^{-\kappa_n r}, \quad \kappa_n = \sqrt{-\frac{2m E_n}{\hbar^2}}. \tag{2}$$

Wavefunctions (1) and (2) are normalized to unity. Using Eq. (3) of the previous problem, (1) and (2) give us ($\chi = rR$)

$$\frac{E_n - E_n^{(0)}}{\kappa_n^{(0)} + \kappa_n} = \frac{\hbar^2}{2m\alpha_0}(\tilde{R}_n^{(0)}(0))^2. \tag{3}$$

To calculate the integral we should note that its value depends strongly on the region of large r, where the function under the integral sign is $\propto \exp\left\{-(\kappa_n^{(0)} + \kappa_n)r\right\}$. Finally, we obtain the relation for the level energy given in the problem statement: $\kappa_n = \kappa_n^{(0)} - \tilde{R}_n^{(0)2}(0)/\alpha_0$.

For $\kappa_n^{(0)} \gg |\tilde{R}_n^{(0)2}(0)/\alpha_0|$ the level shift is small, and this result turns into Problem 4.29. As a final note, see that the state considered is *truly bound* only if $\kappa_n > 0$. Otherwise, it is a *virtual* level.[48]

Problem 4.31

A particle is in a field $U(r)$ (with $rU \to 0$ for $r \to 0$) and a zero-range potential located at the point $\mathbf{r} = \mathbf{0}$ (see Problem 4.10). The particle Green function $G_0(\mathbf{r}, \mathbf{r}'; E)$ for the potential $U(\mathbf{r})$ is known. Show that the spectrum of the bound states of this system can be determined from:

$$\left[\frac{\partial}{\partial r}(rG_0(r, 0; E_n))\right]_{r=0} = -\frac{m\alpha_0}{2\pi\hbar^2}. \tag{1}$$

Obtain from this relation the level shift from problems 4.29 and 11.4:

$$\delta E_n \approx \frac{2\pi\hbar^2}{m}|\psi_n^{(0)}(0)|^2 a_0, \tag{2}$$

where $a_0 \equiv 1/\alpha_0$ is the scattering length for the z.r.p., and $\psi_n^{(0)}(r)$ is the wavefunction of the unperturbed level in the potential $U(r)$. ΔE_n is its shift caused by the z.r.p.

Solution

We use the Green function, and the fact that it decreases as $r \to \infty$ and goes as $1/r$ as $r \to 0$, and observe that the bound state wavefunction in the potential $U(\mathbf{r})$ in the presence of a zero-range potential has the form

$$\psi(\mathbf{r}, E_n) = C_n G_0(\mathbf{r}, 0; E_n) \underset{r \to 0}{=} C_n \frac{m}{2\pi\hbar^2}\left\{\frac{1}{r} + A(E_n) + \ldots\right\}, \tag{3}$$

---

[48] The virtual level in potential $U(r)$ could become real, or the real level could become virtual under the influence of the z.r.p., depending on the sign and value of α_0.

The normalization coefficient is C_n. The restriction on the potential is important: $rU \to 0$ for $r \to 0$; otherwise the expansion above is not valid, and using the z.r.p. approximation is impossible. We compare expansion (3) with the expression defining a zero-range potential (see Problem 4.10) and obtain the equation for the level spectrum,

$$A(E_n) = -\alpha_0, \tag{4}$$

which yields Eq. (2) which was to be derived.

For a free particle, $U \equiv 0$, (4) takes the form $\kappa = \alpha_0$. In the case $\alpha_0 > 0$, this describes a bound state in the zero-range potential. See Problem 4.10.

In the case of large enough α_0, the roots of (4) are close to the values $E_n^{(0)}$ of the Green-function poles, which corresponds to the spectrum in the isolated potential, $U(\mathbf{r})$. Using the known expression for the Green function,

$$G_E(\mathbf{r}, \mathbf{r}', E) = \sum_m \frac{\psi_m^{(0)}(\mathbf{r}) \psi_m^{(0)*}(\mathbf{r}')}{E_m^{(0)} - E}, \tag{5}$$

we see for $\mathbf{r}' = 0$, $\mathbf{r} \to 0$, and $E \to E_n^{(0)}$, that it has the form

$$G_{E_n}(\mathbf{r}, 0; E_n) \approx \frac{m}{2\pi\hbar^2} \cdot \frac{1}{r} + \frac{|\psi_n^{(0)}(0)|^2}{E_n^{(0)} - E_n}, \tag{6}$$

Therefore we have $A(E_n) \approx \frac{2\pi\hbar^2}{m} \frac{|\psi_n^{(0)}(0)|^2}{E_n^{(0)} - E_n}$, and from Eq. (4) we directly obtain the relation (2) for the level shift, $\Delta E_n = E_n - E_n^{(0)}$. See Problem 13.20.

Problem 4.32

For a monotonic attractive potential, $U'(r) \geq 0$ and $U(r) \to 0$ as $r \to \infty$, prove that

$$\frac{2}{\pi\hbar} \int_0^\infty \sqrt{-2mU(r)} dr \geq 1$$

is a necessary condition for the existence of a bound state. Compare with Problem 4.21.

Solution

Consider the potential $\tilde{U}(r)$ whose wavefunction at the threshold of emergence of a bound state, *i.e.*, at $E = 0$, is

$$\chi_0(r) = \cos\left(\frac{1}{\hbar} \int_r^\infty p_0(r) dr\right), \quad \text{where} \quad p_0(r) = \sqrt{-2mU(r)}. \tag{2}$$

This wavefunction corresponds to equation $\chi_0'' - (2m/\hbar^2)\tilde{U}(r)\chi_0 = 0$ with the potential

$$\tilde{U}(r) = U(r) - \frac{\hbar U'(r)}{2\sqrt{-2mU(r)}} \tan\left(\frac{1}{\hbar}\int_r^\infty \sqrt{-2mU(r)}dr\right), \quad (3)$$

and the condition of the emergence of a first bound state in this potential is[49]

$$\frac{1}{\hbar}\int_0^\infty \sqrt{-2mU(r)}dr = \frac{\pi}{2}. \quad (4)$$

If we write $U(r) = \tilde{U}(r) + \delta U(r)$, we see that $\delta U(r) \geq 0$, so the potential well $U(r)$ is less deep than $\tilde{U}(r)$. This proves the statement of the problem.

For a rectangular potential well of the depth U_0 and radius a, the considered condition for the existence of a bound state takes the form $U_0 \geq \pi^2\hbar^2/8ma^2$, which coincides with the exact condition. For the exponential well, $U(r) = -U_0 e^{-r/a}$, we obtain $ma^2 U_0/\hbar^2 \geq \pi^2/32 \approx 0.31$. Compare this with the result of Problem 4.21.

4.3 Symmetries of the Coulomb problem

Problem 4.33

Consider an electron in the Coulomb potential of a nucleus with charge Ze (the nucleus is assumed infinitely heavy compared to the electron), i.e., the single-electron Hamiltonian has the form:

$$\hat{H} = \frac{\hat{\mathbf{p}}^2}{2m} - \frac{K}{r},$$

where m is the electron mass and $K = Ze^2$.

Consider the following operators related to the classical Laplace-Runge-Lenz vector, $\mathbf{M} = \frac{\mathbf{p} \times \mathbf{L}}{m} - K\frac{\mathbf{r}}{r}$ (here, $\mathbf{L} = \mathbf{r} \times \mathbf{p}$ is the angular momentum):

1. $\hat{\mathbf{M}}_1 = \frac{\hat{\mathbf{L}} \times \hat{\mathbf{p}}}{m} - \mathbf{r}/r$;
2. $\hat{\mathbf{M}}_2 = \frac{\hat{\mathbf{p}} \times \hat{\mathbf{L}}}{m} - K\mathbf{r}/r$;
3. $\hat{\mathbf{M}} = \frac{1}{2m}\left(\hat{\mathbf{p}} \times \hat{\mathbf{L}} - \hat{\mathbf{L}} \times \hat{\mathbf{p}}\right) - K\mathbf{r}/r$;

Prove explicitly that the operator, $\hat{\mathbf{M}}$, commutes with the Hamiltonian and check whether the operators $\hat{\mathbf{M}}_1$ and $\hat{\mathbf{M}}_2$ do so as well.

---

[49] This relation provides the boundary condition $\tilde{\chi}_0(0) = 0$; and $\chi_0(r) \to 0$ as $r \to \infty$. See Problem 4.25.

Solution

To evaluate $[\hat{\mathbf{M}}, \hat{H}]$, we use the following commutation relations:

$$[r_i, r_j] = 0, \quad [\hat{p}_i, \hat{p}_j] = 0, \quad [r_i, \hat{p}_j] = i\hbar \delta_{ij}, \quad [\hat{L}_i, \hat{L}_j] = i\hbar \epsilon_{ijk} \hat{L}_k,$$

$$[\hat{L}_i, r_j] = \epsilon_{ikl}[r_k \hat{p}_l, r_j] = -i\hbar \epsilon_{ikl} r_k \delta_{lj} = i\hbar \epsilon_{ijk} r_k, \quad (1)$$

$$[\hat{L}_i, \hat{p}_j] = \epsilon_{ikl}[r_k \hat{p}_l, \hat{p}_j] = i\hbar \epsilon_{ikl} \hat{p}_l \delta_{kj} = i\hbar \epsilon_{ijk} \hat{p}_k.$$

Now, using the expressions for $\hat{H}$ and $\hat{\mathbf{M}}$, we obtain

$$[\hat{\mathbf{M}}, \hat{H}] = \left[\frac{1}{2m}(\hat{\mathbf{p}} \times \hat{\mathbf{L}} - \hat{\mathbf{L}} \times \hat{\mathbf{p}}) - K\frac{\mathbf{r}}{r}, \frac{\hat{\mathbf{p}}^2}{2m} - \frac{K}{r} \right].$$

$$= \frac{1}{4m^2} \left[(\hat{\mathbf{p}} \times \hat{\mathbf{L}} - \hat{\mathbf{L}} \times \hat{\mathbf{p}}), \hat{\mathbf{p}}^2 \right] - \frac{K}{2m} \left\{ \left[(\hat{\mathbf{p}} \times \hat{\mathbf{L}} - \hat{\mathbf{L}} \times \hat{\mathbf{p}}), \frac{1}{r} \right] + \left[\frac{\mathbf{r}}{r}, \hat{\mathbf{p}}^2 \right] \right\}. \quad (2)$$

We use Eqs. (1) to evaluate the components of commutators in Eq. (2):

$$\left[(\hat{\mathbf{p}} \times \hat{\mathbf{L}})_i, \hat{\mathbf{p}}^2 \right] = \epsilon_{ijk}[\hat{p}_j \hat{L}_k, \hat{p}_a \hat{p}_a] = \epsilon_{ijk}\epsilon_{klm}[\hat{p}_j r_l \hat{p}_m, \hat{p}_a \hat{p}_a]$$

$$= \epsilon_{ijk}\epsilon_{klm}\hat{p}_j \hat{p}_m([r_l, \hat{p}_a]\hat{p}_a + \hat{p}_a[r_l, \hat{p}_a]) = 2\epsilon_{ijk}\epsilon_{klm}\hat{p}_j \hat{p}_m \delta_{la} \hat{p}_a$$

$$= 2\epsilon_{ijk}\epsilon_{klm}\hat{p}_j \hat{p}_m \hat{p}_l = 0,$$

since ϵ_{klm} is anti-symmetric in l, m and $\hat{p}_m \hat{p}_l$ is symmetric. Similarly, we can show that $[(\hat{\mathbf{L}} \times \hat{\mathbf{p}})_i, \hat{\mathbf{p}}^2] = 0$. Now for the other commutators in Eq. (2), we first need to evaluate the following (for a general function, $f(r)$):

$$\left[\hat{p}_i, \frac{1}{r} \right] f(r) = -i\hbar \left[\frac{\partial}{\partial r_i}, \frac{1}{r} \right] f(r) = -i\hbar \left\{ \frac{\partial}{\partial r_i}\left(\frac{f(r)}{r}\right) - \frac{1}{r}\frac{\partial f(r)}{\partial r_i} \right\}$$

$$= -i\hbar \frac{\partial}{\partial r_i}\left(\frac{1}{r}\right) f(r) = -i\hbar \frac{r_i}{r^3} f(r),$$

$$\left[\frac{r_i}{r}, \hat{\mathbf{p}}^2 \right] f(r) = -\hbar^2 \left[\frac{r_i}{r}, \frac{\partial^2}{\partial r^2} \right] f(r) = -\hbar^2 \left\{ \frac{r_i}{r}\frac{\partial^2 f(r)}{\partial r^2} - \frac{\partial^2}{\partial r^2}\left(\frac{r_i}{r}f(r)\right) \right\}$$

$$= 2\hbar^2 \frac{r_i}{r^3} f(r).$$

Now, in Eq. (2),

$$\left[(\hat{\mathbf{p}} \times \hat{\mathbf{L}})_i, \frac{1}{r} \right] = \epsilon_{ijk}\left[\hat{p}_j \hat{L}_k, \frac{1}{r} \right] = \epsilon_{ijk}\epsilon_{klm}\left[\hat{p}_j r_l \hat{p}_m, \frac{1}{r} \right]$$

$$= \epsilon_{ijk}\epsilon_{klm} r_l \left(\hat{p}_j\left[\hat{p}_m, \frac{1}{r}\right] + \left[\hat{p}_j, \frac{1}{r}\right]\hat{p}_m \right)$$

$$= -i\hbar \epsilon_{ijk}\epsilon_{klm} r_l \frac{\hat{p}_j r_m + r_j \hat{p}_m}{r^3} = (-i\hbar)^2 \frac{r_i}{r^3} = -\hbar^2 \frac{r_i}{r^3},$$

where we have used $\epsilon_{ijk}\epsilon_{klm} = \delta_{il}\delta_{jm} - \delta_{im}\delta_{jl}$ and Eqs. (1). Similarly,

$$\left[(\hat{\mathbf{L}} \times \hat{\mathbf{p}})_i, \frac{1}{r}\right] = -(-i\hbar)^2 \frac{r_i}{r^3} = \hbar^2 \frac{r_i}{r^3}.$$

Hence, we get from Eq. (2):

$$\left[\hat{M}_i, \hat{H}\right] = -\frac{K}{2m}\left(-\hbar^2 \frac{r_i}{r^3} - \hbar^2 \frac{r_i}{r^3} + 2\hbar^2 \frac{r_i}{r^3}\right) = 0.$$

It is also clear from the above that just $\left[\hat{\mathbf{p}} \times \hat{\mathbf{L}}, \frac{1}{r}\right]$ or $\left[\hat{\mathbf{L}} \times \hat{\mathbf{p}}, \frac{1}{r}\right]$ alone does not cancel against $\left[\frac{\mathbf{r}}{r}, \mathbf{p}^2\right]$. Hence, $\hat{M}_1$ and $\hat{M}_2$ do not commute with the Hamiltonian, while the operator $\hat{\mathbf{M}}$ does commute with $\hat{H}$.

Problem 4.34

Consider the angular momentum operator, $\hat{\mathbf{L}}$, and the Laplace-Runge-Lenz operator, $\hat{\mathbf{M}}$, defined in the previous problem and prove that their components satisfy the following commutation relations:

1. $[\hat{L}_\alpha, \hat{M}_\beta] = i\hbar\epsilon_{\alpha\beta\gamma}\hat{M}_\gamma$.
2. $[\hat{M}_\alpha, \hat{M}_\beta] = -\left(\frac{2\hat{H}}{m}\right)i\hbar\epsilon_{\alpha\beta\gamma}\hat{L}_\gamma$.

where $\hat{H}$ is the Hamiltonian for a particle in the Coulomb potential (see, the previous problem) and the indices α, β, and γ take the values x, y, and z, and the usual Einstein rule for summing over repeated indices is used.

Solution

Using the definitions of the operators $\hat{\mathbf{L}}$ and $\hat{\mathbf{M}}$, we have

$$\left[\hat{L}_\alpha, \hat{M}_\beta\right] = \left[\hat{L}_\alpha, \frac{1}{2m}\left(\hat{\mathbf{p}} \times \hat{\mathbf{L}} - \hat{\mathbf{L}} \times \hat{\mathbf{p}}\right)_\beta\right] - \left[\hat{L}_\alpha, K\frac{r_\beta}{r}\right].$$

We can work it out keeping the indices general, as in the previous problem, but it is a bit involved, so we are just going to show it for $[\hat{L}_x, \hat{M}_y]$, and the others will follow by cyclic permutation.

$$[\hat{L}_x, \hat{M}_y] = \left[(y\hat{p}_z - z\hat{p}_y), \frac{1}{2m}(\hat{p}_z\hat{L}_x - \hat{p}_x\hat{L}_z - \hat{L}_z\hat{p}_x + \hat{L}_x\hat{p}_z) - K\frac{y}{r}\right]$$

$$= i\frac{\hbar}{2m}\left(2z(\hat{p}_x^2 + \hat{p}_y^2) - \hat{p}_z(\hat{p}_x x + \hat{p}_y y + x\hat{p}_x + y\hat{p}_y)\right) - i\hbar K\frac{z}{r}$$

$$= i\hbar\hat{M}_z.$$

Similarly,

$$[\hat{M}_x, \hat{M}_y] = \left[\frac{1}{2m}\left(2x(\hat{p}_y^2 + \hat{p}_z^2) - \hat{p}_x(\hat{p}_z z + \hat{p}_y y + z\hat{p}_z + y\hat{p}_y)\right) - K\frac{x}{r},\right.$$

$$\left.\frac{1}{2m}\left(2y(\hat{p}_x^2 + \hat{p}_z^2) - \hat{p}_y(\hat{p}_x x + \hat{p}_z z + x\hat{p}_x + z\hat{p}_z)\right) - K\frac{y}{r}\right]$$

$$= -\frac{i\hbar}{m^2}(\hat{p}_x^2 + \hat{p}_y^2 + \hat{p}_z^2)(x\hat{p}_y - y\hat{p}_x) - \frac{i\hbar K}{r}(x\hat{p}_y - y\hat{p}_x)$$

$$= -\frac{2H}{m}i\hbar \hat{L}_z.$$

Problem 4.35

Since $[\hat{H}, \hat{\mathbf{M}}] = 0$ (see definitions in the previous two problems), the operator $\hat{\mathbf{M}}$ can not change the energy of an eigenstate of the Hamiltonian with an energy, E_n. Restricting the action of the operator to a set of eigenstates with the same energy, define a new operator, $\hat{\mathbf{M}}' = \sqrt{-\frac{m}{2E_n}}\hat{\mathbf{M}}$ and also the two operators, $\hat{\mathbf{A}} = \frac{1}{2}(\hat{\mathbf{L}} + \hat{\mathbf{M}}')$ and $\hat{\mathbf{B}} = \frac{1}{2}(\hat{\mathbf{L}} - \hat{\mathbf{M}}')$, and prove that

1. $[\hat{A}_\alpha, \hat{A}_\beta] = i\epsilon_{\alpha\beta\gamma}\hat{A}_\gamma$.
2. $[\hat{B}_\alpha, \hat{B}_\beta] = i\epsilon_{\alpha\beta\gamma}\hat{B}_\gamma$.
3. The operators $\hat{\mathbf{A}}$ and $\hat{\mathbf{B}}$ are generators of a symmetry group of the Coulomb problem. What is this group?

Solution

From the derivation in the previous problem, we have

$$\left[\hat{L}_\alpha, \hat{M}'_\beta\right] = \sqrt{-\frac{m}{2E_n}}\left[\hat{L}_\alpha, \hat{M}_\beta\right] = i\hbar\epsilon_{\alpha\beta\gamma}\hat{M}'_\gamma,$$

$$[\hat{M}'_\alpha, \hat{M}'_\beta] = -\frac{m}{2E_n}\left[\hat{M}_\alpha, \hat{M}_\beta\right] = \frac{m}{2E_n}\frac{2H}{m}i\hbar\epsilon_{\alpha\beta\gamma}\hat{L}_\gamma = i\hbar\epsilon_{\alpha\beta\gamma}\hat{L}_\gamma.$$

So,

$$\left[\hat{A}_\alpha, \hat{A}_\beta\right] = \frac{1}{4}\left([\hat{L}_\alpha, \hat{L}_\beta] + [\hat{L}_\alpha, \hat{M}'_\beta] + [\hat{M}'_\alpha, \hat{L}_\beta] + [\hat{M}'_\alpha, \hat{M}'_\beta]\right)$$

$$= \frac{1}{4}i\hbar\epsilon_{\alpha\beta\gamma}\left(\hat{L}_\gamma + \hat{M}'_\gamma + \hat{M}'_\gamma + \hat{L}_\gamma\right) = i\hbar\epsilon_{\alpha\beta\gamma}\hat{A}_\gamma.$$

Similarly for $\hat{\mathbf{B}}$.

The operators $\hat{\mathbf{L}}$ and $\hat{\mathbf{M}}'$ are generators of the $SO(3)$ group. Their direct sum gives the operators $\hat{\mathbf{A}}$ and $\hat{\mathbf{B}}$ which are generators of the $SO(4)$ group. This is the symmetry group of the Coulomb problem that is responsible for the "accidental degeneracy" of the energy levels in the hydrogenic atom.

Problem 4.36

This problem focuses on the spectrum-generating group of the hydrogenic atom. Consider again the Coulomb problem as defined above in Problem 4.33, but now focus on the equation for the radial component, $w(r)$ of the wave function, $\psi(\mathbf{r}) = \frac{1}{r} w(r) Y_l^m(\mathbf{n})$, which satisfies the following equation:

$$\left(r\hat{D}^2 + \frac{A}{r} + B + Cr \right) w(r) = 0,$$

where we defined, $A = -l(l+1)$, $B = 2me^2/\hbar^2$, and $C = 2mE/\hbar^2$, and we also introduced an operator notation for the derivative, $\hat{D} \equiv \frac{d}{dr}$. Define the following operators:

$$\hat{\Gamma}_x = \frac{1}{2} \left(r\hat{D}^2 + \frac{A}{r} + r \right),$$

$$\hat{\Gamma}_y = r\hat{D}, \text{ and}$$

$$\hat{\Gamma}_z = \frac{1}{2} \left(r\hat{D}^2 + \frac{A}{r} - r \right),$$

and prove that they satisfy a set of closed commutation relations. Derive those commutation relations explicitly.

Solution

First we note the following commutation relations:

$$[r\hat{D}, r] = r, \quad [r\hat{D}, r\hat{D}^2] = -r\hat{D}^2, \quad [r, r\hat{D}^2] = -2r\hat{D}, \quad \left[\frac{A}{r}, r\hat{D}\right] = \frac{A}{r},$$

which can be easily shown by operating them on a function $f(r)$. Then we have

$$[\hat{\Gamma}_x, \hat{\Gamma}_y] = \frac{1}{2} \left[r\hat{D}^2 + \frac{A}{r} + r, r\hat{D} \right] = \frac{1}{2} \left(r\hat{D}^2 + \frac{A}{r} - r \right) = \hat{\Gamma}_z,$$

$$[\hat{\Gamma}_y, \hat{\Gamma}_z] = \frac{1}{2} \left[r\hat{D}, r\hat{D}^2 + \frac{A}{r} - r \right] = \frac{1}{2} \left(-r\hat{D}^2 - \frac{A}{r} - r \right) = -\hat{\Gamma}_x,$$

$$[\hat{\Gamma}_z, \hat{\Gamma}_x] = \frac{1}{4} \left[r\hat{D}^2 + \frac{A}{r} - r, r\hat{D}^2 + \frac{A}{r} + r \right] = \frac{1}{4} \left([r\hat{D}^2, r] - [r, r\hat{D}^2] \right) = r\hat{D} = \hat{\Gamma}_y.$$

Thus, $\hat{\Gamma}_{x,y,z}$ indeed satisfy a set of closed commutation relations.

Problem 4.37

Express the radial Schrödinger equation for the Coulomb potential through the Γ-operators defined in the previous problem. Note (do not attempt to prove it) that the operators $\hat{\Gamma}_\pm = \hat{\Gamma}_x \pm i\hat{\Gamma}_y$ act "vertically" between the energy levels with different energies (it is in this sense that they "generate" the spectrum of the Coulomb problem), as opposed to the operators $\hat{A}$ and $\hat{B}$ that act "horizontally" connecting different states with the same energy.

Solution

We note that

$$\hat{\Gamma}_x + \hat{\Gamma}_z = r\hat{D}^2 + \frac{A}{r}, \quad \hat{\Gamma}_x - \hat{\Gamma}_z = r.$$

So, the radial Schrödinger equation becomes

$$\left(\hat{\Gamma}_x + \hat{\Gamma}_z\right) + B + C\left(\hat{\Gamma}_x - \hat{\Gamma}_z\right) w(r) = 0.$$

4.4 Systems with axial symmetry

Problem 4.38

Determine the discrete spectrum of a particle in a two-dimensional potential well of the form

a) $U(\rho) = -\alpha\delta(\rho - a)$;
b) $U(\rho) = -U_0$ for $\rho < a$ and $U(\rho) = 0$ for $\rho > a$;

with angular momentum projection $m = 0$. Consider the case of a shallow well. Compare with one-dimensional motion.

Solution

The two-dimensional Schrödinger equation for states with $m = 0$ and energy $E_{n_\rho 0}$ has the form (μ is the mass of the particle)

$$\left[\frac{d^2}{d\rho^2} + \frac{1}{\rho}\frac{d}{d\rho} + \frac{2\mu}{\hbar^2}\left(E_{n_\rho 0} - U(\rho)\right)\right]\psi_{n_\rho 0}(\rho) = 0. \tag{1}$$

a) The solution of Eq. (1) for the δ-potential in the case of $E_{n_\rho 0} = -\hbar^2\kappa^2/2m < 0$, that is finite at the origin and equal to zero at infinity, has the form $\psi(\rho) = c_1 I_0(\kappa\rho)$ for $\rho < a$ and $\psi(\rho) = c_2 K_0(\kappa\rho)$ for $\rho > a$ ($I_0(x)$ and $K_0(x)$ are the modified Bessel functions). The matching conditions at the point $\rho = a$ (the same as in Problem 2.6) give us the relation

$$x\left[K_0'(x)I_0(x) - K_0(x)I_0'(x)\right] = -\frac{2\mu\alpha a}{\hbar^2}K_0(x)I_0(x), \tag{2}$$

with $x = \kappa a$. This determines the spectrum. Using the Wronskian $W(I_0(x), K_0(x)) = -1/x$, we write (2) in the form

$$K_0(\kappa a)I_0(\kappa a) = \frac{1}{\xi}, \quad \xi \equiv \frac{2\mu\alpha a}{\hbar^2}. \tag{3}$$

Let us first consider the case where the level has a small energy ($\kappa a \ll 1$). Using the known asymptotes $I_0(z) \approx 1 + z^2/4$ and $K_0(z) \approx \ln(2/\gamma z)$ as $z \to 0$, we obtain[50]

$$\ln\frac{2}{\gamma a \kappa} \approx \frac{1}{\xi}, \quad \text{or} \quad E_{00} \sim -\frac{2\hbar^2}{\gamma^2 \mu a^2} e^{-2/\xi}. \tag{4}$$

See that $\xi \ll 1$, i.e., this level with a small binding energy could only exist in a shallow well. This means that in the δ-well there is only one level (with m=0), just as in the one-dimensional case.

With an increase in α, the level deepens, and for $\xi \gg 1$ its energy is $E_{00} \approx -\mu\alpha^2/2\hbar^2$, which is easy to obtain from (3) using the asymptotic expressions for the functions $K_0(z)$ and $I_0(z)$ for $z \to \infty$.

b) For a rectangular potential well, the solution of Eq. (1) is

$$\psi_{n_\rho 0}(\rho) = \begin{cases} c_1 J_0(k\rho), & \rho < a, \quad k = \sqrt{2\mu(U_0 - |E_{n_\rho 0}|)/\hbar^2}, \\ c_2 K_0(\kappa a), & \rho > a, \quad \kappa = \sqrt{-2\mu E_{n_\rho 0}/\hbar^2}. \end{cases} \tag{5}$$

From the continuity of $\psi_{n_\rho 0}(\rho)$ and $\psi'_{n_\rho 0}(\rho)$ at $\rho = a$, it follows that

$$\kappa J_0(ka)K'_0(\kappa a) = k J'_0(ka)K_0(\kappa a), \tag{6}$$

which is the equation for the levels with $m = 0$.

In the case of the shallow well, $\xi = \mu\alpha^2 U_0/\hbar^2 \ll 1$, the arguments of the cylindrical functions in (6) are small. From the $x \ll 1$ asymptotes of J_0 and K_0, we can simplify (6):

$$J_0(x) \approx 1, \quad J'_0(x) = -\frac{x}{2}, \quad K_0(x) \approx \ln\left(\frac{2}{\gamma x}\right), \quad K'_0(x) \approx -\frac{1}{x}$$

$$\mu\left(U_0 - |E_{n_\rho 0}|\right)\frac{a^2}{\hbar^2}\ln\sqrt{\frac{2\hbar^2}{\mu\gamma^2 a^2 |E_{n_\rho 0}|}} \approx 1. \tag{7}$$

This equation has only one root,

$$E_{00} \sim -\frac{2\hbar^2}{\gamma^2 \mu a^2}\exp\left(-\frac{2\hbar^2}{\mu a^2 U_0}\right) = -\frac{2U_0}{\gamma^2 \xi}\exp\left(-\frac{2}{\xi}\right), \tag{8}$$

which is easy to find if we remember that $|E_{n_\rho 0}| \ll U_0$ and neglect $|E_{n_\rho 0}|$ in comparison to U_0.

---

[50] Here $\gamma = 1.781\ldots$ is the Euler constant.

In the shallow two-dimensional well, just as in the similar one-dimensional well, there is always one bound state. But the depth of the level occurrence, as is seen from (4) and (8), is exponentially small in comparison with the well depth.

Problem 4.39

Determine the energy spectrum of the bound states of a particle with an arbitrary value m of the angular-momentum projection for the following two-dimensional potentials:

a) $U(\rho) = -\dfrac{\alpha}{\rho}$;

b) $U = 0$ for $\rho < a$ and $U(\rho) = \infty$ for $\rho > a$.

What is the degree of degeneracy of the levels?

Solution

The Schrödinger equation has the form $(\psi_{n_\rho m} = R_{n_\rho |m|} e^{im\varphi}/\sqrt{2\pi})$

$$\left[\frac{d^2}{d\rho^2} + \frac{1}{\rho}\frac{d}{d\rho} - \frac{m^2}{\rho^2} + \frac{2\mu}{\hbar^2}\left(E_{n_\rho|m|} - U(\rho)\right)\right] R_{n_\rho|m|}(\rho) = 0. \tag{1}$$

a) For $U = -\alpha/\rho$, Eq. (1) has a form similar to Eq. (IV.6), and differs only by a change of $l + 1/2 \to |m|$. So, taking into account the known expression for the energy levels in the three-dimensional Coulomb potential and making the substitution $l + 1/2 \to |m|$, we obtain

$$E_{n_\rho|m|} = -\frac{\mu\alpha^2}{2\hbar^2(n_\rho + |m| + 1/2)^2}. \tag{2}$$

We see that in the two-dimensional field, $U = -\alpha/\rho$, as well as in the three-dimensional one, $U = -\alpha/r$, a degeneracy appears because the energy depends only on the combination $n_\rho + |m|$ of the quantum numbers n_ρ and m. If we introduce the quantum number, $N = n_\rho + |m| + 1$, the analog of the principal quantum number n in the Coulomb field, then we can write expression (2) in the form

$$E_N = -\frac{\mu\alpha^2}{2\hbar^2}\left(N - \frac{1}{2}\right)^{-2}; \quad N = 1, 2, \ldots. \tag{3}$$

This level has the degeneracy $g(N) = 2N - 1$.

b) The solution of Eq. (1), in the case of an infinitely deep rectangular potential well for $\rho < a$, that is finite at the origin, has the form

$$R_{n_\rho|m|} = cJ_{|m|}(\kappa\rho), \text{ where } \kappa = \left(\frac{2\mu E_{n_\rho|m|}}{\hbar^2}\right)^{1/2}.$$

The condition that the wavefunction vanishes at the wall determines the energy spectrum:

$$E_{n_\rho|m|} = \frac{\hbar^2}{2\mu a^2}\alpha^2_{n_\rho+1,m}, \qquad (4)$$

where $\alpha_{k,m} > 0$ is the kth zero of the Bessel function $J_m(\alpha)$. In particular, taking into account the values $\alpha_{10} \approx 2.40$ and $\alpha_{11} \approx 3.83$, we obtain the ground-state energy $E_{00} \approx 2.88\hbar^2/\mu a^2$ (with $m = 0$) and the energy $E_{01} \approx 7.33\hbar^2/\mu a^2$ of the lowest level with $|m| = 1$. Finally, note that levels with $m = 0$ are non-degenerate, while those with $|m| \neq 0$ are doubly degenerate.

Problem 4.40

Find the Green function of a free particle in two dimensions with the energy $E < 0$ and which decreases as $\rho \to \infty$.

Solution

The Green function obeys the equation

$$(\hat{H} - E)G_E \equiv \frac{\hbar^2}{2\mu}(-\Delta_2 + \kappa^2)G_E(\boldsymbol{\rho}, \boldsymbol{\rho}') = \delta(\boldsymbol{\rho} - \boldsymbol{\rho}'), \qquad (1)$$

with $\kappa = \sqrt{-2\mu E/\hbar^2} > 0$. From symmetry, we see that it is a function of the form $G_E = f(|\boldsymbol{\rho} - \boldsymbol{\rho}'|)$. Equation (1) for $\boldsymbol{\rho} \neq \boldsymbol{\rho}'$ and its solution take the form

$$\left(\frac{d^2}{d\rho^2} + \frac{1}{\rho}\frac{d}{d\rho} - \kappa^2\right)f(\rho) = 0, \quad f(\rho) = cK_0(\kappa\rho), \qquad (2)$$

where $K_0(z)$ is the Macdonald function (the other independent solution, $\propto I_0(\kappa\rho)$, increases exponentially as $\rho \to \infty$).

To determine the value of c, we integrate (1) over a circle of a small radius ε centered at the point $\boldsymbol{\rho} = \boldsymbol{\rho}'$. On the right side we get unity. Integration of the second term on the left side (with κ^2) gives zero for $\varepsilon \to 0$. We transform the integral of the first term using the Gauss's law:

$$\int \Delta G_E dV = \oint \nabla G_E d\mathbf{s}.$$

In this two-dimensional case we obtain $dV \equiv dS$, $d\mathbf{s} \equiv d\mathbf{l} = \mathbf{n}dl$.[51] Since $K_0(x) \approx \ln(2/\gamma x)$ as $x \to 0$, then $\nabla K_0(\kappa\rho) \approx -\boldsymbol{\rho}/\rho^2$ as $\rho \to 0$. Integration gives the relation $\pi c\hbar^2/\mu = 1$, so the final expression for G_E is

$$G_E(\boldsymbol{\rho}, \boldsymbol{\rho}') = \frac{\mu}{\pi\hbar^2}K_0(\kappa|\boldsymbol{\rho} - \boldsymbol{\rho}'|). \qquad (3)$$

---

[51] We should note that the unit vector $\mathbf{n}$ (the *outer normal*) is perpendicular to the integration contour. See that

$$d\mathbf{l} = \mathbf{n}dl = \frac{\tilde{\boldsymbol{\rho}}}{\tilde{\rho}}\tilde{\rho}d\varphi = \tilde{\boldsymbol{\rho}}d\varphi \quad (\tilde{\rho} = |\boldsymbol{\rho} - \boldsymbol{\rho}'| = \varepsilon).$$

Problem 4.41

The same as for the previous problem, but with $E > 0$. Determine the Green function $G_E^{(\pm)}$ that has the correct asymptotic behavior as $\rho \to \infty$.

Solution

Consideration similar to the previous problem gives the following form for the Green function:

$$G_E^{(\pm)}(\boldsymbol{\rho}, \boldsymbol{\rho}') = \pm \frac{i\mu}{2\hbar^2} H_0^{(1,2)}(\kappa|\boldsymbol{\rho} - \boldsymbol{\rho}'|), \tag{1}$$

where $\kappa = \sqrt{2\mu E/\hbar^2} > 0$, $H_0^{(1,2)}(z)$ are the Hankel functions.

Problem 4.42

Find the Green function $G_E(\varphi, \varphi')$ for a plane rotator (see Problem 3.2). Analyzing it as an analytic function of the complex variable E show that it has singular points and determine a relation between the locations of the poles on the E plane and the rotator energy levels. Compare with Problem 2.26.

Solution

The Green function obeys the equation

$$\hat{H} G_E \equiv -\frac{\hbar^2}{2I}\left(\frac{d^2}{d\varphi^2} + \kappa^2\right) G_E(\varphi, \varphi') = \delta(\varphi - \varphi'), \tag{1}$$

where $\kappa = \sqrt{2IE/\hbar^2}$. From symmetry, G_E is a function of the form $G_E = G_E(|\varphi - \varphi'|)$. From (1) we obtain, for $\varphi \neq \varphi'$:

$$G_E = C \cos(\kappa|\varphi - \varphi'| + \alpha).$$

The value $C = I/\kappa\hbar^2 \sin\alpha$ comes from matching conditions at the point $\varphi = \varphi'$ (compare with Problem 2.6), while the value of α comes from the continuity of the Green function and its derivative with respect to φ at the points $\varphi - \varphi' = \pm\pi$, which correspond to the same point. This value is $\alpha = -\pi\kappa$.

The Green function has the form

$$G_E(\varphi, \varphi') = -\frac{I}{\kappa\hbar^2 \sin\pi\kappa} \cos(\kappa|\varphi - \varphi'| - \pi\kappa). \tag{2}$$

It has poles at the points $\kappa = \pm m$, $m = 0, 1, 2, \ldots$. These are the points $E_m = \hbar^2 m^2/2I$ of the complex variable E. As expected, the positions of these poles coincide with the rotator energy levels.

Problem 4.43

Find the Green function $G_E(\mathbf{n}, \mathbf{n}')$ for a spherical rotator, where $\mathbf{n}$ is the unit vector directed along the rotator axis (see Problem 3.3). Solve the problem in two ways:

a) by directly solving the differential equation for the Green function
b) by expressing it as a series of relevant eigenfunctions

Solution

a) The Green function is the solution of the equation

$$(\hat{H} - E)G_E \equiv \left(\frac{\hbar^2}{2I}\hat{\mathbf{l}}^2 - E\right)G_E = \delta(\mathbf{n} - \mathbf{n}'). \qquad (1)$$

and is a function of the form $G_E \equiv G_E(\mathbf{n} \cdot \mathbf{n}')$; i.e., it depends only on the angle between the directions of vectors $\mathbf{n}$ and $\mathbf{n}'$. So, choosing the direction of the polar axis $\mathbf{n}_0$ along $\mathbf{n}'$ and introducing the notations $z = \mathbf{n} \cdot \mathbf{n}_0 = \cos\theta$ and $E \equiv \hbar^2 \nu(\nu+1)/2I$, we rewrite (1) in the form

$$(1-z^2)G_E''(z) - 2zG_E'(z) + \nu(\nu+1)G_E(z) = -\frac{2I}{\hbar^2}\delta(\mathbf{n} - \mathbf{n}_0). \qquad (2)$$

For $z \neq 1$ (i.e., $\theta \neq 0$), the right part of (2) is equal to zero. The solution of such an equation is

$$G_E(z) = c_1 \mathcal{P}_\nu(-z) + c_2 \mathcal{P}_\nu(z), \qquad (3)$$

where $\mathcal{P}_\nu(z)$ is a spherical Legendre function of the first kind. Since $\mathcal{P}_\nu(1) = 1$ and $\mathcal{P}_\nu(z) \to \infty$ as $z \to -1$, we should set $c_2 = 0$ in (3) and leave c_1 to be determined by the δ-functional term in (2). To find c_1, consider the limit $z \to 1$ and put $z \approx 1 - \theta^2/2$. Equation (2) becomes

$$\left[\frac{d^2}{d\theta^2} + \frac{1}{\theta}\frac{d}{d\theta} + \nu(\nu+1)\right]G_E = -\frac{2I}{\hbar^2}\delta(\mathbf{n} - \mathbf{n}_0), \qquad (4)$$

which is the equation for the free particle Green function in the two-dimensional case[52] (see Problem 4.40). We get[53]

$$\mathcal{P}_\nu(-z) \approx \frac{2}{\pi}\sin(\pi\nu)\ln\theta \quad \text{at } z = 1 - \frac{1}{2}\theta^2 \to 1,$$

---

[52] This correspondence is not accidental, since the operator $-\hat{\mathbf{l}}^2$ is the Laplace operator on the unit sphere, and a small part of a sphere is almost indistinguishable locally from the tangent plane. The vector $\mathbf{n} - \mathbf{n}_0$ is approximately perpendicular to $\mathbf{n}_0$, and $|\mathbf{n} - \mathbf{n}_0| \approx \theta$ is analogous to the variable ρ.

[53] This result can be obtained from the following equation

$$\mathcal{P}_\nu(\cos\theta) = \frac{2}{\pi}\int_0^\theta \frac{\cos(\nu + 1/2)\beta}{\sqrt{2(\cos\beta - \cos\theta)}}d\beta,$$

if we notice that for $\theta \to \pi$, the integral diverges in the upper limit, and calculate the diverging part.

Taking into account the result of Problem 4.40, we find $c_1 = -I/2\hbar^2 \sin \pi \nu$ and obtain the final expression for the Green function:

$$G_E(\mathbf{n}, \mathbf{n}') = -\frac{I}{2\hbar^2 \sin \pi \nu} P_\nu(-\mathbf{n} \cdot \mathbf{n}'), \quad \nu = -\frac{1}{2} + \left(\frac{2IE}{\hbar^2} + \frac{1}{4}\right)^{1/2}. \tag{5}$$

b) The general method of constructing the Green function as a series involving the eigenfunctions of the corresponding operator gives rise to the following expression in our case:

$$G_E(\mathbf{n}, \mathbf{n}') = \sum_{l,m} \frac{Y_{lm}(\mathbf{n}) Y_{lm}^*(\mathbf{n}')}{E_l - E}, \tag{6}$$

where $E_l = \hbar^2 l(l+1)/2I$ are the rotator energy levels, and Y_{lm} are the spherical harmonics that are the eigenfunctions of the Hamiltonian. Using the spherical harmonic addition theorem (see Eq. (III.6)), (6) may be written as

$$G_E = \frac{I}{2\pi\hbar^2} \sum_l \frac{(2l+1) P_l(\mathbf{n} \cdot \mathbf{n}')}{l(l+1) - \nu(\nu+1)}. \tag{7}$$

Using the relation

$$\sum_{l=0}^{\infty} \left(\frac{1}{\nu - l} - \frac{1}{\nu + l + 1}\right) P_l(z) = \frac{\pi}{\sin \pi \nu} P_\nu(-z), \tag{8}$$

we see that (6) and (7) coincide with (5).

5
Spin

1) The wavefunction of a particle with spin s has $(2s+1)$ components. In the s_z-representation it is described by the column

$$\Psi = \begin{pmatrix} \psi(\mathbf{r}, s) \\ \psi(\mathbf{r}, s-1) \\ \ldots \\ \psi(\mathbf{r}, \sigma) \\ \ldots \\ \psi(\mathbf{r}, -s) \end{pmatrix}. \quad (V.1)$$

$\psi(\mathbf{r}, \sigma)$ is the amplitude of a state with a given value of the spin z-projection σ, with $\sigma = s, s-1, \ldots, -s$. Physically, spin is an intrinsic angular momentum. In this representation the operators of spin-vector projections are the matrices $\hat{s}_x$, $\hat{s}_y$, $\hat{s}_z$, whose elements are given by the general formulae (III.9) and (III.10) with $l = s$ and $m = \sigma$. The matrix $\hat{s}_z$ is diagonal and $(\hat{s}_z)_{\sigma\sigma'} = \sigma\delta_{\sigma\sigma'}$.

For spin $s = 1/2$ these operators are expressed in terms of the Pauli matrices, $\hat{\mathbf{s}} = \frac{1}{2}\hat{\boldsymbol{\sigma}}$:

$$\hat{\sigma}_x = \begin{pmatrix} 0 & 1 \\ 1 & 0 \end{pmatrix}, \ \hat{\sigma}_y = \begin{pmatrix} 0 & -i \\ i & 0 \end{pmatrix}, \ \hat{\sigma}_z = \begin{pmatrix} 1 & 0 \\ 0 & -1 \end{pmatrix}. \quad (V.2)$$

The Pauli matrices have the following property:[54]

$$\hat{\sigma}_i \hat{\sigma}_k = \delta_{ik} + i\varepsilon_{ikl}\hat{\sigma}_l, \quad (V.3)$$

Here $i, k = 1, 2, 3$ and $\hat{\sigma}_1 \equiv \hat{\sigma}_x$, $\hat{\sigma}_2 \equiv \hat{\sigma}_y$, $\hat{\sigma}_3 \equiv \hat{\sigma}_z$. In this particular case of $s = 1/2$, the notation $\psi_1 \equiv \psi(\sigma = +1/2)$ and $\psi_2 \equiv \psi(\sigma = -1/2)$ for the spin components is often used, so that $\Psi = \begin{pmatrix} \psi_1 \\ \psi_2 \end{pmatrix}$. The inner product in spin space is given by

$$\langle \Phi | \Psi \rangle \equiv \Phi^* \Psi = \phi_1^* \psi_1 + \phi_2^* \psi_2.$$

---

[54] In particular, $\hat{\sigma}_x^2 = \hat{\sigma}_y^2 = \hat{\sigma}_z^2 = 1$, $\hat{\sigma}_x \hat{\sigma}_y = i\hat{\sigma}_z$ etc. From (V.3) we can see the anticommutativity of different Pauli matrices: $\hat{\sigma}_i \hat{\sigma}_k + \hat{\sigma}_k \hat{\sigma}_i = 0$ for $i \neq k$.

Note the characteristic property of spin $s = 1/2$ for an arbitrary spin state: it is always possible to find a vector **n** along which the spin projection has a definite value (equal to $\pm 1/2$ of course). We can write down the spin function, normalized to unity, for any state, in the form:

$$\Psi = \begin{pmatrix} \cos(\theta/2) \\ e^{i\varphi} \sin(\theta/2) \end{pmatrix}. \quad (V.4)$$

θ and φ ($0 \leq \theta \leq \pi$, $0 \leq \phi \leq 2\pi$) determine the polar and azimuthal angles of an axis along which the spin projection is $s_n = +1/2$. See Problems 5.3 and 5.9.

2) It is often necessary to deal with the *spin density matrix* $\hat{\rho}$. The elements $\rho_{\sigma\sigma'}$ of this matrix are equal to an average of the bilinear combination of spin wavefunctions $\psi(\sigma)\psi^*(\sigma')$. The averaging is performed over an ensemble of possible states,[55] labelled by a parameter λ,

$$\rho_{\sigma\sigma'} = \overline{\psi(\sigma, \lambda)\psi^*(\sigma', \lambda)}. \quad (V.5)$$

The mean value of operator $\hat{f}$ in the spin space is given by

$$\bar{f} = \text{Tr}\,(\hat{\rho}\hat{f}) = \text{Tr}\,(\hat{f}\hat{\rho}). \quad (V.6)$$

A spin $s = 1/2$ density matrix can be written as

$$\hat{\rho} = \frac{1}{2}(1 + \mathbf{P} \cdot \hat{\boldsymbol{\sigma}}). \quad (V.7)$$

$\mathbf{P} = 2\bar{\mathbf{s}}$ is known as the *polarization vector*. The case with $\mathbf{P} = 0$ corresponds to an entirely unpolarized collection of states. A state in the case $|\mathbf{P}| = 1$ is *pure* and is described by the spin function (V.4) with the choice of the corresponding axis **n** along the vector **P**.

5.1 Spin $s = 1/2$

Problem 5.1

For a particle with the spin $s = 1/2$, find the eigenvalues and eigenfunctions of the operators $\hat{s}_x$, $\hat{s}_y$, $\hat{s}_z$.

---

[55] The density matrix is normalized by the condition $\text{Tr}\hat{\rho} = 1$. Its diagonal elements, $\rho_{\sigma\sigma}$, determine the probabilities of measuring the corresponding values of the spin z-projection, σ. In the case $\hat{\rho}^2 = \hat{\rho}$, the density matrix has the form $\rho_{\sigma\sigma'} = \psi(\sigma)\psi^*(\sigma')$ and characterizes a *pure* state that is described by the wavefunction $\psi(\sigma)$.

Solution

The eigenfunctions $\psi_{s_x} = \begin{pmatrix} a \\ b \end{pmatrix}$ and the eigenvalues s_x of the operator $\hat{s}_x = \frac{1}{2}\hat{\sigma}_x$ are found from the equation $\hat{s}_x \psi_{s_x} = s_x \psi_{s_x}$:

$$\frac{1}{2}\begin{pmatrix} 0 & 1 \\ 1 & 0 \end{pmatrix}\begin{pmatrix} a \\ b \end{pmatrix} = \frac{1}{2}\begin{pmatrix} b \\ a \end{pmatrix} = s_x \begin{pmatrix} a \\ b \end{pmatrix}.$$

So $b = 2s_x a$, $a = 2s_x b$. This system has non-trivial solutions when $4s_x^2 = 1$. We find that $a = b$ for $s_x = +1/2$ and $a = -b$ for $s_x = -1/2$. The eigenfunctions, ψ_{s_x}, that are normalized to unity so that $\langle \psi_{s_x} | \psi_{s_x} \rangle = |a|^2 + |b|^2 = 1$ are:

$$\psi_{s_x=+1/2} = \frac{1}{\sqrt{2}}\begin{pmatrix} 1 \\ 1 \end{pmatrix}, \quad \psi_{s_x=-1/2} = \frac{1}{\sqrt{2}}\begin{pmatrix} 1 \\ -1 \end{pmatrix}.$$

Analagously,

$$\psi_{s_y=+1/2} = \frac{1}{\sqrt{2}}\begin{pmatrix} 1 \\ i \end{pmatrix}, \quad \psi_{s_y=-1/2} = \frac{1}{\sqrt{2}}\begin{pmatrix} 1 \\ -i \end{pmatrix},$$

$$\psi_{s_z=+1/2} = \begin{pmatrix} 1 \\ 0 \end{pmatrix}, \quad \psi_{s_z=-1/2} = \begin{pmatrix} 0 \\ 1 \end{pmatrix}.$$

Problem 5.2

Determine explicitly the form of the operator $\hat{s}_\mathbf{n}$, corresponding to the spin projection in the direction determined by a unit vector $\mathbf{n}$. For a state with a definite value of the spin z-projection, determine the value of $\bar{s}_\mathbf{n}$ and find the probability of the spin $\pm 1/2$ n-projection.

Solution

The spin operator $\hat{\mathbf{s}} = \frac{1}{2}\hat{\boldsymbol{\sigma}}$ is a vector (or, more precisely, pseudo-vector) operator. The n-projection operator $\hat{s}_\mathbf{n}$ may be expressed in terms of its components $\hat{s}_x$, $\hat{s}_y$, $\hat{s}_z$ in the same manner as for an ordinary (non-operator) vectors, $i.e.$

$$\hat{s}_\mathbf{n} = \mathbf{n} \cdot \hat{\mathbf{s}} = \frac{1}{2}\mathbf{n} \cdot \hat{\boldsymbol{\sigma}} = \frac{1}{2}(\sin\theta\cos\varphi \cdot \hat{\sigma}_x + \sin\theta\sin\varphi \cdot \hat{\sigma}_y + \cos\theta \cdot \hat{\sigma}_z),$$

where θ, φ are the polar and azimuthal angles of the $\mathbf{n}$-direction. Using the explicit form of the Pauli matrices, Eq. (V.2), we obtain

$$\hat{s}_\mathbf{n} = \frac{1}{2}\begin{pmatrix} \cos\theta & e^{-i\varphi}\sin\theta \\ e^{i\varphi}\sin\theta & -\cos\theta \end{pmatrix}. \qquad (1)$$

The value of $\bar{s}_\mathbf{n}$ in the state $\psi_{s_z=+1/2} = \begin{pmatrix} 1 \\ 0 \end{pmatrix}$ is equal to

168 Exploring Quantum Mechanics

$$\bar{s}_n = \langle \psi | \hat{s}_n | \psi \rangle = \frac{1}{2} \begin{pmatrix} 1 & 0 \end{pmatrix} \begin{pmatrix} \cos\theta & e^{-i\varphi}\sin\theta \\ e^{i\varphi}\sin\theta & -\cos\theta \end{pmatrix} \begin{pmatrix} 1 \\ 0 \end{pmatrix} = \frac{1}{2}\cos\theta.$$

Analagously, we find $\bar{s}_n = -\frac{1}{2}\cos\theta$ for a state with $s_z = -1/2$. Note that the relation $\bar{s}_n = s_z \cos\theta$ is analogous to the result of Problem 3.11.

Let us denote w_+ for the probability of the value $s_n = +1/2$ and $w_- = 1 - w_+$ for the probability of $s_n = -1/2$. Taking into account that $\bar{s}_n = s_z \cos\theta$, we find that

$$s_z \cos\theta = \bar{s}_n = w_+ \cdot \frac{1}{2} + w_- \cdot \left(-\frac{1}{2}\right) = \frac{1}{2}[2w_+ - 1],$$

$$w_+ = \frac{1}{2}(1 + 2s_z \cos\theta), \quad w_- = \frac{1}{2}(1 - 2s_z \cos\theta). \tag{2}$$

Problem 5.3

For spin $s = 1/2$, the normalized wavefunction of the most general spin state is[56]
$\Psi = \begin{pmatrix} \cos\alpha \\ e^{i\beta}\sin\alpha \end{pmatrix}$, with $0 \leq \alpha \leq \frac{\pi}{2}$, $0 \leq \beta \leq 2\pi$. Find the polar and azimuthal angles of an axis **n** along which spin projection has a definite value equal to $+1/2$.

Solve Problem 5.1 using this result.

Solution

We first find the wavefunction $\Psi_{s_n=1/2} = \begin{pmatrix} a \\ b \end{pmatrix}$ of the spin n-projection operator. Using the form of the operator $\hat{s}_n$ determined in the previous problem and the eigenvalue equation

$$\hat{s}_n \Psi_{s_n=1/2} = \frac{1}{2}\Psi_{s_n=1/2},$$

we obtain the equation $a\sin(\theta/2) = be^{-i\varphi}\cos(\theta/2)$. So if we choose (for normalization to unity) $a = \cos(\theta/2)$, we find $b = e^{i\varphi}\sin(\theta/2)$. The spin function $\Psi_{s_n=1/2}$ takes the form given in the problem statement. Here $\theta = 2\alpha$ and $\varphi = \beta$ determine the polar and azimuth angles.

Choosing $\theta = 2\alpha = \pi/2$ and $\varphi = \beta = 0$, we find the eigenfunction $\psi_{s_x=1/2}$. In the case $\theta = \pi/2$ and $\varphi = \pi$ we obtain the eigenfunction $\psi_{s_x=-1/2}$, and so on; compare with the results of Problem 5.1.

Problem 5.4

An arbitrary rank 2 square matrix, $\hat{A}$, may be expanded in terms of the full system of matrices $\hat{1}$, $\hat{\sigma}_x$, $\hat{\sigma}_y$, $\hat{\sigma}_z$:

$$\hat{A} = a_0\hat{1} + a_x\hat{\sigma}_x + a_y\hat{\sigma}_y + a_z\hat{\sigma}_z \equiv a_0 + \mathbf{a} \cdot \hat{\boldsymbol{\sigma}}. \tag{1}$$

---

[56] Except for phase factor $e^{i\gamma}$.

Prove that the coefficients are given by $2a_0 = \text{Tr } \hat{A}$, $2\mathbf{a} = \text{Tr } (\hat{\boldsymbol{\sigma}} \hat{A})$.

Solution

If we take the *trace* of the both sides of (1) and use $\text{Tr}\hat{\sigma}_i = 0$, $\text{Tr}\hat{1} = 2$, we obtain

$$a_0 = \frac{1}{2}\text{Tr}\hat{A}.$$

If we multiply both sides of (1) by $\hat{\sigma}_k$ on the right and calculate the trace, then we obtain

$$a_k = \frac{1}{2}\text{Tr}\left(\hat{A}\hat{\sigma}_k\right) = \frac{1}{2}\text{Tr}\left(\hat{\sigma}_k\hat{A}\right).$$

Problem 5.5

Simplify the expression $(\mathbf{a} \cdot \hat{\boldsymbol{\sigma}})^n$, where $\mathbf{a}$ is a real numerical vector, n is an integer, and $\hat{\boldsymbol{\sigma}}$ are the Pauli matrices.

Solution

$(\mathbf{a} \cdot \hat{\boldsymbol{\sigma}})^n = |a|^n\hat{1}$ if n is even and $(\mathbf{a} \cdot \hat{\boldsymbol{\sigma}})^n = |a|^{n-1}(\mathbf{a} \cdot \hat{\boldsymbol{\sigma}})$ if n is odd.

Problem 5.6

Find

a) the eigenvalues and eigenfunctions of operator function $\hat{f} = a + \mathbf{b} \cdot \hat{\boldsymbol{\sigma}}$;
b) the explicit form for the operator $\hat{F} = F(a + \mathbf{b} \cdot \hat{\boldsymbol{\sigma}})$.
a and $\mathbf{b}$ are real scalar and vector parameters, and $F(z)$ is some function.

As an illustration, (c) consider the operator $\hat{R}(\boldsymbol{\varphi}_0) = \exp(i\boldsymbol{\varphi}_0 \cdot \hat{\boldsymbol{\sigma}}/2)$ that describes the transformation of the spin wavefunction, $\Psi' = \hat{R}(\boldsymbol{\phi}_0)\Psi$, under rotation of the coordinate system through the angle $\boldsymbol{\phi}_0$. Find the eigenfunctions, $\Psi_{s_n=\pm 1/2}$, of the spin projection operator along the vector $\mathbf{n}$. Compare with Problem 5.3.

Solution

a) The operator has only two eigenvalues, equal to $f_{1,\,2} = a \pm b$. The corresponding eigenfunctions are given in Problem 5.3 with $\mathbf{n} = \pm \mathbf{b}/b$.

b) As a result of Problem 1.22, the form of the operator $\hat{F} = F(\hat{f})$ is

$$\hat{F} = \frac{1}{2}\left[F(a+b) + F(a-b) + (F(a+b) - F(a-b))\frac{\mathbf{b} \cdot \hat{\boldsymbol{\sigma}}}{b}\right].$$

c) In particular, $\hat{R}(\boldsymbol{\varphi}_0) = \cos(\varphi_0/2) + i\sin(\varphi_0/2) \cdot \frac{1}{\varphi_0}(\boldsymbol{\varphi}_0 \cdot \hat{\boldsymbol{\sigma}})$. To transform the spin functions $\Psi_{s_z=1/2}$ into the form $\Psi_{s_n=1/2}$ by rotating the coordinate system, we should choose $\boldsymbol{\varphi}_0$ in the form $\boldsymbol{\varphi}_0 = (\theta\sin\varphi, -\theta\cos\varphi, 0)$ (see Problem 3.24), where θ and φ are the polar and azimuthal angles of the vector $\mathbf{n}$. In this case, $\hat{R}(\phi_0) = \cos(\theta/2) + i\sin(\theta/2)(\sin\varphi \cdot \hat{\sigma}_x - \cos\varphi \cdot \hat{\sigma}_y)$, so we have

$$\Psi_{s_n=+1/2} = \hat{R}\begin{pmatrix}1\\0\end{pmatrix} = \begin{pmatrix}\cos(\theta/2)\\e^{i\varphi}\sin(\theta/2)\end{pmatrix},$$

in accordance with Problem 5.3.

Problem 5.7

Using the transformation law for the spin functions $\Psi = \begin{pmatrix}\psi_1\\\psi_2\end{pmatrix}$ under rotation of the coordinate system, $\Psi' = \hat{R}(\boldsymbol{\varphi}_0)\Psi$ (see the previous problem), show that the quantities

$$S = \Phi^*\Psi = \varphi_1^*\psi_1 + \varphi_2^*\psi_2$$

do not change, i.e., are scalars, while the quantities of the form

$$\mathbf{V} = \Phi^*\hat{\boldsymbol{\sigma}}\Psi \quad \left(\text{or } V_i = \sum_{\alpha,\beta}\phi_\alpha^*(\hat{\sigma}_i)_{\alpha\beta}\psi_\beta\right)$$

transform as vectors.

Solution
From the transformation of the form ($\mathbf{n}_0 = \boldsymbol{\varphi}_0/\varphi_0$)

$$\Psi' \equiv \begin{pmatrix}\psi_1'\\\psi_2'\end{pmatrix} = \exp\left(\frac{i}{2}\boldsymbol{\varphi}_0\cdot\hat{\boldsymbol{\sigma}}\right)\Psi = \left(\cos\frac{\varphi_0}{2} + i\sin\frac{\varphi_0}{2}\mathbf{n}_0\cdot\hat{\boldsymbol{\sigma}}\right)\Psi, \quad (1)$$

we have the transformation law for a complex conjugate spin function $\Phi^* = (\varphi_1^*, \varphi_2^*)$:

$$\Phi^{*\prime} \equiv (\varphi_1^{*\prime}, \varphi_2^{*\prime}) = \Phi^*\left(\cos\frac{\varphi_0}{2} - i\sin\frac{\varphi_0}{2}\mathbf{n}_0\cdot\hat{\boldsymbol{\sigma}}\right). \quad (2)$$

It follows that $\Phi^{*\prime}\Psi' = \Phi^*\Psi$.

Using (1), (2), and relation (V.3), we obtain

$$\mathbf{V}' \equiv \Phi^{*\prime}\hat{\boldsymbol{\sigma}}\Psi' = \cos\varphi_0\cdot\mathbf{V} - \sin\varphi_0\cdot\mathbf{n}_0\times\mathbf{V} + 2\sin^2\frac{\varphi_0}{2}\cdot\mathbf{n}_0(\mathbf{n}_0\cdot\mathbf{V}), \quad (3)$$

which is the vector transformation law for a coordinate system rotation by the angle $\boldsymbol{\phi}_0$. In particular for $\mathbf{n}_0 = (0,0,1)$ (rotation about z), from (3) it follows that $V_z' = V_z$ and

$$V_x' = \cos\varphi_0\cdot V_x + \sin\varphi_0\cdot V_y, \quad V_y' = -\sin\varphi_0\cdot V_x + \cos\varphi_0\cdot V_y. \quad (4)$$

Problem 5.8

Consider the matrix element of the form[57]

$$\langle \Psi^{(2)}|\hat{A}|\Phi^{(1)}\rangle\langle \Phi^{(2)}|\hat{B}|\Psi^{(1)}\rangle \equiv \psi_\alpha^{(2)*} A_{\alpha\beta}\varphi_\beta^{(1)} \varphi_\gamma^{(2)*} B_{\gamma\delta}\psi_\delta^{(1)}.$$

$\hat{A}$, $\hat{B}$ are some 2×2 matrices, and $\Psi^{(1,2)}$, $\Phi^{(1,2)}$ are spin functions. Show that it is possible to rearrange the spin functions:

$$\sum_{i,k=0}^{3} C_{ik}\langle \Psi^{(2)}|\hat{\sigma}_i|\Psi^{(1)}\rangle\langle \Phi^{(2)}|\hat{\sigma}_k|\Phi^{(1)}\rangle.$$

For ease of notation, we put $\hat{1} \equiv \hat{\sigma}_0$. Also, determine the form of the scalar matrix elements:

$$\langle \Psi^{(2)}|\Phi^{(1)}\rangle\langle \Phi^{(2)}|\Psi^{(1)}\rangle \text{ and } \langle \Psi^{(2)}|\hat{\boldsymbol{\sigma}}|\Phi^{(1)}\rangle \cdot \langle \Phi^{(2)}|\hat{\boldsymbol{\sigma}}|\Psi^{(1)}\rangle.$$

Solution

We can regard the expression of the form $A_{\alpha\beta}B_{\gamma\delta}$ as a matrix $M_{\alpha\delta}(\beta,\gamma)$ that depends on β and γ as parameters, and write it in the form of expansion

$$M_{\alpha\delta}(\beta,\gamma) = \sum_{i=0}^{3} C_i(\gamma\beta)(\hat{\sigma}_i)_{\alpha\delta}.$$

The coefficients $C_i(\gamma\beta)$ are determined by the result of Problem 5.4. We can expand C_i in an analogous way. So we obtain

$$A_{\alpha\beta}B_{\gamma\delta} = \sum_{i,k=0}^{3} C_{ik}(\hat{\sigma}_i)_{\alpha\delta}(\hat{\sigma}_k)_{\gamma\beta}, C_{ik} = \frac{1}{4}\sum_{\alpha\beta\gamma\delta}(\hat{\sigma}_i)_{\delta\alpha}(\hat{\sigma}_k)_{\beta\gamma}A_{\alpha\beta}B_{\gamma\delta}. \quad (1)$$

Using (1) with $A_{\alpha\beta} = \delta_{\alpha\beta}, B_{\gamma\delta} = \delta_{\gamma\delta}$ and $C_{ik} = \frac{1}{2}\delta_{ik}$, it is easy to obtain

$$\langle \Psi^{(2)}|\Phi^{(1)}\rangle\langle \Phi^{(2)}|\Psi^{(1)}\rangle = \frac{1}{2}\{\langle \Psi^{(2)}|\Psi^{(1)}\rangle\langle \Phi^{(2)}|\Phi^{(1)}\rangle$$
$$+ \langle \Psi^{(2)}|\hat{\boldsymbol{\sigma}}|\Psi^{(1)}\rangle \cdot \langle \Phi^{(2)}|\hat{\boldsymbol{\sigma}}|\Phi^{(1)}\rangle\}.$$

Using $A_{\alpha\beta} = (\hat{\sigma}_l)_{\alpha\beta}, B_{\gamma\delta} = (\hat{\sigma}_l)_{\gamma\beta}, C_{00} = 3/2, C_{ik} = -(1/2)\delta_{ik}$ for $i, k = 1, 2, 3$, and $C_{ik} = 0$ if $i \neq k$, we obtain

$$\langle \Psi^{(2)}|\hat{\boldsymbol{\sigma}}|\Phi^{(1)}\rangle \cdot \langle \Phi^{(2)}|\hat{\boldsymbol{\sigma}}|\Psi^{(1)}\rangle = \frac{1}{2}\{3\langle \Psi^{(2)}|\Psi^{(1)}\rangle\langle \Phi^{(2)}|\Phi^{(1)}\rangle$$
$$- \langle \Psi^{(2)}|\hat{\boldsymbol{\sigma}}|\Psi^{(1)}\rangle \cdot \langle \Phi^{(2)}|\hat{\boldsymbol{\sigma}}|\Phi^{(1)}\rangle\}.$$

---

[57] Carefully note that the indices 1 and 2 enumerate the different spinors, not the different components of the same spin function.

Pay attention to the *scalar* nature of all the terms in these relations. Compare with Problem 5.7.

Problem 5.9

Determine the projection operators $\hat{P}_{s_z=\pm 1/2}$ that project onto states with a definite value of the spin z-projection.

Generalize the form of $\hat{P}_{s_n=\pm 1/2}$ to a state with a definite spin projection along the direction determined by the unit vector $\mathbf{n}$. Using these projection operators, find the spin function $\Psi_{s_n=\pm 1/2}$ and compare with Eqn. (V.4) and Problem 5.3.

Solution

The desired projection operators follow from 1.35:

$$\hat{P}_{s_z=\pm 1/2} = \frac{1}{2}(1 \pm \hat{\sigma}_z), \quad \text{and} \quad \hat{P}_{s_n=\pm 1/2} = \frac{1}{2}(1 \pm \mathbf{n} \cdot \hat{\boldsymbol{\sigma}}).$$

If we act with the operator $\hat{P}_{s_n=1/2}$ on an arbitrary spin function Ψ, we obtain the eigenfunction of the operator $\hat{s}_n$ that corresponds to the eigenvalue $s_n = 1/2$. For convenience, we take $\Psi = \begin{pmatrix} 1 \\ 0 \end{pmatrix}$, and we find

$$\Psi_{s_n=1/2} = C \hat{P}_{s_n=1/2} \Psi = \frac{C}{2}(1 + \mathbf{n} \cdot \hat{\boldsymbol{\sigma}}) \begin{pmatrix} 1 \\ 0 \end{pmatrix} = \begin{pmatrix} \cos(\theta/2) \\ e^{i\varphi} \sin(\theta/2) \end{pmatrix} \quad (1)$$

where θ and φ are the polar and azimuth angles of the $\mathbf{n}$-direction. We have choosen $C = \cos^{-1}(\theta/2)$ to normalize the spin function to unity.

Problem 5.10

For a system of two identical spins with $s = 1/2$, find the eigenfunctions Ψ_{SS_z} of the total spin square and its z-projection operators. Indicate the characteristic symmetry of these functions with respect to the interchange of particles, and the symmetry's dependence on S.

Solution

The form of the spin functions for for $S = 1$, $S_z = \pm 1$ is obvious:

$$\psi_{11} = \begin{pmatrix} 1 \\ 0 \end{pmatrix}_1 \begin{pmatrix} 1 \\ 0 \end{pmatrix}_2 \quad \text{and} \quad \psi_{1,-1} = \begin{pmatrix} 0 \\ 1 \end{pmatrix}_1 \begin{pmatrix} 0 \\ 1 \end{pmatrix}_2.$$

The spin functions with $S_z = 0$ have the form

$$\psi_{S,S_z=0} = C_S^{(1)} \begin{pmatrix} 1 \\ 0 \end{pmatrix}_1 \begin{pmatrix} 0 \\ 1 \end{pmatrix}_2 + C_S^{(2)} \begin{pmatrix} 0 \\ 1 \end{pmatrix}_1 \begin{pmatrix} 1 \\ 0 \end{pmatrix}_2. \quad (1)$$

From the equation $\hat{\mathbf{S}}^2\psi_{00} = 0$, it follows that $\hat{S}_x\psi_{00} = 0$, i.e.,

$$\hat{S}_x\psi_{00} \equiv \frac{1}{2}(\hat{\sigma}_{x1}+\hat{\sigma}_{x2})\left\{C_0^{(1)}\begin{pmatrix}1\\0\end{pmatrix}_1\begin{pmatrix}0\\1\end{pmatrix}_2 + C_0^{(2)}\begin{pmatrix}0\\1\end{pmatrix}_1\begin{pmatrix}1\\0\end{pmatrix}_2\right\}$$

$$= (C_0^{(1)} + C_0^{(2)})\left\{\begin{pmatrix}1\\0\end{pmatrix}_1\begin{pmatrix}0\\1\end{pmatrix}_2 + \begin{pmatrix}0\\1\end{pmatrix}_1\begin{pmatrix}1\\0\end{pmatrix}_2\right\} = 0.$$

This means that $C_0^{(1)} = -C_0^{(2)}$, and from normalization we have $C_0^{(1)} = 1/\sqrt{2}$. To find the $C_1^{(1,2)}$ in (1) for the case of $S = 1$ we use the eigenfunction orthogonality condition $\langle\psi_{00}|\psi_{10}\rangle = 0$, which gives $C_1^{(1)} = C_1^{(2)}$.

The normalized eigenfunctions ψ_{S0} therefore have the form

$$\psi_{1(0),0} = \frac{1}{\sqrt{2}}\left\{\begin{pmatrix}1\\0\end{pmatrix}_1\begin{pmatrix}0\\1\end{pmatrix}_2 \pm \begin{pmatrix}0\\1\end{pmatrix}_1\begin{pmatrix}1\\0\end{pmatrix}_2\right\}. \tag{2}$$

Spin functions have a well-defined symmetry with respect to interchange of the particles' spin variables: they are symmetric in the case $S = 1$ and antisymmetric for $S = 0$. This falls in accordance with the result of Problem 3.30. Notice that we can use this result to write expressions (2) for the functions $\psi_{SS_z=0}$ without calculations.

Problem 5.11

A system of two spins with $s = 1/2$ is in a state described by the spin function of the form $\Psi_{\alpha\beta} = \varphi_\alpha\chi_\beta$. The multiplicative form of $\Psi_{\alpha\beta}$ points to the fact that there is no correlation between particles' spin states.

Determine the probabilities of the different possible values of S in such a state. Find the mean value $\overline{\mathbf{S}^2}$. Consider the case of $\varphi_\alpha = \chi_\alpha$.

Solution

We write the spin function as a superposition of the symmetric and antisymmetric terms:

$$\Psi_{\alpha\beta} = \frac{1}{2}(\varphi_\alpha\chi_\beta + \chi_\alpha\varphi_\beta) + \frac{1}{2}(\varphi_\alpha\chi_\beta - \chi_\alpha\varphi_\beta). \tag{1}$$

Taking into account the nature of the symmetry of functions ψ_{SS_z} (see the previous problem), we can state that the first, symmetric, term in (1) corresponds to $S = 1$, while the second, antisymmeric, term corresponds to $S = 0$. Assume φ_α, χ_β, and $\Psi_{\alpha\beta}$ are normalized to unity. The normalization of each of these terms determines the probability of the corresponding value of S, so that

$$w(S = 0,\ 1) = \frac{1}{4}(\varphi_\alpha^*\chi_\beta^* \mp \chi_\alpha^*\varphi_\beta^*)(\varphi_\alpha\chi_\beta \mp \chi_\alpha\varphi_\beta) = \frac{1}{2}(1 \mp |\langle\varphi|\chi\rangle|^2).$$

The + sign relates to $S = 1$. Finally, $\overline{\mathbf{S}^2} = 2w(S = 1)$.

174 *Exploring Quantum Mechanics*

Problem 5.12

For a system of two particles with spins $s = 1/2$ show that

a) in states with a definite value of the total spin, the operator $\hat{\boldsymbol{\sigma}}_1 \cdot \hat{\boldsymbol{\sigma}}_2$ also has a definite value;
b) the operator $(\hat{\boldsymbol{\sigma}}_1 \cdot \hat{\boldsymbol{\sigma}}_2)^2$ can be reconstructed in a form that contains the Pauli matrices $\hat{\boldsymbol{\sigma}}_{1,2}$ in powers not higher than 1.

Solution

a) Since $\hat{\mathbf{S}}^2 = \frac{1}{4}(\hat{\boldsymbol{\sigma}}_1 + \hat{\boldsymbol{\sigma}}_2)^2$ and $S_{1,2}^2 = \frac{3}{4}$, then $\hat{\boldsymbol{\sigma}}_1 \cdot \hat{\boldsymbol{\sigma}}_2 = -3 + 2\hat{\mathbf{S}}^2$, and eigenfunctions of $\hat{\mathbf{S}}^2$ are simultaneous eigenfunctions of $\hat{\boldsymbol{\sigma}}_1 \cdot \hat{\boldsymbol{\sigma}}_2$, corresponding to eigenvalues equal to -3 (if $S = 0$) and $+1$ (if $S = 1$).

b) Since the Hermitian operator $\hat{\boldsymbol{\sigma}}_1 \cdot \hat{\boldsymbol{\sigma}}_2$ has only two different eigenvalues, it obeys the equation $(\hat{\boldsymbol{\sigma}}_1 \cdot \hat{\boldsymbol{\sigma}}_2 - 1)(\hat{\boldsymbol{\sigma}}_1 \cdot \hat{\boldsymbol{\sigma}}_2 + 3) = 0$ (compare with Problem 1.21). It follows that $(\hat{\boldsymbol{\sigma}}_1 \cdot \hat{\boldsymbol{\sigma}}_2)^2 = 3 - 2\hat{\boldsymbol{\sigma}}_1 \cdot \hat{\boldsymbol{\sigma}}_2$.

Problem 5.13

For a system of two particles with spins $s = 1/2$, find the form of the spin exchange operator, $\hat{C}$, which acts on the spin function $\Psi_{\alpha\beta}$ to give $\hat{C}\Psi_{\alpha\beta} \equiv \Psi_{\beta\alpha}$, i.e., it interchanges the spin variables of the two particles. Express $\hat{C}$ in terms of the Pauli matrices.

Solution

Let us write the spin function in the form

$$\psi_{\alpha\beta} = \psi^+_{\alpha\beta} + \psi^-_{\alpha\beta}, \text{ where } \psi^{\pm}_{\alpha\beta} = \frac{1}{2}(\psi_{\alpha\beta} \pm \psi_{\beta\alpha}).$$

If we take into account the symmetry nature of spin functions ψ_{SS_z}, we notice that the function $\psi^+_{\alpha\beta}$ corresponds to total spin $S = 1$, while $\psi^-_{\alpha\beta}$ corresponds to $S = 0$ (see Problem 5.11). We can see that $\hat{\mathbf{S}}^2 \psi_{\alpha\beta} = 2\psi^+_{\alpha\beta}$, and accordingly,

$$\psi^-_{\alpha\beta} = \psi_{\alpha\beta} - \frac{1}{2}\hat{\mathbf{S}}^2 \psi_{\alpha\beta}.$$

According to the definition of the operator $\hat{C}$ we have $\hat{C}\psi_{\alpha\beta} \equiv \psi_{\beta\alpha} = \psi^+_{\alpha\beta} - \psi^-_{\alpha\beta}$. It follows from the above that

$$\hat{C} = \hat{\mathbf{S}}^2 - 1 = \frac{1}{2}(1 + \hat{\boldsymbol{\sigma}}_1 \cdot \hat{\boldsymbol{\sigma}}_2). \qquad (2)$$

For the relation between $\hat{\mathbf{S}}^2$ and $\hat{\boldsymbol{\sigma}}_1 \cdot \hat{\boldsymbol{\sigma}}_2$, see the previous problem.

We will describe the properties of the operator $\hat{C}$. It is an Hermitian operator. Spin functions Ψ_S are its eigenfunctions, with corresponding eigenvalues $+1$ for $S = 1$ and -1 for $S = 0$. Obviously, $\hat{C}^2 = 1$.

Problem 5.14

For a system of two particles with spins $s = 1/2$, find the eigenvalues and eigenfunctions of the following operators:

a) $\hat{V}_1 = F(a + b\hat{\boldsymbol{\sigma}}_1 \cdot \hat{\boldsymbol{\sigma}}_2)$ where $F(x)$ is some function of x;
b) $\hat{V}_2 = a(\hat{\sigma}_{1z} + \hat{\sigma}_{2z}) + b\hat{\boldsymbol{\sigma}}_1 \cdot \hat{\boldsymbol{\sigma}}_2$;
c) $\hat{V}_3 = a\hat{\sigma}_{1z}\hat{\sigma}_{2z} + b\hat{\boldsymbol{\sigma}}_1 \cdot \hat{\boldsymbol{\sigma}}_2$;
d) $\hat{V}_4 = a_1\hat{\sigma}_{1z} + a_2\hat{\sigma}_{2z} + b\hat{\boldsymbol{\sigma}}_1 \cdot \hat{\boldsymbol{\sigma}}_2$

a, b are some real parameters, so all the operators, $\hat{V}_i$, are Hermitian.

Solution

First, recall that $\hat{\boldsymbol{\sigma}}_1 \cdot \hat{\boldsymbol{\sigma}}_2 = -3 + 2\hat{\mathbf{S}}^2$, $\hat{S}_z = \frac{1}{2}(\hat{\sigma}_{1z} + \hat{\sigma}_{2z})$.

a) Spin functions, ψ_S, are also eigenfunctions of the operator $\hat{f} = a + b\hat{\boldsymbol{\sigma}}_1 \cdot \hat{\boldsymbol{\sigma}}_2$ that correspond to the eigenvalues $f_S = a - 3b + 2bS(S+1)$. So the eigenvalues of the operator $\hat{V}_1$ are equal to $(V_1)_S = F(f_S)$.

b) Spin functions, ψ_{SS_z}, are also eigenfunctions of the operator $\hat{V}_2$ corresponding to the eigenvalues $(V_2)_{SS_z} = 2aS_z - 3b + 2bS(S+1)$.

c) Since $\hat{\sigma}_{1z}\hat{\sigma}_{2z} = 2\hat{S}_z^2 - 1$, then functions ψ_{SS_z} are the eigenfunctions of the operator $\hat{V}_3$, and the corresponding eigenvalues are equal to $(V_3)_{SS_z} = -a + 2aS_z^2 - 3b + 2bS(S+1)$.

d) Let us find a form of the operator

$$\hat{V}_4 = \frac{1}{2}(a_1 + a_2)(\hat{\sigma}_{1z} + \hat{\sigma}_{2z}) + \frac{1}{2}(a_1 - a_2)(\hat{\sigma}_{1z} - \hat{\sigma}_{2z}) + b\hat{\boldsymbol{\sigma}} \cdot \hat{\boldsymbol{\sigma}}_2,$$

using SS_z-representation where it is described by a matrix with matrix elements $\langle S'S_z'|\hat{V}_4|SS_z\rangle$. Use, for the matrix elements, the following numeration of the states defined by the quantum numbers S, S_z:

$$(S = 1, S_z = 1) \to 1, \ (1, -1) \to 2, \ (1, 0) \to 3, \ (0, 0) \to 4$$

Take into account the explicit form of the spin functions ψ_{SS_z} (see Problem 5.10). We obtain

$$\hat{V}_4 = \begin{pmatrix} A & 0 & 0 & 0 \\ 0 & B & 0 & 0 \\ 0 & 0 & C & E \\ 0 & 0 & E^* & D \end{pmatrix}, \quad \begin{array}{l} A = a_1 + a_2 + b, \\ B = -a_1 - a_2 + b, \\ C = b, \ D = -3b, \\ E = E^* = a_1 - a_2. \end{array} \quad (1)$$

By the unitary transformation this Hermitian matrix can be led to the diagonal form which defines directly its eigenvalues. From the form of the matrix $\hat{V}_4$ it follows that two of its eigenvalues are $(V_4)_1 = A$ and $(V_4)_2 = B$, corresponding to the eigenfunctions $\psi_{S=1, S_z=1}$ and $\psi_{1,-1}$, respectively. The unitary operator that diagonalizes $\hat{V}_4$ "mixes" only the states with quantum numbers $(1, 0)$ and $(0, 0)$. To find the two different eigenvalues we should diagonalize the 2×2 matrix of the form $\begin{pmatrix} C & E \\ E^* & D \end{pmatrix}$. We can easily find these eigenvalues if we remember that under a unitary transformation, the trace and determinant of the matrix are invariant (see Problem 1.51). We have finally:

$$(V_4)_{3,4} = -b \pm \sqrt{(a_1 - a_2)^2 + 4b^2}.$$

Problem 5.15

Spins of N particles equal to s each are added into the total spin $S = Ns$. What is the total spin of any n particles? Does the spin function have a definite symmetry with respect to interchange of the spin variables of any two of the particles?

Solution

The total spin of any n particles has a definite value equal to ns. The spin function is symmetric with respect to the interchange of the spin variables of any two particles (compare with Problem 3.30). If $S < Ns$, then for $N > 2$ the spin function already does not have a definite symmetry with respect to the interchange of any two particles (see, for example, Problem 5.19).

Problem 5.16

A spin function of a system consisting of N spins with $s = 1/2$ is given by

$$\Psi = \begin{pmatrix} 1 \\ 0 \end{pmatrix}_1 \begin{pmatrix} 1 \\ 0 \end{pmatrix}_2 \cdots \begin{pmatrix} 1 \\ 0 \end{pmatrix}_n \begin{pmatrix} 0 \\ 1 \end{pmatrix}_{n+1} \cdots \begin{pmatrix} 0 \\ 1 \end{pmatrix}_N.$$

Determine $\overline{\hat{S}^2}$. For the special cases of $n = 1$ and $n = N - 1$, find also the probabilities of the total spin S possible values.

Solution

See that

$$\hat{S}^2 = \frac{1}{4}\left(\sum_{a=1}^{N} \hat{\sigma}_a\right)^2 = \frac{1}{4}\left(\sum_a \hat{\sigma}_a^2 + \sum_{a \neq b} \hat{\sigma}_a \cdot \hat{\sigma}_b\right).$$

Also, we have that $\hat{\sigma}_a^2 = 3$, that in the state considered there is no correlation between spins of particles, and that the mean values of $\hat{\sigma}_a$ give

$$\overline{(\hat{\sigma}_a)_i} = \begin{cases} 0, & i = 1, 2, \\ 1, & i = 3, a \leq n, \\ -1, & i = 3, a \geq n+1. \end{cases}$$

We easily obtain $\overline{\mathbf{S}^2} = \frac{1}{4}(N^2 - 4nN + 2N + 4n^2)$.

In the cases where $n = 1$ or $n = N - 1$, the total spin could take only two values: $S_1 = N/2$ and $S_2 = (N-2)/2$ ($N \geq 2$). By taking into account the value of $\overline{S^2}$, we can easily find their probabilities:

$$w(S = N/2) = 1/N, \quad w(S = N/2 - 1) = 1 - 1/N$$

Compare with Problem 3.29.

Problem 5.17

The state of a particle with the spin $s = 1/2$ is characterized by the values of the quantum numbers l, m, s_z. Determine the probabilities of the possible values, j, of the total angular momentum, $\mathbf{j} = \mathbf{l} + \mathbf{s}$. Use the results of Problem 3.29.

Solution
We have $\overline{\mathbf{j}^2} = l(l+1) + 3/4 + 2ms_z$. Since only two values of total angular momentum are possible, $j = l \pm 1/2$, then the value of $\overline{\mathbf{j}^2}$ allows us to obtain easily their probabilities:

$$w(l + 1/2) = \frac{l + 2ms_z + 1}{2l + 1}, \quad w(l - 1/2) = \frac{l - 2ms_z}{2l + 1}.$$

Problem 5.18

The angular momenta of two weakly interacting subsystems, equal to 1 and 1/2, are added into the total angular momentum, J. For the states of such a complex system, characterized by the values J and J_z, find the probabilities for different possible values of the z-projection of the added angular momenta and their mean values. For solving the problem, use the operators $\hat{J}_\pm$ without using the Clebsch–Gordan coefficients.

Solution
The "spin" function of the state with $J = 3/2$ and $J_z = 3/2$ has the form

$$\psi_{3/2,\,3/2} = \begin{pmatrix} 1 \\ 0 \\ 0 \end{pmatrix}_1 \begin{pmatrix} 1 \\ 0 \end{pmatrix}_2,$$

and the projections l_z and s_z have values equal to 1 and 1/2. We act with operator $\hat{J}_- \equiv \hat{J}_x - i\hat{J}_y = \hat{j}_{1,-} + \hat{j}_2$ on this function to obtain (for the form of the operator $\hat{j}_-$ for $j=1$, see Problem 3.22):

$$\psi_{3/2,\,1/2} = C\hat{J}_-\psi_{3/2,\,3/2} = \frac{1}{\sqrt{3}}\left\{\begin{pmatrix}0&0&0\\\sqrt{2}&0&0\\0&\sqrt{2}&0\end{pmatrix} + \begin{pmatrix}0&0\\1&0\end{pmatrix}\right\}\begin{pmatrix}1\\0\\0\end{pmatrix}\begin{pmatrix}1\\0\end{pmatrix}$$

$$= \sqrt{\frac{2}{3}}\begin{pmatrix}0\\1\\0\end{pmatrix}\begin{pmatrix}1\\0\end{pmatrix} + \frac{1}{\sqrt{3}}\begin{pmatrix}1\\0\\0\end{pmatrix}\begin{pmatrix}0\\1\end{pmatrix}. \tag{1}$$

The multiplier $C = 3^{-1/2}$ is introduced for the normalization. From (1) we have the desired probabilities for a state with $J = 3/2$ and $J_z = 1/2$:

$$w(l_z = 1) = w\left(s_z = -\frac{1}{2}\right) = \frac{1}{3}, \quad w(l_z = 0) = w\left(s_z = \frac{1}{2}\right) = \frac{2}{3}$$

and the mean values $\overline{l_z} = 1/3$, $\overline{s_z} = 1/6$ ($J_z = \overline{l_z} + \overline{s_z}$).

We write the "spin" function for a state with $J = 1/2$ and $J_z = 1/2$ in the form

$$\Psi_{1/2,1/2} = C_1\begin{pmatrix}1\\0\\0\end{pmatrix}\begin{pmatrix}0\\1\end{pmatrix} + C_2\begin{pmatrix}0\\1\\0\end{pmatrix}\begin{pmatrix}1\\0\end{pmatrix}.$$

Using its orthogonality to $\Psi_{3/2,1/2}$, we find $C_1 = \sqrt{2/3}$, $C_2 = -\sqrt{1/3}$. We obtain the desired probabilities for the state with $J = 1/2$, $J_z = 1/2$:

$$w(l_z = 1) = w\left(s_z = -\frac{1}{2}\right) = \frac{2}{3}, \quad w(l_z = 0) = w\left(s_z = \frac{1}{2}\right) = \frac{1}{3},$$

and the mean values $\overline{l_z} = 2/3$, $\overline{s_z} = -1/6$.

The results for the states with $J_z < 0$ are obtained following the same steps.

Problem 5.19

For a system of three particles with the spin $s = 1/2$ there are eight independent spin states. Classify them by the values of system's total spin. Obtain the complete system of the spin functions, Ψ_{SS_z}, that describe the states with definite values of S and S_z of total spin. Discuss the symmetry of these functions with respect to the interchange of the spin variables. Compare with the case of two particles.

Solution

The possible values of the total spin are $S = 3/2$ and $1/2$.

Now the set of the eigenvalues with $S = 1/2$ is degenerate, since for $S = 1/2$ and a given value of S_z there are two independent spin states. Indeed, the value $S = 1/2$

could be obtained by two independent (for a given S_z) ways: 1) by combining the first two particle spins into a total $S_{12} = 0$, where the system's total spin is determined by the spin of a third particle; 2) by combining the spins into a total $S_{12} = 1$, then combining the spin of the third particle into the total spin $S = 1/2$. Since the number of independent spin states with a given S (without the degeneracy in S, S_z) is equal to $2S + 1$, then the total number of independent spin states is equal to $(2 \cdot 3/2 + 1) + 2(2 \cdot 1/2 + 1) = 8$.

The form of the spin functions $\psi_{S=3/2,\, S_z=\pm 3/2}$ is obvious:

$$\psi_{3/2,3/2} = \begin{pmatrix} 1 \\ 0 \end{pmatrix}_1 \begin{pmatrix} 1 \\ 0 \end{pmatrix}_2 \begin{pmatrix} 1 \\ 0 \end{pmatrix}_3, \quad \psi_{3/2,\,-3/2} = \begin{pmatrix} 0 \\ 1 \end{pmatrix}_1 \begin{pmatrix} 0 \\ 1 \end{pmatrix}_2 \begin{pmatrix} 0 \\ 1 \end{pmatrix}_3. \tag{1}$$

We can also write, without any calculations, the spin functions corresponding to $S = 3/2$, $S_z = \pm 1/2$, by using their symmetry with respect to the interchange of the spin variables of any two particles, whose total angular momentum is equal to 1 (compare with problem 5.10):

$$\psi_{3/2,1/2} = \frac{1}{\sqrt{3}} \left\{ \begin{pmatrix} 1 \\ 0 \end{pmatrix}_1 \begin{pmatrix} 1 \\ 0 \end{pmatrix}_2 \begin{pmatrix} 0 \\ 1 \end{pmatrix}_3 + \begin{pmatrix} 1 \\ 0 \end{pmatrix}_1 \begin{pmatrix} 0 \\ 1 \end{pmatrix}_2 \begin{pmatrix} 1 \\ 0 \end{pmatrix}_3 \right.$$
$$\left. + \begin{pmatrix} 0 \\ 1 \end{pmatrix}_1 \begin{pmatrix} 1 \\ 0 \end{pmatrix}_2 \begin{pmatrix} 1 \\ 0 \end{pmatrix}_3 \right\}$$

$$\psi_{3/2,-1/2} = \frac{1}{\sqrt{3}} \left\{ \begin{pmatrix} 0 \\ 1 \end{pmatrix}_1 \begin{pmatrix} 0 \\ 1 \end{pmatrix}_2 \begin{pmatrix} 1 \\ 0 \end{pmatrix}_3 + \begin{pmatrix} 0 \\ 1 \end{pmatrix}_1 \begin{pmatrix} 1 \\ 0 \end{pmatrix}_2 \begin{pmatrix} 0 \\ 1 \end{pmatrix}_3 \right.$$
$$\left. + \begin{pmatrix} 1 \\ 0 \end{pmatrix}_1 \begin{pmatrix} 0 \\ 1 \end{pmatrix}_2 \begin{pmatrix} 0 \\ 1 \end{pmatrix}_3 \right\}.$$

If a spin function has a total spin of the first two particles equal to $S_{12} = 0$, then it describes the state with $S = 1/2$. We have

$$\psi^{(1)}_{1/2,1/2} = \frac{1}{\sqrt{2}} \left\{ \begin{pmatrix} 1 \\ 0 \end{pmatrix}_1 \begin{pmatrix} 0 \\ 1 \end{pmatrix}_2 - \begin{pmatrix} 0 \\ 1 \end{pmatrix}_1 \begin{pmatrix} 1 \\ 0 \end{pmatrix}_2 \right\} \begin{pmatrix} 1 \\ 0 \end{pmatrix}_3, \tag{3}$$

$$\psi^{(1)}_{1/2,-1/2} = \frac{1}{\sqrt{2}} \left\{ \begin{pmatrix} 1 \\ 0 \end{pmatrix}_1 \begin{pmatrix} 0 \\ 1 \end{pmatrix}_2 - \begin{pmatrix} 0 \\ 1 \end{pmatrix}_1 \begin{pmatrix} 1 \\ 0 \end{pmatrix}_2 \right\} \begin{pmatrix} 0 \\ 1 \end{pmatrix}_3.$$

We can find the second pair of the functions, $\psi^{(2)}_{1/2,\pm 1/2}$, that are linearly independent with respect to (3), by considering the states with a total spin $S_{23} = 0$ for the second and third particles:

$$\psi^{(2)}_{1/2,1/2} = \frac{1}{\sqrt{2}} \begin{pmatrix} 1 \\ 0 \end{pmatrix}_1 \left\{ \begin{pmatrix} 1 \\ 0 \end{pmatrix}_2 \begin{pmatrix} 0 \\ 1 \end{pmatrix}_3 - \begin{pmatrix} 0 \\ 1 \end{pmatrix}_2 \begin{pmatrix} 1 \\ 0 \end{pmatrix}_3 \right\}, \tag{4}$$

$$\psi^{(2)}_{1/2,-1/2} = \frac{1}{\sqrt{2}} \begin{pmatrix} 0 \\ 1 \end{pmatrix}_1 \left\{ \begin{pmatrix} 1 \\ 0 \end{pmatrix}_2 \begin{pmatrix} 0 \\ 1 \end{pmatrix}_3 - \begin{pmatrix} 0 \\ 1 \end{pmatrix}_2 \begin{pmatrix} 1 \\ 0 \end{pmatrix}_3 \right\}.$$

We should note that though functions (3) and (4) are linearly independent, for the same values of S_z they are not orthogonal. The most general spin function of a state with $S = 1/2$ is a superposition of functions (3) and (4). The reader should consider a state with the total spin $S_{13} = 0$ that also corresponds to $S = 1/2$, and ensure that the corresponding function could be expressed in terms of the functions (3) and (4).

In conclusion, we notice that the spin functions of states with the total spin $S = 1/2$ do not have a definite symmetry with respect to the interchange of spin variables of any two particles. Although the first of the functions in (3) is antisymmetric with respect to the interchange of spin variables of the first and second particles, with the interchange of first and third particles it becomes a completely different function $(-\psi^{(2)}_{1/2,1/2})$.

5.2 Spin-orbital states with spin $s = 1/2$; Higher spins

Problem 5.20

The states of a particle with a definite value of λ for the spin projection along its momentum direction are called *helical*[58] states. For a particle with spin $s = 1/2$, find the wavefunctions, $\Psi_{\mathbf{p}_0 \lambda}$, of states with definite momentum $\mathbf{p}_0$ and *helicity* $\lambda = \pm 1/2$.

Solution

It is easy to find the functions if we take into account the result of Problem 5.3 (see also Eq. (V.4)):

$$\psi_{\mathbf{p}_0,\, \lambda=+1/2} = \frac{e^{i\mathbf{p}_0 \mathbf{r}/\hbar}}{(2\pi\hbar)^{3/2}} \begin{pmatrix} \cos(\tilde{\theta}/2) \\ e^{i\tilde{\varphi}} \sin(\tilde{\theta}/2) \end{pmatrix},$$

$$\psi_{\mathbf{p}_0,\, \lambda=-1/2} = \frac{e^{i\mathbf{p}_0 \mathbf{r}/\hbar}}{(2\pi\hbar)^{3/2}} \begin{pmatrix} \sin(\tilde{\theta}/2) \\ -e^{i\tilde{\varphi}} \cos(\tilde{\theta}/2) \end{pmatrix},$$

where $\tilde{\theta}$ and $\tilde{\varphi}$ are the polar and azimuthal angles of the vector $\mathbf{p}_0$.

Problem 5.21

For a particle with spin $s = 1/2$, show that the most general expression for the spin-angular dependence of a wavefunction of the $p_{1/2}$-state (the state with orbital angular momentum $l = 1$ and total angular momentum $j = 1/2$) is given by

$$\Psi = (\hat{\boldsymbol{\sigma}} \cdot \mathbf{n})\chi, \text{ or } \Psi_\alpha = (\hat{\boldsymbol{\sigma}}_{\alpha\beta} \cdot \mathbf{n})\chi_\beta.$$

---

[58] Note that the helicity is a pseudoscalar quantity and changes its sign under coordinate inversion.

We have that $\chi = \begin{pmatrix} a \\ b \end{pmatrix}$ is some spinor which does not depend on the direction of the vector $\mathbf{n}$ ($\mathbf{n} = \mathbf{r}/r$ or $\mathbf{n} = \mathbf{p}/p$, in accordance with the representation used).

Normalize this wavefunction to unity. Find the distribution (averaged over the spin) of particle momenta directions in the given states.

Calculate the mean value of $\mathbf{j}$. Find the way by which it depends on the definite choice of spinor χ.

Determine the form of functions that describe $p_{1/2}$-states with a given value of $j_z = \pm 1/2$.

Solution

1) If we act by the operator $\hat{\mathbf{j}}^2$ on the given function, we encounter the expression $(\hat{\mathbf{l}} + \hat{\boldsymbol{\sigma}}/2)^2 (\hat{\boldsymbol{\sigma}} \cdot \mathbf{n}) \chi$. Using the relation[59] $[\hat{j}_i, (\hat{\boldsymbol{\sigma}} \cdot \mathbf{n})] = 0$ and equation $\hat{\mathbf{l}} \chi = 0$, we can transform this relation to $(\hat{\boldsymbol{\sigma}} \cdot \mathbf{n}) \cdot \frac{1}{4} \hat{\boldsymbol{\sigma}}^2 \chi$ or $\frac{3}{4} (\hat{\boldsymbol{\sigma}} \cdot \mathbf{n}) \chi$. It follows that $\hat{\mathbf{j}}^2 \psi = (3/4)\psi$, i.e., j has a definite value equal to $1/2$. The fact that the function given corresponds to the value $l = 1$ follows from its linear dependence on the vector $\mathbf{n}$ (compare with Eq. (III.7) and Problem 3.42).

2) $\Psi^* \Psi = \chi^* (\hat{\boldsymbol{\sigma}} \cdot \mathbf{n})^2 \chi = \chi^* \chi = \text{const}$ and does not depend on $\mathbf{n}$ means that the distribution over the directions of momentum (or position) is isotropic, as the s-state. So the normalization condition $\int \Psi^* \Psi d\Omega = 1$ would be fulfilled for $\chi^* \chi = |a|^2 + |b|^2 = 1/4\pi$.

3) At last we have the relation

$$\bar{\mathbf{j}} = \langle \Psi | \hat{\mathbf{j}} | \Psi \rangle = \int \chi^* (\hat{\boldsymbol{\sigma}} \cdot \mathbf{n}) \left(\hat{\mathbf{l}} + \frac{1}{2} \hat{\boldsymbol{\sigma}} \right) (\hat{\boldsymbol{\sigma}} \cdot \mathbf{n}) \chi d\Omega = 4\pi \chi^* \frac{\hat{\boldsymbol{\sigma}}}{2} \chi.$$

It follows that the total angular momentum vector in the state considered is the same as the spin vector in the state described by spin function χ. Therefore, if we choose $\sqrt{4\pi} \chi$ of the form $\begin{pmatrix} 1 \\ 0 \end{pmatrix}$ and $\begin{pmatrix} 0 \\ 1 \end{pmatrix}$, we obtain normalized functions of $p_{1/2}$-states with $j_z = +1/2$ and $j_z = -1/2$:

$$\Psi_{j=1/2, l=1, j_z=1/2} = \frac{1}{\sqrt{4\pi}} (\hat{\boldsymbol{\sigma}} \cdot \mathbf{n}) \begin{pmatrix} 1 \\ 0 \end{pmatrix} = \frac{1}{\sqrt{4\pi}} \begin{pmatrix} \cos\theta \\ \sin\theta \cdot e^{i\varphi} \end{pmatrix}.$$

θ, φ are the polar and azimuthal angles of the vector $\mathbf{n}$. The spin-angular function of the state with $j_z = -1/2$ is similar. Compare to Problem 5.24.

Problem 5.22

A particle with the spin $s = 1/2$ has the spin-angular dependence of the wavefunction in the form $\psi_\pm = (1 \pm \hat{\boldsymbol{\sigma}} \cdot \mathbf{n}) \chi$ (spinor χ does not depend on $\mathbf{n}$). Analyze the states of

---

[59] See also Problems 3.5 and 3.28 about the given commutator value.

this particle with respect to the following quantum numbers: j, l, I (parity), and also λ (eigenvalues of the operator $\hat{\lambda} = \hat{\boldsymbol{\sigma}} \cdot \mathbf{n}/2$, spin projection along vector $\mathbf{n}$). If $\mathbf{n} = \mathbf{p}/p$, then λ is helicity.

Find the way by which functions $\psi_\pm$ are transformed with the coordinate inversion.

Solution

These wavefunctions are superpositions of the $s_{1/2}$-state function χ, and $p_{1/2}$-state function $(\hat{\boldsymbol{\sigma}} \cdot \mathbf{n})\chi$ (see the previous problem). Total angular momentum has a well defined value $j = 1/2$. Orbital angular momentum l and parity I do not have definite values. Since the wavefunctions, $\psi_\pm$, are normalized in the same way we come to the conclusion that l can take the two possible values, 0 and 1, with the same probability $1/2$. Then we notice that $\hat{\lambda}\psi_\pm = \pm\psi_\pm/2$, so the spin n-projection has a definite value equal to $\pm 1/2$. Under coordinate inversion, $\psi_\pm$ switch into each other.

Problem 5.23

For a particle with spin $s = 1/2$, show that the spin-angular wavefunction of the form $\Psi = \{2(\mathbf{c} \cdot \mathbf{n}) + i[\mathbf{c} \times \mathbf{n}] \cdot \hat{\boldsymbol{\sigma}}\}\chi$, describes the $p_{3/2}$-state. Assume that vector $\mathbf{c}$ and spinor $\chi = \begin{pmatrix} a \\ b \end{pmatrix}$ do not depend on $\mathbf{n} = \mathbf{r}/r$.

Find the concrete values of $\mathbf{c}$ and χ for which the given function describes the $p_{3/2}$-state with a definite value ($j_z = \pm 1/2, \pm 3/2$) of total angular momentum z-projection.

Solution

Consider the spin-angular wavefunction of the state with $l = 1$: $\Psi_{l=1} = (\mathbf{c} \cdot \mathbf{n})\chi$, where $\mathbf{c}$ and χ do not depend on $\mathbf{n}$ (compare with Problem 3.42). This function does not correspond to a definite value of j, but instead represents the superposition of states with $j = 1/2$ and $j = 3/2$. To select the part that corresponds to $j = 3/2$, we use the projection operator

$$\hat{P}_{j=3/2} = \frac{1}{3}(2 + \hat{\mathbf{l}} \cdot \hat{\boldsymbol{\sigma}}).$$

Compare[60] with Problem 3.36. We easily find

$$\Psi_{j=3/2} = \hat{P}_{j=3/2}\Psi_{l=1} = \frac{1}{3}(2\mathbf{c} \cdot \mathbf{n} + i[\mathbf{c} \times \mathbf{n}] \cdot \hat{\boldsymbol{\sigma}}])\chi \tag{1}$$

in accordance with the problem statement. The reader should normalize the wavefunction and ensure that it is orthogonal to the $p_{1/2}$-state wavefunction from Problem 5.21. We should note that the number of the independent functions, $\Psi_{l=1}$, is equal to six (there are three independent choices of vector $\mathbf{c}$ and two of the spinor χ). There are only four independent functions of the form (1), since they are deduced by excluding the two independent functions corresponding to $j = 1/2$ from $\Psi_{l=1}$.

---

[60] See also Problem 5.24.

Function $\Psi_{l=1}$ with the choice of $\mathbf{c} = (0, 0, 1)$ and $\chi = \begin{pmatrix} 1 \\ 0 \end{pmatrix}$ describes the state with $l_z = 0$ (see Problem 3.18), $s_z = 1/2$ and $j_z = 1/2$; in this case j does not have a definite value. Function (1) for such $\mathbf{c}$ and χ has the form

$$\Psi_{3/2,1,1/2} = \frac{1}{3} \begin{pmatrix} 2\cos\theta \\ -e^{i\varphi}\sin\theta \end{pmatrix}.$$

This describes the state with $j = 3/2$, $l = 1$ and $j_z = 1/2$. Since $\hat{P}_j$ commutes with the operators $\hat{l}^2$ and $\hat{j}_z$, then the eigenfunctions of these operators remain eigenfunctions under the action of operator $\hat{P}_l$.

In the same way, we can find the wavefunctions of other $p_{3/2}$-states. For example, if we choose $\mathbf{c} = (1, i, 0)$ and $\chi = \begin{pmatrix} 1 \\ 0 \end{pmatrix}$ then we will obtain $\Psi_{3/2,1,3/2}$, etc. Compare with the result of Problem 5.24.

Problem 5.24

For a particle with spin $s = 1/2$, find the spin-angular wavefunctions, Ψ_{jlj_z}, of the states with definite values of l, j_z and $j = l \pm 1/2$ (here l and j are the orbital and total angular momenta).

Solve this problem in two ways without using the Clebsch–Gordan coefficients:

a) by using the projection operators $\hat{P}_j$;
b) by using the raising (lowering) operators $\hat{j}_\pm$.

Solution

a) Consider a spin-angle function of the form $\Psi = \begin{pmatrix} Y_{lm}(\theta, \varphi) \\ 0 \end{pmatrix}$ which describes the state of a particle with definite values of l, m, $s_z = 1/2$ and $j_z = m + 1/2$. This function is not the eigenfunction of $\hat{\mathbf{j}}^2$ (except when $m = l$ and $j = l + 1/2$), but describes some superposition of states with $j = l + 1/2$ and $j = l - 1/2$. Using the projection operator[61] $\hat{P}_j$ for the states with given j,

$$\hat{P}_{j=l\pm 1/2} = \frac{1}{2l+1}\left(l + \frac{1}{2} \pm \frac{1}{2} \pm \hat{\boldsymbol{\sigma}} \cdot \hat{\mathbf{l}}\right),$$

and using the relations analogous to Eq. (III.8):

$$\hat{\boldsymbol{\sigma}} \cdot \hat{\mathbf{l}} = 2\hat{\mathbf{s}} \cdot \hat{\mathbf{l}} = 2\hat{s}_z\hat{l}_z + \hat{s}_+\hat{l}_- + \hat{s}_-\hat{l}_+,$$

$$\hat{j}_+\psi_{jj_z} = \sqrt{(j-j_z)(j+j_z+1)}\psi_{j,j_z+1}, \qquad (1)$$

$$\hat{j}_-\psi_{jj_z} = \sqrt{(j-j_z+1)(j+j_z)}\psi_{j,j_z-1},$$

---
[61] Its form follows from Problem 1.35 if we also take into account Problem 3.27.

we easily find the explicit form of the desired spin-angle functions:

$$\psi_{l+1/2,l,m+1/2} = C_1 \hat{P}_{l+1/2} \Psi = \frac{1}{\sqrt{2l+1}} \begin{pmatrix} \sqrt{l+m+1}\, Y_{lm} \\ \sqrt{l-m}\, Y_{l,m+1} \end{pmatrix}, \quad (2)$$

$$\psi_{l-1/2,l,m+1/2} = C_2 \hat{P}_{l-1/2} \Psi = \frac{1}{\sqrt{2l+1}} \begin{pmatrix} \sqrt{l-m}\, Y_{lm} \\ -\sqrt{l+m+1}\, Y_{l,m+1} \end{pmatrix}.$$

The coefficients $C_{1,\,2}$ are chosen for the wavefunction normalization of unity. From (2) we see the orthogonality of the functions considered:

$$\langle j_1, l, j_z | j_2, l, j_z \rangle = 0, \text{ where } j_{1,2} = l \pm 1/2.$$

b) We see that $\psi_{l+1/2,l,l+1/2} = \begin{pmatrix} Y_{ll} \\ 0 \end{pmatrix}$. Acting by the operator $\hat{j}_-^n$ with $n = l + 1/2 - j_z$ on this function, we obtain the $\psi_{l+1/2,l,j_z}$. Taking into account that $\hat{j}_- = \hat{l}_- + \hat{s}_-$, $\hat{s}_-^2 = 0$ and therefore

$$\hat{j}_-^n = \hat{l}_-^n + n \hat{l}_-^{n-1} \hat{s}_-,$$

and using relation (1), we find again the function known from (2):

$$\psi_{l+1/2,l,m+1/2} = C \hat{j}_-^{l-m} \begin{pmatrix} Y_{ll} \\ 0 \end{pmatrix} = \frac{1}{\sqrt{2l+1}} \begin{pmatrix} \sqrt{l+m+1}\, Y_{lm} \\ \sqrt{l-m}\, Y_{l,m+1} \end{pmatrix}.$$

Problem 5.25

Show that the functions Ψ_{jlj_z} considered in the previous problem are related by

$$\Psi_{jl_1 j_z} = \hat{\boldsymbol{\sigma}} \cdot \mathbf{n}\, \Psi_{jl_2 j_z}, \quad l_{1,2} = j \pm 1/2 \ (\mathbf{n} = \mathbf{r}/r \text{ or } \mathbf{p}/p).$$

Determine the spin-angular dependence of the wavefunctions $\Psi_{j\lambda j_z}$ (in the momentum representation) of particle states with definite values of j, j_z, and helicity $\lambda = \pm 1/2$.

Solution

1) We can check the problem statement by direct calculation, but it is easier to verify it by the following argument that depends upon commutativity of operators[62] $\hat{j}_z$ and $(\hat{\boldsymbol{\sigma}} \cdot \mathbf{n})$ and upon the pseudo-scalar nature of the latter.

---

[62] Compare with Problems 3.5 and 3.28.

Let $\Psi_{jl_2j_z}$ be the eigenfunctions of operators $\hat{j}^2$, $\hat{l}^2$, $\hat{j}_z$, where $l_2 = j - 1/2$. This function has a definite parity equal to $I_2 = (-1)^{l_2}$. Consider the function

$$\tilde{\Psi} = (\hat{\boldsymbol{\sigma}} \cdot \mathbf{n})\Psi_{jl_2j_z}.$$

For this function, we easily find

$$\hat{j}_z\tilde{\Psi} = \hat{j}_z(\hat{\boldsymbol{\sigma}} \cdot \mathbf{n})\Psi_{jl_2j_z} = (\hat{\boldsymbol{\sigma}} \cdot \mathbf{n})\hat{j}_z\Psi_{jl_2j_z} = j_z(\hat{\boldsymbol{\sigma}} \cdot \mathbf{n})\Psi_{jl_2j_z} = j_z\tilde{\Psi},$$

$$\hat{j}^2\tilde{\Psi} = \hat{j}^2(\hat{\boldsymbol{\sigma}} \cdot \mathbf{n})\Psi_{jl_2j_z} = (\hat{\boldsymbol{\sigma}} \cdot \mathbf{n})\hat{j}^2\Psi_{jl_2j_z} = j(j+1)\tilde{\Psi},$$

$$\hat{I}\tilde{\Psi} = \hat{I}(\hat{\boldsymbol{\sigma}} \cdot \mathbf{n})\Psi_{jl_2j_z} = -(\hat{\boldsymbol{\sigma}} \cdot \mathbf{n})\hat{I}\Psi_{jl_2j_z} = (-1)^{l_2+1}\tilde{\Psi}.$$

It follows that $\tilde{\Psi}$ is also the eigenfunction of the operators $\hat{j}^2$, $\hat{j}_z$, $\hat{I}$, and its parity is opposite to the parity of wavefunction $\Psi_{jl_2j_z}$. Because there are only two possible values of l for the given j, $l_{1,2} = j \pm 1/2$, and the parity is equal to $(-1)^l$, then $\tilde{\Psi}$ corresponds to to the value $l_1 = j + 1/2$. So the function $\tilde{\Psi}$ is the eigenfunction of the operators $\hat{j}^2$, $\hat{l}^2$, $\hat{j}_z$, so that $\Psi_{jl_1j_z} \equiv \tilde{\Psi} = (\hat{\boldsymbol{\sigma}} \cdot \mathbf{n})\Psi_{jl_2j_z}$. We should note that after averaging over the particle spin state, the relation appears:

$$\langle \tilde{\Psi}|\tilde{\Psi}\rangle = \langle \Psi_{jl_2j_z}|(\hat{\boldsymbol{\sigma}} \cdot \mathbf{n})^2|\Psi_{jl_2j_z}\rangle = \langle \Psi_{jl_2j_z}|\Psi_{jl_2j_z}\rangle.$$

This new function expresses the same characteristics of the angle distributions (over the directions of $\mathbf{n}$) as the original.

2) Now it is easily seen that

$$\Psi_{jj_z,\lambda=\pm 1/2} = \frac{1}{\sqrt{2}}(1 \pm (\hat{\boldsymbol{\sigma}} \cdot \mathbf{n}))\Psi_{jlj_z}.$$

Taking into account the explicit form of the functions Ψ_{jlj_z} from the previous problem, we can find

$$\Psi_{jj_z,\lambda=1/2} = \sqrt{\frac{2}{(2l+1)}}\left\{\sqrt{l+m+1}\cos\left(\frac{\theta}{2}\right)Y_{lm}\right.$$

$$\left. + \sqrt{l-m}\sin\left(\frac{\theta}{2}\right)e^{-i\varphi}Y_{l,m+1}\right\} \times \begin{pmatrix} \cos\left(\frac{\theta}{2}\right) \\ e^{i\varphi}\sin\left(\frac{\theta}{2}\right) \end{pmatrix}$$

$$\Psi_{jj_z,\lambda=-1/2} = \sqrt{\frac{2}{(2l+1)}}\left\{\sqrt{l+m+1}\sin\left(\frac{\theta}{2}\right)Y_{lm}\right.$$

$$\left. - \sqrt{l-m}\cos\left(\frac{\theta}{2}\right)e^{-i\varphi}Y_{l,m+1}\right\} \times \begin{pmatrix} \sin\left(\frac{\theta}{2}\right) \\ -e^{i\varphi}\cos\left(\frac{\theta}{2}\right) \end{pmatrix}.$$

Here $m = j_z - 1/2$. Note that the spin part of the wavefunction is the same as in Eq. (V.4) and Problem 5.3.

Problem 5.26

A particle with the spin $s = 1$ can be described both by a symmetric spinor[63] of the second rank $\psi^{\alpha\beta}(\mathbf{r})$ (the spinor representation), and by a vector function, $\mathbf{V}(\mathbf{r})$ (the vector representation).

Indicate

1) the form of the spin operator in these representations;
2) the connection of such wavefunctions with the wavefunction $\psi(\mathbf{r}, \sigma)$ in the s_z-representation;
3) the explicit form of these wavefunctions for a particle state with orbital angular momentum $l = 1$ and total angular momentum $j = 0$.

Solution

For $s = 1$, the description of a particle spin properties in the terms of the symmetric spinor $\psi^{\alpha\beta}(\mathbf{r})$ is analogous to the description of states with total spin 1 in a system of two spins with $s = 1/2$ (compare with Problems 5.10 and 5.11). By analogy with expression $\hat{\mathbf{S}} = \frac{1}{2}(\hat{\boldsymbol{\sigma}}_1 + \hat{\boldsymbol{\sigma}}_2)$, the form of the spin component operators in the spinor representation immediately follows:

$$\hat{s}\psi^{\alpha\beta} \equiv s^{\alpha\beta}_{\mu\nu}\psi^{\mu\nu}, \quad s^{\alpha\beta}_{\mu\nu} = \frac{1}{2}(\sigma^{\alpha}{}_{\mu} \cdot \delta^{\beta}_{\nu} + \delta^{\alpha}_{\mu} \cdot \sigma^{\beta}{}_{\nu}). \tag{1}$$

The relation between wavefunctions $\psi^{\alpha\beta}$ in the spinor representation and the wavefunctions in the s_z-representation, $\psi(s_z)$ is[64]

$$\psi^{\alpha\beta} = \sum_{s_z} \psi(s_z) \psi^{\alpha\beta}_{s_z}, \tag{2}$$

where the spinor $\psi^{\alpha\beta}_{s_z}$ is an eigenfunction of the operator $\hat{s}_z$. These spinors that correspond to different values of s_z have the form (compare to Problem 5.10)

$$\psi^{\alpha\beta}_{s_z=1} = \delta^{\alpha}_1 \delta^{\beta}_1, \quad \psi^{\alpha\beta}_{s_z=0} = \frac{1}{\sqrt{2}}(\delta^{\alpha}_1 \delta^{\beta}_2 + \delta^{\alpha}_2 \delta^{\beta}_1), \quad \psi^{\alpha\beta}_{s_z=-1} = \delta^{\alpha}_2 \delta^{\beta}_2. \tag{3}$$

Spinor components δ^{α}_{β} are equal to $\delta^1_1 = \delta^2_2 = 1$, $\delta^1_2 = \delta^2_1 = 0$. From (2) and (3), it follows that

$$\psi^{11} = \psi(1), \quad \psi^{22} = \psi(-1), \quad \psi^{12} = \psi^{21} = \frac{1}{\sqrt{2}}\psi(0). \tag{4}$$

---

[63] To solve Problems 5.26 and 5.27 it is necessary to know the basics of the spinor algebra and the relation between spinors and tensors. Here we should distinguish co-variant and contra-variant spinor components. We should write the simple Pauli matrices (V.2) not in the form $(\sigma_i)_{\alpha\beta}$ but as $(\sigma_i)^{\alpha}{}_{\beta}$; here i is vector index, while α and β are contra-variant and co-variant spinor indexes; pay attention to the order of such indexes in Pauli matrices. In Problems 5.26 and 5.27, vector indexes are represented by Latin characters, and the spinor indexes by Greek characters.

[64] Pay attention to the dual meaning of the variable s_z in relation (2): as an argument of the wavefunction $\psi(s_z)$ it is a variable of s_z-representation, while in the case of spinor $\psi^{\alpha\beta}_{s_z}$ it is an eigenvalue of the operator $\hat{s}_z$.

A description of a spin $s = 1$ particle in the vector representation is analogous to that used in Problem 3.44 for $l = 1$. In this representation,

$$\hat{s}_i V_k \equiv s_{i,kl} V_l, \quad s_{i,kl} = -i\varepsilon_{ikl}. \tag{5}$$

The generalization of (2) for the vector representation has the form

$$\mathbf{V} = \sum_\sigma \psi(\sigma) \mathbf{v}_\sigma.$$

The vectors $\mathbf{v}_\sigma$, which are the eigenfunctions of $\hat{s}_z$, have the form[65]

$$\mathbf{v}_{s_z=\pm 1} = \mp \frac{1}{\sqrt{2}}(1, \pm i, 0), \quad \mathbf{v}_{s_z=0} = (0, 0, 1) \tag{6}$$

(compare with Problem 3.41). Due to the mutual orthogonality of these vectors it follows that $\psi(\sigma) = \mathbf{v}_\sigma^* \mathbf{V}$, so we have the relation between the wavefunctions in vector and s_z-representations

$$\psi(\pm 1) = \frac{1}{\sqrt{2}}(\mp V_x + iV_y), \quad \psi(0) = V_z, \tag{7}$$

$$V_x = \frac{1}{\sqrt{2}}(\psi(-1) - \psi(1)), \quad V_y = -\frac{i}{\sqrt{2}}(\psi(1) + \psi(-1)).$$

From (4) and (7) we obtain the relations between the spinor and vector wavefunctions

$$V_x = \frac{1}{\sqrt{2}}(\psi^{22} - \psi^{11}), \quad V_y = -\frac{i}{\sqrt{2}}(\psi^{11} + \psi^{22}), \quad V_z = \sqrt{2}\psi^{12}. \tag{8}$$

These relations may be written more clearly with

$$\mathbf{V} = C\boldsymbol{\sigma}^\gamma{}_\alpha \psi^{\alpha\beta} g_{\gamma\beta} \equiv C\boldsymbol{\sigma}^\gamma{}_\alpha \psi^\alpha_\gamma. \tag{9}$$

$g_{\beta\gamma}$ is an antisymmetric unit spinor of rank 2, whose components are of the form $g_{12} = -g_{21} = 1$, $g_{11} = g_{22} = 0$ and $C = 1/\sqrt{2}$, while $\psi^\alpha_\beta = \psi^{\alpha\gamma} g_{\beta\gamma}$ (and $\psi^\alpha_\alpha = 0$ due to the symmetric property of the spinor ψ^α_β). In this form, the relation of wavefunctions $\mathbf{V}$ and $\psi^{\alpha\beta}$ with each other is immediately obvious, since $\boldsymbol{\sigma}^\gamma{}_\alpha \psi^\alpha_\gamma$ is the only vector (up to a factor) that could be correlated to the spinor $\psi^{\alpha\beta}$. Using the relation

$$\boldsymbol{\sigma}^\alpha{}_\beta \cdot \boldsymbol{\sigma}^\mu{}_\nu = 2\delta^\alpha_\nu \delta^\mu_\beta - \delta^\alpha_\beta \delta^\mu_\nu,$$

expression (9) can be reversed:

$$\psi^\alpha_\beta = \frac{1}{\sqrt{2}} \mathbf{V} \cdot \boldsymbol{\sigma}^\alpha{}_\beta, \quad \mathbf{V} = \frac{1}{\sqrt{2}} \boldsymbol{\sigma}^\alpha_\beta \psi^\beta_\alpha,$$

---

[65] We should note that the choice of phase factors in the relations (3), (6) for eigenfunctions v_{s_z}, $\psi^{\alpha\beta}_{s_z}$ for different values of s_z is a choice based on momentum theory.

$$\psi^{\alpha\beta} = \frac{1}{2}(\psi_\mu^\alpha g^{\mu\beta} + \psi_\mu^\beta g^{\mu\alpha}) \tag{10}$$

$$\psi_1^1 = -\psi_2^2 = \psi^{12}, \quad \psi_2^1 = \psi^{11}, \quad \psi_1^2 = \psi^{22}.$$

Values of anti-symmetric spinor contra-variant components $g^{\alpha\beta}$ coincide with those of $g_{\alpha\beta}$.

Let us consider the spin-orbital eigenfunctions $|\psi_{jlj_z}\rangle$ for a particle with the spin $s = 1$ in different representations. In the general form we have to use the relation from the theory of angular momenta addition (the orbital l, spin $s = 1$ and total j):

$$|\psi_{jlj_z}\rangle = \sum_\sigma C_{lm1\sigma}^{jj_z} Y_{lm}(\mathbf{n})|1,\sigma\rangle, \tag{11}$$

where $|l,\sigma\rangle$ is the purely spin (*i.e.*, it does not depend on coordinates) eigenfunction of the operator $\hat{s}_z$ corresponding to the eigenvalue $s_z \equiv \sigma$; the Clebsch–Gordan coefficients in (11) are not equal to zero only for $j_z = m + \sigma$. From the form of the relation (11), the coefficient written in front of $|1,\sigma\rangle$ is the spin wavefunction of the state considered in the s_z-representation, *i.e.*,

$$\psi_{jlj_z}(\mathbf{n},\sigma) = C_{lm1\sigma}^{jj_z} Y_{lm}(\mathbf{n}), \quad j_z = m + \sigma. \tag{12}$$

On the other hand if we consider $|1,\sigma\rangle$ in (11) to be the basis vectors $\mathbf{v}_\sigma$ from (6) it would describe the spin-angular part of a particle wavefunction in vector representation. Or by replacing $|1,\sigma\rangle$ by spinors from (3), we come to the wavefunctions in spinor representation.

It is instructive, however, to consider the states with lower values of j without using (11). We will try to consider them from the general consideration, using the transformational properties of states wavefunctions corresponding to different values of angular momentum (compare with the problem from sec. 4 in Chapter 3).

We determine the form of the particle state wavefunction with $l = 1$ and $j = 0$. We see that it must depend linearly on vector $\mathbf{n}$ (because l=1; see Problem 3.41). Due to the spherical symmetry of the state with $j = 0$, it should not contain any "external" vector and spinor values. So we have the form of the wavefunction in the vector representation,[66] $V_i = cn_i$ or $\mathbf{V} = c\mathbf{n}$, where $|c| = 1/\sqrt{4\pi}$ from normalization. Of course, this result could be obtained from (11) (Clebsch–Gordan coefficients for this case are found in Problem 3.39). And now by Eq. (10) we easily find the wavefunctions of this state with $l = 1$, $j = 0$ in the spinor representation

$$\psi^{11} = -\frac{c}{\sqrt{2}}e^{-i\phi}\sin\theta, \quad \psi^{12} = \frac{c}{\sqrt{2}}\cos\theta, \quad \psi^{22} = \frac{c}{\sqrt{2}}e^{i\phi}\sin\theta. \tag{13}$$

---

[66] The classic example of a particle with the spin $s = 1$ is a photon. Here, due to the specific property of a photon that is connected with the fact that electromagnetic field is *transversal*, its wavefunction, which is the vector potential $\mathbf{A}(\mathbf{p})$ (in the momentum space), must satisfy the additional condition of the form $\mathbf{pA}(\mathbf{p}) = 0$ (or $\text{div}\mathbf{A}(\mathbf{r}) = 0$, see Chapter 14). The function $\mathbf{A} = \mathbf{f}(\mathbf{p})\mathbf{p}$ of a state with $\mathbf{j} = \mathbf{0}$ does not correspond to that condition. This means that the states of a photon with $\mathbf{j} = \mathbf{0}$ do not exist, so it is impossible for a system to emit it if its total moment both in initial and final states is equal to zero: "$0 - 0$" *transitions are forbidden*.

Angles θ, φ are the polar and azimuth angles of vector n direction, and from Eqs. (4) and (7) we find the wavefunction's s_z-representation,

$$\psi(1) = \frac{1}{\sqrt{3}} Y_{1,-1}(\mathbf{n}), \quad \psi(0) = -\frac{1}{\sqrt{3}} Y_{10}(\mathbf{n}), \quad \psi(-1) = \frac{1}{\sqrt{3}} Y_{11}(\mathbf{n})$$

in accordance with (12).

Finally we will make a remark on the form of the spin-angular wavefunctions in the case $j \neq 0$ by the example of a particle state with $l = 1$. Now the wavefunction contains "external" tensors that characterize the states with the angular momentum j not equal to zero (compare with Problem 3.41). Specifically in the vector representation the wavefunctions desired have the form

$$\mathbf{V}_{j=1}(\mathbf{n}) = \left(\frac{3}{\sqrt{8\pi}}\right)^{1/2} \boldsymbol{\epsilon} \times \mathbf{n}, \quad V_{j=2,i}(\mathbf{n}) = \left(\frac{3}{\sqrt{4\pi}}\right)^{1/2} \varepsilon_{ik} n_k, \quad (14)$$

and from the normalization condition, $\boldsymbol{\epsilon}^* \cdot \boldsymbol{\epsilon} = 1$, $\epsilon^*_{ik}\epsilon_{ik} = 1$. The concrete choice of $\boldsymbol{\epsilon}(j_z)$, $\epsilon_{ik}(j_z)$, for which vector functions (14) describe the states with the definite value of j_z, is determined by the result from Problem 3.41. The form of wavefunctions in other representations could be determined as above in the case of $j = 0$.

Problem 5.27

A particle with spin $s = 3/2$ can be described both by a symmetric spinor of the third rank, $\psi^{\alpha\beta\gamma}(\mathbf{r})$, and by a spinor-vector function $V_k^\alpha(\mathbf{r})$ which satisfies the additional condition $(\hat{\sigma}_k)_\beta^\alpha V_k^\beta = 0$. Indicate the form of the spin operator and the relations both between the wavefunctions in these representations themselves and between these wavefunctions and wavefunction $\psi(\mathbf{r}, \sigma)$ of the s_z-representation.

Obtain the explicit form of the wavefunctions for the particle states with orbital angular momentum $l = 1$ and total angular momentum $j = 1/2$.

Solution

For a particle with $s = 3/2$, describing its spin properties in terms of the symmetric spinor $\psi^{\alpha\beta\gamma}$ is analogous to describing the states with total spin $S = 3/2$ in a system consisting of three spins with $s = 1/2$. Compare with Problem 5.19. Similar to the previous problem, we have:

$$\mathbf{s}^{\alpha\beta\gamma}_{\mu\nu\tau} = \frac{1}{2}(\sigma^\alpha_{\ \mu} \cdot \delta^\beta_\nu \cdot \delta^\gamma_\tau + \delta^\alpha_\mu \cdot \sigma^\beta_{\ \nu} \cdot \delta^\gamma_\tau + \delta^\alpha_\mu \cdot \delta^\beta_\nu \cdot \sigma^\gamma_{\ \tau}), \quad (1)$$

$$\psi^{111} = \psi\left(\frac{3}{2}\right), \quad \psi^{112} = \frac{1}{\sqrt{3}}\psi\left(\frac{1}{2}\right), \quad \psi^{122} = \frac{1}{\sqrt{3}}\psi\left(-\frac{1}{2}\right), \quad \psi^{222} = \psi\left(-\frac{3}{2}\right).$$

(Here we list only the independent spinor components.)

Going to the spinor-vector representation is carried out by using the relation between the spinor of rank 2 and the vector as mentioned in the previous problem (Eqn. (9)):

$$\mathbf{V}^\alpha = \frac{1}{\sqrt{2}} \sigma^\nu{}_\beta g_{\nu\gamma} \psi^{\alpha\beta\gamma} \equiv \frac{1}{\sqrt{2}} \sigma^\nu{}_\beta \psi^{\alpha\beta}_\nu. \tag{2}$$

See that the additional condition[67] $\sigma^\alpha{}_\beta \cdot \mathbf{V}^\beta = 0$ follows automatically. The form of spin component operators in this representation is given by

$$(s_i)^\alpha_{\mu, kl} = -i\varepsilon_{ikl}\delta^\alpha_\mu + \frac{1}{2}\delta_{kl}\sigma_{i,\ \mu}{}^\alpha. \tag{3}$$

Compare with the previous problem.

To determine the form of the state wavefunctions with $l = 1$ and $j = 1/2$ we should remember that they depend linearly on the vector $\mathbf{n}$ and the "external" spinor χ^α that gives the state of a system with momentum $j = 1/2$. We pick the spinor χ^α so that δ^α_1 and δ^α_2 correspond to the system states with $j_z = 1/2$ and $-1/2$. The wavefunction of a particle state with $j = 1/2$, $l = 1$ in the spinor representation has the simplest form for the "mixed" (with co-variant and contra-variant indexes) spinor components:

$$\psi^{\alpha\beta}_\gamma = C\left\{\mathbf{n}\cdot\boldsymbol{\sigma}^\alpha_\gamma \chi^\beta + \mathbf{n}\cdot\boldsymbol{\sigma}^\beta_\gamma \chi^\alpha - \frac{1}{3}(\mathbf{n}\cdot\boldsymbol{\sigma}^\alpha_\mu \chi^\mu \delta^\beta_\gamma + \mathbf{n}\cdot\boldsymbol{\sigma}^\beta_\mu \chi^\mu \delta^\alpha_\gamma)\right\}. \tag{4}$$

We have used both the spinor symmetry over upper indexes and the relation $\psi^{\alpha\beta}_\alpha = 0$ for the spinor $\psi^{\alpha\beta}_\gamma = g_{\gamma\nu}\psi^{\alpha\beta\nu}$ to obtain this relation. Without detailed analysis of the state in the s_z-representation, we consider only the component

$$\psi\left(\sigma = +\frac{3}{2}\right) = \psi^{111} = -\psi^{11}_2 = -2C\sin\theta e^{-i\varphi}\chi^l.$$

It is obvious from the physical considerations that this is proportional to the spherical function $Y_{1,-l}(\mathbf{n})$ and the spinor component, χ^1, which corresponds to the value $j_z = \pm 1/2$, since $j = 1/2$, $l = 1$, $j_z = m + \sigma$, and $\sigma = +3/2$.

The spin-vector wavefunction V^α of a state with $j = 1/2$, $l = 1$ could be found by (2) and (4). Using the relation

$$\sigma^\alpha_{i,\beta}\sigma^\beta_{i,\alpha} = \delta_{ik}\delta^\alpha_\gamma + i\epsilon_{ikl}\sigma^\alpha_{l,\gamma}$$

(compare with Eq. (V.3)), we obtain

$$\mathbf{V}^\alpha = \frac{1}{\sqrt{2}}\sigma^\gamma_\beta \psi^{\alpha\beta}_\gamma = \frac{2\sqrt{2}}{3}C(2\mathbf{n} - i[\mathbf{n}\times\hat{\boldsymbol{\sigma}}])\chi^\alpha. \tag{5}$$

We can obtain this expression in different way. Note that the most general form of a spin-vector that depends linearly on $\mathbf{n}$ and χ^α is

---

[67] To prove it we should use the equality $\psi^{\alpha\beta}_\alpha = 0$ and the relation

$$\sigma^\alpha_\beta \cdot \sigma^\gamma_\nu = 2\delta^\alpha_\nu \delta^\gamma_\beta - \delta^\alpha_\beta \delta^\gamma_\nu.$$

$$\mathbf{V}^\alpha = c_1 \mathbf{n} \chi^\alpha + c_2 [\mathbf{n} \times \boldsymbol{\sigma}_\beta^\alpha] \chi^\beta.$$

The additional condition, $\boldsymbol{\sigma}_\gamma^\alpha \cdot \mathbf{V}^\gamma = 0$, leads to the relation $c_1 = 2ic_2$. We obtain (5). Compare with Problem 5.23.

5.3 Spin density matrix; Angular distributions in decays

Problem 5.28

A system of two particles with $s = 1/2$ is in a state with definite values of S and S_z (S is the total spin). Find density spin matrices for each of the particles in these states for a case when the averaging is performed over the spin state of the other particle.

Solution

The spin density matrix of particle 1, $\rho_{\sigma\sigma'}^{(1)}$, is expressed in terms of spin function $\psi_{SS_z}(\sigma_1, \sigma_2)$ of the system from the general equation (V.5):

$$\rho_{\sigma\sigma'}^{(1)} = \sum_{\sigma_2} \psi_{SS_z}(\sigma, \sigma_2) \psi_{SS_z}^*(\sigma', \sigma_2).$$

Using the known expressions for ψ_{SS_z} from Problem 5.10, we obtain

$$\rho_{\sigma\sigma'}^{(1)}(S=1, S_z=1) = \begin{pmatrix} 1 & 0 \\ 0 & 0 \end{pmatrix} \equiv \frac{1}{2}(1 + \hat{\sigma}_z)_{\sigma\sigma'}, \quad \rho_{\sigma\sigma'}^{(1)}(1,0) = \frac{1}{2}\begin{pmatrix} 1 & 0 \\ 0 & 1 \end{pmatrix}, \quad (1)$$

$$\rho_{\sigma\sigma'}^{(1)}(1,-1) = \begin{pmatrix} 0 & 0 \\ 0 & 1 \end{pmatrix} = \frac{1}{2}(1 - \hat{\sigma}_z)_{\sigma\sigma'}, \quad \rho_{\sigma\sigma'}^{(1)}(0,0) = \frac{1}{2}\begin{pmatrix} 1 & 0 \\ 0 & 1 \end{pmatrix}.$$

Comparing (1) with the general equation (V.7) for $\hat\rho$, we see that in the states with $S=1$, $S_z=0$, and $S=0$, the polarization vector $\mathbf{P}=0$. We have completely unpolarized states. For the case of $S_z = \pm 1$ we have $\mathbf{P} = (0,0,\pm 1)$, so $|\mathbf{P}|=1$. We have completely polarized states. Here the spin state is pure and $\hat\rho^2 = \hat\rho$ (this is connected with multiplicative form of the spin function for $S_z = \pm 1$). The density matrix for the second particle has the same form as for the first one.

Problem 5.29

A particle with spin $s = 1/2$ is in a state with definite values of j, l, and j_z. Find the spin density matrix describing the spin state of the particle irrespective of its position in space.

Solution

The spin density matrix has the form $\hat\rho = \frac{1}{2}(1 + \mathbf{P} \cdot \hat{\boldsymbol{\sigma}})$ where $\mathbf{P} = 2\bar{\mathbf{s}}$ is the polarization vector. This can easily be found by using the result of Problem 3.40 a:

$$\mathbf{P} = 2\bar{\mathbf{s}} = \frac{j(j+1) - l(l+1) + s(s+1)}{j(j+1)} \bar{\hat{\mathbf{j}}}$$

so[68] $\mathbf{P} = (0, 0, \pm\frac{j_z}{l+1/2})$, where the signs $\pm$ correspond to the values $j = l \pm 1/2$. In the cases $j_z = \pm j$ for $j = l + 1/2$ we have $|\mathbf{P}| = 1$ and the spin state is pure.

Problem 5.30

Indicate the restrictions on the quantum numbers,[69] the spin J and the intrinsic parity P, of a neutral particle A^0. Use the existence of the decay $A^0 \to \pi^+\pi^-$ with parity conservation. The quantum numbers of a *pion* are $J^P_{\pi^\pm} = 0^-$.

Determine the angular distribution in the system of a resting particle A^0 if before the decay it was in a state with a definite value of J_z. See also Problem 5.32.

Solution

Since the spin of a pion is $J_\pi = 0$, the total angular momentum J of two-pion system in the center-of-mass system (which is the rest frame for the particle A^0) coincides with the orbital angular momentum L of their relative motion, i.e., $J = L$, and at the same time (due to the angular momentum conservation) it is equal to spin J_A of A^0, i.e., $L = J_A$. Furthermore, the parity (in the center-of-mass system) of the two-pion system is $P_{2\pi} = (-1)^L P_\pi P_\pi = (-1)^{J_A}$ ($(-1)^L$ is the orbital parity of the pion pair). Under the assumption that in the decay considered, the parity is conserved, we see that the intrinsic parity P_A of the particle A^0 must be equal to $P_A = P_{2\pi} = (-1)^{J_A}$. Thus the possible quantum numbers of the particle A^0 are: $J^P_A = 0^+, 1^-, 2^+, \ldots$.

If the particle, A^0, is in a state with a definite value of J_z, then $L_z = J_z$ for the decay pions. Fixing of $L = J$ and $L_z = J_z$ determines uniquely the wavefunction angular dependence for two pions in the form $Y_{JJ_z}(\mathbf{n})$, where $\mathbf{n} = \mathbf{p}/p$, $\mathbf{p}$ is their relative momentum. So the angular distribution of a decay product has the form $\frac{dw}{d\Omega_\mathbf{n}} = |Y_{JJ_z}(\mathbf{n})|^2$. See also Problem 5.32.

Problem 5.31

Show that an existence for the K-meson (spin $J_K = 0$) of two decay chanels – into two pions, K$\to 2\pi$, and into three pions, K$\to 3\pi$ – indicates the non-conservation of parity for its decays,[70]

---

[68] Compare $\bar{\hat{\mathbf{j}}}$ with $\bar{\hat{\mathbf{l}}}$ from Problem 3.10 a.
[69] If we apply the parity conservation law for the decays, then when particles decay and form it is necessary to account for the *intrinsic parities*. See Problem 10.5 for the decay $A^0 \to 2\pi^0$.
[70] Before the discovery of the parity non-conservation it was believed that the two decay channels corresponded to different particles θ and τ. The solution of the $\theta - \tau$-problem stimulated the experiments by which non-conservation of parity for the *weak interaction* was discovered.

Solution

Due to the angular momentum conservation for a system of two pions from the decay, $L = 0$. The two-pion system has the positive parity (compare with the previous problem). At the same time, for a system of three pions with $L = 0$ (in the center-of-mass system) the orbital parity is positive (see Problem 3.47), while the intrinsic, and so the total, parity is negative. The phenomenon of the same particle having two decay channels with different parities of the final states indicates its non-conservation.

Problem 5.32

A resting particle, X, with spin J decays into two spinless particles (for example, into two pions). Determine the angle distribution of decay products if the initial particle

a) has a definite value J_z;
b) is in a state described by a spin density matrix $\rho_{mm'}$, where m is the spin z-projection.

As an illustration, consider the angle distribution of pions from the decay of a *vector* particle $V \to 2\pi$ ($J_V^P = 1^-$).

Solution

a) Due to the particles-decay products' angular momentum conservation, the orbital angular momentum of their relative motion is $L = J$ and $L_z = J_z$, which uniquely determines the wavefunction angular dependence in the form $Y_{JJ_z}(\mathbf{n})$ ($\mathbf{n} = \mathbf{p}/p$, where $\mathbf{p}$ is the momentum of the relative motion of the decay products), and the angular distribution of the decay products has the form $\frac{dw}{d\Omega_n} = |Y_{JJ_z}(\mathbf{n})|^2$.

b) Let $c(m)$ ($m = J, J-1, \ldots, -J$) be the normalized spin wavefunction of the decaying particle in the J_z-representation. Due to the angular momentum conservation across the decay, it also describes the state of the angular products. The angular part of the wavefunction has the form

$$\psi(\mathbf{n}) = \sum_m c(m) Y_{Jm}(\mathbf{n}),$$

and their particle angle distribution for the decay is described by

$$\frac{dw}{d\Omega_n} = \left| \sum_m c(m) Y_{Jm}(\mathbf{n}) \right|^2 = \sum_{mm'} c(m) c^*(m') Y_{Jm}(\mathbf{n}) Y^*_{Jm'}(\mathbf{n}). \tag{1}$$

The desired angle distribution is obtained from (1) by the substitution $c(m)c^*(m') \to \rho_{mm'} = \overline{c(m)c^*(m')}$, where $\hat{\rho}$ is the polarization density matrix of the decaying particle. In particular, we will consider the case of $J = 1$ using the form of spherical functions (III.7). We obtain

$$\frac{dw}{d\Omega_n} = \frac{3}{8\pi} \left\{ (\rho_{11} + \rho_{-1,-1}) \sin^2\theta + 2\rho_{00} \cdot \cos^2\theta - 2\mathrm{Re}\, \rho_{1,-1} \cdot \cos 2\varphi \cdot \sin^2\theta \right.$$

$$+2\mathrm{Im}\,\rho_{1,-1}\cdot\sin 2\varphi\cdot\sin^2\theta - \sqrt{2}\mathrm{Re}\,\rho_{1,0}\cdot\cos\varphi\cdot\sin 2\theta + \sqrt{2}\mathrm{Im}\,\rho_{1,0}\cdot\sin\varphi\cdot\sin 2\theta$$

$$+\sqrt{2}\mathrm{Re}\,\rho_{-1,0}\cdot\cos\varphi\cdot\sin 2\theta + \sqrt{2}\mathrm{Im}\,\rho_{-1,0}\cdot\sin\varphi\cdot\sin 2\theta\Big\}. \qquad (2)$$

The polar and azimuthal angles of the vector **n** are θ, φ, and $\rho_{11}+\rho_{00}+\rho_{-1,-1}=1$. Note that in the case of a completely unpolarized state we have $\rho_{ik}=\delta_{ik}/3$. (The angular distribution is isotropic).

Problem 5.33

Determine the angle distribution of decay $B \to \pi N$ products of some unstable particle B with spin $J_B = 1/2$, if

a) parity is conserved in the decay and parity of B is negative;
b) parity is conserved in the decay but parity of B is positive;
c) the decay takes place without parity conservation.

Assume that the spin state of the nucleon produced is not detected. Remember that the quantum numbers of nucleon and pion are $J_N^P=(1/2)^+$, $J_\pi^P=0^-$.

Solution

Let us determine the form of wavefunction spin-angular dependence properties of the πN-system with total angular momentum $J=1/2$. Since for the pion we have $J_\pi^P=0^-$ while for a nucleon $J_N^P=(1/2)^+$, we see that for a given value of J, the orbital angular momentum L could take only two values: $L = J \pm 1/2$. The parity of the πN-system is

$$P_{\pi N} = (-1)^L P_\pi P_N = (-1)^{L+1}.$$

Fixing the values J_B and P_B uniquely determines L. Taking into account everything mentioned above, we find:

a) $L=0$ for $P_B=-1$, so the wavefunction of the πN-system does not depend on angles, and its spin-angular dependence has the trivial form $\psi_{\pi N}=\chi^{(N)}$. Due to angular momentum conservation, spinor $\chi^{(N)}$ coincides with the spinor $\chi^{(B)}$ that describes the spin state of the particle B. Since the wavefunction does not depend on angles, the decay angular distribution is isotropic.

b) Now $L=1$, and the spin-angular part of the wavefunction has the form $\psi_{\pi N}=C(\hat{\boldsymbol{\sigma}}\cdot\mathbf{n})\chi$, where $\mathbf{n}=\mathbf{p}/p$ and $\chi=\chi^{(B)}$. See Problem 5.21. The angle distribution of the pions is given by[71]

---

[71] After the summation over the independent spin states of a nucleon. If we fix the spin state of the nucleon in the decay, which is described by spinor $\chi_\lambda^{(N)}$, then

$$\frac{dw_\lambda}{d\Omega_n} \propto \left|\chi_\lambda^{(N)*}(\hat{\boldsymbol{\sigma}}\cdot\mathbf{n})\chi^{(B)}\right|^2.$$

$$\frac{dw}{d\Omega_n} = (\psi_{\pi N})^*\psi_{\pi N} = |C|^2 \chi^{(B)*}(\hat{\boldsymbol{\sigma}}\cdot\mathbf{n})^2\chi^{(B)} = |C|^2\chi^{(B)*}\chi^{(B)} = \mathrm{const}.$$

As in case a, it is isotropic (for the same angular dependence as distributions with $L = J \pm 1/2$; see Problem 5.25).

c) Here, due to the non-conservation of parity, parity of the πN-system does not have a definite value. The wavefunction spin-angular dependence has the form of the superposition of wavefunctions considered in a and b:

$$\psi_{\pi N} = (a + b(\hat{\boldsymbol{\sigma}}\cdot\mathbf{n}))\chi^{(B)}.$$

The angle distribution of the decay products is described by

$$\frac{dw}{d\Omega_n} \propto (\psi_{\pi N})^*\psi_{\pi N} = \chi^{(B)*}(a^* + b^*(\hat{\boldsymbol{\sigma}}\cdot\mathbf{n}))(a + b(\hat{\boldsymbol{\sigma}}\cdot\mathbf{n}))\chi^{(B)}$$

$$= \chi^{(B)*}(|a|^2 + |b|^2 + 2(\mathrm{Re}\,ab^*)(\hat{\boldsymbol{\sigma}}\cdot\mathbf{n}))\chi^{(B)} \propto \left(1 + \frac{2\mathrm{Re}\,ab^*}{|a|^2 + |b|^2}\langle\boldsymbol{\sigma}\rangle_B\cdot\mathbf{n}\right), \quad (1)$$

Here as well as in case b, the summation was performed over the independent spin states of the nuclei that appears in the decay.

The characteristic feature of this distribution is connected with its "back and forth" asymmetry of pion escape with respect to the polarization vector $\mathbf{P} \equiv \langle\boldsymbol{\sigma}\rangle_B$ of decaying particle B. The existence of such a correlation between the directions of the polar vector $\mathbf{n}$ and the axial vector $\langle\boldsymbol{\sigma}\rangle_B$ that is not invariant with respect to the coordinate reflection is a smoking gun for parity non-conservation in the process considered.[72]

Problem 5.34

Consider the decay X→ a + B, where X and a are spinless, while the particle B has spin j. Find the polarization density matrix of particle B in the case of

a) fixed space quantization axis z;
b) choice of direction of such an axis along the direction of decay product motion (in the system where particle X at rest).

In the case of a, also determine the elements of density matrix, averaged over the directions of decay product motion.

---

[72] An example of a decay of this type, which proceeds without conservation of parity, is the decay of hyperon into nucleon and pion: for example, $\Lambda^0 \to p\pi^-$. The reader should prove that for such decays of unpolarized particles, a nucleon polarization appears which equals

$$\mathbf{P} = \frac{2\,\mathrm{Re}\,ab^*}{|a|^2 + |b|^2}\mathbf{n}.$$

Solution

a) Since the total angular momentum is $J = 0$, the spin-angle part of the final state wavefunction has the form

$$\psi_{J=0} = \sum_{m=-J}^{J} C_{j,m,j,-m}^{00} Y_{j,-m}(\mathbf{n}) \chi_m \equiv \sum_m c(\mathbf{n}, m) \chi_m. \tag{1}$$

The orbital angular momentum of relative motion is equal to the spin of particle B: j. $\mathbf{n}$ is the unit vector directed along the momentum of relative motion.

$$C_{j,m,j,-m}^{00} = (-1)^{j-m} \frac{1}{\sqrt{2j+1}^{1/2}}$$

are the Clebsch–Gordan coefficients (see Problem 3.39). χ_m is the eigenfunction component j_z operator of the particle B spin.

The quantities $c(\mathbf{n}, m)$ for fixed $\mathbf{n}$ can be considered as the spin wavefunction of particle B in the j_z-representation, so the density matrix has the form

$$\rho_{mm'} = Nc(\mathbf{n}, m)c^*(\mathbf{n}, m') = \frac{4\pi}{2j+1} (-1)^{m-m'} Y_{j,-m}(\mathbf{n}) Y_{j,-m'}^*(\mathbf{n}). \tag{2}$$

$N = 4\pi$ is the normalization coefficient. We should note that for fixed $\mathbf{n}$, the spin state of particle B is pure, since in that case $\hat{\rho}^2 = \hat{\rho}$. If we average $\rho_{mm'}$ over all the directions, $\mathbf{n}$, of the particles' escape after the decay, then, using the spherical functions' orthogonality, we obtain $\overline{\rho}_{mm'} = \delta_{m,m'}/(2j+1)$, which describes the density matrix of absolutely unpolarized states.

b) In (2) we should consider $\mathbf{n}$ to be directed along the z-axis. Since

$$|Y_{lm}(\theta = 0, \varphi)| = \left(\frac{2l+1}{4\pi}\right)^{1/2} \delta_{m,0}$$

then we find

$$\rho_{m,m'} = \delta_{m,0} \delta_{m',0}.$$

This result has a simple interpretation. It means that the particle spin projection along the $\mathbf{n}$-direction has a definite value equal to zero. This immediately follows from the angular momentum: in the problem statement we say that in any direction the projection of J is equal to zero, and since the angular momentum n-projection is equal to zero then the spin n-projection is also equal to zero.

Problem 5.35

The same conditions of previous problem, but now the particle B in turn decays into two spinless particles: B$\to$ b + c. Determine the distribution function for values of the angle γ between vectors $\mathbf{p}_a$ (particle a momentum in the rest frame of particle X) and $\mathbf{p}_b$ (particle b momentum in the rest frame of particle B[73]) that describes the correlation between the particles' escape directions.

Solution

The particle B spin z-projection directed along the vector $\mathbf{n}_0 = \mathbf{p}_a/p_a$ is equal to zero. Accordingly, particles b and c have, in the rest system of the particle B, the orbital angular momentum $l = j$ and its projection $l_z = 0$, so that their angle distribution is given by

$$\frac{dw}{d\Omega_\mathbf{n}} = |Y_{l0}(\mathbf{n})|^2 = \frac{2j+1}{4\pi} P_j^2(\mathbf{n}_0 \cdot \mathbf{n}).$$

Such a distribution also follows from the result of Problem 5.32 if for the density matrix $\rho_{mm'}$ we use its form given in b of the previous problem. Since $|P_l(\cos\theta)| \leq 1$ and $|P_l(\pm 1)| = 1$, then in the decay B→ b + c, particles fly out mainly along (or opposite to) the direction of the particle a momentum, if $j \neq 0$.

Problem 5.36

Find the relation between the spin density matrices, $\hat{\rho}^{(a,b)}(\mathbf{n})$, of the particles a and b that have spin 1/2 and are formed in the decay X→ a + b of some spinless particle X (the vector $\mathbf{n}$ is directed along the momentum of the particles' a and b relative motion). Consider the cases when in the decay:

a) parity is conserved,
b) parity is not conserved.

Solution

The spin density matrices have the form

$$\hat{\rho}^{(a,b)} = \frac{1}{2}(1 + \mathbf{P}_{a,b} \cdot \hat{\boldsymbol{\sigma}}).$$

These describe a spin state of one of the particles after averaging over the spin states of the other particle. Due to the spherical symmetry of the system considered ($J = 0$), the polarization vectors $\mathbf{P}_{a(b)} = \xi_{a(b)} \mathbf{n}$ may depend on the vector $\mathbf{n}$ only. Such a relation between the directions of the polar vector $\mathbf{n}$ and axial vector $\mathbf{P}$ which is not invariant with respect to coordinate inversion, could take place only if the parity is not conserved in the decay. If the parity is conserved, then $\xi_{a,b} = 0$ and the spin state of each particle is completely non-polarized.

When the parity in decay is not conserved, parameters $\xi_{a,b} \neq 0$, though there is a relation $\xi_a = -\xi_b$. Indeed, since $J = 0$, then the projection J over any direction is equal to zero. Now consider the averaged value of the total angular momentum projection,

$$\overline{J_\mathbf{n}} = \overline{s_\mathbf{n}^{(a)}} + \overline{s_\mathbf{n}^{(b)}} + \overline{l_\mathbf{n}} = 0,$$

---

[73] Carefully note that vectors $\mathbf{p}_a$ and $\mathbf{p}_b$ are defined with respect to the different frames of reference.

on the direction **n**. Since $\overline{l_n} \equiv 0$, then $\overline{s_n^{(b)}} = -\overline{s_n^{(b)}}$, and accordingly, $\mathbf{P}_a = -\mathbf{P}_b$ (since $\mathbf{P}s = 2\overline{\mathbf{s}}$).

An example of such a decay without parity conservation is π_{μ_2}-decay: $\pi^+ \to \mu^+ + \nu$. In this decay, muon and neutrino are totally polarized, $P = 1$, anti-parallel to their momenta.

5.4 Bound states of spin-orbit-coupled particles

Problem 5.37

The Hamiltonian of a spin-orbit-coupled particle with spin $s = 1/2$, moving freely in two spatial dimensions, can be written as ($\hbar = 1$):[74]

$$\hat{H}_{so} = -\frac{(\nabla_x^2 + \nabla_y^2)}{2m}\hat{\sigma}_0 + i\alpha(\hat{\sigma}_x \nabla_y - \hat{\sigma}_y \nabla_x) + i\beta(\hat{\sigma}_x \nabla_x - \hat{\sigma}_y \nabla_y). \quad (1)$$

Here, m is the mass of the particle, $\hat{\sigma}_0$ is the 2×2 identity matrix, $\hat{\sigma}_i$ are the Pauli matrices, and the hats denote matrices acting in spin space of the particle. α and β are real constants representing the strengths of the respective spin-orbit interaction. Consider now a system of two such spin-orbit coupled fermions which are identical and satisfy the Pauli exclusion principle.

a) Obtain the relevant two-particle Schrödinger equation in terms of a four-component spinor in momentum space for an arbitrary interaction between the fermions. Express the spinor in terms of the momenta of the two fermions, $\mathbf{k}_1$ and $\mathbf{k}_2$.
b) Reduce the interaction term to a simpler form by assuming that it is an attractive δ-function potential, that it is axially symmetric, and that we may ignore all higher harmonics of the scattering potential except for the lowest s-wave component.
c) Solve for the four-component spinor wavefunction with the potential obtained in b). Write the final expression for the wavefunction as a function of the relative and center-of-mass momenta of the two particles, i.e., $\mathbf{k} = (\mathbf{k}_1 - \mathbf{k}_2)/2$ and $\mathbf{Q} = \mathbf{k}_1 + \mathbf{k}_2$.
d) Find a self-consistency equation satisfied by the wavefunction.
e) Find an approximate analytical expression for the bound state energy of the molecule with zero center-of-mass momentum, i.e., $\mathbf{Q} = \mathbf{0}$. Assume here that $\beta > 0$, and clearly state the regime of validity of the result. During the calculation,

---

[74] The Hamiltonian, H_{so}, describes, for instance, the motion of an electron in a two-dimensional electron gas in semiconductor heterostructures with structural inversion asymmetry and bulk inversion asymmetry. In the so-called III-V (e.g. GaAs) and II-VI (e.g. ZnSe) semiconductors, bulk inversion symmetry is broken due to the existence of two distinct atoms in the Bravais lattice. This gives rise to Dresselhaus spin-orbit interaction, which is modeled by the term proportional to β in H_{so}. Structural inversion asymmetry arises in the presence of an external or built-in electric field that makes the conduction band energy profile inversion-asymmetric in the direction of the electric field. This gives rise to so-called Rashba spin-orbit interaction modeled by the term proportional to α in H_{so}.

an ultraviolet-divergent momentum integral may be encountered. In that case, introduce a cutoff $k_c = 1/R_e$, where R_e physically corresponds to the characteristic spatial range of the interaction.

f) Find the energy spectrum of the bound molecule for small $\mathbf{Q}$, and extract the effective mass of the molecule.

g) Repeat d), e), and f) for the case $\beta = 0$ (the 'purely Rashba' case).

Solution

a) We may write the Schrödinger equation for this two-fermion system in terms of a four-component wavefunction, $|\Phi(\mathbf{k}_1, \mathbf{k}_2)\rangle$, written in the basis, $|\uparrow, \uparrow\rangle$, $|\uparrow, \downarrow\rangle$, $|\downarrow, \uparrow\rangle$, and $|\downarrow, \downarrow\rangle$, where the two arrows represent the spin projections of the two particles. In momentum space we then have

$$\left[\hat{H}_{so}(\mathbf{k}_1) \otimes \hat{\sigma}_0 + \hat{\sigma}_0 \otimes \hat{H}_{so}(\mathbf{k}_2) + \check{V}_{12}\right] |\Phi(\mathbf{k}_1, \mathbf{k}_2)\rangle = E|\Phi(\mathbf{k}_1, \mathbf{k}_2)\rangle.$$

$\check{V}_{12}$ is some arbitrary interaction between the two particles, and the check denotes 4×4 matrices that act in the above-mentioned four-component basis. $\hat{H}_{so}(\mathbf{k}_i)$ is the Hamiltonian (1) for particle i written in momentum space.

b) In momentum space, the interaction term has the form

$$\check{V}_{12}|\Phi(\mathbf{k}_1, \mathbf{k}_2)\rangle = \int \frac{d^2\mathbf{k}'}{(2\pi)^2} \check{V}(\mathbf{k} - \mathbf{k}')|\tilde{\Phi}(\mathbf{k}, \mathbf{Q})\rangle,$$

where $|\tilde{\Phi}(\mathbf{k}, \mathbf{Q})\rangle \equiv |\Phi(\mathbf{Q}/2 + \mathbf{k}, \mathbf{Q}/2 - \mathbf{k})\rangle = |\Phi(\mathbf{k}_1, \mathbf{k}_2)\rangle$. For an axially symmetric interaction potential one can write $\check{V}(\mathbf{k} - \mathbf{k}') = \sum_{l=-\infty}^{\infty} \check{V}_l(k, k') e^{il(\varphi_\mathbf{k} - \varphi_{\mathbf{k}'})}$, where $\varphi_\mathbf{k}$ is the angle between $\mathbf{k}$ and the x-axis. Assuming a short-ranged (i.e., δ-function) attractive interaction, keeping only the s-wave component (i.e., $l = 0$), and imposing anti-symmetry of the wavefunction (since we have two *fermions*), we may replace $\check{V}(\mathbf{k} - \mathbf{k}') \to V_0 \mathcal{P}^{(s)}$, where $V_0 \langle 0$ is the s-wave attractive interaction strength, and $\mathcal{P}^{(s)}$ is a projection operator which projects out the singlet component of the wavefunction. The interaction term then reduces to

$$\check{V}_{12}|\Phi(\mathbf{k}_1, \mathbf{k}_2)\rangle \to V_0 \int \frac{d^2\mathbf{k}'}{(2\pi)^2} \mathcal{P}^{(s)}|\tilde{\Phi}(\mathbf{k}', \mathbf{Q})\rangle.$$

c) The interaction potential from b) can be rewritten as

$$V_0 \int \frac{d^2\mathbf{k}'}{(2\pi)^2} \mathcal{P}^{(s)}|\tilde{\Phi}(\mathbf{k}', \mathbf{Q})\rangle = c_\mathbf{Q} V_0 |0, 0\rangle, \qquad (2)$$

where $|0, 0\rangle \equiv (|\uparrow, \downarrow\rangle - |\downarrow, \uparrow\rangle)/\sqrt{2}$ denotes the singlet state, and $c_\mathbf{Q}$ is a normalization constant. The Schrödinger equation can then be rewritten as

$$\check{G}_2^{-1}|\Phi(\mathbf{k}_1, \mathbf{k}_2)\rangle \equiv \left[\hat{G}^{-1}(\mathbf{k}_1) \otimes \hat{\sigma}_0 + \hat{\sigma}_0 \otimes \hat{G}^{-1}(\mathbf{k}_2)\right] |\Phi(\mathbf{k}_1, \mathbf{k}_2)\rangle = -c_\mathbf{Q} V_0 |0, 0\rangle,$$

with $\hat{G}^{-1}(\mathbf{k}_i) = s_i \hat{\sigma}_0 + \alpha \hat{\boldsymbol{\sigma}} \cdot (\mathbf{b}_i \times \mathbf{e}_z)$. Here we have defined $s_i = k_i^2/2m - (E/2)$, $\mathbf{b}_i = (k_{ix} + \gamma k_{iy}, \gamma k_{ix} + k_{iy})$, and $\gamma = \beta/\alpha$ is the ratio of the Dresselhaus to Rashba interaction strengths. The single-particle inverse Green function, $\hat{G}^{-1}$, can be diagonalized by the unitary matrix,

$$\hat{U}_i = \exp\left[-i\frac{\pi}{4}\left(\frac{\mathbf{b}_i}{|\mathbf{b}_i|}\cdot\hat{\boldsymbol{\sigma}}\right)\right]$$

(a $\pi/2$-rotation around, $\mathbf{b}_i$), as $\hat{U}_i^\dagger \hat{G}^{-1}(\mathbf{k}_i)\hat{U}_i = s_i\hat{\sigma}_0 + \alpha b_i\hat{\sigma}_z$. Therefore, the two-particle Green function, $\check{G}_2$, can be diagonalized by the unitary matrix, $\check{U} = \hat{U}_1 \otimes \hat{U}_2$ as $\check{G}_2 = \check{U}\check{D}\check{U}^\dagger$ with $\check{D} = \operatorname{diag}\{d_1, d_2, d_3, d_4\} \equiv \operatorname{diag}\{(s + \alpha b_1 + \alpha b_2)^{-1}, (s + \alpha b_1 - \alpha b_2)^{-1}, (s - \alpha b_1 + \alpha b_2)^{-1}, (s - \alpha b_1 - \alpha b_2)^{-1}\}$, and $s = s_1 + s_2$. Finally, we obtain the wavefunction

$$|\Phi(\mathbf{k}_1, \mathbf{k}_2)\rangle = -V_0 c_\mathbf{Q} \check{G}_2 |0, 0\rangle$$

$$= -\frac{V_0 c_\mathbf{Q}}{4\sqrt{2}} \begin{pmatrix} ie^{-i\theta_2}(+-+-) - ie^{-i\theta_1}(++--) \\ (++++) - e^{i(\theta_2-\theta_1)}(+--+) \\ e^{-i(\theta_2-\theta_1)}(+--+) - (++++) \\ -ie^{i\theta_1}(++--) + ie^{i\theta_2}(+-+-) \end{pmatrix},$$

where $(p_1 p_2 p_3 p_4) = \sum_{i=1}^4 p_i d_i$ with $p_i = \pm 1$, and $\theta_i = \tan^{-1}(b_{iy}/b_{ix})$. After some algebra, the wavefunction becomes

$$|\tilde{\Phi}(\mathbf{k}, \mathbf{Q})\rangle = -\frac{c_\mathbf{Q} V_0}{d(\mathbf{k}, \mathbf{Q})}$$

$$\times \begin{pmatrix} i\sqrt{2}(s^2(\mathbf{k},\mathbf{Q})\alpha b(\mathbf{k})e^{-i\theta_b} - \alpha^3 B(\mathbf{Q})(\mathbf{B}(\mathbf{Q})\cdot\mathbf{b}(\mathbf{k}))e^{-i\theta_B}) \\ -2is(\mathbf{k},\mathbf{Q})\alpha^2(\mathbf{b}(\mathbf{k})\times\mathbf{B}(\mathbf{Q}))_z \\ i\sqrt{2}(s^2(\mathbf{k},\mathbf{Q})\alpha b(\mathbf{k})e^{i\theta_b} - \alpha^3 B(\mathbf{Q})(\mathbf{B}(\mathbf{Q})\cdot\mathbf{b}(\mathbf{k}))e^{i\theta_B}) \\ s(\mathbf{k},\mathbf{Q})(s^2(\mathbf{k},\mathbf{Q}) - \alpha^2 B^2(\mathbf{Q})) \end{pmatrix}, \quad (3)$$

where the four-component wavefunction is now written in the basis of the three-triplet and one-singlet states: $|1,1\rangle, |1,0\rangle, |1,-1\rangle, |0,0\rangle$. Here, $\mathbf{b}(\mathbf{k}) = (k_x + \gamma k_y, \gamma k_x + k_y)$, $\mathbf{B}(\mathbf{Q}) = (Q_x + \gamma Q_y, \gamma Q_x + Q_y)$, $s(\mathbf{k},\mathbf{Q}) = (k^2/m) + (Q^2/4m) - E(\mathbf{Q})$, and $d(\mathbf{k},\mathbf{Q}) = d_1(\mathbf{k},\mathbf{Q})d_2(\mathbf{k},\mathbf{Q})d_3(\mathbf{k},\mathbf{Q})d_4(\mathbf{k},\mathbf{Q})$. The angles are given by $\theta_b = \tan^{-1}(b_y/b_x)$ and $\theta_B = \tan^{-1}(B_y/B_x)$.

d) Using relation (2) and noting that all components of the wavefunction (3) are odd under $\mathbf{k} \to -\mathbf{k}$ except for the singlet component, we find

$$\int \frac{d^2 k'}{(2\pi)^2} |\tilde{\Phi}(\mathbf{k}', \mathbf{Q})\rangle = c_\mathbf{Q}|0,0\rangle. \quad (4)$$

e) Using the wavefunction (3), the self-consistency condition (4) can be rewritten as

$$\frac{1}{|V_0|} = \int \frac{d^2 k}{(2\pi)^2} \frac{s(\mathbf{k},\mathbf{Q})[s^2(\mathbf{k},\mathbf{Q})) - \alpha^2 B^2(\mathbf{Q})]}{s^4(\mathbf{k},\mathbf{Q}) - 4\alpha^2 s^2(\mathbf{k},\mathbf{Q})[b^2(\mathbf{k}) - B^2(\mathbf{Q})/4] + 4\alpha^4 (\mathbf{B}(\mathbf{Q})\cdot\mathbf{b}(\mathbf{k}))^2}. \quad (5)$$

At $\mathbf{Q} = \mathbf{0}$, the self-consistency equation (5) reduces to

$$\int \frac{d^2\mathbf{k}}{(2\pi)^2} \frac{s(\mathbf{k},0)}{s^2(\mathbf{k},0) - 4\alpha^2 b^2(\mathbf{k})} = \frac{1}{|V_0|}, \qquad (6)$$

Noticing that the threshold energy for molecular formation is $E_{th} = -m\alpha^2(1+\gamma)^2$, the dimensionless binding energy, $\delta > 0$, can be defined via $E(0) = -m\alpha^2(1+\gamma)^2(1+\delta)$. Doing the angular integration in (6), we arrive at

$$\frac{4\pi}{|v_0|} = \int_0^{\xi_c} \frac{[\xi + (1+\gamma)^2(1+\delta)]d\xi}{\sqrt{(\xi + (1+\gamma)^2(1+\delta))^2 - 4\xi(1-\gamma)^2}}$$

$$\times \frac{1}{\sqrt{(\xi + [1+\gamma]^2(1+\delta)]^2 - 4\xi(1+\gamma)^2}}, \qquad (7)$$

where $\xi = (k/(m\alpha))^2$, and $v_0 = V_0 m$ is the dimensionless attractive interaction strength. $\xi_c = 1/(m\alpha R_e)^2$ is a dimensionless ultraviolet cutoff where R_e is the characteristic radius of the interaction. Assuming a small binding energy (i.e., $\delta \ll \gamma \sim 1$) the integrand in (7) is strongly peaked for $\xi = \xi_0 \equiv (1+\gamma)^2(1+\delta)$ due to the near-vanishing of the second square-root factor in the denominator. Setting $\xi = \xi_0$ in all of the other factors (since they are regular at $\xi = \xi_0$), the integral can be done, and we obtain the binding energy

$$\delta \approx \frac{\xi_c}{(1+\gamma)^2} e^{-\frac{8\pi\sqrt{\gamma}}{|v_0|(1+\gamma)}},$$

where we have assumed $\xi_c \gg 1 + \gamma$. Therefore, the bound state energy is given by

$$E(0) = -m(\alpha+\beta)^2 \left[1 + \frac{\xi_c}{(1+\gamma)^2} e^{-\frac{8\pi\sqrt{\gamma}}{|v_0|(1+\gamma)}} \right].$$

f) The energy spectrum of the spin-orbit coupled molecule can again be obtained from the self-consistency condition (5) evaluated at $\mathbf{Q} \neq \mathbf{0}$. The integrals can be done analogously to part e), and we obtain

$$E(\mathbf{Q}) \approx -m(\alpha+\beta)^2 - \Lambda e^{-\frac{8\pi\sqrt{\gamma}}{|v_0|(1+\gamma)}}$$

$$+ \frac{1}{4(1+\gamma)^2} \left[(1+6\gamma+\gamma^2)\frac{Q^2}{2m} + 2(1-\gamma)^2 \frac{Q_x Q_y}{2m} \right],$$

where $\Lambda = m\alpha^2 \xi_c$ is the ultraviolet cutoff, and we have assumed small momenta $Q \ll m\alpha$. Therefore, the effective mass of the molecule, m^*, reads

$$\frac{1}{m^*} = \frac{1}{m} \frac{(1+6\gamma+\gamma^2) + (1-\gamma)^2 \sin 2\theta}{4(1+\gamma)^2},$$

where $\theta = \tan^{-1}(Q_y/Q_x)$.

g) The purely Rashba case can be considered by setting $\gamma = 0$ in the self-consistency condition (5). At $\mathbf{Q} = \mathbf{0}$, analogous calculation as part e) gives the dimensionless binding energy (keeping in mind that the threshold energy for molecular formation for this case is $E_{th} = -m\alpha^2$):

$$\delta \approx \frac{|v_0|^2}{16}.$$

The bound state energy is then

$$E(\mathbf{0}) \approx -m\alpha^2 \left(1 + \frac{|v_0|^2}{16}\right).$$

The spectrum of the Rashba molecule is

$$E(\mathbf{Q}) \approx -m\alpha^2 - m\alpha^2 \frac{|v_0|^2}{16} + \frac{Q^2}{8m},$$

where we once again expanded for $Q \ll m\alpha$. The effective mass here is then $m^* = 4m$.

5.5 Coherent-state spin path-integral

Problem 5.38

A useful quantity for statistical–mechanical calculations is the partition function, which given a Hamiltonian, $\hat{H}$, is defined as follows (for simplicity in this problem we take the Planck constant and the Boltzmann constant $\hbar = k_B = 1$):

$$\mathcal{Z} = \operatorname{tr} e^{-\beta \hat{H}},$$

where $\beta = 1/T$ is the inverse temperature. Sometimes the *path integral* representation of quantum mechanics is used to calculate the partition function.[75] The path integral is constructed by introducing a complete set of states parametrized by a continuous parameter such as what are known as *coherent states* (see below).

Consider an ensemble of spin-s particles with the Hamiltonian $\hat{H}$ expressed in terms of the usual operators $\hat{S}_x$, $\hat{S}_y$, and $\hat{S}_z$ satisfying the standard commutation relations, $[\hat{S}_i, \hat{S}_j] = i\epsilon_{ijk}\hat{S}_k$. With this, we define the *maximal* state $|s\rangle$ by $\hat{S}_z |s\rangle = s |s\rangle$. Coherent states are then defined (in one of many equivalent ways) as $|\mathbf{n}\rangle = e^{-i\phi \hat{S}_z} e^{-i\theta \hat{S}_y} |s\rangle$ (where $\mathbf{n}$ is a point on a unit sphere specified by two angles, θ and ϕ).

For this problem we consider the path integral for $\mathcal{Z} = \operatorname{tr} e^{-\beta \hat{H}}$, following these steps:

---

[75] The general applicability of the path-integral construction is an open mathematical question still under debate, and this section illustrates both its correct usage and its limitations.

1) Consider the *resolution of identity* for these coherent states
$$\frac{2s+1}{4\pi}\int d\mathbf{n}\,|\mathbf{n}\rangle\langle\mathbf{n}| = \hat{1}. \qquad (1)$$

Rewrite the trace of $e^{-\beta\hat{H}}$ (i.e., the partition function $\mathcal{Z}$) using this identity.

2) Consider the operator identity $e^{\hat{A}} = \lim_{N\to\infty}\left(1+\frac{\hat{A}}{N}\right)^N$. Between each multiplicative element, insert a *resolution of identity*. Then, recalling that $1+x \approx e^x$ when x is small, write the partition function as

$$\mathcal{Z} = \lim_{N\to\infty}\left(\frac{2s+1}{4\pi}\right)^N \int \prod_{i=0}^{N-1} d\mathbf{n}_i \exp(-S[\{\mathbf{n}_i\}]).$$

Find the explicit form of $S[\{\mathbf{n}_i\}]$ (S is called the *action*).

3) Assume continuity by letting $|\mathbf{n}_i\rangle - |\mathbf{n}_{i-1}\rangle \to 0$ as $N\to\infty$, and find the new form of S in the continuum limit (i.e., $N\to\infty$ and the index i becomes a continuous variable τ). With this limit, the partition function is now a path integral

$$\mathcal{Z} = \int \mathcal{D}\mathbf{n}(\tau)\, e^{-S[\{\mathbf{n}(\tau)\}]},$$

where loosely, $\mathcal{D}\mathbf{n}(\tau) \equiv \lim_{N\to\infty}\prod_{i=0}^{N-1}\frac{2s+1}{4\pi}d\mathbf{n}_i$. What are the conditions on the paths $\mathbf{n}(\tau)$?

4) Find the explicit form of $S[\{\mathbf{n}(\tau)\}]$ for $\hat{H}=0$ and $\hat{H}=\hat{\mathbf{S}}\cdot\mathbf{x}$, where $\mathbf{x}\cdot\mathbf{x}=1$.

Solution

Most of the derivation does not depend on the particular form of the coherent states, so for this derivation we will use an arbitrary set of states labeled with a continuous variable l and resolution of the identity

$$\int dl\, |l\rangle\langle l| = \hat{1}.$$

1) With the resolution of identity we can immediately write the partition function as

$$\mathcal{Z} = \int dl\, \langle l|e^{-\beta\hat{H}}|l\rangle. \qquad (2)$$

In fact, this just a particular way to take trace.

2) If we then break up the exponential with the identity $e^{\hat{A}} = \left(1+\frac{\hat{A}}{N}\right)^N$ and insert a resolution of the identity between each element $\left(1+\frac{\hat{A}}{N}\right)$, we obtain (letting $l \to l_0$ and $l_N \equiv l_0$)

$$\mathcal{Z} = \lim_{N\to\infty}\int \prod_{i=0}^{N-1} dl_i \prod_{j=1}^{N}\langle l_j|1-\frac{\beta\hat{H}}{N}|l_{j-1}\rangle.$$

At this point we define $\Delta\tau \equiv \beta/N$ and focus on the matrix element $\langle l_j | 1 - H\Delta\tau | l_{j-1}\rangle$. Before we apply the continuity condition, we first attempt to derive the partition function in terms of an action. Note that so far we have made no approximations.

Our first move is to factor out $\langle l_j | l_{j-1}\rangle$ and approximate the resulting term it multiplies as an exponential:

$$\langle l_j \left| 1 - \hat{H}\Delta\tau \right| l_{j-1}\rangle \approx \langle l_j | l_{j-1}\rangle e^{-H(l_j, l_{j-1})\Delta\tau},$$

where $H(l_j, l_{j-1}) \equiv \langle l_j | H | l_{j-1}\rangle / \langle l_j | l_{j-1}\rangle$. This move is justified, since N is large and the correction to the exponential will go as N^{-2}. With only N exponentials multiplying one another, $(e^{1/N^2})^N = e^{1/N} \to 1$. Just to get the other object into the exponential, we merely use the identity $x = \exp[\log(x)]$. At this point we have our first version of a (discrete) path integral, with no continuity assumption yet imposed:

$$Z = \lim_{N\to\infty} \int \prod_{i=0}^{N-1} dl_i \, \exp(-S[\{l_j\}]),$$

where

$$S[\{l_j\}] = -\sum_{j=1}^{N} \left[-\log(\langle l_j | l_{j-1}\rangle) + H(l_j, l_{j-1})\Delta\tau \right]. \tag{3}$$

3) At this point we impose the condition of continuity (which is a key *assumption* in the construction of the coherent-state path-integral). This condition states that $|l_j\rangle = |l_{j-1}\rangle + |\delta l_j\rangle$, where $|\delta l_j\rangle \to 0$ as $N \to \infty$.

We first handle the log term in the action given in (3) to obtain

$$\log(\langle l_j | l_{j-1}\rangle) \approx -\langle l_j | \delta l_j\rangle, \tag{4}$$

where we have explicitly used continuity to obtain this expression. For the term $H(l_j, l_{j-1})\Delta\tau$ we can simply set $l_{j-1} = l_j$, since it already multiplies $\Delta\tau$. Letting $H(l_j) = \langle l_j | \hat{H} | l_j\rangle = H(l_j, l_j)$, we then obtain the expression for the action:

$$S[\{l_j\}] = \sum_{j=1}^{N} \left[\frac{\langle l_j | \delta l_j\rangle}{\Delta\tau} + H(l_j) \right] \Delta\tau.$$

If we now let $N \to \infty$, the index j becomes a continuous variable τ, the term $\frac{|\delta l_j\rangle}{\Delta\tau} = \frac{|l_j\rangle - |l_{j-1}\rangle}{\Delta\tau} \to \partial_\tau |l(\tau)\rangle$, $\Delta\tau = d\tau$, and we obtain the Riemann integral:

$$S[\{l(\tau)\}] = \int_0^\beta d\tau \left\{ \langle l(\tau) | \partial_\tau | l(\tau)\rangle + H[l(\tau)] \right\}.$$

Remember that $l_N = l_0$, and so we have implicitly imposed $l(0) = l(\beta)$ (*i.e.*, the paths are closed). We further define the measure:

$$\mathcal{D}l(\tau) = \lim_{N \to \infty} \prod_{i=0}^{N-1} dl_i.$$

This leads to the final form of our path integral:

$$\mathcal{Z} = \int \mathcal{D}l(\tau) \exp\left(-S[\{l(\tau)\}]\right).$$

At this point we concentrate on the spin coherent-state path integral. The most interesting term is the first one, so we evaluate it:

$$\langle \mathbf{n}(\tau) | \partial_\tau | \mathbf{n}(\tau) \rangle = \langle s | e^{i\theta \hat{S}_y} e^{i\phi \hat{S}_z} \partial_\tau e^{-i\phi \hat{S}_z} e^{-i\theta \hat{S}_y} | s \rangle$$

$$= -i\partial_\tau \phi \, \langle s | \, e^{i\theta \hat{S}_y} \hat{S}_z e^{-i\theta \hat{S}_y} | s \rangle.$$

For clarity, we have dropped θ and ϕ's dependence on τ. There are a number of ways to solve the matrix element remaining, but we will solve it by noting that $\hat{S}_y$ rotates our spherical coordinates and that state $|s\rangle$ is at the top of a sphere of radius s. $\hat{S}_z$ measures the new z-component of the rotated state, so the matrix element will be given by $s \cos\theta$. Our action is thus given by

$$S[\{\mathbf{n}(\tau)\}] = \int_0^\beta d\tau \, \{-is \cos\theta \, \partial_\tau \phi + H(\theta, \phi)\}. \tag{5}$$

Sometimes, different derivations give the first term as $is(1 - \cos\theta)\partial_\tau \phi$, which is only different from ours by a total derivative (something that is usually inconsequential). This term, is sometimes called the Berry phase term, since it has a strong connection to what is known as the Berry phase. The next problem discusses this term in more detail (see also Problem 6.45).

4) If $\hat{H} = 0$, then $H(\theta, \phi) = 0$, and by (5) we can easily get our action. Note that it is *not* zero. There is still a term in the action.

If $\hat{H} = \hat{\mathbf{S}} \cdot \mathbf{x}$, then after a bit of work we have that $H(\mathbf{n}(\tau)) = \mathbf{n}(\tau) \cdot \mathbf{x}$. Note that we can rotate our z-axis onto $\mathbf{x}$ to make the action simple when we turn to coordinates θ, ϕ. In fact, this rotation is one way to easily solve this problem. (See the following problem for how changing this axis affects or does not affect the first term in the action).

Problem 5.39

Consider the Berry action for a free spin-s system given by

$$S_B = is \int_0^\beta d\tau \, [1 - \cos\theta(\tau)]\partial_\tau \phi(\tau), \tag{1}$$

which shows up in physical applications as e^{-S_B}. The integrand is derived from the form $\langle \mathbf{n}(\tau)|\partial_\tau|\mathbf{n}(\tau)\rangle$, where $|\mathbf{n}(\tau)\rangle$ is defined the previous problem. In addition, assume a closed path such that $\mathbf{n}(\beta) = \mathbf{n}(0)$. Show that this action has a geometric interpretation in terms of areas on the sphere, and further show its invariance with respect to the axis of quantization in the context of physical applications. (In the above case we quantized on the positive z axis, and a coordinate transformation can move this axis around the sphere.)

Solution

First we show that S_B is calculating a physical quantity that by its very nature is coordinate-invariant – the area on a sphere enclosed by a curve given by $\mathbf{n}(\tau)$. If we write (1) in terms of $\partial_\tau \mathbf{n}$, then we obtain

$$S_B = is \int_0^\beta d\tau\, \partial_\tau \mathbf{n} \cdot \mathbf{A},$$

where we have defined $\mathbf{A} = \frac{1-\cos\theta}{\sin\theta}\hat\phi$ and $\hat\phi$ is in the direction of the coordinate ϕ. This integral is actually a line integral of the vector potential $\mathbf{A}$ about the closed curve $\mathbf{n}(\tau)$. By Stokes' theorem we can the write expression in terms of a surface integral of the curl of $\mathbf{A}$,

$$S_B = is \int_{\Omega^+} dS\, \mathbf{r} \cdot (\nabla \times \mathbf{A}) = is \int_{\Omega^+} dS,$$

where $\mathbf{r}$ is the radial direction away from the center of the sphere, and Ω^+ is the region enclosed by the curve $\mathbf{n}(\tau)$ that includes the north pole.

Note the ambiguity when calculating the area enclosed by a curve on a sphere. Do we take the area that includes the north pole or the south pole? In this case, our choice of coordinates forced the north pole upon us – $\mathbf{A}$ is singular at the south pole, so many equations we have written six could not apply if we enclosed it. In fact, this is related to the so-called Dirac string. The singularity in $\nabla \times \mathbf{A}$ causes a flux at the south pole (one point!) that will balance the flux through the rest of the sphere. So our coordinate-choice problem boils down to the choice to include the north or south pole in our calculation of the area.

However, we could have equally chosen the north pole to carry the string. This would correspond to a change of coordines $\cos\theta \to -\cos\theta$ and $\partial\phi \to -\partial\phi$. And our new vector potential is $\mathbf{A}' = -\frac{1+\cos\theta}{\sin\theta}\hat\phi = \mathbf{A} - 2\nabla\phi$. This new vector potential differs from the old only by a *gauge transformation*, and is singular at the north pole. So when we perform a similar analysis to the above, we obtain for this transformed action

$$S'_B = -is \int_{\Omega^-} dS,$$

where Ω^- is the enclosed region including the south pole. A minus sign is picked up because if we assume the same orientation of our enclosing curve $\mathbf{n}(\tau)$, then the outward normal to our surface is actual in the negative orientation (recall that Stokes' theorem is sensitive to orientation).

If we remember that $\Omega^- = S^2 - \Omega^+$ (S^2 is the sphere), we can write

$$S'_B = -is \int_{S^2} dS + is \int_{\Omega^+} dS = -4\pi i s + S_B$$

and we see that the transformed S'_B differs from S_B by $-4\pi i s$. This difference is crucial, since recall that s is a half integer *and* only terms like e^{-S_B} are physically relevant. Thus, this difference in the actions actually goes away when calculating physical quantities ($e^{2\pi i n} = 1$ for integers n), and hence S_B is invariant with choice of quantization axis.

This procedure can also be used to produce the result that s must be a half integer, for otherwise the action ceases to make sense physically (it would depend on the choice of "gauge" which is related to the quantization axis).

Problem 5.40

Consider the resolution of the identity (Eq. (1) in Problem 5.38) for spin-1 coherent states ($s = 1$). Relating states and operators to the appropriate matrix representation (3×3 matrices acting on three element vectors), prove the resolution of the identity for this particular case.

Solution

To build coherent states for a spin-1 particle, we consider the standard matrices which make up the SU(2) (spin) algebra:

$$S_x = \frac{1}{\sqrt{2}} \begin{pmatrix} 0 & 1 & 0 \\ 1 & 0 & 1 \\ 0 & 1 & 0 \end{pmatrix}, \quad S_y = \frac{1}{\sqrt{2}} \begin{pmatrix} 0 & -i & 0 \\ i & 0 & -i \\ 0 & i & 0 \end{pmatrix}, \quad S_z = \begin{pmatrix} 1 & 0 & 0 \\ 0 & 0 & 0 \\ 0 & 0 & -1 \end{pmatrix}.$$

These matrices satisfy the commutation relations $[S_i, S_j] = i\epsilon_{ijk} S_k$. To construct the coherent states, we need to figure out how to evaluate the following (which is a coherent state as defined in the previously):

$$|\mathbf{n}\rangle := e^{-i\phi S_z} e^{-i\theta S_y} |1\rangle,$$

where $|1\rangle$ is the eigenstate of S_z with eigenvalue 1. The exponential of S_z is rather easy – as a diagonal matrix. For S_y it is rather more complicated, and we will use these facts (easily calculated):

$$S_y^2 = \frac{1}{2} \begin{pmatrix} 1 & 0 & -1 \\ 0 & 2 & 0 \\ -1 & 0 & 1 \end{pmatrix}, \quad S_y^3 = S_y.$$

Using these facts, we can evaluate the exponential of S_y (odd exponents are S_y, even are S_y^2):

$$e^{-i\theta S_y} = \sum_{n=0}^{\infty} \frac{(-i\theta)^n}{n!} S_y^n = 1 - S_y^2 + \sum_{n=0}^{\infty} \frac{(-i\theta)^{2n+1}}{(2n+1)!} S_y + \sum_{n=0}^{\infty} \frac{(-i\theta)^{2n}}{(2n)!} S_y^2$$

$$= 1 - S_y^2 - i\sin\theta S_y + \cos\theta S_y^2$$

$$= \begin{pmatrix} \frac{1}{2}(1+\cos\theta) & -\frac{1}{\sqrt{2}}\sin\theta & \frac{1}{2}(1-\cos\theta) \\ \frac{1}{\sqrt{2}}\sin\theta & \cos\theta & -\frac{1}{\sqrt{2}}\sin\theta \\ \frac{1}{2}(1-\cos\theta) & \frac{1}{\sqrt{2}}\sin\theta & \frac{1}{2}(1+\cos\theta) \end{pmatrix}.$$

For completeness we write down the other exponential:

$$e^{-i\phi S_z} = \begin{pmatrix} e^{-i\phi} & 0 & 0 \\ 0 & 1 & 0 \\ 0 & 0 & e^{i\phi} \end{pmatrix}.$$

Taking these together it is clear that the coherent states take the form

$$|\mathbf{n}\rangle = \tfrac{1}{2}(1+\cos\theta)e^{-i\phi}|1\rangle + \tfrac{1}{\sqrt{2}}\sin\theta|0\rangle + \tfrac{1}{2}(1-\cos\theta)e^{i\phi}|-1\rangle. \tag{1}$$

Now that we have derived the coherent states as defined, we move to calculating the resolution of the identity (also known as overcompleteness in this case).

All we need to do is show that we get a resolution of the identity from integrating over the outerproduct of these coherent states as so:

$$1 = \frac{3}{4\pi} \int_{S^2} d\mathbf{n} \, |\mathbf{n}\rangle\langle\mathbf{n}|.$$

Recall that the measure on a sphere is $d\mathbf{n} = d\phi d(\cos\theta)$. The matrix to be intregrated is

$$|\mathbf{n}\rangle\langle\mathbf{n}| = \begin{pmatrix} \frac{1}{4}(1+\cos\theta)^2 & \frac{1}{2\sqrt{2}}(1+\cos\theta)\sin\theta e^{-i\phi} & \frac{1}{4}(1-\cos^2\theta)e^{-2i\phi} \\ \frac{1}{2\sqrt{2}}(1+\cos\theta)\sin\theta e^{i\phi} & \frac{1}{2}\sin^2\theta & \frac{1}{2\sqrt{2}}(1-\cos\theta)\sin\theta e^{-i\phi} \\ \frac{1}{4}(1-\cos^2\theta)e^{2i\phi} & \frac{1}{2\sqrt{2}}(1-\cos\theta)\sin\theta e^{i\phi} & \frac{1}{4}(1-\cos\theta)^2 \end{pmatrix}.$$

Clearly, upon integration over ϕ, the off-diagonal terms drop out due to their periodic dependence on ϕ. The other terms are independent of ϕ, so they become multiplied by 2π upon integration. All that is left is the integral over $\cos\theta$ for which we make the assignment $x \equiv \cos\theta$:

$$\frac{3}{4\pi}\int_{S^2} d(\cos\theta) d\phi \, |\mathbf{n}\rangle\langle\mathbf{n}| = \frac{3}{2}\begin{pmatrix} \frac{1}{4}\int_{-1}^{1}(1+x)^2 dx & 0 & 0 \\ 0 & \frac{1}{2}\int_{-1}^{1}(1-x^2)dx & 0 \\ 0 & 0 & \frac{1}{4}\int_{-1}^{1}(1-x)^2 dx \end{pmatrix}$$
$$= 1.$$

Thus, we have proven the resolution of identity for this case.

This allows us to decompose any state into a sum of coherent states by just inserting the identity

$$|\psi\rangle = 1|\psi\rangle = \frac{3}{4\pi}\int_{S^2} d\mathbf{n} \, \langle\mathbf{n}|\psi\rangle|\mathbf{n}\rangle.$$

Problem 5.41

Consider a spin-s system and the following (alternative, yet equivalent) definition of the coherent states:

$$|\mathbf{n}\rangle = e^{\frac{1}{2}\theta[S_-e^{i\phi}-S_+e^{-i\phi}]}|s\rangle$$

with $|s\rangle$ still being the maximal state for S_z, and $S_\pm = S_x \pm iS_y$. Using this form of the state, derive an expansion of $|\mathbf{n}\rangle$ in terms of eigenstates of S_z so that $|\mathbf{n}\rangle = \sum_{m=-s}^{s} A_m(\mathbf{n})|m\rangle$.

Solution

The answer is

$$|\mathbf{n}\rangle = \sum_{m=-s}^{s} \binom{2s}{m+s} [\cos(\theta/2)]^{s+m}[\sin(\theta/2)]^{s-m} e^{i(s-m)\phi}|m\rangle.$$

Problem 5.42

For every operator $\hat{A}$, there are two ways of representing the function with coherent states.[76] One way is given by a function

$$a(\mathbf{n}) = \langle\mathbf{n}|\hat{A}|\mathbf{n}\rangle,$$

and the other is by a function $A(\mathbf{n})$ defined by the property

$$\hat{A} = \frac{2s+1}{4\pi}\int d\mathbf{n}\, A(\mathbf{n})|\mathbf{n}\rangle\langle\mathbf{n}|.$$

---

[76] See E. Lieb, *Commun. Math. Phys.* **31**, 327–340 (1973), for a detailed discussion of the mathematical aspects of this construction.

210 Exploring Quantum Mechanics

In general, these two quantities differ. Calculate these two functions for the following operators: $\hat{S}_x$, $\hat{S}_y$, $\hat{S}_z$, $\hat{S}_x^2$, $\hat{S}_y^2$, $\hat{S}_z^2$.

In general, there is an infinite set of functions $A(\mathbf{n})$, so to find a particular $A(\mathbf{n})$ calculate $a(\mathbf{n})$, then assume that $A(\mathbf{n})$ has a functional form similar to $a(\mathbf{n})$.

Solution

Symmetries are useful in constructing these, since $\hat{S}_z$ can be rotated into the other two operators. The result of this yields the following:

Operator	$a(\mathbf{n})$	$A(\mathbf{n})$
$\hat{S}_x$	$s\cos\theta$	$(s+1)\cos\theta$
$\hat{S}_y$	$s\sin\theta\cos\phi$	$(s+1)\sin\theta\cos\phi$
$\hat{S}_z$	$s\sin\theta\sin\phi$	$(s+1)\sin\theta\sin\phi$
$\hat{S}_x^2$	$s(s-\frac{1}{2})\sin^2\theta\cos^2\phi + s/2$	$(s+1)(s+3/2)\sin^2\theta\cos^2\phi - \frac{1}{2}(s+1)$
$\hat{S}_y^2$	$s(s-\frac{1}{2})\sin^2\theta\sin^2\phi + s/2$	$(s+1)(s+3/2)\sin^2\theta\sin^2\phi - \frac{1}{2}(s+1)$
$\hat{S}_z^2$	$s(s-\frac{1}{2})\cos^2\theta + s/2$	$(s+1)(s+3/2)\cos^2\theta - \frac{1}{2}(s+1)$

Note how the two columns are quite different from each other.

Problem 5.43

Consider a spin-1 system with the following two Hamiltonians:

a) $\hat{H} = \hat{S}_z$;
b) $\hat{H} = \hat{S}_z^2$.

For both Hamiltonians, calculate the partition function using the usual states and operators for a spin-1 system, and compare that with the result from calculating the path integral derived in the first problem of this section:

$$\mathcal{Z} = \int \mathcal{D}\mathbf{n}(\tau) \exp\left\{-\int_0^\beta \left[-\langle \mathbf{n}(\tau)|\partial_\tau|\mathbf{n}(\tau)\rangle + \langle \mathbf{n}(\tau)|\hat{H}|\mathbf{n}(\tau)\rangle\right] d\tau\right\}. \qquad (1)$$

The path integral itself, represented by the $\int \mathcal{D}\mathbf{n}(\tau)$ in (1), is over all closed paths (i.e., $\mathbf{n}(0) = \mathbf{n}(\beta)$).

To evaluate this path integral it is useful to note a couple of facts for path integrals:

1) The condition that a path is closed means that certain angles can be increased by an integer multiple of 2π. For instance, $\varphi(\beta) = \varphi(0) + 2\pi n$ where n is an integer still represents a closed path on the sphere. In fact, n is a *winding number* and can be used to separate *topologically distinct* paths by introducing a sum over their winding numbers, n.

2) By analogy with regular integrals, $\int \mathcal{D}x(\tau)\, e^{-i \int_0^\beta f(\tau)x(\tau)d\tau} = \delta(f)$.

   Note The results for the path integral representation and the operator representation may not agree. Speculate on why each case may or may not agree.

Solution

First, it is a straightforward exercise to evaluate $\mathcal{Z} = \text{tr}\, e^{-\beta \hat{H}}$ in the operator/matrix representation of quantum mechanics, and that results in the following (the subscripts a and b represent the two Hamiltonians):

$$\mathcal{Z}_a = e^{-\beta} + 1 + e^{\beta} \tag{2a}$$

$$\mathcal{Z}_b = 2e^{-\beta} + 1 \tag{2b}$$

The path integral is rather more difficult to calculate, and we will complete both a) and b) at the same time by noting that in both cases $\langle \mathbf{n}(\tau)|\, \hat{H}\, |\mathbf{n}(\tau)\rangle = H(\cos\theta(\tau))$, a function of $\cos\theta(\tau)$. Also, it can shown that $\langle \mathbf{n}(\tau)|\, \partial_\tau\, |\mathbf{n}(\tau)\rangle = -i(1-\cos\theta(\tau))\partial_\tau\varphi(\tau)$. If we write $\mathcal{Z} = \int \mathcal{D}\mathbf{n}(\tau) e^{-S[\mathbf{n}]}$, then we note that after performing an integration by parts, we obtain

$$S[\mathbf{n}] = -i(1-\cos\theta(0))(\varphi(\beta)-\varphi(0)) + i\int_0^\beta \varphi(\tau)\partial_\tau \cos\theta(\tau)\, d\tau$$

$$+ \int_0^\beta H(\cos\theta(\tau))\, d\tau.$$

At this point we can separate the paths by their winding numbers with $\varphi(\beta) - \varphi(0) = 2\pi n$, for winding number n. If we do this, and call $x(\tau) \equiv \cos\theta(\tau)$, then we can say that $\mathcal{D}\mathbf{n}(\tau) = \mathcal{D}x(\tau)\mathcal{D}\varphi(\tau)$, and

$$\mathcal{Z} = \sum_n \int \mathcal{D}x(\tau)\mathcal{D}\varphi(\tau)\, e^{-2\pi i n(1-x(0))} e^{i\int_0^\beta \varphi(\tau)\partial_\tau x(\tau)\, d\tau - \int_0^\beta H(x(\tau))\, d\tau}.$$

Using the path integral fact given in the problem statement, do the integral over φ to obtain $\delta(\partial_\tau x(\tau))$, and then do the integral over; $x(\tau)$ up to a constant that we call x_0 (which still must be integrated over; and recall that x_0 can only range from -1 to $+1$, due to its being the cosine of an angle). This considerably simplifies the path integral to just the expression

$$\mathcal{Z} = \sum_n \int_{-1}^1 dx_0\, e^{-2\pi i n(1-x_0)} e^{-\beta H(x_0)}.$$

The sum over n can be evaluated to a sum of δ-functions

$$\mathcal{Z} = \sum_k \int_{-1}^1 dx_0\, \delta(1-x_0-k) e^{-\beta H(x_0)},$$

where the sum over k is over all integers. Now the sum over x_0 will pick up only integers of x_0 in the interval $[-1, +1]$, so that the sum over k will only be non-zero for $x_0 = -1, 0, +1$, and we obtain

$$\mathcal{Z} = e^{-\beta H(-1)} + e^{-\beta H(0)} + e^{-\beta H(1)}. \tag{3}$$

We can now apply (3) to a) and b) to obtain new expressions for $\mathcal{Z}_a$ and $\mathcal{Z}_b$. For this, we note that for a), $H(\cos\theta) = \cos\theta$, and for b), $H(\cos\theta) = \frac{1}{2}(1 + \cos^2\theta)$. Using these appropriately, we obtain

$$\mathcal{Z}_a = e^{-\beta} + 1 + e^{\beta}, \tag{4a}$$

$$\mathcal{Z}_b = 2e^{-\beta} + e^{-\beta/2}. \tag{4b}$$

If we compare these results with (2a) and (2b), we first see that (4a) agrees with (2a), but then we notice that (4a) does *not* agree with (4b). In fact, they are different expressions. This is currently (as of 2011) an unresolved problem with evaluating the path integral.

Interestingly, part a) does actually have agreement between the two methods, and it is a linear sum of what are called *generators* of the Lie group SU(2) (whose generators are $\hat{S}_x$, $\hat{S}_y$, and $\hat{S}_z$). In fact, the path integral gives the correct result when the Hamiltonian is a linear sum of generators, and the coherent states are constructed using those generators. On the other hand, the Hamiltonian for part b) is quadratic in a generator, and it fails to produce correct results. This observation can be generalized to a number of other cases.

6
Time-dependent quantum mechanics

There are several ways (representations) that are used to describe time-dependent phenomena in quantum mechanics.

In *the Schrödinger representation* a wavefunction (a state vector) changes in time according to the Schrödinger equation

$$i\hbar\frac{\partial}{\partial t}\psi(q,t) = \hat{H}\psi(q,t), \tag{VI.1}$$

while operators of dynamic variables – e.g. coordinates $\hat{q}_i$, momenta $\hat{p}_i$ and spin $\hat{s}_i$ – do not depend on time.

For a time-independent Hamiltonian of a system, the wavefunction can be written in the form of an expansion:

$$\psi(q,t) = \sum_n c(E_n)e^{-iE_n t/\hbar}\psi_{E_n}(q), \tag{VI.2}$$

where $\psi_{E_n}(q)$ are a complete set of eigenfunctions for the Hamiltonian (that describe the *stationary states*). The coefficients in this expansion are uniquely defined by the initial value of the wavefunction:

$$c(E_n) = \int \psi_{E_n}^*(q)\psi(q,t=0)d\tau_q. \tag{VI.3}$$

If a quantum-mechanical operator $\hat{f} \equiv f(\hat{p},\hat{q},t)$ is correlated with some physical quantity f, then the operator that corresponds to the physical quantity $\dot{f} \equiv df/dt$ (a time derivative) is defined by

$$\hat{\dot{f}} \equiv \dot{\hat{f}} = \frac{\partial \hat{f}}{\partial t} + \frac{i}{\hbar}[\hat{H},\,\hat{f}]. \tag{VI.4}$$

Physical quantities for which $\hat{\dot{f}} = 0$ are called *constants of motion*.[77]

The time-dependent Green function $G(q,t;q',t')$ obeying the Schrödinger equation with respect to the variables q, t, and an initial condition of the form $G(q,t';q',t') = \delta(q-q')$, allows one to write the general solution to Eq. (VI.1) in the form

---

[77] For such quantities, the spectrum of eigenvalues and probability distribution for an arbitrary state do not depend on time.

$$\psi(q,t) = \int G(q,t;q',0)\psi_0(q')d\tau_{q'}, \qquad (\text{VI.5})$$

where $\psi_0(q) \equiv \psi(q,t=0)$. In particular, for a time-independent Hamiltonian, the Green function is

$$G(q,t;q',t') = \sum_n e^{-iE_n(t-t')/\hbar}\psi_{E_n}(q)\psi^*_{E_n}(q'). \qquad (\text{VI.6})$$

For a free particle, whose Hamiltonian is $\hat{H} = \frac{1}{2m}\hat{\mathbf{p}}^2$, the Green function has the form

$$G(\mathbf{r},t;\mathbf{r}',t') = \left(-\frac{im}{2\pi\hbar(t-t')}\right)^{3/2} \exp\left[\frac{im(\mathbf{r}-\mathbf{r}')^2}{2\hbar(t-t')}\right]. \qquad (\text{VI.7})$$

In *the Heisenberg representation*, unlike the Schrödinger representation, the wavefunction of a system does not depend on time, while the time-dependence of the operators obeys the equations[78]

$$\frac{d}{dt}\hat{q}_i(t) = \frac{i}{\hbar}[\hat{H}(t),\hat{q}_i(t)], \quad \frac{d}{dt}\hat{p}_i(t) = \frac{i}{\hbar}[\hat{H}(t),\hat{p}_i(t)], \qquad (\text{VI.8})$$

where the Hamiltonian, $\hat{H}(\hat{p}_i(t),\hat{q}_i(t),t)$, is expressed in terms of the Heisenberg operators $\hat{q}(t),\hat{p}(t)$ that satisfy the canonical commutation relations $[\hat{p}_i(t),\hat{q}_k(t)] = -i\hbar\,\delta_{ik}$. In this representation, the relation (VI.4) is not a definition of $\hat{f}$ but a derived relation from (VI.8).

The Schrödinger and Heisenberg representations are connected with a unitary transformation: $\psi(q,t) = \hat{U}(t)\psi_0(q)$. If the Hamiltonian is not explicitly dependent on time, then $\hat{U}(t) = \exp\{-i\hat{H}t/\hbar\}$ and the relation between the operators in these representations has the form

$$\hat{f}_H = e^{i\hat{H}t/\hbar}\hat{f}_S e^{-i\hat{H}t/\hbar}. \qquad (\text{VI.9})$$

There exists another frequently-used representation: the *interaction representation*; it is considered in Problem 6.30.

6.1 The Schrödinger representation; The motion of wave packets

Problem 6.1

Consider the following systems and their wavefunctions ψ_0 at an initial time $(t=0)$:

---

[78] In order to distinguish between representations: in the Heisenberg representation we indicate explicitly the time-dependence of the operators as: $\hat{q}(t)$, $\hat{p}(t)$. We keep $\hat{q}$, $\hat{p}$ for the corresponding operators in the Schrödinger representation. These representations of wavefunctions and operators coincinde at $t=0$; compare with Eq. (VI.9).

1) a particle in an infinitely deep potential well of width a with $\psi_0(x) = A \sin^3\left(\frac{\pi x}{a}\right)$ for $0 \leq x \leq a$;
2) a plane rotator with $\psi_0(\varphi) = A \sin^2 \varphi$;
3) a spherical rotator with $\psi_0(\vartheta, \varphi) = A \cos^2 \vartheta$

Find the wavefunctions for these systems at an arbitrary moment of time. Prove that after a time T the systems return to the initial state.

Solution

The stationary states of these systems were considered in Problems 2.1, 3.2, and 3.3. By using Eq. (VI.2) we easily obtain:

1)
$$\psi(x,t) = \frac{A}{4} e^{-i\omega t}\left(3 \sin\frac{\pi x}{a} - e^{-8i\omega t}\sin\frac{3\pi x}{a}\right), \quad \omega = \frac{\pi^2 \hbar}{2ma^2},$$

2)
$$\psi(\varphi, t) = \frac{A}{2}\left[1 - e^{-2i\hbar t/I}\cos 2\varphi\right],$$

3)
$$\psi(\vartheta, t) = \frac{A}{3}\left[1 + e^{-3i\hbar t/I}(3\cos^2\vartheta - 1)\right].$$

(To determine the expansion of the wavefunction ψ_0 in terms of the Hamiltonian's eigenfunctions it is convenient to use known trigonometric equations without using Eq. (VI.3).)

After time T equal to 1) $ma^2/2\pi\hbar$, 2) $\pi I/\hbar$, 3) $2\pi I/3\hbar$, the systems return to their initial states.

Problem 6.2

The state of a free particle at $t = 0$ is described by the wavefunction

$$\psi_0(x) = A \exp\left[i\frac{p_0 x}{\hbar} - \frac{x^2}{2a^2}\right].$$

Determine the time dependence of the state and the following mean values: $\overline{x(t)}$, $\overline{p(t)}$, $\overline{(\Delta x(t))^2}$, $\overline{(\Delta p(t))^2}$. (See also Problem 6.21.)

Show that the width of the wave packet, $\sqrt{\overline{(\Delta x(t))^2}}$, independently of the parameters that determine the wavefunction $\psi_0(x)$, cannot be arbitrarily small.

Solution

Let us expand the wavefunction $\psi_0(x)$ in the momentum eigenfunctions, which are also the eigenfunctions of the free particle Hamiltonian:[79]

$$\psi_0(x) = \int c(p)\psi_p(x)dp, \quad \psi_p(x) = \frac{1}{\sqrt{2\pi\hbar}}e^{ipx/\hbar}.$$

Evaluating the resulting Gaussian integral, we find

$$c(p) = \int \psi_0(x)\psi_p^*(x)dx = \frac{aA}{\sqrt{\hbar}} \exp\left\{-\frac{(p-p_0)^2 a^2}{2\hbar^2}\right\}. \tag{1}$$

Now, by using Eq. (VI.2) we obtain

$$\psi(x,t) = \int c(p) \exp\left(-i\frac{p^2 t}{2m\hbar}\right) \psi_p(x) dp =$$

$$A\left(1 + \frac{i\hbar t}{ma^2}\right)^{-1/2} \exp\left\{\frac{i\hbar^3 x^2 t + i\hbar m^2 a^4 v_0(2x - v_0 t) - \hbar^2 m a^2 (x - v_0 t)^2}{2m(\hbar^2 a^4 + \hbar^4 t^2/m^2)}\right\}, \tag{2}$$

where $v_0 = p_0/m$. Hence it follows that

$$|\psi(x,t)|^2 = |A|^2 \frac{a}{a(t)} \exp\left\{-\frac{(x - v_0 t)^2}{a(t)}\right\}, \tag{3}$$

$$a^2(t) = a^2(1 + \hbar^2 t^2/(m^2 a^4)).$$

Choosing $|A|^2 = (\pi a^2)^{-1/2}$ to normalize the wavefunction to unity, we find

$$\overline{x(t)} = \int x|\psi(x,t)|^2 dx = v_0 t, \quad \overline{(\Delta x(t))^2} = \frac{1}{2}a^2(1 + \hbar^2 t^2/(m^2 a^4)). \tag{4}$$

Since $c(p,t) = \exp\left(-ip^2 t/2m\hbar\right) c(p)$ is the wavefunction in the momentum representation, then using (1) we obtain

$$\overline{p(t)} = \int p|c(p,t)|^2 dp = p_0, \quad \overline{(\Delta p(t))^2} = \frac{\hbar^2}{2a^2}. \tag{5}$$

(independence of momentum on time reflects the fact that for a free particle momentum is the constant of motion.)

The results (3)–(5) have a simple meaning: the coordinate distribution (3) is the Gaussian wave packet, the center of which, $\overline{x(t)}$, moves with the constant velocity v_0 (equal to $\overline{p}/m = \overline{v}$); the width of the packet here $\sim \sqrt{\overline{(\Delta x(t))^2}}$, increases (i.e., the packet spreads out). The spreading out of the packet is connected with the fact that the momentum (and velocity) of the particle has no definite value.

---

[79] All of the integrals in this problem are calculated within infinite limits.

The width of the packet doubles within a time $t_0 = \sqrt{3}ma^2/\hbar$. We evaluate this "spreading-out" time for two cases. 1) For a microscopic particle with the mass ($m = 10^{-27}$ g) (electron) and $a = 10^{-8}$ cm (atomic dimensions), we have $t_0 \sim 10^{-16}$ sec. 2) For a small but macroscopic particle with $m = 10^{-6}$ g and $a = 10^{-2}$ cm, we have $t_0 \sim 10^{17}$ sec, or $\sim 10^{10}$ years. We should note that when a packet is "tightly" localized with a small $\Delta x(0)$, the spreading of the wave packet is fast; this is connected with the uncertainty relation $\Delta p \Delta x \geq \hbar$ (and since the spreading is determined by the uncertainty Δv of velocity, then it especially clearly seen with the decrease of particle mass). As we see from (4) the width of the packet satisfies the relations $\overline{(\Delta x(t))^2} \geq (a^2/2,\ \hbar t/m,\ \hbar^2 t^2/2ma^2)$.

Problem 6.3

Let us consider a normalized wave packet at $t = 0$:

$$\psi(x, t=0) = \int c(E)\psi_E(x)dE, \quad \int |\psi|^2 dx = 1,$$

which is some superposition of the eigenfunctions of the Hamiltonian corresponding to the continuous part of energy spectrum. Show that the probability density at any point tends to zero as $t \to \infty$. Explain why this fact does not contradict conservation of wavefunction normalization.

Solution

At an arbitrary moment in time,

$$\psi(x, t) = \int c(E) \exp\left(-i\frac{Et}{\hbar}\right) \psi_E(x) dE. \tag{1}$$

The decrease of $|\psi(x,t)|^2$ with $t \to \infty$ occurs due to the rapid oscillations of the expression under the integral sign (its real and imaginary parts), neighboring domains of integration cancel each other. The decrease in $|\psi(x,t)|^2$ means that the particle moves to infinity as $t \to \infty$. This corresponds to the notion in classical mechanics that a particle in motion will continue the motion indefinitely, moving to infinity. We should note that when the probability density of the wave packet decreases, the width of the packet increases – it spreads out and keeps the wavefunction normalized to unity. (As an illustration, see (3) from Problem 6.2).

Problem 6.4

The state of a particle in a δ-potential (see Problem 2.7) at $t = 0$ is described by the wavefunction $\tilde{\psi}_0(x) = A\exp(-\beta|x|)$, $\beta > 0$. are the probability $W(x)dx$ to find the particle inside the coordinate interval $(x, x + dx)$ as $t \to \infty$. Find the value of the integral $\int W(x)dx$ and compare with its initial value. Explain the result obtained.

Solution

Expanding the wavefunction $\tilde{\psi}_0(x)$ in the eigenfunctions of the Hamiltonian, we have

$$\psi(x,t) = c_0 \exp\left(-i\frac{E_0 t}{\hbar}\right)\psi_0(x) + \int_0^\infty c(E)\exp\left(-i\frac{Et}{\hbar}\right)\psi_E(x)dE, \quad (1)$$

where

$$\psi_0(x) = \sqrt{\kappa}e^{-\kappa|x|}, \quad E_0 = -\frac{\hbar^2\kappa^2}{2m}, \quad \kappa = \frac{m\alpha}{\hbar^2}$$

Are the wavefunction and the energy of the unique discrete spectrum state in the δ-potential, see Problem 2.7.

The second term in (1) describes a contribution of continuous spectrum and vanishes as $t \to \infty$, so

$$|\psi(x, t = \infty)|^2 = |c_0\psi_0(x)|^2 = \kappa|c_0|^2 e^{-2\kappa|x|}. \quad (2)$$

Choosing $A = \sqrt{\beta}$ to normalize the wavefunction $\tilde{\psi}_0(x)$ to unity, we then calculate

$$c_0 = \int \tilde{\psi}_0(x)\psi_0^*(x)dx = \frac{2\sqrt{\kappa\beta}}{\kappa + \beta},$$

and can rewrite (2) in the form

$$W(x) \equiv |\psi(x, t = \infty)|^2 = \frac{4\kappa^2\beta}{(\kappa+\beta)^2}e^{-2\kappa|x|}, \quad (3)$$

which determines the distribution over the coordinates of the particle as $t \to \infty$. It is normalized to the value

$$w \equiv \int_{-\infty}^\infty W(x)dx = \frac{4\kappa\beta}{(\kappa+\beta)^2} \leq 1, \quad (4)$$

whose difference from unity means that the particle propagates to infinity with the probability equal to $(1-w)$. To explain the result obtained, we should note that in general,

$$\int \lim_{t\to\infty} |\psi(x,t)|^2 dx \neq \lim_{t\to\infty}\int |\psi(x,t)|^2 dx = 1.$$

Problem 6.5

At $t = 0$, the state of a free particle is determined by its wavefunction $\phi_0(p)$ normalized to unity in the momentum representation. Find the asymptotic behavior of its wavefunction $\psi(x,t)$ as $t \to \infty$. Verify that the normalization of the wavefunctions is preserved under time evolution. As an illustration of the result obtained, consider the wave packet from Problem 6.2.

Solution

The wavefunction has the form

$$\psi(x,t) = \frac{1}{\sqrt{2\pi\hbar}} \int \exp\left[-\frac{i}{\hbar}\left(\frac{p^2 t}{2m} - px\right)\right] \phi_0(p) dp. \qquad (1)$$

As $t \to \infty$ (and $x \to \pm\infty$) the phase factor in the exponential changes considerably, even in the case of small p which leads to rapid oscillations and therefore canceling of the contributions of neighboring integration domains. Here the dominant contribution is given by those integration domains whose phase as a function of p is an extremum, and changes slowly. In our case, the only extremum point is $p_0 = mx/t$. Approximating the value of "shallow" function ϕ_0 by its value at p_0, we can factor it out of the integral, and we find the desired asymptotic form of the wavefunction as $t \to \infty$:

$$\psi(x,t) \approx \frac{\phi_0(mx/t)}{\sqrt{2\pi\hbar}} \int_{-\infty}^{\infty} \exp\left[-\frac{i}{\hbar}\left(\frac{p^2 t}{2m} - px\right)\right] dp$$

$$= \sqrt{\frac{m}{it}} \phi_0\left(\frac{mx}{t}\right) \exp\left(\frac{imx^2}{2\hbar t}\right). \qquad (2)$$

(Note that this function is normalized to unity since $\phi_0(p)$ is normalized.) Let us note the meaning of (2): it is the value of the wavefunction as $t \to \infty$ at the point $x \to \pm\infty$ (so that $x/t = const \equiv v_0 = p_0/m$), and it is given by the wavefunction $\phi_0(p)$ at $p = p_0$, i.e., with a momentum that the free particle in classical mechanics must have in order to cover the distance x in time t. Actually, the exponent of (2) is $iS(x,t)/\hbar$, where S is the classical "action" of such a particle.

Problem 6.6

Consider a wave packet reflecting off of an impenetrable potential wall, i.e., a wave packet in the potential $U(x) = 0$ for $x < 0$ and $U(x) = \infty$ for $x > 0$. Initially,

$$\psi_0(x, t=0) = A \exp\left[i\frac{p_0 x}{\hbar} - \frac{(x+x_0)^2}{2a^2}\right],$$

where $p_0 > 0$, $x_0 > 0$ and it is assumed that $x_0 \gg a$, since we need $\psi(x,0) = 0$ for $x \geq 0$.

Solution

To calculate $\psi(x,t)$ we use the Green function $G(x,t,x',t')$ which obeys the Schrödinger equation for a free particle and the boundary condition $G(x=0,t;x',t') = 0$. Taking into account Eq. (VI.7) we see (in analogy with the "method of images" in the electrostatics) that

$$G(x,t;x',t') = \sqrt{\frac{im}{2\pi\hbar(t-t')}} \left\{ \exp\frac{im(x-x')^2}{2\hbar(t-t')} - \exp\frac{im(x+x')^2}{2\hbar(t-t')} \right\}.$$

Substituting this expression and ψ_0 from the problem statement into, Eq. (VI.5), and calculating the integral, we obtain

$$\psi(x,t) = A\sqrt{\frac{ma^2}{ma^2+i\hbar t}} \left\{ \exp\left\{ \frac{1}{2(1+\hbar^2 t^2/m^2 a^4)a^2} \left(-\left(x+x_0-\frac{p_0 t}{m}\right)^2 \right.\right.\right.$$
$$\left.\left.\left. + \frac{i\hbar t(x+x_0)^2}{ma^2} + \frac{2ip_0 a^2(x+x_0)}{\hbar} - \frac{ip_0 a^2 t}{m\hbar} - \frac{2ip_0 x_0(a^2+\hbar^2 t^2/m^2 a^2)}{\hbar} \right) \right\}$$
$$- \exp\{(x \to -x)\}\}, \qquad (1)$$

where $\exp\{(x \to -x)\}$ denotes the same expression from the first exponential, but with the change $x \to -x$.

The wavefunction (1) is a superposition of two wave packets, the first of which describes the wave packet moving towards the wall, while the second corresponds to the reflected wave packet. At $t=0$, the first term predominates in (1), while the second term predominates for $t > mx_0/p_0$ when $p_0 \gg \hbar/a$.[80]

Problem 6.7

Discuss the reflection of a wave packet on a potential step of the form: $U(x) = 0$ for $x < 0$ and $U(x) = U_0 > 0$ for $x > 0$ (Fig. 2.9). The incident wave packet includes momenta with similar amplitudes from the interval $p_0 \pm \Delta p$ where $\Delta p \ll p_0$ and $E_0 < U_0$. Find the reflection *delay time* at the barrier in comparison with the case of the classical particle.

Solution

Let us first consider a free particle wave packet normalized to unity and summed over momenta $p = \hbar k$ from the interval $(k_0 - \kappa_0, k_0 + \kappa_0)$, with the amplitudes distributed by a Gaussian:

---

[80] Let us emphasize that for the case $p_0 \gg \hbar/a$, we may interpret the initial wave packet as describing the particle falling towards the wall. Otherwise (for $p_0 \leq \hbar/a$), even in the initial state, the particle has, with noticeable probability (~ 1), the negative sign of momentum (compare with Problem 1.37), as compared with the reflected particle; therefore, it does not make sense to characterize the first term in (1) as corresponding to the particles falling onto the wall. The distinction between these cases makes itself evident in the fact that in the case $p_0 \leq \hbar/a$ the first term in (1) is not negligibly small in comparison with the second one, even as $t \to \infty$.

$$\psi_{\text{free}}(x,t) = \frac{1}{\sqrt{2\pi}} \int_{-\kappa_0}^{\kappa_0} \exp\left\{i(k_0+\kappa)x - i\frac{\hbar t}{2m}(k_0+\kappa)^2\right\} \frac{d\kappa}{\sqrt{2\kappa_0}}. \tag{1}$$

For $\kappa_0 \ll k_0$ and for $|t| \ll T = m/\hbar\kappa_0^2$, we can omit the terms $\propto \kappa^2$ in the exponent (i.e., all contributing momenta have the same amplitude) and obtain

$$\psi_{\text{free}}(x,t) \approx \sqrt{\frac{a}{\pi}} \exp\left\{ik_0 x - i\frac{\hbar k_0^2 t}{2m}\right\} \frac{\sin[(x-v_0 t)/a]}{x - v_0 t}, \tag{2}$$

where $v_0 = \hbar k_0/m$ and $a = 1/\kappa_0$. According to (2), $|\psi_{\text{free}}(x,t)|^2$ describes the wave packet whose center moves with constant velocity v_0 (being at the point $x=0$ at the moment $t=0$); the packet width is $\Delta x = |x - v_0 t| \sim a$ and it does not spread out (but only during the time-interval considered.)

Now we consider the reflection of the initial wave packet at the potential step, and the eigenfunction for the Hamiltonian at the energy $E = \hbar^2 k^2/2m$ is

$$\psi_k^+(x) = \frac{1}{\sqrt{2\pi}}\left(e^{ikx} - e^{i\varphi(k)}e^{-ikx}\right), \quad \text{for } x < 0, \tag{3}$$

where

$$\varphi(k) = 2\arctan\sqrt{\frac{E}{U_0 - E}}.$$

(It follows from the matching of function (3) with the wave-function $\psi_k^+(x) = A\exp\{-\sqrt{2m(U_0-E)}x/\hbar\}$.) Let us construct the wave packet $\psi(x,t)$ from the functions (3) and the free variant (1). As a result, we obtain the wavefunction of the particle (with the potential step) for $x \leq 0$ in the form

$$\psi(x,t) = \psi_{\text{in}}(x,t) + \psi_{\text{out}}(x,t). \tag{4}$$

Here $\psi_{\text{in}}(x,t)$ is a part of the wave packet connected with the first term in (3); it coincides with (1) and (2) and describes the particle moving towards the barrier and differs from zero (for $x < 0$, the domain of validity) only for $t \leq a/v_0$, i.e., until the packet reaches the step. The second part of the packet ψ_{out} corresponds to the particle that is reflected at the step (it differs from zero only for $t > 0$). If while integrating over κ we neglect the change in phase $\varphi(k)$ of the reflected wave, i.e., put $\varphi(k) \approx \varphi(k_0)$, then for ψ_{out} we obtain the relation that follows from (2) by changing x to $-x$ and multiplying by $-e^{i\varphi(k_0)}$. In this approximation there is no effect of particle *delay* for reflection. By using a more accurate approximation, $\varphi(k_0 + \kappa) \approx \varphi(k_0) + \varphi'(k_0)\kappa$, we obtain[81]

$$\psi_{\text{out}}(x,t) \approx -\sqrt{\frac{a}{\pi}} e^{i\alpha(x,t)} \frac{\sin[(x - \varphi'(k_0) + v_0 t)/a]}{x - \varphi'(k_0) + v_0 t}. \tag{5}$$

---

[81] We do not indicate the value of the phase $\alpha(x,t)$ in (5).

Hence it follows that the relation for the time delay ($E_0 = p_0^2/2m$),

$$\tau = \frac{\varphi'(k_0)}{v_0} = \frac{\hbar}{\sqrt{E_0(U_0 - E_0)}}. \tag{6}$$

Since $\tau > 0$, there is a delay in the reflection of the particle at the barrier. We can understand this by taking into account that, unlike in classical mechanics, the particle penetrates into the barrier.[82] Note that τ decreases when U_0 increases, so that $\tau = 0$ as $U_0 = \infty$. Concluding, we should note that $\tau \ll a/v_0$, i.e., the time delay is small in comparison with the time it takes the particle to cross a distance comparable to its packet size.

Problem 6.8

Discuss the reflection of a particle by a short-range potential $U(x)$. The state of the particle moving toward the non-zero part of the potential is described by a wave packet normalized to unity. Assuming that the particle momenta from the interval $p_0 \pm \triangle p$ are present in the packet with the same amplitude, find the restrictions on $\triangle p$ for which the values of the reflection and transmission coefficients do not depend on its size and is defined by the simple expressions of stationary theory; see Eq. (II.4).

Solution

Eigenfunctions of the Hamiltonian that correspond to the particles coming from the left, where the potential is zero, have the form

$$\psi_k^+ = \begin{cases} \frac{1}{\sqrt{2\pi}}\left(e^{ikx} + A(k)e^{-ikx}\right), & x < -d, \\ \frac{1}{\sqrt{2\pi}}B(k)e^{ikx}, & x > d, \end{cases} \tag{1}$$

where $k = \sqrt{2mE/\hbar^2}$ and d is the radius of a potential, so we can suggest that $U = 0$ for $|x| > d$. Let us consider the wave packet, normalized to unity (compare with the previous problem):

$$\psi(x,t) = \int_{-\kappa_0}^{\kappa_0} \psi_{k_0+\kappa}^+(x) \exp\left\{-i\frac{\hbar t}{2m}(k_0 + \kappa)^2\right\} \frac{d\kappa}{\sqrt{2\kappa_0}}. \tag{2}$$

For the case of $\kappa_0 \ll k_0$ and so small that we can suggest that

$$A(k_0 \pm \kappa_0) \approx A(k_0) \quad \text{and} \quad B(k_0 \pm \kappa_0) \approx B(k_0),$$

---

[82] Actually, $\tau \sim \delta x / v_0$, where $\delta x \sim \hbar/\sqrt{m(U_0 - E_0)}$ is the distance of particle penetration inside the barrier.

and neglect the term $\propto \kappa^2$ in the exponent (for $|t| \ll T = m/\hbar\kappa_0$). The wavefunction outside the potential range, using the asymptote (1), takes the form

$$\psi(x,t) = \begin{cases} \psi_{\text{in}}(x,t) + \psi_{\text{refl}}(x,t), & x < -d, \\ \psi_{\text{trans}}(x,t), & x > d, \end{cases} \quad (3)$$

where[83]

$$\psi_{\text{in}}(x,t) \approx \sqrt{\frac{a}{\pi}} e^{i\alpha(x,t)} \frac{\sin[(x - v_0 t)/a]}{x - v_0 t},$$

$$\psi_{\text{refl}}(x,t) \approx A(k_0)\sqrt{\frac{a}{\pi}} e^{i\alpha(-x,t)} \frac{\sin[(x + v_0 t)/a]}{x + v_0 t}, \quad (4)$$

$$\psi_{\text{trans}}(x,t) \approx B(k_0)\sqrt{\frac{a}{\pi}} e^{i\alpha(x,t)} \frac{\sin[(x - v_0 t)/a]}{x - v_0 t},$$

while $v_0 = \hbar k_0/m$ and $a = 1/\kappa_0$.

The interpretation of the expressions (3) and (4) is as follows. For $t < -d/v_0$ only ψ_{in} is essentially different from zero, and it describes the particle incident from the left which has not reached the region $|x| \leq d$ where the potential is non-zero. On the other hand, for $t > d/v_0$ only the parts of the wavefunction that describe the particle after interaction with the potential, ψ_{refl} and ψ_{trans}, are essentially different from zero. Here the probabilities of finding the particle in the reflected and transmitted wave packets are equal to $R = |A(k_0)|^2$ and $D = |B(k_0)|^2$ respectively, in accordance with the result of the stationary approach.

Problem 6.9

Consider a *two-level system*[84] (its levels are non-degenerate and have the energies $\varepsilon_1^{(0)}$ and $\varepsilon_2^{(0)}$), that is in one of its stationary states; when $t > 0$, an external field begins to act. The interaction $\hat{V}$ of the system with the field is characterized by the matrix elements V_{11}, V_{22}, $V_{12} = V_{21}^*$ between the unperturbed states $|1\rangle$ and $|2\rangle$, where V_{ab} is time-independent (for $t > 0$). Determine the wavefunction of the system for $t > 0$ and the probabilities of finding the system in eigenstates of unperturbed Hamiltonian $\hat{H}_0$.

Solution

We will describe the wavefunction of the system by the two-component column vector $\Psi(t) = \begin{pmatrix} \psi_1(t) \\ \psi_2(t) \end{pmatrix}$, where the functions $\psi_{1(2)}(t)$ would be the amplitudes of the 1st and

---

[83] See (2) from the previous problem: $\alpha(x,t) = k_0 x - \hbar k_0^2 t/2m$.
[84] The two-level system models the behavior of a system whose energy spectrum has two close levels. Transitions between these and other levels of the system are small for low-energy interactions.

2nd eigenstates of unperturbed Hamiltonian $\hat{H}_0$ in the absence of a perturbation (i.e., $\psi_{1,2}^{(0)} = C_{1,2} \exp\{-i\varepsilon_{1,2}^{(0)} t/\hbar\}$). When the external field acts on the system, its new stationary states (their energies and wavefunctions) are determined by the solution to the Schrödinger equation $(\hat{H}_0 + \hat{V})\Psi_\varepsilon = \varepsilon \Psi_\varepsilon$, which, by substituting $\Psi_\varepsilon = \begin{pmatrix} a_1 \\ a_2 \end{pmatrix}$, takes the form of two algebraic equations

$$\left(\varepsilon_1^{(0)} + V_{11} - \varepsilon\right) a_1 + V_{12} a_2 = 0, \quad V_{21} a_1 + \left(\varepsilon_2^{(0)} + V_{22} - \varepsilon\right) a_2 = 0.$$

The solution gives the "perturbed" energy levels $\varepsilon_{1,2}$ and their corresponding eigenfunctions:

$$\varepsilon_{1,2} = \frac{1}{2}\left[\varepsilon_1^{(0)} + \varepsilon_2^{(0)} + V_{11} + V_{22} \mp \sqrt{\left(\varepsilon_1^{(0)} - \varepsilon_2^{(0)} + V_{11} - V_{22}\right)^2 + 4|V_{12}|^2}\right], \quad (1)$$

$$\Psi_{\varepsilon_1} = N \begin{pmatrix} 1 \\ b \end{pmatrix}, \quad \Psi_{\varepsilon_2} = N \begin{pmatrix} -b^* \\ 1 \end{pmatrix},$$

$$b = \frac{\varepsilon_1 - \varepsilon_1^{(0)} - V_{11}}{V_{12}},$$
$$N = \left(1 + |b|^2\right)^{-1/2} \quad (2)$$

(note that the wavefunctions Ψ_{ε_1} and Ψ_{ε_2} are mutually orthogonal, as expected).

For $t > 0$, the wavefunction of the system has the form

$$\Psi(t) = c_1 \Psi_{\varepsilon_1} e^{-i\varepsilon_1 t/\hbar} + c_2 \Psi_{\varepsilon_2} e^{-i\varepsilon_2 t/\hbar},$$

where the values of $c_{1,2}$ are determined by the initial conditions. In the case under consideration we have $\Psi(0) = \begin{pmatrix} 1 \\ 0 \end{pmatrix}$, and obtain

$$\Psi(t) \equiv \begin{pmatrix} \psi_1(t) \\ \psi_2(t) \end{pmatrix} = \frac{1}{1+|b|^2} \begin{pmatrix} e^{-i\varepsilon_1 t/\hbar} + |b|^2 e^{-i\varepsilon_2 t/\hbar} \\ b\left(e^{-i\varepsilon_1 t/\hbar} - e^{-i\varepsilon_2 t/\hbar}\right) \end{pmatrix}, \quad (3)$$

so that the probability of a transition of the system to the other eigenstate of the unperturbed Hamiltonian is given by

$$w_2(t) = |\psi_2(t)|^2 = \frac{4|b|^2}{(1+|b|^2)^2} \sin^2\left[\frac{1}{2\hbar}(\varepsilon_2 - \varepsilon_1)t\right]. \quad (4)$$

As is seen, its value oscillates between 0 and $w_{\max} = 4|b|^2/(1+|b|^2)^2$. The value of $w_{\max}$ can be close to 1 if the matrix element V_{12} is sufficiently large. Such a situation takes place when[85] $\varepsilon_1^{(0)} = \varepsilon_2^{(0)}$ (the unperturbed energy levels) and $V_{11} = V_{22} = 0$. As

---

[85] We should note the existence of the almost degenerate $2s-$ and $2p-$ states of hydrogen atom, significant transitions between which appear even in a weak field. This leads to an important effect of the electric field on the life time of the metastable $2s-$state; see Problem 11.62.

can be seen from (1), the eigenstates of the perturbed system have equal probability to be in either of the unperturbed energy eigenstates in this case.

6.2 Time-dependent observables; Constants of motion

Problem 6.10

For a charged spinless particle moving in an external electromagnetic field,[86] find the velocity $\hat{\mathbf{v}}$ and acceleration $\hat{\mathbf{w}}$ operators. Compare the results with the expressions for classical theory.

Solution

Taking into account the Hamiltonian (VII.1), from Eq. (VI.4) we find

$$\hat{\mathbf{v}} \equiv \hat{\dot{\mathbf{r}}} = \frac{1}{\mu}\left(\hat{\mathbf{p}} - \frac{e}{c}\mathbf{A}\right), \tag{1}$$

$$\hat{\mathbf{w}} \equiv \hat{\dot{\mathbf{v}}} = \frac{\partial \hat{\mathbf{v}}}{\partial t} + \frac{i}{\hbar}[\hat{H}, \hat{\mathbf{v}}] = \frac{e}{\mu}\boldsymbol{\mathcal{E}} + \frac{e}{2\mu c}(\hat{\mathbf{v}} \times \boldsymbol{\mathcal{H}} - \boldsymbol{\mathcal{H}} \times \hat{\mathbf{v}}), \tag{2}$$

where $\boldsymbol{\mathcal{E}} = -\nabla\varphi - \frac{\partial \mathbf{A}}{c\partial t}$, $\boldsymbol{\mathcal{H}} = \mathrm{rot}\,\mathbf{A}$, which gives the natural quantum-mechanical generalization of the corresponding classical theory expressions. (Here the right-hand part of (2) determines the Lorentz force operator $\hat{\mathbf{F}}_{\mathrm{Lor}}$.)

Problem 6.11

For a neutral particle with a spin s that has its own magnetic moment μ_0 and moves in an electromagnetic field,[87] find the operators for its velocity $\hat{\mathbf{v}}$, acceleration $\hat{\mathbf{w}}$, and time derivative of the spin operator $\hat{\dot{\mathbf{s}}}$.

Solution

Similar to the previous problem, we obtain

$$\hat{\mathbf{v}} = \frac{\hat{\mathbf{p}}}{\mu}, \quad \hat{\mathbf{w}} = \frac{\mu_0}{s\mu}\nabla\left(\hat{\mathbf{s}} \cdot \boldsymbol{\mathcal{H}}(\mathbf{r}, t)\right), \quad \hat{\dot{\mathbf{s}}} = \frac{i}{\hbar}[\hat{H}, \hat{\mathbf{s}}] = \frac{\mu_0}{s\hbar}\hat{\mathbf{s}} \times \boldsymbol{\mathcal{H}}(\hat{\mathbf{r}}, t).$$

(Compare these with the corresponding classical expressions for a neutral particle which has magnetic moment $\boldsymbol{\mu}$ and angular momentum $\mathbf{M} = \kappa\boldsymbol{\mu}$; its interaction with magnetic field is described by the potential $U = -\boldsymbol{\mu} \cdot \boldsymbol{\mathcal{H}}(\mathbf{r}, t)$, and $d\mathbf{M}/dt = \boldsymbol{\mu} \times \boldsymbol{\mathcal{H}}$.)

---

[86] The Hamiltonian is given by Eq. (VII.1).
[87] The Hamiltonian of the particle is again given by Eq. (VII.1).

Problem 6.12

Show that the expectation value of the time derivate of any observable that does not explicitly depend on time is zero when the state is stationary. Using this result,[88] prove that by averaging the operator $\frac{d}{dt}(\hat{\mathbf{p}} \cdot \hat{\mathbf{r}})$ in an attractive potential $U = \alpha r^\nu$, we obtain the *virial theorem*.

Solution

1) By averaging the operator $\hat{\dot{f}}$, we find

$$\overline{\dot{f}} = \int \psi_n^* \hat{\dot{f}} \psi_n d\tau = \frac{i}{\hbar} \int \psi_n^* \left(\hat{H}\hat{f} - \hat{f}\hat{H} \right) \psi_n d\tau = 0.$$

(Here we took into account that $\hat{H}\psi_n = E_n \psi_n$ and under the integral sign $\int \psi_n^* \hat{H} \cdots = \int \left(\hat{H}\psi_n \right)^* \cdots$ due to the Hermitian character of $\hat{H}$.)

2) Taking into account that $d\hat{\mathbf{r}}/dt = \hat{\mathbf{p}}/m$ and $d\hat{\mathbf{p}}/dt = -\nabla U$, for $U = \alpha r^\nu$ we have

$$\frac{d(\hat{\mathbf{p}} \cdot \hat{\mathbf{r}})}{dt} = \dot{\hat{\mathbf{p}}} \cdot \hat{\mathbf{r}} + \hat{\mathbf{p}} \cdot \dot{\hat{\mathbf{r}}} = 2\hat{T} - \nu \hat{U}, \qquad (2)$$

and according to (1), the expectation values for stationary states satisfy $\overline{T}_{nn} = \nu \overline{U}_{nn}/2$, which is the quantum-mechanical generalization of the virial theorem in classical mechanics (where averaging is conducted over time).

Note Another derivation of the virial theorem is obtained by the application of Eq. (I.6). We should note that the power law potential $U = \pm \alpha r^\nu$ is characterized one parameter: α. We can then use the three parameters m, α, and $\hbar$ to obtain a quantity with units of energy: $\varepsilon_0 = \alpha \left(\hbar^2/m\alpha \right)^{\nu/(\nu+2)}$. We cannot form a dimensionless parameter from the same parameters, so just using dimensional considerations, the discrete energy spectrum from the Hamiltonian $\hat{H} = -\left(\hbar^2/2m\right)\triangle + U$ has the form $E_n = C(n,\nu)\varepsilon_0$. Since $U \equiv \alpha \partial \hat{H}/\partial \alpha$ according to Eq.(II.6) we obtain

$$\overline{U}_{nn} = \alpha \frac{\partial E_n}{\partial \alpha} = \frac{2}{\nu+2} E_n.$$

From this relation and the equality $E_n = \overline{T}_{nn} + \overline{U}_{nn}$, the virial theorem follows immediately.

Note that the relation obtained is also valid for systems consisting of an arbitrary number of particles if the interaction between them and the external field is described by the a power-law potential with the same exponent ν.

---

[88] In some cases, given a potential and convenient choice of the operator $\hat{f} = \hat{f}(r, \hat{p})$, we can use the condition

$$\overline{\dot{f}} = \frac{i}{\hbar}\overline{[\hat{H}, \hat{f}]} = 0$$

to calculate different averages.

Problem 6.13

For a system consisting of N identical charged particles in the nth eigenstate, show that the following equality is valid (the so-called *sum rule:*, see Problem 14.11):

$$\frac{2\mu}{e^2\hbar^2} \sum_k (E_k - E_n)|(d_i)_{kn}|^2 = N, \quad (i = 1,\ 2,\ 3),$$

where $(\hat{d}_i)_{kn}$ are the matrix elements of the system's dipole moments. The summation is taken over all the independent stationary states, and μ and e are the mass and charge of the particles.

Solution

If we take the expectation value of $[\hat{p}_{ai},\ \hat{x}_{bk}] = -i\hbar\delta_{ab}\delta_{ik}$ and $d(\hat{x}_{ai}\hat{x}_{bk})/dt = (i/\hbar\mu)(\hat{p}_{ai}\hat{x}_{bk} + \hat{x}_{ai}\hat{p}_{bk})$ (indices a, b enumerate the particles) with respect to an eigenfunction of the Hamiltonian, ψ_n, and we take into account that the second average is equal to zero for $i = k$ (without summation!), then we find

$$\sum_m \{\langle n|\hat{x}_{ai}|m\rangle \langle m|\hat{p}_{bi}|n\rangle + \langle n|\hat{x}_{bi}|m\rangle \langle m|\hat{p}_{ai}|n\rangle\} = i\hbar\delta_{ab} \tag{1}$$

(where we have used the completeness condition $\sum_m |m\rangle \langle m| = 1$). Now we take into account that $\hat{p}_{ai} = (i\mu/\hbar)[\hat{H}, \hat{x}_{ai}]$, so that we obtain

$$\langle m|\hat{p}_{ai}|n\rangle = i\frac{\mu}{\hbar}(E_m - E_n)\langle m|\hat{x}_{ui}|n\rangle. \tag{2}$$

Since $\hat{\mathbf{d}} = e \sum_a \hat{\mathbf{x}}_a$, then after multiplying (1) by e^2, substituting (2), and performing a sum over a and b (over all particles), we obtain the sum rule given in the problem statement.

Problem 6.14

Show that if some time-independent unitary operator $\hat{U}$ leaves a Hamiltonian of a system unchanged, i.e., $\hat{U}\hat{H}\hat{U}^+ = \hat{H}$, then the Hermitian operator $\hat{F}$ related to $\hat{U}$ by $\hat{U} = \exp(i\hat{F})$ (see Problem 1.50) describes a *constant of motion*. Indicate the physical meaning of the constants of motion for a system of N particles that are associated with invariance of its Hamiltonian with respect to the following coordinate transformations:

a) a translation $\mathbf{r}_n \to \mathbf{r}'_n = \mathbf{r}_n + \mathbf{a}$;
b) rotation through an angle $\phi_0 = \varphi_0 \mathbf{n}_0$;
c) an inversion $\mathbf{r}_n \to \mathbf{r}'_n = -\mathbf{r}_n$; $n = 1, 2, \ldots, N$.

Solution

From the conditions of the Hamiltonian being invariant with respect to this unitary transformation, it follows that $\hat{U}\hat{H} - \hat{H}\hat{U} = 0$, and if we write the operator in the form $\hat{U} = \exp(i\hat{F})$, where $\hat{F}$ is an Hermitian operator, then we have $[\hat{F}, \hat{H}] = 0$. Therefore, if $\partial \hat{F}/\partial t = 0$ then $\hat{F}$ is constant in time – the constant of motion. We should note that the existence of such constants of motion is connected with the symmetry of the interaction (Hamiltonian) – its invariance under coordinate transformation – and does not depend on the concrete form of the interaction.

a) The translation operator has the form $\hat{U} = \exp\left\{(i/\hbar)\mathbf{a}\cdot\hat{\mathbf{P}}\right\}$, see Problem 1.7; its commutativity with $\hat{H}$ is equivalent to the condition $[\hat{\mathbf{P}}, \hat{H}] = 0$, which corresponds to the conservation of the total momentum $\hat{\mathbf{P}} = \sum_n \hat{\mathbf{p}}_n$,

b) The operator of coordinate rotation $\hat{U} = \exp\{i\boldsymbol{\varphi}_0 \cdot \hat{\mathbf{J}}\}$, where $\hat{\mathbf{J}} = \hat{\mathbf{L}} + \hat{\mathbf{S}}$ is the angular momentum operator. Commutativity between $\hat{U}$ and $\hat{H}$ is equivalent to condition $[\hat{\mathbf{J}}, \hat{H}] = 0$, which corresponds to conservation of total angular momentum. (If the interaction does not depend on spin, then the Hamiltonian invariance under coordinate rotation leads to the conservation of both orbital and spin momenta separately.)

c) For coordinate inversion, $\hat{U}\psi(\mathbf{r}_n) \equiv \psi(-\mathbf{r}_n)$, and from the commutativity of $\hat{U}$ and $\hat{H}$ we obtain the conservation of parity.

The Hamiltonian of any closed system of particle is invariant with respect to the transformations considered above, which is related to the properties of the free space: its homogeneity, isotropy, and equivalence of "right" and "left" (the last invariance and hence the law of parity conservation are broken by so-called *weak* interactions). External fields change these properties of space. So, the Hamiltonian of a system in an external field does not have the same degree of symmetry. However, some elements of symmetry and their corresponding constants of motion can appear in these cases; see the following problems.

Problem 6.15

Indicate the constants of motion for a system of N spinless particles in the following fields:

1) for free motion;
2) in a field created by an infinite, uniform plane;
3) in a field created by a uniform sphere;
4) in a field created by two points;
5) in a uniform field alternating with the time;
6) in a field created by homogeneously charged straight wire;
7) in a field created by an infinite cylindrical helix with pitch a.

Solution

The Hamiltonian $\hat{H}$ of the system without any external field has the highest degree of symmetry: it is invariant with respect to arbitrary coordinate translation, rotation, or inversion. External fields break this symmetry. The potential energy in external fields has the form:

$$U_{ext}(r_1, \ldots, r_N) = \sum_n U_n(r_n),$$

and it determines the symmetry of the system Hamiltonian $\hat{H} = \hat{H}_0 + U_{ext}$ in general.

The explicit form of the momentum and angular momentum operators are

$$\hat{P}_z = -i\hbar \sum_n \frac{\partial}{\partial z_n}, \quad \hat{L}_z = -i \sum_n \frac{\partial}{\partial \varphi_n},$$

and we have conservation of system energy in the case of $\partial U_{ext}/\partial t = 0$. Considering this alongside parity, we obtain the following constants of motion.

1) The constants of motion are E, **P**, **L**, I (energy, momentum, angular momentum, and parity, respectively).
2) The system is translationally invariant in any direction parallel to the plane x, y that creates the field. It also has azimuthal symmetry with respect to any axis z that is perpendicular to the plane x, y and reflective symmetry with respect to this plane; so

$$U_{\text{ext}} = \sum_n U_n(|z_n|).$$

The constants of motion are E, P_x, P_y, L_z, I.

3) This system has spherical symmetry; the constants of motion are E, **L**, I.
4) The system has axial symmetry around the axis that penetrates the points – our sources of the field; $U_{\text{ext}} = \sum_n U_n(\rho_n, z_n)$ and the constants of motion are E, L_z.

    In the case when points carry same "charges" we have $U_n = U_n(\rho_n, |z_n|)$,[89] and therefore the parity I is conserved too.
5) If the direction of forces that act on the particles depends on time, then there exist no non-trivial constants of motion. If only the values of the forces depend on time (but not the direction) then, choosing the z-axis along this direction, we have

$$U_{\text{ext}} = -\sum_n F_n(t) z_n.$$

---

[89] Let the plane $z = 0$ pass through the middle of the line connecting the points.

and the constants of motion are P_x, P_y, L_z (for $F_n(t) = $ const energy, E is also the constant of motion).

6) The system has axial symmetry with respect to the z-axis, directed along the wire, and translational invariance in this direction also, so that $U_{ext} = \sum_n U_n(\rho_n)$. The constants of motion are E, P_z, L_z, I (conservation of parity I expresses the reflective symmetry with respect to the plane perpendicular to axis z).

7) If the axis of the helix is the z-axis, the angle of rotation about this axis is φ, the pitch of the helix is a, then the potential function is

$$U_{ext} = \sum_n U_n(\rho_n, \varphi_n, z_n),$$

and is invariant with the respect to the transformation of the form: $\varphi_n \to \varphi_n + \delta\alpha$, $z_n = z_n + \frac{a}{2\pi}\delta\alpha$, for a fixed value of ρ_n. This means that the operator U_{ext}, and therefore the Hamiltonian $\hat{H}$, commutes with the operator

$$\sum_n \frac{\partial}{\partial \varphi_n} + \frac{a}{2\pi}\sum_n \frac{\partial}{\partial z_n} = \frac{i}{\hbar}\left(\hbar \hat{L}_z + \frac{a}{2\pi}\hat{P}_z\right).$$

Hence the combination

$$L_z + \frac{a}{2\pi\hbar}P_z$$

is a constant of motion, and due to $\frac{\partial \hat{U}_{ext}}{\partial t} = 0$, the energy E is conserved too.

Problem 6.16

For a spin-1/2 particle, consider the following interactions with an external field:[90]

a) $\hat{U} = U_0(r) + U_1(r)(\hat{\boldsymbol{\sigma}} \cdot \hat{\mathbf{l}})$,
b) $\hat{U} = U_0(r) + U_1(r)(\hat{\boldsymbol{\sigma}} \cdot \hat{\mathbf{n}})$, $\mathbf{n} = \mathbf{r}/r$.

Determine the constants of the motion and the spin-angular dependence of the energy eigenfunctions.

Solution

In both cases the constants of motion are energy E, total angular momentum $\mathbf{j} = \mathbf{l} + \mathbf{s}$, and $\mathbf{j}^2$, of course. Moreover, in case a) the square of the orbital angular momentum $\mathbf{l}^2$ and parity I are constants of motion too.

---

[90] See also Problem 12.5.

Since constants of the motion commute with the Hamiltonian, the energy eigenfunctions can be chosen to simultaneously diagonalize the Hamiltonian and the (commuting) constants of motion. In the case of a) we have the form

$$\psi_{Ejl_{1(2)}j_z} = f(r)\psi_{jl_{1(2)}j_z},$$

where ψ_{jlj_z} are the spin-angular wavefunctions for a particle with spin, $s = 1/2$ (see Problem 5.24), $l = l_{1(2)} = j \pm 1/2$, $I = (-1)^{l_{1,2}}$. The Schrödinger equation reduces to a one-dimensional equation for the function $f(r)$.

For the cases b) the eigenfunctions have the form

$$\psi_{Ejj_z} = f_1(r)\psi_{jl_1j_z} + f_2(r)\psi_{jl_2j_z}$$

and the Schrödinger equation reduces to a system of two linear differential equation for functions f_1 and f_2: Compare with Problem 12.5.

Problem 6.17

Show that if $\hat{f}_1$ and $\hat{f}_2$ are constants of motion of some system, then $\hat{g}_1 = (\hat{f}_1\hat{f}_2 + \hat{f}_2\hat{f}_1)$ and $\hat{g}_2 = i(\hat{f}_1\hat{f}_2 - \hat{f}_2\hat{f}_1)$ are also constants of motion.

As an illustration of this result, indicate another constant of the motion for a system where a) P_x and J_z, b) J_x and J_y are conserved; explain the results obtained from the symmetry properties of this system.

Solution

From the condition $d\hat{f}_{1,2}/dt = 0$, it follows that $d\left(\hat{f}_1\hat{f}_2\right)/dt = 0$ and $d\left(\hat{f}_2\hat{f}_1\right)/dt = 0$, so that $\hat{f}_1\hat{f}_2$ and $\hat{f}_2\hat{f}_1$, as well as their Hermitian combinations indicated in the condition of the problem (if $\hat{f}_{1,2}$ are Hermitian), are the constants of motion.

a) Since $\hat{P}_y = i\left(\hat{P}_x\hat{J}_z - \hat{J}_z\hat{P}_x\right)$ – see Eq. (III.2) – then from the conservation of P_x and J_z, it follows that P_y is also conserved. The conservation of P_x means that system has the translation invariance along the axis x, while conservation of J_z indicates its axial symmetry with respect to the z-axis. The existence of these two symmetries results in the translation symmetry along the axis y (and in xy-plane in general).

b) $\hat{J}_z = -i\left(\hat{J}_x\hat{J}_y - \hat{J}_y\hat{J}_x\right)$ is also a constant of motion. The conservation of J_x and J_y indicates axial symmetry of the system with respect to the x-axis, and y-axis, which leads to the axial symmetry with respect to axis having the common point of intersection with the axes mentioned above, i.e., the system has spherical symmetry.

Problem 6.18

For a particle acted on by a uniform constant force, show that the operator $\hat{\mathbf{G}} = \hat{\mathbf{p}} - \mathbf{F}_0 t$ is the operator of a conserved quantity ($\mathbf{F}_0$ is the force acting on the particle). Compare this with the result of classical mechanics.

Solution

We begin with the Hamiltonian $\hat{H} = \hat{\mathbf{p}}^2/2m - \mathbf{F}_0 \mathbf{r}$, and we obtain

$$\frac{d}{dt}\hat{\mathbf{G}} = \frac{\partial}{\partial t}\hat{\mathbf{G}} + \frac{i}{\hbar}[\hat{H}, \hat{\mathbf{G}}] = -\mathbf{F}_0 - \frac{i}{\hbar}[(\mathbf{F}_0 \mathbf{r}), \hat{\mathbf{p}}] = 0,$$

so that the mean value is

$$\overline{\mathbf{G}} = \overline{\mathbf{p}}(t) - \mathbf{F}_0 t = \text{const}.$$

This is the natural quantum-mechanical generalization of the classical result. The motion of a particle in an homogeneous vector field can be characterized by the constant of the motion $\mathbf{p}_0 = \mathbf{p}(t) - \mathbf{F}_0 t$ (since $\mathbf{v}(t) = \mathbf{v}(0) + \mathbf{F}_0 t/m$), which is equal to the momentum of particle at $t = 0$.

6.3 Time-dependent unitary transformations; The Heisenberg picture of motion

Problem 6.19

Verify the following relation

$$e^{\hat{A}}\hat{B}e^{-\hat{A}} = \hat{B} + \frac{1}{1!}[\hat{A}, \hat{B}] + \frac{1}{2!}[\hat{A}, [\hat{A}, \hat{B}]] + \ldots.$$

Solution

Let us first introduce the operator $\hat{f}(\lambda) = e^{\lambda\hat{A}}\hat{B}e^{-\lambda\hat{A}}$. Differentiating $\hat{f}$ with respect to λ gives

$$\frac{d\hat{f}}{d\lambda} = \hat{A}e^{\lambda\hat{A}}\hat{B}e^{-\lambda\hat{A}} - e^{\lambda\hat{A}}\hat{B}\hat{A}e^{-\lambda\hat{A}} = e^{\lambda\hat{A}}[\hat{A}, \hat{B}]e^{-\lambda\hat{A}}.$$

In a similar manner we find derivatives of the second and higher orders: $\hat{f}''(\lambda) = e^{\lambda\hat{A}}[\hat{A}, [\hat{A}, \hat{B}]]e^{-\lambda\hat{A}}$, etc. Now using the expansion in Taylor series we obtain the required relation:

$$e^{\hat{A}}\hat{B}e^{-\hat{A}} = \hat{f}(\lambda = 1) = \sum_n \frac{1}{n!}\left(\frac{d^n\hat{f}}{d\lambda^n}\right)_{\lambda=0} = \hat{B} + \frac{1}{1!}[\hat{A}, \hat{B}] + \frac{1}{2!}[\hat{A}, [\hat{A}, \hat{B}]] + \ldots.$$

Problem 6.20

For the following systems:

1) a free particle;
2) a particle acted on by a constant force, $U(x) = -F_0 x$;
3) a linear harmonic oscillator,

find the Heisenberg operators for the coordinate and momentum of the particle by

1) using the unitary transformation that connects the Schrödinger and Heisenberg pictures of motion;
2) solving equations of motion for Heisenberg operators.

Solution

1) The form of the Heisenberg operators, according to Eq. (VI.9), is easy to find if we use the result of the previous problem. We will give the answer:

   a) for a free particle,
   $$\hat{x}(t) = \hat{x} + \frac{t}{m}\hat{p}, \quad \hat{p}(t) = \hat{p};$$

   b) for a particle acted on by a constant force,
   $$\hat{x}(t) = \hat{x} + \frac{t}{m}\hat{p} + \frac{F_0 t^2}{2m}, \quad \hat{p}(t) = \hat{p} + \frac{F_0 t}{m};$$

   c) for a linear oscillator,
   $$\hat{x}(t) = \hat{x}\cos\omega t + \hat{p}\frac{\sin\omega t}{m\omega}, \quad \hat{p}(t) = \hat{p}\cos\omega t - \hat{x}m\omega\sin\omega t.$$

   Here $\hat{x}$ and $\hat{p}$ are the ordinary Schrödinger operators, which coincide with Heisenberg operators at $t = 0$.

2) Let us now demonstrate how the Heisenberg operators can be derived from the equations of motion by considering the linear oscillator. Its Hamiltonian is
$$\hat{H} = \frac{1}{2m}\hat{p}^2(t) + \frac{1}{2}k\hat{x}^2(t).$$

Using the value of the commutator $[\hat{p}(t), \hat{x}(t)] = -i\hbar$, the equations of motion take the form:
$$\frac{d\hat{x}(t)}{dt} = \frac{i}{\hbar}[\hat{H}, \hat{x}(t)] = \frac{\hat{p}(t)}{m}, \quad \frac{d\hat{p}(t)}{dt} = \frac{i}{\hbar}[\hat{H}, \hat{p}(t)] = -k\hat{x}(t).$$

The solution of this system of equations is given by ($\omega = \sqrt{k/m}$):
$$\hat{x}(t) = \hat{C}_1\cos\omega t + \hat{C}_2\sin\omega t, \hat{p}(t) = -m\omega(\hat{C}_1\sin\omega t - \hat{C}_2\cos\omega t).$$

If we take into account that Schrödinger and Heisenberg operators coincide at the point $t = 0$, we have $\hat{C}_1 = \hat{x}$, $\hat{C}_2 = \hat{p}/m\omega$.

Problem 6.21

Using the Heisenberg operators for position and momentum, find the expectation values $\overline{x(t)}$, $\overline{p(t)}$, $\overline{(\Delta x(t))^2}$, $\overline{(\Delta p(t))^2}$ for the systems given in the previous problem and for the initial wavefunction of the form

$$\psi_0(x) = A \, \exp\left[i\frac{p_0 x}{\hbar} - \frac{(x-x_0)^2}{2a^2}\right].$$

Solution

The time-dependence of the physical expectation values can be determined using their corresponding Heisenberg operators and taking into account the expectation values in the initial state

$$\overline{\hat{x}} = x_0, \ \overline{\hat{x}^2} = x_0^2 + \frac{1}{2}a^2, \ \overline{\hat{p}} = p_0, \ \overline{\hat{p}^2} = p_0^2 + \frac{\hbar^2}{2a^2}, \ \overline{\hat{x}\hat{p} + \hat{p}\hat{x}} = 2x_0 p_0.$$

Using Problem 6.20 we find:

a) for a free particle,

$$\overline{x(t)} \equiv \overline{\hat{x}(t)} = x_0 + \frac{p_0}{m}t, \ \overline{p(t)} = p_0, \ \overline{(\Delta x(t))^2} = \frac{a^2}{2}\left(1 + \frac{\hbar^2}{m^2 a^4}t^2\right),$$

$$\overline{(\Delta p(t))^2} = \frac{\hbar^2}{2a^2};$$

b) For a particle acted on by a constant force,

$$\overline{x(t)} = x_0 + \frac{p_0}{m}t + \frac{F_0}{2m}t^2, \ \overline{p(t)} = p_0 + \frac{F_0}{m}t, \ \overline{(\Delta x(t))^2}$$

$$= \frac{a^2}{2}\left(1 + \frac{\hbar^2}{m^2 a^4}t^2\right), \overline{(\Delta p(t))^2} = \frac{\hbar^2}{2a^2}.$$

c) For a linear oscillator,

$$\overline{x(t)} = x_0 \cos \omega t + \frac{p_0}{m\omega}\sin \omega t, \ \overline{p(t)} = p_0 \cos \omega t - m\omega x_0 \sin \omega t,$$

$$\overline{(\Delta x(t))^2} = \frac{1}{2}a^2 \left(\cos^2 \omega t + \frac{\hbar^2}{m^2 \omega^2 a^4}\sin^2 \omega t\right),$$

$$\overline{(\Delta p(t))^2} = \frac{\hbar^2}{2a^2}\left(\cos^2 \omega t + \frac{m^2 \omega^2 a^4}{\hbar^2}\sin^2 \omega t\right).$$

We should note that for the oscillator in the case $a^2 = \hbar/m\omega$, both the position and momentum dispersions do not depend on time while their product takes the minimum possible value that is allowed by the uncertainty relation $\Delta x \Delta p = \hbar/2$. Such states of an oscillator are called the *coherent states*, see Problem 10.15.

Determining the desired expectation values in the Schrödinger representation is more laborious as can be seen by comparing this solution with that of Problem 6.2.

Problem 6.22

Using equations of motion for the Heisenberg operators, show that $[\hat{p}_i(t), \hat{x}_k(t)] = -i\hbar\delta_{ik}$.

Solution

Using equations of motion (VI.4) for Heisenberg operators, it is easy to find that

$$\frac{d}{dt}[\hat{p}_i(t), \hat{x}_k(t)] = 0,$$

i.e., the value of the commutator does not depend on the time, and since at $t = 0$ it is equal to $-i\hbar\delta_{ik}$, it remains such at all times. These are sometimes called the *equal time commutation relations*.

Problem 6.23

Find the value of the unequal time commutation relation $[\hat{p}(t), \hat{x}(t')]$ for the systems considered in Problem 6.20.

Solution

Using the form of the position and momentum Heisenberg operators found in problem 6.20, we find

a) $[\hat{p}(t), \hat{x}(t')] = -i\hbar$;
b) $[\hat{p}(t), \hat{x}(t')] = -i\hbar$;
c) $[\hat{p}(t), \hat{x}(t')] = -i\hbar\cos\omega(t-t')$,

for a free particle, a particle acted on by a constant force, and a linear oscillator, respectively.

The commutator for part c) goes to zero for the values $t - t' = \pi(n + 1/2)/\omega$ (n is integer). If at the initial time, $t = 0$, an oscillator state is characterized by a small position dispersion (the limit $\Delta x \to 0$), so that position takes (almost) definite value, then at the moments of time $t' = \pi(n + 1/2)/\omega$ the momentum takes (almost) a definite value. (Compare with the result of Problem 1.30 and expressions for oscillator position and momentum dispersion relations from Problem 6.21.)

Problem 6.24

A particle (described by some normalized wave packet) is acted on by an homogeneous time-dependent force, $\mathbf{F}(t)$, and $F(t) \to 0$ for $t \to \pm\infty$. Determine the change of the expectation value of the particle energy caused by the action of the force. Compare the result obtained with the classical mechanics analog.

Solution

The interaction of the particle with external field is described by the expression $\hat{U} = -\mathbf{F}(t) \cdot \hat{\mathbf{r}}(t)$. And here,

$$\frac{d}{dt}\hat{\mathbf{p}}(t) = \frac{i}{\hbar}[\hat{H},\ \hat{\mathbf{p}}(t)] = \mathbf{F}(t).$$

Hence it follows (in analogy with the *classical* case) that

$$\hat{\mathbf{p}}(t) = \hat{\mathbf{p}}(-\infty) + \int_{-\infty}^{t} \mathbf{F}(t')dt'. \tag{1}$$

Since as $t \to \pm\infty$, the Hamiltonian of the particle has the form

$$\hat{H}(\pm\infty) = \frac{1}{2m}\hat{\mathbf{p}}^2(\pm\infty),$$

and according to (1) we obtain

$$\overline{E}(+\infty) = \overline{E}(-\infty) + \frac{1}{m}\overline{\mathbf{p}}(-\infty)\int_{-\infty}^{\infty}\mathbf{F}(t)dt + \frac{1}{2m}\left[\int_{-\infty}^{\infty}\mathbf{F}(t)dt\right]^2. \tag{2}$$

This relation, as well as (1), is in a form analogous to the classical equivalent, which is obtained from (2) by replacing the quantum-mechanical expectation values by their *definite* classical values and, of course, $E(-\infty) = p^2(-\infty)/2m$. If we consider an *ensemble* of classical particles with some distribution over momenta, then for the averages, even in the classical case, expression (2) is suitable.

Problem 6.25

Consider a one-dimensional harmonic oscillator that at $t \to -\infty$ is in its ground state. It is further subjected to the action of an external force $F(t)$ and $F(t) \to 0$ as $t \to \pm\infty$. Obtain the probabilities that the oscillator is excited into its various excited states and the expectation value of its energy as $t \to +\infty$. When solving the problem, use the Heisenberg picture, and begin from the equations of motion for the *creation* $\hat{a}^+(t)$ and *annihilation* $\hat{a}(t)$ operators.

Solution

Using the Heisenberg operators,

$$\hat{a}(t) = \frac{1}{\sqrt{2\hbar}}\left(\lambda\hat{x}(t) + \frac{i}{\lambda}\hat{p}(t)\right), \quad \hat{a}^+(t) = \frac{1}{\sqrt{2\hbar}}\left(\lambda\hat{x}(t) - \frac{i}{\lambda}\hat{p}(t)\right), \quad (1)$$

where $\lambda = \sqrt{m\omega}$ and $[\hat{a}(t), \hat{a}^+(t)] = 1$, we can write the Hamiltonian of the system in the form

$$\hat{H}(t) = \hbar\omega\left(\hat{a}^+(t)\hat{a}(t) + \frac{1}{2}\right) - \sqrt{\frac{\hbar}{2m\omega}}F(t)(\hat{a}(t) + \hat{a}^+(t)).$$

The equations of motion for the operators are

$$\frac{d}{dt}\hat{a}(t) = \frac{i}{\hbar}[\hat{H}(t), \hat{a}(t)] = -i\omega\hat{a}(t) + i\frac{F(t)}{\sqrt{2m\hbar\omega}}, \quad (2)$$

$$\frac{d}{dt}\hat{a}^+(t) = i\omega\hat{a}^+(t) - i\frac{F(t)}{\sqrt{2m\hbar\omega}}$$

and they can easily be integrated[91]

$$\hat{a}(t) = e^{-i\omega t}\left\{\hat{a}_{\text{in}} + \frac{i}{\sqrt{2m\hbar\omega}}\int_{-\infty}^{t}F(t')e^{i\omega t'}dt'\right\}. \quad (3)$$

Hence as $t \to \pm\infty$, we have

$$\begin{aligned}\hat{a}(t) &= e^{-i\omega t}\hat{a}_{\text{in}} & \text{as } t \to -\infty, \\ \hat{a}(t) &= e^{-i\omega t}\hat{a}_{\text{f}}, \quad \hat{a}_{\text{f}} = \hat{a}_{\text{in}} + \alpha & \text{as } t \to +\infty.\end{aligned} \quad (4)$$

In (4) we use the definition

$$\alpha = \frac{i}{\sqrt{2m\hbar\omega}}\int_{-\infty}^{\infty}F(t)e^{i\omega t}dt.$$

For the Hamiltonian of the system, we obtain

$$\hat{H}(-\infty) \equiv \hat{H}_{\text{in}} = \hbar\omega\left(\hat{a}_{\text{in}}^+\hat{a}_{\text{in}} + \frac{1}{2}\right),$$

$$\hat{H}(\infty) = \hbar\omega\left(\hat{a}_{\text{f}}^+\hat{a}_{\text{f}} + \frac{1}{2}\right). \quad (5)$$

The time-independent state vector $|\psi\rangle$ in Heisenberg representation is determined by the condition that as $t \to -\infty$ the oscillator was in its ground state, *i.e.*, for this vector

---

[91] The operator $\hat{a}^+(t)$ is the Hermitian-conjugate to the operator $\hat{a}(t)$. The time-independent operator $\hat{a}_{\text{in}}$ plays the role of the "initial" condition and the commutation relation $[\hat{a}_{\text{in}}, \hat{a}_{\text{in}}^+] = 1$ must be fulfilled.

we have $\hat{H}_{\text{in}}|\psi\rangle = (\hbar\omega/2)|\psi\rangle$. According to (5), this state is the *vacuum* with respect to the operators $\hat{a}_{\text{in}}$, $\hat{a}_{\text{in}}^+$, i.e., $|\psi\rangle \equiv |0, \text{ in}\rangle$ so that $\hat{a}_{\text{in}}|0, \text{ in}\rangle = 0$.

The oscillator eigenvectors for $t \to +\infty$ are given by

$$|n, \text{ f}\rangle = \frac{1}{\sqrt{n!}} (\hat{a}_{\text{f}}^+)^n |0, \text{ f}\rangle,$$

so that the coefficients in the expansion $|0, \text{ in}\rangle = \sum_n c_n |n, \text{ f}\rangle$ determine the desired transition probability of the oscillator: $w(0 \to n) = |c_n|^2$. Acting by the operator $\hat{a}_{\text{in}}$ on both sides of the expansion, and taking into account its relation with $\hat{a}_{\text{f}}$ given in (4) as well as the identity

$$\hat{a}_{\text{f}}|n, \text{ f}\rangle = \sqrt{n}|n-1, \text{ f}\rangle,$$

we obtain a recurrence relation $c_n = (\alpha/\sqrt{n})c_{n-1}$. From this relation it follows that $c_n = (\alpha^n/\sqrt{n!})c_0$, and the normalization condition

$$\langle 0, \text{ in}|0, \text{ in}\rangle = \sum_n |c_n|^2 = 1$$

gives $|c_0|^2 = \exp\{-|\alpha|^2\}$. For the transition probability to any excited state, we obtain

$$w(0 \to n) = \frac{|\alpha|^{2n}}{n!} e^{-|\alpha|^2} \tag{6}$$

(the *Poisson distribution*). Using the expectation value $\bar{n} = |\alpha|^2$, we find

$$\overline{E}(+\infty) = \hbar\omega \left(\bar{n} + \frac{1}{2}\right) = \hbar\omega \left(|\alpha|^2 + \frac{1}{2}\right).$$

Let us give another way of calculating $\overline{E}(+\infty)$ that does not require calculation of the transition probabilities. According to (4) and (5) we have

$$\hat{H}(+\infty) = \hat{H}_{\text{in}} + \hbar\omega(|\alpha|^2 + \alpha\hat{a}_{\text{in}}^+ + \alpha^*\hat{a}_{\text{in}}).$$

Hence if at $t \to -\infty$ the oscillator was in its kth excited state, i.e., $|\psi\rangle \equiv |k, \text{ in}\rangle$ (in the case of the problem $\langle\psi|\hat{a}_{\text{in}}|\psi\rangle = \langle\psi|\hat{a}_{\text{in}}^+|\psi\rangle = 0$), then

$$\overline{E}(+\infty) = E(-\infty) + \hbar\omega|\alpha|^2, \quad E(-\infty) = \hbar\omega\left(k + \frac{1}{2}\right). \tag{7}$$

(We should note that the expectation value of the energy acquired by the oscillator, which is equal to $\hbar\omega|\alpha|^2$, does not contain the Plank constant, and coincides with the result of classical mechanics.)

Problem 6.26

Find the unitary operator that corresponds to a *Galilean transformation*, i.e., a transformation to a new inertial frame of reference. Prove that the Schrödinger equation is invariant with respect to this transformation. How would the particle wavefunction transform in the position and momentum representations respectively?

Solution

Let a frame of reference K' move relative to a system K with a velocity V along the x-axis, so that $x = x' + Vt$, $t = t'$. Using this transformation we obtain how potentials must transform:

$$U'(x', t') \equiv U'(x - Vt, t) = U(x, t).$$

The unitary operator[92] $\hat{U}$, corresponding to a Galilean transformation, is found from the condition that if the wavefunction $\psi(x, t)$ satisfies the Schrödinger equation,

$$i\hbar \frac{\partial}{\partial t} \psi(x, t) = \hat{H} \psi \equiv \left[\frac{1}{2m} \hat{p}^2 + U(x, t) \right] \psi(x, t), \tag{1}$$

in the frame of reference K, then the wavefunction $\psi'(x', t) = \hat{U} \psi(x, t)$ has to satisfy the Schrödinger equation in the frame of reference K' (and vice versa):

$$i\hbar \frac{\partial}{\partial t} \psi'(x', t) = \hat{H}' \psi' \equiv \left[\frac{1}{2m} (\hat{p}')^2 + U'(x', t) \right] \psi'(x', t). \tag{2}$$

Since both the functions, $\psi(x, t)$ and $\psi'(x', t)$, describe the same physical state (but with respect to different frames of reference), they must satisfy the condition

$$|\psi'(x', t)|^2 \equiv |\psi'(x - Vt, t)|^2 = |\psi(x, t)|^2, \tag{3}$$

which expresses the independence of the probability density from the choice of the frame of reference. Hence, from (3) it follows that the desired unitary operator has the form

$$\hat{U} = e^{iS(x,t)},$$

where $S(x, t)$ is some real function. If we substitute

$$\psi'(x', t) = e^{iS(x,t)} \psi(x, t) \tag{4}$$

---

[92] Do not confuse this operator with the potential energy $U(x, t)$. Also, for simplicity we restrict ourselves to the case of one-dimensional motion.

into (2) and transform to the variables x, t, we obtain the following equation:

$$i\hbar\frac{\partial}{\partial t}\psi(x,t) = -\frac{\hbar^2}{2m}\frac{\partial^2}{\partial x^2}\psi(x,t) + i\hbar\left(-\frac{\hbar}{m}\frac{\partial S}{\partial x} - V\right)\frac{\partial\psi}{\partial x} +$$

$$\left[U(x,t) - \frac{i\hbar^2}{2m}\frac{\partial^2 S}{\partial x^2} + \frac{\hbar^2}{2m}\left(\frac{\partial S}{\partial x}\right)^2 + \hbar V\frac{\partial S}{\partial x} + \hbar\frac{\partial S}{\partial t}\right]\psi(x,t).$$

If we then require that this equation is identical to (1), we will obtain the system of equations

$$\frac{\hbar}{m}\frac{\partial S}{\partial x} + V = 0, \quad -\frac{i\hbar}{2m}\frac{\partial^2 S}{\partial x^2} + \frac{\hbar}{2m}\left(\frac{\partial S}{\partial x}\right)^2 + V\frac{\partial S}{\partial x} + \frac{\partial S}{\partial t} = 0.$$

From the first of them we have $S = -mVx/\hbar + f(t)$, while the second equation allows us to find $f(t)$ and to obtain

$$S(x,t) = -\frac{mVx}{\hbar} + \frac{mV^2 t}{2\hbar} + C. \tag{5}$$

(The unimportant real constant C may be omited, since it is just an overall phase to our wavefunction.)

Let us now find the transformation of the wavefunction in the momentum representation. Multiplying (4) by $\psi_{p'}^*(x')$ (eigenfunctions of the momentum operator) and integrating over x', we obtain (using (5)),

$$\phi'(p',t)' = \exp\left\{-i\frac{mV^2 t}{2\hbar} + i\frac{pVt}{\hbar}\right\}\phi(p,t), \quad p = p' + mV. \tag{6}$$

Hence we obtain the relation

$$w'(p - mV, t) = w(p, t)$$

between the probablity distributions over particle momenta in systems K' and K.

Problem 6.27

Determine the form of the unitary transformation corresponding to a gauge transformation of the electromagnetic field potentials. Show gauge invariance of the Schrödinger equation.

Solution

Let the wavefunction $\psi(\mathbf{r}, t)$ be a solution of the Schrödinger equation:

$$i\hbar\frac{\partial}{\partial t}\psi = \frac{1}{2m}\left(\hat{\mathbf{p}} - \frac{e}{c}\mathbf{A}(\mathbf{r},t)\right)^2\psi + e\varphi(\mathbf{r},t)\psi, \tag{1}$$

where $\mathbf{A}(\mathbf{r},t)$ and $\varphi(\mathbf{r},t)$ are the potentials of an external electromagnetic field. The gauge invariance of the Schrödinger equation means that if we introduce new potentials according to

$$\mathbf{A}' = \mathbf{A} + \nabla f(\mathbf{r},t), \quad \varphi' = \varphi - \frac{\partial}{c\partial t} f(\mathbf{r},t),$$

then there must exist a unitary operator $\hat{U}$ such that the wavefunction $\psi'(\mathbf{r},t) = \hat{U}\psi(\mathbf{r},t)$ describes the same physical state as the initial wavefunction ψ (but with the different choice of the potentials), and hence satisfies the relation $|\psi'(r,t)|^2 = |\psi(r,t)|^2$ and solves the Schrödinger equation:

$$i\hbar\frac{\partial}{\partial t}\psi' = \frac{1}{2m}\left(\hat{\mathbf{p}} - \frac{e}{c}\mathbf{A}'(\mathbf{r},t)\right)^2 \psi' + e\varphi'(\mathbf{r},t)\psi'. \tag{2}$$

Due to the invariance of the probability density, the desired operator must be of the form

$$\hat{U} = e^{iS(\mathbf{r},t)},$$

where $S(\mathbf{r},t)$ is some real function (compare with the previous problem). Substituting the function

$$\psi'(\mathbf{r},t) = e^{iS(\mathbf{r},t)}\psi(\mathbf{r},t)$$

into equation (2), and demanding that the relation obtained coincides with relation (1), we obtain the following equations:

$$c\hbar\nabla S(\mathbf{r},t) = e\nabla f(\mathbf{r},t), \quad c\hbar\frac{\partial}{\partial t}S(\mathbf{r},t) = e\frac{\partial}{\partial t}f(\mathbf{r},t).$$

Hence we find that $S(\mathbf{r},t) = \frac{e}{\hbar c}f(\mathbf{r},t) + C$, which gives the solution of the problem (the unimportant real constant C may be omitted). Thus, for a system of charged particles,

$$\hat{U} = \exp\left\{\frac{i}{\hbar c}\sum_a e_a f(\mathbf{r}_a,t)\right\}.$$

Problem 6.28

Find a time-dependent unitary transformation of the Hamiltonian that leaves the Schrödinger equation invariant. Compare with the canonical transformations in classical mechanics.

Solution

1) In the case of unitary transformations, wavefunctions and operators are transformed in the following way:

$$\psi \to \psi' = \hat{U}\psi, \quad \hat{f} \to f = \hat{U}\hat{f}\hat{U}^+.$$

In particular, for the Hamiltonian of a system we have

$$\hat{H}' = \hat{U}\hat{H}\hat{U}^+. \tag{1}$$

Let us determine the form of the Schrödinger equation for time-dependent unitary transformation. Applying the operator $\hat{U}(t)$ to the both parts of Schrödinger equation, we obtain the equation

$$i\hbar \hat{U}\frac{\partial \psi}{\partial t} \equiv i\hbar \frac{\partial}{\partial t}(\hat{U}\psi) - i\hbar \left(\frac{\partial \hat{U}}{\partial t}\right)\psi = \hat{U}\hat{H}\psi.$$

Substituting $\psi' = \hat{U}(t)\psi$ and taking into account that $\hat{U}^+\hat{U} = 1$, the equation becomes

$$i\hbar \frac{\partial \psi'}{\partial t} = \left[\hat{U}\hat{H}\hat{U}^+ + i\hbar \left(\frac{\partial \hat{U}}{\partial t}\right)\hat{U}^+\right]\psi',$$

which gives the Schrödinger equation with the Hamiltonian:

$$\hat{H}'' = \hat{U}\hat{H}\hat{U}^+ + i\hbar \left(\frac{\partial \hat{U}}{\partial t}\right)\hat{U}^+ = \hat{H}' + i\hbar \left(\frac{\partial \hat{U}}{\partial t}\right)\hat{U}^+. \tag{2}$$

This is the desired solution (here H'' is a Hermitian operator).

2) So, we could assign the system $\hat{H}$ the two different operators after a time-dependent unitary transformation: $\hat{H}'$ and $\hat{H}''$. To understand the result obtained we have to take into account the fact that the Hamiltonian of a system plays, generally speaking, two roles: 1) it determines the wavefunction time evolution in accordance with the Schrödinger equation; and 2) in the case when it is time-independent it is a constant of motion, and its eigenvalues have a definite physical interpretation, i.e., they determine the system's energy spectrum.

If the initial Hamiltonian $\hat{H}$ does not depend on time, then in the case of the time-dependent unitary transformation its roles mentioned above are distributed among between the operators $\hat{H}'$ and $\hat{H}''$ The first of them is the constant of motion (the eigenvalues of $\hat{H}$ and $\hat{H}'(t)$ coincide), while the second determines the time evolution of the wavefunction.

If $\hat{H}(t)$ is time-dependent, then the Hamiltonian is no longer a constant of motion (energy is not conserved). Here the eigenvalues of "instant" Hamiltonians $\hat{H}(t)$ and $\hat{H}'(t)$ have no physical content in general. However, the operator $\hat{H}''$ still determines the time evolution.[93]

As an illustration, consider the transformation from the Schrödinger picture to the Heisenberg picture for the case $\partial \hat{H}/\partial t = 0$, which is given by $\hat{U}(t) = \exp(\frac{i}{\hbar})\hat{H}t$. Here $\hat{H}' = \hat{H}$, $\hat{H}'' = 0$, and from the Schrödinger equation it follows, as is expected, that the system's wavefunctions are time-independent (since it is the Hiesenberg picture).

---

[93] If $\hat{H}''$, unlike $\hat{H}(t)$, does not depend on time, then it fulfills both roles mentioned. see Problem 6.29.

Unitary transformations in quantum mechanics are the analogues of canonical transformations in classical mechanics. Here the tranformation given by (2) for the Hamiltonian is the quantum-mechanical generalization of the equation

$$H''(P,Q,t) = H(P,Q,t) + \frac{\partial f}{\partial t}$$

from classical mechanics which gives a canonical transformation for the Hamiltonian function determined by a time-dependent function $f(t)$.

Problem 6.29

Determine the form of the unitary transformation describing transition to a uniformly rotating coordinate system. Show how the operators for position, momentum, velocity, and energy (the Hamiltonian) are transformed. Compare your results with their classical analogs. As an illustration, consider a charged particle moving in a uniform circular electric field of the form:

$$\mathcal{E}_x = \mathcal{E}_0 \cos \omega t, \quad \mathcal{E}_y(t) = \mathcal{E}_0 \sin \omega t, \quad \mathcal{E}_z = 0.$$

Solution

The unitary operator which describes a rotating reference frame is given by $\hat{U} = e^{i\hat{\mathbf{L}} \cdot \boldsymbol{\omega} t}$, where $\boldsymbol{\omega}$ is the angular velocity of the rotating coordinate system with respect to the initial inertial system. The wavefunction transformation law has the form

$$\psi'(\mathbf{r},t) = e^{i\hat{\mathbf{L}} \cdot \boldsymbol{\omega}} \psi(\mathbf{r},t) \equiv \psi(\mathbf{r}',t), \tag{1}$$

where $\psi'(\mathbf{r},t)$ is the wavefunction in a rotating system, while $\psi(\mathbf{r}',t)$ is the initial wavefunction in a static coordinate system. $\mathbf{r}$ is the position vector in the rotating system, and $\mathbf{r}' = \hat{R}\mathbf{r}$ is the position vector in the static coordinate system, that at the moment t, $\mathbf{r}'$ coincides with $\mathbf{r}$ ($\hat{R}(t=0) = \hat{1}$). Eq (1) implies that the wavefunction does not depend on the coordinate system nor on the variables that describe it (stationary $\mathbf{r}'$ or rotating $\mathbf{r}$).[94]

Let us find the form of our physical operators in the new representation, i.e., in the rotating coordinate system, and their connection with the operators in the initial stationary frame of reference. First we determine the form of the position operator $\hat{\mathbf{r}}_{\text{stat}}(t)$ and momentum operator $\hat{\mathbf{p}}_{\text{stat}}(t)$ with respect to the inertial frame of reference. Using the result of Problem 6.19, the known values of the commutators for the operators $\hat{\mathbf{r}}$, $\hat{\mathbf{p}}$, and $\hat{\mathbf{L}}$, see, Eq. (III.2), and the general formula for operator transformation, $\hat{f}' = \hat{U}\hat{f}\hat{U}^+$, we obtain

---

[94] Compare this with Problems 6.26 and 6.27.

$$\hat{x}_{\text{stat}}(t) = \hat{U}\hat{x}\hat{U}^+ = \hat{U}x\hat{U}^+ = x\cos\,\omega t - y\sin\,\omega t,$$
$$\hat{y}_{\text{stat}}(t) = x\sin\,\omega t + y\cos\,\omega t,\ \hat{z}_{\text{stat}}(t) = z, \qquad (2)$$

as well as

$$\hat{p}_{x,\text{stat}}(t) = \hat{U}\left(-i\hbar\frac{\partial}{\partial x}\right)\hat{U}^+ = \hat{p}_x\cos\,\omega t - \hat{p}_y\sin\,\omega t,$$
$$\hat{p}_{y,\text{stat}}(t) = \hat{p}_x\sin\,\omega t + \hat{p}_y\cos\,\omega t,\ \hat{p}_{z\text{stat}}(t) = \hat{p}_z, \qquad (3)$$

where $\hat{p}_x = -i\hbar\partial/\partial x$ etc., while the z-axis is directed along the vector $\boldsymbol{\omega}$.

The equations in (2) imply that the position operator transforms as a simple vector under the rotation. In this way we can conclude that the position operator in the rotated frame, after being rotated into the stationary coordinate system, is just that – the position operator in the stationary coordinate system.[95] In an analogous way, equations (3) are also exactly the vector transformation of the components of the momentum, which implies that the momentum operator in both systems is the same and has the form $\hat{\mathbf{p}} = -i\hbar\nabla$. Similarly, in the rotating and stationary systems the angular momentum operator has the form

$$\hbar\hat{\mathbf{L}} = \hat{\mathbf{r}} \times \hat{\mathbf{p}} = \hat{\mathbf{r}}_{\text{stat}} \times \hat{\mathbf{p}}_{\text{stat}}.$$

The particle Hamiltonian, $\hat{H} = \hat{\mathbf{p}}^2/2m + U(\mathbf{r},t)$, transitioning to the rotating coordinate system and according to Eq. (2) from Problem 6.28 has the form

$$\hat{H}_{rot} = \hat{U}\hat{H}\hat{U}^+ - \hbar\boldsymbol{\omega}\hat{\mathbf{L}} = \frac{\hat{\mathbf{p}}^2}{2m} + U'(\mathbf{r},t) - \hbar\boldsymbol{\omega}\hat{\mathbf{L}}, \qquad (4)$$

where $U'(\mathbf{r},t) = U(\mathbf{r}',t)$ is the potential in the rotating coordinate system. Now, taking into account the relation $\hat{\mathbf{v}} = \dot{\hat{\mathbf{r}}} = i[\hat{H},\hat{\mathbf{r}}]/\hbar$, it is easy to find the velocity operator in both the stationary and rotating coordinate systems:

$$\hat{\mathbf{v}}_{stat} = \frac{\hat{\mathbf{p}}}{m},\ \hat{v}_{rot} = \frac{\hat{\mathbf{p}}}{m} - [\boldsymbol{\omega} \times \mathbf{r}] = \hat{\mathbf{v}}_{stat} - [\boldsymbol{\omega} \times \mathbf{r}] \qquad (5)$$

Finally, let us consider a particle in an external field whose sources rotate with a constant angular velocity with respect to some axis, so that the potential energy in the initial coordinate system $U(\mathbf{r},t)$ depends explicitly on time. Changing coordinate systems by rotating and using (4), we obtain

$$\hat{H}_{rot} = -\frac{\hbar^2}{2m}\Delta + U'(r) - \hbar\boldsymbol{\omega}\cdot\hat{\mathbf{L}}. \qquad (6)$$

It is important that now $U'(\mathbf{r})$ as well as the Hamiltonian in general do not explicitly depend on time, so the energy in a rotating system is a constant of motion; so,

---

[95] We should note that in the initial rotating frame, r is the position vector and $\hat{\mathbf{r}} = \mathbf{r}$ (the operator is just multiplication by the coordinates). When we transform, $\mathbf{r}$ transforms like a vector into the stationary frame, but the operator transforms in the common-operator way, $\hat{\mathbf{r}}' = \hat{U}\hat{\mathbf{r}}\hat{U}^+$. This statement is also true for momentum.

for a particle moving in an electric field of circular-polarized monochromatic wave, $U'(x) = -e\mathcal{E}_0 x$.

In conclusion, note that the results of the given problem are the natural quantum-mechanical generalizations of the corresponding results of classical mechanics.

Problem 6.30

The Hamiltonian of some system has the form $\hat{H} = \hat{H}_0 + \hat{V}$ where the *unperturbed* Hamiltonian $\hat{H}_0$ does not depend explicitly on time. Consider the unitary transformation from the Schrödinger representation to so-called *interaction representation* that is carried out by the unitary operator,[96] $\hat{U} = \exp\left(\frac{i}{\hbar}\hat{H}_0(t-t_0)\right)$. (In the case $\hat{V} \equiv 0$ and $t_0 = 0$, this transformation gives rise to the Heisenberg representation.)

Determine the time-dependence of the wavefunction and operators of the system in the interaction representation.

As an illustration, consider the excitation of a linear harmonic oscillator, which was in the ground state at $t \to -\infty$, by an external force $F(t)$, where $F(t) \to 0$ as $t \to \pm\infty$. Consider the interaction $\hat{V} = -F(t)\hat{x}$ to be small. Compare with the exact result. See Problem 6.25.

Solution

The relation between wavefunctions and operators in the interaction representation (they have indexes "*int*") with those in the Schrödinger representation has the form

$$\psi_{\text{int}}(t) = \exp\left\{\frac{i}{\hbar}\hat{H}_0(t-t_0)\right\}\psi(t), \quad (1)$$

$$\hat{f}_{\text{int}}(t) = \exp\left\{\frac{i}{\hbar}\hat{H}_0(t-t_0)\right\}\hat{f}\exp\left\{-\frac{i}{\hbar}\hat{H}_0(t-t_0)\right\}.$$

If $\hat{f} = f(\hat{p}, \hat{q}, t)$, then $\hat{f}_{\text{int}} \equiv f(\hat{p}_{\text{int}}, \hat{q}_{\text{int}}, t)$.

Differentiating (1) with respect to time we obtain the equation of motion for the corresponding operator

$$\frac{d}{dt}\hat{f}_{\text{int}} = \frac{\partial}{\partial t}\hat{f}_{\text{int}} + \frac{i}{\hbar}[\hat{H}_0, \hat{f}_{\text{int}}], \quad (2)$$

where

$$\hat{H}_0(\hat{p}, \hat{q}) = \hat{H}_{0,\text{int}} = \hat{H}_0(\hat{p}_{\text{int}}, \hat{q}_{\text{int}}), \ \hat{p}_{\text{int}}(t_0) = \hat{p}, \ \hat{q}_{\text{int}}(t_0) = \hat{q}.$$

---

[96] The value of time t_0 is chosen so that it precedes the moment when interaction $\hat{V}(t)$ is turned on.

The time-evolution of the wavefunction is determined by the equation

$$i\hbar \frac{\partial}{\partial t} \psi_{\text{int}} = \hat{H}'_{\text{int}} \psi_{\text{int}}, \quad \hat{H}'_{\text{int}} = \hat{V}_{\text{int}} \tag{3}$$

(such a form of the Hamiltonian $\hat{H}'_{\text{int}}$ in the interaction picture follows from Eq. (2) of Problem 6.28.)

Let us illustrate the application of the interaction picture for the harmonic oscillator in a uniform time-dependent external field. Here,

$$\hat{H}_0 = \frac{1}{2m}\hat{p}^2 + \frac{1}{2}m\omega^2 \hat{x}^2, \quad \hat{V} = -F(t)x.$$

The coordinate and momentum operators' time-dependence in the interaction representation is the same as in the Heisenberg representation for a free oscillator (*i.e.*, with no interaction). Using the result of Problem 6.20 for $x(t)$, we obtain

$$\hat{V}_{\text{int}}(t) = -F(t)\hat{x}_{\text{int}}(t) = -F(t)\left[x \cos \omega(t-t_0) - \frac{i\hbar}{m\omega} \sin \omega(t-t_0) \frac{\partial}{\partial x}\right]. \tag{4}$$

To solve (3) with $\hat{V}_{\text{int}}$ in (4), consider consecutive iterations that take into account the smallness of force $F(t)$, and we substitute into (3) the wavefunction

$$\psi_{\text{int}}(x,t) = \psi_0(x) + \psi^{(1)}(x,t),$$

where, in accordance with problem statement, $\psi_0(x)$ is the wavefunction of the ground state of the oscillator and $\psi^{(1)}(x, t = -\infty) = 0$ (before the force was turned on, the oscillator was in the ground state). Now taking into account the form of the wavefunction, See Eq. (II.2),

$$\psi_0(x) = \left(\frac{1}{\pi a^2}\right)^{1/4} \exp\left(-\frac{x^2}{2a^2}\right), \quad a = \sqrt{\frac{\hbar}{m\omega}},$$

and using (3) and (4) we obtain

$$i\hbar \frac{\partial}{\partial t} \psi^{(1)}(x,t) = -F(t) e^{i\omega(t-t_0)} x \psi_0(x).$$

Hence it follows that $\psi^{(1)} = \sqrt{2} C_1(t) x \psi_0(x)/a$, where

$$C_1(t) = \frac{i}{\sqrt{2m\hbar\omega}} e^{-i\omega t_0} \int_{-\infty}^{t} F(t') e^{i\omega t'} dt'.$$

Considering that $\psi_1 = \sqrt{2} x \psi_0(x)/a$ is the wavefunction of the first excited state of the oscillator, we note that under the action of uniform field and to the first order in F, there are only transitions to the first excited state from the ground state. The transition probability as $t \to \infty$ is equal to[97]

$$w^{(1)}(0 \to 1) = |C_1(+\infty)|^2 = \frac{1}{2m\hbar\omega} \left| \int_{-\infty}^{\infty} F(t)e^{i\omega t} dt \right|^2,$$

which in the case of a weak force agrees with the exact result. See Problem 6.25.

6.4 The time-dependent Green function

Problem 6.31

Show that for the case of a time-independent Hamiltonian, the time-dependent Green function obeys the equation[98]

$$\hat{q}(-t)G(q,t;q',t'=0) = q'G(q,t;q',0), \tag{1}$$

where $\hat{q}(t)$ is the Heisenberg operator.

Using this relation, find the Green function for a free particle in the coordinate and momentum representation. Obtain it also using Eq. (VI.6). With the help of the Green function, solve Problem 6.2.

Solution

1) Let us first show that if a wavefunction in the Schrödinger representation at $t=0$ is an eigenfunction of a time-independent operator $\hat{f}$, then $\psi(q,t)$ is the eigenfunction of the Heisenberg operator $\hat{f}(-t)$. Indeed, applying the operator $e^{-i\hat{H}t/\hbar}$ to the equation $\hat{f}\psi(q,0) = f\psi(q,0)$, and using the known connection between wavefunctions and operators in the Schrödinger and Heisenberg representations, we obtain

$$\hat{f}(-t)\psi(q,t) = f\psi(q,t).$$

Furthermore, if we choose $\hat{f} \equiv \hat{q} = q$, $f = q'$, then the function $\psi(q,t)$ coincides with Green function $G(q,t:q',0)$[99] and we arrive at the equation for it which is given in the condition of the problem.

2) For a free particle, $\hat{\mathbf{p}}(t) = \hat{\mathbf{p}}$ (the operator does not depend on time). In the momentum representation $\hat{\mathbf{p}} = \mathbf{p}$, and the equation for Green function takes the form $\mathbf{p}G = \mathbf{p}'G$. Hence, it follows that

$$G(\mathbf{p},t;\mathbf{p}',0) = c(\mathbf{p},t)\delta(\mathbf{p}'-\mathbf{p}). \tag{2}$$

---

[97] Note that this probability does not depend on the choice of t_0, as expected.
[98] For the case of several degrees of the freedom this would be a system of equations whose number would correspond to those degrees of freedom.
[99] Indeed, the function $\psi(q,t)$ satisfies the Schrödinger equation in the variables q, t, and at $t=0$ is equal to $\delta(q-q')$, as the Green function demands.

Using the Schrödinger equation (in the variables **p**, t) we find that

$$c(\mathbf{p}, t) = c_0(\mathbf{p}) \exp\left\{-i\frac{p^2 t}{2m\hbar}\right\},$$

while from the initial condition at $t = 0$, it follows that $c_0(\mathbf{p}) = 1$. This unambiguously determines the form of time-dependent Green function in the momentum representation.

In an analogous way, using the relation $\hat{\mathbf{r}}(t) = \hat{\mathbf{r}} + \hat{\mathbf{p}}t/m$ in coordinate representation, we have $(\mathbf{r} - t\hat{\mathbf{p}}/m)G = \mathbf{r}'G$, which is a system of three differential equations. Using the x-component as an example of how they look,

$$\left(x_i + i\frac{\hbar t}{m}\frac{\partial}{\partial x_i}\right)G = x'G.$$

The solution of this system of equations is given by

$$G(\mathbf{r}, t; \mathbf{r}', 0) = a(\mathbf{r}', t) \exp\left\{i\frac{m(\mathbf{r} - \mathbf{r}')^2}{2\hbar t}\right\}. \tag{3}$$

Substituting this function into the Schrödinger equation we obtain $\dot{a} = -3a/2t$. Hence, it follows that

$$a(\mathbf{r}', t) = a_0(\mathbf{r}') t^{-3/2}.$$

In order to determine $a_0(\mathbf{r}')$, we note that according to the initial condition, $\int G dV \to 1$ as $t \to 0$. Calculating the integral we find

$$a(\mathbf{r}', t) = a(t) = \left(-\frac{im}{2\pi\hbar t}\right)^{3/2}. \tag{3}$$

Taking into account (3), we obtain the final expression for the Green function.

Let us note that this expression could also be obtained in the following two ways: 1) by the direct calculation of integral (VI.6), choosing the eigenfunction $\psi_E(r)$ in the form of plain waves and performing the integration over p, and 2) using (2) and the general equation that connects the operator kernel in the coordinate and momentum representations, as was done in Problem 2.20 for the stationary Schrödinger equation.

To conclude, we note that the exponential factor in (2) is equal to $iS/\hbar$, where

$$S(\mathbf{r}, \mathbf{r}', t) = \frac{1}{2}mv^2 t = \frac{m(\mathbf{r} - \mathbf{r}')^2}{2t}$$

is the *action* of a free classical particle which moves from the point $\mathbf{r}'$ at $t = 0$ to the point $\mathbf{r}$ at the time t. Its velocity is equal to $(\mathbf{r} - \mathbf{r}')/t$.

Problem 6.32

Repeat the previous problem but for a particle acted on by a uniform force $U(\mathbf{r}) = -\mathbf{F}_0 \mathbf{r}$.

Solution

Since $\hat{\mathbf{r}}(t) = \hat{\mathbf{r}} + \hat{\mathbf{p}}t/m + \mathbf{F}_0 t^2/2m$ – see Problem 6.20, the equation for the Green function, $\hat{\mathbf{r}}(-t)G = \mathbf{r}'G$, differs from the one considered in the previous problem by only the change $\mathbf{r}' \to \mathbf{r}' - \mathbf{F}_0 t^2/2m$. This is also valid for its solution (3) from Problem 6.31. But now

$$\dot{a} = \left\{ -\frac{3}{2t} + \frac{i}{\hbar} \mathbf{F}_0 \left(\mathbf{r}' - \frac{\mathbf{F}_0 t^2}{2m} \right) \right\} a.$$

Hence we can find $a(\mathbf{r}', t)$ and obtain the final expression for the Green function in the coordinate representation:

$$G(\mathbf{r}, t; \mathbf{r}', 0) = \left(\frac{m}{2\pi i \hbar t} \right)^{3/2} \exp \left\{ \frac{i}{\hbar} \left[\frac{1}{2mt} \left(\mathbf{r} - \mathbf{r}' - \frac{\mathbf{F}_0 t^2}{2m} \right)^2 + \mathbf{F}_0 \mathbf{r} t - \frac{F_0^2 t^3}{6m} \right] \right\}. \quad (1)$$

(Just as in Problem 6.31, the exponent is the *action* for a classical particle acted on by a constant force.)

In the momentum representation,

$$G(\mathbf{p}, t; \mathbf{p}', 0) = \exp \left\{ -\frac{it}{2m\hbar} \left(p^2 - \mathbf{F}_0 \mathbf{p} t + \frac{1}{3} F_0^2 t^2 \right) \right\} \delta(\mathbf{p} - \mathbf{p}' - \mathbf{F}_0 t).$$

Problem 6.33

Repeat Problem 6.30 but for a linear harmonic oscillator.

Solution

1) First we find the Green function by using Eq. (VI.6). For an oscillator we have

$$\psi_n = \left(2^n \sqrt{\pi} a n! \right)^{-1/2} \exp \left(-\frac{x^2}{2a^2} \right) H_n \left(\frac{x}{a} \right), \quad E_n = \hbar \omega \left(n + \frac{1}{2} \right),$$

where $a = \sqrt{\hbar/m\omega}$ The summation in Eq. (VI.6) is performed by using the following identity

$$(1 - z^2)^{-1/2} \exp \left\{ \frac{2xyz - (x^2 + y^2)z^2}{1 - z^2} \right\} = \sum_{n=0}^{\infty} \frac{H_n(x) H_n(y)}{2^n n!} z^n,$$

and this leads to the following result:

$$G(x,t;x',0) = \frac{\exp\left\{i\cot(\omega t)\cdot\left(x^2 - 2xx'\sec(\omega t) + (x')^2\right)/2a^2\right\}}{\sqrt{2\pi i a^2 \sin(\omega t)}}. \tag{1}$$

2) Determining the form of the Green function by using Eq. (1) in Problem 6.31, we obtain (with the Heisenberg position operator in Problem 6.20):

$$\hat{x}(-t)G \equiv \left(\hat{x}\cos\omega t + ia^2 \sin(\omega t)\frac{\partial}{\partial x}\right)G = x'G.$$

Hence, it follows that

$$G = c(x',t)\exp\left\{\frac{i}{2a^2}\left(x^2\cot(\omega t) - 2xx'\operatorname{cosec}(\omega t)\right)\right\}.$$

Substituting this expression into the Schrödinger equation, we obtain

$$\dot{c} + \frac{1}{2}\left(\omega\cot(\omega t) + i\frac{\omega(x')^2}{a^2\sin^2(\omega t)}\right)c = 0.$$

The solution of this equation is

$$c = c_0(\sin(\omega t))^{-1/2}\exp\left\{i\frac{(x')^2\cot(\omega t)}{2a^2}\right\}.$$

From the initial condition we have $c_0 = (2\pi i a^2)^{-1/2}$, and for the Green function we again obtain the expression (1).

The Schrödinger equation for the harmonic oscillator in the momentum representation has the same form as in the coordinate representation, so the expression for the Green function $G(p,t;p',0)$ may be obtained from (1) by just changing $x \to p$ and $a \to \sqrt{\hbar m\omega}$.

Problem 6.34

Find the time-dependent Green function for a charged particle moving in a circular electric field in a rotating coordinate system; see Problem 6.29.

Solution
The Hamiltonian of a particle in this rotating system has the form (see Problem 6.29):

$$\hat{H}_{\text{rot}} = \frac{\hat{\mathbf{p}}^2}{2m} - \omega\hat{L}_z - F\hat{x}, \quad F = e\mathcal{E}_0.$$

The time-dependent Green function can be found by using equation (1) from Problem 6.31. For that we should first find the Heisenberg operators $\hat{\mathbf{r}}(t)$ and $\hat{\mathbf{p}}(t)$. For brevity, we shall use the units $\hbar = m = 1$. The equations of motion for the Heisenberg operators have the form

$$\dot{\hat{x}} = \hat{p}_x + \omega\hat{y}, \quad \dot{\hat{y}} = \hat{p}_y + \omega\hat{x}, \quad \dot{\hat{z}} = \hat{p}_z,$$

$$\dot{\hat{p}}_x = \omega\hat{p}_y + F, \quad \dot{\hat{p}}_y = -\omega\hat{p}_x, \quad \dot{\hat{p}}_z = 0.$$

Due to the linearity of this system of equations we can solve them by looking at the case of non-operator functions. Using the combinations $\hat{p}_x \pm i\hat{p}_y$ and $\hat{x} \pm i\hat{y}$, we find

$$\hat{p}_x + i\hat{p}_y = \hat{A}_p e^{-i\omega t} - \frac{iF}{\omega}, \quad \hat{x} + i\hat{y} = (\hat{A}_x + t\hat{A}_p)e^{-i\omega t} - \frac{F}{\omega^2},$$

where $\hat{A}_x, \hat{A}_p$ are time-independent non-Hermitian operators. Their explicit form is determined by the equivalence of Heisenberg and Schrödinger operators at $t = 0$, which gives

$$\hat{A}_z = x + iy + \frac{F}{\omega^2}, \quad \hat{A}_p = -i\frac{\partial}{\partial x} + \frac{\partial}{\partial y} + \frac{iF}{\omega}.$$

The Green function can now be found up to a factor $c(r', t)$ by the system of equations $\hat{\mathbf{r}}(-t) = \mathbf{r}'G$ (compare with the previous problems). To solve this set of equations it is convenient to introduce compex variables $u = x + iy$ and $v = x - iy$. We determine $c(r', t)$ in the same way as in the previous problems,[100] and arrive at the final expression for the Green function:

$$G(\mathbf{r}, t; \mathbf{r}', 0) = (2\pi i t)^{-3/2} exp\left\{\frac{i}{t}\left[\frac{1}{2}(\mathbf{r} - \mathbf{r}')^2 + \boldsymbol{\rho} \cdot \boldsymbol{\rho}'(1 - \cos\omega t) + \right.\right.$$

$$(x'y - xy')\sin\omega t + \frac{1}{\omega^2}F(1 - \cos\omega t)(x + x') - \frac{1}{\omega^2}F(\omega t - \sin\omega t)(y - y') +$$

$$\left.\left. + \frac{1}{\omega^4}F^2(1 - \cos\omega t) - \frac{1}{2\omega^2}F^2 t^2\right]\right\},$$

where $\boldsymbol{\rho}$ is the radius in the x, y-plane. As $\omega \to 0$ this expression turns into the Green's function from Problem 6.32.

Problem 6.35

Find the time-dependent Green function for a charged particle moving in a uniform magnetic field.

---

[100] Considering the Hamiltonian's time-independence, when substituting the Green function into the Schrödinger equation, it is convenient to express the Hamiltonian in terms of particle coordinate and momentum Schrödinger operators.

Solution

Let us first consider the transverse motion of a particle in a magnetic field, using the vector potential $\mathbf{A} = (0, \mathcal{H}x, 0)$. Taking into account (from Problem 7.1a) the explicit form of the eigenfunctions $\psi_{np_y}(\rho)$, and the spectrum $E_{t,n}$, we can use Eq. (VI.6) as well as the result of Problem 6.33 to obtain

$$G(\rho, t; \rho_0, 0) = \sum_n \int e^{-i\omega_{\mathcal{H}}(n+1/2)t} \psi_{np_y}(\rho) \psi^*_{np_y}(\rho_0) dp_y =$$

$$\frac{1}{2\pi\hbar} \int G_{\text{osc}}(\tilde{x}, t; \tilde{x}_0, 0) e^{ip_y(y-y_0)/\hbar} dp_y. \tag{1}$$

In (1), G_{osc} is the Green function of a one-dimensional harmonic oscillator found in Problem 6.33 with the frequency $\omega_{\mathcal{H}} = |e|\mathcal{H}/mc$, $\tilde{x} = x - cp_y/e\mathcal{H}$, and similarly for $\tilde{x}_0$. Calculating the integral in (1) and multiplying the expression[101] obtained by the Green function of free particle $G_0(z, t; z_0, 0)$ – see Problem 6.31 – we obtain the desired Green function ($a^2 = \hbar/m\omega_{\mathcal{H}}$):

$$G(\mathbf{r}, t; \mathbf{r}_0, 0) = \left(\frac{m}{2i\pi\hbar t}\right)^{3/2} \cdot \frac{\omega t}{2\sin(\omega t/2)} e^{iS/\hbar}, \tag{2}$$

$$\hbar^{-1} S = \frac{m(z-z_0)^2}{2\hbar t} + \frac{1}{4a^2} \left\{\cot\frac{\omega_{\mathcal{H}} t}{2} \cdot (\rho - \rho_0)^2 + 2\frac{e}{|e|}(x + x_0)(y - y_0)\right\}, \tag{3}$$

where S is the action of a classical charged particle moving in a magnetic field. Compare with Problems 6.31 and 6.32.

Note that when we perform a gauge transformation on the vector potential, i.e., the $\mathbf{A} \to \mathbf{A}' = \mathbf{A} + \nabla f(\mathbf{r})$, the Green function in the new gauge is obtained from (2) by multiplying it with

$$e^{ie(f(\mathbf{r})-f(\mathbf{r}_0))/\hbar c}.$$

Compare this result with Problem 6.27. In particular, for the transformation to the symmetric gauge $\mathbf{A}' = (1/2)\mathcal{H} \times \mathbf{r}$ we should choose $f(\mathbf{r}) = -\mathcal{H}xy/2$.

6.5 Quasistationary and quasi-energy states; Berry phase

The general ideas of quasi-energy states are presented in Problems 6.40 and 8.41–8.43.

---

[101] The multiplicative form of the Green function is connected with *separation of variables* of the transverse and longitudinal motion.

Problem 6.36

Calculate the level shift and the decay width of a particle in the ground state of a one-dimensional δ-well (see Problem 2.7) due to the application of a constant force described by the potential $V = -F_0 x$. The field is assumed to be weak, so that $aF_0 \ll \hbar^2/ma^2$, where $a = 1/\kappa_0 = \hbar^2/m\alpha$ determines the typical localization length of the particle in the well.

Solution

The Hamiltonian of the particle has the form

$$\hat{H} = -\frac{\hbar^2}{2m}\frac{d^2}{dx^2} - \alpha\delta(x) - Fx. \tag{1}$$

The level E_0 of the discrete spectrum is isolated in the δ-well; see Problem 2.7. After applying the constant force we obtain a decay width and the state becomes quasistationary. This width, Γ, determines the *lifetime* $\tau = \frac{\hbar}{\Gamma}$ of the state, and is connected with the process where the particle tunnels through the barrier; see Fig. 6.1. To determine the properties of the quasistationary state, it is necessary to find the solution of the Schrödinger equation that has the correct asymptotic behavior for $x \to \pm\infty$: wave moving rightwards for $x \to +\infty$ (the radiation condition), and a decaying wave in the classically inaccessible region for $x \to -\infty$.

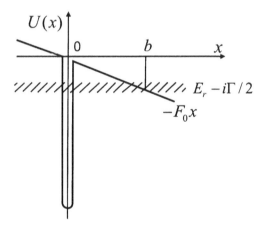

Fig. 6.1

To solve the Schrödinger equation, we see that for $x \neq 0$, we can make the following change of variables:

$$z = \xi\left(x + \frac{E}{F}\right), \quad \xi = \left(\frac{2mF}{\hbar^2}\right)^{1/3}$$

which brings the Schrödinger equation the form $\psi''(z) + z\psi(z) = 0$. Taking into account the asymptotic behavior mentioned above, the solution must be chosen in the form of the following combinations of the Airy functions:

$$\psi = C_1[\text{Ai}(-z) - i\text{Bi}(-z)] \propto_{z \to +\infty} z^{-1/4} \exp\left\{i\frac{2}{3}z^{3/2}\right\}, \quad x > 0, \tag{2}$$

$$\psi = C_2 \text{Ai}(-z) \propto_{z \to -\infty} (-z)^{-1/4} \exp\left\{-\frac{2}{3}(-z)^{3/2}\right\}, \quad x < 0.$$

The matching conditions at the point $x = 0$, according to Problem 2.6, lead to the equation[102]

$$\text{Ai}(-z_0)\text{Bi}(-z_0) + i\text{Ai}^2(-z_0) = \frac{\hbar^2 \xi}{2\pi m\alpha}, \quad z_0 = \frac{\xi E}{F}, \tag{3}$$

determining the spectrum of the quasi-discrete levels.

In the case of a weak field,[103] $\xi\hbar^2/m\alpha \ll 1$, the right-hand side of equation (3) is small. Hence it follows that $|\text{Ai}(-z_0)|$ is small and $\text{Re}(-z_0) \gg 1$. By using the known asymptotics of the Airy functions and Eq. (3), we obtain

$$\left(-\frac{\hbar^2}{2mE}\right)^{1/2}\left[\left(1 + \frac{5}{72\nu^2} + \ldots\right) + \frac{i}{2}\left(1 + O\left(\frac{1}{\nu}\right)\right)e^{-2\nu}\right] = \frac{1}{\kappa_0}, \tag{4}$$

where $\nu = (2/3)(-z_0)^{3/2}$ and $\kappa_0 = m\alpha/\hbar^2$.

Solving (4) order-by-order in $1/\nu$, (Re $\nu \gg 1$), we obtain the zeroth-order approximation (replacing the expression within square bracket by 1), $E \approx E_0 = -\hbar^2\kappa_0^2/2m$, i.e., the unperturbed level in a δ-well. Futhermore, substituting

$$E = E_0 + \triangle E - \frac{i}{2}\Gamma$$

and substituting ν within square brackets in (4) by its unperturbed value $\nu_0 = \hbar^2\kappa_0^3/3mF$, we find both a shift of the level $\triangle E$ and its width Γ that appear due the constant force:

$$\triangle E = -\frac{5}{8}\frac{mF^2}{\hbar^2\kappa_0^4}, \quad \Gamma = \frac{\hbar^2\kappa_0^2}{m}\exp\left(-\frac{2\hbar^2\kappa_0^3}{3mF}\right). \tag{5}$$

The level shift is quadratic in the force applied, and it determines the polarization of ground state of the particle in the δ-well to be $\beta_0 = 5me^2/4\hbar^2\kappa_0^4$ (we put $F = eE$). This can also be calculated by using the perturbation theory; see Problem 8.12.

The exponential smallness of the decay width is due to a small transmittance of the barrier, and may be obtained using the classical expression for the transmittance; see Eq. (IX.7) and Problem 9.28. Such exponential dependence of the level width on

---

[102] The value of the Wronskian $W\{\text{Ai}(z), \text{Bi}(z)\} = 1/\pi$ was used.
[103] When this condition is violated (in the strong-field case), strong widening of level occurs and the specific properties of quasi-stationary state disappear.

Problem 6.37

Find quasi-discrete energy levels (their position and decay widths) for s-states of a particle moving in the potential $U(r) = \alpha\delta(r-a)$; see Fig. 6.2. Consider specifically the case of low-lying states captured by a weakly penetrable barrier $m\alpha a/\hbar^2 \gg 1$. Relate the decay width with the penetrability of the δ-barrier; see Problem 2.30.

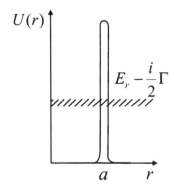

Fig. 6.2

Solution

The wavefunction of the quasi-stationary s-states, $\psi_{k,l=0}(r) = \chi_k(r)/r$, satisfies the equation

$$-\chi_k'' + \tilde{\alpha}\delta(r-a)\chi_k = k^2\chi_k, \quad k = \sqrt{\frac{2mE}{\hbar^2}}, \quad \tilde{\alpha} = \frac{2m\alpha}{\hbar^2}, \tag{1}$$

the boundary condition $\chi_k(0) = 0$, and has the asymptotic behavior, $\chi_k(r) \propto e^{ikr}$ as $r \to \infty$. In such a problem the solution exists only for some complex values $k = k_1 - ik_2$, and in this case

$$E = E_r - \frac{i}{2}\Gamma, \quad E_r = \frac{\hbar^2(k_1^2 - k_2^2)}{2m}, \quad \Gamma = \frac{2\hbar^2 k_1 k_2}{m},$$

where E_r and Γ are the energy and the decay width of the quasi-stationary state (we should note that $k_{1,2} > 0$).

The required solution of Eq. (1) has the form

$$\chi_k(r) = \begin{cases} C_1 \sin kr, & r < a, \\ C_2 e^{ikr}, & r > a. \end{cases} \tag{2}$$

The matching conditions of the wavefunction at the point $r = a$, according to Problem 2.6, give

$$ika - ka \cot ka = \tilde{\alpha}a, \qquad (3)$$

which determines the energy spectrum of quasi-discrete s-levels.

In the case $\tilde{\alpha}a \gg 1$, from (3) it follows that the values of ka for the low-lying levels (such that $|ka| \ll \tilde{\alpha}a$) are close to $(n+1)\pi$. Writing it as

$$k_n a = (n+1)\pi + \varepsilon_1 - i\varepsilon_2, \; n = 0, \, 1, \ldots, \; |\varepsilon_{1,2}| \ll 1$$

and substituting it in (3), we can obtain the approximate values of $\varepsilon_{1,2}$:

$$\varepsilon_1 \approx -\frac{(n+1)\pi}{\tilde{\alpha}a}, \; \varepsilon_2 \approx \varepsilon_1^2,$$

as well as the spectrum of the lowest quasi-discrete s-levels:

$$E_{r,n} \approx \left(1 - \frac{2}{\tilde{\alpha}a}\right) E_n^{(0)}, \; \Gamma_n \approx \frac{4\pi(n+1)}{(\tilde{\alpha}a)^2} E_n^{(0)}, \; E_n^{(0)} = \frac{\pi^2 \hbar^2 (n+1)^2}{2ma^2}. \qquad (4)$$

Note that the decay width of the levels is much smaller than the level spacing between neighboring levels.

As expected, the positions of the quasi-discrete levels are close to the s-levels $E_n^{(0)}$ of a particle in an infinitely deep potential well of radius a, and turn into them as $\tilde{\alpha} \to \infty$, i.e., when the potential barrier becomes impenetrable.

The width of the quasi-stationary state, $\Gamma = \hbar w$, determines the probability w of its *decay* per unit of time (or *life time* of state $\tau = 1/w$). Equation (4) for the level width allows us to illustrate this correspondence if we write it in the form

$$w_n = \frac{\Gamma_n}{\hbar} = \left(\frac{2\pi(n+1)}{\tilde{\alpha}a}\right)^2 \cdot \frac{\pi\hbar(n+1)}{2ma^2} \equiv D \cdot N, \qquad (5)$$

where $D(E_{r,n})$ is the penetration probability (transmission coefficient) through the δ-barrier in the case of a single collision with the barrier – see Problem 2.30 while

$$N = \frac{\pi\hbar(n+1)}{2ma^2} = \frac{v}{2a}$$

determines the number of collisions (of bounces of the particle off of the barrier wall) per unit of time.

In conclusion we note that the results obtained could be applied equally well to the case of $\alpha < 0$ (δ-well), which at first sight looks counterintuitive. In fact this is a peculiarity of quantum mechanics; particles can be reflected by a potential well when it has pronounced discontinuities and kinks. The transmission coefficient in the well's case can be small, $D \ll 1$, even for a large particle energy. On the other hand, if we change it to a smooth well ("spread out" the δ-function), then $D \sim 1$, and the quasi-stationary state disappears (the lifetime will be of the same order as the time of particle motion through the localization domain).

Problem 6.38

Explore the question of the wavefunction normalization in the case when the potential energy of a particle is the complex function, $U(\mathbf{r}) = U_0(\mathbf{r}) - iU_1(\mathbf{r})$, where $U_{0,1}$ are real functions (so-called *optical potential*). The change of wavefunction normalization in time could be interpreted as "absorption" or "creation" of the particle during the interaction process. How is the sign of the imaginary part of the potential related to the character of such processes?

Specifically, consider the one-dimensional δ-well with $U = -(\alpha_0 + i\alpha_1)\delta(x)$, and find the shift and decay width of its ground state, connected with the possibility of particle "absorption" (see also the following problem).

Solution

1) In the usual way, we obtain a generalized continuity equation

$$\frac{\partial \rho(\mathbf{r},t)}{\partial t} + \text{div } \mathbf{j}(\mathbf{r},t) = -\frac{2}{\hbar}U_1(\mathbf{r})|\psi(\mathbf{r},t)|^2, \tag{1}$$

$$\rho = |\psi(\mathbf{r},t)|^2, \quad \mathbf{j} = -\frac{i\hbar}{2m}(\psi^*\nabla\psi - \psi\nabla\psi^*).$$

Integrating (1) over an arbitrary volume gives

$$\frac{d}{dt}\int_V |\psi(\mathbf{r},t)|^2 dV = -\oint_S \mathbf{j} d\mathbf{s} - \frac{2}{\hbar}\int_V U_1(\mathbf{r})(|\psi(\mathbf{r},t)|^2 dV. \tag{2}$$

In the case of $U_1 \equiv 0$, (2) constitutes, the law of the probability conservation: the change of the probability of finding a particle in some volume V per unit time is equal to (with a minus sign) the flow of it through the surface S surrounding this volume. In the case of $U_1 \neq 0$, the second term in the right-hand side of (2) disturbs this balance, and thus it presents an additional mechanism for the probability to change with time, and therefore the normalization of the wavefunction which can be interpreted as a change in the number of particles. In the case of $U_1 > 0$, it is "absorption", and for $U_1 < 0$, "creation" of a particle. We note that the optical potential is usually used for description of a specific channel in a many-channel system. And here the processes of "absorption" and "creation" reflect the interplay between the channels. See the following problem.

2) In the case of a complex δ-well, as in Problem 2.7, we obtain ($\alpha_{0,1} > 0$):

$$\psi_E(x) = Ae^{-\kappa|x|}, \quad \kappa = \left(-\frac{2mE}{\hbar^2}\right)^{1/2} = \frac{m}{\hbar^2}(\alpha_0 + i\alpha_1).$$

Hence it follows that $E = E_0 - i\Gamma/2$, where

$$E_0 = -\frac{m\left(\alpha_0^2 - \alpha_1^2\right)}{2\hbar^2}, \quad \Gamma = \frac{2m\alpha_0\alpha_1}{\hbar^2}$$

determine the position and decay width of the desired quasi-discrete level.

Problem 6.39

Consider the following model of a system with two *channels*. The system consists of two particles in one-dimensional motion. One of them is a structureless particle, while the other is a composite system that has two independent states of "internal" motion with the energy difference Q_0. The wavefunction of such a system in the center-of-mass frame can be presented as a column of the form $\Psi = \begin{pmatrix} \psi_1(x,t) \\ \psi_2(x,t) \end{pmatrix}$, where $x = x_2 - x_1$ is the relative coordinate and $\psi_{1(2)}$ are the amplitudes of finding the composite particle in the 1st or 2nd internal states channels. The interaction between the particles is local and is described by the following operator

$$\hat{U} = -\hat{\alpha}\delta(x) = -\begin{pmatrix} \alpha & \beta \\ \beta & \alpha \end{pmatrix}\delta(x),$$

where α and β are real parameters and $\alpha > 0$.

Find the spectrum of discrete and quasi-discrete levels of such a system. Show that for the energies close to the threshold of the second channel, the dynamics can be described in terms of a one-channel optical potential, and find its form.

Solution

The Hamiltonian of the system considered has the form

$$\hat{H} = -\frac{\hbar^2}{2m}\frac{d^2}{dx^2} - \begin{pmatrix} \alpha & \beta \\ \beta & \alpha \end{pmatrix}\delta(x) + \begin{pmatrix} 0 & 0 \\ 0 & Q_0 \end{pmatrix},$$

where m is the reduced mass of the particles; furthermore, we will consider $Q_0 > 0$, so that the 1st state of the composed particle, which corresponds to the upper component of the wavefunction (see the problem statement), is the *ground state*. The Schrödinger equation becomes a system of two equations and for $x \neq 0$, and we have

$$\psi_1''(x) = -k_1^2\psi_1(x), \quad \psi_2''(x) = -k_2^2\psi_2(x), \tag{1}$$

where $k_1 = \sqrt{2mE/\hbar^2}$ and $k_2 = \sqrt{2m(E - Q_0)/\hbar^2}$. The solution of this system of equations is of the form $\psi_{1,2}(x) = C_{1,2}\exp\{ik_{1,2}|x|\}$, where we have used both the wavefunction continuity at the point $x = 0$ and the required asymptotic behavior of the expanding wave[104] as $x \to \pm\infty$. Now using the matching conditions

---

[104] Here the *closed* channel wavefunction decreases at the large distances.

for the derivatives of the wavefunction at the point $x = 0$, as in Problem 2.6, we have

$$(ik_1 + \tilde{\alpha})C_1 + \tilde{\beta}C_2 = 0, \quad \tilde{\beta}C_1 + (ik_2 + \tilde{\alpha})C_2 = 0, \tag{2}$$

where $\tilde{\alpha} = m\alpha/\hbar^2$ and $\tilde{\beta} = m\beta/\hbar^2$. The condition for the existence of a non-trivial solution leads to the relation

$$(ik_1 + \tilde{\alpha})(ik_2 + \tilde{\alpha}) = \tilde{\beta}^2, \tag{3}$$

which determines the energy spectrum of the discrete and quasi-discrete levels.

First we note that in the case $\beta = 0$, there is no interaction between the channels (i.e., interaction between the particles does not affect the internal motion of the composite particle), and the system considered has two discrete levels: one for each channel. These are the ordinary discrete spectrum levels in a δ-well. Particularly, $E_2^{(0)} = Q_0 + E_0^{(0)} > 0$, where $E_1^{(0)} = E_0^{(0)} = -m\alpha^2/2\hbar^2$, and the discrete level in the second channel lies in the background of the continuous spectrum of the first channel. In this situation, turning on even a weak interaction between the channels, $\beta \ll \alpha$, leads to the appearance of a width of this level, and the corresponding state becomes quasi-stationary. Then from (3) we can find the energy E_2 of this state:

$$E_2 \equiv E_r - \frac{i}{2}\Gamma \approx Q_0 + E_0^{(0)} - 2\frac{\beta^2}{\alpha^2}|E_0^{(0)}|\left\{\frac{E_0^{(0)}}{Q_0} + i\sqrt{\frac{\left(Q_0 - |E_0^{(0)}|\right)|E_0^{(0)}|}{Q_0^2}}\right\}. \tag{4}$$

Note that the level moves upwards and acquires a width (the discrete level $E_1^{(0)}$ of the first channel undergoes only the small shift downwards).

2) To conclude, let us show, in the framework of the given problem, how an *optical potential* can be introduced. If we consider the system appearing in the second channel only as the result of a transition from the first channel, we can write the stationary state wavefunction of the system with (real) energy E in the form $\Psi = \begin{pmatrix} \psi_1(x) \\ C_2\exp(ik_2|x|) \end{pmatrix}$. The matching conditions at the point x_0 give (compare with (2))

$$-\frac{1}{2}\delta\psi_1'(0) = \tilde{\alpha}\psi_1(0) + \tilde{\beta}C_2, \quad -ik_2C_2 = \tilde{\alpha}C_2 + \tilde{\beta}\psi_1(0), \tag{5}$$

where $\delta\psi_1'(0)$ is the jump of the function derivative at the point $x = 0$. Eliminating C_2 from these equations we obtain

$$-\frac{1}{2}\delta\psi_1'(0) = \left(\tilde{\alpha} - \frac{\tilde{\beta}^2}{\tilde{\alpha} + ik_2}\right)\psi_1(0). \tag{6}$$

Since for $x \neq 0$, $\psi_1(x)$ still satisfies Eq. (1), the matching condition (6) (with the condition of $\psi_1(x)$ continuity) implies that this function is the solution of stationary Schrödinger equation with a "potential" of the form

$$U_{\text{opt}}(x, E) = -\alpha_{\text{opt}}(E)\delta(x), \tag{7}$$

where

$$\alpha_{\text{opt}}(E) = \alpha - \frac{m\beta^2}{m\alpha + i\hbar^2 k_2(E)}. \tag{8}$$

So the dynamics in this particular channel of the initially two-channel system can be considered in terms of the wavefunction of only this channel,[105] and the corresponding equation has the form of the stationary Schrödinger equation with the one-channel Hamiltonian. We should note the following properties of the effective (optical) potential in such a Hamiltonian[106].

1) It depends on the energy, and therefore the corresponding "Hamiltonian" is not a self-adjoint operator.
2) As seen from (7) and (8), if the energy of the particle in the channel considered exceeds the threshold of the other channel (*i.e.*, in the case $E > Q_0$), then the optical potential acquires an imaginary part with a negative sign. This corresponds to the fact that, with respect to the initial channel, the transition of the particle into the other, open channel may be considered *absorption*. Compare with Problem 6.38.

Problem 6.40

A charged particle is moving in a uniform electric field that is periodic in time (*i.e.*, $E(t+T) = E(t)$) and whose average over a period is equal to zero. Find the quasi-energy spectrum and the form of the wavefunctions for the *quasi-energy states*.

Specifically consider the cases:

a) $\mathcal{E}(t) = \mathcal{E}_0 \cos \omega t$;
b) $\mathcal{E}_x(t) = \mathcal{E}_0 \cos \omega t$, $\mathcal{E}_y(t) = \mathcal{E}_0 \sin \omega t$, $\mathcal{E}_z = 0$,

(electric fields of the linearly and circularly polarized monochromatic wave).

Solution

1) We consider the *quasi-energy states* (QES) and *quasi-energies* of quantum systems, whose Hamiltonians are periodic (with the period $T = 2\pi/\omega$) functions of time.

---

[105] In spite of interaction transitions between the channels.
[106] Such a simple form of the effective (optical) potential (7), (8) is due to the locality of the interaction. In the general case such an effective potential is a non-local operator that depends on the energy.

QES are defined as states whose wavefunctions are solutions to the time-dependent Schrödinger equation and satisfy the condition

$$\psi_\varepsilon(q, t + T) = e^{-i\varepsilon T/\hbar}\psi_\varepsilon(q, t), \tag{1}$$

where ε is called the quasi-energy (compare with the notion of quasi-momentum and Bloch functions for a particle in a spatially-periodic potential; see Problem 2.53).

The wavefunction of a QES can be written in the form

$$\psi_\varepsilon(q, t) = e^{-i\varepsilon t/\hbar} u_\varepsilon(q, t), \tag{2}$$

where $u_\varepsilon(q, t)$ is a periodic function of time. Its Fourier series expansion is

$$u_\varepsilon(q, t) = \sum_{k=-\infty}^{\infty} e^{-ik\omega t} C_{\varepsilon,k} \varphi_{\varepsilon,k}(q), \tag{3}$$

and defines the *quasi-energy harmonics* $\varphi_{\varepsilon,k}(q)$.

The quasi-energy (as well as quasi-momentum) is uniquely defined modulo an integer multiple of $\pm\hbar\omega$. To get rid of this ambiguity we can restrict our quasi-energies to an interval; for example, $-\hbar\omega/2 < \varepsilon \leq \hbar\omega/2$ is often used, or sometimes the interval such that the quasi-energy and the corresponding value of energy $E_n^{(0)}$ of the stationary Hamiltonian coincide when adiabatically turning off the time-dependent part of the Hamiltonian.

The notion of QES is the natural generalization of a stationary state: the system of QES wavefunctions has properties that, to a large extent, are analogous to the stationary Hamiltonian's eigenfunctions. So, QES wavefunctions with different quasi-energies are orthogonal, and at different moments of time they form a complete set of functions. And the analog to Eq. (VI.2) is an expansion over the QES wavefunctions with constant coefficients, and this determines the general solution of the time-dependent Schrödinger equation.

An important difference between the QES and stationary states manifests itself in the emission processes and also in the resonance effect of perturbations. If we have a system with stationary Hamiltonian $\hat{H}_0$, the frequencies of photon emission (corresponding to spontaneous transitions between the stationary states) and the frequencies of the harmonic perturbation that causes the resonance transitions in the system are determined entirely by the transition frequencies, $\hbar\omega_{f_i} = E_f^{(0)} - E_i^{(0)}$, but the QES situation is different. The corresponding frequencies for QESs are defined by the relation

$$\hbar\omega_{kns} = \varepsilon_k - \varepsilon_n \pm s\hbar\omega, \ s = 0, 1, 2 \ldots,$$

where $\varepsilon_{k,n}$ are quasi-energy levels and the values $n = k$ are allowed. The intensity of the transition for different values of s depends on the amplitudes of quasi-energy harmonics in (3).

2) Now let us return to the actual problem. We will describe a uniform electric field in terms of a vector potential: $\boldsymbol{\mathcal{E}}(t) = -\dot{\mathbf{A}}(t)/c$, which for the case of $\overline{\boldsymbol{\mathcal{E}}}(t) = 0$ is a periodic function in time.[107] The Hamiltonian of the particle takes the form

$$\hat{H}(t) = \frac{1}{2m}\left(\hat{\mathbf{p}} - \frac{e}{c}\mathbf{A}(t)\right)^2. \tag{4}$$

Since the momentum operator commutes with the Hamiltonian, the generalized momentum is a constant of motion. Its eigenfunction is $\propto \exp(i\mathbf{p}\cdot\mathbf{r}/\hbar)$, and is also the eigenfunction of the *instantaneous* Hamiltonian. This function corresponds to the eigenvalues $E(t)$, which are obtained from $\hat{H}(t)$ by simply $\hat{\mathbf{p}} \to \mathbf{p}$. This allows us to write the solution of the Schrödinger equation:

$$\psi_{\mathbf{p},\varepsilon(p)}(\mathbf{r},t) = \frac{1}{(2\pi\hbar)^{3/2}}\exp\left\{\frac{i}{\hbar}\left(\mathbf{p}\cdot\mathbf{r} - \int_0^t E(t')dt'\right)\right\},$$

$$E(t) = \frac{1}{2m}\left(\mathbf{p} - \frac{e}{c}\mathbf{A}(t)\right)^2. \tag{5}$$

For a periodic dependence of the vector potential $\mathbf{A}(t)$ on time, we can describe the QES with the quasi-energy (further, we take $\overline{\mathbf{A}(t)} = 0$):

$$\varepsilon(p) = \overline{E}(t) \equiv \frac{1}{T}\int_0^T E(t)dt = \frac{\mathbf{p}^2 + \mathbf{p}_0^2}{2m}, \quad \mathbf{p}_0^2 = \frac{e^2}{c^2}\overline{\mathbf{A}^2}(t), \tag{6}$$

which is equal to the average value of the particle energy over a period.

To elucidate the physical meaning of $\mathbf{p}$ and $\mathbf{p}_0^2$, we should note that since

$$\mathbf{p} - \frac{e}{c}\mathbf{A}(t) = m\mathbf{v}(t) \text{ and } \overline{\mathbf{A}}(t) = 0,$$

we have that $\mathbf{p} = m\overline{\mathbf{v}}(t)$, i.e., the conserved momentum of the particle determines the average velocity of the particle. Furthermore, from the above equations it follows that

$$\frac{e}{c}\mathbf{A}(t) = m(\overline{\mathbf{v}} - \mathbf{v}(t)), \quad \frac{e^2}{c^2}\overline{\mathbf{A}^2}(t) = m^2\overline{(\overline{\mathbf{v}} - \mathbf{v}(t))^2}$$

Therefore, the value of $p_0^2/2m$ defines the average value of the kinetic energy of the oscillatory motion of the particle in the electric field (on the background of its uniform motion with the velocity $\overline{\mathbf{v}}$).

In the case of a linearly polarized monochromatic wave we have

$$\varepsilon(p) - \frac{p^2}{2m} = \frac{e^2\mathcal{E}_0^2}{4m\omega^2}, \tag{7}$$

---

[107] For different gauge choices of the potentials, see Problem 6.27.

(for the motion of a particle in the field of a circularly polarized wave we should multiply the right-hand side by 2). The divergence of ε_p as $\omega \to 0$ corresponds to the unlimited increase in velocity of a particle moving in a constant uniform electric field.

Problem 6.41

Analyze the quasi-energy states (QES) that appear from a doubly-degenerate level of the Hamiltonian $\hat{H}_0$ due to a periodic perturbation $\hat{V}(t)$ whose matrix elements between the two degenerate states considered are given by[108] $V_{11} = V_{22} = 0, V_{12} = V_{21} = V_0 \sin \omega t$, and where $V_0 = V_0^*$. Expand the QES wavefunctions over the quasi-energy harmonics; see Problem 6.40. Neglect the other states, and compare the resulting model with the two-level system in Problem 6.9.

Solution

We write the wavefunction of the system in the form $\Psi(t) = \begin{pmatrix} \psi_1(t) \\ \psi_2(t) \end{pmatrix} e^{-i\varepsilon_0 t/\hbar}$, where the functions $\psi_{1(2)}(t)$ are the amplitudes of the 1st (2nd) independent stationary state of the unperturbed Hamiltonian $\hat{H}_0$ with the energy ε_0. Then the Schrödinger equation, $i\hbar \partial \Psi / \partial t = \left(\hat{H}_0 + \hat{V} \right) \Psi$, takes the form

$$i\hbar \dot{\psi}_1 = V_0 \sin(\omega t)\psi_2, \ i\hbar \dot{\psi}_2 = V_0 \sin(\omega t)\psi_1,$$

from which we obtain

$$\psi_1 \pm \psi_2 = A_\pm e^{\pm i f(t)}, \ f(t) = \frac{V_0}{\hbar \omega} \cos \omega t,$$

and so

$$\Psi(t) = \frac{C_1}{\sqrt{2}} e^{i(f(t) - \varepsilon_0 t/\hbar)} \begin{pmatrix} 1 \\ 1 \end{pmatrix} + \frac{C_2}{\sqrt{2}} e^{-i(f(t) + \varepsilon_0 t/\hbar)} \begin{pmatrix} 1 \\ -1 \end{pmatrix}. \qquad (1)$$

Each of the terms in the wavefunction (1) describes an independent QES, and here the quasi-energies of both states are the same and equal to ε_0. We now use the expansion

$$e^{iz \cos \omega t} = \sum_{-\infty}^{\infty} i^k J_k(z) e^{ik\omega t}, \ z = \pm \frac{V_0}{\hbar \omega}, \qquad (2)$$

---

[108] This problem models, for example, the influence of electric field $\mathcal{E}(t) = \mathcal{E}_0 \sin \omega t$ on a charged particle in a potential with degenerate $s-$ and $p-$levels ($2s-$ and $2p-$ states with $l_z = 0$ in the hydrogen atom). When the field density is much lower the atomic density, the distortion of wavefunctions is small, and their mixing by the external field is an essential effect.

where J_k is a Bessel function, and this allows us to define, in accordance with (3) from Problem 6.40, the amplitudes of the quasi-energy harmonics of the QES. Their intensities are $\propto J_k^2(z)$, and oscillate as V_0 is raised.

Problem 6.42

Prove that the analysis of the quasi-energy states and calculation of the quasi-energy spectrum for a system in an electric field of a circularly polarized monochromatic wave of the form

$$\mathcal{E}_x(t) = \mathcal{E}_0 \cos \omega t, \quad \mathcal{E}_y(t) = \mathcal{E}_0 \sin \omega t, \quad \mathcal{E}_z = 0$$

can be reduced to solving a time-independent Schrödinger equation.

Solution

To prove the problem statement, consider a coordinate system that is rotating with the same angular velocity as the electric field, ω. The Hamiltonian in this frame of reference, $\hat{H}_{rot}$, does not depend on time and energy is conserved; see Problem 6.29. The stationary state and its energy in the rotating system are exactly the QES and quasi-energy with respect to the initial coordinate system. Furthermore, the corresponding wavefunctions are related by Eq. (1) from Problem 6.29.

Problem 6.43

A localized spin-1/2 (quantum two-level system) is subjected to a periodic-in-time time-dependent magnetic field

$$\mathbf{b}(t) = (b_1 \cos[\omega t], b_1 \sin[\omega t], b_0),$$

where b_0 is a constant component of the field, b_1 is the amplitude of the rotating component, and ω is its frequency. The corresponding Schrödinger equation is $i\partial_t \Psi(t) = \mu \mathbf{b}(t) \cdot \hat{\boldsymbol{\sigma}} \Psi(t)$, where $\Psi(t)$ is a two-component spinor wavefunction, $\hat{\boldsymbol{\sigma}} = (\hat{\sigma}_x, \hat{\sigma}_y, \hat{\sigma}_z)$ is a vector of Pauli matrices, and μ is a magnetic moment. Use the units where $\hbar = 1$.

Initially, the spin is pointing "up" i.e., the initial condition for the Schrödinger equation is $\Psi(0) = \binom{1}{0}$. The evolution operator, $\hat{U}(t)$, can be defined by the relation:

$$\Psi(t) = \hat{U}(t)\Psi(0)$$

(a) "Construct" the exact evolution operator for the given problem.
(b) Find the spin-flip probability, $w_{\mathrm{sf}}(t)$, directly from the evolution operator.
(c) Consider the initial condition corresponding to the spin pointing in the x-direction at $t = 0$, and find the corresponding exact time-dependent wavefunction, $\Psi(t)$.

(d) Calculate the determinant of the evolution operator, $\det \hat{U}(t)$, and provide a mathematical and physical interpretation of your result.

(e) Evaluate explicitly the "velocity" $i\frac{\partial \hat{U}(t)}{\partial t}\big|_{t=0}$, and interpret your result in the context of the original Schrödinger equation.

Solution

(a) First we find the exact solution of the problem.[109] The state of the spin is described by the spinor,

$$\Psi(t) = \begin{pmatrix} \psi_\uparrow(t) \\ \psi_\downarrow(t) \end{pmatrix}.$$

We simplify the time-dependent Schrödinger equation by the following unitary transformation, $\Psi(t) = e^{-i\omega t \sigma_z/2}\tilde{\Psi}(t)$:

$$i\frac{\partial \tilde{\Psi}}{\partial t} + \frac{\omega \sigma_z}{2}\tilde{\Psi} = e^{i\omega t \sigma_z/2}\hat{H}_0 e^{-i\omega t \sigma_z/2}\tilde{\Psi} = \mu\begin{pmatrix} b_0 & b_1 \\ b_1 & -b_0 \end{pmatrix}\tilde{\Psi},$$

which reduces to

$$i\frac{\partial \tilde{\Psi}}{\partial t} = \begin{pmatrix} \Omega & \mu b_1 \\ \mu b_1 & -\Omega \end{pmatrix}\tilde{\Psi} = \hat{H}\tilde{\Psi},$$

where $\Omega = -\omega/2 + \mu b_0$. If we define

$$\mathcal{E} = \sqrt{\Omega^2 + \mu^2 b_1^2}, \quad \cos\theta = \frac{\Omega}{\mathcal{E}},$$

then $H = \mathcal{E}(\cos\theta \sigma_z + \sin\theta \sigma_x)$. The time-evolution operator for $\tilde{\Psi}(t)$ is given by

$$\hat{U}(t) = e^{-i\hat{H}t} = \exp\left[-i\mathcal{E}t(\cos\theta\sigma_z + \sin\theta\sigma_x)\right] = \cos\mathcal{E}t - i\sin\mathcal{E}t(\cos\theta\sigma_z + \sin\theta\sigma_x).$$

Returning to the original wavefunction, $\Psi(t)$, we obtain

$$\Psi(t) = e^{-i\omega t \sigma_z/2}\hat{U}(t)\tilde{\Psi}(0) = \hat{U}(t)\Psi(0).$$

Hence,

$$\hat{U}(t) = \begin{pmatrix} \cos\mathcal{E}t - i\frac{\Omega}{\mathcal{E}}\sin\mathcal{E}t & -i\frac{\mu b_1}{\mathcal{E}}\sin\mathcal{E}t \\ -ie^{i\omega t}\frac{\mu b_1}{\mathcal{E}}\sin\mathcal{E}t & e^{i\omega t}\left(\cos\mathcal{E}t + i\frac{\Omega}{\mathcal{E}}\sin\mathcal{E}t\right) \end{pmatrix}.$$

---

[109] This problem provides a *rare* example of exactly solvable driven two-level-system dynamics. Such a generic problem reduces to a differential equation of the Riccati type, and its solution generally cannot be expressed in terms of standard elementary or special functions.

(b) Given the initial condition, $\Psi(0) = \begin{pmatrix} 1 \\ 0 \end{pmatrix}$, the wavefunction at a later time, t, is

$$\Psi(t) = \hat{U}(t)\Psi(0) = \begin{pmatrix} e^{-i\omega t/2}\left(\cos \mathcal{E}t - i\frac{\Omega}{\mathcal{E}}\sin \mathcal{E}t.\right) \\ -ie^{i\omega t/2}\frac{\mu b_1}{\mathcal{E}}\sin \mathcal{E}t \end{pmatrix}.$$

Therefore, the spin-flip probability is

$$w_{\text{sf}}(t) = |\langle \downarrow | \Psi(t) \rangle|^2 = \frac{\mu^2 b_1^2}{\Omega^2 + \mu^2 b_1^2}\sin^2\left(\sqrt{\Omega^2 + \mu^2 b_1^2}\,t\right).$$

(c) We assume that at $t = 0$ the spin points in the positive x-direction, which implies that $\sigma_x \Psi(0) = \Psi(0)$. So the initial condition is as follows:

$$\Psi(0) = \frac{1}{\sqrt{2}}\begin{pmatrix} 1 \\ 1 \end{pmatrix}.$$

Consequently, the exact wavefunction at time, t, is

$$\Psi(t) = \hat{U}(t)\Psi(0).$$

(d) A simple calculation shows that $\det \hat{U} = 1$. Notice that this is not implied just by the unitarity of timeevolution. The unitarity condition, $\hat{U}^\dagger \hat{U} = \hat{U}\hat{U}^\dagger = \hat{1}$, leads to $|\det \hat{U}|^2 = 1$, so $|\det \hat{U}| = 1$. Therefore, the unitarity implies that $\det \hat{U}$ may give rise to a complex phase: $\det \hat{U} = e^{i\alpha}$. This is consistent with the fact that the time-evolution of a quantum system must be unitary in order to preserve the norm of the wavefunction, up to a global phase. In mathematical language, $\hat{U}$ is an element of unitary group $\mathbb{U}(N)$, where N is the dimension of Hilbert space.

(e) A straightforward calculation shows that

$$i\frac{\partial \hat{U}}{\partial t}\bigg|_{t=0} = \hat{H}(0).$$

Actually, this is a general result. We substitute $\Psi(t) = \hat{U}(t)\Psi(0)$ into the Schrödinger equation, $i\partial_t \Psi(t) = \hat{H}(t)\Psi(t)$, and noting that $\Psi(0)$ is arbitrary, one must have

$$i\partial_t \hat{U}(t) = \hat{H}(t)\hat{U}(t).$$

Now, take the limit $t \to 0$ and noting that $\hat{U}(0) = \hat{1}$, we obtain

$$i\partial_t \hat{U}(t)|_{t=0} = \hat{H}(0).$$

Problem 6.44

Consider a spin-1/2 particle in an arbitrary time-dependent magnetic field, $\mathbf{b}(t)$. The evolution operator, $\hat{U}(t)$, satisfies the equation, $i\left[\partial_t \hat{U}(t)\right]\hat{U}^\dagger(t) = \mu \mathbf{b}(t)\cdot \hat{\boldsymbol{\sigma}}$. This operator is a unitary 2×2 matrix, and any such matrix can be represented in the form $\hat{U}(t) = \exp[-i\boldsymbol{\theta}(t) \cdot \hat{\boldsymbol{\sigma}}]$, where $\boldsymbol{\theta}(t) = (\theta_x(t), \theta_y(t), \theta_z(t)) \equiv \theta(t)\mathbf{n}(t)$ is a time-dependent 3D-vector, $\theta(t) \equiv |\boldsymbol{\theta}(t)|$ is its modulus, and $|\mathbf{n}(t)| \equiv 1$ is its direction. Use the algebraic identities for the Pauli matrices to express explicitly the magnetic field, $\mathbf{b}(t)$, through $\boldsymbol{\theta}(t)$ and its derivatives.

Solution

The time-evolution operator satisfies the standard Schrödinger equation:

$$i\partial_t \hat{U}(t) = \hat{H}(t)\hat{U}(t).$$

Multiplying, both sides by $\hat{U}^\dagger(t)$ from right,

$$\hat{H}(t) = \mu \mathbf{b}(t) \cdot \hat{\boldsymbol{\sigma}} = i\left[\partial_t \hat{U}(t)\right] \hat{U}^\dagger(t).$$

The fact that $\hat{U}(t) \in \mathbb{SU}(2)$ allows us to write $\hat{U}(t) = \exp[-i\boldsymbol{\theta}(t) \cdot \hat{\boldsymbol{\sigma}}] = \exp[-i\theta(t)\mathbf{n}(t) \cdot \hat{\boldsymbol{\sigma}}] = \cos\theta(t) - i\hat{\boldsymbol{\sigma}} \cdot \mathbf{n}(t) \sin\theta(t)$. So

$$i\left(\partial_t \hat{U}(t)\right) \hat{U}^\dagger(t) = \left[\hat{\boldsymbol{\sigma}} \cdot \mathbf{n}\dot{\theta}\cos\theta - i(\dot{\theta} + i\hat{\boldsymbol{\sigma}} \cdot \dot{\mathbf{n}})\sin\theta\right](\cos\theta + i\hat{\boldsymbol{\sigma}} \cdot \mathbf{n}\sin\theta)$$

$$= \dot{\theta}\hat{\boldsymbol{\sigma}} \cdot \mathbf{n} + \hat{\boldsymbol{\sigma}} \cdot \dot{\mathbf{n}}\sin\theta\cos\theta + i(\hat{\boldsymbol{\sigma}} \cdot \dot{\mathbf{n}})(\boldsymbol{\sigma} \cdot \mathbf{n})\sin^2\theta.$$

Now we invoke the following algebraic identity for Pauli matrices:

$$(\hat{\boldsymbol{\sigma}} \cdot \mathbf{a})(\hat{\boldsymbol{\sigma}} \cdot \mathbf{b}) = \mathbf{a} \cdot \mathbf{b} + i\hat{\boldsymbol{\sigma}} \cdot (\mathbf{a} \times \mathbf{b}).$$

Thus we can simplify the last term in the expression as

$$(\hat{\boldsymbol{\sigma}} \cdot \dot{\mathbf{n}})(\hat{\boldsymbol{\sigma}} \cdot \mathbf{n}) = \dot{\mathbf{n}} \cdot \mathbf{n} + i\hat{\boldsymbol{\sigma}} \cdot (\dot{\mathbf{n}} \times \mathbf{n}),$$

where $\dot{\mathbf{n}} \cdot \mathbf{n} = 0$, following the fact that $\mathbf{n}$ is a unit vector: $|\mathbf{n}|^2 = 1$. Putting everything together we find

$$\mu \mathbf{b} \cdot \hat{\boldsymbol{\sigma}} = \dot{\theta}\hat{\boldsymbol{\sigma}} \cdot \mathbf{n} + \hat{\boldsymbol{\sigma}} \cdot \dot{\mathbf{n}}\sin\theta\cos\theta - \sin^2\theta\hat{\boldsymbol{\sigma}} \cdot (\dot{\mathbf{n}} \times \mathbf{n}),$$

which yields the expression of $\mathbf{b}$ in terms of θ and $\mathbf{n}$:

$$\mu \mathbf{b}(t) = \dot{\theta}\mathbf{n} + \sin\theta\cos\theta\dot{\mathbf{n}} - \sin^2\theta\dot{\mathbf{n}} \times \mathbf{n}.$$

Note that the result obtained in this problem can be used to construct exact solutions of quantum dynamics of a spin-1/2 in a time-dependent magnetic field. A common mistake is to write the following: $\partial_t \hat{U}(t) = \partial_t \left(e^{i\boldsymbol{\theta}(t) \cdot \hat{\boldsymbol{\sigma}}}\right) = i\left(\partial_t \boldsymbol{\theta} \cdot \hat{\boldsymbol{\sigma}}\right) e^{i\boldsymbol{\theta}(t) \cdot \hat{\boldsymbol{\sigma}}}$. This is, however, *not* correct, because $\partial_t \boldsymbol{\theta} \cdot \hat{\boldsymbol{\sigma}}$ and $\boldsymbol{\theta} \cdot \hat{\boldsymbol{\sigma}}$ do not commute in general. Let us show this in a general setting. Consider the operator, $e^{\hat{A}(t)}$. It is defined by its Taylor series:

$$e^{\hat{A}} = \sum_{n=0}^{\infty} \frac{\hat{A}^n}{n!}.$$

Taking the derivative with respect to t, we find

$$\frac{\partial}{\partial t} e^{\hat{A}(t)} = \sum_{n=0}^{\infty} \frac{1}{n!} \frac{\partial}{\partial t}\left[\hat{A}^n(t)\right] = \sum_{n=0}^{\infty} \frac{1}{n!} \sum_{m=0}^{n} \hat{A}^m \frac{d\hat{A}}{dt} \hat{A}^{n-m}.$$

Only if $\partial \hat{A}/\partial t$ commutes with $\hat{A}$ (which is generally not the case), this sum can be simplified to $\frac{d\hat{A}}{dt} e^{\hat{A}}$.

Problem 6.45

Consider the same setup as in Problem 6.43: a localized spin-1/2 driven by a periodic time-dependent magnetic field, $\mathbf{b}(t) = (b_1 \cos[\omega t], b_1 \sin[\omega t], b_0)$ with the period, $T = 2\pi/\omega$. But now assume that the perturbation is sufficiently slow, and calculate the dynamic phase and the Berry phase within the adiabatic approximation.[110] Compare your results with the exact results obtained earlier.

Solution

Let us first calculate the Berry phase using the adiabaticity of time-evolution. The instantaneous eigenstates of the time-dependent Hamiltonian $|\mathbf{b}(t)\rangle$ are found from the eigenvalue problem $\hat{\boldsymbol{\sigma}} \cdot \mathbf{b}(t) |\psi(t)\rangle = E(t) |\psi(t)\rangle$ (the Hamiltonian is $H = \mu \hat{\boldsymbol{\sigma}} \cdot \mathbf{b}(t)$ and we set $\mu = 1$ for simplicity). The instantaneous ground state is

$$|\psi_-(t)\rangle = \begin{pmatrix} \cos\left(\frac{\theta}{2}\right) \\ -\sin\left(\frac{\theta}{2}\right) e^{i\omega t} \end{pmatrix}, \text{ with } E_- = -\sqrt{b_0^2 + b_1^2},$$

and where $\tan(\theta/2) = \frac{b_1}{\sqrt{b_0^2 + b_1^2} - b_0}$.

The Berry phase of the ground state is calculated as follows:

$$\gamma = i \int_0^T \partial t \, \langle \psi_-(t) | \partial_t | \psi_-(t) \rangle$$

$$= i \int_0^T \partial t \left(\cos(\theta/2), -\sin(\theta/2) e^{-i\omega t} \right) \begin{pmatrix} 0 \\ -i\omega \sin(\theta/2) e^{i\omega t} \end{pmatrix}$$

$$= -\omega T \sin^2(\theta/2) = -2\pi \sin^2(\theta/2)$$

$$= -\pi \left(1 + \frac{b_0}{\sqrt{b_0^2 + b_1^2}} \right).$$

The dynamical phase is simply

$$-\int_0^T \partial t \, E_-(t) = \sqrt{b_0^2 + b_1^2} \, T.$$

---

[110] If a quantum system is subjected to a slow periodic-in-time perturbation, then upon the completion of a single evolution cycle, the wavefunction does not necessarily return to its original initial value (even though there are no transitions for adiabatic evolution), but acquires a phase that consists of a dynamic phase and a topological Berry phase.

Therefore, the total phase, ϕ, acquired by the wavefunction in the course of a single evolution cycle, is found to be

$$\phi = \sqrt{b_0^2 + b_1^2}\, T - \pi \left(1 + \frac{b_0}{\sqrt{b_0^2 + b_1^2}}\right).$$

Now we use the exact solution obtained in Problem 6.43. The exact time-evolution operator is

$$U(t) = \begin{pmatrix} e^{-i\omega t/2}\cos \mathcal{E} t - i\frac{\Omega}{\mathcal{E}}\sin \mathcal{E} t & -ie^{-i\omega t/2}\frac{b_1}{\mathcal{E}}\sin \mathcal{E} t \\ -ie^{i\omega t/2}\frac{b_1}{\mathcal{E}}\sin \mathcal{E} t & e^{i\omega t/2}\left(\cos \mathcal{E} t + i\frac{\Omega}{\mathcal{E}}\sin \mathcal{E} t\right) \end{pmatrix},$$

where $\mathcal{E} = \sqrt{\Omega^2 + b_1^2}$ and $\Omega = b_0 - \omega/2$.

The initial state is as follows:

$$|\psi(0)\rangle = |\psi_-(0)\rangle = \begin{pmatrix} \cos\left(\frac{\theta}{2}\right) \\ -\sin\left(\frac{\theta}{2}\right) \end{pmatrix}.$$

Therefore, the final state reads:

$$|\psi(T)\rangle = U(T)|\psi(0)\rangle = \begin{pmatrix} e^{-i\omega t/2}\cos(\theta/2)\left[\cos \mathcal{E} T + i\frac{\sqrt{b_0^2+b_1^2}+\omega/2}{\mathcal{E}}\sin \mathcal{E} T\right] \\ -e^{i\omega t/2}\sin(\theta/2)\left[\cos \mathcal{E} T + i\frac{\sqrt{b_0^2+b_1^2}-\omega/2}{\mathcal{E}}\sin \mathcal{E} T\right] \end{pmatrix}.$$

Now we want to compare this exact result with the "adiabatic" Berry phase calculation. Notice that

$$\cos \mathcal{E} T + i\frac{\sqrt{b_0^2+b_1^2} \pm \omega/2}{\mathcal{E}}\sin \mathcal{E} T \approx e^{i\mathcal{E} T}.$$

This is justified by expanding in the adiabatic limit:

$$\frac{\sqrt{b_0^2+b_1^2} \pm \omega/2}{\mathcal{E}} = 1 \mp \frac{1}{\sqrt{b_0^2+b_1^2}}\left(\frac{b_0}{\sqrt{b_0^2+b_1^2}} - 1\right)\omega + O(\omega^2).$$

Hence, the difference between the phase of the exact wavefunction, $|\psi(T)\rangle$, and the phase of the initial condition, $|\psi(0)\rangle$, is

$$\pi + \mathcal{E} T = \pi + \sqrt{\left(b_0 - \frac{\omega}{2}\right)^2 + b_1^2}\, T \approx \sqrt{b_0^2 + b_1^2}\, T + \pi - \frac{\pi b_0}{\sqrt{b_0^2 + b_1^2}} + O(\omega),$$

which indeed agrees with our previous Berry phase result ϕ. (In fact, our calculation gives phases that differ by 2π, but since a quantal phase is defined modulo 2π, the two phases are in fact identical.) Notice that the π-phase originates from the factor, $e^{\pm i\omega T/2}$, in the time-evolution operator. This calculation confirms that the Berry phase indeed arises as the leading-order term in an adiabatic perturbative expansion.

7

Motion in a magnetic field

The Hamiltonian of a charged particle with spin s and (intrinsic) magnetic moment μ_0 moving in an external magnetic field is given by[111]

$$\hat{H} = \frac{1}{2m}\left(\hat{\mathbf{p}} - \frac{e}{c}\mathbf{A}\right)^2 + e\varphi - \frac{\mu_0}{s}\boldsymbol{\mathcal{H}}\cdot\hat{\mathbf{s}}, \qquad \text{(VII.1)}$$

where $\boldsymbol{\mathcal{H}} = \operatorname{curl}\mathbf{A}$ and $\hat{\mathbf{p}} = -i\hbar\boldsymbol{\nabla}$. It is called the *Pauli Hamiltonian*.

The velocity operator is (see Problem 6.10) $\hat{\mathbf{v}} = \frac{1}{m}\left(\hat{\mathbf{p}} - \frac{e}{c}\mathbf{A}(\mathbf{r},t)\right)$, and its components obey the commutation relations,

$$[\hat{v}_i,\ \hat{v}_k] = i\frac{e\hbar}{m^2 c}\varepsilon_{ikl}\mathcal{H}_l, \qquad \text{(VII.2)}$$

or when written in the vector form, $\hat{\mathbf{v}}\times\hat{\mathbf{v}} = i\frac{e\hbar}{m^2 c}\boldsymbol{\mathcal{H}}$.

In a uniform magnetic field, $\mathcal{H}_0$, along the z direction, the "transverse" motion (in the plane perpendicular to the magnetic field) of spinless particles yields the discrete energy spectrum,

$$E_{tr,n} = \hbar\omega_H\left(n + \frac{1}{2}\right),\ n = 0,\ 1,\ 2,\ldots\text{(the *Landau levels*)}, \qquad \text{(VII.3)}$$

where $\omega_H = \frac{|e|}{mc}\mathcal{H}_0$. The explicit form of the corresponding eigenfunctions depends on a particular gauge chosen for the vector potential; see Problem 7.1. If the particle has a non-zero spin s, then an additional term, $-\frac{1}{s}\mu_0\mathcal{H}s_z$, must be included in Eq. (VII.3), where s_z is the z-component of the particle spin.

The electric current density in the presence of a magnetic field has the following form:

$$\mathbf{j} = \mathbf{j}_{\text{orb}} + \mathbf{j}_{\text{sp}}, \qquad \text{(VII.4)}$$

---

[111] The operator of spin magnetic moment is $\hat{\boldsymbol{\mu}} = \mu_0\hat{\mathbf{s}}/s$, so the last term in Eq. (VII.1) has the form $-\hat{\boldsymbol{\mu}}\cdot\boldsymbol{\mathcal{H}}$; for a particle with $s = 1/2$ it is equal to $-\mu_0\hat{\boldsymbol{\sigma}}\cdot\boldsymbol{\mathcal{H}}$.

In this chapter we use the coordinate representation and the Schrödinger picture of motion, and hence we omit, as usual, the operator symbol over quantities that depend only on the coordinate, $\mathbf{r}$; see Problem 7.15.

We should also mention that in this chapter the following conflicting notations are used: m is used to denote either a particle mass or magnetic quantum numbers, and μ corresponds to either a magnetic moment or a reduced mass. The meaning of a particular notation should be clear from the context.

where the first term $\mathbf{j}_{\text{orb}}$ is related to the orbital motion of the particle:

$$\mathbf{j}_{\text{orb}} = \frac{ie\hbar}{2m}\{(\nabla\psi^*)\psi - \psi^*\nabla\psi\} - \frac{e^2}{mc}\mathbf{A}\psi^*\psi, \tag{VII.5}$$

and the second term $\mathbf{j}_{\text{sp}}$ is related to the spin magnetic moment:

$$\mathbf{j}_{\text{sp}} = \frac{\mu}{s}c\ \text{curl}(\psi^*\hat{\mathbf{s}}\psi). \tag{VII.6}$$

7.1 Stationary states in a magnetic field

Problem 7.1

Find the energy levels and the normalized wavefunctions of the stationary states of a charged spinless particle moving in a uniform magnetic field directed along the z-axis, using the following gauges of the vector potential $\mathbf{A}$:

a) $A_x = 0$, $A_y = \mathcal{H}_0 x$, $A_z = 0$; b) $\mathbf{A} = \frac{1}{2}\mathcal{H}_0 \times \mathbf{r}$.

Pay attention to the quantization of the transverse motion and different forms of normalization of the "transverse" part of the eigenfunction. Explain the reason for these differences. Compare your results with the case of a discrete spectrum of a particle in a potential field, $U(r)$.

Solution

a) The fact that the operators $\hat{p}_y$, $\hat{p}_z$ commute with both each other and the Hamiltonian of the system

$$\hat{H} = \frac{1}{2\mu}\left\{\hat{p}_x^2 + \left(\hat{p}_y - \frac{e}{c}\mathcal{H}_0 x\right)^2 + \hat{p}_z^2\right\},$$

ensures that $\hat{p}_y$ and $\hat{p}_z$ are conserved quantities, and accordingly the eigenfunctions of the Hamiltonian can be chosen in the form

$$\psi_{E p_y p_z}(x, y, z) = \frac{1}{2\pi\hbar}\exp\left\{\frac{i}{\hbar}(p_y y + p_z z)\right\}\psi(x). \tag{1}$$

The Schrödinger equation then becomes

$$-\frac{\hbar^2}{2\mu}\psi''(x) + \frac{1}{2\mu}\left(p_y - \frac{e}{c}\mathcal{H}_0 x\right)^2 \psi(x) = E_t\psi(x),$$

where we have introduced the energy of the *transverse* motion $E_t = E - p_z^2/2\mu$. This equation describes an harmonic oscillator with the fundamental frequency, $\omega_H = |e|\mathcal{H}_0/\mu c$; see Eq. (II.2). Hence, the solution is given by:

$$\psi_{n p_y p_z}(x, y, z) = \frac{1}{2\pi\hbar}\exp\left\{\frac{i}{\hbar}(p_y y + p_z z)\right\}\psi_n^{(\text{osc})}\left(x - \frac{cp_y}{e\mathcal{H}_0}\right)$$

and

$$E_{np_z} = E_{t,n} + \frac{p_z^2}{2\mu}, \quad E_{t,n} = \hbar\omega_H\left(n + \frac{1}{2}\right), \quad n = 0, 1, 2, \ldots \tag{2}$$

Let us emphasize that the energy levels of the transverse motion, $E_{t,n}$, i.e., the *Landau levels*, are discrete and have infinite degeneracy (the energy levels do not depend on p_y, which takes its value in $-\infty < p_y < \infty$). The *transverse* part of the eigenfunctions, $\psi_{np_yp_z}$, cannot be normalized to unity, since $|\psi|^2$ does not depend on y.

b) In this gauge, the Hamiltonian takes the form

$$\hat{H} = \frac{\hat{\mathbf{p}}^2}{2\mu} - \frac{e\hbar}{2\mu c}\mathcal{H}_0 \hat{l}_z + \frac{e^2\mathcal{H}_0^2}{8\mu c^2}\rho^2 \equiv \hat{H}_{tr,\,2} + \frac{p_z^2}{2\mu}. \tag{3}$$

Now $\hat{l}_z$, $\hat{p}_z$, and $\hat{H}$ form a set of mutually commuting operators, and we can write the eigenfunction in the form (by using cylindrical coordinates):

$$\psi_{Emp_z}(\rho, z, \varphi) = \frac{1}{2\pi\sqrt{\hbar}}\exp\left\{i\left(m\varphi + \frac{p_z z}{\hbar}\right)\right\}\sqrt{\rho}f(\rho), \tag{4}$$

where an unknown radial part is determined by the following equation:

$$f'' + \frac{2}{\rho}f' + \left[\frac{2\mu E_t}{\hbar^2} - \frac{m^2 - 1/4}{\rho^2} + \frac{em\mathcal{H}_0}{\hbar c} - \frac{e^2\mathcal{H}_0^2}{4\hbar^2 c^2}\rho^2\right]f = 0.$$

This equation has been studied in Problem 4.5. Using the results there, we obtain the solution

$$\sqrt{\rho}f(\rho) = Ce^{-x/2}x^{|m|/2}F\left(\frac{1}{2}\left(-\frac{2E_t}{\hbar\omega_H} + |m| + 1 - \frac{em}{|e|}\right), |m| + 1, x\right), \tag{5}$$

where the following notations are used:

$$x = \frac{\rho^2}{2a_H^2}, \quad a_H = \sqrt{\frac{\hbar}{\mu\omega_H}}, \quad \omega_H = \frac{|e|\mathcal{H}_0}{\mu c}.$$

Demanding that the hyper-geometric function is reduced to a polynomial, we obtain the condition

$$\frac{1}{2}\left(-\frac{2E_t}{\hbar\omega_H} + |m| + 1 - \frac{em}{|e|}\right) = -n_\rho, \quad n_\rho = 0, 1, 2, \ldots,$$

which determines the energy spectrum of the transverse motion. Hence, it follows:

$$E_{t,n} = \hbar\omega_H\left(n + \frac{1}{2}\right), \quad n = 0, 1, 2, \ldots \quad n = n_\rho + \frac{1}{2}\left(|m| - \frac{e}{|e|}m\right) \tag{6}$$

in accordance with Eq. (2).

Here for the energy level, $E_{t,n}$, with a given n, the degenerate eigenstates can take the following values of m (the quantum number that physically corresponds to the z-component of the angular momentum):[112]

$$m = -n, \ -n+1, \ldots, \ 0, \ +1, \ldots, \ +\infty \text{ in the case } e > 0,$$

or

$$m = -\infty, \ldots, \ -1, \ 0 \ldots, \ n \text{ for } e < 0,$$

where the infinite number of possible m implies the infinite degeneracy of the Landau levels.

It should be noticed that now eigenfunctions (4) and (5) of the Hamiltonian are localized in the transverse (perpendicular to $\mathcal{H}_0$) directions (*i.e.*, in the xy-plane). The "transverse" part $\psi_{nm}(\rho, \varphi)$ of the eigenfunction can be normalized to unity, with the value of the normalization coefficient in Eq. (5) being

$$|C|^2 = \frac{(|m| + n_\rho)!}{n_\rho!(|m|!)^2 a_H^2}.$$

Concerning the normalization of the stationary states corresponding to a discrete spectrum, we should note that for particles moving in a potential $U(r)$, they are always localized in a bounded spatial domain. The unusual properties in an homogeneous magnetic field are related to the fact that the discrete Landau levels of the transverse motion are infinitely degenerate. Let us consider a wave-packet that consists of the eigenfunctions, ψ_{np_y} (we omit the z-dependence of the wavefunction), of the form

$$\psi_n(x, y) = \int C(p_y)\psi_{n,p_y}(x, y)dp_y.$$

This wavefunction also belongs to the Landau level, $E_{t,n}$, and if $\int |C(p_y)|^2 dp_y = 1$, then it is normalized and thus describes a localized particle state, unlike the non-normalizable wavefunction, ψ_{n,p_y}. On the contrary, from the normalizable eigenfunctions ψ_{nm}, we can construct a wavefunction of the form $\tilde{\psi}_n = \sum_m C_m \psi_{nm}$. They also belong to the Landau level $E_{t,n}$, but in this case $\sum_m |C_m|^2 = \infty$ and they do not describe states localized in the plane (x, y).

Problem 7.2

For a charged spinless particle moving in a uniform magnetic field, find the operators for 1) the coordinate of the orbital center, $\hat{\boldsymbol{\rho}}_0$, of the "transverse" motion (perpendicular to the magnetic field; *i.e.*, directed along the z-axis); 2) the square of the orbital center coordinate, $\hat{\rho}_0^2$; and 3) the square of the Larmor orbit radius, $\hat{\rho}_L^2$.

---

[112] Note that for the lowest Landau level with $n = 0$, the following condition holds: $e \cdot m \geq 0$.

274 *Exploring Quantum Mechanics*

Obtain the commutation relations of the operators with each other and the Hamiltonian.

Determine the spectrum of the eigenvalues of the operators $\hat{\rho}_0^2$ and $\hat{\rho}_L^2$.

Describe the transverse spatial distribution of the particle in the stationary states ψ_{nmp_z} considered in the previous problem, for the following cases:

a) $m = -\frac{e}{|e|}n$ (consider, in particular, the case $n \gg 1$ and compare with the result of classical mechanics);

b) $n = 0$ and $|m| \gg 1$.

Solution

1) In classical mechanics, the orbit of a charged particle in the plane perpendicular to the direction of the magnetic field is a circle (Larmor's orbit) whose radius square is given by

$$\rho_L^2 = \frac{v_\perp^2}{\omega_H^2} \equiv \frac{v_x^2 + v_y^2}{\omega_H^2}, \quad \omega_H = \frac{|e|\mathcal{H}}{\mu c}. \tag{1}$$

Here the vectors $\boldsymbol{\rho}$, $\boldsymbol{\rho}_0$, $\mathbf{v}_\perp$ represent the position of the particle, the orbital (circle) center, and the velocity, respectively. They obey the relation

$$\boldsymbol{\omega} \times (\boldsymbol{\rho} - \boldsymbol{\rho}_0) = \mathbf{v}_\perp, \quad \boldsymbol{\omega} = \left(0, 0, -\frac{e\omega_H}{|e|}\right).$$

This formula describes a uniform circular motion of the particle in the (xy)-plane, and the sign of ω determines the direction of the motion (clockwise or counterclockwise; here, the z-axis is directed along the magnetic field). From this relation, it follows that

$$x_0 = x - \frac{v_y}{\omega}, \quad y_0 = y + \frac{v_x}{\omega}; \quad \rho_0^2 = x_0^2 + y_0^2. \tag{2}$$

The quantum-mechanical generalizations of Eqs. (1) and (2) of classical mechanics are the corresponding Hermitian operators:

$$\hat{x}_0 = \hat{x} - \frac{\hat{v}_y}{\omega} = x - \frac{-i\hbar \partial/\partial y - eA_y}{\mu \omega}, \quad \hat{y}_0 = \hat{y} + \frac{\hat{v}_x}{\omega}, \tag{3}$$

$$\hat{\rho}_0^2 = \hat{x}_0^2 + \hat{y}_0^2, \quad \hat{\rho}_L^2 = \frac{\hat{v}_x^2 + \hat{v}_y^2}{\omega^2},$$

(here, $\mu \hat{\mathbf{v}} = \hat{\mathbf{p}} - e\mathbf{A}/c$, $\hat{\mathbf{p}} = -i\hbar \nabla$), which yields the following commutation relations:[113]

$$[\hat{H}, \hat{x}_0] = [\hat{H}, \hat{y}_0] = [\hat{H}, \hat{\rho}_0^2] = [\hat{H}, \hat{\rho}_L^2] = 0, \quad [\hat{x}_0, \hat{y}_0] = -i\frac{\hbar c}{e\mathcal{H}}, \quad [\hat{\rho}_0^2, \hat{\rho}_L^2] = 0. \tag{4}$$

---

[113] We can obtain them without choosing a particular gauge for the vector potential, but calculations may be made easier if we do fix the gauge as follows: $\mathbf{A} = (0, \mathcal{H}x, 0)$.

Since these operators commute with the Hamiltonian, the corresponding physical quantities are constants of motion, just as in classical mechanics.

2) Since $\hat{\rho}_L^2 = 2\hat{H}_{tr}/\mu\omega_H^2$, where $\hat{H}_{tr} = \hat{H} - \hat{p}_z^2/2\mu$ is the Hamiltonian of the transverse motion of the particle, by using the results of Problem 7.1, we obtain the eigenvalues of the square of the orbit radius:

$$(\rho_L^2)_n = (2n+1)a_H^2, \quad n = 0, 1, 2, \ldots, \quad a_H^2 = \frac{c\hbar}{|e|\mathcal{H}}. \tag{5}$$

Furthermore, after fixing the gauge of the vector potential to be[114] $\mathbf{A} = (1/2)\mathcal{H} \times \mathbf{r}$, we find that

$$\mu\omega_H^2 \hat{\rho}_0^2 = 2\hat{H}_{tr} + \frac{e}{|e|}\hat{l}_z \omega_H.$$

Therefore, the wavefunctions (4) from Problem 7.1 are also the eigenfunctions of the operator $\hat{\rho}_0^2$, and the spectrum[115] can be read off directly:

$$(\rho_0^2)_k = (2k+1)a_H^2, \quad k = n + \frac{em}{|e|} = 0, 1, 2, \ldots. \tag{6}$$

As for the eigenvalues of $\hat{x}_0$ and $\hat{y}_0$, they both have a continuous spectrum. We should note that the wavefunctions, $\psi_{np_y p_z}$ from Problem 7.1, are the eigenfunctions of the operator, $\hat{x}_0$, with the corresponding eigenvalues: $x_0 = -p_y/\mu\omega$. However, since these operators do not commute with each other, the location of the orbit center is constrained by the uncertainty relation, $\Delta x_0 \cdot \Delta y_0 \geqq a_H^2/2$, and hence is not well-defined (see Problem 1.30).

3a) By using the expressions (4) and (5) from Problem 7.1 for the wavefunctions of the eigenstates with $m = -en/|e|$, we obtain the probability distribution

$$dw(\rho) = |\psi_{nmp_z}|^2 \cdot 2\pi\rho d\rho = \frac{1}{n!a_H^2}\left(\frac{\rho^2}{2a_H^2}\right)^n e^{-\rho^2/2a_H^2}\rho d\rho, \tag{7}$$

which gives the mean values

$$\bar{\rho} = \int \rho dw = \sqrt{2}\frac{\Gamma(n+3/2)}{n!}a_H, \quad \overline{\rho^2} = 2(n+1)a_H^2, \quad \rho_{m.p.} = \sqrt{2n+1}a_H. \tag{8}$$

Here $\rho_{m.p.}$ is the most probable value of ρ, which corresponds to the maximum of the distribution.

We should also note that the operators $\hat{\rho}_0^2$ and $\hat{\rho}_L^2$ have definite values in the states considered. According to Eqs. (5) and (6), they are given by

$$(\rho_0^2)_0 = a_H^2, \quad (\rho_L^2)_n = (2n+1)a_H^2. \tag{9}$$

---

[114] See also Problem 6.27 concerning the behavior of the wavefunction under a gauge transformation.
[115] The spectrum of the operators $\hat{\rho}_L^2$s and $\hat{\rho}_0^2$ can be obtained by identifying their commutation relations (3) and the commutators, $[\hat{x}_0, \hat{y}_0]$ and $[\hat{v}_x, \hat{v}_y]$, with the corresponding commutation relations for the harmonic oscillator: $\hat{H} = \hat{p}^2/2m + m\omega^2\hat{x}^2/2$ and $[\hat{p}, \hat{x}] = -i\hbar$. This algebraic equivalence determines the familiar spectrum, $E_n = \hbar\omega(n+1/2)$.

In the case $n \gg 1$ (*quasi-classical* limit), using the known asymptote of the Γ–function,

$$\Gamma(x) \approx \sqrt{2\pi} x^{x-1/2} e^{-x}, \quad x \to \infty,$$

we obtain, from Eq. (8), $\bar{\rho} \approx \sqrt{2n} a_H$, and therefore,

$$\rho_{m.p.} = \sqrt{(\rho_L^2)_n} \approx \sqrt{\overline{\rho^2}} \approx \bar{\rho} \approx \sqrt{2n} a_H \gg a_H. \tag{10}$$

These relations indicate that if $n \gg 1$, the radial probability distribution has a sharp peak near $\rho_{m.p.}$. Also, relation (7) in the most relevant domain of ρ can be rewritten as

$$dw \approx \left(\pi a_H^2\right)^{-1/2} \exp\left\{-\frac{(\rho - \rho_{m.p})^2}{a_H^2}\right\} d\rho. \tag{11}$$

Consequently, we have $\Delta\rho \equiv \sqrt{\overline{\rho^2} - \bar{\rho}^2} \approx a_H/\sqrt{2} \ll \bar{\rho}$.

Thus, the probability of finding the particle is noticeably different from zero only in a narrow ring-shaped region with the radius, $\sqrt{2n} a_H$, and the width of order a_H. It corresponds to the transition to the classical picture of motion, averaged over the period of the circular orbit. The relation $m = -en/|e|$ for $n \gg 1$, after the substitutions, $m = M_z/\hbar$, together with $n \approx E_t/\hbar\omega_H$, becomes the standard classical relation between the angular momentum with respect to the orbital center and the energy of transverse motion:

$$E_t = \frac{|e|\mathcal{H}|M_z|}{\mu c} = \omega M.$$

b) Now, for $n = 0$ and $m = e|m|/|e|$, we have the probability distribution as follows:

$$dw(\rho) = \frac{1}{|m|! a_H^2} \left(\frac{\rho^2}{2a_H^2}\right)^{|m|} e^{-\rho^2/2a_H^2} \rho d\rho, \tag{12}$$

which differs from Eq. (7) only by the replacement: $n \to |m|$. The analogous changes in Eqs. (8), (10), and (11) define other characteristics of the radial distribution in this case.

However, the interpretation of the states under consideration differs greatly from the previous case. Here, the energy of the transverse motion for $n = 0$ is $\hbar\omega_H/2$, and such states are essentially "non-classical". But taking into account the fact that instead of Eq. (9), we have

$$(\rho_L^2)_0 = a_H^2, \quad \rho_{(0)|m|}^2 = (2|m|+1)a_H^2$$

if $|m| \gg 1$, we can view the states in accordance with the following classical picture: a uniform distribution of the "orbits" of the minimal "radius" equal to $\sqrt{(\rho_L^2)_0} a_H = a_H$ over a narrow ring-shaped region of the radius, $R \approx \sqrt{2|m|} \gg a_H$, with the width, a_H.

Problem 7.3

Find the stationary states and the corresponding energy levels of a charged spinless particle moving in uniform electric and magnetic fields perpendicular to each other.

Solution

Directing the z-axis along the magnetic field and the x-axis along the electric field, and using the vector potential of the form

$$A_x = 0, \quad A_y = \mathcal{H}x, \quad A_z = 0,$$

we obtain the single-particle Hamiltonian as follows:

$$\hat{H} = \frac{1}{2\mu}\left\{\hat{p}_x^2 + \left(\hat{p}_y - \frac{e}{c}\mathcal{H}x\right)^2 + \hat{p}_z^2\right\} - e\mathcal{E}x. \tag{1}$$

Since the operators $\hat{p}_y$, $\hat{p}_z$, and $\hat{H}$ are mutually commuting, the eigenfunctions could be chosen in the following way:

$$\psi_{E p_y p_z}(x,y,z) = \frac{1}{2\pi\hbar}\exp\left\{\frac{i}{\hbar}(p_y y + p_z z)\right\}\psi(x). \tag{2}$$

Hence the Schrödinger equation can be written as

$$f''(x) + \frac{1}{\hbar^2}\left[2\mu E_{tr} + 2\mu e\mathcal{E}x - \left(p_y - \frac{e}{c}\mathcal{H}x\right)^2\right]f(x) = 0, \tag{3}$$

where $E_{tr} = E - p_z^2/2\mu$. This equation can be viewed as the Schrödinger equation of an harmonic oscillator with the frequency, $\omega_H = |e|\mathcal{H}/\mu c$. Using its solution (see Eq. (II.2)), we obtain the eigenfunctions and the spectrum of the Hamiltonian (1) as follows:

$$\psi_{n p_y p_z}(x,y,z) = \frac{1}{2\pi\hbar}\exp\left\{\frac{i}{\hbar}(p_y y + p_z z)\right\}\psi_n^{(\text{osc})}\left(x - \frac{c p_y}{e\mathcal{H}} - \frac{\mu c^2 \mathcal{E}}{e\mathcal{H}^2}\right) \tag{4}$$

and

$$E_{n p_y p_z} = \hbar\omega_H\left(n + \frac{1}{2}\right) - \frac{c\mathcal{E}p_y}{\mathcal{H}} - \frac{\mu c^2 \mathcal{E}^2}{2\mathcal{H}^2} + \frac{p_z^2}{2\mu}, \quad n = 0, 1, 2, \ldots$$

A few remarks on the properties of solution (4) are in order:

1) The energy spectrum of the particle is continuous and is not bounded from below. Consequently, the discrete levels of a particle in a potential, which vanishes ($U \to 0$) at large distances, acquire a finite *width* and become *quasi-stationary* under the influence of an electric field; see also Problem 6.36.
2) The eigenfunctions of the Hamiltonian (4) describe states in which the particle is localized in the x-direction. In the case $\mathcal{E} < \mathcal{H}$, it corresponds to the finite classical motion of the particle in the x-direction.
3) The derivatives

$$\frac{\partial E_{np_y p_z}}{\partial p_z} = \frac{p_z}{\mu}, \quad \frac{\partial E_{np_y p_z}}{\partial p_y} = -\frac{\mathcal{E}}{\mathcal{H}}c$$

determine the z-component of the velocity of the particle and the *drift* velocity[116] of the particle along the y-direction.

Problem 7.4

Consider the same setting as in the previous problem, except that the electric and magnetic fields are parallel.

Solution

Taking the z-axis as the direction of both the electric and magnetic fields, we see that the Hamiltonian of the particle differs from that in Problem 7.1 only by an additional term, $-e\mathcal{E}z$. Therefore the *transverse* and *longitudinal* motions remain separated, and can be treated independently. The longitudinal motion is equivalent to a particle in an homogeneous field, which is different from the free particle motion of Problem 7.1. The solution of the present problem may be obtained from the equations of Problem 7.1 by changing the term $p_z^2/2\mu$ there to the energy E_t of the longitudinal motion, and the plane wave, $\psi_{p_z}(z)$, to the wavefunction, $\psi_{E_t}(z)$, of the particle in an homogeneous field; see also Problem 2.41. We also note that the energy spectrum is continuous and unbounded from below, as in Problem 7.1.

Problem 7.5

Find the energy levels and normalized wavefunctions of the stationary states for a charged spherical oscillator (*i.e.*, a charged particle in the potential, $U(r) = \frac{1}{2}kr^2$) in an external uniform magnetic field. Examine the limiting cases of a weak and strong magnetic field.

Solution

If we choose the vector potential in the form $\mathbf{A} = \frac{1}{2}\mathcal{H} \times \mathbf{r}$, the Hamiltonian of the particle in the cylindrical coordinates (with the z-axis directed along the magnetic field) takes the form

---

[116] Due to the non-relativistic condition, $v \ll c$, solution (4) is valid if $E \ll H$.

$$\hat{H} = -\frac{\hbar^2}{2\mu}\left[\frac{1}{\rho}\frac{\partial}{\partial\rho}\rho\frac{\partial}{\partial\rho} + \frac{1}{\rho^2}\frac{\partial^2}{\partial\varphi^2} + \frac{e\mathcal{H}}{c\hbar}\hat{l}_z\right] + \left(\frac{k}{2} + \frac{e^2\mathcal{H}^2}{8\mu c^2}\right)\rho^2 + \hat{H}_l,$$

where

$$\hat{H}_l = -\frac{\hbar^2}{2\mu}\frac{\partial^2}{\partial z^2} + \frac{k}{2}z^2.$$

Due to the mutual commutativity of the operators $\hat{l}_z$, $\hat{H}_l$, and $\hat{H}$, the eigenfunctions can be chosen as (the operator, $\hat{H}_l$, describes a harmonic oscillator):

$$\psi_{Emn_2}(\rho, z, \varphi) = \frac{e^{im\varphi}}{\sqrt{2\pi}}\psi_{n_2}^{(osc)}(z)\sqrt{\rho}f(\rho), \quad n_2 = 0, 1, 2, \ldots \tag{1}$$

Here the Schrödinger equation leads to

$$f'' + \frac{2}{\rho}f' + \left[\frac{2\mu E_{tr}}{\hbar^2} - \frac{m^2 - 1/4}{\rho^2} + \frac{em\mathcal{H}}{\hbar c} - \frac{\rho^2}{4a^2}\right]f = 0,$$

where the following notations are used:

$$E_{tr} = E - \hbar\omega\left(n_2 + \frac{1}{2}\right), \quad \omega = \sqrt{\frac{k}{\mu}}, \quad \omega_H = \frac{|e|\mathcal{H}}{\mu c}, \quad a = \left(\frac{\hbar}{\mu}\right)^{1/2}(4\omega^2 + \omega_H^2)^{-1/4}.$$

The equation is almost identical to the Schrödinger equation considered in Problem 4.5, apart from a redefinition of some variables. Therefore, using the results of Problem 4.5, we obtain

$$\sqrt{\rho}f_{n_1 m} = C\left(\frac{\rho}{a}\right)^{|m|}\exp\left\{-\frac{\rho^2}{4a^2}\right\}F\left(-n_1, |m|+1, \frac{\rho^2}{2a^2}\right) \tag{2}$$

and

$$E_{tr; n_1 m} = \frac{1}{2}\hbar\sqrt{\omega_H^2 + 4\omega^2}(2n_1 + |m| + 1) - \frac{e\hbar\omega_H}{2|e|}m, \quad n_1 = 0, 1, 2, \ldots.$$

Expressions (1) and (2) determine the eigenfunctions and the spectrum of the Hamiltonian of the oscillator in a magnetic field:

$$E_{n_1 n_2 m} = \frac{1}{2}\hbar\sqrt{\omega_H^2 + 4\omega^2}(2n_1 + |m| + 1) + \hbar\omega\left(n_2 + \frac{1}{2}\right) - \frac{e\hbar\omega_H}{2|e|}m. \tag{3}$$

In the case of a weak magnetic field, $\omega_H \ll \omega$, we have

$$E_{n_1 n_2 m} \approx E_N^{(0)} - \frac{e\hbar}{2\mu c}m\mathcal{H} + \frac{e^2\hbar}{8\mu^2\omega c^2}(2n_1 + |m| + 1)\mathcal{H}^2. \tag{4}$$

Here, $E_N^{(0)} = \hbar\omega(N + 3/2)$ describes the unperturbed oscillator levels (see Problem 4.5), where $N = 2n_1 + |m| + n_2$. The linear-in-$\mathcal{H}$ term in the level shift corresponds to an interaction of the magnetic moment of the oscillator with the magnetic field described by the expression $\hat{V} = -\hat{\boldsymbol{\mu}} \cdot \mathcal{H}$, where $\hat{\boldsymbol{\mu}} = (e\hbar/2\mu c)\hat{\mathbf{l}}$ is the angular magnetic moment of the charged particle. The term in Eq. (4) quadratic in $\mathcal{H}$ determines the *diamagnetic* part of the level shift. In particular for the ground state, the linear term vanishes, and

$$\Delta E_0 \approx -\frac{1}{2}\chi_0 \mathcal{H}^2, \quad \text{where } \chi_0 = -\frac{e^2 \hbar}{4\mu^2 c^2 \omega}$$

determines the *magnetic susceptibility* of the oscillator.

In the case of a strong magnetic field, when $\omega_H \gg \omega$, Eq. (3) yields

$$E_{n_1 n_2 m} \approx E_{tr,\,n} + \frac{\hbar \omega^2}{\omega_H}(2n_1 + |m| + 1) + E_{l;\,n_2}. \tag{5}$$

In this case, the "transverse" part of the energy spectrum is determined mainly by the magnetic field. $E_{tr,\,n} = \hbar \omega_H (n + 1/2)$ describes the spectrum of the Landau levels with $n = n_1 + |m|/2 - em/2|e|$. The second term in (5) gives a correction that originates from the effect of the elastic force (the harmonic potential) on the transverse motion of the particle. The last term, $E_{l,n_2} = \hbar \omega (n_2 + 1/2)$, can be interpreted as the energy of the free oscillation along the magnetic field.

Problem 7.6

Show that a magnetic field, $\mathcal{H}(\mathbf{r})$, that is non-zero in some bounded domain, cannot "bind" a charged spinless particle, *i.e.*, there are no localized stationary states. Why does not this result contradict the existence of *magnetic traps* for charged particles in classical mechanics?

Solution

Indeed, the eigenvalues of the Hamiltonian $\hat{H} = \frac{1}{2m}\left(\hat{\mathbf{p}} - \frac{e}{c}\mathbf{A}\right)^2$ are always positive, and for positive energy $E > 0$ the particle is essentially free, whenever it is far from the region where magnetic field is non-zero. Hence there are no solutions of the Schrödinger equation that are decreasing with r as $r \to \infty$. However, even though true bound states in magnetic fields do not exist, there may exist *quasi-stationary* states (see Chapter 6, sec. 5), whose *lifetime* could be considered as infinitely long for macroscopic (classical) systems.

Problem 7.7

As it is known in one and two dimensions, a particle in any attractive potential well always has bound state(s), where it is localized in a bounded region of space. In three dimensions, no such states exist in the case of a shallow potential well.

Show that in the presence of a uniform magnetic field, a charged particle in a attractive potential always does have bound states in which the particle is localized (and not only in the "transverse" direction), provided that the potential satisfies the conditions $U(r) \leq 0$ and $U(r) \to 0$ as $r \to \infty$. That is to say, in the presence of a magnetic field, any well can "bind" a quantum particle.

In the case of a shallow well $U_0 \ll \hbar^2/ma^2$ (U_0 and a being the characteristic strength and radius of potential), obtain approximate expressions for the binding energy; see also Problem 8.61.

Solution

Let us use the variational method (as in Problem 2.3). Taking the form of the vector potential as $\mathbf{A} = \frac{1}{2}\mathcal{H} \times \mathbf{r}$, we can write the Hamiltonian as

$$\hat{H} = \hat{H}_{tr} + \frac{\hat{p}_z^2}{2\mu} + U(r),$$

where $\hat{H}_{tr}$ is the transverse part of the Hamiltonian with a magnetic field directed along the z-axis; see Eq. (3) from Problem 7.1. Let us now consider the following normalized wavefunctions

$$\psi_m = \sqrt{\kappa} e^{-\kappa|z|} \psi_{n=0,m}(\rho, \varphi), \qquad (1)$$

where ψ_{nm} is the "transverse" part of the wavefunction (4) from Problem 7.1. For this part we have $\hat{H}_{tr}\psi_{0m} = (\hbar\omega_H/2)\psi_{0m}$, where $\omega_H = \frac{|e|\mathcal{H}}{\mu c}$; so the mean value of the energy in the state represented by the wavefunction (1) is equal to

$$\overline{E}_m(\kappa) = \int \psi_m^* \hat{H} \psi_m dV = \frac{\hbar\omega_H}{2} + \frac{\hbar^2\kappa^2}{2\mu} + \kappa \int U(r)e^{-2\kappa|t|}|\psi_{0m}|^2 dV.$$

Since $U(r) \leq 0$, it is always possible to choose a small value of κ to ensure that $\overline{E}_m(\kappa) < \hbar\omega_H/2$. Therefore, the Hamiltonian under consideration has eigenvalues less than $\hbar\omega_H/2$, which is the minimal energy of the particle in a uniform magnetic field. Therefore, there does exist a bound state, where the particle cannot "escape" to infinity. We should note that there is an infinite number of such independent states for different values of the angular moment component, m (see Problem 7.1).

The emergence of the bound states in the problem under consideration, even in the case of a shallow potential, has a simple interpretation: the particle is already bounded by the magnetic field in the transverse direction – see Problem 7.1 – while the existence of the potential well leads to its *binding* in the longitudinal direction, as in the one-dimensional case (see Problem 2.3).

Let us focus on the case of a shallow well. The wavefunctions of the bound states have the approximate form as follows(see Eq. (1)):

$$\psi_{Em} \approx \psi_{0m}(\rho, \varphi)\psi_{\varepsilon m}(z), \quad E_m = \frac{\hbar\omega_H}{2} + \varepsilon. \qquad (2)$$

Here we took into account that the dependence of the eigenfunctions on the *transverse* coordinates is mainly determined by the magnetic field. Substituting this wavefunction into the Schrödinger equation, $\hat{H}\psi_{Em} \approx E\psi_{Em}$, multiplying the latter by $\psi^*_{0m}(\rho)$ from the left, and integrating over the coordinates of the transverse motion,[117] we obtain the one-dimensional Schrödinger equation

$$-\frac{\hbar^2}{2\mu}\psi''_{\varepsilon m}(z) + [U_{eff,\,m}(z) - \varepsilon_m]\psi_{\varepsilon m}(z) = 0 \tag{3}$$

with the *effective potential*

$$U_{eff,\,m}(z) = \int U(r)|\psi^*_{0m}(\rho,\varphi)|^2 d^2\rho = \frac{2}{|m|!}\int_0^\infty U(r)\tilde{\rho}^{2|m|+1}\exp\{-\tilde{\rho}^2\}d\tilde{\rho}. \tag{4}$$

Here,

$$\tilde{\rho} = \frac{\rho}{\sqrt{2}a_H},\quad a_H = \sqrt{\frac{\hbar}{\mu\omega_H}},\quad r = \sqrt{\rho^2 + z^2},$$

and in this procedure we have used the explicit form of the wavefunction as follows (see also Problem 7.1)

$$\psi_{0m} = \tilde{\rho}^{|m|}\frac{\exp\{im\phi - \tilde{\rho}^2/2\}}{\sqrt{2\pi|m|!}a_H}.$$

In the case of a shallow potential well, $U(r)$, the effective potential (4) is also shallow, and the energy level ε_m can be determined using the result of Problem 2.22:

$$\varepsilon_m \approx -\frac{\mu\alpha_m^2}{2\hbar^2},\quad \alpha_m = -\int_{-\infty}^\infty U_{m,eff}(z)dz. \tag{5}$$

(Here the dependence of the wavefunction (2) on z is the same as in Eq. (1) with $\kappa = \mu\alpha_m/\hbar^2$.)

A simple calculation shows that the binding energy of the bound states is small: $|\varepsilon_m| \ll \hbar\omega_H$. Here the binding energies of the states with different angular momentum components, m, depend strongly on the relation between the "magnetic length", a_H, and the radius of the potential well, R. The dependence of ε_m on m is particularly strong for $R \ll a_H \propto \mathcal{H}^{-1/2}$. In this case, the term $\exp\{-\tilde{\rho}^2\}$ in the integral in Eq. (4) can approximated by 1, and from Eqs. (4) and (5) we obtain

$$\varepsilon_m \propto \left(\frac{R}{a_H}\right)^{4|m|+4} \propto \mathcal{H}^{2|m|+2},$$

so the binding energy decreases fast with increasing $|m|$.

---

[117] A similar method for the Schrödinger equation is often used in the framework of the *adiabatic approximation*; see Problem 8.61.

A generalization of the above results to the case where the potential well $U(r)$ is not shallow is given in Problem 8.61.

Problem 7.8

Find the stationary states and the corresponding energy levels for a neutral particle with spin $s = 1/2$ and magnetic moment μ_0 (so that $\hat{\boldsymbol{\mu}} = \mu_0 \hat{\boldsymbol{\sigma}}$) in a uniform magnetic field.

Solution

Choosing the z-axis along the magnetic field, the Hamiltonian of the particle reads $\hat{H} = \frac{1}{2m}\hat{\mathbf{p}}^2 - \mu_0 \mathcal{H} \hat{\sigma}_z$; see Eq. (VII.1). Due to the mutual commutativity of the operators $\hat{H}$, $\hat{\mathbf{p}}$, and $\hat{\sigma}_z$, we can immediately determine the eigenfunctions and the corresponding energy eigenvalues:

$$\psi_{\mathbf{p}s_z} = \frac{1}{(2\pi\hbar)^{3/2}} e^{i\mathbf{p}\cdot\mathbf{r}/\hbar}\chi_{s_z} \ , \ E_{\mathbf{p}s_z} = \frac{\mathbf{p}^2}{2m} - 2\mu_0 \mathcal{H} s_z,$$

where χ_{s_z} are the eigenstates of the operator, $\hat{s}_z$, with eigenvalues $s_z = \pm 1/2$; see Problem 5.1.

Problem 7.9

Consider the same conditions as in the previous problem, but for a charged particle with spin $s = \frac{1}{2}$. Compare with the results of Problems 7.1 and 7.8.

Pay attention to the appearance of an additional degeneracy[118] of the energy levels of the transverse motion with the magnetic moment, $\mu_0 = \frac{e\hbar}{2mc}$ (e and m are the charge and mass of the particle, respectively. Note that this value of μ_0 follows from the Dirac equation describing electrons, muons, and their anti-particles).

Solution

The Hamiltonian of the particle differs from the spinless case by an additional term that has the form $-\mu_0\mathcal{H}\hat{\sigma}_z$ (see Eq. (VII.I), with the z-axis directed along the magnetic field). Since it does not depend on the spatial coordinates, the orbital and the spin degrees of freedom separate. This consideration, combined with the result of Problem 7.1 and the *conservation* of s_z, allows us to write the eigenfunctions of the Hamiltonian as

$$\psi_{n\nu p_z s_z} = \psi_{n p_z \nu}(\mathbf{r})\chi_{s_z}, \tag{1}$$

where $\psi_{np_z\nu}(\mathbf{r})$ are the eigenfunctions of the spinless Hamiltonian. Their explicit form depends on a specific gauge choice (here $\nu \equiv p_y$ and $\nu \equiv m$ correspond to cases a) and b) in Problem 7.1). Eigenfunctions given in (1) have the following energies:

---

[118] This degeneracy is related to the *supersymmetric* character of the Hamiltonian. Some properties and consequences of supersymmetry are studied in Problems 10.2 and 10.27.

$$E_{ns_z p_z} = E_{tr, ns_z} + \frac{p_z^2}{2m}, \quad E_{tr, ns_z} = \hbar \omega_H \left(n + \frac{1}{2}\right) - 2\mu_0 \mathcal{H} s_z, \tag{2}$$

with $n = 0, 1, 2, \ldots$, $\omega_H = |e|\mathcal{H}/mc$. Here the discrete part of the spectrum E_{tr,ns_z} is associated with the transverse motion of the particle.

Considering the spin magnetic moment, $\mu_0 = -|e|\hbar/2mc$, Eq. (2) gives $E_{t,ns_z} \equiv E_N = \hbar \omega_H N$, where $N = n + s_z + \frac{1}{2} = 0, 1, \ldots$. This spectrum has the following property. The ground state with $N = 0$ is "non-degenerate"[119] and its energy is $E_0 = 0$ ($n = 0$, $s_z = -1/2$), while the levels with $N \neq 0$ are "two-fold degenerate". The two-fold degenerate states are those with $n = N$, $s_z = -1/2$, and $n = N - 1$, $s_z = +1/2$.

These properties of the energy spectrum can be understood from the *supersymmetric* nature of the transverse part of the Hamiltonian. Indeed, this Hamiltonian can be reduced to a supersymmetric oscillator (considered also in Problem 10.26), since it can be written as

$$\hat{H}_{tr} = \frac{1}{2}(\hat{\mathbf{p}}_\perp + |e|\mathbf{A}_\perp/c)^2 + \frac{1}{2}|e|\hbar \mathcal{H} \hat{\sigma}_z \equiv \hbar \omega_H \left(\hat{b}^+ \hat{b} + \frac{1}{2}\right) + \hbar \omega_H \left(\hat{f}^+ \hat{f} - \frac{1}{2}\right). \tag{3}$$

Here, $\hat{f} \equiv (\hat{\sigma}_x - i\hat{\sigma}_y) = \begin{pmatrix} 0 & 0 \\ 1 & 0 \end{pmatrix}$ and $\hat{f}^+ \hat{f} = (1 + \hat{\sigma}_z)/2$ and $\{\hat{f}, \hat{f}^+\}_+ = 1$, while $\hat{b} = (\hat{\pi}_y + i\hat{\pi}_x)/\sqrt{2m\hbar\omega_H}$, with $\hat{\boldsymbol{\pi}} \equiv m\hat{\mathbf{v}} = (\hat{\mathbf{p}} + |e|\mathbf{A}/c)$ and $[\hat{b}, \hat{b}^+] = 1$. So, the spin corresponds to a fermionic degree of freedom with $n_F = s_z + 1/2$, while the orbital motion corresponds to a bosonic degree of freedom with $n_B = n$. The spectrum (3) takes the form: $E_{tr, n_B n_F} = \hbar \omega_H (n_B + n_F)$. It coincides with the spectrum, $E_{tr, N}$, and explains the properties mentioned above.

In conclusion, we present the expressions for the operators of supersymmetry transformation, $\hat{Q}$ (see also Problem 10.26) for the given problem:

$$\hat{Q}^+ \equiv \frac{1}{2}\left(\hat{Q}_1 + i\hat{Q}_2\right) = q\hat{b}\hat{f}^+ = \frac{(\hat{\pi}_y + i\hat{\pi}_x)}{\sqrt{2m}} \cdot \begin{pmatrix} 0 & 1 \\ 0 & 0 \end{pmatrix}, \quad \hat{Q} = (\hat{Q}^+)^+,$$

$$\hat{Q}_1 = \hat{Q}^+ + \hat{Q} = \frac{(\hat{\sigma}_x \hat{\pi}_y - \hat{\sigma}_y \hat{\pi}_x)}{\sqrt{2m}}, \quad \hat{Q}_2 = -i(\hat{Q}^+ - \hat{Q}) = \frac{(\hat{\sigma}_x \hat{\pi}_x + \hat{\sigma}_y \hat{\pi}_y)}{\sqrt{2m}},$$

where $q = \sqrt{\hbar \omega_H}$. In terms of the superoperators, the Hamiltonian (3) can be written in the following equivalent forms:

$$\hat{H}_{tr} = \hat{Q}_1^2 = \hat{Q}_2^2 = \{\hat{Q}, \hat{Q}^+\}_+.$$

---

[119] Here we do not count the degeneracy of the levels $E_{t,n}$ of the transverse orbital motion in a magnetic field.

Problem 7.10

Show that the Pauli Hamiltonian (VII.1) for an electron moving in an electromagnetic field can be written in the form,

$$\hat{H} = \frac{1}{2m}\left[\hat{\boldsymbol{\sigma}} \cdot \left(\hat{\mathbf{p}} - \frac{e}{c}\mathbf{A}\right)\right]^2 + e\varphi. \qquad (1)$$

Using this expression, show that

a) For an electron in a stationary uniform magnetic field, the projection of its spin along the direction of its velocity is a *constant of motion*.
b) A magnetic field, $\mathcal{H}(\mathbf{r})$, which is non-zero only in a bounded region of space, cannot "bind" an electron (see also Problem 7.6).

Do these results generalize for other spin-1/2 particles (proton, neutron, and others)?

Solution

1) With the help of relations (V.3) and (VII.2), we can rewrite the Hamiltonian given in the problem as follows:

$$\hat{H} = \frac{1}{2m}\left(\hat{\boldsymbol{\sigma}} \cdot \left(\hat{\mathbf{p}} - \frac{e}{c}\mathbf{A}\right)\right)^2 + e\varphi \equiv \frac{m}{2}\hat{\sigma}_i\hat{\sigma}_k\hat{v}_i\hat{v}_k + e\varphi$$

$$= \frac{m}{2}(\delta_{ik} + i\varepsilon_{ikl}\hat{\sigma}_l)\hat{v}_i\hat{v}_k + e\varphi = \frac{m}{2}\hat{v}^2 + e\varphi - \frac{e\hbar}{2mc}\mathcal{H}\cdot\hat{\boldsymbol{\sigma}}. \qquad (2)$$

Hence, this is the Pauli Hamiltonian for a particle with spin $s = 1/2$, charge e, and the spin magnetic moment $\mu_0 = e\hbar/2mc$ moving in the electromagnetic field. (Electron, muon, and their antiparticles have this value of μ_0, which follows from the Dirac equation).

2) When the particle moves in a stationary magnetic field (in the absence of an electric field), one can choose, $\varphi = 0$, while the vector potential, $\mathbf{A}(\mathbf{r})$, does not depend on time. Hence, the velocity of the particle and the operator, $\hat{\boldsymbol{\sigma}} \cdot \hat{\mathbf{v}}$, do not depend explicitly on time. Furthermore, taking into account the fact that the Hamiltonian, $\hat{H} = m(\hat{\boldsymbol{\sigma}} \cdot \hat{\mathbf{v}})^2/2$, obviously commutes with $\hat{\boldsymbol{\sigma}} \cdot \hat{\mathbf{v}}$, and recalling the equation of motion given in Eq. (VI.4),[120] we come to the conclusion that $\hat{\boldsymbol{\sigma}} \cdot \hat{\mathbf{v}}$ is conserved (constant of motion). Similarly, for a particle moving in a uniform magnetic field, $\hat{\mathbf{v}}^2$ is also a constant of motion.[121] The fact that both $\hat{\boldsymbol{\sigma}} \cdot \hat{\mathbf{v}}$ and $\hat{\mathbf{v}}^2$ are conserved means that the projection of the spin onto the direction of the velocity is also a constant of motion.

This result has an intuitive interpretation: the variation of the velocity and the spin with time are "synchronized" when the magnetic field is uniform: *i.e.*, both involve a precession with the frequency, ω_H; see Problem 7.15. But if the magnetic moment of a particle is different from $e\hbar/2mc$ (for $s = 1/2$), then the relative angle between the velocity and the spin varies with time. This fact forms the basis of

---

[120] We should note that $\partial\hat{\mathbf{v}}/\partial t = -(e/mc)\partial\mathbf{A}/\partial t = 0$ and hence $d\hat{\boldsymbol{\sigma}}\cdot\hat{\mathbf{v}}/dt = (i/\hbar)[\hat{H}, \hat{\boldsymbol{\sigma}}\cdot\hat{\mathbf{v}}] = 0$.
[121] For a spinless particle, $\hat{v}^2$ is also conserved even in an inhomogeneous magnetic field.

an experimental method (see also Problem 7.15 and a relevant footnote there) for measuring the *anomalous* part of the magnetic moment

$$\mu_0' \equiv \mu_0 - \frac{e\hbar}{2mc},$$

when it is small.

3) The argument used in Problem 7.6 to show that there are no bound states for a charged spinless particle in a magnetic field can be applied here for our Hamiltonian (1) (with $\varphi = 0$).

7.2 Time-dependent quantum mechanics in a magnetic field

Problem 7.11

Show that for the motion of a charged particle with a non-zero spin (and spin magnetic moment) in a time-dependent, uniform magnetic field, $\mathcal{H}(t)$, the spin and spatial dependence of the wavefunction are separated (the electric field can be arbitrary).

Solution

The first two terms in the Pauli Hamiltonian (VII.1) do not depend on the spin, while the last term does not depend on the spatial coordinates. Hence we can find a particular set of solutions to the Schrödinger equation in the following form, $\Psi(\mathbf{r}, t) = \psi(\mathbf{r}, t)\chi(t)$, where the functions ψ and χ satisfy the corresponding Schrödinger equations:

$$i\hbar \frac{\partial \psi}{\partial t} = \left[\frac{1}{2m}\left(\hat{\mathbf{p}} - \frac{e}{c}\mathbf{A}\right)^2 + e\varphi\right]\psi, \quad i\hbar \frac{\partial \chi}{\partial t} = -\frac{\mu}{s}\mathcal{H}(t) \cdot \hat{\mathbf{s}}\chi. \qquad (1)$$

A general wavefunction satisfying the Schrödinger equation is a superposition of $(2s+1)$ (corresponding to the number of spin states) solutions whose spin and spatial dependence can be factorized. To illustrate this point, we write the wavefunction of an arbitrary initial state at $t = 0$ in the form (for simplicity, we choose $s = 1/2$)

$$\Psi(\mathbf{r}, t=0) = \begin{pmatrix} \psi_1(\mathbf{r}, 0) \\ \psi_2(\mathbf{r}, 0) \end{pmatrix} = \psi_1(\mathbf{r}, 0)\begin{pmatrix} 1 \\ 0 \end{pmatrix} + \psi_2(\mathbf{r}, 0)\begin{pmatrix} 0 \\ 1 \end{pmatrix}.$$

Due to the linearity of the Schrödinger equation, we have

$$\Psi(\mathbf{r}, t) = \psi_1(\mathbf{r}, t)\chi_1(t) + \psi_2(\mathbf{r}, t)\chi_2(t),$$

where the functions $\psi_{1,2}$ and $\chi_{1,2}$ satisfy the equations given in (1) and the corresponding initial conditions.

Problem 7.12

For a particle with spin $s = 1/2$ and a magnetic moment, μ, moving in a uniform stationary magnetic field, determine the time evolution of the spin wavefunction and the mean values of the spin components (given that the spin and spatial variables are already separated, as shown in the previous problem).

Solution

Directing the z-axis along the magnetic field, the spin sector of the Hamiltonian takes the following form: $\hat{H} = -\mu \mathcal{H} \hat{\sigma}_z$. The Schrödinger equation, $(i\hbar)\partial \psi(t)/\partial t = \hat{H}\psi(t)$, for the spin wavefunction, $\Psi(t) = \begin{pmatrix} C_1(t) \\ C_2(t) \end{pmatrix}$, is reduced to the following set of equations:

$$\dot{C}_1(t) = i\omega C_1(t), \quad \dot{C}_2(t) = -i\omega C_2(t), \text{ where } \omega = \frac{\mu \mathcal{H}}{\hbar}.$$

Hence

$$C_1(t) = e^{i\omega t} C_1(0), \quad C_2(t) = e^{-i\omega t} C_2(0),$$

where the constants, $C_1(0)$ and $C_2(0)$, are determined by the initial conditions, and to normalize the wavefunction we use the condition $|C_1|^2 + |C_2|^2 = 1$.

The mean values of the spin vector components can be evaluated as follows:

$$\bar{s}(t) = \frac{1}{2} \Psi^*(t) \hat{\sigma} \Psi(t), \quad \bar{s}_x(t) = \bar{s}_x(0) \cos 2\omega t + \bar{s}_y(0) \sin 2\omega t, \tag{1}$$

$$\bar{s}_y(t) = \bar{s}_y(0) \cos 2\omega t - \bar{s}_x(0) \sin 2\omega t, \quad \bar{s}_z(t) = \bar{s}_z(0) = \text{const},$$

i.e., the vector $\bar{s}(t)$ precesses around the magnetic field with the angular velocity 2ω.

Problem 7.13

Generalize the results of the previous problem to the case of a time-dependent magnetic field, whose direction does not change with time; i.e., $\mathcal{H}(t) = \mathcal{H}(t)\mathbf{n}_0$.

Solution

The results of the previous problem could be directly generalized to the present case of magnetic field $\mathcal{H}(t) = (0, 0, \mathcal{H}(t))$. Now the Schrödinger equation becomes

$$i\hbar \dot{C}_1 = -\mu \mathcal{H}(t) C_1, \quad i\hbar \dot{C}_2 = \mu \mathcal{H}(t) C_2,$$

and its solution is

$$C_1(t) = e^{i\xi(t)} C_1(0), \quad C_2(t) = e^{-i\xi(t)} C_2(0), \quad \xi(t) = \frac{\mu}{\hbar} \int_0^t \mathcal{H}(t) dt.$$

The mean values of the spin vector components, $\bar{\mathbf{s}}(t)$, are given by equation (1) of the previous problem, after replacing ωt by $\xi(t)$. Hence the vector $\bar{\mathbf{s}}(t)$ still precesses (generally non-uniformly) around the magnetic field direction.

Problem 7.14

Consider a particle with spin $s = 1/2$ and magnetic moment, μ, in the uniform magnetic field $\mathcal{H}(t)$,

$$\mathcal{H}_x = \mathcal{H}_1 \cos \omega_0 t, \quad \mathcal{H}_y = \mathcal{H}_1 \sin \omega_0 t, \quad \mathcal{H}_z = \mathcal{H}_0,$$

where $H_{0,1}$ and ω_0 are constants.

At the moment $t = 0$, the particle is in a state with $s_z = 1/2$. Determine the probabilities of different values of s_z at the moment, $t > 0$. Pay attention to the resonance of the spin-flip probability in its dependence on the frequency ω_0 when $|\mathcal{H}_1/\mathcal{H}_0| \ll 1$.

Solution

The spin part of the Hamiltonian has the form

$$\hat{H}(t) = -\mu \mathcal{H}(t) \cdot \hat{\boldsymbol{\sigma}} = -\mu \begin{pmatrix} \mathcal{H}_0 & \mathcal{H}_1 \exp\{-i\omega_0 t\} \\ \mathcal{H}_1 \exp\{i\omega_0 t\} & -\mathcal{H}_0 \end{pmatrix}.$$

The Schrödinger equation for the spin wavefunction $\Psi(t) = \begin{pmatrix} a(t) \\ b(t) \end{pmatrix}$ is reduced to the following set of equations:

$$\begin{aligned} i\hbar \dot{a} &= -\mu \mathcal{H}_0 a - \mu \mathcal{H}_1 \exp\{-i\omega_0 t\} b, \\ i\hbar \dot{b} &= -\mu \mathcal{H}_1 \exp\{i\omega_0 t\} a + \mu \mathcal{H}_0 b. \end{aligned} \quad (1)$$

By using the substitutions

$$a(t) = e^{-i\omega_0 t/2} \tilde{a}(t), \quad b(t) = e^{i\omega_0 t/2} \tilde{b}(t),$$

the time-dependence of the coefficients of the differential equations is eliminated, allowing us to find its solution (compare with Problem 6.9):

$$\tilde{a}(t) = C_1 e^{i\omega t} + C_2 e^{-i\omega t},$$
$$\tilde{b}(t) = (\omega - \gamma_1)\gamma_2^{-1} C_1 e^{i\omega t} - (\omega + \gamma_1)\gamma_2^{-1} C_2 e^{-i\omega t},$$

where we have introduced the following notations:

$$\gamma_1 = \frac{\mu \mathcal{H}_0}{\hbar} + \frac{\omega_0}{2}, \quad \gamma_2 = \frac{\mu \mathcal{H}_1}{\hbar}, \quad \omega = \sqrt{\gamma_1^2 + \gamma_2^2}.$$

Using the initial conditions, $a(0) = 1$ and $b(0) = 0$, we obtain

$$\Psi(t) = \frac{1}{2\omega} \begin{pmatrix} \left[(\omega + \gamma_1)e^{i\omega t} + (\omega - \gamma_1)e^{-i\omega t}\right]e^{-i\omega_0 t/2} \\ 2i\gamma_2 \sin \omega t \cdot e^{i\omega_0 t/2} \end{pmatrix}. \quad (2)$$

The spin-flip probability (*i.e.*, the probability to find the particle in the spin state with $s_z = -1/2$ at the moment of time, t) is

$$W(s_z = -1/2, t) = \left(\frac{\gamma_2}{\omega}\right)^2 \sin^2 \omega t \equiv g \sin^2 \omega t, \quad (3)$$

where

$$g = \left(\frac{\gamma_2}{\omega}\right)^2 = \frac{\mathcal{H}_1^2}{\mathcal{H}_1^2 + (\mathcal{H}_0 + \hbar\omega_0/2\mu)^2}.$$

If $\mathcal{H}_1 \ll \mathcal{H}_0$, the spin-flip probability (as well as the value of the parameter, g) is small except in the narrow frequency-range $\omega \sim \omega_{0\ \text{res}} = -2\mu\mathcal{H}_0/\hbar$, with the width of the order of $\Delta\omega_0 \sim \mu\mathcal{H}_1/\hbar$. The resonance in the frequency-dependence of the spin-flip probability forms the basis of experimental techniques used to measure magnetic moments.

Problem 7.15

For a charged particle with spin $s = 1/2$ and spin magnetic moment μ, moving in a uniform stationary magnetic field, find the operators of position, velocity, momentum, and spin in the Heisenberg picture. Choose the vector potential in the form,[122] $\mathbf{A} = (0, \mathcal{H}_0 x, 0)$. Solve the problem using a method presented in Problem 6.20.

Compare the time-dependencies of the mean values of the velocity, $\overline{\mathbf{v}(t)}$, and spin, $\overline{\mathbf{s}(t)}$; see also Problem 7.10.

Solution

The problem will be solved similarly to Problem 6.20. We give the expressions for Heisenberg operators of the position, momentum, and particle spin components:

$$\hat{x}(t) = \hat{x} \cos \omega_0 t + \frac{\hat{p}_x}{m\omega_0} \sin \omega_0 t + \frac{\hat{p}_y}{m\omega_0}(1 - \cos \omega_0 t),$$
$$\hat{y}(t) = \hat{y} - \hat{x} \sin \omega_0 t + \frac{\hat{p}_x}{m\omega_0}(\cos \omega_0 t - 1) + \frac{\hat{p}_y}{m\omega_0} \sin \omega_0 t,$$
$$\hat{z}(t) = \hat{z} + \frac{\hat{p}_z}{m}t,$$

$$\begin{aligned}
\hat{p}_x(t) &= \hat{p}_x \cos \omega_0 t + \hat{p}_y \sin \omega_0 t - m\omega_0 \hat{x} \sin \omega_0 t, & \hat{s}_x(t) &= \hat{s}_x \cos \omega t + \hat{s}_y \sin \omega t, \\
\hat{p}_y(t) &= \hat{p}_y, & \hat{s}_y(t) &= \hat{s}_y \cos \omega t - \hat{s}_x \sin \omega t, \\
\hat{p}_z(t) &= \hat{p}_z, & \hat{s}_z(t) &= \hat{s}_z,
\end{aligned}$$

---

[122] Note that this classical expression must be modified in the quantum-mechanical Heisenberg picture: $\hat{\mathbf{A}} = (0, \mathcal{H}_0 \hat{x}(t), 0)$.

$$\omega_H = \frac{e\mathcal{H}_0}{mc}, \quad \omega = \frac{2\mu\mathcal{H}_0}{\hbar},$$

where $\hat{x}_i$, $\hat{p}_i$, $\hat{s}_i$ are the corresponding operators in the Schrödinger picture, and e, m are the charge and mass of the particle, respectively.

The velocity $\hat{\mathbf{v}}(t) = d\hat{\mathbf{r}}(t)/dt$ could be obtained by directly differentiating $\hat{\mathbf{r}}(t)$. The time evolutions of the mean values of $\overline{\mathbf{v}}(t)$ and $\overline{\mathbf{s}}(t)$ describe the precession of these vectors with angular velocities, ω_0 and ω, respectively. In the case $\omega_0 = \omega$, the angle between them does not change with time (this case corresponds to $\mu = e\hbar/2mc$; see Problem 7.10). If $\omega_0 \neq \omega$, then the angle between the two vectors $\overline{\mathbf{v}}(t)$ and $\overline{\mathbf{s}}(t)$ in the azimuthal plane (perpendicular to the magnetic field) varies with time: $\Delta\varphi(t) = (g-2)\omega_0 t/2$, and $g = 2\mu/\mu_0$.

This fact forms the basis of the experimental measurement of the g-factor,[123] when it differs slightly from the value, $g_0 = 2$, given by the Dirac equation. Experiments show that small deviations from the value $g_D = 2$ occur in the g-factors of the electrons and muons, which is in agreement with the prediction of quantum electrodynamics.

Problem 7.16

Under the conditions of Problem 7.12, find the time-dependent Green function, $G_{\alpha\beta}(t, t')$, for the spin degrees of freedom of the particle ($\alpha, \beta = 1$ and 2 are the spin variables).

Solution

The Green function, $G_{\alpha\beta}(t, t')$, satisfies the Schrödinger equation of the spin Hamiltonian, $\hat{H} = -\mu\mathcal{H}\hat{\sigma}_z$ (the z-axis is directed along the magnetic field). At $t = t'$ it is given by $G_{\alpha\beta} = \delta_{\alpha\beta}$. Its explicit form is ($\omega = \mu\mathcal{H}/\hbar$):

$$\hat{G}(t, t') = \begin{pmatrix} \exp\{i\omega(t-t')\} & 0 \\ 0 & \exp\{-i\omega(t-t')\} \end{pmatrix}.$$

Problem 7.17

The same as in the previous problem, but for the conditions of Problem 7.13.

Result

The Green function is obtained from the expression of the previous problem by changing $\omega(t-t') \to \xi(t,t') = (\mu/\hbar)\int_{t'}^{t} \mathcal{H}(t)dt$.

---

[123] The angle, $\Delta\varphi(t)$, is an "accumulative effect". After a sufficiently long measurement time, $\Delta\varphi$ becomes of the order of 1, which allows determination of $(g-2)$, even if this difference is very small.

Problem 7.18

Find the time-dependent Green function, $G_{\alpha\beta}(\mathbf{r},t;\mathbf{r}',t')$, of a neutral particle with spin $s=1/2$ and magnetic moment μ in a uniform stationary magnetic field.

Solution

In accordance with the results of Problem 7.11 about the separation of the spatial and spin variables for a particle moving in a uniform magnetic field, the required Green function is given by the product:

$$G_{\alpha\beta}(\mathbf{r},t;\mathbf{r}',t') = G(\mathbf{r},t;\mathbf{r}',t') \cdot G_{\alpha\beta}(t,t').$$

Here, $G(\mathbf{r},t;\mathbf{r}',t')$ is the Green function of a free spinless particle, – see Eq. (VI.7) – and $G_{\alpha\beta}(t,t')$ is the spin Green function from Problem 7.16.

Problem 7.19

Generalize the result of the previous problem to the case of a uniform but non-stationary magnetic field, whose direction does not vary with time; *i.e.*, $\mathcal{H}(t) = \mathcal{H}(t)\mathbf{n}_0$.

Result

The Green function has the same form as in the previous problem, but now the spin Green function is given by the result from Problem 7.17.

7.3 Magnetic field of the orbital currents and spin magnetic moment

Problem 7.20

Find the mean values of the current density for a charged spinless particle in a uniform magnetic field in the stationary state, ψ_{nmp_z} (see Problem 7.1 b).

Solution

Using the explicit form of the wavefunctions, ψ_{nmp_z} (see Eqs. (4)–(6) from Problem 7.1), with the vector potential chosen as $\mathbf{A} = \frac{1}{2}\mathcal{H}_0 \times \mathbf{r}$, we obtain the current density of the charged spinless particle in the magnetic field from Eq. (VII.5):

$$j_\rho = 0, \quad j_z = \frac{ep_z}{\mu}|\psi_{nmp_z}|^2, \quad j_\varphi = \left(\frac{e\hbar m}{\mu\rho} - \frac{e^2\mathcal{H}_0\rho}{2\mu c}\right)|\psi_{nmp_z}|^2,$$

where cylindrical coordinates are used. We should emphasize that $|\psi|^2$ for the states considered depends only on the radius, ρ.

Problem 7.21

The same as in previous problem, but for a charged particle with spin $s = 1/2$ and magnetic moment μ_0 in the stationary state $\psi_{nmp_z s_z}$ from Problem 7.9.

Solution

The wavefunctions of the states under consideration have the form

$$\psi \equiv \psi_{nmp_z s_z} = \psi_{nmp_z} \chi_{s_z},$$

where χ_{s_z} are the spin eigenfunctions of the operator $\hat{s}_z$ (see also Problem 7.9 and the previous problem). In this case,

$$\mu_0 \psi^* \hat{\boldsymbol{\sigma}} \psi = (0,\ 0,\ 2\mu_0 s_z |\psi_{nmp_z}|^2).$$

Noting that $|\psi_{nmp_z}|^2$ depends only on ρ and according to Eq. (VII.6), we find the components (in cylindrical coordinates) of the current density originating from the spin magnetic moment:

$$j_{\text{sp},\rho} = j_{\text{sp},z} = 0, \quad j_{\text{sp},\varphi} = -2\mu_0 c s_z \frac{\partial}{\partial \rho} |\psi_{nmp_z}|^2. \tag{1}$$

The total current density is given by the sum of the contribution (1) and the corresponding components of the orbital current calculated in the previous problem.

Problem 7.22

Determine the mean value of the magnetic field $\mathcal{H}(0)$ at the origin created by a charged spinless particle moving in the Coulomb field of a nucleus: $U(r) = -Ze^2/r$. Consider the 1s- and 2p-states.

Solution

We use the following equation well known from classical electrodynamics:

$$\mathcal{H}(\mathbf{R}) = \frac{1}{c} \int \frac{\mathbf{j}(\mathbf{r}) \times (\mathbf{R} - \mathbf{r})}{|\mathbf{R} - \mathbf{r}|^3} dV. \tag{1}$$

In the absence of any external magnetic field, the current density is given by Eq. (VII.5) with $\mathbf{A} = 0$. For the stationary s-state, the wavefunction is real, so we have[124] $\mathbf{j} = 0$ and $\mathcal{H} = 0$.

The wavefunction of a $2p$ state has the form:

$$\psi_{2p} = (32\pi a^5)^{-1/2} (\boldsymbol{\varepsilon} \cdot \mathbf{r}) e^{-r/2a}, \quad a = \frac{\hbar^2}{Ze^2 \mu}, \quad |\varepsilon|^2 = 1$$

---

[124] The physical reason for the vanishing magnetic field in the s-states is related to the spherical symmetry of the state.

(For the angular dependence, see Problem 3.42.) Using Eqs. (1) and (VII.5) for $A = 0$, we obtain (we set the charge of the particle to be $-e$)

$$\mathcal{H}(0) = -\frac{ie\hbar}{64\pi\mu a^5 c} \int \frac{e^{-r/a}}{r^3} [\mathbf{r} \times (\boldsymbol{\varepsilon}^*(\boldsymbol{\varepsilon} \cdot \mathbf{r}) - \boldsymbol{\varepsilon}(\boldsymbol{\varepsilon}^* \cdot \mathbf{r}))]dV. \tag{2}$$

(Note that to calculate the current we need to apply the operator ∇ only to the terms $\boldsymbol{\varepsilon} \cdot \mathbf{r}$ and $\boldsymbol{\varepsilon}^* \cdot \mathbf{r}$ in the wavefunctions, because $[\mathbf{r} \times \nabla f(r)] = 0$ and $\nabla(\boldsymbol{\varepsilon} \cdot \mathbf{r}) = \boldsymbol{\varepsilon}$. Introducing the vector, $\mathbf{b} = [\boldsymbol{\varepsilon}^* \times \boldsymbol{\varepsilon}]$, we see that the integral in (2) takes the form

$$\int \frac{1}{r^3} e^{-r/a} \{\mathbf{r}(\mathbf{r} \cdot \mathbf{b}) - \mathbf{b}r^2\} dV \equiv \mathbf{I}. \tag{3}$$

In order to calculate this integral, let us first consider the following integral:

$$\int \frac{1}{r^3} e^{-r/a} x_i x_k dV = C\delta_{ik}. \tag{4}$$

If we perform convolution over the indices i and k, we obtain

$$3C = \int \frac{1}{r} e^{-r/a} dV = 4\pi a^2. \tag{5}$$

From Eqs. (3)–(5) it follows that $\mathbf{I} = -8\pi a^2 \mathbf{b}/3$, and as a result we find the magnetic field "on the nuclei" to be

$$\mathcal{H}(0) = -\frac{ie\hbar}{24\mu a^3 c} [\boldsymbol{\varepsilon}^* \times \boldsymbol{\varepsilon}]. \tag{6}$$

Hence, using the explicit form of $\varepsilon(m)$ – see Problem 3.42 – we obtain

$$\mathcal{H}_{m=0}(0) = 0, \quad \mathcal{H}_{m=\pm 1}(0) = \left(0, 0, \mp \frac{e\hbar}{24\mu a^3 c}\right). \tag{7}$$

In conclusion, we should note that if a nucleus has a non-zero magnetic moment, then its interaction with the magnetic field, $\mathcal{H}(0)$, leads to a *hyperfine structure splitting* of the atomic levels; see Problem 11.2.

Problem 7.23

Find the mean magnetic field created by an electron in the ground state in the Coulomb field of a nucleus with the charge Ze.

Solution

The wavefunction of the state of the electron is $\psi = (\pi a^3)^{-1/2} e^{-r/a} \chi$, where χ is its spin function and $a = \hbar^2/Ze^2\mu$. The orbital current density is zero, so the current is determined solely by the spin magnetic moment. According to Eq. (VII.6) we have

294   *Exploring Quantum Mechanics*

$$\mathbf{j} = \mathbf{j}_{\text{sp}} = -\mu_0 c[\overline{\boldsymbol{\sigma}} \times \nabla \rho], \quad \overline{\boldsymbol{\sigma}} = \chi^* \boldsymbol{\sigma} \chi, \quad \rho(r) = \frac{1}{\pi a^3} e^{-2r/a}.$$

Using the known expression for the vector potential

$$\mathbf{A}(\mathbf{R}) = \frac{1}{c} \int \frac{\mathbf{j}(\mathbf{r})}{|\mathbf{R} - \mathbf{r}|} dV = -\mu_0 \left[\overline{\boldsymbol{\sigma}} \times \int \frac{\nabla \rho(r)}{|\mathbf{R} - \mathbf{r}|} dV \right], \quad (1)$$

we can make here the following transformations:

$$\int \frac{\nabla \rho(r)}{|\mathbf{R} - \mathbf{r}|} dV = \int \left\{ \nabla \frac{\rho(r)}{|\mathbf{R} - \mathbf{r}|} - \rho(r) \nabla \frac{1}{|\mathbf{R} - \mathbf{r}|} \right\} dV = \nabla_R \int \frac{\rho(r)}{|\mathbf{R} - \mathbf{r}|} dV. \quad (2)$$

Here we have applied Gauss's theorem, and the relation $\nabla_r g(\mathbf{R} - \mathbf{r}) \equiv -\nabla_R g(\mathbf{R} - \mathbf{r})$. The integral in Eq. (2) yields[125]

$$\int \frac{\rho(r)}{|\mathbf{R} - \mathbf{r}|} dV = \frac{1}{R} - \left(\frac{1}{R} + \frac{1}{a} \right) e^{-2R/a} \equiv f(R). \quad (3)$$

Hence we obtain $\mathbf{A}(\mathbf{r}) = -\mu_0 [\overline{\boldsymbol{\sigma}} \times \nabla f(r)]$. Let us also give the asymptotic expressions for the magnetic field, $\boldsymbol{H} = \nabla \times \mathbf{A}(\mathbf{r})$:

$$\mathcal{H}(0) = \frac{8}{3a^3} \boldsymbol{\mu}, \quad \mathcal{H}(\mathbf{r}) \approx \frac{3(\boldsymbol{\mu} \cdot \mathbf{r})\mathbf{r} - \boldsymbol{\mu} r^2}{r^5}; \quad \text{for } r \to \infty; \boldsymbol{\mu} = \mu_0 \overline{\boldsymbol{\sigma}}.$$

(At large distances this corresponds to the field of a magnetic dipole.)

Problem 7.24

Find the mean magnetic field produced at the origin by a particle with the spin $s = 1/2$, and magnetic moment μ_0, in the stationary s-state of an arbitrary central potential.

Solution

From Eqs. (1) and (2) of the previous problem, we find

$$\mathbf{A}(\mathbf{R}) = -\mu_0 [\overline{\boldsymbol{\sigma}} \times \nabla_R f(R)], \quad f(R) = \int \frac{|\psi(r)|^2}{|\mathbf{R} - \mathbf{r}|} dV, \quad (1)$$

where $\psi(r)$ is the wavefunction of the s-state. The magnetic field has the form ($\boldsymbol{\mu} = \mu_0 \overline{\boldsymbol{\sigma}}$):

$$\mathcal{H}(\mathbf{R}) = \nabla \times \mathbf{A}(\mathbf{R}) = -\boldsymbol{\mu} \triangle f(R) + (\boldsymbol{\mu} \cdot \nabla) \nabla f(R).$$

---

[125]   This integral describes a contribution to the effective potential created by the electron "cloud". Its value can be found in Problem 4.6.

Therefore, its components are found to be

$$\mathcal{H}_i(\mathbf{R}) = -\mu_k \left(\delta_{ik} \triangle - \frac{\partial^2}{\partial X_i \partial X_k} \right) f(R). \tag{2}$$

Now consider the expression

$$\left. \frac{\partial^2}{\partial X_i \partial X_k} f(R) \right|_{R=0} = C \delta_{ik}. \tag{3}$$

Taking the convolution over i and k, we obtain $3C = \triangle f(R)|_{R=0}$, and using Eq. (1) for $f(R)$ and the equation $\triangle |\mathbf{R} - \mathbf{r}|^{-1} = -4\pi \delta(\mathbf{R} - \mathbf{r})$, we find: $3C = -4\pi |\psi(0)|^2$. Hence, from Eqs. (2) and (3), it follows that

$$\boldsymbol{\mathcal{H}}(0) = \frac{8\pi}{3} |\psi(0)|^2 \boldsymbol{\mu}, \tag{4}$$

which can be compared with the results of the previous problem and Problem 7.22.

8

Perturbation theory; Variational method; Sudden and adiabatic theory

---

The methods of the perturbation theory are based on introducing the system Hamiltonian in the form of $\hat{H} = \hat{H}_0 + \hat{V}$, where the *perturbation* $\hat{V}$ is a small correction. It is assumed that the solutions of the Schrödinger equation for the unperturbed Hamiltonian $\hat{H}_0$ are known and that the specific form of $\hat{V}$ is also known. The perturbation theory methods enable us to consider the effects of perturbation by iterative approximations.

1) In the case when $\hat{H}_0$ and $\hat{V}$ and so the total Hamiltonian $\hat{H}$ do not depend on time, its egenvalues and eigenfunctions of the discrete spectrum can be written as an expansion in powers of the perturbation. Succinctly:

$$E_n = E_n^{(0)} + E_n^{(1)} + E_n^{(2)} + \ldots;$$

$$\psi_n = \sum_m c_{nm} \psi_m^{(0)}, \ c_{nm} = c_{nm}^{(0)} + c_{nm}^{(1)} + \ldots,$$

where $E_n^{(0)}$ and $\psi_n^{(0)}$ are the spectrum and eigenfunctions of the unperturbed Hamiltonian. Here if the unperturbed level $E_n^{(0)}$ is not degenerate, then

$$E_n^{(1)} = \langle \psi_n^{(0)} | \hat{V} | \psi_n^{(0)} \rangle \equiv \langle n | \hat{V} | n \rangle \quad E_n^{(2)} = \sum_m{}' \frac{|\langle m | \hat{V} | n \rangle|^2}{E_n^{(0)} - E_m^{(0)}} \tag{VIII.1}$$

(the sum has no term with $m = n$), while for the eigenfunctions

$$c_{nk}^{(0)} = \delta_{nk}, \ c_{nn}^{(1)} = 0, \ c_{nk}^{(1)} = \frac{\langle k | \hat{V} | n \rangle}{E_n^{(0)} - E_k^{(0)}} \quad \text{for} \ k \neq n. \tag{VIII.2}$$

The condition for the applicability of these results is given by ($n \neq k$):

$$|\langle k | \hat{V} | n \rangle| \ll |E_n^{(0)} - E_k^{(0)}|. \tag{VIII.3}$$

If an unperturbed level $E_n^{(0)}$ is s-fold degenerate and corresponds to mutually orthogonal eigenfunctions $\tilde{\psi}_{n\alpha}^{(0)}$ with $\alpha = 1, 2, \ldots, s$, then the *correct* eigenfunctions in zeroth approximation are $\psi_n = \sum_\alpha c_\alpha^{(0)} \psi_{n,\alpha}^{(0)}$, and corresponding energy level shifts, $E_n^{(1)}$, are determined by solving the system of equations:

$$\sum_\beta (\langle n\alpha | \hat{V} | n\beta \rangle - E_n^{(1)} \delta_{\alpha\beta}) c_\beta^{(0)} = 0. \qquad \text{(VIII.4)}$$

A non-trivial solution to this system exists only if the following determinant is equal to zero:

$$|\langle n\alpha | \hat{V} | n\beta \rangle - E_n^{(1)} \delta_{\alpha\beta}| = 0. \qquad \text{(VIII.5)}$$

The corresponding roots $E_n^{(1)}$ (their number is s) determine the energy level splitting,[126] and their substitution in (VIII.4) determines the wavefunctions of the corresponding sublevels (in the zeroth approximation).

2) In the case of a time-dependent perturbation, $\hat{V}(t)$, for the wavefunctions

$$\psi(t) = \sum_k a_k(t) e^{-\frac{i}{\hbar} E_k^{(0)} t} \psi_k^{(0)}(q) \qquad \text{(VIII.6)}$$

from the Schrödinger equation $i\hbar \frac{\partial}{\partial t} \psi = (\hat{H}_0 + \hat{V}(t)) \psi$, it follows that[127]

$$i\hbar \frac{da_m(t)}{dt} = \sum_k V_{mk}(t) e^{i\omega_{mk} t} a_k(t), \qquad \text{(VIII.7)}$$

where $V_{mk}(t) = \int \psi_m^{(0)*}(q) \hat{V}(t) \psi_k^{(0)}(q) d\tau_q$, $\omega_{mk} = \frac{1}{\hbar}(E_m^{(0)} - E_k^{(0)})$.

The solution of (VIII.7) by successive iterations, $a_k(t) = a_k^{(0)}(t) + a_k^{(1)}(t) + \ldots$, first gives $a_k^{(0)}(t) = const$. Then if we take $\hat{V}(t) \to 0$ for $t \to -\infty$ and assume that before the perturbation was turned on, the system was in the nth discrete state, $\psi_n^{(0)}$, we have $a_k(t - \infty) \to \delta_{nk}$, and we choose $a_k^{(0)} \equiv a_{kn}^{(0)} = \delta_{nk}$ (instead of $a_k(t)$ we now write $a_{kn}(t)$ to emphasize the particular setup). For the first-order correction from (VIII.7), using $a_{kn}^{(1)}(t = -\infty) = 0$, we obtain

$$a_{kn}^{(1)}(t) = -\frac{i}{\hbar} \int_{-\infty}^{t} V_{kn}(t) e^{i\omega_{kn} t} dt. \qquad \text{(VIII.8)}$$

---

[126] If all the roots $E_n^{(1)}$ are different then the degeneracy is removed completely. Otherwise, if degenerate roots exit, the level degeneracy is lifted only partially and an ambiguity exists in determining the corresponding eigenfunctions of zeroth order.

[127] We should note that the time-dependence of the matrix elements $V_{k,n}(t)$ is determined only by the operator $\hat{V}(t)$. The factors $\exp\{-iE_n^{(0)} t/\hbar\}$ have already been separated out.

If the perturbation $\hat{V}(t)$ vanishes as $t \to +\infty$, then to first order in the perturbation, $a_{kn}^{(1)}(t = +\infty)$ determines the system transition probability from its initial nth state into the final kth state ($k \neq n$):

$$W^{(1)}(n \to k) = \frac{1}{\hbar^2} \left| \int_{-\infty}^{\infty} V_{kn}(t) e^{i\omega_{kn}t} dt \right|^2. \quad \text{(VIII.9)}$$

3) The transition probability (per unit time) from the initial i-th state[128] into the final close f-state of the continuous spectrum under the action of a time-independent perturbation, $\hat{V}$ is

$$dw(i \to f) = \frac{2\pi}{\hbar} |V_{fi}|^2 \delta(E_i - E_f) d\nu_f, \quad \text{(VIII.10)}$$

where $d\nu_f$ characterizes the number of final states. Integration over their energies gives the *Fermi's golden rule* for the transition probability:

$$w(i \to f) = \frac{2\pi}{\hbar} |V_{fi}|^2 \rho_f(E_i), \quad \text{(VIII.11)}$$

where $\rho_f(E_i)$ is the final density of states.

An important generalization of (VIII.10) appears for a periodic-in-time perturbation of the form

$$\hat{V}(t) = \hat{F} e^{-i\omega t} + \hat{F}^+ e^{i\omega t}, \quad \text{(VIII.12)}$$

where $\hat{F}$ is a time-independent operator. In this case, the transition probability is given by

$$dw(i \to f) = \frac{2\pi}{\hbar} |F_{fi}|^2 \delta(E_i - E_f - \hbar\omega) d\nu_f. \quad \text{(VIII.13)}$$

8.1 Stationary perturbation theory (discrete spectrum)

Problem 8.1

Consider a particle in an infinitely-deep one-dimensional potential well in the presence of a rather arbitrary perturbing potential, $V(x)$. Prove that the first-order correction to the energy, $E_n^{(1)}$, of the highly excited levels in the well (with $n \gg 1$) is n-independent.

Solution

The eigenfunctions of the unperturbed Hamiltonian have the form $\psi_n^{(0)}(x) = \sqrt{2/a}\sin(\pi(n+1)x/a)$ (for $0 \leq x \leq a$). In the matrix element $<n|V(x)|n>$, we can

---

[128] It can belong either to the discrete or continuous spectrum.

replace the fast oscillating $\sin^2$-term by its mean value of $1/2$ for $n \gg 1$.[129] We obtain:

$$E_n^{(1)} = \langle n|V(\hat{x})|n\rangle \approx \frac{1}{a}\int_0^a V(x)dx.$$

Problem 8.2

For a charged linear oscillator in a uniform electric field directed along the oscillation axis, find a shift of the energy levels in the first two orders of perturbation theory and determine the *polarizability* of different states. Compare with the exact results.

Solution

The perturbation is $V(x) = -e\mathcal{E}x$, and it is obvious that $E_n^{(1)} = 0$. In order to calculate the corrections to second order of perturbation theory, according to (VIII.1) we should use the coordinate matrix elements. See Eq. (II.3). Taking into account the form of the spectrum $E_n^{(0)}$ of the unperturbed oscillator, we obtain

$$E_n \approx E_n^{(0)} + E_n^{(1)} + E_n^{(2)} = \hbar\omega\left(n + \frac{1}{2}\right) - \frac{1}{2}\frac{e^2}{m\omega^2}\mathcal{E}^2, \quad (1)$$

so that the polarizability for all the oscillator states is the same and is equal to $\beta_0 = e^2/m\omega^2$. This result coincides with the exact one (see Problem 2.2). The corrections of the third and higher orders of perturbation theory are equal to zero.

Problem 8.3

The same as in previous problem, but for the ground state of a charged particle moving within a one-dimensional infinitely deep potential well.

Solution

The perturbation is $V(x) = -e\mathcal{E}x$. Using the form of the unperturbed wavefunction (see 8.1) and the symmetry of $|\psi_n^{(0)}(x)|^2$ with respect to the center of the well, we find that $<n|x|n> = a/2$. So in the first order of perturbation theory for all the levels, $E_n^{(1)} = -ea\mathcal{E}/2$. Then we calculate the coordinate matrix elements (for $n \neq 0$):

$$x_{n0} = \frac{2}{a}\int_0^a x\sin\frac{\pi x}{a}\cdot\sin\frac{\pi(n+1)x}{a}dx = \frac{4\left[(-1)^n - 1\right](n+1)a}{\pi^2 n^2 (n+2)^2}.$$

---

[129] Quantum states with $n \gg 1$ are *quasi-classical*. Perturbation theory for such states is considered in Problems 9.10–9.12.

This differs from zero for odd values of n only. Using the results of Problem 2.1 and Eq. (VIII.1), we find the second-order correction $E_0^{(2)} \equiv -\beta_0 \mathcal{E}^2/2$ that gives the polarizability of the ground state:

$$\beta_0 = \frac{1024}{\pi^6} \frac{e^2 m a^4}{\hbar^2} \sum_{k=0}^{\infty} \frac{(k+1)^2}{(2k+1)^5 (2k+3)^5}. \quad (1)$$

This series converges rapidly, and its value is determined mainly by the first term, so $\beta_0 \approx 4.39 \cdot 10^{-3} e^2 m a^4 / \hbar^2$.

Let us comment on the small numerical value of the coefficient. Dimensional estimate of polarizability has the form $\beta \sim e^2/m\omega^2$, where ω is the characteristic frequency (compare with the polarizability of an oscillator from Problem 8.2). This frequency is determined from the relation $\hbar\omega = \Delta E$, where ΔE is the distance to the neighboring level (of opposite parity). In the problem considered, see that

$$\omega = \frac{E_1^{(0)} - E_0^{(0)}}{\hbar} = \frac{3\pi^2 \hbar}{2ma^2}.$$

And, in accordance with (1), we obtain $\beta_0 \approx 0.96 e^2/m\omega^2$.

Problem 8.4

For a isotropic planar oscillator under the action of the perturbation $V = \alpha xy$, determine the shift of the ground-state energy in the first non-vanishing order of perturbation theory. Indicate the conditions for applicability of the result and compare it with the exact one. See Problem 2.49.

Solution

The eigenfunctions and spectrum of the unperturbed Hamiltonian are considered in Problem 2.48, and have the form:

$$\psi_{n_1 n_2}^{(0)} = \psi_{n_1}^{(osc)}(x) \cdot \psi_{n_2}^{(osc)}(y), \quad E_{n_1 n_2}^{(0)} \equiv E_N^{(0)} = \hbar\omega(N+1), \quad (1)$$

$$N = n_1 + n_2 = 0, 1, 2, \ldots.$$

In the first order of perturbation theory, the ground state does not shift: $E_0^{(1)} = 0$. When using (VIII.1) to calculate the second-order corrections, we should take m to be a set of two numbers (n_1, n_2) that define the unperturbed eigenfunctions (1). Using the coordinate matrix elements (see Eq. (II.3)), we find that $<n_1 n_2|V|00>$ differs from zero only for $n_1 = n_2 = 1$, and that $<11|V|00> = \alpha\hbar/2m\omega$. We obtain $E_0^{(2)} = -\alpha^2\hbar/8m^2\omega^2$. The condition for applicability, Eq. (VIII.3), in this problem takes the form of $|\alpha| \ll m\omega^2 = k$.

According to Problem 2.49, the exact value of the ground-state energy is equal to $E_0 = \hbar\omega(\sqrt{1+\alpha/k} + \sqrt{1-\alpha/k})/2$. Its expansion in powers of α/k corresponds to the perturbation theory series. In the case $|\alpha/k| \ll 1$, this series converges very rapidly. On the other hand, in the case of $|\alpha/k| \geq 1$ there is no energy quantization, the series diverges, and perturbation theory breaks down.

Problem 8.5

Under conditions of the previous problem, find the splitting: a) of the first excited and b) of the second excited levels of the oscillator. Indicate the correct eigenfunctions of the zeroth approximation.

Solution

a) The unperturbed oscillator level with $N = 1$ is two-fold degenerate. We will denote the unperturbed eigenfunctions that correspond to it $\psi_{n_1 n_2}^{(0)}$ (see the previous problem) as $\psi_1^{(0)} \equiv \psi_{10}^{(0)}$ and $\psi_2^{(0)} \equiv \psi_{01}^{(0)}$. The matrix elements of the perturbation with respect to these eigenfunctions (using Eq. (II.3)) are $V_{11} = V_{22} = 0$, $V_{12} = V_{21} = \alpha\hbar/2m\omega$. The secular equation (VIII.5) and its solution take the form:

$$\begin{vmatrix} -E_1^{(1)} & \frac{\alpha\hbar}{2m\omega} \\ \frac{\alpha\hbar}{2m\omega} & -E_1^{(1)} \end{vmatrix} = 0, \quad E_{1,1(2)}^{(1)} = \mp \frac{\alpha\hbar}{2m\omega}. \tag{1}$$

The degeneracy is lifted. The *correct* eigenfunctions in the zeroth approximation have the form: $\psi_{1,1(2)}^{(0)} = \frac{1}{\sqrt{2}} \left(\psi_1^{(0)} \mp \psi_2^{(0)} \right)$.

b) The level with $N = 2$ is three-fold degenerate. Its eigenfunctions are $\psi_1^{(0)} \equiv \psi_{20}^{(0)}$, $\psi_2^{(0)} \equiv \psi_{11}^{(0)}$, $\psi_3^{(0)} \equiv \psi_{02}^{(0)}$. The non-vanishing matrix elements of the perturbation are $V_{12} = V_{21} = V_{23} = V_{32} = \alpha\hbar/\sqrt{2}m\omega$. The solution of the secular equation gives the following values for the first-order perturbative corrections:

$$E_{2,1}^{(1)} = -\frac{\alpha\hbar}{m\omega}, \quad E_{2,2}^{(1)} = 0, \quad E_{2,3}^{(1)} = \frac{\alpha\hbar}{m\omega}, \tag{2}$$

so that the level splits into three sub-levels and therefore its degeneracy is completely lifted. The correct functions of the zero approximation that correspond to the splitted levels (2) have the form:

$$\psi_{2,1(3)}^{(0)} = \frac{1}{2} \left(\psi_1^{(0)} \mp \sqrt{2}\psi_2^{(0)} + \psi_3^{(0)} \right), \quad \psi_{2,2}^{(0)} = \frac{1}{\sqrt{2}} \left(\psi_1^{(0)} - \psi_3^{(0)} \right).$$

Compare this result, obtained with the help of perturbation theory, with the exact one. See Problem 2.49.

Problem 8.6

A two-level system (the levels are non-degenerate, with the energies ε_1 and ε_2) is under the action of a perturbation characterized by matrix elements V_{11}, V_{22}, $V_{12} = V_{21}^*$ between initial unperturbed states 1 and 2. Obtain the energy level shifts in the first two orders of perturbation theory. Indicate the conditions for applicability of the results obtained, and compare to exact results.

Solution

1) Energy level shifts in the first two orders of perturbation theory are equal to

$$E_1^{(1)} = V_{11}, \quad E_2^{(1)} = V_{22}, \quad E_1^{(2)} = -E_2^{(2)} = \frac{|V_{12}|^2}{\varepsilon_1 - \varepsilon_2}. \tag{1}$$

The conditions for perturbation theory applicability have the form $|V_{11}|, |V_{22}|, |V_{12}| \ll \varepsilon_2 - \varepsilon_1$.

2) It is useful to compare these results with the exact solution of the problem, which can be obtained by diagonalization of the operator:

$$\hat{H} = \hat{H}_0 + \hat{V} = \begin{pmatrix} \varepsilon_1 + V_{11} & V_{12} \\ V_{21} & \varepsilon_2 + V_{22} \end{pmatrix}.$$

This matrix is the Hamiltonian of the perturbed two-level system in the *energy representation* of the unperturbed Hamiltonian. Its eigenvalues are equal to[130]

$$E_{1(2)} = \frac{1}{2}\left(\varepsilon_1 + \varepsilon_2 + V_{11} + V_{22} \mp \sqrt{(\varepsilon_1 - \varepsilon_2 + V_{11} - V_{22})^2 + 4|V_{12}|^2}\right). \tag{2}$$

In the case of the small values of matrix elements V_{ab}, the expansion of the radical in (2) in powers of the parameter $\sim V/(\varepsilon_2 - \varepsilon_1)$ corresponds to a series of the perturbation theory, whose first terms coincide with the expressions in (1).

For the case of $\varepsilon_1 = \varepsilon_2$, the result (2) follows immediately from the secular equation (VIII.5) for the two-fold degenerate level. And for the case $\varepsilon_1 \neq \varepsilon_2$, equation (2) gives a generalization of perturbation theory for the case of two close levels. The interaction between them is taken into account exactly, while we neglect their interaction with other levels.

Problem 8.7

A Hamiltonian depends on some real parameter λ so that $\hat{H}(\lambda) = \hat{h} + \lambda \hat{W}$, where $\hat{h}$ and $\hat{W}$ do not depend on λ. For the ground level, $E_0(\lambda)$, of such a Hamiltonian, prove that $d^2 E_0(\lambda)/d\lambda^2 < 0$. As an illustration, apply this result to a linear oscillator and to a particle in Coulomb potential.

Solution

Let us consider the values of λ which are close to some λ_0 and write down the Hamiltonian in the form:

$$\hat{H}(\lambda) = \hat{H}(\lambda_0) + (\lambda - \lambda_0)\hat{W}.$$

---

[130] Compare with Problem 6.9.

Using the perturbation theory for $\lambda \to \lambda_0$, we find that

$$E_0(\lambda) = E_0(\lambda_0) + A(\lambda_0)(\lambda - \lambda_0) + B(\lambda_0)(\lambda - \lambda_0)^2 + \ldots,$$

where $B(\lambda_0) < 0$, since the correction of the second approximation to the ground level is always negative. The problem statement follows, since

$$\frac{d^2 E_0(\lambda)}{d\lambda^2} = 2B(\lambda) < 0.$$

Let illustrate this property of $E_0(\lambda)$ on the example of a linear oscillator. Here the Hamiltonian is $\hat{H} = \hat{p}^2/2m + kx^2/2$ and $E_0 = \hbar\omega/2$, where $\omega = \sqrt{k/m}$. Choosing $\lambda = k$, we immediately obtain the inequality $E_0'' < 0$ by a direct differentiation. The Coulomb potential is treated analogously.

Problem 8.8

A planar (two-dimensional) rotor with the moment of inertia I and electric dipole moment $\mathbf{d}$ is placed in an homogeneous electric field, $\boldsymbol{\mathcal{E}}$, that lies in the plane of the rotation. Considering the electric field as a perturbation, find the polarizability of the ground state.

Solution

For the unperturbed rotor, we have (see Problem 3.2):

$$\psi_m^{(0)} = \frac{1}{\sqrt{2\pi}} e^{im\varphi}, \quad E_{|m|}^{(0)} = \frac{\hbar^2 m^2}{2I}, \quad m = 0, \pm 1, \pm 2, \ldots. \tag{1}$$

The perturbation is $V = -\mathbf{d} \cdot \boldsymbol{\mathcal{E}} = -d\mathcal{E} \cos\varphi$. Using the expression $\cos\varphi = (e^{i\varphi} + e^{-i\varphi})/2$, we find that the matrix elements $V_{mm'}$ differ from zero only in the case $m' = m \pm 1$, and then are equal to $-d\mathcal{E}/2$. Then, according to Eqs. (VIII.1) and (1) for the ground state of the rotor, we obtain

$$E_0 \approx E_0^{(0)} + E_0^{(1)} + E_0^{(2)} = -\frac{d^2 I}{\hbar^2} \mathcal{E}^2. \tag{2}$$

So the polarizability of the ground state is equal to $\beta_0 = 2d^2 I/\hbar^2$. Compare with Problems 8.2 and 8.3.

Problem 8.9

Under the conditions of the previous problem, determine up to second order in perturbation theory the shifts, splittings, and polarizabilities of the excited states of the quantum rotor. Indicate the correct eigenfunctions in zeroth approximation. Pay attention to properties of the first excited level.

Solution

Although the excited levels of the rotor are twofold degenerate, we can apply the non-degenerate perturbation theory to calculate the level shifts in a uniform electric field if we take into account the following fact: the perturbation $V = -d\mathcal{E}\cos\varphi$, as well as the Hamiltonian $\hat{H}_0$, are invariant with respect to coordinate reflection along the electric field axis. That is, they are invariant with respect to the transformation $\varphi \to -\varphi$. Therefore, we can classify the Hamiltonian eigenfunctions by the value of *parity* $P = \pm 1$ and consider the corresponding states separately. We immediately determine the correct eigenfunctions in the zeroth approximation (compare with Problem 8.8):

$$\psi_{\mu,+}^{(0)} = \frac{1}{\sqrt{\pi}}\cos\mu\varphi, \quad \psi_{\mu,-}^{(0)} = \frac{1}{\sqrt{\pi}}\sin\mu\varphi,$$

$$E_{\mu,\pm}^{(0)} = \frac{\hbar^2\mu^2}{2I}, \quad \mu = |m| = 1, 2, \ldots$$

and $\psi_{0,+}^{(0)} = 1/\sqrt{2\pi}$, $E_{0,+}^{(0)} = 0$ (ground level).

We begin by calculating the *even* energy levels. For these, we find that only the following perturbation matrix elements are different from zero:

$$V_{\mu+,\mu'+} = \begin{cases} -\frac{1}{2}d\mathcal{E} \ ; & \mu' = \mu \pm 1, \ \mu' \neq 0, \ \mu \neq 0, \\ -\frac{1}{\sqrt{2}}d\mathcal{E} \ ; & \mu' = 1, \ \mu = 0 \text{ or } \mu' = 0, \ \mu = 1. \end{cases}$$

By using the equations from (VIII.1), we obtain

$$E_{\mu,+}^{(1)} = 0; \quad E_{1,+}^{(2)} = \frac{5}{6}\frac{d^2 I}{\hbar^2}\mathcal{E}^2, \quad E_{\mu,+}^{(2)} = \frac{1}{4\mu^2 - 1}\frac{d^2 I}{\hbar^2}\mathcal{E}^2 \text{ for } \mu \geq 2. \quad (1)$$

For the *odd* energy levels, $V_{\mu-,\mu'-} = -d\mathcal{E}/2$ with $\mu' = \mu \pm 1$ (other matrix elements are equal to zero), and their energy shifts are given by

$$E_{\mu,-}^{(1)} = 0; \quad E_{1,-}^{(2)} = -\frac{1}{6}\frac{d^2 I}{\hbar^2}\mathcal{E}^2, \quad E_{\mu,-}^{(2)} = \frac{1}{4\mu^2 - 1}\frac{d^2 I}{\hbar^2}\mathcal{E}^2 \text{ for } \mu \geq 2. \quad (2)$$

Comparing Eq. (1) and (2), we see that in the second order of perturbation theory the energy level with $|m| = 1$ splits and the degeneracy is removed, while for the values $|m| \geq 2$ only a shift appears. The splitting of the mth energy level appears only in the $2|m|$th order of perturbation theory.

Problem 8.10

A three-dimensional rotor with a moment of inertia I and dipole moment **d** is placed into a uniform electric field, considered as a perturbation. Find the polarizability of the ground state.

Solution

The eigenfunctions and eigenvalues of the unperturbed Hamiltonian are (see Problem 3.3)

$$\psi_{lm}^{(0)} = Y_{lm}(\vartheta, \varphi), \quad E_{lm}^{(0)} \equiv E_l^{(0)} = \frac{\hbar^2 l(l+1)}{2I}, \tag{1}$$

where the perturbation is $V = -d\mathcal{E}\cos\vartheta$, the z-axis directed along the electric field. Remembering the relation $\cos\vartheta \cdot Y_{00} = -iY_{10}/\sqrt{3}$ (see Eq. (III.7) and the condition of orthogonality of the spherical harmonics), we find that the perturbation matrix element $V_{lm,00}$ is different from zero only for $l = 1$, $m = 0$, and $V_{10,00} = id\mathcal{E}/\sqrt{3}$. From Eq. (VIII.1) we obtain

$$E_0 \approx E_0^{(0)} + E_0^{(1)} + E_0^{(2)} = -\frac{I}{3}\frac{d^2\mathcal{E}^2}{\hbar^2}. \tag{2}$$

The polarizability of the ground state of the rotor is $\beta_0 = 2d^2 I/3\hbar^2$.

Problem 8.11

Under the conditions of the previous problem, calculate the shifts of the excited energy levels of the rotor in the first non-vanishing order of perturbation theory. When does the level degeneracy get lifted? Is there a further degeneracy lifting in the higher orders of perturbation theory?

Solution

In order to calculate the perturbation matrix elements (see the previous problem) we use the following relation:

$$\cos\vartheta \cdot Y_{lm} = a_{lm}Y_{l+1,m} - a_{l-1,m}Y_{l-1,m}, \quad l \geq 1, \tag{1}$$

$$a_{lm} = -i\sqrt{\frac{(l-m+1)(l+m+1)}{(2l+1)(2l+3)}}.$$

In order to obtain this relation we have used the connection between the spherical harmonics Y_{lm} and the associated Legendre polynomials $P_l^{|m|}$, and recursive relations for the Legendre polynomials.

From Eq. (1) we see that only the perturbation matrix elements for the adjacent energy levels are different from zero:

$$V_{l+1,m;lm} = -V_{lm;l+1,m} = -a_{lm}d\mathcal{E}.$$

Although the energy levels of the unperturbed rotor are degenerate for $l \neq 0$, the calculation of their shift and splitting in electric field does not require the use of degenerate perturbation theory. Due to the conservation of l_z, the action of perturbation on the states with different values of m can be considered separately. The equations of non-degenerate perturbation theory apply. Taking this into account and using equations (VIII.1), we obtain:

$$E^{(1)}_{lm} = 0; \quad E^{(2)}_{lm} = \frac{l(l+1) - 3m^2}{l(l+1)(2l-1)(2l+3)} \frac{d^2 I}{\hbar^2} \mathcal{E}^2 \quad \text{for } l \geq 1. \tag{2}$$

As is seen, the $(2l+1)$-fold degeneracy of the unperturbed rotor level is partially removed. The l-th level splits into $l+1$ sub-levels including the non-degenerate $m=0$ level and l twofold degenerate ($\pm|m|$) levels. Further degeneracy lifting does not occur in the higher levels of perturbation theory. This is connected with the fact that the value $m \equiv l_z$ is the constant of motion and has a definite value together with the energy, and that the Hamiltonian is invariant with respect to coordinate reflection in all planes that include the z-axis, so the energy of the states that differ only by the sign of the projection of the angular momentum on the direction of the electric field must be the same.

Problem 8.12

Find the energy shift and polarizability of the ground state of a charged particle in the one-dimensional δ-well, $U(x) = -\alpha \delta(x)$.

Solution

For the ground level of the particle in the δ-well we have (see Problem 2.7):

$$E_0^{(0)} = -\frac{\hbar^2 \kappa^2}{2m}, \quad \psi_0^{(0)}(x) = \sqrt{\kappa} e^{-\kappa|x|}, \quad \kappa = \frac{m\alpha}{\hbar^2}.$$

We must calculate its shift under a perturbation of the form $V = -e\mathcal{E}x$ in the second-order perturbation theory (it is obvious that $E_0^{(1)} = 0$). For the unperturbed eigenfunctions of the continuous spectrum, it is convenient to use the functions $\psi_{kI}^{(0)}(x)$ which correspond to a definite parity $I = \pm 1$. The even Hamiltonian eigenfunctions which are distorted by the δ-potential give rise to a vanishing perturbation matrix element, and thus their explicit form is not important. The odd wavefunctions are not distorted by the δ-potential and thus coincide with the wavefunctions of a free particle $\psi_{k,-1}^{(0)} = (1/\sqrt{\pi}) \sin(kx)$. So according to Eq. (VIII.1) we have ($E_k = \hbar^2 k^2 / 2m$)

$$E_0^{(2)} \equiv -\frac{1}{2} \beta_0 \mathcal{E}^2 = \sum_I \int_0^\infty \frac{|\langle kI | e\mathcal{E}x | 0 \rangle|^2}{E_0^{(0)} - E_k} dk =$$

$$-\frac{2m\kappa e^2 \mathcal{E}^2}{\pi \hbar^2} \int_0^\infty \frac{dk}{\kappa^2 + k^2} \left| \int_{-\infty}^\infty x \sin kx \cdot e^{-\kappa|x|} dx \right|^2.$$

Using the value of the integral,

$$\int_{-\infty}^{\infty} x \sin kx \cdot e^{-\kappa|x|} dx = \frac{4\kappa k}{(\kappa^2 + k^2)^2}$$

and the integral (App.1.5) we find the level shift and the polarizability:

$$E_0^{(2)} = -\frac{1}{2}\beta_0 \mathcal{E}^2, \quad \beta_0 = \frac{5}{4}\frac{me^2}{\hbar^2 \kappa^4}.$$

Compare with Problem 6.36.

Problem 8.13

For a two-dimensional rotor with a dipole moment **d** placed in a strong electric field, $\mathcal{E}$ ($d\mathcal{E} \gg \hbar^2/I$), find the approximate form of the wavefunctions and energies of the low-lying states.

Solution

In a strong electric field, the wavefunctions of the low-lying states are localized in the region of small angles $|\varphi| \ll 1$, since the potential energy $U(\varphi) = -\mathbf{d}\cdot\boldsymbol{\mathcal{E}} = -d\mathcal{E}\cos\varphi$ has a deep minimum at $\varphi = 0$. See Fig. 8.1.

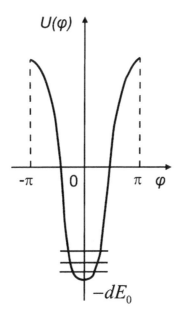

Fig. 8.1

We can expand $U(\varphi)$ in a series and keep only the first two terms of the expansion:

$$U(\varphi) = -d\mathcal{E}\cos\varphi \approx -d\mathcal{E} + \frac{1}{2}d\mathcal{E}\varphi^2. \tag{1}$$

In the zeroth approximation we reduce the rotor Hamiltonian to that of a linear oscillator. Using Eq. (II.3) we obtain[131]

$$E_n^{(0)} = -d\mathcal{E} + \hbar\sqrt{\frac{d\mathcal{E}}{I}}\left(n + \frac{1}{2}\right), \quad n = 0, 1, 2, \ldots, \tag{2}$$

$$\psi_n^{(0)}(\varphi) = \frac{1}{\sqrt{2^n\sqrt{\pi}\varphi_0 n!}}\exp\left\{-\frac{1}{2}\left(\frac{\varphi}{\varphi_0}\right)^2\right\}H_n\left(\frac{\varphi}{\varphi_0}\right), \quad \varphi_0 = \left(\frac{\hbar^2}{Id\mathcal{E}}\right)^{1/4}.$$

The validity of this result relies on the eigenfunction (2) being small for $|\varphi| \sim 1$. Since the wavefunctions, $\psi_n^{(0)}(\varphi)$, are essentially different from zero only around the angles allowed for a classical rotator,

$$\frac{1}{2}d\mathcal{E}\varphi^2 \leq E_n^{(0)} + d\mathcal{E}, \quad \text{or } \varphi^2 \leq \varphi_0^2\left(n + \frac{1}{2}\right),$$

then the required condition is $\mathcal{E} \gg \hbar^2(n+1)^2/Id$. In conclusion, we note that by using the next terms in the expansion (1) in φ^2 (anharmonic corrections), we can determine the value of E_n in (2).

Problem 8.14

The same as the previous problem, but for a three-dimensional rotor (see Problem 8.10).

Solution

The Hamiltonian of the system has the form

$$\hat{H} = \frac{\hbar^2}{2I}\hat{\mathbf{l}}^2 - \mathbf{d}\cdot\boldsymbol{\mathcal{E}} = -\frac{\hbar^2}{2I}\triangle_{\theta,\varphi} - d\mathcal{E}\cos\theta. \tag{1}$$

The polar axis z is directed along the electric field $\mathcal{E}$. For a strong field, the Hamiltonian wavefunctions of the low-lying levels are localized to $\theta \ll 1$ due to the deep minimum of potential energy $U(\theta) = -d\mathcal{E}\cos\theta$ at $\theta_0 = 0$ (see the previous problem). Note that the operator $\triangle_{\theta,\varphi}$ (the Laplacian on the sphere of unit radius) in the case of the small θ could be approximated as the Laplacian in two-dimensional space (on the plane tangential to the unit sphere at the point $\theta_0 = 0$, so that θ is considered as a "radial" variable). The Hamiltonian can be written in the form:

$$\hat{H} \approx -\frac{\hbar^2}{2I}\left(\frac{\partial^2}{\partial x^2} + \frac{\partial^2}{\partial y^2}\right) - d\mathcal{E} + \frac{1}{2}d\mathcal{E}(x^2 + y^2). \tag{2}$$

---

[131] In this limit, we can not view the rotor as a rotating particle.

Here $x = \theta\cos\varphi$, $y = \theta\sin\varphi$, $\theta = \sqrt{x^2 + y^2}$. This Hamiltonian describes a planar oscillator, which allows us (using Problem 2.48) to obtain the spectrum and eigenfunctions of the initial Hamiltonian (1) in the *zeroth* approximation:

$$E^{(0)} = -d\mathcal{E} + \hbar\omega(N+1), \quad \omega = \left(\frac{d\mathcal{E}}{I}\right)^{1/2}, \quad N = 0, 1, 2, \ldots, \tag{3}$$

$$\psi^{(0)}_{n_1 n_2} = C_{n_1} C_{n_2} \exp\left\{-\frac{\theta^2}{2\theta_0^2}\right\} H_{n1}\left(\frac{x}{\theta_0}\right) H_{n2}\left(\frac{y}{\theta_0}\right), \quad n_1 + n_2 = N,$$

where $\theta_0 = (\hbar^2/Id\mathcal{E})^{1/4}$. Since the oscillator eigenfunctions (2) are localized in the domain $\theta^2 \leq (N+1)\theta_0^2$, the condition $\theta \ll 1$ used above gives the applicability condition for relation (3) as $\theta_0^2 \ll 1/(N+1)$, or $\mathcal{E} \gg \hbar^2(N+1)^2/dI$.

In the approximation considered, the N-th level $E_N^{(0)}$ has the degeneracy $g(N) = N+1$. If in the expansion of $\cos\theta$ we consider the next term, $O(\theta^4)$, and if we use a more accurate expression for the Laplacian, $\Delta_{\theta,\varphi}$, we can find a splitting of levels $E_N^{(0)}$.

Problem 8.15

A particle moves in a central potential of the form ($a > 0$):

$$\text{a) } U(r) = -\frac{U_0}{e^{r/a} - 1}, \quad \text{b) } U(r) = -U_0\frac{a}{r} e^{-r/a},$$

where $U_0 \gg \hbar^2/ma^2$. Find the difference between the low-lying energy levels and the energy levels of the Coulomb potential, $U_C(r) = -U_0\frac{a}{r}$, in first-order perturbation theory. Pay attention to a lifting of the *accidental* Coulomb level degeneracy.

Solution

For $r \ll a$, the potentials considered reduce to the form $U(r) \approx -U_0 a/r$. In such a Coulomb field, the wavefunctions of low-energy states are localized at distances of order $r_n \sim a_0 n^2$ from the origin; here $a_0 = \hbar^2/mU_0 a$ and n is the principal quantum number. If $r_n \ll a$ (i.e. $\xi \equiv mU_0 a^2/\hbar^2 \gg n^2$), it is obvious that in the zeroth approximation the low-energy states and the corresponding eigenfunctions will have the same form as in the Coulomb field:

$$E^{(0)}_{n_r l} = -\frac{mU_0^2 a^2}{2\hbar^2 n^2}, \quad \psi^{(0)}_{n_r l m} = \psi^{\text{Coul}}_{n_r l m}. \tag{1}$$

See Eq. (IV.3). Here the difference between the potentials considered and the Coulomb potential acts as the perturbation $V(r) = U(r) + U_0\frac{a}{r}$. Since the particle angular momentum is a constant of the motion, and since under the perturbation, eigenfunctions (1) are the *correct* functions in the zeroth approximation, then the first-order correction is determined by Eq. (VIII.1):

$$E_{n_r l}^{(1)} = \int V(r)|\psi_{n_r lm}^{(0)}|^2 d^3r. \tag{2}$$

a) Expanding the perturbation potential into a series in the power r/a, we have:

$$V(r) = -U_0 \left[-\frac{1}{2} + \frac{r}{12a} + O\left(\frac{r^3}{a^3}\right) \right]. \tag{3}$$

We also note the value of the integral:

$$\int r|\psi_{n_r lm}^{(0)}|^2 d^3r = \frac{1}{2}[3n^2 - l(l+1)]\frac{\hbar^2}{mU_0 a}. \tag{4}$$

According to (2)–(4) we find:

$$E_{n_r l}^{(1)} = U_0 \left\{ \frac{1}{2} - \frac{3n^2 - l(l+1)}{24\xi} \right\}, \quad n = n_r + l + 1. \tag{5}$$

We should note that if in expansion (3) we take into account the term $\sim r^3/a^3$, then we would exceed the accuracy of the approximation used. Since its contribution to the level shift $\sim U_0/\xi^3$ has the same order of magnitude as a second-order perturbation, and we do not account for it, inclusion of this term is unnecessary.

b) We have $V(r) \approx U_0(1 - r/2a + r^2/6a^2)$ and

$$E_{n_r l}^{(1)} = U_0 \left\{ 1 - \frac{3n^2 - l(l+1)}{4\xi} + \frac{n^2}{12\xi^2}[5n^2 + 1 - 3l(l+1)] \right\}. \tag{6}$$

As seen from (5) and (6), the accidental Coulomb degeneracy of the levels with respect to l is lifted.

We should note that in the case of the Hulthen potential, the Schrödinger equation for the s-states has an exact solution (see Problem 4.8). In this case, ($l = 0$), $E_{n,l}^{(0)}$ together with (5) gives the exact result.

Problem 8.16

For a particle in the central attractive potential $U(r) = -\alpha/r^\nu$ with $0 < \nu < 2$ and $\alpha > 0$, find the energy levels, $E_{n_r l}$, with large value of the angular momentum $l \gg 1$ and the radial quantum number $n_r \sim 1$. For the Coulomb potential, i.e., $\nu = 1$, compare this result with the exact one.

Solution

The Schrödinger equation for the radial part, $\chi(r) = rR(r)$, the wavefunction has the form:

$$-\frac{\hbar^2}{2m}\chi'' - \frac{\alpha}{r^\nu}\chi + \frac{\hbar^2 l(l+1)}{2mr^2}\chi = E_{n_r l}\chi.$$

The effective potential energy in this equation,

$$U_{\text{eff}} = -\frac{\alpha}{r^\nu} + \frac{\hbar^2 l(l+1)}{2mr^2},$$

has a minimum at the point

$$r_0 = \left(\frac{\hbar^2 l(l+1)}{m\alpha\nu}\right)^{1/(2-\nu)}.$$

In the case of $\nu < 2$ and $l \to \infty$, the wavefunctions of the low-energy radial states are localized near the minimum point in a domain, where we can restrict ourselves to a few first terms in the expansion of the effective potential:

$$U_{\text{eff}} \approx -\frac{1}{2}\alpha(2-\nu)r_0^{-\nu} + \frac{1}{2}\alpha(2\nu-\nu^2)r_0^{-(\nu+2)}(r-r_0)^2 + \ldots. \quad (1)$$

Compare with Problems, 8.13–8.15. In the *zeroth* approximation we essentially return to the problem of a linear oscillator with equilibrium point $r = r_0$ and elasticity $k = U''_{eff}(r_0)$. This allows us to obtain[132] (compare with Eq. (II.2)):

$$\chi^{(0)}_{n_r l} = \left(2^{n_r}\sqrt{\pi}n_r!a\right)^{-1/2}\exp\left(-\frac{(r-r_0)^2}{2a^2}\right)H_{n_r}\left(\frac{r-r_0}{a}\right), \quad (2)$$

$$E^{(0)}_{n_r l} = -\frac{1}{2}\frac{\alpha(2-\nu)}{r_0^\nu} + \hbar\sqrt{\frac{1}{m}\alpha(2\nu-\nu^2)r_0^{-(2+\nu)}}\left(n_r + \frac{1}{2}\right), \quad (3)$$

where $a = \left(\hbar^2 r_0^{2+\nu}/\alpha m(2\nu-\nu^2)\right)^{1/4}$.

For the applicability of these results, the condition we used in the derivation must be fulfilled: namely, the radial function (2) must be localized within region $|r - r_0| \ll r_0$. So we have $l \gg (n_r + 1/2)/\sqrt{2-\nu}$ (compare with Problems 8.13 and 8.14).

As an illustration we will apply the result obtained to the case of the Coulomb potential, $\nu = 1$. Here we have $E_n = -m\alpha^2/2\hbar^2 n^2$. Writing $n \equiv l + 1/2 + n_r + 1/2$, and performing the expansion

$$E_n = -\frac{m\alpha^2}{2\hbar^2}\frac{1}{(l+1/2+n_r+1/2)^2} \approx -\frac{m\alpha^2}{2\hbar^2(l+1/2)^2}$$
$$+ \frac{m\alpha^2}{\hbar^2(l+1/2)^3}\left(n_r + \frac{1}{2}\right), \quad (4)$$

we see that (3) represents the first two terms of the expansion (4) for E_n over the small parameter $(n_r + 1/2)/(l + 1/2)$. Note that since $l \gg 1$, in the approximation considered, $l(l+1) \approx (l+1/2)^2$.

---

[132] Taking into account the following terms in expansion (1) (anharmonic corrections) allows us, with the help of perturbation theory, to obtain more accurate expressions for the radial wavefunction and energy spectrum. Compare with Chapter 9, Sec. 4.

Problem 8.17

The same as in the previous problem, but for the potential $U(r) = \alpha r^\nu$ with $\alpha, \nu > 0$ (now $E_{n_r l} > 0$). For a spherical oscillator, $\nu = 2$, compare the result obtained with the exact one.

Solution

The solution of this problem can be obtained by changing $-\alpha \to \alpha$ and $-\nu \to \nu$ in the equations of the previous problem. Now $E_{n_r l} > 0$, and there are no restrictions for the value $\nu > 0$.

In the case of a spherical oscillator, the exact spectrum has the form $E_N = \hbar\omega(N + 3/2)$, where $N = 2n_r + l$ and $\omega = \sqrt{2\alpha/m}$ (see Problem 4.5). Equation (3) in the previous problem reproduces this spectrum with the only difference $l + 1/2 \to \sqrt{l(l+1)}$, which is inconsequential, since $l \gg 1$.

Problem 8.18

A particle is inside an impenetrable ellipsoid of rotation, so the potential has form:

$$U(x,y,z) = \begin{cases} 0, & \frac{x^2+y^2}{a^2} + \frac{z^2}{b^2} < 1, \\ \infty, & \frac{x^2+y^2}{a^2} + \frac{z^2}{b^2} \geq 1, \end{cases}$$

and $|a - b| \ll a$. Find the energy shift of the ground level with respect to the ground level of the particle in a spherical well of the same volume, to first order in perturbation theory.

Solution

The substitutions $x = x'$, $y = y'$, $z' = az/b$ yield the following form of the Schrödinger equation and the boundary condition:

$$-\frac{\hbar^2}{2m}\left(\frac{\partial^2}{\partial x'^2} + \frac{\partial^2}{\partial y'^2} + \left(\frac{a}{b}\right)^2 \frac{\partial^2}{\partial z'^2}\right)\psi = E\psi, \quad \psi(r' = a) = 0.$$

Since $|a - b| \ll a$, then writing $a = (1 + \varepsilon)b$ with $|\varepsilon| \ll 1$, we can write the Hamiltonian in the form $\hat{H} = \hat{H}_0 + \hat{V}$ with

$$\hat{H}_0 = -\frac{\hbar^2}{2m}\Delta', \quad \hat{V} = -\frac{\hbar^2}{2m}(2\varepsilon + \varepsilon^2)\frac{\partial^2}{\partial z'^2}. \tag{1}$$

Here $\hat{V}$ is the perturbation operator. For the ground state of the unperturbed Hamiltonian, we have (compare with Problem 4.1):

$$\psi_0^{(0)} = \frac{1}{\sqrt{2\pi a r}} \sin\frac{\pi r}{a}, \quad r \leq a; \quad E_0^{(0)} = \frac{\pi^2 \hbar^2}{2ma^2}. \tag{2}$$

In what follows in this problem, we shall omit the primes on variables.

From (1) we see that to obtain the correction of the first order in small parameter ε, we have to calculate the mean value of $\partial^2/\partial z^2$ for the ground state. Because of the spherical symmetry of the wavefunction, we have

$$\overline{\frac{\partial^2}{\partial x^2}} = \overline{\frac{\partial^2}{\partial y^2}} = \overline{\frac{\partial^2}{\partial z^2}} = \frac{1}{3}\overline{\Delta}.$$

By using Eqs. (1) and (2), we obtain:

$$E_0^{(1)} = \frac{\varepsilon \pi^2 \hbar^2}{3ma^2}, \quad E_0 \approx E_0^{(0)} + E_0^{(1)} = \left(1 + \frac{2}{3}\varepsilon\right) \frac{\pi^2 \hbar^2}{2ma^2}. \tag{3}$$

As the volume of the ellipsoid is equal to

$$\frac{4\pi}{3}a^2 b \approx \frac{4\pi}{3}a^3(1-\varepsilon) \equiv \frac{4\pi}{3}R^3,$$

then according to (3) we see that $E_0 \approx \pi^2\hbar^2/2mR^2$, where R is the radius of the sphere of the same volume. Hence, in the first approximation with respect to the deformation parameter ε, the ground-state energy depends on the volume of ellipsoid only. Using the fact that in the s-states, the particle exerts equal surface pressure onto all parts of the walls of the spherical well, and also using the expression for the work done under changes of volume, $-PdV$, we see that the result obtained remains valid in the case of small surface deformations of sufficiently arbitrary shape, which preserve the volume.

Problem 8.19

Solve the previous problem for the excited states of the particle. Explain why the degeneracy with respect to the projection of the angular momentum is lifted in the first and higher orders of perturbation theory.

Solution
The problem can be solved similarly to the previous one, but now

$$\psi_{n_r lm}^{(0)} = \frac{\tilde{C}}{\sqrt{r}} J_{l+1/2}\left(\alpha_{n_r+1,l}\frac{r}{a}\right) Y_{lm}, \quad E_{n_r l}^{(0)} = \frac{\alpha_{n_r+1,l}^2 \hbar^2}{2\mu a^2}.$$

See Problem 4.9. Note that the unperturbed lth level is $(2l+1)$-fold degenerate. Since l_z is a constant of motion, the above wavefunctions are the correct functions in the zeroth approximation, so the energy level shifts are given by

$$E_{n_r lm}^{(1)} = -\varepsilon \frac{\hbar^2}{\mu} \int \psi_{n_r lm}^{(0)*} \frac{\partial^2}{\partial z^2} \psi_{n_r lm}^{(0)} dv. \tag{1}$$

To evaluate this integral, let us first consider the more general matrix element of the form

$$\left\langle n_r l m' \left| \frac{\partial^2}{\partial x_i \partial x_k} \right| n_r l m \right\rangle.$$

After integrating over the coordinates, it becomes $\left\langle m' \left| \hat{T}_{ik} \right| m \right\rangle$, where $\hat{T}_{ik}$ is now an ordinary matrix, operating in the space of the state vectors with the angular momentum, l. The vectors $|m>$ define a basis in this space. The most general form of the tensor, operator T_{ik}, is (compare with Problem 3.40, with $\hat{l}_i$ as the vector-matrices of the angular momentum components):

$$\hat{T}_{ik} = A\delta_{ik} + B\varepsilon_{ikl}\hat{l}_l + C\left(\hat{l}_i\hat{l}_k + \hat{l}_k\hat{l}_i\right). \tag{2}$$

Due to the symmetry of $\hat{T}_{ik}$, we have $B = 0$. From the condition

$$\hat{l}_i \hat{T}_{ik} = \overline{\hat{l}_i \frac{\partial^2}{\partial x_i \partial x_k}} \equiv 0,$$

we obtain

$$A + [2l(l+1) - 1]C = 0. \tag{3}$$

The convolution in (2) over the indexes i and k gives

$$3A + 2l(l+1)C = \hat{T}_{ii} = \overline{\triangle}. \tag{4}$$

Determining A and C from (3) and (4), and taking into account $-(\hbar^2/2\mu)\overline{\triangle} = E^{(0)}_{n_r l}$, we obtain

$$E^{(1)}_{n_r l m} = 2\varepsilon \frac{2l^2 + 2l - 1 - 2m^2}{(2l-1)(2l+3)} E^{(0)}_{n_r l}. \tag{5}$$

Hence it follows that the level degeneracy is lifted: it splits into $(l+1)$ sublevels, one of which ($m = 0$) is non-degenerate, while the others sublevels ($m = \pm|m|$) are two-fold degenerate. Note that this remaining degeneracy is preserved in higher orders of perturbation theory.

We should note that the average value of the first-order correction over all the sublevels is

$$\overline{E^{(1)}_{n_r l m}} = \frac{1}{2l+1} \sum_m E^{(1)}_{n_r l m} = \frac{2\varepsilon}{3} E^{(0)}_{n_r l}.$$

See the calculation of the sum in Problem 3.1. So the value $\overline{E_{n_r l m}}$ is determined by the volume of the ellipsoid only, just as in the case of the ground state (see the remark on this matter in the previous problem).

Problem 8.20

Using perturbation theory, obtain the quantization condition for the angular momentum squared and find the form of spherical harmonics for the case $l \approx |m| \gg 1$.

Solution

The equation $\hat{\mathbf{l}}^2 \psi_{lm} = L^2 \psi_{lm}$ by the substitution

$$\psi_{lm} = \frac{1}{\sqrt{2\pi \sin \vartheta}} e^{im\varphi} \chi(\vartheta)$$

takes the form of the one-dimensional Schrödinger equation,

$$\chi'' + \left(L^2 + \frac{1}{4} - \frac{m^2 - 1/4}{\sin^2 \vartheta} \right) \chi = 0 \tag{1}$$

with $\hbar = 1$, "mass" $\mu = 1/2$, "potential energy" $U(\vartheta) = (m^2 - 1/4)/\sin^2 \vartheta$, and "energy" equal to $L^2 + 1/4$. In the case of $m^2 \gg 1$, the "potential" has a deep minimum at $\vartheta = \pi/2$, and so the wavefunctions of lower "energy levels" are localized in the proximity of this point. Expanding $U(\vartheta)$ into a series

$$U(\vartheta) = \left(m^2 - \frac{1}{4} \right) \left[1 + \left(\vartheta - \frac{\pi}{2} \right)^2 + \frac{2}{3} \left(\vartheta - \frac{\pi}{2} \right)^4 + \ldots \right], \tag{2}$$

we see that the problem considered reduces to the Schrödinger equation for a linear oscillator. In the *zeroth* approximation, omitting the term $\sim \left(\vartheta - \frac{\pi}{2} \right)^4$ in (2), we obtain (with $x = \vartheta - \pi/2$)

$$\chi_n^{(0)} = C \frac{|m|^{1/4}}{\sqrt{2^n \sqrt{\pi} n!}} e^{-|m|x^2/2} H_n(\sqrt{|m|} x), \quad n = 0, 1, \ldots, \tag{3}$$

$$\left(L_n^2 \right)^{(0)} = m^2 + 2|m|(n + 1/2) - 1/2,$$

where C is a phase factor (see below). In the wavefunction and expression for L_n^2, the terms of the order $1/|m|$ are omitted.

In order to refine this value of L^2, we include the next anharmonic term $\sim (\vartheta - \pi/2)^4$ in (2). It is equal to

$$\left(L_n^2 \right)^{(1)} = \frac{2}{3} m^2 \overline{\left(\vartheta - \frac{\pi}{2} \right)^4} = n^2 + n + \frac{1}{2}.$$

Adding this to the value of zeroth approximation from (3), we obtain

$$L^2 = \left(L_n^2 \right)^{(0)} + \left(L_n^2 \right)^{(1)} = (|m| + n)(|m| + n + 1), \tag{4}$$

which coincides with the exact value of $\hat{L}^2 = l(l+1)$, where $l = |m| + n$.

We should note that the localization condition (3) in the region $|\vartheta - \pi/2| \ll 1$ demands that $n \ll |m|$ and $|m| \gg 1$ (compare with Problem 8.13). Finally, the spherical function phase factor is fixed (see Eqs. (III.8) and (III.5)), so that in (3), we should have

$$C = i^l(-1)^{(m+|m|)/2}.$$

8.2 Variational method

Problem 8.21

For a particle in a potential from Problem 2.8, $U(x) = F_0 x$ for $x \geq 0$ and $U(x) = \infty$ for $x < 0$, find the energy of the ground state by the variational method, using the trial functions of the form

$$\text{a)}\ \psi = Ax \exp(-\alpha x), \quad \text{b)}\ \psi = Bx \exp\left(-\frac{1}{2}\alpha x^2\right)$$

for $x \geq 0$. Compare with the exact result.

Solution

Let us calculate the mean energy $\overline{E}(\alpha)$ and find its minimum value $\min \overline{E} = \overline{E}(\alpha_0)$. According to the main idea of the variational method, the minimal value on the class of trial functions approximates the ground-state energy, $E_{0,\text{var}} = \overline{E}(\alpha_0)$. The trial functions we use reflect the characteristic properties of the exact wavefunction of the ground state, $\psi_0(x)$, which can be seen from the general considerations: 1) $\psi_0(x) \propto x$ as $x \to 0$; 2) the rapid (exponential) decrease on large distances; 3) the absence of zeros (except boundary values); 4) its smooth behavior on the both sides of the extremum.

To calculate the mean value $\overline{E} = \overline{T} + \overline{U}$, it is convenient to normalize the trial function and use the following relation

$$\overline{T} = \frac{\hbar^2}{2m} \int |\psi'(x)|^2 dx,$$

and taking into account the integral (App.1.6).

a) From the normalization condition for the wavefunctions, we have $A^2 = 4\alpha^3$; then we find

$$\overline{T} = \frac{\hbar^2 \alpha^2}{2m}, \quad \overline{U} = \frac{3}{2}\frac{F_0}{\alpha}.$$

Minimization of $\overline{E}(\alpha)$ gives (at $\alpha_0 = (3mF_0/2\hbar^2)^{1/3}$)

$$E_{0,\text{var}} = \left(\frac{243}{32}\right)^{1/3} \left(\frac{\hbar^2 F_0^2}{m}\right)^{1/3} \approx 1.966 \left(\frac{\hbar^2 F_0^2}{m}\right)^{1/3}. \tag{1}$$

b) In an analogous way we obtain

$$B^2 = 4\sqrt{\frac{\alpha^3}{\pi}}, \quad \overline{T} = \frac{3}{4}\frac{\hbar^2\alpha}{m}, \quad \overline{U} = \frac{2F_0}{\sqrt{\pi\alpha}}, \quad \alpha_0 = \left(\frac{16m^2 F_0^2}{9\pi\hbar^4}\right)^{1/3},$$

$$E_{0,\text{var}} = \left(\frac{81}{4\pi}\right)^{1/3}\left(\frac{\hbar^2 F_0^2}{m}\right)^{1/3} \approx 1.861\left(\frac{\hbar^2 F_0^2}{m}\right)^{1/3}. \qquad (2)$$

Since the variational calculation gives a restriction from above on the exact value of the ground-state energy, we can confirm that (2) is more accurate than (1); note that the exact value is $E_0 = 1.856(\hbar^2 F_0^2/m)^{1/3}$. We should also note that in both cases, the relation $2\overline{T}(\alpha_0) = \overline{U}(\alpha_0)$ is valid due to the virial theorem (see Problem 6.12).

Problem 8.22

The same as in the previous problem, but for

a) a δ-well (see Problem 2.7) and trial function $\psi(x) = A(a + |x|)^{-\nu}$;
b) a linear oscillator and $\psi(x) = A(a^2 + x^2)^{-\nu/2}$ (ν is integer);
c) the Coulomb potential and $\psi(r) = A(a + r)^{-\nu}$,

where a and ν are variational parameters.

Solution

Making the same calculation as in the previous problem, we obtain the following results.

a)

$$A^2 = \frac{1}{2}(2\nu - 1)a^{2\nu-1}, \quad \overline{T} = \frac{\hbar^2\nu^2(2\nu-1)}{2m(2\nu+1)a^2}, \quad \overline{U} = -\frac{(2\nu-1)\alpha}{2a}.$$

Minimizing $\overline{E}(a, \nu)$ first with respect to the parameter a, we obtain

$$\overline{E}(a_0, \nu) = -\frac{m\alpha^2}{2\hbar^2}\left(1 - \frac{1}{4\nu^2}\right), \quad a_0 = \frac{2\hbar^2\nu^2}{(2\nu+1)m\alpha}. \qquad (1)$$

Minimizing with respect to ν (it is obvious that $\nu_0 = \infty$) gives the value $E_{0,\text{var}} = -m\alpha^2/2\hbar^2$, which coincides with the exact result (see Problem 2.7). This is due to the fact that $\lim(1 - z/\nu)^\nu = e^{-z}$ as $\nu \to \infty$; the trial function for the values of $a_0(\nu)$ from (1) for $\nu \to \infty$ coincides with the exact wavefunction of the ground state.

As is seen from (1), the values of $\overline{E}(a_0, \nu)$ are close to E_0, even for the values of ν different from $\nu_0 = \infty$:

ν	1	2	3	5	7	10
$\overline{E}(\nu)/E_0$	0.750	0.938	0.972	0.990	0.995	0.998

Strong deviations occur only for small values of ν, where the slow decrease at large distances begins to manifest itself.

b) Using the value of the integral (App.1.5), we find:

$$A^2 = \frac{2^{\nu-1}(\nu-1)! a^{2\nu-1}}{\pi(2\nu-3)!!}, \quad \overline{T} = \frac{\nu(2\nu-1)}{8(\nu+1)} \frac{\hbar^2}{ma^2}, \quad \overline{U} = \frac{\overline{kx^2}}{2} = \frac{ka^2}{2(2\nu-3)}.$$

Minimizing over a we obtain ($\omega = \sqrt{k/m}$):

$$\overline{E}(a_0, \nu) = \frac{1}{2}\sqrt{1 + \frac{3}{2\nu^2 - \nu - 3}}\,\hbar\omega, \quad a_0 = \left(\frac{\nu(2\nu-3)(2\nu-1)\hbar^2}{4(\nu+1)mk}\right)^{1/4}, \quad (2)$$

while the following minimization over ν gives ($\nu = \nu_0 = \infty$) the exact value $E_0 = \hbar\omega/2$. Explanation of this fact is the same as in the case of the δ-well, and the statement of $\overline{E}(a_0, \nu)$ being close to E_0 at $\nu \neq \nu_0$, which we made before, remains valid for the oscillator as well.

c) For the Coulomb potential, $U = -\alpha/r$, we obtain

$$4\pi A^2 = (\nu-1)(2\nu-3)(2\nu-1)a^{2\nu-3}, \quad \overline{T} = \frac{\nu(\nu-1)(2\nu-3)}{2(2\nu+1)} \frac{\hbar^2}{ma^2},$$

$$\overline{U} = -\frac{2\nu-3}{2}\frac{\alpha}{a}.$$

Minimization over α gives

$$\overline{E}(a_0, \nu) = -\left(1 - \frac{3}{8\nu(\nu-1)}\right)\frac{m\alpha^2}{2\hbar^2}, \quad a_0 = \frac{2\nu(\nu+1)\hbar^2}{(2\nu+1)m\alpha}, \quad (3)$$

while the following minimization over ν gives the exact result, $E_0 = -m\alpha^2/2\hbar^2$. The situation here is the same as for the potentials used above.

Finally, we should note that if the trial function used gives the accuracy of variational calculation of E_0 high enough so that

$$\left|\frac{E_{0,\text{var}}}{E_0} - 1\right| \equiv \gamma^2 \ll 1,$$

then a variational calculation of another generic property, f, of the ground state (such as $|\psi|^2$, $\overline{(\Delta x)^2}$, etc.) has usually a much lower accuracy. Generally speaking $|f_{var}/f_{ex} - 1| \sim \gamma$. For the δ-well, the trial function considered for $\nu = 10$ gives the value of E_0 with an error 0.2%. The mean value calculated with its help,

$$\overline{x^2} = \frac{8\nu^4}{(2\nu-3)(2\nu-2)(2\nu+1)^2}\left(\frac{\hbar^2}{ma}\right)^2,$$

differs from the exact value by 19%. For the Coulomb potential, with $\nu = 10$, according to (3), $E_{0,var}$ differs from the exact value by 0.4%. While

$$|\psi(0)|^2 = \frac{(2\nu-3)(2\nu-1)(2\nu+1)^3}{32\pi\nu^3(\nu-1)^2}\left(\frac{ma}{\hbar^2}\right)^3$$

differs from the exact value by 15%. This reduction in accuracy is connected with the fact that the trial functions used for finite values of the parameter ν decrease as a power law at large distances, while the exact wavefunctions decrease exponentially.

Problem 8.23

Find the energy a) of the ground state and b) of the first excited state for a particle moving within a one-dimensional infinitely deep potential well, approximating the Hamiltonian eigenfunctions by the simplest polynomials that satisfy the required conditions.

Solution

To better determine the form of polynomials that faithfully approximate the wavefunctions we should first take into account the boundary conditions $\psi(0) = \psi(a) = 0$ and the absence of ground-state wavefunction zeros (except at the boundaries). Furthermore, the trial function for the first excited level must be orthogonal to the ground-state wavefunction; only if this condition is fulfilled will the value of $\overline{E}$ be restricted from above for the excited level energy (compare to Problem 8.28). In problems with one-dimensional symmetric potentials it is easy to fulfill this orthogonality condition due different parities of the ground and first excited state for coordinate inversion with respect to the well center.

a) For the ground state we choose $\psi = Ax(x-a)$ for $0 \leq x \leq a$; this trial function, as well as the exact one, is even for coordinate inversion with respect to the well center $x = a/2$. By normalizing the wavefunction we find (note that $\overline{U} = 0$ means $\overline{E} = \overline{T}$):

$$A^2 = \frac{30}{a^5}, \quad \overline{E} = \frac{\hbar^2}{2m}\int |\psi'(x)|^2 dx = \frac{5\hbar^2}{ma^2} \approx 1.013 E_0. \tag{1}$$

b) Now, for the first excited level we choose $\psi = Bx(x-a/2)(x-a)$; here the multiplier $(x - a/2)$ gives rise to the proper symmetry of the wavefunction. We find

$$B^2 = \frac{840}{a^7}, \quad \overline{E} = \overline{T} = 21\frac{\hbar^2}{ma^2} \approx 1.064 E_1. \tag{2}$$

The variational results (1) and (2) for the quantity $\overline{E}$ determine the approximate values of energy levels E_0 and E_1. They turn out to be rather close to the exact values, $E_n = \pi\hbar^2(n+1)^2/2ma^2$, which is connected with the fact that the trial functions correctly represent the key properties of the exact wavefunctions $\psi_0(x)$ and $\psi_1(x)$. Compare with Problem 8.21.

Problem 8.24

For two particles of the same mass m moving in a one-dimensional infinitely deep potential well and interacting with each other as impenetrable points, evaluate the energy of the ground level by approximating the ground state by the simplest polynomial that fulfills the conditions demanded. Compare with the exact value (see Problem 2.51).

Solution

Due to the mutual impenetrability of particles, the two-particle wavefunction satisfies the condition $\psi(x_1, x_2) = 0$ at $x_1 = x_2$. So, taking into account the boundary conditions on the well walls and assuming that particle 1 is to the left of the particle 2, we approximate the exact ground-state wavefunction by

$$\psi(x_1, x_2) = Ax_1(x_1 - x_2)(a - x_2), \quad 0 \leq x_1 \leq x_2 \leq a.$$

This plays the role of the trial function for the variational calculation of the ground-level energy E_0. Normalizing the wavefunction, which gives $A^2 = 5040a^{-8}$, and taking into account $\overline{U} = 0$ we find

$$E_{0,\text{var}} = \overline{T_1} + \overline{T_2} = -\frac{\hbar^2}{2m} \int_0^a \int_0^{x_2} \psi^* \left(\frac{\partial^2}{\partial x_1^2} + \frac{\partial^2}{\partial x_2^2} \right) \psi dx_1 dx_2 = 28\frac{\hbar^2}{ma^2}, \quad \overline{T_1} = \overline{T_2}.$$

This value differs from the exact value $E_0 = 5\pi^2\hbar^2/2ma^2$ by 13%. See Problem 2.51 and a discussion about the level degeneracy.

Problem 8.25

Find the energy of the lower p-level of a particle in an infinitely deep spherical potential well of radius a by the variational method. Use the trial radial function of the form $R(r) = Ar(a^\nu - r^\nu)$ for $r \leq a$, where ν is a variational parameter ($\psi_{l=1,m}(\mathbf{r}) = R(r)Y_{1m}(\mathbf{n})$). Compare with the exact result.

Solution

In this problem there are no complications connected with the fact that the trial function must be chosen to be orthogonal to the wavefunctions with the angular momentum $l = 0$; the angular dependence in the spherical functions Y_{lm} enforces this orthogonality automatically. By using the value $\overline{U} = 0$ and normalizing the trial function given in the problem statement, we find

$$A^2 = \frac{5(5+\nu)(5+2\nu)}{2\nu^2 a^{2\nu+5}},$$

and

$$\overline{E}(\nu) = \overline{T} = \frac{\hbar^2}{2m} \int_0^a R^*(r) \left[\left(-\frac{1}{r}\frac{d^2}{dr^2} + \frac{2}{r^2} \right) R(r) \right] r^2 dr$$

$$+ \frac{5(5+\nu)(5+2\nu)}{4(3+2\nu)} \frac{\hbar^2}{ma^2}.$$

The minimum of $\overline{E}(\nu)$ is $E_{01var} \approx 10.30\hbar^2/ma^2$, with the minimum point $\nu_0 \approx 0.37$. But other values of $\nu \sim 1$ also give a close result, as we can see from the following:

ν	0	0.4	1	2
$\frac{\overline{E}ma^2}{\hbar^2}$	10.42	10.30	10.50	11.25

We should note that the exact value, $E_{01} = 10.10\hbar^2/ma^2$, follows from Problem 4.9 if we use the value $x_0 = 4.4934$ of the first zero of the Bessel function $J_{3/2}(x)$.

Problem 8.26

For the ground level of a particle in the one-dimensional δ-well, $U_0(x) = -\alpha\delta(x)$, find the level shift in a weak uniform field, i.e. for the perturbation $V = -F_0 x$, using the variational method and the trial function of the form:

$$\psi(x) = C\psi_0^{(0)}(x)\left(1 + \varepsilon F_0\, xe^{-\gamma|x|}\right), \quad \psi_0^{(0)}(x) = \sqrt{\kappa_0}\, e^{-\kappa_0|x|},$$

where $\psi_0^{(0)}(x)$ is the wavefunction of the unperturbed state (see Problem 2.7) and ε and γ are variational parameters. Compare with the exact result from Problem 8.12.

This problem illustrates the possibility of calculating terms of the perturbation theory series for Hamiltonian $\hat{H} = \hat{H}_0 + \hat{V}$ using variational method.

Solution

Let us calculate the mean value $\overline{E}(\varepsilon, \gamma)$ in accordance with the main idea of the variational method; that is by minimizing this value over the free variational parameters. First we normalize the trial function.

$$C^2 = \left(1 + \frac{\kappa_0 \varepsilon^2 F^2}{2(\kappa+\gamma)^3}\right)^{-1}, \quad \text{i.e. } C^2 \approx 1 - \frac{\varepsilon^2 F^2}{2(1+\gamma)^3}. \tag{1}$$

We use fact that the external field is weak and also introduce the units $\hbar = m = \alpha = 1$; in this case $\kappa_0 = 1$ and $E_0^{(0)} = -1/2$ (the level energy in the absence of a field). Now we have

$$\overline{T} = \frac{1}{2}\int |\psi'(x)|^2 dx = \frac{1}{2}C^2\left(1 + \frac{\varepsilon^2 F_0^2}{2(1+\gamma)}\right), \qquad (2)$$

$$\overline{U} = \overline{-\delta(x) - F_0 x} = -C^2\left(1 + \frac{8\varepsilon F_0^2}{(2+\gamma)^3}\right).$$

Using (1) we obtain

$$\overline{E}(\varepsilon, \gamma) = -\frac{1}{2} - \frac{8\varepsilon F_0^2}{(2+\gamma)^3} + \frac{(1 + (1+\gamma)^2)\varepsilon^2 F_0^2}{4(1+\gamma)^3}. \qquad (3)$$

Minimizing this expression as a function of ε, we obtain

$$\overline{E}(\varepsilon_0, \gamma) = -\frac{1}{2} - \frac{1}{2}\beta_0(\gamma)F_0^2, \quad \beta_0(\gamma) = \frac{128(1+\gamma)^3}{(2+\gamma)^6[1+(1+\gamma)^2]}. \qquad (4)$$

Now minimization over the parameter γ allows us to find the level shift. The minimum is at $\gamma = \gamma_0 \approx 0.34$. Here $\beta_{0,\text{var}} = 1.225$, while the exact value is $\beta_0 = 5/4 = 1.250$.

Problem 8.27

Using the variational method and a trial function of the form $\psi(r) = Ce^{-\kappa r}$, with $\kappa > 0$ as a variational parameter, obtain a *sufficient* condition for the existence of a bound state in a central potential $U(r) < 0$ ($U(r) \to 0$ as $r \to \infty$).

Apply your result to the potentials considered in Problem 4.8 and compare with both the exact result and the *necessary* condition of the bound state existence from Problem 4.21.

Solution

Let us calculate the mean energy of the particle[133] (with normalization $C^2 = \kappa^3/\pi$):

$$\overline{T} = \frac{\hbar^2 \kappa^2}{2m}, \quad \overline{U} = 4\kappa^3 \int_0^\infty r^2 U(r) e^{-2\kappa r} dr, \quad \overline{E}(\kappa) = \overline{T} + \overline{U}.$$

$E_0 \leq \overline{E}(\kappa)$ (E_0 is the ground state energy), so if for some value of $\kappa \geq 0$ we would have $E_0(\kappa) \leq 0$, then in the potential considered there certainly exists at least one discrete spectrum state (the potential *binds* the particle). Therefore, the desired condition takes the form:

---

[133] Only the sign of $\overline{E}(\kappa)$ is important, so normalizing the trial function is unnecessary for our purposes.

$$\max\left\{\kappa\int_0^\infty |U(r)|r^2 e^{-2\kappa r} dr\right\} \geq \frac{\hbar^2}{8m} \tag{1}$$

The use of the maximum value corresponds to an optimal choice of the parameter κ.

For potential $U = -\alpha\delta(r-a)$, condition (1) takes the form

$$\max(\alpha\kappa a^2 e^{-2\kappa a}) \geq \frac{\hbar^2}{8m}, \text{ or } \xi \equiv \frac{m\alpha a}{\hbar^2} \geq \frac{e}{4} \approx 0.68,$$

while the exact and necessary condition (in the given case, they coincide with each other) is $\xi \geq 1/2$.

For the potential from Problem 4.8 b), condition (1) gives

$$\xi \equiv \frac{ma^2 U_0}{\hbar^2} \geq \frac{27}{32} \approx 0.84.$$

The exact condition is $\xi \geq 0.72$, while the necessary one is $\xi \geq 1/2$.

Problem 8.28

Let ψ_a with $a = 0, 1, \ldots, N-1$ be some system of N mutually orthogonal wavefunctions normalized to unity, and let $\bar{E}_a$ be the mean values of the Hamiltonian, $\hat{H}$, for these states. Prove that

$$\sum_{a=0}^{N-1} \bar{E}_a \geq \sum_{n=0}^{N-1} E_n^{(0)}, \tag{1}$$

where $E_n^{(0)}$ are the exact discrete energy levels of the Hamiltonian and the summation over them accounts for their degeneracy.

Solution

Let us consider the following sum:

$$\frac{1}{N}\sum_{a=0}^{N-1} <\psi_a|\hat{H}|\psi_a> = \frac{1}{N}\sum_{a=0}^{N-1} \bar{E}_a. \tag{2}$$

Now we expand the wavefunctions $\psi_a = \sum_n C_{an}\psi_n^{(0)}$ into a series of the Hamiltonian $\hat{H}$ eigenfunctions and obtain the relation,

$$\frac{1}{N}\sum_{a=0}^{N-1}\sum_{n=0}^{\infty} |C_{an}|^2 E_n^{(0)} = \frac{1}{N}\sum_{a=0}^{N-1} \bar{E}_a. \tag{3}$$

Coefficients C_{an} have the following properties:

$$\sum_n |C_{an}|^2 = 1 \text{ and } \sum_{a=0}^{N-1} |C_{an}|^2 \leq 1. \tag{4}$$

The first of them is obvious, while the second property is a consequence of the mutual orthogonality of the wavefunctions ψ_a. Indeed, from the condition $<\psi_b|\psi_a>= \delta_{ab}$ it follows that

$$\sum_n C_{an} C_{bn}^* = \delta_{ab}.$$

Multiplying this relation by $C_{an'}^* C_{bn'}$ and summing over a and b, we obtain

$$\sum_{a=0}^{N-1} |C_{an'}|^2 = \sum_{n=0}^{\infty} \left| \sum_{a=0}^{N-1} C_{an} C_{an'}^* \right|^2 = \left(\sum_{a=0}^{N-1} |C_{an'}|^2 \right)^2 + \sum_{n=0}^{'} \left| \sum_{a=0}^{N-1} C_{an} C_{an'}^* \right|^2, \tag{5}$$

where the prime on the sum indicates the absence of the term with $n = n'$.

Now we note that the left-hand side of Eq. (3) can be written in the form:

$$\overline{E} \equiv \sum_{n=0}^{\infty} w_n E_n^{(0)}, \text{ where } w_n = \frac{1}{N} \sum_{a=0}^{N-1} |C_{an}|^2. \tag{6}$$

It is proper to think of w_n as some probability distribution normalized to 1 ($\sum_n w_n = 1$), for which from (4) we have $w_n \leq 1/N$. In this case we see that

$$\overline{E} \geq \frac{1}{N} \sum_{n=0}^{N-1} E_n^{(0)}, \tag{7}$$

where $E_n^{(0)}$ with $n = 0, 1, \ldots, N-1$ are the energy values of the N states from (1). From (3) and (7) the inequality[134] follows.

In conclusion, we note that the result obtained is the generalization of the inequality $\overline{E} \geq E_0$, and constitutes the basis for variational calculation involving excited states. It is important that the trial functions are mutually orthogonal, but there is no need to enforce orthogonality to the eigenfunctions that correspond to the lower energy levels.

---

[134] The equality is fulfilled only for ψ_a coinciding with the exact wavefunctions $\psi_n^{(0)}$.

8.3 Stationary perturbation theory (continuous spectrum)

Problem 8.29

Using the result of Problem 2.42, find the amplitude of the *reflected wave* in a short-range potential $U(x)$ ($U(x) \to 0$ for $x \to \pm\infty$) in the first two orders of perturbation theory. Discuss the conditions for applicability of perturbation theory for calculating the particle reflection coefficient. Consider the application of this result to the following specific potentials:

a) $U(x) = \alpha \delta(x)$,

b) $U(x) = \begin{cases} U_0 e^{-x/a} & \text{for } x > 0, \\ 0 & \text{for } x \leq 0, \end{cases}$

c) $U(x) = \dfrac{U_0}{\cosh^2(x/a)}$,

d) $U(x) = U_0 e^{-x^2/a^2}$,

and compare with the exact amplitudes.

Solution

Using an integral form of the Shrödinger equation,

$$\psi_p^+(x) = e^{ipx/\hbar} - i\frac{m}{\hbar|p|} \int_{-\infty}^{\infty} e^{i|p(x-x')|/\hbar} U(x') \psi_p^+(x') dx', \tag{1}$$

we obtain the expression for the reflected wave amplitude:

$$A(p) = -i\frac{m}{\hbar|p|} \int_{-\infty}^{\infty} e^{ipx/\hbar} U(x) \psi_p^+(x) dx. \tag{2}$$

(See Problem 2.42). This gives the reflection coefficient for the particles, $R(p) = |A(p)|^2$.

The solution of Eq.(1) as an expansion in powers of the potential

$$\psi_p^+(x) = \psi_p^{(0)}(x) + \psi_p^{(1)}(x) + \ldots,$$

gives

$$\psi_p^{(0)}(x) = e^{ipx/\hbar},$$

$$\psi_p^{(1)}(x) = -i\frac{m}{\hbar|p|} \int_{-\infty}^{\infty} \exp\left\{\frac{i}{\hbar}(|p||x-x'| + px')\right\} U(x') dx'. \tag{3}$$

We have the analogous expansion for the amplitude $A(p) = A^{(1)}(p) + A^{(2)}(p) + \ldots$, where

$$A^{(1)}(p) = -i\frac{m}{\hbar|p|}\int_{-\infty}^{\infty} e^{2ipx/\hbar}U(x)dx, \tag{4}$$

$$A^{(2)}(p) = -\frac{m^2}{\hbar^2 p^2}\int\int_{-\infty}^{\infty}\exp\left\{\frac{i}{\hbar}(px' + px + |p||x - x'|)\right\}U(x')U(x)dx'dx.$$

Using the Fourier transform of the potential,

$$\tilde{U}(p) = \int_{-\infty}^{\infty} e^{ipx/\hbar}U(x)dx,$$

and relation (E.1.3), it is convenient to write Eq. (4) in the form,

$$A^{(1)}(p) = -\frac{im}{\hbar|p|}\tilde{U}(2p), \tag{5}$$

$$A^{(2)}(p) = \frac{im^2}{\pi\hbar^2|p|}\int_{-\infty}^{\infty}\frac{\tilde{U}(p-q)\tilde{U}(p+q)}{q^2 - p^2 - i\gamma}dq,$$

where $\epsilon > 0$ is infinitesimal. Using these equations gives the structure of a perturbation theory series for $A(p)$, and we can make the following conclusions for its applicability.

1) Perturbation theory is not applicable for $p \to 0$, i.e., in the case of slow particles. This is not surprising, since according to Problem 2.39 we have $R(p) \to 1$ for $p \to 0$, while the applicability of perturbation theory demands the fulfillment of the condition $R \ll 1$.
2) Denoting by U_0 and a the typical values of the potential and its radius correspondingly, we see that in the case of $pa/\hbar \leq 1$ (i.e., for the particles that are not fast, here $\tilde{U}(p) \sim U_0 a$), the applicability of perturbation theory demands fulfillment of the condition

$$U_0 \ll \frac{\hbar|p|}{ma} \quad \text{for} \quad \frac{|p|a}{\hbar} \leq 1. \tag{6}$$

In this case, both $|\psi_p^{(1)}| \ll |\psi_p^{(0)}|$ and $|A^{(2)}| \ll |A^{(1)}|$.

3) In the case of fast particles, $p \to \infty$, though the distortion of the wavefunction, $\psi_p^+(x)$, by comparison with the unperturbed one, $\psi_p^{(0)}$, is always small, the applicability of perturbation theory (and the convergence of the series for $A(p)$ in general) depends on the nature of $\tilde{U}(p)$ decrease for $p \to \infty$.

As is seen from Eq. (5), if the asymptotic behavior of $\tilde{U}(p)$ is

$$|\tilde{U}(p)| > C\exp\{-\alpha|p|^\nu\}, \quad |p| \to \infty$$

with $\nu < 1$ (*i.e.*, the slower than exponential $\propto e^{-\alpha|p|}$), the integral in Eq. (5) decreases faster than $\tilde{U}(2p)$, and therefore $|A^{(2)}|/|A^{(1)}| \to 0$ as $|p| \to \infty$ (the same condition takes applies to the higher-order $A^{(n)}$). This means that for $p \to \infty$, the perturbation theory is applicable and the series for $A(p)$ converges for any value of U_0.

In the case of a decrease law of the form

$$|\tilde{U}(p)| < B\exp\{-\alpha|p|^\nu\}, \quad \nu > 1, \quad |p| \to \infty,$$

i.e., faster than exponential $\propto e^{-\alpha|p|}$, the integral in Eq. (5) decreases slower than $\tilde{U}(2p)$ (see below in d), so that $|A^{(2)}|/|A^{(1)}| \to \infty$ and the perturbation series diverges.

We should note that a transitional situation occurs in the case of a decrease in the form $\tilde{U}(p) \propto |p|^\gamma e^{-\alpha|p|}$ with $\gamma = 1$: for $\gamma > 1$ the perturbation theory is applicable, while for $\gamma < 1$ it is non-applicable. In the case of $\gamma = 1$, the integral in Eq. (5) decreases in the same way as $\tilde{U}(2p)$. Here, perturbation theory applicability and the convergence of series $A(p)$ depends on the value of parameter $ma^2 U_0/\hbar^2$.

Let us consider some concrete applications of perturbation theory.

a) For the δ-potential, by using Eq. (4), we obtain

$$A^{(1)}(p) = -\frac{im\alpha}{\hbar|p|}, \quad A^{(2)}(p) = -\frac{m\alpha^2}{\hbar^2 p^2}, \tag{7}$$

which is useful to compare with an expansion of the exact expression for the amplitude (see Problem 2.30):

$$A(p) = -\frac{im\alpha}{\hbar|p| + im\alpha} = -\frac{im\alpha}{\hbar|p|} - \frac{m\alpha^2}{\hbar^2 p^2} + \ldots$$

In accordance with Eq. (6), perturbation theory is valid in the case of $m\alpha \ll \hbar p$ (for δ-potential $a = 0$, but $U_0 a \sim \alpha$); the perturbation theory series for $A(p)$ converges if the condition $m\alpha/\hbar p < 1$ is fulfilled.

For the calculation of $A(p)$ in the other cases, we shall consider $p > 0$ and take into account the symmetry of the function under the integrand in Eq. (4), rewriting it in the form:

$$A^{(2)}(p) = -\frac{2m^2}{\hbar^2 p^2}\int_{-\infty}^{\infty} e^{2ipx/\hbar} U(x) \int_{-\infty}^{x} U(x')dx' dx. \tag{8}$$

b) For the potential under consideration, the simple integration in Eqs. (4) and (8) gives

$$A^{(1)}(p) = -\frac{imaU_0}{\hbar p(1 - 2ipa/\hbar)}, \quad A^{(2)}(p) = -\frac{m^2 a^2 U_0^2}{\hbar^2 p^2 (1 - ipa/\hbar)(1 - 2ipa/\hbar)}. \tag{9}$$

As is seen for $pa/\hbar \sim 1$, the condition of perturbation theory applicability, $|A^{(2)}|/|A^{(1)}| \ll 1$, coincides with Eq. (6). Due to the power-law decrease of $\hat{U}(p)$, perturbation theory is applicable in the case of $p \to \infty$. The parameter of expansion is mU_0/p^2. For the potential considered, the exact value of the amplitude is

$$A(p) = \frac{J_{1-2ika}(2i\xi)}{J_{-1-2ika}(2i\xi)}, \tag{10}$$

where $k = p/\hbar$, $\xi = (2ma^2U_0/\hbar^2)^{1/2}$, $J_\nu(x)$ is the Bessel function of νth order.

c) For the potential $U = U_0 \text{ch}^{-2}(x/a)$ in first-order perturbation theory,

$$A^{(1)} = -\frac{imU_0}{\hbar} \int_{-\infty}^{\infty} \frac{e^{eipx/\hbar}}{\text{ch}^2(x/a)} dx. \tag{11}$$

The integral is calculated with the help of residue theory by closing the integration contour into the upper half-plane of the complex variable x. The singular points of the function under the integral, the poles of the second order, are $x_n = ia(\pi n + \pi/2)$, where $n = 0, 1, \ldots$, $a > 0$. Since for $x \to x_n$ we have

$$\text{ch}^2\left(\frac{x}{a}\right) \approx -\frac{(x-x_n)^2}{a^2} + O\left((x-x_n)^4\right)$$

and hence here

$$\frac{e^{2ipx/\hbar}}{\text{ch}^2(x/a)} \approx -e^{2ipx_n/\hbar}\left[\frac{a^2}{(x-x_n)^2} + \frac{2ipa^2}{\hbar}\frac{1}{(x-x_n)}\right],$$

then the total contribution of all of the poles is

$$A^{(1)} = -\frac{imU_0}{\hbar p} 2\pi i \sum_{n=0}^{\infty} \left(-\frac{2ipa^2}{\hbar} e^{2ipx_n/\hbar}\right) = -\frac{2\pi i m a^2 U_0}{\hbar^2 \text{sh}(\pi pa/\hbar)}. \tag{12}$$

The series is geometric.

The integral for $A^{(2)}$, according to Eq. (8), takes the form

$$A^{(2)} = -\frac{2maU_0}{\hbar^2 p^2} \int_{-\infty}^{\infty} \frac{1}{\text{ch}^3(x/a)} \exp\left\{\frac{2ipx}{\hbar} + \frac{x}{a}\right\} dx, \tag{13}$$

and after some simple transformations it can be expressed in terms of integral (11), which allows us to obtain

$$A^{(2)} = -4\pi \left(1 + \frac{ipa}{\hbar}\right) \frac{m^2 a^3 U_0^2}{\hbar^3 p\,\text{sh}(\pi pa/\hbar)}. \tag{14}$$

Comparing Eqs. (12) and (14), we see that for $pa/\hbar \sim 1$, the applicability of perturbation theory demands the fulfillment of (6). As for the case of $pa/\hbar \gg 1$, the perturbation theory is applicable only with the condition $U_0 \ll \hbar^2/ma^2$. This situation differs from the one for the two previous potentials, and is connected with the exponential decrease of $\hat{U}(p) \propto p e^{-\pi p a/\hbar}$ for $p \to \infty$. We should note that this potential allows an exact calculation of $A(p)$. With this result it is easy to reach the conclusion that when $|U_0| > \hbar^2/8ma^2$, the perturbation theory series diverges for $p \to \infty$.

d) For the potential $U(x) = U_0 e^{-x^2/a^2}$, we obtain

$$A^{(1)}(p) = -i\sqrt{\pi}\frac{maU_0}{\hbar p}\exp\left\{-\frac{p^2 a^2}{\hbar^2}\right\}. \tag{15}$$

In calculating the amplitudes of second order in the perturbation in the case of a fast particle, with Eq. (5) and with $p \gg \hbar/a$, we see that in the integral over q, the domain $q \leq \hbar/a$ plays the main role (contribution of the rest of the space is not important due to the exponential decrease of the function under the integral). As a consequence, we can put $q = 0$. Then it is easy to calculate the integral, which allows us to obtain

$$A^{(2)}(p) \approx -i\sqrt{2\pi}\frac{m^2 a U_0^2}{\hbar p^3}\exp\left\{-\frac{p^2 a^2}{2\hbar^2}\right\}, \text{ for } \frac{pa}{\hbar} \gg 1. \tag{16}$$

A comparison of this and Eq. (15) shows that $|A^{(2)}|/|A^{(1)}| \to \infty$ as $|p| \to \infty$ and the perturbation theory for fast particles is not applicable (the series for $A(p)$ is divergent).

In conclusion, we should mention that the different roles of higher orders of perturbation theory for $p \to \infty$, and how they depend on the $\tilde{U}(p)$ decrease, correctly reflect the actual physical situation per the following arguments. For the "slow" decrease of $\tilde{U}(p)$ (i.e., roughly speaking, for the case $|\tilde{U}| > Ce^{-\alpha p}$), a sharp change in particle momentum happens due to a single interaction. For the "fast" decrease, a noticeable change in momentum requires a large number of interactions, each giving only a small change. Compare with the results of Problems 4.18 and 13.84.

Problem 8.30

Determine the reflection coefficient, $R(E)$, for a fast particle in the case of a potential, $U(x)$, having a jump at the point $x = 0$ (see Fig. 8.2). Generalize this result to the case of a potential that has gaps at several points. Apply the result to the potential in Problem 8.29 b and to the square barrier from Problem 2.31. Compare with the asymptote for $E \to \infty$ of the exact result for $R(E)$. See also Problem 9.27.

Solution

From Eq. (4) of the previous problem we can determine amplitude of the reflected wave, $A^{(1)}$, from the Fourier transform of the potential.

$$\tilde{U}(k) = \int_{-\infty}^{\infty} U(x) e^{ikx} dx = \frac{i}{k}\int_{-\infty}^{\infty} e^{ikx}\frac{\partial U(x)}{\partial x}dx \tag{2}$$

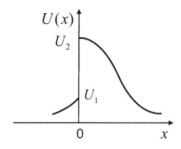

Fig. 8.2

(Here we should use the relation $ike^{ikx} = \partial e^{ikx}/\partial x$ and perform the integration by parts.) For the potential, which has a jump at the point $x = 0$, its derivative $U'(x)$ includes the term $(U_2 - U_1)\delta(x)$, and this δ-function determines the asymptotic behavior:

$$\tilde{U} \approx \frac{i(U_2 - U_1)}{k} \text{ for } k \to \infty.$$

The contribution from the other part of the integration region is inessential due to the fast oscillation of the function under the integral. Thus

$$R(p) \approx \frac{m^2(U_2 - U_1)^2}{4p^4}, \quad p \to \infty. \tag{2}$$

The generalization of this formula for the case of potential which contains jumps at several points, x_n, is given by

$$R(p) = \frac{m^2}{4p^4} \left| \sum_n \Delta U_n e^{2ipx_n/\hbar} \right|^2, \quad p \to \infty, \tag{3}$$

where ΔU_n is the potential jump at the corresponding point x_n. Note that for discontinuous potentials, $R(p) \propto 1/p^4$ as $p \to \infty$.

According to (2) and (3) we find the reflection coefficients

$$1) \ R(p) \approx \frac{m^2 U_0^2}{4p^4}, \quad 2) \ R(p) \approx \frac{m^2 U_0^2}{p^4} \sin^2\left(\frac{pa}{\hbar}\right)$$

for the potential from Problem 8.29 b and for a square potential, respectively, which coincide with the asymptotes of the exact expressions for $R(p)$ as $p \to \infty$. See also Problem 8.31.

Problem 8.31

The same as in the previous problems but for a jump in the derivative of the potentials at the point $x = 0$. Apply the result obtained to the parabolic barrier of the form: $U(x) = U_0(1 - x^2/a^2)$ for $|x| < a$ and $U(x) = 0$ for $|x| > a$.

Solution

As in Eq. (1) from the previous problem, we obtain

$$\tilde{U}(k) = -\frac{1}{k^2} \int_{-\infty}^{\infty} e^{ikx} U''(x) dx. \tag{1}$$

In this case, the potential has a derivative $U'(x)$ jump, so that $U''(x)$ contains a δ-function term of the form $-\Delta F \delta(x)$. $-\Delta F = U'(0+) - U'(0-)$ is the jump of the derivative, which determines the relevant asymptote of $A^{(1)}$. The reflection coefficient for the potential whose derivative has discontinuities at several points x_n for $p \to \infty$ is equal to

$$R(p) \approx \frac{m^2 \hbar^2}{16 p^6} \left| \sum_n \Delta F_n e^{2ipx_n/\hbar} \right|^2, \tag{2}$$

thus $R(p) \propto 1/p^6$. For the parabolic potential, (2) gives

$$R(p) = \frac{m^2 \hbar^2 U_0^2}{a^2 p^6} \sin^2 \left(\frac{2pa}{\hbar} \right).$$

In conclusion, we should note that that the reflection coefficient for $p \to \infty$ and the analytical properties of the potential energy, $U(x)$, as a function of x are closely related. If the potential has critical points (*singularities*) on the real axis x, then $R(p)$ decreases as a power law. Here, the weaker the singularity, the faster the decrease; compare with the results of the previous problem. But if $U(x)$ has no critical points on the real axis x (an infinitely differentiable function), then $R(p)$ decreases exponentially. See also Problem 4.18.

Problem 8.32

As is known, the particle energy spectrum in a periodic potential consists of bands. For such a one-dimensional potential, $U(x + a) = U(x)$, regarded as a small perturbation, find the energy spectrum $E_n(q)$, where n is the band label and $\hbar q$ is the *quasi-momentum* (here $-\pi/a \leq q \leq \pi/a$).

Determine a relation between the momentum, $\hbar k$, of a free particle and the quasi-momentum, $\hbar q$, and find the correct eigenfunctions in the zeroth approximation, $\psi_{n,q}^{(0)}(x)$. Find the energy gap between the neighboring energy bands. Apply the results obtained to the potential from Problem 2.53.

Solution

Using the known relation for the Bloch functions,

$$\psi_{n,q}(x+a) = e^{iqa}\psi_{n,q}(x), \quad -\frac{\pi}{a} < q < \frac{\pi}{a}, \tag{1}$$

it will be sufficient to solve the Schrödinger equation on the interval $0 < x < a$:

$$-\frac{\hbar^2}{2m}\psi_{n,q}''(x) + U(x)\psi_{n,q}(x) = E_n(q)\psi_{n,q}(x). \tag{2}$$

Here, relation (1) can be viewed mathematically as a condition that gives a self-adjoint extension[135] of the Hermitian operator $\hat{p}^2/2m + U(x)$ in Eq. (2) defined on this interval, to all values of q (see Problem 1.29). For a given value, q, the spectrum, $E_n(q)$, is discrete, while the continuous dependence on q leads to a band structure of the energy spectrum in general.

Neglecting $U(x)$ in Eq. (2) we obtain the following unperturbed eigenfunctions and eigenvalues:

$$\psi_k^{(0)} = \frac{1}{\sqrt{a}}e^{ikx}, \quad E_k^{(0)} = \frac{\hbar^2 k^2}{2m}.$$

Taking into account relation (1), we can classify these expressions in terms of the quasi-momentum q (and the band number n). Writing $k = 2\pi n/a + q$, where $n = 0, \pm 1, \ldots$, we obtain

$$E_n^{(0)}(q) = \begin{cases} \frac{\hbar^2}{2m}\left(\frac{\pi n}{a} + |q|\right)^2, & n = 0, 2, 4, \ldots, \\ \frac{\hbar^2}{2m}\left(\frac{\pi(n+1)}{a} - |q|\right)^2, & n = 1, 3, 5, \ldots. \end{cases} \tag{3}$$

This spectrum consists of the unperturbed spectrum $E_k^{(0)}$ (see dashed lines on Fig. 8.4), so the neighboring bands touch one another (there are no forbidden values of the

---

[135] It actually implies imposing the following two boundary conditions,

$$\psi(a) = e^{iqa}\psi(0), \quad \psi'(a) = e^{iqa}\psi'(0),$$

which corresponds to the $(2,2)$ deficiency index of this operator (see, Problem 1.29 for the relevant definition and discussion).

energy). The relation between the momentum and the quasi-momentum of a free particle is given by

$$k = \begin{cases} \frac{\pi n}{a} + q, & q > 0, \\ -\frac{\pi n}{a} + q, & q < 0, \end{cases} \quad n = 0, 2, 4, \ldots, \tag{4a}$$

$$k = \begin{cases} -\frac{\pi(n+1)}{a} + q, & q > 0, \\ \frac{\pi(n+1)}{a} + q, & q < 0. \end{cases} \quad n = 1, 3, 5, \ldots, \tag{4b}$$

from which we have the explicit form of the unperturbed Hamiltonian eigenfunctions, $\psi_{n,q}^{(0)}(x) = e^{ikx}/\sqrt{a}$, which satisfy condition (1).

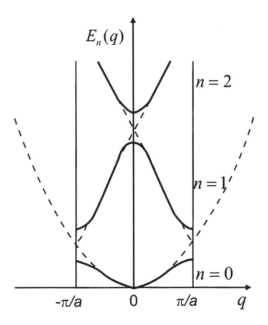

Fig. 8.3

From relations (3) and (4) (see also Fig. 8.3), we find that for a given value of q, the unperturbed levels, $E_n^{(0)}(q)$, are separated by a finite interval, generally speaking. So for the calculation of level shifts under the influence of the potential $U(x)$, we can use Eq. (VIII.1) of non-degenerate perturbation theory. For the first-order correction, we obtain

$$E_n^{(1)}(q) = \frac{1}{a} \int_0^a U(x) dx \equiv \overline{U}. \tag{5}$$

(In this approximation, the shift is the same for all values of q and n.) The condition for applicability of (5) is given by

$$\left|E_n^{(0)}(q) - E_{n+1}^{(0)}(q)\right| \gg E_n^{(1)}(q).$$

As we see from Eq. (3) and Fig. 8.4, this condition is broken for 1) $q \approx 0$, and 2) $q \approx \pm \pi/a$, when the neighboring energy bands touch each other for values of energy equal to

1) $E_n^{(0)}(0) = E_{n+1}^{(0)}(0)$, and 2) $E_n^{(0)}(\pm\pi/a) = E_{n-1}^{(0)}(\pm\pi/a)$.

Here $n = 1, 3, 5, \ldots$ in both cases.

For the values of q given above to calculate the shifts $E_n^{(0)}(q)$ we should use perturbation theory valid for the closely-positioned levels. Compare to Problem 8.6. Just as in the case of strongly degenerated levels, the perturbation mixes the unperturbed eigenfunctions with the close energies, so

$$\tilde{\psi} = C_1 \psi_{n,q}^{(0)}(x) + C_2 \psi_{n+1,q}^{(0)}(x), \quad n = 0, 1, 2, \ldots \tag{6}$$

The perturbed levels and the coefficients $C_{1,2}$ which determine the correct eigenfunctions in the zeroth approximation are found in the same way as in the degenerate perturbation theory. The secular equation takes the form

$$\begin{vmatrix} \overline{U} + E_n^{(0)}(q) - E & U_{n,n+1} \\ U_{n+1,n} & \overline{U} + E_{n+1}^{(0)}(q) - E \end{vmatrix} = 0, \tag{7}$$

where we take into account the value (5) for U_{nn}. In the matrix element $U_{n+1,n}$ we can use the eigenfunctions $\psi_{n(n+1),q}^{(0)}(x)$ for the values of q that correspond to the levels joining together. The values of the corresponding momenta are $\hbar k = \pm(n+1)h/a$ (in the case where nth and $(n+1)$-th zones meet), so that in Eq. (7),

$$|U_{n+1,n}| = \frac{1}{a}\left|\int_0^a \exp\left\{\pm 2i\pi(n+1)\frac{x}{a}\right\} U(x)dx\right| \equiv \triangle_n. \tag{8}$$

See that $\triangle_n$ does not depend on the choice of the sign $\pm$.

The solution of Eq. (7) gives

$$E_{n(n+1)}(q) = \tilde{U} + \frac{1}{2}\left[E_n^{(0)}(q) + E_{n+1}^{(0)}(q) \mp \sqrt{\left(E_{n+1}^{(0)}(q) - E_n^{(0)}(q)\right)^2 + 4\triangle_n^2}\right]. \tag{9}$$

The signs $(-)$ and $(+)$ correspond to the (lower) n and (upper) $n+1$ bands respectively. From relations (3), (5), and (9), the plot for $E_n(q)$ is shown in Fig. 8.4 (solid line, and we put $\overline{U} = 0$). Taking into account the interaction, we obtain the band structure with an energy gap present in the spectrum. The gap width – the distance between the neighboring n and $n+1$ zones – is equal to $2\Delta_n$. Coefficients $C_{1,2}$ in expression (6) are equal to

$$C_1 = \frac{U_{n,n+1}}{E - \overline{U} - E_n^{(0)}(q)} C_2. \tag{10}$$

The two values of E are given by Eq. (9). Particularly, at the point of an avoided crossing of the unperturbed levels, we obtain $C_1 = \pm C_2 = 1/\sqrt{2}$.

As we deviate (with a change of q) from the avoided crossing, we have

$$\delta E_n^{(0)}(q) \equiv E_{n+1}^{(0)}(q) - E_n^{(0)}(q) \gg \Delta_n,$$

from Eq. (9) it follows that

$$E_{n(n+1)}(q) \approx E_{n(n+1)}^{(0)}(q) + \tilde{U} \mp \frac{\Delta_n^2}{\delta E_n^{(0)}(q)}. \tag{11}$$

Here the last term is a part of the second-order correction, corresponding to the term associated with the closest level in Eq. (VIII.1).

Using the explicit form for $E_n^{(0)}$, Eq. (9) may be rewritten in a more informative form. For example, when in Eq. (9) for the lowest band the value of n is odd (the avoided crossing of levels is for $q = 0$), we have

$$E_{n(n+1)}(q) \approx \frac{\hbar^2}{2m}(k_n^2 + q^2) + \tilde{U} \mp \sqrt{\frac{\hbar^4 k_n^2 q^2}{m^2} + \Delta_n^2}, \tag{12}$$

where $k_n = \pi(n+1)/a$. For $|q| \ll m\Delta_n/\hbar^2 k_n$ from here it follows[136]

$$E_{n(n+1)}(q) \approx \frac{\hbar^2}{2m}k_n^2 + \tilde{U} \mp \Delta_n \mp \frac{1}{2}\left[\frac{\hbar^2 k_n^2}{m^2 \Delta_n} \mp \frac{1}{m}\right]\hbar^2 q^2. \tag{13}$$

In the case when for the lower band the value of n is even and the avoided crossing takes place for $q = \pm\pi/a$, we obtain the expression analogous to Eqs. (12) and (13), but with the change $|q| \to \pi/a - |q|$.

In conclusion, we should note that if we apply our results to the potential from Problem 2.53 we have $\overline{U} = \alpha$ and $\Delta_n = |\alpha|$, so the energy gap between the neighboring zones has a constant value independent of n. This is specific to the δ-potential; for any other potential, $\Delta_n \to 0$ for $n \to \infty$. At the same time, the width of the allowed bands increases linearly with n.

---

[136] We should pay attention to the quadratic dependence of the energy on the quasi-momentum close to the band edge: $E(q) - E(0) \propto q^2$.

8.4 Non-stationary perturbation theory; Transitions in continuous spectrum

Problem 8.33

A charged linear oscillator is under the effect of an homogeneous electric field that changes in time by the law:

a) $\mathcal{E}(t) = \mathcal{E}_0 \exp\{-t^2/\tau^2\}$;
b) $\mathcal{E}(t) = \mathcal{E}_0(1 + t^2/\tau^2)^{-1}$;
c) $\mathcal{E}(t) = \mathcal{E}_0 \exp\{-t^2/\tau^2\}\cos\omega_0 t$.

Assuming the oscillator was in the nth quantum state before the field was turned on (for $t \to -\infty$), find the transition probabilities into different states for $t \to +\infty$ in the first order of perturbation theory. For $n = 0$, compare the result obtained to the exact one (see Problem 6.25).

Solution

The oscillator perturbation has the form $V = -ex\mathcal{E}(t)$. Its matrix elements are $V_{kn}(t)$, and hence the probabilities for the oscillator transitions in the first order of perturbation theory are non-zero only for $k = n \pm 1$ (see Eq. (II.3); transitions appear only between the neigboring levels). Using Eq. (VIII.9) we obtain

$$W^{(1)}(n \to k) = \frac{e^2|I|^2}{2m\hbar\omega} \cdot \begin{cases} (n+1), & k = n+1, \\ n, & k = n-1, \end{cases} \tag{1}$$

where $I(\omega) = \int\limits_{-\infty}^{\infty} e^{i\omega t}\mathcal{E}(t)dt$. We should note that $|I(\omega)|^2$ does not depend on the sign of ω. For the forms of $\mathcal{E}(t)$ given above, we find ($\tau > 0$):

a) $I(\omega) = \mathcal{E}_0 \int\limits_{-\infty}^{\infty} \exp\{i\omega - t^2/\tau^2\}dt = \sqrt{\pi}\tau\mathcal{E}_0 \exp\{-\frac{\omega^2\tau^2}{4}\}$,

b) $I(\omega) = \mathcal{E}_0 \int\limits_{-\infty}^{\infty} e^{i\omega t}\frac{dt}{1+(t/\tau)^2} = \pi\tau\mathcal{E}_0 e^{-|\omega\tau|}$,

c) $I(\omega) = \mathcal{E}_0 \int\limits_{-\infty}^{\infty} \exp\{i\omega t - t^2/\tau^2\}\cos(\omega_0 t)dt =$

$$\frac{1}{2}\sqrt{\pi}\tau\mathcal{E}_0\left\{\exp\left[-\frac{1}{4}(\omega-\omega_0)^2\tau^2\right] + \exp\left[-\frac{1}{4}(\omega+\omega_0)^2\tau^2\right]\right\}.$$

The main condition for the applicability of this result obtained by perturbation theory is the fulfillment of inequality (VIII.3), which here takes the form:

$$e\mathcal{E}_0\sqrt{\frac{(n+1)\hbar}{m\omega}} \ll \hbar\omega.$$

For a non-resonant perturbation this condition implies small transition probabilities, $W^{(1)}(n \to k) \ll 1$. In the case of a weak resonance interaction (see c), this condition also restricts the duration of the perturbation.

We should note that for the slow constant field turning on and off, i.e., for $\tau \to \infty$, the transition probabilities go to zero,[137] except for the case c, where for $\omega \to \omega_0$ the probabilities increase with τ, which is connected with the resonant nature of the field.

Problem 8.34

An homogeneous electric field that changes in time as $\boldsymbol{\mathcal{E}}(t) = \mathcal{E}(t)\mathbf{n}_0$ is applied to a two-dimensional rotor with the dipole moment d. Before the field is applied, the rotor has a definite value of energy and a definite projection of the angular momentum, m. Calculate the probabilities of different values of the projection of the angular momentum and rotor energies for $t \to +\infty$ in the first order of perturbation theory. Consider in particular the forms of $\mathcal{E}(t)$ given in the previous problem.

Solution

The perturbation has the form $V = -d\mathcal{E}(t)\cos\varphi$. φ is the angle between the rotor axis and the direction of the electric field. Perturbation matrix elements are different from zero only for $m' = m \pm 1$, and are equal to

$$V_{m'm} = -\frac{d\mathcal{E}(t)}{2}.$$

See Problem 8.8. In the first order of perturbation theory, only transitions between the neighboring levels occur, and their probabilities are

$$W^{(1)}(m \to m') = \frac{d^2}{4\hbar^2} \left| \int_{-\infty}^{\infty} \mathcal{E}(t) e^{i\omega_{m'm}t} dt \right|^2, \quad m' = m \pm 1. \tag{1}$$

The values of the integral in (1) are given in the previous problem, but it is necessary to take into account the fact that now the values of the transition frequencies $\omega_{m'm}$ are equal to $(1 \pm 2m)\hbar/2I$ for $m' = m \pm 1$. The applicability condition is also given in the previous problem.

Problem 8.35

The same as in the previous problem, but for a spherical (three-dimensional) rotor. Before the electric field was turned on, the rotor was in the state with quantum numbers $l, l_z = m$. The field is directed along the z-axis.

---

[137] This property, which reflects the adiabaticity of the slow perturbations, remains even in the case of strong fields. See Problems 8.54 and 8.55.

Solution

The perturbation has the form $V = -d\mathcal{E}(t)\cos\theta$. Using the values of matrix elements V_{kn} from Problem 8.11 (where we considered a stationary field), from Eq. (VIII.9) we obtain

$$W^{(1)}(l,m \to l'm) = \frac{d^2}{\hbar^2}\left|\int_{-\infty}^{\infty}\mathcal{E}(t)e^{i\omega_{l'l}t}dt\right|^2 \times \begin{cases} |a_{lm}|^2, & l' = l+1, \\ |a_{l-1,m}|^2, & l' = l-1. \end{cases}$$

In the first order of perturbation theory, transitions appear only to the neighboring energy levels. Due to conservation of l_z, the value of m does not change. The transition frequencies are equal to

$$\omega_{l+1,l} = \frac{\hbar(l+1)}{I} \quad \text{and} \quad \omega_{l-1,l} = -\frac{\hbar l}{I}.$$

The values of the integral in Eq. (1) are given in Problem 8.33.

Problem 8.36

For the conditions of Problem 8.34, determine the transition probabilities for the case when the direction of the electric field rotates in-plane with the rotor and with the angular velocity ω_0, i.e., $\mathcal{E}_x = \mathcal{E}(t)\cos\omega_0 t$, $\mathcal{E}_y = \mathcal{E}(t)\sin\omega_0 t$. Pay attention to the possibility of a resonant increase of the transition probabilities even for a slow dependence, $E(t)$, of the form a and b from Problem 8.33.

Solution

Considering interaction between the rotator and the field as a perturbation,

$$V = -d\mathcal{E}(t) = -d\mathcal{E}(t)\cos(\varphi - \omega_0 t),$$

we can see that the results for the rotor transition probabilities in the rotating field are obtained from Eqn. (1) of Problem 8.34 by changing the corresponding integral to

$$\int_{-\infty}^{\infty}\mathcal{E}(t)\exp\{i(\omega_{m'm} + m\omega_0 - m'\omega_0)t\}dt.$$

This means that the transition probability is determined by the Fourier component of the field, with the frequency

$$\tilde{\omega}_{m'm} = \omega_{m'm} + (m - m')\omega_0,$$

which could be small for the corresponding value of ω_0. The probability of such a transition could increase sharply if the duration of the perturbation is long enough. This resonance situation can be understood if we go into a rotating reference frame that rotates together with the field (see Problem 6.29). In this coordinate system the unperturbed energy levels, $E_{rot,m} = E_m^{(0)} - \hbar\omega_0 m$, with the different values of the projection of the angular momentum m can be degenerate, and the perturbation slowly changing in time could sharply enhance (if its duration is long enough) transitions between the corresponding states (see Problem 8.40). The value $\tilde{\omega}_{m'm}$ is the frequency of the transition for rotor states considered in the rotating frame of reference.

Problem 8.37

Obtain a relation between the wavefunction and the transition amplitude from the initial nth (for $t \to -\infty$) to finite kth (for $t \to +\infty$) states in the second order of non-stationary perturbation theory. Assume that for $t \to \pm\infty$, the perturbation is absent.

Solution
Using Eqs. (VIII.6)–(VIII.9), we find

$$a_{kn}(t) = \delta_{kn} + a_{kn}^{(1)}(t) + a_{kn}^{(2)}(t) + \ldots,$$

where the values of the first-approximation amplitudes $a_{kn}^{(1)}(t)$ are determined by Eq. (VIII.8). Substituting them into Eq. (VIII.7), we obtain

$$\dot{a}_{kn}^{(2)}(t) = -\frac{i}{\hbar} \sum_m V_{km} a_{mn}^{(1)}(t) e^{i\omega_{km}t}.$$

Hence, using Eq. (VIII.8) and the fact that $a_{kn}^{(2)}(-\infty) = 0$, we find

$$a_{kn}^{(2)}(t) = -\frac{1}{\hbar^2} \sum_m \int_{-\infty}^{t} V_{km}(t') e^{i\omega_{km}t'} \int_{-\infty}^{t'} V_{mn}(t'') e^{i\omega_{mn}t''} dt'' dt'. \tag{1}$$

The transition probability for the system from the initial nth to the finite kth (for $t \to +\infty$) state is equal to ($k \neq n$):

$$W(n \to k) = |a_{kn}(t = +\infty)|^2 = |a_{kn}^{(1)}(\infty) + a_{kn}^{(2)}(\infty) + \ldots|^2.$$

If $a_{kn}^{(1)}(\infty) = 0$, then

$$W^{(2)}(n \to k) = |a_{kn}^{(2)}(\infty)|^2, \quad k \neq n,$$

defines the probability of a transition that is forbidden in the first order of perturbation theory.

Problem 8.38

In the conditions of Problem 8.33, find in second-order perturbation theory the oscillator transition probabilities that are *forbidden* in the first order.[138] Compare these second-order transition probabilities with $W^{(1)}(n \to k)$.

Solution

From relation (1) of the previous problem and the values of the perturbation ($V = -e\mathcal{E}(t)x$) matrix elements for the oscillator (see Eq. (II.3)), it follows that in second order, oscillator transitions from the initial states with quantum numbers n to the final states with $n \pm 2$ appear and those are forbidden in first-order perturbation theory (see Problem 8.33). Here, the sum in the relation is reduced to the single term with $m = n \pm 1$. The transition frequencies in this term coincide, $\omega_{km} = \omega_{mn} = \pm\omega$ (due to the equal level spacing between the oscillator levels), which allows to simplify the integration to

$$\int_{-\infty}^{\infty} f(t)e^{i\omega t} dt \int_{-\infty}^{t} f(t')e^{i\omega t'} dt' = \frac{1}{2} \left[\int_{-\infty}^{\infty} f(t)e^{i\omega t} dt \right]^2.$$

After the transformations above, we obtain

$$a_{n\pm2,n}^{(2)}(\infty) = -\frac{e^2}{4m\hbar\omega} g_n^{(\pm)} \left[\int_{-\infty}^{\infty} \mathcal{E}(t)e^{\pm i\omega t} dt \right]^2,$$

where

$$g_n^{(+)} = \sqrt{(n+1)(n+2)} \text{ and } g_n^{(-)} = \sqrt{(n-1)n}.$$

The probabilities of the transitions considered are

$$W^{(2)}(n \to k) = |a_{n\pm2,n}^{(2)}(\infty)|^2.$$

If we compare them with the transition probabilities from first-order perturbation theory (see Problem 8.33), we see that $W^{(2)} \sim [W^{(1)}]^2$; hence $W^{(2)}/W^{(1)} \sim W^{(1)} \ll 1$, as long as perturbation theory applies.

Problem 8.39

If we use relation (VIII.8), then we obtain the unreasonable result that $W_n = |a_{nn}(+\infty)|^2 > 1$ for the probability of the system to remain in the initial nth state. Resolve this paradox and recover the conservation law of the wave-function normalization, taking into account the transitions in first-order perturbation theory.

---

[138]   I.e., transitions for which $W^{(1)}(n \to k) = 0$.

Solution

To resolve the paradox, we notice that to calculate the square of modulus of a quantity such as $a = 1 + a^{(1)} + a^{(2)} + \ldots$, that is the expansion in some small parameter $V \ll 1$ (so that $|a^{(n)}| \sim V^n$), to an accuracy up to the second order in V, it is necessary to know the real part[139] of a with the same accuracy, since

$$|a|^2 = 1 + 2\text{Re } a^{(1)} + |a^{(1)}|^2 + 2\text{Re } a^{(2)} + O(V^3), \quad |a^{(1)}|^2 \sim \text{Re } a^{(2)} \sim V^2.$$

(According to Eq. (VIII.8), $a_{mn}^{(1)}$ is an imaginary quantity.)

Now we should discuss the conservation of the wavefunction normalization, taking into account transitions in first-order perturbation theory. According to Eq. (1) from Problem 8.37, we have

$$a_{mn}(t = +\infty) \approx 1 - \frac{i}{\hbar} \int_{-\infty}^{\infty} V_{mn}(t) dt$$

$$- \frac{1}{\hbar^2} \sum_m \int_{-\infty}^{\infty} V_{nm}(t) e^{i\omega_{nm} t} \int_{-\infty}^{t} V_{mn}(t') e^{i\omega_{mn} t'} dt' dt,$$

so that the probability for the system to remain in the initial nth state as $t \to +\infty$ to an accuracy of the second order in perturbation is

$$W_n^{(2)} = |a_{nn}(\infty)|^2 = 1 + \frac{1}{\hbar^2} \left[\int_{-\infty}^{\infty} V_{nn}(t) dt \right]^2$$

$$- \frac{1}{\hbar^2} \left\{ \sum_m \int_{-\infty}^{\infty} V_{nm}(t) e^{i\omega_{nm} t} \int_{-\infty}^{t} V_{mn}(t') e^{i\omega_{mn} t'} dt' dt + \text{c.c.} \right\},$$

where "c.c." denotes the complex conjugate. Taking into account $\omega_{nm} = -\omega_{mn}$, $V_{nm}^* = V_{mn}(t)$, and using the relation

$$\int_{-\infty}^{\infty} dt \int_{-\infty}^{t} f(t,t') dt' = \int_{-\infty}^{\infty} dt' \int_{t'}^{\infty} f(t,t') dt,$$

expression (1) could be transformed to the desired form:

$$W_n^{(2)} = 1 - \frac{1}{\hbar^2} {\sum_m}' \left| \int_{-\infty}^{\infty} V_{mn}(t) e^{i\omega_{mn} t} dt \right|^2 = 1 - {\sum_m}' W^{(1)}(n \to m),$$

where the prime after the sum means the absence of a term with $m = n$.

---

[139] It is enough to have the imaginary part only in the first order.

Problem 8.40

For a system that for $t \to -\infty$ was in the n_1th quantum state corresponding to a two-fold degenerate level $E_n^{(0)}$ of a Hamiltonian $\hat{H}_0$, a time-dependent perturbation $\hat{V}(t)$ is applied. Find the wavefunction in the zeroth approximation at an arbitrary moment in time. Assume that the diagonal matrix elements for the degenerate states satisfy the condition $V_{n_1 n_1} = V_{n_2 n_2} = 0$ (which takes place, for example, if the states $|n_{1,2}>$ have definite and opposite parities, and the perturbation is proportional to the system dipole moment). How should we modify Eq. (VIII.8) for transition amplitudes between states with different energies?

Solution

While transitions between states with different energies are small, transitions between the states that correspond to the degenerate level can be significant if the duration of the perturbation is long enough. The situation is analogous to the case of a resonant perturbation action and may be viewed formally as the case of an exact resonance with the frequency $\omega = 0$. Taking into account the probability of transitions between the degenerate states, and without assuming that the probabilities are small, we write down the system wavefunction in *zeroth* approximation (compare with Eq. (VIII.6)):

$$\psi(t) = [a_1(t)\psi_1^{(0)} + a_2(t)\psi_2^{(0)}]e^{-i\omega_n t}.$$

We write $1, 2$ instead of $n_{1,2}$. As usual, we obtain the equations

$$i\hbar \dot{a}_1 = f(t) a_2, \quad i\hbar \dot{a}_2 = f(t) a_1. \tag{1}$$

Here we took into account that $V_{nn} = 0$, and denote $V_{12} = f(t)$, considering $f(t)$ to be real. From (1), it follows that

$$\frac{d}{dt}(a_1 \pm a_2) = \mp \frac{i}{\hbar} f(t)(a_1 \pm a_2).$$

So, taking into account the initial conditions, we find

$$a_1(t) = \cos \xi(t), \quad a_2(t) = -i \sin \xi(t), \quad \xi(t) = \frac{1}{\hbar} \int_{-\infty}^{t} f(t') dt'. \tag{2}$$

These results for $\xi \ll 1$ correspond to Eq. (VIII.8).

The generalization of Eq. (VIII.8) for the transition amplitudes between states with different energies is

$$a_k^{(1)}(t) = -\frac{i}{\hbar} \sum_{n=1,2} \int_{-\infty}^{t} V_{kn}(t') a_n^{(0)}(t') e^{i\omega_{kn} t'} dt'.$$

In conclusion, we should note the oscillating character of the time-dependence of the amplitude $a_{1,2}(t)$ and the transition probabilities, which appears even for the case of a weak but long-lasting perturbation. We should also mention that according to Eq. (1), superpositions $|1> \pm |2>$ of the initial states are *diagonal* (there are no transitions between them). The appearance of such independent states is connected with the restrictions on the values of the perturbation matrix elements V_{ab}.

Problem 8.41

For a periodic in time perturbation, $\hat{V}(q, t+T) = \hat{V}(q,t)$, find the wavefunctions of quasi-energy states[140] (QES) in the zeroth approximation and the quasi-energy spectrum in the first order of perturbation theory. Consider the energy spectrum of the unperturbed Hamiltonian to be discrete and assume that it does not contain level spacings corresponding to the resonant transitions, $E_n^{(0)} - E_k^{(0)} \neq \pm \hbar \omega$, with $\omega = 2\pi/T$ (see 8.43).

Solution

We write the QES wavefunctions in the form of an expansion

$$\psi_{\varepsilon_n}(q,t) = \exp\{-i(E_n^{(0)} + \varepsilon_n^{(1)} + \ldots)t/\hbar\}$$
$$\times \left\{ c_n(t)\psi_n^{(0)}(q) + \sum_k{}' c_{nk}(t)\psi_k^{(0)}(q) \right\} \quad (1)$$

(where prime means the absence of the term with $k=n$ in the sum). Here $E_n^{(0)}$ and $\psi_n^{(0)}$ are the eigenvalues and eigenfunctions of the unperturbed Hamiltonian, which in this case coincides with the quasi-energy and wavefunction of QES in the *zeroth* approximation. The coefficients in the expansion are periodic functions: for example, $c_n(t+T) = c_n(t)$. Substituting (1) into the Schrödinger equation, multiplying it by $\psi_n^{(0)*}(t)$ from the left, and integrating over the coordinates, we obtain (taking into account only the first-order terms):

$$i\hbar \dot{c}_n(t) + \varepsilon_n^{(1)} c_n(t) = V_{nn}(t) c_n(t). \quad (2)$$

Therefore,

$$c_n(t) = c_0 \exp\left\{ \frac{i}{\hbar}\left(\varepsilon_n^{(1)} t - \int_0^t V_{nn}(t')dt' \right) \right\}. \quad (3)$$

---

[140] These quasi-energy states are also called "Floquet states" in the literature. They represent a time analogue to the Bloch states that appear in a periodic spatial potential. The existence of both the Bloch states and the quasi-energy (Floquet) states are related to the general mathematical fact (Floquet's theorem) that a fundamental solution to a general (matrix) first-order differential equation with periodic coefficients can be written as a product of a periodic part and a simple exponential dependence typical for the problem with constant coefficients.

The value of $\varepsilon_n^{(1)}$ (of the first-order correction to the quasi-energy) is determined by the periodicity condition on $c_n(t)$. Introducing $\overline{V_{nn}(t)}$ – the mean value of the perturbation matrix element – we rewrite the exponential factor from (3) in the form:

$$\frac{i}{\hbar}\left[(\varepsilon_n^{(1)} - \overline{V}_{nn})t - \int_0^t (V_{nn}(t') - \overline{V}_{nn})dt'\right].$$

Since the integral term here is a periodic function, periodicity condition on the function, $c_n(t)$ gives

$$\varepsilon_n^{(1)} = \overline{V_{nn}(t)}. \tag{4}$$

So the value of the quasi-energy in first-order perturbation theory, $\varepsilon_n \approx E_n^{(0)} + \varepsilon_n^{(1)}$, coincides with the average instantaneous energy over the period, $E_n(t) = E_n^{(0)} + V_{nn}(t)$, Compare to the result of Problem 8.56 for the case of an adiabatic change.

Problem 8.42

Under the conditions of the previous problem, find the second perturbative correction to the quasi-energy for the case of $V_{nn}(t) \equiv 0$. Consider specifically a perturbation of the form $\hat{V} = \hat{V}(q)\cos\omega t$ and the limiting cases $\omega \to 0$ and $\omega \to \infty$. Obtain the dynamic polarizability of the levels in the field of a linearly polarized monochromatic wave, $\hat{V} = -\hat{\mathbf{d}} \cdot \boldsymbol{\mathcal{E}}_0 \cos\omega t$, and then in particular for an oscillator.

Hint

For a system in the electric field of a linearly polarized wave, the dynamic polarizability, $\beta_n(\omega)$, is connected with the second-order perturbative correction to quasi-energy by the relation[141]

$$\varepsilon_n^{(2)} = -\frac{1}{4}\beta_n(\omega)\mathcal{E}_0^2.$$

Solution

1) First we write the wavefunction of the QES in the form (compare to the previous problem; hereafter we put $\hbar = 1$):

$$\psi_{\varepsilon_n}(q,t) = e^{-i(E_n^{(0)} + \varepsilon_n^{(1)} + \varepsilon_n^{(2)} + \ldots)t}\left\{c_n(t)\psi_n^{(0)}(q) + \sum_k{}' c_{nk}(t)\psi_k^{(0)}(q)\right\}. \tag{1}$$

Here, in the context of perturbation theory,

$$c_n = 1 + c_n^{(1)} + c_n^{(2)} + \ldots, \quad c_{nk} = c_{nk}^{(1)} + \ldots.$$

---

[141] An additional term $1/2$ in comparison to the static case is connected with the mean value $\overline{\cos^2\omega t} = 1/2$. Compare to Problem 8.56.

Substituting (1) into the Schrödinger equation and multiplying it from the left[142] by $\langle\psi_k^{(0)}|$ with $k \neq n$, as usual, we find in the first approximation ($\omega_{kn} = E_k^{(0)} - E_n^{(0)}$):

$$i\dot{c}_{nk}^{(1)}(t) = \omega_{kn}c_{nk}^{(1)}(t) + V_{kn}(t).$$

The general solution of this equation is:

$$c_{nk}^{(1)}(t) = e^{-i\omega_{kn}t}\left(c_{nk}^{(1)}(0) - i\int_0^t V_{kn}(t')e^{i\omega_{kn}t'}dt'\right). \tag{2}$$

The value of the constant $c_{nk}^{(1)}(0)$ is found from the condition of $c_{nk}^{(1)}(t+T) = c_{nk}^{(1)}(t)$ periodicity, and is equal to

$$c_{nk}^{(1)}(0) = -\frac{i}{e^{i\omega_{kn}T} - 1}\int_0^T e^{i\omega_{kn}t}V_{kn}(t)dt. \tag{3}$$

(Relations (2) and (3) are used below to determine the corrections of the second order for quasi-energy.)

Now multiplying the Schrödinger equation by $\langle\psi_n^{(0)}|$, we find (for terms of the first order in perturbation)

$$i\dot{c}_n^{(1)}(t) = -\varepsilon_n^{(1)} + V_{nn}(t).$$

Further calculations are performed under the assumption that $V_{nn} \equiv 0$. Here $c_n^{(1)} = i\varepsilon_n^{(1)}t + c_n^{(1)}(0)$, and from periodicity we have $\varepsilon_n^{(1)} = 0$ (compare to the previous problem). We can choose the value of the constant as $c_n^{(1)}(0) = 0$ (this choice does not effect the q- and t-dependence of the wavefunction), so that $c_n^{(1)}(t) = 0$.

For terms of the second order in perturbation theory that appear when we multiply the Schrödinger equation by $\langle\psi_n^{(0)}|$, we find

$$i\dot{c}_n^{(2)}(t) = -\varepsilon_n^{(2)} + \sum_k{}' V_{nk}(t)c_{nk}^{(1)}(t), \tag{4}$$

where $c_{nk}^{(1)}$ are determined by Eqs. (2) and (3). Thus

$$c_n^{(2)}(t) = i\varepsilon_n^{(2)}t - i\sum_k{}'\int_0^t V_{nk}(t')c_{nk}^{(1)}(t')dt' + c_n^{(2)}(0). \tag{5}$$

The constant $c_n^{(2)}(0)$ could be omitted as was $c_n^{(1)}(0)$ above.

---

[142] This symbolic notation implies a multiplication by $\psi_k^{(0)*}(q)$ and integration over coordinates.

The value of $\varepsilon_n^{(2)}$ is found from the periodicity of the coefficient $c_n^{(2)}(t)$. Taking into account the fact that both $V_{nk}(t)$ and $c_n^{(1)}(t)$, as well as their product in (5), are periodic functions with the period T, we find the correction:

$$\varepsilon_n^{(2)} = \frac{1}{T}{\sum_k}' \int_0^T V_{nk}(t) c_{nk}^{(1)}(t) dt$$

$$= -\frac{i}{T}{\sum_k}' \left\{ \frac{1}{e^{i\omega_{kn}T} - 1} \left| \int_0^T V_{nk}(t) e^{i\omega_{kn}t} dt \right|^2 \right.$$

$$\left. + \int_0^T V_{nk}(t) e^{-i\omega_{kn}t} \int_0^t V_{nk}(t') e^{i\omega_{kn}t'} dt' dt \right\}. \tag{6}$$

Using the fact that the perturbation is a Hermitian operator, we see (for $\omega_{kn} \neq 2\pi N/T$) that the relation inside the brackets is imaginary and $\varepsilon_n^{(2)}$ is real. If $V_{nk}(q,t) \equiv V_{nk}(q)$ (i.e., the perturbation does not depend on time) then (6) becomes a standard equation of stationary second-order perturbation theory (VIII.1) for the level shift.

2) In the case of an harmonic perturbation of the form $\hat{V} = \hat{V}(q)\cos\omega t$ with $\omega = 2\pi/T$, relation (6) is simplified:

$$\varepsilon_n^{(2)} = \frac{1}{2}{\sum_k}' |V_{nk}(q)|^2 \frac{\omega_{kn}}{\omega^2 - \omega_{kn}^2}. \tag{7}$$

We analyze this result below.

First, note that as $\omega \to 0$, relation (7) differs from the standard perturbation theory equation for $\hat{V} = \hat{V}(q)$ only by the factor $1/2$. This corresponds to the fact that $\varepsilon_n^{(2)}$ is obtained as a result of an averaging, $\overline{E_n^{(2)}(t)} \propto \overline{\cos^2\omega t}$, for an "instantaneous" perturbation $\hat{V}(q)\cos\omega t$, and $\overline{\cos^2(\omega t)} = 1/2$ (these results also follow from the adiabatic approximation; see Problem 8.56). In the opposite limiting case, $\omega \to \infty$, from (7) it follows that $\varepsilon_n^{(2)} \propto 1/\omega^2$ (see Eq. (10) below).

In the important case of a system of charged particles interacting with an electromagnetic wave, where $\hat{V} = -\mathcal{E}_0 \cdot \hat{\mathbf{d}} \cos\omega t$, from (7) it follows that the *dynamic polarizability* of the system is (the z-axis is directed along $\mathcal{E}_0$):

$$\beta_n(\omega) = 2 {\sum_k}' \frac{\omega_{kn} |\langle k|\hat{d}_z|n\rangle|^2}{\omega_{kn}^2 - \omega^2}, \quad \varepsilon_n^{(2)} = -\frac{1}{4}\beta_n(\omega)\mathcal{E}_0^2. \tag{8}$$

For $\omega = 0$, the dynamic polarizability coincides with the static polarizability. Specifically for a linear oscillator, as well as in the case of a stationary electric field in Problem 8.2, we find

$$\beta_n(\omega) = \frac{e^2}{m(\omega_0^2 - \omega^2)}, \qquad (9)$$

where ω_0 is its fundamental frequency.

Using the relation $p_{kn} = im\omega_{kn}r_{kn}$ and the sum rule from Problem 6.13, we see that for a system of N charged particles with the same mass m and charge e, Eq. (8) can be transformed to the form

$$\beta_n(\omega) = -\frac{1}{\omega^2}\left\{\frac{e^2}{m}N - \sum_k{}' \frac{2\omega_{kn}^3|\langle k|\hat{d}_z|n\rangle|^2}{\omega_{kn}^2 - \omega^2}\right\}. \qquad (10)$$

The first term, $\beta_0(\omega) = -e^2N/m\omega^2$, corresponds to a level shift in the wave field, as if the particle were free (compare to Problem 6.40, and also with (9) for $\omega_0 = 0$); this term is dominant for $\omega \to \infty$. The correction term in (10) is reduced to a diagonal matrix element, if we take into account the sum rule in Problem 14.11.

Eqs. (7)–(10) are not applicable for $\omega \to \omega_{kn}$, where $\omega_{kn} = (E_k^{(0)} - E_n^{(0)})/\hbar$ is the frequency of transitions between discrete levels, which are caused by the perturbation, $\hat{V}$. In this resonance situation, even a weak perturbation leads to a strong mutual interaction between resonance levels. Compare to the previous problem.

A different scenario is realized if the resonant states with $k \equiv \nu$ are in the continuous spectrum. Here, the application of perturbation results in 'ionization' of the system and a QES attenuation in time. To determine its lifetime, according to Eq. (7), we should make the substitution $E_n^{(0)} \to E_n^{(0)} + i\gamma$ or $\omega_{kn}^2 \to \omega_{kn}^2 - i\gamma$, where $\gamma > 0$ is infinitesimal.

Writing

$$\varepsilon_n^{(2)} = \Delta\varepsilon_n^{(2)} - \frac{i}{2}\Gamma_n$$

and replacing the summation over k by the integration over ν, we find the *width* of the QES,

$$\Gamma_n = -2\mathrm{Im}\varepsilon_n^{(2)} = \frac{\pi}{2}\int |V_{\nu n}|^2 \delta(E_\nu - E_n^{(0)} - \omega)d\nu, \qquad (11)$$

in accordance with the general equation for transition probability (remember that $\Gamma_n = \hbar/\tau_n = \hbar\omega_n$).

In conclusion, we note that for a system in a linearly-polarized electric field, $\hat{V} = -\boldsymbol{\mathcal{E}}_0 \cdot \hat{\mathbf{d}}\cos(\omega t)$, the level width and the imaginary part of the dynamic polarizability are connected with the *photo-ionization*[143] cross-section for this state by the relation

---

[143] This relation follows automatically from the comparison of the $V_{\nu n}$ and single-photon transition operator Eqs. (XIV.12) and (XIV.13), which determine the photo-effect cross-section. See Problems 14.18–14.20.

Problem 8.43

Analyze the quasi-energy states that appear when a periodic resonant perturbation of the form $\hat{V} = \hat{V}_0 \cos \omega t$ is applied to a two-level system with the energies $E_{1,2}^{(0)}$, where $|\omega - \omega_0| \ll \omega_0$, $\hbar \omega_0 = E_2^{(0)} - E_1^{(0)}$. Operator $\hat{V}_0$ does not depend on time, and its diagonal matrix elements are equal to zero, while the off-diagonal elements are given by $(\hat{V}_0)_{12} = V_0$, $V_0 = V_0^*$, where $V_0 \ll \hbar \omega_0$. Analyze the quasi-energy harmonics of QES and compare to Problem 6.41.

Solution

We write the wavefunction in the form

$$\psi(t) = \begin{pmatrix} \psi_1(t) \\ \psi_2(t) \end{pmatrix}, \quad \psi_{1,2} = a_{1,2}(t) e^{-i E_{1,2}^{(0)} t / \hbar}$$

(compare to Problem 6.41). The Schrödinger equation, $i\hbar \partial \psi / \partial t = (\hat{H}_0 + \hat{V})\psi$, is reduced to a system of two equations ($\hbar \omega_0 = E_2^{(0)} - E_1^{(0)}$):

$$i\hbar \dot{a}_1 = V_0 e^{-i\omega_0 t} \cos(\omega t) a_2, \quad i\hbar \dot{a}_2 = V_0 e^{i\omega_0 t} \cos(\omega t) a_1. \tag{1}$$

Consider the time factors:

$$e^{\pm i \omega_0 t} \cos \omega t = \frac{1}{2}[e^{\pm i(\omega_0 - \omega)t} + e^{\pm i(\omega_0 + \omega)t}].$$

The first term is a slowly changing function, while the second one is a rapidly changing function of time. In the case of a weak perturbation, $V_0 \ll \hbar \omega$, terms in Eqs. (1) that contain a fast-changing factor could be omitted, since they do not play an important role in transitions (compare to Problem 8.40). Taking into account this fact and making the substitution

$$a_{1,2}(t) = \tilde{a}_{1,2}(t) e^{\mp i \gamma t / 2}, \quad \gamma = \omega_0 - \omega,$$

we transform the system (1) to the form ($v_0 = V_0/\hbar$):

$$2i\dot{\tilde{a}}_1 = -\gamma \tilde{a}_1 + v_0 \tilde{a}_2, \quad 2i\dot{\tilde{a}}_2 = v_0 \tilde{a}_1 + \gamma \tilde{a}_2. \tag{2}$$

We find two independent solutions of this system of equations with constant coefficients by the substitution $\tilde{a}_{1,2}(t) = C_{1,2}e^{-i\lambda t}$:

$$\lambda_1 = \frac{1}{2}\sqrt{\gamma^2 + v_0^2}, \quad C_1^{(1)} = \frac{v_0}{\gamma + \sqrt{\gamma^2 + v_0^2}} C_2^{(1)},$$

$$\lambda_2 = -\frac{1}{2}\sqrt{\gamma^2 + v_0^2}, \quad C_1^{(2)} = -\frac{1}{v_0}\left(\gamma + \sqrt{\gamma^2 + v_0^2}\right) C_2^{(2)}.$$

So the general solution of the Schrödinger equation has the form

$$\psi(t) = \frac{A_1}{\sqrt{1+\beta^2}} \begin{pmatrix} \beta \\ -e^{-i\omega t} \end{pmatrix} e^{-i\varepsilon_1 t/\hbar} + \frac{A_2}{\sqrt{1+\beta^2}} \begin{pmatrix} e^{i\omega t} \\ \beta \end{pmatrix} e^{-i\varepsilon_2 t/\hbar}, \quad (3)$$

where

$$\varepsilon_1 = E_1^{(0)} + \frac{1}{2}\gamma + \lambda, \quad \lambda = \frac{1}{2}\sqrt{\gamma^2 + v_0^2}, \quad (4)$$

$$\varepsilon_2 = E_2^{(0)} - \frac{1}{2}\gamma - \lambda, \quad \beta = \frac{\gamma + 2\lambda}{v_0}.$$

We should note that for exact resonance, i.e., $\omega = \omega_0$, we have $\gamma = 0$ and $\beta = |v_0|/v_0 = \pm 1$.

Each of the two terms in wavefunction (3) describes an independent QES, where $\varepsilon_{1,2}$ are the quasi-energies of these states. As is seen, there are only two quasi-energy harmonics (see Problem 6.40), which correspond to the states $\begin{pmatrix} 1 \\ 0 \end{pmatrix}$ and $\begin{pmatrix} 0 \\ 1 \end{pmatrix}$, i.e., they are the eigenfunctions of the unperturbed Hamiltonian, $\hat{H}_0$. Higher harmonics have amplitudes proportional to powers of the small parameter, $V_0/\hbar\omega_0 \ll 1$, and so do not appear in the approximation considered. The disappearance of these terms is connected with neglecting fast-changing terms in the system of equations.

Problem 8.44

For the system with two channels considered in Problem 6.39, and in the case of a weak coupling between the channels ($\beta \ll \alpha$), find the width of the quasi-stationary state in the channel with the excited composite particle, by perturbation theory. Consider the non-interacting (a) and interacting (b) cases separately, and compare your results in these two cases with each other and the exact result.

Solution

The transition probability per unit of time from a discrete state to a continuous state under the influence of a constant perturbation is,

$$w = \frac{2\pi}{\hbar} \int |V_{\nu n}|^2 \delta(E_\nu - E_n^{(0)}) d\nu. \quad (1)$$

The perturbation in our case, $\hat{V} = -\begin{pmatrix} 0 & \beta \\ \beta & 0 \end{pmatrix}\delta(x)$, is the part of the interaction that corresponds to the coupling between the channels. For the wavefunction, $\psi_n^{(0)}$, of the initial state,[144] we take the wavefunction $\psi_n^{(0)} = \begin{pmatrix} 0 \\ \psi_0(x) \end{pmatrix}$, of the bound state in the channel with the excited composite particle (for definitions, see Problem 6.39). Here

$$\psi_0(x) = \sqrt{\kappa_0}e^{-\kappa_0|x|}, \quad \kappa_0 = \frac{m\alpha}{\hbar^2}, \quad E_n^{(0)} = Q_0 - \frac{\hbar^2\kappa_0^2}{2m}.$$

The state corresponds to a particle in the ground state of the δ-well (see Problem 2.7), with the bottom of the continuous spectrum shifted by Q_0 in this channel. Finally, we have the wavefunction $\psi_\nu^{(0)} = \begin{pmatrix} \psi_\nu(x) \\ 0 \end{pmatrix}$, with the function $\psi_\nu(x)$ describing a continuum state in the main channel with energy $E_\nu = E_n^{(0)}$; see below for a specific choice of ψ_ν.

a) Neglecting interaction in the main channel, we can use for ψ_ν the wavefunction of free motion, i.e., $\psi_\nu = (2\pi)^{-1/2}e^{ikx}$; here $\nu \equiv k$, $-\infty < k < \infty$, and $E_\nu = \hbar^2 k^2/2m$. Now the perturbation matrix element is

$$V_{\nu n} = \langle k|\hat{V}|0\rangle = -\beta \int \psi_k^*(x)\delta(x)\psi_0(x)dx = -\beta\sqrt{\frac{\kappa_0}{2\pi}}$$

and according to (1) we obtain

$$\Gamma = \hbar\omega = \frac{\beta^2 \kappa_0}{\hbar}\sqrt{\frac{2m}{E_n^{(0)}}}. \tag{2}$$

b) To determine Γ more accurately, we should also take into account the interaction (the δ-potential) in the final state. For the wavefunction ψ_ν, it is convenient to choose a state $\psi_{k,I}$ with a definite parity I and $k = \sqrt{2mE_\nu/\hbar^2} > 0$. For the δ-potential these wavefunctions have the form

$$\psi_{k,-1} = \frac{1}{\sqrt{\pi}}\sin kx, \quad \psi_{k,+1} = \frac{1}{\sqrt{\pi}}\cos(k|x| + \delta),$$

using the matching conditions at $x = 0$ (see Problem 2.6), we find $\tan\delta = m\alpha/\hbar^2 k$. The matrix element $V_{\nu n}$, where $\nu \equiv (k, I)$, differs from zero only for even states, with $I = +1$, and there is equal to

$$V_{\nu n} = -\beta\sqrt{\frac{\kappa_0}{\pi}}\cos\delta.$$

---

[144] This state is truly bound only if we neglect the perturbation. Under the influence of the perturbation it becomes quasi-stationary with the width $\Gamma = \hbar\omega$. Coupling to the opened channels plays the role of a finite transparency of the barrier for quasi-stationary states of a one-channel system. Compare to Problems 6.36 and 6.37.

Taking into account the value of δ, according to (1), we obtain

$$\Gamma = \hbar\omega = \frac{\beta^2 \kappa_0 k}{Q_0}, \quad k = \sqrt{\frac{2m E_n^{(0)}}{\hbar^2}}. \tag{3}$$

Let us now compare our results (2) and (3). In case a), equation (2) is applicable only for $Q_0 \gg |E_0| = \hbar^2 \kappa_0^2/2m$, i.e., if the kinetic energy in the ground channel is much larger than the binding energy. Indeed, in this case Eqs. (2) and (3) almost coincide. For $Q_0 \sim |E_0|$, Eq. (2) is not applicable (for particles with energy $E \sim |E_0|$, the reflection coefficient $R \sim 1$ and we cannot consider them as free). Equation (3), which was derived under the assumption of a weak inter-channel coupling, $\beta \ll \alpha$, remains valid even in this case. Comparing (3) to the exact solution, Eq. (4) from Problem 6.39, confirms this fact.

Problem 8.45

Find the probability of "ionization" per unit of time for a particle in the ground state of a one-dimensional δ-well up to first order in perturbation theory (see Problem 2.7), if the particle is under the influence of a uniform, periodic-in-time field, $V(x,t) = -xF_0 \cos\omega_0 t$. Solve the problem both neglecting the interaction in the final state and taking it into account.

Solution

The transition probability into a continuum state due to a periodic perturbation is

$$w = \frac{2\pi}{\hbar} \int |F_{\nu n}|^2 \delta(E_\nu - E_n^{(0)} - \hbar\omega_0) d\nu. \tag{1}$$

In this problem, $\hat{V} = -F_0 x \cos(\omega_0 t)$ and hence $\hat{F} = -F_0 x/2$. Then (compare to the solution of the previous problem) $\psi_n^{(0)} = \sqrt{\kappa} e^{-\kappa|x|}$ is the wavefunction of the ground state of the δ-well, $\kappa = m\alpha/\hbar^2$, and $E_n^{(0)} \equiv E_0 = -\hbar^2 \kappa^2/2m$ is the ground-state energy.

When neglecting the effect of the δ-potential on the particle in the final state, we should choose $\psi_\nu = (2\pi)^{-1/2} e^{ikx}$, with $\nu \equiv k$, $-\infty < k < \infty$, $E_\nu = \hbar^2 k^2/2m$. Calculating the matrix element

$$F_{\nu n} = -\frac{F_0 \sqrt{\kappa}}{2\sqrt{2\pi}} \int_{-\infty}^{\infty} x \exp\{-(\kappa|x| + ikx)\} dx = i\frac{\sqrt{2} k \kappa^{3/2} F_0}{\sqrt{\pi}(k^2 + \kappa^2)^2},$$

according to (1) we find

$$w = \frac{2\hbar F_0^2 |E_0|^{3/2} \sqrt{\hbar\omega_0 - |E_0|}}{m(\hbar\omega_0)^4}. \tag{2}$$

Since we neglected the interaction in the final state, our results are formally valid if the condition $\hbar\omega_0 \gg |E_0|$ is fulfilled. However, the results are actually applicable for $\hbar\omega_0 > |E_0|$ (and in the vicinity of the threshold). To see this, we include the interaction. We choose for ψ_ν the exact wavefunctions of the unperturbed Hamiltonian, $\psi_{k,I}$, that correspond to a definite parity I (compare to the previous problem). Now notice that the matrix element $F_{\nu n}$ is different from zero only for the odd states, whose wavefunctions are not distorted by the δ-potential and coincide with the free-particle wavefunction. So Eq. (2) holds even if we take into account interaction in the final state, where there are no restrictions on the energy of the outgoing particle.

In conclusion, note that $w = 0$ for $\hbar\omega_0 < |E_0|$ in the first-order perturbation theory. Particle transitions to continuous spectrum appear in higher orders of perturbation theory ("multi photon ionization") and therefore has lower probability. Compare to tunnel ionization in a static field, in the limiting case $\omega_0 \to 0$. See Problem 6.39.

Problem 8.46

A particle is in a one-dimensional short-range potential, $U(x)$, such that $U(x) \to 0$ for $x \to \pm\infty$. Considering this potential as a perturbation, find the reflection coefficient using perturbation theory for transitions in the continuous spectrum. Determine applicability conditions for your result and compare it to Problem 8.29.

Solution

For transitions in the continuous spectrum,

$$dw_{\nu_0\nu} = \frac{2\pi}{\hbar}|V_{\nu_0\nu}|^2 \delta(E_\nu - E_{\nu_0})d\nu, \quad \nu \neq \nu_0. \tag{1}$$

ν_0 and ν are the wave "vectors" (one-dimensional motion) of free particles, and the corresponding wavefunctions are

$$\psi_k = \sqrt{\frac{m}{\hbar k}}e^{ikx}, \quad \psi_{k'} = \frac{1}{\sqrt{2\pi}}e^{ik'x}.$$

Let us point out a few caveats about normalization. While the final wavefunctions are normalized in the usual way, the initial states are to be normalized according to the following considerations. If we consider the system to be inside a "box" of a large dimension, L, the initial wavefunction should be chosen in the form $\psi_{\nu_0} = e^{ikx}/\sqrt{L}$ (normalized to unity). The transition probability w has the right physical dimension of inverse time, T^{-1}. But for the states of the continuous spectrum, we usually consider not a single particle but a particle current, with the current density $j = \rho v = v/L$. An important quantity in this case is not the probability, but a "cross-section" defined by the formula $\sigma = w/j$. It does not depend on a choice of L (unlike the transition probability). In the one-dimensional case, the "cross-section" is dimensional and has the physical meaning of a reflection coefficient.

If we perform integration over ν (i.e. over k') in Eq. (1), we obtain

$$R = w(k \to k' = -k) = \frac{m^2}{\hbar^4 k^2} \left| \int_{-\infty}^{\infty} U(x) e^{2ikx} dx \right|^2.$$

Transitions appear to the states with $k' = -k$ that correspond to reflected particles.

8.5 Sudden perturbations

Problem 8.47

A system described by Hamiltonian $\hat{H}_0$ is in the nth discrete state. At $t = 0$, the Hamiltonian changes abruptly and becomes $\hat{H}_f = \hat{H}_0 + \hat{V}_0$, where neither $\hat{V}_0$ nor $\hat{H}_0$ depend on time. Find the probabilities of different stationary states for $t > 0$. Find the mean energy and show that for a small perturbation, $\hat{V}_0$, these results can also be obtained by time-dependent perturbation theory.

Solution

1) The wave-function does not have time to change during the sudden change and therefore coincides with the initial wavefunction (the n-th eigenstate, $\psi_{n,i}$, of the original Hamiltonian) right after the perturbation is turned on.

The change of the expectation value of the system energy after the perturbation is turned on therefore reads[145]

$$\overline{\Delta E} = \langle n, i | \hat{V}_0 | n, i \rangle. \qquad (1)$$

The expansion coefficients of wavefunction $\psi_{n,i}$ in the eigenfunctions $\psi_{k,f}$ of the final Hamiltonian ($\hat{H}_f = \hat{H}_0 + \hat{V}_0$),

$$\psi_{n,i} = \sum_k c_{kn} \psi_{k,f},$$

determine the desired transition probabilities:

$$w(n_i \to k_f) = |c_{kn}|^2 = |\langle k, f | n, i \rangle|^2. \qquad (2)$$

2) Considering $\hat{V}_0$ as a small perturbation, we can use the known expansion of perturbation theory (see Eq. (VIII.2)) and find, analogously to (2),

$$w^{(1)}(n_i \to k_f) = \frac{|\langle k, f | \hat{V}_0 | n, i \rangle|^2}{(E_{n,i} - E_{k,i})^2}, \quad k \neq n. \qquad (3)$$

---

[145] See also Problem 9.22, where sudden perturbations are considered within the quasi-classical approximation.

The same result could be obtained by non-stationary perturbation theory. Integration in Eq. (VIII.8) by parts gives

$$a^{(1)}_{kn}(t) = \frac{1}{\hbar\omega_{kn}} \int_{-\infty}^{t} e^{i\omega_{kn}t} \frac{\partial V_{kn}}{\partial t} dt - \frac{V_{kn}(t)}{\hbar\omega_{kn}} e^{i\omega_{kn}t}. \tag{4}$$

In this problem $\hat{V} = \hat{V}_0 \eta(t)$, where $\eta(t)$ is the step function ($\eta(t) = 1$ for $t > 0$ and $\eta(t) = 0$ for $t < 0$). Since $d\eta(t)/dt = \delta(t)$, we see that for $t > 0$, the first term in the right-hand side of (4) that determines transition probability reproduces Eq. (3). The second term in (4) describes a distortion of the wavefunction of the nth state for $t > 0$ by perturbation $\hat{V}_0$, and is unrelated to system transitions.

Problem 8.48

A system experiences an impulse perturbation $\hat{V} = \hat{W}_0 \delta(t)$, so its Hamiltonian has the form $\hat{H} = \hat{H}_0 + \hat{W}_0 \delta(t)$. For $t < 0$ the system was in the nth state of discrete spectrum. Find the probabilities of different quantum states for $t > 0$. For small perturbation $\hat{V}$, compare these probabilities to the result of non-stationary perturbation theory.

Illustrate your general result on the example of the specific petrubation with $W_0 = -P_0 x$ (here, P_0 is constant parameter that has the physical dimension of momentum and x is the particle coordinate).

Solution

To determine how the wavefunction changes due to the impulse, it is convenient to consider it a $\tau \to 0$ limit of the interaction $\hat{V}(t, \tau) = \hat{W}_0 f(t)$, where the function $f(t)$ differs from zero only for $|t| < \tau$, and its integral in the limits from $-\tau$ to τ is equal to 1. The Schrödinger equation for $|t| < \tau$ takes the form $i\hbar\dot{\psi} = \hat{W}_0 f(t)\psi$ (the term $\hat{H}_0$ in the Hamiltonian is omitted since for the infinitely small period of time τ, it does not yield a noticeable change; we only need $f(t) \sim 1/\tau \to \infty$). The solution of this equation gives

$$\psi(t) = \exp\left\{-\frac{i}{\hbar}\int_{-\tau}^{t} f(t')dt' \hat{W}_0\right\} \psi(-\tau). \tag{1}$$

If we put $t = \tau$ and take the limit $\tau \to 0$ we find

$$\psi(t = 0+) = \exp\left\{-\frac{i\hat{W}_0}{\hbar}\right\} \psi(t = 0-). \tag{2}$$

Taking into account that, $\psi(0-) = \psi_n^{(0)}$, we find the desired probabilities:

$$w(n \to k) = \left|\int \psi_k^{(0)*} \exp\left\{-\frac{i}{\hbar}\hat{W}_0\right\} \psi_n^{(0)} d\tau_q\right|^2. \tag{3}$$

Expanding the exponential in the case of a small perturbation, where $|(\hat{W}_0)_{kn}| \ll \hbar$, gives

$$w(n \to k) \approx w^{(1)}(n \to k) = \frac{1}{\hbar^2}|(\hat{W}_0)_{kn}|^2, \quad k \neq n,$$

which coincides with the result of non-stationary perturbation theory that is obtained by integration (due to the δ-function) in Eq. (VIII.8).

The impulse of the form $V(x,t) = -xP_0\delta(t)$ exerted on a classical particle results in giving the particle a momentum kick, $P_0 = \int F(t)dt$. This interpretation remains valid in quantum mechanics, which follows (compare to Problem 6.26) from (2). Indeed, wavefunctions in the momentum representation just before ($a_i(p) = \langle p|t = 0-\rangle$) and just after ($a_f(p) = \langle p|t = 0+\rangle$) the impulse are related by expression $a_f(p) = a_i(p - P_0)$.

Problem 8.49

A particle is in the ground state of an infinitely deep potential well with the width a ($0 < x < a$). At some moment, the well's right wall moves suddenly (in a short period of time, τ) to the point $b > a$. Find probabilities of different quantum states after the wall stops.

Solution

From equation (2) from Problem 8.47, we find

$$w(0 \to k) = \frac{4ab^3}{\pi^2[a^2(k+1)^2 - b^2]^2} \sin^2 \frac{\pi(k+1)a}{b}.$$

The applicability condition is $\tau\omega_{k0} = \pi^2(k+1)^2\hbar\tau/ma^2 \ll 1$.

Problem 8.50

A particle is in the ground state of a δ-well, so that $U = -\alpha\delta(x)$. The parameter α that characterizes the "depth" of the well changes suddenly to $\tilde{\alpha}$ (physically, this may occur if the charge of an atomic nucleus changes suddenly, for example, for β-decay). Find

a) the probability of the particle to remain in the ground state,
b) the momentum distribution of the particle escaped from the well.

Solution
a) Use the relation

$$\psi_0(x, \alpha) = \sqrt{\kappa}e^{-\kappa|x|}, \quad \text{where} \quad \kappa = \frac{m\alpha}{\hbar^2},$$

for the ground state wavefunction in a δ-well (see Problem 2.7). According to general equation (2) from Problem 8.47 for transition probabilities, we find the probabilities for the particle to remain bound by the well:

$$w_0 \equiv w(0_i \to 0_f) = \left| \int \psi_0^*(x, \tilde{\alpha}) \psi_0(x, \alpha) dx \right|^2 = \frac{4\alpha\tilde{\alpha}}{(\alpha + \tilde{\alpha})^2}. \tag{1}$$

b) To consider transitions into the continuous spectrum, it is convenient to choose the states, $\psi_{k,I}(k)$ corresponding to states with definite parity I as the basis classifying the final states. Such functions, normalized to δ-functions and with $k = \sqrt{2mE/\hbar^2} > 0$, were obtained in Problem 8.44. Using the expression for them and according to a generalization of Eq. (2) from Problem 8.47 to the case of continuous spectrum, we find

$$dw(k) = \left| \int \psi_{k,I=+1}^*(x, \tilde{\alpha}) \psi_0(x, \alpha) dx \right|^2 dk = \frac{4}{\pi} \frac{\kappa(\kappa - \tilde{\kappa})^2 k^2 dk}{(\kappa^2 + k^2)^2 (\tilde{\kappa}^2 + k^2)}. \tag{2}$$

This distribution is normalized to $1 - w_0$, where w_0 is determined by Eq. (1). Transitions appear only to the even final states, and therefore the probabilities of the momentum values $p = \pm \hbar k$ are the same. As is seen from Eqs. (1) and (2), in the case of $\alpha \approx \tilde{\alpha}$, the escape probability is small, while for the values $\tilde{\alpha} \ll \alpha$ and $\tilde{\alpha} \gg \alpha$, the probability for the particle to remain in the bound state is small.

Problem 8.51

A particle is in the ground state of a δ-well, so that $U = -\alpha \delta(x)$. At $t = 0$, the well begins to move with a constant speed V. Find the probability that the well will carry the particle with it. Consider the limiting cases of small and large velocities.

Solution
To calculate the probability, we turn to the coordinate system K' that moves together with the well, where $x' = x - Vt$. The particle wavefunctions immediately after the motion commences in the initial, $\psi_0(x)$, and moving, $\tilde{\psi}_0(x')$, coordinate systems have the form

$$\tilde{\psi}_0(x') = \exp\left\{-\frac{i}{\hbar} mVx'\right\} \psi_0(x'), \quad \psi_0(x) = \sqrt{\kappa} e^{-\kappa|x|}.$$

$\kappa = m\alpha/\hbar^2$ (see Problem 2.7). The relation between the wavefunctions reflects the fact that the Galilean transformation is just the substitution $p \to p' = p - mV$ (compare to 6.26). Since the wavefunction of the particle bound state in K' is obtained from $\psi_0(x)$ by changing $x \to x'$, then the probability given from Eq. (2) in Problem 8.47 becomes

$$w_0 = \left| \int \psi_0^*(x') \tilde{\psi}_0(x') dx' \right|^2 = \frac{1}{(1 + V^2/4v_0^2)^2},$$

where $v_0^2 = \alpha^2/\hbar^2$ (we should note that v_0^2 coincides with $\overline{v^2}$, which is the mean value of the velocity square in the ground state of the particle in a δ-well). In the case $V \ll v_0$, we have $w_0 \approx 1 - V^2/2v_0^2 \approx 1$, which means that the particle remains confined to the well with the large probability. On the contrary, in the limiting case of $V \gg v_0$, $w_0 \approx (2v_0/V)^4 \ll 1$, the particle will most probably escape the well.

Problem 8.52

For a charged oscillator in the ground state, an homogeneous electric field is suddenly applied along the oscillation axis. Find the excitation probability into different oscillator states after the field turns on. Compare to the result of Problem 6.25.

Solution

The probabilities are $w(0 \to n) = |\langle n, f|0, i\rangle|^2$ (see Eq. (2) in Problem 8.47). The easiest way to calculate these matrix elements is to use the creation and annihilation operator formalism (compare to Problem 6.25). For an unperturbed oscillator, $\hat{a}_i = (2\hbar)^{-1/2}(\lambda x + i\lambda^{-1}\hat{p})$, where $\lambda = \sqrt{m\omega}$; the ground state is determined by the relation $\hat{a}_i|0, i\rangle = 0$. Application of an electric field is equivalent to a shift of the oscillator equilibrium point to $x_0 = e\mathcal{E}/m\omega^2$, so that here $\hat{a}_f = \hat{a}_i - \lambda(2\hbar)^{-1/2}x_0$. Meanwhile, the final (for $t > 0$, after the field turns on) stationary states are determined by the relations

$$|n, f\rangle = \frac{1}{\sqrt{n!}}(\hat{a}_f^+)^n|0, f\rangle, \quad \hat{a}_f|0, f\rangle = 0.$$

The expansion coefficients $|0, i\rangle = \sum_n c_n|n, f\rangle$ were calculated in Problem 6.25. Using these values we find the desired transition probabilities:

$$w(0 \to n) = |c_n|^2 = \frac{1}{n!}\alpha^{2n}e^{-\alpha^2}, \quad \alpha = -\frac{\lambda x_0}{\sqrt{2\hbar}}. \tag{1}$$

As is seen, the dependence of the transition probabilities on the quantum state n of the oscillator is described by the Poisson distribution.

Problem 8.53

For a linear oscillator in the ground state at the moment $t = 0$, the equilibrium point of the oscillator begins to move with velocity V. Find the probability of transitions into different oscillator states for $t > 0$.

Solution

We should act as in the previous problem. The transformation to the frame of reference moving with the velocity V corresponds to the change of particle momentum by $-mV$, so that now $\hat{a}_f = \hat{a}_i - imV/\sqrt{2\hbar\lambda}$. The probability distributions are determined by Eq. (1) of the previous problem, where we should put $|\alpha| = mV/\sqrt{2\hbar\lambda}$.

8.6 Adiabatic approximation

a) Adiabatic approximation in non-stationary problems

Problem 8.54

A Hamiltonian, $\hat{H}(\hat{p}, q, \lambda(t))$, explicitly depends on time. For each moment of time, t, a discrete energy spectrum, $E_n(\lambda(t))$, of the "instantaneous" Hamiltonian is assumed to be known. A complete system of the corresponding orthonormal eigenfunctions, $\psi_n(q, \lambda(t))$, is also known.

Write the wave equation for the system in the basis of functions $\psi_n(q, \lambda(t))$. Prove that for an adiabatic change (in the limit $\dot{\lambda} \to 0$), the distribution of quantum states does not depend on time. What is the classical analog of this result?

Solution

We write the system wavefunction in the form of expansion[146]

$$\psi(q, t) = \sum_n C_n(t) \psi_n(q, t) \exp\left\{ -\frac{i}{\hbar} \int_0^t E_n(t') dt' \right\}. \tag{1}$$

Here

$$\hat{H}(\hat{p}, q, \lambda(t)) \psi_n(q, \lambda(t)) = E_n(\lambda(t)) \psi_n(q, \lambda(t)), \tag{2}$$

while coefficients $C_n(t)$ are the wavefunctions in the representation used. Substituting (1) into the Schrödinger equation, multiplying both its sides by $\psi_k^*(t)$ from the left, and integrating over coordinates q using wavefunction $\psi_n(q, t)$ orthogonality, we find

$$\dot{C}_k(t) = -\sum_n C_n(t) \exp\left\{ \frac{i}{\hbar} \int_0^t (E_k - E_n) dt' \right\} \int \psi_k^*(t) \dot{\psi}_n(t) dq. \tag{3}$$

This is the sought-after equation.[147] It is convenient to use it for Hamiltonians slowly-varying with time. Indeed, since $\dot{\psi}_n = \dot{\lambda} \partial \psi_n / \partial \lambda$, then $\langle \psi_k | \dot{\psi}_n \rangle \propto \dot{\lambda} \ll 1$, and in the

---

[146] For brevity below we write $\psi_n(q, t)$ instead of $\psi_n(q, \lambda(t))$, and often do not show the coordinate q dependence.

[147] We can write it in the form of the Schrödinger equation, $i\hbar \dot{C}_k = \sum_n H''_{kn} C_n \equiv \hat{H}'' C_k$. Here the operator (matrix) $\hat{H}''$ is Hermitian and describes the Hamiltonian of the system in the energy representation of the instantaneous Hamiltonian. However, its connection with the initial Hamiltonian due to a non-trivial time-dependence of the unitary transformation is not obvious. Compare to Problem 6.28.

zeroth approximation we can put the right side of equation (3) equal to zero and obtain[148]

$$C_n(t) \approx C_n^{(0)} = \text{const.} \qquad (4)$$

This can be characterized as the (approximate) conservation of the quantum state number during the adiabatic change. This result is the quantum-mechanical analog of the adiabatic invariance of the value

$$I = \frac{1}{2\pi} \oint p dq,$$

in classical mechanics. This analogy becomes more straightforward in the quasi-classical case (see the next chapter) if we take into account the *Bohr–Sommerfeld quantization rule*.

In conclusion, we should mention the following circumstance. Despite the slow evolution of the Hamiltonain, it can change significantly during a long-enough time interval (its final form may be nothing like the initial Hamiltonian). Nevertheless, if at the initial moment of time the system was in the nth quantum state, then it would most likely remain in the same quantum state but with the wavefunction $\psi_n(q, \lambda(t))$.

Problem 8.55

In the conditions of the previous problem, consider the system to be in the nth non-degenerate quantum state at $t = t_0$, and find its wavefunction for $t > t_0$ in the first order of adiabatic perturbation theory.

Using the results obtained, consider the excitation of a charged linear oscillator, that was in the ground state at $t \to -\infty$, under the action of an homogeneous electric field $\mathcal{E}(t)$. Compare to the exact solution (see Problem 6.25). Analyze the forms of $\mathcal{E}(t)$ given in Problem 8.33.

Solution

1) Let us first clarify the conditions for result (4) of the previous problem to be valid, by transforming the matrix element $\langle \psi_k | \dot\psi_n \rangle$ from Eq. (3). We differentiate both parts of Eq. (2) from Problem 8.54 with respect to λ, then multiply it by ψ_k^* from the left and integrate over the coordinates. Taking into account the hermiticity of $\hat{H}$, we obtain

$$\int \psi_k^* \dot\psi_n dq = \frac{1}{E_n(t) - E_k(t)} \int \psi_k^* \left(\frac{\partial \hat{H}}{\partial t}\right) \psi_n dq, \quad k \neq n. \qquad (1)$$

---

[148] For an applicability condition, see the following problem.

More precisely, for $E_k \neq E_n$. In the case of $n = k$, we can always make $\langle \psi_n | \dot\psi_n \rangle$ equal to zero[149] by a change of the phase factor. So relation (3) from Problem 8.54 takes the form

$$\dot C_k(t) = \sum_n{}' \frac{1}{\hbar \omega_{kn}(t)} \left(\frac{\partial \hat H}{\partial t}\right)_{kn} \exp\left\{ i \int_0^t \omega_{kn}(t') dt' \right\} C_n(t). \tag{2}$$

The term with $n = k$ in the sum is absent, and $\hbar \omega_{kn} = E_k - E_n$. If the derivative, $\partial \hat H/\partial t$, is small enough, then $\dot C_k \approx 0$ and

$$C_k \approx C_{kn}^{(0)} = \text{const} = \delta_{kn}$$

from the problem condition. In the next approximation of adiabatic perturbation theory, according to (2), for $k \neq n$ we have

$$\dot C_{kn}^{(1)}(t) = \frac{1}{\hbar \omega_{kn}(t)} \left(\frac{\partial \hat H}{\partial t}\right)_{kn} \exp\left\{ i \int_0^t \omega_{kn}(t') dt' \right\}.$$

Integrating with the given initial condition, we obtain

$$C_{kn}^{(1)}(t) = \int_{t_0}^t \frac{1}{\hbar \omega_{kn}(t')} \left(\frac{\partial \hat H}{\partial t'}\right)_{kn} \exp\left\{ i \int_0^{t'} \omega_{kn}(t'') dt'' \right\} dt'. \tag{3}$$

An order-of-magnitude estimation of $C_{kn}^{(1)}$ has the form

$$|C_{kn}^{(1)}| \sim \frac{\left|\frac{\partial \hat H}{\partial t} \frac{1}{\omega_{kn}}\right|}{|E_k - E_n|}.$$

In the right-hand we have the ratio of the Hamiltonian change during the time of Bohr period, ω_{kn}^{-1}, and the energy difference between the corresponding levels. If this ration is small then the evolution of the Hamiltonian can be considered slow (*adiabatic*). We should mention that if while $\hat H(t)$ is changing with time, some levels come closer to each other (so that $E_k(t') \approx E_n(t')$), then the adiabaticity is broken, and in this case the transitions between the nth and kth states are most intensive.

2) For the oscillator in an electric field, we have

$$\hat H = \frac{\hat p^2}{2m} + \frac{kx^2}{2} - e\mathcal{E}(t)x.$$

---

[149] Since for a real eigenfunction we have

$$\langle \psi_n | \dot\psi_n \rangle = \frac{1}{2}\frac{\partial}{\partial t} \int \psi_n^2 dq = 0.$$

The eigenfunctions and eigenvalues of the instantaneous Hamiltonian are given in Problem 2.2. Here $\partial \hat{H}/\partial t = -e\dot{\mathcal{E}}x$, while the matrix element $(\partial \hat{H}/\partial t)_{k0}$ for the levels $k \neq 0$ is different from zero only for $k = 1$, where it is equal to $-ea\dot{\mathcal{E}}/\sqrt{2}$, $a = \sqrt{\hbar/(m\omega)}$. From Eq. (3) we obtain (we put $t_0 = -\infty$)

$$C_{10}^{(1)}(t = +\infty) = -\frac{ea}{\sqrt{2}\hbar\omega} \int_{-\infty}^{\infty} \frac{\partial \mathcal{E}}{\partial t} e^{i\omega t} dt. \tag{4}$$

So the probability of a single allowed oscillator transition in the first order of adiabatic perturbation theory is

$$W(0 \to 1) = |C_{10}^{(1)}|^2 = \frac{e^2 a^2}{2\hbar^2} \left| \int_{-\infty}^{\infty} \mathcal{E}(t) e^{i\omega t} dt \right|^2. \tag{5}$$

Assuming that the electric field is turned off for $t \to +\infty$, we make substitution (4) and integrate by parts. This result coincides with that obtained in Problem 8.33 using the methods of usual non-stationary perturbation theory[150] and differs from the exact result only slightly (see Problem 6.25) for $W \ll 1$. The transition probabilities for the forms of $\mathcal{E}(t)$ given in the problem statement coincide with those given in Problem 8.33.

Problem 8.56

Analyze the quasi-energy states[151] in the adiabatic approximation. Consider the energy spectrum of the instantaneous Hamiltonian to be discrete and non-degenerate.

Solution

The solution of the Schrödinger equation in the "zeroth" approximation of adiabatic approximation theory for a periodic dependence on time (see the previous problem in the case when $\lambda(t + T) = \lambda(t)$) is

$$\psi_{\varepsilon_n}(q, t) = \exp\left\{-\frac{i}{\hbar} \int_0^t E_n(\lambda(t)) dt\right\} \psi_n(q, \lambda(t)). \tag{1}$$

---

[150] The reason that the result of adiabatic approximation (whose applicability condition is $\tau \ll \omega^{-1}$, $ea\mathcal{E}/\hbar\omega^2\tau \ll 1$) coincides with the result of perturbation theory (applicability condition is $eaE \ll \hbar\omega$) is that the specific action of the homogeneous field on the oscillator is reduced to a shift of the equilibrium point. This causes the perturbation matrix elements to differ from zero only for transitions between the neighboring oscillator levels.

[151] See Problem 6.40.

This describes the QES. It is assumed that the phase factor in the eigenfunctions ψ_n themselves is chosen so that $\dot\psi_n \sim \dot\lambda$. Transforming the exponent with

$$\int_0^t E_n dt = \int_0^t (E_n(t) - \overline{E}_n) dt + \overline{E}_n t,$$

we see that the value of quasi-energy for this state in the zeroth approximation is equal to

$$\varepsilon_n = \overline{E}_n = \frac{1}{T} \int_0^T E_n(\lambda(t)) dt, \qquad (2)$$

i.e., it coincides with the mean value over the period of Hamiltonian change of $E_n(t)$. The expansion of the periodic function

$$\exp\left\{-\frac{i}{\hbar} \int_0^t (E_n(t) - \overline{E}_n) dt\right\} \psi_n(q, \lambda(t))$$

into a Fourier series determines the quasi-energy harmonics of QES. See Problem 6.40.

Problem 8.57

A particle is in the field of two δ-wells approaching each other, so that

$$U(x,t) = -\alpha\left[\delta\left(x - \frac{L(t)}{2}\right) + \delta\left(x + \frac{L(t)}{2}\right)\right].$$

For $t \to -\infty$, the wells were infinitely separated and the particle was bound by one of them. The distance $L(t)$ between wells decreases slowly, and at some moment of time the wells collide: $U(x) = -2\alpha\delta(x)$. Find the probability for a particle to remain in a bound state.

Solution

The probability for the particle to remain bound is equal to $1/2$.

Taking into account parity conservation, it is convenient to analyze the time-dependence of even and odd wavefunction components separately. Denoting the bound state wavefunction in a δ-well by $\psi_0(x)$, and assuming for concreteness that for $t \to -\infty$ the particle was bound by the right well, we write the wavefunction of the initial state in the form $\psi(t = -\infty) = (\psi_+ + \psi_-)/\sqrt{2}$, where

$$\psi_\pm = \frac{1}{\sqrt{2}}\left\{\psi_0\left[x - \frac{L(-\infty)}{2}\right] \pm \psi_0\left[x + \frac{L(+\infty)}{2}\right]\right\}.$$

For a large distance between the wells, $L(-\infty) = \infty$, both the even ψ_+ and odd ψ_- wavefunction components describe the bound particle with energy equal to the energy in the field of one well (the level is two-fold degenerate).

Now we note that regardless how exactly the wells approach each other, we can predict that when the wells run into one another, the odd wavefunction component would describe an "unbinding" of the particle. That is due to the fact that in the field of a single δ-well, there is only one even discrete state. (We should note that when the two wells are close, the discrete odd level and continuum merge together, and the adiabatic approximation for the odd part of the wavefunction is not applicable.)

The time-dependence of the even component of the wavefunction depends strongly on a specific motion of the wells, but if it has adiabatic nature then the particle remains in the ground state. Since for the initial state the probability of the particle being in an odd state is $1/2$, then the probability for it to remain bound for a slow collision of the wells is also $1/2$, which we mention at the beginning of this solution.

The applicability condition of this result is $|\dot{L}| \ll \alpha/\hbar$. But this condition must be fulfilled only when the wells approach to a distance of the order of the particle localization length in the δ-potential ground state, i.e., $L \leq \hbar^2/m\alpha$. At large distances there is not such a strict restriction on the well velocity, since in this case the particle localized in one of the wells does not "feel" the other well, while for the well motion with an arbitrary (but constant) velocity, in accordance with the principle of relativity, there are no transitions. (On distances $L \ll \hbar^2/m\alpha$ it is only required that the acceleration $\ddot{L}$ is not too large.)

b) **Adiabatic approximation in stationary problems**

Problem 8.58

The Hamiltonian of a system consisting of two subsystems has the form

$$\hat{H} = \hat{H}_1(x) + V(x,\xi) + \hat{H}_2(\xi),$$

where x, ξ are the coordinates of the first and second subsystem, and $V(x,\xi)$ describes an interaction between them. Considering the characteristic frequencies of the first ("fast") subsystem to be much larger than the frequencies of the second ("slow") subsystem, reduce the problem of finding the energy levels and corresponding wavefunctions of the full system to the solution of the Schrödinger equation for separate subsystems.

Use these results to analyze the states in the lower part of the energy spectrum in a two-dimensional potential of the form

$$U(x,y) = \begin{cases} 0, & \frac{x^2}{a^2} + \frac{y^2}{b^2} \leq 1, \\ \infty, & \frac{x^2}{a^2} + \frac{y^2}{b^2} > 1, \end{cases}$$

for the case $b \gg a$.

Solution

Let us introduce $\psi_{n_1}(x,\xi)$ and $E_{n_1}(\xi)$, to be the eigenfunctions and the eigenvalues of the operator $\hat{H}' = \hat{H}_1(x) + V(x,\xi)$ for the "slow" subsystem with fixed values of coordinates ξ, so that

$$[\hat{H}_1(x) + \hat{V}(x,\xi)]\psi_{n_1}(x,\xi) = E_{n_1}(\xi)\psi_{n_1}(x,\xi). \tag{1}$$

These play a role analogous to eigenfunctions and eigenvalues of the instantaneous Hamiltonian in adiabatically-driven systems (see Problems 8.54 and 8.55). For the exact eigenfunctions of the total system Hamiltonian, an expansion of the form

$$\Psi_N(x,\xi) = \sum_{n_1} \phi_{N n_1}(\xi) \psi_{n_1}(x,\xi)$$

is valid.

The important thing is that for the "fast" subsystem, the change of "slow" system state works like an adiabatic perturbation, where the quantum numbers are conserved (see Problem 8.54). Neglecting transitions, we obtain an approximate expression for the wavefunction, which corresponds to taking into account only one term in the sum given above:

$$\Psi_N \approx \psi_{n_1 n_2}(x,\xi) = \phi_{n_1 n_2}(\xi)\psi_{n_1}(x,\xi). \tag{2}$$

Using the Schrödinger equation,

$$[\hat{H}_1 + V(x,\xi) + \hat{H}_2]\phi_{n_1 n_2}(\xi)\psi_{n_1}(x,\xi) \approx E_{n_1 n_2}\phi_{n_1 n_2}(\xi)\psi_{n_1}(x,\xi),$$

and taking into account relation (1), we make the following transformations. We multiply both parts of the equation obtained by $\psi_{n_1}^*$ from the left, integrate over the coordinates x of the "fast" subsystem, and neglect the action of operator[152] $\hat{H}_2(\xi)$ on the variable ξ from wavefunction $\psi_{n_1}(x,\xi)$ (*i.e.*, put $\hat{H}_2\phi\psi \approx \psi\hat{H}_2\phi$; here again we take advantage of the difference between the typical motion times); as a result, we obtain the Schrödinger equation for the "slow" subsystem:

$$[\hat{H}_2(\xi) + E_{n_1}(\xi)]\phi_{n_1 n_2}(\xi) = E_{n_1 n_2}\phi_{n_1 n_2}(\xi). \tag{3}$$

As is seen in the approximation considered, interaction with the "fast" subsystem is characterized by an effective potential, here as $U_{eff}(\xi) = E_{n_1}(\xi)$.

Eqs. (1)–(3) are the basis of *adiabatic approximation* for stationary states. For example, with the potential given in the problem condition, since $b \gg a$, we have motion along x as a "fast" subsystem, and motion along y as a "slow" subsystem. For a fixed y, motion along x is a motion in an infinitely deep well with the width $a(y) = 2a\sqrt{1 - y^2/b^2}$, so that

---

[152] An analogy with the case of adiabatic approximation considered in Problems 8.54 and 8.55, is seen in the fact that there, while calculating the derivative $\partial\psi/\partial t$, we could omit the term obtained by differentiating the instantaneous wavefunction $\psi_n(q, \lambda(t))$ over time.

$$\psi_{n_1} = \sqrt{\frac{2}{a(y)}} \sin\left[\frac{\pi(n_1+1)}{a(y)}\left(x + \frac{a(y)}{2}\right)\right], \quad E_{n_1}(y) = \frac{\hbar^2\pi^2(n_1+1)^2}{2ma^2(y)}.$$

According to Eq. (3), along y the particle moves in an effective potential:

$$U(y) = E_{n_1}(y) = \frac{\hbar^2\pi^2(n_1+1)^2 b^2}{8ma^2(b^2-y^2)}, \quad |y| < b.$$

For such a potential, wavefunctions of the energy levels that are not too high are localized on the distances $|y| \ll b$, where the effective potential can be expanded in a series:

$$U(y) \approx \frac{\hbar^2\pi^2(n_1+1)^2}{8ma^2} + \frac{\hbar^2\pi^2(n_1+1)^2}{8ma^2 b^2} y^2.$$

The calculation of the wavefunction, $\phi_{n_1 n_2}(y)$, and levels, $E_{n_1 n_2}$, can be reduced to the problem of an harmonic oscillator, which allows us to obtain

$$E_{n_1 n_2} = \frac{\hbar^2\pi^2(n_1+1)^2}{8ma^2} + \frac{\hbar^2\pi(n_1+1)}{2mab}\left(n_2 + \frac{1}{2}\right),$$

$$\psi_{n_1 n_2} = (2^{n_2}\sqrt{\pi}y_0 n_2!)^{-1/2} \exp\left\{-\frac{y^2}{2y_0^2}\right\} H_{n_2}\left(\frac{y}{y_0}\right) \psi_{n_1}(x,y), \quad n_{1,2} = 0,1,\ldots, \quad (4)$$

$$\left(y_0 = \sqrt{\frac{2ab}{\pi(n_1+1)}}, \quad y_0^2\left(n_2+\frac{1}{2}\right) \ll b^2\right).$$

Problem 8.59

The Hamiltonian has the form

$$\hat{H} = \frac{1}{2m}\hat{p}_x^2 + \frac{1}{2M}\hat{p}_y^2 + \frac{k(x^2+y^2)}{2} + \alpha xy, \quad |\alpha| < k,$$

where $M \gg m$ (two bound oscillators with different masses). Find the energy levels and corresponding wavefunctions using the adiabatic approximation. Compare the result obtained with the exact solution (see Problem 2.50).

Solution

The fast subsystem is the light oscillator with the mass m which is characterized by the coordinate x. For a fixed y, the coordinate of the slow subsystem, we have ($\omega = \sqrt{k/m}$):

$$\psi_{n_1} \equiv \psi_{n_1}^{osc}\left(x + \frac{\alpha y}{k}\right), \quad E_{n_1}(y) = \hbar\omega\left(n_1 + \frac{1}{2}\right) - \frac{\alpha^2}{2k}y^2.$$

The wavefunctions and energy levels of the slow subsystem are determined according to Eq. (3) of the previous problem. In this case,

$$E_{n_1 n_2} = \hbar\omega\left(n_1 + \frac{1}{2}\right) + \hbar\omega\sqrt{\frac{m}{M}\left(1 - \frac{\alpha^2}{k^2}\right)}\left(n_2 + \frac{1}{2}\right);$$

compare to the exact result for the spectrum from Problem 2.50 by making an expansion over the small parameter $\sqrt{m/M}$.

Problem 8.60

Two particles with sharply different masses, $M \gg m$, are in an infinitely deep potential well of width a, and interact with each other as mutually impenetrable points. Find the low-lying energy levels and the corresponding wavefunctions.

Solution

We use the adiabatic approximation. See Problem 8.58. The fast subsystem is the light particle (coordinate x_1). Its levels and wavefunctions for a fixed value of x_2, the coordinate of the heavy particle (slow subsystem), have the form

$$\psi_{n_1} = \sqrt{\frac{2}{a - x_2}} \sin\frac{\pi(n_1 + 1)(x_1 - x_2)}{a - x_2}, \quad E_{n_1}(x_2) = \frac{\hbar^2 \pi^2 (n_1 + 1)^2}{2m(a - x_2)^2}.$$

For concreteness, we put $x_2 < x_1$; the wavefunction is equal to zero for the values $x_1 > a$ and $x_1 < x_2$. Compare to Problem 2.51.

The energy, $E_{n_1}(x_2)$, acts as an effective potential, $U(x_2)$, for the heavy particle with $0 < x_2 < a$ (outside this interval, $U = \infty$). For the lower energy states, the typical position of the heavy particle, $x_2 \ll a$, and $U(x_2)$ could be expanded in series (compare to Problem 8.58):

$$U(x_2) \approx \frac{\hbar^2 \pi^2 (n_1 + 1)^2}{2ma^2} + \frac{\hbar^2 \pi^2 (n_1 + 1)^2}{ma^3} x_2, \quad x_2 \geq 0.$$

Now calculation of the spectrum, $E_{n_1 n_2}$, and the eigenfunctions, $\phi_{n_1 n_2}(x_2)$, is reduced to the one considered in Problem 2.8. Using the result of this problem, we obtain

$$E_{n_1 n_2} = \frac{\hbar^2 \pi^2 (n_1 + 1)^2}{2ma^2} + \left[\frac{\hbar^6 \pi^4 (n_1 + 1)^4}{2m^2 M a^6}\right]^{1/3} \alpha_{n_2 + 1}; \quad n_{1,2} = 0, 1, \ldots.$$

Here $-\alpha_k$, where $k = 1, 2, \ldots$, is the sequence of the Airy function zeros increasing order.

Problem 8.61

Using the adiabatic approximation, analyze the energy spectrum and the corresponding wavefunctions of bound states in a central attractive potential, $U(r)$, and in the presence of a strong, homogeneous magnetic field.

Use your general considerations to find (a) the shift of Landau levels by a short-range potential and (b) the ground state of the hydrogen atom in a strong magnetic field.

Solution

The particle Hamiltonian has the form

$$\hat{H} = \hat{H}_t + \frac{1}{2\mu}\hat{p}_z^2 + U(\sqrt{\rho^2 + z^2}),$$

where

$$\hat{H}_t = -\frac{\hbar^2}{2\mu}\left(\frac{1}{\rho}\frac{\partial}{\partial\rho}\rho\frac{\partial}{\partial\rho} + \frac{1}{\rho^2}\frac{\partial^2}{\partial\varphi^2}\right) - \frac{e\hbar}{2\mu c}\mathcal{H}\hat{l}_z + \frac{e^2\mathcal{H}^2}{8\mu c^2}\rho^2$$

is the Hamiltonian of the transverse motion in an homogeneous magnetic field, directed along the z-axis with the vector potential $\mathbf{A} = [\mathcal{H} \times \mathbf{r}]/2$ (see Problem 7.1).

In the case of a strong enough magnetic field, the particle motion in the transverse direction is determined generally by this field. Here $\omega_H = |e|\mathcal{H}/\mu c$, which is the typical frequency of such motion, exceeds strongly the frequency of the longitudinal motion. So, we can use the *adiabatic approximation* (see Problem 8.58). Here as a "fast" subsystem, we take the motion of the particle in the transverse direction. For this, the potential, $U(r)$, can be considered as a perturbation. And the wavefunctions of the "fast" subsystem (for a fixed value of the coordinate z of the longitudinal motion that characterizes the "slow" subsystem) in the "zeroth" approximation have the form

$$\psi_{n_1} \equiv \psi_{nm}(\mathbf{r}) \approx \psi_{nm}^{(0)}(\boldsymbol{\rho}) = \frac{e^{im\varphi}}{\sqrt{2\pi}}\frac{1}{a_H|m|!}\left[\frac{(n_\rho + |m|)!}{n_\rho!}\right]^{-1/2} x^{|m|/2}$$

$$\times e^{-x/2} F(-n_\rho, |m|+1, x) \quad (1)$$

(they do not depend on z), where

$$x = \frac{\rho^2}{2a_H^2}, \quad a_H = \sqrt{\frac{\hbar}{\mu\omega_H}}, \quad n = n_\rho + \frac{|m| - em/|e|}{2}.$$

Energy levels of the "fast" subsystem in the first order of perturbation theory are described by the expressions

$$E_{n_1}(z) \equiv E_{nm}(z) \approx E_n^{(0)} + E_{mn}^{(1)}(z); \quad E_n^{(0)} = \hbar\omega_H\left(n + \frac{1}{2}\right), \tag{2}$$

$$E_{mn}^{(1)}(z) = \int_0^\infty U(\sqrt{\rho^2 + z^2})|\psi_{nm}^{(0)}(\boldsymbol{\rho})|^2 2\pi\rho d\rho,$$

here $E_n^{(0)}$ determine the Landau levels.

Now, according to Problem 8.58, solving for the wavefunctions $\psi_{nm}^{(0)}(\boldsymbol{\rho})\psi_{nn_2}(z)$ and the energy spectrum $E_{n_1n_2} \equiv E_{nmn_2}$ of the bound states is reduced to a solution of a one-dimensional Schrödinger equation in an effective potential that coincides with $E_{nm}^{(1)}(z)$. The properties of such bound states depend essentially both on the form of potential $U(r)$ and quantum numbers n, m that characterize the transverse motion (the "fast" subsystem).

Let us consider some special cases.

1) It is possible to analyze the case of a "shallow" spherical well $U(r)$ (of an arbitrary form) with radius R and characteristic depth U_0, for which $\mu R^2 U_0/\hbar^2 \ll 1$, and without particle bound states in the absence of a magnetic field. The effective potential $U_{eff} \equiv E_{nm}^{(1)}(z)$ also defines a "shallow" well, but one-dimensionally. In such a well, for each pair of quantum numbers n, m there is one and only one bound state, for which (compare, for example, to Problem 2.22)

$$E_{nm0} \approx E_n^{(0)} - \frac{\mu\alpha_{nm}^2}{2\hbar^2}; \quad \psi_0(z) \approx \sqrt{\frac{\mu\alpha_{nm}}{\hbar^2}}\exp\left\{-\frac{\mu\alpha_{nm}}{\hbar^2}|z|\right\},$$

$$\alpha_{nm} = -2\int_0^\infty\int_0^\infty U(\sqrt{\rho^2 + z^2})|\psi_{nm}^{(0)}(\boldsymbol{\rho})|^2 2\pi\rho d\rho dz > 0. \tag{3}$$

So, for example, in the case of a weak magnetic field,[153] for which $a_H \gg R$, it follows that

$$\alpha_{nm} \propto -a_H^{-2|m|-2}\int\int U(r)\rho^{2|m|+1}d\rho dz \sim RU_0\left(\frac{R}{a_H}\right)^{2|m|+2} \propto \mathcal{H}^{|m|+1}. \tag{4}$$

In the integral, values $\rho \leq R \ll a_H$ are important, for which, according to (1), $\psi_{nm}^{(0)} \propto \rho^{|m|}/a_H^{|m|+1}$. We see that the binding energy, equal to $\mu\alpha_{nm}^2/2\hbar^2$, decreases sharply with the increase of $|m|$. We should emphasize that the value $l_\parallel \sim \hbar^2/\mu\alpha_{nm}$ determines a localization domain in the z-direction. Also, we can see $l_\parallel \gg l_\perp$, where $l_\perp \sim a_H$ is the size of the particle's localization domain in the plane perpendicular to the magnetic field. Thus the particle localization domain has a needle-shaped form. This property remains even with an increase of the magnetic field, when condition $a_H \gg R$ is not fulfilled. The needle-shaped form of the particle's wavefunction

---

[153] We should mention that the results obtained for a "shallow" well do not put any restrictions on the value of the magnetic field. The field could also be weak. Compare to Problem 7.7.

localization domain is connected with the difference in periods for motion along and across the magnetic field in the adiabatic approximation.

2) Now let the attractive potential $U(r)$ be strong, so that $\mu R^2 U_0/\hbar^2 \geq 1$. To consider it as perturbation on the background of the magnetic field, the latter must be so strong that the condition $R \gg a_H \propto 1/\sqrt{\mathcal{H}}$ is fulfilled. For the states with quantum numbers n, $m \sim 1$ (or $\sqrt{n}$, $\sqrt{|m|} \ll R/a_H$; compare to Problem 7.2) in relation (2), we can factor $U(|z|)$ outside the integral sign and obtain

$$E_{nm}^{(1)}(z) \approx U(|z|), \tag{5}$$

so that the effective one-dimensional potential has the same form as the initial central potential, $U(r)$. See Problems 4.1 and 2.5 for the relation between the energy spectrum in a symmetric one-dimensional potential, $U(|z|)$, and the spectrum of s-levels in central field $U(r)$.

3) Substitution of $U(r) \to U(|z|)$ in (2) that gives (5) is not valid at short distances, $|z| \leq a_H$, in potentials with Coulomb (or stronger) attraction, since the particle can now fall into the center. We will analyze the position of levels with $n_2 = 0$ (low-lying levels of the longitudinal motion) for the hydrogen-like atom in a strong magnetic field. In this case we write $E_{nm}^{(1)} \approx -Ze^2/(|z| + a_H)$; in comparison with (5), here a "cutoff" of the Coulomb potential on small distances is used. Using the variational method with a trial wavefunction of the form $\psi_0 = \sqrt{\kappa} e^{-\kappa|z|}$, where κ is the variational parameter, we find

$$E_{nm0} \approx E_n^{(0)} + \overline{\hat{H}(z)} = E_n^{(0)} + \frac{\hbar^2 \kappa^2}{2\mu} - \kappa \int_{-\infty}^{\infty} \frac{Ze^2}{|z| + a_H} e^{-2\kappa|z|} dz$$

$$\approx E_n^{(0)} + \frac{\hbar^2 \kappa^2}{2\mu} - 2\kappa Ze^2 \ln \frac{1}{2\kappa a_H}. \tag{6}$$

To approximately calculate this integral to logarithmic accuracy, we can replace the exponent by 1, while replacing the limits of integration $\pm\infty$ by the values $\sim (\pm 1/\kappa)$.

Minimization of this relation for the hydrogen atom with $a_H = 10^{-2}$ (we use atomic units) gives a binding energy of 12.5 (for $\kappa = 3.4$).

We emphasize that just as in the previous case, according to Eq. (3), the depth of the longitudinal levels (for fixed quantum numbers n, m) is much smaller than the distance $\hbar\omega_H$ between the neighboring Landau levels.

4) Now, we make a few concluding statements. The first is connected with the problem of generalizing the results obtained above in (1) to include the impact of a "shallow" well on Landau levels in the case of a weak magnetic field, when $a_H \gg R$ (R is the radius of an arbitrary short-range potential) with the help of *perturbation theory with respect to the scattering length* (see Problems 4.29, 4.31, and 4.11). We should note, being limited by states with $m = 0$ (for $m \neq 0$ see Problems 13.36 and 13.37), that according to (3) and for $a_H \gg R$, the integral in relation

$$a_{n0} \approx -\frac{1}{a_H^2} \int_{-\infty}^{\infty} \int_0^{\infty} U(r) \rho d\rho dz$$

differs from the s-scattering length a_0^B in the Born approximation only by the coefficient $\mu/\hbar^2$. So, replacing a_0^B by the scattering length a_0 in potential $U(r)$ in the case of $a_0 < 0$ gives the desired generalization of (3). Now

$$\alpha_{n0} \approx -\frac{\hbar^2 a_0}{\mu a_H^2}, \quad E_{n00} \approx E_n^{(0)} - \frac{1}{2}\mu\omega_H^2 a_0^2. \tag{7}$$

For $a_0 > 0$ there are no bound states, just as in the case of a repulsive potential in the conditions of Eq. (1). This equation becomes inapplicable in the case of $|a_0| \geq a_H$, when in the (isolated) potential $U(r)$, there exists a "shallow" s-level with the energy $\sim \hbar\omega_H$. In this case there is an important reconstruction of the Landau level spectrum with $m = 0$ (i.e., their shifts become $\sim \hbar\omega_H$; compare to Problems 11.4 and 9.3). We should emphasize that the rest of the "deep" levels with both angular momentum $l = 0$ and $l \neq 0$, if they exist, undergo only a small shift under the weak magnetic field.

Now we note that the levels considered, E_{nmn_2}, under the combined action of the magnetic field and potential, are truly bound only for the values of quantum number $n_\rho = 0$ (for each m). For the levels $n_\rho \geq 1$ they correspond to quasi-stationary states, so that under the potential $U(r)$ (which results in the formation of a bound state in the longitudinal direction) transitions to even lower levels, $E_{n'}^{(0)}$, of transversal motion with $n'_\rho < n_\rho$ are possible. In the longitudinal direction, the particle is not bound and has the energy $E_t \approx \hbar\omega_H(n_\rho - n'_\rho)$, Here we take into account the smallness of the depth where the levels are lying. In the case of a "shallow" well and a weak magnetic field, as considered in Eq. (1), the width such quasi-stationary states with $m = 0$

$$\Gamma_{n00} \approx \sum_{n'=0}^{n-1} \frac{\mu}{\hbar^3}\sqrt{\frac{2\mu}{\hbar\omega_H(n-n')}} a_0^3, \quad a_0 = -\frac{1}{a_H^2}\int_{-\infty}^{\infty}\int_0^{\infty} U(r)\rho d\rho dz, \tag{8}$$

could be obtained in the same way[154] as Eq. (3) from Problem 8.4. With the help of substituting α_0 by the scattering length a_0, as mentioned above, expression (8) could be generalized to the case of a "strong" short-range potential.

5) Finally, we discuss the peculiarities of the quantum-mechanical problem of particle motion in the one-dimensional Coulomb potential $U = -\alpha/|x|$ on the whole axis $-\infty < x < +\infty$. As noted in section (3) above, here the particles falls into the center, the point $x = 0$; from Eq. (6) for the "cutoff" of the potential, $a_H \to 0$, it follows that $E_0 \to -\infty$. The central point is that the particle Hamiltonian

$$\hat{H} = \frac{1}{2m}\hat{p}^2 - \frac{\alpha}{|x|}, \quad -\infty < x < +\infty \tag{9}$$

---

[154] For parameter β, which determines the relation between the two channels in the conditions of Problem 8.44, we now have

$$\beta_{n_1 n_2 m} = -\int U(r)\psi_{n_1 m}^*(\rho)\psi_{n_2 m}(\rho)dV.$$

Compare to α_{nm} from Eq. (3).

for motion on the whole axis is Hermitian, but is not a self-conjugate operator. This is connected with the fact that Hamiltonian eigenfunctions for $x = \pm|x| \to 0$ have the form[155] (on the right and on the left):

$$\psi_\pm(x) = C_{1,\pm}\left\{1 - \frac{2|x|}{a_B}\ln\frac{|x|}{a_B} + O\left(x^2\ln\frac{|x|}{a_B}\right)\right\} + C_{2,\pm}\left[|x| + O\left(\frac{x^2}{a_B}\right)\right],$$

$$a_B = \frac{\hbar^2}{m\alpha^2}, \tag{10}$$

and the usual continuity condition for the regular potential wavefunction and its derivative cannot be fulfilled at the point $x = 0$, since $\psi'_\pm(x)$ becomes infinite for $x \to 0$.

But Hermitian operator (9) allows a self-adjoint extension. For a discussion about the additional conditions that give such an operator extension, see Problem 1.29. We note that for functions that satisfy conditions (10) for $|x| \to 0$, the following relation is valid:

$$\int_\epsilon^\infty \psi_2^* \hat{H}\psi_1 dx + \int_{-\infty}^{-\epsilon} \psi_2^* \hat{H}\psi_1 dx = \int_\epsilon^\infty (\hat{H}\psi_2)^* \psi_1 dx + \int_{-\infty}^{-\epsilon} (\hat{H}\psi_2)^* \psi_1 dx$$

$$+ \frac{\hbar^2}{2m}\left\{\psi_2^*(\epsilon)\psi_1'(\epsilon) - \psi_2^{*'}(\epsilon)\psi_1(\epsilon) - \psi_2^*(-\epsilon)\psi_1'(-\epsilon) + \psi_2^{*'}(-\epsilon)\psi_1(-\epsilon)\right\}. \tag{11}$$

Here $\epsilon > 0$, and the term outside the integral for $\epsilon \to 0$ is equal to

$$\frac{\hbar^2}{2m}\left\{C_{1,+}^{(2)*}C_{2,+}^{(1)} - C_{2,+}^{(2)*}C_{1,+}^{(1)} + C_{1,-}^{(2)*}C_{2,-}^{(1)} - C_{2,-}^{(2)*}C_{1,-}^{(1)}\right\}, \tag{12}$$

where the upper indices 1 and 2 correspond to the wavefunctions $\psi_{1,2}$. If for the self-adjoint extension of the operator (9), we keep the same wavefunctions as (10), then the continuity conditions at the point $x = 0$ are

$$C_{1,+}^{(1,2)} = C_{1,-}^{(1,2)}. \tag{13}$$

In addition to that relation, from the condition of term (12) outside the integral turning into zero, we obtain

$$\frac{C_{2,+}^{(1)} + C_{2,-}^{(1)}}{C_{1,\pm}^{(1)}} = \left\{\frac{C_{2,+}^{(2)} + C_{2,-}^{(2)}}{C_{1,\pm}^{(2)}}\right\}^* = \beta = \text{const.} \tag{14}$$

The real parameter β, which determines the matching conditions at the point $x = 0$, gives the self-adjoint Hamiltonian extension of Eq. (9). From a physical point of view, different choices of the parameter β correspond to different "cutoffs" of the potential on small distances. Compare to the case considered in paragraph

---

[155] Pay attention to the logarithmic term and its energy independence; an energy dependence appears only in the subleading correction terms in the asymptotic expression (10). Compare to Problem 9.14.

(3). There, the value of β could be uniquely found from the position of the ground state. We should also emphasize that the parameter β determines not only the energy spectrum of bound states but also of continuous states (particle reflection by the potential).

Now we obtain the discrete spectrum of Hamiltonian (9) using the matching conditions at the point $x = 0$, (13) and (14). The solution for the Schrödinger equation in the one-dimensional Coulomb potential that exponentially decreases as $|x| \to \infty$, with the energy $E = -m\alpha^2/2\hbar^2\nu^2$, is expressed in terms of the Whittaker function, $W_{\nu,1/2}(x)$:

$$\psi_\pm(x) = C_\pm W_{\nu,1/2}\left(\frac{2|x|}{\nu a_B}\right) = \frac{C_\pm}{\Gamma(1-\nu)} \left\{ 1 - \frac{2|x|}{a_B} \ln \frac{2|x|}{\nu a_B} \right.$$
$$\left. - \left[\frac{1}{\nu} + 2\psi(1-\nu) - 2 + 4\gamma\right] \frac{|x|}{a_B} + O\left(\frac{x^2}{a_B^2} \ln \frac{|x|}{a_B}\right) \right\}. \quad (15)$$

Here $\nu > 0$, $\psi(z) = \Gamma'(z)/\Gamma(z)$ is the logarithmic derivative of the Γ-function, and $\gamma = 0.5772\ldots$ is the Euler constant.

The energy levels have definite parity. For odd states, $\psi_\pm(0) = 0$. To satisfy this, the expression in square brackets in (15) must be infinite. Since $\psi(z)$ becomes infinite only at the points $z = -k$, where $k = 0, 1, 2, \ldots$, and since

$$\psi(z) \approx -\frac{1}{(z+k)} \quad \text{for } z \to -k, \quad (16)$$

we see that for the odd levels, ν takes the values $\nu_n^- = n$ with $n = 1, 2, \ldots$. Therefore, the spectrum of such levels

$$E_n^- = -\frac{m\alpha^2}{2\hbar^2 n^2} \quad (17)$$

coincides with the spectrum of s-levels in the central field $U = -\alpha/r$. This is expected; see Problems 4.1 and 2.5.

For the even particle levels, according to relations (14) and (15), we obtain[156]

$$-\frac{1}{\nu} - 2\psi(1-\nu) + 2 - 4\gamma + 2\ln\frac{\nu}{2} = \beta a_B. \quad (18)$$

Their spectrum depends strongly on the value of parameter β. We will analyze two limiting cases. Let $\beta > 0$ and $\beta a_B \gg 1$. Taking into account relation (16), we see that in this case, the even levels are only slightly shifted with respect to the odd levels. Writing $\nu_n^+ = n + \Delta\nu_n$, according to Eq. (18) we obtain

$$E_n^+ = -\frac{m\alpha^2}{2\hbar^2(n + \Delta\nu_n)^2}, \quad \Delta\nu_n \approx -\frac{2}{\beta a_B}, \quad n = 1, 2, \ldots, \quad (19)$$

---

[156] Pay attention to the difference of the arguments in the logarithm in Eqs. (10) and (15).

and here $|\Delta\nu_n| \ll 1$. This formula is valid also in the physically more interesting case of $\beta < 0$ with $|\beta|a_B \gg 1$. See the "cutoff" of the Coulomb potential, considered in paragraph (3). Now the even levels (19) move up slightly with respect to the odd levels. But in addition to these, there appears an additional deep-lying level, E_0^+, for which, from relation (18), we find

$$\nu_0^+ \equiv \nu_0 \approx -\frac{1}{\beta a_B + 2\ln(|\beta|a_B)} \ll 1, \qquad (20)$$

and thus

$$E_0^+ = -\frac{m\alpha^2}{2\hbar^2 \nu_0^2}. \qquad (21)$$

This level (without $E_n^{(0)}$) is described by Eq. (6). By determining the value of β according to Eq. (19), we can obtain the spectrum of the even the excited levels of the longitudinal motion for a hydrogen-like atom in a magnetic field.

9

Quasi-classical approximation; $1/N$-expansion in quantum mechanics

The quasi-classical approximation is also known as the *WKB (Wentzel–Kramers–Brillouin) method*. Within this approximation, two independent solutions of the Schrödinger equation[157] (II.1) can be written as

$$\psi_E^{(\pm)}(x) = \frac{1}{\sqrt{p(x)}} \exp\left\{\pm \frac{i}{\hbar} \int_c^x p(x)dx\right\}, \qquad \text{(IX.1)}$$

with

$$p(x) = \sqrt{2m[E - U(x)]},$$

being the classical momentum.

This approximation is valid if the *quasi-classical condition* is fulfilled:

$$\left|\frac{d\lambda}{dx}\right| \equiv \hbar \frac{d}{dx}\frac{1}{p(x)} = m\hbar \left|\frac{U'(x)}{p^3(x)}\right| \ll 1. \qquad \text{(IX.2)}$$

A general solution of the Schrödinger equation in the quasi-classical approximation is expressed as a superposition of the wavefunctions (IX.1):

$$\psi_E(x) = C_1 \psi_E^+(x) + C_2 \psi_E^-(x).$$

Usually, there are regions of x where the quasi-classical condition (IX.2) breaks down (for example, in the vicinity of a classical turning point). This raises the problem[158] of matching the quasi-classical functions, where we must link the solution of the Schrödinger equation on opposite sides of such regions.

---

[157] We remind the reader that the Schrödinger equation for a particle in a central potential reduces to a one-dimensional form (see Eq. (IV.5)). Some difficulties, however, appear here due a term corresponding to centrifugal energy, $U_{\text{cf}}(r) = \hbar^2 l(l+1)/2mr^2$, in the effective potential. The term makes the quasi-classical approximation inapplicable for $r \to 0$ since, in this case, $d\lambda/dr \propto r^{-1/2} \to \infty$. An effective way to overcome this difficulty is to use the *Langer transformation* which we discuss later.

[158] Its solution is needed, for instance, in order to take account of the boundary condition.

The matching conditions based on a linearization of the potential near a classical turning point can often be used:

$$U(x) \approx U(x_0) - F(x_0)(x - x_0), \quad F(x_0) = -U'(x_0), \quad p(x_0) = 0.$$

Here it is assumed that x belongs to the region close enough to the turning point, x_0, where the truncated Taylor expansion is reliable, but at the same time far enough from it, so that the quasi-classical condition (IX.2) remains reliable. Near the right turning point (corresponding to $x = b$ in Fig. 9.1), the solutions have the form

$$\psi(x) = \frac{C}{2\sqrt{|p(x)|}} \exp\left\{-\frac{1}{\hbar}\int_b^x |p(x)|dx\right\}, \quad x > b, \quad \text{(IX.3a)}$$

$$\psi(x) = \frac{C}{\sqrt{p(x)}} \sin\left\{\frac{1}{\hbar}\int_x^b p(x)dx + \frac{\pi}{4}\right\}, \quad x < b. \quad \text{(IX.3b)}$$

Near the left turning point (corresponding to $x = a$ in Fig. 9.1), we have

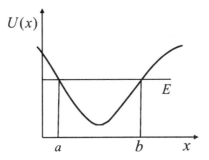

Fig. 9.1

$$\psi(x) = \frac{C_1}{2\sqrt{|p(x)|}} \exp\left\{-\frac{1}{\hbar}\int_x^a |p(x)|dx\right\}, \quad x < a; \quad \text{(IX.4a)}$$

$$\psi(x) = \frac{C_1}{\sqrt{p(x)}} \sin\left\{\frac{1}{\hbar}\int_a^x p(x)dx + \frac{\pi}{4}\right\}, \quad x > a. \quad \text{(IX.4b)}$$

For the potential shown in Fig. 9.1 and for the discrete levels, $E = E_n$, the *Bohr–Sommerfeld quantization rule* follows from the condition that expressions (IX.3 b) and (IX.4 b) (which describe the same solution of the Schrödinger equation) coincide, i.e., the sum of the phases in the sine terms is a multiple of π:[159]

---

[159] In a more general case, when the conditions of matching Eqs. (IX.3) and (XI.4) are not applicable, the right-hand side in the quantization rule is equal to $\pi(n + \alpha)$, where the quasi-classical

$$\frac{1}{\hbar}\int_a^b \sqrt{2m(E_n - U(x))}dx = \pi\left(n + \frac{1}{2}\right), \quad n = 0, 1, 2, \ldots, \tag{IX.5}$$

Although strictly speaking, the quasi-classical quantization rules accurately determine the spectrum, E_n, only for $n \gg 1$, a sufficient accuracy even for $n \sim 1$ is often achieved in the case of a smooth potential.

Differentiating (IX.5) with respect to n gives the level spacing between the neighboring levels

$$\delta E_n \equiv E_{n+1} - E_n \approx \frac{\partial}{\partial n}E_n = \hbar\omega(E_n),$$

where $\omega(E) = \frac{2\pi}{T(E)}$ is the frequency of quasi-classical motion for a particle with energy E_n, and T is its period (see Eq. (IX.7)).

For a bound-state wavefunction, we can usually use the following expression (see Eq. (IX.3,4))

$$\psi_n(x) \approx \begin{cases} \frac{C_n}{\sqrt{p(x)}}\sin\left(\frac{1}{\hbar}\int_a^x p(x)dx + \frac{\pi}{4}\right), & a < x < b, \\ 0, & x < a, \ x > b. \end{cases} \tag{IX.6}$$

Here we have neglected the possibility of the particle penetrating the classically forbidden region, where the wavefunction is suppressed exponentially. In order to normalize the wavefunction to unity we should choose

$$C_n^2 = \frac{2m\omega(E_n)}{\pi}, \quad T(E_n) = \frac{2\pi}{\omega(E_n)} = 2m\int_a^b \frac{dx}{p(x, E_n)}. \tag{IX.7}$$

The quantum mechanical probability density $|\psi_n(x)|^2$ oscillates rapidly as a function of x, due to $n \gg 1$. After averaging[160] over a small interval of values x these oscillations disappear, and the probability density takes the form

$$\overline{|\psi_n(x)|^2} = \frac{2m}{T(E_n)p(x)} = \frac{2}{T(E_n)v(x)}.$$

This corresponds to the *classical probability*

$$\omega_{class}(x) = \frac{2}{T}dt = \frac{2}{T(E)}\frac{dx}{v(x)}, \quad a < x < b, \tag{IX.8}$$

which is determined by the time interval dt the particle takes to pass through spatial interval dx, divided by one half of its period.

---

correction, $\alpha \sim 1$. In this case, the domain of quasi-classical applicability usually persists down to $n \sim 1$.

[160] It reduces to the substitution of $\sin^2\{\ldots\}$ by its mean value $1/2$.

We shall note here that whenever a derivative of the wavefunction needs to be computed, the leading contribution comes from differentiating only the trigonometric factor (sines and cosines), since it is the most rapidly-varying part of the wavefunction.

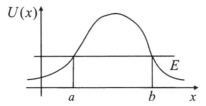

Fig. 9.2

One of the most important applications of the quasi-classical method is in calculating transmission coefficients (penetrabilities) of potential barriers. The penetrability of a generic barrier shown in Fig. 9.2 in the quasi-classical approximation is described by the expression

$$D(E) = \exp\left\{-\frac{2}{\hbar}\int_a^b |p(x,E)|dx\right\}. \qquad (IX.9)$$

This expression is applicable for a large negative argument of the exponential function such that $D \ll 1$. Equation (IX.9) as well as (IX.5) assume the possibility of quasi-classical solution-matching in the vicinity of turning points based on the linear approximation of the potential. If this condition is broken, the quasi-classical result (IX.9) is valid only to within the accuracy of the factor in front of the exponent (although the main feature – namely, the exponential suppression of the barrier penetration coefficient – is still captured).

Langer transformation

1) Let us recall the Schrödinger Eq. (IV.5) for the radial component $\chi_{El}(r)$. One sees that the centrifugal energy in the effective potential (here and below $\hbar = m = 1$, $E = k^2/2$),

$$U_{eff}(r) = U(r) + \frac{l(l+1)}{2r^2}, \qquad (1)$$

dominates for small distances, i.e., $r \to 0$. In this case, $d\lambda(r)/dr \propto r^{-1/2} \to \infty$, and the quasi-classical condition is not satisfied. To overcome this difficulty Langer proposed using the change of variables and wavefunction transform as follows:

$$r = e^x, \quad \psi(x) = e^{-x/2}\chi_{El}(e^x), \quad -\infty < x < \infty. \qquad (2)$$

As a result, the radial Schrödinger equation becomes

$$\frac{d^2\psi(x)}{dx^2} + p^2(x)\psi(x) = 0, \quad p^2(x) = [k^2 - 2U(e^x)]e^{2x} - \nu^2, \quad \nu = l + \frac{1}{2}. \quad (3)$$

This then has the form of the one-dimensional Schrödinger equation for which the barrier is located at $x \to -\infty$ (corresponding to $r \to 0$ in the original variables). The equation thus provides a quasi-classical solution for the radial function having the correct dependence on angular momentum for small distances. Indeed, applying Eq. (IX.4a) to Eq. (3) using Eq. (2) gives

$$\chi^{WKB}_{El}(r) = \frac{C}{\sqrt{|p_L(r)|}} \exp\left\{-\int_r^{r_-} |p_L(r)|dr\right\} = c^{WKB}_{El}(r^{l+1} + \ldots), \quad r \to 0. \quad (4)$$

where $r_- = e^a$ is the left turning point and the quasi-classical momentum is described by the relation

$$p_L(r) = \left[k^2 - 2U(r) - \frac{(l+\frac{1}{2})^2}{r^2}\right]^{1/2}. \quad (5)$$

Application of Bohr–Sommerfeld quantization (IX.5) to Eq. (3), and writing it in terms of the variable r, gives the quantization condition:

$$\int_{r_-}^{r_+} p_L(r)dr = \pi\hbar\left(n_r + \frac{1}{2}\right), \quad (6)$$

with the quasi-classical momentum $p_L(r)$.

We see that although the quasi-classical condition breaks down at small distances for the potential in (1), the quantization condition of its spectrum is still described by the usual expression (IX.5), where for centrifugal potential the following substitution is fulfilled:[161] $l(l+1) \to (l+1/2)^2$. This sometimes is referred to as the Langer correction.

Note that the application of Kramers matching conditions to Eq. (3) (which are used for deriving the quantization rule (IX.5)) is strictly-speaking not justified. The issue is that in the case of $\nu \sim 1$ we cannot use the matching condition (IX.4) for this equation in the vicinity of the left turning point $x = a$, based on the linear expansion of the potential: the non-quasi-classical domain here is much wider. It is easier to see this for free motion (i.e., $U = 0$) where, according to Eq. (5), $p(x) = \nu\sqrt{e^{2(x-a)} - 1}$, $a = \ln(\nu/k)$ is the turning point. For $|x - a| \ll 1$ we have $p(x) \approx \nu\sqrt{2(x-a)}$, and the quasi-classical condition (IX.2) takes the form

$$\nu^{-2/3} \ll |x - a| \ll 1. \quad (6)$$

---

[161] We note that the Langer method is also applicable for s-states.

These inequalities are fulfilled simultaneously only in the case of $\nu \gg 1$ (*i.e.*, $l \gg 1$), when the centrifugal potential becomes quasi-classical.

2) We discuss the modification of Kramers matching condition for quasi-classical solutions of Eq. (3) in the vicinity of the left turning point $x = a$ in the two following cases.

Matching at large energies

As the energy grows at the left turning point $a \to -\infty$, there $r_- \to 0$ and, in relation (3), we can omit the term with potential, $U(e^z)$. The equation can then be solved in terms of the Bessel functions, and allows one to match the quasi-classical asymptotes, $\psi(x)$. It gives the following quasi-classical relations for $\chi_{El}(r)$

$$\chi_{El}(r) = \begin{cases} \dfrac{C}{\sqrt{p_L(r)}} \cos\left(\int\limits_{r_-}^{r} p_L(r) dr - \dfrac{\pi}{4}\right), & r > r_-, \\ \dfrac{C'}{\sqrt{|p_L(r)|}} \exp\left(-\int\limits_{r}^{r_-} |p_L(r)| dr\right), & r < r_-, \end{cases} \qquad (7)$$

where

$$\frac{C'}{C} = \frac{1}{2}\xi(\nu), \quad \xi(\nu) = \sqrt{2\pi} \nu^{\nu-1/2} \frac{e^{-\nu}}{\Gamma(\nu)}. \qquad (8)$$

This can be compared to Eqs. (IX.4a) and (IX.4b). We see that the quasi-classical phase, $-\pi/4$, has not changed, unlike the relation between the coefficients C and C'. We note that $\xi(1/2) = \sqrt{2/e} = 0.8578$ and $\xi(1) = \sqrt{2\pi}e^{-1} = 0.9221$, while $\xi(x) \approx 1 - 1/12x$ for $x \to \infty$, so that for $l \gg 1$, relations (8) become Kramers' matching conditions (IX.4).

The modification made for the matching conditions influences the values of the radial wavefunction in the region under the barrier on small distances.[162] Let us now see its role by calculating the "asymptotic coefficient at zero", $c_{n_r l}$, on the example of a spherical oscillator $U(r) = \omega^2 r^2/2$ (remembering that we are using the units, where $\hbar = m = 1$). This coefficient determines the behavior of a bound-state normalized wavefunction, $\int\limits_0^\infty \chi_{n_r l}^2(r) dr = 1$, at small distances,

$$\chi_{n_r l}(r) \approx c_{n_r l} r^{l+1}, \quad r \to 0, \qquad \text{see Eq. (4)}. \qquad (9)$$

The exact value of $c_{n_r l}^{(ex)}$ for the oscillator, according to Problem 4.5, is equal to

$$c_{n_r l}^{(ex)} = \sqrt{2} \frac{\Gamma(l + n_r + 3/2)}{\sqrt{n_r!} \Gamma(l + 3/2)} \omega^{l/2 + 3/4}, \qquad (10)$$

---

[162] We should emphasize that Langer quantization – *i.e.*, according to Eq. (6) – in the cases of a harmonic oscillator and a particle in the Coulomb potential actually gives the exact energy spectrum.

while the quasi-classical values are defined by

$$c_{n_r l}^{(q)} = \xi(\nu) c_{n_r l}^{(L)}, \quad c_{n_r l}^{(L)} = \left\{ \frac{1}{\pi} \frac{n^n e^\nu}{(n_r + 1/2)^{n_r+1/2} \nu^{2l+2}} \right\}^{1/2} \omega^{l/2+3/4}, \quad n = n_r + l + 1, \tag{11}$$

where $c_{n_r l}^{(L)}$ are obtained using the Kramers matching conditions. In the table below we compare the specific values of these coefficients. Below, we provide the values of the ratios

$$\eta_{n_r l}^{(L)} = \frac{c_{n_r l}^{(L)}}{c_{n_r l}^{(ex)}}, \quad \eta_{n_r l}^{(q)} = \frac{c_{n_r l}^{(q)}}{c_{n_r l}^{(ex)}}, \tag{12}$$

which characterize an accuracy of the corresponding approximation.

l	η	$n_r = 0$	$n_r = 1$	$n_r = 2$	$n_r = 5$	$n_r = \infty$
0	$\eta^{(L)}$	1.1469	1.1620	1.1642	1.1655	1.1658
	$\eta^{(q)}$	0.9838	0.9967	0.9986	0.9997	1
1	$\eta^{(L)}$	1.0293	1.0493	1.0531	1.0555	1.0563
	$\eta^{(q)}$	0.9744	0.9934	0.9969	0.9992	1
2	$\eta^{(L)}$	1.0039	1.0251	1.0295	1.0325	1.0337
	$\eta^{(q)}$	0.9711	0.9917	0.9959	0.9988	1
5	$\eta^{(L)}$	0.9826	1.0045	1.0095	1.0133	1.0153
	$\eta^{(q)}$	0.9678	0.9895	0.9943	0.9981	1

As we see, the modification of the matching conditions, given in Eqs. (7) and (8), not only ensures the asymptotic accuracy of the quasi-classical approximation in calculating $c_{n_r l}$ for $n_r \to \infty$, but also gives satisfactory results for the states with small values of the radial quantum number, n_r, and even for the ground state. These properties persist also in the case of other smooth potentials.

Matching conditions in the case of level concentration, $E_n \to -0$

Let us now consider the case of attractive potentials, $U(r) = -gr^\alpha$, with $-2 < \alpha < 0$ (for $\alpha = -1$, it is the Coulomb potential), where there is an infinite number of levels accumulating towards the point $E = -0$. In Eq. (3), in the vicinity of the left turning point, we can neglect the term with energy $E = k^2/2$, which allows us to obtain the exact (not quasi-classical) solution. Using this solution we can match the quasi-classical asymptotes. This leads to the previous relations (7), where now

$$\frac{C'}{C} = \frac{1}{2}\xi(\mu), \quad \mu = \frac{2l+1}{2+\alpha}. \tag{13}$$

We will illustrate their role in calculating the asymptotic coefficient at zero in the case of the Coulomb potential (for which $\mu = 2l + 1$). According to Eq. (IV.3) the exact value is

$$c_{n_r l}^{(ex)} = \frac{2^{l+2}}{(2l+1)! n^{l+2}} \left\{ \frac{(n+l+1)!}{n_r!} \right\}^{1/2} g^{l+3/2}, \quad n = n_r + l + 1, \tag{14}$$

while the quasi-classical relations for them are (see Eq. (11))

$$c_{n_r l}^{(q)} = \xi(\mu) c_{n_r l}^{(L)}, \quad c_{n_r l}^{(L)} = \frac{\nu^{2l+3/2} e^\nu}{2^{l+1} \sqrt{\pi} n^{l+2}} \frac{(n+\nu)^{(n+\nu)/2}}{(n-\nu)^{(n-\nu)/2}} g^{l+3/2}. \tag{15}$$

Comparisons to the exact values are given in the following table.

l	η	$n_r = 0$	$n_r = 1$	$n_r = 2$	$n_r = 5$	$n_r = \infty$
0	$\eta^{(L)}$	1.0602	1.0787	1.0819	1.0838	1.0844
	$\eta^{(q)}$	0.9776	0.9947	0.9977	0.9994	1
1	$\eta^{(L)}$	0.9974	1.0189	1.0235	1.0267	1.0281
	$\eta^{(q)}$	0.9702	0.9911	0.9955	0.9987	1
2	$\eta^{(L)}$	0.9844	1.0063	1.0112	1.0150	1.0162
	$\eta^{(q)}$	0.9681	0.9897	0.9945	0.9982	1

We see that the modified matching conditions ensure a good asymptotic accuracy of the quasi-classical asymptotic coefficients.

9.1 Quasi-classical energy quantization

Problem 9.1

Using the quasi-classical approximation find the energy spectra in the following two cases:

a) linear oscillator,
b) bound states in the potential, $U(x) = -U_0 \cosh^{-2}(x/a)$.

Solution

a) For a linear oscillator the integration in Eq. (IX.5) gives $E_n = \hbar\omega(n + 1/2)$, which coincides with the exact result.
b) For the given potential, the integration in Eq. (IX.5) can be done using the substitution

$$\sinh\frac{x}{a} = \kappa\sin t, \quad \kappa = \sqrt{\frac{U_0}{|E_n|} - 1},$$

which leads to

$$E_n = -\frac{\hbar^2}{2ma^2}\left[\sqrt{\frac{2ma^2U_0}{\hbar^2}} - \left(n+\frac{1}{2}\right)\right]^2. \tag{1}$$

We note that this problem can be solved exactly, and the exact result has the form of Eq. (1) but with $2ma^2U_0/\hbar^2 + 1/4$ inside the square root sign, so for the values of the quasi-classical parameter $\xi \equiv \sqrt{2ma^2U_0/\hbar^2} \gg 1$ (when there exist many bound states in the potential), the quasi-classical and exact results are close for $n \sim 1$. The quasi-classical result reproduces the exact spectrum even when there are only three to four discrete levels in the potential. Indeed, the maximum value n is determined by the condition $n + 1/2 \leq \xi$, i.e. $n_{max} \approx \xi$ for $\xi \gg 1$. Here, the difference between the exact and the quasi-classical results is given by $\sqrt{\xi^2 + 1/4} - \xi \approx 1/8\xi \ll 1$.

Problem 9.2

Obtain an energy quantization rule and find the corresponding quasi-classical wavefunctions for a potential of the form[163] shown in Fig. 9.3. Apply the result obtained

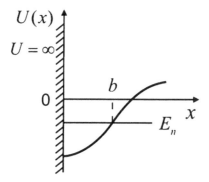

Fig. 9.3

to the potential considered in Problem 2.8. Pay particular attention to the closeness of the quasi-classical and exact values E_n even for $n \sim 1$.

---

[163] The result of this problem can be straightforwardly transplanted to the case of s-states of a particle in a central potential; see Problem 4.1.

Solution

The matching of quasi-classical solutions in the vicinity of the right turning point, $x = b$, is done in the usual way by Eqs. (IX.3). Now the expression for a wavefunction for $x < b$ is valid, generally speaking, for the values of x close to the left turning point $x = 0$ as well (which however is not a "stopping" point here!). Using the boundary condition $\psi(0) = 0$ allows us to obtain the quantization rule:

$$\frac{1}{\hbar} \int_0^b \sqrt{2m(E_n - U(x))}\, dx = \pi \left(n + \frac{3}{4} \right), \quad n = 0, 1, 2, \ldots \quad (1)$$

We should note that the change in the matching conditions affects the value of the quasi-classical correction only; we have $n + 3/4$ instead of $n + 1/2$ in the quantization rule (IX.5). We should also note that relation (1) could be obtained from the Bohr–Sommerfeld quantization rule, applied to odd levels in the symmetric potential $U = U(|x|)$ (i.e., by substituting n by $2n + 1$; see Problem 2.5).

For the potential, $U = Fx$ for $x > 0$, we find in accord with (1) that

$$E_n = \left(\frac{9\pi^2}{8} \right)^{1/3} \left(n + \frac{3}{4} \right)^{2/3} \varepsilon_0, \quad \varepsilon_0 = \left(\frac{\hbar^2 F^2}{m} \right)^{1/3}. \quad (2)$$

This quasi-classical result differs only slightly from the exact one for all values of n (not only for $n \gg 1$). So, the values of E_n/ε_0 in (2) for $n = 0$ and 1 are equal to 1.842 and 3.240; the exact results are 1.856 and 3.245 (note that using the quantization rule Eq. (IX.5) would have resulted in a loss in accuracy for $n \sim 1$: giving, for example, 1.405 and 2.923 instead of the values given above).

Problem 9.3

A particle moves in a central field which is a superposition of a long-range potential, $U(r)$, shown in Fig. 9.3 (with the substitution $x \to r$ so that there is no potential barrier for small distances) and a short-range potential approximated by the zero-range potential (z.r.p.) (see Problems 4.10 and 4.31). Obtain a quantization rule for the s-levels and discuss the level shift in the potential $U(r)$ under the influence of the z.r.p. Pay special attention to the possibility of spectrum reconstruction, i.e., the large shifts, comparable to the level spacing of the unperturbed potential $U(r)$.

Solution

For the radial wavefunction of an s-state, $\chi_{n_r}(r) = rR_{n_r 0}$ (see Eq. (IV.5)), we have for $r < b$:

$$\chi_{n_r}(r) = \frac{C}{\sqrt{p(r)}} \sin \left\{ \frac{1}{\hbar} \int_0^r p(r)\, dr + \gamma \right\}, \quad (1)$$

$$p = \sqrt{2m[E_{n_r,0} - U(r)]},$$

($\gamma = 0$ in the absence of the z.r.p.), so that as $r \to 0$,

$$\chi_{n_r}(r) \approx \tilde{C}\left\{\sin\gamma + \left(\frac{p(0)}{\hbar}\cos\gamma\right)r\right\}.$$

Note that due to the quasi-classical property it is sufficient to take account of the dependence on r only inside the sine and replace $p(r)$ by $p(0)$. Comparing this expansion to the boundary condition from Problem 4.10, which determines the z.r.p., we find[164] $\gamma = -\arctan\frac{p(0)a_0}{\hbar}$. Matching function (1), in accordance with Eq. (IX.3), with the solution decreasing in the classically forbidden region, we obtain the required quantization rule:

$$\frac{1}{\hbar}\int_0^b \sqrt{2m[E_{n_r,0} - U(r)]}\,dr = \pi\left(n_r + \frac{3}{4} + \frac{1}{\pi}\arctan\frac{p_{n_r}(0)a_0}{\hbar}\right). \tag{2}$$

In the case of $a_0 = 0$ this relation determines the spectrum $E_{n_r,0}^{(0)}$ for s-levels in a potential $U(r)$ (without the z.r.p). In the case of $|p_{n_r}(0)a_0/\hbar| \ll 1$ the value of the arctan in (2) is small, hence the energy level shift is also small. Writing $E_{n_r,0} = E_{n_r,0}^{(0)} + \Delta E_{n_r,0}$ and performing an expansion of the square root in (2),

$$\int_0^b \sqrt{E^{(0)} + \Delta E - U}\,dr \approx \int_0^{b_0}\left\{\sqrt{E^{(0)} - U} + \frac{\Delta E}{2\sqrt{E^{(0)} - U}}\right\}dr,$$

we find the level shift under the action of the z.r.p.:

$$\Delta E_{n_r,0} = \frac{1}{\pi}\omega_{n_r}^{(0)}p_{n_r}^{(0)}a_0, \tag{3}$$

where $\omega_{n_r}^{(0)} = \left[\frac{1}{\pi}\int_0^{b_0}\frac{m\,dr}{p_{n_r}^{(0)}(0)}\right]^{-1}$ is a frequency of the radial motion of a classical particle with zero angular momentum in the potential $U(r)$. Using the quasi-classical arguments for the wavefunction normalization (see Eqn. (IX.7)) this result may be rewritten in the form ($\psi = \chi/\sqrt{4\pi}r$):

$$\Delta E_{n_r,0} = \frac{2\pi\hbar^2}{m}\left|\psi_{n_r00}^{(0)}(0)\right|^2 a_0, \tag{4}$$

which corresponds to a level shift that is obtained using perturbation theory with respect to the scattering length (see Problem 4.29).

On the contrary, for $|p_{n_r}(0)a_0/\hbar| \geq 1$ the level shifts are large and are comparable to the distance between the unperturbed levels in the potential $U(r)$. For example, let

---

[164] Here $a_0 = \alpha_0^{-1}$ is the scattering length for the z.r.p. The results, being expressed in terms of the scattering length, are applicable for arbitrary short-range potentials.

us consider the case $a_0 = \infty$. The distinguishing feature of this case is that in the z.r.p. there exists a "shallow" *real* ($a_0 > 0$) or *virtual* ($a_0 < 0$) level with the energy $E_0 \approx -\hbar^2/2ma_0^2$ (see Problems 4.11 and 13.49), which is of the same order of magnitude as the levels in the potential $U(r)$ (which leads to a resonant regime in some sense).

In conclusion we should note that the wavefunction (1) corresponds to the case $E > U(0)$, which could exist only without the z.r.p. With the z.r.p. we must also consider the values $E < U(0)$. Here, the wavefunction for $r > 0$ is described by a decreasing "quasi-classical exponent" instead of (1). We can see that such a solution exists only for the values, $a_0 > 0$, and corresponds to $E_0 = -\hbar^2/2ma_0^2 + U(0)$, describing a discrete level, shifted by $U(0)$, that exists in an isolated zero range potential.

Problem 9.4

In the quasi-classical approximation, analyze the energy spectrum for a particle moving in a symmetric potential well $U_0(x)$ separated by a δ-barrier $\alpha\delta(x)$ so that $U(x) = U_0(x) + \alpha\delta(x)$.

Consider the limiting cases of

1) the weakly reflecting barrier;
2) the weakly penetrable barrier.

Solution

It is convenient to analyze the spectra of the even and odd levels separately. For the odd states, $\psi(0) = 0$. Here, taking into account the matching conditions for the δ-potential from Problem 2.6, we see that the wavefunction derivative is continuous at $x = 0$. Hence the particle in the odd states does not "feel" the δ-potential, while the odd-level spectrum is determined by quantization rule (IX.5) for values $n = 2k + 1$ (where $k + 1$ is the number of an odd level):

$$\frac{1}{\hbar}\int_0^{b_-} \sqrt{2m(E_k^- - U_0(x))}dx = \pi\left(k + \frac{3}{4}\right). \tag{1}$$

For the even levels, the condition for the wavefunction derivative jump takes the form $\psi'(0+) = (m\alpha/\hbar^2)\psi(0)$. For $x > 0$ we use Eq. (IX.3) to find

$$-p(0)\cos\left[\frac{1}{\hbar}\int_0^b p(x)dx + \frac{\pi}{4}\right] = \frac{m\alpha}{\hbar}\sin\left[\frac{1}{\hbar}\int_0^b p(x)dx + \frac{\pi}{4}\right].$$

(Due to the quasi-classical property it is sufficient to differentiate with respect to x only the sine term.) Hence, introducing $\tan\gamma = m\alpha/p(0)\hbar$ we obtain the quantization rule for the even levels with $n = 2k$:

$$\int_0^{b_0} \sqrt{2m[E_k^+ - U_0(x)]}dx = \pi\hbar\left(k + \frac{1}{4} + \frac{1}{\pi}\arctan\frac{m\alpha}{p_k^+(0)\hbar}\right). \tag{2}$$

For $\alpha = 0$ it reproduces the even-level spectrum $E_{0,k}^+$ for potential $U_0(x)$. In the case $m\alpha/\hbar p_k^+(0) \ll 1$ the level shifts are small (much smaller than the distance between the unperturbed levels). Writing $E_k^+ = E_{0,k}^+ + \Delta E_k^+$ and performing an expansion under the integral sign (compare to the previous problem), we obtain

$$\Delta E_k^+ = \frac{4m\alpha}{T(E_{0,k}^+)\sqrt{2m[E_{0,k}^+ - U_0(0)]}}, \tag{3}$$

which coincides with $\alpha|\psi_{0,k}^+(0)|^2$ and corresponds to the result of perturbation theory for $V = \alpha\delta(x)$; here $T(E_{0,k}^+)$ is the period of motion, and $\psi_{0,k}^+(0)$ is the wavefunction of an odd state at $x = 0$ in the potential $U_0(x)$ (see Eq. (IX.7)).

The case considered above corresponds to weak reflection ($R \ll 1$) from the δ-potential whose transmission coefficient is equal to (see Problem 2.30)

$$D(p) = \left(1 + \frac{m^2\alpha^2}{\hbar^2 p^2}\right)^{-1}.$$

With increasing α, even level shifts increase. For $m\alpha/\hbar p_k^+(0) \gg 1$, they move closer to the neighboring odd levels above. This is the case of the weakly penetrable barrier, $\alpha > 0$. Substituting

$$p_k^+(0) \approx p_k^-(0) = \sqrt{2m[E_k^- - U_0(x)]}$$

into the right-hand side of Eq. (2) and using the expansion $\arctan x \approx \pi/2 - 1/x$ for $x \gg 1$, we find, according to Eqs. (1) and (2), the distance between neighboring even and odd levels in this case:

$$\delta E_k \equiv E_k^- - E_k^+ \approx \frac{4\hbar^2}{m\alpha T(E_k^-)}\sqrt{2m[E_k^- - U_0(x)]}, \tag{4}$$

or, taking into account the relation for $D(p)$,

$$\delta E_k = \frac{2\hbar}{\tau(E_k^-)}D^{1/2}(p_k^-(0)). \tag{5}$$

Here, $\tau(E_k^-) = T(E_k^-)/2$ is the period of classical motion of a particle with the energy E_k^- in the potential $U_0(x)$ divided by the potential barrier (at $x = 0$). Note that this Eq. (5) for a level splitting in a symmetric potential is quite general and remains valid for a wide range of weakly-penetrable barriers of a rather arbitrary shape.

The results obtained above are valid even in the case of $\alpha < 0$, i.e., for the δ-well. But now the even levels move down, and in the case of $m|\alpha|/\hbar p_k^+(0) \gg 1$ an even level (with the number $k + 1$) approaches the lower neighboring odd level (with the number

k). Here, the ground-state level $E_0^+(\alpha)$ with increase in $|\alpha|$ has the following behavior. It decreases, first reaching the value, $E_0^+(\alpha_0) = U_0(0)$. With the subsequent increase in $|\alpha|$ it approaches the ground-state energy in the δ-well, shifted by $U_0(0)$ (compare this to the discussion at the end of the previous problem).

Problem 9.5

For a particle in a potential, which behaves as the attractive Coulomb potential, $U(r) \approx -\alpha/r$, at small distances, obtain the wavefunctions and quantization rule for the s-levels with the energy $|E| \ll m\alpha^2/\hbar^2$ in the quasi-classical approximation. Apply these results to the pure Coulomb potential, $U(r) = -\alpha/r$, and the Hulthen potential, $U(r) = -U_0 \left(e^{r/a} - 1\right)^{-1}$. Compare your results with the exact solution for the spectrum (see Problem 4.8).

Solution

A complication of this problem is due to the fact that the quasiclassical conditions for the radial Schrödinger equation for the function $\chi = rR$ (see Eq. (IV.5)) are broken at small distances, $r \to 0$. Indeed, here $p(r) \approx \sqrt{2m\alpha/r}$ and $d\lambda/dr \propto r^{-1/2} \to \infty$. Hence, to take account of the boundary condition $\chi(0) = 0$ in the quantization rule we should find an exact (non-quasi-classical) solution of the Schrödinger equation for small distances, and match it with the quasi-classical solution (IX.3).

The Schrödinger equation for small distances takes the form $\chi'' + (\tilde{\alpha}/r)\chi = 0$, where $\tilde{\alpha} = 2m\alpha/\hbar^2$. With the help of substitutions, $\chi = \sqrt{r}\varphi$ and $z = 2\sqrt{\tilde{\alpha}r}$, we obtain the Bessel equation with $\nu = 1$ for the function $\varphi(z)$. Using the boundary condition at zero, $\chi(0) = 0$, we find

$$\chi(r) = A\sqrt{r}J_1(2\sqrt{\tilde{\alpha}r}). \tag{1}$$

The asymptotic behavior of this function for $\sqrt{\tilde{\alpha}r} \gg 1$ has the form

$$\chi(r) \approx A\left(\frac{r}{\pi^2\tilde{\alpha}}\right)^{1/4} \sin\left(2\sqrt{\tilde{\alpha}r} - \frac{\pi}{4}\right). \tag{2}$$

We now see that for values of r satisfying

$$r \ll \frac{\alpha}{|E|},$$

we have $\frac{1}{\sqrt{\tilde{\alpha}}} \ll \sqrt{r}$, and $|d\lambda/dr| \sim (\tilde{\alpha}r)^{-1/2} \ll 1$, i.e., the quasi-classical method is applicable.[165] The quasi-classical solution of the Schrödinger equation

$$\chi_{quas} = \frac{C}{\sqrt{p(r)}} \sin\left(\frac{1}{\hbar}\int_0^r p(r)dr + \gamma\right), \quad p(r) = \sqrt{2m(E - U(r))} \tag{3}$$

---

[165] Here we assume that on such distances the potential still behaves as $U \approx -\alpha/r$ and E corresponds to the *upper* levels with $n \gg 1$ in such a potential.

for such distances takes the form (here $p(r) \approx \hbar\sqrt{\tilde{\alpha}/r}$):

$$\chi_{quas}(r) \approx C \left(\frac{r}{\tilde{\alpha}\hbar^2}\right)^{1/4} \sin(2\sqrt{\tilde{\alpha}r} + \gamma). \tag{4}$$

By matching it with the exact solution (2) we find $\gamma = -\pi/4$. (Compare this to the value $\gamma = \pi/4$ that is obtained from the matching condition within the quasi-classical method in the vicinity of the left turning point $x = a$ in Eq. (IX.4), based on the linear approximation of the potential, and also with $\gamma = 0$ using the conditions of Problem 9.2.)

Finally, the matching condition between the quasi-classical solution (3) for $\gamma = -\pi/4$ and the solution (IX.3 b) (with substitution $\psi(x) \to \chi(x)$), which fulfills the boundary condition at $r \to \infty$, gives the quantization rule:

$$\frac{1}{\hbar} \int_0^b \sqrt{2m(E_{n_r,0} - U(r))}\, dr = \pi(n_r + 1). \tag{5}$$

For the Coulomb potential, $U = -\alpha/r$, we therefore find

$$E_{n_r,0} = -\frac{m\alpha^2}{2\hbar^2(n_r + 1)^2},$$

which coincides with the exact result. This also can be proven for the Hulthen potential, where the exact result for the s-level spectrum is also available (see Problem 4.8).

Problem 9.6

For a central potential of the form shown in Fig. 9.3. (restricted as $r \to 0$), find the radial wavefunctions of the stationary states of a particle with the angular momentum[166] $l \sim 1$ in the domain of its classical motion and in the quasi-classical approximation. Using the obtained result, analyze the modified quantization rule (IX.5), and find the energy spectra for a) the spherical oscillator $U = m\omega^2 r^2/2$; and b) the particle in one dimension in the potential $U(x) = U_0 \tan^2 \frac{\pi x}{a}$ for $|x| < \frac{a}{2}$ and $U = \infty$ for $|x| > a/2$. Compare this to the result of quantization by Eq. (IX.5) and the exact solution for the spectrum.

Solution

Note that for the potential $U(r) \approx \alpha/r^2$, the quasi-classical property is broken in the case of $2m\alpha/\hbar^2 = l(l+1) \lesssim 1$ and at small distances ($r \to 0$). Hence, we should use the exact (non-quasi-classical) solution of the Schrödinger equation that satisfies the boundary condition $\chi(0) = 0$; we obtain

---

[166] This case is of particular interest because for $l \gg 1$ the centrifugal barrier $\hbar^2 l(l+1)/2mr^2$ is already quasi-classical, and we can use the matching conditions (IX.4); see also the previous problem.

$$\chi'' + \left[k_0^2 - \frac{\tilde{\alpha}}{r^2}\right]\chi = 0, \quad \chi = A\sqrt{r}J_\nu(k_0 r). \tag{1}$$

Here, $k_0^2 = 2m(E - U_0(0))/\hbar^2$, and potential[167] $U_0(r)$ is replaced by its value at the origin $U_0(0)$, $\tilde{\alpha} = 2m\alpha/\hbar^2$, and $\nu = \sqrt{\tilde{\alpha} + 1/4}$. Using the asymptotic behavior of the Bessel function we see that over such distances where $k_0 r \gg 1$ (for values $\nu \sim 1$) the exact solution of the Schrödinger equation (1) takes the form

$$\chi(r) \approx \sqrt{\frac{2}{\pi k_0}} A \sin\left(k_0 r - \frac{\pi \nu}{2} + \frac{\pi}{4}\right), \tag{2}$$

and coincides with the quasi-classical expression

$$\chi_{quas} = \sqrt{\frac{2\hbar}{\pi p(r)}} A \sin\left\{\frac{1}{\hbar}\int_a^r p \, dr + \gamma\right\}, \tag{3}$$

$$\gamma = \frac{\pi}{2}\left[\sqrt{\tilde{\alpha}} - \sqrt{\tilde{\alpha} + \frac{1}{4}} + \frac{1}{2}\right],$$

where $p(a) = 0$. Indeed, for the distances considered here, we have

$$p(r) \approx \hbar k_0 \sqrt{1 - \frac{\tilde{\alpha}}{k_0^2 r^2}}.$$

The integral in (3) is easy to calculate using integration by parts with the substitution $z = 1/r$. For $r \gg \sqrt{\tilde{\alpha}}/k_0$ it becomes $\hbar(k_0 r - \pi\sqrt{\tilde{\alpha}}/2)$. (We should note that for a free particle, $U_0(r) \equiv 0$, the quasiclassical phase of the radial wavefunction, according to (3), is equal at large distances to $kr - \pi l/2$ and coincides with the exact relation.)

Eq. (3) solves the problem of incorporating the boundary condition, $\chi(0) = 0$, in the quasi-classical solution in the classically allowed region of motion. As $\tilde{\alpha} \to 0$ we have $\gamma \to 0$ and, from Eq. (3), we have the result of Problem 9.2. On the contrary, for $\tilde{\alpha} \gg 1$, the barrier, α/r^2, is already quasi-classical, $|d\lambda/dr| \propto \tilde{\alpha}^{-1/2} \ll 1$; here, $\gamma \approx \pi/4$ in accordance with Eq. (IX.4 b).

Using Eq. (3) and the usual matching condition (IX.3) for the right turning point, we obtain the following quantization rule for levels with the angular momentum l in a central potential $U(r)$, which is bounded at the origin:

---

[167] Eq. (1) corresponds to the one-dimensional Schrödinger equation with the potential $U(r) = \alpha/r^2 + U_0(r)$, where $U_0(r)$ is a smooth function of r. For a centrifugal potential, $\tilde{\alpha} = l(l+1)$, and $\nu = l + 1/2$.

$$\frac{1}{\hbar}\int_a^b \sqrt{2m\left[E_{n_r l} - \frac{\hbar^2 l(l+1)}{2mr^2} - U(r)\right]}\,dr =$$
$$\pi\left(n_r + \frac{l}{2} - \frac{1}{2}\sqrt{l(l+1)} + \frac{3}{4}\right). \tag{4}$$

2) Hence, for the spherical oscillator, $U = m\omega^2 r^2/2$, using

$$\int_a^b \frac{1}{x}\sqrt{(b-x)(x-a)}\,dx = \frac{\pi}{2}(a+b-2\sqrt{ab}),$$

we find $E_{n_r l} = \hbar\omega(2n_r + l + 3/2)$, which coincides with the exact relation for the spectrum (see Problem 4.5).

Let us now consider the potential $U = U_0 \tan^2(\pi x/a)$. For $x \to \pm a/2$ it takes the form $U \approx \alpha/r^2$, where $r = a/2 - |x|$ and $\alpha = U_0 a^2/\pi^2$. Therefore, in order to take account of the boundary conditions $\psi(\pm a/2) = 0$ and derive the quantization rule, we should use relations analogous to Eq. (3) for the quasi-classical wavefunction in the vicinity of both turning points. We then obtain

$$\frac{1}{\hbar}\int_{-b}^{b}\sqrt{2m\left[E_n - U_0 \tan^2\left(\frac{\pi x}{a}\right)\right]}\,dx =$$
$$\pi\left[n + \frac{1}{2} + \sqrt{\frac{2mU_0 a^2}{\pi^2 \hbar^2} + \frac{1}{4}} - \sqrt{\frac{2mU_0 a^2}{\pi^2 \hbar^2}}\right].$$

This can be compared to $\pi(n + 1/2)$ on the right-hand side of Eq. (IX.5). Calculating the integral using the substitution

$$\sin\frac{\pi x}{a} = \kappa \sin t, \quad \kappa = \left(1 + \frac{U_0}{E_n}\right)^{1/2},$$

we find

$$E_n = \frac{\pi^2 \hbar^2}{2ma^2}\left\{\left(n + \frac{1}{2}\right)^2 + 2\left(n + \frac{1}{2}\right)\sqrt{\frac{2mU_0 a^2}{\pi^2 \hbar^2} + \frac{1}{4}} + \frac{1}{4}\right\},$$

which gives the exact spectrum. In view of the result obtained here, we note that quantization according to Eq. (IX.5) for the values $m\alpha/\hbar^2 \leq 1$ leads to a loss of accuracy even for large enough values of n. For $U_0 = 0$ (i.e., for an infinitely deep potential well), we have $E_n \propto (n+1)^2$ for the exact spectrum, while $E_n \propto (n+1/2)^2$ follows from Eq. (IX.5). Therefore, for $n = 10$, the deviation is approximately 10%.

Problem 9.7

In the previous problem we obtained the quantization rule for states of a particle with the angular momentum l in a central potential $U(r)$, restricted in the limit $r \to 0$:[168]

$$\frac{1}{\hbar}\int_a^b \left\{2m\left[E_{n_r l} - \frac{\hbar^2 l(l+1)}{2mr^2} - U(r)\right]\right\}^{1/2} dr =$$

$$\pi\left[n_r + \frac{1}{2} + \frac{1}{2}\left(l + \frac{1}{2} - \sqrt{l(l+1)}\right)\right]. \qquad (1)$$

Prove that within the quasi-classical accuracy it is equivalent to the Langer correction:

$$\frac{1}{\hbar}\int_a^b \left\{2m\left[E_{n_r l} - \frac{\hbar^2 (l+1/2)^2}{2mr^2} - U(r)\right]\right\}^{1/2} dr = \pi\left(n_r + \frac{1}{2}\right). \qquad (2)$$

Show also that both these relations, within the same accuracy, are equivalent to the more simple quantization rule:

$$\frac{1}{\hbar}\int_0^b \tilde{p}(r)dr \equiv \frac{1}{\hbar}\int_0^b \sqrt{2m[E_{n_r l} - U(r)]}\, dr = \pi\left(n_r + \frac{l}{2} + \frac{3}{4}\right), \qquad (3)$$

where the radial momentum $\tilde{p}(r)$ does not contain the centrifugal potential, and the value of the quasi-classical correction to n_r depends only on the value of the angular momentum l.

Using the quantization rules, calculate the energy spectra for: a) a spherical oscillator $U = m\omega^2 r^2/2$; and b) a particle in an infinitely deep spherical well of radius R (take into account the change in the matching condition for the right turning point $r = R$). Compare to the exact spectrum.

Solution

Let us begin by noting that for quasi-classical states with $n_r \gg 1$, and for classical motion over distances, $r \lesssim b$, the kinetic energy, $E_{n_r l} - U(r)$, has the following order-of-magnitude estimate

$$E_{n_r l} - U(r) \sim \frac{\hbar^2 n_r^2}{mb^2},$$

which follows from the quantization rules. In the integrand associated with the quantization rules, the centrifugal potential for $l \sim 1$ is a correction of the order $(l + 1/2)^2/n_r^2$ (which is outside the level of accuracy that we are working at) and

---

[168] We should emphasize that all three quantization conditions the standard usual quasi-classical matching condition at the right turning point according to Eq. (IX.3). The Langer quantization scheme is valid even for a potential which, for $r \to 0$, has the form $U(r) \approx \alpha/r^\nu$ with $0 \le \nu < 2$. A generalization of the quantization rule for this case is given in the solution.

can be omitted. The exception is the region of small distances near the left turning point. The difference in the contributions to the integral in this region determines the difference between the quasi-classical corrections on the right-hand sides of the quantization rules given in the problem.

To prove this statement we divide the integration domain into two: from a (or 0) to c and from c to b. We choose c[169] in such a way that: 1) we can also neglect a change in $U(r)$ in the region $r < c$ and replace the potential by $U(0)$; and 2) $\hbar^2/mc^2 \ll E_{n_r l} - U(0)$ is satisfied. According to 2) the centrifugal potential can be omitted while integrating between c and b. Therefore, the integrals in all three cases are the same (for the same value $E_{n_r l}$). Using

$$\int_a^c \sqrt{1 - \frac{\alpha}{r^2}}\, dr = \left[r\sqrt{1 - \frac{\alpha}{r^2}} + \sqrt{\alpha}\arcsin \frac{\sqrt{\alpha}}{r} \right]\Bigg|_a^c \approx c - \frac{\pi\sqrt{\alpha}}{2}, \qquad (4)$$

where $a = \sqrt{\alpha}$ and $c \gg a$ (see the previous problem for its calculation) ensure that indeed the difference in the right-hand sides of the quantization rules is compensated by the difference in their left-hand side integrals. Hence the equivalence follows (with quasi-classical accuracy) for the three quantization rules given in the problem.

2) For a spherical oscillator, $U = m\omega^2 r^2/2$, they all lead to the same result that coincides with the exact expression for the spectrum $E_{n_r l} = \hbar\omega(2n_r + l + 3/2)$. (Here the calculation is especially easy for the last quantization rule (3) not involving the centrifugal potential. For the values of the integrals for the other two quantization rules, see the previous problem.)

For a particle in an infinitely deep potential well we should make the following modification: replace $n_r + 1/2$ by $n_r + 3/4$ (as in Problem 9.3, it is now connected with the right turning point $x = R$). Here, the last of the quantization rules (3) immediately gives

$$\sqrt{\frac{E_{n_r l}}{\varepsilon_0}} = \pi\left(n_r + \frac{l}{2} + 1\right), \quad i.e. \quad E_{n_r l} = \pi^2 \varepsilon_0 \left(n_r + \frac{l}{2} + 1\right)^2, \qquad (5)$$

where $\varepsilon_0 = \hbar^2/2mR^2$ ($E_0 = \pi^2 \varepsilon$ is the ground state). From the Langer quantization rule and from Eq. (4), it follows that[170]

$$\sqrt{\frac{E_{n_r l}}{\varepsilon_0} - \left(l + \frac{1}{2}\right)^2} + \left(l + \frac{1}{2}\right)\arcsin\left[\left(l + \frac{1}{2}\right)\sqrt{\frac{\varepsilon_0}{E_{n_r l}}}\right] = \pi\left(n_r + \frac{l}{2} + 1\right). \qquad (6)$$

---

[169] The possibility to choose such c is provided by the quasi-classical condition, $|d\lambda/dr| \ll 1$, over small distances subject to potential $U(r)$ (without the centrifugal potential).

[170] For the values $n_r \gg (l + 1/2)$

$$\delta^2 \equiv \left(l + \frac{1}{2}\right)^2 \frac{\varepsilon_0}{E_{n_r l}} \ll 1$$

and omitting the correction connected with δ from (6) we obtain (5).

For an illustration of the accuracy of the quasi-classical result, we write, using $E_{n_r l} = g_{n_r l} \varepsilon_0$, the table of values $g_{n_r l}$ for a set of quantum numbers n_r and l calculated according to Eq. (5) and the exact values (in brackets). For the exact spectrum $g_{n_r l} = x_{n_r l}^2$,

l	$n_r = 0$	$n_r = 1$	$n_r = 2$	$n_r = 3$	$n_r = 4$
1	22.2	61.7	120.9	199.9	298.6
	(20.2)	(59.7)	(118.9)	(197.9)	(296.5)
2	39.5	88.8	157.9	246.7	355.5
	(33.2)	(82.7)	(151.9)	(240.7)	(349.3)
3	61.7	120.9	199.9	298.6	417.0
	(48.8)	(108.5)	(187.6)	(286.4)	(404.9)
4	88.8	157.9	246.7	355.3	483.6
	(67.0)	(137.0)	(226.2)	(334.9)	(463.3)

where $x_{n_r l}$ is the $(n_r + 1)$-th zero (without $x = 0$) of the Bessel function $J_\nu(x)$ with $\nu = l + 1/2$. We confine ourselves to small values of n_r only when the quasi-classical approach is not applicable to emphasize that in this case the quasi-classical result is close to the exact one. It is important here that the choice of quasi-classical corrections in the quantization conditions is consistent with the boundary conditions.[171] There are no s-states in the table; for them the quasi-classical spectrum coincides with the exact one, $g_{n_r 0} = \pi^2 (n_r + 1)^2$. As is seen in the table, the difference between the exact and quasi-classical values for $E_{n_r l}$ with a fixed l weakly depends on n_r. Since the subleading quasi-classical correction on the right-hand side of the quantization rule is $\sim 1/n_r$ for $n_r \gg 1$, this property persists down to $n_r \sim 1$.

3) Let us make two concluding remarks regarding the quantization rule without the centrifugal potential. Since n_r and l appear here only in the combination $2n_r + l$, we have an unexpected (approximate) level degeneracy for arbitrary quasi-classical states, similar to the degeneracy well-known from the exact solution of a spherical oscillator. The degeneracy is removed however if we include the subleading quasi-classical corrections.

The quantization rule (3) corresponds to a potential restricted at small distances. However, we can generalize it to the case where $U(r) \propto 1/r^\nu$ for $r \to 0$ and $0 \leq \nu < 2$ if we replace the right hand side[172] with the expression

---

[171] Note that quantization according to Eq. (IX.5) leads to a noticeable loss in accuracy (see Problem 9.6).
[172] Such a modification of the quantization rule could be obtained if, at small distances, we use the exact solution of the Schrödinger equation $\chi_l'' - (l(l+1)/r^2)\chi_l - (2m\alpha/\hbar^2)r^{-\nu}\chi_l = 0$ and then match it with the quasi-classical solution; see also the Problem 9.9.

$$\pi\left[n_r + \frac{1}{2} + \frac{2l+1}{2(2-2\nu)}\right]$$

(for a bounded potential $\nu = 0$; compare also with Problem 9.5). Here, for the Coulomb potential, $U = -\alpha/r$, the quasi-classical result obtained gives the exact spectrum for all values of the angular momentum l.

Problem 9.8

For a one-dimensional attractive potential which, for $|x| \to \infty$, has the form $U(x) \propto x^{-4}$, determine within the quasi-classical approximation the condition for the appearance of a new bound state as the depth of the well increases. Apply the result obtained to the potential from Problem 2.40 and compare to the exact result.

Solution

The condition can be obtained from the quantization rule (IX.5) using $n = N - 1$ (N being the level number) and by taking the value of E_{N-1} to zero (hence $a \to -\infty$ and $b \to +\infty$). However, if we take into account the quasi-classical correction of $1/2$ subleading to n, we would go beyond its regime of accuracy. It is connected with the fact that the matching condition for quasi-classical solutions based on the linear potential approximation is not applicable near the turning points in this problem: turning points move to infinity where for $E = 0$ the quasi-classical property is broken, since $|d\lambda/dx| \propto |x| \to \infty$.

To obtain a quasi-classical correction we should find the exact solution of the Schrödinger equation at large distances and then match it with the quasi-classical solution. The Schrödinger equation takes the form:

$$\psi'' + \frac{\tilde{\alpha}}{x^4}\psi = 0, \quad U \approx -\frac{\alpha}{x^4}, \quad \tilde{\alpha} = \frac{2m\alpha}{\hbar^2}.$$

Using the substitution, $\psi = \varphi(z)/z$, where $z = \sqrt{\tilde{\alpha}}/|x|$, we obtain $\varphi''(z) + \varphi(z) = 0$. The existence of a zero-energy solution to the Schrödinger equation which is bounded as $x \to \pm\infty$, corresponds to the moment a new discrete state appears as the potential well deepens (see Problem 2.13). Then the solution for $\varphi(z)$ must be chosen in the form $\varphi = A \sin z$. Hence, the exact wavefunction for large distances, just when the level appears, has the form

$$\psi(x) \approx A_{\pm} x \sin\left(\frac{\sqrt{\tilde{\alpha}_{\pm}}}{x}\right), \quad x \to \pm\infty. \tag{1}$$

For such large distances, but $|x| \ll \sqrt{\tilde{\alpha}}$, the quasi-classical condition, $|d\lambda/dx| \ll 1$, is already satisfied and the quasi-classical method is applicable (but it is of course assumed that on such distances potential is still described by an asymptotic expression, i.e., $U \propto x^{-4}$). In the quasi-classical approximation we have

$$\psi = \frac{C_1}{\sqrt{p}} \sin\left[\frac{1}{\hbar}\int_{-\infty}^{\pi} pdx + \gamma_1\right] = \frac{C_2}{\sqrt{p}} \sin\left[\frac{1}{\hbar}\int_x^{\infty} pdx + \gamma_2\right]. \qquad (2)$$

We find the values for parameters $\gamma_{1,2}$ from the matching conditions of the quasi-classical solutions with the exact ones found in Eq. (1). Here, $p/\hbar \approx \sqrt{\tilde{\alpha}_{\pm}}/x^2$, so we see that $\gamma_1 = \gamma_2 = 0$. Now, from condition (2) (the phases in the sines sum to πN) we obtain the relation desired:

$$\frac{1}{\hbar}\int_{-\infty}^{\infty} \sqrt{-2mU(x)}dx = \pi N. \qquad (3)$$

(The direct use of Eq. (IX.5), as was mentioned earlier in this solution, gives the value of the right-hand side equal to $\pi(N-1/2)$). Hence for the potential $U = -U_0(x^2/a^2 + 1)^{-2}$, we find

$$\xi \equiv \frac{2mU_0 a^2}{\hbar^2} = N^2. \qquad (4)$$

The exact result (see Problem 2.40) is obtained by the substitution $N^2 \to N^2 - 1$. We see that conditions (3) and (4) yield a higher accuracy[173] than those obtained by Eq. (XI.5). We find that ξ_N for $N = 10$ (according to (4)) differs from the exact result by 1%, while the "simplified" conditions give a difference of approximately 10%.

Problem 9.9

Using the quasi-classical approximation, obtain the condition for the appearance of new bound states with angular momentum $l \sim 1$, as one increases the well depth of an attractive potential with the following properties: $U \approx -\alpha_2 r^{-\nu_2}$ with $\nu_2 > 2$ for large distances, and $U \approx -\alpha_1 r^{-\nu_1}$ with $0 \leq \nu_2 < 2$ for $r \to 0$. Illustrate your results on the examples of the Hulthen, Yukawa, and Tietz potentials.

Solution

As in the previous problem we first write the quasi-classical expression for the radial function ($\psi = Y_{lm}\chi_l(r)/r$) for $E = 0$ in two ways:

$$\chi_l(r) = \frac{C_1}{\sqrt{\tilde{p}}}\sin\left[\frac{1}{\hbar}\int_0^r \tilde{p}\,dr + \gamma_{1\ell}\right] = \frac{C_2}{\sqrt{\tilde{p}}}\sin\left[\frac{1}{\hbar}\int_r^{\infty}\tilde{p}\,dr + \gamma_{2\ell}\right]. \qquad (1)$$

Note that $\chi_l \propto r^{l+1}$ for $r \to 0$ and $\chi_l \propto 1/r^l$ for $r \to \infty$ when a bound state appears (see Problem 4.26). In (1), $\tilde{p} = \sqrt{-2mU(r)}$ is the radial momentum without the centrifugal potential (for a reasoning behind this approximation for states with $l \sim 1$

---

[173] Here the difference between the classical and exact results appears as $\sim 1/N^2$.

in the quasi-classical domain, see Problem 9.7). In this problem, however, the quasi-classical property is not obeyed over both small and large distances. We should, therefore, use the exact solution to the Schrödinger equation which is matched with the quasi-classical solutions (1) in order to find $\gamma_{1(2)l}$. We then can obtain the threshold condition for the appearance of a bound state using (1).

The Schrödinger equation both for $r \to 0$ and $r \to \infty$ takes the form

$$\chi_l'' + \left[\frac{\tilde{\alpha}}{r^\nu} - \frac{l(l+1)}{r^2}\right]\chi_l = 0, \quad \tilde{\alpha} = \frac{2m\alpha}{\hbar^2}. \tag{2}$$

As is known, solutions of this equation are expressed in terms of cylindrical functions:

$$\sqrt{r} Z_s(\beta r^\mu); \quad s = \frac{2l+1}{2-\nu}, \quad \mu = \frac{2-\nu}{2}, \quad \beta = \frac{2\sqrt{\tilde{\alpha}}}{2-\nu}.$$

In order to satisfy the boundary condition at small distances ($\chi_l \propto r^{l+1}$) the solution should be chosen in the form

$$\chi_l = C\sqrt{r} J_{s_1}(\beta_1 r^{\mu_1}).$$

Using the asymptotic expression for the Bessel function, J_s, as $z \to \infty$, we obtain

$$\chi_l \propto r^{\nu_1/2} \sin\left[\frac{2\sqrt{\tilde{\alpha}_1}}{2-\nu_1} r^{\mu_1} - \frac{\pi s_1}{2} + \frac{\pi}{4}\right]. \tag{3}$$

This corresponds to the quasi-classical solution applicable for short distances such that $U \approx -\alpha_1/r^{\nu_1}$, but in which the contribution from the centrifugal potential can already be neglected.[174] Comparing with the first expression for χ_1 in (1) gives

$$\frac{1}{\hbar}\int_0^r \tilde{p}\,dr = \frac{2\sqrt{\tilde{\alpha}_1}}{2-\nu_1} r^{\mu_1} \quad \text{and} \quad \gamma_{1l} = \pi\left[\frac{2l+1}{2(\nu_1 - 2)} + \frac{1}{4}\right]. \tag{4}$$

For large distances we choose the solution of Eq. (2) in the form

$$\chi_l = \tilde{C}\sqrt{r} J_{-s_2}(-\beta_2 r^{\mu_2}),$$

and find the parameter value

$$\gamma_{2l} = \pi\left[\frac{2l+1}{2(2-\nu_2)} + \frac{1}{4}\right]. \tag{5}$$

Now, from the equality in Eq. (1) (the phases in the sines must sum to πN_1), we obtain the condition for the appearance of the N_lth bound state for a particle with the angular momentum l:

---

[174] We should emphasize that the part of the quasi-classical phase in Eq. (3) which depends on r does not contain l, because taking account of the centrifugal potential in the principle domain of quasi-classical motion is outside the limits of accuracy for the considered approximation. see Problem 9.7.

$$\frac{1}{\hbar}\int_0^\infty \sqrt{-2mU(r)}\,dr = \pi\left[N_l + \frac{2l+1}{2(2-\nu_1)} + \frac{2l+1}{2(\nu_2-2)} - \frac{1}{2}\right]. \qquad (6)$$

We note that the case $\nu_2 \to \infty$ corresponds to a potential which exponentially decreases for large distances.

Let us now consider applications of the result in (6).

1) For s-states in the Hulthen potential, $U(r) = -U_0/e^{r/a-1}$, we should choose $\nu_1 = 1$, $\nu_2 = \infty$ and $l = 0$ in (6). After integration, we obtain

$$\frac{2ma^2 U_0}{\hbar^2} = N_s^2,$$

which coincides with the exact result (see Problem 4.8).

2) For the Yukawa potential, $U = -\alpha e^{-r/a}/r$, we obtain (according to (6))

$$\xi \equiv \frac{2m\alpha}{\hbar^2 a} = \frac{\pi N_s^2}{2}.$$

Hence the values of ξ equal to 1.57, 6.28, and 14.14 correspond to the appearance of the lower 1s-, 2s-, and 3s-states, respectively. The exact values obtained by numerical integration of the Schrödinger equation are 1.68, 6.45, and 14.34.

3) Let us now consider the Tietz potential, $U = -\frac{a}{r(r+1)^2}$, for which $\nu_1 = 1$ and $\nu_2 = 3$. According to (6) we have

$$\frac{2m\alpha}{\hbar^2 a} = \left(N_1 + 2l + \frac{1}{2}\right)^2.$$

For this potential we can find the exact condition for the appearance of the first bound state, $N_l = 1$, for an arbitrary value of l. The wavefunction when the bound state appears has the form

$$\chi_l = C r^{l+1}(r+a)^{-2l-1},$$

where

$$\frac{2m\alpha}{\hbar^2 a} = (2l+1)(2l+2).$$

We see that even for $N_l = 1$ the quasi-classical result is quite close to the exact condition for all values of the particle's angular momentum, l.

Problem 9.10

Using the Bohr–Sommerfeld quantization rule, obtain an expression for the energy levels shifts caused by a small change in a potential, $\delta U(x)$, and compare it with the result of first-order perturbation theory. What is the interpretation of the result obtained in the framework of the classical theory? As an illustration consider the energy-level shifts for a linear oscillator caused by an anharmonicity, $\delta U(x) = \beta x^4$, and compare this to the exact result obtained from first-order perturbation theory.

Solution

Let us denote the energy levels in the potential fields $U_0(x)$ and $U_0(x) + \delta U(x)$ by $E_n^{(0)}$ and $E_n^{(0)} + \delta E_n$, respectively. Expanding the function under the integral in the quantization rule,

$$\int_{a+\delta a}^{b+\delta b} \sqrt{2m\left[E_n^{(0)} + \delta E_n - U_0(x) - \delta U(x)\right]}\, dx = \pi \hbar \left(n + \frac{1}{2}\right). \tag{1}$$

Taking into account the smallness of δE_n and δU we find the desired level shift[175]

$$\delta E_n \approx E_n^{(1)} = \frac{2}{T}\int_a^b \frac{\delta U(x)}{v_0(x)}\, dx, \tag{2}$$

$$v_0(x) = \sqrt{\frac{2}{m}\left[E_n^{(0)} - U_0(x)\right]},$$

where $T(E_n^{(0)})$ is the period of classical motion in the potential $U_0(x)$.

The result corresponds to the equation of first-order perturbation theory, $E_n^{(1)} = (\delta U(x))_{nn}$ (see Eq. (VIII.1)). It can be calculated from it using the quasi-classical expression (IX.6) for unperturbed wavefunction and replacing the rapidly oscillating sine-squared with its mean value $1/2$.

The classical interpretation of Eq. (2) can be given in terms of the adiabatic invariant. It describes the change in the energy, $E_n^{(0)}$ (which in classical theory is not quantized!) of a particle's finite motion inside the potential $U_0(x)$ due to an *adiabatic* change in the potential, $\delta U(x)$. Here $I = \frac{1}{2\pi}\oint p\, dx = \text{const}$, and in the case of a small potential perturbation, δU, we have Eq. (2). Compare this with the case of a sudden perturbation discussed in Problem 9.22.

Let us apply Eq. (2) to the case of an oscillator with a weak anharmonicity:

$$U_0(x) = \frac{1}{2}m\omega^2 x^2, \quad E_n^{(0)} = \hbar\omega\left(n + \frac{1}{2}\right), \quad \delta U(x) = \beta x^4.$$

$$\delta E_n = \frac{\omega}{\pi}\int_{-b}^{b} \frac{\beta x^4\, dx}{\sqrt{2E_n^{(0)}/m - \omega^2 x^2}} = \frac{3}{2}\beta\left(\frac{\hbar}{m\omega}\right)^2\left(n^2 + n + \frac{1}{4}\right). \tag{3}$$

---

[175] See Problem 9.3. We note that the change in the left-hand side of Eq. (1) due to a shift of the turning points vanishes in the first order. If $\delta U(x)$ differs from zero only outside the domain of classical motion, then the level shift is $\delta E_n = 0$ (according to Eq. (2)). Of course, a small level shift does occur in this case. However, it is, generally speaking, exponentially small, and its calculation requires a separate analysis; see Problem 9.4.

We should emphasize that taking into account 1/4, which is subleading to $n^2 + n$, is, strictly speaking, beyond the accuracy of the quasi-classical approximation (see Problem 9.13). Indeed, the true expression for $E_n^{(1)}$ differs from (3) by 1/2 instead of 1/4. We see that the quasi-classical result satisfactory reproduces the exact one for $n \sim 1$ (except for the case $n = 0$).

Problem 9.11

For a particle in a symmetric potential well, $U(x)$, obtain the energy level shifts in a weak uniform electric field and the polarizabilities of its stationary states using the Bohr–Sommerfeld quantization rule. What is the interpretation of the result in the framework of the classical theory? Find the polarizabilities for a linear oscillator and a particle in an infinitely deep potential well.

Solution

Let us denote the unperturbed energy levels by $E_n^{(0)}$, and their shifts by $\Delta E_n(\mathcal{E})$. The shifts are determined by the quantization rule:

$$\int_{a_1}^{a_2} \sqrt{2m[E_n^{(0)} + \Delta E_n - U(x) + e\mathcal{E}x]}\, dx = \pi\hbar\left(n + \frac{1}{2}\right). \tag{1}$$

Here, $V = -e\mathcal{E}x$ is the potential energy of a particle with charge e in an homogeneous electric field $\mathcal{E}$. Since $\Delta E_n \propto \mathcal{E}^2$ an expansion with respect to E in the left-hand side of Eq. (1) should take account of the second-order terms. To this end we transform the left-hand side of (1) in the following way:

$$\frac{3}{2}\frac{\partial}{\partial E_n^{(0)}} \int_{a_1}^{a_2} \frac{1}{2m}[2m(E_n^{(0)} + \Delta E_n - U(x) + e\mathcal{E}x)]^{3/2}\, dx \approx$$

$$\frac{3}{2}\frac{\partial}{\partial E_n^{(0)}} \int_{-a}^{a} \left\{\frac{p_n^3(x)}{2m} + \frac{3}{2}p_n(x)(\Delta E_n + e\mathcal{E}x) + \frac{3m}{4p_n(x)}(e\mathcal{E}x)^2\right\} dx, \tag{2}$$

where $p_n = \sqrt{2m(E_n^{(0)} - U(x))}$ and $\pm a$ are the turning points for the unperturbed motion. The first term with $p_n^3(x)$ reproduces the right-hand side in (1); the second term is equal to $T(E_n^{(0)})\Delta E_n/2$, while the third term determines the desired level shift,

$$\Delta E_n \equiv -\frac{1}{2}\beta_n E^2 = -\frac{me^2\mathcal{E}^2}{T(E_n^{(0)})}\frac{\partial}{\partial E_n^{(0)}} \int_{-a}^{a} \frac{x^2\, dx}{\sqrt{2m[E_n^{(0)} - U(x)]}}. \tag{3}$$

Here $T(E_n^{(0)})$ is the period of unperturbed motion, and β_n is the polarizability of the nth state, and it in turn determines the mean dipole moment of the system,

$d_{nn} = \beta_n \mathcal{E}$, induced by the external field. According to (3) the polarizability is equal to (for $E = E_n^{(0)}$):

$$\beta(E) = \frac{e^2}{T(E)} \frac{\partial}{\partial \mathcal{E}} (T(\mathcal{E})\overline{x^2}), \tag{4}$$

where $\overline{x^2}$ is the mean value of x^2 in potential $U(x)$ averaged over a period of classical motion. The classical interpretation of this relation is based on considering the mean dipole moment over the period:

$$\overline{d}_{class}(\mathcal{E}) = \frac{1}{T(\mathcal{E})} \oint \frac{ex}{v(x,\mathcal{E})} dx, \quad v = \pm\sqrt{\frac{2(E - U(x) + e\mathcal{E}x)}{m}}.$$

Expanding the integral here in powers of E as in Eq. (2) we obtain the relation $\overline{d}_{class} = \beta \mathcal{E}$, where β is defined via Eq. (4). We should note also that the polarizability determines a change in energy of the classical system when the field is turned on slowly: in accordance with (3) $\Delta E = -\beta \mathcal{E}^2/2$ (see the previous problem).

For the oscillator $T(\mathcal{E}) = $ const, while $\overline{x^2} = \mathcal{E}/m\omega^2$ (as follows, for example, from the virial law). Here, $\beta_n = e^2/m\omega^2$, which coincides with the exact quantum-mechanical result (see Problem 2.2).

For a particle in an infinitely deep potential well, $\beta_n = -e^2 a^2/24 E_n^{(0)} < 0$, where a is the well width.

Problem 9.12

Using the Bohr–Sommerfeld quantization rule, obtain quasi-classical expressions for the first- and second-order shifts in energy levels caused by a potential $V(x)$. As an illustration determine the shifts for a linear oscillator due to the anharmonicity, $V(x) = \alpha x^3$, and compare them with the exact result from second-order perturbation theory.

Solution

Writing $E_n = E_n^{(0)} + E_n^{(1)} + E_n^{(2)}$, $U(x) = U_0(x) + V(x)$, and performing the expansion in Eq. (IX.5) in a way similar to the previous problem, we find

$$E_n^{(1)} = \overline{V(x)},$$

$$E_n^{(2)} = \frac{1}{2T\left(E_n^{(0)}\right)} \frac{\partial}{\partial E_n^{(0)}} \left\{ T\left(E_n^{(0)}\right) \left[\left(\overline{V(x)}\right)^2 - \overline{(V(x))^2} \right] \right\}. \tag{1}$$

Here an overbar denotes the average of the corresponding quantity over one period of classical motion in the potential, $U_0(x)$, with the energy, $E_n^{(0)}$. More specifically, it is defined by[176]

---

[176] I.e., by averaging over the classical probability, see Eq. (IX.8).

$$\overline{f(x)} \equiv \frac{2}{T\left(E_n^{(0)}\right)} \int_a^b \frac{f(x)dx}{v_n(x)}, \quad v_n = \sqrt{\frac{2}{m}\left[E_n^{(0)} - U_0(x)\right]}. \tag{2}$$

For an oscillator with $U_0(x) = m\omega^2 x^2/2$ and anharmonicity $V(x) = \alpha x^3$, we find $\overline{V(x)} = 0$ and

$$\overline{(V(x))^2} = \frac{\omega}{\pi} \int_{-b}^{b} \frac{\alpha^2 x^6 dx}{\sqrt{2E_n^{(0)}/m - \omega^2 x^2}} = \frac{5}{2}\alpha^2 \left(\frac{E_n^{(0)}}{m\omega^2}\right)^3.$$

Hence $E_n^{(1)} = 0$, and from Eq. (1) we have

$$E_n^{(2)} = -\frac{15}{4}\frac{\alpha^2}{m\omega^2}\left(\frac{\hbar}{m\omega}\right)^2\left(n^2 + n + \frac{1}{4}\right). \tag{3}$$

This differs from the exact value only by the factor $1/4$; the exact result is equal to $11/30$ (such a difference is within the error bars allowed by the quasi-classical accuracy of the Bohr–Sommerfeld quantization rule; see Problem 9.13). Here the quasi-classical result faithfully reproduces the exact one even for states with $n \sim 1$ (except for the case $n = 0$).

Problem 9.13

Find a quasi-classical correction to the Bohr–Sommerfeld quantization rule to next up to the next-to-leading order in $\hbar$. Prove that if such a correction is taken into account, the level energy becomes

$$E_n = E_n^{BS} + \Delta E_n, \quad \Delta E_n = \frac{\hbar^2}{24mT(E)}\frac{\partial^2}{\partial E^2}(\overline{(U'(x))^2}T(E)),$$

where the overline means average with the classical probability, $dw = 2dx/T(E)v(x)$, for an energy, $E = E_n^{BS}$, determined by the Bohr–Sommerfeld quantization rule. Calculate the next-order correction with respect to $\hbar$ for the results in Problems 9.10 and 9.12, where anharmonic oscillators with $V(x) = \beta x^4$ and $V(x) = \alpha x^3$ were considered.

Solution

1) The desired quantization rule correction can be obtained if we use more precise quasi-classical wavefunctions that take into account the quasi-classical correction to next order in $\hbar$, while obtaining the Bohr–Sommerfeld rule.

The wavefunction to the right of the right turning point,[177] $x = b$, now has the form

$$\psi = \frac{C}{2\sqrt{-ip}} \exp\left\{ \frac{i}{\hbar} \int_b^x p\, dx - \frac{im\hbar}{4} \frac{F}{p^3} - \frac{i\hbar}{24} \frac{\partial^2}{\partial E^2} \int_b^x \frac{F^2}{p} dx \right\}, \tag{1}$$

where $F = -dU/dx$. Here we used the next-to-leading order results for the wavefunction available in the literature (see, for example, vol. III of the theoretical physics course by Landau and Lifshitz) and also the following identity

$$\frac{i\hbar m^2}{8} \int \frac{F^2}{p^5} dx = \frac{i\hbar}{24} \frac{\partial^2}{\partial E^2} \int \frac{F^2}{p} dx.$$

(On the real axis for $x > b$ the value $p(x)$ is imaginary and $ip < 0$.) If we move into the region of classical motion on the contour in the complex plane, x, we obtain the wavefunction for[178] $x < b$:

$$\psi = \frac{C}{\sqrt{p}} \sin\left\{ \frac{1}{\hbar} \int_x^b p\, dx + \frac{m\hbar}{4} \frac{F}{p^3} - \frac{\hbar}{24} \frac{\partial^2}{\partial E^2} \int_x^b \frac{F^2}{p} dx + \frac{\pi}{4} \right\}. \tag{2}$$

By matching with the decreasing solution for $x \to -\infty$, we find the wavefunction in the region $x > a$:

$$\psi = \frac{C'}{\sqrt{p}} \sin\left\{ \frac{1}{\hbar} \int_a^x p\, dx - \frac{m\hbar}{4} \frac{F}{p^3} - \frac{\hbar}{24} \frac{\partial^2}{\partial E^2} \int_a^x \frac{F^2}{p} dx + \frac{\pi}{4} \right\}, \tag{3}$$

and identifying the expressions (2) and (3), we obtain the quantization rule:

$$\frac{1}{\hbar} \int_a^b p\, dx = \pi\left(n + \frac{1}{2}\right) + \frac{\hbar}{24} \frac{\partial^2}{\partial E^2} \int_a^b \frac{F^2}{p} dx. \tag{4}$$

The last term here is the sought-after correction. Therefore, by writing $E_n = E_n^{BS} + \Delta E_n$, where ΔE_n is the energy level shift of interest, and performing an expansion with respect to ΔE_n (see Problem 9.3) we obtain

$$\Delta E_n = \frac{\hbar^2}{12T(E)} \frac{\partial^2}{\partial E^2} \int_a^b \frac{(U'(x))^2 dx}{\sqrt{2m(E - U(x))}} \bigg|_{E=E_n^{BS}}, \tag{5}$$

(on the right-hand side of Eq. (4) we neglect the change in the energy levels), which coincides with the relation given in the problem statement.

---

[177] Here we denote the left and right turning points by a and b respectively.
[178] The upper limit for the integral in (2) must be chosen to be the turning point $x = b$.

Let us now consider some applications of Eq. (5). For an oscillator we have $U' = m\omega^2 x$; here, $\overline{(U')^2} = m\omega^2 E$, and since $T = 2\pi/\omega = const$ (independent of energy), we find $\Delta E_n = 0$. This is a natural result, since for an oscillator, the Bohr–Sommerfeld law reproduces the exact spectrum and there are no higher-order corrections with respect to $\hbar$.

For an oscillator with the anharmonicity, βx^4, performing the corresponding expansions in Eq. (5), we find the part of the shift linear in β,

$$\Delta E_n = \frac{3\beta \hbar^2}{8m^2\omega^2}.$$

If we combine this with Eq. (3) from Problem 9.10 we obtain the result from first-order perturbation theory, $E_n^{(1)}$, which coincides with the exact one.

In an analogous way, for the anharmonicity $V = \alpha x^3$ we obtain using Eq. (5):

$$\Delta E_n = -\frac{7\alpha^2 \hbar^2}{16m^3\omega^4}$$

for the correction quadratic in α, which along with Eq. (3) from Problem 9.12, reproduces the exact result from second-order perturbation theory.

2) Finally, we would like to present another method of obtaining quasi-classical corrections to the quantization rule to higher order in $\hbar$; the method is based on analyzing the non-linear equation equivalent to the Schrödinger equation,

$$\chi' = \frac{2m}{\hbar^2}(U - E_n) - \chi^2, \tag{6}$$

which is written in terms of the logarithmic derivative $\chi = \psi'/\psi$. Here,

$$\chi = -\sqrt{\frac{2m(U(x) - E_n)}{\hbar^2} - \chi'(x)} \tag{7}$$

(for the choice of a sign, see below).

We integrate (7) using the residue theorem over a contour, which in the complex plane of variable x surrounds the part of the real axis between the turning points a and b:

$$\oint_C \chi dx = -\frac{1}{\hbar} \oint_C \sqrt{2m(U - E_n) - \hbar^2 \chi'} dx = 2\pi i n. \tag{8}$$

Here we have taken into account the fact that the zeros of the wavefunction are the poles of the function $\chi(x)$. The number of these poles is equal to n, and the residue for each of them is 1.

Relation (8) is exact (for analytic potentials) and is valid for an arbitrary choice of the contour C. But for further transformations it is convenient to choose the contour such that it is not too close to the interval on the real axis between the turning points. In this case, according to Eqs. (7) and (8), χ' is a subleading correction on the integration contour. Indeed, near the interval (a, b) on the real axis, $\psi(x)$ has an oscillatory behavior, $\psi \propto \sin((i/\hbar) \int p dx + \gamma)$, and the values χ' and χ^2 are of the same order. If we move into the complex plane there exists only one exponentially-increasing term in the wavefunction. Here, $\chi = \psi'/\psi$ does not contain a rapidly changing factor, and the derivative χ' appears to be a small quantity, of the order of $d\lambda/dx \ll 1$, in comparison to χ^2. Hence, in this case Eq. (7) can be solved by successive iterations

$$\chi = \chi^{(0)} + \chi^{(1)} + \chi^{(2)} + \ldots,$$

and we find

$$\chi^{(0)} = -\frac{i}{\hbar} p_n(x),$$

$$\chi^{(1)} = -\frac{\chi^{(0)\prime}}{2\chi^{(0)}} = \frac{mU'}{2p_n^2}, \qquad (9)$$

$$\chi^{(2)} = \chi^{(0)} \left[\frac{\hbar^2 \chi^{(1)\prime}}{2p_n^2} - \frac{1}{8} \left(\frac{\hbar^2 \chi^{(0)\prime}}{p_n^2} \right)^2 \right] \quad etc.$$

Here, as usual, $p_n = \sqrt{2m(E_n - U(x))}$. The turning points a, b are the branch points for the function, $p_n(x)$. Here, we introduce a branch cut along the interval (a, b) on the real axis x. Above the cut, $p_n > 0$ ($p_n < 0$ below the cut). We should note that $p_n = -i\sqrt{2m(U(x) - E_n)}$, where the phase, $U(x) - E_n$, to the right of the right turning point, b, is chosen equal to zero; it is in accordance with the choice of sign in Eq. (7), so that $\chi < 0$ for $x > b$. We should emphasize that the turning points are neither critical points nor branch points for the exact solution $\chi(x)$.

Substituting the expansion of χ into Eq. (8), we obtain[179]

$$\frac{i}{\hbar} \oint p_n dx = 2\pi \left[n + \frac{1}{2} + \frac{\hbar}{24\pi} \frac{\partial^2}{\partial E_n^2} \int_a^b \frac{(U'(x))^2}{p_n(x)} dx + \ldots \right\}. \qquad (10)$$

Here we took into account that[180]

$$\oint \chi^{(1)} dx = -\frac{1}{2} \oint d \ln p(x) = -i\pi,$$

---

[179] The integration contour could now be deformed so that it includes the interval (a, b) along the real axis. We emphasize that analytic properties of the exact solution and its quasi-classical expansion are different!

[180] Here we use the equation $\ln x = \ln |z| + i \arctan z$, and take into account that the phase $p(x)$ changes by 2π while going around the integration contour (at each branch point a phase equal to π is accumulated).

while for an integral transformation from $\chi^{(2)}$ we integrated by parts in the term with $\chi^{(1)'}$ and used the relation

$$\oint \frac{(U')^2}{p^5} dx = \frac{1}{3m^2} \frac{\partial^2}{\partial E^2} \oint \frac{(U')^2}{p} dx = -\frac{2}{3m^2} \frac{\partial^2}{\partial E^2} \int_a^b \frac{(U')^2}{p} dx.$$

The quantization rule (10) coincides with expression (4) obtained above in another way.

Problem 9.14

Let us consider a singular attractive potential $U = -\alpha/r^\nu$ with $\nu > 2$ as $r \to 0$, which classically gives rise to trajectories where a particle "falls into the center." In this case, both independent solutions of the radial Schrödinger equation behave in a similar way at small distances (compare this to $R_l \propto r^l$ and $\propto r^{-l-1}$ for a regular potential with $\nu < 2$) and, at first sight, there is no energy quantization, since a single condition for a decreasing wavefunction in the classically forbidden region for $r \to \infty$ can always be fulfilled.

Using a regularization (small-distance cut-off) of the potential by an impenetrable sphere of radius r_0 surrounding the origin, prove that energy quantization does occur, but for $r_0 \to 0$ it is necessary to fix[181] a position of one of the levels (for each value of l) in order to uniquely determine the full spectrum. Obtain a quantization rule and find the corresponding boundary condition on the wavefunction for $r \to 0$. Find also the energy spectrum for the potential $U = -\alpha/r^2$ in the conditions for "falling into the origin".

Solution

1) The most general solution of the radial Schrödinger equation within the quasi-classical approximation in the region of finite particle motion has the form ($\psi_{Elm} = Y_{lm}\chi_{El}/r$)

$$\chi_{El} = \frac{C}{\sqrt{p(r)}} \sin\left(\frac{1}{\hbar}\int_a^r p\,dr + \gamma\right), \quad p = \hbar\sqrt{\frac{\tilde\alpha}{r^\nu} - \frac{l(l+1)}{r^2} - \kappa^2}. \qquad (1)$$

Here[182] $\tilde\alpha = 2m\alpha/\hbar^2$ and $E = -\hbar^2\kappa^2/2m$. Since the integral in (1) for $r \to 0$ diverges ($\nu > 2$), both independent solutions behave in a similar way – with infinite sine oscillations – and ensure convergence at the normalization integral at small

---

[181] It is important to note here that the Hamiltonian is Hermitian but not self-adjoint (see Problem 1.29). Extending to the case where it is also self-adjoint requires an additional condition that is equivalent to fixing a position of one of the levels.

[182] For concreteness, we assume that the potential has the form $U = -\alpha/r^\nu$ in the entire space (and $\nu > 2$). We emphasize that the value of the lower integration limit a in Eq. (1) is not connected with a turning point, and can be chosen in an arbitrary way.

distances. Hence, for an arbitrary value of E and by a proper choice of γ, we can make the wavefunction decrease in the classically forbidden region for $r \to \infty$. Hence, naively, there is no quantization of the energy spectrum. A mathematical subtlety however exists due to the fact that the condition for the *self-adjoint extension* for the Hamiltonian operator here constrains values of parameter γ. This is seen by first "cutting" the potential with an impenetrable sphere at the origin of radius r_0 (as is given in the problem statement), and then proceeding to the limit of $r_0 \to 0$.

For small but finite r_0 we have the boundary condition $\chi(r_0) = 0$. Here, in Eq. (1) we should put $a = r_0$ and $\gamma = 0$. Then, as usual, we obtain the quantization rule:

$$\int_{r_0}^{b} \sqrt{2m\left[E_{n_r l} + \frac{\alpha}{r^\nu} - \frac{\hbar^2 l(l+1)}{2mr^2}\right]}\, dr = \pi\hbar\left(n_r + \frac{3}{4}\right) \qquad (2)$$

(see Problem 9.2).

The energy-level spectrum that follows from the quantization rule (2) and depends on a particular choice of r_0, and with the decrease of r_0 it has two features. First, the level with a given fixed value of n_r shifts down, and the corresponding $E_{n_r l} \to -\infty$ as $r_0 \to 0$. Secondly, in some energy domain $\tilde{E} < E < \tilde{E} + \Delta E$ (with $E < 0$), new levels appear with increasingly greater values of n_r, so that $n_r \to \infty$ as $r_0 \to 0$. Although the positions of these levels depend on r_0 and there is no limit for them as $r_0 \to 0$, the distance between the neighboring levels (inside the given domain) is fixed to $\hbar\omega(E_{n_r l})$, as is typical in the quasi-classical approximation.

Hence the energy spectrum in the problem for $E < 0$ is discrete. However, there must be additional conditions which uniquely fix positions of the levels (the potential itself does not provide that, in contrast to the case of $\nu < 2$). We can see that a position of a single level (for each value of the angular momentum, l) entirely determines the entire spectrum. Indeed, writing down an expression analogous to Eq. (2) with another value of the radial quantum number $\tilde{n}_r$ and subtracting one from the other, we obtain the relation (we can put $r_0 = 0$ and hence find a specific quantization rule determining the spectrum ($\varepsilon \to 0$)):

$$\int_{\varepsilon}^{b} \sqrt{2m\left[E_{nl} + \frac{\alpha}{r^\nu} - \frac{\hbar^2 l(l+1)}{2mr^2}\right]}\, dr - \int_{\varepsilon}^{\tilde{b}} \sqrt{2m\left[E_{0l} + \frac{\alpha}{r^\nu} - \frac{\hbar^2 l(l+1)}{2mr^2}\right]}\, dr =$$

$$= \pi\hbar n; \quad n = 0, \pm 1, \ldots. \qquad (3)$$

Here we have introduced the following notations: $E_{n_r l}$ is replaced by E_{nl}, and $E_{\tilde{n}_r l}$ by E_{0l}. We also used the quantum number $n = n_r - \tilde{n}_r$ that characterizes an ordering of the levels with respect to the fixed level E_{0l}. The fixed value of E_{0l} determines the entire spectrum. Note that the number of levels is infinite, because the values of n are not restricted from below and $E_{nl} \to -\infty$ as $n \to -\infty$, which corresponds to the case of "falling into the origin".

2) Let us discuss a connection of the quantization rule (3) with the corresponding restrictions on the wavefunctions (1). For this we emphasize that relation (3) follows from the wavefunction of the form

$$\chi_{El} = \frac{C}{\sqrt{p_n(r)}} \sin\left(\frac{1}{\hbar}\int_\varepsilon^b p_n dr - \frac{1}{\hbar}\int_\varepsilon^{\tilde{b}} p_0 dr - \frac{\pi}{4}\right) \approx_{r\to 0}$$

$$\tilde{C} r^{\nu/4} \sin\left[\frac{2\sqrt{\tilde\alpha}}{(\nu-2)r^{(\nu-2)/2}} + \tilde\gamma_l\right],$$

with $\varepsilon \to 0$. It is important here that the value of the phase, $\tilde\gamma_l$, does not depend on energy. Here the wavefunctions of all the states (for a given l) for $r \to 0$ have the same radial dependence which does not depend on energy and makes the term outside the integral sign in

$$\int_0^\infty \chi_2^* \hat H_l \chi_1 dr = \int_0^\infty (\hat H_l \chi_2)^* \chi_1 dr - \frac{\hbar^2}{2m}[\chi_2^* \chi' - \chi_2^{*\prime}\chi_1]\Big|_{r=\varepsilon\to 0}^\infty$$

equal to zero. This means that fixing[183] a value of $\tilde\gamma_l$ defines a self-adjoint extension of the Hermitian operator $\hat H$ (in the states with a definite angular momentum, l; see Problem 1.29).

3) The analysis given above can be applied to the case of $\nu = 2$ if we put the *Langer correction* into the centrifugal potential. Using

$$\int \sqrt{\frac{c^2}{r^2} - a^2}\, dr = \sqrt{c^2 - a^2 r^2} - \frac{c}{2}\ln\frac{c + \sqrt{c^2 - a^2 r^2}}{c - \sqrt{c^2 - a^2 r^2}}, \tag{5}$$

we obtain the quantization rule (3) in the form

$$a\ln\frac{E_{0l}}{E_{nl}} = 2\pi n, \quad a_l = \sqrt{\frac{2m\alpha}{\hbar^2} - \left(l + \frac{1}{2}\right)^2} > 0.$$

Hence, we have the explicit expression for the spectrum:

$$E_{nl} = E_{0l}\exp\left\{-\frac{2\pi n}{\alpha_l}\right\}, \quad n = 0, \pm 1, \pm 2, \ldots. \tag{6}$$

We should note that the number of levels is infinite both due to the existence of states with an arbitrarily large binding energy ($n \to -\infty$) and due to the condensation of levels for $E = 0$ (for $n \to +\infty$); the latter does not occur in the case

---

[183] Fixing $\tilde\gamma_l$ is equivalent to fixing a position of one of the levels. We emphasize that the same parameter, $\tilde\gamma_l$, determines the properties of both the discrete spectrum for $E < 0$ and the continuous spectrum for $E > 0$. We should note also that if we consider a smooth regularization of the potential for $r < r_0$ of the form $U(r) = U(r_0)$ (instead of the impenetrable wall), then we obtain $\tilde\gamma_l = \tilde\gamma_0 + \pi l/2$ for the $\tilde\gamma_l$ dependence on l; see Problem 9.7.

of $\nu > 2$. We also write the expression for the wavefunctions for small distances. Using Eqs. (5) and (6), and according to Eq. (4), we find

$$\chi_{E_n l}(r) \approx C_n \sqrt{r} \sin\left\{\alpha_l \left[\ln\left(\frac{\kappa_{0l} r}{2\alpha_l}\right) + 1\right] - \frac{\pi}{4}\right\}. \tag{7}$$

There is no phase-dependence on energy for $r \to 0$.

Finally, for the potential, $U = -\alpha/r^2$, the Schrödinger equation has an exact solution. Here,

$$\chi_{El} \approx C\sqrt{r} K_{s\alpha_l}(\kappa r), \tag{8}$$

where K_ν is the MacDonald function and the exact spectrum coincides with the quasi-classical one (6). Indeed, for $r \to 0$ the radial function (8) takes the form

$$\chi_{El}(r) \approx C|\Gamma(i\alpha_l)|\sqrt{r}\cos\left\{\alpha_l \ln\frac{kr}{2} + \arg\Gamma(-i\alpha_l)\right\}. \quad k = \frac{1}{\hbar}\sqrt{-2mE}.$$

From condition (7) it follows that

$$\alpha_l \ln\frac{k_{nl}}{k_{0l}} = -\pi n, \quad \text{or} \quad E_{nl} = E_{0l} \exp\left(\frac{2\pi n}{\alpha_l}\right), \quad n = 0, \pm 1, \pm 2, \ldots,$$

which coincides with (6).

9.2 Quasi-classical wavefunctions, probabilities, and mean values

Problem 9.15

Obtain the quasi-classical wavefunction in the momentum representation in the regime, where the particle momentum takes its typical values. Find the momentum distribution of a particle in a discrete stationary state. Determine a classical interpretation of the result.

Solution

1) Let us first consider the quasi-classical wavefunction in the form of a traveling wave:

$$\psi^{(\pm)}(x) = \frac{C}{\sqrt{p(x)}} \exp\left\{\pm\frac{i}{\hbar}\int_a^x p(x')dx'\right\},$$

$$p(x) = \sqrt{2m(E - U(x))}.$$

The corresponding wavefunctions in momentum representation are given by

$$\phi^{(\pm)}(p) = \frac{C}{\sqrt{2\pi\hbar}} \int \exp\left\{\frac{i}{\hbar}\left[\pm\int_a^x p\,dx' - px\right]\right\} \frac{dx}{\sqrt{p(x)}}. \tag{1}$$

(Do not confuse p as the variable of representation with $\pm p(x)$ – momentum of the classical particle.) The characteristic property of the integral in (1) for the quasi-classical states is that the phase, $\varphi(x)$, in the exponential varies rapidly as a function of x, and the value of the integral is determined mainly by the contribution near the stationary points of the phase. Let us denote the positions of these points by[184] $x_s^{\pm}(p)$. They are determined from the condition

$$\frac{\partial \varphi^{\pm}(x)}{\partial x} = 0, \quad \text{or} \quad \pm p(x_s^{\pm}) = p. \tag{2}$$

Expanding $\varphi^{\pm}(x)$ in the vicinities of the points, $x_s^{\pm}(p)$, we have

$$\varphi^{\pm}(x) \approx \varphi^{\pm}(x_s^{\pm}) \pm \frac{m\omega(x_s^{\pm})}{2\hbar v(x_s^{\pm})}(x - x_s^{\pm})^2, \tag{3}$$

where $\omega = -U'/m$ and v are the acceleration and velocity of the classical particle at the corresponding points, $x_s^{\pm}$. Factoring out the slowly varying $p^{-1/2}(x)$ outside the integral sign (i.e., replacing it by its value at the corresponding stationary point) and calculating the integrals using the expansions[185] (3) we find:

$$\phi^{\pm}(p) \approx \sum_s \frac{C}{m\sqrt{\mp i\omega(x_s)}} \exp\left\{\frac{i}{\hbar}\left[\pm \int_a^{x_s} p\,dx - px_s\right]\right\}. \tag{4}$$

We should emphasize that all the points on a classical trajectory with momentum p contribute to the sum (here $\phi^+ = 0$ for the values $p < 0$, while $\phi^- = 0$ for $p < 0$).
2) Let us now consider the wavefunction, $\psi_n(x)$, of a stationary state in the potential with a single minimum, as shown in Fig. 9.1. Writing the sine function in Eq. (IX.6) in terms of exponents, and using (4), we find for $p > 0$,

$$\phi_n(p) = \sqrt{\frac{-i}{mT(E_n)}} \left\{ \frac{\exp\{i\varphi(x_1(p))\}}{\sqrt{\omega(x_1(p))}} + \frac{\exp\{i\varphi(x_2(p))\}}{\sqrt{\omega(x_2(p))}} \right\}, \tag{5}$$

while $\phi_n(p)$ for $p < 0$ can be obtained by complex conjugation of (5), calculated with the momentum fixed to $|p|$. Wavefunction (5) is different from zero only for the following values of momentum:

$$0 \leq p \leq p_0 = \sqrt{2m(E_n - \min U)}.$$

Here we have taken into account that relation (2) has two roots on both sides of the minimum points of $U(x)$, which merge together as $p \to p_0$ (at the minima of $U(x)$, Eq. (2) has no roots for the larger values of p).

---

[184] If there are no such points, then $\phi^{\pm}(p) = 0$. In the classically forbidden domain, where the wavefunction decreases exponentially, we can consider $\psi = 0$.

[185] Although the integration is performed over narrow regions around the points, $x_s^{\pm}$, due to its fast convergence we can extend it to infinity limits; the integrals obtained are the Poisson integral.

The momentum distribution, $dW_n(p) = |\phi_n(p)|^2 dp$, due to the existence of the rapidly oscillating exponents in (5), rapidly oscillates as well (compare to the osculations of $|\psi_n(x)|^2$). But if we average this distribution over a small momentum interval, the interference term disappears and we have

$$d\overline{W}_n(p) = \left(\frac{1}{\omega(x_1(p))} + \frac{1}{\omega(x_2(p))}\right) \frac{dp}{mT(E_n)}, \quad -p_0 \le p \le p_0 \tag{6}$$

(here $\omega > 0$). This expression has a simple classical interpretation. Indeed, substituting into $dW_{cl} = dt/T(E)$, the time of motion along the trajectory dt by the particle momentum change

$$dt = \left(\frac{dt}{dp}\right) dp = \pm \frac{dp}{m\omega(p)},$$

and taking into account the two-valuedness of $\omega(p)$ with respect to p, we obtain the classical momentum distribution that coincides with relation (6) (compare this to the classical probability $dW_{cl} = 2dx/Tv(x)$ for the particle coordinates).

Problem 9.16

For the stationary states of a discrete spectrum, determine the probability of finding the particle in the classically forbidden region. Apply the result obtained to a linear oscillator.

Solution

The main contribution to the desired probability comes from the regions that adjoin the turning points where the quasi-classical property is broken. Let us consider the exact solution of the Schrödinger equation based on the linear approximation of the potential near these points. In the vicinity of the right turning point, $x = b$, the Schrödinger equation takes the form:

$$\psi''(x) - \frac{2m|F(b)|}{\hbar^2}(x-b)\psi(x) = 0, \tag{1}$$

where $|F(b)| = U'(b)$ and $E = U(b)$. With the substitution

$$z = \left(\frac{2m|F(b)|}{\hbar^2}\right)^{1/3}(x-b)$$

we transform (1) into the form $\psi''(z) - z\psi(z) = 0$. The solution of this equation which decreases as one goes deeper into the classically forbidden region is expressed in terms of the Airy function, i.e., $\psi = c\mathrm{Ai}(z)$. Its asymptotes are[186]

---

[186] Since in this case $\zeta = \hbar^{-1}\left|\int_b^x p(x)dx\right|$, (2) reproduces the matching conditions given in Eqs. (IX.3,4) (the region $|\zeta| \gg 1$ is already the quasi-classical region).

$$\text{Ai}(z) \approx \begin{cases} \frac{1}{2\pi^{1/2}z^{1/4}}e^{-\zeta}, & z \gg 1 \\ \frac{1}{\pi^{1/2}|z|^{1/4}}\sin(\zeta+\frac{\pi}{4}), & -z \gg 1 \end{cases} \quad (2)$$

where $\zeta = 2|z|^{3/2}/3$. By matching the solution with the quasi-classical wavefunction (IX.6) normalized to unity, we find

$$c^2 = 2\pi \left(\frac{4m^2}{|F|\hbar}\right)^{1/3} T^{-1}(E_n),$$

and for the probability of the particle being in the classically forbidden region on the right of the turning point, $x = b$, we have

$$w_1 = \int_b^\infty |\psi(x)|^2 dx \approx \frac{2\pi}{T(E_n)} \left(\frac{2m\hbar}{F^2(b)}\right)^{1/3} \int_0^\infty (\text{Ai}(z))^2 dz.$$

Now using a relation between the Airy and MacDonald functions, $\text{Ai}(z) = \sqrt{z/3\pi^2}K_{1/3}(\zeta)$, where $\zeta = 2|z|^{3/2}/3$ and using $\int_0^\infty z^{-\nu}K_\nu^2(z)dz$, we find the desired probability

$$w = w_1 + w_2 = \frac{3^{1/3}\Gamma^2(2/3)}{2\pi T(E_n)} \left[\left(\frac{2m\hbar}{F^2(a)}\right)^{1/3} + \left(\frac{2m\hbar}{F^2(b)}\right)^{1/3}\right], \quad (3)$$

where for the potential given in Fig. 9.1, both of the classically forbidden regions are taken into account; $\Gamma(z)$ is the Γ-function, and $\Gamma(2/3) \approx 1.354$.

For the oscillator $U = m\omega^2 x^2/2$, $E_n = \hbar\omega(n+1/2)$, and we have

$$F^2(a) = F^2(b) = 2m\hbar\omega^3\left(n+\frac{1}{2}\right),$$

and according to (3) we obtain

$$w_n \approx 0.134\left(n+\frac{1}{2}\right)^{-1/3}. \quad (4)$$

As usual, a quasi-classical result remains accurate for $n \sim 1$ as well. For the ground state and the first excited state, it follows from (4) that $w_0 \approx 0.169$ and $w_1 \approx 0.117$, while the true values of these probabilities are equal to 0.157 and 0.112.

Problem 9.17

Using the quasi-classical approximation, find the mean value of a physical quantity $F(x)$ that is a function only of the coordinate of a particle in the nth stationary state of a discrete spectrum. As an illustration of the result, calculate the mean values $\overline{x^2}$ and $\overline{x^4}$ for the linear oscillator and compare them with the exact values.

Solution

Using the approximation (IX.6) for the wavefunction, we obtain

$$F_{nn} = \frac{2}{T(E_n)} \int_a^b \frac{F(x)dx}{v(x)} = \frac{\sqrt{2m}}{T(E_n)} \int_a^b \frac{F(x)dx}{\sqrt{E_n - U(x)}}, \quad (1)$$

so the quantum mechanical mean value coincides with the corresponding mean value over a period of motion in classical mechanics.

In the case of the linear oscillator, $U(x) = m\omega^2 x^2/2$. By using (1), we then obtain

$$\overline{x^2} = a^2\left(n + \frac{1}{2}\right), \quad \overline{x^4} = \frac{3}{2}a^4\left(n^2 + n + \frac{1}{4}\right), \quad (2)$$

where $a^2 = \hbar/m\omega$ (the integrals are calculated by using the substitution $x = \sqrt{2E_n/m\omega^2}\sin\varphi$). The value of $\overline{x^2}$ coincides with the exact result; the quasi-classical mean value, $\overline{x^4}$, differs from the true result by the last term $1/4$ (in the brackets) instead of $1/2$. These results are within the quasi-classical accuracy of relation (1) that provides the correct values of the first two terms in the expansion in $1/n$ of the considered physical quantity; see Problems 9.10 and 9.13.

Problem 9.18

In the quasi-classical approximation, find the mean value of a physical quantity $F(p)$ that is a function only of the momentum of a particle in the nth stationary state of a discrete spectrum. As an illustration of the result, calculate the mean values $\overline{p^2}$ and $\overline{p^4}$ for the linear oscillator and compare with the exact results.

Solution

Using the approximation for the wavefunction (IX.6) we express the sine function in terms of exponents. We see that

$$\hat{p}^k \frac{1}{\sqrt{p(x)}} \sin\left(\frac{1}{\hbar}\int_a^x p(x)dx\right) \approx$$

$$\frac{1}{2i\sqrt{p(x)}}\left[p^k(x)\exp\left(\frac{i}{\hbar}\int_a^x p(x)dx\right) - (-p(x))^k \exp\left(-\frac{i}{\hbar}\int_a^x p(x)dx\right)\right]. \quad (1)$$

Due to the quasi-classical property it is necessary to differentiate only the exponents, since they are the most rapidly-varying terms. By a relation analogous to (1), with the substitution $(\pm p(x))^k \to F(\pm p(x))$, action of the operator, $\hat{F} \equiv F(\hat{p})$, on the wavefunction is described. Now the calculation of the mean value of F_{nn} is reduced to the calculation of four integrals. Two of them include rapidly oscillating terms,

$\exp[(\pm 2i/\hbar) \int p dx]$, hence they are small and can be neglected. As a result we obtain

$$F_{nn} = \frac{2}{T(E_n)} \int_a^b \frac{F(p(x)) - F(-p(x))}{v(x)} dx = \frac{1}{T(E_n)} \oint \frac{F(p(x))dx}{v(x)}, \quad (2)$$

which coincides with the mean value over the period of classical motion.[187]

For the oscillator, according to (2), we have

$$\overline{p^2} = m\hbar\omega \left(n + \frac{1}{2}\right), \quad \overline{p^4} = \frac{3}{2}(m\hbar\omega)^2 \left(n^2 + n + \frac{1}{4}\right). \quad (3)$$

The value of $\overline{p^2}$ coincides with the exact one, while $\overline{p^4}$ differs from exact result only by corrections which are subleading to $n^2 + n$: in the exact result it is equal to $1/2$ instead of $1/4$; see the comment in the previous problem for the accuracy of the quasi-classical approximation while calculating mean values.

Problem 9.19

For the quasi-classical states of a discrete spectrum, indicate the order of magnitude for the product of the uncertainties, $\Delta p \cdot \Delta x$. Compare this estimate to the exact value of $\sqrt{\overline{(\Delta p)^2} \cdot \overline{(\Delta x)^2}}$ for the linear oscillator and for a particle in an infinitely deep potential well.

Solution

The estimation can be obtained by using the quantization rule (IX.5). Taking into account the fact that the integral in this expression is order-of-magnitude equal to $(b-a)p_{ch}$, where p_{ch} is the characteristic value of the particle momentum, and that $\Delta x \sim (b-a)/2$ and $\Delta p \sim p_{ch}$ (as $\overline{p} = 0$), we obtain

$$\Delta p \cdot \Delta x \sim \frac{\pi}{2}\hbar \left(n + \frac{1}{2}\right). \quad (1)$$

(Here we keep $1/2$ on top of the leading term n to use this estimation for $n \sim 1$, and also for $n = 0$.) The exact value of $\sqrt{\overline{(\Delta p)^2} \cdot \overline{(\Delta x)^2}}$ for the linear oscillator is equal to $(n+1/2)\hbar$ (for all values of n), while for a particle in an infinitely deep potential well the value is equal to $\pi n\hbar/2\sqrt{3}$ in the case $n \gg 1$.

To avoid misunderstandings it should be noted that the obtained estimate, $\hbar^{-1}\Delta x \cdot \Delta p \sim n$, for $n \gg 1$ corresponds to the stationary quasi-classical states whose wavefunctions are essentially different from zero in the whole region of classical motion. For localized wave packets created from a large number of quasi-classical states there is only a lower bound for the product, $\Delta x \cdot \Delta p$, that is determined by the *uncertainty*

---

[187] We can write it in the form, $F_{nn} = \int F(p)dW_n(p)$, where the momentum probability distribution is determined by the result from 9.15.

relation. Compare, for example, to the case of the coherent states of the harmonic oscillator. See Problem 6.21.

Problem 9.20

Using the quasi-classical approximation obtain the matrix elements, F_{mn}, for an operator of the form $\hat{F} = F(x)$ in the case where $|m - n| \sim 1$, *i.e.*, for closely-positioned levels of a discrete spectru. Find a relation between them and the Fourier components, $\tilde{F}_s$, of the function $F(x(t))$ in classical mechanics:

$$F(x(t)) = \sum_{s=-\infty}^{\infty} \tilde{F}_s e^{is\omega t}, \quad \tilde{F}_s = \frac{1}{T} \int_0^T F(x(t)) e^{-is\omega t} dt,$$

where $T(E) = \frac{2\pi}{\omega}$ is the period of motion of a classical particle with energy $E = \frac{1}{2}(E_m + E_n)$. As an illustration of the result, compute the matrix elements, x_{mn} and $(x^2)_{mn}$, for the linear oscillator and compare with the exact ones.

Solution

The main contributions to the matrix elements come from the integration domain between the turning points. We can use Eq. (XI.6) for the wavefunction in this calculation. Due to the expected proximity of the states, n and m, locations of the turning points and the frequency of motion for them can be assumed the same and calculated as follows:

$$F_{mn} = \int \psi_m^* \hat{F} \psi_n dx \approx \frac{\omega}{\pi} \int_a^b \frac{F(x)}{v(x)} \cos\left[\frac{1}{\hbar} \int_a^x (p_m - p_n) dx'\right] dx -$$

$$\frac{\omega}{\pi} \int_a^b \frac{F(x)}{v(x)} \cos\left[\frac{1}{\hbar} \int_a^x (p_m + p_n) dx'\right] dx. \quad (1)$$

The function under the integral in the second term oscillates rapidly (due to $m, n \gg 1$), hence its value is negligible. Taking into account the fact that in the quasi-classical approximation the difference between the neighboring energy levels is equal to $\hbar\omega$, we find

$$p_m(x) - p_n(x) \approx \hbar\omega \frac{m-n}{v(x)},$$

and then we obtain

$$F_{mn} \approx \frac{\omega}{\pi} \int_a^x \frac{F(x)}{v(x)} \cos\left[\omega(m-n) \int_a^x \frac{dx'}{v(x')}\right] dx. \quad (2)$$

Due to the time-dependence of the coordinates for a classical particle, $x = x(t)$, which moves with an energy[188] $E = (E_m + E_n)/2$, we make a substitution $t = t(x)$ in (2) and obtain

$$F_{mn} = \frac{2}{T} \int_0^{T/2} F(x(t)) \cos[\omega(m-n)t] dt,$$

$$F_{mn} = \frac{1}{T} \int_0^T F(x(t)) e^{-i\omega(m-n)t} dt = \tilde{F}_{s=m-n}. \qquad (3)$$

(The initial moment of time is chosen so that $x(0) = a$; and the change in x from a to b and back corresponds to the change of t from 0 to $T(E) = 2\pi/\omega$.) (3) gives the desired relation between the quantum-mechanical matrix elements and Fourier components in classical mechanics.

Let us analyze the accuracy of relation (3) for the case of the linear oscillator coordinates. Here $x(t) = A\cos\omega t$, so that, according to (3), only the Fourier components, $\tilde{x}_1 = \tilde{x}_{-1} = A/2$, differ from zero. Since the classical oscillator energy is equal to $E = m\omega^2 A^2/2$, then equating it to $(E_n + E_m)/2$ and taking into account that $E_n = \hbar\omega(n+1/2)$, we find A^2 and the non-vanishing matrix elements of the oscillator coordinates in the classical approximation,

$$x_{n+1,n} = x_{n,n+1} = \sqrt{\frac{\hbar}{2m\omega}(n+1)},$$

which coincides with the exact result. In the same way for the square of the coordinate, $x^2(t) = A^2 \cos^2(\omega t)$, we find that only the following Fourier components are different from zero: $(\tilde{x}^2)_0 = 2(\tilde{x}^2)_2 = 2(\tilde{x}^2)_{-2} = A^2/2$. Using the relation between A^2 and $E_{m,n}$ mentioned above (note that the values of A^2 for the cases $m = n$ and $m \neq n$ are different) we find that, according to (3), the non-vanishing matrix elements are

$$(x^2)_{nn} = a^2\left(n + \frac{1}{2}\right), \quad (x^2)_{n,n+2} = (x^2)_{n+2,n} = \frac{1}{2}a^2\left(n + \frac{3}{2}\right), \qquad (4)$$

where $a^2 = \hbar/2m\omega$. The true value, $(x^2)_{nn}$, coincides with (4), while for $(x^2)_{n,n+2}$ it differs from (4) by the substitution, $n + 3/2 \to [(n+1)(n+2)]^{1/2}$. We see that the quasi-classical formula gives a high enough accuracy even for $n, m \sim 1$.

Problem 9.21

Generalize the result of the previous problem for the case of an operator of the form, $\hat{F} = F(\hat{p})$. Apply it to calculate the matrix elements of the operators $\hat{p}$ and $\hat{p}^2$ for an oscillator.

---

[188] The use of the physically meaningful quantity describing the average energy ensures the hermiticity of the corresponding matrix elements, $F_{mn} = F_{nm}^*$.

Solution

The problem can be solved in a way similar to the previous one. We should take into account that the action of the operator, $\hat{p}$, on a quasi-classical function it is sufficient to differentiate the latter over x only in the argument of the sine (see Problem 9.18). After simple transformations we obtain

$$(p^k)_{mn} = \frac{2}{T} \int_0^{T/2} dt \, p^k(t) \cdot \begin{cases} \cos[\omega(m-n)t], & k \text{ is even,} \\ -i\sin[\omega(m-n)t], & k \text{ is odd.} \end{cases} \quad (1)$$

Since signs of the particle momenta for $0 < t < T/2$ and $T/2 < t < T$ are opposite, both relations (1) could be combined into one:

$$(p^k)_{mn} = \frac{1}{T} \int_0^T p^k(t) e^{-i\omega(m-n)t} dt \equiv \left(\widetilde{p^k}(t)\right)_{s=m-n}. \quad (2)$$

We see that the analogous relation, $F_{mn} = \overline{F}_{s=m-n}$, between the quasi-classical matrix elements and the classical Fourier components exists for an arbitrary physical quantity[189] $F(p)$, Eq. (3) in see Problem 9.20.

The matrix elements, p_{mn} and $(p^2)_{mn}$, for the oscillator momentum are calculated in the same way as in the previous problem for the coordinate, and are described by the same expressions with the substitution, $\hbar/m\omega \to \hbar m\omega$ (there also appear some unessential phase factors). Such analogy is not accidental, reflects the fact that the Schrödinger equation and the wavefunction for an oscillator in coordinate and momentum representations have similar forms.

Problem 9.22

A particle is in the nth stationary state in a potential $U(x)$. Suddenly (at $t=0$), potential energy becomes equal to $U(x) + V(x)$. What are the particle's mean energy and its fluctuations at $t > 0$?

Assuming that $n \gg 1$ and the potential change is large enough[190] so that $|V'_{char}(x)| \times (b-a) \gg \hbar\omega_n$, where $T_n = 2\pi/\omega_n$ is the period of classical particle motion in the initial state, find the probabilities of its transitions to new stationary states. In what case can the system "ionize"? Provide an interpretation of the results using classical mechanics. As an illustration, consider the linear oscillator, $U = m\omega^2 x^2/2$, with a sudden appearance of an homogeneous field $V = -F_0 x$.

---

[189] It is fulfilled in the quasi-classical approximation for an arbitrary physical quantity $F(x,p)$.
[190] Physically, this condition means that in the initial state represents a large number of states of the final Hamiltonian H_f.

Solution

1) Due to the instantaneous nature of the potential change, wavefunction coincides with $\psi_{i,n}$ immediately after the potential, $V(x)$, turns on. (Here and below, i and f correspond to the stationary states of the particle in the potentials $U(x)$ and $U(x) + V(x)$; if it does not lead to misunderstandings we will omit them.) For the desired mean values we obtain (for $t > 0$)

$$\overline{E} = E_n + \overline{V(x)}, \quad \overline{(\Delta E)^2} = \overline{V^2(x)} - \overline{(V(x))}^2. \tag{1}$$

(After $V(x)$ turns on, the Hamiltonian does not depend on time again, so the energy is conserved.) In the general case, the quantum-mechanical averaging of (1) is performed over the state with the wavefunction, $\psi_n(x)$, but in the quasi-classical approximation it can be replaced by a more simple averaging over a period of finite motion of a classical particle with energy E_n in potential $U(x)$ (see Problem 9.17).

2) The transition probability after the sudden change in the potential is (see Problem 8.47)

$$w(n \to k) = \left| \int \psi_{f,k}^*(x) \psi_{s,n}(x) dx \right|^2. \tag{2}$$

Here, using the relations of the form (IX.6) for the wavefunctions (for the potentials given above), we transform the matrix elements in the following way:

$$\langle k, f | n, i \rangle = \frac{m}{\sqrt{T_f(E_k) T_i(E_n)}} \int_a^b \left\{ \exp\left[\frac{i}{\hbar} \left(\int_{a_f}^x p_k dx' - \int_{a_i}^x p_n dx' \right) \right] - \right.$$

$$\left. \exp\left[\frac{i}{\hbar} \left(\int_{a_f}^x p_k dx' + \int_{a_i}^x p_n dx' \right) + i\frac{\pi}{2} \right] + c.c. \right\} \frac{dx}{\sqrt{p_k(x) p_n(x)}}, \tag{3}$$

where c.c. means the complex conjugate, $a = \max(a_i, a_f)$, $b = \min(b_i, b_f)$; $a_{i,f}$ and $b_{i,f}$ are the turning points in the quasi-classical states, $\psi_{i,n}$, $\psi_{f,k}$.

A notable feature of the integrals in (3) are the large and rapidly changing values of the phases in the exponent. Such integrals are essentially different from zero only when the phases as functions of x have stationary points, and the integration in the vicinity of them gives the main contribution to the integrals. We can see that there are no such points in the second exponent in (3) and its complex conjugate (by definition, $p_{n,k} > 0$), and these terms can then be omitted. For the first exponent, the saddle-point condition, $\partial\varphi/\partial x = 0$, takes the form[191] $p_{f,k}(x_s) = p_{i,n}(x_s)$, i.e.,

$$\sqrt{E_{i,n} - U(x_s)} = \sqrt{E_{f,k} - U(x_s) - V(x_s)}, \text{ or } E_{f,k} = E_{i,n} + V(x_s). \quad (4)$$

Hence we see that the energy interval of allowed transitions is bounded by conditions

$$V_{min} < E_{f,k} - E_{i,n} < V_{max},$$

where $V_{max(min)}$ is the maximum (minimum) value of $V(x)$ on the interval $a_i < x < b_i$ for a classical particle in the initial state. We should note that condition (4) has a clear physical meaning. Due to the instantaneous potential change, the classical particle momentum does not change. Therefore, (4) gives the trajectory points in phase space (p, x) which are the same for the initial and final states.

To calculate the integral we expand the phase of the exponent in the vicinity of a stationary point:

$$\frac{1}{\hbar}\left(\int_{a_f}^{x} p_k dx - \int_{a_i}^{x} p_n dx\right) \approx \varphi_k(x_s) + \frac{1}{2\hbar}\left(\frac{F_f(x_s)}{v_k(x_s)} - \frac{F_i(x_s)}{v_n(x_s)}\right)(x - x_s)^2 =$$

$$\varphi_k(x_s) - \frac{mV'(x_s)}{2\hbar p_n(x_s)}(x - x_s)^2. \quad (5)$$

Now we should note that due to the assumed condition, $|V'|(b-a) \gg \hbar\omega_n$, the range of values of $(x - x_s)$, where the phase change is of order 1 and which gives the main contribution to the integral, is small with respect to the characteristic region of particle motion, $\sim (b-a)$. Hence in (5) we can restrict ourselves to the quadratic term in the expansion and expand the integration in the vicinity of each stationary point, x, to the whole domain (due to rapid integral convergence) and obtain

$$\int \psi_{f,k}^*(x)\psi_{i,n}(x)dx = \sqrt{\frac{2\pi m\hbar}{T_s T_f}} \sum_s \left\{\frac{\exp\{i\varphi_k(x_s)\}}{\sqrt{iV'(x_s)p_n(x_s)}} + c.c.\right\}. \quad (6)$$

Due to the large value of phases $\varphi_k(x_s)$ in (6), probability values (2), as well as the value of matrix element (6) itself, are subject to strong oscillation even for a small change in the number k of the final state. But after the averaging over a small interval of final states, such oscillations disappear and we obtain the following expression

$$\bar{w}(n \to k) = \frac{4\pi m\hbar}{T_s(E_n)T_f(E_k)} \sum_s \frac{1}{p_{i,n}(x_s)|V'(x_s)|}. \quad (7)$$

---

[191] A number of stationary points x depends on both the form of the potentials and the values of $E_{n,k}$. If $V(x)$ is a monotonic function of x, then there is one point or there are no points at all; in the case when $U(x)$ and $V(x)$ are symmetric functions (with one minimum), there are two or no such points. The absence of stationary points implies that the probability of the corresponding transition is negligible.

3) Let us now find the probability of excitations of final states k that correspond to some energy interval, dE_f. Since the distance between the neighboring levels in the quasi-classical approximation is equal to $\Delta E_k = \hbar\omega_f(E_k)$, the desired probability is obtained by multiplying (7) by $dk = dE_f/\Delta E_k$ – the number of states in this interval – and is described by the expression

$$dw_f = \frac{2m}{T_i(E_n)} \sum_s \frac{dE_f}{p_{i,n}(x_s)|V'(x_s)|}. \qquad (8)$$

There is no Planck's constant in this equation and it can be interpreted within the framework of classical mechanics. We see that the classical particle's energy after $V(x)$ has turned on depends on the coordinate, x, of the particle, and is equal to $E_f = E_i + V(x)$. Hence the probability distribution for E_f is determined by the coordinate distribution for a particle, in the initial state, which has the form

$$dw = \frac{2dx}{T(E_i)v_i(x)}.$$

Now, if we turn from variable x to energies E_f, and take into account the multi-valued nature of x as a function of E_f, we obtain (8), which confirms that this distribution is normalized to unity and hence the same for the probability distribution (7).

If, for a particle's energy $E \geq \tilde{E}_0$, the motion in the potential $U(x) + V(x)$ is unbounded, then equation (8) for $E_f > \overline{E_0}$ gives an energy distribution for the particles escaping from the well, while the integration over the energy E_f within the limits, $\overline{E_0}$ and $E_{max} = E_{i,n} + \max V$, gives the full escape probability after the change in potential (system ionization).[192]

4) When applying the results to an oscillator we have

$$E_{i,n} = \hbar\omega\left(n + \frac{1}{2}\right), \quad E_{f,k} = \hbar\omega\left(k + \frac{1}{2}\right) - \frac{F_0^2}{2m\omega^2}.$$

According to Eq. (4) we find

$$x_s = \frac{[\hbar\omega(n-k) + F_0^2/2m\omega^2]}{F_0}.$$

Then, according to Eq. (7), we obtain the probability of transitions:[193]

$$\bar{w}(n \to k) = \frac{\hbar\omega}{\pi}\left\{\frac{2\hbar F_0^2(n+1/2)}{m\omega} - \left[\hbar\omega(n-k) + \frac{F_0^2}{2m\omega^2}\right]^2\right\}^{-1/2}.$$

---

[192] This fact becomes evident if we take into account the analogy with classical mechanics, mentioned above. If we consider it quasi-classically we should take into account the change of the wavefunction, $\psi_{f,k}$, which for $E_k > \tilde{E}_0$ describes states of the continuous spectrum.

[193] We see that $\bar{w}(n \to k) = \bar{w}(k \to n)$.

(They are different from zero only for the values of k for which the expression in the square root is positive.) Eq. (8) here can be written in a more illuminating way:

$$dw_f = \frac{dE_f}{\pi\left[(E_f - E_{f,min})(E_{f,max} - E_f)\right]^{1/2}},$$

where $E_{f,max(min)} = E_{i,n} \pm a_n F_0$ is the maximum (minimum) energy that the oscillator can have for $t > 0$ within the stated conditions of the problem. Here, $a_n = \sqrt{2\hbar(n + 1/2)/m\omega}$ is the amplitude of a classical oscillation for $t < 0$.

9.3 Penetration through potential barriers

Problem 9.23

Using the quasi-classical approximation, determine the penetrabilities of the following potential barriers:

a) a triangular barrier (see Problem 2.36);
b) $U(x) = U_0 \cosh^{-2}(x/a)$;
c) the barrier from Problem 2.35.

Indicate the conditions for the applicability of these results, and compare them to the exact values of $D(E)$.

Solution

By using Eq. (IX.9) we obtain the following results.

a)
$$D(E) = \exp\left\{-\frac{8\sqrt{2m|E|^3}}{3\hbar F_0}\right\}, \tag{1}$$

which for large values in the exponent, when $D \ll 1$, coincides with the exact result for the penetrability through a triangular barrier (see Problem 2.36).

b) In the case of potential, $U(x) = U_0/\cosh^2(x/a)$, using the substitution,

$$\sinh(x/a) = \kappa \sin t, \quad \kappa = \sqrt{U_0/E - 1},$$

for the integral, we find the barrier penetrability in the quasi-classical approximation:

$$D(E) = \exp\left\{-\frac{2\pi}{\hbar}\sqrt{2ma^2}\left(\sqrt{U_0} - \sqrt{E}\right)\right\}; \tag{2}$$

while its exact value is

$$D(E) = \frac{\sinh^2(\pi ka)}{\sinh^2(\pi ka) + \cosh^2\left(\pi\sqrt{2mU_0a^2/\hbar^2 - 1/4}\right)}, \quad k = \frac{1}{\hbar}\sqrt{2mE}.$$

As we see, while the conditions

$$2\pi\sqrt{2ma^2/\hbar^2}\left(\sqrt{U_0}-\sqrt{E}\right)\gg 1 \text{ and } E\gg \hbar^2/ma^2 \tag{3}$$

are fulfilled, the quasi-classical result and the exact one are in close agreement. The first of the conditions in (3), which provides the value $D(E)\ll 1$, is a usual condition for the applicability of the quasi-classical approximation when computing barrier penetrability. As for the second one, we should make the following remark. Its origin lies in the fact that while deriving Eq. (IX.9) we used the quasi-classical matching conditions in the vicinity of classical turning points, based on the linear approximation of the potential. But in problems similar to this one (when $U \to 0$ for large distances), the use of such conditions is not valid for small particle energies.[194] This is seen from the case of $E=0$, where for large distances the quasi-classical method is not applicable; see Problems 9.8 and 9.9. The change in the solution matching conditions leads to a modification of the classical equation (IX.9), and causes the appearance of an additional factor before the exponent, which is of order unity. For this problem, as is seen from the relations for $D(E)$, the factor is equal to $4\sinh^2(\pi k\alpha)e^{-2\pi k\alpha}$ and is especially important for $E\to 0$, since there it is $\propto E$ (but with the increase in energy it approaches unity).

c) For the given potential,

$$D(E)\sim \exp\left\{-\frac{4a\sqrt{2m}}{3\hbar U_0}(U_0-E)^{3/2}\right\}. \tag{4}$$

This expression determines only an order of magnitude for transmission coefficient. However, it shows an important fact of its exponential smallness. In the exact value there appears an additional factor, equal to $4\sqrt{E(U_0-E)}/U_0$, in front of the exponent. Its appearance is connected with the change in the classical matching conditions near the turning point $x=0$ in comparison with those which lead to Eq. (IX.9) (compare to the barrier penetrability from b) for $E\to 0$).

A calculation of the factor in front of the exponent in the quasi-classical relations for barrier penetrability, when the matching conditions (based on the linear potential approximation at the turning points) are not applicable, are discussed in Problems 9.24–9.26.

Problem 9.24

Obtain a pre-exponential multiplicative factor in the quasi-classical expression for the transmission coefficient in the case of a potential barrier of the form shown in Fig. 9.4a) for the particle energy given there. Apply the result obtained to the potential barrier considered in Problem 2.35, and compare with the exact one (see also Problem 9.23c)).

---

[194] Except for the case of slowly decreasing potentials, $U\propto |x|^{-\nu}$ with $\nu < 2$ for $x\to \pm\infty$.

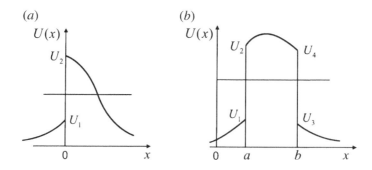

Fig. 9.4

Solution

Proceeding in the same way as when deriving the quasi-classical equation (IX.9) for the barrier penetrability, and considering the falling particle moving from the left to the right, we can write the wavefunction for $x > b$ in the form

$$\psi(x) = \frac{C}{\sqrt{p(x)}} \exp\left(\frac{i}{\hbar}\int_b^x p(x')dx' + \frac{i\pi}{4}\right). \tag{1}$$

The wavefunction in the barrier region is given by

$$\psi(x) = \frac{C}{\sqrt{|p(x)|}} \exp\left(\frac{1}{\hbar}\int_x^b |p(x')|dx'\right), \quad 0 < x < b. \tag{2}$$

(Here we omit the term that dies out inside the barrier.) It is important here that this quasi-classical expression is valid up to the turning point,[195] $x = 0$, where the potential has a jump. The wavefunction for $x \leq 0$ has the form

$$\psi(x) = \frac{1}{\sqrt{p(x)}} \left\{ \exp\left(\frac{i}{\hbar}\int_0^x p(x')dx'\right) + A\exp\left(-\frac{i}{\hbar}\int_0^x p(x')dx'\right) \right\}. \tag{3}$$

For such a normalization, the transmission coefficient for the barrier is $D = |C|^2$.

Matching the wavefunctions (2) and (3) at the point $x = 0$ (using the continuity of the wavefunction and its derivative, and due to the quasi-classical property, it is enough to differentiate only the exponential factors as they are changing rapidly), we obtain the relations

---

[195] Here, E is should not be too close to U_2 and U_1 (so that in the vicinity of the point, $x = 0$, the quasi-classical property is not broken), but if $U(x) = U_1 = $ const for $x < 0$ then there is no second restriction.

$$(1+A)p_1^{-1/2}(0) = CD_0(E)|p_2(0)|)^{-1/2},$$
$$(1-A)p_1^{1/2}(0) = iC|p_2(0)|^{1/2}D_0^{-1/2}(E),$$

where

$$p_1(0) = \sqrt{2m(E-U_1)}, \quad |p_2(0)| = \sqrt{2m(U_2-E)},$$

while $D_0(E)$ is given by the quasi-classical expression (IX.9) for the barrier penetrability. Hence we can find the value of C and also the barrier transmission coefficient:

$$D(E) = \frac{4\sqrt{(E-U_1)(U_2-E)}}{U_2-U_1} \exp\left(-\frac{2}{\hbar}\int_0^b |p(x')|dx'\right). \tag{4}$$

The appearance of the additional factor multiplying the exponential (whose value is of order unity) in comparison to Eq. (IX.9), is related to a change in the quasi-classical matching conditions in the vicinity of the turning point, $x = 0$.

For the potential from Problem 2.35 we have $U_2 = U_0$, $U_1 = 0$, and the pre-exponential factor is equal to $4\sqrt{E_0(U_0-E)}/U_0$. Here (4) coincides with the exact result (for large values of the exponent, when $D \ll 1$; see Problem 9.23 c).

Problem 9.25

Repeat the previous problem for a potential barrier of the form shown in Fig. 9.4b. Apply the result to the case of a rectangular potential barrier (see Problem 2.31).

Solution

We solve the problem similarly to the previous one, but taking into account the fact that now the quasi-classical matching conditions, based on the linear approximation of the potential, change at both turning points a, b. Obtaining the final result is straightforward without calculations. First we should note that the modification of the quasi-classical equation (IX.9), that is connected with the change in the wavefunction matching condition, is reduced to the appearance of an additional pre-exponential factor, $\alpha(E)$, which depends on the character of how the quasi-classical conditions break near the turning points. An important property is

$$\alpha(E) = \alpha_1(E) \cdot \alpha_2(E), \tag{1}$$

where, independent of one another, multipliers $\alpha_{1,2}(E)$ are connected with each of the turning points. (At first sight such a relation is not obvious, due to the different forms of the solutions that are matched at the left and right turning points. If, however, we take into account that the transmission coefficient is independent of the direction in which the particle falls towards the barrier for a given energy, it becomes more evident; see Problem 9.26). Using the matching conditions from Eq. (IX.9), we have $\alpha_{1,2} = 1$. And specifically for the conditions of the previous problem:

$$\alpha_1 = \frac{4\sqrt{(E-U_1)(U_2-E)}}{U_2 - U_1}, \quad \alpha_2 = 1. \qquad (2)$$

According to (1), for our problem, α_1 is described by relation (2), while $\alpha_2(E)$ differs from $\alpha_1(E)$ only by the substitutions $U_1 \to U_3$ and $U_2 \to U_4$. This statement solves the problem of barrier penetrability,

$$D(E) = \alpha_1(E)\alpha_2(E)D_0(E),$$

where $D_0(E)$ is described by Eq. (IX.9). Hence, for a rectangular barrier of height U_0 and width a we obtain

$$D(E) = 16 \frac{E(U_0 - E)}{U_0^2} \exp\left\{-\frac{2}{\hbar}\sqrt{2m(U_0 - E)a^2}\right\} \quad \text{for}$$

$$\frac{m(U_0 - E)a^2}{\hbar^2} \gg 1. \qquad (3)$$

Compare this to the exact result given in Problem 2.31.

Problem 9.26

In the quasi-classical approximation, find the transmission coefficient of slow particles, $E \to 0$, for a potential barrier that decreases in a power-law fashion as $x \to \pm\infty$, $U(x) \approx U_{1,2}\left(\frac{a}{|x|}\right)^{\nu_{1,2}}$ with $\nu_{1,2} > 2$. Extend the result to the case of a potential barrier decreasing exponentially for large distances.

Solution

Eq. (IX.9), for $E \to 0$, is not applicable.[196] According to the equation, $D(E=0) \neq 0$, while the exact result gives $D(E) \approx bE \to 0$. To calculate the coefficient b (see Problem 2.39), we should find the solution to the Schrödinger equation for $E = 0$ which satisfies the boundary condition, $\psi(x) \to 1$, as $x \to +\infty$. Since, for $E = 0$, the quasi-classical property over large distances is broken, we should use the exact solution to the Schrödinger equation in this region, and match it with the quasi-classical result at finite distances (compare this to Problems 9.8 and 9.9).

The Schrödinger equation for large distances, where $U(x) \approx U_{1,2}(a/|x|)^{\nu_{1,2}}$, takes the form

$$\psi'' - \frac{\alpha}{|x|^\nu} = 0, \quad \alpha = \frac{2mUa^\nu}{\hbar^2}.$$

The solutions of this equation are to be expressed in terms of the cylindrical functions:

$$\sqrt{|x|} Z_s(2i\sqrt{\alpha}s|x|^{1/2s}), \quad s = \frac{1}{2-\nu}.$$

---

[196] see Problem 9.23 b).

Using the boundary condition, $\psi(+\infty) = 1$, we choose the solution to the Schrödinger equation (for large distances on the right) in the form:

$$\psi = C\sqrt{|x|}J_{-s_1}(2i\sqrt{\alpha_1}s_1|x|^{1/2s_1}), \quad C = (i\sqrt{\alpha_1}s_1)^{s_1}\Gamma(1-s_1), \tag{1}$$

where $\Gamma(z)$ the Γ-function. For $x \ll \alpha_1^{-s_1}$, using the asymptotic form of the Bessel function

$$J_\nu(z) \approx \left(\frac{2}{\pi z}\right)^{1/2}\sin\left(z - \frac{\pi\nu}{2} + \frac{\pi}{4}\right), \quad z \to \infty,$$

we obtain

$$\psi(x) \approx Ci^{s_1}\left(\frac{\nu_1 - 2}{4\pi\sqrt{\alpha_1}}\right)^{1/2} x^{\nu_1/4} \exp\left\{\frac{2\sqrt{\alpha_1}}{\nu_1 - 2}x^{1/2s_1}\right\}. \tag{2}$$

(Here we have taken into account the fact that the Bessel function argument in (1) is imaginary for $\nu > 2$; here the term in (2), which exponentially decreases as x decreases, is omitted). This solution has a quasi-classical form, which determines the wavefunction, $\psi(x)$, in the whole quasi-classical domain at finite distances (where the potential is not described by its asymptotic form):

$$\psi \approx \frac{C_1}{\sqrt{|p(x)|}}\exp\left\{\frac{1}{\hbar}\int_x^\infty |p|dx\right\} \equiv \frac{C_2}{\sqrt{|p(x)|}}\exp\left\{-\frac{1}{\hbar}\int_{-\infty}^x |p|dx\right\}, \tag{3}$$

where

$$p = \sqrt{-2mU(x)},$$

$$C_1 = C_2 \exp\left\{-\frac{1}{\hbar}\int_{-\infty}^\infty |p|dx\right\} = \tag{4}$$

$$\frac{1}{2}\sqrt{\frac{\hbar}{\pi}}(\nu_1 - 2)^{\nu_1/2(\nu_1-2)}\alpha_1^{-1/2(\nu_1-2)}\Gamma(1-s_1).$$

Now we see that the solution to the Schrödinger equation for large negative distances (i.e., for $x \to -\infty$) should be chosen in the form:

$$\psi(x) = \overline{C}\sqrt{-x}H^{(2)}_{-s_2}(2i\sqrt{\alpha_2}s_2(-x)^{1/2s_2}), \tag{5}$$

where $H^{(2)}_\nu(z)$ is the Hankel function. Using the asymptotic form for $H^{(2)}_\nu(z)$ as $z \to \infty$ we obtain

$$\overline{C} = -i\sqrt{\frac{\pi}{(\nu_2-2)\hbar}}\exp\left\{\frac{i}{2}\pi s_2\right\}C_2. \tag{6}$$

Now using the relations between the Hankel and Bessel functions, we find the asymptotic behavior of solution (5) for $x \to -\infty$ (here the argument of the Hankel function goes to zero). It has the form $\psi \approx -Bx$, and using conditions (4) and (6) we obtain

$$B = \frac{1}{\beta_1 \beta_2} \exp\left\{-\frac{1}{\hbar} \int_{-\infty}^{+\infty} |p| dx\right\},$$

where $\beta_{1,2}$ are determined by the relation

$$\beta = \frac{\sqrt{2\pi}}{\Gamma(1-s)} \left[\frac{\alpha}{(\nu-2)^\nu}\right]^{1/2(\nu-2)} \tag{7}$$

(Indices 1, 2 are omitted for brevity).

Finally, using the result of Problem 2.39 we obtain the transmission coefficient for slow particles in the form:

$$D(E) = \gamma_1(E)\gamma_2(E) \exp\left\{-\frac{2}{\hbar}\int_{-\infty}^{+\infty}\sqrt{2mU(x)}dx\right\} \propto E, \tag{8}$$

where

$$\gamma_{1,2}(E) = 2k\beta_{1,2}^2. \tag{9}$$

The modification to the quasi-classical formula (IX.9), *i.e.*, the appearance of an additional multiplicative factor in front of the exponent, $\gamma(E)$, and the form of the factor

$$\gamma(E) = \gamma_1(E) \cdot \gamma_2(E)$$

reflect a general pattern, as mentioned in the previous problem.

In an analogous way, we can consider the case of exponentially decreasing potentials, $U \propto e^{-|x|/R}$, for large distances. (For the exact solution of the Schrödinger equation, see, for example, Problem 4.8 b.) But if we take into account Problem 9.23 b, then the final result is evident without a calculation: in this case

$$\gamma_{1,2} = 2\sinh(2\pi k R_{1,2})\exp\{-2\pi k R_{1,2}\} \tag{10}$$

and for a slow particle,

$$\gamma_{1,2} \approx 4\pi k R_{1,2}. \tag{11}$$

It is instructive to obtain the last relation from Eqs. (7) and (9) by considering the exponential potential as a limiting case of a power-law potential in the limit of $\nu \to \infty$. Here we first introduce $x_0 = \nu R$ and consider the values of x close to x_0. Here,

$$\left(\frac{x_0}{x}\right)^\nu \equiv \left(\frac{x_0}{x_0 + (x-x_0)}\right)^\nu \approx_{\nu\to\infty} e^{-\tilde{x}/R}, \quad \tilde{x} = x - x_0.$$

For the transition from the power-law potential, $U = \tilde{\alpha}/x^\nu$, to the exponential one, $U = U_0 e^{-x/R}$, we must put

$$\frac{\tilde{\alpha}}{x_0^\nu} \equiv \frac{\tilde{\alpha}}{(\nu R)^\nu} = U_0, \quad i.e. \quad \tilde{\alpha} = (\nu R)^\nu U_0. \tag{12}$$

Substituting (12) into (7) (here $\alpha = 2m\tilde{\alpha}/\hbar^2$) and taking $\nu \to \infty$, we obtain $\beta = \sqrt{2\pi R}$ and, according to (9), we obtain relation (11) for the potential with an exponential tail.

Problem 9.27

Using the quasi-classical approximation, determine the coefficient of the particle to be reflected above a barrier in the case where the potential has a jump at point $x = 0$ (see, for example, Fig. 8.2). Compare this with perturbation theory result derived in Problem 8.30.

Solution

Particle reflection is determined mainly by the fact that the potential has a singularity at the point $x = 0$. The wavefunction that describes the reflection (and transmission) process of a particle, which impinges on the barrier from the left, has the form (within the quasi-classical approximation):

$$\psi(x) = \begin{cases} \frac{1}{\sqrt{p(x)}} \left\{ \exp\left(\frac{i}{\hbar} \int\limits_0^x p dx'\right) + A \exp\left(-\frac{i}{\hbar} \int\limits_0^x p dx'\right) \right\}, & x < 0, \\ \frac{C}{\sqrt{p(x)}} \exp\left(\frac{i}{\hbar} \int\limits_b^x p dx'\right) & x > 0. \end{cases} \tag{1}$$

where $p = \sqrt{2m(E - U(x))} > 0$; here the reflection coefficient is equal to $R(E) = |A|^2$. From the conditions of the wavefunction and its derivative being continuous at $x = 0$ (as it usually is in the case of the quasi-classical approximation, it is enough to differentiate the exponential factors only, since they are the most rapidly varying components), we obtain

$$\sqrt{p_2}(1 + A) = \sqrt{p_1} B, \quad \sqrt{p_1}(1 - A) = \sqrt{p_2} B,$$

where $p_{1,2} = \sqrt{2m(E - U_{1,2})}$. Hence it follows that $A = (p_1 - p_2)/(p_1 + p_2)$ and the reflection coefficient is given by

$$R(E) = |A|^2 = \frac{\left(\sqrt{E - U_1} - \sqrt{E - U_2}\right)^2}{\left(\sqrt{E - U_1} + \sqrt{E - U_2}\right)^2}. \tag{2}$$

We should note that for (2) to be satisfied for finite values of energy it is not necessary for $R(E)$ to be small. But for $E \to \infty$, we have (from (2))

$$R(E) \approx \frac{(U_2 - U_1)^2}{16E^2} \to 0,$$

which coincides with the result of perturbation theory (see Problem 8.30).

Problem 9.28

In the quasi-classical approximation find the shift and width of the ground-state energy level in a δ-potential well, $U = -\alpha\delta(x)$, arising under an action of a weak uniform field, $V = -F_0 x$. Compare with the results of Problems 6.36 and 8.12.

Solution

The energy level considered under the action of the field corresponds to a quasi-stationary state. The position, E_0, and width, Γ, of the quasi-discrete energy levels are determined by the conditions for the existence of a solution of the Schrödinger equation which has a complex-valued energy, $E = E_0 - i\Gamma/2$, and the form of an *outgoing* wave as $x \to \pm\infty$ (if in some direction of motion $U(x) > E_0$ then, in this direction, the solution is exponentially decreasing), see Problem 6.36.

In this problem the solution to the Schrödinger equation in the quasi-classical approximation has the form[197] (see Fig. 6.1)

$$\psi(x) = \begin{cases} \dfrac{A}{\sqrt{p(x)}} \exp\left\{\dfrac{i}{\hbar} \int_x^0 p(x')dx'\right\}, & x < 0, & (1.1) \\[2mm] \dfrac{C}{\sqrt{p(x)}} \exp\left\{-\dfrac{i}{\hbar} \int_b^x p(x')dx'\right\}, & 0 < x < b, & (1.2) \\[2mm] \dfrac{C}{\sqrt{p(x)}} \exp\left\{\dfrac{i}{\hbar} \int_b^x p(x')dx' + i\dfrac{\pi}{4}\right\}, & x > b, & (1.3) \end{cases}$$

where $p(x) = \sqrt{2m(E + F_0 x)}$ (We should emphasize that in the classically forbidden regions and when the width is neglected, $p(x)$ is purely imaginary, and $ip(x) < 0$.) Here we used the known matching condition for the solutions in the vicinity of turning point $x = b$. In Problem (1.2) we only kept the term that increases exponentially inside the barrier.

---

[197] Strictly speaking, the turning points are now complex; but because of the exponentially small Γ this fact is not reflected in the matching conditions.

From the matching conditions at point $x = 0$ (see Problem 2.6) we obtain

$$A = C \exp\left\{-\frac{i}{\hbar}\int_0^b p(x)dx\right\},$$

$$A + C\exp\left\{-\frac{i}{\hbar}\int_0^b p(x)dx\right\} = \left(\frac{2im\alpha}{\hbar p(0)}\right)A. \tag{2}$$

(To calculate the wavefunction derivatives we differentiate only the exponential factors, since they are the most rapidly varying factors within the quasi-classical conditions, see Eq. (6) below). Hence, we obtain $\hbar p(0) = im\alpha$ or $E = E_0^{(0)} = -m\alpha^2/2\hbar^2$, which coincides with the energy level without perturbation, $V = -F_0 x$.

We see in the considered approximation that neither the shift nor the level width can be obtained at this level of approximation. To obtain the shift we should have used more accurate quasi-classical relations for the wavefunction that take into account the next-to-leading order corrections in $\hbar$ (see below). As to the level width, it has not appeared in the calculation above because we have omitted the function exponentially-decaying inside the barrier, $0 < x < b$. Direct calculations using this function are very cumbersome, but one can circumvent these technical complications by using the following considerations.

Taking into account the physical meaning of Γ as the value that determines the decay probability per unit of time, $w = \Gamma/\hbar$, in the state considered we focus on the current density impinging on the right-hand side of the turning point, $x = b$. Using Eq. (1.3) we obtain

$$j = -\frac{i\hbar}{2m}\left(\psi^*\frac{\partial}{\partial x}\psi - \psi\frac{\partial}{\partial x}\psi^*\right) = \frac{|C|^2}{m}. \tag{3}$$

Here, if the wavefunction is normalized so that the particle is in the vicinity of the well with probability ≈ 1, then this current density gives the decay probability per unit of time, $w = j$. As is seen from Eqs. (1.1) and (1.2), the probability density, $|\psi^2|$, is essentially different from zero only for the values

$$|x| \leq \sqrt{\frac{\hbar^2}{2m|E|}} \approx \frac{\hbar^2}{m\alpha^2},$$

and if

$$F_0 \cdot \frac{\hbar^2}{m\alpha} \ll |E_0|, \quad \text{i.e.,} \quad F_0 \ll \frac{m^2\alpha^3}{\hbar^4}, \tag{4}$$

then in $p(x)$ we can omit the term with $F_0 x$. As a result, using the wavefunction normalized to unity in the region of particle localization and near the well, we obtain

$$|\psi|^2 \approx \frac{1}{|p(0)|} A^2 \exp\left\{-\frac{2m\alpha|x|}{\hbar^2}\right\}, \quad A^2 = \frac{m^2\alpha^2}{\hbar^3}.$$

(In the approximation considered, the wavefunction in this region coincides with the wavefunction of bound states in the δ-well that is unperturbed by the field.) According to Eq. (2) we obtain

$$|C|^2 \exp\left\{\frac{2}{\hbar} \int_0^b \sqrt{2m(|E_0| - F_0 x)} dx\right\} = A^2 = \frac{m^2\alpha^2}{\hbar^3}.$$

After the integration we obtain the value of $|C|^2$ and the level width

$$\Gamma = \frac{\hbar |C|^2}{m} = \frac{m\alpha^2}{\hbar^2} \exp\left\{-\frac{2}{3} \frac{m^2\alpha^3}{\hbar^4 F_0}\right\}. \tag{5}$$

Now we discuss the level shift. Using the more accurate relations for the quasi-classical wavefunction in Eq. (1),

$$\psi = \frac{C}{\sqrt{p(x)}} \left(1 \mp \frac{im\hbar}{4} \frac{F}{p^3} \mp \frac{im^2\hbar}{8} \int_0^x \frac{F^2}{p^5} dx\right) \exp\left\{\pm \frac{i}{\hbar} \int_0^x p dx\right\},$$

and differentiating the factors in front of the exponents while matching the solutions at the point $x = 0$, we obtain a correction to Eq. (2) and find

$$ip(0) + \frac{5i}{8} \frac{m^2 \hbar^2 F_0^2}{p^5(0)} = -\frac{m\alpha}{\hbar}.$$

Substituting $p(0)$ in the second correction term by its unperturbed value, $im\alpha/\hbar$, we find

$$E_0 = -\frac{m\alpha^2}{2\hbar^2} - \frac{5}{8} \frac{\hbar^6 F_0^2}{m^3 \alpha^4}. \tag{6}$$

The second term here gives the level shift (and the polarizability), and coincides with the exact result in the second-order perturbation theory (see Problems 8.12 and 6.36), where the level shift and the level width are obtained from the exact solution of the Schrödinger equation.

Problem 9.29

Using the quasi-classical approximation, obtain the energies, E_{0n}, and widths, Γ_n, of the quasi-stationary states in the one-dimensional potential shown in Fig. 9.5. Extend the results to the case where the barrier on the left and on the right of the well has a

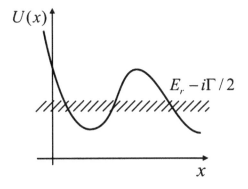

Fig. 9.5

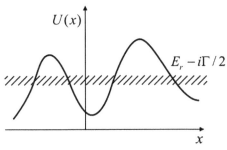

Fig. 9.6

finite penetrability (Fig. 9.6). Apply the result to the case of a linear oscillator subject to a weak anharmonicity, $V(x) = -\lambda x^3$.

Solution

1) The quasi-classical solution to the Schrödinger equation for quasi-stationary states has the form:[198]

$$\psi(x) = \begin{cases} \dfrac{\sqrt{i}C}{2\sqrt{p(x)}} \exp\left(\dfrac{i}{\hbar} \int\limits_{x}^{a_1} p(x')dx'\right), & x < a_1, \quad (1.1) \\[6pt] \dfrac{C}{\sqrt{p(x)}} \sin\left(\dfrac{i}{\hbar} \int\limits_{a_1}^{x} p(x')dx' + \dfrac{\pi}{4}\right), & a_1 < x < b_1, \quad (1.2) \\[6pt] \dfrac{C_1}{\sqrt{p(x)}} \exp\left(-\dfrac{i}{\hbar} \int\limits_{x}^{a_2} p(x')dx'\right) + \\[2pt] \dfrac{C_2}{\sqrt{p(x)}} \exp\left(\dfrac{i}{\hbar} \int\limits_{x}^{a_2} p(x')dx'\right), & b_1 < x < a_2, \quad (1.3) \\[6pt] \dfrac{C_1}{\sqrt{p(x)}} \exp\left(\dfrac{i}{\hbar} \int\limits_{a_2}^{x} p(x')dx'\right), & x > a_2. \quad (1.4) \end{cases}$$

[198] In Eq. (1), a_1, b_1, a_2 are the turning points.

Here (1.1) decreases exponentially as $x \to -\infty$; (1.2) is written by using the matching conditions for solution (IX.4). Then in relations (1.3) and (1.4) we obtain the same coefficient C_1 as it follows from the matching condition for the solution in the vicinity of point, $x = a_2$. Neglecting in (1.3) the second term, which exponentially decreases inside the barrier, we can use conditions (IX.3) in order to match the solution near the point $x = b_1$. Hence we obtain the Bohr–Sommerfeld quantization rules for E_{0n}, which determine the position of the quasi-discrete levels, and the relation between coefficients

$$C_1 = \frac{\sqrt{i}}{2}(-1)^n \exp\left(-\frac{1}{\hbar}\int_{b_1}^{a_2} |p(x')|dx'\right) C.$$

Now using $|C|^2 = 4m/T(E_{0n})$, which provides the normalized wavefunction in the region of the particle's classical motion, $a_1 < x < b_1$, and calculating the flow of probability for $x > a_2$ (compare this to the previous problem), we obtain the following expression for the width of the quasi-stationary states:

$$\Gamma_n = \hbar\omega_n = \frac{\hbar D(E_{0n})}{T(E_{0n})} = \frac{\hbar\omega(E_{0n})}{2\pi}\exp\left\{-\frac{2}{\hbar}\int_{b_1}^{a_2}|p|dx\right\}. \tag{2}$$

Here, the intuitive meaning of ω_n is given as follows: the probability for a particle to leave the well per unit of time is equal to the number of times the classical particle hits the barrier $1/T$ per unit of time, multiplied by the quantum-mechanical probability to pass through the barrier in a single collision. Then the expression for the level width has a larger range of applicability, since it is not connected with the quasi-classical formula for the barrier penetrability; compare this, for example, to Problem 6.37. If the barrier has a finite penetrability on both sides of the well, we have

$$\Gamma_n = \frac{\hbar\omega}{2\pi}(D_1 + D_2).$$

2) We now apply our results to the case of the linear oscillator with anharmonicity, $U = m\omega^2 x^2/2 - \lambda x^3$. Level shifts for the unperturbed oscillator were calculated in Problems 9.10 and 9.13. The level width calculation, according to (2), is reduced to the calculation of the integral:

$$\int_{x_1}^{x_2} |p(x)|dx = \int_{x_1}^{x_2} \sqrt{2m\left[\frac{1}{2}m\omega^2 x^2 - \lambda x^3 - E_{0n}\right]}dx. \tag{3}$$

To approximately calculate it, we divide the integration domain into two: from x_1 to d and from d to x_2, where the value of d is assumed to satisfy the condition

$$\sqrt{\frac{E_{0n}}{m\omega^2}} \ll d \ll \frac{m\omega^2}{\lambda}. \tag{4}$$

In the first integral we could consider λx^3 as a small correction, and in the second one the same can be done for the term with E_{0n}. Expanding with respect to these parameters we find

$$\frac{1}{\hbar}\int_{x_1}^{d}|p|dx \approx \frac{1}{\hbar}\int_{x_{10}}^{d}\left\{\sqrt{(m\omega x)^2 - 2mE_{0n}} - \frac{\lambda m x^3}{\sqrt{(m\omega x)^2 - 2mE_{0n}}}\right\}dx$$

$$\approx \frac{m\omega d^2}{2\hbar} - \frac{E_{0n}}{2\hbar\omega} - \frac{E_{0n}}{2\hbar\omega}\ln\frac{2m\omega^2 d^2}{E_{0n}} - \frac{\lambda d^3}{3\hbar\omega}, \tag{5}$$

and also

$$\frac{1}{\hbar}\int_{d}^{x_2}|p|dx \approx \frac{1}{\hbar}\int_{d}^{x_{20}}\left\{\sqrt{(m\omega x)^2 - 2m\lambda x^3} - \frac{mE_{0n}}{\sqrt{(m\omega x)^2 - 2m\lambda x^3}}\right\}dx$$

$$\approx \frac{m^3\omega^5}{15\hbar\lambda^2} - \frac{m\omega d^2}{2\hbar} + \frac{\lambda d^3}{3\hbar\omega} - \frac{E_{0n}}{\hbar\omega}\ln\frac{2m\omega^2}{d\lambda}. \tag{6}$$

The sum of (5) and (6) gives the exponent in Eq. (2) and also the relation for the sought-after level width (the auxiliary variable d used to calculate the integral drops out of the final result):

$$\Gamma_n = \frac{\hbar\omega}{2\pi}\left[\frac{8m^3\omega^5}{\lambda^2\hbar(n+1/2)}\right]^{n+1/2}\exp\left\{-\frac{2m^3\omega^5}{15\hbar\lambda^2} + \left(n+\frac{1}{2}\right)\right\}. \tag{7}$$

Here, E_{0n} is replaced by the unperturbed value, $\hbar\omega(n+1/2)$.

Problem 9.30

Evaluate the penetrability of a centrifugal barrier and the particle *lifetime* in the quasi-stationary state (related to the level width by $\tau = \hbar/\Gamma$) for a short-range potential, $U_S(r)$, with a radius r_S; the state energy is $E \ll \hbar^2/mr_S^2$.

Solution

The qualitative form of the effective potential,

$$U_{\text{eff}}(r) = U_S(r) + \frac{\hbar^2 l(l+1)}{2mr^2},$$

is shown in Fig. 9.7; at small distances, $r \to 0$, and in the region, $r > r_S$, the centrifugal potential dominates. Here, $U_0 \geq \hbar^2/mr_S^2$, since, in the other case of a "shallow" well, there exists neither a truly bound state (with $E < 0$) nor a quasi-stationary state (with $E > 0$). For the radial wavefunction, $\chi(r) = rR$ (see Eq. (IV.5)), we can use one-

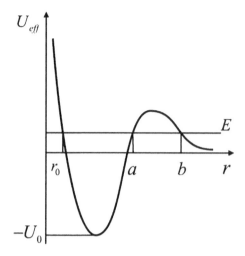

Fig. 9.7

dimensional quasi-classical approximation. Calculating the integral ($E = \hbar^2 k^2/2m$),

$$\frac{1}{\hbar}\int_a^b |p_r(r)|dr \approx \int_a^b \sqrt{\frac{(l+1/2)^2}{r^2} - k^2}\,dr$$

$$\approx \int_a^b \frac{(2l+1)dr}{2r} \approx \frac{2l+1}{2}\ln\frac{2l+1}{2kr_S}. \tag{1}$$

(Here the turning points are $a \sim r_s$ and $b \approx (l+1/2)/k$; for the centrifugal potential the Langer correction is included, i.e., using the substitution, $l(l+1) \to (l+1/2)^2$), we obtain the following estimation for the penetrability through the centrifugal barrier using Eq. (IX.9)

$$D \sim \left(\frac{2kr_S}{2l+1}\right)^{2l+1} \propto (kr_S)^{2l+1}, \tag{2}$$

(pay attention to the energy dependence of this expression).

We note that by using a relation for the barrier analogous to (1), which is applicable at the coordinate origin (with the change of b by r_0 and r_S by $r < b$), we obtain (for $r \to 0$)

$$\chi_{quas} = \frac{C}{\sqrt{|p_r|}}\exp\left(-\frac{1}{\hbar}\int_r^{r_0}|p_r|dr\right) \propto r^{l+1},$$

in accordance with the exact result, $R_l = \chi/r \propto r^l$.

To estimate the lifetime,[199] τ, we find the probability, w, for the particle to leave the well per unit time. This probability is obtained by multiplying the number of times the particle hits the barrier per unit time (which is approximately equal to[200] $v_S/r_S \sim \hbar/mr_S^2$) by the single collision probability for it to pass through the barrier (which coincides with D). We then have

$$\tau = \frac{1}{w}, \quad w \sim \frac{\hbar}{mr_S^2}\left(\frac{2kr_S}{2l+1}\right)^{2l+1}. \tag{3}$$

(For a more exact relation for τ, see Eq. (XIII.17).) Compare this to the energy dependence, $w \propto k^{2l+1}$, for a centrifugal barrier in the case of slow particles with an exponential dependence for the Coulomb barrier (see the next problem).

Problem 9.31

Extend the result of the previous problem to the case of a particle in the attractive Coulomb potential outside the well:[201] $U_C = -\zeta e^2/r$ and $r_S \ll a_S = \hbar^2/m\zeta e^2$; consider $E \ll \zeta e^2/a_S$.

Solution

The form of the effective potential is shown in Fig. 9.8. For the states with small energy ($E \to 0$):

$$\frac{1}{\hbar}\int_a^b |p_r|dr \approx \int_a^b \left\{\frac{(l+1/2)^2}{r^2} - \frac{2}{a_B r}\right\}^{1/2} dr$$

$$\approx \frac{2l+1}{2}\left\{\ln\frac{(2l+1)^2 a_B}{2r_S} - 2\right\}. \tag{1}$$

Here, $a \approx r_S$, $b = (2l+1)^2 a_B/8$, and we took into account that $r_S \ll a_B$. Hence,

$$D \sim \left(\frac{2e^2 r_S}{(2l+1)^2 a_B}\right)^{2l+1} \sim \frac{1}{[(2l+1)!]^2}\left(\frac{2r_S}{a_B}\right)^{2l+1} \tag{2}$$

($e = 2.718\ldots$; in the last transformation we used the Stirling equation). Note that Coulomb attraction "shortens" the centrifugal barrier and, for small energies ($E \to 0$), increases its penetrability; compare this to the previous problem.

---

[199] In the quasi-classical approximation the quasi-stationary state energy, as well as the energy of the exact bound state, is determined by the Bohr–Sommerfeld quantization rule; compare this to the previous problem.

[200] Here $v_s = p_S/m$ is the characteristic velocity of the particle in the well, $p_S \sim \sqrt{mU_0} \geq \hbar/r_S$ (do not confuse p_S with the momentum, $p = \hbar k$, of the particle leaving the well, $p_S \gg \hbar k$).

[201] Such a problem appears in the theory of hadron atoms (see Problem 11.4). There, the quantity of interest is the penetrability of the barrier that separates a region of nuclear attraction from a region with the Coulomb attraction.

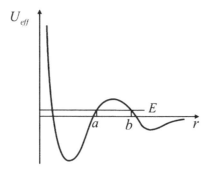

Fig. 9.8

2) On the contrary, in the case of the repulsive Coulomb potential, the barrier penetrability for slow particles decreases sharply. Here the dominant contribution to the integral ($E = mv^2/2$, $U_C = \alpha/r$),

$$\frac{1}{\hbar}\int_a^b |p_r| dr \approx \frac{1}{\hbar}\int_0^{\alpha/E} \sqrt{\frac{2m\alpha}{r} - (mv)^2}\, dr = \frac{\pi \alpha}{\hbar v}$$

comes from large distances, where the centrifugal potential is small. Taking this into account we arrive at the relation

$$D(E) \sim e^{-2\pi\alpha/\hbar v} \tag{3}$$

for the barrier penetrability; compare Eqs. (3) to (2) and also to Eq. (2) from the previous problem.

9.4 1/N-expansion in quantum mechanics

1/N-expansion is a computational method used in different fields of theoretical physics. The idea of the method is to "construct" a parameter N such that for $N \gg 1$ the solution simplifies and allows an expansion in powers of $1/N$. For a suitable choice of parameter N, the region of applicability of such an expansion may extend down to $N \sim 1$, which characterizes the initial system. (Compare this to the high accuracy of quasi-classical results for E_n even for $n \sim 1$, despite the fact that the formal condition of its applicability is $n \gg 1$.) The auxiliary parameter N is often connected with increasing the number of degrees of freedom (the number of states, spatial dimensionality, *etc.*) of the considered system. Below, we discuss several elementary applications of this method for solutions to the one-dimensional Schrödinger equation.

Problem 9.32

Analyze the discrete spectrum of a particle in a one-dimensional potential, $U(x) = -U_0 f(x/a)$, which is a well with a single minimum (at the point $x_0 = 0$), using $1/N$-expansion. Here, choose

$$N = \left(\frac{ma^2 U_0}{\hbar^2}\right)^{1/2}. \tag{1}$$

To illustrate the accuracy of the expansions for the energy levels and the wavefunction, consider the following potentials:

a) $U(x) = -U_0 \cosh^{-2}(x/a)$;
b) $U(x) = U_0(a/x - x/a)^2$;
c) $U(x) = U_0(e^{-2x/a} - be^{-x/a})$.

Solution

As N increases, the potential well becomes deeper and the number of bound states grows as well. In the case $N \gg 1$, the wavefunctions of the low-lying energy levels are localized around the potential minimum point, where $U'(0) = 0$. Therefore we can expand the potential energy in terms of x/a (below we put $\hbar = m = a = 1$):

$$U(x) = U(0) + \frac{1}{2}\omega^2 x^2 + \alpha x^3 + \beta x^4 + \ldots,$$

where

$$\omega^2 = U''(0), \quad \alpha = \frac{1}{6}U'''(0), \quad \beta = \frac{1}{24}U^{\mathrm{IV}}(0). \tag{2}$$

Here the third and higher order terms are considered as a perturbation; the unperturbed system is a linear oscillator with frequency ω. The perturbation series with respect to such anharmonic corrections,

$$E_n = U(0) + \omega\left(n + \frac{1}{2}\right) - \frac{15}{4}\frac{\alpha^2}{\omega^4}\left(n^2 + n + \frac{11}{30}\right)$$

$$+ \frac{3}{2}\frac{\beta}{\omega^2}\left(n^2 + n + \frac{1}{2}\right) + \ldots, \tag{3}$$

may be regarded as a $1/N$-expansion for the particle energy levels; here $U(0) \sim N^2$, $\omega \sim N$, $\frac{\alpha^2}{\omega^4} \sim \frac{\beta}{\omega^2} \sim N^0 = 1$, and so on. The result obtained for E_n is asymptotically exact as, $N \to \infty$. However, in the case of smooth potentials it has a high enough accuracy even for $N \geq 1$ as well (to increase the accuracy we should consider higher-order corrections). As an illustration, let us consider some specific potentials.

a) For the potential

$$U(x) = -U_0 \cosh^{-2}(x/a)$$

the expansion (3) takes the form, $N = \sqrt{U_0}$, and $\omega = \sqrt{2U_0}$,

$$E_n(1/N) = -N^2 + \sqrt{2}N\left(n+\frac{1}{2}\right) - \frac{1}{2}\left(n^2+n+\frac{1}{2}\right) + O\left(N^{-1}\right)$$

$$\approx -U_0 + \sqrt{2U_0}\left(n+\frac{1}{2}\right) - \frac{1}{2}\left(n^2+n+\frac{1}{2}\right). \tag{4}$$

Let us compare this both with the exact energy spectrum,

$$E_n = -\frac{1}{8}\left[\sqrt{8U_0+1} - (2n+1)\right]^2, \tag{5}$$

and with the quasi-classical expression, $E_n^{(q)}$, that differs from (5) by, $8U_0 + 1 \to 8U_0$. For the ground state for different values of N we have

$N = \sqrt{U_0}$	1	1.5	2	3	4
$-E_0(1/N)$	0.543	1.439	2.836	7.129	13.422
$-E_0$	0.500	1.410	2.814	7.114	13.411
$-E_0^{(q)}$	0.418	1.314	2.711	7.004	13.297

For the first excited state, $n = 1$, we have

$N = \sqrt{U_0}$	1.5	2	3	4
$-E_1(1/N)$	0.318	1.007	3.886	8.765
$-E_1$	0.231	0.942	3.842	8.732
$-E_1^{(q)}$	0.193	0.882	3.761	8.640

We should note that the level appears at the value $U_0 = 1$. (According to (5), the total number of discrete levels in the considered potential for $N \gg 1$ is $N_{\text{lev}} \approx \sqrt{2}N$.)

Let us discuss the wavefunctions within the $1/N$-expansion. From the perturbation theory equations (VIII.2), we obtain

$$\psi_0(x) \approx \left[\psi_0^{(0)}(x) + c_{02}^{(1)}\psi_2^{(0)}(x) + c_{04}^{(1)}\psi_4^{(0)}(x)\right], \tag{6}$$

where $\psi_n^{(0)}(x)$ are the eigenfunctions of a linear oscillator with frequency $\omega = \sqrt{2U_0}$. Here the linear oscillator perturbation has the form, $V = -2U_0 x^4/3$, so that the non-zero expansion coefficients, $c_{02}^{(1)} = \langle k|V|0\rangle$ are equal to

$$c_{02}^{(1)} = \frac{1}{4\sqrt{U_0}}, \quad c_{04}^{(1)} = \frac{1}{8\sqrt{3U_0}}.$$

Let us emphasize that the wavefunction[202] (6) is normalized to unity up to within the quadratic order in $1/N$.

The exact wavefunction for the ground state has the form

$$\tilde{\psi}_0(x) = A \cosh^{-s} x, \quad s = \frac{1}{2}(\sqrt{8U_0 + 1} - 1).$$

It is easy to find the normalization coefficient for the values, $U_0 = 1$ (here $s = 1$) and $U_0 = 3$ ($s = 2$): $A(1) = 1/\sqrt{2}$ and $A(3) = \sqrt{3}/2$.

Let us compare the wavefunction values at zero. Here, $\tilde{\psi}_0(0) = A$ while

$$\psi_0(0) = \psi_0^{(0)}(0)\left(1 - \frac{3}{16\sqrt{2U_0}}\right) = \left(\frac{\sqrt{2U_0}}{\pi}\right)^{1/4}\left(1 - \frac{3}{16\sqrt{2U_0}}\right),$$

and for the relation $R = \psi_0(0)/\tilde{\psi}_0(0)$ we obtain

$$R = 1.0048 \quad \text{for} \quad U_0 = 1 \quad \text{and} \quad R = 1.0020 \quad \text{for} \quad U_0 = 3.$$

As we see, $1/N$-expansion gives accurate results for both the energy levels and the wavefunction in the region where the particle is localized even in the case of relatively small values of N, when there are only several discrete levels in the potential. This remains true for more general potentials as long as they are "smooth" enough.

b) For the potential, $U(x) = U_0(a/x - x/a)^2$, we obtain ($N = \sqrt{U_0}$, we put $a = 1$)

$$E_n(1/N) = \sqrt{8U_0}\left(n + \frac{1}{2}\right) + \frac{1}{8} + \ldots \quad (7)$$

A comparison to the exact expression for the spectrum

$$\tilde{E}_n = \sqrt{8U_0}\left[n + \frac{1}{2} + \frac{1}{4}(\sqrt{8U_0 + 1} - \sqrt{8U_0})\right] \quad (8)$$

shows that we again obtain highly accurate results within the $1/N$-expansion even for $N \geq 1$; so, for $N = 1$ for the ground level, $E_0/\tilde{E}_n = 1.026$.

c) For the potential ($U_0 > 0$, $a > 0$, $b > 0$),

$$U(x) = U_0(e^{-2x/a} - be^{-x/a}),$$

and the $1/N$-expansion gives ($N = \sqrt{U_0}$, $a = 1$):

$$E_n(1/N) = -\frac{1}{4}U_0 b^2 + \sqrt{\frac{1}{2}U_0 b^2}\left(n + \frac{1}{2}\right) - \frac{1}{2}\left(n + \frac{1}{2}\right)^2 + \ldots \quad (9)$$

The three expansion terms coincide with the exact result for the spectrum, $E_n = -[\sqrt{2U_0 b^2} - (2n + 1)]^2/8$ (see Problem 2.9).

---

[202] The relation for the wavefunction is valid only in the region where the particle is localized (and is not applicable for large distances in the classically forbidden domain; in the latter region we can use the quasi-classical expression for the wavefunction).

Problem 9.33

Analyze the energy spectrum for bound s-states of a particle moving in a spherically-symmetric potential, $U(r)$, using the $1/N$-expansion with spatial *dimensionality*,[203] D, as the parameter N, i.e., considering $N \equiv D \gg 1$. To illustrate this method, consider the potentials a) $U(r) = Fr$, b) $U(r) = \alpha r^4$, and c) $U(r) = -\alpha/r$.

Solution

In the N-dimensional space and for spherically symmetric functions we have[204]

$$\triangle f(r) = f'' + \frac{N-1}{r} f', \quad r^2 = \sum_{i=1}^{N} x_i^2.$$

Hence, the Schrödinger equation for the s-states, using the substitution $\psi(r) = \chi(r)/r^\nu$ with $\nu = (N-1)/2$, takes the form of the ordinary one-dimensional Schrödinger equation (below we consider $\hbar = m = 1$):

$$-\frac{1}{2}\chi'' + (U_{\text{eff}}(r) - E)\chi = 0, \quad U_{\text{eff}}(r) = \frac{(N-1)(N-3)}{8r^2} + U(r). \tag{1}$$

In the case of large N, due to the quasi-centrifugal barrier $\propto N^2/r^2$, the wavefunctions and the energy spectrum around the minimum of the effective potential have the properties[205] similar to those noted in the previous problem for the potential, $U = U_0 f(x/a)$ for $U_0 \to \infty$, and can be calculated along the same lines.

To solve Eq. (1) for $N \gg 1$ using the $1/N$-expansion it is convenient to obtain the solution in the form of an expansion in powers of $1/\tilde{N}$, where $\tilde{N}^2 = (N-1)(N-3)$, and then perform an expansion in $1/N$. Let us consider application of this method to some specific potentials.

a) For the potential, $U(r) = Fr$, equation (1) takes the form:

$$-\frac{1}{2}\chi'' + \left[\frac{\tilde{N}^2}{8r^2} + Fr\right]\chi = E\chi.$$

Performing an expansion of U_{eff} in the vicinity of the point, $r_0 = \left(\tilde{N}^2/4F\right)^{1/3}$, where the effective potential minimum is located,

$$U_{\text{eff}} = \frac{3}{2}Fg^{-1/3} + \frac{3}{2}Fg^{1/3}x^2 - 2Fg^{2/3}x^3 + \frac{5}{2}Fgx^4 + \ldots,$$

---

[203] It is also called the $1/D$-expansion; for real space, $D = 3$, but sometimes this method can be used even in one- and two-dimensional systems.

[204] This result can be obtained by using the following relations: $\Delta = \text{div}\nabla$, $\nabla f(r) = f'\mathbf{r}/r$, and $\text{div}\mathbf{r} = N$.

[205] This statement is valid only for potentials with an infinite number of bound states, and below we restrict ourselves to this case. For short-range attractive potentials with a finite number of discrete states we should use the method from the following problem, which involves writing the potential in the form, $U(r) = N^2 gv(r/R)$.

where $x = r - r_0$ and $g = 4F/\tilde{N}^2$. We then obtain (similarly to the previous problem):

$$E_0(1/\tilde{N}) = \frac{3}{2}Fg^{-1/3} + \frac{1}{2}\sqrt{3F}g^{1/3} + \frac{1}{72}g^{2/3} + \ldots \quad (2)$$

Taking into account that

$$g = \frac{4F}{N^2 - 4N + 3} \approx 4FN^{-2}\left(1 + \frac{4}{N} + \frac{13}{N^2}\right), \quad (3)$$

we obtain the following $1/N$-expansion for the ground level:

$$E_0(1/N) = \frac{3}{2}\left(\frac{FN}{2}\right)^{2/3}\left\{1 + \frac{1}{N}\left(-\frac{4}{3} + \frac{2}{\sqrt{3}}\right) + \frac{1}{N^2}\left(-\frac{20}{27} + \frac{4}{3\sqrt{3}}\right) + \ldots\right\}$$

$$= \frac{3}{2}\left(\frac{FN}{2}\right)^{2/3} \cdot (1 - 0.17863N^{-1} + 0.02906N^{-2} + \ldots). \quad (4)$$

For $N = 3$ we obtain $E_0 = 1.8549F^{2/3}$, which differs from the exact value, $\tilde{E}_0 = \beta_1\left(F^2/2\right)^{1/3} = 1.8558F^{2/3}$ (here $-\beta_1 = -2.3381$ is the first zero of the Airy function, $\mathrm{Ai}(-\beta_1) = 0$; see Problem 2.8) by merely 0.05%. For the one-dimensional potential, $U = F|x|$ and $N = 1$, and it follows from (4) that $E_0 = 0.8036F^{2/3}$, while the exact value is equal to $\tilde{E}_0 = 0.8086F^{2/3}$. The high accuracy of Eq. (4) is due to the serendipitous fact that the actual expansion parameter is $\approx 1/5N$.

b) For the potential, $U(r) = \alpha r^4$, we obtain the following expansions:

$$E_0(1/\tilde{N}) = 3\alpha^{1/3}\left(\frac{\tilde{N}^2}{16}\right)^{2/3}\left\{1 + \frac{4\sqrt{6}}{3\tilde{N}} + \frac{44}{27\tilde{N}^2} + \ldots\right\},$$

and in terms of the expansion parameter, $1/N$, we obtain

$$E_0(1/N) = 3\alpha^{1/3}\left(\frac{N^2}{16}\right)^{2/3}\left\{1 + \frac{8(\sqrt{3/2} - 1)}{3N} + \frac{50 - 24\sqrt{6}}{27N^2} + \ldots\right\}$$

$$= 3\alpha^{1/3}\left(\frac{N^2}{16}\right)^{2/3}(1 + 0.5993N^{-1} - 0.3255N^{-2} + \ldots). \quad (5)$$

(the expansion parameter is $\approx 1/2N$). Hence, for $N = 3$ we have $E_0 = 2.379\alpha^{1/3}$, which differs from the exact numerical result of the Schrödinger equation, $\tilde{E}_0 = 2.394\alpha^{1/3}$, by merely 0.4 %.

c) For the Coulomb potential, $U(r) = -\alpha/r$, we obtain

$$E_0(1/\tilde{N}) = -\frac{2\alpha^2}{\tilde{N}^2}\left(1 - \frac{2}{\tilde{N}} + \frac{2}{\tilde{N}^2} + \ldots\right)$$

and the $1/N$-expansion takes the form

$$E_0(1/N) = -\frac{2\alpha^2}{N^2}\left(1 + \frac{2}{N} + \frac{3}{N^2} + \ldots\right). \tag{6}$$

Here, in comparison to (4) and (5), the expansion convergence is worse. We can "improve" this if we use the expansion parameter $1/(N-1)$. Then relation (6) takes the form, $E_0(1/(N-1)) = -2\alpha^2/(N-1)^2$, which coincides with the exact result. The use of this slightly modified expansion parameter is connected with the fact that for $N = 1$ it involve the "falling into the origin" behavior (gives $E_0 = -\infty$) that appears in the one-dimensional Coulomb potential,[206] (see Problem 8.61).

Problem 9.34

Analyze the energy spectrum of bound states in a spherically symmetric attractive potential, $U(r) = U_0 v(r/R)$, using the $1/N$-expansion with the parameter

$$N \equiv n = l + n_r + 1. \tag{1}$$

Consider the radial quantum number, n_r, to be fixed, and consider a large angular momentum, $l \to \infty$. Illustrate the accuracy of the obtained energy levels by comparing them with the exact solutions of the Schrödinger equation.

Solution

Let us write the potential in the form, $U(r) = n^2 g v(r/R)$, i.e., put $U_0 \equiv gn^2$, and function $v(z)$ determines the shape of the potential. Then the effective potential energy appearing in Eq. (IV.5) takes the form that is suitable for subsequent use in the $1/n$-expansion,

$$U_{\text{eff}}(r) = n^2 \left\{ gv(r) + \frac{1}{2r^2} - \frac{2n_r + 1}{2nr^2} + \frac{n_r(n_r+1)}{2n^2 r^2} \right\} \tag{2}$$

(here and below, $\hbar = m = R = 1$). In an analogous way we write the $1/n$-expansion for the energy levels:

$$E_{nn_r} = n^2 \left\{ \varepsilon^{(0)} + \frac{1}{n}\varepsilon^{(1)}_{n_r} + \frac{1}{n^2}\varepsilon^{(2)}_{n_r} + \ldots \right\}. \tag{3}$$

In the limiting case, $n \to \infty$, the particle is localized in the vicinity of the effective potential minimum, r_0, for which

$$gr_0^3 v'(r_0) = 1, \quad \varepsilon^{(0)} = \min U_{\text{eff}} = \frac{1}{2r_0^2} + gv(r_0). \tag{4}$$

---

[206] The interpolation between the regions $N \to \infty$ and $N \sim 1$ is often used in other versions of the $1/N$-expansion method.

Restricting ourselves to the vicinity of this point by expanding to quadratic order in $(r - r_0)$, we obtain using (2),[207]

$$\chi''_{nn_r} - \left[n^2\omega^2(r-r_0)^2 - \frac{n(2n_r+1)}{r_0^2} - 2n\varepsilon_{n_r}^{(1)}\right]\chi_{nn_r}(r) = 0, \quad (5)$$

$$\omega = \left(gv''(r_0) + \frac{3}{r_0^4}\right)^{1/2}.$$

In this approximation we have

$$\varepsilon_{n_r}^{(1)} = \left(\omega - \frac{1}{r_0^2}\right)\left(n_r + \frac{1}{2}\right), \quad (6)$$

$$\chi_{nn_r}^{(0)} = \left\{\frac{m\omega}{2^{2n_r}\pi(n_r!)^2}\right\}^{1/4} H_{n_r}(\xi)e^{-\xi^2/2}, \quad \xi = \sqrt{n\omega}(r - r_0). \quad (7)$$

In an analogous way we can find terms of higher order in the $1/n$-expansion. These higher-order results are quite cumbersome however, and so we illustrate the accuracy of the $1/n$-expansion up to the leading approximation only.

For power-law attractive potentials ($G > 0$, $\nu > -2$),

$$U(r) = G\frac{1}{\nu}r^\nu \equiv n^2\left(\frac{\hbar^2}{mR^2}\right)\frac{1}{\nu}\left(\frac{r}{R}\right)^\nu, \quad R = \left(\frac{n^2\hbar^2}{mG}\right)^{1/(2+\nu)}. \quad (8)$$

$1/n$-expansion for the energy levels takes the form[208]

$$E_{nn_r} = n^2\left(\frac{\hbar^2}{mR^2}\right)\varepsilon_{nn_r} = n^2\left(\frac{\hbar^2}{mR^2}\right)\left\{\varepsilon^{(0)} + \frac{1}{n}\varepsilon_{n_r}^{(1)} + \ldots\right\}$$

$$= n^2\left(\frac{\hbar^2}{mR^2}\right)\left\{\frac{2+\nu}{2\nu} + \frac{1}{n}(\sqrt{2+\nu} - 1)\left(n_r + \frac{1}{2}\right) + \ldots\right\}. \quad (9)$$

Hence for the Coulomb potential, $\nu = -1$, and the spherical oscillator, $\nu = 2$, the exact results for the energy spectra follow.

Let us also consider the case of the linear potential, $\nu = 1$. For this potential, from (9), the $1/n$-expansion for the energy levels has the form:

---

[207] Let us note that here the expansion in $1/n$ is connected with the n-dependence of various terms in (2), and with the subsequent expansion in $(r - r_0)$ that has the smallness of the order of $n^{-1/2}$. Here this expansion in E_{nn_r} contains only integer powers of $1/n$, as in the conditions of Problem 9.33.

[208] In this case, $\omega = \sqrt{2+\nu}$. The singularity appearing here for $\nu = -2$ (compare to Eq. (12) below) manifests the appearance of the "falling into the origin" type of behavior see Problem 9.14.

$$\varepsilon_{nn_r}(1/n) = \frac{2}{3} + \frac{1}{n}(\sqrt{3}-1)\left(n_r + \frac{1}{2}\right) + \ldots \tag{10}$$

A comparison of this result to the true values of $\varepsilon_{nn_r}^{(\text{exact})}$, obtained from numerically solving the radial Schrödinger equation, is given in the table below, where we also give the values of error

$$\delta_{nn_r}^{(0(1))} = \frac{E_{nn_r}(1/n)}{E_{nn_r}^{(\text{exact})}} - 1, \tag{11}$$

in the zeroth- and first-order approximations in the $1/n$-expansion.

	$n_r = 0$			$n_r = 1$		
l	0	1	2	0	1	2
$\varepsilon_{nn_r}^{(\text{exact})}$	1.85576	2.66783	3.37178	3.24461	3.87679	4.46830
$\delta_{nn_r}^{(0)}$	-0.19	-0.11	-0.075	-0.27	-0.20	-0.15
$\delta_{nn_r}^{(1)}$	$5.5 \cdot 10^{-3}$	$1.4 \cdot 10^{-3}$	$6.3 \cdot 10^{-4}$	$2.5 \cdot 10^{-3}$	$1.2 \cdot 10^{-3}$	$7.2 \cdot 10^{-3}$

In conclusion, we present the result of the second-order approximation in $1/n$ for power-law potentials:

$$\varepsilon_{n_r}^{(2)} = \frac{2-\nu}{144(2+\nu)}\{\nu^2 - 15\nu - 52 + 36\sqrt{2+\nu} + 6n_r(n_r+1)\}$$
$$\times (\nu^2 - 9\nu - 34 + 24\sqrt{2+\nu}\}, \tag{12}$$

which increases the accuracy of the approximation.

Problem 9.35

For a short-range attractive potential, $U(r) = U_0 v(r/R)$, $(U(r) \to 0$ for $r \to \infty)$, find, using $1/n$-expansion (see the previous problem), the critical values, $\xi_{nn_r,cr}$, of a parameter in the potential

$$\xi \equiv \frac{mU_0 R^2}{\hbar^2},$$

where the nn_rth level appears as the potential well deepens.

Solution
The problem is solved in a way analogous to the previous one, but now, for the $1/n$-expansion, we determine not the energy level but the value of the coupling constant:

$$U_0 \equiv g_{nn_r} n^2, \quad g_{nn_r} = g_0 + \frac{1}{n}g_{n_r}^{(1)} + \frac{1}{n^2}g_{n_r}^{(2)} + \ldots \tag{1}$$

Here, the expansion coefficients, $g_{n_r}^{(k)}$, are to be found from the condition that, when a level appears, its energy $E = 0$ so $\varepsilon_{n_r}^{(k)} = 0$ at all orders in the $1/n$-expansion.

The position of the classical equilibrium point, r_0, and the value of the coupling constant, g_0, in the zeroth-order approximation are determined from the following relations (see Eq. (4) from the previous problem):

$$r_0 v'(r_0) = -2v(r_0), \quad g_0 = -\frac{1}{2r_0^2 v(r_0)}. \tag{1}$$

For the first-order correction we obtain

$$g_{n_r}^{(1)} = -\frac{2n_r + 1}{2v(r_0)}\left(\omega - \frac{1}{r_0^2}\right) = (2n_r + 1)(\omega r_0^2 - 1)g_0, \tag{2}$$

where the frequency, ω, is given by Eq. (5) from the previous problem with the substitution, $g \to g_0$.

We illustrate the application of the results obtained on the example of the Yukawa potential,[209] $U(r) = -U_0(R/r)\exp(-r/R)$. In this case,

$$g_{nn_r} = e\left\{\frac{1}{2} - \frac{\sqrt{2}-1}{2\sqrt{2}}(2n_r + 1)\frac{1}{n} + \ldots\right\}, \tag{3}$$

where $e = 2.718\ldots$. A comparison of this relation to the exact values, $g_{nn_r}^{(\text{exact})}$, obtained by numerically solving the radial Schrödinger equation, is given in the following tables.

	$n_r = 0$			
l	0	1	2	5
$g_{nn_r}^{(\text{exact})}$	0.8399	4.5410	10.947	46.459
$\delta_{nn_r}^{(0)}$	0.62	0.20	0.12	0.053
$\delta_{nn_r}^{(1)}$	0.14	0.022	$8.3 \cdot 10^{-3}$	$1.8 \cdot 10^{-3}$

	$n_r = 1$			
l	0	1	2	5
$g_{nn_r}^{(\text{exact})}$	3.2236	8.8723	17.210	58.496
$\delta_{nn_r}^{(0)}$	0.69	0.38	0.26	0.14
$\delta_{nn_r}^{(1)}$	-0.054	-0.025	-0.014	$-5.6 \cdot 10^{-3}$

---

[209] Here the role of the second-order correction, $g_{n_r}^{(2)}$, is more important than in the previous problem.

Here, we also give the values of the errors:

$$\delta_{nn_r}^{(0(1))} = \frac{g_{nn_r}(1/n)}{g_{nn_r}^{(\text{exact})}} - 1, \qquad (4)$$

for the corresponding approximation in the $1/n$-expansion (compare this to the previous problem).

10

Identical particles; Second quantization

The wavefunction of a system that contains *identical* particles has a definite symmetry with respect to particle permutations, so that

$$\psi(\ldots,\xi_\alpha,\ldots,\xi_b,\ldots) = \pm\psi(\ldots,\xi_b,\ldots,\xi_\alpha,\ldots).$$

Here $\xi_n \equiv (\mathbf{r}_n, \sigma_n)$ denotes a set of variables (position and spin) for each of the corresponding particles. The wavefunction is symmetric with respect to permutations of particles with integer spin – *bosons* – and it is anti-symmetric for particles with half-integer spin – *fermions*. Therefore, when the system is in a definite quantum state its wavefunction in general can be obtained by symmetrizing products of single-particle wavefunctions for bosonic particles and anti-symmetrization of such wavefunctions for fermions. The total state of the system can also be described by designating occupation numbers of the single-particle states.

It is convenient to use the many-particle second-quantization formalism that automatically ensures the proper quantum-mechanical symmetry of identical particles. This formalism relies on the *occupation number (Fock) representation* with operators $\hat{a}_i^+$, $\hat{a}_i$ – *the particle creation and annihilation operators* (in corresponding discrete quantum states characterized by i). In the case of bosons, these operators obey the canonical commutation relations:

$$[\hat{a}_i, \hat{a}_k] = [\hat{a}_i^+, \hat{a}_k^+] = 0, \quad [\hat{a}_i, \hat{a}_k^+] \equiv \hat{a}_i\hat{a}_k^+ - \hat{a}_k^+\hat{a}_i = \delta_{ik}, \qquad (\text{X.1})$$

while in the case of fermions, they obey the canonical anti-commutation relations:

$$\{\hat{a}_i, \hat{a}_k\} = \{\hat{a}_i^+, \hat{a}_k^+\} = 0, \quad \{\hat{a}_i, \hat{a}_k^+\} \equiv \hat{a}_i\hat{a}_k^+ + \hat{a}_k^+\hat{a}_i = \delta_{ik}, \qquad (\text{X.2})$$

(so that for fermion operators, $\hat{a}_i^2 = (\hat{a}_i^+)^2 = 0$).

The operator $\hat{n}_i = \hat{a}_i^+ \hat{a}_i$ yields the particle number in the corresponding quantum state i. For states normalized to unity with well-defined occupation numbers we have

$$\hat{a}_i|\ldots,n_i,\ldots\rangle = \sqrt{n_i}|\ldots,n_i-1,\ldots\rangle;$$

$$\hat{a}_i^+|\ldots,n_i,\ldots\rangle = \sqrt{n_i+1}|\ldots,n_i+1,\ldots\rangle$$

(for fermions $n_i = 0$ or 1, for bosons $n_i = 0, 1, 2, \ldots$).

448 *Exploring Quantum Mechanics*

For a system of identical particles (bosons or fermions), the operator for an additive one-particle physical quantity, $\hat{F}^{(1)} = \sum_a \hat{f}_a^{(1)}$, is expressed in the occupation-number representation in terms of the $\psi(\xi)$-operators as follows:[210]

$$\hat{F}^{(1)} = \int \hat{\psi}^+(\xi) \hat{f}^{(1)} \hat{\psi}(\xi) d\xi, \qquad (X.3)$$

where $\hat{f}^{(1)}$ is the usual single-particle operator in the coordinate representation, while the operators $\hat{\psi}$, $\hat{\psi}^+$ act in the space of occupation-number functions.

Similarly, an additive two-particle quantity $F^{(2)} = \sum_{a<b} f_{ab}^{(2)}$ in the occupation number representation has the form

$$\hat{F}^{(2)} = \frac{1}{2} \int \hat{\psi}^+(\xi) \hat{\psi}^+(\xi') \hat{f}^{(2)} \hat{\psi}(\xi') \hat{\psi}(\xi) d\xi d\xi'. \qquad (X.4)$$

10.1 Quantum statistics; Symmetry of wavefunctions

Problem 10.1

Consider a system of two identical particles with spin s. Find the number of different spin states symmetric (anti-symmetric) with respect to the interchange of the spin variables.

Solution

Spin functions (not symmetrized in the spin variables) have the form[211] $\chi_{s_z}(\sigma_1)\chi_{s'_z}(\sigma_2)$. $(2s+1)^2$ functions of this form are possible. The following combinations of these functions:

---

[210] Operators $\hat{\psi}(\xi)$, $\hat{\psi}^+(\xi)$ represent an important special case of the operators $\hat{a}_i^+$, $\hat{a}_i$ that create and destroy particles in position and spin states designated by $i = \mathbf{r}, \sigma$. Another common choice is when $\hat{a}_i^+$, $\hat{a}_i$ correspond to states denoted by $i = \mathbf{p}, \sigma$. To obtain a discrete spectrum of momentum eigenvalues, the system can be assumed to be in a "box" of large but finite volume $V = L^3$. An expansion of the ψ-operator in plane waves takes the form:

$$\hat{\psi}(\xi) = \frac{1}{\sqrt{V}} \sum_{\mathbf{k},\sigma} e^{i\mathbf{k}\cdot\mathbf{r}} \hat{a}_{\mathbf{k},\sigma}, \quad \mathbf{p} = \hbar\mathbf{k}.$$

The limit $V \to \infty$ is taken using the substitution $\sum_{\mathbf{k}} \cdots \to (V/(2\pi)^3) \int d^3k \ldots$.

Generalization of Eqs. (X.3) and (X.4) to the case of some arbitrary a_i-operator (instead of $\hat{\psi}(\xi)$) reduces to using the i-representation for operators $\hat{f}^{(1)}$, $\hat{f}^{(2)}$.

[211] Here $\chi_{s_z}(\sigma)$ is a normalized single-particle spin wavefunction with a definite value of spin projection s_z. In the s_z-representation it has the form $\chi_{s_z}(\sigma) = \delta_{\sigma,s_z}$.

$$\psi^+_{s_z s_z} = \chi_{s_z}(\sigma_1)\chi_{s_z}(\sigma_2),$$

$$\psi^{\pm}_{s_z s'_z} = \frac{1}{\sqrt{2}}\{\chi_{s_z}(\sigma_1)\chi_{s'_z}(\sigma_2) \pm \chi_{s'_z}(\sigma_1)\chi_{s_z}(\sigma_2)\}, \quad s_z \neq s'_z, \tag{1}$$

normalized to unity, have a definite symmetry with respect to particle permutation: ψ^+ are symmetric, while ψ^- are anti-symmetric functions. The number of independent symmetric and anti-symmetric states are $(s+1)(2s+1)$ and $s(2s+1)$ respectively.

The wavefunctions given above do not correspond to definite values of the total spin S (except for the case of $s = 1/2$; see Problem 5.10). However, using the result of Problem 3.30 we can state that in the symmetric states the total spin values are $S = 2s, \ 2s - 2, \ 2s - 4, \ldots$, while in the anti-symmetric states, the values are $S = 2s - 1, \ 2s - 3, \ldots$ (here $S \geq |s_z + s'_z|$).

Problem 10.2

Show that if n identical particles with spin s are in different orbital states $\varphi_1(\mathbf{r}), \varphi_2(\mathbf{r}), \ldots, \varphi_n(\mathbf{r})$ (here $\langle \varphi_i | \varphi_k \rangle = \delta_{ik}$), then the total number of independent states, including the spin degrees of freedom, is equal to $G = (2s+1)^n$, and it does not depend on the particle quantum statistics.

What is the number of possible states in the case where the particles are distinguishable?

Solution

Taking into account the spin, the wavefunctions of single-particle states have the form:

$$\psi_{i,s_{z,i}} = \varphi_i(\mathbf{r})\chi_{s_{z,i}}(\sigma),$$

where χ_{s_z} is the spin function (the index i in $s_{z,i}$ emphasizes that different orbital states s_z have generally different values for $s_{z,i}$). The set of numbers $s_{z,i}$ uniquely determines n different single-particle states:

$$\psi_{1,s_{z,1}}, \psi_{2,s_{z,2}}, \ldots, \psi_{n,s_{z,n}} \tag{1}$$

as well as a unique state of the total system of n identical particles. The system wavefunction is obtained by symmetrization (or anti-symmetrization) of products of functions (1). Any change of the set of values $s_{z,i}$ gives another distinct set of single-particle states (1) (here it is important that all wavefunctions φ_i are different). Since each $s_{z,i}$ can take $(2s+1)$ different values, the total number of different single-particle sets (1) and hence the number of independent system states is equal to $(2s+1)^n$.

For a system of distinguishable particles, the manner in which particles are assigned to different single-particle states is important. The number of different orbital states here is equal to $n!$, while the total number of system states, with the spin degrees of freedom taken into account, is equal to $(2s+1)^n n!$ (here there is no symmetrization (nor anti-symmetrization) of wavefunctions).

Problem 10.3

Let $\psi_{f_i}(\xi)$ be the wavefunctions of single particle states normalized to unity (f_i are the quantum numbers of some complete set). Write normalized wavefunctions for states of a system of three identical weakly interacting particles (consider both the bosonic and fermionic cases) which occupy single particle states with given quantum numbers f_1, f_2, f_3.

Solution

For bosons, the form of the system wavefunction depends on whether or not the occupied single-particle states coincide. Here we should consider three cases.

1) All the particle are in the same state, $f_1 = f_2 = f_3 = f$. The system wavefunction, normalized to unity, is $\psi = \psi_f(1)\psi_f(2)\psi_f(3)$ (here and below instead of particle variables we give only the quantum number, so $\psi(1) \equiv \psi(\mathbf{r}_1, \sigma_1)$, etc.).

2) Two of three occupied states coincide. Now,

$$\psi = \frac{1}{\sqrt{3}}\{\psi_1(1)\psi_2(2)\psi_2(3) + \psi_2(1)\psi_1(2)\psi_2(3) + \psi_2(1)\psi_2(2)\psi_1(3)\} \quad (1)$$

(we put $f_1 \neq f_2 = f_3$ for concreteness and write the index a instead of f_a for brevity). The expression inside the curly brackets is derived from the wavefunction symmetry condition with respect to the permutation for any two particles. The coefficient $\frac{1}{\sqrt{3}}$ is determined by the normalization condition for the wavefunction ψ:

$$\int |\psi|^2 d\xi_1 d\xi_2 d\xi_3 = 1, \quad \int \psi_a^*(\xi)\psi_b(\xi)d\xi = \delta_{ab} \quad (2)$$

(integration over ξ includes also a summation over the spin variable); while calculating the normalization integral, only three of the nine terms in $|\psi^2|$ give a non-vanishing contribution due to the orthogonality of the single-particle wavefunctions given in (2).

3) If all three occupied states are different, then

$$\psi = \frac{1}{\sqrt{6}}\{\psi_1(1)\psi_2(2)\psi_3(3) + \psi_2(1)\psi_3(2)\psi_1(3) + \psi_3(1)\psi_1(2)\psi_2(3)\pm$$
$$\pm \psi_1(1)\psi_3(2)\psi_2(3) \pm \psi_3(1)\psi_2(2)\psi_1(3) \pm \psi_2(1)\psi_1(2)\psi_3(3)\}, \quad (3)$$

where the upper signs correspond to bosons.

In the case of fermions, all three occupied states must be different, and the anti-symmetric wavefunction is determined by relation (3) with the choice of lower signs.

Problem 10.4

Three identical bosons with spin $s = 1$ are in the same orbital states described by the wavefunction $\varphi(\mathbf{r})$. Write the normalized spin functions for the total system. How many independent states are there? What are the possible values of the total spin of the system?

Solution

Since the coordinate part of the system wavefunction, $\varphi(\mathbf{r}_1)\varphi(\mathbf{r}_2)\varphi(\mathbf{r}_3)$, is symmetrical, the spin part must also be symmetrical with respect to the particles' mutual permutation. Non-symmetrized products of spins functions have the form:

$$\chi_{s_{z,1}}(\sigma_1)\chi_{s_{z,2}}(\sigma_2)\chi_{s_{z,3}}(\sigma_3),$$

where $\chi_{s_{z,i}}(\sigma)$ is the spin function of a single particle with a definite value of the spin projection s_z. The number of these products is $3 \times 3 \times 3 = 27$, but the symmetry condition reduces the number of independent states. Such states correspond to different (not connected to each other by permutations) sets $\{s_{z,a}\}$ of values s_z for single particles. The corresponding spin functions of the system are obtained by symmetrization of the one-particle spin functions; compare to the previous problem.

As an example, for the set of values $s_{z,1} = s_{z,2} = s_{z,3} = 1$ we have

$$\psi_{111} = \chi_1(1)\chi_1(2)\chi_1(3) = \begin{pmatrix}1\\0\\0\end{pmatrix}_1 \begin{pmatrix}1\\0\\0\end{pmatrix}_2 \begin{pmatrix}1\\0\\0\end{pmatrix}_3, \qquad (1)$$

while for the set $s_{z,1} = s_{z,2} = 1$, $s_{z,3} = 0$,

$$\psi_{110} = \frac{1}{\sqrt{3}}\{\chi_1(1)\chi_1(2)\chi_0(3) + \chi_1(1)\chi_0(2)\chi_1(3) + \chi_0(1)\chi_1(2)\chi_1(3)\} =$$

$$= \frac{1}{\sqrt{3}}\left\{\begin{pmatrix}1\\0\\0\end{pmatrix}_1\begin{pmatrix}1\\0\\0\end{pmatrix}_2\begin{pmatrix}0\\1\\0\end{pmatrix}_3 + \begin{pmatrix}1\\0\\0\end{pmatrix}_1\begin{pmatrix}0\\1\\0\end{pmatrix}_2\begin{pmatrix}1\\0\\0\end{pmatrix}_3 + \right.$$

$$\left. + \begin{pmatrix}0\\1\\0\end{pmatrix}_1\begin{pmatrix}1\\0\\0\end{pmatrix}_2\begin{pmatrix}1\\0\\0\end{pmatrix}_3\right\}. \qquad (2)$$

We will also write the other sets of values $(s_{z,1}, s_{z,2}, s_{z,3})$ that lead to new states of the system: $(1, 1, -1)$, $(1, 0, 0)$, $(1, 0, -1)$, $(1, -1, -1)$, $(0, 0, 0)$, $(0, 0, -1)$, $(0, -1, -1)$, $(-1, -1, -1)$. The total number of independent spin states is equal to 10.

From these ten states, seven correspond to the value $S = 3$ of total system spin, while the remaining three correspond to the value $S = 1$. Indeed spin function (1) corresponds to $S = 3$. Function (2) also corresponds to $S = 3$ with projection $S_z = 2$. Two states, $(1, 1, -1)$ and $(1, 0, 0)$, correspond to the value $S_z = 1$. Both these states represent a superposition of states with the total spin $S = 3$ and $S = 1$.

Problem 10.5

Indicate the restrictions on the quantum numbers (spin J_A and intrinsic parity P_A) of some neutral particle A^0 that follow from the fact that it decays to the two π^0-mesons: $A^0 \to 2\pi^0$ (for pion $J_\pi^P = 0^-$). Compare to Problem 5.30.

Solution

The total angular momentum of two pions in their center-of-inertia frame (it is also a rest frame for the particle A_0) coincides with the orbital momentum L of their relative motion which, due to the angular momentum conservation, is equal to the spin of the particle A^0, so that $J_A = L$. However, the condition of symmetry on the two π^0-mesons wavefunctions requires that L take only even values. Thus $J_A = 0, 2, 4$ (indeed, the permutation of pions is equivalent to coordinate inversion in their center-of-inertia frame. Since $\mathbf{r} = \mathbf{r}_1 - \mathbf{r}_2$, the permutation multiplies the wavefunction by $(-1)^L$). Here the two-pion system parity is even, and if it is conserved during the decay, then $P_A = +1$.

Problem 10.6

It is known that the reaction $\pi^- + d \to n + n$, corresponding to the capture of a slow π^--meson (its spin is $J_\pi = 0$), occurs from the ground state of a meso-deuterium with the parity conservation.

Taking into account the fact that proton's and neutron's intrinsic parities are the same and the quantum numbers of deuteron $J_A^P = 1^+$, find the pion intrinsic parity.

Solution

For this problem, the quantum numbers of the $\pi^- d$-system are the following:[212] $J = J_d = 1$ is the total angular momentum, and $P = P_\pi$ is the parity that coincides with the pion's intrinsic parity. Due to the angular momentum conservation, the two neutrons in the final state also have total angular momentum $J = 1$ (in the center-of-inertia system), and since for them $\mathbf{J} = \mathbf{L} + \mathbf{S}$ (L is the orbital momentum of the relative motion, S is the total spin, the neutron's spin is $s = 1/2$), only the following values of L and S are possible:

$$1)\ L = 0,\ S = 1,\quad 2)\ L = 1,\ S = 0; \tag{1}$$
$$3)\ L = 1,\ S = 1,\quad 4)\ L = 2,\ S = 1.$$

It is seen that the anti-symmetric condition for the two-neutron wavefunction forbids them from having the states with quantum number sets 1), 2), and 4). In this case, we should take into account the fact that spin functions with $S = 1$ and $S = 0$ are symmetric and anti-symmetric respectively (see Problem 5.10), while the symmetry of the coordinate functions with a given value of the orbital momentum L coincides with the parity of these functions $(-1)^L$ (since the coordinate permutation is equivalent to

---

[212] For the ground state of a meso-deuterium (i.e., a $\pi^- d$-atom), the angular moment is $l = 0$.

a reflection with respect to the center-of-mass, $\mathbf{r} = \mathbf{r}_1 - \mathbf{r}_2$). Therefore, for the total angular momentum, $J = 1$, the two-neutron system could have only the quantum numbers $L = 1$ and $S = 1$ and hence negative parity. Thus it follows that $P_\pi = -1$ and the pion is a so-called *pseudo-scalar* particle, *i.e.*, it has $J_\pi^P = 0^-$.

Problem 10.7

Three identical bosonic particles with zero spin weakly interacting with each other are in stationary states with the same quantum numbers n_r and l, where $l = 1$, in some central potential. Show that a total angular momentum L of the system cannot take on a zero value.

Solution

The problem of possible states for a system of three bosons is equivalent to Problem 10.4, and wavefunctions could be obtained by the substitution of the spin functions χ_{s_z} in Problem 10.4 for the spherical functions $Y_{1m}(\mathbf{n})$. Hence it follows that the total angular moment could take the values $L = 3$ and $L = 1$.

The states of the system with the total angular momentum $L = 0$ could be much more easily obtained from the form of the wavefunction $\psi_{L=0}$. For this problem, three independent single-particle states with $l = 1$ could be chosen in the form $\psi_i = x_i f(r)$, where x_i are the components of the radius-vector, $\mathbf{r}$; see Problem 3.21. The wavefunctions (non-symmetrized) for a system of three particles in $l = 1$ states have the form:

$$\psi_{ikl} = x_{1i} x_{2k} x_{3l} f(r_1) f(r_2) f(r_3). \tag{1}$$

The wavefunction of a state with total angular momentum $L = 0$ does not change under the rotation of the coordinate system. From the wavefunctions (1) it is possible to make only one scalar (or pseudo-scalar) function that has the property:

$$\psi_{L=0} = (\mathbf{r}_1 \cdot [\mathbf{r}_2 \times \mathbf{r}_3]) \, f(\mathbf{r}_1) f(\mathbf{r}_2) f(\mathbf{r}_3), \tag{2}$$

but it is anti-symmetric with respect to particle permutation and could not describe a bosonic system.

Problem 10.8

Indicate possible values of a total spin S of two identical Bose particles with spin s in a state with the total angular momentum L (L here is the angular momentum in the center-of-inertia system), *i.e.*, what states ^{2S+1}L of the system are possible? Consider, for example, the case of $s = 0$.

Solution

A coordinate permutation of two particles is equivalent to their reflection with respect to the center of mass (since $\mathbf{r} = \mathbf{r}_1 - \mathbf{r}_2$). Hence the symmetry of the coordinate wavefunction for a state with a given value of angular momentum L coincides with the parity of the state, which is equal to $(-1)^L$. The bosonic symmetry property requires that in the even states, spin variable permutation does not change the wavefunction, while in the odd states it yields a sign change for the wavefunction. Hence, taking into account the result of Problem 3.30 for the wavefunction symmetry over addition of two identical angular momenta, we see that for the states with angular momenta $L = 0, 2, 4, \ldots$, only the values of total spin $S = 2s, 2s-2, \ldots, 0$ are possible, while for the states with $L = 1, 3, 5, \ldots$, only $S = 2s-1, 2s-3, \ldots, 1$ are possible.

As an example, for spinless bosons $(s=0)$ only the even values of L are possible. Consequently, we obtain an exclusion rule that prohibits a of a neutral particle with spin $S_v = 1$ (*vector* meson) into two π^0-mesons; see also Problem 10.5.

Problem 10.9

The same as the previous problem, but considering identical Fermi particles.

Result

In a system of two identical fermions with an even angular momentum, L, the total spin could take even values $S = 2s-1, 2s-3, \ldots, 0$, while for the odd L only the odd $S = 2s, 2s-2, \ldots, 1$ are possible. Compare to the previous problem, and also to Problem 10.6.

Problem 10.10[213]

A system consists of two identical spinless Bose particles in states described by mutually orthogonal wavefunctions, normalized to unity, $\psi_{1,2}(\mathbf{r})$. Indicate the probability of finding both particles in the same small volume dV. Compare it to the case of distinguishable particles.

Solution

The system's wavefunction normalized to unity has the form

$$\psi(\mathbf{r}_1, \mathbf{r}_2) = \frac{1}{\sqrt{2}}\{\psi_1(\mathbf{r}_1)\psi_2(\mathbf{r}_2) + \psi_2(\mathbf{r}_1)\psi_1(\mathbf{r}_2)\},$$

$$\iint |\psi|^2 dV_1 dV_2 = 1. \qquad (1)$$

---

[213] Problems 10.10 and 10.11 give examples of important interference effects between states of identical particles. This is in contrast to the interference effects in single-particle quantum mechanics, where the interference occurs between states describing the same particle.

Identical particles; Second quantization 455

The probabilities of two particles to be in the same volume dV simultaneously is equal to

$$dw_{bos} = |\psi(\mathbf{r},\mathbf{r})|^2 dV\, dV = |\psi_1(\mathbf{r})|^2 dV \cdot |\psi_2(\mathbf{r})|^2 dV + |\psi_1(\mathbf{r})\psi_2(\mathbf{r})|^2 (dV)^2, \qquad (2)$$

which is greater than the analogous probability,

$$dw_{dist} = |\psi_1(\mathbf{r})|^2 dV \cdot |\psi_2(\mathbf{r})|^2 dV, \qquad (3)$$

for the case of distinguishable particles. This result shows the existence of interference between identical particles. We can characterize it as a tendency for bosons to mutually approach. Different aspects of this interference phenomenon are discussed in the following problem.

Such an interference between identical particles, in the same spin states, also takes place in the case of fermions. Here $dw_{ferm} = 0$, and we can describe the interference as a tendency for fermions to mutually repel. For particles (both bosons and fermions) in different (orthogonal) spin states, there is no such interference.

Problem 10.11

Two identical Bose particles with zero spin occupy two states described by wave-functions, $\psi_{1,2}(\mathbf{r})$, normalized to unity. Find the (average) particle density for such a system, and compare it to the case of distinguishable particles.

Solution

$$\psi(\mathbf{r}_1,\ \mathbf{r}_2) = C\{\psi_1(\mathbf{r}_1)\psi_2(\mathbf{r}_2) + \psi_2(\mathbf{r}_1)\psi_1(\mathbf{r}_2)\}, \qquad (1)$$

where $2C^2 = (1 + |\langle \psi_1|\psi_2\rangle|^2)^{-1}$. The mean particle density is obtained by averaging the corresponding operator,

$$\hat{n}(\mathbf{r}) \equiv n(\mathbf{r}) = \sum_a \delta(\mathbf{r} - \mathbf{r}_a), \qquad (2)$$

where the summation is performed over all particles. Such an operator form is connected with the fact that the analogous classical quantity depends only on the particle coordinates (but not on their momenta; compare to the operators of potential energy $\hat{U}(\mathbf{r}_a) = U(\mathbf{r}_a)$ in the coordinate representation), where $\mathbf{r}$ acts as an "external" parameter. We see that

$$\overline{n}(\mathbf{r}) \equiv \langle \psi|\hat{n}(\mathbf{r})|\psi\rangle = 2C^2\{|\psi_1(\mathbf{r})|^2 + |\psi_2(\mathbf{r})|^2 + \Delta(\mathbf{r})\}, \qquad (3)$$

where

$$\Delta(\mathbf{r}) = \psi_1(\mathbf{r})\psi_2^*(\mathbf{r})\langle\psi_1|\psi_2\rangle + \psi_2(\mathbf{r})\psi_1^*(\mathbf{r})\langle\psi_2|\psi_1\rangle.$$

Let us discuss the result obtained for $\overline{n}(\mathbf{r})$. First we should note that $\int \overline{n}(\mathbf{r})dV = 2$ is independent of the form of the functions, $\psi_{1,2}(\mathbf{r})$. If these functions are orthogonal, so that $\langle \psi_1|\psi_2\rangle = 0$, then $C^2 = 1/2$, $\Delta(\mathbf{r}) = 0$ and

$$\overline{n}(\mathbf{r}) = |\psi_1(\mathbf{r})|^2 + |\psi_2(\mathbf{r})|^2 \equiv \overline{n}_{dist}(\mathbf{r}), \tag{4}$$

as in the case of distinguishable particles. But if $\langle\psi_1|\psi_2\rangle \neq 0$, then $\overline{n}(\mathbf{r})$ differs from $\overline{n}_{dist}(\mathbf{r})$. There appears an interference between the different (but identical) particles, as discussed in the previous problem. Since $2C^2 < 1$, in the areas of space where the wavefunctions $\psi_{1,2}(\mathbf{r})$ do not "overlap" (so that $\psi_1^*(\mathbf{r})\psi_2(\mathbf{r}) \approx 0$), we have $\overline{n} < \overline{n}_{dist}$. Taking into account the normalization of $\overline{n}(\mathbf{r})$, we see that in the essential region of wavefunction overlap it is $\overline{n} > \overline{n}_{dist}$, in accordance with the tendency of bosons to mutually approach. For fermions in the same spin states, the change of sign in terms that include $|\langle\psi_1|\psi_2\rangle|^2$ and Δ corresponds (here $2C^2 \leq 1$) to the opposite interference effect – mutual repulsion; compare to Problem 10.10.

10.2 Elements of the second quantization formalism (the occupation-number representation)

Problem 10.12

Find the commutation relation for the Hermitian and anti-Hermitian parts of a bosonic annihilation operator $\hat{a}$ (or creation operator $\hat{a}^+$).

Solution

Writing

$$\hat{a} = \frac{1}{2}(\hat{a} + \hat{a}^+) + i\frac{1}{2i}(\hat{a} - \hat{a}^+) \equiv \hat{A} + i\hat{B},$$

(here $\hat{A} = (\hat{a} + \hat{a}^+)/2$), we find $[\hat{A}, \hat{B}] = i/2$; compare to $[\hat{p}, \hat{x}] = -i\hbar$, and see the following problem.

Problem 10.13

In the terms of operators for position $\hat{x}$ and momentum $\hat{p}$ of a particle, construct the operators $\hat{a}$ and $\hat{a}^+$ having the properties of boson annihilation and creation operators.

Determine the wavefunction $\psi_0(x)$ of the vacuum state.

Solution
Writing $\hat{a} = \alpha\hat{x} + \beta\hat{p}$ and $\hat{a}^+ = \alpha^*\hat{x} + \beta^*\hat{p}$, we have

$$[\hat{a}, \hat{a}^+] = i\hbar(\alpha\beta^* - \alpha^*\beta) = 1.$$

The choice of parameters α, β is not unique. We can choose, for example,

$$\alpha = \frac{1}{\sqrt{2L}}, \quad \beta = \frac{iL}{\sqrt{2\hbar}}.$$

(L is a real parameter with dimensions of length.)

From the condition $\hat{a}|0\rangle = 0$ or

$$\left(\frac{x}{L} + L\frac{\partial}{\partial x}\right)\psi_0(x) = 0,$$

we find the wavefunction of the vacuum state,

$$\psi_0(x) = (\pi L^2)^{-1/4} \exp\left\{-\frac{x^2}{2L^2}\right\}. \tag{1}$$

This wavefunction has the form of linear oscillator ground state, see Problem (11.2). Such a state does not change with time (it is stationary). In the case of a real free particle, the Gaussian wave packet (1) *spreads out*; see Problem (6.2) and (6.21), which in terms of pseudo-particles (corresponding to the operators $\hat{a}$ and $\hat{a}^+$) could be interpreted as their creation and annihilation with time.

Problem 10.14

Can we consider $\hat{a}'^\dagger$, $\hat{a}'$ as creation and annihilation operators of some new particles corresponding to the transformation of the form $\hat{a}' = \hat{a}^+$, $\hat{a}'^+ = \hat{a}$? Analyze the states $|n'\rangle$ (*i.e.*, states with a definite number n of new particles) in the basis of initial particles states. Write the unitary operator $\hat{U}$ that accomplishes the transformation considered.

Solution

For the bosonic operators we cannot do this, since in this case $[\hat{a}', \hat{a}'^+] = -1$ (unlike $[\hat{a}, \hat{a}^+] = 1$).

In the case of fermion operators, the transformation is legitimate, since we still have $\hat{a}'^2 = 0$ and $\{\hat{a}', \hat{a}'^+\} = 1$. Here the vacuum state of "new" particles $|0'\rangle$ is the one-particle state $|1\rangle$ in the initial particle basis, *i.e.*, $|0'\rangle = |1\rangle$ and $|1'\rangle = |0\rangle$. Such "new" particles are called *holes* (on the background of initial particles). For fermions the transformation considered is unitary and is provided by the operator $\hat{U} = \hat{a} + \hat{a}^+$, since here, $\hat{a}' = \hat{U}\hat{a}\hat{U}^+ = \hat{a}^+$.

Problem 10.15

Find the eigenfunctions and eigenvalues of the annihilation and creation operators. For these states, obtain a distribution function over the number of particles. Consider the cases of bosonic and fermionic operators.

Prove that for a linear oscillator, the eigenfunctions of the annihilation operator $\hat{a} = (m\omega\hat{x} + i\hat{p})/\sqrt{2m\hbar\omega}$ describe the *coherent states*; see also Problem 6.21.

Solution

1) First we remind ourselves that operators $\hat{a}$ and $\hat{a}^+$ act in the space of state vectors,

$$|\psi\rangle = \sum c_n |n\rangle = c_0|0\rangle + c_1|1\rangle + \ldots,$$

where symbol $|n\rangle$ corresponds to the n-particle state. Here,

$$\hat{a}|n\rangle = \sqrt{n}|n-1\rangle, \quad \hat{a}^+|n\rangle = \sqrt{n+1}|n+1\rangle,$$

for fermions $\hat{a}^\dagger|1\rangle = 0$.

Eigenfunctions $|\alpha\rangle \equiv \sum c_n|n\rangle$ and eigenvalues α of the boson operator $\hat{a}$ are determined from the relation $\hat{a}|\alpha\rangle = \alpha|\alpha\rangle$. Since[214]

$$\hat{a}|\alpha\rangle = \hat{a}\sum c_n|n\rangle = \sum c_n \sqrt{n}|n-1\rangle = \sum c_{n+1}\sqrt{n+1}|n\rangle,$$

then the relation takes the form:

$$\sum_n (c_{n+1}\sqrt{n+1} - \alpha c_n)|n\rangle = 0. \tag{1}$$

Therefore, taking into account the independence of states $|n\rangle$, it follows that

$$c_{n+1} = \frac{\alpha}{\sqrt{n+1}} c_n = \frac{\alpha}{\sqrt{n+1}} \cdot \frac{\alpha}{\sqrt{n}} c_{n-1} = \cdots = \frac{\alpha^{n+1}}{\sqrt{(n+1)!}} c_0. \tag{2}$$

It can be seen that the eigenvalue of a bosonic operator $\hat{a}$ can be any complex number α (operator $\hat{a}$ is non-Hermitian!), while the corresponding eigenfunction $|\alpha\rangle$ could be normalized to unity. The condition $\langle\alpha|\alpha\rangle = 1$ gives

$$\langle\alpha|\alpha\rangle = \sum_n |c_n|^2 = |c_0|^2 \sum_n \frac{|\alpha|^{2n}}{n!} = 1, \quad \text{i.e.} \quad |c_0|^2 = e^{-|\alpha|^2}, \tag{3}$$

so that the distribution over the number of particles in state $|\alpha\rangle$ is determined by the relation,

$$w_n = |c_n|^2 = \exp\{-|\alpha|^2\}\frac{|\alpha|^{2n}}{n!}, \tag{4}$$

and is a Poisson distribution with $\bar{n} = |\alpha|^2$.

The equation for eigenfunctions and eigenvalues of the bosonic operator $\hat{a}^+$ has no solution. The fermion operators $\hat{a}$, $\hat{a}^+$ have one eigenfunction each: $|0\rangle$ is the eigenfunction for $\hat{a}$, while $|1\rangle$ is the eigenfunction for $\hat{a}^+$; the corresponding eigenvalues are equal to 0.

---

[214] Here and below, the summation over n in all the equations is performed from $n = 0$ to $n = \infty$.

2) Let us now consider the linear oscillator and find the form of eigenfunctions $\psi_n(x)$ of the operator

$$\hat{a} = \frac{1}{\sqrt{2m\hbar\omega}}(m\omega\hat{x} + i\hat{p}) \tag{5}$$

in the coordinate representation. From the relation $\hat{a}\psi_\alpha = \alpha\psi_\alpha$ it follows that

$$\psi_\alpha(x) = C\exp\left\{-\frac{m\omega}{2\hbar}\left(x - \sqrt{\frac{2\hbar}{m\omega}}\alpha\right)^2\right\}, \tag{6}$$

which is the wavefunction of a coherent state considered in Problem 6.21. If we put (compare the relation for eigenvalues with the form of operator (5))

$$\alpha = \frac{1}{\sqrt{2m\hbar\omega}}(m\omega x_0 + ip_0),$$

then we see that the change of the coherent state in time happens so that its wavefunction at any time t remains an eigenfunction of the operator $\hat{a}$, but with a time-dependent eigenvalue $\alpha(t)$, where

$$\alpha(t) = \alpha e^{-i\omega t}. \tag{7}$$

This fact is evident in the Heisenberg representation, where $\hat{a} = e^{-i\omega t}\hat{a}$ (compare to Problem 6.25), but the wavefunction ψ_α does not depend on time. It also follows from the expansion given above that $|\alpha\rangle = \sum c_n|n\rangle$, if we take into account that for an oscillator,

$$|n,t\rangle = \exp\left\{-\frac{i}{\hbar}E_n t\right\}|n\rangle, \quad E_n = \hbar\omega\left(n + \frac{1}{2}\right),$$

and use relation (2) for the expansion coefficients.

Problem 10.16

Is the transformation from the operators $\hat{a}$, $\hat{a}^+$ to new operators $\hat{a}' = \hat{a} + \alpha$, $\hat{a}'^+ = \hat{a}^+ + \alpha^*$ (α is a complex number) a unitary? What is the form of the unitary operator here? Consider the cases of bosonic and fermionic operators $\hat{a}$, $\hat{a}^+$.

Analyze the vacuum states for "new" particles $|0'\rangle$ in the basis of initial particles states $|n\rangle$, and find the number distribution over the initial particle states.

Solution
For the fermionic operators, $(\hat{a}')^2 = 2\alpha\hat{a} + \alpha^2 \neq 0$ (for $\alpha \neq 0$), and therefore the transformation considered is not unitary.

For the boson operators, it is still $[\hat{a}', \hat{a}'^+] = 1$, and the transformation is unitary. Taking into account the result from Problem 6.19, we find an explicit form of the unitary operator,

$$\hat{U} = \exp\{\alpha^*\hat{a} - \alpha\hat{a}^+\}, \tag{1}$$

that performs such a transformation – here, $\hat{a}' = \hat{U}\hat{a}\hat{U}^+$. Then, using the equation from the condition for Problem 1.10, we can write operator (1) in the form:

$$\hat{U} = e^{-|\alpha|^2/2}e^{-\alpha\hat{a}^+}e^{-\alpha^*\hat{a}}.$$

Now it is easy to find a state for the "new" vacuum in the initial basis,

$$|0'\rangle = \hat{U}|0\rangle, \quad e^{-|\alpha|^2/2}e^{-\alpha\hat{a}^+}e^{-\alpha^*\hat{a}}|0\rangle = e^{-|\alpha|^2/2}\sum_n \frac{(-\alpha)^n}{\sqrt{n!}}|n\rangle, \tag{2}$$

if we expand the exponents in terms of the operators $\hat{a}$ and $\hat{a}^+$.

See another way to find the states $|0'\rangle$ from the equation $\hat{a}|0'\rangle = 0$ in the previous problem. In this problem, the distribution over the number of initial particles in state $|0'\rangle$ that follows from (2) coincides with expression (4) (Poisson distribution) from the previous problem.

Problem 10.17

Repeat the same analysis as in the previous problem, but for a transformation of the form $\hat{a}' = \alpha\hat{a} + \beta\hat{a}^+$, $\hat{a}'^+ = \alpha\hat{a}^+ + \beta\hat{a}$ (here α, β are real numbers).

Solution

For the fermionic operators, the transformation considered is unitary if the following conditions are fulfilled:

$$(\hat{a}')^2 = (\alpha\hat{a} + \beta\hat{a}^+)^2 = \alpha\beta = 0, \quad \{\hat{a}', \hat{a}'^+\}_+ = \alpha^2 + \beta^2 = 1,$$

i.e., only in the trivial case of $\alpha = \pm 1$, $\beta = 0$, and also the case $\alpha = 0$, $\beta = \pm 1$, which corresponds to a transformation from particles to *holes*; see Problem 10.14.

For bosonic operators the transition is unitary if the condition $\alpha^2 - \beta^2 = 1$ is fulfilled. If we write $|0'\rangle = \sum_n c_n|n\rangle$ and repeat the solution to Problem 10.15, from the relation $\hat{a}'|0'\rangle = 0$ we can find

$$c_{2k} = \left(-\frac{\beta}{\alpha}\right)^k \sqrt{\frac{(2k-1)!!}{2^k k!}} c_0, \tag{1}$$

where $c_n = 0$ for the odd values of n. From the normalization condition $\langle 0'|0'\rangle = 1$ we obtain $|c_0^2| = |\alpha|^{-1}$; the distribution over the initial particle numbers in the "new" vacuum is $w_n = |c_n|^2$.

Problem 10.18

An arbitrary single-particle state $|1\rangle$ can be written in the form $|1\rangle = \sum_i C_{f_i} \hat{a}^+_{f_i} |0\rangle$, where $\hat{a}^+_{f_i}$ is a creation operator for a particle in state $\psi_{f_i}(\xi)$ (f_i is a set of quantum numbers of some complete set). Clarify the quantum-mechanical interpretation of the coefficients C_{f_i}.

As an illustration, consider a single-particle state of a spinless particle of the form:

$$|1\rangle = \int \varphi(\mathbf{r}) \hat{\psi}(\mathbf{r}) dV |0\rangle.$$

Normalize this state vector to unity and calculate the mean value of a physical quantity, f, using the second quantization operator (X.3).

Solution

The relation between the state vectors,

$$|1\rangle = \sum_n C_{f_n} \hat{a}^+_{f_n} |0\rangle, \tag{1}$$

is equivalent to the expansion[215] $\psi = \sum_f C_f \psi_f$ of the wavefunction ψ of an arbitrary single-particle state over a complete set of eigenfunctions ψ_f. Hence C_f is the wavefunction of the state considered in the f-representation.

Since the operator $\psi^\dagger(\mathbf{r})$ "creates" a particle at the point $\mathbf{r}$, $\varphi(\mathbf{r})$ (see the condition) is the wavefunction in the coordinate representation ($f \equiv \mathbf{r}$). Of course, the calculation of the mean value of any additive physical quantity whose operator in the occupation number representation is defined by the relation (X.3) gives the usual quantum-mechanical equation (I.5)

$$\overline{F} = \langle 1 | \hat{F} | 1 \rangle = \int \varphi^*(\mathbf{r}) \hat{f} \varphi(\mathbf{r}) dV,$$

while the normalization condition $\langle 1 | 1 \rangle = 1$ takes the form $\int |\varphi|^2 dV = 1$. These relations can also be derived using the general properties of creation and annihilation operators. Compare to the solution of Problem 10.23.

Problem 10.19

Operators $\hat{a}_{f_k}$, $\hat{a}^+_{f_k}$, and $\hat{a}_{g_k}$, $\hat{a}^+_{g_k}$ are the annihilation and creation operators for a particle in states defined by quantum numbers f_k and g_k of two different complete sets. Indicate the relations between these operators.

---

[215] The derivation of this relation is accomplished by projecting the state vectors in (1) onto the vectors $|\xi\rangle$: here $\langle \xi | 1 \rangle = \psi(\xi)$, $\langle \xi | \hat{a}^\dagger_f | 0 \rangle = \psi_f(\xi)$.

Solution

The operators are related in the following way:

$$\hat{a}^+_{f_i} = \sum_k C(f_i, g_k) \hat{a}^+_{g_k}, \quad \hat{a}_{f_i} = \sum_k C^*(f_i, g_k) \hat{a}_{g_k}. \quad (1)$$

To define $C(f_i, g_k)$ we apply (1) to the vacuum state:

$$\hat{a}^+_{f_i} |0\rangle = \sum_k C(f_i, g_k) \hat{a}^+_{g_k} |0\rangle.$$

This relation is equivalent to the one for the eigenfunction

$$\psi_{f_i} = \sum_k C(f_i, g_k) \psi_{g_k} \quad (2)$$

in the previous problem. Hence it follows that

$$C(f_i, g_k) = \int \psi^*_{g_k} \psi_{f_i} d\xi,$$

so that $C(f_i, g_k)$ is the eigenfunction of ψ_{f_i} in the g-representation.

Problem 10.20

A two-particle state of identical bosons (or fermions) is described by the state vector $|2\rangle = \hat{a}^+_{f_1} \hat{a}^+_{f_2} |0\rangle$. Normalize it to unity. Find the form of normalized wavefunction in the coordinate representation. Consider both the case of same and different quantum numbers $f_{1,2}$.

Solution

If $f_1 \neq f_2$ then the state vector $|2\rangle = \hat{a}^+_1 \hat{a}^+_2 |0\rangle$ is normalized to unity; indeed

$$\langle 2|2\rangle = \langle 0|\hat{a}_2 \hat{a}_1 \hat{a}^+_1 \hat{a}^+_2 |0\rangle = \langle 0|\hat{a}_2(1 \pm \hat{a}^+_1 \hat{a}_1)\hat{a}^+_2 |0\rangle$$
$$= \langle 0|1 \pm \hat{a}^+_2 \hat{a}_2 \pm \hat{a}_2 \hat{a}^+_1 (\pm \hat{a}^+_2 \hat{a}_1)|0\rangle = \langle 0|0\rangle = 1$$

(signs "+" and "−" correspond to bosons and fermions, respectively).

In the case $f_1 = f_2 = f$, the normalized two-boson state has the form

$$|2\rangle = \frac{1}{\sqrt{2}} (\hat{a}^+_f)^2 |0\rangle,$$

while the analogous two-fermion state does not exist.

Wavefunctions of two-particle states considered in the coordinate representation have the form:

$$\psi(\xi_1, \xi_2) = \frac{1}{\sqrt{2}} \{\psi_1(\xi_1)\psi_2(\xi_2) \pm \psi_2(\xi_1)\psi_1(\xi_2)\}, \quad f_1 \ne f_2;$$

$$\psi(\xi_1, \xi_2) = \psi_f(\xi_1)\psi_f(\xi_2), \quad f_1 = f_2 \equiv f.$$

Problem 10.21

Perform the same analysis as in the previous problem, but for the case of a three-particle state, $|3\rangle = \hat{a}^+_{f_1} \hat{a}^+_{f_2} \hat{a}^+_{f_3} |0\rangle$.

Solution

In the case of different values of all quantum numbers f_a for the state vector given, we have $\langle 3|3\rangle = 1$ (both for bosons and fermions). Here, the system's wavefunction in the coordinate representation is described by Eq. (3) from Problem 10.3.

If all three f_a are the same, then to ensure normalization we should include the multiplier $1/\sqrt{3!} = 1/\sqrt{6}$; the wavefunction of the corresponding three-boson state is $\psi = \psi_f(\xi_1)\psi_f(\xi_2)\psi_f(\xi_3)$. If only two of quantum numbers f_a coincide, then the normalization factor should be $1/\sqrt{2}$, while the corresponding wavefunction is described by Eq. (1) from Problem 10.3.

Problem 10.22

For a system consisting of identical particles, determine a form of the particle density operator $\hat{n}(\mathbf{r})$ (at the point $\mathbf{r}$) and of the particle number operator $\hat{N}(v)$ for some volume v in the occupation-number representation.

Solution

The following operator corresponds to the density of particles with the spin projection, s_z at the point $\mathbf{r}$ in space

$$\hat{n}(\mathbf{r}, s_z) = \sum_a \delta(\mathbf{r} - \mathbf{r}_a)\delta_{\sigma_a, s_z}\delta_{\sigma'_a, s_z}. \tag{1}$$

(Compare to Problem 10.11; the operator is written in the coordinate representation for orbital variables and in the s_z-representation for spin variables). It is a sum of single-particle operators (particle density is an additive quantity), so that its form in the occupation number representation is given by Eq. (X.3), from which we obtain

$$\hat{n}(\mathbf{r}, s_z) = \hat{\psi}^*(\mathbf{r}, s_z)\hat{\psi}(\mathbf{r}, s_z), \quad \hat{n}(\mathbf{r}) = \sum_{s_z} \hat{n}(\mathbf{r}, s_z). \tag{2}$$

Here, $\hat{n}(\mathbf{r})$ is the particle density operator independent of the particles' spin projections. Operators $\hat{N}(v, s_z)$ and $\hat{N}(v)$ are obtained from $\hat{n}$ by integration over the corresponding volume; see also Problems 10.28–31.

Problem 10.23

Prove the following commutation relations:

$$[\hat{\mathbf{P}}, \hat{\psi}(\xi)] = i\hbar \frac{\partial}{\partial \mathbf{r}} \hat{\psi}(\xi), \quad [\hat{\mathbf{P}}, \hat{\psi}^+(\xi)] = i\hbar \frac{\partial}{\partial \mathbf{r}} \hat{\psi}^+(\xi),$$

where $\hat{\mathbf{P}}$ and $\hat{\psi}(\xi)$ are momentum and field (ψ-operators) operators in the occupation-number representation for a system of identical bosons and fermions.

Solution

The momentum operator $\hat{\mathbf{P}}$ (an additive physical quantity) for an identical particles system, according to Eq. (X.3), has the form

$$\hat{\mathbf{P}} = -i\hbar \int \hat{\psi}^+(\xi') \frac{\partial}{\partial \mathbf{r}'} \hat{\psi}(\xi') d\xi'.$$

Using commutation relations for the bosonic ψ-operators

$$[\hat{\psi}(\xi), \hat{\psi}(\xi')] = [\hat{\psi}^+(\xi), \hat{\psi}^+(\xi')] = 0; \quad [\hat{\psi}(\xi), \hat{\psi}^+(\xi')] = \delta(\xi - \xi'),$$

we find

$$[\hat{\mathbf{P}} \hat{\psi}(\xi)] = \hat{\mathbf{P}} \hat{\psi}(\xi) - \hat{\psi}(\xi) \hat{\mathbf{P}}$$

$$= -i\hbar \int \hat{\psi}^+(\xi') \frac{\partial}{\partial \mathbf{r}'} \hat{\psi}(\xi') d\xi' \hat{\psi}(\xi) + i\hbar \hat{\psi}(\xi) \int \hat{\psi}^+(\xi') \frac{\partial}{\partial \mathbf{r}'} \hat{\psi}(\xi') d\xi'$$

$$= -i\hbar \int \{\hat{\psi}^+(\xi') \hat{\psi}(\xi) - \hat{\psi}(\xi) \hat{\psi}^+(\xi')\} \frac{\partial}{\partial \mathbf{r}'} \hat{\psi}(\xi') d\xi'$$

$$= i\hbar \int \delta(\xi - \xi') \frac{\partial}{\partial \mathbf{r}'} \hat{\psi}(\xi') d\xi' = i\hbar \frac{\partial}{\partial \mathbf{r}} \hat{\psi}(\xi). \tag{1}$$

In an analogous way we obtain

$$[\hat{\mathbf{P}}, \hat{\psi}^+(\xi)] = -i\hbar \int \hat{\psi}^+(\xi') \frac{\partial}{\partial \mathbf{r}'} \delta(\xi - \xi') d\xi' = i\hbar \frac{\partial}{\partial \mathbf{r}} \hat{\psi}^+(\xi). \tag{2}$$

Replacing the bosonic ψ-operator commutation relations by fermionic anti-commutation relations does not change the results obtained.

Problem 10.24

Investigate the stationary states (energy spectrum and wavefunctions) for the "transverse" motion of a charged spinless particle in a uniform magnetic field using the creation and annihilation operators.[216] We fix a gauge and use the following expression $\mathbf{A} = [\mathcal{H} \times \mathbf{r}]/2$ for the vector potential.

Solution

1) The Hamiltonian of the transverse motion in a magnetic field has the form (see Problem 7.1):

$$\hat{H}_t = \frac{1}{2m}[\hat{p}_x^2 + \hat{p}_y^2 + m^2\omega^2(x^2 - y^2)] - \frac{e}{|e|}\hbar\omega\hat{l}_z, \quad \hbar\hat{l}_z = x\hat{p}_y - y\hat{p}_x, \quad (1)$$

where $\omega = |e|H/2mc$. We can transform it using annihilation (and creation) operators of "oscillation quanta" along the x and y axes:

$$\hat{a}_x = \frac{1}{\sqrt{2m\hbar\omega}}(m\omega x + i\hat{p}_x), \quad \hat{a}_y = \frac{1}{\sqrt{2m\hbar\omega}}(m\omega y + i\hat{p}_y),$$

to the form

$$\hat{H}_t = \hbar\omega\left\{\hat{a}_x^+\hat{a}_x + \hat{a}_y^+\hat{a}_y + 1 + i\frac{e}{|e|}(\hat{a}_y\hat{a}_x^+ - \hat{a}_y^+\hat{a}_x)\right\}. \quad (2)$$

Instead of the operators $\hat{a}_{x,y}$, it is convenient to use their linear combinations

$$\hat{a}_1 = \frac{1}{\sqrt{2}}\left(\hat{a}_x + i\frac{e}{|e|}\hat{a}_y\right), \quad \hat{a}_2 = \frac{1}{\sqrt{2}}\left(\hat{a}_x - i\frac{e}{|e|}\hat{a}_y\right), \quad (3)$$

which are also independent annihilation operators, since for them we have

$$[\hat{a}_i, \hat{a}_k^+] = \delta_{ik}, \quad [\hat{a}_i, \hat{a}_k] = [\hat{a}_i^+, \hat{a}_k^+] = 0; \quad i = 1, 2.$$

Now we have ($\omega_H = 2\omega$),

$$\hat{H}_t = \hbar\omega_H\left(\hat{a}_1^+\hat{a}_1 + \frac{1}{2}\right), \quad \hat{l}_z = \frac{e}{|e|}(\hat{a}_2^+\hat{a}_2 - \hat{a}_1^+\hat{a}_1). \quad (4)$$

Since these operators are expressed only in terms of number operators $\hat{n}_{1,2}$, the eigenvectors $|n_1, n_2\rangle$ of the latter ones are also the eigenvectors of Hamiltonian $\hat{H}_t$ and operator $\hat{l}_z$. Hence, we obtain the Landau level spectrum

$$E_{t,n_1} = \hbar\omega_H\left(n_1 + \frac{1}{2}\right),$$

---

[216] The creation/annihilation operators should be chosen in the same way as for of a linear oscillator, where $\hat{a} = (m\omega\hat{x} - i\hat{p})/\sqrt{2\hbar m\omega}$ that yields the Hamiltonian $\hat{H} = \hbar\omega(\hat{a}^+\hat{a} + 1/2)$. For the problem considered, we need to introduce two pairs of creation and annihilation operators, but the Hamiltonian can be chosen in such a way that explicitly includes only one such pair.

and find that the levels are infinitely-degenerate, since they do not depend on n_2. Here $l_z = (e/|e|)(n_2 - n_1)$. Compare to Problem 7.1.

2) Let us now derive the form of the eigenfunctions $\psi_{n_1 n_2}$ in the coordinate representation. First we obtain the wavefunction ψ_{00} of the vacuum state. From the solution of equations $\hat{a}_1 \psi_{00} = 0$ and $\hat{a}_2 \psi_{00} = 0$ we have

$$\psi_{00} = \sqrt{\frac{m\omega}{\pi \hbar}} \exp\left\{-\frac{1}{2\Delta^2}(x^2 + y^2)\right\}, \quad \Delta^2 = \frac{\hbar}{m\omega}. \tag{5}$$

The wavefunctions

$$\psi_{n_1 n_2} = \frac{1}{\sqrt{n_1! n_2!}} (\hat{a}_1^+)^{n_1} (\hat{a}_2^+)^{n_2} \psi_{00}$$

are obtained from ψ_{00} by differentiation. Here, instead of x, y it is convenient to use the variables

$$\xi = \frac{1}{\sqrt{2}\Delta}(x + iy), \quad \xi^* = \frac{1}{\sqrt{2}\Delta}(x - iy).$$

In the case of $e > 0$ we find

$$\psi_{n_1 n_2} = \left[\frac{2^{n_2 - n_1} m\omega}{\pi \hbar (n_1!)(n_2!)}\right]^{1/2} \left(\xi^* - \frac{\partial}{\partial \xi}\right)^{n_1} \xi^{n_2} e^{-\xi^* \xi}. \tag{6}$$

Eigenfunctions for $e < 0$ are obtained by complex conjugation of (6).

Problem 10.25

The energy spectrum $E_{n_1 n_2} = \hbar \omega_1 \left(n_1 + \frac{1}{2}\right) + \hbar \omega_2 \left(n_2 + \frac{1}{2}\right)$ of a two-dimensional (planar) oscillator with $U = \frac{1}{2} m(\omega_1^2 x^2 + \omega_2^2 y^2)$ in the case of commensurate frequencies $\omega_{1,2}$ contains degenerate levels. For special cases where a) $\omega_1 = \omega_2$, and b) $\omega_1 = 2\omega_2$ identify a symmetry of the Hamiltonian responsible for this degeneracy. Find explicitly the symmetry operators.

Solution

1) The planar oscillator Hamiltonian has the form $\hat{H} = \hat{H}_x + \hat{H}_y$, where $\hat{H}_{x,y}$ are linear oscillator Hamiltonians. From the commutativity of the operators $\hat{H}_{x,y}$ with each other and with the Hamiltonian $\hat{H}$, we can find the spectrum

$$E_{n_1 n_2} = \hbar \omega_1 \left(n_1 + \frac{1}{2}\right) + \hbar \omega_2 \left(n_2 + \frac{1}{2}\right); \quad n_{1,2} = 0, 1, 2, \ldots \tag{1}$$

and its eigenfunctions in the form of oscillator wavefunction products. Compare to Problem 2.48. For incommensurate frequencies, the oscillator levels are non-degenerate.

2) In the case of commensurate frequencies, an additional symmetry appears.[217] We find this symmetry in the existence of new operators that commute with $\hat{H}$ (and do not commute with $\hat{H}_{x,y}$), which gives rise to the degeneracy of the energy levels. Here symmetry operators that act on the eigenfunctions of the Hamiltonian that correspond to a given eigenvalue transform them into each other. In particular, in the case of frequency $\omega_1 = \omega_2$ (i.e., for isotropic oscillator), the Hamiltonian is rotationally-invariant that implies angular momentum conservation.

Using the energy spectrum from (1) we can relate symmetry operators to creation and annihilation operators for "oscillation quanta" along the axes x and y. If the relation between frequencies $k\omega_1 = s\omega_2$ holds where k, s are integers, such "additional" symmetry operators could be chosen to have the form

$$\hat{Q} = (\hat{a}_x^+)^k (\hat{a}_y)^s, \quad \hat{Q}^+ = (\hat{a}_x)^k (\hat{a}_y^+)^s. \qquad (2)$$

Here Hermitian linear combinations of these operators

$$\hat{Q}_1 = \hat{Q} + \hat{Q}^+, \quad \hat{Q}_2 = i(\hat{Q} - \hat{Q}^+) \qquad (3)$$

describe conserved quantities – integrals of motion for the system considered. For example, if $\omega_1 = \omega_2$ (when $k = s = 1$) we have (an expression for $\hat{a}$, can be found for example in Problem 10.5)

$$\hat{l}_z = -i(\hat{Q} - \hat{Q}^+), \quad \hat{p}_x \hat{p}_y + m^2 \omega^2 xy = m\hbar\omega(\hat{Q} + \hat{Q}^+). \qquad (4)$$

The conservation of l_z for the isotropic oscillator is evident.

Problem 10.26

Consider a *supersymmetric* oscillator described by the Hamiltonian

$$\hat{H} \equiv \hat{H}_B + \hat{H}_F = \hbar\omega(\hat{b}^+\hat{b} + \hat{f}^+\hat{f}), \qquad (1)$$

$$\hat{H}_B = \hbar\omega(\hat{b}^+\hat{b} + 1/2), \quad \hat{H}_F = \hbar\omega(\hat{f}^+\hat{f} - 1/2),$$

where $\hat{b}(\hat{b}^+)$ and $\hat{f}(\hat{f}^+)$ are the bosonic and fermionic annihilation (creation) operators respectively. The energy spectrum of this Hamiltonian is of the form $E_N = \hbar\omega N$, $N = n_B + n_F$, and the state vectors are $|n_B, n_F\rangle$. The spectrum has the following properties: $E_N \geq 0$, the levels with $E_N > 0$ are two-fold degenerate, the ground state $E_0 = 0$ is non-degenerate.

Indicate the form of the symmetry operators, and prove that the Hamiltonian could be expressed in terms of the anti-commutator of these operators. Explain the spectrum properties using the symmetry operators.

---

[217] In classical physics such a symmetry manifests itself in that the trajectories are closed curves.

Comment

The symmetry that appears in transformations that replace bosons by fermions, and *vice versa*, is called *supersymmetry*. It has a set of attractive properties, and yields hope for the construction of a unified theory of elementary particles. Here and in the following problem we consider the elementary properties of supersymmetry and its application to the simplest quantum-mechanical systems.

Solution

Since, for the system considered, the energy spectrum $E_N = \hbar\omega(n_B + n_F)$ depends only on the total number of particles $N = n_B + n_F$, we see that

$$\hat{Q} = q\hat{b}^\dagger \hat{f}, \quad \hat{Q}^\dagger = q\hat{b}\hat{f}^\dagger, \quad \text{here } \hat{Q}^2 = (\hat{Q}^\dagger)^2 = 0 \tag{2}$$

(q is a real parameter. It is convenient to choose $q = \sqrt{\hbar\omega}$) is the operator that replaces fermions by bosons $(\hat{Q})$, and *vice versa*, $(\hat{Q}^\dagger)$. These operators commute with the Hamiltonian so that they are the symmetry operators of the system. Compare to the previous problem.

The important property of the symmetry operators considered is that the system Hamiltonian $\hat{H}$ is equal to their anti-commutator, so that[218]

$$\{\hat{Q}, \hat{Q}^+\} \equiv \hat{Q}\hat{Q}^+ + \hat{Q}^+\hat{Q} = \hat{H}, \quad \{\hat{Q},\hat{Q}\} = \{\hat{Q}^+, \hat{Q}^+\} = 0,$$

$$[\hat{H}, \hat{Q}] = [\hat{H}, \hat{Q}^+] = 0. \tag{3}$$

If instead of operators $\hat{Q}$, $\hat{Q}^+$ we introduce their Hermitian combinations,

$$\hat{Q}_1 = \hat{Q} + \hat{Q}^+, \quad \hat{Q}_2 = i(\hat{Q} - \hat{Q}^+),$$

then relation (3) takes the more compact form:

$$\{\hat{Q}_i, \hat{Q}_k\} = 2\delta_{ik}\hat{H}, \quad [\hat{Q}_i, \hat{H}] = 0; \quad i, k = 1, 2. \tag{4}$$

The anti-commutators in (3) and (4), along with the commutators which appear in the Hamiltonian, are the characteristic property of supersymmetry. The algebra of operators (3) and (4) provides useful information about the spectrum (without using the concrete form of operators $\hat{Q}$, $\hat{Q}^+$ and $\hat{H}$):

1) The eigenvalues are non-negative, *i.e.*, $E \geq 0$. Indeed, in any state, mean values $\overline{\hat{Q}\hat{Q}^+} \geq 0$ and $\overline{\hat{Q}^+\hat{Q}} \geq 0$ (compare to Problem 1.15), so that $\overline{\hat{H}} \geq 0$ and hence $E \geq 0$.

2) Levels with $E \neq 0$ are two-fold degenerate. First we should note that the Hermitian operator $\hat{S} = \hat{Q}^+\hat{Q}$ commutes with the Hamiltonian. Thus, there is a complete set of functions ψ_{ES} that are the eigenfunctions both of $\hat{H}$ and $\hat{S}$. Then from the equation $\hat{S}\psi_{ES} = S\psi_{ES}$ and $(\hat{Q}^+)^2 = 0$, it follows that $S\hat{Q}^+\psi_{ES} = 0$. We surmise that either the eigenvalues vanish, $S = 0$, or $\hat{Q}^+\psi_{ES} = 0$. However, in the second

---

[218] Recall that $[\hat{b}, \hat{b}^+] = 1$, $\{\hat{f}, \hat{f}^+\} = 1$, $\hat{f}^2 = (\hat{f}^\dagger)^2 = 0$.

case the relation $\hat{H}\psi_{ES} = E\psi_{ES}$ tells us that the eigenvalues are $S = E$. Operator $\hat{S}$ has no other eigenvalues (for the states with given E). It is important to note that for $E \neq 0$ there are two values $S_1 = 0$ and $S_2 = E$, and the corresponding eigenfunctions pass into each other under the operators $\hat{Q}$ and $\hat{Q}^+$. This explains the two-fold degeneracy of levels with $E \neq 0$.[219] Indeed, let $\psi_{E0} \neq 0$ (here $\hat{Q}\psi_{E0} = 0$), then we have $\psi_{EE} \propto \hat{Q}^+\psi_{E0} \neq 0$ (since $\hat{Q}(\hat{Q}^+\psi_{E0}) = E\psi_{E0} \neq 0$); in the same way, $\psi_{E0} \propto \hat{Q}\psi_{EE}$ (here $\hat{Q}^+\psi_{EE} = 0$).

3) The absence of degeneracy for a level with[220] $E = 0$ (if it exists). From the equation $\hat{H}\psi_0 = 0$ it follows that $\hat{Q}\psi_0 = 0$ and $\hat{Q}^+\psi_0 = 0$, since no new states appear under the action of the symmetry operators on ψ_0.

The spectrum and eigenvectors of the supersymmetric oscillator illustrate the general properties mentioned above. Let us add one more trivial realization of the operator (3) algebra that is connected with the choice $\hat{Q} = \hat{f}$ and $\hat{Q}^+ = \hat{f}^+$, where $\hat{H} = 1$. The spectrum of such a "Hamiltonian" consists of a single "level" with $E = 1$ (here $E > 0$!), which is two-fold degenerate with eigenvectors $|n_F\rangle$, where $n_F = 0$ or 1 and there is no level with $E = 0$. See more interesting examples in the following problem and also in Problem 7.9.

Problem 10.27

Let us introduce the operators

$$\hat{Q} = \hat{A}^+\hat{f}, \quad \hat{Q}^+ = \hat{A}\hat{f}^+, \quad \hat{H} = \hat{Q}\hat{Q}^+ + \hat{Q}^+\hat{Q},$$

where $\hat{f}$, $\hat{f}^+$ are the fermion annihilation and creation operators, while $\hat{A}$ and $\hat{A}^+$ are some operators commuting with them.

Verify that the system[221] considered is supersymmetric (see the previous problem).

Show that for a proper choice of the "coordinate" $\hat{A}$, $\hat{A}^+$ and spin $\hat{f}$, $\hat{f}^+$ operators, the supersymmetric Hamiltonian considered characterizes one-dimensional motion of a particle with spin $s = 1/2$. What are the consequences of supersymmetry?

Solution

1) It is easy to prove that

$$\{\hat{Q}, \hat{Q}^+\} = \hat{H}, \quad [\hat{Q}, \hat{H}] = [\hat{Q}^+, \hat{H}] = 0, \quad \hat{Q}^2 = (\hat{Q}^+)^2 = 0,$$

---

[219] Here the degenerate states are called *super-partners*. We emphasize that we mean the level degeneracy that is connected with the Hamiltonian supersymmetry.

[220] The absence of a state with $E = 0$ is called *spontaneous supersymmetry breaking*.

[221] Here $\hat{H}$ is its Hamiltonian, and the system vector state space is defined by the space where operators $\hat{A}$, $\hat{f}$ are defined and hence also $\hat{Q}$, $\hat{Q}^+$, $\hat{H}$.

as it is supposed to be for a supersymmetric system. If we write $\hat{A} = \hat{A}_1 + i\hat{A}_2$, where $\hat{A}_{1,2}$ are Hermitian operators, we transform the system Hamiltonian to

$$\hat{H} = \hat{A}_1^\dagger - \hat{A}_2^\dagger - i[\hat{A}_1, \hat{A}_2](\hat{f}^+\hat{f} - \hat{f}\hat{f}^+). \qquad (1)$$

Now we see that if we choose

$$\hat{A}_1 = \frac{1}{\sqrt{2m}}\hat{p}, \quad \hat{A}_2 = W(x),$$

where $W(x)$ is a real function, and if we identify the operators $\hat{f}$, $\hat{f}^+$ with spin operators $\hat{f} = \begin{pmatrix} 0 & 0 \\ 1 & 0 \end{pmatrix}$ and $\hat{f}^+ = \begin{pmatrix} 0 & 1 \\ 0 & 0 \end{pmatrix}$, so that

$$\hat{f}^+\hat{f} - \hat{f}\hat{f}^+ = \begin{pmatrix} 1 & 0 \\ 0 & -1 \end{pmatrix} = \hat{\sigma}_z,$$

then Hamiltonian (1) takes the form of the Pauli Hamiltonian (VII.1) for one-dimensional motion with spin $s = 1/2$.

$$\hat{H} = \frac{1}{2m}\hat{p}^2 + U(x) - \mu\mathcal{H}(x)\sigma_z, \qquad (2)$$

where

$$U(x) = W^2(x), \quad \mu\mathcal{H}(x) = \frac{\hbar}{\sqrt{2m}}W'(x).$$

Using the general properties of eigenfunctions and eigenvalues of a supersymmetric Hamiltonian given in the previous problem, we can make some interesting statements about the spectrum of the ordinary one-dimensional Hamiltonian (spinless particle). Indeed, eigenfunctions $\psi_{E\sigma}$ of Hamiltonian (2) and the operator $\hat{\sigma}_z$ that commutes with it have the form

$$\psi_{E,+1} = \begin{pmatrix} \psi_E^+ \\ 0 \end{pmatrix}, \quad \psi_{E,-1} = \begin{pmatrix} 0 \\ \psi_E^- \end{pmatrix}.$$

Here, from the Schrödinger equation, $\hat{H}\psi_E = E\psi_E$, it follows that

$$\left(\frac{1}{2m}\hat{p}^2 + U_\pm(x)\right)\psi_E^\pm = E_\pm\psi_E^\pm, \qquad (3)$$

where

$$U_\pm(x) = W^2(x) \mp \frac{\hbar}{\sqrt{2m}}W'(x). \qquad (4)$$

Formally, equations (3) are the Schrödinger equations for one-dimensional motion in two different potentials $U_\pm(x)$. From supersymmetry of the Hamiltonian in Eq. (2), it follows that their discrete spectra of the two potentials coincide.[222] $E_\pm$ coincide, except for one level $E_0 = 0$ which could exist only in one of these fields. Such a conclusion is based on the fact that the functions $\psi_{E\sigma}$ coincide with the functions ψ_{ES} given in the previous problem, and here $S = (1 \pm \sigma)E/2$. Taking into account the relation between eigenfunctions ψ_{ES} for different values of S we find relations for eigenfunctions $\psi_E^\pm(x)$ from (3) (for the states with $E_+ = E_-$):

$$\psi_E^\pm(x) \propto \left(\frac{1}{\sqrt{2m}}\hat{p} \mp iW(x)\right)\psi_E^\mp(x). \tag{5}$$

2) The fact that $E_+ = E_-$, in some cases allows us to find the spectrum algebraically, without solving the Schrödinger equation. For example, let us consider the linear oscillator with $U = m\omega^2 x^2/2$ and denote its spectrum by E_n. Let us now choose the superpotential

$$W = \sqrt{\frac{m\omega^2}{2}}x, \quad \text{then } U_\pm = \frac{1}{2}m\omega^2 x^2 \mp \frac{1}{2}\hbar\omega. \tag{6}$$

The corresponding spectra are $E_n^\pm = E_n \mp \hbar\omega/2$. In one of them (evidently among E^+) the value $E_0^+ = 0$ appears, while coincidence of the other levels means that $E_{n+1}^+ = E_n^-$. Hence it follows that $E_0 = \hbar\omega/2$ and $E_{n+1} = E_n + \hbar\omega$. This immediately gives the spectrum of the linear oscillator as $E_n = \hbar\omega(n + 1/2)$. The spectrum for the field $U = -U_0 \cosh^{-2}(x/a)$ can be found analogously, choosing the superpotential $W \propto \tanh(x/a)$.

3) Let us consider the existence of a zero-energy bound state. From the equation $\hat{H}\psi_0 = 0$ it follows that

$$\hat{Q}\psi_0 = 0, \quad \hat{Q}^+\psi_0 = 0$$

(compare to the previous problem), and hence applying it to Hamiltonian (2) we obtain

$$\left(\frac{\hat{p}}{\sqrt{2m}} \pm iW(x)\right)\psi_0^\pm = 0, \quad \psi_0^\mp = B^\mp \exp\left\{\pm\frac{\sqrt{2m}}{\hbar}\int_0^x W\,dx\right\}.$$

Taking into account the fact that $W(\pm\infty) \neq 0$ (see the footnote), we see that one of the functions $\psi_0^\pm$ increases over large distances, and we should choose $B = 0$ for it (in accordance with the general result for a level with $E = 0$ being non-degenerate). For the other function satisfying boundary conditions $\psi_0(\pm\infty) = 0$, a zero-energy state ψ_0 exists only if the signs of the superpotentials at infinity $W(\pm\infty)$ are opposite. For example, for $W(x) = W_0 = \text{const}$ in Eqs. (3) we have $U_\pm = W_0^2 = \text{const}$, and there is no state with $E_0 = 0$. In the case of $W(x) = W_0 x/|x|$ we have

---

[222] Since $E_\pm \geq 0$, the existence of a discrete spectrum states means that the limit $U_\pm(x) \to C_\pm > 0$ for $x \to \pm\infty$ (hence here $\lim W(x) \neq 0$). If $W(x)$ is an even function, then potentials $U_\pm(x)$ are obtained from one another by mirror reflection and their spectra coincide, $E_+ = E_-$. There is no level with $E_0 = 0$.

$$U_{\pm}(x) = W_0^2 \mp \alpha\delta(x), \quad \alpha = \sqrt{\frac{2}{m}\hbar W_0},$$

and the existence of the zero-energy state is connected to the presence of a single discrete level in a δ-well; see Problem 2.7.

10.3 The simplest systems with a large number of particles ($N \gg 1$)

Problem 10.28

For the ground state of a *Bose*-gas consisting of N identical non-interacting particles with the spin $s = 0$ in a volume V, calculate the mean particle number density, the mean number of particles in a volume v, and the fluctuation of this particle number.

Solve this problem by averaging the physical operators in the occupation-number representation.

Solution

Expressing $\hat{\psi}(\mathbf{r})$-operators in terms of creators $\hat{a}_{\mathbf{k}}^+$ and annihilators $\hat{a}_{\mathbf{k}}$ of a particle with a given momentum $\mathbf{p} = \hbar\mathbf{k}$, the operator for particle number density $\hat{n}(\mathbf{r}) = \hat{\psi}^+(\mathbf{r})\hat{\psi}(\mathbf{r})$ (see Problem 10.22) (here $\hat{\psi}(\mathbf{r}) = \sum_{\mathbf{k}}(1/\sqrt{V})e^{i\mathbf{k}\cdot\mathbf{r}}\hat{a}_{\mathbf{k}}$) becomes

$$\hat{n}(r) = \frac{1}{V}\sum_{\mathbf{k}_1\mathbf{k}_2} e^{i(\mathbf{k}_2 - \mathbf{k}_1)\mathbf{r}} \hat{a}_{\mathbf{k}_1}^+ \hat{a}_{\mathbf{k}_2}. \tag{1}$$

The mean particle number density $\overline{n(\mathbf{r})}$ is obtained from operator (1) by averaging over the ground state $|\psi_0\rangle = |N_{k=0}0_{k\neq 0}\rangle$ (all the particles have zero momentum $p = 0$). Since

$$\langle\psi_0|\hat{a}_{\mathbf{k}_1}^+ \hat{a}_{\mathbf{k}_2}|\psi_0\rangle = \begin{cases} N, & k_1 = k_2 = 0, \\ 0, & \text{in all other cases,} \end{cases}$$

we have the natural result $\overline{n} = N/V$.

The mean particle number in the volume v is obtained by averaging the operator $\hat{N}(v) = \int_v \hat{n}(\mathbf{r})d^3r$, and is equal to $\overline{N(v)} = Nv/V$.

To calculate the fluctuations of particle number, we first average the operator $\overline{N^2(v)}$ over the state $|\psi_0\rangle$. Since

$$N^2(v) = \frac{1}{V^2}\int_v\int_v \sum_{\mathbf{k}_1\mathbf{k}_2\mathbf{k}_3\mathbf{k}_4} \exp\{i[(\mathbf{k}_2 - \mathbf{k}_1)\cdot\mathbf{r} + (\mathbf{k}_4 - \mathbf{k}_3)\cdot\mathbf{r}']\}\hat{a}_{\mathbf{k}_1}^+\hat{a}_{\mathbf{k}_2}\hat{a}_{\mathbf{k}_3}^+\hat{a}_{\mathbf{k}_4}\, d^3r d^3r',$$

we must first find the matrix elements

$$\langle \psi_0 | \hat{a}_{\mathbf{k}_1}^+ \hat{a}_{\mathbf{k}_2} \hat{a}_{\mathbf{k}_3}^+ \hat{a}_{\mathbf{k}_4} | \psi_0 \rangle$$

to calculate $\overline{N^2(v)}$. Using the explicit form of $|\psi_0\rangle$, we see that the matrix elements are non-vanishing only when conditions $k_1 = k_4 = 0$, $\mathbf{k}_2 = \mathbf{k}_3$ are fulfilled. These elements are equal to N^2 for $\mathbf{k}_2 = \mathbf{k}_3 = 0$ and to N for $\mathbf{k}_2 = \mathbf{k}_3 \equiv \mathbf{k} \neq 0$. Taking this into account we obtain

$$\overline{N^2(v)} = \frac{1}{V^2} \int_v \int_v \left\{ N^2 + N \sum_{k \neq 0} e^{i\mathbf{k}\cdot(\mathbf{r}-\mathbf{r}')} \right\} d^3r\, d^3r'. \tag{2}$$

Because the functions $\psi_{\mathbf{k}} = e^{i\mathbf{k}\cdot\mathbf{r}}/\sqrt{V}$ form a complete set, the sum here is equal to $V\delta(\mathbf{r}-\mathbf{r}') - 1$, and integration gives

$$\overline{N^2(v)} = \left(\frac{Nv}{V}\right)^2 + \frac{Nv}{V} - \frac{Nv^2}{V^2}. \tag{3}$$

Hence

$$\overline{(\Delta N(v))^2} = \overline{N^2(v)} - \overline{N(v)}^2 = \frac{Nv}{V}\left(1 - \frac{v}{V}\right). \tag{4}$$

For $v = V$ we have $\overline{(\Delta N(v))^2} = 0$, which is obvious, since the total number of particle in the system is equal to N and does not fluctuate. When $v \ll V$, according to (4), we have

$$\overline{(\Delta N(v))^2} \approx \frac{Nv}{V} = \overline{N(v)}.$$

Let us note that for a system of N non-interacting classical particles in a volume V, the distribution of the number of particles N_v in the volume v has the form

$$W(N_v) = \frac{N!}{N_v!(N-N_v)!} \left(\frac{v}{V}\right)^{N_v} \left(1 - \frac{v}{V}\right)^{N-N_v}$$

(binomial distribution). For such a distribution, the calculation of mean values $\overline{N_v}$, $\overline{N_v^2}$, $\overline{(\Delta N_v)^2}$ gives results that coincide with those obtained above (see also a remark in the next problem).

Problem 10.29

Under the same conditions as in the previous problem, consider the spatial correlation of density fluctuations. For a homogeneous system it is characterized by a *correlation function* $\nu(\mathbf{r})$ ($\mathbf{r} = \mathbf{r}_1 - \mathbf{r}_2$) equal to

$$\nu(\mathbf{r}) = \frac{\overline{n_1 n_2} - \overline{n}^2}{\overline{n}}, \quad n_{1,2} \equiv n(\mathbf{r}_{1,2}),$$

where $\bar{n}$ is the mean particle number density.

Compare this result to the corresponding result for a system of classical particles.

Solution

Since particle density operators in different points of space commute with each other, the operator of the form $\hat{n}_1 \hat{n}_2$ can be written:

$$\hat{n}_1 \hat{n}_2 = \hat{\psi}^+(\mathbf{r}_1)\hat{\psi}(\mathbf{r}_1)\hat{\psi}^+(\mathbf{r}_2)\hat{\psi}(\mathbf{r}_2)$$

$$= \frac{1}{V^2} \sum_{k_1 k_2 k_3 k_4} \exp\{i[(\mathbf{k}_2 - \mathbf{k}_1)\mathbf{r}_1 + (\mathbf{k}_4 - \mathbf{k}_3)\mathbf{r}_2]\}\hat{a}^+_{k_1}\hat{a}_{k_2}\hat{a}^+_{k_3}\hat{a}_{k_4},$$

and its mean value in the Bose-gas ground state is

$$\overline{n_1 n_2} = \frac{1}{V^2}\left\{N^2 + N\sum_{k\neq 0} e^{ik(\mathbf{r}_1-\mathbf{r}_2)}\right\} = \frac{N}{V}\delta(\mathbf{r}_1 - \mathbf{r}_2) + \frac{N^2}{V^2} - \frac{N}{V^2}. \quad (1)$$

Compare this to the derivation of Eqs. (2) and (3) in the previous problem. Hence[223]

$$\frac{1}{\bar{n}}\{\overline{n_1 n_2} - \bar{n}^2\} = \delta(r) - \frac{\bar{n}}{N},$$

and the correlation function becomes equal to

$$\nu = -\frac{\bar{n}}{N}. \quad (2)$$

In order to understand the results obtained, we derive correlators similar to (1), (2) for the case of non-interacting classical particles. Taking into account the fact that the probability distribution of particle coordinates is described in terms of a product $d^3 r_a/V$ and $n(\mathbf{r}) = \sum_a \delta(\mathbf{r} - \mathbf{r}_a)$, we find

$$\overline{n(\mathbf{r}'_1)n(\mathbf{r}'_2)} = \int \ldots \int \sum_{a,b} \delta(\mathbf{r}'_1 - \mathbf{r}_a)\delta(\mathbf{r}'_2 - \mathbf{r}_b)\frac{d^3 r_1}{V}\ldots\frac{d^3 r_N}{V}$$

$$= \frac{N}{V}\delta(\mathbf{r}'_1 - \mathbf{r}'_2) + \frac{N(N-1)}{V^2}, \quad (3)$$

which coincides with Eq. (1). (Note that the term with a δ-function in (3) corresponds to the terms with $a = b$. The number of such terms is N. The second term corresponds to the terms with $a \neq b$, and their number is $N(N-1)$.)

For macroscopic systems, the value of N is extremely large, and therefore the last term in (1) can be omitted and we have $\nu = 0$ (there is no correlation) in (2). On the other hand, for finite values of N we have $\nu \neq 0$. Here the characteristic properties of ν – its independence from $\mathbf{r}$ and its sign $\nu < 0$ – have an intuitive explanation for

---

[223] The term with the δ-function, that goes to zero as $r \neq 0$, has a universal character and does not depend on the form of the distribution function.

classical particles. Indeed, the value of $\overline{n_1 n_2}$ is lower than $\overline{n}^2$, since a single particle cannot contribute to the particle number density at different points of space at the same time, regardless of the distance between them (in the case where $N \gg 1$ the density at different points in space is determined mainly by the contribution of different particles).

Many characteristics of a Bose-gas in the ground state, considered here and in the previous problem, are the same as for a gas of classical particles. It is not accidental. Indeed, the wavefunction of the ground state for the Bose gas has the form

$$\Psi_0 = \psi_0(\mathbf{r}_1)\psi_0(\mathbf{r}_2)\ldots\psi_0(\mathbf{r}_N), \quad \psi_0(r) = \frac{1}{\sqrt{V}},$$

i.e., is a product of single-particle wavefunctions. This is similar to a gas of distinguishable particles. The particles therefore do not interfere with each other, and for each one of them $|\psi_0|^2 = 1/V$, which corresponds to a constant probability distribution over the volume.

Problem 10.30

In the ground state of an ideal *Fermi*-gas of N particles in volume V, find the mean particle number density and the mean particle number in some volume v.

This problem should be solved by averaging the physical operators in the occupation-number representation.

Solution

The particle density operator $\hat{n}(\mathbf{r})$ has the form

$$\hat{n}(\mathbf{r}) = \sum_\sigma \hat{n}(\mathbf{r}, \sigma) = \frac{1}{V} \sum_\sigma \sum_{\mathbf{k}_1 \mathbf{k}_2} e^{i(\mathbf{k}_2 - \mathbf{k}_1)\mathbf{r}} \hat{a}^+_{\mathbf{k}_1 \sigma} \hat{a}_{\mathbf{k}_2 \sigma}. \tag{1}$$

Compare to Problem 10.28 and 10.22, $\sigma \equiv s_z$. The ground state of the Fermi-gas is determined by occupation numbers $n_{\mathbf{k}\sigma}$, equal to 1 for $|\mathbf{k}| < k_F$ and 0 for $|\mathbf{k}| > k_F$, so that

$$|\psi_0\rangle = \prod \hat{a}^+_{\mathbf{k}\sigma}|0\rangle, \tag{2}$$

where the product includes operators $\hat{a}^+_{\mathbf{k}\sigma}$ with quantum numbers $(\mathbf{k}\sigma)$ of occupied states. Here the Fermi momentum, $p_F = \hbar k_F$, is found from the condition

$$\sum_\sigma \sum_{\mathbf{k}(k<k_F)} 1 = (2s+1) \int_{k<k_F} \frac{V\, d^3k}{(2\pi)^3} = \frac{(2s+1)V k_F^3}{6\pi^2} = N,$$

i.e., we have

$$k_F = \left[\frac{6\pi^2 N}{(2s+1)V}\right]^{1/3}.$$

Since we see that the matrix element $\langle\psi_0|\hat{a}^+_{\mathbf{k}_1\sigma}\hat{a}_{\mathbf{k}_2\sigma}|\psi_0\rangle$ is different from zero (and equal to 1) only for $\mathbf{k}_1 = \mathbf{k}_2 \equiv \mathbf{k}$, and $|\mathbf{k}| \leq k_F$, from Eqs. (1) and (2) we obtain

$$\bar{n} = \langle\psi_0|\hat{n}(\mathbf{r})|\psi_0\rangle = \frac{N}{V}$$

(which is expected), and $\bar{n}(\sigma) = \bar{n}/(2s+1)$, while $\overline{N(v)} = \bar{n}v = Nv/V$.

Problem 10.31

Under the conditions of the previous problem, consider the correlation of particle number densities with definite values of spin z-projection at different points in space: find $\overline{n(\mathbf{r}_1, s_{z1})n(\mathbf{r}_2, s_{z2})}$, and compare to the product $\overline{n(\mathbf{r}_1, s_{z1})} \cdot \overline{n(\mathbf{r}_2, s_{z2})}$. Consider the cases of different and identical values of s_{z1} and s_{z2}.

Find the density-density correlation function (see Problem 10.29).

Solution
The operator $n(\mathbf{r}_1, \sigma_1)n(\mathbf{r}_2, \sigma_2)$ has the form

$$\hat{n}(\xi_1)\hat{n}(\xi_2) = \frac{1}{V}\sum_{\{\mathbf{k}\}} \exp\{i[(\mathbf{k}_2 - \mathbf{k}_1)\mathbf{r}_1 + (\mathbf{k}_4 - \mathbf{k}_3)\mathbf{r}_2]\}\hat{a}^+_{\mathbf{k}_1\sigma_1}\hat{a}_{\mathbf{k}_2\sigma_1}\hat{a}^+_{\mathbf{k}_3\sigma_2}\hat{a}_{\mathbf{k}_4\sigma_2}. \quad (1)$$

(Compare to Problems 10.29 and 10.30). Letting $|\psi_0\rangle$ be the wavefunction of the Fermi-gas ground state given in the previous problem, it is easy to see that the matrix element obtained by averaging

$$\langle\psi_0|\hat{a}^+_{\mathbf{k}_1\sigma_1}\hat{a}_{\mathbf{k}_2\sigma_1}\hat{a}^+_{\mathbf{k}_3\sigma_2}\hat{a}_{\mathbf{k}_4\sigma_2}|\psi_0\rangle,$$

for the case $\sigma_1 \neq \sigma_2$, is different from zero (and equal to 1) only for $\mathbf{k}_1 = \mathbf{k}_2$, $\mathbf{k}_3 = \mathbf{k}_4$ and $|\mathbf{k}_{1,3}| \leq k_F$. Taking this into account, for $\sigma_1 \neq \sigma_2$ we find

$$\langle\psi_0|\hat{n}(\xi_1)\hat{n}(\xi_2)|\psi_0\rangle = \frac{1}{V^2}\sum_{|\mathbf{k}_{1,2}|\leq k_F} 1 = \frac{\bar{n}^2}{(2s+1)^2}, \quad \bar{n} = \frac{N}{V}. \quad (3)$$

(For the calculation of the sum over $\mathbf{k}_{1,2}$, see the previous problem.) Since $\bar{n}(\sigma) = \bar{n}/(2s+1)$, the result (3) means that $\overline{n_1 n_2} = \overline{n_1} \cdot \overline{n_2}$, *i.e*, in the case of different values of the spin projections $\sigma_1 \neq \sigma_2$ there is no correlation between the particle densities at different points in space.

In the case of $\sigma_1 = \sigma_2$, the situation is different. Now the matrix element (2) is different from zero and equal to 1 in the following cases:

1) $\mathbf{k}_1 = \mathbf{k}_2$, $|\mathbf{k}_2| \leq k_F$, $\mathbf{k}_3 = \mathbf{k}_4$, $|\mathbf{k}_4| \leq k_F$;
2) $\mathbf{k}_1 = \mathbf{k}_4$, $|\mathbf{k}_4| \leq k_F$, $\mathbf{k}_2 = \mathbf{k}_3$, $|\mathbf{k}_3| > k_F$.

Using this fact, we find

$$\overline{n(\mathbf{r}_1, \sigma) n(\mathbf{r}_2, \sigma)} = \frac{1}{V^2} \left\{ \sum_{|\mathbf{k}_{1,2}| \leq k_F} 1 + \sum_{|\mathbf{k}_1| \leq k_F} \sum_{|\mathbf{k}_2| \geq k_F} e^{i(\mathbf{k}_2 - \mathbf{k}_1)(\mathbf{r}_1 - \mathbf{r}_2)} \right\}. \quad (4)$$

Then, using the relation

$$\frac{1}{V} \sum_{|\mathbf{k}| \geq k_F} e^{i\mathbf{k}\cdot\mathbf{r}} = \frac{1}{V} \left\{ \sum_{\mathbf{k}} e^{i\mathbf{k}\cdot\mathbf{r}} - \sum_{|\mathbf{k}| < k_F} e^{i\mathbf{k}\cdot\mathbf{r}} \right\} = \delta(\mathbf{r}) - \frac{1}{V} \sum_{|\mathbf{k}| < k_F} e^{i\mathbf{k}\cdot\mathbf{r}}$$

and calculating the integral (in spherical coordinates with the polar axis directed along the vector $\mathbf{r}$) we obtain:

$$\frac{1}{V} \sum_{|\mathbf{k}| \leq k_F} e^{i\mathbf{k}\cdot\mathbf{r}} = \frac{V}{(2\pi)^3} \int_{k \leq k_F} e^{i\mathbf{k}\cdot\mathbf{r}} d^3k = \frac{V}{2\pi^2 r^2} \left\{ \frac{\sin k_F r}{r} - k_F \cos k_F r \right\},$$

We transform (4) using ($\mathbf{r} = \mathbf{r}_1 - \mathbf{r}_2$, $\overline{n}(\sigma) = \overline{n}/(2s+1)$):

$$\overline{n(\mathbf{r}_1, \sigma) n(\mathbf{r}_2, \sigma)} = \overline{n(\sigma)}^2 - \frac{1}{4\pi^4 r^4} \left\{ \frac{\sin k_F r}{r} - k_F \cos k_F r \right\}^2.$$

Hence we obtain the correlation function:

$$\nu(\mathbf{r}, \sigma) = -\frac{[\sin k_F r - k_F r \cos k_F r]^2}{4\pi^4 \overline{n}(\sigma) r^6}. \quad (5)$$

Let us discuss this result. The character of the particle density correlations can be made physically clear. Identical particles with different values of spin projection behave like distinguishable ones, so there is no correlation between them. The sign of the correlation function $\nu(\mathbf{r}, \sigma) < 0$ in the case of the same spin projections is also natural. It shows the "repulsive" character of the fermion exchange interaction. For the values $r = |\mathbf{r}_1 - \mathbf{r}_2| \to \infty$, the correlation disappears.

In conclusion, we should note that the full correlation function for particle number density $\nu(\mathbf{r})$ coincides (independent of the spin projection) with $\nu(\mathbf{r}, \sigma)$.

Problem 10.32

Considering the interaction between particles as a perturbation, find the ground state energy of the Bose-gas (consisting of N spinless particles in volume V) in the first order of the perturbation theory (the interaction between particles is described by a short-range repulsive potential $U(\mathbf{r}) \geq 0$, $\mathbf{r} = \mathbf{r}_a - \mathbf{r}_b$).

Solution

The particle interaction operator in the occupation numbers representation, according to Eq. (X.4), has the form:

$$\hat{U} = \frac{1}{2} \iint \hat{\psi}^+(\mathbf{r}_1)\hat{\psi}^+(\mathbf{r}_2) U(|\mathbf{r}_1 - \mathbf{r}_2|)\hat{\psi}(\mathbf{r}_2)\hat{\psi}(\mathbf{r}_1) d^3r_1 d^3r_2. \quad (1)$$

Using the expansion of the ψ-operator over plane waves (given in Problem 10.28) we transform relation (1) to the form

$$\hat{U} = \frac{1}{2} \sum_{\mathbf{k}_1 \mathbf{k}_2 \mathbf{k}_3 \mathbf{k}_4} U_{\mathbf{k}_1 \mathbf{k}_2}^{\mathbf{k}_3 \mathbf{k}_4} \hat{a}_{\mathbf{k}_3}^+ \hat{a}_{\mathbf{k}_4}^+ \hat{a}_{\mathbf{k}_2} \hat{a}_{\mathbf{k}_1}, \quad (2)$$

where the matrix elements of the interaction potential have the form:

$$U_{\mathbf{k}_1 \mathbf{k}_2}^{\mathbf{k}_3 \mathbf{k}_4} = \frac{1}{V^2} \int_V \int_V U(|\mathbf{r}_1 - \mathbf{r}_2|) \exp\{i[(\mathbf{k}_1 - \mathbf{k}_3)\mathbf{r}_1 + (\mathbf{k}_2 - \mathbf{k}_4)\mathbf{r}_2]\} d^3r_1 d^3r_2.$$

In the first order of the perturbation theory we have

$$E_0 \approx E_0^{(0)} + E_0^{(1)} = \langle \psi_0 | \hat{U} | \psi_0 \rangle, \quad (3)$$

where $|\psi_0\rangle$ is the wavefunction of the non-interacting ground state with all particles having the momentum $\mathbf{p} = \hbar \mathbf{k} = 0$. Here $E_0^{(0)} = 0$, while the matrix element

$$\langle \psi_0 | \hat{a}_{\mathbf{k}_3}^+ \hat{a}_{\mathbf{k}_4}^+ \hat{a}_{\mathbf{k}_2} \hat{a}_{\mathbf{k}_1} | \psi_0 \rangle$$

differs from zero only when $k_a = 0$, in which case it is equal to $N(N-1) \approx N^2$ (due to $N \gg 1$). Hence,

$$E_0 \approx E_0^{(1)} = \frac{U_{00}^{00} N^2}{2}.$$

Since the radius R of the potential $U(r)$ is assumed to be small, so that $R \ll L \sim V^{1/3}$, the integral

$$\int_V U(\mathbf{r}_1 - \mathbf{r}_2) d^3r_2 = \iiint_{-\infty}^{\infty} U(\mathbf{r}_1 - \mathbf{r}_2) d^3r_2 = \int U(\mathbf{r}) d^3r \equiv \tilde{U}_0 \quad (3')$$

could be integrated over all the space and appears not to depend on r_1. As a result we obtain $U_{00}^{00} = \tilde{U}_0/V$, and the final expression for the energy is

$$E_0 \approx \frac{1}{2}\tilde{U}_0 \frac{N^2}{V} = \frac{1}{2}\bar{n}\tilde{U}_0 N, \quad \bar{n} = \frac{N}{V}. \quad (4)$$

It has a simple meaning. It is a product $\tilde{U}_0/V$ of any two particles' mean interaction energy over the volume ($w = 1/V$) with the number of total pairs $N^2/2$.

In conclusion, let us mention the following. Result (4), obtained by using perturbation theory over the potential, demands that the interaction between the particles is weak, so that the condition[224] $\tilde{U}_0 \ll \hbar^2 R/m$ is fulfilled. However, it could be generalized to the case of a strong repulsive[225] potential with a small range of action using *perturbation theory in terms of the scattering length* (see Problem 4.29, 4.31 and 11.4). Using the method of short-range potentials (mentioned in Problem 11.4), and the fact that $\tilde{U}_0$ ($\tilde{U}_0 = 2\pi\hbar^2 a_0/\mu$, $\mu = m/2$ is the particles reduced mass) differs from the scattering length a_0 only by a factor in the Born approximation, we can state that Eq. (4) for E_0 remains valid in the case of strong short-range potential when expressed in terms of the exact scattering length a_0. Also, it is valid when $a_0 \ll (\overline{n})^{-1/3}$.

Problem 10.33

Repeat the analysis from the previous problem for a Fermi-gas of spinful particles with $s = 1/2$. It is assumed that the interaction potential does not depend on spin and satisfies the condition $k_F R_0 \ll 1$, where R_0 is the potential radius and $\hbar k_F$ is the Fermi momentum.

Solution

The problem is to be solved in a way analogous to the previous one. The interaction operator $\hat{U}$ is determined by Eqs. (1) and (2), where we only need to add a spin index $\sigma \equiv s_z$ to the operators $\hat{a}_{\mathbf{k}_a}$, $\hat{a}^+_{\mathbf{k}_a}$ (i.e, to replace $\hat{a}_{\mathbf{k}_a}$ by $\hat{a}_{\mathbf{k}_a \sigma_a}$, etc., and add a summation over σ_a to the summation over $\mathbf{k}_a$. Here $\sigma_3 = \sigma_1$ and $\sigma_4 = \sigma_2$).

While calculating the first-order correction $E_0^{(1)} = \langle \psi_0 | \hat{U} | \psi_0 \rangle$ to the Fermi-gas ground state energy $E_0^{(0)}$ (assuming no interaction), we should take into account the occupation numbers $n_{k\sigma}$ in the state $|\psi_0\rangle$. They are equal to 1 for $k < k_F$ and 0 for $k > k_F$. Averaging the operator $\hat{U}$ (see Eq. (2) from Problem 10.32) we see that in the sums over k_a, all the terms for which at least one of the k_a is higher than k_F go to zero, so that only the matrix elements $U^{\mathbf{k}_3 \mathbf{k}_4}_{\mathbf{k}_1 \mathbf{k}_2}$ with $k_a < k_F$ occur in $E_0^{(1)}$. Turning to the center-of-mass variables,

$$\mathbf{r} = \mathbf{r}_1 - \mathbf{r}_2, \quad \mathbf{R} = \frac{\mathbf{r}_1 + \mathbf{r}_2}{2},$$

we transform the matrix element considered into the form:

$$U^{\mathbf{k}_3 \mathbf{k}_4}_{\mathbf{k}_1 \mathbf{k}_2} = \frac{1}{V^2} \int_V U(r) \exp\left\{\frac{i}{2}(\mathbf{k}_1 - \mathbf{k}_2 - \mathbf{k}_3 + \mathbf{k}_4)\mathbf{r}\right\} d^3 r$$

$$\times \int \exp\left\{i(\mathbf{k}_1 + \mathbf{k}_2 - \mathbf{k}_3 - \mathbf{k}_4)\mathbf{R}\right\} d^3 R.$$

---

[224] Note that $\tilde{U}_0 \sim U_0 R^3$, where U_0, R are the characteristic value and radius of particles' pair-interaction potential.

[225] The repulsive character of the interaction ensures stability of the gaseous phase at the absolute zero of temperature $T = 0$.

Using the condition $R_0 \ll V^{1/3}$ on the effective radius of potential $U(r)$, the integration over $\mathbf{r}$ can be extended to all the space (compare to the previous problem), and due to the inequality $k_F R_0 \ll 1$ the exponential factor can be replaced by unity. We obtain

$$U_{\mathbf{k}_1 \mathbf{k}_2}^{\mathbf{k}_3 \mathbf{k}_4} \approx \frac{1}{V} \delta_{\mathbf{k}_1+\mathbf{k}_2, \mathbf{k}_3+\mathbf{k}_4} \tilde{U}_0, \quad \tilde{U}_0 = \int U(r) dr^3,$$

(the factor $\delta_{\mathbf{k}_1+\mathbf{k}_2, \mathbf{k}_3+\mathbf{k}_4}$ corresponds to momentum conservation), and hence

$$E_0^{(1)} = \frac{\tilde{U}_0}{2V} \langle \psi_0 | \sum_{\sigma_{1,2}} \sum_{\{k_a\}} \delta_{\mathbf{k}_1+\mathbf{k}_2, \mathbf{k}_3+\mathbf{k}_4} \hat{a}_{\mathbf{k}_3 \sigma_1}^{+} \hat{a}_{\mathbf{k}_4 \sigma_2}^{+} \hat{a}_{\mathbf{k}_2 \sigma_2} \hat{a}_{\mathbf{k}_1 \sigma_1} | \psi_0 \rangle. \tag{1}$$

Using the explicit form of the wavefunctions $|\psi_0\rangle$ we see that of all the terms only those for which $\mathbf{k}_3 = \mathbf{k}_1$, $\mathbf{k}_4 = \mathbf{k}_2$, or $\mathbf{k}_4 = \mathbf{k}_1$, $\mathbf{k}_3 = \mathbf{k}_2$ differ from zero, so that (1) can be transformed to

$$E_0^{(1)} = \frac{\tilde{U}_0}{2V} \langle \psi_0 | \sum_{\sigma_{1,2}} \sum_{\mathbf{k}_{1,2}} (\hat{a}_{\mathbf{k}_1 \sigma_1}^{+} \hat{a}_{\mathbf{k}_2 \sigma_2}^{+} + \hat{a}_{\mathbf{k}_2 \sigma_1}^{+} \hat{a}_{\mathbf{k}_1 \sigma_2}^{+}) \hat{a}_{\mathbf{k}_2 \sigma_2} \hat{a}_{\mathbf{k}_1 \sigma_1} | \psi_0 \rangle. \tag{2}$$

Here the double sum over $\mathbf{k}_{1,2}$ for $\sigma_1 = \sigma_2$ is equal to zero because

$$\hat{a}_{\mathbf{k}_1 \sigma}^{+} \hat{a}_{\mathbf{k}_2 \sigma}^{+} + \hat{a}_{\mathbf{k}_2 \sigma}^{+} \hat{a}_{\mathbf{k}_1 \sigma}^{+} = \{\hat{a}_{\mathbf{k}_1 \sigma}^{+}, \hat{a}_{\mathbf{k}_2 \sigma}^{+}\} = 0$$

for fermion operators. For $\sigma_1 \neq \sigma_2$, we may use the anti-commutation properties of the operators $\hat{a}$, $\hat{a}^+$ to write the first part of sum (2) in the form

$$\sideset{}{'}\sum_{\sigma_1 \sigma_2} \sum_{\mathbf{k}_1 \mathbf{k}_2} \hat{a}_{\mathbf{k}_1 \sigma_1}^{+} \hat{a}_{\mathbf{k}_2 \sigma_2}^{+} \hat{a}_{\mathbf{k}_2 \sigma_2} \hat{a}_{\mathbf{k}_1 \sigma_1} = \sideset{}{'}\sum_{\sigma_{1,2}} \sum_{\mathbf{k}_{1,2}} \hat{n}_{\mathbf{k}_2 \sigma_2} \hat{n}_{\mathbf{k}_1 \sigma_1}, \quad \sigma_1 \neq \sigma_2, \tag{3}$$

where $\hat{n}_{\mathbf{k},\sigma}$ are the occupation-number operators. Since $\hat{N}(\sigma) = \sum_{\mathbf{k}} \hat{n}_{\mathbf{k}\sigma}$ is the total number of particles in a state σ and since in the ground state these numbers have definite values equal to $N/2$, the average of relation (3) over the state $|\psi_0\rangle$ yields $2(N/2)^2 = N^2/2$. The second part of the sum in (2) for $\sigma_1 \neq \sigma_2$ is different from zero only for the values $\mathbf{k}_1 = \mathbf{k}_2$. We can see that this second part is equal to N, i.e., is negligibly small (since $N \gg 1$) compared to the first part of the sum. Hence we find

$$E_0^{(1)} = \frac{1}{4} \tilde{U}_0 \frac{N^2}{V} \tag{4}$$

and the ground state energy is

$$E_0 \cong E_0^{(0)} + E_0^{(1)} = \frac{3}{10} \left(\frac{3\pi^2 N}{V}\right)^{2/3} \frac{\hbar^2}{m} N + \frac{1}{4} \tilde{U}_0 \frac{N^2}{V}. \tag{5}$$

An additional factor of $1/2$ appears in the relation for $E_0^{(1)}$ as compared to the Bose-gas (see the previous problem) for a simple reason. In the approximation considered,

$k_F R_0 \ll 1$, fermions with the same value of spin projection do not interact with each other due to the Pauli principle. Compare to Problem 10.31.

In conclusion, let us note that the generalization of the result from our perturbation theory to the case of strong short-range potential, mentioned in the previous problem, is also valid for a Fermi-gas.

Problem 10.34

An ideal Fermi-gas of neutral particles with the spin $s = 1/2$ and the spin magnetic moment μ_0 (so that $\hat{\boldsymbol{\mu}} = \mu_0 \hat{\boldsymbol{\sigma}}$) is placed in an external homogeneous magnetic field. For the ground state of this system, find:

1) the single-particle states' occupation numbers;
2) the magnetic susceptibility (in a weak field).

Assume that the interaction between the magnetic moments is negligibly small.

Solution

The energy of a particle with momentum $\mathbf{p}$ and spin projection (onto the direction of the magnetic field) s_z is equal to (for $s = 1/2$)

$$\varepsilon_{ps_z} = \frac{1}{2m}p^2 - 2\mu_0 s_z \mathcal{H}.$$

Denote the number of particles in the ground state of the Fermi-gas with $s_z = \pm 1/2$ by $N^{\pm}$. Since the ground state is the state wherein the system energy attains its minimum value, it corresponds to the occupation number

$$n_{\mathbf{p},\pm} = 1 \quad \text{for} \quad |\mathbf{p}| \leq p_{F,\pm} \quad \text{and} \quad n_{\mathbf{p},\pm} = 0 \quad \text{for} \quad |\mathbf{p}| > p_{F,\pm}$$

and

$$\frac{1}{2m}p_{F,+}^2 - \mu_0 \mathcal{H} = \frac{1}{2m}p_{F,-}^2 + \mu_0 \mathcal{H}. \tag{1}$$

This relation means that equally filling states with $s_z = \pm 1/2$ to their maximum energies allowed by the Fermi momentum provides the energy minimum for the whole system.

Expressing $p_{F,\pm}$ in terms of $N_\pm$ in the usual way,

$$N_\pm = \sum_{\mathbf{p}} n_{\mathbf{p},\pm} = \int_{p<p_{F,\pm}} \frac{V d^3 p}{(2\pi\hbar)^3} = \frac{V p_{F,\pm}^3}{6\pi^2 \hbar^3}, \quad p_{F,\pm} = \left(\frac{6\pi^2 \hbar^3 N_\pm}{V}\right)^{1/3},$$

according to (1), we find the equation that determines $N_\pm$:

$$N_+^{2/3} - N_-^{2/3} = 4m\mu_0 \mathcal{H} \left(\frac{V}{6\pi^2\hbar^3}\right)^{2/3}, \quad N_+ + N_- = N. \tag{2}$$

Here, the magnetic moment of the gas is equal to

$$M_z = \mu_0(N_+ - N_-), \quad M_x = M_y = 0. \tag{3}$$

In the absence of a magnetic field, $N_+ = N_- = N/2$. In the case of a weak field, $N_\pm$ only slightly differs from $N/2$. Writing them in the form $N_\pm = N/2 \pm n$ and performing an expansion over the small parameter n/N in (2) (using $(N/2 \pm n)^{2/3} \approx (N/2)^{2/3}(1 \pm 4n/3N)$), we find:

$$2n = (N_+ - N_-) = m\mu_0 N \mathcal{H} \left(\frac{\sqrt{3}V}{\pi^2\hbar^3 N}\right)^{2/3}. \tag{4}$$

(The condition $n/N \ll 1$ equivalently defines the case of a weak field.)

From (3) and (4) in a weak field we have

$$M = \chi_0 V \mathcal{H}, \quad \chi_0 = \frac{3m\mu_0^2}{(3\pi^2\hbar^3)^{2/3}} \left(\frac{N}{V}\right)^{1/3}. \tag{5}$$

Since the *magnetic susceptibility* $\chi_0 > 0$, the Fermi-gas is a *paramagnet*.

With the increase of the field, the value $|N_+ - N_-|$ increases monotonically for $\mathcal{H} \geq \mathcal{H}_{cr}$, where

$$\mathcal{H}_{cr} = \frac{1}{4m|\mu_0|} \left(\frac{6\pi^2\hbar^3 N}{V}\right)^{2/3}. \tag{6}$$

When the spins of all the particles array in the same direction, the system's magnetic moment saturates at

$$M = |\mu_0|N, \tag{7}$$

and is directed along the magnetic field.

Problem 10.35

Describe the screening of a point charge q placed inside a conductor by the conduction electrons. Use statistical methods,[226] considering conduction electrons as a Fermi-gas (on the background of a positively charged uniform distribution that ensures overall electroneutrality) at temperature $T = 0$. Electron density distortions in the vicinity of the charge can be considered small.

---

[226] This is similar to the Thomas–Fermi method; see Chapter 2, sec. 3.

Solution

At $T = 0$, electrons occupy the lowest energy states. The electron density $n(\mathbf{r})$, the maximum kinetic energy of electrons at point $\mathbf{r}$ (equal to $\varepsilon_0(\mathbf{r}) = p_0^2(\mathbf{r})/2m$), and the electrostatic potential $\varphi(\mathbf{r})$ inside the conductor,[227] are connected by the relations

$$\Delta\varphi = -4\pi\rho = -4\pi q\delta(\mathbf{r}) + 4\pi e\delta n(\mathbf{r}), \tag{1}$$

where

$$\frac{1}{2m}p_0^2(\mathbf{r}) - e\varphi(\mathbf{r}) = \varepsilon_F = \frac{1}{2m}p_F^2 = \text{const}, \tag{2}$$

$$n(\mathbf{r}) = \frac{1}{3\pi^2\hbar^3}p_0^3(\mathbf{r}), \quad \delta n = n - n_0 = \frac{1}{3\pi^2\hbar^3}(p_0^3(\mathbf{r}) - p_F^3).$$

Here $e > 0$ is the electron charge, $\delta n(r)$ is a change of electron density caused by the charge q; relation (2) provides the minimum of energy for the whole electron system. The parameter ε_F characterizes the unperturbed system. (This remains so if a charge is held at a large distance from it.)

Due to the assumed smallness of $|\delta n| \ll n_0$ we have $|p_0 - p_F| \ll p_F$, and from relations (2) we obtain

$$e\varphi \approx \frac{1}{m}p_F(p_0(\mathbf{r}) - p_F), \quad \delta n = \frac{1}{\pi^2\hbar^3}p_F^2(p_0(\mathbf{r}) - p_F).$$

Now Eq. (1) takes the form

$$\Delta\varphi = -4\pi q\delta(\mathbf{r}) + \kappa^2\varphi, \quad \kappa = \left(\frac{4me^2 p_F}{\pi\hbar^3}\right)^{1/2}, \tag{3}$$

and its solution[228]

$$\varphi(r) = \frac{q}{r}e^{-\kappa r} \tag{4}$$

shows that the electric field inside the conductor is screened over distances $r \sim 1/\kappa$.

Problem 10.36

Find the charge distribution in the vicinity of a charged surface of a conductor (with the "surface" charge density σ). Use statistical considerations assuming that the conduction electrons form a degenerate Fermi-gas. Consider fluctuations in the electron density in the vicinity of the conductor surface to be small, and compare your results to the previous problem.

---

[227] The quantities considered are spherically symmetric with respect to the point $r = 0$, where the charge q is placed. Note that here, unlike in the case of an atom, $n(r) \to n_0 \neq 0$ for $r \to \infty$.
[228] Compare, for example, to Problem 4.20.

Solution

Since the width of the near-surface region inside the conductor, where the charge is distributed (*i.e.*, $\rho \neq 0$), has a microscopic size, we can neglect the curvature of the surface and consider only effectively the one-dimensional case. Considering the surface as a plane at $z = 0$, the electron charge and potential distribution in the vicinity of the surface are described by the equation

$$\varphi''(z) = -4\pi\rho(z) = 4\pi e\, \delta n(z) \approx \kappa^2 \varphi(z), \tag{1}$$

where

$$\kappa = \left(\frac{4m^2 p_F}{\pi\hbar^3}\right)^{1/2}.$$

Compare to the previous problem. The solution of the equation that satisfies the boundary condition $\varphi(z) \to \varphi_0 = 0$ as $z \to \infty$ (we consider that the conductor corresponds to the values $z > 0$) has the form:

$$\varphi(z) = Ae^{-\kappa z}, \quad z \geq 0, \tag{2}$$

while the volume charge density is

$$\rho(z) = -\frac{\kappa^2}{4\pi}\varphi(z) = -\frac{\kappa^2 A}{4\pi}e^{-\kappa z}, \quad z \geq 0. \tag{3}$$

The value of A is determined by the "surface" charge density:

$$\sigma_0 = \int_0^\infty \rho(z)dz, \quad A = -\frac{4\pi\sigma_0}{\kappa}. \tag{4}$$

The same value of A follows from the boundary condition $E_n(0) = \varphi'(0) = 4\pi\sigma_0$, familiar from electrostatics.

11

Atoms and molecules

Most calculations for atomic systems are based on the assumption that individual electrons composing the atom occupy certain single-electron quantum states within the atom. In the framework of this approximation, the atomic wavefunction can be written as an anti-symmetric combination of products of the single-electron states. The most accurate approach within this approximation involves numerical solution of the Hartree–Fock equations, obtained using the self-consistent-field method for the one-electron states.

For systems with a large number of electrons, the self-consistent field idea develops into the Thomas–Fermi method, where the average electron density, $n(r)$, in the ground-state of a neutral atom or a charged ion, is connected with the electrostatic potential, $\varphi(r)$, by the relation[229]

$$n(r) = \frac{1}{3\pi^2}[2(\varphi(r) - \varphi_0)]^{3/2}. \tag{XI.1}$$

For a neutral atom, $\varphi_0 = 0$, and the Poisson equation, $\Delta \varphi = -4\pi\rho$, leads to the Thomas–Fermi equation as follows (below, $r \neq 0$):

$$\Delta \varphi = 4\pi n = \frac{8\sqrt{2}}{3\pi}\varphi^{3/2}. \tag{XI.2}$$

We introduce the more convenient quantities x and $\chi(x)$, according to

$$r = xbZ^{-1/3}, \quad \varphi(r) = \frac{Z}{r}\chi(x) = \frac{Z^{4/3}}{b}\frac{\chi(x)}{x}, \tag{XI.3}$$

where $b = (3\pi/8\sqrt{2})^{2/3} \approx 0.885$, and Z is the number of electrons (the charge of the nucleus). We transform relation Eq. (XI.2) to the form

$$\sqrt{x}\chi''(x) = \chi^{3/2}(x), \tag{XI.4}$$

---

[229] The electrons are assumed to form a many-particle state, a degenerate Fermi-gas at absolute zero, in the presence of a slowly varying field with the potential energy, $U(r) = -e\varphi(r)$. Here $-e\varphi_0$ determines the maximum total energy of the occupied electron states, and $e(\varphi(r) - \varphi_0)$ determines the maximum kinetic energy $(1/2m_e)p_F^2(r)$. The case of $\varphi_0 > 0$ corresponds to *positively-charged* ions.

Here and below we use *atomic units* (a.u.), where $e = \hbar = m_e = 1$.

with boundary conditions $\chi(0) = 1$ and $\chi(\infty) = 0$. The function $\chi(x)$ is *universal* in the Thomas–Fermi method (that is, it remains the same independent of the atomic system at hand). Numerical calculations in the Thomas–Fermi model give the full ionization energy of an atom as follows: $E_0 = -(3\chi'(0)/7b) \, Z^{7/3} = 0.769 \, Z^{7/3} = 20.9 \, Z^{7/3}$ eV, with $\chi'(0) = -1.589$ (see also Problem 11.21).

This chapter presents, in particular, a set of problems that focus on the properties of weakly-bound particles in external electric and magnetic fields (it is assumed that the particle is bound by a short-range neutral potential of radius, r_S, and that the binding energy is small: $\varepsilon = \kappa^2/2$, $\kappa r_S \ll 1$). Such models[230] are used in atomic physics to study negatively charged ions, where the outer weakly-bound electron is considered to be in a short-range collective potential produced by the neutral atom.

In this class of problems, large distances play a dominant role. In the case of a short-range potential, the unperturbed wavefunction has the following asymptotic behavior in this limit:

$$\psi^{(0)}_{nlm}(\mathbf{r}) \approx C_{\kappa l} \sqrt{\frac{\kappa}{2\pi}} \frac{e^{-\kappa r}}{r} Y_{lm}(\mathbf{n}), \quad r \gg r_S \qquad (XI.5)$$

where $C_{\kappa l}$ is the so-called asymptotic coefficient (at infinity).

The Schrödinger equation that determines the bound states of a one-electron atom (an atom with nucleus charge Ze and a single electron) is given in Chapter 4. The radial wavefunction for the lower levels is given by Eqs. (IV.4) with $a = \hbar^2/(m_e Z e^2)$.

11.1 Stationary states of one-electron and two-electron atoms

Problem 11.1

Find relativistic corrections arising due to the velocity-dependence of the electron mass to the energy levels of a *hydrogen-like* atom[231] in the first-order perturbation theory with respect to the electron velocity.[232]

Solution

The relativistic correction to the classical Hamiltonian function of a charged particle in a static electric field follows from the classical equation (we assume that $-e$ is the charge of the particle and $p \ll mc$):

---

[230] Many properties of states with a small binding energy, $E \ll \hbar^2/mr_S^2$, are determined by just two parameters: the binding energy and a so-called "asymptotic coefficient", C_κ, (see Problems 11.36 and 11.37). They in turn are connected with the parameters of low-energy scattering: the *scattering length* a_l and the *effective radius* r_l see; Eq. (XIII.15).

[231] A hydrogen-like/helium-like atom or ion is a system consisting of a nucleus with an arbitrary charge Ze and one/two electrons.

[232] For the model of a spinless charged particle (as considered in the problem), the result will determine a *fine structure* of the hydrogen-like atom. In the case of an electron with spin, as follows from the Dirac equation, there exists an additional correction arising from the *spin-orbit interaction*. Its contribution to the energy levels has the same order of magnitude as the correction calculated in this problem.

$$H = \sqrt{p^2c^2 + m^2c^4} - e\varphi(r) - mc^2 \simeq \frac{p^2}{2m} - \frac{p^4}{8m^3c^2} - e\varphi(r), \tag{1}$$

and is equal to $-p^4/8m^3c^2$. The quantum-mechanical generalization of this correction corresponds to the operator, $\hat{V} = -\hat{\mathbf{p}}^4/8m^3c^2$, which can be viewed as a perturbation. Since it commutes with the operators, $\hat{\mathbf{l}}^2$ and $\hat{l}_z$, the eigenfunctions $\psi_{n_rlm}^{(0)}$ of the unperturbed Hamiltonian ($U = -e\varphi = -Ze^2/r$; see also Eq. IV.3) are the correct zeroth-order approximation. Thus, according to Eq. (VIII.1), the shift of the energy level is

$$E_{n_rl}^{(1)} = \int \psi_{n_rlm}^{(0)*} \hat{V} \psi_{n_rlm}^{(0)} d^3r = -\frac{1}{2mc^2} \langle n_rlm | \left(\frac{\hat{\mathbf{p}}^2}{2m}\right)^2 | n_rlm \rangle \equiv,$$

$$-\frac{1}{2mc^2} \langle n_rlm | \left(\hat{H}_0 + \frac{Ze^2}{r}\right)^2 | n_rlm \rangle =$$

$$-\frac{1}{2mc^2} \langle n_rlm | \hat{H}_0^2 + \hat{H}_0 \frac{Ze^2}{r} + \frac{Ze^2}{r} \hat{H}_0 + \left(\frac{Ze^2}{r}\right)^2 | n_rlm \rangle, \tag{2}$$

where $\hat{H}_0$ is the unperturbed Hamiltonian of the hydrogen-like atom.

Since $\psi_{n_rlm}^{(0)}$ is an eigenfunction of the operator, $\hat{H}_0$, all the operator terms in Eq. (2) involving $\hat{H}_0$ can be replaced by the corresponding eigenvalue below:[233]

$$E_{n_rl}^{(0)} = E_n^{(0)} = -\frac{m(Ze^2)^2}{2\hbar^2 n^2}, \quad n = n_r + l + 1. \tag{3}$$

Then, the first perturbative correction is related to the two matrix elements

$$\langle n_rlm | \frac{Ze^2}{r} | n_rlm \rangle = -E_n^{(0)}, \quad \langle n_rlm | \frac{1}{r^2} | n_rlm \rangle = -\frac{4mE_n^{(0)}}{(2l+1)\hbar^2 n}. \tag{4}$$

The easiest way to calculate these matrix elements is to use relation (I.6). Let us note that (3) can be viewed as eigenvalues of the operator

$$\hat{H}_l = -\frac{\hbar^2}{2m} \frac{1}{r^2} \frac{d}{dr} r^2 \frac{d}{dr} - \frac{\hbar^2 l(l+1)}{2mr^2} - \frac{Ze^2}{r}.$$

Now, differentiating Eq. (3) with respect to the parameters, Z and l, and using (I.6), we obtain Eq. (4) and

$$E_{n_rl}^{(1)} = \left(\frac{Ze^2}{\hbar c}\right)^2 \left\{-\frac{3}{4n^2} + \frac{2}{(2l+1)n}\right\} E_n^{(0)}. \tag{5}$$

---

[233] In particular,

$$\int \psi_n^{(0)*} \hat{H}_0 \frac{Ze^2}{r} \psi_n^{(0)} d^3r = \int (\hat{H}_0 \psi_n^{(0)})^* \frac{Ze^2}{r} \psi_n^{(0)} d^3r = E_n^{(0)} \int \frac{Ze^2}{r} |\psi_n^{(0)}|^2 d^3r,$$

where we used the fact that $\hat{H}_0$ is an Hermitian operator.

We see that the relativistic correction lifts the accidental degeneracy of the levels in the Coulomb field. The level with a given n splits into n components corresponding to the possible values of the angular momentum, $l = 0, 1, \ldots, (n-1)$ of the unperturbed level. The relative shift of the levels decreases as the angular momentum increases. The width of the *fine structure* interval, the distance between the energies of the levels corresponding to $l = 0$ and $l = (n-1)$,

$$\Delta E_{FS} = 4(Z\alpha)^2 \frac{n-1}{n(2n-1)} |E_n^{(0)}|, \tag{6}$$

also decreases with the increase in n. Here $\alpha = e^2/\hbar c = 1/137$ is the *fine-structure constant*. Let us note that for the $n = 2$ level, the splitting is $\Delta E_{FS}(2) = 1.21 \cdot 10^{-4} Z^4$ eV.

Generally, for the levels with $n \sim 1$, Eq. (5) gives

$$\left| \frac{E_{n_r l}^{(1)}}{E_n^{(0)}} \right| \sim (Z\alpha)^2 \sim 10^{-4} Z^2.$$

Note that perturbation theory is reliable if $Z \ll 137$. The atomic number is a natural parameter that controls the applicability of the theory here, because the typical electron velocity inside the atom is $v \sim Ze^2/\hbar = Z\alpha c$, and for the values $Z \sim 100$ the motion becomes essentially relativistic.

Let us make a few further comments on the applicability of the results. As emphasized in the problem and as is obvious from its solution, the correction we found does not take into account the effects of spin-orbit interaction, but rather describes a spinless charged particle. The solution correctly describes the energy shift within this model in the first order of perturbation theory in $(Z\alpha)^2$. However, note that this naïve Hamiltonian perturbation theory does not capture properly the relativistic effects in higher orders of perturbation theory, which are properly described by the fundamental equation of quantum field theory that describes such particles: the *Klein–Gordon equation* (see also Problems 15.13–15.15). Furthermore, while the "spinless model" is mathematically well-defined, it does not have an obvious physical realization. Real spinless particles, such as the π- and K-mesons for example, involve the *strong interaction*, which has a far higher impact on the atomic levels than the relativistic corrections discussed here (see Problem 11.4).

On the other hand, the inclusion of spin-orbit interaction to describe the real electron involves solving the *Dirac equation* for the model of a hydrogen-like atom with a point nucleus. The fine structure of the model has the following property. A level with a given n, which is $2n^2$-fold degenerate according to the non-relativistic theory (2 due to the spin), splits into n components due to the relativistic effects (as in the case of a spinless particle). Each sublevel corresponds to some value, j, of the total electron angular momentum $j = 1/2, 3/2, \ldots, n - 1/2$. In this case, the accidental degeneracy is not totally removed, since the levels with the same values of j but different $l = j \pm 1/2$ remain degenerate. Hence, the $n = 2$ level, for example, splits into two components, one of which corresponds to $2p_{3/2}$-states, and the other to degenerate $2s_{1/2}$ and $2p_{1/2}$-states. Here the width of the fine-structure split level is equal to $\Delta_{FS} \approx 4.5 \cdot 10^{-5} Z^4$ eV. A further lifting of the degeneracy in l (the *Lamb shift*) appears due to so-called radiative corrections which are discussed in Problem 11.62.

Problem 11.2

Consider the *hyperfine structure* of the s-levels of a hydrogen-like atom arising from the interaction between the electron and nuclear magnetic moments. Take the model of a point nucleus with spin, I, and the magnetic moment, μ_0, so that $\hat{\boldsymbol{\mu}} = \mu_0 \hat{\mathbf{I}}/I$. Estimate the hyperfine splitting and compare it with the fine structure discussed in the previous problem.

The nuclear magnetic moment is of the order $e\hbar/m_p c$, with m_p being the proton mass. In the case of the hydrogen atom, compare the result obtained with the experimentally observed hyperfine splitting of the ground-state levels: $\Delta\nu_{HFS} \equiv \Delta E_{HFS}/2\pi\hbar \approx 1420$ MHz.[234] The proton magnetic moment is equal to $\mu_p = 1.396 \, e\hbar/m_p c$.

Solution

1) According to classical electrodynamics, the mutual interaction of two magnetic moments has the form[235] $V = -\boldsymbol{\mu}_1 \cdot \mathcal{H}_2(\mathbf{r})$, where

$$\mathcal{H}_2(\mathbf{r}) = \nabla \times \mathbf{A}_2 = \nabla \times \frac{[\boldsymbol{\mu}_2 \times \mathbf{r}]}{r^3} \equiv \{(\boldsymbol{\mu}_2 \cdot \nabla)\nabla - \boldsymbol{\mu}_2 \Delta\}\frac{1}{r}$$

is the magnetic field produced by the second moment. The quantum-mechanical generalization of this interaction to the case of spin magnetic moments is the operator $\hat{V}$, obtained by the change of classical variables to the corresponding operators (the convention here and below in this problem is that the electron, corresponds to the first particle):

$$\boldsymbol{\mu}_1 \to \hat{\boldsymbol{\mu}}_e = -\frac{e\hbar}{m_e c}\hat{\mathbf{s}}, \quad \boldsymbol{\mu}_2 \to \hat{\boldsymbol{\mu}}_{nuc} = \frac{\mu_0}{I}\hat{\mathbf{I}}.$$

The Hermitian operator, $\hat{V}$, can be written in the form

$$\hat{V} = \frac{e\hbar\mu_0}{m_e c I}\hat{I}_i\hat{s}_k\left(\frac{\partial}{\partial x_i}\frac{\partial}{\partial x_k} - \delta_{ik}\Delta\right)\frac{1}{r}. \tag{1}$$

We now consider it as a perturbation and take into account the form of the unperturbed Hamiltonian. The wavefunction of the s-states can be written as $\psi^{(0)} = \psi_{ns}^{(0)}(r)\chi$ (where χ is the spin part of the wavefunction). Now we average

---

[234] The energy corresponding to frequency $\nu = 1$ MHz is equal to $\varepsilon_0 \approx 4.136 \cdot 10^{-9}$ eV.
[235] Note that particle permutation does not change the form of the interaction.

the operator $\hat{V}$ (VIII.5) over the electron coordinates using the secular equation of degenerate perturbation theory.[236] We have

$$\int |\psi_{ns}^{(0)}(r)|^2 \left(\frac{\partial}{\partial x_i} \frac{\partial}{\partial x_k} - \delta_{ik}\Delta \right) \frac{1}{r} d^3r = C\delta_{ik}.$$

The tensor δ_{ik} appears in the right-hand side, because no other tensors are possible that are consistent with the spherical symmetry of the s-state. To determine the value of C, we trace over the indices i and k. Taking into account that $\delta_{ii} = 3$ and

$$\frac{\partial}{\partial x_i} \frac{\partial}{\partial x_i} \frac{1}{r} \equiv \Delta \frac{1}{r} = -4\pi\delta(\mathbf{r}),$$

we find that $C = (8\pi/3)|\psi_{ns}^{(0)}(0)|^2$. Thus, we obtain

$$\overline{\hat{V}} = \frac{8\pi e\hbar\mu_0}{3m_e cI} |\psi_{ns}^{(0)}(0)|^2 (\hat{\mathbf{I}} \cdot \hat{\mathbf{s}}); \quad |\psi_{ns}^{(0)}(0)|^2 = \frac{Z^3 e^6 m_e^3}{\pi \hbar^6 n^3}. \tag{2}$$

Here $\overline{\hat{V}}$ remains an operator in the spin space. The eigenfunctions of this operator are spinors that correspond to a definite value $J = I \pm 1/2$ of the total angular momentum of the atom and determine the *hyperfine splitting* of the ns-level:

$$E_{HFS} = E_{n,J}^{(1)} = \frac{4\pi e\hbar\mu_0}{3m_e cI} |\psi_{ns}^{(0)}(0)|^2 \cdot \begin{cases} I, & J = I + \frac{1}{2}, \\ -(I+1), & J = I - \frac{1}{2}. \end{cases} \tag{3}$$

The level splits into two sublevels in accordance with two possible values of the total angular momentum. Compare to Problem 3.27.

The value of the hyperfine splitting is

$$\Delta E_{HFS}(ns) = E_{HFS}\left(J = I + \frac{1}{2}\right) - E_{HFS}\left(J = I - \frac{1}{2}\right)$$

$$= \frac{4\pi(2I+1)e\hbar\mu_0}{3m_e cI} |\psi_{ns}^{(0)}(0)|^2. \tag{4}$$

A comparison (for $n \sim 1$ and $Z \sim 1$) with the fine-structure interval (6) from Problem 11.1 gives $\Delta E_{HFS}/\Delta E_{FS} \sim m_e/m_p \sim 10^{-3}$. That is, the hyperfine splitting is actually much smaller than fine structure effects.

2) For the ground-state of the hydrogen atom, $n = 1$, we estimate from Eq. (4): $\Delta \nu_{HFS}(1s) = \Delta E_{HFS}/2\pi\hbar \approx 1420$ MHz. But a more accurate comparison shows that the theoretical result, 1418.6 MHz, differs from the experimental value of 1420.4 MHz by ≈ 0.1 %. This difference exceeds by an order of magnitude the relativistic correction ($\sim \alpha^2 \sim 10^{-4}$, and we have taken into account the difference

---

[236] Note that the zeroth-order perturbation theory wavefunctions have a definite value of the angular momentum ($l = 0$ in this problem), which is related to the accidental Coulomb degeneracy of atomic levels in the non-relativistic hydrogen atom. This degeneracy, however, is removed by relativistic effects which give rise to the fine structure of the atomic spectrum, as discussed in the previous problem. Note also that the interaction, $\hat{V}$, does not mix the $s_{1/2}$- and $p_{1/2}$-states.

between reduced mass of the *ep*-system and the electron mass, m_e). The inconsistency is due to the *anomalous magnetic moment* of the electron, whose value is equal to $\mu_0(1 + \alpha/2\pi)$, where $\mu_0 = -e\hbar/2m_e c$.

Finally, we note that there exist other reasons behind the actual hyperfine structure apart from the magnetic interaction considered here. They include interaction of the electron orbital current with the nuclear magnetic field, ($\mathbf{j}_{orb} = 0$ for $l = 0$). Furthermore, note that in general a distortion of the potential due to the quadrupole moment of the nucleus with $I \geq 1$ also contributes to the splitting. This effect, however, is less important than the contribution of the magnetic interactions, and is absent in the case of a spherically symmetric *s*-state.

Problem 11.3

Calculate the shift of the *s*-levels of a hydrogen-like atom due to nuclear finite-size effects. Use first-order perturbation theory. Consider the nuclear charge distribution to be spherically symmetric. Provide a numerical estimate of the correction using the model of a nucleus as a uniformly charged sphere with radius $R \approx 1.2 \times 10^{-13} A^{1/3}$ cm, $A \approx 2Z$, where A is the atomic mass number. Compare the magnitude of the finite-size corrections with those of the fine-structure and hyperfine-structure effects discussed in the two previous problems.

How important are the finite-size effects for a μ-atom? The interaction of the muon with the nucleus has the electrostatic character.

Solution

We denote the nuclear potential as $\varphi(r)$. Outside the nucleus, $\varphi = Ze/r$. The distortion of the Coulomb potential, $\delta\varphi = \varphi - Ze/r$, near the nucleus determines the perturbation $V = -e\delta\varphi(r)$. Therefore, the shift of the *ns*-levels due to this Coulomb perturbation is

$$\Delta E_{ns} = \int V(r)|\psi_{ns}^{(0)}(r)|^2 d^3r \approx -e|\psi_{ns}^{(0)}(0)| \int \delta\varphi(r) d^3r. \tag{1}$$

We approximated $|\psi_{ns}^{(0)}(r)|^2 \approx |\psi_{ns}^{(0)}(0)|^2$, because the electron wavefunction $\psi_{ns}^{(0)}(r)$ is almost constant on the length scales of the nuclear radius ($r \leq R \approx 10^{-13} - 10^{-12}$ cm).[237]

The integral in Eq. (1) depends on the charge distribution in the nucleus. Taking into account that $\Delta r^2 = 6$, the integral in Eq. (1) takes the form:[238]

---

[237] Note that for states with a non-zero angular momentum, $l \neq 0$, the wavefunction scales as $\psi \propto r^l$, at small distances. Consequently, the finite-size effects in the level shifts decrease sharply with the increase in l.

[238] Here we used the Poisson equation in the form, $\Delta(\delta\varphi) = -4\pi[\rho - Ze\delta(\mathbf{r})]$, and the fact that the contribution of the δ-function term in the integral in Eq. (2) is equal to zero.

$$\int \delta\varphi(r)dV \equiv \frac{1}{6}\int \delta\varphi\Delta r^2 dV = \frac{1}{6}\int r^2 \Delta(\delta\varphi(r))dV =$$
$$= -\frac{2\pi}{3}\int r^2\rho(r)dV = -\frac{2\pi Ze}{3}\langle r^2\rangle. \tag{2}$$

Here $\rho(r)$ is the charge density inside the nucleus, and

$$\langle r^2\rangle \equiv r_e^2 = \frac{1}{Ze}\int r^2 \rho(r)dV,$$

where r_e is the nuclear charge radius. The quantity, $\langle r^2\rangle$, also determines the behavior of nuclear electric form-factor $F(q)$ for $q\to 0$:

$$F(q) \approx Ze\left(1 - \frac{1}{6}\langle r^2\rangle q^2\right).$$

For a uniform charge distribution in the sphere with radius R, we find $\langle r^2\rangle = 3R^2/5$, and according to Eqs. (1) and (2) we obtain[239]

$$\Delta E_{ns} = \frac{2\pi}{5}Ze^2 R^2 |\psi_{ns}^{(0)}(0)|^2, \quad |\psi_{ns}^{(0)}(0)|^2 = \frac{Z^3}{\pi a_B^3 n^3}. \tag{3}$$

The numerical value of the ratio

$$\left|\frac{\Delta E_{ns}}{E_n^{(0)}}\right| = \frac{4}{5n}Z^2\left(\frac{R}{a_B}\right)^2 \approx 8\cdot 10^{-10}\frac{Z^{8/3}}{n} \tag{4}$$

(where we used $R \approx 1.5 \cdot 10^{-11}Z^{1/3}$ cm). It is substantially smaller than both the relativistic correction (see Problem 11.1) and the hyperfine splitting (see Problem 11.2). For $Z \sim 1$, the finite size correction is just about 10^{-5} of the former and 10^{-2} of the latter.

It is interesting to estimate the contribution of the proton size, whose charge radius is $r_e \approx 0.8 \cdot 10^{-13}$ cm, to the Lamb shift of the hydrogenic levels $2s_{1/2}$ and $2p_{1/2}$ (see Problem 11.1). According to Eqs. (1) and (2), we obtain

$$\Delta E_{2s} = \frac{1}{12}\left(\frac{r_e}{a_B}\right)^2 \frac{e^2}{a_B} \approx 5.2\cdot 10^{-10} \; eV \approx 0.12 \; MHz.$$

Remarkably, the shift of the $2p_{1/2}$-level is $(a_B/r_e)^2 \sim 10^{10}$ times smaller. Since the experimental value of the Lamb shift is $\Delta_{LS} \approx 1058$ MHz, the contribution of the proton size is not important, and is of the order or smaller than the experimental error in Δ_{LS}.

Let us note, however, that the finite-size effects are clearly seen in the μ-atoms. This is related to the fact that the muon Bohr radius, $a_{\mu,B} = (m_e/m_\mu)a_B$, is 207 times

---

[239] This result could also be obtained from Eq. (1) and using the relation $\varphi = \frac{Ze(3R^2-r^2)}{2R^3}$ for the electrostatic potential inside a uniformly charged ball.

smaller than the electron Bohr radius. So, we should add an additional factor, equal to $(m_\mu/m_e)^2 \approx 4.3 \cdot 10^4$, to the estimate (4). For the ground level, $n = 1$, and the value $Z = 27$, we find $|\Delta E_{1s}/E_{1s}^{(0)}| \approx 0.2$ for muons.

Problem 11.4

Consider the shifts of the ns-levels in *pion* and *hadron* atoms[240] caused by short-range *strong nuclear interaction* between the pion and nucleus. Prove that the corresponding shift is described by the following perturbation theory equation

$$\Delta E_{ns} = \frac{2\pi\hbar^2}{m}|\psi_{ns}^{(0)}(0)|^2 a_s,$$

where $\psi_{ns}^{(0)}(0)$ is the value of the unperturbed Coulomb wavefunction, a_s is the pion *scattering length*, describing scattering off of the nuclear potential (see Problem 4.29, and also Problem 13.36, where a generalization for $l \neq 0$ is discussed).

Comment

The effect of a short-range potential, $U_s(r)$, with the radius r_s on the levels, $E_n^{(0)} \ll \hbar^2/mr_s^2$, of a long-range potential, U_L, with the radius $r_L \gg r_s$ could be taken into account as a change in the boundary conditions as follows: the wavefunction at small distances is constrained to have the asymptotic behavior, $\psi_n \propto (1 - a_s/r)$ for $r \ll r_L$. Thus the shift of the energy levels is determined entirely by the scattering length, a_s, of the short-range potential, U_s, and does not depend on its detailed form.

Solution

1) Consider the effect of the short-range potential, $U_s(r)$, with the radius r_s (*i.e.*, we assume that $U_s(r) \approx 0$, for $r \geq r_s$) on the levels of a particle with the energy, $|E_n^{(0)}| \ll \hbar^2/mr_s^2$, determined by a "long-range" potential, $U_L(r)$, with the radius, $r_L \gg r_s$. The latter is assumed to be small at distances of order $r \sim r_s$, so that $U_L \ll \hbar^2/mr_s^2$; however, for $r \geq r_L$ there are no restrictions on its value.

For these hadron atoms, the Coulomb attractive potential plays the role of U_L, while U_s describes their *strong* nuclear interaction. For example, for the pion atoms, $r_L \sim a_{\pi,B} = \hbar^2/m_\pi e^2 \approx 2 \cdot 10^{-11}$ cm and $r_s \approx 2 \cdot 10^{-13}$ cm. Considering an arbitrary interaction, U_s, the unperturbed levels, $E_n^{(0)}$, shift only weakly, and these shifts can be determined using perturbation theory in terms of the scattering length developed below.

Consider first the case where $U_L(r)$ is a spherically-symmetric potential, and focus on the s-levels. The conventional perturbation theory gives, in the first approximation,

---

[240] Hadron atoms are systems of two hadrons bound by Coulomb interaction, such as pion–proton ($\pi^- p$) or proton–anti-proton ($p\bar{p}$) atoms, for example. The Coulomb levels of such a system are $E_n^{(0)} = -\frac{m(\zeta e^2)^2}{2\hbar^2 n^2}$, where $m = \frac{m_1 m_2}{m_1 + m_2}$ is the reduced mass. $\zeta = -Z_1 Z_2 > 0$, $Z_{1,2}$ are the hadron charges, and $a_B = \frac{\hbar^2}{\zeta m e^2}$ is the Bohr radius.

$$\Delta E_{ns} = \int U_s(r)|\psi_{ns}^{(0)}(r)|^2 d^3r \approx |\psi_{ns}^{(0)}(0)|^2 \int U_s(r)d^3r. \tag{1}$$

Note that only small distances, $r \leq r_s$, are important in the integral above, and due to the smallness of r_s, the value of the unperturbed wavefunction hardly changes within this integration domain (see Eq. (1) from the previous problem).

The integral above is proportional to the Born approximation *scattering amplitude* of a particle with zero energy $E = 0$ in the potential $U_s(r)$ see; Eq. (XIII.6). Since $a_s = -f(E = 0)$ is the s-wave *scattering length* of the potential, $U_s(r)$, Eq. (1) takes the form

$$\Delta E_{ns} = \frac{2\pi\hbar^2}{m}|\psi_{ns}^{(0)}(0)|^2 a_s^B, \quad a_s^B = \frac{m}{2\pi\hbar^2}\int U_s(r)d^3r, \tag{2}$$

where a_s^B is the scattering length in the Born approximation.

2) Eqs. (1) and (2) are applicable only in the case of a weak potential, $U_s \ll \hbar^2/mr_s^2$, where $|a_s^B| \ll r_s$. If the potential U_s is sufficiently strong, naïve perturbation theory breaks down. However, in this case a correction to the energy can still be straightforwardly obtained from Eq. (2) via its simple modification: by replacing the Born scattering length, a_s^B, by the exact scattering length, a_s, of the potential $U(r)$. Since quite generally, $a_s \sim r_s$, the level shift will remain small. For example, for the long-range attractive potential $U_L(r)$ of the radius r_L, the unperturbed spectrum can be estimated as $E_n^{(0)} \sim \hbar^2/mr_L^2$ and the wavefunction near the origin is $|\psi_{ns}^{(0)}(0)|^2 \sim r_L^{-3}$. The s-levels shift according to Eq. (2), $\Delta E_{ns} \sim (a_s/r_L)E_{sn}^{(0)}$, which remains small as long as $a_s \ll r_L$ (the opposite case, $a_s \gtrsim r_L$, is considered below).

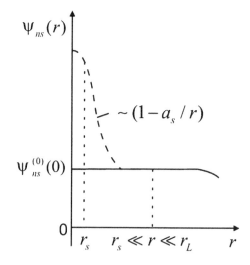

Fig. 11.1

To clarify the method introduced here, consider Fig. 11.1, which illustrates a qualitative change of the wavefunction. The solid line shows the unperturbed wavefunction, $\psi_{ns}^{(0)}(r)$, and the dashed line corresponds to the wavefunction, $\psi_{ns}(r)$, which includes the effects of the short-range potential, $U_s(r)$. The latter wavefunction has the following properties: 1) it remains finite at $r=0$; 2) its explicit form at $r \leq r_s$ depends strongly on the potential, $U_s(r)$; 3) outside the range of the potential, U_s,

$$r_s \ll r \ll r_L, \quad \left(\frac{m|E_n|}{\hbar^2}\right)^{-1/2}, \tag{3}$$

it has the form $\psi_{ns} \approx A(1 - a_s/r)$, where a_s is the scattering length of the potential, U_s, and in this region the wavefunction depends on the long-range potential, U_L, only weakly. This is connected with the fact that in this domain the Schrödinger equation can be approximated by $\Delta \psi_{ns}(r) = 0$ or $(r\psi_{ns}(r))'' = 0$.
Now let us note that for the values of scattering length that satisfy the conditions

$$|a_s| \ll r_L \quad \text{and} \quad |a_s| \ll \left(\frac{m|E_n^{(0)}|}{\hbar^2}\right)^{-1/2}, \tag{4}$$

the exact wavefunction $\psi_{ns}(r)$ is quite different from the unperturbed wavefunction $\psi_{ns}^{(0)}(r)$ at small distances, but they almost coincide at large distances, $r \gg |a_s|$. We conclude that the level shift is small and its value depends on the scattering length of the short-range potential only, but not on the detailed shape of the potential, $U_s(r)$. Therefore, $U_s(r)$ can be replaced by a fictitious potential, the *pseudo-potential* $\tilde{U}_s(r)$, which is within the domain of validity of perturbation theory. A requirement here is that the Born scattering length, $\tilde{a}_s^B$, of the pseudo-potential[241] coincides with a_s, and its radius is consistent with Eq. (4) for a_s.

3) Now we obtain the equation for the shift of the energy level from the Schrödinger equation. We start from the following equations:

$$\left[-\frac{\hbar^2}{2m}\Delta + U_L - E_{ns}^{(0)}\right]\psi_{ns}^{(0)} = 0, \quad \left[-\frac{\hbar^2}{2m}\Delta + U_L + U_s - E_{ns}\right]\psi_{ns} = 0.$$

Multiplying the first equation by $\psi_{ns}^{(0)}$ and the second by $\psi_{ns}^{(0)*}$, subtracting one from the other and integrating over space, except for a spherical domain of radius d in the vicinity of the origin (we assume, $r_s \ll d \ll r_L$), we find

---

[241] The choice of potential $\tilde{U}_s(r)$ is not unique and is quite arbitrary. In the first order of approximation theory we can use the simplest relation:

$$\tilde{U}_s = \alpha \delta(\mathbf{r}), \quad \text{where} \quad \alpha = \frac{2\pi\hbar^2 a_s}{m}.$$

Note that such a "potential", unlike the one-dimensional δ-potential and the potential of a three-dimensional δ-sphere, has only a formal meaning and faithfully represents the perturbation only in the first order.

496 *Exploring Quantum Mechanics*

$$\Delta E_{ns} \cdot \int_{r \geq d} \psi_{ns}^{(0)*} \psi_{ns} dV = \frac{\hbar^2}{2m} \oint_{r=d} (\psi_{ns}^{(0)*} \boldsymbol{\nabla} \psi_{ns} - \psi_{ns}^* \boldsymbol{\nabla} \psi_{ns}^{(0)}) \cdot d\mathbf{s} \quad (5)$$

The phases of the wavefunctions are chosen so that $\psi_{ns}^{(0)}/\psi_{ns}$ is real and the terms containing U_L cancel each other.

Since the level is perturbed only weakly and $\psi_{ns}^{(0)} \sim \psi_{ns}$ for $r \gg r_s, a_s$, which is the only domain that essentially contributes to the integral, the integral on the left-hand side of Eq. (5) can be simply replaced by unity. In the right-hand side, we use the asymptotic form of the wavefunctions at the distances $r_s \ll r \ll r_L$, as follows:

$$\psi_{ns}^{(0)}(r) \approx \psi_{ns}^{(0)}(0), \quad \psi_{ns}(r) \approx \left(1 - \frac{a_s}{r}\right) \psi_{ns}^{(0)}(0),$$

and obtain

$$\Delta E_{ns} = E_{ns} - E_{ns}^{(0)} = \frac{2\pi\hbar^2}{m} |\psi_{ns}^{(0)}(0)|^2 a_s, \quad (6)$$

which agrees with Eq. (2) above.

Let us comment on Eq. (6).

1. The equation remains quantitatively reliable as long as relation (4) holds. The latter breaks down only in the case of a resonance, where $|a_s| \geq r_L$, and the potential $U_s(r)$ features a shallow (real or virtual) level[242] with $l = 0$ and the energy

$$E_s \sim E_{ns}^{(0)} \ll \frac{\hbar^2}{mr_s^2}.$$

   In this case, the levels can shift strongly and are not described by Eq. (6).[243] A complete reconstruction of the spectrum of the long-range potential under the action of a short-range center is possible (the *Zeldovich effect*; see Problems 4.11 and 9.3).

2. Equation (6) remains applicable to non-central long-range potentials, $U_L(\mathbf{r})$, as well.

3. Note that the sign of the shift in the energy levels is determined by the sign of the scattering length, a_s (and not by the bare form of the potential, $U_s(r)$). For a repulsive potential, the scattering length is always positive, and the levels shift upward, "pushed out". For an attractive potential, either negative or positive a_s is possible (in the case of "shallow" well, $a_s < 0$, and the levels move downwards, "sucked in").

---

[242] The existence of "deep" levels with the energy $|E_s| \sim \hbar^2/mr_s^2$ in the potential $U_s(r)$ does not affect the applicability of Eq. (6).

[243] If an attractive part of the potential $U_s(r)$ is present, but is separated by a weakly penetrable barrier, then the shift of the levels of the long-range potential is guaranteed to remain small. This situation is realized for states with non-zero angular momenta, where the centrifugal force gives rise to such a barrier; see also Problem 11.74.

4. The method of dealing with a "strong" short-range potential, $U_s(r)$, described above, involves replacing the perturbative Born scattering length by an exact value, a_s. This approach is, in fact, more general than the problem considered here (see Problems 4.27 and 8.61, for example).
5. A generalization of Eq. (6) to the case of non-zero angular momenta is given in Problem 13.36, and an extension of perturbation theory in terms of the scattering length to the continuous spectrum is considered in Problem 13.37.
6. Let us emphasize that Eq. (6) is valid without any restrictions on the form of the short-range interaction. The interaction may involve inelastic processes (such as, for example, in the case of a $p\bar{p}$-atom, where annihilation into pions is possible). In this case, the scattering length a_s becomes a complex quantity, and the level shifts also become complex. The imaginary part describes the width of the levels corresponding to the *quasi-stationary* states.

Problem 11.5

Use first-order perturbation theory to calculate the ground-state energy of a two-electron atom/ion, considering the Coulomb interaction between electrons as a perturbation. Calculate the atomic *ionization energy* and compare it to the following experimental data for the helium atom and the lithium, beryllium, carbon, and oxygen ions: $I(\text{He}) = 24.6 \ eV$; $I(\text{Li}^+) = 75.6 \ eV$; $I(\text{Be}^{++}) = 154 \ eV$; $I(\text{C}^{4+}) = 392 \ eV$; $I(\text{O}^{6-}) = 739 \ eV$.

Solution

1) If we neglect the interaction between electrons, the ground-state energy is $E_0^{(0)} = -Z^2$, while the ground-state wavefunction has the form:

$$\psi_0^{(0)} = \frac{1}{\pi} Z^3 \exp\{-Z(r_1 + r_2)\}.$$

The first-order correction to the energy due to electrons interaction is

$$E_0^{(1)} = \int \frac{|\psi_0^{(0)}(\mathbf{r}_1, \mathbf{r}_2)|^2}{|\mathbf{r}_1 - \mathbf{r}_2|} dV_1 dV_2 = \frac{Z^6}{\pi^2} \int e^{-2Z(r_1+r_2)} \frac{r_1^2 r_2^2 dr_1 dr_2 d\Omega_1 d\Omega_2}{\sqrt{r_1^2 + r_2^2 - 2\mathbf{r}_1 \cdot \mathbf{r}_2}}. \quad (1)$$

To calculate the integrals below,

$$I = \int f(r_1) g(r_2) \frac{r_1^2 r_2^2 dr_1 dr_2 d\Omega_1 d\Omega_2}{\sqrt{r_1^2 + r_2^2 - 2\mathbf{r}_1 \cdot \mathbf{r}_2}}, \quad (2)$$

it is convenient to first integrate over the angles of the vector, $\mathbf{r}_2$, for a fixed $\mathbf{r}_1$ in spherical coordinates, choosing the polar axis along $\mathbf{r}_1$. Here,

$$\int \frac{d\Omega_2}{|\mathbf{r}_1 - \mathbf{r}_2|} = \int \frac{\sin\theta \, d\theta d\varphi}{\sqrt{r_1^2 + r_2^2 - 2r_1 r_2 \cos\theta}} = \begin{cases} \frac{4\pi}{r_1}, & r_1 > r_2, \\ \frac{4\pi}{r_2}, & r_2 > r_1, \end{cases}$$

and Eq. (2) takes the form:

$$I = 16\pi^2 \left\{ \int_0^\infty f(r_1)r_1 \int_0^{r_1} g(r_2)r_2^2 dr_2 dr_1 + \int_0^\infty g(r_2)r_2 \int_0^{r_2} f(r_1)r_1^2 dr_1 dr_2 \right\}. \tag{2'}$$

From Eqs. (2) and (2), we find

$$\int \frac{e^{-\alpha r_1 - \beta r_2}}{|\mathbf{r}_1 - \mathbf{r}_2|} dV_1 dV_2 = 32\pi^2 \frac{\alpha^2 + \beta^2 + 3\alpha\beta}{\alpha^2 \beta^2 (\alpha + \beta)^3} \tag{3}$$

and, according to Eq. (1), obtain (here, we put $\alpha = \beta = 2Z$ in Eq. (3))

$$E_0 \approx E_0^{(0)} + E_0^{(1)} = -Z^2 + \frac{5}{8} Z. \tag{4}$$

Therefore, the *ionization energy* is

$$I_0 = E_{0,H} - E_0 = \frac{1}{2} Z^2 - \frac{5}{8} Z, \tag{5}$$

with $E_{0,H} = -Z^2/2$ as the ground-state energy of the corresponding one-electron ion.

2) Interestingly, the first-order perturbation theory results obtained above may be improved further almost without any additional calculations. Let us represent the system Hamiltonian as follows (introducing an effective nuclear charge, Z_{eff}):

$$\hat{H} = \hat{H}_0 + V_{eff}, \quad \hat{H}_0 = -\frac{1}{2}(\Delta_1 + \Delta_2) - Z_{eff}\left(\frac{1}{r_1} + \frac{1}{r_2}\right),$$

$$V_{eff} = \frac{1}{|\mathbf{r}_1 - \mathbf{r}_2|} - (Z - Z_{eff})\left(\frac{1}{r_1} + \frac{1}{r_2}\right).$$

Here the choice of $Z_{eff} < Z$, which gives the mutual screening of the nucleus by electrons, leads to a decrease of the perturbation V_{eff} with respect to the case $V = 1/|\mathbf{r}_1 - \mathbf{r}_2|$, used in Eq. (1). From a physical point of view, it seems natural to choose Z_{eff} such that the first-order correction vanishes, i.e., $\overline{V}_{eff} = 0$. Since now $E_0^{(0)} = -Z_{eff}^2$, while

$$\langle\psi_{eff}^{(0)}|\frac{1}{|\mathbf{r}_1 - \mathbf{r}_2|}|\psi_{eff}^{(0)}\rangle = \frac{5}{8}Z_{eff}, \quad \langle\psi_{eff}^{(0)}|\left(\frac{1}{r_1} + \frac{1}{r_2}\right)|\psi_{eff}^{(0)}\rangle = -\frac{2E_0^{(0)}}{Z_{eff}}.$$

(The last relation can be obtained using the arguments of Problem 11.1, for example.) Then from the condition, $\overline{V}_{eff} = 0$, we find $Z_{eff} = Z - 5/16$ and obtain

$$E_0 = -\left(Z - \frac{5}{16}\right)^2, \quad I_0 = \frac{1}{2}Z^2 - \frac{5}{8}Z + \frac{25}{256}. \tag{6}$$

See, Eqs. (4) and (5). Interestingly, the "new" value of E_0 coincides with the result of a variational calculation in Problem 11.6.

The table below presents a comparison of the ionization energies, I_0, with the experimental data (energies are expressed in electron-Volts, eV, and the atomic energy unit is equal to 27.2 eV).

Ion	He	Li$^+$	Be^{++}	C^{4+}	O^{6+}
According to (5)	20.4	71.4	150	388	735
According to (6) and 11.6	23.1	74.0	152	390	737
Exper. value I_0	24.6	75.6	154	392	739

We see that second-order corrections to the results (4), (5), and (6) do not depend on Z. This explains the weak dependence on Z of the difference between the experimental and calculated values of I_0, as seen in the table.

Problem 11.6

Find the ground-state energy and ionization energy of a two-electron atom/ion using the variational method. Take the product of hydrogen functions with some effective charge Z_{eff}, which plays the role of a variational parameter, as a trial function. Compare your results with Problem 11.5. Discuss the stability of a hydrogen ion H^-.

Solution

The expectation value of the energy of the two-electron ion in the trial state,

$$\psi_{test} = \frac{\alpha^3}{\pi} \exp\{-\alpha(r_1 + r_2)\}, \quad \alpha \equiv Z_{eff},$$

can be calculated if we follow the following steps:

1) write the Hamiltonian in the form

$$\hat{H} = -\left(\frac{1}{2}\Delta_1 + \frac{\alpha}{r_1}\right) - \left(\frac{1}{2}\Delta_2 + \frac{\alpha}{r_2}\right) - (Z - \alpha)\left(\frac{1}{r_1} + \frac{1}{r_2}\right) + \frac{1}{|\mathbf{r}_1 - \mathbf{r}_2|},$$

2) take into account the relation

$$\left(\frac{1}{2}\Delta + \frac{\alpha}{r}\right) e^{-\alpha r} = \frac{\alpha^2}{2} e^{-\alpha r},$$

and

3) use the values of integrals

$$\int \frac{1}{r} e^{-2\alpha r} dV = \frac{\pi}{\alpha^2}, \quad \int |\psi_{test}|^2 \frac{dV_1 dV_2}{|\mathbf{r}_1 - \mathbf{r}_2|} = \frac{5\alpha}{8}.$$

(For the second calculation, see Problem 11.5.)

We find

$$\overline{E}(\alpha) = \alpha\left(\alpha - 2Z + \frac{5}{8}\right).$$

Minimization over the parameter α gives (here $\alpha = Z_{\text{eff}} = Z - 5/16$) the variational value of the ground-state energy as follows:

$$E_{0,var} = \min \overline{E}(\alpha) = -\left(Z - \frac{5}{16}\right)^2. \tag{1}$$

The ionization energy is

$$I_0 = -\frac{1}{2}Z^2 - E_0 = \frac{1}{2}Z^2 - \frac{5}{8}Z + \frac{25}{256}. \tag{2}$$

For comparison to experimental data, see the table in the previous problem.

For the hydrogen ion, H^-, we have from Eq. (1) $E_{0,var} = -0.47$, which is higher than the ground-state energy of the hydrogen atom (equal to -0.50). The H^- ion is not stable against auto-ionization according to this calculation and Eq. (1) (see Problem 11.8).

Problem 11.7

Prove that the trial wavefunction with the effective charge, $Z_{\text{eff}} = Z - 5/16$, from the previous problem, is the most optimal among all trial functions of the form, $\psi_{\text{trial}} = \psi(r_1 + r_2)/4\pi$ (i.e., within the class of functions that depend only on variable $u = r_1 + r_2$).

Solution

To prove this statement, we write $\overline{E} = \langle \psi_{test} | \hat{H} | \psi_{test} \rangle$ as an integral over the variable, u, and calculate the first variation of the functional. It is convenient to use spherical coordinates, go back to the variable, $r_2 = u - r_1$, and integrate over r_1.

The normalization integral takes the form:

$$N \equiv \int |\psi_{\text{trial}}|^2 r_1^2 dr_1 d\Omega_1 r_2^2 dr_2 d\Omega_2 = \int_0^\infty \frac{u^5}{30} |\psi(u)|^2 du. \tag{1}$$

Using

$$\Delta_2 \psi(r_1 + r_2) = \frac{1}{(u-r_1)^2} \frac{\partial}{\partial u}(u-r_1)^2 \frac{\partial}{\partial u}\psi(u),$$

we find the average kinetic energy as follows:

$$\overline{T} = 2\overline{T}_2 = -\int_0^\infty \psi^*(u) \int_0^u \frac{\partial}{\partial u}(u-r_1)^2 \frac{\partial}{\partial u}\psi(u) r_1^2 dr_1 du =$$

$$-\int_0^\infty \psi^*(u) \left[\frac{u^5}{30}\frac{d^2}{du^2} + \frac{u^4}{6}\frac{d}{du}\right]\psi(u) du. \qquad (2)$$

The interaction energy between the electrons and the nucleus is given by:

$$\overline{U}_{en} = 2\overline{U}_2 = -2Z\int_0^\infty |\psi(u)|^2 \int_0^u (u-r_1)r_1^2 dr_1 du = -\int_0^\infty \frac{Zu^4}{6}|\psi(u)|^2 du, \qquad (3)$$

while the inter-electron interaction reads

$$\overline{U}_{ee} = 2\iint_{r_1 \leq r_2} |\psi_{test}|^2 \frac{dV_1 dV_2}{|\mathbf{r}_1 - \mathbf{r}_2|} = 2\int_0^\infty |\psi(u)|^2 \int_0^{u/2}(u-r_1)r_1^2 dr_1 du$$

$$= \int_0^\infty \frac{5u^4}{96}|\psi(u)|^2 du. \qquad (4)$$

Note that since $\psi(r_1 + r_2)$ does not depend on the angles, the integral over the angles in Eq. (4) is readily evaluated.

To determine an optimal trial function, $\psi(u)$, normalized to unity, for which $\overline{E} = \overline{T} + \overline{U}_{un} + \overline{U}_{ee}$ is minimal, it is convenient to examine the functional $\overline{E} - E_{0,var}N$ and look for its extrema. Here $E_{0,var}$ acts as a *Lagrange factor*. The presence of the term, $E_{0,var}N$, allows us not to "worry" about the wavefunction normalization. Using relations (1)–(4) and setting the first variation of the functional to zero, we obtain the equation (compare this to the variational principle for the Schrödinger equation):

$$\left[\frac{d^2}{du^2} + \frac{5}{u}\frac{d}{du} + \left(5Z - \frac{25}{16}\right)\frac{1}{u} + E_{0,var}\right]\psi(u) = 0.$$

By replacing the function $\psi(u) = u^{-5/2}\chi(u)$, so that $\chi(0) = 0$, we obtain the equation:

$$\chi'' - \frac{15}{4u^2}\chi + \left(5Z - \frac{25}{16}\right)\frac{1}{u}\chi + E_{0,var}\chi = 0. \qquad (5)$$

It has the form of the Schrödinger Eq. (IV.5) (with $\hbar = m = 1$) for a particle in the Coulomb potential $-\alpha/r$ with $\alpha = \frac{5}{2}(Z - \frac{5}{16})$ and with a fictitious "angular momentum", $l = 3/2$ (or equivalently $l(l+1) = 15/4$).[244] Here, $E_{n,l} = \frac{1}{2}E_{0,var}$. The spectrum reads

$$E_{n,l} = -\frac{\alpha^2}{2(n_r + l + 1)^2} = -\frac{25(Z - 5/16)^2}{8(n_r + 5/2)^2},$$

and the $n_r = 0$ term determines the variational ground-state energy on the class of trial functions considered:

$$E_{0,var} = 2E_{0,3/2} = -\left(Z - \frac{5}{16}\right)^2. \tag{6}$$

The optimal variational wavefunction that minimizes $\overline{E}$ is, according to Eq. (5), the Coulomb wavefunction with $l = 3/2$ and $n = n_r + l + 1 = 5/2$:

$$\chi = Cu^{l+1}e^{-au/n}, \quad \text{or} \quad \psi_0 = \frac{1}{\pi}Z_{eff}^3 \exp\{-Z_{eff}(r_1 + r_2)\}, \tag{7}$$

where $Z_{eff} = Z - 5/16$, in accordance with the statement in the problem.

Problem 11.8

Find the expectation value of the energy of a two-electron ion with the nuclear charge, Z, in the following state:

$$\psi(r_1, r_2) = C[\exp\{-\alpha r_1 - \beta r_2\} + \exp\{-\beta r_1 - \alpha r_2\}].$$

Set the parameters to be $\alpha = 1$ and $\beta = 0.25$, and prove the stability of the hydrogen ion H^-.

Solution

The calculation of $\overline{E} = \iint \psi(r_1, r_2) \hat{H} \psi(r_1, r_2) dV_1 dV_2$ reduces to the following integrals:

$$\int e^{-\gamma r} dV = \frac{8\pi}{\gamma^3}, \quad \int \frac{1}{r} e^{-\gamma r} dV = \frac{4\pi}{\gamma^2},$$

$$\int e^{-\alpha r} \Delta e^{-\beta r} dV = -\int (\nabla e^{-\alpha r})(\nabla e^{-\beta r}) dV = \frac{8\pi\alpha\beta}{(\alpha + \beta)^3}$$

and the integral (3) from Problem 11.5.

---

[244] If we set $l = -5/2$, the wavefunction $\chi \propto u^{l+1}$ does not satisfy the proper boundary condition in the origin.

Evaluating those, we obtain

$$\overline{E}(\alpha, \beta) = 2C^2 \left\{ -Z(\alpha + \beta) + \frac{1}{2}(\alpha^2 + \beta^2) + \frac{\alpha\beta}{\alpha + \beta} + \frac{\alpha^2\beta^2}{(\alpha + \beta)^3} \right.$$
$$\left. + 20\frac{\alpha^3\beta^3}{(\alpha + \beta)^5} + 64[\alpha\beta - Z(\alpha + \beta)]\frac{\alpha^3\beta^3}{(\alpha + \beta)^6} \right\}, \quad (1)$$

where $2C^2 = (1 + 64\alpha^3\beta^3/(\alpha + \beta)^6)^{-1}$ follows from the normalization condition.

If $\alpha = \beta = Z - 5/16$, from Eq. (1) we recover the result of Problem 11.6. But the possibility of an independent variation of parameters α and β allows us to improve the result. For the example of the ion H^-, choosing $\alpha = 1$ and $\beta = 0.25$, we find

$$E_{0,var} = \overline{E} = -0.512, \quad (2)$$

which is below the ground-state energy of the hydrogen atom, equal to $-1/2$, and hence we prove that a stable H^- ion does exist (a more sensitive variational calculation gives $E_0 = -0.528$).

Problem 11.9

Estimate the energy and ionization energy of a helium-like atom in an excited state. Use the approximation where the motion of the excited electron is viewed as occurring in the nuclear field screened by the ground-state electron. Compare the results to the relevant experimental data (the details are presented in the solution).

Solution

In the model considered, the energy of a helium-like atom in an nL-state, which is the same for para- and ortho-states, is a sum of terms corresponding to a hydrogen-like atom with charge Z (ground-state energy, $E_{n=1} = -Z^2/2$) and the energy, $\overline{E}_n = -(Z-1)^2/2n^2$, of the n-level state in the field of a screened charge, $Z - 1$. The energy and ionization energy in this approximation are

$$E_{nL}(Z) = -\frac{Z^2}{2} - \frac{(Z-1)^2}{2n^2}, \quad I_{nL}(Z) = \frac{(Z-1)^2}{2n^2}. \quad (1)$$

For the $2L$-states of the helium atom we have $I_{2L}(\text{He}) = 0.125$, and a comparison to the experimental values is given in the following table.

Term	2^3S	2^1S	2^3P	2^1P
I_{exp}	0.175	0.146	0.133	0.124
(1) with Rydberg correction	0.172	0.145	0.134	0.124

For the $3L$-state of the helium atom, the model gives $I_{3L}(\text{He}) = 556 \cdot 10^{-4}$, while the experimental values are

Term	3^3S	3^1S	3^3P	3^1P	3^3D	3^1P
$I_{exp} \cdot 10^4$	685	612	580	550	560	555
(1) with Rydberg correction	684	611	582	551	557	556

The approximation here neglects an "overlap" of the 1s- and nl-electron wavefunctions (with $l \equiv L$). If $n \gg 1$, this is justified by the large size of the nl-electron orbit (recall that $r_n \propto n^2$) and hence by the small probability of finding the electron in the "ground-state region".

To improve the result, taking into account the quasi-classical property of electron motion for $n \gg 1$, we may use the *Rydberg correction* Δ_l (that does not depend on n), which implies replacing n by $n + \Delta_l$ in Eq. (1). As is seen in the tables, the Rydberg correction[245] improves the accuracy, especially for[246] $n = 2$ and $n = 3$. As for the states with angular momentum $l \neq 0$, the model considered reproduces experimental values quite well, even without the Rydberg correction. This is connected to the fact that with the increase in l, the probability to find the excited electron at small distances decreases fast, so that here, $\psi_l \propto r^l$. Hence the value of the Rydberg correction decreases fast with the increase in l.

Problem 11.10

Find the energy and ionization energy of a helium-like atom in the 2^3S-state by the variational method. Use a antisymmetrized product of the hydrogen 1s- and 2s-states as a trial function, and introduce an effective nuclear charge, Z_{eff}, as a variational parameter. In the case of the helium atom and lithium ion Li$^+$, compare the results obtained to the following experimental data (in atomic units): $I_{He}(2^3S) = 0.175$ and $I_{Li^+}(2^3S) = 0.610$.

Solution

The normalized wavefunction of the 2^3S-state has the form ($\alpha \equiv Z_{eff}$):

$$\psi(r_1, r_2) = \frac{1}{\sqrt{2}} \{\psi_1(r_1)\psi_2(r_2) - \psi_2(r_1)\psi_1(r_2)\}, \tag{1}$$

$$\psi_1 = \sqrt{\frac{\alpha^3}{\pi}} e^{-\alpha r}, \quad \psi_2(r) = \sqrt{\frac{\alpha^3}{8\pi}} \left(1 - \frac{1}{2}\alpha r\right) e^{-\alpha r}. \tag{2}$$

---

[245] The experimental values of the Rydberg correction Δ_{Sl} for the helium atom (S is the electrons' total spin) are equal to

$$\Delta_{0s} = -0.140, \quad \Delta_{0p} = +0.012, \quad \Delta_{0d} = -0.0022 \text{ (for } S = 0\text{)},$$
$$\Delta_{1s} = -0.269, \quad \Delta_{1p} = -0.068, \quad \Delta_{1d} = -0.0029 \text{ (for } S = 1\text{)}.$$

[246] Here the peculiarity of quasi-classical approximation, mentioned in Chapter 9, manifests itself in that it remains reasonable even for $n \sim 1$ and improves the accuracy of the results.

To calculate the energy of the state (1), we write the Hamiltonian as a sum of three terms:

$$\hat{H} = \left\{-\frac{1}{2}(\Delta_1 + \Delta_2) - \frac{\alpha}{r_1} - \frac{\alpha}{r_2}\right\} - (Z-\alpha)\left(\frac{1}{r_1} + \frac{1}{r_2}\right) + \frac{1}{|\mathbf{r}_1 - \mathbf{r}_2|}$$
$$\equiv \hat{H}_1 + \hat{H}_2 + \hat{H}_3. \tag{3}$$

Since the wavefunction (1) is an eigenfunction of the Hamiltonian $\hat{H}_1$, we easily find $\overline{\hat{H}_1} = -5\alpha^2/8$. And also, using virial theorem, we determine the mean value of the second term:

$$\overline{\hat{H}_2} = \frac{Z-\alpha}{\alpha}\overline{\left(-\frac{\alpha}{r_1} - \frac{\alpha}{r_2}\right)} = -\frac{5}{4}\alpha(Z-\alpha). \tag{4}$$

At last we find $\overline{\hat{H}_3} = K - J$, where

$$K = \iint \psi_1^2(r_1)\psi_2^2(r_2)\frac{dV_1 dV_2}{|\mathbf{r}_1 - \mathbf{r}_2|}, \tag{5}$$

$$J = \iint \psi_1(r_1)\psi_2(r_1)\psi_1(r_2)\psi_2(r_2)\frac{dV_1 dV_2}{|\mathbf{r}_1 - \mathbf{r}_2|}, \tag{6}$$

characterizes Coulomb interaction between the electrons. Its first part, connected to the integral, K, has the classical interpretation of interacting charge densities corresponding to the 1s- and 2s-electrons. The integral, J, determines a so-called *exchange interaction*. It is a part of the electron Coulomb interaction and has a purely quantum origin, connected to the symmetry properties of the fermion wavefunction describing identical particles.

Using relations (2) for functions $\psi_{1,2}(r)$, we see that the calculation of K and J is reduced to the following integrals:

$$I_{k,n} = \iint \frac{r_1^k r_2^n}{|\mathbf{r}_1 - \mathbf{r}_2|} e^{-\alpha r_1 - \beta r_2} dV_1 dV_2,$$

where n, k are integers. We find

$$I_{k,n} = (-1)^{k+n}\frac{\partial^k}{\partial \alpha^k}\frac{\partial^n}{\partial \beta^n}\iint \frac{e^{-\alpha r_1 - \beta r_2}}{|\mathbf{r}_1 - \mathbf{r}_2|} dV_1 dV_2 \equiv (-1)^{k+n}\frac{\partial^k}{\partial \alpha^k}\frac{\partial^n}{\partial \beta^n} I_{0,0}.$$

The integral $I(0,0)$ was calculated before (see Eq. (3) from Problem 11.5). A more cumbersome but straightforward calculation gives $K = 17\alpha/81$ and $J = 16\alpha/729$ (let us note that the exchange part of the Coulomb interaction is ten times smaller than the direct Coulomb interaction).

Finally, we obtain

$$\overline{E}(\alpha) = \overline{\hat{H}_1} + \overline{\hat{H}_2} + K - J = \frac{5}{8}\alpha^2 - \frac{5}{4}\alpha Z + \frac{137}{729}\alpha,$$

and a minimization over the parameter, $\alpha = Z_{eff}$, gives the variational energy

$$E(2^3 S) = \min \overline{E} = -\frac{5}{8}\left(Z - \frac{548}{3645}\right)^5 \approx -\frac{5}{8}(Z - 0.150)^2, \quad (7)$$

where $Z_{eff} = Z - 0.150$, and the ionization energy is

$$I(2^3 S) = \frac{5}{8}(Z - 0.150)^2 - \frac{1}{2}Z^2. \quad (8)$$

According to Eq. (8) we have $I = 0.139$ for the helium atom and $I = 0.576$ for the lithium ion, Li$^+$. (Note that the variational result for I is rather far from the experimental value for the helium atom, which is connected to the fact that our trial function has used the same values of effective charge for both the "inner" 1s- and "outer" 2s-electrons; see Problems 11.6 and 11.8.)

Problem 11.11

Calculate the hyperfine splitting of the triplet $2^3 S$-state of a helium atom with the He^3 nucleus. The nuclear spin is $I = 1/2$ and its magnetic moment is $\mu = -1.064\, e\hbar/m_p c$. For calculations, use an approximate form of the $2^3 S$-state wavefunction that neglects electron–electron interaction. Compare the results with the following experimental values of the hyperfine splitting: $\Delta \nu_{HFS} \equiv \Delta E_{HFS}/2\pi\hbar = 6740$ MHz.

Solution

We follow here the same strategy as in Problem 11.2. We choose the perturbation to be of the form $\hat{V} = \hat{V}_1 + \hat{V}_2$, where $\hat{V}_{1,2}$ describes interaction between each electron and the nuclear magnetic field (see Eq. (1) from Problem 11.2). Due to the spherical symmetry of the s-state, there is no interaction between electrons' orbital currents and the magnetic field in the leading order. Averaging the operator, $\hat{V}$, with the wavefunction $\psi(r_1, r_2)$, we find

$$\overline{\hat{V}} = \frac{8\pi e\hbar\mu}{3m_e cI} C(\hat{\mathbf{I}} \cdot \hat{\mathbf{S}}), \quad (1)$$

where $\hat{\mathbf{S}} = \hat{\mathbf{s}}_1 + \hat{\mathbf{s}}_2$ is the total electron spin, and

$$C = \int |\psi(0, r_2)|^2 dV_2 = \int |\psi(r_1, 0)|^2 dV_1. \quad (2)$$

The eigenvalues of the operator $\overline{\hat{V}}$ (acting in the spin space) determine the hyperfine structure of the levels. Since $S = 1$ and $I = 1/2$, then

$$E_{HFS} = \frac{8\pi e\hbar\mu C}{3m_e c} \cdot \begin{cases} 1, & J = \frac{3}{2}, \\ -2, & J = \frac{1}{2} \end{cases} \quad (3)$$

where J is the total angular momentum, and the value of hyperfine splitting is equal to

$$\Delta E_{HFS} = \left| E_{HFS}\left(J = \frac{3}{2}\right) - E_{HFS}\left(J = \frac{1}{2}\right)\right| = \frac{8\pi e\hbar\mu C}{m_e c}. \tag{4}$$

Using the approximate expression for the wavefunction,

$$\psi(r_1, r_2) = \frac{1}{\sqrt{2}}\{\psi_1(r_1)\psi_2(r_2) - \psi_2(r_1)\psi_1(r_2)\},$$

where $\psi_{1,2}$ are the wavefunctions of a hydrogen-like atom with $Z = 2$ for the 1s- and 2s-states, we find

$$C = \frac{1}{2}(|\psi_1(0)|^2 + |\psi_2(0)|^2) = \frac{9}{2\pi}\frac{1}{a_B^3}, \quad a_B = \frac{\hbar^2}{m_e e^2}. \tag{5}$$

The numerical value of E_{HFS}, according to Eqs. (4) and (5), is equal to

$$\Delta\nu_{HFS} = \frac{\Delta E_{HFS}}{2\pi\hbar} \approx 7340 \text{ MHz}. \tag{6}$$

It follows from Eq. (5) that the dominant contribution to the hyperfine structure comes from the 1s-electron. The 2s-electron is eight times lower. In fact, its actual contribution is even smaller, since the probability of finding the 2s-electron wavefunction at $r = 0$ is further reduced due to a screening of the nuclear charge by the 1s-electron. If we use the value of the screened charge $Z = 1$, then the contribution of the 2s-electron becomes 64 times smaller, and instead of Eq. (6) we obtain $\Delta\nu_{HFS} \approx 6630$ MHz, which is, of course, much closer to the experimental value.

Problem 11.12

Prove that among all possible stationary states of a helium-like atom, only the electron configurations of type $1snl$ (*i.e.*, one of the electrons must be in the ground 1s-state) are stable with respect to decay (auto-ionization) into a hydrogen-like ion and a free electron.

Solution

If we neglect the electron–electron interaction, the energy of the state where both electrons are excited is

$$E_{n_1 n_2} = -\frac{1}{2}Z^2\left(\frac{1}{n_1^2} + \frac{1}{n_2^2}\right).$$

If $n_{1,2} \geq 2$, this energy is higher than the ground-state energy, $E_0 = -Z^2/2$, of the one-electron ion ground-state. This simple energetic argument indicates the instability of the state. Interaction between electrons gives rise to a non-zero probability of (auto)-ionization, where a transition to the continuous spectrum occurs for one of the electrons, while the other electron occupies the 1s-orbit yielding the ground-state of the ion.

Clearly, the repulsive electron–electron interactions can only increase the energy of the two-electron atom, so the conclusion above remains valid. This can be proven more formally using Eq. (1.6). Specifically, for the eigenvalues, $E_n(\beta)$, of the Hamiltonian,

$$\hat{H}(\beta) = -\frac{1}{2}(\Delta_1 + \Delta_2) - \frac{Z}{r_1} - \frac{Z}{r_2} + \frac{\beta}{|\mathbf{r}_1 - \mathbf{r}_2|},$$

we find

$$\frac{\partial E_n}{\partial \beta} = \overline{\frac{\partial \hat{H}}{\partial \beta}} = \overline{\frac{1}{|\mathbf{r}_1 - \mathbf{r}_2|}} > 0,$$

so "turning on" the ee-interaction indeed raises the energy.

Note that the above conclusion about instability of excited two-electron states against auto-ionization is based entirely on energy arguments. The complete picture is in fact more complicated, as apart from the energetic analysis, the selection rules have to be considered as well, which may prohibit certain transitions that appear energetically favorable within a naïve analysis such as above. Connected to electron configuration $2pnl$ with $n \geq 2$, we have the set of states stable with respect to ionization, so that their decay is forbidden by the angular momentum and parity conservation laws. These are the states with the angular momentum $L \geq 1$ and parity equal to $(-1)^{L+1}$ (for example, the $3P^+$-term for the electron configuration $(2p)^2$, see Problem 11.72).

Problem 11.13

Estimate the ionization energies of the ground ^2S-(electron configuration $(1s)^22s$) and first excited ^2P-states (electron configuration is $(1s)^22p$) of a lithium-like atom, assuming that the electronic interaction between the ground state and the excited state is effectively reduced to a screening of the nuclei charges (for the excited electron).

In the case of lithium compare your results to experimental values: $I(^2S) = 5.39$ eV and $I(^2P) = 3.54$ eV.

Solution

In this model, we can write the energy of the three-electron atom (or ion) state with electron configuration $(1s)^2nl$, $n \geq 2$, as a sum of the energy of the ground-state of the corresponding two-electron ion $\tilde{E}_0$, which corresponds to the two 1s-electrons, and the energy of outer nl-electron, E_n in the nuclear Coulomb field, whose charge is screened by two 1s-electrons. This determines the ionization energy

$$I_n = |E_n| = \frac{(Z-2)^2}{2n^2}.$$

Its numerical value for lower $2S$- and $2P$-states of lithium atom is the same (in the model considered) and is equal to $I_{2L} = 0.125$ a.u. ≈ 3.40 eV.

In conclusion, we note that the general considerations about the applicability of the model and corrections to it are similar to those presented in the solution for problem 11.9. One of these corrections is the Rydberg correction: $\Delta_0 = -0.400$ for S-states, using which we obtain $I_{2S} = 5.31$ eV (close to the experimental value); for P-states, $\Delta_1 = -0.047$.

11.2 Many-electron atoms; Statistical atomic model

Problem 11.14

Find possible atomic terms consistent with the following electron configurations (above the closed shells):

a) np; b) $(np)^2$; c) $(np)^3$; d) $(np)^4$; e) $(np)^5$; f) $(np)^6$.

Find their parities. Using the Hund law, find the corresponding ground states.

Solution

Atomic spectra that correspond to $(np)^L$ electron configuration above the closed shells are given in table

Configuration	$np, (np)^5$	$(np)^2, (np)^4$	$(np)^3$	$(np)^6$
Terms	$^2P_{1/2, 3/2}$	$^1S_0, {}^1D_2,$	$^2P_{1/2, 3/2}$	$^1S_0$
		$^3P_{0,1,2}$	$^2D_{3/2, 5/2},$	
			$^4S_{3/2}$	
Parity I	-1	+1	-1	+1

Since the parity of an electron state with the angular momentum, l is equal to $(-1)^l$ and is a multiplicative quantity (for the electrons that form a closed shell, it is positive), then $I = (-1)^k$.

The ground states according to Hund law, are $^2P_{1/2}, {}^3P_0, {}^4S_{3/2}, {}^3P_2, {}^2P_{3/2}$ in the cases a)–f) (in the order of k increase). See Problems 11.17 and 11.18.

Problem 11.15

Find atomic terms allowed for the electron configuration $(nl)^2$.

Solution

The wavefunction of two electrons must be anti-symmetric with respect to the interchange of spin and spatial variables. Since 1) the radial dependence of two equivalent electrons' wavefunctions is symmetric with respect to the interchange of r_1 and r_2, 2) the spin part of the wavefunction is symmetric for the total electron spin, $S = 1$ and anti-symmetric for $S = 0$, and 3) the nature of the angular symmetry is determined by the value of the total angular momentum, L (see Problem 3.30), then we obtain for the electron configuration $(nl)^2$:

$$S = 0, \ L = 2l, 2l - 2, \ldots, 0 \quad \text{(singlet terms)},$$
$$S = 1, \ L = 2l - 1, 2l - 3, \ldots, 1 \quad \text{(triplet terms}, \ l \neq 0).$$

Compare to Problems 10.8 and 10.9.

Problem 11.16

Two terms, ^1L and ^3L (L is the total angular momentum $L = l$), correspond to the atomic states that have the electron configuration $nsn'l$ above the closed shells. Considering interaction between electrons as a perturbation, prove that the energy of the triplet term is lower than the energy of the singlet. Do not specify the form of the ns- and $n'l$-electrons radial functions.

Solution

The spatial part of the wavefunction has the form

$$\psi_\pm = \frac{1}{\sqrt{2}}\{\psi_1(\mathbf{r}_1)\psi_2(\mathbf{r}_2) \pm \psi_2(\mathbf{r}_1)\psi_1(\mathbf{r}_2)\}. \tag{1}$$

Signs $+$ and $-$ correspond to the singlet and triplet terms, and ψ_1 and ψ_2 are the wavefunctions of the ns- and $n'l$-electrons. When we neglect electron interaction, wavefunctions (1) correspond to the same energy. The difference due to electron interaction to first order of perturbation theory is

$$\Delta E_\pm = \langle \psi_\pm | \frac{1}{|\mathbf{r}_1 - \mathbf{r}_2|} | \psi_\pm \rangle \equiv K \pm J.$$

The *exchange* integral (compare to Problem 11.10),

$$J = \iint \psi_1^*(\mathbf{r}_1)\psi_2^*(\mathbf{r}_1) \frac{1}{|\mathbf{r}_1 - \mathbf{r}_2|} \psi_1(\mathbf{r}_2)\psi_2(\mathbf{r}_2) dV_1 dV_2, \tag{2}$$

determines the energy splitting.

Let us show that $J > 0$. We see that without loss of generality we can consider the wavefunction ψ_1 of a (non-degenerate) ns-state to be real, and write[247] ψ_2 in the form $\psi_2 = \varphi_1 + i\varphi_2$, where $\varphi_{1,2}$ are real functions. We rewrite expression (2) as

$$J = \iint \frac{1}{|\mathbf{r}_1 - \mathbf{r}_2|} \psi_1(\mathbf{r}_1)\psi_1(\mathbf{r}_2)\{\varphi_1(\mathbf{r}_1)\varphi_1(\mathbf{r}_2) + \varphi_2(\mathbf{r}_1)\varphi_2(\mathbf{r}_2)\} dV_1 dV_2.$$

We see that this expression is positive if we compare it to the known equations for the energy of an electrostatic field, produced by a charge distribution with the charge density $\rho(\mathbf{r})$:

$$W = \int \frac{\mathbf{E}^2(\mathbf{r})}{8\pi} dV = \frac{1}{2} \iint \frac{\rho(\mathbf{r}_1)\rho(\mathbf{r}_2)}{|\mathbf{r}_1 - \mathbf{r}_2|} dV_1 dV_2 > 0.$$

So $J > 0$ and the singlet term is above the triplet. The physical explanation of the fact is that in the triplet state, which is anti-symmetric in electron's coordinates, the probability density becomes zero for $\mathbf{r}_1 = \mathbf{r}_2$, which leads to a decrease in the electron's Coulomb interaction energy (and the system energy in general) in comparison to the singlet state. Compare this to the condition of S being maximum for the ground term of an atom according to Hund's law.

In conclusion, let us mention that in this approximation there are no electrons in closed shells. Their existence appears indirectly in the form of a self-consistent field that determines the one-electron wavefunctions $\psi_{1,2}$ of the "outer" electrons. For a discussion about the accuracy of this approximation, see Problems 11.17 and 11.18.

Problem 11.17

An atom contains two equivalent np-electrons above closed shells. Considering interaction between electrons as a perturbation, find the energy order of the ^1S, ^1D, ^3P terms. Make sure that the values of quantum levels S and L in the ground state satisfy the Hund law. Show that these terms satisfy the expression

$$\Delta \equiv \frac{E(^1S) - E(^1D)}{E(^1D) - E(^3P)} = \frac{3}{2}.$$

In this approximation, the expression also holds true for the configuration $(np)^4$. Do not specify the explicit form of the np-electron radial wavefunction.

Comment

To construct the correct wave function of zeroth approximation that corresponds to a definite value of the angular momentum, L, it is convenient to use tensor formalism (see the Problems in Chapter 3, sec. 4).

---

[247] One-electron nl-levels with $l \neq 0$, as well the terms $^{1,3}L$ with $L = 1$ considered, are degenerate with respect to the projection of the angular momentum on the z-axis. Since the energy does not depend on the value of l_z, then considering the state with $L_z = l_z = 0$ and taking into account the fact that wavefunction ψ_2 here is real, we can conclude that $J > 0$ from relation (2).

Solution

The wavefunction of a single p-electron has the form

$$\psi(\mathbf{r}) = \sqrt{\frac{3}{4\pi}}(\mathbf{a} \cdot \mathbf{n})\varphi(r); \quad \mathbf{n} = \frac{\mathbf{r}}{r}, \quad |a|^2 = 1, \quad \int_0^\infty \varphi^2 r^2 dr = 1.$$

The coordinate part of the wavefunction for a system of two np-electrons with a definite value of the total angular momentum L is described by the expression

$$\psi_L(\mathbf{r}_1, \mathbf{r}_2) = a_{ik}(L) n_{1i} n_{2k} \varphi(r_1) \varphi(r_2). \tag{1}$$

Compare to Problem 3.45. Depending on the value of L, tensor $a_{ik}(L)$ has the following properties:

$$a_{ik} = \begin{cases} \sqrt{\frac{3}{16\pi^2}} \delta_{ik}, & L = 0 \text{ (term } S\text{)}, \\ \sqrt{\frac{9}{32\pi^2}} \varepsilon_{ikl} b_l, \ |\mathbf{b}|^2 = 1 & L = 1 \text{ (term } P\text{)}, \\ a_{ki}, \ a_{ii} = 0, \ a_{ik} a_{ik}^* = \frac{9}{16\pi^2}, & L = 2 \text{ (term } D\text{)}. \end{cases} \tag{2}$$

As is seen, wavefunctions of S- and D-terms are symmetric with respect to the interchange of $\mathbf{r}_1$ and $\mathbf{r}_2$, since these are singlet terms, $S = 0$. The P-term, asymmetric in coordinates, is a triplet, $S = 1$.

If we neglect the interaction of np-electrons, all terms have the same energy. The energy difference due to their interaction in the first order of perturbation theory

$$E_L^{(1)} = \iint |\psi_L(\mathbf{r}_1, \mathbf{r}_2)|^2 \frac{1}{|\mathbf{r}_1 - \mathbf{r}_2|} dV_1 dV_2 \tag{3}$$

could be written in the form

$$E_L^{(1)} = 2 a_{ik} a_{lm}^* \int_0^\infty dr_2 \int_0^{r_2} dr_1 r_1^2 r_2^2 \varphi^2(r_1) \varphi^2(r_2) \iint \frac{n_{1i} n_{2k} n_{1l} n_{2m} d\Omega_1 d\Omega_2}{\sqrt{r_1^2 + r_2^2 - 2 r_1 r_2}}. \tag{4}$$

The contribution of the integration regions $r_1 > r_2$ and $r_1 < r_2$ in relation (3) is the same.

First, perform integration over the angles of the coordinate of the first electron in (4). Recall the following constraint on the tensors involved

$$\int \frac{n_{1i} n_{1l} d\Omega_1}{\sqrt{r_1^2 + r_2^2 - 2 r_1 r_2 \mathbf{n}_1 \mathbf{n}_2}} = A \delta_{il} + B n_{2i} n_{2l}. \tag{5}$$

First performing convolution over i and l, then multiplying (5) by $n_{2i}n_{2l}$, we obtain two relations (for $r_2 > r_1$):

$$3A + B = \int_{-1}^{1} \frac{2\pi dz}{\sqrt{r_1^2 + r_2^2 - 2r_1 r_2 z}} = \frac{4\pi}{r_2},$$

$$A + B = \int_{-1}^{1} \frac{2\pi z^2 dz}{\sqrt{r_1^2 + r_2^2 - 2r_1 r_2 z}} = \frac{\pi}{3r_2^3}\left(\frac{8}{5}r_1^2 + 4r_2^2\right). \tag{6}$$

These allow us to determine A and B (it is convenient to direct the polar axis along the vector $\mathbf{n}_2$, here $\mathbf{n}_1 \cdot \mathbf{n}_2 = \cos\theta_1 \equiv z$).

Now we integrate over the angles of the coordinate of the second electron, using the known relations:

$$\int n_i n_k d\Omega = \frac{4\pi}{3}\delta_{ik}, \quad \int n_i n_k n_l n_m d\Omega = \frac{4\pi}{15}(\delta_{ik}\delta_{lm} + \delta_{il}\delta_{km} + \delta_{im}\delta_{kl}).$$

As the result of integration over the angles, Eq. (4) takes the form

$$E_L^{(1)} = 2a_{ik}(L)a_{lm}^*(L)\int_0^\infty dr_2 \int_0^{r_2} dr_1 r_1^2 r_2^2 \varphi^2(r_1)\varphi^2(r_2)$$

$$\times \left[\frac{4\pi}{3}\delta_{il}\delta_{km}A + \frac{4\pi}{15}(\delta_{ik}\delta_{lm} + \delta_{il}\delta_{km} + \delta_{im}\delta_{kl})B\right]. \tag{7}$$

Using relation (2) and the values of A and B that follow from (6), we find

$$E_L^{(1)} = 2\int_0^\infty dr_2 \int_0^{r_2} dr_1 r_1^2 r_2^2 \varphi^2(r_1)\varphi^2(r_2)\left[1 + \frac{8}{5}b_L \frac{r_1^2}{r_2^2}\right], \tag{8}$$

where

$$b_L = \begin{cases} \frac{1}{4}, & L = 0, \\ -\frac{1}{8}, & L = 1, \\ \frac{1}{40}, & L = 2. \end{cases}$$

We obtain the order of terms

$$E(^3P) < E(^1D) < E(^1S),$$

so that ^3P is the ground state in accordance with Hund's. Let us emphasize that this relation does not depend on a specific form of the wave-function of the np-electron, which is determined by a self-consistent field of closed shells electrons. We could use this as a check on the accuracy of the approximation considered. We compare the obtained value of the ratio $\Delta = 3/2$ to its experimental values for some atoms and ions that have electron configuration $(np)^2$, and also $(np)^4$ (*i.e.*, two *holes* in the np-shell). These values are given in the following table:

Atom	C	N^+	O^{2+}	O	Si	Ge	Sn	Te
Config.	$2p^2$	$2p^2$	$2p^2$	$2p^4$	$3p^2$	$4p^2$	$5p^2$	$5p^4$
Δ_{exp}	1.13	1.14	1.14	1.14	1.48	1.50	1.39	1.50

Problem 11.18

The same as in the previous problem, but for the atom with three equivalent np-electrons. Prove the following relation

$$\Delta = \frac{E(^2P) - E(^2D)}{E(^2D) - E(^4S)} = \frac{2}{3}.$$

Solution

The possible terms are ^4S, ^2P, ^2D. The calculation of their energies shifts due to np-electrons' mutual interaction goes along the strategy outlined in the previous problem. But now it is more tedious, which is connected to a more complex structure of the three-electron wavefunction.

1) We start from the ^4S term that corresponds to the maximum possible value of the total spin, $S = 3/2$. Here the spin part $\chi_{\alpha\beta\gamma}$ of the wavefunction is symmetric with respect to the interchage of the spin variables of any two electrons. So, the spatial part of the wavefunction (actually just the angular part, since the radial wavefunction dependence is the same for all electrons) must be anti-symmetric, which gives the angular dependence of wavefunction in the form

$$\mathbf{n}_1 \cdot [\mathbf{n}_2 \times \mathbf{n}_3] \equiv \varepsilon_{ikl} n_{1i} n_{2k} n_{3l},$$

so that

$$\psi(^4S) = C_1 \varepsilon_{ikl} n_{1i} n_{2k} n_{3l} \varphi(r_1)\varphi(r_2)\varphi(r_3)\chi_{\alpha\beta\gamma}. \tag{1}$$

Since the coordinate part of the wavefunction is pseudo-scalar, and its form does not change under rotation, then it really corresponds to the angular momentum $L = 0$ and describes the S-term (compare to Problem 3.47). Wavefunction normalization of (1) gives $C_1^2 = 9/128\pi^3$ (to calculate the normalization integral, take into account the values of the "angular" integrals given in the previous problem and the known relation $\varepsilon_{ikl}^2 = 6$).

The ^4S-term shifts due to electrons' mutual interaction, and the shift in the first order of perturbation theory is equal to

$$\Delta E(^4S) = \overline{V_{12}} + \overline{V_{13}} + \overline{V_{23}} = 3\overline{V_{12}} = 3\iiint |\psi(^4S)|^2 \frac{dV_1 dV_2 dV_3}{|\mathbf{r}_1 - \mathbf{r}_2|}. \tag{2}$$

After elementary integration over $\mathbf{r}_3$, this relation takes the form of Eq. (4) from the previous problem. It is only necessary to replace $a_{ik}a_{lm}^*$ by

$$4\pi C_1^2 (\delta_{il}\delta_{km} - \delta_{im}\delta_{kl}). \tag{3}$$

Here we used the relation

$$\varepsilon_{ikl}\varepsilon_{sll} = \delta_{is}\delta_{kl} - \delta_{il}\delta_{ks}.$$

With this, all equations from Problem 11.7 up to Eq. (7) could be used for our problem. Moreover, after the substitution of Eq. (3) into Eq. (7) from Problem 11.17, we obtain an expression for the shift (2) of ^4S-term which differs from the result (8) of Problem 11.17 only by a factor of 3, and the value of parameter b_L becomes equal to $b(^4S) = -1/8$.

2) Now consider ^2P. Since it corresponds to the angular momentum, $L = 1$, the coordinate part of the wavefunction must be expressed in terms of vector components. Such a vector, since it is linear in all the three vectors $\mathbf{n}_a$, where $a = 1, 2, 3$, could be made in three independent ways:

$$(\mathbf{n}_1 \cdot \mathbf{n}_2)\mathbf{n}_3, \quad (\mathbf{n}_1 \cdot \mathbf{n}_3)\mathbf{n}_2, \quad (\mathbf{n}_2 \cdot \mathbf{n}_3)\mathbf{n}_1.$$

The condition that the wave-function is antisymmetric uniquely determines its form as follows:

$$\psi(^2P) = \varphi(r_1)\varphi(r_2)\varphi(r_3)\mathbf{C}_{2i}\{(\mathbf{n}_1 \cdot \mathbf{n}_2)n_3 \varepsilon_{\alpha\beta}\chi_\gamma - (\mathbf{n}_2 \cdot \mathbf{n}_3)n_1 \varepsilon_{\gamma\beta}\chi_\alpha$$
$$-(\mathbf{n}_1 \cdot \mathbf{n}_3)n_2 \varepsilon_{\alpha\gamma}\chi_\beta\}. \tag{4}$$

α, β, γ are spin variables of 1, 2, and 3 electrons, anti-symmetric spin function $\varepsilon_{\alpha\beta} = -\varepsilon_{\beta\alpha} = \begin{pmatrix} 0 & 1 \\ -1 & 0 \end{pmatrix}$ describes the state of the two electrons with the total spin equal to zero (it is normalized to 2), $\chi = \begin{pmatrix} a \\ b \end{pmatrix}$ is the spinor, normalized to unity, that determines both the spin state of the last electron and of the whole system, with $S = 1/2$. Vector $\mathbf{C}_2$ characterizes the orbital state (compare to Problem 3.45). Further calculations are analogous to those made for the ^4S-term. Wavefunction normalization gives $|\mathbf{C}_2|^2 = 9/256\pi^3$. Applying the same consideration as above for the S-term now to the ^2P-term. We have instead of expression (3) the factor

$$4\pi[2|C_2|^2\delta_{ik}\delta_{lm} + 2C_{2i}C_{2l}^*\delta_{km} + 2C_{2l}C_{2i}^*\delta_{km} - 3C_{2m}C_{2l}^*\delta_{ik} - 3C_{2l}C_{2m}^*\delta_{ik}], \tag{5}$$

while Eq. (8) from Problem 11.17, multiplied by 3, determines term shift $\Delta E(^2P)$, if we put b_L to be equal to $b(^2P) = 0$.

3) Finally, the term ^2D corresponds to the angular momentum $L = 2$, and the coordinate part of the corresponding wavefunction must be expressed in terms the components of a symmetric tensor of the second rank, which in the conditions

of the problem is linear in all vectors n_a, with the trace equal to zero. This tensor is a superposition of tensors of the following type

$$[\mathbf{n}_1 \times \mathbf{n}_2]_i n_{3k} \equiv \varepsilon_{isl} n_{is} n_{2l} n_{3k}. \tag{6}$$

The form of this superposition is determined by the antisymmetry of the wavefunction of the three-electron system:[248]

$$\psi(^2D) = C_{ik}\varepsilon_{ipl}\{n_{1k}n_{2p}n_{3l}(\varepsilon_{\alpha\beta}\chi_\gamma + \varepsilon_{\alpha\gamma}\chi_\beta) + n_{1p}n_{2k}n_{3l}(\varepsilon_{\alpha\beta}\chi_\gamma + \varepsilon_{\beta\gamma}\chi_\alpha)$$
$$+ n_{1l}n_{2p}n_{3k}(\varepsilon_{\gamma\beta}\chi_\alpha + \varepsilon_{\alpha\gamma}\chi_\beta)\}\varphi(r_1)\varphi(r_2)\varphi(r_3). \tag{7}$$

Here, $C_{ik} = C_{ki}$ and $C_{ii} = 0$. This tensor determines the orbital states; see Problem 3.45. The meaning of functions $\varepsilon_{\alpha\beta}$ and χ_γ is the same as in relation (4). Normalization of this wavefunction gives

$$C_{ik}C_{ki}^* = \frac{1}{128\pi^3}.$$

Calculating the shift $\Delta E(^2D)$ gives us expression (8) from Problem 11.17, multiplied by 3, and here the value b_L must be put equal to $b(^2D) = -1/20$.

From the values b_L given above, we obtain the order of terms:

$$E(^4S) < E(^2D) < E(^2P).$$

The standard term is ^4S, according to the Hund law with respect to the value of the total spin S.[249] Also, the ratio of energy differences given in the problem statement is $\Delta = 2/3$. Experimental values of this ratio for atoms with the electron configuration $(np)^3$ are given in the following table:

Atom	N	O$^+$	S$^+$	As	Sb	Bt
Config.	$2p^3$	$2p^3$	$3p^3$	$4p^3$	$5p^3$	$6p^3$
Δ_{exp}	0.500	0.509	0.651	0.715	0.908	1.121

Compare with the results of the previous problem.

Problem 11.19

Consider the statistical model of the ground state of a neutral atom with the nuclear charge, $Z \gg 1$, neglecting electron mutual interaction. Using this model, find:

a) electron density $n(r)$ and $\overline{r^k}$ for a single electron;

---

[248] Note that expression (6) is anti-symmetric with respect to permutation of $\mathbf{n}_1$ and $\mathbf{n}_2$. In the same way, relation (7) is anti-symmetric with respect to permutation, inside the curly brackets, of indices p and l, *convoluted* with indices of the anti-symmetric tensor ε_{ipl}.

[249] In the case considered, $S = 3/2$. Here, for the system of three equivalent p-electrons, the angular momentum is $L = 0$.

b) electron momentum distribution $\bar{n}(p)$ and also $\bar{p}$ and $\overline{p^2}$;
c) the typical value of electron's angular moment;
d) total ionization energy of atom $E_{tot.ion.} = -E_0$.

Pay attention to the independence of these values from Z. Compare to the results of the Thomas–Fermi model.

Neglecting electron–electron interaction, obtain the exact ground-state energy E_0, and for $Z \gg 1$ compare it to the result of the statistical model.

Solution

The ground state can be built by filling up the lowest-lying single-electron states one by one. In the quasi-classical approximation, the electron density $n(r)$ is connected to their momentum maximum $p_0(r)$ by the relation:

$$n(r) = \frac{1}{3\pi^2} p_0^3(r) = \frac{2\sqrt{2}}{3\pi^2} \frac{Z^{3/2}}{R^{3/2}} \frac{(1-\tilde{x})^{3/2}}{\tilde{x}^{3/2}}, \quad \tilde{x} = \frac{r}{R}. \tag{1}$$

We have used the fact that the total electronic energy is

$$\varepsilon_0 = \frac{1}{2} p_0^2 - \frac{Z}{r} = \text{const} \equiv -\frac{Z}{R}, \tag{2}$$

and does not depend on r (this makes system energy minimum). For $r > R$ we have[250] $n = 0$, while the value R is determined from the normalization condition for the neutral atom, $\int n(r) dV = Z$, and is equal to $R = (18/Z)^{1/3}$.

It is interesting to compare this electron density to the result of the Thomas–Fermi model. If we write (1) in the form of Eqs. (XI.1) and (XI.3), then for $\chi(x)$ we obtain (with $r \leq R$)

$$\chi(x) = 1 - \frac{r}{R} = 1 - 0.338x; \quad x = Z^{1/3} \frac{r}{b}, \quad b = 0.885.$$

Comparison of the results obtained using the two different models is presented in the following table:

x	0	0.5	1.0	1.5	2.0	2.5	3.0
$\chi_{TF}(x)$	1	0.607	0.424	0.314	0.243	0.193	0.157
$\chi(x)$	1	0.831	0.662	0.439	0.324	0.155	0

The higher electron density in the vicinity of nuclei in the model reflects the absence of nuclei charge screening, which is connected to neglecting the electrons' mutual repulsive interaction.

---

[250] In the model considered, the atom has a well-defined radius. But for peripheral electrons, because of the screening effect on the nuclei, the model predictions are unreliable.

We will now discuss the main conclusions that follow from the model considered.

a) Since the density $n(r)$ is normalized to the total electron number, equal to Z, then the function $w(r) = n(r)/Z$ is the probability distribution density for the coordinates of a single electron. It is evident that $\overline{r^n} \propto Z^{-n/3}$, and it is easy to obtain that $\overline{r} \approx 0.98\, Z^{-1/3}$. See that the average distance of the electrons from the nucleus decreases with the increase of Z as $Z^{-1/3}$.

b) The electron number density in momentum space is

$$\tilde{n}(p) = \frac{1}{4\pi^3} V_q(p) = \frac{8Z^3}{3\pi^2(p^2 + 2Z/R)^3}. \tag{3}$$

Compare to Eq. (1). Here, $V_q(p)$ is the volume in r-space where an electron is still "allowed" to have momentum p. As is seen from relation (2), this is the volume of a full sphere with radius

$$r(p) = \frac{2Z}{(p^2 + 2Z/R)}. \tag{4}$$

This density is also normalized to the electron number Z, so the relation $\tilde{w}(p) = \tilde{n}(p)/Z$ has the meaning of the probability distribution function of a single electron's momenta. Now, it is easy to obtain

$$\overline{p} \approx 1.11 Z^{2/3} \quad \text{and} \quad \overline{p^2} = (12)^{1/3} Z^{4/3} \approx 2.29\, Z^{4/3}.$$

So, the characteristic value of electron's momentum increases as $Z^{2/3}$.

c) Taking into account a) and b) for the typical values of electrons' angular momentum, we can estimate

$$l_{char} \sim r_{char} \cdot p_{char} \sim Z^{1/3}. \tag{5}$$

d) Using the virial theorem, according to which for Coulomb interaction $E = \overline{U}/2$, we find the energy of total atom ionization as

$$E_{tot.ion.} = \frac{1}{2} \int_0^R \frac{Z}{r} n(r) 4\pi r^2 dr = \left(\frac{3}{2}\right)^{1/3} Z^{7/3} \approx 1.14\, Z^{7/3}. \tag{6}$$

The integral here is calculated by substitution $r/R = \sin^2 u$. The same result follows from the relation $-E = \overline{T} = Z\overline{p^2}/2m$, but in the Thomas–Fermi model $E_{tot.ion.} = 0.77\, Z^{2/3}$ (the higher value of (6) is connected with neglecting the electrons' mutual interaction, which decreases the total energy).

Let us note that if we neglect electron–electron interaction, the atomic energy is equal to the sum of single electron energies, $\varepsilon_n = -Z^2/2n^2$. Placing them on the lowest levels and taking into account the Pauli principle and the Coulomb level degeneracy, equal to $2n^2$, we have

$$E_0 = \sum_{n=1}^{n_{max}} 2n^2 \varepsilon_n = -Z^2 n_{max}, \quad \sum_{n=1}^{n_{max}} 2n^2 = Z, \tag{7}$$

where n_{max} is the maximum value of the principal quantum number where there are still electrons. In the case $n_{max} \gg 1$ we can replace summation by integration in the second of the sums in (7), to find $n_{max} \approx (3Z/2)^{1/3}$. Energies of the atomic ground-state, according to the first of the sums in (7), coincides with the result of the statistical model in (6).

In conclusion, we emphasize that the results of the simple atom model considered here differ from the results of the Thomas–Fermi approximation only by a numerical coefficient ~ 1, and correctly determine the dependence on Z.

See also Problem 11.39.

Problem 11.20

Find the number of s-electrons as a function of the nuclear charge, Z, using the Thomas-Fermi approximation.

Solution

Single-electron s-levels are determined by the quasi-classical quantization rule for an electron in a self-consistent field ($U = -\varphi(r)$):

$$\int_0^{r_0} \sqrt{2[E_n + \varphi(r)]} dr = \pi(n + \gamma). \tag{1}$$

From the assumption of the Thomas–Fermi distribution, the total number of s-electrons in the atom is two times larger than the number of occupied levels (with spin taken into account), for which $E_n \leq E_{max}$, where E_{max} is the maximum value of energy of the Thomas–Fermi electrons. For a neutral atom $E_{max} = 0$ and for the corresponding value of n_{max} in (1) we should put $r_0 = \infty$. Using Thomas–Fermi units (XI.3) and omitting quasi-classical correction $\gamma \sim 1$ in relation (1), we obtain the total number of s-electrons in the atom,

$$N(l = 0) = 2n_{max} \approx \frac{2}{\pi}\sqrt{2b}Z^{1/3} \int_0^\infty \sqrt{\frac{\chi(x)}{x}} dx = aZ^{1/3}, \tag{2}$$

with the numerical factor $a \approx 3.5$ (it could be estimated using the simple approximation from Problem 11.22 for $\chi(x)$).

According to (2), for $Z = 27$ we have $N \approx 10$, while for the atom $_{27}$Co the number of s-electrons is actually equal to 8. For $Z = 64$, according to (2), $N = 14$, while the number of s-electrons in atom $_{64}$Gd is actually 12.

Problem 11.21

In the Thomas–Fermi model, express electron's kinetic energy, their interaction energy, the electron–nucleus interaction energy, and the total energy of the atom, $E[n(r)]$, in terms of the electron density $n(r)$.

Prove that the function $n_0(r)$ which minimizes the functional $E[n(r)]$ is the solution of the Thomas–Fermi equation (XI.2) with $\varphi = (1/2)(3\pi^2 n_0(r))^{2/3}$. Using variational analysis prove in the framework of the Thomas–Fermi model: a) $U_{en} = -7U_{ee}$ is the relation between the electron interaction energy U_{ee} and the electron–nuclei interaction U_{en}; b) the virial theorem.

Using a trial function of the form[251]

$$n_{\text{trial}}(r) = \frac{\alpha \lambda^3 Z^{3/2}}{16\pi r^{3/2}} \exp\left\{-\lambda \sqrt{rZ^{1/3}}\right\}, \quad \int n_{test} dV = \alpha Z,$$

where α, λ are variational parameters, find the energy E of the ground-state of a neutral atom with the nuclear charge Z by the variational method. Compare to the result of the Thomas–Fermi model.

Solution

Electron interaction energy and the electron-nucleus interaction energy is determined by the electrostatic equations:

$$U_{en} = -Z \int \frac{n(r)}{r} dV, \quad U_{ee} = \frac{1}{2} \iint \frac{\rho(r)\rho(r')}{|\mathbf{r}-\mathbf{r}'|} dV dV'$$

$$= \frac{1}{2} \iint \frac{n(r)n(r')}{|\mathbf{r}-\mathbf{r}'|} dV dV'. \quad (1)$$

Electron kinetic energy is determined from the condition that they are distributed (with occupation numbers $n_k = 1$) over the lower energy levels in the self-consistent atomic field, and is equal to

$$T = \frac{3}{10}(3\pi^2)^{2/3} \int n^{5/3}(r) dV. \quad (2)$$

This expression follows from the quasi-classical equation for the number of available quantum states:

$$\Delta N = \frac{2\Delta \Gamma}{(2\pi)^3} = \frac{2\Delta V_q \Delta V_p}{(2\pi)^3},$$

which for the values $\Delta V_q = 1$ and $\Delta V_p = 4\pi p_{max}^3/3$ connect the electron density $n = \Delta N$ with p_{max}, here $\overline{p^2} = 3\, p_{max}^2/5$.

---

[251] Let us emphasize that here we imply an unconstrained minimum of functional $E[n(r)]$, without an additional condition for $n(r)$ normalization. The exact function, $n_0(r)$, becomes automatically normalized to the electron number Z. The approximate trial function is not supposed to satisfy such a condition.

Let us note an interesting property of the energy functional. In the conditions of this problem, $E[n(r)]$ for a neutral atom takes the minimum value. On the contrarty, if we introduce the functional $E[\varphi(r)]$ (see Problem 11.22), then for a neutral atom it takes maximum value. The results of Problems 11.21 and 11.22 give both the upper and lower bound for the atomic energy in the Thomas–Fermi model.

The atomic (ionic) energy in the quasi-classical approximation is expressed in terms of the electron density by

$$E[n(r)] = \frac{3(3\pi^2)^{2/3}}{10}\int n^{5/3}dV - Z\int \frac{n}{r}dV + \frac{1}{2}\iint \frac{n(r)n(r')}{|\mathbf{r}-\mathbf{r}'|}dV dV'. \tag{3}$$

Variation of the functional $E[n(r)]$ gives

$$\delta E = \int \delta n(r)\left\{\frac{(3\pi^2)^{2/3}}{2}n^{2/3}(r) - \frac{Z}{r} + \int \frac{n(r')dV'}{|\mathbf{r}-\mathbf{r}'|}\right\}dV.$$

The extremal value condition gives the equation for the function $n(r)$ which minimizes the atomic energy:

$$\frac{1}{2}(3\pi^2)^{2/3}n^{2/3}(r) - \frac{Z}{r} + \int \frac{n(r')dV'}{|\mathbf{r}-\mathbf{r}'|} = 0. \tag{4}$$

If we apply the Laplace operator to both sides of this equation and use the relation

$$\Delta \frac{1}{|\mathbf{r}-\mathbf{r}'|} = -4\pi\delta(\mathbf{r}-\mathbf{r}'),$$

we obtain the differential form of Eq. (4):

$$\Delta\left[\frac{1}{2}(3\pi^2)^{2/3}n^{2/3}(r)\right] = -4\pi[Z\delta(\mathbf{r}) - n(r)]. \tag{5}$$

So by taking onto account the electrostatic Poisson equation, $\Delta\varphi = -4\pi\rho$, we conclude that the value $\varphi = \frac{1}{2}(3\pi^2)^{2/3}n^{2/3}$ describes the atom's electrostatic potential, while Eq. (5) here

$$\Delta\varphi = \frac{8\sqrt{2}}{2\pi}\varphi^{3/2}$$

(for the values $r \neq 0$) coincides with the Thomas–Fermi equation (XI.2).

From Eq. (4) for $r \to \infty$ it follows that $\int n(r)dV = Z$, i.e., the neutral atom has minimum energy, but the ion does not. This proves the atom stability in the Thomas–Fermi model, and means that the statistical model could not explain the existence of stable *negative*[252] ions.

a) Examine the relations between quantities T, U_{ee}, U_{ne} for a neutral atom. Denote the electron spatial density by $n_0(r)$ in the Thomas–Fermi model. The change of the functional $E[n_0(r)]$ value, which determines the atomic energy, with the substitution $n_0(r) \to n(r) = (1+\lambda)n_0(r)$, $|\lambda| \ll 1$ is equal to

$$\delta E = E[n(r)] - E[n_0(r)] \approx \left(\frac{5}{3}T + 2U_{ee} + U_{en}\right)\lambda,$$

---

[252] The existence of such ions is connected to the properties of external electron shells, the consideration of which in the statistical model is inadequate.

and the condition $\delta E = 0$, gives

$$5T + 6U_{ee} + 3U_{en} = 0. \tag{6}$$

In an analogous way, considering the transformation of the form $n(r) = n_0\left((1+\lambda)r\right)$ with $|\lambda| \ll 1$, we obtain

$$3T + 5U_{ee} + 2U_{en} = 0. \tag{7}$$

From (6) and (7) we have both relation $U_{en} = -7U_{ee}$ and the virial theorem $2T = -(U_{ee} + U_{en}) \equiv -U$.

b) Finally, we will calculate the ground-state energy through the variational method for $E[n_0(r)]$. For the trial function given in problem condition, according to Eq. (3) we obtain

$$E(\alpha, \lambda) = \frac{9}{400}\left(\frac{3\pi}{2}\right)^{2/3}\alpha^{5/3}\lambda^4 Z^{7/3} - \frac{1}{2}\alpha\lambda^2 Z^{7/3} + \frac{1}{16}\alpha^2\lambda^2 Z^{7/3}. \tag{8}$$

(While calculating U_{ee} it is convenient to use the value of integral (2) from Problem 11.15.) Minimization of relation (8) over λ gives

$$E(\alpha) = \min_{(\lambda = \lambda_0)} E(\alpha, \lambda) = -\frac{25}{576}\left(\frac{2}{3\pi}\right)^{2/3}\alpha^{1/3}(\alpha - 8)^2 Z^{7/3}.$$

where

$$\lambda_0^2(\alpha) = \frac{25(8-\alpha)}{18}\left(\frac{2}{3\pi\alpha}\right)^{2/3}.$$

The minimization over α gives $\alpha = \alpha_0 = 8/7$ and the ground-state energy

$$E_{0,var} \approx -0.759\ Z^{7/3}, \tag{10}$$

and the value of the parameter $\lambda_0(\alpha_0) = 1.761$. Compare to the exact result $E_0 = -0.769\ Z^{7/3}$ for the Thomas–Fermi model.

In conclusion, let us comment on the properties of the trial function $n_{\text{trial}}(r)$. Formally, it is normalized to the number of electrons, equal to αZ, but the value of $E_{0,var}$ obtained using this corresponds to a neutral atom with the number of electrons equal to Z. (The choice of $\alpha = 1$ gives the less accurate value of E_0, though the difference is not essential: instead of 0.759 in Eq. (10) we have 0.757). After minimization in λ, the trial function considered corresponds to a choice of the universal function $\chi(x)$ in the Thomas–Fermi model in the form ($b = 0.885$):

$$\chi_{\text{trial}}(x) = \frac{25(8-\alpha)}{144}\exp\{-\bar{\lambda}\sqrt{x}\}; \quad x = Z^{1/3}\frac{r}{b}, \quad \bar{\lambda} = \frac{2}{3}\lambda_0(\alpha)\sqrt{b}.$$

It is easy to see that the difference of χ_{trial} from the exact function, χ_{T-F}, is more significant than in the case of energy values E_0. For example, for $\alpha = 8/7$ we have $\chi_{\text{trial}}(0) = 1.19$, while $\chi_{T-F}(0) = 1$. Compare to Problem 8.22.

Problem 11.22

In the framework of the statistical model of a neutral atom, write its energy, $E[\varphi(r)]$ in terms of the potential $\varphi(r)$, in such a form that the Thomas–Fermi equation (XI.2) follows from the extremum of the functional $E[\varphi(r)]$.

Using the trial functions,

$$\varphi(r) = \frac{Z}{r}\chi(r), \quad \chi(r) = \frac{1}{(1+\alpha Z^{1/3}r)^2},$$

where α is a variational parameter, find the ground-state energy using the variational method. Compare to the previous problem and to the exact result of the Thomas–Fermi model.

Solution

We write the atomic electrostatic potential in the form $\varphi = Z/r + \varphi_{el}$, where $\varphi_{el}(r)$ is the potential produced by the electrons. Using the relation

$$n(r) = -\rho_{el}(r) = \frac{(2\varphi)^{3/2}}{3\pi^2},$$

(see Eq. (XI.1)) and the known equations from electrostatics, we find

$$\int \rho_{el}(r)\varphi(r)dV = -\frac{2\sqrt{2}}{3\pi^2}\int \varphi^{5/2}(r)dV = 2U_{ee} + U_{en},$$

$$\frac{1}{2}\int \rho_{el}(r)\varphi_{el}(r)dV = -\frac{1}{8\pi}\int \left(\varphi - \frac{Z}{r}\right)\Delta\left(\varphi - \frac{Z}{r}\right)dV = U_{ee}, \quad (1)$$

$$T = \frac{2\sqrt{2}}{5\pi^2}\int \varphi^{5/2}dV.$$

For the electron kinetic energy, see Eq. (2) from previous problem. Now,

$$E[\varphi(r)] = T + U_{en} + U_{ee} = -\frac{4\sqrt{2}}{15\pi^2}\int \varphi^{5/2}(r)dV$$

$$+ \frac{1}{8\pi}\int \left(\varphi - \frac{Z}{r}\right)\Delta\left(\varphi - \frac{Z}{r}\right)dV. \quad (2)$$

If we vary $\varphi(r)$ to extremize (in this case, maximize; see the previous problem) the functional $E[\varphi(r)]$, we indeed obtain the Thomas–Fermi equation for the potential. Now consider the potential[253]

$$\varphi = (1+\lambda)\varphi_0\left((1+\lambda)r\right),$$

---

[253] As we vary the potential, the condition, $\varphi(r) \approx Z/r$ must be fulfilled for $r \to 0$; otherwise, as is seen from expression (1), the value of U_{ee} becomes infinite.

where $\varphi_0(r)$ is the solution of the Thomas–Fermi equation and $|\lambda| \ll 1$. From the maximum condition on $E[\varphi(r)]$, we obtain the relation

$$T + 4U_{ee} + U_{en} = 0, \tag{3}$$

and from Eq. (1) we have

$$5T + 6U_{ee} + 3U_{en} = 0. \tag{4}$$

We obtain both equation $U_{en} = -7U_{ee}$ and the virial theorem for Coulomb interaction in an atom.

For the variational calculation of the atomic ground-state energy E_0, it is convenient to transform Eq. (2). Since the function, $\varphi - Z/r$, has no singularities at the point $r = 0$, then in the second of the integrals in Eq. (2) we can replace Δ by $\frac{1}{r}\frac{\partial^2}{\partial r^2}r$. Integration by parts gives us

$$E[\varphi(r)] = -\frac{16\sqrt{2}}{15\pi}\int_0^\infty r^2 \varphi^{5/2}(r)dr - \frac{1}{2}\int_0^\infty \left[\frac{\partial}{\partial r}(r\varphi(r))\right]^2 dr. \tag{5}$$

For the trial function given in the problem condition, we obtain (the first integral by substitution $x = \sqrt{r}$ is reduced to E 1.5):

$$E(\alpha) = -\frac{7\sqrt{2}}{24}Z^{7/3}\frac{1}{\sqrt{\alpha}} - \frac{2}{5}Z^{7/3}\alpha. \tag{6}$$

The maximum value $E(\alpha_0)$ of this quantity gives the energy of the atom's ground-state:

$$E_{0,var} = E(\alpha_0) = -0.771\, Z^{7/3}, \tag{7}$$

where

$$\alpha_0 = \left(\frac{35}{48}\sqrt{2}\right)^{2/3} \approx 0.643.$$

In the conditions of the problem, (7) is a lower bound for the true value of E_0 in the Thomas–Fermi model, equal to $E_0 = -0.769\, Z^{7/3}$. Compare to the result of the previous problem.

In conclusion, let us note that the trial function considered reproduces with high accuracy not only the value of E_0, but also the universal function $\chi(x)$ of the Thomas–Fermi model. Comparison of $\chi_{\text{trial}} = (1 + \tilde{\alpha}x)^{-2}$, $x = Z^{1/3}r/b$, $\tilde{\alpha} = \alpha_0 b \approx 0.569$ to the exact function χ_{T-F} is given in the following table:

x	0	0.5	1.0	2.0	5.0
$\chi_{T-F}(x)$	1	0.607	0.424	0.243	0.079
$\chi_{\text{trial}}(x)$	1	0.606	0.406	0.219	0.068

Their difference is especially small in the region $x \leq 1$, where most electrons are; with the increase of x, relation χ_{test}/χ_{T-F} decreases. It is connected to the fact that the trial function considered, $n = (2\varphi)^{3/2}/(3\pi^2)$, is normalized to the electron number, equal to

$$\frac{\sqrt{2}}{3\alpha_0^{3/2}} Z = \frac{32}{35} Z.$$

This is lower than Z. Compare to the previous problem.

11.3 Principles of two-atom-molecule theory

Problem 11.23

Classify possible terms of the hydrogen molecular ion H_2^+. Give possible values of the electron angular momentum, L with respect to the symmetry center for different ion terms.

Solution

For terms of a hydrogen molecular ion H_2^+ with the quantum number Λ, the electron angular momentum projection on the direction of the axis, passing through the nuclei, could only take the values $m = \pm\Lambda$. So, wavefunctions of such terms could be written in the form of an expansion over the spherical functions:

$$\psi_m(r, \theta, \varphi) = \sum_{L \geq \Lambda} R_{L\Lambda} Y_{Lm}(\theta, \varphi) = R(r, \theta)e^{im\varphi}, \quad (1)$$

where r, θ, φ are the spherical coordinates. (The polar axis is directed along the ionic symmetry axis, and the system origin is at the center of the segment that connects the nuclei.)

For the Σ-terms (for which $m = \Lambda = 0$) the wavefunction (1) does not change under reflection of the electron coordinates through the plane which passes through the ion symmetry axis. For such transformation of coordinates, r and θ remain the same, and the wavefunction does not depend on φ, since $m = 0$. This means that Σ-states are Σ^+-terms, while Σ^--terms for H_2^+ do not exist, which is a specific property of the one-electron system.

Wavefunction (1) could be chosen as an eigenfunction of the inversion operator $\hat{I}$ which inverts electron coordinates through the point $r = 0$ and commutes with the system Hamiltonian. Since here $\hat{I} Y_{Lm} = (-1)^L Y_{Lm}$, the sum in (1) includes either only even or only odd values of L. In the former case, wavefunction (1) corresponds to *even* terms with quantum numbers Λ_e, while in the latter case it corresponds to *odd* terms Λ_o. Remember that classification of two-atom molecular terms as even and odd arises for identical molecular nuclei and is connected with the behavior of the wavefunction under coordinate inversion.

The possible ion terms:

$$^2\Sigma_g^+,\ ^2\Sigma_u^+,\ ^2\Pi_g^+,\ ^2\Pi_u^+,\ ^2\Delta_g^+,\ ^2\Delta_u^+,\ \ldots.$$

Problem 11.24

The state of a system consisting of two electrons is described by wavefunction $\psi = \psi(\mathbf{r}_1, \mathbf{r}_2)\chi_{\alpha\beta}$, where $\chi_{\alpha\beta}$ is the spin function, and $\psi(\mathbf{r}_1, \mathbf{r}_2)$ has the form:

a) $\psi = f(r_1, r_2)$;
b) $\psi = (\mathbf{r}_1 \cdot \mathbf{n}_0 + \mathbf{r}_2 \cdot \mathbf{n}_0)f(r_1, r_2)$;
c) $\psi = ([\mathbf{r}_1 \times \mathbf{r}_2] \cdot \mathbf{n}_0)f(r_1, r_2)$;
d) $\psi = (\mathbf{r}_1 \cdot \mathbf{n}_0 + \mathbf{r}_2 \cdot \mathbf{n}_0)([\mathbf{r}_1 \times \mathbf{r}_2] \cdot \mathbf{n}_0)f(r_1, r_2)$.

In accordance with the standard classification of two-atom molecules, classify these states, considering the vector $\mathbf{n}_0$ as the radius-vector connecting the nuclei.

Solution

None of the four wavefunctions given in the problem condition change under rotation around the coordinate axis that is parallel to the vector $\mathbf{n}_0$ and passes through $r = 0$. So all of them describe states with projection $m = \Lambda = 0$ of total electrons' angular momentum onto this axis, *i.e.*, the Σ-states.

Now, the wavefunctions have a definite parity with respect to electron coordinate inversion: wavefunctions a) and c) are even, *i.e.*, describe the Σ_e-states, while odd wavefunctions b) and d) describe the Σ_o-states (compare to the previous problems).

Finally, under electron coordinate inversion in a plane that includes the axis mentioned above, wavefunctions a) and b) do not change, *i.e.*, correspond to the Σ^+-states, while c) and d) change sign and are described by the Σ^--states.

Therefore, we have the following classification of states considered:

a) Σ_g^+; b) Σ_u^+; c) Σ_g^-; d) Σ_u^-

Their multiplicity, 1 and 3, is defined by the value, S, of the total electron spin that depends on the symmetry of coordinate wavefunctions $\psi(\mathbf{r}_1, \mathbf{r}_2)$ with respect to electron permutation.

Problem 11.25

For a two-atom molecule, estimate ratios of the following quantities:

a) level spacings between electronic vibrational and rotational levels;
b) distance between the nuclei and amplitude of nuclear oscillations;
c) typical periods and velocities of electronic and atomic motion.

Solution

The main physical fact in molecular quantum mechanics is the smallness of the ratio

$$\frac{m}{M} \sim 10^{-4} - 10^{-3}$$

where m is the electron mass and M is the reduced nuclear mass. This provides different orders of magnitudes for the quantities given in the problem statement.

a) Molecule's linear dimensions a_{mol} and distances a_{nn} between the nuclei have the same order of magnitude as the localization length of valence (outer) electrons in the atom, a_e:

$$a_{mol} \sim a_{nn} \sim a_e \sim a_B = \frac{\hbar^2}{me^2}.$$

Characteristic values of valence electron's energy in the molecule, as well as the level spacing between the neighboring electron terms of the molecule for "fixed" nuclei, are equal by the order of magnitude: $E_{el} \sim \hbar^2/ma_B^2$. The characteristic values of the intervals between oscillation and rotational levels of a molecule for the same electron term are much smaller

$$E_{osc} \sim \hbar\omega_{osc} \sim \sqrt{\frac{m}{M}} E_{el}, \quad E_{rot} \sim \frac{\hbar^2}{I} \sim \frac{\hbar^2}{Na_B^2} \sim \sqrt{\frac{m}{M}} E_{osc} \sim \frac{m}{M} E_{el}.$$

Oscillation levels of a molecule, $E_{osc,v} = \hbar\omega_{osc}(v + 1/2)$, are the levels of an oscillator with mass M and elastic coefficient k, whose order of magnitude from dimensional considerations is determined by the relation $ka_B^2 \sim \hbar^2/ma_B^2$, while $\omega_{osc} = \sqrt{k/M}$. Molecule's rotational levels, $E_{rot,K} = \hbar^2 K(K+1)/2I$, are the levels of a spherical rotator with the moment of inertia $I = MR_0^2$, where R_0 is an equilibrium distance between nuclei, $R_0 \sim a_B$ (in practice, for terms with $\Lambda \neq 0$, molecular rotation is simulated by a symmetric spinning top).

b) Evaluation of a nuclear oscillations amplitude from the relation $E_{osc} \sim k a_{osc}^2$ gives

$$a_{osc} \sim \left(\frac{m}{M}\right)^{1/4} a_B \ll a_B.$$

c) The characteristic periods of different type of motion in the molecule are

$$\tau_{el} \sim \frac{a_B}{v_{el}} \sim \frac{ma_B^2}{\hbar}, \quad \tau_{osc} \sim \frac{1}{\omega_{osc}} \sim \left(\frac{M}{m}\right)^{1/2} \tau_{el},$$

$$\tau_{rot} \sim \frac{a_B}{v_{nuc\ rot}} \sim a_B \left(\frac{M}{E_{rot}}\right) \sim \frac{M}{m}\tau_{el}.$$

The different orders of magnitudes of these time-scales ensure the applicability of the adiabatic approximation (see Chapter 8, sec. 6), according to which, the energy levels of a molecule are given in the form:

$$E = E_{el} + E_{osc} + E_{rot},$$

and $E_{rot} \ll E_{osc} \ll E_{el}$.

Problem 11.26

Considering the following properties of the hydrogen molecule H_2 to be known:

1) dissociation energy of the molecular ground-state into two unperturbed hydrogen atoms: $I_0 = 4.46$ eV;
2) oscillation frequency, ω_e, of the molecule: $\hbar\omega_e = 0.54$ eV;
3) rotational constant: $B_e = 7.6 \cdot 10^{-3}$ eV;
   find the corresponding values for molecules HD and D_2, where one or both proton-nuclei are replaced by a deuteron.

Compare the isotope shift effects for a hydrogen atom and a molecular hydrogen.

Solution

The energy of the molecular ground-state is equal to

$$E_0 = E_{el,0} + E_{osc,0} = E_0(R_0) + \frac{1}{2}\hbar\omega_{osc}, \tag{1}$$

where $E_0(R)$ is the ground-state energy, R_0 is the equilibrium distance between nuclei, $\omega_{osc} \equiv \omega_e = \sqrt{E_0''(R_0)/M}$, M is the nuclei reduced mass. The rotational constant of the molecule is $B_e = \hbar^2/2MR_0^2$.

Since after replacing the molecular nuclei by their isotopes, the function $E_0(R)$ and the value R_0 remain the same, then taking into account the relation $m_d \approx 2m_p$, we find:

$$(\hbar\omega_e)_{HD} \approx \frac{\sqrt{3}}{2}(\hbar\omega_e)_{H_2} = 0.46 \text{ eV}, \quad (B_e)_{HD} \approx \frac{3}{4}(B_e)_{H_2} = 5.7 \cdot 10^{-3} \text{ eV};$$

$$(\hbar\omega_e)_{D_2} \approx \frac{1}{\sqrt{2}}(\hbar\omega_e)_{H_2} = 0.38 \text{ eV}, \quad (B_e)_{O_2} \approx \frac{1}{2}(B_e)_{H_2} = 3.8 \cdot 10^{-3} \text{ eV}.$$

The molecular dissociation energy is

$$I_0 = E_1^{(0)} + E_2^{(0)} - E_0,$$

where $E_{1,2}^{(0)}$ are the ground levels of the corresponding hydrogen atom, taking into account the finiteness of their nuclei masses:[254]

$$E^{(0)} \approx -\frac{m_e e^4}{2\hbar^2}\left(1 - \frac{m_e}{M_{nuc}}\right) = -13.60\left(1 - \frac{m_e}{M_{nuc}}\right) \text{ eV}.$$

From the given equations we obtain

$$(I_0)_{HD} \approx 4.50 \text{ eV}, \quad (I_0)_{D_2} \approx 4.54 \text{ eV}.$$

The effect of the isotopic shift in a hydrogen atom has the magnitude of the order $(\Delta E/E)_{at} \sim m_e/M_{nuc} \sim 10^{-3}$, while in a molecule it is

$$\left(\frac{\Delta E}{E}\right)_{mol} \sim \sqrt{\frac{m_e}{M_{nuc}}} \sim \frac{1}{40},$$

and therefore manifests itself more clearly as a change in the frequency of nuclear oscillations ω_{osc}.

Problem 11.27

Find possible rotational states of the molecules H_2, deuterium D_2, and HD, which are in the ground-state Σ_g^+, as a function of the total nucleus spin (the deuterium spin is equal to 1).

How does the parity of the term depend on the value of the molecule's angular momentum?

Solution

1) Restrictions on possible values of the molecular angular momentum K for a fixed value of the total nuclear spin are due to the fact that the wavefunction of a system with identical particles (in this case the nuclear subsystem in molecules H_2 and D_2) must have a definite symmetry with respect to permutation of the variables (spin and coordinate) of any two such particles. Remember that the spin wavefunction of a two-spin system, where[255] i is the spin of each system, is symmetric with respect to the permutation of spin variables if it has $I = 2i, 2i - 2, \ldots$ for the total spin, and is asymmetric if it has $I = 2i - 1, 2i - 3, \ldots$ (see Problem 3.30), and also there are of course constraints on the wavefunction symmetry for identical bosons and fermions. Using this information, we see that under permutation of the spatial variables, the nuclear wavefunction of the molecule does not change sign for an even value of the nuclear spin ($I = 0$ for H_2 and $I = 0; 2$ for D_2) and changes sign

---

[254] See the footnote for the solution of Problem 11.30.
[255] Nuclear spins are usually denoted by i and I.

for odd I. We should emphasize that this conclusion, as well as the possible values of the angular momentum K, corresponds to any two-atom molecule with $\Lambda = 0$ and identical nuclei. We can examine what restrictions on K this imposes.

For the molecules considered with $\Lambda = 0$, the wavefunction dependence on the nuclear coordinates is defined by the relation

$$\psi_{KMn\Lambda=0} = \psi_{n\Lambda=0}(\mathbf{r}_1, \mathbf{r}_2, \mathbf{R})\psi_{osc}(R)Y_{KM}(\theta, \varphi). \tag{1}$$

Here Y_{KM} are spherical functions; θ, φ – polar and azimuth angles of the radius-vector $\mathbf{R} = \mathbf{R}_2 - \mathbf{R}_1 \equiv R\mathbf{n}$ connecting the nuclei; $\psi_{n\Lambda=0}$ is the wavefunction of the molecule's electron Σ-term, and $\psi_{osc}(R)$ is the wavefunction of the nuclei's oscillation. For the permutation of nuclei coordinates, i.e., for a transformation of the form $\mathbf{R} \to -\mathbf{R}$, the oscillation part of the wavefunction does not change, while the spherical function is multiplied by $(-1)^K$.

More delicate is a question concerning the electron term wavefunction $\psi_{n\Lambda=0}$. This function is a scalar (or pseudo-scalar, depending on term quantum numbers) that depend only on vectors $\mathbf{r}_1$, $\mathbf{r}_2$, and $\mathbf{R}$. The most general form of such a function is

$$\psi_{n\Lambda=0} = \psi(r_1, r_2, R, \mathbf{r}_1\cdot\mathbf{R}, \mathbf{r}_2\cdot\mathbf{R}, \mathbf{r}_1\cdot\mathbf{r}_2, [\mathbf{r}_1\times\mathbf{r}_2]\cdot\mathbf{R}). \tag{2}$$

This function does not change under rotation of the electron subsystem around the axis that passes through the vector $\mathbf{R}$, as is needed for the term with $\Lambda = 0$. Under coordinate inversion in a plane that includes the vector $\mathbf{R}$, we have

$$\hat{P}_1\psi_{n\Lambda=0} \equiv \psi(r_1, r_2, R, \mathbf{r}_1\cdot\mathbf{R}, \mathbf{r}_2\cdot\mathbf{R}, \mathbf{r}_1\cdot\mathbf{r}_2, -[\mathbf{r}_1\times\mathbf{r}_2]\cdot\mathbf{R}) =$$
$$= \sigma_1\psi(r_1, r_2, R, \mathbf{r}_1\cdot\mathbf{R}, \mathbf{r}_2\cdot\mathbf{R}, \mathbf{r}_1\cdot\mathbf{r}_2, [\mathbf{r}_1\times\mathbf{r}_2]\cdot\mathbf{R}), \tag{3}$$

where σ_1 is equal to $+1$ and -1 for the Σ^+ and Σ^--terms. In the same way, for electron's coordinates reflection with respect to the center of the segment that connects the nuclei, we obtain

$$\hat{P}_2\psi_{n\Lambda=0} \equiv \psi(r_1, r_2, R, -\mathbf{r}_1\cdot\mathbf{R}, -\mathbf{r}_2\cdot\mathbf{R}, \mathbf{r}_1\cdot\mathbf{r}_2, [\mathbf{r}_1\times\mathbf{r}_2]\cdot\mathbf{R}) =$$
$$= \sigma_2\psi(r_1, r_2, R, \mathbf{r}_1\cdot\mathbf{R}, \mathbf{r}_2\cdot\mathbf{R}, \mathbf{r}_1\cdot\mathbf{r}_2, [\mathbf{r}_1\times\mathbf{r}_2]\cdot\mathbf{R}), \tag{4}$$

where σ_2 is equal to $+1$ for *even* Σ_g and -1 for *odd* Σ_u-terms.

Now we see that the transformation $\mathbf{R} \to -\mathbf{R}$ is equivalent to a product of transformations performed in Eq. (3) and (4), i.e., $\hat{P}(\mathbf{R} \to -\mathbf{R}) = \hat{P}_1\hat{P}_2$, so that

$$\hat{P}\psi_{n\Lambda=0} = \sigma_1\sigma_2\psi_{n\Lambda=0},$$

and for the wavefunction of the molecule,

$$\hat{P}(\mathbf{R} \to -\mathbf{R})\psi_{KMn\Lambda=0} = (-1)^K\sigma_1\sigma_2\psi_{KMn\Lambda=0}. \tag{5}$$

In accordance with the symmetry properties discussed at the beginning, we have

$$(-1)^K\sigma_1\sigma_2 = (-1)^I. \tag{6}$$

This expression gives a relation between possible values of the quantum numbers K, I, σ_1, σ_2 for a two-atom molecule with identical nuclei. In particular, for the hydrogen molecule, the ground term is Σ_g^+. Here $\sigma_1 = \sigma_2 = +1$, so according to (6) for molecule H_2 with the total nuclear spin $I = 0$ and for molecule D_2 with $I = 0, 2$, only even values of the angular momentum $K = 0, 2, 4, \ldots$ are possible, while for $I = 1$ only the odd K are possible (the possible values of K for molecule HD with different nuclei do not depend on the total spin of its nuclei).

2) Now note that from relations (3) and (4) it follows that for wavefunction (2) the product of transformations $\hat{P}(\mathbf{R} \to -\mathbf{R})\hat{P}_2$, which corresponds to inversion of coordinates for both electrons and nuclei, is equivalent to transformation of $\hat{P}_1$, and wavefunction (1) under inversion is multiplied by $(-1)^K \sigma_1$. This factor determines the parity of the molecule in the $\Lambda = 0$ state (but with not necessarily identical nuclei). So for the Σ^+ (Σ^-) terms, the states of the molecule with even (odd) angular momentum K are *positive*, while those with odd (even) K are *negative*.[256]

In conclusion, let us first discuss the values of the nuclear angular momentum. It does not coincide with the value of molecule rotational angular momentum K and, furthermore nor does it have a definite value, just as the electrons' angular momentum does not. But for the Σ-terms of molecules with identical nuclei, all its possible values, L_{nuc}, have the same parity, so that $(-1)^{L_{nuc}} = (-1)^I$; here, the relation (6) connects the values of L_{nuc} with quantum numbers K, σ_1, σ_2. For example, for the Σ_g^+- and Σ_g^--terms, we have $L_{nuc} = K, K \pm 2, \ldots$.

Molecule states are classified as *symmetric with respect to nuclei* for $(-1)^{L_{nuc}} = (-1)^K \sigma_1 \sigma_2 = +1$ and as *anti-symmetric* for $(-1)^{L_{nuc}} = (-1)^K \sigma_1 \sigma_2 = -1$ (remembering that the factor $(-1)^K \sigma_1 = (-1)^I \sigma_2$ determines the parity of a molecular level). This classification is connected to definite values of the quantity $(-1)^{L_{nuc}}$ (for a two-atom molecule with identical nuclei) that corresponds to Hamiltonian's invariance with respect to nuclei's coordinate permutation.

The restriction, see Eq. (6), on the possible values of the rotational moment K for different parity of the total nuclear spin I leads to a dependence of the molecular levels on I due to different values of the rotational energy (even in the absence of spin terms in Hamiltonian). A nuclear exchange interaction appears. But due to the smallness of rotational energy ($\sim m_e/M_{nuc}$), it is much weaker than the exchange interaction of electrons in the atom. So for the corresponding levels $E_{I,K}$ of the *ortho-hydrogen* ($I = 1$) and *para-hydrogen* ($I = 0$), we have:

$$E_{I=1,K+1} - E_{I=0,K} = 2B_c(K+1) = 0.015(K+1) \ eV,$$

where $K = 0, 2, 4, \ldots$ (a value of the rotational constant for the molecule H_2: see Problem 11.26).

---

[256] Do not confuse the parity of a molecular level with signs $+$ and $-$ of electron terms $\Sigma^{\pm}$!

Problem 11.28

Find electron terms $E(R)$ of a negative molecular ion $(AB)^-$ in the framework of the model where interaction between the external electron and atoms A and B is approximated by a zero-range potential (see Problem 4.10).

Determine

a) if a stable ion $(AB)^-$ can exist, in the case when stable ions A^- and B^- do not exist;

b) level spacing between the even and odd terms for $R \to \infty$ (in the case of identical atoms, $A \equiv B$).

Solution

The solution of the Schrödinger equation for a particle in the combined field of two zero-range potentials which are localized at the points $\mathbf{r}_{1,2}$ for energy values $E = -\hbar^2 \kappa^2 / 2m < 0$ has the form:[257]

$$\psi_E(\mathbf{r}) = \frac{c_1}{|\mathbf{r} - \mathbf{r}_1|} e^{-\kappa|\mathbf{r} - \mathbf{r}_1|} + \frac{c_2}{|\mathbf{r} - \mathbf{r}_2|} e^{-\kappa|\mathbf{r} - \mathbf{r}_2|}. \tag{1}$$

The boundary conditions for $\mathbf{r} \to \mathbf{r}_{1,2}$ associated with the zero-range potentials give (see Problem 4.10)

$$(\kappa - \alpha_1) R c_1 = e^{-\kappa R} c_2, \quad e^{-\kappa R} c_1 = (\kappa - \alpha_2) R c_2, \tag{2}$$

where $\mathbf{R} = \mathbf{r}_1 - \mathbf{r}_2$. A consistency condition for this system of equations with respect to $c_{1,2}$ gives an equation for spectrum $E(R)$ that models electron terms of the molecular ion:

$$(\kappa - \alpha_1)(\kappa - \alpha_2) = \frac{1}{R^2} e^{-2\kappa R}. \tag{3}$$

We will analyze this equation.

1) In the case of $\alpha_{1,2} > 0$, there exist bound states in both potentials (negative atom ions) with energies $E_{1,2}^{(0)} = -\hbar^2 \alpha_{1,2}^2 / 2m$. These states represent Σ^+-terms, (since wavefunction (1) does not change under rotation around the axis that passes through the vector $\mathbf{R}$), have the following properties. For $R \to \infty$ there are two separate states, and for them, $E_{1,2}(R) \to E_{1,2}^{(0)}$, while their wavefunctions in the case of $\alpha_1 \neq \alpha_2$ are localized on each of the centers individually. As R decreases, the lower term deepens, and for this lower term, $E(R) \to -\infty$ for $R \to 0$, while the other term moves upward and for $R = R_c = (\alpha_1 \alpha_2)^{-1/2}$ enters the continuous spectrum (see Fig. 11.2). Deepening of the term with decrease of R (attraction) shows that there is a stable ion $(AB)^-$. In this model, the equilibrium distance

---

[257] Compare to Problem 4.10. Note that analogously to (1), a similar linear combination of exponentials determines the form of the solution for particle bound states in an arbitrary number of zero-range potentials (it follows from integral form of Schrödinger equation; see Problem 4.20).

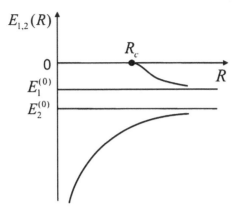

Fig. 11.2

between the nuclei is $R_0 = 0$ and $E(R_0) = -\infty$. It means that the real values of these quantities depend essentially on the form of the potential on atomic distances. So, we emphasize that approximation of the interaction between the electron and atom by a zero-range potential is only reasonable if the electron binding energy in the atom ion is $|E_0| \ll \hbar^2/ma_B^2$, while the domain of the wavefunction localization is $\sim \kappa^{-1} = \alpha^{-1} \gg a_B$. The second electron term, which increases with decrease in R, corresponds to repulsive interaction and does not lead to the appearance of a stable molecular ion.

In the case of $\alpha_1 = \alpha_2 \equiv \alpha > 0$, it is seen from Eqs. (2) and (3) that for the terms considered, $c_1 = \pm c_2$, so that they have a definite parity, i.e., are Σ_g^+- and Σ_u^+-terms. For the difference between their energies for $R \to \infty$ from (3), we obtain

$$E_u(R) - E_g(R) \approx \frac{2\hbar^2\alpha}{mR} e^{-\alpha R}. \tag{4}$$

In the right-hand side of Eq. (3) we can put $\kappa = \alpha$, so that $(\kappa_{g,u} - \alpha) \approx \pm e^{-\alpha R}/R$, while $\Delta E \approx \hbar^2 \alpha \Delta \kappa / m$.

2) In the case of $\alpha_1 > 0$, $\alpha_2 < 0$, there is only one stable atomic ion. According to (3), there is only one term with $E(R) < 0$, and for it, as well as in the previous case, $E(R) \to E_1^{(0)}$ for $R \to \infty$ and $E(R) \to -\infty$ for $R \to 0$.

3) Finally, in the case of $\alpha_{1,2} < 0$ there are no stable atomic ions. But for an electron in the field of the two atoms that are separated by a distance $R < R_c = (\alpha_1\alpha_2)^{-1/2}$, a bound state appears. We have that $E(R_c) = 0$ and $E(R)$ decreases with decrease in R, and $E(R) \to -\infty$ for $R \to 0$. This gives the possibility of a stable state in a molecular ion. But this conclusion is valid only in the case of $R_c \gg a_B$. If $R_c \leq a_B$, then the study based on zero-range potentials is not valid. (For example, interaction of the particle and an impenetrable sphere of radius a for the values of energy $E \ll \hbar^2/ma^2$ could be approximated by a zero-range potential with $\alpha = -1/a$; while we know that there no bound states in the field of two impenetrable spheres.)

Problem 11.29

Find the ground term, $E_0(R)$, of a molecular hydrogen ion H_2^+ by variational method, approximating the wavefunction of the state by a "hydrogen" function of the form:

$$\psi_{\text{trial}}(r) = \sqrt{\frac{\alpha^3}{\pi R^3}} \exp\left\{-\frac{\alpha r}{R}\right\},$$

where r is the distance between the electron and the center of the segment that connects nuclei-protons, and α is a variational parameter.

Calculate the minimum energy of the term E_0, equilibrium distance between the nuclei R_0, and energy of nuclear zero-point oscillations $E_{osc,0}$, and compare them to experimental values: $E_0 \approx -0.60$ a.u., $R_0 \approx 2.0$ a.u., $E_{osc,0} \approx 0.0044$ a.u.

Is it possible, using this result, to make a conclusion about stability of the ion, H_2^+?

Solution

The mean value of the electron Hamiltonian for "fixed" ion nuclei

$$\hat{H}_{el} = -\frac{1}{2}\Delta - \frac{1}{|\mathbf{r} - \mathbf{R}/2|} - \frac{1}{|\mathbf{r} + \mathbf{R}/2|} + \frac{1}{R}$$

in the state described by wavefunction $\psi_{\text{trial}}(r)$ is equal to

$$E_0(R, \alpha) = \overline{\hat{H}_{el}} = \frac{\alpha^2}{2R^2} - [3 - 2(2+\alpha)e^{-\alpha}]\frac{1}{R}. \quad (1)$$

Since the wavefunction has the form of a "hydrogen" function, then for mean values $\overline{T}$ and $\overline{|\mathbf{r} \pm \mathbf{R}/2|^{-1}}$ we can use the known expressions; in particular:

$$\overline{\left|\mathbf{r} \pm \frac{\mathbf{R}}{2}\right|^{-1}} = [2 - (2+\alpha)e^{-\alpha}]\frac{1}{R},$$

as it follows from the Eq. (4) of Problem 4.6, if we put $r = R/2$, $a = R/\alpha$, $e = 1$, and subtract $2/R$.

In the framework of the variational method, expression (1) could be considered as some approximate value of the true energy[258] $E_0(R)$ of the ground term, and the best approximation is obtained by minimizing (1) over parameter α. Optimal value $\alpha(R)$ follows from the condition $\partial E_0(R, \alpha)/\partial \alpha = 0$ and we have

$$\frac{\alpha e^\alpha}{(\alpha + 1)} = 2R. \quad (2)$$

---

[258] More precisely, we should add to $\overline{\hat{H}_{el}}$ the small term $\hat{l}_{el}^2/2MR^2 \propto m_e/M_{\text{nuc}}$, which is obtained after averaging over the electron state of the centrifugal energy of the nuclei.

Here expressions (1) and (2) determine the dependence of $E_{0,var}$, that in the vicinity of the absolute minimum R_0 has the form

$$E_{0,var}(R) = E_0(R_0) + \frac{1}{2} E_0''(R_0)(R - R_0)^2. \tag{3}$$

A simple numerical calculation gives

$$E_{0,var}(R) = -0.470 + 0.078(R - 1.78)^2. \tag{4}$$

According to (4) the ground-state properties are

$$E_0 = -0.47, \quad R_0 = 1.78, \quad E_{osc,0} = \frac{1}{2}\omega_e = \sqrt{\frac{m_e E''(R_0)}{2m_p}} = 0.0065.$$

These results differ significantly from the experimental values, which is connected to the simple choice of the trial function.

Problem 11.30

Estimate the typical distance between the nuclei in a μ-mesomolecular ion of hydrogen,[259] and also the quantities ω_μ and B_μ for the ion in the adiabatic approximation, using results from the previous problem for the simple ion H_2^+.

Solution

Since $m_\mu/m_p \approx 1/9 \ll 1$, characteristic nuclei velocities are much lower than muon velocities, and we can use adiabatic approximation. The nuclei motion occurs in the effective potential $U(R) = E_0(R)$, defined by the energy of the ground muon term for "fixed" nuclei, and has the same form as in the case of the usual molecular ion H_2^+, if we use the muon atom units $m_\mu = e = \hbar = 1$.

In this approximation, ion characteristic ion size is determined by the distance R_0 that corresponds to the minimum of $E_0(R)$, and for the mesomolecular ion it is $R_0 \approx 2$ a.u.$= 2\hbar^2/m_\mu e^2 \approx 5 \cdot 10^{-11}$ cm. The values of ω_μ and B_μ (in the muon atomic units):

$$\omega_\mu \approx 0.27\sqrt{\frac{m_\mu}{M}}, \quad B_\mu = \frac{m_\mu}{2MR_0^2} = \frac{m_\mu}{8M}, \tag{1}$$

where M is the nuclei reduced mass. Compare to Problem 11.26.

We should mention the subtleties of the adiabatic approximation applied for mesomolecular systems, due to the value $\chi \sim 1/10$ of the adiabatic parameter, compared to $\chi \sim 10^{-4} - 10^{-3}$ for common molecules (see Problem 11.25). Although the approximation of "fixed" nuclei for a calculation of $E_0(R)$ is still valid (but with

---

[259] Due to a small size of the mesomolecule ions, Coulomb barrier penetrability, that separates nuclei, increases significantly. So, in the case when ion nuclei are heavy isotopes of hydrogen (d and t), the muon acts as a catalyst for the nuclear fusion reaction (for example, $dt \to n\alpha + 17.6$ MeV); see also Problems 11.59 and 11.74.

much lower accuracy), there is no further separation between the nuclei motion into independent oscillatory motion and rotational motion. The oscillations have a strongly anharmonic character and depend strongly on K, so that the use of adiabatic approximation equations for the spectrum,

$$E_{Kv} = E_0 + \omega_\mu \left(v + \frac{1}{2}\right) + B_\mu K(K+1), \qquad (2)$$

is not valid. But for states with $v = 0$, $K = 0$ and 1 it gives reasonable values of energy. For example, for the $dt\mu$-system, we have the following values of the binding energy[260]: $\varepsilon_{Kv} = -1/2 - E_{Kv}$, equal to $\varepsilon_{00} \approx 0.059$ a.u. ≈ 330 eV and $\varepsilon_{10} \approx 0.036$ a.u. ≈ 200 eV; compared to the exact values 319 eV and 232 eV. Let us note that, as is seen from exact calculations, for the mesolecular hydrogen ions there are two bound states ($v = 0$ and 1) with $K = 0$ and $K = 1$, one ($v = 0$) with $K = 2$, and there are no stable states with $K \geq 3$.

11.4 Atoms and molecules in external fields; Interaction of atomic systems

Problem 11.31

Calculate the polarizability of the hydrogen atom in the ground state using variational method, with the following trial functions:

a) $\psi(\mathbf{r}) = C\psi_0(r)(1 + \alpha\boldsymbol{\mathcal{E}}_0 \cdot \mathbf{r}) = C\pi^{-1/2}e^{-r}(1 + \alpha\mathcal{E}_0 r\cos\theta);$

b) $\psi(\mathbf{r}) = C\pi^{-1/2}[e^{-r} + \alpha\gamma^{5/2}\mathcal{E}_0 r e^{-\gamma r}\cos\theta],$

where α, γ are variational parameters, $\psi_0 = e^{-r}/\sqrt{\pi}$ is the wavefunction of the unperturbed ground-state, and $\mathcal{E}_0$ is the external electric field. Compare to the exact value $\beta_0 = 9/2$ (we use atomic units).

Solution

First we write the wavefunction in the form $\psi_{test} = C(\psi_0 + \alpha\mathcal{E}\psi_1)$, where

$$\psi_1 = \sqrt{\frac{3}{4\pi}}\cos\theta \cdot \frac{(2\gamma)^{5/2}}{2\sqrt{6}} r e^{-(2\gamma r)/2}.$$

---

[260] Here in the energy of the mesoatomic ground-state, equal to $-1/2$, we did not make a correction due to a finite mass (heavier) nucleus, which leads to an increase in the binding energy. Its value is of the same order, $\sim m_\mu/M_{nuc}$, as a contribution to the energy from the term $\vec{l}_\mu^2/2MR^2$ that appears after averaging over the muon state of the centrifugal nuclei energy. This term in the mesomolecule is more important than in common molecules, and leads to an increase in term's energy (and decrease in the binding energy). Hence, the unaccounted terms have opposite signs and partly compensate each other.

Note that the muon atom energy unit is ≈ 5.63 keV.

Let us note that the functions ψ_1, normalized to unity, coincides with the wavefunction of the state with quantum numbers $n = 2$, $l = 1$, $m = 0$ of a hydrogen-like atom with the nuclear charge $Z = 2\gamma$ ($\gamma = 1$ in case a), and satisfies the equation:

$$\left(-\frac{1}{2}\Delta - \frac{2\gamma}{r}\right)\psi_1 = -\frac{Z^2}{2n^2}\psi_1 = -\frac{\gamma^2}{2}\psi_1.$$

Writing the system Hamiltonian in the form

$$\hat{H} = -\frac{1}{2}\Delta - \frac{1}{r} + \mathcal{E}z \equiv \hat{H}_0 + \mathcal{E}r\cos\theta, \tag{1}$$

with axis z along the electric field direction, we find

$$\langle\psi_1|\hat{H}_0|\psi_1\rangle \equiv \langle\psi_1| -\frac{1}{2}\Delta - \frac{2\gamma}{r} + \frac{2\gamma-1}{r}|\psi_1\rangle = \frac{\gamma^2-1}{2}. \tag{2}$$

The value $\langle\psi_1|\frac{1}{r}|\psi_1\rangle \equiv \frac{1}{2\gamma}\langle\psi_1|\frac{2\gamma}{r}|\psi_1\rangle = \frac{\gamma}{2}$ follows from the virial theorem. Then it is evident that

$$\langle\psi_0|\hat{H}_0|\psi_0\rangle = -\frac{1}{2}, \quad \langle\psi_0|z|\psi_0\rangle = \langle\psi_1|z|\psi_1\rangle = 0,$$

$$\langle\psi_1|\hat{H}_0|\psi_0\rangle = \langle\psi_0|\hat{H}_0|\psi_1\rangle = -\frac{1}{2}\langle\psi_1|\psi_0\rangle = 0, \tag{3}$$

$$C^2 = \frac{1}{1+\alpha^2 E^2} \approx 1 - \alpha^2 \mathcal{E}^2.$$

Note that wavefunctions ψ_0 and ψ_1 are mutually orthogonal, as they correspond to different values of the angular momentum. Finally,

$$\langle\psi_1|z|\psi_0\rangle = \langle\psi_0|z|\psi_1\rangle = \frac{\gamma^{5/2}}{\pi}\int e^{-(1+\gamma)r}r^4\cos^2\theta \, dr d\Omega = \frac{32\gamma^{5/2}}{(1+\gamma)^5}. \tag{4}$$

Expressions (1)–(4) allow us to find $\overline{E}(\alpha,\gamma) = \langle\psi|\hat{H}|\psi\rangle$ with accuracy up to the terms $\sim \mathcal{E}^2$. In case a) we have

$$\overline{E}(\alpha) = -\frac{1}{2} + \frac{1}{2}\alpha^2\mathcal{E}^2 + 2\alpha\mathcal{E}^2.$$

Minimization over the parameter α gives

$$E_{0,var} = \min \overline{E}(\alpha) = \overline{E}(\alpha_0) = -\frac{1}{2} - 2\mathcal{E}^2, \quad \alpha_0 = -2. \tag{5}$$

Comparing this approximate value of the hydrogen atom ground-state energy in a weak electric field to the exact result, $-1/2 - \beta_0 \mathcal{E}^2/2$, where β_0 is the ground-state polarizability, we find its approximate (variational) value:

$$\beta_{0,var} = 4. \tag{6}$$

In case b), we have

$$\overline{E}(\alpha, \gamma) = -\frac{1}{2} + \frac{1}{2}(1 - \gamma + \gamma^2)\alpha^2 \mathcal{E}^2 + \frac{64\gamma^{5/2}}{(1+\gamma)^5}\alpha\mathcal{E}^2.$$

Minimization over the parameter α gives

$$\overline{E}(\gamma) = -\frac{1}{2} - \frac{2^{11}\gamma^5}{(1+\gamma)^{10}(1-\gamma+\gamma^2)}\mathcal{E}^2,$$

and the following minimization over parameter[261] γ allows us to find polarizatility more accurately than in case a):

$$\beta_{0,var} = \frac{2^{12}\gamma_0^5}{(1+\gamma_0)^{10}(1-\gamma_0+\gamma_0^2)} = 4.475, \quad \gamma_0 = 0.797. \tag{7}$$

This differs from exact $\beta_0 = 9/2$ by only 0.6 %.

In conclusion, we should note that if the trial function is chosen so that in the absence of an electric field it coincides with the exact wavefunction of the unperturbed Hamiltonian, then variational calculation of ground-state polarizability gives restriction from below on its exact value (so even without knowing the exact value of polarizability, we can conclude that result (7) is more accurate than (6)).

Problem 11.32

Using the known value $\beta_0 = 9/2$ a.u. of the hydrogen atom ground-state polarizability, obtain an approximate value of the ground 1^1S-state polarizability for a two-electron atom or ion:

a) neglecting electron-electron interaction:
b) taking it into account as mutual partial screening of the nuclear charge, choosing the effective charge to be equal to $Z_{eff} = Z - 5/16$ (see Problem 11.6).

Compare these results to the experimental data in this solution.

Solution

The Hamiltonian and ground-state energy of the hydrogen atom in an homogeneous electric field have the form:

$$\hat{H} = \frac{\hat{p}^2}{2m} - \frac{e^2}{r} + e\mathcal{E}z, \quad E_0 \approx E_0^{(0)} + E_0^{(2)}, \quad E_0^{(2)} = -\frac{1}{2}\beta_0\mathcal{E}^2,$$

---

[261] Optimal value of γ_0 is obtained from the condition $\partial \overline{E}/\partial \gamma = 0$, which is an algebraic equation of third degree with respect to γ. From it we have $\gamma_0 \approx 0.797$ (the two other equation roots are complex).

and $E_0^{(0)} = -\frac{1}{2}(me^4/\hbar^2)$, while $\beta_0 = \frac{9}{2}(\hbar^2/me^2)^3$. For a hydrogen-like atom with the nuclear charge Ze, the Hamiltonian is

$$\hat{H} = \frac{\hat{p}^2}{2m} - \frac{Ze^2}{r} + e\mathcal{E}z. \tag{1}$$

Energy E_0 could be obtained from the relations given above using substitutions $e \to \sqrt{Z}e$, $\mathcal{E} \to \mathcal{E}/\sqrt{Z}$, so that

$$E_0^{(2)} = -\frac{9}{4} \cdot \frac{a_B^3}{Z^4}\mathcal{E}^2,$$

and the polarizability of such an atom is $\beta_0 = 9a_B^3/2Z^4$.

The Hamiltonian of a helium-like atom in an electric field, while neglecting electron-electron interaction, is equal to $\hat{H} = \hat{H}_1 + \hat{H}_2$, where $\hat{H}_{1,2}$ have the form (1). The energy and polarizaility of such a system (in its ground-state) is obtained by multiplying the corresponding quantities for a hydrogen-like atom with the same nuclear charge Z by 2. In this approximation,

$$\beta_{0,2e}^{(0)} = 9\frac{a_B^3}{Z^4} = \frac{9}{Z^4} \text{ a. u.} \tag{2}$$

It seems natural that a more accurate value of two-electron atom (ion) polarizability could be obtained as a result of substitution in (2) of the charge Z with its effective charge:

$$\beta_{0,2e} = \frac{9}{(Z - 5/16)^4}. \tag{3}$$

A comparison of the theoretically-obtained values of polarizability to experimental data for the 1^1S-state of a helium atom and some two-electron ions is given in the following table:

Ion	He	Li$^+$	Be^{2+}	B^{3+}	C^{4+}
According to (2)	0.56	0.111	0.035	$1.4 \cdot 10^{-2}$	$6.9 \cdot 10^{-3}$
According to (3)	1.11	0.173	0.049	$1.9 \cdot 10^{-2}$	$8.6 \cdot 10^{-3}$
Exp. val.	1.36	0.196	0.054	$2 \cdot 10^{-2}$	$8.8 \cdot 10^{-3}$

Problem 11.33

Consider the *Stark effect* for excited states of the hydrogen atom with the principal quantum number, $n = 2$, in the first order of perturbation theory. To solve the problem, use the eigenfunctions of the unperturbed Hamiltonian ψ_{nlm} in spherical coordinates.

Obtain the correct functions of zeroth approximation and discuss applicability conditions for the results obtained.

Solution

The unperturbed level of the hydrogen atoms with $n = 2$ is four-fold degenerate (without taking spin of electron into account). For the calculation of its splitting in an homogeneous electric field, we use the secular equation. The corresponding eigenfunctions ψ_{nlm} of the unperturbed Hamiltonian we renormalize in the following way:

$$\psi_1^{(0)} = \psi_{200}, \quad \psi_2^{(0)} = \psi_{210}, \quad \psi_3^{(0)} = \psi_{211}, \quad \psi_4^{(0)} = \psi_{21,-1}.$$

Here, $\psi_{2lm} = R_{2l}(r)Y_{lm}$, where

$$R_{20} = \frac{1}{\sqrt{2a_B^3}}\left(1 - \frac{r}{2a_B}\right)e^{-r/2a_B}, \quad R_{21} = \frac{r}{\sqrt{24a_B^5}}e^{-r/2a_B},$$

while spherical functions Y_{00} and Y_{1m} are given in Eq. (III.7). It is easy to see that only the following matrix elements of perturbation $V = ez\mathcal{E} = er\cos\theta \cdot \mathcal{E}$ are different from zero:

$$V_{12} = -V_{21} = i\frac{e\mathcal{E}}{16\pi a_B^4}\iint e^{-r/a_B}\left(1 - \frac{r}{2a_B}\right)r^4 \cos^2\theta dr\, d\Omega = -3iea_B\mathcal{E}.$$

The secular equation, $|V_{ik} - E_2^{(1)}\delta_{ik}| = 0$, and its solution take the form:

$$\begin{vmatrix} -E_2^{(1)} & -3iea_B\mathcal{E} & 0 & 0 \\ 3iea_B\mathcal{E} & -E_2^{(1)} & 0 & 0 \\ 0 & 0 & -E_2^{(1)} & 0 \\ 0 & 0 & 0 & -E_2^{(1)} \end{vmatrix} = (E_2^{(1)})^2\left((E_2^{(1)})^2 - 9e^2 a_B^2 \mathcal{E}^2\right) = 0,$$

$$E_{2,1}^{(1)} = 3ea_B\mathcal{E}, \quad E_{2,2}^{(1)} = -3ea_B\mathcal{E}, \quad E_{2,3}^{(1)} = E_{2,4}^{(1)} = 0. \tag{1}$$

Spittings occur on three sublevels, two of which are non-degenerate, while one is two-fold degenerate.

The true functions of zeroth approximation $\psi_\pm^{(0)}$ that correspond to split levels $E_{2,1(2)}^{(1)} = \pm 3ea_B\mathcal{E}$ have the form:

$$\psi_\pm^{(0)} = \frac{1}{\sqrt{2}}(\psi_1^{(0)} \pm i\psi_2^{(0)}) = \frac{1}{8\sqrt{\pi}a_B^{5/2}}(2a_B - r \mp r\cos\theta)e^{-r/2a_B}.$$

In these states, which have no definite parity, electrons have a mean dipole moment different from zero. It is directed along the electric field and is equal to $\mp 3ea_B$. Here we recover accidental degeneracy specific to the Coulomb potential, which gives the linear Stark effect in the hydrogen atom (for excited levels). States $\psi_{3,4}^{(0)}$ with $l = 1$,

$l_z = \pm 1$ correspond to a definite parity and their energy change in the electric field $\propto \mathcal{E}^2$.

The applicability condition for the results obtained (1) is

$$5 \cdot 10^{-5} \text{ eV} \ll 6ea_B \ll 3 \text{ eV}, \quad \text{or} \quad 2 \cdot 10^3 \text{ V/cm} \ll \mathcal{E} \ll 10^8 \text{ V/cm}.$$

That is, the Stark splitting must be much larger than the fine structure splitting (see Problem 11.1) but much smaller than the level spacing between the neighboring unperturbed atomic levels.

Problem 11.34

Calculate the energy shift in an homogeneous electric field and the polarizability of a particle bound state in a zero-range potential by the variational method, using the trial function of the form:[262]

$$\psi_{\text{trial}} = C[\psi_0(r) + \lambda(\boldsymbol{\mathcal{E}} \cdot \mathbf{r})e^{-\gamma r}], \quad \psi_0 = \sqrt{\frac{\kappa_0}{2\pi}} \frac{e^{-\kappa_0 r}}{r},$$

where λ, γ are variational parameters, and ψ_0 is the wavefunction of the unperturbed state. Compare to the exact value (see the following problem).

Solution

Writing the trail function in the form $\psi_{\text{trial}} = C(\psi_0 + \psi_1)$, where $\psi_1 = \lambda(\mathbf{E} \cdot \mathbf{r})e^{-\gamma r}$, we find matrix elements for the unperturbed Hamiltonian:

$$\langle \psi_0 | \hat{H}_0 | \psi_0 \rangle = -\frac{\hbar^2 \kappa_0^2}{2m}, \quad \langle \psi_0 | \hat{H}_0 | \psi_1 \rangle = \langle \psi_1 | \hat{H}_0 | \psi_0 \rangle = 0,$$

$$\langle \psi_1 | \hat{H}_0 | \psi_1 \rangle = \langle \psi_1 | \frac{\hat{p}^2}{2m} | \psi_1 \rangle = \frac{\hbar^2}{2m} \lambda^2 \int (\nabla(\boldsymbol{\mathcal{E}} \cdot \mathbf{r})e^{-\gamma r})^2 dV = \frac{\pi \lambda^2 \hbar^2 \mathcal{E}^2}{2m\gamma^3},$$

and the perturbation $\hat{V} = -e(\boldsymbol{\mathcal{E}} \cdot \mathbf{r})$:

$$\langle \psi_0 | \hat{V} | \psi_0 \rangle = \langle \psi_1 | \hat{V} | \psi_1 \rangle = 0,$$

$$\langle \psi_1 | \hat{V} | \psi_0 \rangle = \langle \psi_0 | \hat{V} | \psi_1 \rangle = -\lambda e \sqrt{\frac{\kappa_0}{2\pi}} \int \frac{1}{r} (\boldsymbol{\mathcal{E}} \cdot \mathbf{r})^2 e^{-(\gamma+\kappa_0)r} dV$$

$$= -4\sqrt{2\pi\kappa_0} \frac{\lambda e \mathcal{E}^2}{(\gamma + \kappa_0)^4}.$$

Now choosing

$$C^2 = \frac{1}{1 + \pi\lambda^2 \gamma^{-5} \mathcal{E}^2} \approx 1 - \pi\lambda^2 \gamma^{-5} \mathcal{E}^2$$

---

[262] Pay attention to the fact that $\psi_{\text{trial}}(r)$, as well as $\psi_0(r)$, for $r \to 0$ satisfies the boundary condition consistent with the zero-range potential (see Problem 4.10).

to satisfy the normalization condition for the trial function, we have

$$\overline{E}(\lambda,\gamma) = \langle\psi_{\text{trial}}|\hat{H}_0 + \hat{V}|\psi_{\text{trial}}\rangle \approx -\frac{\hbar^2\kappa_0^2}{2m} - \frac{8\sqrt{2\pi\kappa_0}\lambda e}{(\gamma+\kappa_0)^4}\mathcal{E}^2$$
$$+ \frac{\pi\lambda^2\hbar^2(\gamma^2+\kappa_0^2)}{2m\gamma^5}\mathcal{E}^2.$$

After minimization over parameter λ, we obtain

$$\overline{E}(\gamma) = \min_\lambda \overline{E}(\lambda,\gamma) = -\frac{\hbar^2\kappa_0^2}{2m} - \frac{64me^2\kappa_0\gamma^5}{\hbar^2(\gamma^2+\kappa_0^2)(\gamma+\kappa_0)^8}\mathcal{E}^2. \tag{1}$$

The subsequent minimization over parameter γ (which implies $\gamma = \kappa_0$) gives a variational energy shift under the action of the electric field

$$E_{0,\text{var}} = \min \overline{E}(\lambda,\gamma) = -\frac{\hbar^2\kappa_0^2}{2m} - \frac{me^2}{8\hbar^2\kappa_0^4}\mathcal{E}^2 \tag{2}$$

and the polarizability

$$\beta_{0,\text{var}} = \frac{me^2}{4\hbar^2\kappa_0^4} \propto \frac{1}{\kappa_0^4},$$

which coincides with the exact result. See the following problem.

Problem 11.35

Find the exact polarizability for a particle bound by a zero-range potential (see Problem 4.10). Apply the result obtained to the H^- ion (compare to Problem 11.36).

Solution

To calculate a ground level shift calculation due to the perturbation of the form $V = -e\mathcal{E}z$, it is convenient to choose wavefunctions with definite values of the angular momentum l and its projection l_z, as unperturbed eigenfunctions. Since the wavefunction of the ground-state

$$\psi_0^{(0)} = \frac{\sqrt{\kappa}e^{-\kappa r}}{\sqrt{2\pi r}} \quad \left(\text{its energy is } E_0^{(0)} = -\frac{\hbar^2\kappa^2}{2m}\right)$$

corresponds to the value $l = 0$, perturbation matrix elements $\langle n|(-e\mathcal{E}z)|0\rangle$ are different from zero only for states $|n\rangle$ with $l = 1$ and $l_z = 0$. The zero-range potential will not affect the particle with $l \neq 0$, hence eigenfunctions of $\hat{H}_0$ for $l \neq 0$ coincide with the wavefunctions of a free particle, and for $l = 1$, $l_z = 0$ they have the form:

$$\psi_{k10}^{(0)} = \sqrt{\frac{k}{r}} J_{3/2}(kr) Y_{10}(\mathbf{n}) = -i\sqrt{\frac{3}{2}} \frac{\cos\theta}{\pi r} \left(\frac{\sin kr}{kr} - \cos kr \right), \qquad (1)$$

normalized so that $\langle k'l'm'|klm\rangle = \delta(k-k')\delta_{ll'}\delta_{mm'}$.

We calculate the perturbation matrix element:

$$\langle k10|(-e\mathcal{E}z)|0\rangle = -i\sqrt{\frac{4\kappa}{3\pi}} \frac{e\mathcal{E}}{k} \int_0^\infty e^{-\kappa r}(\sin kr - kr\cos kr)dr$$

$$= -4i\sqrt{\frac{\kappa}{3\pi}} \frac{ek^2\mathcal{E}}{(k^2+\kappa^2)^2}. \qquad (2)$$

According to Eq. (VIII.1) we obtain (now we should replace $\sum'_m \to \sum_{l,m} \int_0^\infty dk$, and notice that in the sum over l and m, only one term with $l=1$, $m=0$ differs from zero):

$$E_0^{(2)} = -\frac{32me^2\kappa\mathcal{E}^2}{3\pi\hbar^2} \int_0^\infty \frac{k^4 dk}{(k^2+\kappa^2)^5} = -\frac{me^2}{8\hbar^2\kappa^4}\mathcal{E}^2. \qquad (3)$$

The polarizability is

$$\beta_0 = \frac{me^2}{4\hbar^2\kappa^4}. \qquad (4)$$

For an application of this equation to the ion, H^-, see the following problem.

Problem 11.36

Find the polarizability of a weakly-bound state of a charged particle with the angular momentum $l=0$ in a central potential $U_S(r)$ of the radius r_S, so that $\kappa r_S \ll 1$, where $\kappa = \sqrt{-2E_0^{(0)}/\hbar^2}$, $E_0^{(0)}$ is the energy of the unperturbed state. Apply the result obtained to ion H^-.

One can use the following parameters for the ion H^- (a two-electron system): $\kappa = 0.235$ a.u. (binding energy $\varepsilon = 0.754$ eV), asymptotic coefficient square (see Eq. (XI.5)) $C_{\kappa 0}^2 = 2.65$, polarizability $\beta_0 = 206$ a.u.

Solution

Due to the fact that the wavefunction of a state with a small binding energy decreases quite slowly with increase in r, while the perturbation $V = -eEz$ increases, the dominant role in the sum (VIII.1) of second-order perturbation theory, which gives the level shift, is played by the states of the continuous spectrum with a small energy $E \leq \hbar^2\kappa^2/m$. Indeed, for such states, perturbation matrix elements $\langle k|V|0\rangle$, where the dominant role is played by large distances $r \sim \kappa^{-1}$, are especially large. On the other hand, at large distances, wavefunctions of the states considered are straightforwardly related to the wavefunctions of free particles. For example, the wavefunction of an unperturbed bound state with $l = 0$ outside the region of potential action differs from the wavefunction in the zero-range potential only by an additional factor $C_{\kappa 0}$ (see the previous problem). Wavefunctions of the continuous spectrum for slow particles with $l \neq 0$ on distances $r > r_S$ actually coincide (in the absence of a perturbation) with the wavefunction of a free particle. (Due to the centrifugal barrier, which is poorly penetrable for slow particles, they do not "feel" the potential center.) They are described by Eq. (1) from the previous problem.

Therefore, the shift and polarizaility of a shallow s-level in the short-range potential are determined from the equations of the previous problem for a zero-range potential by adding the factor $C_{\kappa 0}^2$. For example, polarizability is described by the expression:

$$\beta_0 = \frac{me^2}{4\hbar^2 \kappa^4} C_{\kappa 0}^2, \quad \kappa r_S \ll 1. \tag{1}$$

Let us emphasize that the dominant role of the large distances for the matrix elements $\langle k|eEz|0\rangle$ for low-energy states (seen from Eqs. (2) and (3) of the previous problem) manifests itself as a divergence in the expressions for $E_0^{(2)}$ and β_0 for $\kappa \to 0$.

Let us make a few additional comments:

1) For the s-state of a particle with a small binding energy in a short-range potential, the parameters κ and $C_{\kappa 0}^2$ determine most physical properties in response to external electric and magnetic fields (see Problems 11.46 and 11.66). They also determine scattering of slow particles, $kr_S \ll 1$, off of this potential. Using effective-range expansion (see, Chapter 13, sec. 3),

$$k \cot \delta_0 = -\frac{1}{a_0} + r_0 \frac{k^2}{2} + \ldots,$$

parameters of the low-energy s-scattering, the scattering length a_0, and the effective interaction radius r_0, are connected to κ and $C_{\kappa 0}^2$ by the relations

$$a_0 = \frac{2C_{\kappa 0}^2}{\kappa(1 + C_{\kappa 0}^2)}, \quad r_0 = \frac{C_{\kappa 0}^2 - 1}{\kappa C_{\kappa 0}^2}. \tag{2}$$

For a zero-radius potential, $C_{\kappa 0}^2 = 1$ and $r_0 = 0$.

2) In negative atomic ions, an external electron with can be considered as experiencing a short-range potential of a neutral atom. Using such a method (without any specification of "internal" electrons states) we can describe properties of the ion, determined by the external electron. For an ion H$^-$, using values $\kappa = 0.235$ a.u. and $C_{\kappa 0}^2 = 2.65$ given in the problem condition, according to Eq. (1) we find $\beta_0 = 216$; compare to the result of the variational calculation, $\beta_0 = 206$. See also Problem 13.40 for the scattering of slow electrons on hydrogen atoms.

3) Note that the dominant role of large distances for the polarizability with a small binding energy is preserved for both $l = 1$ (here $\beta_{l=1} \propto \kappa^{-3}$, see the following problem) and $l = 2$ ($\beta_{l=2} \propto \kappa^{-1}$). For larger values of the angular momentum, the polarizability is determined by $r \leq r_S$, and depends on a specific form of the potential and wavefunction on these distances. It can be estimated as

$$\beta_{l \geq 3} \sim \frac{me^2 r_S^4}{\hbar^2}.$$

Such a dependence of the polarizability on the angular momentum is connected to a decrease of the centrifugal barrier with an increase of l that leads to stronger localization of the bound state with increase of l.

Problem 11.37

The same as in the previous problem, but for a weakly bound state of a particle with the angular momentum $l = 1$.

Solution

Calculation of the polarizability is analogous to the in the two previous problems. Now the wavefunction of the unperturbed bound state outside the potential range, for $r > r_S$, has the form (for $l = 1$):

$$\psi_{\kappa l m}^{(0)} = 2\kappa C_{\kappa l} \frac{1}{\sqrt{\pi r}} K_{l+1/2}(\kappa r) Y_{lm}(\mathbf{n}), \tag{1}$$

where K_ν is the MacDonald function, and $C_{\kappa l}$ is the *asymptotic coefficient*.[263] As in Problem 11.36, we can choose wavefunctions of free particles to describe states of the continuos spectrum:[264]

---

[263] The wavefunction asymptote is

$$\psi_{\kappa l m}^{(0)} \approx \sqrt{2\kappa} C_{\kappa l} Y_{lm} \frac{e^{-\kappa r}}{r}$$

for $r \to \infty$. As in the case of $l = 0$, the asymptotic coefficient is connected to the parameters of low-energy scattering with the angular momentum l, that define the effective-range expansion

$$k^{2l+1} \cot \delta_l \approx -\frac{1}{a_l} + r_l \frac{k^2}{2},$$

by the relation

$$\frac{1}{C_{\kappa l}^2} = -r_l \kappa^{1-2l} + (-1)^l (2l+1) + O((\kappa r_S)^{-3-2l}).$$

Here $r_l < 0$ and $C_{\kappa l}^2 \propto \kappa^{2l-1}$ for $\kappa \to 0$ for $l \geq 1$.

[264] For slow particles, $k r_S \ll 1$, the wavefunction of the continuous spectrum outside of the potential range differs essentially from the wavefunction of a free particle only in a resonant wave; see Chapter 13, sec. 3.

$$\psi_{klm}^{(0)} = \sqrt{\frac{k}{r}} J_{l+1/2}(kr) Y_{lm}(\mathbf{n}). \qquad (2)$$

The wavefunctions given approach the exact ones only at large distances $r > r_S$ (outside of the potential range). But perturbation matrix elements $\langle klm|(-e\mathcal{E}z)|\kappa lm\rangle$ for $k \leq \kappa$ are defined by large distances $r \sim 1/\kappa$. While calculating the matrix element $\langle klm|z|\kappa lm\rangle$, integration over the angles is easily performed.

For a particle with $l = 1$ and the angular momentum projection onto the electric field direction $l_z = \pm 1$, only one perturbation matrix element differs from zero, for which

$$|\langle k, 2, \pm 1|z|\kappa, 1, \pm 1\rangle|^2 = \frac{16 C_{\kappa 1}^2 k^6}{5\pi\kappa(k^2+\kappa^2)^4}.$$

The level shift in second-order perturbation theory becomes equal to

$$E_{1,\pm 1}^{(2)} = -\frac{2m}{\hbar^2} \int_0^\infty \frac{|\langle k, 2, \pm 1|e\mathcal{E}z|\kappa, 1, \pm 1\rangle|^2}{k^2 + \kappa^2} dk = -\frac{me^2 C_{\kappa l}^2}{8\hbar^2 \kappa^4}\mathcal{E}^2,$$

so the polarizabilities are

$$\beta_{1,\pm 1} = \frac{me^2 C_{\kappa l}^2}{4\hbar^2 \kappa^4} \propto \frac{1}{\kappa^3}. \qquad (3)$$

Remember that $C_{\kappa l}^2 \propto \kappa$. See the comment in Problem 11.36.

For the state with $l = 1$ and $l_z = 0$, there are two non-vanishing matrix elements of the perturbation potential corresponding to the angular momenta $l = 2$ and $l = 0$,

$$|\langle k00|z|\kappa 10\rangle|^2 = \frac{4 C_{\kappa l}^2 k^2 (k^2 + 3\kappa^2)^2}{3\pi\kappa(k^2+\kappa^2)^4},$$

$$|\langle k20|z|\kappa 10\rangle|^2 = \frac{64 C_{\kappa l}^2 k^6}{15\pi\kappa(k^2+\kappa^2)^4}.$$

Calculation of the integrals, analogous to $E_{1,\pm 1}^{(2)}$, allows us to find a correction to the energy and the polarizability, which is as follows

$$\beta_{1,0} = 7\beta_{1,\pm 1}. \qquad (4)$$

In the state with $l_z = 0$, the particle localization domain is somewhat extended along the electric field, than in the states with $l_z = \pm 1$. So, the field has a larger effect on such a state.

Problem 11.38

Obtain approximate expressions for the polarizabilities of the excited 2^3S- and 2^1S-states of a two-electron atom (ion). Compare to the experimental values for the helium atom and lithium ion Li^+ (a.u.): $\beta_{He}(2^3S) = 316$, $\beta_{He}(2^1S) = 803$, $\beta_{Li^+}(2^3S) = 47$, $\beta_{Li^+}(2^1S) = 99$.

Take into account the close proximity of the $2S$- and $2P$-levels and the following experimental values: $E_{He}(2^3P) - E_{He}(2^3S) = 1.14$ eV and $E_{He}(2^1P) - E_{He}(2^1S) = 0.602$ eV, as well as the corresponding values 2.26 eV and 1.29 eV for the ion Li^+. Treat an excited electron, as moving in the field of nucleus charge screened to unity by the $1s$-electron, and neglect exchange effects.

Solution

An important property of the states that gives rise to the large values of their polarizabilities, is the closeness of helium-like atoms' (ions) $2S$- and $2P$-levels. So in the sum (VIII.1) for a level shift under the influence of the perturbation $V = (z_1 + z_2)E$, the dominant role is played by the term that contains the small energy difference in the denominator, so

$$E_{2S}^{(2)} = -\frac{1}{2}\beta(2S)E^2 \approx |\langle 2P, L_z = 0|(z_1 + z_2)|2S\rangle|^2 \frac{E^2}{E_{2S}^{(0)} - E_{2P}^{(0)}}. \quad (1)$$

The multiplicity symbol is omitted.

In the states $2S$ and $2P$, one of the electrons is in the ground $1s$-state, while the other is in the excited $2s$- or $2p$-state. Since the "excited" electron is on the average much further from the nucleus, then to calculate the perturbation matrix element in Eq. (1) we can use the following approximate expressions for the wavefunctions (the same for both singlet and triplet states):

$$\psi_{2S} \approx \psi_{1s}(r_1, Z)\psi_{2s}(r_2, Z-1), \quad \psi_{2P} \approx \psi_{1s}(r_1, Z)\psi_{2p}(\mathbf{r}_2, Z-1). \quad (2)$$

These wavefunctions have no definite symmetry, due to the neglecting of exchange effects. The one-particle wavefunctions in expressions (2) correspond to Coulomb wavefunctions, and for the "excited" electron, the nucleus charge is chosen to be equal to $Z - 1$, which gives its partial screening by the $1s$-electron (compare to Problem 11.9). Using known expressions for the Coulomb wavefunction, we find the perturbation matrix element:

$$|\langle 2P, L_z = 0|z_1 + z_2|2S\rangle| \approx \frac{3}{Z-1}.$$

Only the second, "excited" electron contributes. According to Eq. (1) we obtain:

$$\beta(2S) = \frac{18}{(Z-1)^2(E_{2P}^{(0)} - E_{2S}^{(0)})}. \quad (3)$$

Using the experimental values for the energy difference of 2S- and 2P-states, given in problem condition, we find the polarizabilities:

$$\beta_{\text{He}}(2^3S) = 428, \quad \beta_{\text{He}}(2^1S) = 813;$$

$$\beta_{Li^+}(2^3S) = 54, \quad \beta_{Li^+}(2^1S) = 53.$$

In conclusion, let us note that analogous closeness of energy levels is typical for many atomic systems, and explains the large numerical values of their polarizabilities. For example, for the ground 2^2S-state of the lithium atom, $E_{2S} = 5.39$ eV, while for the excited 2^2P-state, $E_{2P} = 3.54$eV. Estimation of polarizability, according to Eq. (3), with replacing $Z - 1$ by $Z - 2$ (due to the presence of two 1s-electrons), gives $\beta_{1s} \approx 265$ a.u.(The experimental value is 162; the main reason for the over-estimated value of β according to Eq. (3) is due to the approximation assuming that the "excited" electron is moving in the field of a nuclear charge screened by two 1s-electrons; for a lithium atom, screening is not so strong, as seen from the Rydberg correction $\Delta_s = -0.40$.)

Problem 11.39

Provide an order-of-magnitude estimate of the polarizability of an atom and its dependence on the nuclear charge Z in the Thomas–Fermi model. Neglecting the interaction between electrons (see Problem 11.19), calculate explicitly the numerical coefficient in the leading Z-dependence. Compare your results to the contribution of valence electrons to polarizability.

Solution

We aim to estimate the polarizability of Thomas–Fermi electrons (T-F) β_{T-F} that determines their dipole moment $\mathbf{d} = \beta_{T-F}\boldsymbol{\mathcal{E}}$ induced by an electric field. We note that although linear relation is valid only for a weak field, nevertheless it gives the right order of magnitude of d also in the case of strong fields. Call $\mathcal{E}_{T-F} \sim Ze/r_{T-F}^2$ the characteristic value of nuclear electric field in the region $r \sim r_{T-F} \sim a_B/Z^{1/3}$, where the electrons are predominately located. For $\mathcal{E} \sim \mathcal{E}_{T-F}$, the electron shift under the action of the external field would be of the order of r_{T-F}, so that here $d \sim Zer_{T-F}$ and

$$\beta_{T-F} = \frac{d}{E} \sim \frac{Zer_{T-F}}{Ze/r_{T-F}^2} = r_{T-F}^3 \sim \frac{a_B^3}{Z} = \frac{1}{Z} \text{ a. u.} \tag{1}$$

To find the numerical coefficient in this scaling relation $\beta_{T-F} = \gamma/Z$, consider the statistical model of an atom neglecting interaction between electrons (see Problem 11.19). Now, with the weak electric field, we have

$$p_0(\mathbf{r}) = \left[2\left(\frac{Z}{r} - \boldsymbol{\mathcal{E}}\cdot\mathbf{r} - \frac{Z}{R}\right)\right]^{1/2} \approx \left[2\left(\frac{Z}{r} - \frac{Z}{R}\right)\right]\left(1 - \frac{3r R\mathbf{r}\cdot\boldsymbol{\mathcal{E}}}{2Z(R-r)}\right).$$

Thus
$$n(\mathbf{r}) = p_0^3(r)/3\pi^2 \approx n_0(r)\left(1 - \frac{3r R \mathbf{r} \cdot \boldsymbol{\mathcal{E}}}{2Z(R-r)}\right),$$

where $n_0(r) = \frac{2Z^2(1-z)^{3/2}}{9\pi^2 x^{1/2}}$ is the density in an unperturbed atom, $x = r/R$, and the value $R = (18/Z)^{1/3}$ remains the same as before. The dipole moment is[265]

$$\mathbf{d} = -\int \mathbf{r} n(\mathbf{r}) dV \approx \frac{3R}{2Z}\int \mathbf{r}(\mathbf{r}\cdot\boldsymbol{\mathcal{E}})\frac{rn_0(r)}{R-r}dV =$$
$$= \left\{\frac{R}{2Z}\int \frac{r^3 n_0(r)}{R-r}dV\right\}\boldsymbol{\mathcal{E}} = \frac{63}{16Z}\boldsymbol{\mathcal{E}}$$

gives the polarizability in the model considered:

$$\beta_{T-F} = \frac{63}{16Z}. \tag{2}$$

Similarly to Eq. (1), an estimate of the contribution of the external (valence) electrons that are on the periphery of the atom ($r \sim a_B$), where $\mathcal{E}_{at} \sim e/a_B^2 \ll \mathcal{E}_{T-F}$, gives $\beta_{val} \sim a_B^3 = 1$ a.u. Therefore, we conclude that the atomic polarizability is determined mostly by the valence electrons. Note also that the atomic polarizabilities are usually large, $\sim 10 - 100$ a.u. See also Problem 11.38.

Problem 11.40

Find the Stark splitting of rotational components for a two-atom molecule that has a constant dipole moment (in a coordinate system rigidly connected to the mocululular symmetry axis). The Stark splitting is assumed to be small with respect to the level spacing between neighboring rotational levels for the molecule electron term $^1\Sigma$. Compare to the result of Problem 8.11 for a spherical rotor.

Solution

The stationary states of the molecule in the absence of a field are described by wavefunctions (compare to Problem 11.27, $\Lambda = S = 0$):

$$\psi_{nvKM}^{(0)} = \psi_{el,n}(\mathbf{R}, \xi_1, \xi_2, \ldots)\psi_{osc,v}(R)\psi_{rot,KM}(\theta, \varphi), \tag{1}$$

(ξ_a label both the electron coordinates and spins), while their energy is

$$E_{nvK}^{(0)} = E_n^{(0)} + \hbar\omega_{osc}\left(v + \frac{1}{2}\right) + B_e K(K+1).$$

It is convenient to calculate matrix elements of the perturbation $\hat{V} = -\hat{\mathbf{d}}\cdot\mathbf{E}$ in two steps: first, integrate over the electron coordinates and the relative distance $\mathbf{R}$ between

---

[265] Compare, for example, to the calculation of integral 1 in Problem 7.22.

the nuclei for a fixed molecular axis orientation, and then perform integration over the angles θ, φ that determine the direction of this axis. In the case of matrix elements that are diagonal in quantum numbers n and v, the first integration gives

$$\int (\psi_{el,n}\psi_{osc,v})^* \hat{\mathbf{d}} \psi_{el,n}\psi_{osc,v} d\tau = \mathbf{d} = d\mathbf{n}_0, \quad \mathbf{n}_0 = \frac{\mathbf{R}}{R}. \tag{2}$$

The direction of the vector $\mathbf{d}$ along the molecule axis is seen from symmetry considerations. Directing the z-axis along the electric field $\mathcal{E}$ and taking into account the fact that the rotational wavefunction $\psi_{rot,KM}$ (for $\Lambda = 0$) is a spherical function $Y_{KM}(\theta,\varphi)$, we find that perturbation matrix elements between wavefunctions (1) which correspond to the same molecule level (i.e., differ only by the values of M) are equal to zero, so that $\int \cos\theta Y^*_{KM'} Y_{KM} d\Omega = 0$. In the first order of perturbation theory, the molecular levels are not shifted. Since in the case of an homogeneous field, the projection of the angular momentum onto the direction of vector $\mathcal{E}$ is a "good" quantum number, wavefunctions (1) are true functions of the zeroth approximation, and we can use perturbation theory for non-degenerate levels. The second-order correction is

$$E^{(2)}_{nvKM} = \sum_{k'}{}' \frac{|\langle k'|\hat{\mathbf{d}} \cdot \mathcal{E}|k\rangle|^2}{E^{(0)}_k - E^{(0)}_{k'}}, \tag{3}$$

where for brevity we have used one index, k, to describe different molecular states.

Now note that in sum (3) we can restrict ourselves to such states $|k'\rangle$ that correspond to the initial electron term and differ only by the rotational quantum number K. In Eq. (3),

$$E^{(0)}_k - E^{(0)}_{k'} = B_e[K(K+1) - K'(K'+1)],$$

while the contribution of states with other quantum numbers is much smaller due to larger energy denominators, since $E_{rot} \ll E_{osc} \ll E_{el}$. Taking this fact and relation (2) into account, we can write Eq. (3) in the form ($K \neq 0$)

$$E^{(2)}_{nvKM} = \frac{d^2 \mathcal{E}^2}{B_e} \sum_{K'} \frac{|\langle K'M|\cos\theta|KM\rangle|^2}{K(K+1) - K'(K'+1)}$$

$$= \frac{d^2 \mathcal{E}^2 [K(K+1) - 3M^2]}{2B_e K(K+1)(2K-1)(2K+3)}. \tag{4}$$

We have used the result of problem 8.11. Since the wavefunction of state $|KM\rangle$ is a spherical function, then sum (4) is analogous to the one calculated in Problem 8.11, using substitutions $l \to K$, $m \to M$, $\hbar^2/2I \to B_e$.

In the case of $K = 0$ we have $E^{(2)}_{nv} = -d^2\mathcal{E}^2/6B_e$ (see Problem 8.10).

Let us finally note that for two-atom molecules with identical nuclei, the dipole moment (2) vanishes, and the influence of an electric field on the molecular levels requires a different analysis.

Atomic systems in an external magnetic field

Problem 11.41

Consider the *Zeeman effect* for the hydrogen atom. Assume that the magnetic field is strong and the Zeeman splitting is much larger than the fine structure splitting (see Problem 11.1). Determine the condition of applicability of your results.

Solution

In the approximation linear over a magnetic field, the perturbation potential has the form[266]

$$\hat{V} = \mu_B \mathcal{H}(\hat{l}_z + 2\hat{s}_z),$$

where μ_B is the Bohr magneton and the z-axis is directed along the vector $\mathcal{H}$. Since operators $\hat{l}_z$ and $\hat{s}_z$ commute with each other and with the Hamiltonian, the eigenfunctions of the unperturbed Hamiltonian $\psi_{nlm}\chi_s$ (see Eq. (IV.3), χ_s is the spin part of the wavefunction) are true functions of the zeroth approximation, and the first-order correction to the energy is

$$E^{(1)}_{nl l_z s_z} = \langle nll_z s_z|\hat{V}|nll_z s_z\rangle = \mu_B \mathcal{H}(l_z + 2s_z). \tag{1}$$

As is seen, the $2n^2$-fold degenerate level splits into $2n + 1$ components (remember that $l_z = 0, \pm 1, \ldots, \pm(n-1)$, while $s_z = \pm 1/2$), the outermost of which are non-degenerate.

The applicability condition of the result in Eq. (1) requires that the level splittings $\Delta E_{Zeem} = 2n\mu_B\mathcal{H}$ are much larger than the fine structure interval ΔE_{FS} (see Problem 11.1), but also much smaller than the level spacing ΔE_n between the neighboring hydrogen levels. For $n = 2$, condition $\Delta_{FS} \ll \Delta E_{Zeem} \ll \Delta E_n$ takes the form

$$5 \cdot 10^{-5} \text{ eV} \ll 4\mu_B\mathcal{H} \ll 2 \text{ eV}, \quad \text{or} \quad 3 \cdot 10^3 \text{ Oe} \ll \mathcal{H} \ll 10^8 \text{ Oe}.$$

Recall that $e/a_B^2 = 5.14 \cdot 10^9$ V/cm $= 1.71 \cdot 10^7$ Oe.

Problem 11.42

Consider the Zeeman effect for the ground state of a hydrogen atom taking into account its hyperfine structure (see Problem 11.2). Pay attention to the level shift dependence on H in the cases of weak ($\mu_B\mathcal{H} \ll \Delta$) and strong ($\Delta \ll \mu_B\mathcal{H}$) external magnetic fields. Here $\Delta \approx 1420$ MHz is the hyperfine level splitting.

Solution

We denote by $\Delta_{0(1)}$ the ground level shift due to interaction between the electron and proton spins for the states with the total spin $S = 0(1)$. They are determined by

---

[266] Electron spin magnetic moment is $\mu_e = -\mu_B = -|e|\hbar/2m_ec$.

Eq. (3) of Problem 11.2, and $\Delta = \Delta_1 - \Delta_0 = 1420$ MHz. In the absence of a magnetic field, wavefunctions of the corresponding unperturbed states have the form $\psi_0(r)\chi_{SS_z}$, where χ_{SS_z} is the spin function and $\psi_0(r)$ is the wavefunction of the hydrogen atom ground-state. The level with $S = 0$ is non-degenerate, while the one with $S = 1$ is three-fold degenerate. To calculate the shift (and splitting) of the levels under the action of the perturbation[267] $\hat{V} = \mu_B(\hat{l}_z + 2\hat{s}_{e,z})\mathcal{H}$, we find its matrix elements. We use the following numeration of states: 1 for $S = 0$, $S_z = 0$; 2 for 1, 0; 3 for 1, +1, and 4 for 1, −1, and also the explicit form of spin functions[268] χ_{SS_z} (see Problem 5.10). We find that only four matrix elements are different from zero:

$$\langle 1|\hat{V}|2\rangle = \langle 2|\hat{V}|1\rangle = \langle 3|\hat{V}|3\rangle = -\langle 4|\hat{V}|4\rangle = \mu_B\mathcal{H}.$$

We also include unperturbed hyperfine shifts $\Delta_{0,1}$ into the perturbation matrix and we come to the secular equation:

$$\begin{vmatrix} \Delta_0 - E^{(1)} & \mu_B\mathcal{H} & 0 & 0 \\ \mu_B\mathcal{H} & \Delta_1 - E^{(1)} & 0 & 0 \\ 0 & 0 & \Delta_1 + \mu_B\mathcal{H} - E^{(1)} & 0 \\ 0 & 0 & 0 & \Delta_1 - \mu_B\mathcal{H} - E^{(1)} \end{vmatrix} = 0, \qquad (1)$$

from which we find

$$E^{(1)}_{1,2} = \frac{1}{2}\left[(\Delta_0 + \Delta_1) \mp \sqrt{(\Delta_1 - \Delta_0)^2 + 4\mu_B^2\mathcal{H}^2}\right], \quad E^{(1)}_{3,4} = \Delta_1 \pm \mu_B\mathcal{H}. \qquad (2)$$

See that for the states with $S = 1$, components of level with $S_z = \pm 1$ have a linear shift in $\mathcal{H}$. For the singlet and triplet unperturbed states with $S_z = 0$, which are "mixed" by the magnetic field, level shifts in a weak field, $\mu_B\mathcal{H} \ll \Delta$ (note that $\mathcal{H}_0 = \Delta/\mu_B \approx 10^3$ Oe), are quadratic in the field:

$$E^{(1)}_{1(2)} \approx \Delta_{0(1)} \mp \frac{\mu_B^2\mathcal{H}^2}{\Delta_1 - \Delta_0}. \qquad (3)$$

But in a strong field, when $\Delta \ll \mu_B\mathcal{H}$, we have

$$E^{(1)}_{1,2} \approx \mp\mu_B\mathcal{H}, \quad \chi_{1(2)} \approx \frac{1}{\sqrt{2}}(\chi_{0,0} \mp \chi_{1,0}). \qquad (4)$$

In the states here, electron and proton spin projection onto the direction of the magnetic field have definite values (while the interaction between the spin magnetic moments that breaks their conservation acts like perturbation).

---

[267] We restrict to the part of it linear in the field $\mathcal{H}$. Proton magnetic moment interaction with the magnetic field is not accounted for, due to its smallness ($\mu_p/\mu_e \sim m_e/m_p \sim 10^{-3}$).

[268] Particularly,

$$\chi_{1(0),0} = \frac{1}{\sqrt{2}}\left\{\begin{pmatrix}1\\0\end{pmatrix}_e \begin{pmatrix}0\\1\end{pmatrix}_p \pm \begin{pmatrix}0\\1\end{pmatrix}_e \begin{pmatrix}1\\0\end{pmatrix}_p\right\}.$$

$$\hat{\sigma}_{e,z}\chi_{1(0),0} = \chi_{0(1),0}$$

Problem 11.43

Consider the Zeeman effect for the ground level of a *positronium*, a bound state of an electron–positron system (similar to the hydrogen atom), taking into account its fine structure.[269] Find the spin functions in the presence of a magnetic field.

Solution

This solution is analogous to the previous solution. Now, interaction with the magnetic field has the form:[270]

$$\hat{V} = 2\mu_B(\hat{s}_{ez} - \hat{s}_{nz})\mathcal{H}.$$

Using the same numeration of states as in the previous problem, replacing the proton spin by the positron spin, we see that only two perturbation matrix elements are different from zero: $\langle 1|\hat{V}|2\rangle = \langle 2|\hat{V}|1\rangle = 2\mu_B\mathcal{H}$. The secular equation takes the form:

$$\begin{vmatrix} \Delta_0 - E^{(1)} & 2\mu_B\mathcal{H} & 0 & 0 \\ 2\mu_B\mathcal{H} & \Delta_1 - E^{(1)} & 0 & 0 \\ 0 & 0 & \Delta_1 - E^{(1)} & 0 \\ 0 & 0 & 0 & \Delta_1 - E^{(1)} \end{vmatrix} = 0,$$

$\Delta_{0(1)}$ are the level shifts of para-(ortho-)positronium that determine its fine structure. Its solution gives level shifts:

$$E^{(1)}_{1,2} = \frac{1}{2}\left[(\Delta_0 + \Delta_1) \mp \sqrt{(\Delta_1 - \Delta_0)^2 + 16\mu_B^2\mathcal{H}^2}\right], \quad E^{(1)}_{3,4} = \Delta_1.$$

Note that the levels of ortho-positronium with $S_z = \pm 1$ in this approximation do not shift. The shift of components with $S_z = 0$ in a weak field $\mu_B\mathcal{H} \ll \Delta$ is quadratic in $\mathcal{H}$, as it is included in the problem add a factor of 4 into the second term. In the case of a strong field, $\Delta_{0(1)} \ll \mu_B\mathcal{H}$, the shifts are linear: $E^{(1)}_{1,2} \approx \mp 2\mu_B\mathcal{H}$.

Spin functions for states with $S_z = 0$ that are "mixed" by the magnetic field have the form:

$$\chi_{1(2)} = C^{(1)}_{1(2)}\chi_{0,0} + C^{(2)}_{1(2)}\chi_{1,0}, \quad C^{(1)}_{1,2} = \frac{2\mu_B\mathcal{H}}{E^{(1)}_{1,2} - \Delta_0}C^{(2)}_{1(2)}.$$

---

[269] The magnetic spin interaction in positronium is of the same order of magnitude (there is no small parameter $\sim m_e/m_p$) as other relativistic corrections to the Hamiltonian. Unlike usual atoms, in positronium it is not meaningful to speak of a hyperfine-level structure. Positronium-level classification with respect to S (0 or 1) of total spin remains if we take into account the relativistic effects (*para-* and *ortho-positronium*). Fine splitting for ground levels of ortho- and para-positronium is $\Delta = \Delta_1 - \Delta_0 \approx 8.2 \cdot 10^{-4}$ eV.

[270] The orbital magnetic moment of positronium is equal to zero, since electron and positron contributions compensate each other.

In particular, in the case of a strong field we have $C^{(1)}_{1(2)} \approx \mp C^{(2)}_{1(2)}$, so that

$$\chi_1 \approx \begin{pmatrix} 0 \\ 1 \end{pmatrix}_e \begin{pmatrix} 1 \\ 0 \end{pmatrix}_p, \quad \chi_2 \approx \begin{pmatrix} 1 \\ 0 \end{pmatrix}_e \begin{pmatrix} 0 \\ 1 \end{pmatrix}_p,$$

i.e., particle spin projections onto the direction of field $\mathcal{H}$ have definite values – a natural physical result. In the case of a weak field, state "mixing" is small, but this fact strongly affects the lifetime of positronium (see Problem 11.61).

Problem 11.44

For the ground-state of the hydrogen atom, find the diamagnetic part of the level shift, connected to electron's orbital motion.

Solution

The level shift considered is equal to $\Delta E = -\frac{1}{2}\chi \mathcal{H}^2$, where χ is determined by the known equation for diamagnetic susceptibility:[271]

$$\chi = -\frac{e^2}{6mc^2}\overline{r^2}, \quad \overline{r^2} = 3a_B^2; \quad \chi = -\frac{1}{2}\alpha^2 a_B^3. \tag{1}$$

The mean value $\overline{r^2}$ is calculated in the ground level of the atom. This shift is the same for all electron spin states. The part of the level shift that depends on the electron spin state (and is dominant) has already been considered in Problem 11.41. See also Problem 11.42.

Problem 11.45

Find the magnetic susceptibility for a helium atom in the ground-state, using the approximate form of the wavefunction from the variational calculation in Problem 11.6. Consider also the magnetic susceptibility of 1 cm^3 of gaseous helium in normal conditions, and compare it to its experimental value, $-8.4 \cdot 10^{-11}$.

Solution

From the known equation for diamagnetic susceptibility, we obtain

$$\chi_{at} = -\frac{e^2}{6mc^2} \sum_{a=1,2} \overline{r_a^2} = -\frac{e^2}{mc^2} \frac{1}{Z_{eff}^2} \equiv -\alpha^2 \frac{1}{Z_{eff}^2} a_B^3. \tag{1}$$

The mean values $\overline{r_a^2}$ are calculated with the wavefunction from 11.6 and $Z_{eff} = 27/16$. The numeric value of susceptibility according to (1) is $\chi_{at} = -2,77 \cdot 10^{-30}$ cm^3 (see

---

[271] Here, $\alpha = e^2/\hbar c \approx 1/137$ is the fine structure constant. We emphasize that smallness $\sim \alpha^2$ in the quantity χ reflects the relativistic nature of the interaction with the magnetic field. Compare this to $\beta \sim a_B^4$ for the polarizability in an electric field.

also the footnote in the previous problem). In normal conditions we can consider all gas atoms to be in the ground-state, so that for magnetic susceptibility of 1 cm^3 of helium gas we obtain

$$\chi_{\text{He}} = n_0 \chi_{at} = -7.5 \cdot 10^{-11},$$

where $n_0 = 2.69 \cdot 10^{19}$ cm^3 is the Loschmidt number.

Problem 11.46

Find the shift and magnetic susceptibility of the ground level for a charged particle in a zero-range potential placed in an external homogeneous magnetic field. Generalize the result to the case of a weakly-bound state of a particle with the angular momentum $l = 0$ in a short-range potential $U_S(r)$ of radius r_S. Compare to Problems 11.35 and 11.36.

Solution

The unperturbed wavefunction of the $l = 0$ state is $\psi_0 = \sqrt{\kappa} e^{-\kappa r}/\sqrt{2\pi} r$ (see Problem 4.10). The diamagnetic shift is equal to

$$\Delta E_0 = \frac{e^2 \mathcal{H}^2}{12 mc^2} \int r^2 \psi_0^2 dV = \frac{e^2}{24 mc^2 \kappa^2} \mathcal{H}^2, \quad (1)$$

so that the magnetic susceptibility is

$$\chi_0 = -\frac{e^2}{12 mc^2 \kappa^2}. \quad (2)$$

The applicability condition for the result obtained is $\Delta E_0 \ll |E_0^{(0)}| = \hbar^2 \kappa^2/2m$.

These equations are valid for particle states with a small binding energy and $l = 0$, in an arbitrary short-range potential (a negative ion) $U_s(r)$ of radius r_s, where the particle localization domain, which is $\sim \kappa^{-1}$, is much larger than r_s. The dominant role of large distances manifests itself as a divergence of ΔE_0 as $\kappa \to 0$ and in the χ_0-dependence on the particle binding energy.

A correction due to a finite radius ($r_s \neq 0$) of the potential is found by including factors $C_{\kappa 0}^2$ to the relations obtained, where $C_{\kappa 0}$ is the asymptotic coefficient (compare to Problem 11.36). In particular, the magnetic susceptibility is

$$\chi_0 = -\frac{e^2 C_{\kappa 0}^2}{12 mc^2 \kappa^2}. \quad (3)$$

Problem 11.47

Find paramagnetic and diamagnetic level shifts for a weakly-bound state of a charged particle with the angular momentum $l = 1$ in a short-range potential, $U_S(r)$.

Solution

The wavefunction of the particle with a small binding energy decreases at large distances quite slowly, while the perturbation (the part that is quadratic in field)

$$\hat{V} = -\frac{e\hbar}{2mc}\boldsymbol{\mathcal{H}}\cdot\hat{\mathbf{l}} + \frac{e^2}{8mc^2}[\mathbf{r}\cdot\boldsymbol{\mathcal{H}}]^2 \tag{1}$$

increases. Due to this, large distances[272] play the dominant role in the integral of matrix element $\langle 1l_z|[\boldsymbol{\mathcal{H}}\times\mathbf{r}]^2|1l_z\rangle$ that determines the diamagnetic part of the level shift. Outside the range of the potential, U_s, the wavefunction of the unperturbed state has the form given in Problem 11.37. To calculate the perturbation matrix element, we can use this expression for the wavefunction for all values of r (the region $r \leq r_s$ is not important). Now, using the value of integral

$$\int_0^\infty x^3 K_{3/2}^2(x)dx = \frac{5\pi}{8}$$

we find the energy shifts:

$$\Delta E_{l=1,l_z=0} = \frac{e^2\kappa^2 C_{\kappa 1}^2 \mathcal{H}^2}{2\pi mc^2}\langle 10|\sin^2\theta|10\rangle \int_0^\infty r^3 K_{3/2}^2(\kappa r)dr = \frac{e^2 C_{\kappa 1}^2}{8mc^2\kappa^2}\mathcal{H}^2,$$

$$\Delta E_{l=1,l_z=\pm 1} = \mp\frac{e\hbar}{2mc}\mathcal{H} + \frac{e^2 C_{\kappa 1}^2}{4mc^2\kappa^2}\mathcal{H}^2. \tag{2}$$

We emphasize that the linear-in-the-field perturbation term (1) gives also a linear-in-the-field paramagnetic level shift (higher corrections to it are equal to zero). Also, remember that $C_{\kappa 1}^2 \propto \kappa$ for $\kappa \to 0$ (see Problem 11.37), so that $\Delta E^{(2)} \propto \kappa^{-1}$.

Problem 11.48

Find the Zeeman splitting of the rotational levels of a two-atom molecule, assuming that the shift is small with respect to the distance between neighboring rotational levels. The electron term of the molecule is $^1\Sigma$.

---

[272] The dominant role of large distances manifests itself as a divergence of susceptibility for $\kappa \to 0$. Note that for the angular momentum values $l \geq 2$, the role of large distances is not special. Diamagnetic susceptibility is determined by the form of the wavefunction on distances $\sim r_s$ and its value

$$\chi_{dia} \sim \frac{e^2 r_S^2}{mc^2}$$

depends on a specific form of the potential U_S. Compare to the properties of polarizability for a weakly bound state, considered in Problems 11.36 and 11.37.

Solution

The perturbation has the form:

$$\hat{V} = \mu_B(\hat{\mathbf{L}} + 2\hat{\mathbf{S}}) \cdot \mathcal{H} + \frac{e^2}{8mc^2}\sum_a \{r_a^2 \mathcal{H}^2 - (\mathbf{r}_a \cdot \mathcal{H})^2\}. \tag{1}$$

Summation is performed over all of the molecule's electrons, **L** and **S** are the total angular and spin momenta, and nuclear magnetism, due to its smallness $\sim m_e/M_{Nuc}$, is not taken into account.

Using the expression for eigenfunctions $\psi^{(0)}_{n v K M}$ of the unperturbed Hamiltonian (given in Problem 11.40), we see that to first order in $\mathcal{H}$, level shift is absent, since $\Lambda = S = 0$ and

$$\int (\psi_n^{el}\psi_v^{osc})\hat{\mathbf{L}}(\psi_n^{el}\psi_v^{osc}) \equiv A\mathbf{n}_0 = \Lambda \mathbf{n}_0 = 0.$$

The level shift due to the second term in expression (1) is equal to

$$E^{(2)}_{KM} = \frac{e^2 \mathcal{H}_i \mathcal{H}_k}{8mc^2} \int \psi^{(0)*}_{nvKM} \sum \{r_a^2 \delta_{ik} - x_{ia} x_{ak}\} \psi^{(0)}_{nvKM} d\tau. \tag{2}$$

Integrating over the coordinates of electrons and the variable R, we obtain

$$E^{(2)}_{KM} = \frac{e^2 \mathcal{H}_i \mathcal{H}_k}{8mc^2} \int Y^*_{KM}\{a\delta_{ik} - bn_{0i}n_{0k}\} Y_{KM} d\Omega_{n_0}, \tag{3}$$

where a and b are some constants that do not depend on the quantum numbers K and M. Using the expression

$$\int Y^*_{KM} n_{03} n_{03} Y_{KM} d\Omega_{n_0} \equiv \int \cos^2\theta |Y_{KM}|^2 d\Omega = \frac{2K^2 + 2K - 1 - 2M^2}{(2K-1)(2K+3)}$$

(see Problem 8.11) and formulas Eqs. (2) and (3), we find

$$E^{(2)}_{KM} = \frac{e^2 \mathcal{H}^2}{8mc^2}\left\{a - b\frac{2K^2 + 2K - 1 - 2M^2}{(2K-1)(2K+3)}\right\}. \tag{4}$$

Note that the contribution of second-order perturbation theory in terms of $\hat{V}' = \mu_B \hat{\mathbf{L}} \cdot \mathcal{H}$ in expression (1) has the same dependence on K and M as in Eq. (4).

Thus Eq. (4) gives the desired Zeeman splitting of rotational levels for a two-atom molecule with electron term $^1\Sigma$, to second order in a magnetic field. Note that the contribution to level shift $E^{(2)}_{KM}$ from the $\mathcal{H}^2$ term in Eq. (1) is positive. Just as in the case of atoms with $L = S = 0$, it corresponds to *diamagnetism*. Input of the second approximation from the linear-in-the-field term in Eq. (1) (which is absent in case of atoms with $L = S = 0$) could have either sign. For example, in the ground term, this part of the shift is negative.

In conclusion, we note that for molecular terms other than $^1\Sigma$, levels shifts in a weak magnetic field are linear in $\mathcal{H}$.

Interactions of atomic systems at large distances

Here we assume that the relative velocities of the colliding composite particles are not too large, so that the adiabatic approximation is applicable.

Problem 11.49

Find the interaction potential of a charged particle (ion, electron, *etc.*) with an unexcited hydrogen atom at large distances from each other.

Solution

The interaction operator for a particle and an atom (the usual electrostatic interaction of a charged system), assuming that the atom size is much smaller than the distance between them, has the form (Ze is the particle charge, atom is at the origin)

$$\hat{U} = Ze\varphi_{at}(\mathbf{R}) = Ze\left\{\frac{\mathbf{d}\cdot\mathbf{R}}{R^3} + \varphi_{at\ quadr}(\mathbf{R}) + \ldots\right\}. \quad (1)$$

Assume we can neglect excited states of the atom, in accordance with the main idea of adiabatic approximation (see Chapter 8, sec. 8 Problem 8.58). The potential $U(R)$ of interaction between the particle and the atom for their slow relative motion is determined by a change in atom energy due to the interaction (1), and is obtained by averaging this operator, considered as a perturbation over the atomic state, *i.e.*, $U(R) = \overline{\hat{U}}$. In the first order of perturbation theory, $\overline{\hat{U}} = 0$, since all multipole moments of the atom are zero due to spherical symmetry (for states with $L = 0$). In second order, we find:

$$U(R) = -\frac{\beta_0}{2}\left(\frac{Ze}{R^2}\right)^2 = -\frac{9Z^2e^2a_B^3}{4R^4}. \quad (2)$$

We have taken into account the fact that the perturbation $\hat{V} = Ze\mathbf{d}\cdot\mathbf{R}/R^3$ is equivalent to atom interaction with an homogeneous electric field $\boldsymbol{\mathcal{E}} = -Ze\mathbf{R}/R^3$, and used the value $\beta_0 = 9a_B^3/2$ for polarizability of the hydrogen atom ground-state. Let us emphasize that interaction (2) is attractive, and decreases $\propto R^{-4}$ with increase in R.

Problem 11.50

Find the interaction energy between a charged particle and a two-atom molecule that are far from each other. It is assumed that the molecule has a constant dipole moment d (in a coordinate system rigidly bound to the molecular axis) and is in the state with the rotational quantum number $K = 0$. The electron term of the molecule is $^1\Sigma$.

Solution

As in the previous problem, interaction is determined by the dipole term $\hat{V} = Ze(\mathbf{d} \cdot \mathbf{R})/R^3$. (Here it is important that the molecule is in a state with the rotational number $K = 0$, since otherwise the quadrupole interaction, which is different from zero even in the first order of perturbation theory and decreases with increase in R as R^{-3}, would be dominant.) So,

$$U(R) = -\frac{1}{2}\beta_0 \left(\frac{Ze}{R^2}\right)^2 = -\frac{Z^2 e^2 d^2}{6 B_c R^4}. \tag{1}$$

In the polarization potential we used the value of the molecular polarizability from Problem 11.40 (for $K = 0$).

Problem 11.51

Find how the interaction between a charged particle and a hydrogen atom in an excited state depends on the distance between them. Compare your results to Problem 11.49.

Solution

Generally speaking, $U(R) \propto R^{-2}$, which corresponds to interaction between a charge and a dipole moment, whose existence for an excited state of unperturbed hydrogen atom is connected to the accidental degeneracy of levels with different values of l and different parity, specific to the Coulomb potential. The proportionality coefficient in that power law depends on a state of the atom, and the corresponding independent states[273] that diagonalize the perturbation are the same as for the atom in an homogeneous electric field. For the level with $n = 2$, these are considered in Problem 11.33.

In conclusion, we note that the mean value of $U(R)$ over all independent states of the hydrogen atom with given n vanishes. The mean value of the next, $\propto R^{-3}$ term in the potential, corresponding to the interaction between the particle and the quadrupole moment of the atom, is also equal to zero. The value $\overline{U}(R) \propto R^{-4}$ originates from polarization effects as discussed in two previous problems.

Problem 11.52

Find the interaction energy between two hydrogen atoms (in the ground-state) at a large distance, R, from one another, using the variational method. For calculation, we Use the trial function of the form:[274]

---

[273] For these, the angular momentum projection on the direction $\mathbf{R}$ has a definite value. Vector $\overline{\mathbf{d}}$ is directed along $\mathbf{R}$ and $\overline{\mathbf{d}} = 0$ for the states with $l_z = \pm(n-1)$. But we should take into account that particle motion gives rise to transitions appear between the states. Compare to Problem 11.55.

[274] By considering the electrons localized in the vicinity of their nuclei corresponds to neglecting the exchange interaction, which decreases exponentially at large distances. The choice of the trial function (and its difference from the unperturbed wavefunction) reflects the dipole–dipole nature of the atom-atom interaction.

$$\psi_{\text{trial}} = C\psi_0(r_1)\psi_0(r_2)[1 + \alpha(x_1 x_2 + y_1 y_2 - 2z_1 z_2)],$$

where $\psi_0(r)$ is the ground-state wavefunction, α is a variational parameter, $\mathbf{r}_1$, $\mathbf{r}_2$ are radius vectors of the electrons in the two atoms relative to their nuclei, and axis z is directed along the axis that connects the nuclei.

Solution

The atom-atom interaction, $U(R)$, is determined by the electron term $E(R)$. For atoms in the ground-state, the term $E_0(R)$ could be calculated by variational method. In the problem considered, the Hamiltonian of the electron subsystem has the form:

$$\hat{H} = \hat{H}_{01} + \hat{H}_{02} + \hat{U} \approx \hat{H}_{01} + \hat{H}_{02} - \frac{e^2(2z_1 z_2 - x_1 x_2 - y_1 y_2)}{R^3},$$

where $\hat{H}_{01(2)}$ are Hamiltonians of isolated hydrogen atoms, and their interaction is taken into account in the dipole–dipole approximation. Compare to Problem 11.49.

It is easy to find the average energy $\overline{E}(\alpha, R)$ using the trial wavefunction ψ_{trial} if we take into account the following expressions (compare to Problem 11.49):

1) $\langle 0|x|0 \rangle \equiv \int x \psi_0^2(r) dV = 0$ due to the function under the integral being odd in x. In the same way, all integrals containing an odd power of any of the of components of the vectors $\mathbf{r}_1$ or $\mathbf{r}_2$ are equal to zero;
2) $\langle 0|\hat{H}_0|0 \rangle = -e^2/2a_B$, because ψ_0 is an eigenfunction of the operator $\hat{H}_0$;
3) $\langle 0|x^2|0 \rangle = \langle 0|y^2|0 \rangle = \langle 0|z^2|0 \rangle = \frac{1}{3}\langle 0|r^2|0 \rangle = a_B^2$;
4) $\langle 0|x\hat{H}_0 x|0 \rangle = \langle 0|y\hat{H}_0 y|0 \rangle = \langle 0|z\hat{H}_0 z|0 \rangle = \int (z\psi_0)\hat{H}_0(z\psi_0)dV = 0$. The easiest way to calculate such matrix elements is given at the beginning of Problem 11.34.

Summarizing, we obtain (here $C^2 \approx 1 - 6\alpha^2 a_B^4$ from the normalization of the trial wavefunction):

$$\overline{E}(\alpha, R) \approx -\frac{e^2}{a_B} + 6\alpha^2 e^2 a_B^3 + 12\alpha e^2 a_B^4 \frac{1}{R^3},$$

and after a minimization over the parameter α, we find the approximate variational value of the ground-state energy for the system of two hydrogen atoms considered, $E_0(R)$, and their interaction energy, $U(R)$:

$$E_0(R) = -\frac{e^2}{a_B} + U(R), \quad U(R) = -6\frac{e^2 a_B^5}{R^6}. \tag{1}$$

The term, $-e^2/a_B$, corresponds to the energy of two isolated atoms.

The power-law decay of the atom-atom interaction energy with distance, $U \propto R^{-6}$, corresponds to the van der Waals force. Let us note that the exact numerical calculation gives the coefficient of order 6.5 (instead of 6 in Eq. (1)).

Problem 11.53

Find the large-distance asymptote of the interaction energy between two molecules with the dipole moments $\mathbf{d}_1$ and $\mathbf{d}_2$ (in coordinate systems rigidly bound to the molecular axes). It is assumed that the molecules are in states with rotational quantum numbers $K_{1,2} = 0$, and that their electron terms are $^1\Sigma$.

Solution

A molecule-molecule interaction appears in the second order of perturbation theory in the dipole–dipole interaction (see the previous problem):

$$\hat{V} = \frac{(\hat{\mathbf{d}}_1 \cdot \hat{\mathbf{d}}_2)R^2 - 3(\hat{\mathbf{d}}_1 \cdot \mathbf{R})(\hat{\mathbf{d}}_2 \cdot \mathbf{R})}{R^5}$$

and is determined by the expression:

$$U(R) = E_0^{(2)}(R) = \sum_{k_1 k_2}{}' \frac{|\langle k_1 k_2 | \hat{\mathbf{d}}_1 \cdot \hat{\mathbf{d}}_2 R^2 - 3(\hat{\mathbf{d}}_1 \cdot \mathbf{R})(\hat{\mathbf{d}}_2 \cdot \mathbf{R})|0,0\rangle|^2}{(E_{0,1} + E_{0,2} - E_{k_1} - E_{k_2})R^{10}}, \qquad (1)$$

where $k_{1,2}$ are the sets of quantum numbers that characterize stationary states of an isolated molecule (see Problem 11.40). Here it is important that rotational quantum numbers of the molecules are $K_1 = K_2 = 0$, because otherwise the interaction appears in the first order in quadrupole interaction and decreases with distance as R^{-3}.

Note that the dominant role in sum (1) is played by terms that correspond to states for which all quantum numbers except for K and M are the same as for the colliding molecules. This is because the energy denominators are anomalously small, due to the smallness of rotational energy (compare to Problem 11.40). Direct axis z along the vector $\mathbf{R}$ and perform the integration in matrix elements over the electron coordinates and the distance between the molecular nuclei. Sum (1) is:

$$U(R) = -\sum_{K_1 M_1 K_2 M_2} \frac{d_1^2 d_2^2 |\langle K_1 M_1, K_2 M_2 | 3n_{1z} n_{2z} - \mathbf{n}_1 \mathbf{n}_2 | 0,0\rangle|^2}{[B_1 K_1 (K_1 + 1) + B_2 K_2 (K_2 + 1)] R^6}. \qquad (2)$$

$\mathbf{n}_{1,2}$ are unit vectors which determine molecular axes' orientations. The wavefunctions, $|KM\rangle$, correspond to spherical functions. Taking into account their forms (III.7) we find that the matrix elements $\langle KM|n_i|00\rangle$ are different from zero only for $K=1$ and are equal to

$$\langle 11|n_x|00\rangle \equiv \langle 11|\sin\theta\cos\varphi|00\rangle = -\langle 1,-1|n_x|00\rangle = \frac{i}{\sqrt{6}},$$

$$\langle 11|n_y|00\rangle = \langle 1,-1|n_y|00\rangle = \frac{1}{\sqrt{6}},$$

$$\langle 10|n_z|00\rangle = -\frac{i}{\sqrt{3}}.$$

The final expression for the energy of interaction between the molecules is

$$U(R) = -\frac{d_1^2 d_2^2}{3(B_1 + B_2)R^6}. \tag{3}$$

This is the van der Waals attraction.

Problem 11.54

Calculate interaction (including the *exchange potential*[275]) of an atom with its negatively-charged ion (system A^-A) at large distances. Consider the valence electron inside the ion as a weakly-bound particle in a short-range potential $U_S(r)$ created by the atom, approximating it with a zero-range potential.

Solution

We consider the electron in the combined field of two identical atoms as moving in the field of two zero-range potentials. The corresponding electronic terms were determined in Problem 11.28, from which it follows that the exchange potential at large distances is equal to

$$\Delta(R) = E_g(R) - E_u(R) = -\frac{2\hbar^2 \alpha}{mR} e^{-\alpha R}. \tag{1}$$

Now writing the interaction potential for even and odd terms in the form $U_{g,u}(R) = U_0(R) \pm \frac{1}{2}\Delta(R)$, we find for their common part:

$$U_0(R) \approx \frac{\hbar^2 \alpha}{mR} e^{-2\alpha R} - \frac{\hbar^2}{2mR^2} e^{-2\alpha R} - \frac{\beta_0 e^2}{2R^4}. \tag{2}$$

Here we have also taken into account the polarization potential according to Problem 11.49, where β_0 is the atom polarizability. Let us also note that to deduce interaction (2) from Eq. (3) of problem 11.28 by successive iterations, we should rewrite it in the form ($\alpha_1 = \alpha_2 = \alpha$):

$$\kappa_{g,u} - \alpha = \pm\frac{1}{R}\exp\{-\kappa_{g,u}R\} \approx \pm\frac{1}{R}e^{-\alpha R} - \frac{1}{R}e^{-2\alpha R}.$$

For $\kappa_{g,u}$ in the exponent, we used the value of its first approximation, $\kappa_{g,u} \approx \alpha \pm e^{-\alpha R}/R$, and correspondingly expanded the exponential factor.

---

[275] The exchange potential, $\Delta(R) = E_g(R) - E_u(R)$, is the energy difference between even E_g and odd E_u electron terms. It determines (with a coefficient 1/2) the interaction matrix element between the states $|1\rangle$ and $|2\rangle$, $\langle 2|\hat{H}(R)|1\rangle$, where the electron in $|i\rangle$ is localized in the vicinity of the ith nuclei. This matrix element (unlike the diagonal $\langle 1(2)|\hat{H}(R)|1(2)\rangle$) characterizes interaction where there is the exchange of an electron between the atom and the ion. See also, Problem 13.88.

Problem 11.55

Consider interaction between an unperturbed hydrogen atom and a hydrogen atom in the excited state with $n = 2$. Determine the correct functions of the zeroth approximation that diagonalize the atoms' dipole–dipole interaction operator.

Solution

Interaction potentials (for different states) are obtained by diagonalizing the dipole-dipole interaction operator:

$$\hat{V} = \frac{(\mathbf{d}_1 \cdot \mathbf{d}_2)R^2 - 3(\mathbf{d}_1 \cdot \mathbf{r})(\mathbf{d}_2 \cdot \mathbf{R})}{R^5}, \quad \mathbf{d}_{1,2} = -\mathbf{r}_{1,2},$$

where $\mathbf{r}_{1,2}$ are the radius vectors of electrons in the hydrogen atoms with respect to their nuclei (compare to Problem 11.49). The diagonalization is performed in the basis of the Hamiltonian $\hat{H}_0 = \hat{H}_{01} + \hat{H}_{02}$ for two non-interacting hydrogen atoms corresponding to the degenerate unperturbed level

$$E^{(0)} = -\frac{1}{2n_1^2} - \frac{1}{2n_2^2} = -\frac{5}{8},$$

where $n_1 = 1$ and $n_2 = 2$. For degenerate states, both matrix elements of the operators $\mathbf{d}_{1,2}$ and perturbation $\hat{V}$ are different from zero.

Note the following:

1) matrix elements of $\mathbf{d}_{1,2}$ are non-zero only for states with different parity;
2) angular momentum projections onto the direction of $\mathbf{R}$ are constants of the motion;
3) there exists a symmetry due to having two identical atoms in the problem.
   With these, the form of unperturbed eigenfunctions that diagonalize the perturbation $\hat{V}$ operator, i.e., the correct functions of the zeroth approximation, is straightforward to find as follows:

$$\psi_{1,2} = \frac{1}{\sqrt{2}}(|1s, 2s\rangle \pm |2s, 1s\rangle), \quad \psi_{5,6} = \frac{1}{\sqrt{2}}(|1s, 2p1\rangle \pm |2p1, 1s\rangle); \quad (1)$$

$$\psi_{3,4} = \frac{1}{\sqrt{2}}(|1s, 2p0\rangle \pm |2p0, 1s\rangle); \quad \psi_{7,8} = \frac{1}{\sqrt{2}}(|1s, 2p, -1\rangle \pm |2p, -1, 1s\rangle).$$

Here the first symbols $1s$, $2s$, $2pm$ in the state vectors $|...\rangle$ characterize electron states in the first atom, and the second symbols characterize states in the second atom. See that each of the atoms has probability $1/2$ to be in the ground, $1s$-state, and probability $1/2$ to be in the excited state with $n = 2$.

Using known expressions for "hydrogen" wavefunctions, $\psi_{nlm} = R_{nl}Y_{lm}$ (see Eqs. (IV.3) and (III.7)), we find the matrix elements $\langle 2lm|\mathbf{r}|1s\rangle$:

$$\langle 2p0|z|1s\rangle = -\frac{i}{\sqrt{3}}d_0, \quad \langle 2p1|x+iy|1s\rangle = -\langle 2p,-1|x-iy|1s\rangle = i\sqrt{\frac{2}{3}}d_0,$$

$$d_0 = \frac{256}{81\sqrt{6}}. \tag{2}$$

Now writing

$$\mathbf{d}_1 \cdot \mathbf{d}_2 = \frac{1}{2}(x_1+iy_1)(x_2-iy_2) + \frac{1}{2}(x_1-iy_1)(x_2+iy_2) + z_1 z_2,$$

it is easy to find the dipole–dipole interaction potential $U_a(R) = \langle a|\hat{V}|a\rangle$ in the states given above, $a = 1, 2 \ldots 8$. They have the form:

$$U_a(R) = g_a \frac{d_0^2}{R^3}, \quad \text{where} \quad g_{1,2} = 0, \quad g_{3,4} = \mp\frac{2}{3}, \quad g_{5,6} = g_{7,8} = \pm\frac{1}{3}.$$

In conclusion, we note that as the atoms move, the direction of $\mathbf{R}$ changes. This will induce transitions between the almost-degenerate states in Eq. (1). See Problem 13.89.

Problem 11.56

Find a large-distance asymptote of the interaction potential of two atoms in the case when the valence electron of one of the atoms is weakly bound, so that $|E_0| \ll \hbar^2/ma_B^2$. Use scattering length perturbation theory developed in Problem 11.4 and determine the condition of applicability of your results.

Solution

The interaction potential $U(R) = \Delta E(R)$ is determined by the energy change ΔE of the valence electron, caused by its additional interaction with the other atom. This interaction could be described by a short-range potential $U_S(r)$ with a radius of action of the order of the atomic size. Due to the fact that the valence electron's localization domain is large enough that $L \approx \kappa_0^{-1} \gg a_B$, the level shift is determined by the equation from scattering length perturbation theory (see, Problem 11.4):

$$U(R) = \Delta E(R) \approx \frac{2\pi\hbar^2}{m} \left|\psi^{(0)}(R)\right|^2 a_s, \tag{1}$$

where $\psi^{(0)}(r)$ is an unperturbed wavefunction of the valence s-electron, and a_s is the scattering length from "another" atom (see Problems 11.4 and 4.29).

We will make several remarks regarding applicability of expression (1):

1) It is assumed that $|a_s| \ll L$ (see 11.4). This means that there exists no weakly-bound negative ion for the "other" atom.

2) This formula is valid for distances $R \sim L$. For larger distances, it gives an exponentially small shift. At these distances, the atoms' interaction would be determined by van der Waals forces (it is assumed that both atoms are in S-states).

3) Eq. (1) needs clarification in the case when the angular momentum l of the weakly-bound electron is different from zero. Now the interaction will depend on the value of the projection of the angular momentum onto the direction of $\mathbf{R}$. If $l_z = 0$, then the potential is still described by Eq. (1). If $l_z \neq 0$, then Eq. (1) gives $U(R) = 0$. Here, the level shift would be determined by the interaction between the electron and the atom in the state with angular momentum equal to $|l_z|$.

4) Let us note that Eq. (1) also describes interaction between a negative ion with a weakly bound outer-shell electron and the "other" atom.

11.5 Non-stationary phenomena in atomic systems

Problem 11.57

A tritium atom (a superheavy hydrogen isotope) is in its ground-state. As a result of β-decay, the *triton* turns into helium: $^3\text{H} \to e + \tilde{\nu} + ^3\text{He}$. Find:

a) the mean energy acquired by the electron in the β-decay;
b) the probability that a ground-state helium ion He^+ forms from β-decay;
c) the formation probability of an excited state of a Helium atom with the principal quantum number, $n = 2$.

Take into account that the electron in the β-decay is relativistic (the energy release in the decay is ≈ 17 keV).

Solution

a) The β-electron flies through the atom very fast and the problem could be solved using the sudden approximation. The electron Hamiltonian before the decay $\hat{H}_1$, immediately turns into $\hat{H}_2$ after the decay, where[276]

$$\hat{H}_1 = -\frac{1}{2}\Delta - \frac{1}{r}, \quad \hat{H}_2 = -\frac{1}{2}\Delta - \frac{2}{r}.$$

The electron wavefunction immediately after the nuclear decay (at the moment $t = 0$), as well as before decay, has the form:

$$\psi_{1s}(r, Z) = \sqrt{\frac{Z^3}{\pi}} e^{-Zr} \quad \text{with} \quad Z = 1.$$

---

[276] The effect of nuclei recoil is unessential (compare to Problem 11.58). Thus the nucleus can be considered at rest at all stages of the decay process.

A change in the atomic electron energy appears precisely at the moment of nucleus decay, but for $t > 0$ its mean value does not depend on time, and is equal to

$$\overline{E} = \langle 1s, Z = 1|\hat{H}_2|1s, Z = 1\rangle \equiv \langle 1s, 1|\hat{H}_1 - \frac{1}{r}|1s, 1\rangle = -\frac{3}{2}. \tag{1}$$

So, mean atom energy acquired as a result of the decay (due to the β-electron) is

$$\overline{E}_{rot} = \overline{E} - E_0 = -1 \text{ a.u.} = -27.2 \text{ eV}.$$

See that $\overline{E}_{rot} < 0$, i.e., the atomic (ion) energy decreases.

b) The probability for the electron to remain in the ground-state of the helium ion that appears as a result of the decays, according to main equation of the theory of sudden perturbations (see Problem 8.47), is given by the following overlap integral

$$w_{1s \to 1s} = \left|\int \psi_{1s}(r, Z=2)\psi_{1s}(r, Z=1)dV\right|^2 = \frac{512}{729} \approx 0.70. \tag{2}$$

In the same way, and taking into account the form of the wavefunction of the $2S$ state of a hyrdogen-like atom (helium ion $Z = 2$),

$$\psi_{2s}(r, Z) = \sqrt{\frac{Z^3}{8\pi}}\left(1 - \frac{Zr}{2}\right)e^{-Zr/2},$$

we find that the probability of transition into this state is

$$w_{1s \to 2s} = \left|\int \psi_{2s}(r, Z=2)\psi_{1s}(r, Z=1)dV\right|^2 = \frac{1}{4}. \tag{3}$$

Let us note that due to the spherical symmetry of the Hamiltonian $\hat{H}_{1,2}$ and in the approximation of instantaneous escape of the β-electron, the angular momentum of the atomic electron is conserved, and since it is equal to zero in the initial state, then after the nucleus decays, only electron transitions in the s-states are possible. With nuclear decay, atom (ion) ionization can occur, i.e., atomic electron transitions into the states of the continuous spectrum with $E > 0$ are possible. The angular distribution of such escaping electrons is isotropic. It should be noted that the ionization probability is small, as is seen from the calculated probabilities, see Eqs. (2) and (3).

Problem 11.58

A nucleus of an atom (the atom is in the stationary state, ψ_0) experiences an instant impulse and acquires a momentum **P**. Express in a general form the probability of atomic transition into a stationary state ψ_n as the result of such a "shaking." Consider specifically the case of a hydrogen atom and calculate the total probability of its excitation and ionization.

Solution

Usually, for the wavefunction of an atom we take a wavefunction of the shell electrons $\psi(\mathbf{r}_1, \mathbf{r}_2, \ldots, \mathbf{r}_N)$ (spin variables are omitted for the sake of brevity), while the atomic nucleus is considered to be fixed at the coordinate origin. Free motion of the nucleus (whose position almost coincides with the system's center of mass) does not affect the electron state. To determine how the atomic state changes as a result of an instant transfer of momentum $\mathbf{P} = M\mathbf{V}$ to the nucleus (M is the nucleus mass), we see that due to Galilean invariance, it is equivalent to a change to $\mathbf{p} = -m\mathbf{V}$ of all electrons' momenta (and the initial nuclear momentum). The wavefunction of the electron shell immediately after the "shaking" takes the form:

$$\tilde{\psi}_0 = \exp\left\{-\frac{i}{\hbar}m\mathbf{V}\cdot\sum_a \mathbf{r}_a\right\} \psi_0(\mathbf{r}_1, \mathbf{r}_2, \ldots, \mathbf{r}_N), \tag{1}$$

where ψ_0 is the wavefunction immediately before the "shaking" (compare to Problem 6.26). Further evolution of the electron wavefunction is determined by the Schrödinger equation, and the excitation probabilities for different stationary states

$$w_{0\to n} = |\langle \psi_n | \tilde{\psi}_0 \rangle|^2 = \left| \int \psi_n^* \exp\left\{-\frac{i}{\hbar}m\mathbf{V}\cdot\sum_a \mathbf{r}_a\right\} \psi_0 d\tau \right|^2 \tag{2}$$

do not depend on time (compare to Problem 8.47).

For example, for a hydrogen-like atom in the ground, $1s$ state, the probability of remaining in the initial state according to Eq. (2) is equal to ($\mathbf{q} = m\mathbf{P}/\hbar M$):

$$w_0 = \left| \frac{Z^3}{\pi a_B^3} \int \exp\left\{-\frac{2Zr}{a_B} - i\mathbf{q}\cdot\mathbf{r}\right\} dV \right|^2 = \frac{1}{(1+q^2 a_B^2/4Z^2)^4}. \tag{3}$$

Here, the total probability of excitation and ionization of the atom is $w = 1 - w_0$. For the limiting cases of weak, $qa_B/Z \ll 1$, and strong, $qa_B/Z \gg 1$, "shaking," we have:

$$w = \begin{cases} \frac{q^2 a_B^2}{Z^2} \ll 1, & \frac{qa_B}{Z} \ll 1, \\ 1 - \left(\frac{2Z}{qa_B}\right)^8 \approx 1, & \frac{qa_B}{Z} \gg 1 \end{cases}.$$

Note that $qa_B/Z \equiv V/v_a$, where $v_a = Z\hbar/ma_B$ is the characteristic electron velocity in the initial state.

Problem 11.59

For a mesomolecular $dt\mu$-system that is in its ground-state ($K = v = 0$), estimate the probability of the α-particle, that appears in the nuclear fusion reaction $dt \to n\alpha + 17.6$ MeV, "catching" the muon.[277]

Solution

Using adiabatic approximation for the $dt\mu$-system (see Problem 11.30), we see that when the deuteron and triton move closer down to nuclear lengthscales, where the fusion reaction occurs $dt \to n\alpha$, the muon wavefunction is reduced to that of a hydrogen-like atom in the ground-state with the nuclear charge $Z = 2$. Then, as a result of the reaction, a fast α-particle (nucleus ^{4}He) appears, and the probability for it to "pick up" the muon, i.e., to form a muon-atom ion $(\mu\alpha)^+$, is determined by Eq. (2) of the previous problem. The dominant mechanism here is associated with the process where the muon is caught in the ground state whose probability w_0 is calculated by Eq. (3) of the previous problem, if we put $Z = 2$ and take for a_B the muon Bohr radius. The value of q^2 here is determined by the 17.6 MeV energy release in the reaction $dt \to n\alpha$, 3.52 MeV of which belongs to the α-particle. Taking into account the fact that $m_\mu = 207 m_e$, $m_\alpha = 7286 m_e$, we find $q^2 a_B^2 \approx 35.5$ MeV and hence $w_0 \approx 9.3 \cdot 10^{-3}$. Thus, one muon could cause about 100 events of the reaction $dt \to n\alpha$. A more accurate calculation that takes into account adiabatic corrections to the wavefunction, the finiteness of the ratio m_μ/m_α, and also transitions into other bound states of the system $(\mu\alpha)^+$, gives ≈ 160 events of the fusion reaction.

Problem 11.60

Generalize the result of Problem 11.58 to the case of a two-atom molecule, i.e., obtain the general expression for the probability of the molecular transitions from a stationary state ψ_0 to an excited state ψ_n as the result of a sudden "shaking," when a momentum $\mathbf{P}$ is given to one of the molecule's nuclei (for example, the recoil momentum for emission of a quantum by an excited nuclei). Apply the result obtained to calculate the probability for the molecule to remain in the initial state if the nuclear velocity change is much smaller than the characteristic velocities of the electrons in the molecule. The molecular term is $^1\Sigma$, and it is in a state with quantum numbers $K = v = 0$. Discuss the conditions under which excitation of the rotational and oscillatory degrees of freedom occurs in the molecule.

Solution

The wavefunction of the system immediately before the "shaking" has the form:

$$\psi_{syst} = \psi(\mathbf{R}_{c.m.}) \cdot \psi_0(\boldsymbol{\rho}_1 - \mathbf{R}_{c.m.}, \ldots, \boldsymbol{\rho}_N - \mathbf{R}_{c.m.}; \mathbf{R}_1 - \mathbf{R}_2).$$

---

[277] When bound in a mesoatomic ion μHe, the muon stops acting as a catalyst of the fusion reaction. This fact (and not the finiteness of the muon lifetime) places restrictions on the number of reaction events initiated by one muon, and so places restrictions on the energy effectiveness of the μ-catalysis. See also Problem 11.74.

The wavefunction $\psi(\mathbf{R}_{c.m.})$ describes the motion of the center of mass, while ψ_0 is the wavefunction of a molecular "internal" state. Here (compare to Problem 11.40), we use $\mathbf{r}_a = \boldsymbol{\rho}_a - \mathbf{R}_{c.m.}$ as the radius vectors of the electrons with respect to the center of mass.

If a momentum $\mathbf{P}$ is imparted on nucleus 1, with the radius-vector $\mathbf{R}_1$, then the wavefunction of the system immediately after the "shaking" is

$$\tilde{\psi}_{syst} = \exp\left\{\frac{i}{\hbar}\mathbf{P}\cdot\mathbf{R}_1\right\}\psi_{syst}. \tag{1}$$

Note the relation ($\mathbf{R} = \mathbf{R}_1 - \mathbf{R}_2$, $M = M_1 + M_2$):

$$\mathbf{R}_1 = \mathbf{R}_{c.m.} + \frac{M_2}{M}\mathbf{R} - \frac{m}{M}\sum_a(\boldsymbol{\rho}_a - \mathbf{R}_{c.m.}).$$

From Eq. (1) we find that the change of the molecular wavefunction as a result of the "shaking" is (compare to Eq. (1) from Problem 11.58 for an atom)

$$\tilde{\psi}_0 = \exp\left\{\frac{i}{\hbar}\cdot\frac{M_2}{M}\mathbf{P}\cdot\mathbf{R} - \frac{i}{\hbar}\cdot\frac{m}{M}\mathbf{P}\cdot\sum_a\mathbf{r}_a\right\}\psi_0(\mathbf{r}_1,\ldots,\mathbf{r}_N;\mathbf{R}).$$

The transition probabilities in the molecule are calculated by the usual equation: $w(0 \to n) = |\langle\psi_n|\tilde{\psi}_0\rangle|^2$. This matrix element contains an integration over the coordinates of all electrons, the relative distance between the nuclei $\mathbf{R}$, and the angles that determine the orientation of the molecular axis. A summation over the spin indices is also performed.

2) For a molecule with electron term $^1\Sigma$, the probability of remaining in the initial state with quantum numbers $K = v = 0$, using the form of the wavefunction from Problem 11.40, is determined by expression:

$$w_0 = \left|\int\exp\left\{\frac{iM_2\mathbf{P}\cdot\mathbf{R}}{\hbar M} - \frac{im\mathbf{P}}{\hbar M}\cdot\sum_a\mathbf{r}_a\right\}\cdot|\psi_n^{el}\psi_0^{osc}\psi_{00}^{rot}|^2 d\tau_{el}dR\, d\Omega_R\right|^2. \tag{2}$$

From the condition $P/M \ll v_{el} \approx \hbar/ma_B$ and characteristic values for the electron coordinates $r \sim a_B$, we can replace the factor $\exp\{-i(m\mathbf{P}/\hbar M)\sum_a\mathbf{r}_a\}$ by unity, in Eq. (2). We have $\int|\psi_n^{el}|^2 d\tau_{el} = 1$ (which corresponds to the fact that change of the electron term is a low-probability event). Performing an integration over the angles

$$\frac{1}{4\pi}\int\exp\left\{i\frac{M_2PR\cos\theta}{\hbar M}\right\}d\Omega_R = \frac{\sin\alpha R}{\alpha R}, \quad \alpha = \frac{M_2P}{\hbar M} = \frac{\mu V}{\hbar}$$

(μ is the nuclei's reduced mass), we transform Eq. (2) to the form:

$$w_0 = \left|\int\frac{1}{2\alpha R}(e^{i\alpha R} - e^{-i\alpha R})|\psi_0^{osc}(R)|^2 dR\right|^2. \tag{3}$$

The wavefunction

$$\psi_0^{osc} = \left(\frac{\mu\omega_e}{\pi\hbar}\right)^{1/4} \exp\left\{-\frac{\mu\omega_e(R-R_0)^2}{2\hbar}\right\}$$

is essentially different from zero only in the region, where

$$|R - R_0| \leq \left(\frac{\hbar}{\mu\omega_e}\right)^{1/2} \ll R_0 \sim a_B$$

(R_0 is the equilibrium distance between molecule nuclei, and w_e is the oscillation frequency). So, to calculate the integral in Eq. (3), the value of R in the denominator could be replaced by R_0. The integral is easily calculated, since

$$\int |\psi_0^{osc}(R)|^2 e^{\pm i\alpha R} dR = \left(\frac{\mu\omega_e}{\pi\hbar}\right)^{1/2} e^{\pm i\alpha R_0} \exp\left\{-\frac{\mu\omega_e}{\hbar}(R-R_0)^2 \pm \right.$$

$$\left. \pm i\alpha(R-R_0)\right\} dR = \exp\left\{\pm i\alpha R_0 - \frac{\hbar\alpha^2}{4\mu\omega_e}\right\}.$$

Due to the fast convergence, we can integrate in infinite limits.

As a result, the final relation for the probability that the molecule remains in the initial state with the quantum numbers $K = v = 0$ is

$$w_0 = \left(\frac{\sin\alpha R_0}{\alpha R_0}\right)^2 \cdot \exp\left(-\frac{\hbar\alpha^2}{2\mu\omega_e}\right). \tag{4}$$

The physical meaning of the two factors in this expression is easily explained. The first of them, $[(\sin\alpha R_0)/\alpha R_0]^2$, gives the probability that rotational degrees of freedom are not excited. It becomes essentially different from unity only for $\alpha R_0 \geq 1$ or $\mu V R_0 \geq \hbar$. This is a natural physical result if we recall the angular momentum quantization and that $\mu V R_0$ characterizes the value of the angular momentum imparted onto the molecule. The second factor in Eq. (4) corresponds to the probability that the oscillatory degrees of freedom are not excited.

Problem 11.61

Find the change of lifetime for the ground-states of ortho- and para-positronium (see Problem 11.43) upon application of an homogeneous magnetic field.

Comment

The finiteness of positronium lifetime is connected into the annihilation of the electron-positronium pair into photons. In the absence of external fields, the ortho- and para-positronium lifetimes have different values: $\tau_1 \approx 1.4 \cdot 10^{-7}$ s for the ortho-positronium, and $\tau_0 \approx 1.2 \cdot 10^{-10}$ s for the para-positronium, which is connected to the difference of their decay channels: they decay into three and two photons respectively. Also note that in the presence of several decay channels, the total probability of decay ω (the quantity inverse to the lifetime τ) is equal to the sum of the partial probabilities.

Solution

A magnetic field strongly affects the positronium lifetime because it "mixes" the ortho- and para-states (see Problem 11.43) that have sharply different lifetimes. The lifetimes of the quasi-stationary states that appear are determined by the relation:

$$\frac{1}{\tilde{\tau}_1} = \omega_0^{(1)} \cdot \frac{1}{\tau_0} + \omega_1^{(1)} \cdot \frac{1}{\tau_1}, \tag{1}$$

where $\omega_{0(1)}^{(1)} = |C_{0(1)}^{(1)}|^2$ are the probabilities of positronium being in para- (ortho-) states. Spin functions for the positronium ground-state in a magnetic field were found in Problem 11.43. Since states of ortho-positronium with projection of spin $S_z = \pm 1$ in the direction of the magnetic field remain (quasi)stationary and are not distorted by the weak magnetic field, their lifetimes are not changed.

A completely different situation takes place for states with $S_z = 0$. Now the magnetic field "mixes" the ortho- and para-states. Using the results of Problem 11.43 for coefficients $C_{0(1)}^{(1)}$, and according to Eq. (1), we find

$$\frac{1}{\tilde{\tau}_{1(2)}} = \frac{y \pm 1}{2y} \cdot \frac{1}{\tau_0} + \frac{y \mp 1}{2y} \cdot \frac{1}{\tau_1}. \tag{2}$$

We have used $y = \sqrt{1 + (4\mu_B \mathcal{H}/\Delta)^2}$ (Δ is the fine splitting of the positronium ground level), and the signs + and − correspond to states 1 and 2. The former corresponds to para-positronium and the latter to orthopositronium. Since $\tau_1 \gg \tau_0$, then from Eq. (2) we see that even a weak magnetic field strongly affects the lifetime of orthopositronium, decreasing it as

$$\tilde{\tau}_2 \approx \left(1 + \frac{4\mu_B^2 \mathcal{H}^2}{\Delta^2} \cdot \frac{\tau_1}{\tau_0}\right)^{-1} \tau_1, \quad \mu_B \mathcal{H} \ll \Delta. \tag{3}$$

This result follows from expression (1) if we use coefficients $C_1^{(2)} \approx 1$ and $C_2^{(2)} \approx 2\mu_B \mathcal{H}/\Delta$, according to Eq. (VIII.2) of perturbation theory.

With an increase of magnetic field, the value $\tilde{\tau}_1$ increases, while $\tilde{\tau}_2$ decreases. In a strong magnetic field, when $\mu_B \mathcal{H} \gg \Delta$, these lifetimes approach each other $\tilde{\tau}_1 \approx \tilde{\tau}_2 \approx \tau_0/2$. In this case, the states 1 and 2 considered contain the ortho- and para-states of positronium with equal probability.

Problem 11.62

Find the change in lifetime of the metastable 2s-states of the hydrogen atom with a weak homogeneous electric field applied.

Comment

Excited states of atomic systems are, properly speaking, quasi-stationary states, since they have a finite lifetime, connected to electrons' transition to lower levels with photon emission. Usually, the characteristic lifetime is $\tau \sim 10^{-9}$ s. For example, for the $2p$-state of the hydrogen atom, $\tau_{2p} = 1.6 \cdot 10^{-9}$ s. But the lifetime of $2s$-state is much larger, $\tau_{2s} \approx 0.1$ s, and is determined by transition to ground-state with a two-photon emission (see, for this, Problem 14.6 and 14.8). While solving the problem we should take into account the anomalous closeness between the energies of $2s_{1/2}$- and $2p_{1/2}$-states, the difference of which is $\Delta E/2\pi\hbar \approx 1058$ MHz (the so-called *Lamb shift*).

Solution

The physical cause for the strong influence of even a weak electric field on the lifetime of the hydrogen atom $2s$-state lies in the fact that it "mixes" it with $2p$-state, and the latter quickly emits photons, passing to the ground-state (compare to the previous problem). The special role of the $2p$- (and especially $2p_{1/2}$-) states is due to their near-degeneracy in energy with the $2s$-state. The level shift, appearing for the non-relativistic level $E_{n=2}$ of the hydrogen atom due to relativistic and so-called *radiational* corrections, is shown in Fig. 11.3. $\Delta_{LS}/2\pi\hbar = 1058$ MHz is the Lamb shift, while $\Delta_{HS}/2\pi\hbar = 1.1 \cdot 10^4$ MHz is due to fine structure.

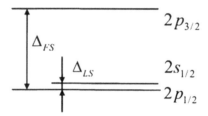

Fig. 11.3

Upon applying a weak electric field, instead of a "pure" $2s_{1/2}$-state, a superposition appears:

$$\psi_E \approx C_1 \psi_{2s_{1/2}} + C_2 \psi_{2p_{1/2}},$$

where we restrict ourselves only to the $2p_{1/2}$ state due to the smallness of Δ_{LS} with respect to Δ_{FS} (see below). The lifetime of such a state is

$$\frac{1}{\tau} = |C_1|^2 \cdot \frac{1}{\tau_{2s}} + |C_2|^2 \cdot \frac{1}{\tau_{2p}}.$$

As usual, from the Schrödinger equation $\hat{H}|\psi_E\rangle = E|\psi_E\rangle$, by successively multiplying by $\langle 2s_{1/2}|$, $\langle 2p_{1/2}|$, we obtain

$$V^* C_2 = E^{(1)} C_1, \quad V C_1 = (E^{(1)} + \Delta_{FS}) C_2, \tag{1}$$

where $E^{(1)}$ is the difference of energy between the 2s state and the unexcited level $E_{2s}^{(0)}$, while

$$V = \langle 2p_{1/2}|e\mathcal{E}z|2s_{1/2}\rangle = i\sqrt{3}ea_B\mathcal{E}.$$

This perturbation matrix element is calculated for states with similar values of the angular momentum projection $j_z = \pm 1/2$ onto the direction of the electric field, and differs from the matrix element $\langle 2p0|e\mathcal{E}z|2s\rangle$ only by the factor[278] $1/\sqrt{3}$, calculated from Problem 11.33, neglecting spin-orbital interaction.

We will not present the general solution[279] of the system of equations (1) that gives the shifts of the $2s_{1/2}$- and $2p_{1/2}$-levels and values of the corresponding coefficients C_i, but instead consider only limiting cases.

To satisfy the conditions $ea_B\mathcal{E} \ll \Delta_{LS}$ for coefficients C_i that describe the distortion of the wavefunction in the 2s-state by an electric field, we can use the result of perturbation theory (VIII.2), according to which $C_{2s} \approx 1$ and $C_{2p_{1/2}} \approx V/\Delta_{LS}$. We obtain an expression for the lifetime of the state considered:

$$\frac{1}{\tau} = \frac{1}{\tau_{2s}} + \frac{3e^2 a_B^2 \mathcal{E}^2}{\Delta_{LS}^2} \cdot \frac{1}{\tau_{2p}}. \tag{2}$$

Let us note that $C_{2p_{3/2}} = \sqrt{2}V/(\Delta_{LS} - \Delta_{FS})$ and the correction involving the $2p_{3/2}$-state is approximately 4% of the second term on the right-hand side of (2).

In the case of $ea_B\mathcal{E} \gg \Delta_{LS}$ (we note that $E_0 = \Delta_{LS}/ea_B \approx 800$ V/cm) in the superposition considered, the 2s- and 2p-states are given with the same probability, and now $\tau \approx \tau_{2p}/2$.

In conclusion, we note that here, as well as in the previous problem, the widths of the states considered, $\Gamma = \hbar/\tau$, are small with respect to the level spacing, ΔE, between the unexcited levels. In analogous problems with $\Gamma \geq \Delta E$, it is necessary to take into account the attenuation of states.

Problem 11.63

To first order in perturbation theory, find the probability for a charged particle subjected to an electrostatic field of a monochromatic electromagnetic wave to be pulled out of a bound state of a zero-range potential. The particle momentum λ is considered to be much larger than the particle's localization domain, so that the electric field could be considered homogeneous, changing in time in an harmonic way.

Consider the most general case of a wave with an elliptical polarization. Generalize the result obtained to the case of a weakly-bound state of a particle in a short-range potential $U_S(r)$ of a finite radius.

---

[278] It is a Clebsch–Gordan coefficient. Its value follows, for example, from the result of Problem 5.18.
[279] It could be obtained as a result of simple substitutions from equations that give the solution for Problems 11.61 and 11.43 (for states with $S_z = 0$).

Comment

For a monochromatic wave of an elliptical polarization that propagates in the direction of the z axis, the electric field at a fixed point in space changes in time by the law:

$$\boldsymbol{\mathcal{E}}(t) = (\mathcal{E}_0 \cos \omega t, \zeta \mathcal{E}_0 \sin \omega t, 0),$$

where ζ is the degree of ellipticity ($|\zeta| \leq 1$), and $\mathcal{E}_0$ is the amplitude of the field intensity. The sign of ζ gives the direction of rotation for the vector $\boldsymbol{\mathcal{E}}(t)$. For $\zeta = 0$, the wave is linearly-polarized, while for $\zeta = \pm 1$ it has a circular polarization.

Solution

We use the equation for transition probability per unit of time from a state of the discrete spectrum to a state of the continuous spectrum under the action of a periodic perturbation:

$$dw_\nu = \frac{2\pi}{\hbar} |F_{\nu n}|^2 \delta(E_\nu - E_n^{(0)} - \hbar\omega) d\nu. \tag{1}$$

In the problem considered, the perturbation has the form

$$\hat{V} = -e\boldsymbol{\mathcal{E}}(t) \cdot \mathbf{r} \equiv \hat{F} e^{-i\omega t} + \hat{F}^+ e^{i\omega t},$$

and

$$\hat{F} = \frac{-e\mathcal{E}_0(x + i\zeta y)}{2}.$$

In the perturbation matrix element $F_{\nu n} = \langle \psi_\nu^{(0)} | \hat{F} | \psi_n^{(0)} \rangle$, $\psi_n^{(0)}$ is the wave-function of the initial state as follows,

$$\psi_0 = \sqrt{\frac{\kappa_0}{2\pi}} \frac{e^{-\kappa_0 r}}{r},$$

which describes the particle ground-state in a zero-range potential (see Problem 4.10). We choose[280] ν to be the wave vector $\mathbf{k}$ of the particle leaving to infinity after ionization, and correspondingly for $\psi_\nu^{(0)}$, we take wavefunctions $\psi_\mathbf{k}^{(-)}$. But in the conditions of the problem, instead of $\psi_\mathbf{k}^{(-)}$ we should use the wavefunctions of a free particle: $\psi_\mathbf{k}^{(0)} = (2\pi)^{-3/2} e^{i\mathbf{k}\cdot\mathbf{r}}$. This is connected with the fact that in the case of a zero-range potential, $\psi_\mathbf{k}^{(-)}$ differs from $\psi_\mathbf{k}^{(0)}$ only by terms that correspond to the angular momentum $l = 0$ (a zero-range potential does not affect a particle with $l \neq 0$), and

---

[280] Remember that ν is a set of quantum numbers that describe a continuous spectrum of unexcited states. Another convenient set for ν is $\nu = (k, l, m)$, where l is the angular momentum of the leaving particle. The perturbation matrix element is different from zero only for the value $l = 1$, while the corresponding wavefunctions $\psi_\nu^{(0)}$ coincide with the wavefunctions $\psi_{klm}^{(0)}$ of a free particle. See Problem 11.35.

the input of these terms into the matrix element $\langle \psi_{\mathbf{k}}^{(-)} | \hat{F} | \psi_0 \rangle$ is equal to zero. We obtain:

$$F_{\nu n} = -\frac{e\sqrt{\kappa_0}\mathcal{E}_0}{8\pi^2} \int \frac{x + i\zeta y}{r} e^{-\kappa_0 r - i\mathbf{k}\cdot\mathbf{r}} dV. \tag{2}$$

The integral here is equal to

$$i\left(\frac{\partial}{\partial k_z} + i\zeta\frac{\partial}{\partial k_y}\right) \int \frac{1}{r} e^{-\kappa_0 r - i\mathbf{k}\cdot\mathbf{r}} dV = i\left(\frac{\partial}{\partial k_z} + i\zeta\frac{\partial}{\partial k_y}\right) \frac{4\pi}{\kappa^2 + k^2}. \tag{3}$$

It is calculated in spherical coordinates, with the polar axis directed along vector $\mathbf{k}$. Using

$$d\nu \equiv d^3k = k^2 dk \, d\Omega_k = \frac{mk \, d\Omega_k \, dE_k}{\hbar^2},$$

and performing integration over $E_k \equiv E_\nu$ in Eq. (1), we obtain

$$\frac{dw}{d\Omega_k} = \frac{2e^2 m\kappa_0 k^3 \mathcal{E}_0^2}{\pi\hbar^3(\kappa_0^2 + k^2)^4} \sin^2\theta \cdot (\cos^2\varphi + \zeta^2 \sin^2\varphi). \tag{4}$$

The polar axis is chosen along axis z, so that $k_z = k\sin\theta\cdot\cos\varphi$. Since $E_0^{(0)} = -\hbar^2\kappa_0^2/2m$, then for the escaping particles, $\hbar k = \sqrt{2m\hbar\omega - \hbar^2\kappa_0^2}$.

Note the following properties of the angular distribution of escaping particles. For the values $\zeta = \pm 1$ (that correspond to a circular polarization), we have

$$\frac{dw}{d\Omega_k} \propto \sin^2\theta,$$

which corresponds to a particle with angular momentum $l = 1$ and its projection $l_z = \pm 1$. In the case of $\zeta = 0$ (linear polarization of the wave along axis x), the angular distribution is

$$\frac{dw}{d\Omega_k} \propto \cos^2\theta',$$

where θ' is the angle between vector $\mathbf{k}$ and axis x. This corresponds to a particle with $l = 1$ and $l_x = 0$.

Performing integration over the angles in expression (4), we obtain the total probability of ionization per unit of time:

$$w = \frac{2e^2(1 + \zeta^2)}{3\hbar m} \cdot \frac{\sqrt{\omega_0}(\omega - \omega_0)^{3/2}}{\omega^4} \mathcal{E}_0^2. \tag{5}$$

where $\omega_0 = \hbar\kappa_0^2/2m$ is the ionization threshold frequency. Let us note that the dependence

$$w \propto (\omega - \omega_0)^{3/2} \propto k^3 \quad \text{for} \quad \omega \to \omega_0$$

is determined by the angular momentum ($w \propto k^{2l+1}$) of the leaving particle – in this case equal to $l = 1$ – and is connected to the penetrability of the centrifugal barrier for slow particles. Compare to Problem 9.30.

Let us make several final remarks:

1) For $\omega < \omega_0$, the ionization probability according to Eqs. (1) and (5) is equal to zero. In this case, it is determined by higher orders of perturbation theory.
2) The generalization of the results obtained, Eqs. (4) and (5), to the case of ionization of a state with a small binding energy and the angular momentum $l = 0$ in a short-range potential $U_s(r)$ with a finite radius, $r_s \neq 0$, is obtained by inserting the factor $C_{\kappa 0}^2$ into these equations. Compare to Problem 11.36.
3) We emphasize that apart from the usual applicability condition of perturbation theory $|F_{\nu n}| \ll |E_\nu - E_n|$ or $em\mathcal{E}_0/\hbar^2 \kappa_0^3 \ll 1$, validity of expressions (4) and (5) requires satisfaction of two more conditions: $\kappa_0 r_s \ll 1$ and $k r_s \ll 1$. The former reflects the weakly-bound nature of the state considered, while the latter places a restriction on the frequency of the wave, $\hbar\omega \ll \hbar^2/mr_s^2$. If these are not satisfied, then in the integral of matrix element $F_{\nu n}$, an essential role is played by the region $r \leq r_S$ (in the case of $kr_s \geq 1$ this is connected to fast oscillations of $e^{i\mathbf{k}\cdot\mathbf{r}}$), where the wavefunctions $\psi_n^{(0)}$ and $\psi_k^{(-)}$ depend on a concrete form of potential. Then, the replacement of the potential by a zero-range potential is not justified.
4) The considered state in the field of a wave that is periodic in time, strictly speaking, is *a quasi-energy state* (see Chapter 6, sec. 5). In Eq. (8) from Problem 8.42 we calculate the quasi-energy, the imaginary part of which gives the width of this QES, which is connected to ionization probability (5) by relation $\Gamma = \hbar w$. See Problem 11.66.

Problem 11.64

The same as in the previous problem, but now a particle in a short-range potential $U_S(r)$ that is weakly bound with the angular momentum $l = 1$ is subjected to the field of a linearly polarized wave.[281] The frequency of this wave satisfies the condition $\hbar\omega \ll \hbar^2/mr_s^2$, where r_s is the radius of the potential.

Solution

This problem is solved analogously to the previous Problem. Due to the weak-bound nature of the state, $\kappa_0 r_s \ll 1$, and the restriction on the wave frequency $\hbar\omega \ll \hbar^2/mr_s^2$, in perturbation matrix element $\langle \nu | e\mathcal{E}z | n \rangle$ (axis z is directed along the vector $\boldsymbol{\mathcal{E}}(t)$) the essential role is played by large distances[282] $r \gg r_s$. For the wavefunction of the initial state $\psi_n^{(0)} \equiv \psi_{\kappa, l=1, m}$ we can use expression (1) from Problem 11.37. The wavefunction of the final state (as in Problem 11.63) could be taken in the form of

---

[281] Wavefunctions of quasi-stationary states, that appear in the wave-field, are characterized by a definite value of the projection of the angular momentum onto the direction of the electric field. For them, the ionization probability per unit of time is connected to the width of the level $\Gamma = \hbar w$.

[282] Compare to the case of a statistic electric field in the conditions of Problem 11.37.

a plane wave $\psi_{\mathbf{k}}^{(-)} = \psi_{\mathbf{k}}^{(0)}$. (This is connected to the fact that the leaving particle is slow, $kr_s \ll 1$, and so the phase shifts $\delta_l \sim (kr_S)^{2l+1}$, that determine the difference of the wavefunction from that of a free particle for $r > r_s$, are small; see Eq. (XIII.15)). With the expressions $Y_{lm}(\mathbf{n}) = a_i(m)x_i/r$ for spherical functions (see Problem 3.41), the perturbation matrix element could be transformed to the form:

$$\langle k|e\mathcal{E}_0 z|\kappa lm\rangle = -\frac{e\kappa_0 C_{\kappa 1}\mathcal{E}_0}{\sqrt{2\pi^2}} a_i(m) \frac{\partial}{\partial k_z}\frac{\partial}{\partial k_i} \int r^{-3/2+\varepsilon} e^{-i\mathbf{k}\cdot\mathbf{r}} K_{3/2}(\kappa_0 r) dV, \quad (1)$$

with $\varepsilon \to 0$ ($\varepsilon > 0$; we cannot immediately put $\varepsilon = 0$ due to the appearance of a "false" divergence of the integral, which disappears after differentiation over k_i). After performing integration over the angles in spherical coordinates, the integral takes the form:

$$\frac{4\pi}{k} \int r^{-1/2+\varepsilon} \sin kr \cdot K_{3/2}(\kappa_0 r) dr =$$

$$\pi \left(\frac{2}{\kappa_0}\right)^{3/2+\varepsilon} \Gamma\left(\frac{\varepsilon}{2}\right) \Gamma\left(\frac{3}{2}+\frac{\varepsilon}{2}\right) F\left(\frac{3+\varepsilon}{2},\frac{\varepsilon}{2},\frac{3}{2},-\frac{k^2}{\kappa_0^2}\right), \quad (2)$$

where F is an hypergeometric function:

$$F(\alpha,\beta,\gamma,z) = 1 + \frac{\alpha \cdot \beta}{\gamma} \cdot \frac{z}{1!} + \frac{\alpha(\alpha+1)\cdot\beta(\beta+1)}{\gamma(\gamma+1)} \cdot \frac{z^2}{2!} + \ldots, \quad |z|<1.$$

Using the fact that $\varepsilon \to 0$,

$$F(\gamma+\varepsilon,\varepsilon,\gamma,z) \approx 1 + \varepsilon \sum_{n=1}^{\infty} \frac{z^n}{n} = 1 - \varepsilon \ln(1-z),$$

and $\Gamma(\varepsilon) \approx \varepsilon^{-1}$. The integral in expression (1), according to Eq. (2), is written in the form:

$$B(\kappa_0,\varepsilon) - \sqrt{\frac{2\pi^3}{\kappa_0^3}} \ln\left(\frac{k^2+\kappa_0^2}{\kappa_0^2}\right).$$

The first term diverges $\propto 1/\varepsilon$ for $\varepsilon \to 0$, but does not depend on k and so does not contribute to matrix element (1).

Now, using the connection between $a_i(m)$ and $Y_{lm}(\mathbf{n})$, we find:

$$a_i(m) \frac{\partial}{\partial k_z}\frac{\partial}{\partial k_i} \ln(k^2+\kappa_0^2) = -\frac{-4k\cos\theta \cdot Y_{lm}(\mathbf{k}/k) + 2i\sqrt{3/4\pi}(k^2+\kappa_0^2)\delta_{m,0}}{(k^2+\kappa_0^2)^2}. \quad (3)$$

This determines the perturbation matrix element (1), and we obtain the angular distribution of the particle after ionization:

$$\frac{dw}{d\Omega} = \frac{e^2\mathcal{E}_0^2 C_{\kappa 1}^2 \sqrt{\omega-\omega_0}}{2\hbar m\omega^4 \sqrt{\omega_0}} \left|\sqrt{\frac{3}{4\pi}}\omega\delta_{l_z,0} + 2i(\omega-\omega_0)\cos\theta Y_{1l_z}\left(\frac{\mathbf{k}}{k}\right)\right|^2. \quad (4)$$

Compare to the derivation of Eq. (4) from Problem 11.63.

We will discuss the result obtained, Eq. (4).

For the ionization from states with the projections of the angular momentum, $l_z = \pm 1$, the angular distribution of the leaving particle is described by the expression:

$$\frac{dw}{d\Omega} \propto \cos^2\theta \cdot \sin^2\theta \propto |Y_{21}|^2,$$

which corresponds to a particle that has the angular momentum $l = 2$ and $l_z = \pm 1$. At the threshold of ionization, i.e., for $\omega \to \omega_0 = \hbar\kappa_0^2/2m$, we have $dw/d\Omega \propto (\omega - \omega_0)^{5/2}$, as it is supposed to be ($\propto k^{2l+1}$) for the angular momentum $l = 2$. Compare to the previous problem.

In the case of $l_z = 0$ for the bound particle, the angular distribution (4) describes interference of the s- and d-waves. Here on the threshold, the s-wave dominates and

$$\frac{dw}{d\Omega} \propto (\omega - \omega_0)^{1/2} \quad \text{for} \quad \omega \to \omega_0.$$

Performing in expression (4) an integration over the angles, we find the total probability of ionization per unit of time. For a particle that in the initial state has the angular momentum $l = 1$ and projection $l_z = \pm 1$, we obtain

$$w_{l_z = \pm 1} = \frac{2e^2 C_{\kappa 1}^2 \mathcal{E}_0^2 (\omega - \omega_0)^{5/2}}{5m\hbar \omega^4 \sqrt{\omega_0}}, \tag{5}$$

for a particle with $l = 1$ and $l_z = 0$:

$$w_{l_z = 0} = \frac{e^2 C_{\kappa 1}^2 \mathcal{E}_0^2 \sqrt{\omega - \omega_0}(7w^2 - 4\omega\omega_0 + 12\omega_0^2)}{10m\hbar \omega^4 \sqrt{\omega_0}}. \tag{6}$$

Problem 11.65

The same as in Problem 11.63, but for a hydrogen-like atom in the ground-state. Consider only the case of an external field with a large frequency, $\hbar\omega \gg m(Ze^2)^2/\hbar^2$, so that the leaving electron is has a high velocity.

Comment

To solve the problem, use the operator of the form $\hat{V} = -(e/mc)\mathbf{A}(t) \cdot \hat{\mathbf{p}}$ to describe the interaction between charged particle and an homogeneous electric field. This operator corresponds to the choice of potential, $\varphi = 0$. Here $\boldsymbol{\mathcal{E}}(t) = -\partial \mathbf{A}(t)/c\partial t$, so for an elliptically polarized monochromatic wave,

$$\mathbf{A}(t) = -\frac{c\mathcal{E}_0}{\omega}(\sin\omega t, -\zeta \cos\omega t, 0).$$

Such interaction is equivalent to the one used in the two previous problems, $\hat{V}' = -e\boldsymbol{\mathcal{E}}(t) \cdot \mathbf{r}$ ($\mathbf{A}' = 0$, $\varphi' = -\boldsymbol{\mathcal{E}}(t) \cdot \mathbf{r}$), and is obtained by a gauge transformation with $\chi = -c\mathbf{r} \cdot \int \boldsymbol{\mathcal{E}}(t)dt$ (see, for example, Problem 6.27). Its advantage, in comparison to $\hat{V}'$, is the fact that in the case of large frequency ω, when the leaving particle is fast, while calculating the perturbation matrix element, we can take the wavefunction of the final state, $\psi_\mathbf{k}^{(-)}$, to be a plane wave.

Solution

As in the two previous problems, we calculate the ionization probability using Eq. (1) from Problem 11.63. Now, though,

$$\hat{F} = -\frac{i\mathcal{E}_0}{2\omega}(\hat{p}_x + i\zeta\hat{p}_y), \quad \psi_n^{(0)} \equiv \psi_{1s} = \sqrt{\frac{Z^3}{\pi}} e^{-Zr}$$

(we use atomic units), but still $\psi_\nu^{(0)} \approx (2\pi)^{-3/2} e^{i\mathbf{k}\cdot\mathbf{r}}$.

The perturbation matrix element is

$$F_{\nu n} = -\frac{i\sqrt{Z^3}\mathcal{E}_0}{4\sqrt{2}\pi^2\omega} \int e^{-i\mathbf{k}\cdot\mathbf{r}}(\hat{p}_x + i\zeta\hat{p}_y) e^{-Zr} dV = -\frac{i\sqrt{2Z^5}\mathcal{E}_0}{4\pi\omega^3}(k_x + i\zeta k_y).$$

Under the integral, we can replace operators $\hat{p}_{x,y}$ by $k_{x,y}$. Then it is easily calculated in spherical coordinates with polar axis along $\mathbf{k}$. In the expression given we have taken into account that $k \approx \sqrt{2\omega} \gg Z$. So, the probability of ionization per unit of time with the given direction of electron escape becomes equal to

$$\frac{dw}{d\Omega} = \frac{Z^5 \mathcal{E}_0^2}{\sqrt{2}\pi\omega^{9/2}} \sin^2\theta \cdot (\cos^2\varphi + \zeta^2 \sin^2\varphi). \tag{1}$$

The angular distribution has the same form as in the conditions of Problem 11.63. See Eq. (4) of that problem and the comment connected to it.

Integration of expression (1) over the angles gives the total probability of atom (ion) ionization per unit of time:

$$w(\omega) = \frac{64(1+\zeta^2)}{3} Z^5 \left(\frac{a_B^2 \mathcal{E}_0}{e}\right)^2 \left(\frac{\omega_0}{\omega}\right)^{9/2} \omega_0, \tag{2}$$

$$\omega \gg Z^2 \omega_0, \quad \omega_0 = me^4/\hbar^3.$$

Problem 11.66

Find the *dynamic polarizability* $\beta_0(\omega)$ for a particle in a zero-range potential. Generalize the result obtained to the weakly-bound state of a particle with the angular momentum $l = 0$ in a short-range potential $U_S(r)$, and in the corresponding limiting cases, compare it to Problems 11.36 and 11.63. To solve the problem, use the perturbation theory for quasi-energy states from Problem 8.42.

Solution

Dynamic polarizability $\beta_n(\omega)$ determines the change of *quasi-energy* (the quadratic in the field term) of a system in the quasi-energy state ψ_{ε_n}, that is subjected to a monochromatic uniform electric field. In the case of a linearly polarized wave, *i.e.*, an electric field of the form $\mathcal{E}(t) = \mathcal{E}_0 \cos \omega t$, QES are characterized by a definite value of the projection of the angular momentum onto the direction of the electric field (chosen along the axis z), so we can use the theory for non-degenerate (in the absence of perturbation) levels given in Problem 8.42. Note that for states with an angular momentum different from zero, in the field of an elliptically polarized wave, difficulties appear, connected to the problem of finding true zeroth approximation functions. But for initial states with the angular momentum $l = 0$, there are no such problems, and here the change of quasi-energy is described by the expression:

$$E_0^{(2)} = -\frac{1}{4}(1 + \zeta^2)\beta_0(\omega)\mathcal{E}_0^2,$$

where ζ is the degree of ellipticity.

The initial relation for the calculation of $\beta_0(\omega)$ is Eq. (8) from Problem 8.42, which now has the form:

$$\beta_0(\omega) = 2 \int \frac{\omega_{k0} |\langle \psi_{\mathbf{k}}^{(-)} | z | \psi_0 \rangle|^2 d^3 k}{\omega_{k0}^2 - \omega^2 - i\gamma}. \tag{1}$$

$\gamma > 0$ is an infinitely small quantity, wavefunctions ψ and $\psi_{\mathbf{k}}^{(-)}$ are the same as in Problem 11.63, $E_0 = -\kappa_0^2/2$ is the energy of the unperturbed state, $E_k = k^2/2$ and $\omega_{k0} = (k^2 + \kappa_0^2)/2$, and here we use the system of units $e = \hbar = m = 1$.

The matrix element of coordinate z in Eq. (1) was calculated in Problem 11.63. Its angular part $\propto k_z = k \cos \theta$. After an elementary integration over the angles ($d^3 k = k^2 dk \, d\Omega$), Eq. (1) becomes equal to

$$\beta_0(\omega) = \frac{32\kappa_0}{3\pi} \int_{-\infty}^{\infty} \frac{k^4 dk}{(\kappa_0^2 + k^2)^3[(\kappa_0^2 + k^2)^2 - 4\omega^2 - i\gamma]}. \tag{2}$$

The integral is calculated using residues by closing the integration contour, for example, in the upper half-plane of the complex variable k. Calculation could be simplified if we first perform some simple algebraic transformations and write the function under the integral sign in (2) in the form:

$$\frac{k^4}{4\omega^2(\kappa_0^2 + k^2)} \left[-\frac{1}{(\kappa_0^2 + k^2)^2} + \frac{1}{(\kappa_0^2 + k^2)^2 - 4\omega^2 - i\gamma} \right]. \tag{3}$$

The integral of the first term in the brackets is easily calculated (especially if we write $(z + k^2)^{-3}$ with $z = \kappa_0^2$ as $d^2(z + k^2)^{-1}/2dz^2$ and take differentiation over z outside the integral sign). Its contribution to $\beta_0(\omega)$ is $-1/\omega^2$. Then, writing the term that corresponds to the second term in the brackets in the form

$$\frac{k^4}{32\omega^4} \left(-\frac{2}{\kappa_0^2 + k^2} + \frac{1}{\kappa_0^2 + k^2 - 2\omega - i\gamma} + \frac{1}{\kappa_0^2 + k^2 + 2\omega + i\gamma} \right) \tag{4}$$

($\omega > 0$), we obtain the final result:

$$\beta_0(\omega) = C_{\kappa 0}^2 \left[-\frac{1}{\omega^2} - \frac{2\kappa_0^4}{\omega^4} + \frac{\kappa_0}{3\omega^4}(\kappa_0 - 2\omega)^{3/2} + \frac{\kappa_0}{3\omega^4}(\kappa_0^2 + 2\omega)^{3/2} \right]. \quad (5)$$

Here we introduced the factor $C_{\kappa 0}^2$ (the square of the *asymptotic coefficient*), which corresponds to a generalization of the result for a zero-range potential to the case of a weakly-bound state of a particle with the angular momentum $l = 0$ in a short-range potential $U_s(r)$. Compare to Problems 11.36 and 11.63.

Let us discuss some properties of the dynamic polarizability in (5):

1) For small frequencies, $\omega \ll \kappa_0^2$, we have

$$\beta_0(\omega) \approx \beta_0 \left(1 + \frac{7\omega^2}{6\kappa_0^4}\right), \quad \text{where} \quad \beta_0 = \frac{1}{4\kappa_0^4} C_{\kappa 0}^2, \quad (6)$$

and where β_0 reproduces the value of the static polarizability from Problem 11.36.

2) For values of frequency $\omega > \kappa_0^2/2$ (above the ionization threshold), an imaginary part appears in $\beta_0(\omega)$ that determines a width Γ of the state, $E_0^{(2)} \equiv \Delta E - i\Gamma/2$,

$$\Gamma(\omega) = \frac{\sqrt{2}(1+\zeta^2)}{3\omega^4} \kappa_0 C_{\kappa 0}^2 \mathcal{E}_0^2 \left(\omega - \frac{\kappa_0^2}{2}\right)^{3/2}. \quad (7)$$

The existence of a quasi-energy level width reflects the possibility of system ionization, the probability (per unit of time) of which $w = \Gamma$ (since $\hbar = 1$) coincides, of course, with the result of Problem 11.63.

3) In the case of a zero-range potential, $C_{\kappa 0}^2 = 1$, and from Eq. (5) we have $\beta_0(\omega) \approx -1/\omega^2$ for $\omega \to \infty$ in accordance with the general result (see Problem 6.40). For a potential of finite radius r_s, Eq. (5) is applicable only for frequencies that are not too large: $\omega \ll r_s^{-2}$. For more about this restriction, see Problem 11.63.

Problem 11.67

Estimate up to exponential accuracy (*i.e.*, do not pay attention to overall coefficients of order one) the probability of escape per unit of time of a charged particle, bound in a central potential $U(r)$ ($U(r) \to 0$ for $r \to \infty$), under the influence of a weak homogeneous electric field.

Solution

The important, exponential factor in the probability is determined by the penetrability of the electrostatic barrier $V = -e\mathcal{E}z$, and is equal to ($c = \hbar = m = 1$):

$$w \sim \frac{v}{r_S} D \sim \frac{1}{r_S^2} \exp\left\{-2 \int_0^b \sqrt{\kappa_0^2 - 2\mathcal{E}z}\, dz \right\} \approx \frac{1}{r_S^2} \exp\left\{-\frac{2\kappa_0^3}{3\mathcal{E}}\right\}. \quad (1)$$

The right turning point is $b = \kappa_0^2/2\mathcal{E}$, the left turning point is $a \sim r_s$, r_s is the potential radius, we put $\kappa_0 r_s \ll 1$ (a weakly-bound particle, its energy is $E_0 = -\kappa_0^2/2$), and v/r_s determines how many times the particle hits the barrier per unit of time (compare to Problems 9.30 and 9.28).

To refine this result, we see that in Eq. (1) the barrier penetrability corresponds to a particle that moves along the field (in the direction of axis z). For a particle that moves at an angle θ with the field, the barrier is less transparent, and this fact could be taken into account if we see that the effective electric field for it is $\mathcal{E}\cos\theta \approx \mathcal{E}(1-\theta^2/2)$. Here, for the probability of particle tunneling ionization from the state with angular momentum l and its projection m onto the direction of the field, we obtain[283]

$$w_{lm} \approx \frac{1}{r_S^2} \int |Y_{lm}(\mathbf{n})|^2 D(\mathcal{E}\cos\theta) d\Omega \approx$$

$$\approx \frac{1}{r_S^2} \int |Y_{lm}|^2 \exp\left\{-\frac{2\kappa_0^3}{3\mathcal{E}}\left(1+\frac{\theta^2}{2}\right)\right\} d\Omega \approx$$

$$\approx \frac{2l+1}{|m|! r_S^2} \frac{(l+|m|)}{(l-|m|)!} \left(\frac{3\mathcal{E}}{4\kappa_0^3}\right)^{|m|+1} \exp\left\{-\frac{2\kappa_0^3}{3\mathcal{E}}\right\}. \tag{2}$$

(Since only small angles $\theta \ll 1$ are important, then in the spherical function it is enough to restrict ourselves to only the factor $\propto \theta^{|m|}$.) For a short-range potential, this equation correctly gives a dependence of the ionization probability on a weak electric field intensity both in the exponential and the overall coefficient.

Problem 11.68

Find the probability of a K-electron ejection from an atom for a dipole transition of the nucleus from an excited state, as a result of direct electrostatic interaction between the electron and the nuclear protons (*inner conversion*). Use the ψ function of a hydrogen-like atom with a K-electron as the wavefunction of the electron's initial state. The velocity of the escaping electron is assumed to be much larger than atomic velocities.

Solution

The Hamiltonian of the "nucleus+electron" system has the form:

$$\hat{H} = \hat{H}_{nuc} - \frac{1}{2}\Delta_e - \sum_p \frac{1}{|\mathbf{r}_e - \mathbf{r}_p|}. \tag{1}$$

---

[283] Strictly speaking, in Eq. (2) we should also include the factor $\sim (\kappa_0 r_S)^{2l+1}$, which takes into account the decrease in probability of particle escape due to the centrifugal barrier. Compare to Problem 9.30.

$\hat{H}_{nuc}$ is the Hamiltonian of the nucleus subsystem in the center-of-inertia system. Summation is performed over all protons. Writing

$$U = -\sum_p \frac{1}{|\mathbf{r}_e - \mathbf{r}_p|} = -\frac{Z}{r_e} - \sum_p \left(\frac{1}{|\mathbf{r}_e - \mathbf{r}_p|} - \frac{1}{r_e} \right) \equiv -\frac{Z}{r_e} + V,$$

we see that V is the relevant part of the interaction that depends on the state of the nucleus subsystem, and is responsible for the transition of interest (in the approximation a point nucleus, $V \equiv 0$). Its probability (per unit of time) is given by the equation

$$dw_\nu = 2\pi |V_{\nu n}|^2 \delta(E_\nu - E_n^{(0)}) d\nu. \tag{2}$$

The wavefunctions that enter the perturbation matrix element $V_{\nu n}$ are

$$\psi_n^{(0)} = \psi_0^{nuc}\psi_0, \quad \psi_0 = \left(\frac{Z^3}{\pi}\right)^{1/2} e^{-Zr_e}, \quad \psi_\nu^{(0)} = \psi_1^{nuc}\psi_\mathbf{k}, \quad \psi_\mathbf{k} = \frac{1}{(2\pi)^{3/2}} e^{i\mathbf{k}\cdot\mathbf{r}_e},$$

where $\psi_{0(1)}^{nuc}$ is the wavefunction of the nucleus subsystem, so that

$$V_{\nu n} = -\frac{Z^{3/2}}{\sqrt{8\pi^2}} \iint \psi_1^{nuc} e^{-i\mathbf{k}\cdot\mathbf{r}_e} \sum_p \left(\frac{1}{|\mathbf{r}_e - \mathbf{r}_p|} - \frac{1}{r_e}\right) e^{-Zr_e} \psi_0^{nuc} d\tau \, dV_c.$$

In this expression, $d\tau$ is the product involving all independent nucleus coordinates (including spin variables).

To calculate the matrix element, we expand the Coulomb potential in a Fourier integral:

$$\sum_p \frac{1}{|\mathbf{r}_e - \mathbf{r}_p|} = \frac{1}{2\pi^2} \int \frac{d^3q}{q^2} e^{i\mathbf{q}\cdot\mathbf{r}_e} \sum_p e^{-i\mathbf{q}\cdot\mathbf{r}_p}.$$

Now

$$V_{\nu n} = -\frac{Z^{3/2}}{2^{5/2}\pi^4} \int \frac{d^3q}{q^2} \int d\tau \psi_1^{nuc} \sum_p (e^{i\mathbf{q}\cdot\mathbf{r}_p} - 1)\psi_0^{nuc} \int dV_c e^{(\mathbf{q}-\mathbf{k})\cdot\mathbf{r}_e - Zr_e}. \tag{3}$$

We can expand the "nucleus" exponent into a series over powers $\mathbf{q}\cdot\mathbf{r}_p$. This corresponds to the expansion of terms $|\mathbf{r}_e - \mathbf{r}_p|^{-1}$ over a small parameter $r_p/r_e \sim ZR_{nuc}/a_B \ll 1$ (but the latter expansion is less convenient for further transformations than the transition to Fourier components). Expanding the exponent, we obtain

$$\int \psi_1^{nuc} \sum_p (e^{-i\mathbf{q}\cdot\mathbf{r}_p} - 1)\psi_0^{nuc} d\tau \approx \int \psi_1^{nuc} \left(-i\mathbf{q}\sum_p \mathbf{r}_p\right)\psi_0^{nuc} d\tau \equiv -i\mathbf{q}\cdot\mathbf{d}_{10}, \tag{4}$$

where $\mathbf{d}_{10}$ denotes the matrix element of the nuclear dipole moment. Now, we integrate over the electron coordinates, which gives

$$\int \exp\{i(\mathbf{q}-\mathbf{k})\cdot\mathbf{r}_c - Zr_e\}dV_c = \frac{8\pi Z}{[Z^2+(\mathbf{q}-\mathbf{k})^2]^2}.$$

We transform matrix element (3) to the form

$$V_{\nu n} = \frac{i\sqrt{2}Z^{5/2}}{\pi^3}\mathbf{d}_{10}\cdot\int\frac{d^3q}{q^2}\frac{\mathbf{q}}{[Z^2+(\mathbf{q}-\mathbf{k})^2]^2}. \tag{5}$$

Since the velocity of the escaping electron is much larger than the atomic velocities[284] i.e., $k \gg Z$ (here the energy of escaping electron is $E_k = k^2/2 \approx E_{nuc,0} - E_{nuc,1}$), then the dominant role in integral (5) is played by the region of $\mathbf{q}$ for which $|\mathbf{q}-\mathbf{k}| \leq Z$, and so

$$\int\frac{d^3q}{q^2}\frac{\mathbf{q}}{[Z^2+(\mathbf{q}-\mathbf{k})^2]^2} \approx \frac{\mathbf{k}}{k^2}\int\frac{d^3q}{[Z^2+(\mathbf{q}-\mathbf{k})^2]^2} = \frac{\pi^2}{Z}\frac{\mathbf{k}}{k^2}.$$

$q \approx k$, so it follows that $qr_{nuc} \ll 1$, which explicitly justifies the aforementioned expansion of the exponential. Now,

$$V_{\nu n} = \frac{i\sqrt{2}Z^{3/2}}{\pi k^2}\mathbf{k}\cdot\mathbf{d}_{10}. \tag{6}$$

Since in Eq. (2), $d\nu \equiv k^2 dk\, d\Omega = k\, dE_k\, d\Omega$, then if we perform an integration in it first over E_k and then over the angles of the escaping electrons, using (6), then we find that

$$dw = \frac{4Z^3}{\pi}\frac{|\mathbf{k}\cdot\mathbf{d}_{10}|^2}{k^3}d\Omega, \quad w = \frac{16Z^3}{3k}|\mathbf{d}_{10}|^2. \tag{7}$$

Since there are two K-electrons in the atom, the total probability of inner conversion (per unit of time) of electrons on the K-shell is equal to

$$w_K = 2w = \frac{32Z^3}{3k}|\mathbf{d}_{10}|^2 = \frac{32m^3e^6}{3\hbar^7}\frac{Z^3e^2}{\hbar\nu}|\mathbf{d}_{10}|^2. \tag{8}$$

In this last expression, we returned to the conventional units.

On the other hand, the probability of a dipole photon emission for a nuclear transition is

$$w_{emis} = \frac{4\omega^3}{3\hbar c^3}|\mathbf{d}_{10}|^2,$$

---

[284] In this case, we can choose the plane wave (the wavefunction of a free electron) as a wavefunction of the escaping electron.

(see Eq. (XIV.10)), $\hbar\omega = mv^2/2$, so that the coefficient of inner conversion is

$$\beta_K \equiv \frac{w_K}{w_{emis}} = \frac{Z^3 \alpha^4}{2}\left(\frac{2mc^2}{\hbar\omega}\right)^{7/2}, \quad \alpha = \frac{e^2}{\hbar c} \approx \frac{1}{137}. \tag{9}$$

As is seen, it increases sharply with increase in Z and decrease of frequency ω.

Problem 11.69

The same as in the previous problem, but when the initial and final states of the nucleus have the angular momentum equal to zero and the same parity (such processes are called *monopole* or *E0 transitions*).

Solution

This problem is solved in a way similar to the previous one, and the initial stages of solutions to both problems are identical. But here, while calculating the nuclear part of the matrix element, we cannot use Eq. (4), since we now have $\mathbf{d}_{10} = 0$. Taking the following term in the expansion of the exponential in powers of $\mathbf{q} \cdot \mathbf{r}_p$, we obtain the nuclear part of the matrix element in the form:

$$-\frac{1}{2}\int \psi_1^{nuc*} \sum_p (\mathbf{q}\cdot\mathbf{r}_p)(\mathbf{q}\cdot\mathbf{r}_p)\psi_0^{nuc} d\tau = -\frac{q_i q_k}{2}\int \psi_1^{nuc*}\sum_p x_{pi}x_{pk}\psi_0^{nuc}d\tau \equiv$$

$$-\frac{1}{6}q_i q_k Q_0 \delta_{ik} = -\frac{1}{6}q^2 Q_0, \tag{1}$$

where

$$Q_0 = \int \psi_1^{nuc*}\left(\sum_p r_p^2\right)\psi_0^{nuc} d\tau. \tag{2}$$

It has been assumed that the parities of the initial and final states of nucleus are the same. Otherwise $Q_0 = 0$.

The perturbation matrix element is equal to

$$V_{\nu n} = \frac{1}{12\pi^2}Q_0 \int e^{i\mathbf{q}\cdot\mathbf{r}_e}\psi_k^*(\mathbf{r}_e)\psi_0(r_e) d^3q\, dV_c,$$

and after the integration over $\mathbf{q}$ that leads to the δ-function $\delta(\mathbf{r}_e)$, we obtain

$$V_{\nu n} = \frac{2\pi}{3}Q_0 \psi_k^*(0)\psi_0(0). \tag{3}$$

Now we easily find the angular distribution of escaping electron and the total probability of its escape (per unit of time):

$$\frac{dw}{d\Omega} = \frac{8\pi^3}{9}k|Q_0|^2|\psi_k^*(0)\psi_0(0)|^2, \quad w = \frac{32\pi^4}{9}k|Q_0|^2|\psi_k^*(0)\psi_0(0)|^2. \tag{4}$$

The isotropic nature of the distribution $dw/d\Omega$ is evident from symmetry.

Taking into account the existence of two K-electrons in the atom, the probability of the $E0$-transition is equal to $w_{E0} = 2w$. In particular, if the escaping electron is fast $k \gg Z$, then $\psi_k(0) \approx (2\pi)^{-3/2}$ and so

$$w_{E0} = \frac{8}{9} Z^3 k |Q_0|^2. \tag{5}$$

In conclusion, we note that from these results, it follows that inner conversion plays a dominant role for the electrons in the K-shell (since $w \propto |\psi(0)|^2$, compare to Problems 14.18 and 14.19).

Problem 11.70

Find the probability of K-electron escape as a result of the *Auger effect* in a μ-mesoatom (the muon in an excited state goes to the lower level, giving energy to the electron). Consider only a dipole or so-called Auger P-transition, for which a change of the muon angular momentum is $|\Delta l| = 1$. Assume that the size of the muon orbit is much smaller than the electron's, and approximate the electron in the final state as a free particle. Use, as an example, the muon transition $2p \to 1s$.

Solution

The Hamiltonian of a system that consists of a muon and an electron (which is one of the K-shell electrons), that are in the Coulomb field of a nucleus with charge Z, has the form:

$$\hat{H} = -\frac{1}{2m_\mu}\Delta_\mu - \frac{Z}{r_\mu} - \frac{1}{2}\Delta_e - \frac{Z}{r_e} + \frac{1}{|\mathbf{r}_e - \mathbf{r}_\mu|}. \tag{1}$$

Since the size of the muon orbit is much lower than the electron orbit ($m_\mu = 207 m_e$ and the Bohr radius $a_B \propto m^{-1}$), this Hamiltonian could be written in the form

$$\hat{H} = \hat{H}_\mu + \hat{H}_e + \hat{V},$$

where

$$\hat{H}_\mu = -\frac{1}{2m_\mu}\Delta_\mu - \frac{Z}{r_\mu}, \quad \hat{H}_e = -\frac{1}{2}\Delta_e - \frac{Z-1}{r_e}$$

and

$$\hat{V} = \frac{1}{|\mathbf{r}_e - \mathbf{r}_\mu|} - \frac{1}{r_e}.$$

If we neglect the perturbation $\hat{V}$, the system considered consists of two independent subsystems: the muon and electron (with the nucleus charge screened to unity).

Note that a calculation of the Auger effect is analogous to the calculation of the probability of inner conversion (see Problem 11.68), since we can consider the muon subsystem as a nucleus. Moreover, equations that describe the dipole (or P-)

Auger effect are obtained from the equations of problem 11.68 by replacing $Z \to Z - 1$ and substituting the initial and final states of the muon wavefunction instead of the nucleus wavefunction $\psi_{0,1}^{nuc}$. The probability (per unit of time) of the process with two K-electrons taken into account has the form:

$$w_P = \frac{32}{3} \frac{(Z-1)^3 e^2}{\hbar v} \frac{m_e^3 e^6}{\hbar^7} |\mathbf{d}_{10}|^2. \tag{2}$$

Compare to Eq. (8) from Problem 11.68. v is the velocity of the escaping electron, and $\mathbf{d}_{10}$ is the matrix element of the muon dipole moment. In particular, for the muon transition $2p \to 1s$ we have

$$|\mathbf{d}_{10}|^2 = \frac{2^{15}}{(3^{10} Z^2 m_\mu^2)}, \quad v^2 = \frac{3}{4} Z^2 m_\mu$$

in atomic units, $m_\mu = 207$, and according to (2) we obtain:

$$w_\mu(2p \to 1s) \approx 4.6 \cdot 10^{11} (Z-1)^3 / Z^3 \text{ s}^{-1}.$$

Compare to the probability of z photon emission $\omega_{emis} \approx 1.3 \cdot 10^{11} Z^4 \text{ s}^{-1}$.

In conclusion, we note that since the size of the muon orbit is $a_{n,\mu} \sim n^2/Zm_\mu$, while the size of the K-electron orbit is $\sim 1/Z$, the condition of applicability of Eq. (2) becomes the inequality $n^2 \ll m_\mu \approx 200$ (and that the escaping electron is fast).

Problem 11.71

The same as in the previous problem, but for $\Delta l = 0$ (Auger S-transitions). Consider the case, where the muon angular momentum in the initial and final states is equal to zero. Consider also the special case of the muon transition $2s \to 1s$.

Solution

If the dipole Auger transition is analogous to the process of inner conversion for a dipole nuclear transition (see Problem 11.68), then an Auger S-transition, in the case where muon has the angular momentum $l = 0$ in the initial and final states, is the analog of conversion for an $E0$-transition in nucleus (see Problem 11.69). In the same way that the solution of the previous problem followed the solution of Problem 11.68, the solution of this one follows Problem 11.69.

The probability of the transition considered (taking into account two K-electrons) is described by the expressions

$$w_S(n_1 s \to n_2 s) = \frac{8}{9}(Z-1)^3 k |Q_0|^2, \quad Q_0 = \int \psi_{n_2 00}^*(r_\mu) r_\mu^2 \psi_{n_1 00}(r_\mu) d^3 r_\mu. \tag{1}$$

Compare to Eqs. (5) and (2) from Problem 11.69.

In particular, using the explicit expression for wavefunction ψ_{n00} of a hydrogen-like mesoatom, we find the value $Q_0^2 = 2^{19}/(3^{10} Z^4 m_\mu^4)$ for the muon transition $2s\,t \to 1s$, and since here $k^2 = 3Z^2 m_\mu/4$, then according to Eq. (1) we obtain

$$w_S(2s \to 1s) = \frac{2^{21}}{\sqrt{3}\,3^{11}} \frac{(Z-1)^3}{m_\mu^{7/2} Z^3}\ \text{a.u.} \approx 2.21 \left(\frac{Z-1}{Z}\right)^3 10^9\ s^{-1}.$$

Problem 11.72

Classify (i.e., find all possible values of the following quantum numbers: total electron spin S, total angular momentum L, and parity I) the lower *autoionization states*[285] (AIS) of a two-electron atom or ion, connected to the electron configuration $nlnl'$, with $n=2$. Considering the interaction between the two electrons as a perturbation, find energy levels for these states in the first order of perturbation theory. Determine the true wavefunctions of the zeroth approximation.

Discuss the width of the AIS levels and their dependence on the nuclear charge, Z.

Solution

1) There are four (remembering spin, eight) independent one-particle states with the principal quantum number $n=2$: one 2s-state and three 2p-states (with $l_z = 0$ and ± 1). For the electron configuration $2s^2$, we have $L=0$, $I=+1$, $S=0$. The wavefunction of such a term $^1S^+$ has the form

$$\psi_{2s^2} = \psi_{2s}(r_1)\psi_{2s}(r_2)\chi_{\alpha\beta}^{(-)}, \tag{1}$$

where $\psi_{2s}(r) = \sqrt{Z^3/8\pi}\,(1 - Zr/2)e^{-Zr/2}$ is the wavefunction of a hydrogen-like atom's 2s-state, while $\chi_{\alpha\beta}^{(-)}$ is the anti-symmetric spin wavefunction, which corresponds to the value $S=0$ (due to the Pauli principle there is no state with $S=1$ for configuration $2s^2$).

For configuration $2s2p$, we have $L=1$, $I=-1$, and both singlet, $S=0$, and triplet, $S=1$, states are possible. The coordinate parts of the wavefunctions for such terms $^{1,3}P^-$ have the form:

$$\psi_{2s2p} = \frac{1}{\sqrt{2}}\{\psi_{2s}(r_1)\psi_{2p}(\mathbf{r}_2) \pm \psi_{2p}(\mathbf{r}_1)\psi_{2s}(\mathbf{r}_2)\}, \tag{2}$$

where $\psi_{2p}(\mathbf{r}) = \sqrt{Z^5/32\pi}\,(\mathbf{a}\cdot\mathbf{r})e^{-Zr/2}$ is the wavefunction of the 2p-states. Here $|\mathbf{a}|=1$ (see Problem 3.42), while signs $+$ and $-$ correspond to the values of S, equal to 0 and 1.

Finally, with an even, $I=+1$, configuration $2p^2$, both singlet $^1S^+$ and $^1D^+$-terms and triplet $^3P^+$-term are involved. The wavefunctions for these terms are

---

[285] By *autoionization states* we mean states of atomic systems with two or more excited electrons that are unstable against ionization (electron escape). So that when an excitation energy is transferred to one of the electrons, the latter leaves the atom (ion). AIS are quasi-stationary states, and usually appear as resonances.

determined by the corresponding equation from Problem 11.17, where for $\varphi(r)$ we should take the radial wavefunction of the $2p$-state. In particular, for the $^1S^+$-term we have

$$\psi_{2p^2,L=0} = \frac{\sqrt{3}}{96\pi} Z^5 (\mathbf{r}_1 \cdot \mathbf{r}_2) e^{-Z(r_1+r_2)/2} \chi_{\alpha\beta}^{(-)}. \qquad (3)$$

If we neglect interaction between electrons, the energies of all these states are the same and equal to $E^{(0)} = -2(Z^2/2n^2) = -Z^2/4$. This is larger than the energy of the corresponding one-electron ion (ground-state), equal to $-Z^2/2$, which shows the instability of such states. As a result of electron–electron interaction, a transition of one of the electrons to the $1s$-state together with a simultaneous escape of the other one is possible. Compare to Problem 11.12.

2) Let us now calculate a change in energy of the states considered due to mutual electron interaction. We begin with the $^1S^+$-terms. Since there are two of them (for configurations $2s^2$ and $2p^2$), we should use the secular equation. A calculation of the relevant matrix elements gives:

$$\langle 2s^2 | \frac{1}{|\mathbf{r}_1-\mathbf{r}_2|} | 2s^2 \rangle = \frac{77}{512} Z, \quad \langle 2p^2 | \frac{1}{|\mathbf{r}_1-\mathbf{r}_2|} | 2p^2 \rangle = \frac{111}{512} Z,$$

$$\langle 2s^2 | \frac{1}{|\mathbf{r}_1-\mathbf{r}_2|} | 2p^2 \rangle = \langle 2p^2 | \frac{1}{|\mathbf{r}_1-\mathbf{r}_2|} | 2s^2 \rangle = -\frac{45}{512\sqrt{3}} Z. \qquad (4)$$

The methods for calculating the corresponding integrals are described in Problems 11.5, 11.10, and 11.17. Now, the solution of the secular equation allows us to find shifts of the $^1S^+$-terms and the true functions of the zeroth approximation:

$$E^{(1)}_{^1S,1(2)} = \frac{47 \mp \sqrt{241}}{256} Z = \begin{cases} 0.123Z = E^{(1)}_{^1S,1}, \\ 0.244Z = E^{(1)}_{^1S,2}, \end{cases} \qquad (5)$$

$$|^1S^+, 1\rangle = 0.880\{|2s^2\rangle + 0.540|2p^2\rangle\}, \qquad (6)$$

$$|^1S^+, 2\rangle = 0.880\{-0.540|2s^2\rangle + |2p^2\rangle\}.$$

As is seen, in the lower of the split $^1S^+$-terms, the configuration $2s^2$ is present with a high probability; for it, we have $w_{2s^2,1} \approx 0.774$. In the other split $^1S^+$-term, the configuration $2p^2$ has a higher probability. Hence, these terms are sometimes classified as $2s^2\ ^1S^+$ and $2p^2\ ^1S^+$. We should, though, take into account that the "mixing" of configuration (6) strongly affects the widths of these auto-ionization states.

Shifts of the other terms are determined by the mean value of the perturbation $V = |\mathbf{r}_1 - \mathbf{r}_2|^{-1}$ in the corresponding states, and are as follows,

$$E^{(1)}(^3P^+) = \frac{21}{128} Z \approx 0.164Z, \quad E^{(1)}(^1D^+) = \frac{237}{1280} Z \approx 0.185Z,$$

$$E^{(1)}(^{1(3)}P^-) = \frac{83 \pm 15}{512} Z = \begin{cases} 0.191Z, & ^1P^-, \\ 0.133Z, & ^3P^-. \end{cases} \qquad (7)$$

Note that the shift for configuration $2p^2$ can be calculated using Eq. (7) from Problem 11.17.

From expressions (5) and (7), we see that one of the $^1S^+$ terms has the lowest energy (the other one has the highest energy). For this AIS, the energy of the escaping electron is equal to $Z^2/4 + 0.123Z$, which for a helium atom is 33.9 eV. The excitation energy for this state from the atom ground-state ($E_0(\text{He}) = -79.0$ eV) is 58.5 eV. (Although $Z = 2$ is not large, these values differ from exact ones by only 0.5 eV, and for other terms the difference is not greater than 2 eV.)

3) Now consider the widths of the AIS. Their values to first order in the perturbation $V = |\mathbf{r}_1 - \mathbf{r}_2|^{-1}$ could be found by the usual equation for the transition probability per unit of time ($\Gamma_i = w_{i \to f}$, $\hbar = 1$):

$$w_{i \to f} = 2\pi \int |\langle f|\hat{V}|i\rangle|^2 \delta(E_f - E_i) d\nu_f. \tag{8}$$

To describe the final states $|f\rangle$ with a set of quantum numbers for the escaping electron, we can choose k, l, m: $E_k = k^2/2$. Since here another electron passes to the 1s-state, but the total spin does not change after ionization, $S_f = S_i$, then the coordinate part of the wavefunction ψ_f can be obtained as a result of the corresponding symmetrization (or anti-symmetrization) of the function

$$Y_{lm}(\mathbf{n}_1) R_{kl}(r_1, Z) \psi_{1s}(r_2), \tag{9}$$

where R_{kl} is the radial wavefunction of an electron in the Coulomb potential,

$$R_{kl} = \frac{2k|\Gamma(l+1-iZ/k)|}{\sqrt{2\pi}(2l+1)!} (2kr)^l e^{-ikr + \pi Z/2k} F\left(\frac{iZ}{k} + l + 1, 2l + 2, 2ikr\right).$$

Here, we have made a modification that takes into account the form of the potential: $-Z/r$ (instead of $-1/r$), and added the factor $(2\pi)^{-1/2}$, corresponding to wavefunction normalization $\psi_{klm} = Y_{lm} R_{kl}$ of the form:

$$\langle klm|k'l'm'\rangle = \delta_{ll'} \delta_{mm'} \delta(k - k').$$

For such normalization, integration $\int d\nu_f \ldots$ in Eq. (8) is reduced to $\sum_{lm} \int dk \ldots$ and takes the form:

$$w_{i \to f} = \frac{2 \cdot 2\pi}{k} \sum_m \left|\langle k, l_f, m, 1s | \frac{1}{|\mathbf{r}_1 - \mathbf{r}_2|} |\psi_i\rangle\right|^2. \tag{10}$$

The angular momentum of the escaping electron, l_f, coincides with the angular momentum L of the state considered. The wavefunction of the final state coincides with Eq. (9), i.e., is not symmetrized; so we put an additional factor of 2 into Eq. (10). It is important that the wavefunctions of the initial states, given above, are symmetrized in the proper way.

Calculation of the radial part of the perturbation matrix element in Eq. (10) can not be done analytically and requires use of numerical methods (integration over the angles is performed in the usual way, as, for example, in Problem 11.17). So, we will limit oneself to only a few remarks about the width of auto-ionization states.

1) If for k we use the unperturbed value $k^2/2 = E_1^{(0)} - E_{1s}$, i.e., $k = Z/\sqrt{2}$, then we easily see that the width $\Gamma_i = w_{i \to f}$, according to Eq. (10), becomes independent from the nuclear charge Z.

2) The numerical values of the width are quite different for different AIS. The lower $^1S^+$- and also $^1D^+$- and $^1P^-$-terms (we can see that the width is two orders of magnitude smaller than the characteristic value $me^4/\hbar^2 \approx 27$ eV, if we note that according to Eqs. (6) and (7) the perturbation matrix element is of order $\sim 0.1 - 0.2$) have the largest width, $\Gamma \sim 0.2$ eV. For the state $^3P^-$ and second term of $^1S^+$, the width is much (more than one order) lower. This is connected to a strong cancellation in the matrix element of the contributions coming from different terms. In the case of the $^3P^-$-state, these are "direct" and "exchange" interactions, while in case of the $^1S^+$-term they are contributions coming from the $2s^2$- and $2p^2$-configurations (see Eq. (6)).

3) Calculation of the width in first order of perturbation theory for small values of Z is not very accurate (the results have a rather qualitative nature). This is mainly connected to the effect of nucleus screening for the escaping electron due to the $1s$-electron.

4) Finally, note an interesting fact about the $^3P^+$ state (for configuration $2p^2$) that has the angular momentum $L = 1$, positive parity, and electron spin $S = 1$. The decay of such a state with the transition of one electron to the $1s$-state and the escape of the other one is forbidden by the laws of angular momentum and parity conservation. Indeed, the escaping electron must have the angular momentum $l_f = 1$, but then the parity of the final state becomes negative. This means that if we take into account only Coulomb interaction, such a state, which exists on the background of a continuous spectrum, remains truly bound. Relativistic corrections to the interaction (specifically the spin-orbital interaction terms) lead to the appearance of an ionization width for such a state.

Problem 11.73

Find the time-dependence of the μ^+-muon polarization vector in the ground-state of a *muonium* that is in a uniform magnetic field perpendicular to the initial muon polarization (*i.e.*, investigate muon spin precession). Consider the case of a weak magnetic field, *i.e.*, $\mu_B^{(e)} \mathcal{H} \ll \Delta$, where $\mu_B^{(e)}$ is the electron Bohr magneton and Δ is the hyperfine splitting of the muonium ground level (compare to Problem 11.2). Assume that the electron is unpolarized at muonium formation, while the muon is entirely polarized.[286]

---

[286] Muonium is a kind of hydrogen atom, whose "nucleus" is the μ^+-muon. We can neglect the change of muon and electron polarization during the process of muonium formation. Recall the main characteristics of a muon: spin $1/2$, mass $m_\mu \approx 207 m_e$, magnetic moment equal to $e\hbar/2m_\mu c$, and

Solution

The Hamiltonian of the spin subsystem for the ground-state of muonium in a magnetic field (directed along the axis z) takes the form:

$$\hat{H} = \frac{\Delta}{4}\hat{\boldsymbol{\sigma}}^\mu \cdot \hat{\boldsymbol{\sigma}}^{(e)} + \frac{e\hbar}{2m_e c}\mathcal{H}\hat{\sigma}_z^{(e)}. \tag{1}$$

The first term here describes the interaction of muon and electron spin magnetic moments, and is defined by Eqs. (2) and (4) of Problem 11.2. Here,

$$\Delta = \frac{32\pi}{3}\mu_B^{(e)}\mu_B^{(\mu)}\psi_{1s}^2(0)$$

is the hyperfine splitting of the ground, $1s$-level of muonium. In the second term in Eq. (1) we neglected the interaction of the muon magnetic moment with the magnetic field due to the smallness (207 times) of the corresponding Bohr magneton in comparison to electron's $\mu_B^{(e)}$.

Considering the interaction with the magnetic field as a perturbation ($\mu_B^{(e)}\mathcal{H} \ll \Delta$), we see that spin functions χ_{SS_z} of the electron–muon system that correspond to the states with different values of the total spin S and its projection S_z (see Problem 5.10) are the true functions of the zeroth approximation for Hamiltonian (1). The corresponding eigenvalues, E_{SS_z}, are determined by Eqs. (VIII.1) in first- and second-order perturbation theory (see also Problem 11.42):

$$E_{1,\pm 1} = \frac{\Delta}{4} \pm \mu_B^{(e)}\mathcal{H}, \quad E_{10} \approx \frac{\Delta}{4} + \frac{(\mu_B^{(e)}\mathcal{H})^2}{\Delta},$$

$$E_{00} = -\frac{3\Delta}{4} - \frac{(\mu_B^{(e)}\mathcal{H})^2}{\Delta}. \tag{2}$$

For the temporal dependence of the spin wavefunction, according to Eq. (VI.2) we have[287] ($n \equiv S, S_z$):

$$\psi(t) = \sum_n c_n e^{-iE_n t/\hbar}\chi_{SS_z}. \tag{3}$$

Coefficients c_n in this expansion are determined by initial conditions. Since for $t = 0$ the muon is polarized in a direction perpendicular to the magnetic field, then (choosing this direction to be axis x) we have (see Problem 5.1):

$$\psi(0) \propto \chi_{s_x=1/2}^{(\mu)}, \quad \text{so that} \quad \psi(0) = \frac{1}{\sqrt{2}}\begin{pmatrix}1\\1\end{pmatrix}_\mu \begin{pmatrix}a_1\\a_2\end{pmatrix}_e. \tag{4}$$

---

lifetime $\tau_\mu \approx 2.2 \cdot 10^{-6}$ s. Since the polarization of the muon could be easily measured by the angular distribution of the positron that appears in its decay $\mu^+ \to e^+\nu\tilde{\nu}$, then the dynamics of muon spin could be used for the investigation of material properties.

[287] First consider a *pure* spin state of a system, and then go to the description using the density matrix.

Using the explicit form of spin functions χ_{SS_z}:

$$\chi_{1(0),0} = \frac{1}{\sqrt{2}}\left\{\begin{pmatrix}1\\0\end{pmatrix}_\mu \begin{pmatrix}0\\1\end{pmatrix}_e \pm \begin{pmatrix}0\\1\end{pmatrix}_\mu \begin{pmatrix}1\\0\end{pmatrix}_e\right\},$$

$$\chi_{11} = \begin{pmatrix}1\\0\end{pmatrix}_\mu \begin{pmatrix}1\\0\end{pmatrix}_e, \quad \chi_{1,-1} = \begin{pmatrix}0\\1\end{pmatrix}_\mu \begin{pmatrix}0\\1\end{pmatrix}_e.$$

From the condition that wavefunction (3) coincides with function (4) at $t = 0$, we find:

$$c_{10} + c_{00} = \sqrt{2}c_{1,-1} = a_1, \quad c_{10} - c_{00} = \sqrt{2}c_{11} = a_2. \tag{5}$$

Therefore,

$$c_{1(0),0} = \frac{a_1 \pm a_2}{2}.$$

The (complex) quantities a_i uniquely define the system's spin function $\psi(t)$. Through these coefficients, in the form of a bilinear combination $a_k^* a_i$, any spin characteristics of the electron–muon system may be expressed at an arbitrary moment of time. As is seen from Eq. (4), these quantities, a_i, determine the electron spin function at the moment of time $t = 0$. If the initial state of the electron is given by a density matrix $\hat{\rho}$, then in the corresponding bilinear combination we should make the substitution:

$$a_i a_k^* \to \overline{a_i a_k^*} = \rho_{ik}.$$

For an entirely unpolarized state (as in the conditions of this problem),

$$\overline{a_1 a_1^*} = \overline{a_2 a_2^*} = \frac{1}{2}, \quad \overline{a_1 a_2^*} = 0. \tag{6}$$

For calculation of the muon polarization vector,

$$\mathbf{P}^\mu(t) = \langle\psi(t)|\hat{\boldsymbol{\sigma}}^\mu|\psi(t)\rangle,$$

we should first consider the action of the operators $\hat{\sigma}_z^\mu$ and $\hat{\sigma}_x^\mu + i\hat{\sigma}_y^\mu$ on the spin functions χ_{SS_z}. We find:

$$\hat{\sigma}_z^\mu|1(0),0\rangle = |0(1),0\rangle, \quad \hat{\sigma}_z^\mu|1,\pm1\rangle = \pm|1,\pm1\rangle,$$

$$(\hat{\sigma}_x^\mu + i\hat{\sigma}_y^\mu)|1(0),0\rangle = \pm\sqrt{2}|1,1\rangle, \quad (\hat{\sigma}_x^\mu + i\hat{\sigma}_y^\mu)|1,1\rangle = 0,$$

$$(\hat{\sigma}_x^\mu + i\hat{\sigma}_y^\mu)|1,-1\rangle = \sqrt{2}(|1,0\rangle + |0,0\rangle).$$

Now, we see that

$$P_x^{(\mu)}(t) + iP_y^{(\mu)}(t) = \frac{1}{4}\sum_k e^{i\omega_k t}, \quad P_z^{(\mu)}(t) = 0, \tag{7}$$

where

$$\hbar\omega_1 = E_{11} - E_{10} = \mu_B^{(e)}\mathcal{H} - \frac{(\mu_B^{(e)}\mathcal{H})^2}{\Delta}, \quad \hbar\omega_3 = E_{00} - E_{1,-1} = \hbar\omega_1 - \Delta,$$

$$\hbar\omega_2 = E_{10} - E_{1,-1} = \mu_B^{(e)}\mathcal{H} + \frac{(\mu_B^{(e)}\mathcal{H})^2}{\Delta}, \quad \hbar\omega_4 = E_{11} - E_{00} = \hbar\omega_2 + \Delta. \quad (8)$$

We now discuss these results:

1) For $t = 0$, we have $P(0) = 1$ – a fully polarized (along axis x) muon in accordance with the problem condition. But in the following moments of time, $P(t) < 1$, so that a *depolarization* occurs.
2) The dependence $P_x + iP_y \propto e^{i\omega t}$ describes a uniform precession of the polarization vector $\mathbf{P}(t)$ around axis z with the frequency ω. According to Eq. (7), we can characterize the time–dependence of $\mathbf{P}(t)$ as a four-frequency precession.
3) The frequencies of precession, ω_k, have essentially different orders of magnitude. Two of them, ω_3 and ω_4, are large. For them,

$$\frac{|\omega|}{2\pi} \approx \frac{\Delta}{2\pi\hbar} \equiv \nu_0 \approx 4.5 \cdot 10^3 \text{ MHz.}$$

Compare to 1420 MHz for hydrogen (see Problem 11.2). Frequencies $\omega_{1,2}$, due to the condition $\mu_B^{(e)}\mathcal{H} \ll \Delta$, have much lower values. If we average the vector $\mathbf{P}(t)$ over the fast oscillations, with the period $T = \nu_0^{-1} \sim 10^{-10}$ s (this time it is much smaller than the muon lifetime), then Eq. (7) takes the form:

$$(P_x + iP_y) = \frac{1}{4}(e^{i\omega_1 t} + e^{i\omega_2 t}) = \frac{1}{2}e^{i\omega_H t}\cos\left(\frac{\hbar\omega_H^2}{\Delta}f\right). \quad (9)$$

This is a two-frequency precession. We have defined $\omega_H = \mu_B^{(e)}\mathcal{H}/\hbar$. Let us note that in the absence of a magnetic field, $\omega_1 = \omega_2 = 0$, so that $\langle \mathbf{P}(t) \rangle = (1/2, 0, 0)$. In this case, the depolarization degree of the muon does not depend on the hyperfine splitting Δ (which is why the dynamics of muon spin is interesting in a magnetic field).

Problem 11.74

Estimate the nuclear fusion reaction rate of $dt \to n\alpha$ for a meso-molecular $dt\mu^-$ ion in a state with the rotational quantum number $K = 0$. How does the rate affect the number of reaction events that are initiated by a single muon? Compare to Problem 11.59.

Hint

Use adiabatic approximation and "perturbation theory in terms of the scattering length" (see Problem 11.4). The scattering length in the resonant s-state for a dt-

system is $a_s \approx -(90 + i \cdot 30)$ Fm (see the comment in the problem solution). Connect the quantity $|\psi^{(0)}(0)|^2$ to the penetrability of the Coulomb barrier that divides the nuclei in a mesomolecular ion.

Solution

The main equation of perturbation theory in terms of the scattering length for the level shift (see, for example, Eq. (5) from Problem 11.4) remains valid even in the presence of *absorption* on small distances inside the system. This absorption, which corresponds to inelastic scattering, appears due to a mixing of different channels,[288] which in the absence of a short-range potential act as independent systems dt and $n\alpha$. The scattering length here is a complex quantity. The change in system energy under the short-range interaction is also complex $\Delta E = \Delta E_r - i\Gamma/2$, and its imaginary part describes the width of the s-level

$$\Gamma = -\frac{4\pi\hbar^2}{m}|\psi^{(0)}(0)|^2 \text{Im } a_s, \tag{1}$$

and determines the lifetime $\tau = \hbar/\Gamma$ and the reaction rate $\lambda_f = \tau^{-1} = \Gamma/\hbar$ in this state.

For the mesomolecular ion, the wavefunction $\psi^{(0)}(r_\alpha)$ in (1) should be taken to be the wavefunction of the nucleus subsystem in the adiabatic approximation for the s-state (since the rotational quantum number $K = 0$). A sketch of the effective potential for this subsystem without the short-range nuclear part is given in Fig. 11.4.

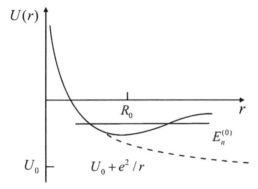

Fig. 11.4

To estimate the quantity $|\psi^{(0)}(0)|^2$, we first note that the typical value of $|\psi^{(0)}(r)|^2$ in the relevant region where the nuclei are localized in the ion is of the order:

$$|\psi_{char}|^2 \sim \frac{3}{4\pi R_0^3}, \tag{2a}$$

---

[288] This includes channels associated with the continuous spectrum (the $n\alpha\mu$-channel in this problem).

where $R_0 = 2L_\mu$ is the characteristic size of the ion (see Problem 11.30), $L_\mu = \hbar^2/m_\mu e^2$, and m_μ is the mass of the muon. In relation with this estimate, note that in the adiabatic approximation, a more natural estimate is the following

$$|\psi_{char}|^2 \sim \frac{1}{4\pi R_0^2 \cdot 2a_{osc}}. \tag{2b}$$

This corresponds to a localization domain in a spherical layer of radius R_0 and a width of the order of twice the amplitude of nuclear oscillations in the ion. Since $a_{osc} \sim (m_\mu/M_{nuc})^{1/4} L_\mu$ (compare to Problems 11.25 and 11.30), then both estimates become essentially the same.

But the quantity $|\psi^{(0)}(0)|^2$ is much smaller, which is connected to the existence of a weakly-penetrable Coulomb barrier, that separates the "molecular" region of nuclear motion and the nuclear region, where the fusion reaction $dt \to n\alpha$ is taking place. Taking this into account for $|\psi^{(0)}(0)|^2$, we have the following estimate:

$$|\psi^{(0)}(0)|^2 \sim P(k)|\psi_{char}|^2 \approx \frac{3}{16ka_B L_\mu^3} e^{-2\pi/ka_B}. \tag{3}$$

We have $a_B = \hbar^2/me^2$, m is the nuclei reduced mass, and $E = \hbar^2 k^2/2m$ is the energy of motion of the dt-system on nuclear distances. The factor

$$P(k) = \frac{2\pi}{ka_B} \frac{1}{\exp\{2\pi/ka_B\} - 1}$$

characterizes a ratio of the wavefunction squared for a particle at $r \to 0$ in the Coulomb repulsive potential, $U = e^2/r$, and that of a free particle; (here the dominant exponential factor $\exp\{-2\pi/ka_B\}$ for $ka_B < 1$ can be calculated according to quasi-classical equation (IX.7); see 9.31.)

Important in this problem is a numerical value of energy E in the expression for $P(k)$. It is defined from the following considerations: the effective potential of the dt-system in the adiabatic approximation (see Fig. 11.4) outside the nuclear region has the form $U = e^2/r + E_\mu(r)$, where $E_\mu(r)$ is the energy of the ground muon term for a fixed distance r from the nuclei. As $r \to 0$ (the essential region of the Coulomb barrier) we have

$$E_\mu(r) \approx -\frac{\mu(2e^2)^2}{2\hbar^2} \equiv U_0,$$

where $\mu = m_\mu M(m_\mu + M)$, $M = m_d + m_t$. This follows from the fact that for $r \ll L_\mu$, the molecular term $E_\mu(r)$ coincides with the ground level of the μ-mesoatom with the nuclear charge $Z = 2$ and mass M. If we denote by $E_{0\nu}^{(0)}$ the mesomolecular levels (with $K = 0$) in the effective potential, then $E = E_{0\nu}^{(0)} - U_0$. Usual mesomolecular levels give binding energies $\varepsilon_{0\nu}^{(0)}$, which are counted from the ground level of the mesoatom with the heaviest of the ion nuclei (this defines the bottom of the continuous spectrum for separated nuclei). In this case,

$$E_{0\nu}^{(0)} = -\frac{\tilde{\mu}e^4}{2\hbar^2} - \varepsilon_{0\nu}^{(0)},$$

where $\tilde{\mu} = m_\mu m_t/(m_\mu + m_t)$, so that

$$E = \frac{2\mu e^4}{\hbar^2} - \frac{\tilde{\mu}e^4}{2\hbar^2} - \varepsilon_{0\nu}^{(0)},$$

or $E \approx 8.3$ keV$-\varepsilon_{0\nu}^{(0)}$.

Using the following values for the dt-system $a_B = 24.0$ fm, $\hbar^2/ma_B^2 = 60.0$ keV, $L_\mu = 2.56 \cdot 10^{-11}$ cm, and $E = 8$ keV so that $ka_B = 0.52$ (for binding energy of the molecular levels, see Problem 11.30), and also using Eqs. (1) and (3) we find:

$$\Gamma \sim 1.5 \cdot 10^{-3} \text{ eV}, \quad \tau \sim 5 \cdot 10^{-13} \text{ s}^{-1}. \tag{4}$$

In conclusion, we make a few remarks

1) Nuclear reaction lead time in the mesomolecule is lower than the muon lifetime by seven orders of magnitude. So, it does not play a noticeable role in the kinetics of μ-catalysis, and does not affect the number of nuclear fusion events,[289] initiated by a single muon. Compare to Problem 11.59.

2) We can notice a large value of scattering length for the dt-system. It is connected to the existence of a quasi-stationary state with a small energy – nucleus ^{5}He($3/2^+$) (the energy of the resonance is $E_R \approx 50$ keV and the width is $\Gamma_R \approx 70$ keV; resonant phenomena for scattering are discussed in Chapter 13, Sec. 3).

   The value of τ given above corresponds to the resonant s-state of the dt-system with the total spin $I = 3/2$. For the non-resonant state with $I = 1/2$, the evaluation of τ (and Γ) differs from Eq. (4), but the conclusion about the fusion reaction lead time being small remains valid for this state.

3) As mentioned in Problems 11.4 and 9.3, if s-states with a small energy exist in a short-range potential then necessarily a reconstruction of the s-levels in the long-range potential occurs. However, no reconstruction of the mesomolecular levels under the influence of the nuclear resonance interaction actually takes place. This is connected to the weak penetrability of the Coulomb barrier, which divides the molecular and nuclear regions of nuclei motion in the mesomolecule. An analogous situation takes place for states with a non-zero angular momentum due to the centrifugal barrier (even in the absence of the Coulomb barrier). The shifts of levels with $l \neq 0$ in conditions of a resonance are also small. See Problem 13.36.

---

[289] Note that $ka_B \approx \sqrt{3m_\mu/m}$ and the exponent in the Coulomb barrier penetrability is ≈ -12.5. If we now go back to the conventional atomic systems (replacing m_μ with m_e), the value of the exponent increases greatly (to approximately -180) and the molecule lifetime becomes so large that the nuclear fusion reaction becomes unobservable.

12

Atomic nucleus

1) Atomic nuclei consist of *nucleons* – protons and neutrons – that are bound together by nuclear forces, which are characterized by a small radius and large strength. Qualitative features of the nucleon-nucleon interactions can be described under the following assumptions about the typical length scales and energy scales in the system: the range of the nuclear forces is about $R \approx 2 \cdot 10^{-13}$ cm, and the typical potential for the same force is $U_0 \approx 40$ MeV.[290] For a proton-neutron system, there exists a bound state (the only possible bound state in a two-nucleon system) – the *deuteron* – whose binding energy is $\varepsilon_0 = 2.23$ MeV and the deuteron has quantum numbers, $J^P = 1^+$.

The striking similarity between the properties of the proton and neutron (both have spin, $s = 1/2$, and very close masses: $m_p = 1836.1 m_e$ and $m_n = 1838.6 m_e$) reflects the concept of *isotopic symmetry*. This symmetry groups elementary particles, *hadrons*, into multiplets according to their quantum numbers in this symmetry group, in analogy to ordinary spin. We refer to the corresponding quantum numbers as *isospin*. It can be viewed as a vector quantity in an abstract three-dimensional space, and is characterized by a definite value of the isospin, T.[291] Possible values of T are connected to the eigenvalues, $T(T+1)$, of the operator, $\hat{\mathbf{T}}^2$, and are equal to $0, 1/2, 1, \ldots$. Particles, that belong to a given multiplet, differ from each other by their value of electric charge that correspond to different values of the T_3 component of the isospin.[292] The number of particles in a multiplet is equal to $(2T+1)$. All of them have the same spin, parity, and close masses.

The isospin of the nucleon is $T_N \equiv \tau = 1/2$. Isospin operators for the nucleon, $\hat{\boldsymbol{\tau}}$ have the form of the standard Pauli matrices for spin $s = 1/2$:

$$\hat{\tau}_1 = \frac{1}{2}\begin{pmatrix} 0 & 1 \\ 1 & 0 \end{pmatrix}, \quad \hat{\tau}_2 = \frac{1}{2}\begin{pmatrix} 0 & -i \\ i & 0 \end{pmatrix}, \quad \hat{\tau}_3 = \frac{1}{2}\begin{pmatrix} 1 & 0 \\ 0 & -1 \end{pmatrix}. \tag{XII.1}$$

---

[290] For comparison, Coulomb interaction on such a distance is $U_{Coul} = e^2/R \approx 0.7$ MeV, and the interaction of the nuclear magnetic moments is $U_{mag} \sim \mu^2/R^3 \sim 10^{-2}$ MeV, where $\mu = e\hbar/m_p c$.

[291] In addition to the doublet of nucleons (p, n), we note the existence of the iso-triplet of pions (π^+, π^0, π^-), here $T_3(\pi^\pm) = \pm 1$, $T_3(\pi^0) = 0$.

[292] The "quantization axis" in the iso-space is usually dictated by isotopic symmetry breaking (which is primarily due to electromagnetic interaction). There exists a relatively small splitting of the masses of particles in multiplets due to breaking of this symmetry.

The physical states of a nucleon – proton or neutron – are described by eigenfunctions of the operator, $\hat{\tau}_3$, so that[293]

$$\psi_p = \psi_{\tau_3=+1/2} = \begin{pmatrix} 1 \\ 0 \end{pmatrix}, \quad \psi_n = \psi_{\tau_3=-1/2} = \begin{pmatrix} 0 \\ 1 \end{pmatrix}, \quad \text{(XII.2)}$$

where the eigenvalues, $\tau_3 = \pm 1/2$, determine the particle charge: $q = e(1 + 2\tau_3)/2$.
2) Classification of states of atomic nuclei can be obtained from the *shell model*. In this model, each nucleon is considered as moving in some self-consistent field (mean field) produced by other nucleons. For the model to produce results consistent with experimental data (*e.g.*, nucleon energy levels, $E_{n_r j l}$), one needs to introduce, apart from a spherically symmetric self-consistent potential, $U(r)$, a spin-orbit interaction of the form, $\hat{U}_{ls} = -f(r)\hat{\mathbf{l}} \cdot \hat{\mathbf{s}}$.

In the shell model, the spin and parity, J^P, magnetic μ and quadrupole Q moments of nuclei are determined only by nucleons outside the filled closed shells. In particular, in the case of a nucleus with one nucleon in the shell nl_j, it has spin $J = j = l \pm 1/2$ and parity $P = (-1)^l$. Here, the operator for the nuclear magnetic moment takes the form $\hat{\boldsymbol{\mu}} = g_l \hat{\mathbf{l}} + g_s \hat{\mathbf{s}}$, where g_l and g_s are the orbital and spin gyromagnetic factors, equal to $g_l = 1$, $g_s = 5.59$ for proton and $g_l = 0$, $g_s = -3.83$ for neutron (μ and g are expressed in units of the nuclear-magneton, equal to $e\hbar/2m_p c$). Averaging this operator (see Problem 3.40) gives the magnetic moment of a nucleon in the nl_j-shell as follows:

$$\mu_j \equiv g_j j = \langle j, l, j_z = j | \hat{\mu}_z | j, l, j_z = j \rangle =$$
$$= \frac{(g_l + g_s)j(j+1) + (g_l - g_s)[l(l+1) - 3/4]}{2(j+1)}. \quad \text{(XII.3)}$$

The values of μ_j and g_j for a series of states are given in the following table:

		$s_{1/2}$	$p_{1/2}$	$p_{3/2}$	$d_{3/2}$	$d_{5/2}$
Proton	μ	2.79	−0.26	3.79	0.12	4.79
	g	5.59	−0.53	2.53	0.08	1.92
Neutron	μ	−1.91	0.64	−1.91	1.15	−1.91
	g	−3.83	1.28	−1.27	0.77	−0.76

(XII.4)

The predictions of the shell model for a nucleus with only one nucleon above a closed shell are the same as for a nucleus with a missing proton or neutron (*i.e.*, a *hole*), and the values of μ and g are also the same. For the nuclei with an empty shell, the predictions of the shell model that employs one-particle nuclear potentials are not unique. The properties of such nuclei depend strongly on *residual interactions* of the nucleons with each other. From an analysis of experimental data,

---

[293] In the literature, a "reversed" classification is also used sometimes, where $\tau_3 = +1/2$ corresponds to the neutron, and $\tau_3 = -1/2$ to the proton.

the residual interactions for nucleons of the same charge suggest a "pairing" into a state with zero total angular momentum. Hence, for an even number of protons and/or neutrons, their total angular momentum in the ground state is equal to zero. Here, the spin and parity of a nucleus with an odd number of nucleons are entirely determined by quantum numbers of the *unpaired* nucleon.

The characteristic size of a nucleus that consists of A nucleons, is $R \approx r_0 A^{1/3}$, where $r_0 = 1.2 \cdot 10^{-13}$ cm.

12.1 Nuclear forces—the fundamentals; The deuteron

Problem 12.1

Approximating the proton–neutron interaction by a spherically symmetric potential well of radius[294] $R = 1.7 \cdot 10^{-13}$ cm, and using the value $\varepsilon_0 = 2.23$ MeV for the deuteron binding energy, evaluate the depth of the potential well, U_0. Also, find the probability of finding nucleons outside the well and the average distance between the nucleons, $\langle r \rangle$.

Solution

For a particle with a mass μ, confined within a square potential well of the radius R and depth U_0, the wavefunction of the s-state with an energy E has the form $\psi_E = \frac{1}{\sqrt{4\pi r}} \chi_E(r)$, with

$$\chi_E = \begin{cases} A \sin\left(\frac{1}{\hbar}\sqrt{2\mu(U_0 - \varepsilon_0)}r\right), & r < R, \\ B e^{-\kappa r}, & r > R, \end{cases}$$

where $\varepsilon_0 = |E| = \hbar^2 \kappa^2 / 2\mu$ is the binding energy. Using the standard matching conditions at $r = R$, we find the following equation for the spectrum:

$$x \cot x = -\kappa R, \quad x = \frac{1}{\hbar}\sqrt{2\mu(U_0 - \varepsilon_0)R^2}. \tag{1}$$

In the case of the deuteron, μ is given by the reduced mass of the proton–neutron system, $\mu \approx (m_p + m_n)/4$. Here,

$$\frac{\hbar^2}{2\mu} = \frac{m_e}{m_p + m_n} \cdot \frac{2\hbar^2}{m_e a_B^2} \cdot a_B^2 \approx 4.15 \cdot 10^{-25} \text{ MeV} \cdot \text{cm}^2,$$

and $\kappa R \approx 0.394$. From Eq. (1) we find $x \approx 1.79$ and $U_0 = 48.1$ MeV.

Now let us perform some estimates using general quantum-mechanical equations that describe a weakly-bound particle in a potential well.

---

[294] Note that for a square well, the *effective interaction radius*, r_0, coincides with its actual radius, R; see Problem 13.43. Here, the value of R is chosen such that r_0 agrees with the experimentally measured value. See also Problem 11.36.

At the threshold where the bound state just appears in a square well, the corresponding "critical" wavefunction reads

$$\chi_{E=0}(r) = \begin{cases} C \sin \lambda r, & r < R, \\ 1, & r > R, \end{cases} \quad (2)$$

where $\lambda = \sqrt{2\mu \tilde{U}_0/\hbar^2}$ and $\tilde{U}_0$ is the threshold value of the well depth corresponding to the appearance of a new s-level (see Problem 4.25). Using the matching conditions again at $r = R$ gives

$$\lambda R = \left(n + \frac{1}{2}\right)\pi \quad \text{and} \quad C = (-1)^n,$$

where $n = 0, 1, 2, \ldots$. Note that $\lambda R = \pi/2$ in the ground state. Hence for the proton–neutron system, we have

$$\tilde{U}_0 = \frac{\pi^2 \hbar^2}{8\mu R^2} = 35.4 \text{ MeV}$$

($R = 1.7 \cdot 10^{-13}$ cm). Here, for deuteron $\varepsilon_0 \ll \tilde{U}_0$, so that it is indeed a weakly-bound system. For the zero-energy level $E = 0$ to move down to the value $E_0 = -\varepsilon_0$, the well must be deepened by δU_0. In accordance with Problem 4.27, we find

$$\delta U_0 \approx \sqrt{\frac{2\hbar^2 \varepsilon_0}{\mu R^2}} = \frac{4}{\pi}\sqrt{\varepsilon \tilde{U}_0} = 11.3 \text{ MeV}$$

and

$$U_0 = \tilde{U}_0 + \delta U_0 \approx 46.7 \text{ MeV.} \quad (3)$$

As shown, the value δU_0 is much larger than ε_0. This is a natural result, because a weakly-bound particle with $l = 0$ is located mainly outside the region, where the attractive forces act. We should also note that for a weakly bound state, strictly speaking, the condition, $\kappa R \ll 1$, must be fulfilled. This value for the deuteron is $\kappa R \approx 0.4$, which is not so small. Nevertheless, Eq. (3) differs from the exact value (48.1 MeV) by only 3%. This high accuracy of the method is due to the fact that the corresponding expansion parameter is $\varepsilon/\tilde{U}_0 \ll 1$.

For a weakly-bound state, the wavefunction has the form

$$\chi_E(r) \approx \sqrt{2\kappa} C_{\kappa 0} e^{-\kappa r} \chi_{E=0}(r), \quad (4)$$

where $C_{\kappa 0}$ is the asymptotic coefficient (see Problem 11.36). Integrating the function, χ_E^2, in the region of the well and taking into account that $\kappa R \ll 1$, we obtain

$$C_{\kappa 0}^2 = (1 - \kappa r_0)^{-1} = (1 - \kappa R)^{-1} \approx 1.65. \quad (5)$$

The probability of finding the nucleons outside the well reads

$$w_{out} = \int_R^\infty \chi_E^2(r)dr = C_{\kappa 0}^2 e^{-2\kappa R} \approx 1.65 e^{-0.79} \approx 0.75,$$

while the mean distance between the nucleons is

$$\langle r \rangle \approx \frac{C_{\kappa 0}^2}{2\kappa} \approx 3.6 \cdot 10^{-13} \text{ cm},$$

which is approximately twice as large as the range of the potential.

Problem 12.2

What is the magnetic moment of a system consisting of a proton and a neutron in the following states:

a) $^1S_0$; b) $^3S_1$; c) $^1P_1$; d) $^3P_0$; e) $^3P_1$; f) $^3D_1$?

Use the values of magnetic moments, $\mu_p = 2.79$ and $\mu_n = -1.91$, of the free nucleons (in nuclear magnetons). Using the fact that the deuteron has spin, $J_d = 1$, and its magnetic moment is $\mu_d = 0.86$, and that it is in a superposition of the $^3S_1$- and $^3D_1$ states, evaluate the magnitude of the admixture of D-wave (compare to Problem 12.3).

Solution

The mean value of the z-component of a magnetic moment of a particle in a state with $J_z = J$, is

$$\mu_0 = \langle J, J_z = J | \hat{\mu}_z | J, J_z = J \rangle. \tag{1}$$

For a system that consists of a proton and a neutron, we have:

$$\hat{\mu} = \hat{\mu}_{orb} + \hat{\mu}_{sp} = \frac{1}{2}\hat{\mathbf{L}} + \mu_p \hat{\sigma}_p + \mu_n \hat{\sigma}_n = \frac{1}{2}\hat{\mathbf{L}} + (\mu_p + \mu_n)\hat{\mathbf{S}} + \frac{1}{2}(\mu_p - \mu_n)(\hat{\sigma}_p - \hat{\sigma}_n). \tag{2}$$

Here, again the magnetic moments are expressed in units of nuclear magnetons, $e\hbar/2m_pc$. The orbital magnetic moment is connected to the proton motion and is given by $\hat{\mathbf{L}}/2$, since the angular momentum of the proton is equal to a half of the total angular momentum in the center-of-mass frame.

In the singlet, ^1L, states ($S = 0$), we have $\overline{\hat{\mathbf{S}}} = \overline{\hat{\sigma}_p} = \overline{\hat{\sigma}_n} = 0$ and $J = L$, so that according to Eqs. (1) and (2), $\mu(^1L) = L/2$. In particular,

$$\mu(^1S_0) = 0; \quad \mu(^1P_1) = \frac{1}{2}; \quad \mu(^1D_2) = 1 \quad etc.$$

For the triplet ^3L_J-state, $\overline{(\hat{\sigma}_p - \hat{\sigma}_n)} = 0$ ($S = 1$ is symmetric, while the operator $(\hat{\sigma}_p - \hat{\sigma}_n)$ is antisymmetric with respect to permutation of the proton and neutron spins), so that from Eqs. (1) and (2) it follows that

$$\mu(^3L_J) = \langle J, J_z = J, L, S | \frac{1}{2}\hat{L}_z + (\mu_p + \mu_n)\hat{S}_z | J, J_z = J, L, S \rangle. \tag{3}$$

Using the results of Problem 3.40, we obtain

$$\mu(^3L_J) = \frac{1}{2(J+1)}\{1.38 J(J+1) - 0.38[L(L+1) - 2]\}. \tag{4}$$

For example, we have

$$\mu(^3S_1) = 0.88; \quad \mu(^3P_1) = 0.69; \quad \mu(^3P_0) = 0; \quad \mu(^3D_1) = 0.31.$$

For the deuteron, the experimental value of its magnetic moment, $\mu_0 = 0.86$, shows that the wavefunction of the deuteron has $J_d = 1$ and exists in a superposition of the S- and D-wave, and that the admixture of the D-wave is small, around $w_D \approx 0.04$ (compare to Problem 12.3).

Problem 12.3

The magnetic moment of the deuteron, described by the superposition of the $(^3S_1 + ^3D_1)$-states, is equal to

$$\mu_d = (1-w)\mu(^3S_1) + w\mu(^3D_1) \approx 0.86 \text{ nuc.mag},$$

where $\mu(^3S_1)$ and $\mu(^3D_1)$ are the magnetic moments of the proton–neutron system in the states $^3S_1$ and $^3D_1$, and $w \approx 0.04$ is the admixture of the D-component.

Explain why no analog of the above relation for the magnetic moment exists for the deuteron's quadrupole moment. Note that the quadrupole moment is zero in the $^3S_1$-state and negative in the $^3D_1$ state. Its experimental value for the deuteron is $Q_d \approx 2.82 \cdot 10^{-27}$ cm$^2 > 0$.

Solution

Writing the deuteron wavefunction as a superposition of S- and D-waves, $\psi_d = \psi_S + \psi_D$, where

$$\langle \psi_d | \psi_d \rangle = 1, \quad \langle \psi_S | \psi_S \rangle = 1 - w_D, \quad \langle \psi_D | \psi_D \rangle = w_D \approx 0.04$$

(a specific expression for the wavefunction, $\psi_{S,D}$, is given in Problem 12.5), we find:

$$\bar{\mu} = \int \psi_S^* \hat{\boldsymbol{\mu}} \, \psi_S dr + \int \psi_D^* \hat{\boldsymbol{\mu}} \, \psi_D dr. \tag{1}$$

Here we have taken into account that the interference terms are equal to zero:

$$\langle \psi_S | \hat{\boldsymbol{\mu}} | \psi_D \rangle = \langle \psi_D | \hat{\boldsymbol{\mu}} | \psi_S \rangle = \mathbf{0}.$$

Indeed, since $\hat{\boldsymbol{\mu}} = \hat{\boldsymbol{\mu}}_{orb} + \hat{\boldsymbol{\mu}}_{sp}$ – see Problem 12.2 – we have $\hat{\boldsymbol{\mu}}_{orb}\psi_S = \frac{1}{2}\hat{\mathbf{L}}\psi_S = \mathbf{0}$ ($L = 0$ for S-wave), and also

$$\langle \psi_S | \hat{\boldsymbol{\mu}}_{sp} | \psi_D \rangle \propto \langle \psi_S | \psi_D \rangle = 0$$

due to orthogonality of the wavefunctions with different values of L. The expression for μ_d follows from Eq. (1).

For the quadrupole moment however, $\hat{Q}_{zz} = e(3z_p^2 - r_p^2)$, the situation is quite different, since the interference term does not vanish:

$$\frac{1}{4}\left\{\int \psi_S^*(3z^2 - r^2)\psi_D d\tau + \int \psi_D^*(3z^2 - r^2)\psi_S d\tau\right\} \neq 0.$$

Moreover, taking into account the small D-wave admixture in the deuteron, we expect that its contribution to the quadrupole moment $\propto w_D$ would be much lower than that of the interference term $\propto \sqrt{w_D}$. This explains the positive value observed experimentally.

Problem 12.4

Discuss experimentally observed properties of the deuteron that (discussed in the previous problems of this Chapter) suggest that the proton–neutron interaction is spin dependent? Consider the following spin-dependent potentials:

a) $\hat{U}_S = V(r)\hat{\boldsymbol{\sigma}}_1 \cdot \hat{\boldsymbol{\sigma}}_2 = V(r)(2\hat{S}^2 - 3)$;
b) $\hat{U} = V(r)\hat{\mathbf{S}} \cdot \hat{\mathbf{L}}$ (spin-orbit interaction);
c) $\hat{U} = V(r)[6(\hat{\mathbf{S}} \cdot \mathbf{n})^2 - 2\hat{S}^2]$ (tensor forces) ($\mathbf{n} = \mathbf{r}/r$, $\mathbf{r} = \mathbf{r}_1 - \mathbf{r}_2$, $\hat{\mathbf{S}} = 1/2(\hat{\boldsymbol{\sigma}}_1 + \hat{\boldsymbol{\sigma}}_2)$ is the operator of the total spin)
and determine which of these potential(s), together with a central potential, could explain the aforementioned properties of the deuteron. Find the integrals of motion for these potentials.

Solution

Experimental data for a deuteron, and for its magnetic and quadrupole moment (see also the previous problems), show that its wavefunction is a superposition of S- and D-wave, so that the orbital angular momentum L does not have a definite value, as it must be for spin-independent central forces. From the potentials given, only the third one, which describes tensor forces, could lead to the given state of the deuteron.

a) Indeed, the potential in a) is central, although the magnitude of the interaction does depend on spin: $\hat{U}_S = -3V(r)$, if $S = 0$, and $\hat{U}_S = V(r)$, if $S = 1$. For this potential, the integrals of motion are the angular momentum, L, and the total spin, S, which are separately conserved.
b) In the presence of spin-orbit interaction, L and S are not conserved separately, and only the total angular momentum $J = L + S$ remains an integral of motion. Nevertheless, this potential cannot lead to a superposition of S- and D-waves. Indeed, although for such potential vectors $\mathbf{L}$ and $\mathbf{S}$ are not conserved, the corresponding operators squared give rise to independent integrals of motion, $\hat{\mathbf{L}}^2$ and $\hat{\mathbf{S}}^2$, which commute with the operator of spin-orbit interaction.

c) The tensor interaction, unlike the spin-orbit interaction, does not conserve the angular orbital momentum, nor its square. It leads to a state, that is a superposition $^3S_1 + {}^3D_1$, which is qualitatively consistent with the experimentally observed properties of the deuteron (see also the following problem).

Finally, note that even though the tensor potential does not conserve **S**, it still conserves the value of $\mathbf{S}^2$. Also note that for all potentials considered in a), b), and c), the total angular momentum J (and hence $\mathbf{J}^2$) and parity are integrals of motion.

Problem 12.5

Show that the deuteron wavefunction can be written as follows:

$$\psi_d = \psi(^3S_1) + \psi(^3D_1) = [f_0(r) + f_2(r)\hat{S}_{12}]\chi_{S=1}.$$

Here, $\hat{S}_{12} = 6(\hat{\mathbf{S}} \cdot \mathbf{n})^2 - 2\hat{\mathbf{S}}^2$, $\hat{\mathbf{S}} = 1/2(\hat{\boldsymbol{\sigma}}_1 + \hat{\boldsymbol{\sigma}}_2)$ is the operator of the total nucleon spin, and $\chi_{S=1}$ is an arbitrary spin function for $S=1$ (see also Problem 5.26).

Use the proton–neutron interaction in the form $\hat{U} = U_S(r) + U_T(r)\hat{S}_{12}$ (a superposition of central and tensor interactions; see the previous problem) to obtain equations for the radial functions, $f_{0,2}(r)$, of the deuteron. Prove also that a potential of tensor forces, considered as a small perturbation, leads to a shift of the ^3S-level that appears only second-order perturbation theory.

Solution

1) A spherically symmetric wavefunction, $\psi(^3S_1) = f_0(r)\chi_{S=1}$, describes the state with $L=0$ and $J=S$. Hence, this wavefunction corresponds to the $^3S_1$-wave.

Let us show now that the wavefunction $\psi(^3D_1) = f_2(r)\hat{S}_{12}\chi_{S=1}$ indeed describes the $^3D_1$-state. We write it in the form:

$$\psi(^3D_1) = f_2(r)\{6n_i n_k - 2\delta_{ik}\}\hat{S}_i\hat{S}_k\chi_{S=1}. \tag{1}$$

The angular part of this wavefunction, $T_{ik} = 6n_i n_k - 2\delta_{ik}$, is a symmetric tensor of the second rank with zero trace. According to Problem 3.41, we confirm that the wavefunction in Eq. (1) indeed describes a state with the angular momentum, $L=2$. Then, since the commutator $[\hat{\mathbf{S}}^2, \hat{S}_i\hat{S}_k] = 0$, we have:

$$\hat{\mathbf{S}}^2\psi(^3D_1) = f_2(r)\{6n_i n_k - 2\delta_{ik}\}\hat{S}_i\hat{S}_k\hat{\mathbf{S}}^2\chi_{S=1} = 2\psi(^3D_1),$$

i.e., the wavefunction, $\psi(^3D_1)$, describes the state with spin $S=1$. In an analogous way, from the commutativity of $\hat{\mathbf{J}}$ with the scalar operator $f_2(r)\hat{S}_{12}$, it follows that

$$\hat{\mathbf{J}}^2\psi(^3D_1) = f_2(r)\{6n_i n_k - 2\delta_{ik}\}\hat{S}_i\hat{S}_k\hat{\mathbf{J}}^2\chi_{S=1} = 2\psi(^3D_1).$$

(since the spin part of the wavefunction, $\chi_{S=1}$, does not depend on the angles, we have $\hat{\mathbf{J}}^2\chi = \hat{\mathbf{S}}^2\chi = 2\chi$). Hence, the wavefunction, $\psi(^3D_1)$, corresponds to a state with $J=1$ and indeed describes a $^3D_1$-wave.

2) Let us discuss now the properties of the operator, $\hat{S}_{12} = 6(\hat{\mathbf{S}} \cdot \mathbf{n})^2 - 2\hat{\mathbf{S}}^2$. Taking into account that $(\hat{\mathbf{S}} \cdot \mathbf{n})^3 = (\hat{\mathbf{S}} \cdot \mathbf{n})$ for spin $S = 1$ (see Problem 1.21), we obtain[295]

$$\hat{S}_{12}^2 = -2\hat{S}_{12} + 4\hat{\mathbf{S}}^2. \tag{2}$$

Then, using the value of the integral

$$\int n_i n_k d\Omega = \frac{4\pi}{3}\delta_{ik},$$

we find

$$\int \hat{S}_{12} d\Omega = 0, \quad \int \hat{S}_{12}^2 d\Omega = 16\pi \hat{\mathbf{S}}^2. \tag{3}$$

Hence, we obtain the normalization condition for the wavefunction of the deuteron as a superposition of $^3S_1$ and $^3D_1$ waves:

$$\langle \psi_d | \psi_d \rangle = 4\pi \int_0^\infty \{|f_0(r)|^2 + 8|f_2(r)|^2\} r^2 dr = 1, \tag{4}$$

where $\langle \chi_{S=1} | \chi_{S=1} \rangle = 1$. We also find that the shift of the $^3S_1$-level vanishes in first-order perturbation theory in the tensor interaction:

$$E^{(1)} = \langle \psi(^3S) | \hat{U}_T | \psi(^3S) \rangle = 0.$$

Now, the Schrödinger equation for the deuteron is

$$\left(-\frac{\hbar}{2m}\Delta + \hat{U}\right)\psi_d = E_d \psi_d$$

(where m is the reduced mass). Together with Eq. (2), the radial functions, $\tilde{f}_{0,2} = r f_{0,2}$, are described by the equations

$$-\frac{\hbar^2}{2m}\tilde{f}_0'' + [U_S(r) - E_d]\tilde{f}_0 + 8U_T(r)\tilde{f}_2 = 0$$

and

$$\left[-\frac{\hbar^2}{2m}\frac{d^2}{dr^2} + \frac{3\hbar^3}{mr^2} + U_S(r) - 2U_T(r) - E_d\right]\tilde{f}_2 + U_T(r)\tilde{f}_0 = 0. \tag{5}$$

Let us note that the tensor interaction causes transitions between states with different values of the orbital angular momentum L (but with the same parity) only in the triplet, $S = 1$, states. In the singlet states, $S = 0$, the tensor forces are absent and the orbital angular momentum is an integral of motion, coinciding with J.

---

[295] Since the possible values of S are 0 and 1, then $(\hat{\mathbf{S}}\mathbf{n})\hat{\mathbf{S}}^2 = 2(\hat{\mathbf{S}}\mathbf{n})$, $(\hat{\mathbf{S}}^2)^2 = 2\hat{\mathbf{S}}^2$ and relation (2) is valid for both $S = 1$ and $S = 0$; that is, for all states of the system of two nucleons.

Finally, note (in connection with the deuteron wavefunction considered here) that the relation

$$\langle \hat{\mathbf{J}} \rangle = \chi^* \hat{\mathbf{S}} \chi$$

holds (see Problem 5.21).

Problem 12.6

For a system of two nucleons:

1) find the eigenfunctions and eigenvalues of the isospin (*i.e.*, its value, T, and projection, T_3);
2) determine the value of the isospin, T, in the ^{2S+1}L-state with a definite values of the total spin S and the angular momentum, L (see Problem 10.9);
3) find the isospin part of the deuteron wavefunction.

Solution

1) Using the formal analogy between the isospin and spin, the wavefunctions ψ_{TT_3} of the two-nucleon system with definite values of the total isospin, T, and its projection, T_3, could be written as in Problem 5.10 (see also Eqs. (XII.1) and (XII.2)):

$$\psi_{11} = \begin{pmatrix} 1 \\ 0 \end{pmatrix}_1 \begin{pmatrix} 1 \\ 0 \end{pmatrix}_2 \equiv \psi_P(1)\psi_P(2);$$

$$\psi_{10} = \frac{1}{\sqrt{2}} \left\{ \begin{pmatrix} 1 \\ 0 \end{pmatrix}_1 \begin{pmatrix} 0 \\ 1 \end{pmatrix}_2 + \begin{pmatrix} 0 \\ 1 \end{pmatrix}_1 \begin{pmatrix} 1 \\ 0 \end{pmatrix}_2 \right\} \equiv$$

$$\equiv \frac{1}{\sqrt{2}} \{\psi_P(1)\psi_n(2) + \psi_n(1)\psi_P(2)\};$$

$$\psi_{1,-1} = \begin{pmatrix} 0 \\ 1 \end{pmatrix}_1 \begin{pmatrix} 0 \\ 1 \end{pmatrix}_2 \equiv \psi_n(1)\psi_n(2); \qquad (1)$$

$$\psi_{00} = \frac{1}{\sqrt{2}} \left\{ \begin{pmatrix} 1 \\ 0 \end{pmatrix}_1 \begin{pmatrix} 0 \\ 1 \end{pmatrix}_2 - \begin{pmatrix} 0 \\ 1 \end{pmatrix}_1 \begin{pmatrix} 1 \\ 0 \end{pmatrix}_2 \right\} \equiv$$

$$\equiv \frac{1}{\sqrt{2}} \{\psi_P(1)\psi_n(2) - \psi_n(1)\psi_P(2)\}.$$

Let us emphasize that for the states with $\psi_{T,T_3=0}$, each of the nucleons is not in a state with a definite charge, but with the 50% probability could be in either the proton or neutron states. This is a reason why an interaction that conserves isospin has, generally speaking, an exchange character.

2) According to isotopic symmetry, the proton and neutron are considered as different charge (or equivalently isospin) states of the same particle – a nucleon. Since a

nucleon is a fermion, a wavefunction of a many-nucleon system must be anti-symmetric with respect to permutation of particles – including the coordinate, spin, and isospin degrees of freedom – which gives rise to a generalized Pauli principle. For a two-nucleon system:

a) permutation of coordinates is equivalent to inversion with respect to the system center of mass, and symmetry of the coordinate part of the wavefunction with a given value of the angular momentum, L, is determined entirely by parity, $(-1)^L$;

b) symmetry of the spinor part of the wavefunction for states with the total spin, S, with respect to permutation of spin variables is determined by the factor $(-1)^{S+1}$;

c) symmetry of the isospin wavefunction is given by the factor $(-1)^{T+1}$.

Therefore, permutation of two nucleons in the state, ψ_{LST}, with definite values of quantum numbers L, S, and T reduces to a multiplication by $(-1)^{L+S+T}$. Using the condition that the wavefunction is anti-symmetric, we have the relation:

$$(-1)^T = (-1)^{L+S+1}. \tag{2}$$

In states with $S = 1$ and even values of L – as, for example, for the deuteron (see Problem 12.5) – the isospin of two nucleons is $T = 0$. (Per the generalized Pauli principle, we find that the $T = 1$ states with two identical nucleons – two protons or two neutrons – are forbidden; see also Problem 10.9.)

3) The isospin part of the deuteron wavefunction with $T = 0$ is described by the function ψ_{00} above.

Problem 12.7

Which properties of the real nuclear forces would have had to be "eliminated" if the deuteron state were a superposition $^1P_1 + {}^3P_1$ (in contrast to $^3S_1 + {}^3D_1$ in the real deuteron)? Discuss a possible interaction that could produce such a state.

Solution

In the ^3P and ^1P states, the proton–neutron system would have had different values of the isospin, $T = 1$ and 0 respectively; see the previous problem. It means that the interaction, which gives the superposition $^1P_1 + {}^3P_1$, does not conserve the isospin, i.e., it is not isoscalar. As an example of such an interaction, we can write:

$$\hat{U} = V(r)(\hat{\boldsymbol{\sigma}}_1 \cdot \hat{\boldsymbol{\tau}}_3^{(1)} + \hat{\boldsymbol{\sigma}}_2 \cdot \hat{\boldsymbol{\tau}}_3^{(2)})\hat{1} = \frac{1}{2}V(r)(\hat{\boldsymbol{\sigma}}_p - \hat{\boldsymbol{\sigma}}_n) \cdot \hat{1}.$$

Compare with Problem 12.8.

Problem 12.8

Assume that nucleon–nucleon interaction has the following isotopic structure: $\hat{U} = \hat{V}_1 + \hat{V}_2 \hat{\tau}_3^{(1)} \hat{\tau}_3^{(2)}$, where $\hat{V}_{1,2}$ are isospin-independent operators (i.e., operators acting in the coordinate and/or spin space, symmetric under nucleon permutation) and find the form of the interaction for a) two protons, b) two neutrons and c) a proton and w neutron.

Does this interaction have 1) isotopic invariance; 2) charge symmetry?

Solution

Since for the proton and neutron, $\tau_{3p} = +1/2$ and $\tau_{3n} = -1/2$, then for the interaction considered and different charge states, we have:

$$\hat{U}_{pp} = \hat{U}_{nn} = \hat{V}_1 + \frac{1}{4}\hat{V}_2, \quad \hat{U}_{pn} = \hat{V}_1 - \frac{1}{4}\hat{V}_2.$$

The equality $\hat{U}_{pp} = \hat{U}_{nn}$ shows the charge-independence of the interaction. But it does not have isotopic invariance, since it is a superposition of the isoscalar $\hat{V}_1$ and the tensor component $\hat{V}_2 \hat{\tau}_3^{(1)} \hat{\tau}_3^{(2)}$. Hence, the interaction in the pn-system differs from the interaction between two identical nucleons.

Problem 12.9

For the system of two nucleons, find the following:

1) the most general form of the isospin-invariant interaction, $\hat{U}$. Express it in terms of the operators, $\hat{U}_T$, of the nucleon-nucleon interaction in states with a definite value of isotopic spin: $T = 0$ and $T = 1$;
2) the isospin structure of the nucleon Coulomb interaction.

Solution

1) The desired form must be a scalar in the isospace, and as such could only contain the following operators:

$$\hat{1}, \ \hat{\tau}_{1,2}^2, \ \hat{\tau}_1^2 \hat{\tau}_2^2, \ldots, \ \hat{\tau}_1 \hat{\tau}_2, \ (\hat{\tau}_1 \cdot \hat{\tau}_2)^2, \ldots,$$

where $\hat{\tau}_{1,2}$ are (iso-vector) operators, acting on the isospin of a single nucleon. Only two among these operators are independent: $\hat{1}$ and $\hat{\tau}_1 \hat{\tau}_2$. Indeed, the operator, $\hat{\tau}_1^2 = \hat{\tau}_2^2 = 3/4$, is proportional to the identity operator, and any power of $\hat{\tau}_1 \hat{\tau}_2$ can be expressed in terms of a linear combination of $\hat{1}$ and $\hat{\tau}_1 \hat{\tau}_2$, because

$$(\hat{\tau}_1 \hat{\tau}_2)^2 = \frac{3}{16} - \frac{1}{2}(\hat{\tau}_1 \hat{\tau}_2)$$

(see Problem 5.12). Hence, a generic isospin-invariant interaction has the form

$$\hat{U} = \hat{V}_1 + \hat{V}_2(\hat{\tau}_1 \hat{\tau}_2), \tag{1}$$

where $\hat{V}_{1,2}$ are isospin-independent operators acting in the coordinate and/or spin space, which are symmetric with respect to nucleon permutation.

610 Exploring Quantum Mechanics

Since $\hat{\boldsymbol{\tau}}_1 \cdot \hat{\boldsymbol{\tau}}_2 = (1/2)\hat{\mathbf{T}}^2 - 3/4$, where $\hat{\mathbf{T}}$ is the total isospin operator of the two-nucleon system, then according to Eq. (1) we obtain for states with a definite value of T:

$$\hat{U}(T=0) = \hat{V}_1 - \frac{3}{4}\hat{V}_2, \quad \hat{U}(T=1) = \hat{V}_1 + \frac{1}{4}\hat{V}_2. \tag{2}$$

Hence, we can express $\hat{V}_{1,2}$ in terms of $\hat{U}(T)$ and write interaction (1) as follows:

$$\hat{U} = \frac{1}{4}(\hat{U}_0 + 3\hat{U}_1) + (\hat{U}_1 - \hat{U}_0)\hat{\boldsymbol{\tau}}_1 \cdot \hat{\boldsymbol{\tau}}_2, \tag{3}$$

where $\hat{U}_{0,1} \equiv \hat{U}(T=0,1)$.

2) In a two-nucleon system, the Coulomb interaction is non-zero only if both nucleons are in the charged, proton, state. Since the operator corresponding to the nucleon charge is

$$\hat{q}_N = \frac{e}{2}(1 + 2\hat{\tau}_3),$$

the Coulomb interaction has the form

$$\hat{U}_{Coul} = \frac{e^2}{4|\mathbf{r}_1 - \mathbf{r}_2|}(1 + 2\hat{\tau}_3^{(1)})(1 + 2\hat{\tau}_3^{(2)}). \tag{4}$$

Clearly, the Coulomb interaction breaks both isospin invariance and charge invariance.

Problem 12.10

Find the mean value of Coulomb interaction energy for protons inside a ^{3}He nucleus and estimate the size of the mirror nuclei, tritium ^{3}H, and helium ^{3}He. Use the fact that in β-decay, ^{3}H $\to$ ^{3}He $+ e^- + \tilde{\nu}$, the maximum kinetic energy of the electron is $\varepsilon_0 = 17$ keV. Also recall that the nuclei considered have no excited states, and $(m_n - m_p) \times c^2 \approx 2.5 m_e c^2 \approx 1.3$ MeV.

Solution

If isospin symmetry were exact, then the tritium and helium, ^{3}He, would have had many identical nuclear properties (masses, energy levels, and their quantum numbers), and β-decay would have been prohibited by energy conservation. The breaking of this symmetry originates primarily from the electromagnetic interaction, and results in a difference between the ^{3}H and ^{3}He masses and hence the corresponding rest energies, Mc^2.

This difference between the nucleon masses is basically due to the[296] masses of the proton and neutron being different and the Coulomb interaction between protons in the nucleus ^{3}He. We have

---

[296] Here we neglect both the interaction of the nucleon magnetic moments (which is much smaller than the Coulomb interaction) and also the influence of electromagnetic interactions on the nuclear potential.

$$[M(^3\text{H}) - M(^3\text{He})]c^2 = (m_n - m_p)c^2 - \int \frac{e^2}{r}|\psi|^2 d\tau,$$

where ψ is the wavefunction of the ^{3}He nucleus, and $\mathbf{r} = \mathbf{r}_1 - \mathbf{r}_2$ is the distance between the protons. Taking into account the expression

$$m_e c^2 + \varepsilon_0 = [M(^3\text{H}) - M(^3\text{He})]c^2,$$

we find

$$\overline{U}_{Coul} = \left\langle \frac{e^2}{r} \right\rangle = (m_n - m_p - m_e)c^2 - \varepsilon_0 \approx 1.5 m_e c^2 \approx 0.77 \text{ MeV}.$$

This result can be used to estimate the mean size, R:

$$R \sim \langle r \rangle \sim \left[\left\langle \frac{1}{r} \right\rangle\right]^{-1} \approx 1.9 \cdot 10^{-13} \text{ cm}$$

(Here we used $e^2/a_B \approx 27$ eV and $a_B \approx 0.53 \cdot 10^{-8}$ cm.)

Problem 12.11

As mentioned before, nuclear sizes are determined by the relation $R = r_0 A^{1/3}$, where A is the number of nucleons in a nucleus.

Estimate the value of r_0 from β^+-decay data for a nucleus that contains $(Z+1)$ protons and Z neutrons, so that $A = 2Z+1$. Express it in terms of the maximum energy of positron decay, ε_0. Take into consideration that the decaying nuclei and its decay product are mirror nuclei in the same states (that is, they have the same quantum numbers, except for the T_3 isospin component). Consider the Coulomb interaction of protons inside the nucleus and setting it equal to the electrostatic energy of a uniformly charged ball with the same charge and radius as the nucleus. Obtain a numerical estimate of r_0 from the data, $\varepsilon_0 = 3.48$ MeV, for the decay $^{27}_{14}\text{Si} \to ^{27}_{13}\text{Al} + e^+ + \nu$.

Solution

Using the mass difference between the mirror nuclei (given in previous problem) and the value $U_{Coul} = 3(Ze)^2/5R$, corresponding to the Coulomb energy of a charged ball (with the charge, Ze, and radius, R), we find

$$\varepsilon_0 = -(m_n - m_p)c^2 + \frac{3(2Z+1)e^2}{5(2Z+1)^{1/3} r_0},$$

where the last term is the difference between the electrostatic energies of two full-spheres with the same radii $R = r_0 A^{1/3}$, but different charges, $(Z+1)e$ and Ze (here, $A = 2Z+1$). Hence,

$$r_0 = \frac{3(2Z+1)}{5(2Z+1)^{1/3}} \frac{e^2}{(\varepsilon_0 + \Delta)},$$

where $\Delta = (m_n - m_p)c^2 \approx 1.29$ MeV, and from the β-decay data for the nucleus ^{27}Si we obtain: $r_0 \approx 1.6 \cdot 10^{-13}$ cm.

12.2 The shell model

Problem 12.12

Assuming that the self-consistent field inside a nucleus can be approximated by the potential, $U(r) = -U_0 + m\omega^2 r^2/2$ (m is the mass of the nucleon), find single-particle energy levels.

What are the values of the *magic* numbers given by such a model of the self-consistent potential?

What are the predictions of the model for the angular momenta and parities of the nuclear ground states?

Estimate the value of the model parameter, $\hbar\omega$, using the nuclei size data.

Solution

To find the single-particle energy levels and the corresponding eigenfunctions, we consider the Schrödinger equation,

$$\left[-\frac{\hbar^2}{2m}\Delta - U_0 + \frac{1}{2}m\omega^2 r^2\right]\psi_N = E_N\psi_N,$$

which was solved before in Problem 4.4. Here,

$$E_N = -U_0 + \hbar\omega\left(N + \frac{3}{2}\right), \quad N = 2n_r + l = 0, 1, 2, \ldots.$$

For every level with a given value of N, there exist one-particle states, $\psi_{n_r l m}$, with the angular momenta, $l = N, N-2, \ldots, 1(0)$. Therefore, the level degeneracy is $G(N) = (N+1)(N+2)/2$ (without taking the nucleon spin into account).

Fig. 12.1 shows the spectrum of the single-particle levels for the potential considered. Displayed at right are the following quantities: $G(N)$ is the level degeneracy; $n(N) = 2G(N)$ is the maximum number of nucleons for each charged state, which could occupy a given level (the doubling of $G(N)$ is due to spin); and $M(N)$ is the maximum number of nucleons in each charged state, which could be placed on all levels, starting from the lowest and ending with $M(N+1) = M(N) + n(N+1)$.

The *magic* numbers, $M(N)$, are equal to 2, 8, 20, 40, 70, ...

The accidental degeneracy of the levels is a specific peculiarity of the oscillator potential, which possesses an additional symmetry. If we slightly deform the potential by $\delta U(r)$, the accidental degeneracy is lifted and the degenerate level splits into as

N		G(N)	n(N)	M(N)
6	—— 4s, 3d, 2g, 1i	28	56	168
5	—— 3p, 2f, 1h	21	42	112
4	—— 3s, 2d, 1g	15	30	70
3	—— 2p, 1f	10	20	40
2	—— 2s, 1d	6	12	20
1	—— 1p	3	6	8
0	—— 1s	1	2	2

Fig. 12.1

many sublevels as there are different values of l that correspond to it.[297] This fact is illustrated in Fig. 12.2, where a splitting of the levels with $N = 3$ and 4 is presented schematically.

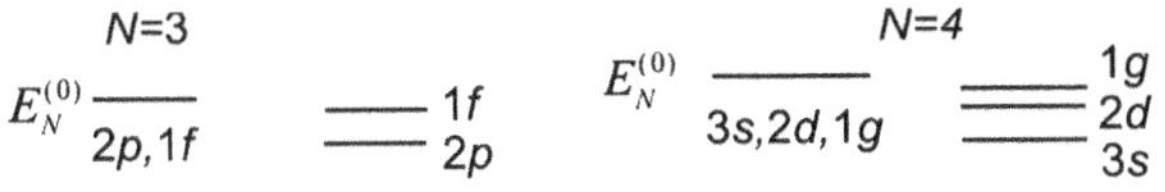

$$E_N^{(0)} \quad \overline{\quad} \quad 2p, 1f \qquad \overline{\quad} \quad 1f \atop \overline{\quad} \quad 2p \qquad\qquad E_N^{(0)} \quad \overline{\quad} \atop 3s, 2d, 1g \qquad \overline{\quad} \quad 1g \atop \overline{\quad} \quad 2d \atop \overline{\quad} \quad 3s$$

N=3 .... N=4

Fig. 12.2

The model does not provide well-defined predictions for the value of the total nuclear spin, J, due to the large degeneracy of the levels. However, it does provide accurate predictions for the parity in the ground states, which is related to the parity, $P = (-1)^l = (-1)^N$, of all single-particle states that correspond to the level with a given N, and the fact that the parity is a multiplicative quantum number. For example, the ground state of the nucleus $^{13}_{6}\text{C}$ has a negative parity, while that for of $^{17}_{8}\text{O}$ is positive.

To estimate the parameter, $\hbar\omega$, for nuclei with $A \approx 2Z \gg 1$, we equate the size of the nucleus, $R = r_0 A^{1/3}$, where $r_0 = 1.2 \cdot 10^{-13}$ cm, to the "radius" $R(N_{max})$ of the quasi-classical orbit for a nucleon in an outer filled level that is given by the relation

$$m\omega^2 R^2(N_{max}) = \hbar\omega(2N_{max} + 3).$$

Taking into account the estimate $N_{max} = (3A/2)^{1/3}$ (as follows from the condition $A = \sum_{N \leq N_{max}} 4G(N) \approx \sum_{N \leq N_{max}} 2N^2 \approx \tfrac{2}{3} N_{max}^2$, with the sum replaced by an integral), we find

$$\hbar\omega \approx \frac{2 N_{max} \hbar^2}{m r_0^2 A^{2/3}} \approx 60 A^{-1/3} \quad \text{MeV}.$$

---

[297] The order of the sublevels l depends on a specific form of the perturbation, $\delta U(r)$.

Problem 12.13

Using the conditions of the previous problem, discuss the effect of adding a spin-orbital interaction of the form $\hat{U}_{1s} = -\alpha\hat{\mathbf{l}}\cdot\hat{\mathbf{s}}$ on the one-nucleon energy spectrum. For $\alpha = \hbar\omega/10$, provide a graphical illustration of the low-lying single-particle levels.[298]

In the framework of the model considered, find the angular momenta (spins) and parities of the ground states of the following nuclei: $^6_2$He, $^6_3$Li, $^{10}_5$B, $^{12}_6$C, $^{13}_6$C, $^{13}_7$N, $^{14}_6$C, $^{16}_8$O, $^{17}_8$O, $^{27}_{13}$Al, and $^{40}_{20}$Ca.

Solution

The energy levels of a particle with spin in a central potential do not depend on the spin and are determined only by quantum numbers n_r, l (but not l_z, s_z). With a spin-orbital interaction present, the energy of a particle (with spin, $s = 1/2$, and a certain value of l) splits into two sub-levels, corresponding to the values $j = l \pm 1/2$ of the total angular momenta (except for s-levels). In this case the "good" quantum numbers are the total angular momenta j, its projection j_z, and parity $P = (-1)^l$. (Even though l itself is not conserved, the orbital angular momentum squared remains an integral of motion; i.e., the operator, $\hat{\mathbf{l}}^2$, commutes with Hamiltonian.) Hence, the eigenfunctions of the Hamiltonian can be chosen in the form:

$$\psi_E(r) = f(r)\psi_{jlj_z}(n),$$

where the spin-orbit functions, ψ_{jlj_z}, were discussed in Problem 5.24 (their explicit form is not germane to this problem).

Since the operator, $\hat{\mathbf{l}}\cdot\hat{\boldsymbol{\sigma}} = \hat{\mathbf{j}}^2 - \hat{\mathbf{l}}^2 - 3/4$, for states with definite j and l, also has definite value:

$$\hat{\mathbf{l}}\cdot\hat{\boldsymbol{\sigma}} = \begin{cases} l, & \text{in the state with } j = l + \frac{1}{2}, \\ -l - 1, & \text{in the state with } j = l - \frac{1}{2}, \end{cases}$$

the spin-orbit interaction gives the following splitting:

$$\Delta E_{jl} = \begin{cases} \alpha(l+1), & j = l - \frac{1}{2}, \\ -\alpha l, & j = l + \frac{1}{2} \end{cases} \quad (1)$$

to the unperturbed level, $E_{n_r l}$, and is independent of the concrete form of the central potential, $U(r)$. Therefore, the energy levels are described by the expression:

$$E_{n_r jl} = -U_0 + \hbar\omega\left(2n_r + l + \frac{3}{2}\right) + \Delta E_{jl}. \quad (2)$$

The width of the level splitting is

$$\Delta E_l = E_{j=l-1/2,l} - E_{j=l+1/2,l} = (2l+1)\alpha, \quad l \neq 0,$$

---

[298] The parameter $\alpha > 0$, because, according to experiment, the level with $j = l + 1/2$ lies below that with $j = l - 1/2$.

and it increases with increasing l. For degenerate levels, the maximum value of the orbital angular momentum is $l_{max} = N$, so for them the splitting is

$$\Delta E_N = (2N+1)\alpha, \quad N \geq 1.$$

Let us also note that the mean value of level shift vanishes as follows:

$$\sum_j (2j+1)\Delta E_{jl} = 0.$$

The left-hand side of Fig. 12.3 shows a level splitting for the unperturbed oscillator with $N = 2$; its right-hand side displays the low-lying single-particle levels of our model.

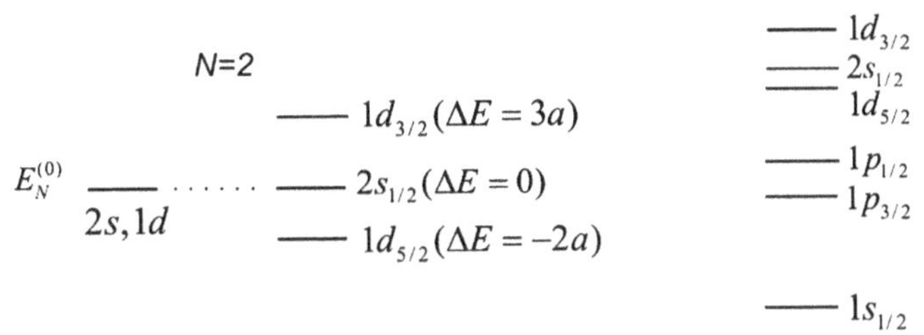

Fig. 12.3

Taking into account that the nuclear ground state in the shell model is defined by putting the nucleons on the lowest one-particle levels with the Pauli principle taken into account (the Aufbau building-up principle), the total angular momentum, J, and the parity, P, of the filled shells are given by $J^P = 0^+$, while the quantum numbers of the "hole-like" states are the same as those of the corresponding single-particle level. This yields the following predictions for the spins and parities of the ground states of the given nuclei:

1) For the nuclei ^{12}C, ^{14}C, ^{16}O, and ^{40}Ca we find $J^P = 0^+$ (these nuclei have only filled shells).
2) The nuclei ^{13}C, ^{13}N, ^{17}O, and ^{27}Al have only one nucleon (proton or neutron) or one hole above the filled shells, whose quantum numbers determine J^P of these nuclei and are equal to $(1/2)^-$, $(1/2)^-$, $(5/2)^+$, and $(5/2)^+$ respectively.
3) Predictions made by the shell model with respect to spin (but not parity) for ^{6}He, ^{6}Li, ^{10}B are not unique. For example, for ^{6}Li, which has a $(1s)^4$ proton and a neutron in the state $1p_{3/2}$ above the filled shell, J^P could take one of the following values: 3^+, 2^+, 1^+, 0^+. In the same way, we obtain the predicted values for ^{10}B, which has one proton and one neutron hole in the shell $1p_{3/2}$. ^{6}He has two neutrons in state $1p_{3/2}$ above the full shell $(1s)^4$, and according to the model the possible values of J^P are 2^+ and 0^+ (quantum numbers 3^+ and 1^+ are forbidden by the Pauli principle, see Problem 12.6). But if we take into account the "pairing effect" due

to the residual interactions of neutrons, then for the nucleus ^{6}He, the prediction is unique: $J^P = 0^+$.

Problem 12.14

Using the shell model, find the spin–isospin dependence of the wavefunctions for the tritium (^{3}H) and helium (^{3}He) nuclear ground-states.

Solution

In the nuclei considered, the three nucleons are in the 1s-shell. Such a configuration could be considered as one hole in the 1s-state, which defines the quantum numbers – spin, parity and isospin – of these nuclei. $J^P = (1/2)^+$, $T = 1/2$ (here $T_3 = -1/2$ and $+1/2$ for the nuclei, ^{3}H and ^{3}He, respectively).

The orbital (coordinate) part of their wavefunction is symmetric with respect to permutation of the nucleon coordinates (which all are in the 1s orbital state). Hence, the spin–isospin part of the wavefunction must be anti-symmetric (see Problem 12.6). Its explicit form is found by anti-symmetrizing the wavefunction of a system of identical fermions with given occupied single-particle states. For brevity, below, we use the following notation for the spin–isospin wavefunctions: $p_\uparrow(1)$ is the wavefunction of the first nucleon in the proton-state with the definite value $s_z = +1/2$ of the spin projection, and so forth.

If the three nucleons occupy the states $p_\uparrow$, $p_\downarrow$, and $n_\uparrow$, then the spin-isospin part of the wavefunction is defined by the determinant below

$$\psi_{T_3=1/2, S_z=1/2} = \frac{1}{\sqrt{6}} \begin{vmatrix} p_\uparrow(1) & p_\downarrow(1) & n_\uparrow(1) \\ p_\uparrow(2) & p_\downarrow(2) & n_\uparrow(2) \\ p_\uparrow(3) & p_\downarrow(3) & n_\uparrow(3) \end{vmatrix}. \tag{1}$$

This wavefunction corresponds to the ^{3}He nucleus in the $J_z = +1/2$ state.

Making the substitution $p \leftrightarrow n$ in Eq. (1), we obtain a spin–isospin wavefunction for ^{3}H nuclei with $J_z = +1/2$. In the same way, substitutions $\uparrow \leftrightarrow \downarrow$ give wavefunctions for the $J_z = -1/2$ states.

Problem 12.15

Find possible values of the total angular momentum, J, and the isospin, T, for nuclei that contain two nucleons in states $p_{1/2}$ with the same n above filled shells. The nuclei with such configurations are $^{14}_6$C, $^{14}_7$N, and $^{14}_8$O (two nucleons above the filled shells: $(1s_{1/2})^4(1p_{3/2})^8$).

Solution

Since J and the isospin of the filled shells are zero, J and T are determined by the nucleons above the closed shells. Here, the possible values of J and T are restricted by the anti-symmetry constraint (in accordance with the generalized Pauli principle; see also Problem 12.6) on the nucleon wavefunction.

Notice that the isospin part of a two-nucleon wavefunction is symmetric with respect to isospin permutation, if $T = 1$, and is antisymmetric if $T = 0$; also, the wavefunction of two nucleons with the same angular momenta, $j_1 = j_2 = 1/2$, is symmetric with respect to permutation of the j_z-variables, if $J = 2j, 2j - 2, \ldots$, and it is antisymmetric, if $J = 2j - 1, 2j - 3, \ldots$. Finally, taking into account the same radial dependence of the wavefunction for the two fermions, we see that:

1) for $T = 1$, only $J = 2j - 1, 2j - 3, \ldots$ are possible;
2) for $T = 0$, only $J = 2j, 2j - 2, \ldots$ are possible.

Problem 12.16

The same as in the previous problem, but for two nucleons in the state $p_{3/2}$.

Solution

For $T = 1$ only $J = 2$ and $J = 0$ are possible; while for $T = 0$, only $J = 3$ and $J = 1$ are allowed.

Problem 12.17

Use the shell model to find the spin and magnetic moments of the following nuclei in their ground states:[299]

$$^3\text{H}\left(J = \frac{1}{2}, \mu = 2.91\right); \quad ^3\text{He}\left(\frac{1}{2}, -2.13\right);$$

$$^{11}_{5}\text{B}\left(\frac{3}{2}, 2.69\right); \quad ^{13}_{6}\text{C}\left(\frac{1}{2}, 0.70\right); \quad ^{15}_{7}\text{N}\left(\frac{1}{2}, -0.28\right);$$

$$^{17}_{8}\text{O}\left(\frac{5}{2}, -1.89\right); \quad ^{29}_{14}\text{Si}\left(\frac{1}{2}, -0.55\right).$$

Use the scheme for one-particle levels from Problem 12.13.

Solution

Taking into account the hierarchy of one-nucleon levels found in Problem 12.13, we see that indeed the nuclei considered contain only either one nucleon or one hole above the filled shells. Therefore, the nuclear spin J and parity P are determined by this single-particle entity, and according to Eq. (XII.4) the same holds for the moment, μ.

---

[299] We indicate the experimental values of the spin J and magnetic moment μ of a nucleus in brackets. Note that the nuclei under consideration in this problem either contain only one nucleon above their filled shells or have one hole in a filled shell.

618 Exploring Quantum Mechanics

Nucleus	J^P	μ	Nucleus	J^P	μ
^{3}H$(p(1s_{1/2}))$	$1/2^+$	2.79	^{13}N$(p(1p_{1/2}))$	$1/2^-$	-0.26
^{3}He$(n(1s_{1/2}))$	$1/2^+$	-1.91	^{17}O$(n(1d_{5/2}))$	$5/2^+$	-1.91
^{11}B$(p(1p_{3/2})^{-1})$	$3/2^-$	3.79	^{29}Si$(n(2s_{1/2}))$	$1/2^+$	-1.91
^{13}C$(n(1p_{1/2}))$	$1/2^-$	0.64			

The table provides a nucleus configuration only above the filled shells; $(nlj)^{-1}$ denotes the existence of a hole in the state nlj.

Agreement between the calculated and experimental values μ is quite good, except for the nuclei ^{11}B and ^{29}Si.

Problem 12.18

In the shell model, find the magnetic moment of a nucleus that contains one proton and one neutron (or the corresponding hole) above filled shells in the states nlj depending on the nuclear spin J.

Compare the results with experimental data for the following nuclei:

$$^2\text{H}\,(J=1,\ \mu=0.86);\quad {}_3^6\text{Li}\,(1,\ 0.82);$$

$$_5^{10}\text{B}\,(3,\ 1.80);\quad {}_7^{14}\text{N}\,(1,\ 0.40).$$

Use the one-nuclei level scheme from Problem 12.13.

Solution

In the shell model, the magnetic moment, spin, and parity of a nucleus are determined by the nucleons above the filled shells. In this case, the magnetic moment operator takes the form (see Problem 12.20):

$$\hat{\mu} = g_p \hat{\mathbf{j}}_p + g_n \hat{\mathbf{j}}_n, \tag{1}$$

where $g_{p,n}(l,j)$ are the gyromagnetic factors for the proton and neutron in the state l_j; see table (XII.4). Averaging this operator and using the results of Problem 3.40, we find the magnetic moment of the nucleus in the configuration $p(nl_j)^1 n(nl_j)^1$ and for the nuclear spin, J (here, $\hat{\mathbf{J}} = \hat{\mathbf{j}}_p + \hat{\mathbf{j}}_n$ and $j_p = j_n = j$), as follows:

$$\mu(J) = \langle J, J_z = J|\hat{\mu}_z|J, J_z = J\rangle = \frac{1}{2}[g_p(l,j) + g_n(l,j)]J. \tag{2}$$

This equation also determines the magnetic moment of a nucleus that has one proton and one neutron hole in the state nl_j. Equation (2) yields:

Nucleus	J	μ
^{2}H$(p(1s)^1, n(1s)^1)$	1	0.88
^{6}Li$(p(1p_{3/2})^1, n(1p_{3/2})^1)$	1	0.63
^{10}B$(p(1p_{3/2})^{-1}, n(1p_{3/2})^{-1})$	3	1.89
^{14}N$(p(1p_{1/2})^1, n(1p_{1/2})^1)$	1	0.38

Problem 12.19

Calculate the magnetic moment of a nucleus that contains one proton and one neutron (or the corresponding holes) above filled shells in the same states within the LS-coupling scheme.[300]

Apply the result obtained to the ground state of $^6_3$Li, which has spin $J = 1$. Assuming that the nucleons above the filled shell $(1s)^4$ are in the $1p$-state, find the nuclear magnetic moment for different possible states L and S, and compare these results to the experimental value of $\mu_{exp} = 0.82$, and also with the results of the previous problem. What is the isospin of the states considered?

Solution

The problem involves the nuclear magnetic moment operator in the LS-coupling scheme,

$$\hat{\boldsymbol{\mu}} = g_L \hat{\mathbf{L}} + g_S \hat{\mathbf{S}},$$

where $g_{L,S}$ are the orbital and spin gyromagnetic factors for the nuclei (unfilled shell). Averaging this operator according to Problem 3.40, we find

$$\mu(L, S, J) = \langle J, J_z = J | \hat{\mu}_z | J, J_z = J \rangle =$$
$$= \frac{1}{2(J+1)} \{ (g_L + g_S) J(J+1) + (g_L - g_S)[L(L+1) - S(S+1)] \}, \quad (1)$$

where L, S are the total angular and spin momenta of the nucleons, which, along with J, characterizes the state of a nucleus in the framework of the LS-coupling scheme.

Let us find $g_{L,S}$ for the proton–neutron system. The orbital magnetic moment is given by (see Problem 12.21)

$$\hat{\boldsymbol{\mu}}_{orb} = g_{l,p} \hat{\mathbf{l}}_p + g_{l,n} \hat{\mathbf{l}}_n,$$

with the gyromagnetic factors $g_{l,p} = 1$ and $g_{l,p} = 0$, and it takes the form $\hat{\boldsymbol{\mu}}_{orb} = g_L \hat{\mathbf{L}}$ only after averaging over the states with well-defined values of L. As well as above, we take advantage of the results of Problem 3.40 and obtain (here $l_p = l_n$):

---

[300] Here, the single-particle levels are characterized by quantum numbers n, l, but not n, l, j, as in the jj-coupling scheme.

$$\mu_L = g_L L, \quad g_L = \frac{1}{2}(g_{l,p} + g_{l,n}) = \frac{1}{2}. \tag{2}$$

Similarly, we find

$$\mu_S = g_S S, \quad g_S = \frac{1}{2}(g_{s,p} + g_{s,n}) = 0.88, \tag{3}$$

and according to Eqs. (1), (2), and (3) we obtain:

$$\mu(L, S, J) = 0.69 J - \frac{0.19}{(J+1)}[L(L+1) - S(S+1)]. \tag{4}$$

If we apply Eq. (4) to the nucleus ^{6}Li with $J = 1$ for different values of L, S (consistent with $J = 1$), it gives

$$\mu(0, 1, 1) = 0.88 \ (T = 0); \quad \mu(2, 1, 1) = 0.31 \ (T = 0),$$
$$\mu(1, 0, 1) = 0.50 \ (T = 0); \quad \mu(1, 1, 1) = 0.69 \ (T = 1),$$

where we provide the values of the isospin T for the corresponding nuclear states (see Problems 12.6 and 12.15). Experimental data for ^{6}Li show a tendency towards the LS-coupling scheme in this nucleus, here $L = 0$, $S = 1$.

Problem 12.20

Use the jj-coupling scheme to find the magnetic moment of a nucleus that has the same number of protons and neutrons above filled shells in the same nlj states depending on the value of J.

Apply the result to the nucleus $^{22}_{11}$Na that has $J = 3$ and the magnetic moment $\mu_{exp} = 1.75$.

Solution

The magnetic moment operator of a nucleus in an nl_j state can be written in the form (see Problem 12.9):

$$\hat{\mu}_N = g_p(l, j)\left(\frac{1}{2} + \hat{\tau}_3\right)\hat{\mathbf{j}} + g_n(l, j)\left(\frac{1}{2} - \hat{\tau}_3\right)\hat{\mathbf{j}},$$

where the gyromagnetic factors, $g_{p,n}$, are determined by Eqs. (XII.3) and (XII.4). Hence, for these nuclei,

$$\hat{\mu} = \frac{1}{2}(g_p + g_n)\sum_a \hat{\mathbf{j}}_a + (g_p - g_n)\sum_a \hat{\tau}_{3,a}\hat{\mathbf{j}}_a, \tag{1}$$

where the sum is taken over all the nucleons in the (unfilled) shell nl_j. If we average this operator over the nuclear states with a definite value of the isospin T, the second sum vanishes due to the fact that $T_3 = 0$; indeed,

$$\langle T, T_3 = 0|\hat{\tau}_{3a}|T, T_3 = 0\rangle \propto \langle T, T_3 = 0|\hat{T}_3|T, T_3 = 0\rangle = 0.$$

Hence, the nuclear magnetic moment is determined by the first sum in Eq. (1), and since $\sum_a \hat{\mathbf{j}}_a = \hat{\mathbf{J}}$, we have

$$\mu(J) = \frac{1}{2}[g_p(l,j) + g_n(l,j)]J. \tag{2}$$

$^{22}_{11}$Na in its ground state has the nucleon configuration p$(1d_{5/2})^3$n$(1d_{5/2})^3$ above the closed shells; see Problem 12.13. Taking into account the values of $g_{p,n}(l,j)$ according to Eq. (2), we find the shell model prediction for the magnetic moment of the nucleus ^{22}Na (that has the spin $J = 3$): $\mu_{shell} = 1.74$, which is near-identical to the experimental value: $\mu_{exp} = 1.75$.

Problem 12.21

The same as in the previous problem, but using the LS-coupling scheme.

Solution

The magnetic moment here takes the form:

$$\hat{\boldsymbol{\mu}}_N = (g_{l,p}\hat{\mathbf{l}} + g_{s,p}\hat{\mathbf{s}})\left(\frac{1}{2} + \hat{\tau}_3\right) + (g_{l,n}\hat{\mathbf{l}} + g_{s,n}\hat{\mathbf{s}})\left(\frac{1}{2} - \hat{\tau}_3\right),$$

where the gyromagnetic factors are $g_{l,p} = 1$, $g_{l,n} = 0$, $g_{s,p} = 5.59$, $g_{s,n} = -3.83$. Hence, for the nucleus we have

$$\hat{\boldsymbol{\mu}} = g_L\hat{\mathbf{L}} + g_S\hat{\mathbf{S}} + \sum_a [(g_{lp} - g_{ln})\hat{\mathbf{l}}_a + (g_{sp} - g_{sn})\hat{\mathbf{s}}_a]\hat{\tau}_{3a}, \tag{1}$$

$$\hat{\mathbf{L}} = \sum_a \hat{\mathbf{l}}_a, \quad \hat{\mathbf{S}} = \sum_a \hat{\mathbf{s}}_a, \quad g_L = \frac{g_{lp} + g_{ln}}{2} = \frac{1}{2},$$

$$g_S = \frac{g_{sp} + g_{sn}}{2} = 0.88,$$

where the sum is taken over the unfilled shell (see Eq. (1) of the previous problem, which determines the magnetic moment in the jj-coupling scheme).

After averaging Eq. (1) over the nuclear states that correspond to a definite value of the isospin T and its projection $T_3 = 0$, the last term vanishes. Hence the magnetic moment of such a nucleus is determined only by the first, *isoscalar* part, $\hat{\boldsymbol{\mu}}_{isosc} = g_L\hat{\mathbf{L}} + g_S\hat{\mathbf{S}}$, of the operator $\hat{\boldsymbol{\mu}}$. Therefore,

$$\mu(L, S, J) = \frac{(g_L + g_S)J(J+1) + (g_L - g_S)[L(L+1) - S(S+1)]}{2(J+1)}. \tag{2}$$

See Problems 12.19 and 12.20.

Problem 12.22

Using the shell model, find a relation between the magnetic moments of the ground states of the mirror nuclei. Consider all nucleons (of both charge states) to be in same states nlj above filled shells.

Solution

The magnetic moment operator, where all the nucleons are above the filled shells and are in the same states, nlj, has the form (see Problem 12.20):

$$\hat{\boldsymbol{\mu}} = \frac{1}{2}[g_p(l,j) + g_n(l,j)]\hat{\mathbf{J}} + [g_p(l,j) - g_n(l,j)]\sum_a \hat{\tau}_{3a}\hat{\mathbf{j}}_a.$$

After taking the expectation value (see also Problems 12.18 and 12.21), we obtain

$$\mu = \mu_{isosc} + \mu_{isovec},$$

where

$$\mu_{isosc} = \frac{1}{2}(g_p + g_n)J, \quad \mu_{isovec} = (g_p - g_n)\left\langle \left|\sum_a \hat{\tau}_{3a}\hat{j}_{za}\right| \right\rangle. \tag{1}$$

Let us denote the mirror nuclei as A and $\overline{A}$. Since they differ from each other only by the sign of the isospin projection T_3, their *isoscalar* parts of the magnetic moment are the same, while the *isovector* parts have the opposite signs, so that

$$\mu(A) + \mu(\overline{A}) = [g_p(l,j) + g_n(l,j)]J. \tag{2}$$

For the mirror nuclei, ^{3}H and ^{3}He, Eq. (2) gives

$$\mu(^3\text{H}) + \mu(^3\text{He}) = 0.88.$$

Note that the experimental value is 0.78 (see also Problem 12.17).

Problem 12.23

Find the quadrupole moment, Q_0, of nuclei that have only one proton above filled shells, for the states: a) $s_{1/2}$, b) $p_{3/2}$, and c) $d_{5/2}$ (express Q_0 in terms of $\langle r_p^2 \rangle$). Consider $A \gg 1$.

Solution

Let us remind ourselves that the following mean value actually defines the nuclear quadrupole moment:

$$Q_0 = \langle J, J_z = J| \sum_p (3z_p^2 - r_p^2)|J, J_z = J\rangle.$$

For the nuclei considered, the dominant contribution to Q_0 comes from the proton above filled shells, hence

$$Q_0 \approx \int \psi^*_{n_j l_j}(3\cos^2\theta - 1)\psi_{n_j l_j} r^2 dV, \tag{1}$$

where $\psi_{n_j l_j}$ is the wavefunction of the proton, $J = j$, and $J_z = j_z$. Since in this problem, $j = l + 1/2$, the wavefunction takes the form:

$$\psi_{n_j l_j} = Y_{ll}(\mathbf{n})\begin{pmatrix}1\\0\end{pmatrix} f(r), \quad |Y_{ll}|^2 = \frac{(2l+1)!!}{2^{l+2}\pi l!}\sin^{2l}\theta.$$

Here, according to Eq. (1), we find

$$Q_0 = \frac{(2l+1)!!}{2^{l+2}\pi l!}\int(3\cos^2\theta - 1)\sin^{2l}\theta d\Omega \langle r_p^2 \rangle, \tag{2}$$

where

$$\langle r_p^2 \rangle = \langle njlj|r^2|njlj\rangle = \int_0^\infty r^4 f^2(r) dr$$

is the mean value of the radius-vector squared for the proton.

An elementary integration over the angles in Eq. (2) gives

a) $Q_0(s_{1/2}) = 0$; b) $Q_0(p_{3/2}) = -\frac{2}{5}\langle r_p^2 \rangle$; c) $Q_0(d_{5/2}) = -\frac{4}{7}\langle r_p^2 \rangle.$

Note the signs of Q_0 (see also the following two problems).

When calculating the quadrupole moment, we can neglect the contribution from protons in filled shells. They give a spherically symmetric charge distribution, whose quadrupole moment with respect to the symmetry center is guaranteed to vanish. Their contribution to Q_0 (defined relative to the nuclear center of mass) is $\sim Z/A^2 \ll 1$ of the value (2) (see Problem 12.26).

Finally, the quadrupole moments of the nuclei that have one proton above filled shell and one proton hole, respectively, differ by a sign (in contrast to the magnetic moments of such nuclei).

Problem 12.24

Generalize the results of the previous problem to the case of a nucleus with a proton outside a filled shell in a state with an arbitrary value of l and $j = l + 1/2$.

Solution

The quadrupole moment is again defined by Eq. (2) from the previous problem. Rewrite it in the form

$$Q_0 = \int (3\cos^2\theta - 1)|Y_{ll}|^2 d\Omega \langle r_p^2 \rangle, \tag{1}$$

and use the expression

$$\sin^2\theta \cdot |Y_{ll}|^2 = \frac{2l+2}{2l+3}|Y_{l+1,l+1}|^2$$

(see the formula for $|Y_{ll}|^2$ given in Problem 12.23). Using Eq. (1) here and performing the integration, we obtain

$$Q_0 = -\frac{2l}{2l+3}\langle r_p^2 \rangle = -\frac{2j-1}{2j+2}\langle r_p^2 \rangle. \tag{2}$$

In conclusion, let us comment on the sign, $Q_0 < 0$, and the limiting value $Q_0 = -\langle r_p^2 \rangle$ for $j \to \infty$. Both these properties become apparent if we note that in the quasi-classical limit, the trajectory of a particle with $J_z = J$ lies in the equatorial plane (here $z = 0$ and from the expression for Q_0 it follows $Q_0 = -\langle r_p^2 \rangle$).

Problem 12.25

Find the quadrupole moment of a nucleus that has only one proton above filled shells in a state with an arbitrary l and $j = l - 1/2$. Compare it to the results of the previous two problems.

Solution

The expression for Q_0 (with the corresponding value of the total proton angular momentum, $j = l - 1/2$) obtained in the previous problem remains valid here as well. This result follows from the same angular distributions for the probability density, described by the wavefunctions of the form $\psi_{jl_{1(2)}j_z}$ with $l_{1,2} = j \pm 1/2$, as discussed in Problem 5.25.

Hence, the same remarks regarding the sign of the nuclear quadrupole moment, made in the two previous problems, can be transplanted to this case as well.

Problem 12.26

Find the quadrupole moment of a nucleus that has only one neutron above filled shells in a state with the orbital angular momentum, l, and the total angular momentum, $j = l \pm 1/2$.

Comment

Consider this nucleus as a combination of two subsystems: a neutron above filled shells and a set of filled shells (as a whole), which moves with respect to the nuclear center of mass.

Solution

In this approximation, where the nuclear center of mass coincides with the system of filled shells, the nuclear quadrupole moment is equal to zero (see Problem 12.23). The center of the nucleon charge distribution for the filled shells coincides with their center of mass, and is located in the point

$$\mathbf{r}_{\text{shell}} = -\frac{\mathbf{r}_n}{A-1},$$

where $\mathbf{r}_n$ is the radius-vector of the neutron, measured with respect to the total center of mass. Here, the quadrupole moment of the nucleus (originating from the protons of the filled shells) is defined by the expression

$$Q_0^{(n)} = \frac{Z}{(A-1)^2} \langle \psi_{njlj} | (3\cos^2\theta - 1)r^2 | \psi_{njlj} \rangle$$

(see Problem 12.23), where ψ_{njlj} is the neutron wavefunction. Hence,

$$Q_0^{(n)} = -\frac{2j-1}{2j+2} \frac{Z}{(A-1)^2} \langle r_n^2 \rangle, \tag{1}$$

which follows directly from the results of the two previous problems: here, the nuclear spin is $J = j$. Note also that $Q_0^{(n)} < 0$.

Problem 12.27

In the shell model with a self-consistent potential of the harmonic-oscillator type (see Problem 12.12), obtain the expression for the nucleon's radial density inside a nucleus with $A \gg 1$. Use quasi-classical considerations. Here, neglect the Coulomb interaction between the protons and consider nuclei with the same number of protons and neutrons. How do these results compare with the experimental data for heavy nuclei?

Solution

If we neglect the Coulomb interactions, the levels for protons and neutrons are the same in the ground state. Here, for the ground-state of the nucleus with $A = 2Z$ the same single-particle proton and neutron levels would be occupied.

Let us denote the maximum energy of occupied states as ε_F (see Fig. 12.4, where a constant energy, $-U_0$, has been omitted in the potential). We see that the volume of phase space (corresponding to the occupied states in volume dV) is equal to

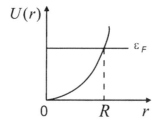

Fig. 12.4

$$d\Gamma = \frac{4\pi p_F^3}{3} dV = \frac{4\pi}{3}(2m\varepsilon_F - m^2\omega^2 r^2)^{3/2} dV.$$

Dividing it by $(2\pi\hbar)^3$, we obtain the number of filled states. In each of them, there are four nucleons (in different charge and spin states), so that the total number of nucleons in the volume, dV, is

$$dN = \frac{2}{3\pi^2\hbar^3}(m\omega R)^3 \left(1 - \frac{r^2}{R^2}\right)^{3/2} dV, \quad r \leq R = \sqrt{\frac{2\varepsilon_F}{m\omega^2}}. \tag{1}$$

Hence the nucleon density inside the nucleus is described by the following expression:

$$n(r) = \frac{2}{3\pi^2\hbar^3}(m\omega R)^3 \left(1 - \frac{r^2}{R^2}\right)^{3/2} dV, \quad r \leq R, \tag{2}$$

where $n_p = n_n = n/2$ and for $r > R$, $n \equiv 0$.

Normalization condition for expressions (1) and (2) is tied to the total number of nucleons, and yields the relation

$$\frac{1}{\hbar} m\omega R^2 = (12A)^{1/3}$$

between the parameters ω and R. Hence, Eq. (2) takes the form

$$n(r) = \frac{8A}{\pi^2 R^3}\left(1 - \frac{r^2}{R^2}\right)^{3/2}, \quad r \leq R \tag{3}$$

(where the normalization integral is calculated using the substitution, $r/R = \sin u$).

Let us note that for heavy nuclei, Eq. (3) (based on the harmonic oscillator self-consistent potential) contradicts experimental data, which indicates that in such nuclei the nucleon density is almost constant except for a narrow region near the boundary (see Problem 12.28).

Problem 12.28

The same as in the previous problem, but for the following self-consistent potential:

$$U(r) = \begin{cases} -U_0, & r < R \\ \infty, & r > R. \end{cases}$$

Choose the model parameter, R, equal to the nuclear radius $R = r_0 A^{1/3}$, $r_0 = 1.2 \cdot 10^{-13}$, and find a threshold momentum of nucleons and their maximum velocity.

Solution

The problem can be solved in the same way as the previous one. Now, the threshold momentum is constant, $p = \sqrt{2m\varepsilon_F}$ and instead of Eq. (2) of the previous problem we obtain:

$$n(r) = \frac{2p_F^3}{3\pi^2 \hbar^3} = \text{const}, \quad r < R. \tag{1}$$

Normalization to the total number of nucleons, A, gives $(p_F R)^3 = 9\pi A \hbar^3 / 8$, or

$$p_F = \frac{1}{2}(9\pi)^{1/3} \frac{\hbar}{r_0} \tag{2}$$

having made use of the expression for R.

The maximum velocity for the nucleons (treated like an ideal Fermi-gas) corresponding to the momentum p_F is

$$v_F = \frac{p_F}{m} = \frac{1}{2}(9\pi)^{1/3} \frac{m_e}{m} \frac{a_B}{r_0} \frac{\hbar}{m_e a_B} \approx \frac{c}{4},$$

where $m/m_e \approx 1840$, $a_B \approx 0.53 \cdot 10^{-8}$ cm, $v_{at} = \hbar / m_e a_B = c/137$, and c is the speed of light, while the maximum kinetic energy is $\varepsilon_F \approx 30$ MeV $\ll mc^2 \approx 940$ MeV, so that nucleons can be considered non-relativistic.

12.3 Isotopic invariance

Problem 12.29

Charges (measured in the units of the proton's charge, e) of different particles in an isospin multiplet are generally expressed in terms of the isospin component, T_3, corresponding to a given particle: $q = (1/2)Y + T_3$, where Y is a so-called *hypercharge* (for example, for a nucleon, $Y_N = 1$, for a pion, $Y_\pi = 0$, etc.).

Prove that isospin conservation in the presence of interactions implies hypercharge conservation.

Solution

Let us label particles in the initial and final states by indices, i and f, so that

$$q_{i,f} = \frac{1}{2}Y_{i,f} + T_{3,i,f}.$$

From the charge conservation law, it follows that $\sum q_i = \sum q_f$, while the isospin conservation[301] yields $\sum T_{3,i} = \sum T_{3,f}$. Hence, we obtain $\sum Y_i = \sum Y_f$, which indeed implies hypercharge conservation.

Note that since the mean value, $\overline{T_3}$, for particles in a given multiplet is equal to zero, the hypercharge is twice the mean electric charge of the particles in the multiplet.

Problem 12.30

Find the most general form of an isospin-invariant pion–nucleon interaction operator.

How are the operators of the πN-interaction in states with a definite value of the isospin ($T = 1/2$, $3/2$) connected to the operator $\hat{U}$? Express $\hat{U}$ in terms of the operators, $\hat{U}(T)$.

Solution

The operator can be expressed in terms of the following operators: the identity operator, $\hat{1}$; the nuclear isospin operators, $\hat{\tau}_i$; and the pion $\hat{t}_k$ components that must be iso-scalars. Since the operators $\hat{\tau}$ and $\hat{t}$ are isovector operators, then we can produce only the following isoscalar operators $\hat{\tau}^2$, $\hat{t}^2$, $(\hat{\tau} \cdot \hat{t})$, and also different combinations of these operators. But they can all be expressed as linear combinations of two operators: $\hat{1}$ and $(\hat{\tau} \cdot \hat{t})$ (see Problem 12.9). Indeed, $\hat{\tau}^2 = 3/4$ and $\hat{t}^2 = 2$ are multiples of the identity operator, while for $\hat{\tau} \cdot \hat{t}$, the expression $(\hat{\tau} \cdot \hat{t})^2 = (1 - \hat{\tau} \cdot \hat{t})/2$ holds. It follows, for example, from Problem 1.21, that the operator $\hat{\tau} \cdot \hat{t} = \hat{\mathbf{T}}^2/2 - 11/8$ has only two different eigenvalues, equal to -1 and $+1/2$, in the states with the total isospin of the πN-system equal to $T = 1/2$ and $3/2$, respectively.

Hence, the sought after operator has the form

$$\hat{U} = \hat{V}_1 + \hat{V}_2(\hat{\tau} \cdot \hat{t}), \tag{1}$$

where $\hat{V}_{1,2}$ are operators in configuration space that do not depend on the isospin. Hence, we find the form of operators $\hat{U}(T)$ for the πN-interaction in states with the definite values of total isospin, $T = 1/2$ and $3/2$, as follows:

$$\hat{U}\left(T = \frac{1}{2}\right) = \hat{V}_1 - \hat{V}_2, \quad \hat{U}\left(T = \frac{3}{2}\right) = \hat{V}_1 + \frac{1}{2}\hat{V}_2. \tag{2}$$

---

[301] In fact, we only require the T_3-components to be conserved.

We can relate $\hat{U}$ with these operators:

$$\hat{U} = \frac{1}{3}\left[\hat{U}\left(\frac{1}{2}\right) + 2\hat{U}\left(\frac{3}{2}\right)\right] - \frac{2}{3}\left[\hat{U}\left(\frac{1}{2}\right) - \hat{U}\left(\frac{3}{2}\right)\right]\hat{\tau}\cdot\hat{t}. \tag{3}$$

Problem 12.31

Repeat the previous problem for a system of two pions. Express the operator $\hat{U}$ in terms of the $\pi\pi$-interaction operators for states with a definite isospin.

Solution

The solution is analogous to the previous one. The difference lies in the fact that now the operator $\hat{t}_1 \cdot \hat{t}_2 = \frac{1}{2}\hat{T}^2 - 2$ has three different eigenvalues equal to -2, -1, and $+1$ corresponding to the values 0, 1, and 2 of the total pion isospin. Hence, the independent isoscalar operators are $\hat{1}$, $\hat{t}_1 \cdot \hat{t}_2$, and $(\hat{t}_1 \cdot \hat{t}_2)^2$ (here $(\hat{t}_1 \cdot \hat{t}_2)^3 = 2 + (\hat{t}_1 \cdot \hat{t}_2) - 2(\hat{t}_1 \cdot \hat{t}_2)^2$), while the isotopic-invariant $\pi\pi$-interaction has the form:

$$\hat{U}_{\pi\pi} = \hat{V}_1 + \hat{V}_2(\hat{t}_1 \cdot \hat{t}_2) + \hat{V}_3(\hat{t}_1 \cdot \hat{t}_2)^2. \tag{1}$$

Operators for the pion interaction in the states with definite values of the isospin read

$$\hat{U}_{\pi\pi}(T=0) = \hat{V}_1 - 2\hat{V}_2 + 4\hat{V}_3; \quad \hat{U}_{\pi\pi}(T=1) = \hat{V}_1 - \hat{V}_2 + \hat{V}_3;$$

$$\hat{U}_{\pi\pi}(T=2) = \hat{V}_1 + \hat{V}_2 + \hat{V}_3.$$

Problem 12.32

For a two-pion system find the isospin structure of the Coulomb interaction operator.

Solution

Since the pion charge operator is connected to the t_3-component of the isospin by $\hat{q}_\pi = e\hat{t}_3$ (compare to Problem 12.29, $e > 0$), then the Coulomb interaction between the pions has the form:

$$\hat{U}_{Coul} = \frac{e^2}{|\mathbf{r}_1 - \mathbf{r}_2|}\hat{t}_3^{(1)}\hat{t}_3^{(2)}.$$

Problem 12.33

Repeat the previous problem for the Coulomb interaction in a πN-system.

Result

The operator for the πN Coulomb interaction is

$$\hat{U}_{Coul} = \frac{e^2}{|\mathbf{r}_\pi - \mathbf{r}_N|}(1 + 2\hat{\tau}_3)\hat{t}_3,$$

see Problems 12.9 and 12.32.

Problem 12.34

What are the possible values of the isospin for a two-pion system in states with a definite value of the orbital angular momentum, L?

Solution

Isotopic invariance suggests that we should consider different particles that correspond to the same multiplet as identical particles but in different charge states. Here the principle of indistinguishability of identical particles is valid for the different particles in the same multiplet. In particular, for mutual permutation of two pions (that are spinless bosons) the wavefunction must remain unchanged (and remain symmetric).

For two pions, permutation of the pion's spatial variables is equivalent to coordinate inversion with respect to the center of mass, so that the symmetry of the wavefunction coincides with the parity, $(-1)^L$. Symmetry of the isospin part of the wavefunction is determined by the factor $(-1)^T$, where T is the total isospin of the system, which follows from the result of Problem 3.30 and taking into account an analogy between the angular momentum and isospin (with, $T_\pi = 1$). Hence, the symmetric wavefunction of two pions leads to $(-1)^{L+T} = 1$ and the possible values L and T are of the same parity: in the states with an even L, only $T = 0$ and $T = 2$ are possible; if L is odd, only $T = 1$ can be realized.

Problem 12.35

For a system that consists of two π^0-mesons, find the probabilities, $w(T)$, to have different values of the total isospin and the mean value, $\overline{\mathbf{T}^2}$.

Solution

For the pion, $T_\pi = 1$ and $T_3(\pi_0) = 0$. Due to the analogy between the angular momentum and isospin, the solution to the current problem follows one-to-one that to Problem 3.32. Hence, we just provide the final results as follows:

$$w(T=2) = \frac{2}{3}, \quad w(T=1) = 0, \quad w(T=0) = \frac{1}{3}, \quad \overline{\mathbf{T}^2} = 4.$$

Problem 12.36

Find the probabilities to have different values of the pion–nucleon total isospin and the mean value $\overline{T^2}$ in the following charge states: $\pi^+ p$, $\pi^+ n$, $\pi^0 p$, $\pi^0 n$, $\pi^- p$, and $\pi^- n$.

Solution

In the charged states of the πN-system considered, the t_3-components of the nucleon and pion isospin have definite values.[302] Hence, using the technique introduced in Problem 3.29, we find $\overline{T^2}$ and the probabilities, $w(T)$, of the total isospin for the πN-system ($T = 1/2$ and $3/2$), as follows:

1) for $\pi^+ p$ and $\pi^- n$-systems $w(3/2) = 1$, $\overline{T^2} = 15/4$;
2) for $\pi^+ n$ and $\pi^- p$-systems $\overline{T^2} = 7/4$, $w(1/2) = 2/3$, $w(3/2) = 1/3$;
3) for $\pi^0 p$ and $\pi^0 n$-systems $\overline{T^2} = 11/4$, $w(1/2) = 1/3$, $w(3/2) = 2/3$.

Problem 12.37

A neutral particle f^0 with isospin $T = 0$ decays into two pions: $f^0 \to 2\pi$. The possible decay channels are $f^0 \to \pi^+ \pi^-$ and $f^0 \to 2\pi^0$. Find the relation between the probabilities of decay via these channels.

Solution

The problem could be solved in different ways. For example, using the algebraic analogy between the angular momentum and isospin properties, we can use Clebsch–Gordan coefficients. The case of addition of angular momentum with the sum equal to zero was discussed in Problem 3.39. In our context, these results lead to

$$\psi_{T=0}(2\pi) = \frac{1}{\sqrt{3}}\{\psi_1(1)\psi_{-1}(2) - \psi_0(1)\psi_0(2) + \psi_{-1}(1)\psi_1(2)\}. \tag{1}$$

Here $\psi_{T=0}(2\pi)$ is the isospin wavefunction of a state with two pions with $T = 0$, while $\psi_{t_3}(1(2))$ are the isospin wavefunctions for pions with a definite value of t_3 ($t_3 = \pm 1, 0$).

The probabilities of f_0 decaying into different charge states of a two-pion system, $\pi^+ \pi^-$ and $2\pi^0$ are proportional to the probabilities of finding pions in the states with a zero total isospin $T = 0$ (due to isospin conservation in decays). According to Eq. (1), the probability of a charged state with two π^0 is equal to $w_{T=0}(2\pi^0) = (1/\sqrt{3})^2 = 1/3$ (recall that $t_3(\pi^0) = 0$); hence, $w_{T=0}(\pi^+\pi^-) = 2/3$, so that

$$\frac{w(f^0 \to \pi^+\pi^-)}{w(f^0 \to 2\pi^0)} = 2. \tag{2}$$

Let us examine alternative ways of obtaining the same result:

1) Consider the decay of N particles f^0. As a result of their decay, there appear $Nw\,(f^0 \to \pi^+\pi^-)$ charged π^+-mesons, the same number, Nw, of π^-, and $2Nw$ ($f^0 \to 2\pi^0$) uncharged pions (in the decay $f^0 \to 2\pi^0$, two π^0 are produced). The initial f^0 system is isospin-symmetric (isotropic in the isospace, since $T_f = 0$). The final state, which includes decayed pions, must also be isospin-symmetric. This symmetry demands in particular that the same number of pions are in different charge states – π^+, π^-, and π^0 – which leads to Eq. (2).

---

[302] Recall that for the pion, $T_\pi = 1$ and $T_3(\pi^\pm) = \pm 1$, and $T_3(\pi^0) = 0$.

The solution given is very intuitive from the physical point of view, and can be generalized to the case of more complicated (with respect to isospin) decays and reactions; see, for example, Problem 12.39.

2) Let us use the result of Problem 1.43, according to which $|\psi_A(B)|^2 = |\psi_B(A)|^2$. We choose the operators $\hat{t}_3^{(1)}$, $\hat{t}_3^{(2)}$ for the pion's isospin components as the "$\hat{B}$-operators", and $\hat{\mathbf{T}}^2$ and $\hat{T}_3$ as the "$\hat{A}$-operators". The aforementioned relation from Problem 1.43 then takes the form (see Problem 3.33):

$$|\psi_{TT_3}(t_3^{(1)}, t_3^{(2)})|^2 = |\psi_{t_3^{(1)}, t_3^{(2)}}(T, T_3)|^2. \tag{3}$$

If we now set $T = 0$ and $t_3^{(1)} = t_3^{(2)} = 0$, we obtain

$$w_{T=0}(2\pi^0) = w_{2\pi^0}(T=0) = \frac{1}{3}, \tag{4}$$

where we have used the probability to have $T = 0$ in the $2\pi^0$-system obtained in Problem 12.35.

Problem 12.38

Prove that the isospin part of the wavefunction for a system of three pions in the state with the total isospin, $T(3\pi) = 0$, has a definite symmetry with respect to interchange of the isospin variables of any two pions, and find the character (symmetric/antisymmetric) of this symmetry.

On the basis of this result, show that the neutral particle, ω^0, with the isotopic spin $T = 0$, cannot decay into three π^0-mesons; i.e., the decay process $\omega^0 \to 3\pi^0$ is forbidden.

Solution

To find the isospin part of the wavefunction with a zero total isospin, we rely again on the formal algebraic analogy between the angular momentum and isospin properties. We also take into account the following considerations and facts:

1) To describe a single pion isospin state with $t_\pi = 1$, we use a vector $\boldsymbol{\phi}$ in the isospace. Here, the relation between the vector representation and the conventional t_3-representation is the same as in the case of the angular momentum (e.g., see Problem 3.44). The pion in the charge state π^0 (i.e., $t_3 = 0$) is described a vector, $\boldsymbol{\phi}_{\pi^0}$, in the isospace with the following components: $\boldsymbol{\phi}_{\pi^0} = (0, 0, 1)$.
2) The wavefunction of the state with isospin $T = 0$ is a scalar (or a pseudo-scalar) in the isospace (does not change under isospin rotations).
3) From the three vectors, $\boldsymbol{\phi}_a$, that describe the isospin states of a single pion, we can make only one isospin scalar (or pseudo-scalar) for the three-pion system:

$$\psi_{T=0}(3\pi) = \boldsymbol{\phi}_1 \cdot [\boldsymbol{\phi}_2 \times \boldsymbol{\phi}_3] = \varepsilon_{ikl}\varphi_{1i}\varphi_{2k}\varphi_{3l}. \tag{1}$$

The isospin function is therefore anti-symmetric with respect to permutation of any two pion isospin variables. Hence, it does not contain the term that have all three pions in the same charge state, which proves that the corresponding decay is impossible (if isospin is conserved).

Problem 12.39

A particle Δ that has isospin $T = 3/2$ and charge states Δ^{++}, Δ^{+}, Δ^{0}, and Δ^{-}, corresponds to the values $+3/2$, $+1/2$, $-1/2$, and $-3/2$ of the isospin projection T_3 and decays into a pion and a nucleon: $\Delta \to \pi N$.

Show the possible decay channels for different charge states of the Δ-particle, and find relations between the probabilities of decay into these channels.

Solution

1) Decays of the particles Δ^{++} anf Δ^{-} involve one charge channel, while decays of particles Δ^{+} and Δ^{0} involve two channels:

$$\Delta^{++} \to \pi^+ p, \qquad \Delta^- \to \pi^- n,$$

$$\Delta^+ \to \begin{cases} \pi^+ n, \; w_1, \\ \pi^0 p, \; w_2, \end{cases} \qquad \Delta^0 \to \begin{cases} \pi^- p, \; w_1, \\ \pi^0 n, \; w_2. \end{cases} \tag{1}$$

Let us note that due to isospin invariance, the decay probabilities per unit of time for all the particles in the same multiplet are the same.[303] The respective probabilities of different decay channels that are mirror reflection of one another in the isospace are also the same (as, for example, the decays $\Delta^+ \to \pi^+ n$ and $\Delta^0 \to \pi^- p$).

For the calculation of the specific probabilities, $w_{1,2}$, for the particles, Δ^+ and Δ^0, we note that they are determined by the probabilities of forming the corresponding charge states in a πN-system in the state with $T = 3/2$ and $T_3 = \pm 1/2$. The isospin wavefunction of the πN-system, that is formed as a decay process of Δ^+, has the form:

$$\psi_{T=3/2, T_3=1/2} = C_1 |\pi^+ n\rangle + C_2 |\pi^0 p\rangle = C_1 \psi_1(\pi) \psi_{-1/2}(N) + C_2 \psi_0(\pi) \psi_{1/2}(N),$$

where $\psi_{t_3}(\pi(N))$ are normalized isospin wavefunctions of a pion (nucleon) in a state with a definite value of t_3. Here, the quantities $|C_1|^2$ and $|C_2|^2$ determine the probabilities of different charge states ($\pi^+ n$ and $\pi^0 p$, respectively) of the πN-system, and hence the decay probabilities of Δ into different channels are $w_1 = |C_1|^2$ and $w_2 = |C_2|^2$. Again taking into account the analogy between the angular momentum and isospin algebras, we see that $C_{1,2}$ here correspond to the Clebsch–Gordan coefficients and could be found using the known equations for these coefficients. Their values, $C_1 = 1/\sqrt{3}$ and $C_2 = \sqrt{2/3}$, were obtained in Problem 5.18. Hence we find the following relations:

---

[303] One of the consequences of isospin invariance breaking is the appearance of a mass difference for particles in a given isomultiplet. It is reflected in the released energy during a decay, and also may significantly affect the relative decay probabilities in various channels.

$$\frac{w(\Delta^+ \to \pi^+ n)}{w(\Delta^+ \to \pi^0 p)} = \frac{w(\Delta^0 \to \pi^- p)}{w(\Delta^0 \to \pi^0 n)} = \frac{1}{2}. \tag{2}$$

2) Let us mention two alternative methods to solve the problem, which do not involve a calculation of $C_{1,2}$.

The first method is based on the following physical considerations. Let us consider a non-polarized (in the isospace) particle "beam" of the Δ particles, where all charged states of Δ come in the same numbers, N_0. Among the decay products – pions and nucleons – different charge states would be in the following proportions:

$$N(\pi^+) = N(\pi^-) = N_0(1 + w_1),$$

$$N(\pi^0) = 2N_0 w_2 = 2N_0(1 - w_1).$$

From these considerations, it is evident that the "beam" of decaying pions must remain unpolarized (in the isospace), and the different charge states of the pion are presented in same numbers: $N(\pi^+) = N(\pi^-) = N(\pi^0)$. Hence we find $w_1 = 1/3$ and obtain the relation (2). Note also that the nucleon beam emitted in this decay is unpolarized as well, i.e., $N(p) = N(n)$.

The second method is based on using Eq. (3) from Problem 12.37. We leave it to the reader to reproduce the results of this problem within this method (see also Problem 12.36).

Problem 12.40

Repeat the previous problem for a particle N^* that has isospin $T = 1/2$ and charge states $N^{*+}(T_3 = 1/2)$, $N^{*0}(T_3 = -1/2)$, and decays into a pion and nucleon: $N^* \to \pi N$.

Solution
Using the same method as in the previous problem, we obtain

$$\frac{w(N^{*+} \to \pi^+ n)}{w(N^{*+} \to \pi^0 p)} = \frac{w(N^{*0} \to \pi^- p)}{w(N^{*0} \to \pi^0 n)} = 2.$$

Problem 12.41

Prove that

$$\frac{d\sigma(p + p \to d + \pi^+)}{d\sigma(n + p \to d + \pi^0)} = 2,$$

where $d\sigma$ are differential cross-sections of the corresponding reactions involving the same energies, angles, and spin orientations.

Solution

Since the isospins $T_d = 0$ and $T_\pi = 1$, the final states of the two reactions considered exist in different isospin states of the same "pion+deuteron" system with $T = 1$. Isospin conservation demands that the reactions can take place only if the initial nucleon–nucleon system has $T = 1$. In the reaction $pp \to d\pi^+$, both nucleons are in a state with $T = 1$ (and $T_3 = 1$); while in the reaction $pn \to d\pi^0$, the isotopic state of the nucleons with $T = 1$ and $T_3 = 0$ is present only with probability $1/2$ (the realization of the nucleon–nucleon system with $T = 0$ is equally probable).

From the viewpoint of coordinate and spin degrees of freedom, the cross-sections of both reactions are the same. However, the isospin invariance constraints the ratio of their cross-sections to be equal to the ratio of the decay probabilities into a desired isospin channel with $T = 1$ in the initial states and is equal to 2; q.e.d.

Problem 12.42

Prove that

$$\frac{d\sigma(p + d \to d + n + \pi^+)}{d\sigma(p + d \to d + p + \pi^0)} = 2,$$

where the meaning of $d\sigma$ is the same as in the previous problem.

Solution

Since $T_d = 0$, the isospin invariance implies that the deuteron plays the role of a "catalyst" in the process of proton "dissociation" into a nucleon and a pion:

$$p \to N\left(T = \frac{1}{2}\right) + \pi(T = 1).$$

At the initial stage of the process, $T = 1/2$ and $T_3 = 1/2$, and due to isospin conservation, the πN-system has to have the same values of T and T_3 in the final state. From the viewpoint of the coordinate and spin degrees of freedom, the reactions are the same, so that the ratio of their cross-sections is equal to the ratio of the charge states, $\pi^+ n$ and $\pi^0 p$, in the pion–nucleon system with $T = 1/2$ and $T_3 = 1/2$. The latter ratio is equal to 2 (see Problems 12.40 and 12.36), which proves the statement.

Problem 12.43

Using the hint that the pions are scattered by nucleons (in some energy interval) mainly through intermediate states of a πN-system with the total isotopic spin, $T = 3/2$ (here, interaction in the state $T = 1/2$ is negligibly small), find the differential cross-sections of the following three relations (assuming the same relative energies, angles, and spin orientations)

$$\pi^+ + p \to \pi^+ + p \quad (\text{I})$$
$$\pi^- + p \to \pi^0 + n \quad (\text{II})$$
$$\pi^- + p \to \pi^- + p \quad (\text{III})$$

Solution

Since the problem condition states that the reactions considered go only through the state with isospin $T = 3/2$ and are the same from the viewpoint of coordinate and spin degrees of freedom, we conclude that their cross-sections are proportional to the probabilities (or "weights") of the isotopic state with $T = 3/2$ required in both the initial and final states of the pion–nucleon system:

$$d\sigma \propto w_i\left(T = \frac{3}{2}\right) \cdot w_f\left(T = \frac{3}{2}\right). \quad (1)$$

These probabilities can be calculated using the same methods as used in Problem 12.36 (see also Problem 12.39), and we find them to be equal to 1 for $\pi^+ p$-; $2/3$ – for $\pi^0 n$-; and $1/3$ – for $\pi^- p$. Hence, it follows that the relation

$$d\sigma(\text{I}) : d\sigma(\text{II}) : d\sigma(\text{III}) = 9 : 2 : 1 \quad (2)$$

between the cross-sections of the reactions considered.

Problem 12.44

Taking into account charge symmetry of the nucleon–nucleon and pion–nucleon interactions, find the relation between differential cross-sections of the processes

$$n + p \to p + p + \pi^-, \quad n + p \to n + n + \pi^+.$$

Solution

The reactions

$$n + p \to p + p + \pi^-, \quad p + n \to n + n + \pi^+$$

are "mirror" reflections of one another in the isospace. Hence, due to isospin invariance, the differential cross-sections of these reactions for the same momenta and spins of the corresponding "mirror" particles (p and n, π^+ and π^-) are identical. Let us emphasize that replacing the particles by their "mirror" iso-partners must be performed in both stages of the process: the initial and final stages. It means that momentum and spin properties of the proton and neutron in the initial states of reactions must be interchanged as well.

13

Particle collisions

1) The quantum scattering problem for a particle with momentum $\mathbf{p}_0 = \hbar \mathbf{k}_0$, interacting with a potential, $U(\mathbf{r})$, involves solving the familiar Schrödinger equation,

$$\left[-\frac{\hbar^2}{2m}\Delta + U(\mathbf{r})\right]\psi^+_{\mathbf{k}_0}(\mathbf{r}) = E\psi^+_{\mathbf{k}_0}(\mathbf{r}). \qquad (\text{XIII.1})$$

A solution to this equation has the following large-distance asymptotic behavior:[304]

$$\psi^+_{\mathbf{k}_0}(\mathbf{r}) \approx_{r\to\infty} e^{i\mathbf{k}_0 r} + \frac{f(\mathbf{k},\mathbf{k}_0)}{r}e^{ikr}, \quad \mathbf{k} = \frac{k_0 \mathbf{r}}{r}, \qquad (\text{XIII.2})$$

where $\mathbf{k}$ is the wavevector of the scattered particle ($k = k_0 = \sqrt{2mE/\hbar^2}$). Here, $f(\mathbf{k}, \mathbf{k}_0)$ is the *scattering amplitude*, which is a key quantity of interest in the quantum theory of scattering. The square of its absolute value gives the *differential scattering cross-section*, $d\sigma/d\Omega$, which integrated over a sphere yields the total scattering cross-section, $\sigma = \int |f(\mathbf{k},\mathbf{k}_0)|^2 d\Omega_k$.

Using the Green function for free particles,

$$G^+_E(\mathbf{r},\mathbf{r}') = \frac{m}{2\pi\hbar^2 |\mathbf{r}-\mathbf{r}'|}e^{ik|\mathbf{r}-\mathbf{r}'|}, \qquad (\text{XIII.3})$$

Eq. (XIII.1), along with boundary condition (XIII.2), can be expressed as an integral equation:

$$\psi^+_{\mathbf{k}_0}(\mathbf{r}) = e^{i\mathbf{k}_0\cdot\mathbf{r}} - \frac{m}{2\pi\hbar^2}\int \frac{e^{ik|\mathbf{r}-\mathbf{r}'|}}{|\mathbf{r}-\mathbf{r}'|}U(\mathbf{r}')\psi^+_{\mathbf{k}_0}(\mathbf{r}')dV'. \qquad (\text{XIII.4})$$

This leads to the following expression for the scattering amplitude $f(\mathbf{k}, \mathbf{k}_0)$:

$$f(\mathbf{k},\mathbf{k}_0) = -\frac{m}{2\pi\hbar^2}\int e^{-i\mathbf{k}\cdot\mathbf{r}}U(r)\psi^+_{\mathbf{k}_0}(\mathbf{r})dV, \qquad (\text{XIII.5})$$

which is very convenient for constructing various perturbative expansions and for approximate calculations.

---

[304] For $U(\mathbf{r})$ that falls faster than $\propto 1/r$ at large distances, $r \to \infty$.

First Born approximation: Set $\psi_{\mathbf{k}_0}^+ = e^{i\mathbf{k}_0\cdot\mathbf{r}}$. Following Eq. (XIII.5), this leads to the following approximate equation for the scattering amplitude:

$$f^B(\mathbf{q}) = -\frac{m}{2\pi\hbar^2}\tilde{U}(\mathbf{q}), \quad \tilde{U}(\mathbf{q}) = \int e^{-i\mathbf{q}\cdot\mathbf{r}} U(\mathbf{r})dV, \quad \text{(XIII.6)}$$

We use $\mathbf{q} = \mathbf{k} - \mathbf{k}_0$ for the change in momentum of the particle after scattering. If the scattering is elastic, there is no change in the magnitude of this momentum, and $q = 2k\sin(\theta/2)$, where θ is the scattering angle. The first Born approximation constitutes the lowest-order expansion of the scattering amplitude in powers of the interaction potential. This approximation is valid if one of the following two inequalities is satisfied:

$$U_0 \ll \frac{\hbar^2}{mR^2} \quad \text{or} \quad U_0 \ll \frac{\hbar v}{R}, \quad \text{(XIII.7)}$$

where U_0 and R are the typical strength and radius of the potential respectively.

For scattering in a central potential, the scattering amplitude depends only on the energy E and the polar angle θ, and in the first Born approximation it depends only on the magnitude of momentum transfer, $\hbar q$. In the latter case, Eq. (XIII.6) can be written as

$$f^B(q) = -\frac{2m}{\hbar^2} \int_0^\infty U(r) \frac{\sin qr}{q} r\, dr. \quad \text{(XIII.8)}$$

2) For a spherically-symmetric potential, the scattering amplitude can be expanded in terms of the *partial waves* as follows:

$$f(k,\theta) = \sum_{l=0}^\infty (2l+1)\varphi_l(E)P_l(\cos\theta),$$

$$\varphi_l = \frac{e^{2i\delta_l}-1}{2ik} = \frac{1}{k(\cot\delta_l - i)}, \quad \text{(XIII.9)}$$

where the *phase shifts* $\delta_l(k)$ are related to the large distance asymptote of the radial wavefunction $R_{k,l}$, by $rR_{k,l} \approx C\sin(kr - \pi l/2 + \delta_l)$, and the sum is over all angular momenta, l. The total scattering cross-section is given by

$$\sigma = \sum_{l=0}^\infty \sigma_l,$$

$$\sigma_l = \frac{4\pi}{k^2}(2l+1)\sin^2\delta_l = \frac{4\pi}{k^2}(2l+1)(\cot^2\delta_l + 1)^{-1}. \quad \text{(XIII.10)}$$

Comparing Eqs. (XIII.9) and (XIII.10), we obtain the following identity (the *optical theorem*):[305]

---

[305] The optical theorem follows from very basic conservation laws and as such has a very general nature. It is valid even for scattering of composite particles, where inelastic processes are possible. In the

$$\operatorname{Im} f(E, \theta = 0) = \frac{k}{4\pi}\sigma(E). \qquad (\text{XIII.11})$$

Below, we list explicit approximate expressions for the phase-shifts under various approximation schemes:

1. In the Born approximation (where $|\delta_l^B| \ll 1$),

$$\delta_l^B(k) = -\frac{\pi m}{\hbar^2}\int_0^\infty U(r) J_{l+1/2}^2(kr) r\, dr. \qquad (\text{XIII.12})$$

2. In the quasi-classical approximation,

$$\delta_l^{(q)} = \int_{r_0}^\infty \left\{ \sqrt{k^2 - \frac{2m}{\hbar^2}U(r) - \frac{(l+1/2)^2}{r^2}} - k \right\} dr +$$

$$+ \frac{1}{2}\pi\left(l + \frac{1}{2}\right) - kr_0, \qquad (\text{XIII.13})$$

where r_0 is the turning point of the motion. If $|U(r)| \ll E$, this expression is further simplified to

$$\delta_l^{(q)} = -\int_{r_0}^\infty \frac{m U(r) dr}{\hbar^2\sqrt{k^2 - (l+1/2)^2/r^2}}, \quad r_0 = \frac{(l+1/2)}{k}. \qquad (\text{XIII.14})$$

3) For the scattering of slow particles ($kR \ll 1$), and if the potential falls rapidly enough (see Problems 13.28–30), an *effective range expansion* holds:

$$k^{2l+1}\cot\delta_l = -\frac{1}{a_l} + \frac{r_l k^2}{2} + \ldots, \qquad (\text{XIII.15})$$

where a_l and r_l are called the *scattering length* and the *effective interaction range*[306] respectively for the state with the orbital angular momentum, l; these parameters determine the low-energy scattering for the corresponding partial wave, in accordance with Eqs. (XIII.9) and (XIII.10).

If there are no shallow *real/virtual* (for $l = 0$) or *quasi-stationary* (for $l \neq 0$) levels in the potential, then the second term involving the effective interaction range can be ignored. In this case,

$$\delta_l \approx -a_l k^{2l+1}, \quad |a_l| \le R^{2l+1}, \quad \text{and} \quad \sigma_l \le 4\pi(2l+1)(kR)^{4l}R^2.$$

---

latter case, we should use in Eq. (XIII.11), the total collision cross-section for $\sigma(E)$, while $f(E,0)$ should be understood as the forward-scattering amplitude of *elastic* scattering of the *composite* particle.

[306] Note, however, that these parameters have the physical dimensions of length only for $l = 0$. In general, their dimensionality is $[a_l] = L^{2l+1}$, $[r_l] = L^{1-2l}$, where L is a unit of length.

On the other hand, if the potential supports a shallow bound state with the energy $|E_l| \ll \hbar^2/mR^2$ and angular momentum, l, the scattering cross-section for the corresponding partial wave, $\sigma_l(E)$, is sharply peaked at $E \approx E_l$ (*resonant scattering*). Specifically, if a shallow s-level ($l = 0$) exists, then it follows that $|a_0| \gg R$ and

$$\sigma(E) \approx \sigma_{l=0}(E) \approx \frac{2\pi\hbar^2}{m} \frac{1 + r_0\kappa_0}{E + \varepsilon_0}, \qquad (\text{XIII.16})$$

where $|E_0| = \varepsilon_0 = \hbar^2 \kappa_0^2/2m$ and κ_0 is given to second order in $1/a_0$ by[307] $\kappa_0 = 1/a_0 + r_0/2a_0^2$. If $a_0 > 0$, it follows that $\kappa_0 > 0$, and ε_0 gives the energy of a real bound state. For $a_0 < 0$, the level is virtual.

For resonant scattering with an angular momentum $l \neq 0$, the type of the energy-dependence and in particular the behavior of $\sigma_l(E)$ depend crucially on the nature of the level specifically on whether the level is real or quasi-stationary ($a_l < 0$). In the latter case, writing E_l as $E_l = E_R - i\Gamma_R/2$ (where E_R and Γ_R are the level energy and its width), from the poles of the scattering amplitude in the l-th harmonic located at $\cot \delta_l(E_l) = i$, we find:[308]

$$E_R \equiv \frac{\hbar^2 k_R^2}{2m} \approx \frac{\hbar^2}{ma_l r_l} > 0 \quad \text{and} \quad \Gamma_R \approx \left(\frac{2\hbar^2}{m|r_l|}\right) k_R^{2l+1}.$$

In this case, the partial-wave cross-section, $\sigma_l(E)$, changes sharply close to E_R and is given by

$$\sigma_l(E) \approx \frac{(2l+1)\pi}{k_R^2} \frac{\Gamma_R^2}{(E - E_R)^2 + \Gamma_R^2/4}. \qquad (\text{XIII.17})$$

4) For scattering of fast particles ($kR \gg 1$ and $E \gg |U(\mathbf{r})|$), the following relation holds for the amplitude of the scattered wave in the region of small scattering angles $\theta \leq 1/kR$ (the *eikonal approximation*):

$$f(\mathbf{k}_0, \mathbf{q}_\perp) = \frac{k}{2\pi i} \int [S(\boldsymbol{\rho}) - 1] e^{-i\mathbf{q}_\perp \cdot \boldsymbol{\rho}} d^2\rho \qquad (\text{XIII.18})$$

Here $\mathbf{q}_\perp$ is the component of $\mathbf{q}$ in a direction perpendicular to the incident momentum, $\hbar \mathbf{k}_0$ (for small-angle scattering, $q_\perp \approx q \approx k\theta$ and $q_\parallel \approx k\theta^2/2$) and

$$S(\boldsymbol{\rho}) = e^{2i\delta(\boldsymbol{\rho})}, \quad \delta(\boldsymbol{\rho}) = -\frac{1}{2\hbar v} \int_{-\infty}^{\infty} U(\boldsymbol{\rho}, z) dz. \qquad (\text{XIII.19})$$

---

[307] The poles of the partial-wave amplitude $\phi_l(E)$ are given by $\cot \delta_l(E) = i$. Combining this with the wave-vector corresponding to a bound state $\kappa = -i\sqrt{2mE/\hbar^2}$ and taking into account Eq. (XIII.15) for $a_0 \gg R(\approx r_0)$, it follows that $\kappa_0 \approx 1/a_0$, $\varepsilon_0 \approx \hbar^2/2ma_0^2$ to first order in $1/a_0$. The second-order term is obtained by substituting the first-order term in (XIII.15).

[308] For a shallow level with $l \neq 0$ to exist, effective interaction range must be negative, $r_l < 0$, so that $E_R, \Gamma_R > 0$ (see Problem 13.44).

Using the optical theorem with Eq. (XIII.18), we obtain the total scattering cross-section:[309]

$$\sigma(E) = 2 \int [1 - \cos 2\delta(\boldsymbol{\rho})] \, d^2\rho. \qquad (XIII.20)$$

In the case of a central potential, Eq. (XIII.19) for $\delta(\rho)$ coincides with the quasi-classical expression (XIII.14) for $l \approx k\rho \gg 1$, and Eq. (XIII.18) could be written in the form:

$$f(k, \theta) = ik \int_0^\infty \left[1 - e^{2i\delta(\rho)}\right] J_0(k\rho\theta) \rho \, d\rho, \qquad (XIII.21)$$

where $J_0(z)$ is the zeroth-order Bessel function.

5) In case of a spin-dependent interaction, the scattering amplitude becomes a matrix, $\hat{f}$, in the spin-basis, whose elements $\chi_f^* \hat{f} \chi_i$ describe the scattering from an initial spin state χ_i into a final state χ_f. For the scattering of spin-1/2 particles off of spinless ones and in the case of a parity-conserving interaction,[310]

$$\hat{f} = A(k, \theta) + iB(k, \theta)\boldsymbol{\nu} \cdot \hat{\boldsymbol{\sigma}}, \quad \boldsymbol{\nu} = \frac{[\mathbf{k}_0 \times \mathbf{k}]}{|[\mathbf{k}_0 \times \mathbf{k}]|}. \qquad (XIII.22)$$

The differential scattering cross-section summed over the spin states of the scattered particles is given by

$$\frac{d\sigma}{d\Omega} = |A|^2 + |B|^2 + 2\mathrm{Im}(AB^*)\boldsymbol{\nu} \cdot \mathbf{P}_0, \qquad (XIII.23)$$

where $\mathbf{P}_0 = 2\chi_i^* \hat{\boldsymbol{\sigma}} \chi_i$ is the polarization vector of the particles before collision. The polarization state of the scattered particles depends on both the interaction and the initial polarization, $\mathbf{P}_0$. If the particles were not polarized before collision, $\mathbf{P}_0 = \mathbf{0}$, then the polarization vector after scattering is

$$\mathbf{P} = \frac{2\mathrm{Im}(AB^*)}{|A|^2 + |B|^2} \boldsymbol{\nu}. \qquad (XIII.24)$$

As before, Eq. (XIII.22) for the amplitude can be expanded in terms of the partial waves:

$$A = \frac{1}{2ik} \sum_{l=0}^\infty [(l+1)(\exp\{2i\delta_l^+\} - 1) + l(\exp\{2i\delta_l^-\} - 1)] P_l(\cos\theta),$$

$$B = \frac{1}{2ik} \sum_{l=1}^\infty (\exp\{2i\delta_l^+\} - \exp\{2i\delta_l^-\}) \sin\theta P_l'(\cos\theta), \qquad (XIII.25)$$

---

[309] For this relation to hold, it is only required that $kR \gg 1$ (see Problem 13.51).
[310] Note that we have factored out the i term in front of B. This step is a matter of convention and different conventions are used in the literature (both with and without separating out the i-term; keep it in mind when comparing your results with the literature).

where $\tilde{\delta}_l^{\pm}$ are the phase shifts for states with the orbital angular momentum l and the total angular momentum $j = l \pm 1/2$.

6) The scattering amplitude has well-defined *analytic* and *unitarity* properties. In particular, it satisfies the following unitarity condition:[311]

$$f(\mathbf{k}, \mathbf{k}_0) - f^*(\mathbf{k}_0, \mathbf{k}) = \frac{ik}{2\pi} \int f(\mathbf{k}', \mathbf{k}_0) f^*(\mathbf{k}', \mathbf{k}) d\Omega'. \quad \text{(XIII.26)}$$

(For $\mathbf{k} = \mathbf{k}_0$, the optical theorem is reproduced.)

The analytic properties of scattering amplitudes are effectively *dispersion relations* for them. The simplest analytic properties exist for the forward-scattering amplitudes in a central potential, which as a function of energy E[312] in the complex plane satisfies a dispersion relation of the form:

$$f(E, 0) = f^B(0) + \sum_n \frac{d_n}{E - E_n} + \frac{1}{\pi} \int_0^\infty \frac{\operatorname{Im} f(E', 0)}{E' - E} dE'. \quad \text{(XIII.27)}$$

Here, $f^B(0)$ is the amplitude of the first Born approximation; see Eq. (XIII.6). The summation is performed over all the discrete states in the potential (each of which corresponds to a pole in the scattering amplitude); d_n is the residue at the pole corresponding to E_n, and is given by

$$d_n = -(-1)^{l_n}(2l_n + 1)\frac{\hbar^2 A_n^2}{2m}, \quad \text{(XIII.28)}$$

where A_n is a normalization coefficient for the large-distance asymptote of the radial wavefunction corresponding to a bound state with the angular momentum l_n; that is, $rR_n \approx A_n \exp\{-\kappa_n r\}$ (compare to Eq. (XI.5)).

13.1 Born approximation

Problem 13.1

Using the Born approximation, find the scattering amplitude and the total scattering cross-section of a particle on the following potentials:

a) $U(r) = \frac{\alpha}{r} e^{-r/R}$; b) $U(r) = \alpha \delta(r - R)$; c) $U(r) = U_0 e^{-r/R}$;

d) $U(r) = \frac{\alpha}{r^2}$; e) $U(r) = \begin{cases} U_0, & r < R, \\ 0, & r > R; \end{cases}$ f) $U(r) = U_0 e^{-r^2/R^2}$.

---

[311] It is consequence of the unitarity of the S-matrix. The same relation holds as a unitarity condition for scattering of composite particles, but only for values of energy for which inelastic processes are absent ("elastic" unitarity).

[312] It has the following singular points: branch points $E = 0$ and $E = \infty$ and poles E_n on real half-axis $E < 0$, which coincide with the positions of discrete levels.

Consider the limiting cases of slow and fast particles. Find the conditions under which the approximations in these limiting cases are valid.

Solution

The scattering amplitude (under the first Born approximation) can be calculated from Eqs. (XIII.6) and (XIII.8). The total scattering cross-section is

$$\sigma(E) = \int |f|^2 d\Omega = 2\pi \int_0^\pi |f|^2 \sin\theta \, d\theta = \frac{\pi\hbar^2}{2mE} \int_0^{8mE/\hbar^2} f^2(q) dq^2, \qquad (1)$$

with $q = 2k\sin(\theta/2)$. This leads to the following results:

a) $$f = -\frac{2m\alpha R^2}{\hbar^2(1+q^2 R^2)}, \quad \sigma(E) = 16\pi \left(\frac{m\alpha R^2}{\hbar^2}\right)^2 \frac{1}{1+4k^2 R^2}. \qquad (2)$$

For R finite, the scattering cross-section also has a finite magnitude. However, as $R \to \infty$, the short-range Yukawa potential considered here crosses over to a long-range Coulomb potential $U = \alpha/r$. For the latter, the differential cross-section is described by the Rutherford formula, $d\sigma/d\Omega = 4m^2\alpha^2/\hbar^4 q^4$, and the total scattering cross-section is infinite.

b) $$f = -\frac{2m\alpha R^2}{\hbar^2} \frac{\sin qR}{qR}, \quad \sigma(E) = \frac{4\pi m\alpha^2 R^2}{\hbar^2 E} \int_0^{\sqrt{8mER^2/\hbar^2}} \frac{\sin^2 x}{x} dx. \qquad (3)$$

In the limiting cases, we have

$$\sigma(E) \approx_{E\to 0} \frac{16\pi (m\alpha R^2)^2}{\hbar^4}, \quad \sigma(E) \approx_{E\to\infty} \frac{\pi m\alpha^2 R^2}{\hbar^2 E} \ln \frac{8mER^2}{\hbar^2}.$$

In the $E \to \infty$ limit, the integral in (3) diverges; to calculate this diverging integral we replace the oscillating factor $\sin^2 x$ by its mean value, $1/2$.

c) $$f = -\frac{4mU_0 R^3}{\hbar^2(1+q^2 R^2)^2},$$

$$\sigma(E) = \frac{8\pi m R^4 U_0^2}{3\hbar^3 E} \left[1 - \frac{1}{(1+8mER^2/\hbar^2)^3}\right]. \qquad (4)$$

d) $$f = -\frac{\pi m\alpha}{\hbar^2 q} = -\frac{\pi m\alpha}{2\hbar^2 k \sin(\theta/2)} \qquad (5)$$

The total scattering cross-section is infinite, which is related to a slow decrease of the potential at large distances. For more on scattering from the potential $U = \alpha/r^2$, see Problem 13.19.

e) $$f = \frac{2mU_0 R}{\hbar^2 q^2} \left(\cos qR - \frac{\sin qR}{qR}\right),$$

$$\sigma(E) = \frac{2\pi}{k^2}\left(\frac{mU_0R^2}{\hbar^2}\right)^2\left[1 - \frac{1}{(2kR)^2} + \frac{\sin 4kR}{(2kR)^3} - \frac{\sin^2 2kR}{(2kR)^4}\right]. \tag{6}$$

In the limiting cases we have

$$\sigma(E) \approx_{E\to 0} \frac{16\pi m^2 U_0^2 R^6}{9\hbar^4}, \quad \sigma(E) \approx_{E\to\infty} \frac{\pi m U_0^2 R^4}{\hbar^2 E}.$$

The limit $E \to \infty$ is also discussed in Problem 13.2.

f) $f = -\dfrac{\sqrt{\pi}mU_0R^3}{2\hbar^2}e^{-q^2R^2/4}$,

$$\sigma(E) = \frac{\pi^2 m U_0^2 R^4}{4\hbar^2 E}(1 - e^{-4mER^2/\hbar^2}). \tag{7}$$

Due to the exponential decrease of $f(q)$, the Born approximation is not applicable for large values of q^2 (see Problem 13.13). Therefore, the inclusion of the exponentially-small term in the equation for $\sigma(E)$ for $E \to \infty$ is beyond the accuracy of the approximation used.

Problem 13.2

Prove that the total scattering cross-section (in the Born approximation) of a high-energy particle in a potential, $U(r)$ with $kR \gg 1$ is described by the expression[313]

$$\sigma(E)\Big|_{E\to\infty} \approx \frac{m}{2\hbar^2 E}\iint\left[\int_{-\infty}^{\infty}U(\rho,z)dz\right]^2 d^2\rho. \tag{1}$$

The momentum of the particle before scattering is along the z-axis; ρ is a two-dimensional radius-vector lying in the plane perpendicular to the z-axis.

Apply this result to the potential $U(r) = U_0\exp\{-r^2/R^2\}$ and to a rectangular potential well (barrier) of strength, U_0, and radius, R. Compare to Problem 13.1.

Prove also that, in this case, the *transport scattering cross-section* has an asymptotic behavior given by

$$\sigma_{tr} = \int(1-\cos\theta)d\sigma\Big|_{E\to\infty} \approx \frac{1}{8E^2}\iint\left[\int_{-\infty}^{\infty}\frac{\rho}{r}\frac{\partial U(r)}{\partial r}dz\right]^2 d^2\rho. \tag{2}$$

State conditions of applicability for these results.

---

[313] For this problem see also Problems 13.14, 13.51, and 13.52.

Solution

1) From Eq. (XIII.6), the scattering amplitude is related to the Fourier transform of the potential, $\tilde{U}(\mathbf{q})$, with

$$f_B(\mathbf{q}) = -\frac{m}{2\pi\hbar^2}\tilde{U}(\mathbf{q}), \quad \tilde{U}(\mathbf{q}) = \int e^{-i\mathbf{q}\cdot\mathbf{r}}U(r)dV.$$

From this, it follows that $f_B(q)$ is essentially different from zero only if $qR \lesssim 1$ (where R is the range of the potential); for $qR \gg 1$, this the integral is small due to the fast oscillating integrand. Since $q^2 = 2k^2(1 - \cos\theta)$, the two conditions, $kR \gg 1$ and $qR \leq 1$, lead to the known conclusions that the scattering of fast particles occurs primarily into small angles, $\theta \sim 1/kR \ll 1$. From this, it follows that $q \approx k\theta \sim R^{-1}$.

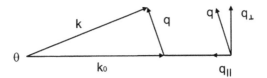

Fig. 13.1

For what follows, we decompose $\mathbf{q}$ into two perpendicular components, $\mathbf{q} = \mathbf{q}_\parallel + \mathbf{q}_\perp$, where $\mathbf{q}_\parallel$ is directed along the incident momentum, $\hbar\mathbf{k}_0$ (see Fig. 13.1). For $\theta \ll 1$, we have

$$q_\parallel = k(1 - \cos\theta) \approx \frac{1}{2}k\theta^2, \quad q_\perp = k\sin\theta \approx k\theta,$$

so that $q_\perp \gg q_\parallel$. It is safe to consider $\mathbf{q} \approx \mathbf{q}_\perp$, so that

$$\tilde{U}(\mathbf{q}) \approx \tilde{U}(\mathbf{q}_\perp) = \iiint e^{-i\mathbf{q}_\perp\cdot\boldsymbol{\rho}}U(\boldsymbol{\rho}, z)dz\, d^2\rho. \quad (3)$$

To find the scattering cross-section we will carry out the q-integration only in a small region of the 4π-sphere around the z-axis. This region can be considered as approximately flat, so that $k^2 d\Omega \approx dS = dq_{\perp x}dq_{\perp y} = d^2 q_\perp$. Therefore,

$$\sigma(E) = \int |f_B|^2 d\Omega \approx \frac{m^2}{4\pi^2\hbar^4 k^2}\iint |\tilde{U}(\mathbf{q}_\perp)|^2 d^2 q_\perp.$$

Using expression (3), we have

$$\sigma(E) = \frac{m}{8\pi^2\hbar^2 E}\iint \left[\iiint e^{-i\mathbf{q}_\perp\cdot\boldsymbol{\rho}}U(\boldsymbol{\rho}, z)dz\, d^2\rho\right] \times \\ \left[\iiint e^{i\mathbf{q}_\perp\cdot\boldsymbol{\rho}'}U(\boldsymbol{\rho}', z')dz'\, d^2\rho'\right] d^2 q_\perp.$$

After performing the $\mathbf{q}_\perp$-integration using

$$\iint \exp\{-i\mathbf{q}_\perp \cdot (\boldsymbol{\rho} - \boldsymbol{\rho}')\}d^2q_\perp = (2\pi)^2\delta(\boldsymbol{\rho} - \boldsymbol{\rho}') \quad (4)$$

and the resulting δ-function integration over $\boldsymbol{\rho}'$, we are immediately led to the following asymptote of the scattering cross-section:

$$\sigma(E)_{E\to\infty} = \frac{m}{2\hbar^2 E}\iint\left[\int_{-\infty}^{\infty} U(\boldsymbol{\rho}, z)dz\right]^2 d^2\rho.$$

For the potential $U(r) = U_0\exp\{-r^2/R^2\}$ and the rectangular well (barrier), using this expression, we obtain

$$\sigma(E) = \frac{\pi^2 m U_0^2 R^4}{4\hbar^2 E} \quad \text{and} \quad \sigma(E) = \frac{\pi m U_0^2 R^4}{\hbar^2 E},$$

respectively (see Problem 13.1 e, f).

2) In an analogous way, we can find the energy-dependence of the *transport cross-section* (2) for $E \to \infty$:

$$\sigma_{tr}(E) = \int (1 - \cos\theta)d\sigma\Big|_{E\to\infty} \approx \frac{m}{8\pi^2\hbar^4 k^4}\iint q_\perp^2 |\tilde{U}(q_\perp)|^2 d^2 q_\perp. \quad (5)$$

Now, instead of Eq. (4), we use the integral

$$\iint q^2 e^{-i\mathbf{q}\cdot\boldsymbol{\rho}} d^2 q = -(2\pi)^2 \Delta_\perp \delta(\boldsymbol{\rho}).$$

Following some transformations in Eq. (5) for the case of a spherically-symmetric potential, we obtain the desired asymptotic form:

$$\sigma_{tr}(E)\Big|_{E\to\infty} \approx \frac{1}{8E^2}\iint\left[\int_{-\infty}^{\infty}\frac{\rho}{r}\frac{\partial U}{\partial r}dz\right]^2 d^2\rho.$$

Note that this expression does not contain Planck's constant, and coincides with σ_{tr} for fast particles $(E \gg U)$ in classical mechanics, defined as

$$\sigma_{tr,class}(E) = \int_0^\infty [1 - \cos\theta(\rho)] \cdot 2\pi\rho\, d\rho \approx \pi\int_0^\infty \theta^2(\rho)\rho\, d\rho.$$

We have used an equation for small-angle scattering from classical mechanics.

The agreement of the result of Born approximation with the classical result at first sight might appear strange. However, it becomes less surprising if we recall that the same result follows from the quasi-classical eikonal approximation (XIII.18) too; Compare to the Rutherford equation.

In conclusion, we note that expression in Eq. (2) assumes that the differential scattering cross-section at large momenta falls faster than $\propto q^{-2}$, so that the upper limit of the q^2-integration could be set to infinity.

Problem 13.3

Using Born approximation, obtain an expression for the scattering amplitude of a particle in an exchange potential;[314] that is, $\hat{U}_{exc}\psi(\mathbf{r}) \equiv U(r)\psi(-\mathbf{r})$.

Find how the scattering amplitude in this case is related to the scattering amplitude of the particle in a regular garden-variety potential, $U(r)$. Obtain the angular distribution of fast particles.

Solution

Substituting $U\psi_{\mathbf{k}_0}^+$ in Eqs. (XIII.4, 5) by $\hat{U}_{ex}\psi_{\mathbf{k}_0}^+$, $\psi_{\mathbf{k}_0}^+(\mathbf{r})$ by $e^{i\mathbf{k}_0 \cdot \mathbf{r}}$, and taking into account the exchange nature of $\hat{U}_{ex}$, we find

$$f_{ex}^B(\mathbf{k}_0, \mathbf{k}) = f_{ex}^B(\boldsymbol{\Delta}) = -\frac{m}{2\pi\hbar^2}\int e^{-i\boldsymbol{\Delta}\cdot\mathbf{r}}U(r)dV, \quad (1)$$

where $\boldsymbol{\Delta} = \mathbf{k} + \mathbf{k}_0$ and $\Delta^2 = 2k^2(1+\cos\theta)$. Thus,

$$f_{ex}^B(E, \theta) = f^B(E, \pi - \theta),$$

where $f^B(E,\theta)$ is the amplitude of scattering in a regular spherically-symmetric potential, $U(r)$. From this relation, it is clear that for fast particles ($kR \gg 1$), the scattering from the exchange potential occurs primarily "backwards," with the relevant angles, $\pi - \theta \lesssim (kR)^{-1}$. For more on scattering from an exchange potential, see Problem 13.56.

Problem 13.4

Find the differential and total cross-section of elastic scattering of fast electrons by a hydrogen atom in its ground state. Neglect a polarization of the atom. See also Problem 13.77.

Solution

The scattering potential is of the form:

$$U(r) = -e^2\left(\frac{1}{r} + \frac{1}{a_B}\right)e^{-2r/a_B},$$

---

[314] For a two-body problem, where $\mathbf{r} = \mathbf{r}_1 - \mathbf{r}_2$, such a potential describes interaction resulting in a permutation of the particles (*exchange*). Such interactions appear naturally in nuclear physics.

(see Problem 4.6). According to Eq. (XIII.8), we obtain

$$f^B(q) = \frac{2(8 + q^2 a_B^2)}{(4 + q^2 a_B^2)^2} a_B. \tag{1}$$

The total scattering cross-section ($d\Omega = \pi k^{-2} dq^2$):

$$\sigma = \frac{\pi}{k^2} \int_0^{4k^2} (f^B)^2 dq^2 = \frac{4\pi}{k^2} \left[\frac{7}{12} - \frac{7 + 9k^2 a_B^2 + 3k^4 a_B^4}{12(1 + k^2 a_B^2)^3} \right] \approx \frac{7\pi a_B^2}{3(ka_B)^2}. \tag{2}$$

Here, we have taken into account the condition of applicability of the Born approximation (XIII.7), viz. $ka_B \gg 1$. (See also Problems 13.77 and 13.78)

Problem 13.5

The same as the previous problem, but for a helium atom. Choose the atomic wavefunction using the variational calculation performed in Problem 11.6.

Solution

In the approximation considered in Problem 11.6, we found that the mean electronic density in the ground state of a helium atom is $n = \frac{2}{\pi a^3} e^{-2r/a}$, where $a = a_B/Z_{eff} = 16 a_B/27$. From this, we calculate the form-factor:

$$F(\mathbf{q}) = \int n(\mathbf{r}) e^{-i\mathbf{q}\cdot\mathbf{r}} dV = \frac{2}{\pi a^3} \int e^{-2r/a - i\mathbf{q}\cdot\mathbf{r}} dV = \frac{32}{(4 + q^2 a^2)^2}.$$

Using the known equation,

$$\frac{d\sigma}{d\Omega} = \frac{4[Z - F(q)]^2}{q^4 a_B^2},$$

we find the differential and total cross-section of elastic scattering from the helium atom:

$$\sigma = \frac{\pi}{k^2} \int_0^{4k^2} \frac{d\sigma}{d\Omega} dq^2 = \frac{4\pi a^2}{k^2 a_B^2} \frac{7k^6 a^6 + 18k^4 a^4 + 12k^2 a^2}{(1 + k^2 a^2)^3} \approx \frac{28\pi a^2}{3(ka_B)^2}.$$

We have $ka_B \gg 1$, and the total cross-section differs from the cross-section for scattering from the hydrogen atom by the factor $4(16/27)^2 \approx 1.40$.

Problem 13.6

Find the Z-dependance of the cross-section for the elastic scattering of fast electrons by a neutral atom with $Z \gg 1$; use the Thomas–Fermi model for the interaction potential.

Solution

Under the Thomas–Fermi approximation, the interaction potential between the scattered electrons and the neutral atom (ignoring an atom polarization) is

$$U(r) = -\varphi(r) = -\frac{Z}{r}\chi\left(\frac{rZ^{1/3}}{b}\right),$$

where $\chi(x)$ is the universal function in the Thomas–Fermi model (see Eq. (XI.3). We use the atomic system of units $e = \hbar = m_e = 1$.) The scattering amplitude in the Born approximation is

$$f^B(Z, q) = -2\int_0^\infty U(r)\frac{\sin qr}{q}r\,dr = Z^{1/3}\Phi\left(\frac{q}{Z^{1/3}}\right), \quad (1)$$

where $\Phi(x)$ is a new universal function (the same for all atoms):

$$\Phi(x) = \frac{2}{x}\int_0^\infty \chi\left(\frac{y}{b}\right)\sin(xy)dy. \quad (2)$$

As $x \to \infty$, only a small region of y in the vicinity of the lower limit contributes to the integral (2). Taking into account that $\chi(0) = 1$ and $\int_0^\infty \sin y\,dy = 1$,[315] we find $\Phi(x) \approx 2/x^2$, as $x \to \infty$. Hence, from Eq. (1), we have $f^B \approx 2Z/q^2$ for $q \gg Z^{1/3}$. It is not unexpected that this corresponds to the amplitude of Rutherford electron scattering in atomic nuclei, since for a large momentum transfer, the screening by atomic electrons can be ignored. As $x \to 0$, the function $\Phi(x)$ takes a finite value.

Eqs. (1) and (2) give the differential cross-section of elastic scattering; the total scattering cross-section is given by

$$\sigma = \frac{\pi}{k^2}\int_0^{4k^2}\left[f^B(q)\right]^2 dq^2 \approx \frac{\pi Z^{2/3}}{k^2}\int_0^\infty \Phi^2\left(\frac{q}{Z^{1/3}}\right)dq^2 = \pi C\frac{Z^{4/3}}{k^2},$$

where $C = 7.14$.

Problem 13.7

Using the Born approximation, find the amplitudes for scattering from two identical potential centers that are at some distance, **a**, from each other (*viz.*, of the form $U(\mathbf{r}) = U_0(r) + U_0(|\mathbf{r} - \mathbf{a}|))$. Express your answer in terms of the scattering amplitude, $f_0^B(q)$, for scattering from $U_0(r)$ alone.

---

[315] To evaluate the integral, the usual procedure is to introduce a "cutoff" factor $e^{-\alpha y}$ with $\alpha > 0$ in the integrand, and in the final result take the limit $\alpha \to 0$.

From this expression, determine a relation between the differential cross-sections for scattering of fast electrons from monoatomic and homonuclear diatomic molecules. Note that this quantity should be averaged over the molecular axis orientations, considering all of them to appear with equal probability.

Find a relation between the total scattering cross-sections from one and two centers for the following limiting cases:

a) $ka \ll 1$ (here the quantity kR — R being the range of potential $U_0(r)$— could be arbitrary);

b) $kR \sim 1$ and $a \gg R$ (i.e., the distance between the centers is much larger than the range of $U_0(r)$).

Solution

The scattering amplitude for two identical force centers separated by a distance, $\mathbf{a}$, is given by

$$f_{2c}^B(\mathbf{q}) = -\frac{m}{2\pi\hbar^2}\int e^{-i\mathbf{q}\cdot\mathbf{r}}[U_0(r) + U_0(|\mathbf{r}-\mathbf{a}|)]dV = f_0^B(q)[1 + e^{-i\mathbf{q}\cdot\mathbf{a}}], \quad (1)$$

from which the differential scattering cross-section follows:

$$\frac{d\sigma_{2c}}{d\Omega} = 2(1 + \cos\mathbf{q}\cdot\mathbf{a})[f_0^B(q)]^2. \quad (2)$$

Although the form of the potential implies two identical force centers, the expressions in (1) and (2) are also applicable to a diatomic molecule. This is because only the valence electrons participate in the molecular formation, leaving behind two inner cores (nuclei screened by inner electrons) that are essentially the same as those of two isolated atoms. However, Eq. (2) must be averaged over all $\mathbf{a}$. Neglecting nuclear vibrations (see Problem 11.25), it is sufficient to average only over the directions of $\mathbf{a}$. For an isotropic distribution, this means that we perform the averaging using a direction-element $dw = d\Omega_n/4\pi$ ($d\Omega_n$ is an infinitesimal element of the solid angle that includes the direction of the vector $\mathbf{a} = a\mathbf{n}$). We obtain (choosing the z-axis along $\mathbf{q}$):

$$\overline{\cos\mathbf{a}\cdot\mathbf{q}} = \frac{1}{4\pi}\int \cos\mathbf{a}\cdot\mathbf{q}\, d\Omega_n = \frac{\sin qa}{qa}.$$

Thus,

$$\frac{d\sigma_{mol}}{d\Omega} = 2\left(1 + \frac{\sin qa}{qa}\right)\frac{d\sigma_{at}}{d\Omega}.$$

As can be seen, the relation between the atomic and the molecular scattering cross-sections changes with the distance between the nuclei, a. (Similar relations can be found for multiatomic molecules, even heterogeneous ones. They form the basis of the diffraction methods used for examining molecular structures.)

Let us discuss the relation between the total scattering cross-sections. For $ka \ll 1$ and $qa \ll 1$, $f_{2c}^B \approx 2f_0^B$, the scattering cross-section on two centers is *four times* larger than on a single center. In the cases $kR \leq 1$ and $a \gg R$, we have $ka \gg 1$, and hence the quantity qa changes essentially, even for a small change of the scattering angle. So, in the integration (2) over the angles, the term with the fast oscillating factor, $\cos(\mathbf{q} \cdot \mathbf{a})$, is much smaller than the first term, so that in this case the cross-section of scattering on two centers is *twice* as large as on one center.

Problem 13.8

Generalize the result of the previous problem for the case of a system with arbitrary number N of identical centers, located at the points $\mathbf{a}_n$, $n = 1, 2, \ldots, N$.

Discuss properties of the angular distribution of the scattered particles for an *ordered* array of a large number ($N \gg 1$) of centers along a straight line with the same distance, b, between closest neighbors.

Solution

In the Born approximation, the scattering amplitude is described by the expression (compare to the previous problem):

$$f_N^B(\mathbf{q}) = f_0^B(q) G_N(\mathbf{q}), \quad G_N(\mathbf{q}) = \sum_n e^{-i\mathbf{q} \cdot \mathbf{a}_n}. \tag{1}$$

Let us emphasize that the factor $G_N(\mathbf{q})$ depends only on the mutual position of the centers and on the vector $\mathbf{q}$ (but not on the form of interaction between the particle and a single center).

In the case of an array of potential centers ordered along a line with the unit vector $\mathbf{j}$ ($\mathbf{a}_n = b(n-1)\mathbf{j}$), we have

$$G_N = \sum_{n=1}^N \exp\{-ib(n-1)\mathbf{q} \cdot \mathbf{j}\} = \frac{1 - \exp\{-ibN\mathbf{q} \cdot \mathbf{j}\}}{1 - \exp\{-ib\mathbf{q} \cdot \mathbf{j}\}},$$

$$|G_N(\mathbf{q})|^2 = \left[\frac{\sin(bN\mathbf{q} \cdot \mathbf{j}/2)}{\sin(b\mathbf{q} \cdot \mathbf{j}/2)}\right]^2. \tag{2}$$

In the case of $N \gg 1$, the quantity $|G_N(\mathbf{q})|^2$ is especially large for specific values, $\mathbf{q} = \mathbf{q}_s$ with[316] $b\mathbf{q}_s \cdot \mathbf{j} = 2\pi s$, where $s = 0, \pm 1, \pm 2, \ldots$. So, the particles are scattered mostly into particular directions, for which

$$\cos \beta_s \approx \frac{\mathbf{k}_0 \cdot \mathbf{j}}{k_0} + \frac{2\pi s}{bk_0},$$

---

[316] Restriction on $|s|_{max}$ is determined by the momentum of scattered particles. In the case of $k < \pi/b$, only the value $s = 0$ is possible. In this case, $q = 0$, so that *coherent* scattering is absent.

where β_s is the angle between the vectors $\mathbf{k}$ and $\mathbf{j}$ (note that $\mathbf{qj} = \mathbf{kj} - \mathbf{k}_0\mathbf{j}$). At maximum, $|G_N|^2 = N^2$. The maxima are sharp; their width is $\Delta\beta_s \propto 1/\sqrt{N}$. The scattering cross-section into such interval of angles is $\propto N$. For other scattering angles, $|G_N|^2 \sim 1$.

These results could be obtained by taking advantage of the large-N limit in the expression for $|G_N|^2$. Using the relation

$$\lim_{N \to \infty} \frac{\sin^2 \alpha N}{\pi \alpha^2 N} = \delta(\alpha),$$

and expanding $\sin(b\mathbf{q} \cdot \mathbf{j}/2)$ in expression (2) in the vicinity of $\mathbf{q}_s$, we find

$$|G_N(\mathbf{q})|^2 \approx \frac{2\pi N}{b} \sum_{s=-\infty}^{\infty} \delta\left(\mathbf{k} \cdot \mathbf{j} - \mathbf{k}_0 \cdot \mathbf{j} - \frac{2\pi s}{b}\right). \tag{3}$$

Such method allows for an easy generalization to the case of a "crystalline" arrangement of the scattering centers. In particular, for a system of centers with $\mathbf{a}_{\{n\}} = n_1\mathbf{b}_1 + n_2\mathbf{b}_2 + n_3\mathbf{b}_3$, $n_a = 0, 1, \ldots, N_a - 1$, we have[317]

$$|G_N(\mathbf{q})|^2 = \left|\sum_{\{n\}} \exp\{-i\mathbf{q} \cdot \mathbf{a}_{\{n\}}\}\right|^2 = \frac{(2\pi)^3 N}{b_1 b_2 b_3} \sum_{\boldsymbol{\tau}} \delta(\mathbf{k} - \mathbf{k}_0 - 2\pi\boldsymbol{\tau}), \tag{4}$$

where $N = N_1 \cdot N_2 \cdot N_3$ is the total number of scattering centers,

$$\boldsymbol{\tau} = s_1\mathbf{a}_1 + s_2\mathbf{a}_2 + s_3\mathbf{a}_3, \quad s_a = 0, \pm 1, \ldots,$$

$$\mathbf{a}_1 = \frac{1}{\Delta}[\mathbf{b}_2 \times \mathbf{b}_3], \quad \mathbf{a}_2 = \frac{1}{\Delta}[\mathbf{b}_3 \times \mathbf{b}_1], \quad \mathbf{a}_3 = \frac{1}{\Delta}[\mathbf{b}_1 \times \mathbf{b}_2], \quad \Delta = (\mathbf{b}_1 \cdot [\mathbf{b}_2 \times \mathbf{b}_3])$$

(where $\mathbf{a}_1 \cdot \mathbf{b}_1 = \mathbf{a}_2 \cdot \mathbf{b}_2 = \mathbf{a}_3 \cdot \mathbf{b}_3 = 1$).

The δ-function terms in Eq. (4) determine the directions, $\mathbf{k} = \mathbf{k}_0 + 2\pi\boldsymbol{\tau}$, of particle elastic scattering in crystals (*Bragg's law*).

Problem 13.9

In the Born approximation, find the scattering amplitude for the collision of two *extended* (non-point-like) particles, interacting with each other via electrostatic forces. The particles are considered to be extended, which means they are characterized by some charge distribution, $\rho_{1,2}(r)$, that is assumed to be spherically symmetric

---

[317] Here we have used the relation:

$$\prod_{a=1}^{3} \delta(\mathbf{q} \cdot \mathbf{b}_a - 2\pi s_a) = \frac{1}{b_1 b_2 b_3} \delta(\mathbf{q} - 2\pi\boldsymbol{\tau}).$$

with respect to the center of mass of the corresponding particle[318] and does not change in the collision process. Express the scattering amplitude in terms of the charge distributions, $\rho_{1,2}(r)$, and the *form-factors* $F_{1,2}(q)$.

Solution

The interaction potential is

$$U(r) = \iint \frac{\rho_1(r_1')\rho_2(r_2')}{|\mathbf{r} + \mathbf{r}_1' - \mathbf{r}_2'|} dV_1' dV_2',$$

where $\mathbf{r} = \mathbf{r}_1 - \mathbf{r}_2$, and $\mathbf{r}_{1,2}$ are the radius vectors of the centers of mass of the colliding particles. Taking into account the relation

$$\int \frac{e^{-i\mathbf{q}\cdot\mathbf{r}} dV}{|\mathbf{r} + \mathbf{r}_1' - \mathbf{r}_2'|} = \frac{4\pi}{q^2} e^{i\mathbf{q}\cdot(\mathbf{r}_1' - \mathbf{r}_2')},$$

and according to Eq. (XIII.6), we obtain

$$f^B(q) = -\frac{2m}{\hbar^2 q^2} F_1(q) F_2(q)$$

(where m is the reduced mass of the particles). Here,

$$F_{1,2}(q) = \int e^{-i\mathbf{q}\cdot\mathbf{r}} \rho_{1,2}(r) dV$$

are the form-factors of the corresponding charge distributions. Note that $F_{1,2}(0) = e_{1,2}$ are particle charges. Some properties of the form-factors are considered in Problems 13.80 and 13.84.

Problem 13.10

Obtain an expression for the scattering amplitude in the n-th order of perturbation theory for a potential $U(r)$.

Hint

First, using Eqs. (XIII.5) and (XIII.4), obtain an integral equation for the scattering amplitude (the *Lippmann-Schwinger equation*).

---

[318] The fact that charge distribution is not point-like denotes the composite character of colliding particles (compare, for example, to Problem 13.4). Since, though, the change of their inner states in a collision process is to be neglected, then the problem is reduced to a common problem of two-body scattering. In connection with this problem, see also Problem 13.80.

Solution

Substituting expression (XIII.4) into Eq. (XIII.5), and using the momentum representation for the free-particle Green function,

$$\frac{e^{ik_0|\mathbf{r}-\mathbf{r}'|}}{|\mathbf{r}-\mathbf{r}'|} = \frac{1}{2\pi^2}\int \frac{e^{i\boldsymbol{\kappa}\cdot(\mathbf{r}-\mathbf{r}')}}{\kappa^2 - k_0^2 - i\varepsilon} d^3\kappa,$$

($\varepsilon > 0$, $\varepsilon \to 0$), we obtain the Lippmann–Schwinger equation:

$$f(\mathbf{k}, \mathbf{k}_0) = -\frac{m}{2\pi\hbar^2}\left[\tilde{U}(\mathbf{k}-\mathbf{k}_0) + \int \frac{\tilde{U}(\mathbf{k}-\boldsymbol{\kappa})f(\boldsymbol{\kappa},\mathbf{k}_0)}{2\pi^2(\kappa^2 - k_0^2 - i\varepsilon)} d^3\kappa\right]. \tag{1}$$

Let us emphasize that for physical real elastic scattering, $k^2 = k_0^2 = 2mE/\hbar^2$. On the other hand, Eq. (1) connects the scattering amplitudes for values $k^2 \neq k_0^2$ (outside the physical sheet or so-called energy surface; see also Eq. (XIII.5)).

Using Eq. (1), we find a recurrence relation for the terms in the expansion, $f(\mathbf{k}, \mathbf{k}_0) = \sum_n f^{(n)}$, of the scattering amplitude in powers of interaction, and then obtain an explicit expression for them $(\mathbf{q} = \mathbf{k} - \mathbf{k}_0)$:

$$f^{(1)}(\mathbf{k}, \mathbf{k}_0) = f^B(\mathbf{q}) = -\frac{m}{2\pi\hbar^2}\tilde{U}(\mathbf{q}), \quad \tilde{U}(\mathbf{q}) = \int e^{-i\mathbf{q}\cdot\mathbf{r}} U(\mathbf{r}) dV;$$

$$f^{(n)}(\mathbf{k},\mathbf{k}_0) = \left(-\frac{m}{2\pi\hbar^2}\right)^n \int \cdots \int \tilde{U}(\mathbf{k}-\boldsymbol{\kappa}_{n-1})\frac{\tilde{U}(\boldsymbol{\kappa}_{n-1} - \boldsymbol{\kappa}_{n-2})d^3\kappa_{n-1}}{2\pi^2(\kappa_{n-1}^2 - k_0^2 - i\varepsilon)} \cdots$$

$$\frac{\tilde{U}(\boldsymbol{\kappa}_1 - \mathbf{k}_0)d^3\kappa_1}{2\pi^2(\kappa_1^2 - k_0^2 - i\varepsilon)}. \tag{2}$$

For applications of these results, see Problems 13.15 and 13.50.

Problem 13.11

In the Born approximation, the forward scattering amplitude (angle $\theta = 0$) is a real quantity and so does not satisfy the optical theorem (XIII.11). Why does this fact not contradict the successful description of differential and total scattering cross-section in the framework of the Born approximation, whenever it applies?

Write an expression for the scattering amplitude in second-order perturbation theory. Find Im $f^{(2)}(E, \theta = 0)$ and explain the result obtained.

Solution

The Born approximation is the first linear term in the expansion of the scattering amplitude in power of the interaction (more specifically, of parameters given in Eq. (XIII.7)). On the other hand, an expansion of the scattering cross-section begins from the term quadratic in interaction, since $\sigma \propto |f|^2$. So, in the relation

$4\pi \,\text{Im}\, f(E,0) = k\sigma(E)$, the left-hand side must have a second-order interaction term, as the right-hand side does. It follows that necessarily $\text{Im}\, f^B(\theta=0) = 0$.

In the second order of perturbation theory, the scattering amplitude is described by expression (2) of the previous problem. In particular, for forward scattering ($\theta=0$, $\mathbf{k} = \mathbf{k}_0$), it gives

$$f^{(2)}(\mathbf{k}_0, \mathbf{k}_0) = \frac{m^2}{8\pi^4 \hbar^4} \int \frac{|\tilde{U}(\boldsymbol{\kappa} - \mathbf{k}_0)|^2}{\kappa^2 - k_0^2 - i\varepsilon} d^3\kappa. \tag{1}$$

We used the fact that $\tilde{U}(\mathbf{q}) = \tilde{U}^*(-\mathbf{q})$. It follows that

$$\text{Im}\, f^{(2)}(\mathbf{k}_0, \mathbf{k}_0) = \frac{m^2}{8\pi^4 \hbar^4} \int \frac{\varepsilon}{(\kappa^2 - k_0^2)^2 + \varepsilon^2} |\tilde{U}(\boldsymbol{\kappa} - \mathbf{k}_0)|^2 d^3\kappa. \tag{2}$$

If we note that[319] $\varepsilon/\pi(x^2 + \varepsilon^2) = \delta(x)$ (for infinitesimal $\varepsilon > 0$), and write $\boldsymbol{\kappa} = \kappa\mathbf{n}$, $d^3\kappa = \frac{1}{2}\kappa d\kappa^2 d\Omega_\mathbf{n}$, we transform expression (2) to the form

$$\text{Im}\, f^{(2)}(\mathbf{k}_0, \mathbf{k}_0) = \frac{m^2}{16\pi^3 \hbar^4} \iint |\tilde{U}(\kappa\mathbf{n} - \mathbf{k}_0)|^2 \kappa \delta(\kappa^2 - k_0^2) d\kappa^2 d\Omega_\mathbf{n} =$$

$$\frac{m^2 k_0}{16\pi^3 \hbar^4} \int |\tilde{U}(k_0 \mathbf{n} - \mathbf{k}_0)|^2 d\Omega_\mathbf{n} \equiv \frac{k_0}{4\pi} \int |f^B(\mathbf{k} - \mathbf{k}_0)|^2 d\Omega_\mathbf{n},$$

i.e., $\text{Im}\, f^{(2)}(E, \theta=0) = k\sigma_B(E)/4\pi$, which expresses the optical theorem in the second order of perturbation theory.

Problem 13.12

In the second order of perturbation theory, find the scattering amplitude of the Yukawa potential, $U(r) = \frac{\alpha}{r} e^{-r/R}$. Compare $f^{(1)}$ to $f^{(2)}$ for different values of energy and scattering angle.

Solution

For the Yukawa potential,

$$f^{(1)}(q) \equiv f^B(q) = -\frac{2m\alpha R^2}{\hbar^2(1 + q^2 R^2)}, \quad \mathbf{q} = \mathbf{k} - \mathbf{k}_0, \tag{1}$$

and according to Eq. (2) from Problem 13.10, in the second order of perturbation theory we have

$$f^{(2)}(\mathbf{k}_0, \mathbf{k}_0) = \frac{2m^2 \alpha^2 R^4}{\pi^2 \hbar^4} \int \frac{d^3\kappa}{[1 + R^2(\mathbf{k}_0 - \boldsymbol{\kappa})^2][1 + R^2(\mathbf{k} - \boldsymbol{\kappa})^2](\kappa^2 - k_0^2 - i\varepsilon)}. \tag{2}$$

---

[319] Indeed, this function differs from from zero only for $|x| \leq \sqrt{\varepsilon} \to 0$, while integrating it in the infinite limits gives unity, as it is supposed to be for the δ-function.

Using relation (with $k^2 = k_0^2$)

$$\frac{1}{1 + R^2(\mathbf{k}_0 - \boldsymbol{\kappa})^2} \cdot \frac{1}{1 + R^2(\mathbf{k} - \boldsymbol{\kappa})^2} =$$

$$\int_0^1 \frac{d\xi}{\{1 + k^2 R^2 + \kappa^2 R^2 - 2R^2 \boldsymbol{\kappa}\cdot[\mathbf{k}_0 \xi + (1-\xi)\mathbf{k}]\}^2},$$

in Eq. (2) it is easy to perform integration over the angles ($d^3\kappa = \kappa^2 d\kappa\, d\Omega$), and obtain

$$f^{(2)}(\mathbf{k}_0, \mathbf{k}_0) = \frac{4m^2\alpha^2 R^4}{\pi\hbar^4} \int_0^1 \int_{-\infty}^{\infty} \frac{\kappa^2}{\kappa^2 - k^2 - i\varepsilon} K(k, q, \kappa, \xi) d\kappa\, d\xi, \qquad (3)$$

where

$$K = \{(1 + k^2 R^2 + \kappa^2 R^2)^2 - 4\kappa^2 R^4 [k^2 - \xi(1-\xi)q^2]\}^{-1}. \qquad (4)$$

Taking into account the parity of the function under the integral, the integration over the variable κ may be extended to the entire axis.

Equation (4) for K could be transformed to

$$K = \frac{1}{(\kappa R - \alpha_1)(\kappa R - \alpha_2)(\kappa R - \alpha_3)(\kappa R - \alpha_4)}.$$

This function, considered as a function of the complex variable κ, is (as is the entire expression under the integral in Eq. (3)) *meromorphic*, and has only simple poles at the points $\kappa_n = \alpha_n/R$. We have

$$\alpha_1 \equiv \alpha = \sqrt{k^2 R^2 - \xi(1-\xi)q^2 R^2} + i\sqrt{1 + \xi(1-\xi)q^2 R^2},$$

(since $0 \leq \xi \leq 1$ and $q^2 \leq 4k^2$, then both radicals are real and positive) and $\alpha_2 = \alpha^*$, $\alpha_3 = -\alpha$, $\alpha_4 = -\alpha^*$.

Now, in Eq. (3), it is easy to perform an integration over κ using residues, closing the integration contour into the upper half-plane of the complex variable κ.

The positions of the poles are shown in Fig. 13.2. A contribution to the integral in Eq. (3) from the pole at the point $\kappa = k + i\varepsilon$ is

$$I_1(\xi) = \frac{i\pi k}{1 + 4k^2 R^2 + 4k^2 q^2 R^4 \xi(1-\xi)}.$$

This part of the integral is imaginary. The contributions from the poles located at the points $\kappa = \alpha_1/R$ and $\kappa = \alpha_4/R$ are real and equal to

$$I_2(\xi) = \frac{\pi}{2R\sqrt{1 + q^2 R^2 \xi(1-\xi)}[1 + 4k^2 R^2(1 + q^2 R^2 \xi(1-\xi))]}.$$

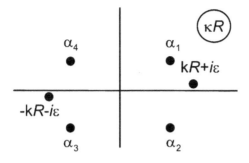

Fig. 13.2

Finally, performing an integration over the variable ξ in

$$f^{(2)}(\mathbf{k},\mathbf{k}_0) = \frac{4m^2\alpha^2 R^4}{\pi\hbar^4}\int_0^1 [I_1(\xi) + I_2(\xi)]d\xi,$$

we obtain the final expression for the scattering amplitude of the Yukawa potential in second-order perturbation theory:

$$f^{(2)}(\mathbf{k},\mathbf{k}_0) = \frac{2m^2\alpha^2 R^3}{\hbar^4}\left\{\frac{2\arctan(qR/2\sqrt{\Delta(k,q)})}{qR\sqrt{\Delta(k,q)}} + \frac{i}{qR\sqrt{\Delta(k,q)}}\ln\frac{\sqrt{\Delta(k,q)} + kqR^2}{\sqrt{\Delta(k,q)} - kqR^2}\right\}, \quad (5)$$

where

$$\Delta(k,q) = 1 + k^2R^2(4 + q^2R^2).$$

Let us discuss these results. We first note that calculating the imaginary part of the amplitude in second-order perturbation theory for $\theta = 0$ (here also $q = 0$) and using the optical theorem, Eq. (XIII.11), we can find the total scattering cross-section in the Born approximation. The result obtained coincides, of course, with the calculation of cross-section by the equation $\sigma_B = \int (f^B)^2 d\Omega$. See Problem 13.1 a.

It is interesting to compare Im $f^{(2)}$ and Re $f^{(2)}$ with each other and with the amplitude of the first approximation for different values of the energy and the scattering angle. In the case of slow particles, when $kR \ll 1$, according to Eqs. (5) and (1), we obtain

$$\frac{\text{Im } f^{(2)}}{\text{Re } f^{(2)}} \approx 2kR \ll 1, \quad \left|\frac{f^{(2)}}{f^{(1)}}\right| \approx \frac{m\alpha R}{\hbar^2}.$$

Taking into account the optical theorem, we can conclude that the smallness of the ratio Im $f^{(2)}/$Re $f^{(2)}$ (for $kR \ll 1$) is a general result for fairly arbitrary short-range potentials. The smallness of $|f^{(2)}/f^{(1)}|$ assumes fulfilment of the condition $m\alpha R/\hbar^2 \ll 1$, which ensures applicability of the Born approximation (see Eq. (XIII.7); for the Yukawa potential, $U_0 \sim \alpha/R$). These estimates are also useful for particles of "moderate" energy, $kR \sim 1$, but in this case, Im $f^{(2)}$ and Re $f^{(2)}$ are quantities of the same order for an arbitrary scattering angle.

Let us consider the case of fast particles, $kR \gg 1$. For small scattering angles, when $qR \lesssim 1$ (this domain provides a dominant contribution into the total scattering cross-section), we find

$$\frac{\text{Im } f^{(2)}}{\text{Re } f^{(2)}} \sim kR \gg 1, \quad \left|\frac{f^{(2)}}{f^{(1)}}\right| \sim \frac{m\alpha}{\hbar^2 k}.$$

The smallness of Re $f^{(2)}$ in comparison to Im $f^{(2)}$ is a general result (see Problem 13.14), while the condition $|f^{(2)}/f^{(1)}| \ll 1$, as expected, assumes that the applicability condition of the Born approximation for fast particles is satisfied, namely the second of conditions (XIII.7). The obtained estimates remain valid for larger value of qR. (Here, some specific properties of the Yukawa potential manifest themselves, in particular, a power-law decrease of $\tilde{U}(q)$ for $q \to \infty$; compare to result of the previous problem.)

Problem 13.13

In the second order of perturbation theory, find the scattering amplitude in the potential $U(r) = U_0 e^{-r^2/R^2}$ for large momenta transferred, $qR \gg 1$. Compare the real and imaginary parts of $f^{(2)}$ with each other and with the Born amplitude.

Solution

The Fourier component of the potential is

$$\tilde{U}(q) = \int U_0 \exp\left\{-i\mathbf{q} \cdot \mathbf{r} - \frac{r^2}{R^2}\right\} dV = \pi^{3/2} U_0 R^3 e^{-q^2 R^2/4},$$

and according to Eq. (2) from Problem 13.10, the scattering amplitude in the second order of perturbation theory is described by the expression:

$$f^{(2)}(\mathbf{k}, \mathbf{k}_0) = \frac{m^2}{8\pi^4 \hbar^4} \int \frac{\tilde{U}(\mathbf{k} - \boldsymbol{\kappa})\tilde{U}(\boldsymbol{\kappa} - \mathbf{k}_0)}{\kappa^2 - k_0^2 - i\varepsilon} d^3\kappa =$$

$$\frac{m^2 U_0^2 R^6}{8\pi \hbar^4} e^{-q^2 R^2/4} \int \frac{\exp\{-\frac{1}{2}R^2[\boldsymbol{\kappa} - \frac{1}{2}(\mathbf{k} + \mathbf{k}_0)]^2\}}{\kappa^2 - k_0^2 - i\varepsilon} d^3\kappa. \quad (1)$$

The dominant role in the integral comes from the region of values $\boldsymbol{\kappa}$, where $|\boldsymbol{\kappa} - (\mathbf{k} + \mathbf{k}_0)/2|R \leq 1$ (outside of this domain, the integrand is exponentially small). For large energies, $kR \gg 1$, and large momenta transferred, $qR \gg 1$, in this domain the denominator of the integrand in Eq. (1) changes only slightly, and it can be factored outside the integral sign at the point $\boldsymbol{\kappa} = \mathbf{k} + \mathbf{k}_0/2$. Then, a simple integration gives (for $k^2 = k_0^2$):

$$f^{(2)}(\mathbf{k}, \mathbf{k}_0) \approx \operatorname{Re} f^{(2)} \approx -\frac{\sqrt{2\pi} m^2 U_0^2 R^3}{\hbar^4 q^2} \exp\left\{-\frac{1}{8} q^2 R^2\right\}. \qquad (2)$$

In this approximation, amplitude $f^{(2)}$ is a real function. Its imaginary part according to Eq. (1) could be easily found in general if we note that

$$\operatorname{Im} \frac{1}{\kappa^2 - k_0^2 - i\varepsilon} = \pi \delta(\kappa^2 - k_0^2)$$

(compare to Problem 13.11). Writing $d^3\kappa = \kappa\, d\kappa^2\, d\Omega/2$ and integrating first over κ^2 and then over the angles (choosing the polar axis along the vector $\mathbf{k} + \mathbf{k}_0$), we find:

$$\operatorname{Im} f^{(2)} = \left(\frac{m U_0 R^3}{4\hbar^2}\right)^2 \iint \kappa \delta(\kappa^2 - k^2) \exp\left\{-\kappa^2 R^2 + \frac{1}{2}\boldsymbol{\kappa} \cdot (\mathbf{k} + \mathbf{k}_0) R^2\right\} d\kappa^2 d\Omega =$$

$$\frac{\pi m^2 U_0^2 R^4}{2\hbar^4 |\mathbf{k} + \mathbf{k}_0|} \sinh\left(\frac{1}{2}|\mathbf{k} + \mathbf{k}_0| k R^2\right) e^{-k^2 R^2}. \qquad (3)$$

(Note that $|\mathbf{k} + \mathbf{k}_0| = (4k^2 - q^2)^{1/2}$.)

For large energies, $kR \gg 1$, and small scattering angles, when $qR \lesssim 1$, the scattering amplitude of the second order is determined mainly by the imaginary part of Eq. (3), as is seen from Problem 13.14 (Eq. (2) for $qR \lesssim 1$ is not applicable). For large changes of momentum, on the contrary, the integral is dominated by the real part of the scattering amplitude, $f^{(2)}$. Moreover, since

$$\operatorname{Re} f^{(2)} \propto e^{-q^2 R^2/8}, \quad \text{while} \quad f^{(1)} = f^a \propto \tilde{U}(q) \propto e^{-q^2 R^2/4},$$

then for the values of qR large enough, $|\operatorname{Re} f^{(2)}| \gg |f^B|$. This signal a break-down of the Born approximation, independently of the parameter U_0 that characterizes the interaction strength.[320] Let us emphasize that the discovered properties are characteristic for potentials with an exponential decrease of the Fourier component $\tilde{U}(q) \propto \exp\{-\alpha q^n\}$ with $n \geq 1$ for $q \to \infty$ (compare to Problem 8.29, and also to the case of a power-law decrease of $\tilde{U}(q)$, considered in the previous problem).

In conclusion, let us note that using Eq. (3) and the optical theorem (XIII.11), we can find the scattering cross-section in the Born approximation; the result, of course, coincides with its calculation by the equation $\sigma_B = \int (f^B)^2\, d\Omega$ (see Problem 13.1 f).

Problem 13.14

Prove that the scattering amplitude of second-order perturbation theory for large energies, $kR \gg 1$, but for relatively low momenta[321] transfered, $qR \lesssim 1$, is described

---

[320] It corresponds to the large values of qR (the contribution of which to the total cross-section is small).

[321] This region of momenta transferred contributes dominantly to the total scattering cross-section. Here $|\mathbf{q}_\perp| \approx q$, where $\mathbf{q}_\perp$ is the component of $\mathbf{q}$ perpendicular to the initial momentum $\mathbf{k}_0$ (directed along the axis z).

by (R is the radius of the potential)

$$f^{(2)} \approx i \frac{m^2}{4\pi \hbar^4 k} \iint \left[\int_{-\infty}^{\infty} U(\rho, z) dz \right]^2 \exp\{-i\mathbf{q}_\perp \cdot \boldsymbol{\rho}\} d^2\rho.$$

Apply this result to the potential $U(r) = U_0 e^{-r^2/R^2}$, and compare to Problem 13.13.

Solution

In the second order of perturbation theory, the scattering amplitude is given by

$$f^{(2)}(\mathbf{k}, \mathbf{k}_0) = \frac{m^2}{8\pi^4 \hbar^4} \int \frac{\tilde{U}(\mathbf{k} - \boldsymbol{\kappa}) \tilde{U}(\boldsymbol{\kappa} - \mathbf{k}_0)}{\kappa^2 - k_0^2 - i\varepsilon} d^3\kappa. \tag{1}$$

See Problem 13.10. For a large energy, $kR \gg 1$, and a small scattering angle, $qR \leq 1$, the vector $\boldsymbol{\kappa}$, as well as $\mathbf{k}$, is "close" to $\mathbf{k}_0$. Let us write these vectors in the form:

$$\mathbf{k} = \mathbf{k}_0 + \mathbf{q} = k_\| \mathbf{n}_0 + \mathbf{q}_\perp, \quad \boldsymbol{\kappa} = \kappa_\| \mathbf{n}_0 + \boldsymbol{\kappa}_\perp,$$

where $\mathbf{n}_0 = \mathbf{k}_0/k_0$, while $\mathbf{q}_\perp, \boldsymbol{\kappa}_\perp \perp \mathbf{n}_0$. See that $|\mathbf{q}_\perp| \approx q$ and $k_\| = k_0 + q_\|$, $q_\| \approx -q^2/2k_0$. Compare to Problem 13.2. After substitution of the explicit expression for Fourier components of the potential (in terms of $U(\mathbf{r})$), the integral in Eq. (1) takes the form:

$$\int \frac{U(\boldsymbol{\rho}_1, z_1) U(\boldsymbol{\rho}_2, z_2) \exp\{-i[\kappa'_\|(z_1 - z_2) + \boldsymbol{\kappa}_\perp \cdot (\boldsymbol{\rho}_1 - \boldsymbol{\rho}_2) + \mathbf{q}_\perp \cdot \boldsymbol{\rho}_2]\}}{2k_0 \kappa'_\| + (\kappa'_\|)^2 + \kappa_\perp^2 - i\varepsilon} \times$$

$$d^3 r_1 d^3 r_2 d\kappa'_\| d^2\kappa_\perp. \tag{2}$$

Here $\kappa'_\| = \kappa_\| - k_0$, and the term $-iq_\| z_2$ in the exponent is omitted, since $|q_\| z_2| \leq q^2 R/k_0 \ll 1$. Let us note that the terms in the perturbative expansion for the scattering amplitude (see Eq. (2) from Problem 13.10) could be interpreted as describing a number of single-particle collisions with the scattering potential, each accompanied by a momentum change for the particle being scattered. In this sense, the radius-vector $\mathbf{r}_1$ in the corresponding integral corresponds to the point of first collision, after which the particle momentum becomes equal to $\boldsymbol{\kappa}$ (after the second collision, it takes the value $\mathbf{k}$).

Closing the contour in the upper half-plane of the complex variable $\kappa'_\|$ in the case of $z_2 > z_1$, and in the lower half-plane for $z_2 < z_1$, we can perform the integration over $\kappa'_\|$ in Eq. (2):

$$\int \frac{\exp\{-i\kappa'_\|(z_1 - z_2)\} d\kappa'_\|}{2k_0 \kappa'_\| + (\kappa'_\|)^2 + \kappa_\perp^2 - i\varepsilon} = \begin{cases} \frac{i\pi}{k_0} \exp\left\{\frac{i\kappa_\perp^2}{2k_0}(z_1 - z_2)\right\}, & z_2 > z_1, \\ \frac{i\pi}{k_0} \exp\{2ik_0(z_1 - z_2)\}, & z_2 < z_1. \end{cases} \tag{3}$$

If $|k_0 z_{1,2}| \sim k_0 R \gg 1$, then in the case of $z_1 > z_2$, the exponent in Eq. (3) is a fast oscillating function. This means that for the subsequent integration over $z_{1,2}$ in Eq. (2), a contribution of such $z_{1,2}$ would be small, and can be neglected. Thus, Eq. (3) could be considered equal to

$$\frac{i\pi}{k_0}\eta(z_2 - z_1)\exp\left\{i\frac{\kappa_\perp^2(z_1 - z_2)}{2k_0}\right\}, \qquad (4)$$

where $\eta(z)$ is a step function,[322] and since typically we have $z_{1,2} \lesssim R$ and $\kappa_\perp \lesssim R^{-1}$, then the exponent can be replaced by 1. The appearance of a step function in Eq. (4) has a physical explanation: for fast particles, each successive "collision" takes place for larger and larger values of z (there is no backward scattering).

Now, in Eq. (2), integration over $\kappa_\perp$ is easily performed, and the δ-function, $\delta(\boldsymbol{\rho}_1 - \boldsymbol{\rho}_2)$, allows integration over $\boldsymbol{\rho}_2$. As a result, we obtain

$$f^{(2)} = i\frac{m^2}{2\pi\hbar^4 k_0}\int\left\{\int_{-\infty}^{\infty} U(\boldsymbol{\rho}, z_1)dz_1 \int_{z_1}^{\infty} U(\boldsymbol{\rho}, z_2)dz_2\right\} e^{-i\mathbf{q}_\perp \cdot \boldsymbol{\rho}} d^2\rho.$$

Finally, replacing the lower limit of integration over z_2 by $-\infty$ and introducing the coefficient $1/2$, we obtain the expression given in the problem condition. Let us note that for a central potential, the scattering amplitude is purely imaginary (a real part of the amplitude $f^{(2)}$ is much smaller than the imaginary part, and in this approximation it does not appear; compare to Problems 13.12 and 13.13).

For the potential $U = U_0 \exp\{-r^2/R^2\}$, we obtain

$$f^{(2)} = i\frac{\pi m^2 U_0^2 R^4}{8\hbar^4 k_0} e^{-q^2 R^2/8},$$

which coincides with Eq. (3) from Problem 13.13 for the values $qR \lesssim 1$.

In conclusion, note that taking $q = 0$ in the equation for $f^{(2)}$ and using the optical theorem (XIII.11), we obtain the scattering cross-section in the Born approximation, which coincides with the result of Problem 13.2.

Problem 13.15

From the solution of the Lippmann–Schwinger equation (see Problem 13.10), find the scattering amplitude for a separable potential with the following integral kernel $U(r, r') = \lambda\chi(r)\chi^*(r')$. Find the angular distribution and total scattering cross-section.

Solution
The Lippmann–Schwinger equation (Eq. (1) from Problem 13.10) remains valid for an arbitrary interaction $\hat{U}$, if we replace the potential Fourier component $\tilde{U}(\mathbf{k} - \mathbf{k}')$ by the corresponding kernel $\tilde{U}(\mathbf{k}, \mathbf{k}')$. For the separable potential,

---

[322] Recall that $\eta(z) = 1$ for $z > 0$, and $\eta(z) = 0$ for $z < 0$.

$$\tilde{U}(\mathbf{k},\mathbf{k}') = \lambda g(k)g^*(k'), \quad g(k) = \int e^{-i\mathbf{k}\cdot\mathbf{r}}\chi(r)dV,$$

and the Lippmann–Schwinger equation reads

$$f(\mathbf{k},\mathbf{k}_0) = -\frac{\lambda m}{2\pi\hbar^2}g(k)\left[g^*(k_0) + \int\frac{g^*(\kappa)f(\mathbf{k},\mathbf{k}_0)d^3\kappa}{2\pi^2(\kappa^2 - k_0^2 - i\varepsilon)}\right]. \quad (1)$$

Denoting the integral here by $F(k_0)$, we have

$$f(\mathbf{k},\mathbf{k}_0) = -\frac{\lambda m}{2\pi\hbar^2}g(k)[g^*(k_0) + F(k_0)],$$

and after substituting this expression back into the integral, we obtain a self-consistent equation for $F(k_0)$ and determine the scattering amplitude $(k = k_0)$:

$$f(\mathbf{k},\mathbf{k}_0) \equiv f(k) = -\frac{\lambda m}{2\pi\hbar^2}\frac{g^2(k)}{1+K(k)},$$

where

$$K(k) = \frac{\lambda m}{4\pi^3\hbar^2}\int\frac{|q(\kappa)|^2 d^3\kappa}{\kappa^2 - k^2 - i\varepsilon}.$$

One of the notable properties of the scattering amplitude is its independence from the scattering angle,[323] so that the angular distribution of the scattered particles is isotropic, and the total scattering cross-section is simply $\sigma(E) = 4\pi|f|^2$ (it is informative to prove its consistence with the optical theorem). Note that in the large-energy limit: $\sigma(E) \propto |g(k)|^4$ for $E \to \infty$.

Problem 13.16

For $E = 0$, compare the values of the exact and Born scattering amplitudes in a potential $U(r)$ for

a) a repulsive potential $U(r) \geq 0$;
b) an attractive potential in which there are no bound states (*i.e.*, the potential well is "shallow" enough).

Prove that the Born approximation in case a) overestimates the scattering cross-section, while in case b), it underestimates it.

Solution

For $E = 0$, we have

$$f^B(0) = -\frac{m}{2\pi\hbar^2}\int U(r)dV, \quad f_{ex}(0) = -\frac{m}{2\pi\hbar^2}\int U(r)\psi_0(r)dV, \quad (1)$$

---

[323] Compare to the scattering from a zero-range potential, considered in Problem 13.20.

where the wavefunction $\psi_0(r)$ satisfies the equation:

$$\psi_0(r) = 1 - \frac{m}{2\pi\hbar^2}\int U(r')\frac{1}{|\mathbf{r}-\mathbf{r}'|}\psi_0(r')dV'. \tag{2}$$

If there are no bound states in the potential $U(r)$, then the wavefunction for $E = 0$ has no zeros, and since $\psi_0(\infty) = 1$, then $\psi_0(r) > 0$. As follows from Eq. (2), for a repulsive potential, $0 \leq \psi_0(r) \leq 1$, and according to Eq. (1), we have the inequality $|f^B(0)| > |f_{ex}(0)|$, i.e., the Born approximation overestimates the scattering cross-section. In the analogous case of an attractive potential, $U(r) \leq 0$, we obtain $|f^B(0)| < |f_{ex}(0)|$, so the Born approximation underestimates the scattering cross-section (in the absence of bound states in the potential). Let us emphasize that the relations between the cross-sections do not assume that the potential is weak, as required for the Born approximation to be formally valid. In connection to this problem, see also Problems 13.69 and 13.70.

Problem 13.17

In the Born approximation, obtain a scattering amplitude for scattering of a charged particle by a magnetic field, $\mathcal{H}(\mathbf{r})$. Prove gauge invariance of the result obtained.[324]

Solution
To first order in the magnetic field, the interaction has the form:

$$\hat{V} = -\frac{e}{2mc}(\hat{\mathbf{p}}\cdot\mathbf{A} + \mathbf{A}\cdot\hat{\mathbf{p}}).$$

Substituting this expression into Eq. (XIII.5) instead of $U(r)$, and replacing the wavefunction $\psi_{\mathbf{k}_0}^+$ by a plane wave, we obtain the scattering amplitude

$$f(\mathbf{k},\mathbf{k}_0) = \frac{e}{4\pi\hbar^2 c}\int e^{-i\mathbf{k}\cdot\mathbf{r}}(\hat{\mathbf{p}}\cdot\mathbf{A} + \mathbf{A}\cdot\hat{\mathbf{p}})e^{i\mathbf{k}_0\cdot\mathbf{r}}d^3r = \frac{e}{4\pi\hbar c}(\mathbf{k}+\mathbf{k}_0)\cdot\tilde{\mathbf{A}}(\mathbf{q}),$$

where $\tilde{\mathbf{A}}(\mathbf{q}) = \int e^{-i\mathbf{q}\cdot\mathbf{r}}\mathbf{A}(\mathbf{r})dV$ is the Fourier transform of the vector potential. Under a gauge transformation, a term $\nabla\chi(\mathbf{r})$ is added to $\mathbf{A}(\mathbf{r})$, and consequently $i\mathbf{q}\tilde{\chi}(\mathbf{q})$ to $\tilde{\mathbf{A}}(\mathbf{q})$. But since $\mathbf{q}\cdot(\mathbf{k}+\mathbf{k}_0) = 0$, then the value of f and the value of the differential scattering cross-section do not change, in accordance with gauge invariance.

13.2 Scattering theory: partial-wave analysis

Problem 13.18

Obtain an expression for phase shifts directly from the partial-wave expansion of the Born scattering amplitude in a spherically-symmetric potential.

---
[324] See also Problem 13.24.

Solution

Let us use the expansion:

$$\frac{\sin qr}{qr} = \frac{\sin\sqrt{2k^2r^2 - 2k^2r^2\cos\theta}}{\sqrt{2k^2r^2 - 2k^2r^2\cos\theta}} =$$

$$\frac{\pi}{2kr}\sum_{l=0}^{\infty}(2l+1)[J_{l+1/2}(kr)]^2 P_l(\cos\theta),$$

which follows from the *addition theorem* for cylindrical functions, if we note that $(\sin z)/z = \sqrt{\pi/2z}\,J_{1/2}(z)$. Substituting this into Eq. (XIII.8), we obtain the Born amplitude in the form of a series:

$$f^B = \sum_{l=0}^{\infty}(2l+1)\left\{-\frac{\pi m}{\hbar^2 k}\int_0^{\infty}U(r)[J_{l+1/2}(kr)]^2 r\, dr\right\}P_l(\cos\theta). \quad (1)$$

Comparing this to expansion (XIII.9) of the scattering amplitude in partial waves, we conclude that the Born approximation corresponds to small scattering phase-shifts,[325] $|\delta_l(k)| \ll 1$, where

$$f = \sum_l (2l+1)\frac{e^{2i\delta_l}-1}{2ik}P_l(\cos\theta) \approx \sum_l (2l+1)\frac{\delta_l(k)}{k}P_l(\cos\theta). \quad (2)$$

From a comparison between Eqs. (1) and (2), we reproduce the known expression for the Born phase shifts - Eq. (XIII.12) - which is a linear-in-the-potential term in the expansion for $\delta_l(k)$.

Problem 13.19

Find phase shifts in the field $U(r) = \alpha/r^2$, $\alpha > 0$.

Perform summation of the series (XIII.9), *i.e.*, of the partial-wave expansion of the scattering amplitude for the following cases:

a) $m\alpha/\hbar^2 \ll 1$ for an arbitrary scattering angle;
b) $m\alpha/\hbar^2 \gtrsim 1$ for small-angle scattering;
c) $m\alpha/\hbar^2 \gg 1$ for the backward scattering ($\theta = \pi$).

Find the differential scattering cross-section and compare it to that in the Born approximation and the scattering cross-section that follows from classical mechanics.

---

[325] The smallness of all phase shifts is a necessary (but not sufficient) condition for the Born approximation to remain valid. This smallness is responsible for the scattering amplitude being real in this approximation.

Solution

Writing the wavefunctions as $\psi_{klm} = u_{kl}(r)Y_{lm}/\sqrt{r}$, for the function $u_{kl}(r)$ we have the following equation (compare to Eq. (IV.6)):

$$u_{kl}'' + \frac{1}{r}u_{kl}' + \left\{k^2 - \frac{1}{r^2}\left[\left(l+\frac{1}{2}\right)^2 + \frac{2m\alpha}{\hbar^2}\right]\right\}u_{kl} = 0.$$

Its solution, satisfying the boundary condition $u(0) = 0$, is $u_{kl} = CJ_\nu(kr)$, where J_ν is the Bessel function with the index

$$\nu = \sqrt{\left(l+\frac{1}{2}\right)^2 + \frac{2m\alpha}{\hbar^2}}.$$

The asymptote of this solution for $r \to \infty$ is

$$u_{kl} \approx C\sqrt{\frac{2}{\pi kr}}\sin\left[kr - \frac{\pi\nu}{2} + \frac{\pi}{4}\right] \equiv C\sqrt{\frac{2}{\pi kr}}\sin\left[kr - \frac{\pi l}{2} + \delta_l\right].$$

The phase shifts are

$$\delta_l = -\frac{\pi}{2}\left[\sqrt{\left(l+\frac{1}{2}\right)^2 + \frac{2m\alpha}{\hbar^2}} - \left(l+\frac{1}{2}\right)\right]. \tag{1}$$

Since δ_l does not depend on k, then according to Eq. (XIII.9) the scattering amplitude has the form $f(k,\theta) = F(\theta)/k$, and the differential scattering cross-section goes as $d\sigma/d\Omega \propto k^{-2} \propto E^{-1}$. The same energy-dependence[326] (but not a dependence on the scattering angle) occurs in classical mechanics (see Eq. (5)).

Let us consider some special cases where it is possible (approximately) to sum the series (XIII.9) with phase shifts (1).

a) In the case $m\alpha/\hbar^2 \ll 1$, according to (1) we have

$$\delta_l \approx -\frac{\pi m\alpha}{(2l+1)\hbar^2}$$

where $|\delta_l| \ll 1$. Expanding $e^{2i\delta_l}$ in Eq. (XIII.9), we obtain[327]

$$f(E,\theta) \approx -\frac{\pi m\alpha}{\hbar^2 k}\sum_{l=0}^{\infty}P_l(\cos\theta) = -\frac{\pi m\alpha}{2\hbar^2 k\sin(\theta/2)}, \tag{3}$$

---

[326] For the potential $U = \alpha/r^2$, the dependence $d\sigma/d\Omega \propto E^{-1}$, both in classical and in quantum mechanics, is easily obtained from a dimensional analysis.

[327] To sum this series, use the generating function of the Legendre polynomials (with $z = \cos\theta$ and $x = 1$):

$$\frac{1}{\sqrt{1 - 2xz + x^2}} = \sum_{l=0}^{\infty}x^l P_l(z).$$

which coincides with the result of the Born approximation (see Problem 13.1 d). The angular dependence of the differential cross-section,

$$\frac{d\sigma}{d\Omega} = |f|^2 = \frac{\pi^2 m \alpha^2}{8\hbar^2 E \sin^2(\theta/2)}, \qquad (4)$$

is dissimilar from the corresponding result of classical mechanics:

$$\left(\frac{d\sigma}{d\Omega}\right)_{cl} = \frac{\pi^2 \alpha (\pi - \theta)}{E\theta^2 (2\pi - \theta)^2 \sin\theta}. \qquad (5)$$

b) In the case $m\alpha/\hbar^2 \gtrsim 1$, it is impossible to perform summation of the partial-wave series for an arbitrary scattering angle. But it is easy to see that for sufficiently small scattering angles, Eqs. (3) and (4) are valid. It is connected to an unbounded increase of the scattering amplitude for $\theta \to 0$. Since each term of series (XIII.9) is bounded, then such a divergence of the amplitude means that there are a lot of terms with large values of l in the sum (generally, the smaller the scattering angles, the larger the values of l). But for values of l large enough, relation (2) is still valid, from which Eq. (3) and (4) are deduced. The fact that for small scattering angles, the Born approximation is applicable independently of the value of $m\alpha/\hbar^2$, is reasonable: In the conditions of the problem, large distances contribute dominantly to the relevant quantities (due to the aforementioned divergence of the scattering amplitude), where $U = \alpha/r^2 \ll \hbar v/r$, and the potential can be considered as a perturbation.

c) Let us consider backward scattering. In this case, we have:

$$\sum_{l=0}^{\infty} (2l+1) P_l(\cos\theta) = 4\delta(1-\cos\theta),$$

i.e., for $\theta \neq 0$ the sum is equal to zero, and $P_l(-1) = (-1)^l$. Now series (XIII.9) takes the form:

$$f(E, \theta = \pi) = \frac{1}{2k} \sum_{l=0}^{\infty} (2l+1) \exp\left\{ -i\pi \sqrt{\left(l + \frac{1}{2}\right)^2 + \frac{2m\alpha}{\hbar^2}} \right\}. \qquad (6)$$

In the case of $m\alpha/\hbar^2 \gg 1$, the dominant role in this sum is played by the terms with large angular momenta[328] $l \lesssim (m\alpha/\hbar^2)^{1/4}$. Here, the neighboring terms differ from one another only slightly, and the summation could be replaced by an integration:

---

[328] For large values of l, terms of the sum (6) oscillate quickly, and as result of their mutual compensation, the corresponding part of the sum becomes small. Let us note that for the calculation of series (6), as well as for the previous sum with the δ-function, we should introduce a cutoff factor, $e^{-\gamma l}$ with $\gamma > 0$, in integral (7), and then the limit take $\gamma \to 0$ in the final result.

$$f(E, \pi) \approx \frac{1}{k} \int_0^\infty l \exp\left\{-i\pi\sqrt{l^2 + \frac{2m\alpha}{\hbar^2}}\right\} dl. \quad (7)$$

Using the substitution $x = \sqrt{l^2 + 2m\alpha/\hbar^2}$, we finally obtain

$$f(E, \pi) = -i\frac{\sqrt{2m\alpha}}{\pi\hbar k} \exp\left\{-i\pi\sqrt{\frac{2m\alpha}{\hbar^2}}\right\}. \quad (8)$$

This differential scattering cross-section for $\theta = \pi$ coincides with result (5) of classical mechanics. This remains so also for the region of scattering angles (which expand with an increase of α), that are close to $\theta = \pi$.

Let us note in conclusion that in the case of an attractive potential, $\alpha < 0$, the phase shifts, (1), could become imaginary. Formally, this coincides with the appearance of "absorption" in the system. This instability can be linked to the classical phenomenon of "fall into the center" (see Problem 9.14). What concerns the small-angle scattering, it is described by Eqs. (3) and (4), independently of the sign of α.

Problem 13.20

Find the wavefunction, $\psi_{\mathbf{k}_0}^+(\mathbf{r})$, scattering amplitude, and scattering cross-section of a particle in a zero-range potential (see Problem 4.10). Find the value of effective radius, r_0.

Solution

Since a zero-range potential affects only particles with the angular momentum $l = 0$, then only an s-scattering phase-shift, $\delta_0(k)$, is non-zero, and the scattering amplitude does not depend on the angle θ. The asymptotic form of the wavefunction,

$$\psi_{\mathbf{k}}^+(\mathbf{r}) = e^{i\mathbf{k}\cdot\mathbf{r}} + \frac{f(E)}{r} e^{ikr}, \quad (1)$$

determines the exact solution of the Schrödinger equation for all $r > 0$. Comparing the wavefunction expansion for $r \to 0$,

$$\psi_{\mathbf{k}}^+ \approx \frac{f}{r} + (1 + ikf),$$

with the boundary condition for the zero-range potential (see Problem 4.10), we find

$$f(E) = -\frac{a_0}{1 + ika_0}, \quad \sigma = 4\pi|f|^2 = \frac{4\pi}{k}\mathrm{Im}\, f(E) = \frac{4\pi a_0^2}{1 + k^2 a_0^2}. \quad (2)$$

Here, $a_0 = 1/\alpha_0$. Note that $f(0) = -a_0$, where a_0 is the scattering length of a z.r.p.

Finally, using the equality $e^{2i\delta} - 1 = 2i/(\cot\delta - i)$, we obtain the s-scatterig phase-shift on the zero-range potential as follows:

$$k \cot \delta_0(k) = -\frac{1}{a_0}. \tag{3}$$

Its comparison with the effective-range expansion, (XIII.15), proves that for scattering from a z.r.p., the effective radius of interaction is $r_0 = 0$.

Problem 13.21

Reconstruct the interaction potential $U(r)$ from the s-scattering phase-shift $\delta_0(k)$, considering it to be known for all energies, and assuming that $|\delta_0(k)| \ll 1$.

To illustrate applications of this result, consider the following dependencies:

a) $\delta_0(k) = \mathrm{const}$;   b) $\delta_0(k) = \dfrac{\alpha k}{1 + \beta k^2}$.

Solution

Since $|\delta_0(k)| \ll 1$ for all energies, we can consider the potential as a perturbation and use expression (XIII.12) for the phase shift of the s-wave:

$$\delta_0(k) = -\frac{2m}{\hbar^2 k} \int_0^\infty U(r) \sin^2 kr \, dr = -\frac{m}{\hbar^2 k} \int_0^\infty U(r)[1 - \cos 2kr] dr. \tag{1}$$

Multiplying both parts of this equality by $(-\hbar^2 k/2m)$ and differentiating by k, we obtain

$$-\frac{\hbar^2}{2m} \frac{d}{dk}[k\delta_0(k)] = \int_0^\infty rU(r) \sin 2kr \, dr \equiv \frac{i}{2} \int_{-\infty}^\infty rU(|r|) e^{-2ikr} dr. \tag{2}$$

Here we have used the *even* continuation of the potential, $U(-|r|) \equiv U(|r|)$, into the region with negative values of r.

Equation (2) determines the Fourier component of the potential (to be more precise, of the function $rU(|r|)$) and allows us to find the potential using the inverse Fourier transform:

$$rU(|r|) = \frac{i\hbar^2}{\pi m} \int_{-\infty}^\infty e^{2ikr} \frac{d}{dk}[k\delta_0(k)] dk. \tag{3}$$

Since according to Eq. (1), $\delta_0(k)$ is an odd function of the variable k, and so $\delta_0'(k)$ is an even function, then Eq. (3) takes the form:

$$U(r) = -\frac{2\hbar^2}{\pi m r} \int_0^\infty \sin(2kr) \frac{d}{dk}[k\delta_0(k)]dk. \tag{4}$$

Let us consider applications of this general formula.

a) In the case of $\delta_0(k) = \text{const} = C$ (for $k > 0$), according to Eq. (4) we obtain:

$$U(r) = -\frac{2\hbar^2 C}{\pi m r} \int_0^\infty \sin 2kr\, dk = -\frac{\hbar^2 C}{\pi m r^2} \equiv \frac{\gamma}{r^2}. \tag{5}$$

To calculate this integral, introduce the "cutoff" factor $e^{-\lambda k}$ with $\lambda > 0$, and in the final result take $\lambda \to 0$.

b) Substituting $[k\delta_0(k)]' = 2\alpha k/(1+\beta k^2)^2$ into Eq. (3) and calculating the integral using residues, we find:

$$U(r) = -\frac{2\alpha\hbar^2}{m\sqrt{\beta^3}} \exp\left\{-\frac{2r}{\sqrt{\beta}}\right\} \equiv U_0 e^{-r/R}. \tag{6}$$

Note that the condition $|\delta_0(k)| \ll 1$ assumes that $|C| \ll 1$ and $|\alpha| \ll \sqrt{\beta}$. The reconstructed potentials (5) and (6), as expected, are consistent with the applicability condition of the Born approximation for any energy.

Problem 13.22

Obtain phase shifts in the Born approximation for an exchange potential (see Problem 13.3).

Solution

Since the action of the operator $\hat{U}_{ex}$ on the wavefunction of a state with a definite value of angular momentum, l, is reduced to its multiplication by $(-1)^l U(r)$, then the expression for the phase shift, $\delta^B_{ex,l}(k)$, differs from the Born expression, (XIII.12), for a regular potential, $U(r)$, only by the factor $(-1)^l$.

Problem 13.23

Develop scattering theory based on a partial-wave analysis for an axially-symmetric potential, $U(\rho)$, in two-dimensional quantum mechanics.

What is a form of the optical theorem in two dimensions?

Solution

A theory of scattering for an axially-symmetric potential in two dimensions can be developed along the same lines as that for a spherically-symmetric potential in three dimensional quantum mechanics.

The Hamiltonian in an axially-symmetric field has the form:

$$\hat{H} = -\frac{\hbar^2}{2\mu}\left(\frac{1}{\rho}\frac{\partial}{\partial\rho}\rho\frac{\partial}{\partial\rho} + \frac{1}{\rho^2}\frac{\partial^2}{\partial\varphi^2}\right) + U(\rho), \tag{1}$$

while the solution of the Schrödinger equation has the following asymptotic behavior

$$\psi_{\mathbf{k}}^{+}(\boldsymbol{\rho}) \approx e^{ik\rho\cos\varphi} + \sqrt{i}\frac{f(k,\varphi)}{\sqrt{\rho}}e^{ik\rho}, \quad \rho \to \infty. \tag{2}$$

Incident particles move in the direction of the positive x axis; here, $x = \rho\cos\varphi$. Note that now the scattered wave is cylindrical (not spherical), and so instead of $1/r$ we have $1/\sqrt{\rho}$. In this two-dimensional case, the scattering amplitude, $f(k,\varphi)$, has the physical dimension of the square root of length, and the physical dimension of the differential scattering cross-section $d\sigma/d\varphi = |f|^2$ is length. Finally, a phase factor $\sqrt{i}$ is used for convenience.

Since the operator $\hat{l}_z = -i\partial/\partial\varphi$ commutes with both the Hamiltonian of free motion and Hamiltonian (1), then the two-dimensional plane wave and the exact wavefunction $\psi_{\mathbf{k}}^{+}(\boldsymbol{\rho})$ can be expanded in the eigenfunctions, $\psi_m = e^{im\varphi}$, of this operator (analogous to the expansion over spherical functions in the case of a central potential):

$$e^{ik\rho\cos\varphi} = \sum_{m=-\infty}^{\infty} A_m J_{|m|}(k\rho)e^{im\varphi}, \quad \psi_{\mathbf{k}}^{+} = \sum_{m=-\infty}^{\infty} B_m R_{km}(\rho)e^{im\varphi}. \tag{3}$$

Here, we have taken into account that the radial function of free motion with an "angular momentum" m is expressed in terms of the Bessel functions $J_{|m|}(k\rho)$, with the coefficients $A_m = i^{|m|}$. The values of the coefficients B_m are determined from the condition that the difference, $\psi_{\mathbf{k}}^{+} - e^{ikx}$, at large distances contains the outgoing waves, $\propto e^{ik\rho}$, for each term of the sum over m. Writing the asymptote of the radial function in the form[329]

$$R_{km}(\rho) = \sqrt{\frac{2}{\pi k\rho}}\sin\left(k\rho - \frac{\pi|m|}{2} + \frac{\pi}{4} + \delta_m\right), \tag{4}$$

we find $B_m = e^{i\delta_m}A_m$. As a result, we obtain a partial-wave decomposition of the scattering amplitude:

---

[329] The normalization factor in this function and the phase δ_m are chosen so that for $\delta_m = 0$, asymptote (4) coincides with the asymptote of a Bessel function, which describes the radial function of free motion.

$$f(k, \varphi) = \frac{1}{i\sqrt{2\pi k}} \sum_{m=-\infty}^{\infty} (e^{2i\delta_m} - 1)e^{im\varphi}. \tag{5}$$

Note that the phase shift δ_m does not depend on the sign of m.

According to Eq. (5), the total scattering cross-section is

$$\sigma = \int_0^{2\pi} |f|^2 d\varphi = \frac{4}{k} \sum_{m=-\infty}^{\infty} \sin^2 \delta_m, \tag{6}$$

and the optical theorem in this two-dimensional case reads

$$\operatorname{Im} f(k, \varphi = 0) = \sqrt{\frac{k}{8\pi}} \sigma(E). \tag{7}$$

Notice that extracting the factor $\sqrt{i}$ in Eq. (2) in the Born approximation is indeed convenient, as this convention ensures that for $|\delta_m| \ll 1$, the scattering amplitude given becomes real, according to Eq. (5).

Problem 13.24

Using the partial-wave decomposition, find the scattering amplitude and differential scattering cross-section of charged particles in an axially symmetric magnetic field, $\mathcal{H}(\rho)$, directed along the z axis and localized at small distances $\rho \leq a$ in the vicinity of this axis.

Hint: Consider the limiting case $a \to 0$, and a finite value, Φ_0, of the magnetic flux. It is convenient to choose the vector potential in the form $A_\varphi = \Phi_0/2\pi\rho$, $A_z = A_\rho = 0$.

Solution

In the polar coordinates, the Hamiltonian of transverse motion in a magnetic field $\hat{H}_\perp = (\hat{\mathbf{p}} - e\mathbf{A}/c)^2/2\mu$, takes the form:[330]

$$\hat{H}_\perp = \frac{\hbar^2}{2\mu} \left\{ -\frac{1}{\rho} \frac{\partial}{\partial \rho} \rho \frac{\partial}{\partial \rho} + \frac{1}{\rho^2} \left[\lambda^2 + 2i\lambda \frac{\partial}{\partial \varphi} - \frac{\partial^2}{\partial \varphi^2} \right] \right\}, \tag{1}$$

where $\lambda = e\Phi_0/2\pi\hbar c$. Since the operator $\hat{H}_\perp$ commutes with $\hat{l}_z$, the two-dimensional theory of scattering developed in the previous problem is also applicable here. Now, though, the phase shifts, δ_m, depend on the sign of m. The radial functions here, as well as for free motion, are expressed in terms of the Bessel functions, $J_\nu(k\rho)$, but with the index $\nu = |m - \lambda|$. Using their known asymptote, we find

---

[330] Note that free motion along the field is not particularly interesting and therefore is not being considered. Also note that for to derive Hamiltonian (1), we used the following azimuthal component of the gradient operator, $\rho^{-1}\nabla_\varphi = \partial/\partial\varphi$.

$$\delta_m = -\frac{\pi}{2}(|m-\lambda| - |m|).$$

The scattering amplitude (see Eq. (5) of the previous problem) becomes

$$f(k,\varphi) = \frac{1}{i\sqrt{2\pi k}} \sum_{m=-\infty}^{\infty} \{e^{i\pi(m-|m-\lambda|)} - 1\}e^{im\varphi}. \tag{2}$$

Denoting the minimum value of m (that is still larger than λ) by m_0, and dividing the sum into two pieces: with $m \geq m_0$ and with $m \leq m_0 - 1$, we can calculate the latter summing up a geometrical progression. This yields a closed equation for the amplitude and differential cross-section:

$$f = i\frac{\exp\{i(m_0 - 1/2)\varphi\}}{\sqrt{2\pi k}} \frac{\sin \pi\lambda}{\sin(\varphi/2)}, \quad \frac{d\sigma}{d\varphi} = \frac{\sin^2(e\Phi_0/2\hbar c)}{2\pi k \sin^2(\varphi/2)}\bigg|_{\phi \to 0} \propto \phi^{-2}.$$

An interesting property of this result is the infinite value of the total scattering cross-section, *i.e.*, particle scattering occurs even for an arbitrarily large impact parameter. From the point of view of classical theory, this conclusion appears counter-intuitive: the magnetic field and the Lorentz force are different from zero only on the axis, and hence have no influence on the particle's classical motion. This paradox is a manifestation of the fact that the phenomenon due to the magnetic flux described in this problem, the *Aharonov–Bohm effect*, is purely quantum, and disappears in the classical limit; for $\hbar \to 0$ and $k^{-1} = \hbar/p \to 0$, the scattering is indeed absent. In quantum mechanics, the interaction between a charged particle and a magnetic field is characterized by a vector potential: as in the Hamiltonian.[331] In this problem, despite the fact that outside the axis, the field vanishes, $\mathcal{H} = 0$, the vector potential can not be turned off ("gauged away") by any gauge transformation ($\oint \mathbf{A} \cdot d\mathbf{l} = \Phi_0$), and its slow, $\propto 1/\rho$, decrease on large distances explains the infinite scattering cross-section.

Problem 13.25

Find the energy dependence of phase shifts, $\delta_l(k)$, with a fixed value of l, in the limit $k \to \infty$. Consider potentials with the following behavior at small distances ($r \to 0$): $U \approx \alpha/r^\nu$ with a) $\nu < 1$; b) $1 < \nu < 2$; c) $\nu = 1$.

Solution

For large energies, the Born approximation is applicable, and we can use expression (XIII.12) for the phase shifts. For $k \to \infty$, the argument of the Bessel function is large,

---

[331] See also an excellent discussion about whether the vector potential is "real" in *The Feynman Lectures on Physics*, vol. 6, Ch. 15.

$x = kr \gg 1$, in the entire integration domain, except for a narrow region of small r. Using the known asymptote for $J_\mu(x)$, we obtain

$$\delta_l(k) \underset{k \to \infty}{=} -\frac{2m}{\hbar^2 k} \int_0^\infty U(r) \sin^2\left(kr - \frac{\pi l}{2}\right) dr \approx$$

$$-\frac{m}{\hbar^2 k} \int_0^\infty U(r)\, dr \propto k^{-1}. \tag{1}$$

The fast-oscillating factor $\sin^2(x - \pi l/2)$ was replaced by its mean value, equal to $1/2$. This result is valid in case a), where $rU(r) \to 0$ for $r \to 0$.

For more singular potentials, Eq. (1) is not applicable, due to a divergence of the integral. Such a divergence means that now the region of small r plays a dominant role, and we cannot replace $J_\mu(x)$ by its asymptote. Dividing the integration domain over r in Eq. (XIII.12) into two: from $r = 0$ to some small but finite R, and from $r = R$ to infinity, we see that the second of these integrals is proportional to k^{-1}, as in Eq. (1). A contribution of the first of these integrals is dominant, and we can put there $U = \alpha/r^\nu$. Making substitution $x = kr$, we obtain the[332]

$$\delta_l \approx -\frac{\pi m \alpha}{\hbar^2} k^{\nu-2} \int_0^\infty \frac{J_{l+1/2}^2(x) dx}{x^{\nu-1}} =$$

$$-\frac{\pi m \alpha}{\hbar^2} \frac{\Gamma(\nu-1)\Gamma(l+(3-\nu)/2)}{2^{\nu-1}\Gamma^2(\nu/2)\Gamma(l+(\nu+1)/2)} k^{\nu-2} \propto k^{\nu-2}. \tag{2}$$

Normally, we would replace the upper limit of integration, equal to kR, for $k \to \infty$ by ∞. For $\nu = 1$ however, such a replacement is not valid due to the divergence of the integral (in the upper limit).

Using the asymptote $J_\mu(x)$ for $x \to \infty$, we calculate the divergent part as follows

$$\delta_l \approx -\frac{m\alpha}{\hbar^2 k} \ln kR, \quad \nu = 1. \tag{3}$$

This equation has the *logarithmic accuracy* in accordance with an uncertainty in the value of R.

Let us note that the ν-dependence of the phase-shifts, $\delta_l(k)$, for $k \to \infty$, affects the scattering of particles with a large momentum transfer. This is connected with the fact that $f^B \propto \tilde{U}(q)$ for large values of q is determined by the *singularities* of the potential $U(r)$ as a function of r. For singular at $r \to 0$ potentials, $U \approx \alpha/r^\nu$, we have $\tilde{U} \propto q^{\nu-3} \propto k^{\nu-3}$ for $q \to \infty$. But such a difference in the energy dependence of phase-

---

[332] For $\nu = 2$, the independence of phase-shifts from k holds for any values of α and l, even if the Born approximation is not applicable (see Problem 13.19). Note also that for $\nu > 2$, the Born approximation breaks down.

shifts with fixed values of l does not affect the total scattering cross-section, $\sigma \propto 1/E$, that is determined by angular momenta $l \sim kR \gg 1$.

Problem 13.26

Prove that for large values of energy and angular momentum, where $kR \sim l \gg 1$, the Born approximation, (XIII.12), for phase-shifts crosses over to the quasi-classical expression, see Eq. (XIII.14).

Solution

In Eq. (XIII.12) for the values $l \sim kR \gg 1$ (R is the potential radius), we divide the integration domain into two parts: from $r = 0$ to $r = r_0 \equiv (l + 1/2)/k$, and from $r = r_0$ to $r = \infty$. In the second of these integrals, we use the *tangent approximation* for the Bessel function, i.e., the following asymptote

$$J_\nu \left(\frac{\nu}{\cos \beta} \right) \approx \sqrt{\frac{2}{\pi \nu \tan \beta}} \cos \left(\nu \tan \beta - \beta \nu - \frac{\pi}{4} \right), \quad \nu \gg 1. \tag{1}$$

Here, $\nu = l + 1/2$ and $\nu/\cos \beta = kr$, so that $\nu \tan \beta = k\sqrt{r^2 - r_0^2}$. After replacing the fast-oscillating factor, $\cos^2[\ldots]$, under the integral by its mean value of $1/2$, we find:

$$\delta_l^B(k) \approx -\frac{m}{\hbar^2 k} \int_{r_0}^{\infty} \frac{rU(r)dr}{\sqrt{r^2 - r_0^2}}. \tag{2}$$

We restrict ourselves to the second of the integrals mentioned, since the contribution of the first one is small. This smallness is connected to the fact that for $\nu \gg 1$, the function $J_\nu(x)$ decreases exponentially with the decrease of x from the value $x = \nu$. Such a decrease of $J_\nu(x)$ is a manifestation of the common quasiclassical behavior, where the quasiclassical wavefunctions decreases rapidly into a classically-forbidden region, inside a barrier (in this case, the centrifugal barrier).

Equation (2) coincides with the known quasi-classical expression, (XIII.14), for the phase shift in the case of $|U(r)| \ll E$. The use of the Born approximation assumes the fulfilment of a stricter condition, $|U(r)| \ll \hbar v/r$, with $|\delta_l^{B''}| \ll 1$.

According to Eq. (2), a dependence of the phase shift on l and k (in the relevant parameter range) is determined by the expression:

$$\delta_l^B(k) = \frac{g(s)}{k}, \quad s = \frac{l + 1/2}{k}. \tag{3}$$

The function $g(s)$ depends on the concrete form of the potential. If the Born approximation applies, by replacing the summation over l with an integration in Eq. (XIII.9), we obtain the known result, $\sigma \propto 1/E$ for $E \to \infty$ (for a short-range potential).

13.3 Low-energy scattering; Resonant scattering

Problem 13.27

Find the energy dependence of the scattering cross-section, $\sigma(E)$, in a field that decreases at large distances as $U(r) \approx \alpha/r^\nu$, $2 < \nu \leq 3$, and in the low-energy limit, $E \to 0$.

Solution

For potentials with this behavior at large distances, the Born amplitude for $q \to 0$ diverges. Since $q = 2k\sin(\theta/2)$, then for $E \to 0$ in the Born approximation, the total scattering cross-section would diverge as well (here, $q \to 0$ for all scattering angles). The divergence of the scattering cross-section signals the importance of large distances, where conditions (XIII.7) are fulfilled, and therefore the Born approximation does work. Here, according to Eq. (XIII.12), we have:

$$f(q) \approx -\frac{2m\alpha}{\hbar^2 q^{3-\nu}} \int_{qR}^{\infty} \frac{\sin x}{x^{\nu-1}} dx. \tag{1}$$

We consider only the diverging part of the amplitude and total scattering cross-section. We do not consider a contribution from distances $r \leq R$, where the potential is not described by its known asymptote.

For $\nu < 3$, in Eq. (1) we set the lower integration limit as $qR = 0$ and obtain:

$$f = \frac{C_\nu}{q^{3-\nu}}, \quad C_\nu = \frac{\pi m \alpha}{\hbar^2 \Gamma(\nu-1)\cos(\pi\nu/2)}, \tag{2}$$

using the integral

$$\int_0^\infty \frac{\sin x}{x^{\nu-1}} dx = \Gamma(2-\nu)\sin\frac{\pi\nu}{2} = -\frac{\pi\sin(\pi\nu/2)}{\Gamma(\nu-1)\sin\pi\nu}.$$

The total scattering cross-section is (for $q^2 = 2k^2(1-\cos\theta)$)

$$\sigma = 2\pi \int_0^\pi f^2 \sin\theta \, d\theta \underset{E\to 0}{\approx}$$

$$2\pi C_\nu^2 \left(\frac{\hbar^2}{4mE}\right)^{3-\nu} \int_{-1}^1 \frac{dz}{(1-z)^{3-\nu}} \propto \frac{1}{E^{3-\nu}} \to \infty. \tag{3}$$

If $\nu = 3$, we cannot replace the lower limit of integration in Eq. (1) by zero, due to integral divergence. Since $\sin x \approx x$ for $x \to 0$, the diverging part of the integral is equal to $\ln(1/qR)$, so that

$$f \approx \frac{2m\alpha}{\hbar^2} \ln qR, \tag{4}$$

and the total scattering cross-section is

$$\sigma = \int f^2 d\Omega \approx \int \frac{4m^2\alpha^2}{\hbar^4} \ln^2 kR \, d\Omega = \frac{16\pi m^2 \alpha^2}{\hbar^4} \ln^2 kR \to \infty. \tag{5}$$

For a calculation of the diverging, $\propto \ln^2 k$, part of the scattering cross-section, we put $\ln qR \approx \ln kR$. An uncertainty in the value of R gives rise to the logarithmic accuracy of Eqs. (4) and (5).

Problem 13.28

Find the k-dependence of the phase shifts, $\delta_l(k)$, for slow particles, and discuss the effective-range expansion in the Born approximation. What are the constraints on the large-distance asymptotic behavior of the potential that ensure the applicability of the expansion (XIII.15)? Introduce the following parameters of low-energy scattering: scattering length, a_l, and an effective range of the interaction, r_l, with the angular momentum, l.

Solution

For $k \to 0$, the argument, $x = kr$, of the Bessel function in Eq. (XIII.12) is small, and we can use the known expansion of $J_\mu(x)$ for $x \to 0$. We consider only the first two terms of the expansion,[333] and obtain

$$\delta_l^B(k) \approx k^{2l+1}[A_l + B_l k^2 + \ldots] \propto k^{2l+1}. \tag{1}$$

Here,

$$A_l = c_l \int_0^\infty U(r) r^{2l+2} dr, \quad B_l = -\frac{c_l}{2l+3} \int_0^\infty U(r) r^{2l+4} dr,$$

$$c_l = -\frac{\pi m}{2^{2l+1}\Gamma^2(l+3/2)\hbar^2}. \tag{2}$$

Taking into account the effective range expansion, Eq. (XIII.15), and also the smallness of the phase shift in the Born approximation, we find:

$$a_l^B = -A_l, \quad r_l^B = -2B_l A_l^{-2}. \tag{3}$$

One of the applicability conditions for these results is a fast decrease of the potential at large distances. This guarantees convergence of the integral in Eq. (2). For an exponential decrease of the potential, no restriction on the values[334] of l appears.

---

[333] These transformations are valid for fixed finite value of k, but $l \gg (kR)^2$, which allows us to make a conclusion for the phase shifts at large values of the angular momentum: $\delta_l \propto (kR/l)^{2l+1}$.

[334] The dependence $\delta_l \propto k^{2l+1}$ is valid for "strong" potentials including those for which the Born approximation is not valid.

For potentials with a power-law decrease, $U \approx \alpha/r^\nu$, we have a different situation. For the values $l < (\nu - 5)/2$, the integrals in Eq. (2) converge in the upper limit, and the notions of the scattering length a_l and effective range r_l are still well-defined.

In the case of $(\nu - 5)/2 \leq l < (\nu - 3)/2$, the second of the integrals in Eq. (2) diverges. Here, the effective range expansion, Eq. (XIII.15), is not valid, but the notion of the scattering length and the dependence $\delta_l \approx -a_l k^{2l+1}$ for slow particles are preserved.

Finally, for the values $l \geq (\nu - 3)/2$, the dependence $\delta_l \propto k^{2l+1}$ is broken for $k \to 0$. Low-energy scattering with an arbitrary angular momentum l in potentials with a power "tail" is considered in the following two problems (see Problem 13.27).

Problem 13.29

For a potential with a power-law decrease at large distances, $U \approx \alpha/r^\nu$ with $\nu > 2$, find the k-dependence of phase-shifts for slow particles with an angular momentum, l.

Solution

For the values $l < (\nu - 3)/2$, $\delta_l = A_l k^{2l+1}$. In the Born approximation, the coefficient A_l in this case is obtained as in the previous problem. In the case $l \geq (\nu - 3)/2$, we can not replace the Bessel function, $J_\mu(x)$, in Eq. (XIII.12) by its first, $\propto x^\mu$, expansion term, as it would have led the dependence $\delta_l \propto k^{2l+1}$, and a divergence in the integral. Now, we divide the integration domain in Eq. (XIII.12) into two parts: from $r = 0$ to $r = R$, and from $r = R$ to $r = \infty$ (the value of R is such that for $r > R$, we can use an asymptotic form of the potential), we see that the first integral, proportional to k^{2l+1}, is less important than the second one. We find:

$$\delta_l \approx -\frac{\pi m \alpha}{\hbar^2} k^{\nu-2} \int_{kR}^{\infty} \frac{1}{x^{\nu-1}} J_{l+1/2}^2(x) dx. \tag{1}$$

For $l > (\nu - 3)/2$, we can replace the lower integration limit by zero and obtain:

$$\delta_l \approx -\frac{\pi m \alpha}{\hbar^2} \frac{\Gamma(\nu - 1)\Gamma(l + (3 - \nu)/2)}{2^{\nu-1}\Gamma^2(\nu/2)\Gamma(l + (\nu + 1)/2)} k^{\nu-2}. \tag{2}$$

In the case of $l = (\nu - 3)/2$, the integral in Eq. (1) for $k \to 0$ diverges in the lower limit. Using the expansion of $J_\mu(x)$ for $x \to 0$, we find the divergent part of the integral and the phase shift:

$$\delta_l \approx \frac{\pi m \alpha}{2^{2l+1}\Gamma^2(l + 3/2)} k^{2l+1} \ln kR. \tag{3}$$

Let us make a few remarks:

1) Although the above results are based on the Born approximation, for $\nu > 2$ they apply more broadly. Indeed, in this case, the large distances are essential, where $|U(r)| \ll \hbar^2/mr^2$, so that the potential can be considered as a perturbation.
2) For a potential that is a superposition of a "strong" short-range and a "weak" long-range (with a power-law asymptote) potential, generally speaking,[335] the contributions to the phase shift are additive: $\delta_l \approx \delta_{l,shor} + \delta_{l,long}$. For abnormally small k, the dominant contribution comes from the long-range potential, while for k not too small it comes from the short-range potential. See Problem 13.37, and also Problem 13.42.

Problem 13.30

For potential with a power-law decrease at large distances, $U \approx \alpha/r^\nu$, discuss how the effective-range expansion, see Eq. (XIII.15), should be modified in order to describe scattering of particles with angular momenta, l, in the range $(\nu - 5)/2 \leq l < (\nu - 3)/2$. As an illustration of this result, consider s-wave scattering of slow particles in a potential with the large-distance asymptote, $U \approx \alpha/r^4$.

Solution

Taking into account the solutions of the two previous problems, it is easy to obtain the following expansions:

$$\delta_l + a_l k^{2l+1} \approx -\frac{\pi m \alpha}{\hbar^2} k^{\nu-2} \int_0^\infty \frac{1}{x^{\nu-1}} \left[J_{l+1/2}^2(x) - \frac{x^{2l+1}}{2^{2l+1}\Gamma^2(l+3/2)} \right] dx,$$

$$\frac{\nu - 5}{2} < l < \frac{\nu - 3}{2}. \tag{1}$$

In the case $l = (\nu - 5)/2$, we can also obtain Eq. (1), but the lower limit of the integral will be equal to kR. Calculating the divergent part, we find:

$$\delta_l + a_l k^{2l+1} \approx -\frac{\pi m \alpha}{2^{2l+1}(2l+3)\Gamma^2(l+3/2)} k^{2l+3} \ln kR. \tag{2}$$

In particular, for a potential that at large distances has the form $U \approx \alpha/r^4$ (i.e., $\nu = 4$), and for $l = 0$, we obtain from Eq. (1):

$$\delta_0 + a_0 k \approx -\frac{2m\alpha}{\hbar^2} k^2 \int_0^\infty \frac{1}{x^3} \left[\frac{\sin^2 x}{x} - x \right] dx = \frac{2\pi m \alpha}{3\hbar^2} k^2.$$

---

[335] Except for the case, where there exists a state with a small binding energy in the short-range potential.

The generalization of the effective range expansion, Eq. (XIII.15), takes the form:

$$k \cot \delta_0(k) \approx -\frac{1}{a_0} + \frac{2\pi m \alpha}{3\hbar^2 a_0^2} k.$$

Problem 13.31

Find the scattering length, a_0, in the following potentials:

a) $U(r) = \begin{cases} -U_0, & r < R, \\ 0, & r > R; \end{cases}$

b) $U(r) = -U_0 R \delta(r - R);$

c) $U(r) = -U_0 e^{-r/R};$

d) $U(r) = -U_0 [1 + (r/R)^2]^{-2};$

e) $U(r) = U_0 (R/r)^4, \quad U_0 > 0.$

What is notable about the potential parameters that correspond to a divergent scattering length?

What is the reason behind a non-analytic dependence of $a_0(U_0)$ on the parameter U_0, for $U_0 \to 0$ in case e)?

Solution

To calculate the scattering length, a_0, we should find a bounded solution of the radial Schrödinger equation with $E = 0$ and $l = 0$. Its asymptote,[336]

$$R_{l=0, E=0} \approx 1 - \frac{a_0}{r} \quad \text{for} \quad r \to \infty,$$

determines the value of a_0. The solutions of the Schrödinger equation for the potentials considered were discussed in Chapter 4, so here we make only several remarks,[337] (below, $\chi = r R_{l=0, E=0}$ and $\lambda = \sqrt{2m U_0 R^2/\hbar^2}$).

a) For the square potential well,

$$\chi = \begin{cases} A \sin \frac{\lambda r}{R}, & r < R, \\ r - a_0, & r > R. \end{cases}$$

From the conditions of χ and χ' continuity at the point $r = R$, we find:

$$a_0 = \left(1 - \frac{1}{\lambda} \tan \lambda\right) R.$$

---

[336] Such an asymptote assumes a decrease of the potential at large distances faster than $\propto 1/r^3$.
[337] Let us reiterate that the boundary condition, $\psi(\infty) = 0$, that is used for states of the discrete spectrum, does not appear here.

b) For the δ-spherical potential,

$$\chi = \begin{cases} Cr, & r < R, \\ a_0 - r, & r > R. \end{cases}$$

Matching the solution at the point $r = R$ (see Problem 4.8) gives

$$a_0 = -\frac{\lambda^2}{1 - \lambda^2} R.$$

c) For an exponential potential, the solution to the Schrödinger equation is

$$\chi = J_0(2\lambda) N_0(x) - N_0(2\lambda) J_0(x), \quad \text{where } x = 2\lambda e^{-r/2R}.$$

Here, J_0 and N_0 are Bessel and Neumann functions, respectively. We have used the boundary condition $\chi(r = 0) = 0$ (compare to Problem 4.8). Since for $r \to \infty$, we have $x \to 0$, then we can use the expression (with $\mathcal{C} = e^\gamma = 1.781\ldots$, $\gamma = 0.5772\ldots$ is Euler's constant):

$$J_0(x) \approx 1 - \frac{x^2}{4}, \quad N_0(x) \approx \frac{2}{\pi} J_0(x) \ln \frac{x}{2} + \frac{2\gamma}{\pi} + \frac{1}{2\pi}(1-\gamma)x^2, \quad |x| \ll 1,$$

and find the scattering length:

$$a_0 = \frac{\pi R}{J_0(2\lambda)} \left[\frac{2}{\pi} J_0(2\lambda) \ln \gamma\lambda - N_0(2\lambda) \right]. \tag{3}$$

d) For the potential given, the Schrödinger equation has the form (compare with Problem 4.25):

$$\psi_{E=0, l=0} = C \sqrt{1 + \frac{R^2}{r^2}} \sin\left(\xi \arctan \frac{r}{R}\right),$$

where $\xi = \sqrt{1 + \lambda^2}$. From its asymptote, we find:

$$a_0 = \xi R \cot\left(\frac{1}{2}\pi\xi\right). \tag{4}$$

e) In analogy to the solution to Problem 4.25, we find the wavefunction,

$$\psi_{E=0, l=0} = A \exp\left\{-\frac{\lambda R}{r}\right\},$$

and the scattering length,

$$a_0 = \lambda R. \tag{5}$$

Now, we discuss the results obtained. We first mention an essential difference in the dependence of the scattering length, a_0, on the parameter $\lambda \propto \sqrt{U_0}$ for $U_0 \to 0$ in case e) in comparison to the other cases a)–d). According to (1)–(4), a_0 for small λ is expanded into a series in powers of λ^2, and is an analytic function of the parameter

U_0. Depending on the sign of U_0, these expressions give a scattering length in either an attractive potential $U_0 > 0$, or a repulsive one $U_0 < 0$. For example, Eq. (1) for $U_0 < 0$ takes the form:

$$a_0 = \left(1 - \frac{1}{|\lambda|}\tanh|\lambda|\right) R$$

and describes the scattering length of a potential barrier. In the case of the potential $U = \alpha/r^4$, with $\alpha = U_0 R^4$, the dependence of a_0 on the parameter α (or, equivalently, U_0) is non-analytic, since $a_0 \propto \sqrt{\alpha}$. This non-analyticity reflects a qualitatively different way in which the potential affects the particle for small values of $\alpha > 0$ versus $\alpha < 0$, that manifests itself in the "fall into the center" phenomenon in the latter case[338] $\alpha < 0$. Here, Eq. (5) is invalid, since the boundary condition, $\psi(0) = 0$, used for its derivation, could not be fulfilled and requires a modification. Compare to Problem 9.14.

Eqs. (1)–(4) reflect the general nature of the dependence of the scattering length, a_0, on the parameters U_0 and R for a regular potential of a fixed sign and of the form $U(r) = -U_0 f(r/R)$, where $f(|z|) \geq 0$ (attraction for $U_0 > 0$ and repulsion for $U_0 < 0$). Its qualitative nature is shown in Fig. 13.3.

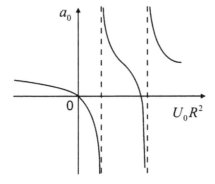

Fig. 13.3

We now list some properties of the solution:

1) For $|\lambda|^2 \ll 1$ (a "weak" potential), the scattering length is also small, since $|a_0| \sim |\lambda|^2 R \ll R$. Writing $a_0 = -\beta\lambda^2 R$ from a) – d), we find that the parameter β, is equal to $1/3$, 1, 2, $\pi/4$ respectively. These results can be obtained from the Born approximation equation:

$$a_0^B = -f^B(E=0) = \frac{m}{2\pi\hbar^2}\int U(r)dV.$$

The sign of the scattering length coincides with the sign of potential.

---

[338] Compare to the analogous property of the scattering phases δ_l, in the case of an attractive potential $U = -|\alpha|/r^2$, which follows from Eq. (1) of Problem 13.19. For $2m|\alpha|/\hbar^2 = (l+1/2)^2$, the particle with the angular momentum, l, "falls into the center."

2) For an increase of $|\lambda|^2$, the value of the scattering length increases. The dependence of a_0 on U_0 is monotonic, except for the *unique* values, λ_n, of the parameter λ, where, in an attractive potential, the scattering length, a_0, becomes infinite. Such values of λ correspond to the appearance of a new bound state in the potential with the angular momentum $l = 0$.[339] Indeed, the wavefunction's asymptote for $r \to \infty$ at the threshold of emergence of a new bound state is $\psi_{E=0} \propto 1/r$ (but not $C_1 + C_2/r$, as in the general case; compare to Problem 4.25). This corresponds to a divergent scattering length, $a_0 \to \infty$.

Let us prove that the dependence of a_0 on U_0 is monotonous if the potential does not change sign. The two Schrödinger equations for $E = 0$ and $l = 0$ are

$$\chi_0'' = -\frac{2m}{\hbar^2} U_0 f\left(\frac{r}{R}\right) \chi_0, \quad \chi'' = -\frac{2m}{\hbar^2}(U_0 + \delta U_0) f\left(\frac{r}{R}\right)\chi.$$

The boundary conditions take the form $\chi_0(0) = \chi(0) = 0$, while the asymptotes of the solutions for $r \to \infty$ are $\chi_0 \approx a_0 - r$ and $\chi \approx a - r$, with $a = a_0 + \delta a_0$ (where δa_0 is a change in the s-wave scattering length due to a change of the potential by $-\delta U_0 f$). Multiplying the first of these equations by χ and the second one by χ_0, subtracting them term by term, and integrating over r from 0 to ∞, we find:

$$\delta a_0 = -(\chi\chi_0' - \chi_0\chi')|_0^\infty \approx -\frac{2m}{\hbar^2}\delta U_0 \int_0^\infty f\left(\frac{r}{R}\right) \chi_0^2(r) dr.$$

In the integral, we put $\chi \approx \chi_0$, which is valid due to the assumed smallness of $\delta U_0 \to 0$ and the condition $a_0 \neq \infty$; here, δa_0 is also small and $\delta a_0/\delta U_0 < 0$, which proves the monotonic $a_0(U_0)$ dependence.

3) Along with the values of λ close to λ_n, for which the scattering amplitude is anomalously large (resonant scattering), there also exist values $\tilde{\lambda}_n$, where the scattering length on the contrary vanishes. For such values, $\tilde{\lambda}_n$, of the parameters of the potential, the scattering cross-section, $\sigma = 4\pi a_0^2$, of particles with energy $E = 0$ vanishes. So for λ close to $\tilde{\lambda}_n$, the scattering cross-section for slow particles, $kR \ll 1$, can be abnormally small, which manifests itself in the *Ramsauer–Townsend effect*. Let us emphasize that this smallness of the scattering cross-section, $\sigma \ll \pi R^2$, for slow particles in a "strong" short-range potential of radius R might appear only if the potential is attractive.

Finally, let us mention the following property of the scattering length in the case of a strong potential, with $|\lambda| \gg 1$. If the potential has a well-defined radius, R, as in the case of a rectangular δ-well (or the corresponding barrier), then $a_0 \approx R$, except for narrow regions of λ in the vicinity of the points[340] λ_n (see Eqs. (1) and (2)). This property exists also for rapidly (exponentially) decreasing

---

[339] In the case of the δ-well, Eq. (2), only one bound state with the angular momentum $l = 0$ appears, as $\lambda \geq \lambda_0 = 1$.

[340] These narrow regions also include the points $\tilde{\lambda}_n$. As is seen from Eqs. (1) and (3), for rapidly decreasing potentials, the points λ_n and $\tilde{\lambda}_n$ move closer as $n \to \infty$; for the step potential (see Eq. (4)) this statement is not valid.

potentials. From Eq. (3), for $\ln \gamma \lambda \gg 1$, we have $a_0 \approx R_{eff} \approx 2R \ln \gamma \lambda$, except when the Bessel function, $J_0(2\lambda)$, is close to zero. These exceptional values of λ correspond to the appearance of new bound states. The difference from the previous case lies in the weak, logarithmic dependence of R_{eff} on λ, which is seen from the expression $|U(R_{eff})| \approx \hbar^2/mR_{eff}^2$. For potentials with a power-law decrease, as seen from Eq. (4), we have a different situation. In this case, $a_0(\lambda)$ is a "lively" function of λ in the entire interval between λ_n and λ_{n+1}.

Problem 13.32

Find the scattering length and cross-section for slow particles scattered by an impenetrable ellipsoid, i.e., by the potential:

$$U(r) = \begin{cases} \infty, & \frac{x^2+y^2}{b^2} + \frac{z^2}{c^2} < 1, \\ 0, & \frac{x^2+y^2}{b^2} + \frac{z^2}{c^2} \geq 1. \end{cases}$$

Assume $c \geq b$. Consider specifically two limiting cases: $c \approx b$ and $c \gg b$.

Solution

This problem is reduced to solving the Schrödinger equation with energy $E = 0$, which takes the form $\Delta \psi(\mathbf{r}) = 0$ with the boundary condition $\psi(\mathbf{r}_0) = 0$ on the surface of the ellipsoid and with the asymptote, $\psi \approx 1 - a_0/r$, at large distances. The scattering length is[341] $a_0 = -f(E=0)$.

For the function $\phi(\mathbf{r}) = 1 - \psi(\mathbf{r})$, the equation, the boundary condition, and the asymptote are:

$$\Delta \varphi = 0, \quad \varphi(\mathbf{r}_0) = 1, \quad \varphi \approx \frac{a_0}{r} \quad \text{for} \quad r \to \infty. \tag{1}$$

According to these relations, the function $\varphi(\mathbf{r})$ can be considered as an electrostatic potential of a charged ellipsoid that on its surface takes the value $\varphi_0 = 1$, and $e = a_0$ is the ellipsoid's total charge. Since $e = C\varphi_0$, where C is the capacitance, the particle's scattering length for an "impenetrable" ellipsoid is formally equivalent to the electrostatic capacitance of a conductor of the same form.[342] The solution of the electrostatic problem, Eq. (1), for an oblong ellipsoid is known and has the form (for $c > b$):

$$\varphi(\mathbf{r}) = \frac{e}{2\sqrt{c^2-b^2}} \ln \frac{z + \sqrt{c^2-b^2} + [(z+\sqrt{c^2-b^2})^2 + x^2 + y^2]^{1/2}}{z - \sqrt{c^2-b^2} + [(z-\sqrt{c^2-b^2})^2 + x^2 + y^2]^{1/2}}. \tag{2}$$

---

[341] Let us emphasize that in the limit of zero energy, particle scattering is isotropic even in the case of a non-spherically-symmetric short-range interaction.

[342] The relation $a_0 = C$ is valid for an "impenetrable" body of an arbitrary form. For example, for particle scattering from an impenetrable disc of radius R, the scattering cross-section is $\sigma(E = 0) = (16/\pi)R^2$ (the disc capacitance is $C = 2R/\pi$).

This result can be obtained by the method of images: the potential created by the ellipsoid coincides with that of a homogeneously-charged segment, whose ends are at the foci of the ellipsoid, with the coordinates $x = y = 0$, $z = \pm\sqrt{c^2 - b^2}$. Determining e from the condition $\varphi(\mathbf{r}_0) = 1$ (here, it is convenient to chose the point on the ellipsoid that lies on the rotation axis, $x = y = 0$, $z = c$), we find the scattering length:

$$a_0 = e = 2\sqrt{c^2 - b^2} \left[\ln \frac{c + \sqrt{c^2 - b^2}}{c - \sqrt{c^2 - b^2}} \right]^{-1}. \tag{3}$$

In the case $c = b$, we have $a_0 = c$ and $\sigma = 4\pi c^2$, the known result for slow particle scattering cross-section for an impenetrable sphere of radius $R = c$. For a strongly prolate ellipsoid, $c \gg b$, Eq. (3) gives:

$$a_0 \approx \frac{c}{\ln(2c/b)}, \quad \sigma(E = 0) \approx \frac{4\pi c^2}{(\ln(2c/b))^2}. \tag{4}$$

The generalization to the case of an oblate ellipsoid, $b > c$, can be obtained using analytic continuation of the scattering length (3). Writing $\sqrt{c^2 - b^2} = i\sqrt{b^2 - c^2} \equiv iv$ and using the relation

$$\ln(c \pm iv) = \frac{1}{2}\ln(c^2 + v^2) \pm i\arctan\frac{v}{c},$$

we obtain

$$\ln\left(\frac{c + iv}{c - iv}\right) = 2i\arctan\frac{v}{c} = 2i\arctan\frac{\sqrt{b^2 - c^2}}{c} = 2i\arccos\frac{c}{b},$$

and for the scattering length, we obtain ($b > c$)

$$a_0 = \frac{\sqrt{b^2 - c^2}}{\arccos(c/b)}. \tag{5}$$

If we put here $c = 0$ and $b = R$, we obtain the scattering length, $a_0 = 2R/\pi$, for an impenetrable disc of radius R.

Problem 13.33

In the quasi-classical approximation, find the scattering length, a_0, for a repulsive potential that has the asymptotic behavior $U \approx \alpha/r^4$ at large distances.

To illustrate the quasi-classical result obtained, apply it to the potentials a) $U = \alpha(R^2 + r^2)^{-2}$ and, b) $U = \alpha(R + r)^{-4}$, where $R > 0$. Compare these with exact results.

Solution

At finite distances, the quasi-classical solution of Eq. (IV.5) for the function $\chi = rR(r)$ with $E = 0$ and $l = 0$ should be chosen as follows

$$\chi_{quas}(r) = \frac{C}{\sqrt{|p|}} \exp\left\{-\frac{1}{\hbar}\int_r^\infty |p|dr\right\}, \quad |p| = \sqrt{2mU(r)}. \quad (1)$$

Here, only the wavefunction that vanishes at the origin is kept. By neglecting the increasing part of the solution, we enforce the boundary condition,[343] $\chi(0) = 0$.

But at large distances, $r \to \infty$, we have $|p(r)| \propto 1/r^2 \to 0$, and the quasiclassical approximation is not applicable. So, we should match solution (1) at large distances, where $U \approx \alpha/r^4$, with the exact solution of the Schrödinger equation, whose asymptote, $\chi(r) \approx a_0 - r$, gives the scattering length. Such an exact solution has the form (see Problem 4.25):

$$\chi = r\left[\frac{a_0}{d}\sinh\frac{d}{r} - \cosh\frac{d}{r}\right] \underset{r\to\infty}{\approx} a_0 - r, \quad d = \sqrt{\frac{2m\alpha}{\hbar^2}}. \quad (2)$$

Matching of the solutions (1) and (2) can be performed only if in Eq. (2), as well as in Eq. (1), there exists no term exponentially increasing towards the origin, $r \to 0$. Hence, we find the scattering length in the quasi-classical approximation:

$$a_{0,quas} = d = \sqrt{\frac{2m\alpha}{\hbar^2}}. \quad (3)$$

This result for $U = \alpha/r^4$ reproduces the exact one.

For the other potentials from the problem condition, the scattering lengths are (see Problem 13.31 d; $\xi = \sqrt{2m\alpha/\hbar^2 R^2} = d/R$):

a) $a_0 = R\sqrt{\xi^2 - 1}\coth\left(\frac{1}{2}\pi\sqrt{\xi^2 - 1}\right)$, b) $a_0 = R\xi\left(\coth\xi - \frac{1}{\xi}\right)$. (4)

As expected, for $\xi \to \infty$ they become Eq. (3). Even for the values $\xi \geq 1$, the quasi-classical result is close to the exact one. As an illustration, below is a table showing the ratios, $\eta = a_{0,quas}/a_0$ for several ξ:

ξ	1	2	4	6
η, a)	1.571	1.145	1.033	1.014
η, b)	3.202	1.862	1.332	1.200

---

[343] If $U(r)$ is a bounded potential, $\chi_{quas}(0)$ is different from zero, but is exponentially small. Such a break-down of the boundary condition gives exponentially small errors in the results for relevant observables. More important are corrections connected to the deviation of the potential from its asymptotic form at large distances. See the following problem.

The lower accuracy in case b is due to a larger difference between the potential $U(r)$ and its asymptote, α/r^4, at large distances. See the following problem.

Problem 13.34

In the conditions of the previous problem, find a quasi-classical correction to the scattering length using the subleading term in the large-distance expansion of the potential: $U = \alpha r^{-4}(1 + b/r + \dots)$.

Solution

If the potential $U(r)$ differs from α/r^4 only at finite distances, $r \leq R$, then for $\xi = \sqrt{2m\alpha/\hbar^2 R^2} \gg 1$, the difference of a_0 from its quasi-classical value $\sqrt{2m\alpha/\hbar^2}$ is exponentially small and is outside the limits of applicability of the quasi-classical approximation.

If the difference, $U(r) - \alpha/r^4$, is different from zero for $r > R$, then power-law corrections in the parameter, $1/\xi \propto \hbar$, appear. To determine the first such correction in the case of $U(r) \approx (\alpha/r^4)(1 + b/r)$ for $r \to \infty$, it is necessary to find a solution of the Schrödinger equation at large distances, whose accuracy allows to determine correctly a correction in the asymptote of the potential. To do this, we note that within the required accuracy, $U = \alpha/(r - b/4)^4$. For this potential, the Schrödinger equation allows an exact solution, which can be obtained by replacing r by $r - b/4$ in Eq. (2) of the previous problem, and has the form:

$$\chi = -\left(r - \frac{b}{4}\right) \exp\left\{-\frac{\sqrt{2m\alpha}}{\hbar(r - b/4)}\right\}. \tag{1}$$

We have omitted here a second independent solution of the Schrödinger equation, which increases exponentially with a decrease in r. From the asymptote of Eq. (1) for $r \to \infty$, we obtain:

$$a_{0,quas} = \sqrt{\frac{2m\alpha}{\hbar^2}}\left(1 + \frac{\hbar b}{\sqrt{32m\alpha}}\right). \tag{2}$$

In the same way, for a potential with the asymptote

$$U(r) = \frac{\alpha}{r^4}\left(1 + \frac{b_1}{r} + \frac{b_2}{r^2} + \dots\right),$$

we can find a more accurate quasi-classical correction to Eq. (2) if we note that, within the required accuracy, it can be written in the form:

$$U = \frac{\alpha}{[(r+b)^2 + R^2]^2}, \quad \text{where } b = -\frac{1}{4}b_1, \quad R^2 = \frac{5}{16}b_1^2 - \frac{1}{2}b_2.$$

For such a potential, the solution of the Schrödinger equation with $E = 0$ can be obtained as in Problem 4.25, and then we find the following correction to Eq. (2):

$$a_{0,quas} = \sqrt{\frac{2m\alpha}{\hbar^2}} \left[1 + \frac{\hbar b_1}{\sqrt{32m\alpha}} + \frac{\hbar^2}{64m\alpha}(8b_2 - 5b_1^2)\right]. \tag{3}$$

The inclusion of these quasi-classical corrections greatly increases the accuracy of the results. For example, according to Eq. (3) instead of the values given in the table in the previous problem, we now have 0.786, 1.002, 1.0005, 1.0001 in case a, and 0, 0.931, 0.999, 0.99998 in case b.

In conclusion, we give quasi-classical expressions for the scattering length, a_0, in the case of repulsive potentials with other asymptotic behaviors at infinity, $r \to \infty$.

In the case of a power-law asymptote, $U = (\alpha/r^\nu)(1 + b/r + \ldots)$ with $\nu > 3$, we find:

$$a_{0,quas} = \frac{\Gamma[(\nu-3)/(\nu-2)]}{\Gamma[((\nu-1)/(\nu-2))]} \left(\frac{2m\alpha}{(\nu-2)^2\hbar^2}\right)^{1/(\nu-2)} + \frac{b}{\nu}, \tag{4}$$

where $\Gamma(z)$ is the Γ-function. For $\nu \to 3$, $a_{0,quas} \to \infty$, which reflects the fact that in the potential $U = \alpha/r^3$, the particle scattering cross-section diverges in the limit $E \to 0$.

For a potential with the exponential decrease, $U \approx U_0 e^{-r/R}$,

$$a_{0,quas} = R\left[\ln\left(\frac{2mU_0R^2}{\hbar^2}\right) + 2\gamma\right], \tag{5}$$

where $\gamma = 0.5772\ldots$ is Euler's constant.

Problem 13.35

For an attractive potential, $U(r) \leq 0$, that has a power-law decrease, $U \approx -\alpha/r^\nu$ with $\nu > 3$, at large distances, find the scattering length, a_0, in the quasi-classical approximation. Assume that the potential has the following behavior at small distances, $U \propto r^{-\beta}$ with $0 \leq \beta < 2$. What is special about the parameters of the potential that correspond to a divergent scattering length?

Consider applications of this result to the potentials a) $U = -\alpha(r+R)^{-4}$, and b) $U = -\alpha(r^2 + R^2)^{-2}$, and compare to the exact solution.

Solution

At finite distances, the quasi-classical solution of the Schrödinger Eq. (IV.5) for $E = 0$ and $l = 0$ has the form (compare to Problem 9.9):

$$\chi_{quas} = rR(r) = \frac{C}{\sqrt{p(r)}} \sin\left(\frac{1}{\hbar}\int_0^r p(r)dr + \gamma_0\right), \quad p = \sqrt{-2mU(r)}, \tag{1}$$

where γ_0 depends on the form of the potential at small distances. In the case $U \approx -A/r^\beta$ for $r \to 0$, we have:

$$\gamma_0 = -\frac{\pi\beta}{4(2-\beta)}.$$

At large distances, $p(r) \to 0$ and the quasi-classical method is not applicable. Here, though, by taking into account the form of potential, $U \approx -\alpha/r^\nu$, we could find the exact solution of the Schrödinger equation:

$$\chi = \sqrt{r}[C_1 J_s(2s\sqrt{\tilde\alpha}r^{-1/2s}) + C_2 J_{-s}(2s\sqrt{\tilde\alpha}r^{-1/2s})], \tag{2}$$

where $s = 1/(\nu - 2)$ and $\tilde\alpha = 2m\alpha/\hbar^2$. On the other hand at distances, where the argument of the Bessel functions in Eq. (2) is large, the quasi-classical approximation is applicable, and both Eqs. (1) and (2) are valid. Using the asymptote of the Bessel function, $J_\nu(z)$, for $z \to \infty$, solution (2) is transformed to

$$\chi = \sqrt{\frac{\pi s \hbar}{p(r)}} \left[C_1 \sin\left(\frac{1}{\hbar}\int_r^\infty p\, dr - \frac{\pi s}{2} + \frac{\pi}{4}\right) + C_2 \sin\left(\frac{1}{\hbar}\int_r^\infty p\, dr + \frac{\pi s}{2} + \frac{\pi}{4}\right)\right]. \tag{3}$$

We use $p(r) = \sqrt{2m\alpha/r^\nu}$, and by matching Eqs. (3) and (1) obtain:

$$\frac{C_1}{C_2} = -\frac{\sin(\tau + \pi s/2 + \pi/4)}{\sin(\tau - \pi s/2 + \pi/4)}, \quad \tau = \frac{1}{\hbar}\int_0^\infty \sqrt{-2mU(r)}\, dr + \gamma_0. \tag{4}$$

Furthermore, using an expansion of $J_{\pm s}(z)$ for $z \to 0$, we find the asymptote of solution (2) at large distances:

$$\chi(r) \approx \frac{C_1}{\Gamma(1+s)}(s\sqrt{\tilde\alpha})^s + \frac{C_2}{\Gamma(1-s)}(s\sqrt{\tilde\alpha})^{-s}, \quad r \to \infty. \tag{5}$$

Since $\chi \propto r - a_0$, and taking into account relation (4), we obtain the quasi-classical scattering amplitude as follows:

$$a_{0,quas}^{(0)} \approx \frac{\Gamma(1-s)}{\Gamma(1+s)} \frac{\sin(\tau + \pi s/2 + \pi/4)}{\sin(\tau - \pi s/2 + \pi/4)} \left(s\sqrt{\frac{2m\alpha}{\hbar^2}}\right)^{2s}. \tag{6}$$

The parameters of the potential for which the scattering length becomes infinite correspond to the appearance of a new discrete state, as the well becomes deeper. According to Eqs. (6) and (4), this happens when the following conditions are fulfilled (with $l = 0$):

$$\frac{1}{\hbar}\int_0^\infty \sqrt{-2mU(r)}\,dr = \pi\left[N + \frac{\beta}{4(2-\beta)} + \frac{1}{2(\nu-2)} - \frac{1}{4}\right], \tag{7}$$

where $N = 1, 2, \ldots$ is the sequence number of the new bound state (compare to Problem 9.9).

For the potentials given in the problem condition, we have $\nu = 4$, $s = 1/2$, $\gamma = \beta = 0$. From Eq. (6), we obtain

$$\text{a) } a_{0,quas}^{(0)} = R\xi \cot \xi, \quad \text{b) } a_{0,quas}^{(0)} = R\xi \cot\left(\frac{\pi}{2}\xi\right), \tag{8}$$

where $\xi = \sqrt{\tilde{\alpha}}/R$. The exact values of the scattering length are equal to, respectively,[344]

$$\text{a) } a_0 = R(\xi \cot \xi - 1), \quad \text{b) } a_0 = R\sqrt{\xi^2 + 1}\cot\left(\pi\frac{\sqrt{\xi^2+1}}{2}\right). \tag{9}$$

As is seen, for $\xi \to \infty$ the quasi-classical and exact expressions for the scattering length coincide. Their difference scales as $1/\xi$ for finite values of ξ, and is due to quasi-classical corrections. A calculation of such corrections is more difficult (compare to Problem 13.34), since it is necessary to take into account the next term in $\hbar$ in the phase of the wavefunction at finite distances, *i.e.*, in the quasi-classical region.

Problem 13.36

Consider a quantum particle moving in the presence of two potentials: a long-range potential, $U_L(r)$, with the radius r_L and a short-range potential, $U_S(r)$, with the radius, $r_S \ll r_L$. Treat the latter short-range potential as a small perturbation and generalize the results of Problem 11.4 for the level shift due to $U_S(r)$ (perturbation theory in terms of the scattering length) for a state with an arbitrary angular momentum, l (as opposed to $l = 0$ in Problem 11.4). Assume that at small distances, $r \lesssim r_S$, the long-range potential is weak ($|U_L| \ll \hbar^2/mr_S^2$) and that

$$\left|E_{n_r l}^{(0)}\right| \ll \frac{\hbar^2}{mr_S^2}.$$

Solution

We use the method described in Problem 11.4. Using ordinary perturbation theory for the potential $U_S(r)$, we obtain

$$\Delta E_{n_r l} = \overline{U_S(r)} = \int U_S(r)|\psi_{n_r lm}^{(0)}(\mathbf{r})|^2 dV. \tag{1}$$

---

[344] This exact solution to the Schrödinger equation could be obtained using substitutions analogous to those in Problem 4.25 b and c.

At the relevant small distances (where the potential, $U_S(r)$, has an effect), we can use the following expression for the unperturbed wavefunction:

$$\psi^{(0)}_{n_r lm}(\mathbf{r}) = R_{n_r l}(r) Y_{lm}(\mathbf{n}), \quad R_{n_r l}(r) \approx Q_{n_r l} r^l \quad \text{for} \quad r \ll r_L.$$

Using this, we can express integral (1) in terms of the scattering length with the angular momentum l for the potential $U_S(r)$, in the Born approximation (see Problem 13.28):

$$a_l^B = \frac{\pi m}{2^{2l+1} \Gamma^2(l + 3/2) \hbar^2} \int_0^\infty r^{2l+2} U_S(r) dr.$$

Finally, by replacing a_l^B with the exact scattering length, a_l^S, in the short-range potential, $U_S(r)$, we obtain the level shift expressed in terms of the scattering length:

$$\Delta E_{n_r l} = \frac{\hbar^2}{2m} [(2l+1)!!]^2 Q_{n_r l}^2 a_l^{(S)}. \tag{2}$$

We used the fact that $2^{l+1} \Gamma(l + 3/2) = \sqrt{\pi}(2l+1)!!$. Also, note a simplification that occurs for the s-states: $Q_{n0}^2 = 4\pi \psi_{n0}^{(2)}(0)$.

We now discuss the main result – Eq. (2).

1) Since, generally speaking, $a_l^{(S)} \propto r_S^{2l+1}$ while $Q_{n_r l}^2 \propto r_L^{-2l-1}$, then with an increase of l, level shifts decrease fast, $\propto (r_S/r_L)^{2l+1}$. This is connected to a decrease in the penetrability of the centrifugal barrier that divides the regions, where the short-range and long-range potentials act.

2) If the short-range interaction gives rise to inelastic processes (for example, annihilation into pions due to nuclear interaction for the $p\bar{p}$-hadron atom; see also Problem 11.74), then the scattering length, $a_l^{(S)}$, acquires an imaginary part. In this case, $\Delta E_{n_r l}$ describes not only a shift in energy, but also a widening of the level. The level becomes quasi-stationary and has a finite lifetime.

Let us note that the level width

$$\Gamma_{n_r l} = -2 \, \text{Im} \, \Delta E_{n_r l} \propto \text{Im} \, a_l^{(S)}$$

is connected with the inelastic scattering cross-section (the reaction cross-section), $\sigma_{rl}^{(S)}$, in the partial wave with the angular momentum, l due to the short-range interaction of slow particles, since

$$-\text{Im} \, a_l^{(S)} = \frac{\sigma_{rl}^{(S)}(k)}{4} (2l+1) \pi k^{2l-1} |_{k \to 0}. \tag{3}$$

Note that the long-range potential can also strongly affect the reaction cross-section for slow particles (see the following problem).

3) As mentioned in Problem 11.4, for the case of angular momentum $l = 0$, if the condition $|a_0^{(S)}| \ll r_L$ is broken, then Eq. (2) is not applicable. In this case, large shifts of the s-levels are possible in the long-range potential, *i.e.*, a reconstruction of the spectrum occurs (see also Problem 9.3). In the case of $l \neq 0$ we have a different situation and there are no large level shifts, which is due to a low-transparency of the centrifugal barrier. But in the case of a large scattering length, $|a_l^{(S)}| \gg r_S^{2l+1}$, when there is a shallow level with the angular momentum l in the potential, Eq. (2) does require modification. Here, we note that the solution of the Schrödinger equation in the potential $U_S(r)$ for slow particles, $kr_S \ll 1$, and for the distances $r_s \ll r \ll (k^{-1}, r_L)$, has the form:

$$R_{kl} \approx C\left[r^l + B_l(k)\frac{1}{r^{l+1}}\right], \qquad (4)$$

where

$$B_l^{-1} = \frac{1}{(2l-1)!!(2l+1)!!}k^{2l+1}\cot\delta_l^{(S)}(k).$$

The substitution $k^{2l+1}\cot\delta_l^{(S)} \to -1/a_l^{(S)}$, which leads to Eq. (2) for the level shift (it could be more formally obtained, as in Problem 11.4 for the angular momentum $l = 0$), corresponds to the non-resonant case, where $a_l^{(S)} \leq r_S^{2l+1}$. A generalization of this equation to the resonant case, with the effective range expansion (XIII.15) taken into account, is obtained by replacing $a_l^{(S)}$ with[345]

$$a_l^{(S)} \to \left[\frac{1}{a_l^{(S)}} - r_l^{(S)}E\right]^{-1}. \qquad (5)$$

Generally speaking, we can put $E = E_{n_r l}^{(0)}$. For $l \neq 0$, the second, singular term in Eq. (4) (which is independent of $a_l^{(S)}$ and scales $\propto 1/r^{l+1}$) is small in the region $r \sim r_L$. This ensures the validity of Eqs. (2) and (5) and yields the corresponding modification of the expression for level shift,

$$\Delta E_{n_r l} = \frac{1}{2}[(2l+1)!!]^2 Q_{n_r l}^2 \left(\frac{1}{a_l^{(S)}} - r_l^{(S)}E_{n_r l}^{(0)}\right)^{-1}. \qquad (6)$$

The importance of the $l \neq 0$ condition to the validity of Eq. (6) in the resonant case is connected to a large effective radius that makes the level shifts small, since

---

[345] Below in this footnote $\hbar = m = 1$, so that $E = k^2/2$. Note that Eq. (4) can be viewed as a boundary condition for the Schrödinger equation with potential $U_L(r)$ at small distances, which is due to the appearance of the short-range potential $U_S(r)$ (compare to Problem 11.4). But in the case $l \neq 0$, to take the limit $r \to 0$ is impossible due to the divergence of the normalization integral (since it gives $\psi \propto 1/r^{l+1}$ at short distances $r \leq r_s$, the wavefunction coincides with that in the short-range potential). In the case $l = 0$, the limit $r_s = 0$ is possible and corresponds to approximating of the short-range potential by a zero-range potential. See Problem 4.10.

$r_l^{(S)} \propto r_S^{1-2l}$ (see Problem 13.44). The substitution $E = E_{n_r l}^{(0)}$ used in Eqs. (5) and (6) is invalid only if

$$E_S^{(0)} \equiv \frac{1}{a_l^{(S)}} r_l^{(S)} \approx E_{n_r l}^{(0)}. \tag{7}$$

This is because $E_S^{(0)}$ describes a level with the angular momentum l that exists in an isolated potential, $U_S(r)$. Condition (7) is fulfilled when there are two close levels, related to both potentials. If we do not substitute $E_{n_r l}^{(0)}$ for E in Eq. (6), we obtain a equation for $\Delta E_{n_r l} = E - E_{n_r l}^{(0)}$, and solve it as follows:

$$E_{1,2} = \frac{1}{2} \left\{ E_S^{(0)} + E_{n_r l}^{(0)} \pm \left[(E_S^{(0)} - E_{n_r l}^{(0)})^2 + \frac{1}{|r_l^{(S)}|} ((2l+1)!!)^2 Q_{n_r l}^2 \right]^{1/2} \right\}. \tag{8}$$

This gives the energy of these levels with their interaction taken into account. Note that this equation describes "quasi-crossing" levels. In the case of $E_S^{(0)} > E_{n_r l}^{(0)}$, one of the roots of Eq. (8) describes a shift of the unperturbed level $E_S^{(0)}$ upwards under the influence of the long-range potential,[346] $U_L(r)$, while the second one describes a level shift downward of $E_{n_r l}^{(0)}$ under the influence of the short-range potential, $U_S(r)$. In the case $E_S^{(0)} < E_{n_r l}^{(0)}$, their roles are switched.

Finally, we discuss the form of wavefunctions in the presence of a level quasi-crossing. For a sufficiently large difference between $E_S^{(0)}$ and $E_{n_r l}^{(0)}$, with $E_S^{(0)} > E_{n_r l}^{(0)}$, the wavefunction that corresponds to the first of the levels in Eq. (8) is approximately $\psi_S^{(0)}$, localized at small distances $r \lesssim r_s$, while the wavefunction that corresponds to the second level in Eq. (8) is approximately $\psi_{n_r l}^{(0)}$ localized at large distances, $r \sim r_L$. In the case $E_S^{(0)} < E_{n_r l}^{(0)}$, wavefunctions corresponding to the two levels (in the essential localization domain) change places. In the case of an exact resonance $E_S^{(0)} = E_{n_r l}^{(0)}$, the quantum particle occurs in the two states with an equal probability (equal to 1/2); *i.e.*, has the same probability, 1/2, to be bound in either the localization domains of the potential $U_S(r)$ (for $r \lesssim r_S$) and or that of $U_L(r)$ (for $r \sim r_L$). Wavefunctions of the corresponding states are the same for $r \lesssim r_s$ and differ by a sign in the region $r \gg r_s$, which gives rise to their orthogonality.

---

[346] Strictly speaking, here we reproduce only the part of the shift that is connected to the action of $U_L(r)$ at large distances, $r \gg r_s$. The influence of U_L on the level shift $E_S^{(0)}$ at short distances manifests itself in a renormalization of the parameters $a_l^{(S)}$, $r_l^{(S)}$. For example, if $U_L(r) \approx U_0$ for $r \leq r_S$, then it would be more accurate to replace E in Eq. (5) by $E - U_0$. Such a replacement corresponds to a renormalization of the scattering length, *i.e.*, replacing $a^{(S)}$ by $a_l^{(SL)}$, $1/a_l^{(SL)} = 1/a_l^{(S)} + r_l^{(S)} U_0$. The effective radius is not renormalized in this case. See also Problem 13.42.

Problem 13.37

A particle is moving in a combined field of two potentials: a strong short-range potential $U_S(r)$ with the radius r_S and a long-range potential $U_L(r)$ with the radius $r_L \gg r_S$, which is weak at small distances $r \lesssim r_S$: $|U_L| \ll \hbar^2/mr_S^2$. Assuming that the solution of the Schrödinger equation for the potential $U_L(r)$ is known, find a change of the phase shift, $\Delta\delta_l^{(S)}$, in this potential under the influence of U_S, in the case of slow particles, so that $kr_S \ll 1$. Express $\Delta\delta_l^{(S)}$ in terms of the scattering length $a_l^{(S)}$ in the potential U_S.

Under what conditions is the phase shift in the field $U = U_S + U_L$ approximately equal to the sum of phase shifts in the potentials U_S and U_L separately?

Apply this result to the long-range Coulomb potential.

Solution

We are going to use the methods developed in the previous problem: (1) find a change of the phase shift, $\Delta\delta_l^{(S)}$, by perturbation theory, (2) express it in terms of the scattering length by a short-range potential in the Born approximation, and (3) replace the scattering length by the exact scattering length $a_l^{(S)}$ in the "strong" potential $U_S(r)$. A formal justification of this method is provided in Problem 11.4.

To determine the shift, $\Delta\delta_l^{(S)}$, under the influence of $U_S(r)$ by perturbation theory, we write two Schrödinger equations for the radial wavefunctions, $\psi_{klm} = \chi_{kl} Y_{lm}/r$, of the continuous spectrum:

$$\chi_{kl}^{(0)\prime\prime} + \left[k^2 - \frac{l(l+1)}{r^2} - \frac{2m}{\hbar^2} U_L(r)\right] \chi_{kl}^{(0)} = 0,$$

$$\chi_{kl}^{\prime\prime} + \left[k^2 - \frac{l(l+1)}{r^2} - \frac{2m}{\hbar^2}(U_L(r) + U_S(r))\right] \chi_{kl} = 0. \qquad (1)$$

The solutions of these equations, which correspond to the boundary condition $\chi(0) = 0$, have the following asymptotes at large distances:

$$\chi_{kl}^{(0)} \approx \sin\left(kr \pm \frac{Z}{ka_B}\ln 2kr - \frac{\pi}{2}l + \delta_l^{(L)}\right),$$

$$\chi_{kl} \approx \sin\left(kr \pm \frac{Z}{ka_B}\ln 2kr - \frac{\pi}{2}l + \delta_l^{(L)} + \Delta\delta_l^{(S)}\right). \qquad (2)$$

For concreteness, we have considered the case where at large distances: $U_L \approx \mp Ze^2/r$, $a_B = \hbar^2/me^2$.

Multiplying the first of the equations in (1) by χ_{kl} and the second by $\chi_{kl}^{(0)}$, subtracting them term by term, and integrating over r from 0 to ∞, we obtain:

$$k \sin \Delta \delta_l^{(S)} = -\frac{2m}{\hbar^2} \int_0^\infty U_S(r) \chi_{kl}(r) \chi_{kl}^{(0)}(r) dr. \tag{3}$$

We can approximate the sine-function by its argument, and χ_{kl} by $\chi_{kl}^{(0)}$. Taking into account the short-range nature of the potential U_S and also the relation $\chi_{kl}^{(0)} \approx Q_{kl} r^{l+1}$ for $r \to 0$, we find:

$$\Delta \delta_l^{(S)} = -\frac{2m}{\hbar^2 k} Q_{kl}^2 \int_0^\infty r^{2l+2} U_S(r) dr. \tag{4}$$

The integral here is expressed in terms of the scattering length with angular momentum l in the Born approximation (see the previous problem). Now, we replace it by the exact scattering length, and obtain the desired result:

$$\Delta \delta_l^{(S)}(k) = -[(2l+1)!!]^2 a_l^{(S)} \frac{Q_{kl}^2}{k}. \tag{5}$$

We now analyze Eq. (5):

1) Let us first note that in the case $U_L \equiv 0$, we have

$$\chi_{kl}^{(0)} = \sqrt{\frac{\pi k r}{2}} J_{l+1/2}(kr).$$

Here,

$$Q_{kl}^2 = \frac{k^{2l+2}}{[(2l+1)!!]^2},$$

so that

$$\Delta \delta_l^{(S)} = -a_l^{(S)} k^{2l+1}, \tag{6}$$

i.e., Eq. (6) coincides with the scattering phase $\delta_l^{(S)}$ from the isolated potential $U_S(r)$, as it should be.

2) The same relation, $\Delta \delta_l^{(S)} = \delta_l^{(S)}$, holds approximately also in the case of a long-range potential $U_L(r)$ that can be considered as perturbation (see conditions (XIII.7)). In this case, the scattering phase in the potential, which is the superposition $U_S(r) + U_L(r)$, is equal to the sum of the scattering phases for each of them separately.

3) Let us finally focus on the case of the long-range Coulomb potential, $U_L = \mp Ze^2/r$. Here,

$$\chi_{kl}^{(0)} = \frac{Zr}{a_B} C_{kl}^{(\mp)} \frac{(2kr)^l}{(2l+1)!} e^{\mp ikr} F\left(i\frac{Z}{ka_B} + l + 1, 2l+2, \pm 2ikr\right),$$

$$\delta_l^{(L)} = \arg \Gamma\left(l + 1 \mp \frac{iZ}{ka_B}\right),$$

$$C_{kl}^{(\mp)} = 2k' e^{\pm \pi/2k'} \left|\Gamma\left(l + 1 \mp \frac{i}{k'}\right)\right| =$$

$$e^{\pm \pi/2k'} \left(\frac{\pi k'}{\sinh \pi k'}\right)^{1/2} \prod_{s=1}^{l} \sqrt{s^2 + \frac{1}{(k')^2}}.$$

We let $k' = ka_B/Z$, so that

$$(Q_{kl}^{(\mp)})^2 = \frac{\pi Z (2k)^{2l+1}}{[(2l+1)!]^2 a_B} \frac{(\mp 1)}{e^{\mp 2\pi Z/ka_B} - 1} \prod_{s=1}^{l} \left[s^2 + \left(\frac{Z}{ka_B}\right)^2\right]. \tag{7}$$

For $l = 0$, the products are replaced by 1.

In the case of "fast" particles, when $ka_B \gg Z$ (but still $kr_s \ll 1$), the Coulomb potential could be considered as a perturbation, and from Eqs. (5) and (7), the results discussed above in 1) and 2) follow.

A completely different situation arises for $ka_B \leq Z$. Here, values of $Q_{kl}^{(\mp)}$ differ greatly from the unperturbed values given in 1). Particularly for $ka_B \ll Z$, according to Eqs. (5) and (7), for an s-wave we obtain

$$\text{a)} \quad \Delta\delta_0^{(S)} = \frac{2\pi Z}{ka_B} \delta_0^{(S)}; \quad \text{b)} \quad \Delta\delta_0^{(S)} = \frac{2\pi Z}{ka_B} e^{-2\pi Z/ka_B} \delta_0^{(S)} \tag{8}$$

for the attractive and repulsive Coulomb potential respectively. The essential change of the phase shift value (its increase by $2\pi Z/ka_B \gg 1$ in the case of attraction, and its exponential decrease in the case of repulsion) has a clear physical underpinning: in the case of slow particles the long-range Coulomb attraction (repulsion) greatly increases (decreases) the probability of finding particles at small distances. This also strongly affects the cross-sections of inelastic processes caused by a short-range interaction, that accompany collisions of slow particles.

In conclusion, let us note that all of this assumes that the scattering from the short-range potential is non-resonant, i.e., there is no "shallow" level in the potential $U_S(r)$. Here $a_l^{(S)} \leq r_S^{2l+1}$ and $\delta_l^{(S)} \approx -a_l^{(S)} k^{2l+1} \ll 1$. However, by substituting the scattering length $a_l^{(S)}$, as in the previous problem, with the expression

$$\left(\frac{1}{a_l^{(S)}} - \frac{1}{2} k^2 r_l^{(S)}\right)^{-1}, \tag{9}$$

Eq. (5) could be generalized to include the resonant case. The condition of its applicability is $\Delta\delta_l^{(S)} \ll 1$ (which was used to transform Eq. (3)). Note that for a repulsive long-range potential, this condition may remain satisfied even if the scattering phase $\delta_l^{(S)}$ for the isolated short-range potential $U_S(r)$ is not small.[347]

Problem 13.38

How should we modify the *Rutherford formula* to describe the differential scattering cross-section in the Coulomb potential $U = \pm Ze^2/r$, which is "distorted" at small distances $r \lesssim r_S$? Assume that the conditions $kr_S \ll 1$ and $Ze^2 \ll \hbar v$ are fulfilled. The distortion of the Coulomb field is described by a potential $U_S(r)$, with a known scattering length, $a_0^{(S)}$.

Solution

The scattering amplitude is described by the expression:

$$f \approx f_{Coul} + f_S; \quad f_{Coul} = \pm \frac{Ze^2 m}{2\hbar^2 k^2 \sin^2(\theta/2)}, \quad f_S = -a_0^{(S)}. \tag{1}$$

This follows from the fact that we only need to consider influence of the short-range potential, $U_S(r)$, on the particles with the angular momentum $l = 0$ (since $kr_s \ll 1$), and that we can consider the Coulomb potential as a perturbation (since $Ze^2 \ll \hbar v$). According to the previous problem, the phase shift in the total potential field (the superposition $U_{Coul} + U_S$) is approximately equal to the sum of phase shifts for each of the potentials separately.

The differential scattering cross-section is described by

$$\frac{d\sigma}{d\Omega} = \left(\frac{Ze^2}{2mv^2}\right)^2 \frac{1}{\sin^4(\theta/2)} + \left(a_0^{(S)}\right)^2 \mp \frac{Ze^2 a_0^{(S)}}{mv^2 \sin^2(\theta/2)}. \tag{2}$$

The last term corresponds to an interference of the scattering amplitudes due to the Coulomb and the short-range interactions. As is seen, its character depends of the sign of the scattering length $a_0^{(S)}$.

Problem 13.39

Find the scattering length, a_l, of a particle with an arbitrary angular momentum, l for the following potentials:

a) an impenetrable sphere with radius R;
b) $U(r) = -\alpha\delta(r - R)$;
c) a square well of radius R and depth U_0.

Compare to the case of $l = 0$ from Problem 13.31.

---

[347] However, if the scattering length is large, a renormalization of the parameters describing low-energy scattering may become significant. See the previous problem, and also Problem 13.42.

Solution

The scattering length, a_l, could be found from the asymptote of the radial wavefunction, $R_{kl}(r)$, for $E = 0$:

$$R_{0l} \approx r^l - \frac{1}{r^{l+1}}\{(2l-1)!!(2l+1)!!a_l\}, \quad \text{for } r \to \infty.$$

This follows, for example, from a comparison of the expression for the wavefunctions of slow particles at distances $d \ll r \ll 1/k$:

$$R_{kl} \approx r^l + B_l(k)\frac{1}{r^{l+1}}, \quad B_l^{-1} = \frac{k^{2l+1}\cot\delta_l}{(2l-1)!!(2l+1)!!},$$

with the effective range expansion (XIII.15).

Below, we present the final results:

a) for scattering from the impenetrable sphere:

$$a_l = \frac{1}{(2l-1)!!(2l+1)!!}R^{2l+1} \equiv a_{sph,l}; \tag{1}$$

b) for scattering from the δ-well:

$$a_l = \frac{\xi}{\xi - 2l - 1}a_{sph,l}, \quad \xi = \frac{2m\alpha R}{\hbar^2}; \tag{2}$$

c) for scattering from the square potential well:

$$a_l = -\frac{J_{l+3/2}(\lambda)}{J_{l-1/2}(\lambda)}a_{sph,l}, \quad \lambda - \sqrt{\frac{2mU_0 R^2}{\hbar^2}}. \tag{3}$$

Note that the properties of the scattering length, a_l, as a function of the parameters of the potentials (in cases b) and c)) are similar to those considered in Problem 13.31 for an s-wave. In particular, when a new bound state appears, the scattering length, a_l, becomes infinite.

Problem 13.40

Estimate the singlet (*i.e.*, the total electron spin is $S = 0$) electron s-scattering length, $a_0(1)$, for scattering off of an unexcited hydrogen atom. Keep in mind that there exists a weakly-bound state, the ion H^-, with the binding energy $\varepsilon_0 = 0.754$ eV $= 0.0277$ a.u.

a) Neglect the finite-size effects for both the atomic size and the interaction range of the external electron with the atom.
b) Consider the external electron as weakly bound in a finite-range potential, and use the following asymptotic coefficient $C^2_{\kappa 0} = 2.65$ (see Problem 11.36).

Compare to the result of a variational calculation: $a_0(1) = 5.97$ a.u.

Solution

a) In this approximation, which corresponds to considering the external electron as in a zero-range potential, we have (see Problems 13.20 and 4.10):

$$a_0 = \kappa_0^{-1} = (2\varepsilon_0)^{-1/2} = 4.25 \tag{1}$$

(in atomic units).

b) Considering the external electron as weakly bound in the potential of finite radius r_S, we first use the relation between the effective interaction radius r_0 and the asymptotic coefficient:

$$r_0 = \frac{C_{\kappa 0}^2 - 1}{\kappa_0 C_{\kappa 0}^2}.$$

Now, taking into account the effective-range expansion, Eq. (XIII.15), and the fact that the scattering amplitude, as a function of energy, has a pole at $E = -\varepsilon_0$ (Note that at the pole, $k \cot \delta_0 = ik = -\kappa_0$), we find:

$$a_0 = \frac{2C_{\kappa 0}^2}{\kappa_0(1 + C_{\kappa 0}^2)} = 6.17. \tag{2}$$

As is seen, the correction to the effective radius strongly affects the value of the scattering length. This is because $\kappa_0 r_s \approx 0.6$ is not that small. In connection to this, let us mention the role of the subleading, $\propto k^4$, term in the effective-range expansion, Eq. (XIII.15). It is usually written as

$$-Pr_0^3 k^4, \tag{3}$$

A typical value of this parameter, P, here is small, $|P| \lesssim 0.1$. Taking this into account, we should see that the scattering length (2) is determined to within a few percent accuracy.

Problem 13.41

In a proton–neutron system, estimate the triplet s-wave scattering length, $a_0(3)$, taking into account the existence of a weakly bound state, the deuteron, with the binding energy $\varepsilon_0 = 2.23$ MeV. Compare to the experimental value, $a_0(3) = 5.39 \cdot 10^{-13}$ cm.

Solution

Neglecting the finiteness of the effective range of the interaction, we have

$$a_0(3) = \kappa_0^{-1} = \left(\frac{\hbar^2}{2\mu\varepsilon_0}\right)^{1/2} = 4.3 \cdot 10^{-13} \text{ cm}, \tag{1}$$

where $\mu \approx m_p/2$ is the reduced mass of the pn-system. Since the range of the nuclear forces is $r_s \sim 10^{-13}$ cm, then $\kappa_0 r_s \sim 0.3$, Equation (1) has the same accuracy: about 20 %. If we include the term with the effective range ($r_0(3) = 1.7 \cdot 10^{-13}$ cm) in expansion (XIII.15), we reproduce the experimental value of the triplet scattering length. Compare with the previous problem.

Problem 13.42

a) Using the experimental value of the singlet s-wave scattering length, $a_0(1) = -23.7 \cdot 10^{-13}$ cm, for a proton–neutron system, estimate the energy of the shallow virtual level[348] in such a system in the state with $S = 0$ and $l = 0$.

b) For a proton–proton system, $a_0(1) = -7.77 \cdot 10^{-13}$ cm. Does the significant difference between the scattering lengths for the pn- and pp-systems contradict the isotopic invariance of the nuclear interaction? The effective range of the interaction $r_0(1)$, is equal to $2.67 \cdot 10^{-13}$ cm and $2.77 \cdot 10^{-13}$ cm, respectively, for the pn- and pp-systems.

Solution

a) Neglecting the effective range of the interaction, we obtain (compare to Problems 13.20 and 4.10, and also Problem 13.40)

$$\varepsilon_{virt} \approx \frac{\hbar^2}{2\mu a_0^2(1)} = 56 \text{ keV}.$$

Here, $\mu = m_p/2$ is the reduced mass of the pn-system.

b) Since the Coulomb interaction at small distances is weaker than the nuclear forces by approximately two orders of magnitude, then naïvely we might assume that the difference of the low-energy scattering parameters for the pp-system from those for the pn-system should remain to within a few percent. This is indeed so specifically for the effective range of the interaction.

The scattering lengths however do differ noticeably: by a factor of three. But this does not imply that isotopic invariance of nuclear interactions is strongly broken, since the difference could be explained by the Coulomb interaction in the pp-system. In the pn-system there is a shallow virtual level, and the scattering length is large (approximately twenty times larger than the interaction radius!). In such conditions, this function depends sharply on the parameters of the potential (see Problem 13.31), and differs greatly for even a small change of those (in this problem, due to the Coulomb interaction).

Let us provide a simple estimate of the *renormalization* of the scattering length. We write the pn-interaction in the form $U_{pn} = U_0(r) + \delta U(r)$, where $U_0(r)$ corresponds to the threshold of the appearance of a new bound state, here $\delta U(r) \geq 0$, since the level in the pn-system is virtual. For the pp-system, considering protons to be point-like, we have $U_{pp} = U_0 + \delta U + e^2/r$. Using the result of Problem 4.27 for a shallow s-level depth, we find:

---

[348] The virtual nature of the level and the sign of $a_0(1) < 0$ both follow from the absence of a real bound state. Accounting for a finite effective range, $r_0(1)$, we obtain $\varepsilon_{virt} = 67$ keV.

$$\kappa_{pp} = -\frac{2\mu}{\hbar^2} \int_0^d \left(\delta U(r) + \frac{e^2}{r} \right) \chi_0^2(r) dr. \tag{1}$$

We use $\chi_0(r)$ as the wavefunction ($\chi = rR$) at the threshold of the appearance of a new bound state, normalized by the condition $\chi_0(r) = 1$ outside of the potential range. The upper integration limit is $d \sim a_B = \hbar^2/m_p e^2 \approx 29 \cdot 10^{-13}$ cm. (At larger distances, a contribution coming from the Coulomb interaction is taken into account independently, and appears in the amplitude of Coulomb scattering of the protons.) The integral of $\delta U(r)$ in Eq. (1) gives κ_{pn} (in both systems $\kappa < 0$, since the levels are virtual). To estimate the integral of the Coulomb potential, we put $\chi_0^2(r) = 1$. Due to a divergence in the lower limit, we introduce a cutoff at $r_s \approx 10^{-13}$ cm (of the order of the nuclear interaction radius). Finally, taking into account the relations $\kappa_{pn} = 1/a_0^{pn}(1)$ and $\kappa_{pp} = 1/a_0^{pp}(1)$, according to Eq. (1), we obtain:

$$\frac{1}{a_0^{pp}(1)} - \frac{1}{a_0^{pn}(1)} \approx -\frac{1}{a_B} \ln\left(\frac{a_B}{r_0}\right). \tag{2}$$

Here, instead of r_s, we put the effective interaction radius, r_0. Let us emphasize that the uncertainty in parameters d and r_s, for which $d/r_s \gg 1$, is "softened" in the final result because of the logarithm. So, we see that accounting for the protons' "weak" Coulomb interaction explains the significant difference of the scattering lengths for the pp- and pn-systems.

Problem 13.43

Prove the following relation for the effective range of the interaction (see, Eq. (XIII.15)):

$$r_0 = 2 \int_0^\infty \left\{ \left(-\frac{r}{a_0} + 1 \right)^2 - \chi_0^2(r) \right\} dr,$$

where $\chi_0(r)$ is the radial wavefunction ($\chi_0 = rR_0$) of a state with $l = 0$ and $E = 0$, normalized by the condition $\chi_0(r) = (-r/a_0 + 1)$ for $r \to \infty$ (a_0 is the scattering length).

Find r_0 for an impenetrable sphere of radius R, and also for a δ-well, $U(r) = -\alpha\delta(r - R)$ and a square well of radius R at the threshold for bound state formation ($a_0 = \infty$) in the two latter cases.

Solution

We denote the radial wavefunctions for the values of energy $E = 0$ and $E = \hbar^2 k^2/2m$ by χ_0 and χ, respectively. They satisfy the following equations:

$$\chi_0'' - \tilde{U}(r)\chi_0 = 0, \quad \chi'' - [\tilde{U}(r) - k^2]\chi = 0; \quad \tilde{U} = \frac{2m}{\hbar^2} U(r). \tag{1}$$

To normalize $\chi_0(r)$, use the condition $\chi_0(r) \approx (1 - r/a_0)$ for $r \gg d$ (d is the potential radius). For wavefunction $\chi(r)$ at distances $d \ll r \ll 1/k$, we have

$$\chi(r) = k \cot \delta_0 \cdot r[1 + O(k^2 r^2)] + [1 + O(k^2 r^2)], \quad (2)$$

where $O(k^2 r^2)$ corresponds to corrections in the asymptote.

Multiplying the first equation of (1) by χ, the second one by χ_0, subtracting them term by term, and integrating in the region where expansion (2) is valid, we obtain:

$$\int_0^r \frac{d}{dr}\{\chi_0' \chi - \chi_0 \chi'\} dr \equiv -\frac{1}{2} r_0 k^2 + O\left(k^2 r, k^2 \frac{r^2}{a_0}, k^2 \frac{r^3}{a_0^2}\right) =$$

$$k^2 \int_0^r \chi(r) \chi_0(r) dr. \quad (3)$$

Here, we have taken into account the asymptotes given above, the boundary condition $\chi(0) = 0$, and the effective range expansion (XII.15). The integral on the right-hand side of the expression can be cast into the form

$$\int_0^r \chi \chi_0 dr \approx \int_0^r \left[\chi_0^2 \pm \left(-\frac{r}{a_0} + 1\right)^2\right] dr =$$

$$= \int_0^r \left[\chi_0^2 - \left(-\frac{r}{a_0} + 1\right)^2\right] dr + \frac{r^3}{3a_0^2} - \frac{r^2}{a_0} + r$$

(here we have used $\chi \approx \chi_0$; the notation $\pm(-r/a_0 + 1)^2$ means an addition or subtraction of the corresponding term).

We have used the fact that $\chi \approx \chi_0$. In the last integral we could extend the integration to infinite limits. Then, leaving only r independent terms in Eq. (3), we obtain

$$r_0 = 2 \int_0^\infty \left\{\left(-\frac{r}{a_0} + 1\right)^2 - \chi_0^2(r)\right\} dr. \quad (4)$$

For the scattering from an impenetrable sphere of radius R we have $\chi_0 = 1 - r/R$ for $r > R$ and $\chi_0 = 0$ for $r < R$. In this case, the scattering length is $a_0 = R$, and according to Eq. (4) we find the effective range of the interaction, $r_0 = \frac{2}{3}R$.

For $a_0 = \infty$ from Eq. (4) the well-known result for effective range, r_0, at the threshold of s-level appearance follows. In particular, for the scattering from a δ-well with $a_0 = \infty$, we have $\chi_0 = 1$ for $r > R$ and $\chi_0 = r/R$ otherwise. According to Eq. (4) with $a_0 = \infty$, we obtain $r_0 = \frac{4}{3}R$. For the scattering from a square potential well $r_0 = R$, and the effective range is the same whenever a new bound state appears (independently of its sequence number). Here, for $r < R$, the wavefunction is

$$\chi_0 = \sin\left[\pi\left(n_r + \frac{1}{2}\right)\frac{r}{R}\right],$$

where $n_r = 0, 1, 2, \ldots$ is the sequence number of bound state that appear with a deepening of the well.

Note that both the value $r_0 \sim R$ and the sign $r_0 > 0$ of the "critical" effective range (at the threshold of appearance of a new s-state) are quite general for attractive potentials, $U(r) \leq 0$. Compare to the case of angular momenta $l \neq 0$ in Problem 13.44, and also to the case $l = 0$ for a well surrounded by a potential barrier, in Problem 13.47.

Problem 13.44

Prove that the effective range of the interaction r_l, in a state with $l \neq 0$ (see Eq. (XIII.15)) and at the threshold of bound-state[349] formation, is

$$r_l = -2[(2l-1)!!]^2 \frac{1}{C_l^2},$$

where C_l is the normalization coefficient in the wavefunction of the zero-energy state $(E = 0)$:

$$\chi_l^{(0)}(r) \approx C_l \frac{1}{r^l} \quad \text{for} \quad r \to \infty \quad \text{and} \quad \int_0^\infty \left(\chi_l^{(0)}(r)\right)^2 dr = 1.$$

Find r_l for a δ-well.

Solution

We will follow the method outlined in the previous problem. Taking into account that the wavefunction when a new bound state is just about to appear has the form $\chi_l^{(0)} = C_l r^{-l}$ for $r \gg d$, and also the relation

$$\chi_{kl} \approx C_l\left[r^{-l} + \frac{k^{2l+1}\cot\delta_l}{(2l-1)!!(2l+1)!!}r^{l+1}\right], \quad d \ll r \ll \frac{1}{k}$$

(compare, for example, to Problem 13.39), we find:

$$-\frac{1}{[(2l-1)!!]^2}C_l^2 k^{2l+1}\cot\delta_l = k^2\int_0^r \chi_{kl}\chi_l^{(0)}, \quad d \ll r \ll \frac{1}{k}.$$

Now, if we replace χ_{kl} by $\chi_l^{(0)}$ in the right-hand side (due to a small value of k^2), extend the integration to infinity (in the case $l = 0$, we cannot do this due to a divergence of

---

[349] This case is interesting, since in the absence of a shallow level in the potential, the term containing the effective range in Eq. (XIII.15) acts as a small correction.

the integral at the upper limit) and use the relation $k^{2l+1}\cot\delta_l = r_l k^2/2$ (since $a_l = \infty$ at the threshold), then we obtain:

$$r_l = -2[(2l-1)!!]^2 C_l^{-2}, \quad l \geq 1. \tag{1}$$

The value of C_l^2 is uniquely determined by the normalization of the radial wavefunction, $\chi_l^{(0)}$, to 1.

Unlike in the case of s-scattering, the effective range of the interaction is negative, $r_l < 0$, and is of the order of $|r_l| \sim R^{1-2l}$, where R is the radius of the potential (which is of the same magnitude as the localization domain of a zero-energy bound-state with the angular momentum $l \neq 0$).

For a δ-well, the wavefunction at the threshold of appearance of a new bound state with the angular momentum, l, has the form:

$$\chi_l^{(0)} = \begin{cases} C_l r^{-l}, & r > R, \\ C_l \frac{r^{l+1}}{R^{2l+1}}, & r < R. \end{cases}$$

From the normalization condition, we find C_l^2 and the effective range of the interaction is:

$$r_l = -\frac{4(2l-3)!!(2l+1)!!}{2l+3} R^{1-2l}, \quad l \geq 1.$$

In conclusion, note that a change in the effective range as a result of a small change in the potential is also small. So, the "critical" value of r_l at the threshold of bound-state formation is applicable also in the case of a shallow level (real and quasi-discrete).

Problem 13.45

Find the phase shift, $\delta_0(k)$, and scattering cross-section for scattering of slow particles in the following potentials:

a) an impenetrable sphere of radius R;
b) a δ-well, $U(r) = -\alpha\delta(r-R)$;
c) a square well of radius R and depth U_0.

Use the effective range expansion.

Solution
a) For the radial function $\chi_l = rR_l$ with $l = 0$, the Schrödinger equation and its solution, which corresponds to the boundary condition $\chi_0(R) = 0$, have the form:

$$\chi_0'' + k^2\chi_0 = 0, \quad \chi_0(r) = A\sin[k(r-R)].$$

So, the s-scattering phase shift is $\delta_0(k) = -kR$, and the scattering cross-section for slow particles, when $kR \ll 1$, is described by the expression:

$$\sigma \approx \sigma_{l=0}(k) = \frac{4\pi}{k^2} \sin^2 \delta_0 \approx 4\pi R^2 \left(1 - \frac{1}{3} k^2 R^2\right).$$

Recall that the first correction to the cross-section is connected with p-wave scattering, and is $\propto k^4$. The contribution of higher waves is less important.

In the cases b) and c), it is also possible to find $\delta_0(k)$ from a exact solution to the Schrödinger equation. But it is easier to consider slow particle scattering using the effective range expansion, Eq. (XIII.15). The parameters of low-energy scattering, a_l and r_l, could be obtained using the zero-energy solution of the Schrödinger equation. See Problems 13.31 and 13.43, specifically for a discussion of the s-wave scattering.

b) The solution of the Schrödinger equation for $E = 0$ and $l = 0$ has the form:

$$\chi_0(r) = \begin{cases} r - a_0, & r > R, \\ Cr, & r < R. \end{cases}$$

Matching the solution at the point $r = R$ (see Problem 2.6) gives:

$$a_0 = \frac{\tilde{\alpha} R}{\tilde{\alpha} - 1}, \quad C = 1 - \frac{a_0}{R}, \quad \tilde{\alpha} = \frac{2m\alpha R}{\hbar^2}.$$

If $\tilde{\alpha}$ is not close to 1, then $|a_0| \leq R$, and for slow particles $\sigma \approx 4\pi a_0^2$ (a correction of the order of $k^2 R^2$ to this expression could be found by calculating the effective range, r_0, according to Problem 13.43). However, if $\tilde{\alpha}$ is close to 1, then $|a_0| \gg R$ and the scattering cross-section has a pronounced energy-dependence that is described by resonance equation (XIII.16). For this, the effective range of the interaction for a δ-well at the threshold of s-state formation is $r_0 = \frac{4}{3}R$ (see Problem 13.43). Note that the relation $\sigma \approx 4\pi a_0^2$ needs to be corrected also for values of the parameter $\tilde{\alpha}$ close to $2l + 1$ with $l \geq 1$ (when a bound state with the angular momentum l appears in the system), due to the resonant nature of scattering in the l-th partial wave (see Problem 13.46).

c) The parameters of low-energy scattering, a_0 and r_0, for the square potential well were found in Problems 13.31 and 13.43.

Problem 13.46

Consider scattering of slow particles in the partial wave with the angular momentum, $l \neq 0$, for the potential $U(r) = -\alpha \delta(r - R)$. Consider specifically the case of angular resonant scattering with a quasi-stationary state of small energy, $E_R \ll \hbar^2/mR^2$, and find its width, Γ_R.

Solution

As in the previous problem, we use the effective range expansion, Eq. (XIII.15). The scattering length,

$$a_l = -\frac{1}{(2l-1)!!(2l+1)!!}\frac{\tilde{\alpha}}{2l+1-\tilde{\alpha}}R^{2l+1}, \quad \tilde{\alpha} = \frac{2m\alpha R}{\hbar^2}, \tag{1}$$

was calculated in Problem 13.39. When $\tilde{\alpha}$ is not close to $2l+1$, the scattering in lth partial wave is non-resonant. For slow particles,

$$\sigma_l \approx 4\pi(2l+1)a_l^2 k^{4l},$$

and the dominant scattering is s-wave.

If $\tilde{\alpha}$ is close to $2l+1$, then in the potential there exists a shallow *real* (for $\tilde{\alpha} > 2l+1$) or *quasi-discrete*[350] (for $\tilde{\alpha} < 2l+1$) level. Scattering has a resonant nature, and the term with the following effective range (according to Problem 13.44)

$$r_l \approx -4(2l-3)!!(2l+1)!!\frac{R^{1-2l}}{2l+3} \tag{2}$$

is essential.

The energy spectrum is determined by the poles of the partial scattering amplitude $f_l(E) = 1/k(\cot\delta_l - i)$. At the pole E_l we have $\cot\delta_l(E_l) = i$, and using the effective range expansion, Eq. (XIII.15), we obtain for "shallow" levels:

$$ik_R^{2l+1} = -\frac{1}{a_l} + \frac{1}{2}r_l k_R^2. \tag{3}$$

In the case $l \geq 1$, the left-hand side of this equation is small in comparison to both terms in the right-hand side, so that it could be solved by successive iterations. In the "zeroth" approximation, neglecting the left part in Eq. (3), we obtain:

$$E_R \approx \frac{\hbar^2}{mr_l a_l} \approx \frac{(2l-1)(2l+3)(2l+1-\tilde{\alpha})}{4(2l+1)}\frac{\hbar^2}{mR^2}. \tag{4}$$

Since $r_l < 0$ (for $l \geq 1$), then for $a_l > 0$ (*i.e.*, for $\tilde{\alpha} > 2l+1$) we have $E_R < 0$, so that the level corresponds to a bound state in the δ-well (compare to Problem 4.9). For the scattering of slow particles in this case, $f_l \approx (k\cot\delta_l)^{-1}$, while the corresponding partial scattering cross-section is

---

[350] Since $l \geq 1$, the s-level for $\tilde{\alpha} < 1$ is *virtual*.

$$\sigma_l(E) = 4\pi(2l+1)|f_l|^2 \approx 4\pi(2l+1)\frac{\hbar^4}{m^2 r_l^2}\frac{k^{4l}}{(E+|E_R|)^2}. \tag{5}$$

As is seen, $\sigma_l \propto (kR)^{4(l-1)}R^2$, i.e., in this case the resonant cross-section in the lth partial wave has the same order of magnitude as the non-resonant cross-section in the lower partial wave, with the angular momentum equal to $l-1$. So, except for the case $l=1$, it does not provide a significant contribution to the scattering cross-section.

In the case of $\alpha_l < 0$ (i.e., for the value $\tilde{\alpha} < 2l+1$, $l \geq 1$), the situation is different. Here, according to Eq. (4), $E_R > 0$ and the left part of Eq. (3) is imaginary. The next iteration allows us to obtain the imaginary part of the amplitude's pole, $E_l = E_R - i\Gamma_R/2$, which determines the width of the quasi-stationary state considered:

$$\Gamma_R \approx \frac{2\hbar^2}{m|r_l|}k_R^{2l+1} \sim (k_R R)^{2l+1}\frac{\hbar^2}{mR^2}, \quad k_R = \frac{1}{\hbar}\sqrt{2mE_R}. \tag{6}$$

The dependence of Γ_R on k_R is determined by the energy-dependence of the transmission coefficient of the centrifugal barrier:

$$\Gamma_R \propto D_l \sim \exp\left\{-2\int_R^b \sqrt{\frac{(l+1/2)^2}{r^2} - k_R^2}\, dr\right\} \sim$$

$$\exp\left\{-(2l+1)\ln\frac{l+1/2}{k_R R}\right\} \propto (k_R R)^{2l+1}.$$

$b = (l+1/2)/k_R$ is the quasi-classical turning point. Compare to Problem 9.30.

Now the partial scattering cross-section, $\sigma_l(E)$, is large for energies in a narrow region of width $\sim \Gamma_R$ around E_R, and is

$$\sigma_l(E) \approx \frac{\pi(2l+1)}{k^2}\frac{\Gamma_R^2}{(E-E_R)^2+\Gamma_R^2/4}. \tag{7}$$

It is large, $\sigma_l \sim k_R^{-2} \gg R^2$, and is much larger than the scattering cross-section in the s-wave ($\sigma_0 \sim \pi R^2$), so the total cross-section is $\sigma \approx \sigma_l$. Outside this energy region, for $|E - E_R| \gg \Gamma_R$, the cross-section is described analogously to Eq. (5), with the change $(E+|E_R|)^2 \to (E-E_R)^2$. Therefore, the estimate of the scattering cross-section given above for $a_l > 0$ is the same as for $a_l < 0$ considered here.

Problem 13.47

On the basis of the specific model, including a potential well of strength U_0 and radius R that is surrounded by a δ-barrier $\overline{U}(r) = \alpha\delta(r-R)$ (Fig. 13.4), investigate how the existence of a weakly-penetrating barrier[351] (in this model, $m\alpha R/\hbar^2 \gg 1$) affects scattering of slow particles in the s-state.

Hint: Use the effective range expansion, paying attention to the effect of the weakly penetrable barrier on the effective range of the interaction, r_0.

Solution

We use the effective range expansion and calculate the parameters of low-energy scattering, a_0 and r_0. The wavefunction $\chi_0 = rR_0(r)$ for $E = 0$ and $l = 0$ has the form:

$$\chi_0(r) = \begin{cases} C\sin\kappa_0 r, \ r < R, \ \kappa_0 = \frac{1}{\hbar}\sqrt{2mU_0}. \\ 1 - \frac{r}{a_0}, \quad r > R. \end{cases}$$

Matching the solution at the point $r = R$ (see Problem 2.6), we obtain:

$$C\sin\lambda = 1 - \frac{R}{a_0}, \quad \frac{R}{a_0} + \lambda C \cos\lambda = \tilde{\alpha}\left(\frac{R}{a_0} - 1\right), \tag{1}$$

where $\lambda = \kappa_0 R$ and $\tilde{\alpha} = 2m\alpha R/\hbar^2$. Hence, we find the scattering length:

$$a_0 = R\left(1 - \frac{1}{\tilde{\alpha} + \lambda\cot\lambda}\right). \tag{2}$$

When the inequality $\tilde{\alpha} \gg 1$ is fulfilled, we generally have $a_0 \approx R$, which corresponds to the scattering length of an impenetrable sphere of radius R (a physically intuitive result due to the small penetrability of the barrier).

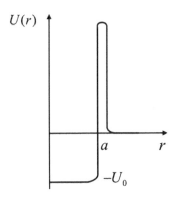

Fig. 13.4

A large difference from the scattering length of an impenetrable sphere appears if λ is close to $n\pi$, with $n = 1, 2, \ldots$. Writing here $\lambda = n\pi + \gamma$, where $|\gamma| \ll 1$, we obtain:

---

[351] For physical applications, especially interesting is the case where such barrier originates from the Coulomb repulsion of charged particles. But due to the slow decrease of the Coulomb potential, this case needs special consideration.

$$a_0 \approx R\left[1 - \frac{1}{\tilde{\alpha} + n\pi/\gamma}\right]. \qquad (3)$$

When $\gamma \approx \gamma_{0,n} = -n\pi/\tilde{\alpha}$, the scattering length is large, $|a_0| \gg R$, and becomes infinite for $\gamma = \gamma_{0,n}$. Such potential parameters correspond to the appearance of a level with energy $E = 0$ (the nth level). In the case of $|a_0| \gg R$, slow particle scattering is resonant.

Now calculate the effective range, r_0. According to Eq. (1), for $a_0 = \infty$ we have $C \approx (-1)^{n+1}\tilde{\alpha}/n\pi$, and from the equation for effective range, r_0, at the threshold of s-level formation,

$$r_0 = 2\int_0^\infty (1 - \chi_0^2(r))dr, \quad \chi_0 = 1 \text{ for } r \to \infty$$

(compare Problem 13.43 for the case $a_0 = \infty$), we find:

$$r_0 = R(2 - C^2) \approx -C^2 R \approx -\left(\frac{\tilde{\alpha}}{n\pi}\right)^2 R. \qquad (4)$$

Unlike the case of s-scattering by a well without a barrier (where $r_0 \sim R$ and $r_0 > 0$; see Problem 13.43), the effective range at the threshold is large, $|r_0| \gg R$, and negative, $r_0 < 0$. Therefore, the term containing the effective range in Eq. (XIII.15) is essential[352] for description of the resonant s-scattering. The situation is analogous to the one that takes place for scattering of particles with non-zero angular momenta, and the physical reason for this similarity is the existence of the weakly-penetrable δ-barrier that for $l = 0$ plays the same role as the centrifugal potential for $l \neq 0$.

Both the amplitude of resonant scattering,

$$f \approx f_0 = \frac{1}{-\frac{1}{a_0} + \frac{1}{2}r_0 k^2 - ik}, \qquad (5)$$

and the scattering cross-section, $\sigma \approx 4\pi|f_0|^2$, have a pronounced energy dependence, whose character depends on the relation among the three small parameters $kR \ll 1$, $R/|a_0| \ll 1$, $R/|r_0| \ll 1$. Let us consider two cases.

1) If $|a_0| \gg |r_0|$, then

$$\sigma \approx \frac{4\pi}{\left(\frac{1}{a_0}\right)^2 + k^2 + \frac{1}{4}r_0^2 k^4}, \qquad (6)$$

so that the scattering cross-section is almost independent of the sign of the scattering length. In the regime with $k \ll 1/|r_0|$, the term containing the effective range, as in the ordinary s-scattering, acts as a correction. It begins to play an important role for $k > 1/|r_0|$.

---

[352] Let us note that $|r_0| \gg R$ holds only of there exists a "shallow" level in the system.

2) In the case of $|a_0| \ll |r_0|$, scattering strongly depends on the sign of the scattering length. For $a_0 > 0$, we have (in the denominator of Eq. (5), we could neglect the $-ik$ term):

$$\sigma(E) \approx 4\pi \left(\frac{\hbar^2}{mr_0}\right)^2 \frac{1}{(E+\varepsilon)^2}, \quad \varepsilon = -\frac{\hbar^2}{ma_0 r_0} > 0. \tag{7}$$

The cross-section is also described by an analogous equation (but with $\varepsilon \equiv -E_R < 0$) if $a_0 < 0$, except for a narrow region of energies[353], $|E - E_R| \sim \Gamma_R$, where

$$\sigma(E) \approx \frac{\pi}{k_R^2} \frac{\Gamma_R^2}{(E-E_R)^2 + \Gamma_R^2/4}, \tag{8}$$

$$\Gamma_R = \frac{2\hbar^2 k_R}{m|r_0|} = \frac{2(n\pi\hbar)^2 k_R}{m\tilde{\alpha}^2 R}. \tag{9}$$

Parameters E_R and Γ_R determine the position and width of the quasi-discrete level; the value of the width is connected to the penetrability of the δ-barrier. Compare to the previous problem, and also with Problem 13.48.

Problem 13.48

Find the partial amplitude of s-wave scattering in the potential $U(r) = \alpha \delta(r - R)$. In the case of a weakly-penetrable δ-barrier, determine, $E_{R,n}$, and the widths, $\Gamma_{R,n}$, of the lowest quasi-discrete s-levels (with $E_{R,n} \sim \hbar^2/mR^2$).

Compare the scattering cross-section from the δ-sphere and impenetrable sphere. What is the difference, $\Delta\sigma(E)$, between these cross-sections if the energy is close to the energy of the quasi-discrete level?

Solution
The Schrödinger equation and its s-wave solution are

$$\chi'' - \tilde{\alpha}\delta(r-R)\chi + k^2\chi = 0, \quad \tilde{\alpha} = \frac{2m\alpha}{\hbar^2},$$

$$\chi(r) = \begin{cases} A \sin kr, & r < R, \\ (S_0 e^{ikr} - e^{-ikr}), & r > R. \end{cases}$$

Matching at the point $r = R$ (see Problem 2.6), we find

$$S_0 = e^{2i\delta_0} = e^{-2ikR} \frac{\tilde{\alpha}R\sin kR + kR\cos kR + ikR\sin kR}{\tilde{\alpha}R\sin kR + kR\cos kR - ikR\sin kR}. \tag{1}$$

If the conditions $\tilde{\alpha}R \gg 1$ and $kR \sim 1$ (or more precisely, $kR \ll \tilde{\alpha}R$) are fulfilled, then from Eq. (1) we generally have $S_0 \approx e^{-2ikR}$, i.e., $\delta_0 \approx -kR$, which corresponds to the scattering from an impenetrable sphere (see Problem 13.45).

---

[353] Where the real part of the denominator in Eq. (5) is close to zero.

The case of particle energies for which $kR \approx n\pi$ with $n = 1, 2, \ldots$ needs special consideration, and in Eq. (1) we cannot simply replace the fractional factor by 1. Writing $kR = n\pi + \gamma$, with $|\gamma| \ll 1$, we have

$$\tilde{\alpha} R \sin kR + kR \cos kR - ikR \sin kR \approx (-1)^n (\tilde{\alpha} R \gamma + n\pi - in\pi\gamma) \approx$$
$$(-1)^n (\tilde{\alpha} R - in\pi)(kR - n\pi + \lambda + i\lambda^2),$$

where $\lambda = n\pi/\tilde{\alpha} R \ll 1$. Equation (1) becomes

$$S_0 \approx e^{-2ikR} \frac{kR - n\pi + \lambda - i\lambda^2}{kR - n\pi + \lambda + i\lambda^2}. \tag{2}$$

Multiplying both the numerator and denominator by

$$(kR + n\pi - \lambda) \frac{\hbar^2}{2mR^2} \approx \frac{n\pi\hbar^2}{mR^2},$$

we transform Eq. (2) to a more convenient form,

$$S_0 \approx e^{2i\delta_0^{(0)}} \frac{E - E_{R,n} - i\Gamma_{R,n}/2}{E - E_{R,n} + i\Gamma_{R,n}/2}, \tag{3}$$

where $\delta_0^{(0)} = -kR$, and

$$E_{R,n} = \frac{\pi^2 n^2 \hbar^2}{2mR^2} \left(1 - \frac{2}{\tilde{\alpha} R}\right),$$

$$\Gamma_{R,n} = \frac{2\pi^2 n^2 \hbar^2}{mR^2} \frac{n\pi}{(\tilde{\alpha} R)^2} \ll E_{R,n}.$$

Equation (3) has a familiar form for resonant scattering from a quasi-discrete level. $\delta_0^{(0)}$ describes the phase of *potential* scattering (*i.e.*, phase far from resonance), while $E_{R,n}$ and $\Gamma_{R,n}$ give the position and the width of the quasi-discrete level. Therefore:

1) the phase of potential scattering coincides with the scattering phase for an impenetrable sphere of radius R;

2) the position, $E_{R,n}$, of the quasi-discrete levels almost coincides with the levels in an infinitely deep well of radius R;

3) the level width, which determines the lifetime of the quasi-stationary state, could be written in the form:

$$\Gamma_{R,n} = \frac{\hbar}{\tau_n} \equiv \hbar D N = \hbar \cdot 4\lambda^2 \cdot \frac{\pi n \hbar}{2mR^2},$$

where D is the transmission coefficient (for a single collision) of the δ-barrier for an energy equal to $E_{R,n}$ (see Problem 2.30), while $N = v/2R = \pi n \hbar / 2mR^2$ gives the number of times the particle hits the "wall" per unit of time.

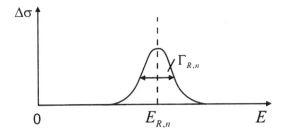

Fig. 13.5

From physical considerations, we should assume that analogous results take place for non-zero values of the angular momentum. So, for $kR \ll \tilde{a}R$, the scattering cross-section on the δ-sphere almost coincides with that on the impenetrable sphere of the same radius, except for narrow regions, ΔE, in the vicinity of quasi-discrete levels. Now, the partial cross-section is described by the expression:

$$\sigma_l = 4\pi(2l+1)\frac{1}{k^2}|S_l - 1|^2.$$

Note that there is no scattering with the angular momentum $l = 0$ from the impenetrable sphere for particles with energy close to the resonant energy $E_{R,n}$ of the s-level (here $\delta_0^{(0)} \approx n\pi$). Hence, we see that the difference between the scattering cross-sections of the δ- and the impenetrable sphere in the vicinity of a quasi-discrete s-level is equal to (Fig. 13.5)

$$\Delta \sigma = \pi R^2 \frac{\Gamma_{R,n}^2}{n^2\pi^2[(E - E_{R,n})^2 + \Gamma_{R,n}^2/4]} > 0.$$

Problem 13.49

Parameters of a potential $U_0(r)$ are chosen so that there exists a zero-energy bound state with the angular momentum l in it (the scattering length is $a_l^{(0)} = \infty$). Find the scattering length, a_l, in this partial wave, for a small change of the potential by $\delta U(r)$.

Using the result obtained, discuss how the dependence of the level position on the perturbation, $\delta U(r)$, differs in the $l = 0$ case from the $l \neq 0$ case. Compare to Problems 4.27 and 4.28.

Solution

Let $\chi_l^{(0)}(r)$ and $\chi_l(r) = rR(r)$ be the zero-energy solutions to the Schrödinger equation with the potentials $U_0(r)$ and $U_0(r) + \delta U(r)$ respectively, normalized by the following conditions (compare to Problem 13.39):

$$\chi_l^{(0)} = r^{-l} \quad \text{and} \quad \chi_l = r^{-l} - r^{l+1}\{(2l-1)!!(2l+1)!!a_l\}^{-1} \quad \text{for} \quad r \to \infty.$$

If we write the Schrödinger equation for $\chi_l^{(0)}$ and χ_l, and multiply the first equation by χ_l and the second one by $\chi_l^{(0)}$, then subtract one from the other term by term, and integrate the result over r from 0 to ∞, we find:

$$\frac{\hbar^2}{2[(2l-1)!!]^2 ma_l} = -\int_0^\infty \delta U \chi_l \chi_l^{(0)} dr \approx -\int_0^\infty \delta U \left(\chi_l^{(0)}\right)^2 dr. \tag{1}$$

Compare this to the analogous transformation in Problem 13.31. We emphasize that relation (1) is valid for any value of angular momentum.

Using the result obtained and the effective range expansion (XIII.15), from the condition $\cot \delta_l(E_l) = i$ we can find the position of the pole in the partial scattering amplitude, which determines the change of the level $E_l^{(0)} = 0$ under the influence of the perturbation. If $l = 0$, we have:

$$\kappa_0 = -i\sqrt{\frac{2mE_0}{\hbar^2}} \approx \frac{1}{a_0}$$

and for a positive scattering length $a_0 > 0$ (for $\delta U < 0$), we obtain the known result:

$$E_0 \approx -\frac{2m}{\hbar^2} \left[\int_0^\infty \delta U \left(\chi_0^{(0)}(r)\right)^2 dr\right]^2. \tag{2}$$

Compare to Problem 4.27. In the case $a_0 < 0$, we also have $\kappa_0 < 0$, so that the level E_0 is virtual. We emphasize that the s-wave scattering cross-section depends weakly on the sign of $\delta U(r)$ (see Eq. (XIII.16)).

Now, consider the case with $l \neq 0$. Using Eq. (1) and the result of Problem 13.44 for the effective range of the interaction, r_l, at the threshold where a new bound state appears, we see that the level shift (determined by the poles of the partial amplitude),

$$E_l \approx -\frac{\hbar^2}{ma_l|r_l|}, \quad l \geq 1, \tag{3}$$

is linear in δU and is described by first-order perturbation theory (compare to Problem 4.28). For $a_l > 0$, the level is real with $E_l < 0$. If $a_l < 0$, then $E_l > 0$ determines the energy of a quasi-stationary state. Its width is

$$\Gamma_l \approx \frac{2\hbar^2}{m|r_l|} \left(\frac{2mE_l}{\hbar^2}\right)^{l+1/2}. \tag{4}$$

In the case of $l \neq 0$, the nature of resonant scattering depends strongly on the sign of $\delta U(r)$. See, for example, Problem 13.46.

13.4 Scattering of fast particles; Eikonal approximation

Problem 13.50

Derive Eq. (XIII.18) for the scattering amplitude of fast particles by summing up the non-perturbative series, $f = \sum_n f^{(n)}$, derived in Problem 13.10 (see Eq. (2) there).

Solution

The Fourier components of the potential, $\tilde{U}(\boldsymbol{\kappa})$ are essentially different from zero only for $\kappa R \lesssim 1$ (R is the radius of the potential). So, for fast particles, $kR \gg 1$, and small scattering angles, $qR = |\mathbf{k} - \mathbf{k}_0| \leq 1$, the integrals over $\boldsymbol{\kappa}_k$ in this equation are dominated by the integration domains where $|\boldsymbol{\kappa}_k - \mathbf{k}_0| \lesssim 1/R$. Writing

$$\boldsymbol{\kappa}_k = \mathbf{k}_0 + \tilde{\kappa}_k^{\|}\mathbf{n}_0 + \boldsymbol{\kappa}_k^{\perp},$$

where $\mathbf{n}_0 = \mathbf{k}_0/k_0$ and $\boldsymbol{\kappa}_k^{\perp} \perp \mathbf{n}_0$, then for the energy denominators we have the approximate expression:

$$\kappa_k^2 - k_0^2 - i\varepsilon \approx 2k_0\tilde{\kappa}_k^{\|} - i\varepsilon. \tag{1}$$

We neglected the terms $(\tilde{\kappa}_k^{\|})^2$ and $(\kappa_k^{\perp})^2$ in comparison to $2k_0\tilde{\kappa}_k^{\|}$, which leads to an error $\sim 1/kR$. Now we substitute the Fourier components of the potential into Eq. (2) from Problem 13.10:

$$\tilde{U}(\boldsymbol{\kappa}_k - \boldsymbol{\kappa}_{k-1}) = \iiint U(\boldsymbol{\rho}_k, z_k) \exp\{-i[(\tilde{\kappa}_k^{\|} - \tilde{\kappa}_{k-1}^{\|})z + (\tilde{\boldsymbol{\kappa}}_k^{\perp} - \tilde{\boldsymbol{\kappa}}_{k-1}^{\perp}) \cdot \boldsymbol{\rho}_k]\} d^2\rho_k dz_k.$$

Recall the following identity for the delta-function:

$$\iint \exp\{i(\boldsymbol{\rho}_{k+1} - \boldsymbol{\rho}_k) \cdot \boldsymbol{\kappa}_k^{\perp}\} d^2\kappa_k^{\perp} = (2\pi)^2 \delta(\boldsymbol{\rho}_{k+1} - \boldsymbol{\rho}_k)$$

Now integrating over $\boldsymbol{\kappa}_k^{\perp}$, and using the δ-function to integrate over $\boldsymbol{\rho}_k$, we see that in all factors of $U(\boldsymbol{\rho}_k, z_k)$, the values $\boldsymbol{\rho}_k$ (with different k) become the same. Finally, using the integral

$$\int \frac{\exp\{i(z_{k+1} - z_k)\kappa_k^{\|}\}}{2k_0\kappa_k'' - i\varepsilon} d\kappa_k^{\|} = \frac{i\pi}{k_0}\eta(z_{k+1} - z_k)$$

($\eta(z)$ is the step function; see Problem 13.14), we find (for $k = k_0$):

$$f^{(n)} \approx \frac{k}{2\pi i}\left(-\frac{im}{\hbar^2 k}\right)^n \int\limits_{-\infty}^{z_2} dz_1 \int\limits_{-\infty}^{z_3} dz_2 \ldots \int\limits_{-\infty}^{\infty} dz_n \times$$

$$\iint U(\boldsymbol{\rho}, z_1) \ldots U(\boldsymbol{\rho}, z_n) e^{-i\mathbf{q}_\perp \cdot \boldsymbol{\rho}} d^2\rho. \tag{2}$$

In the exponent we neglected the term $-iq_\| z_n$, because $q_\| \ll q_\perp$ (compare to Problem 13.2). In this expression we extend integration to infinite limits over all z_k if we introduce the factor $(n!)^{-1}$. As a result, we obtain the scattering amplitude in the eikonal approximation:

$$f \approx \frac{k}{2\pi i} \iint \left\{ \sum_{n=1}^{\infty} \frac{1}{n!} \left(-\frac{im}{\hbar^2 k}\right)^n \left[\int_{-\infty}^{\infty} U(\boldsymbol{\rho}, z) dz\right]^n \right\} e^{-i\mathbf{q}_\perp \cdot \boldsymbol{\rho}} d^2\rho =$$

$$= \frac{ik}{2\pi} \iint \left\{ 1 - \exp\left\{-\frac{i}{\hbar v} \int_{-\infty}^{\infty} U(\boldsymbol{\rho}, z) dz\right\}\right\} e^{-i\mathbf{q}_\perp \cdot \boldsymbol{\rho}} d^2\rho. \quad (3)$$

In conclusion, we discuss the conditions of applicability of Eq. (3), which follow form the above derivation. Omission of the terms $q_\| z \leq q^2 R/k \approx k\theta^2 R \ll 1$ in the exponent of Eq. (2) implies small scattering angles $\theta \ll 1/\sqrt{kR}$. Since $kR \gg 1$, then this region includes scattering angles $\theta \lesssim 1/kR$, which dominate the total cross-section. This, as we noted in the context of Eq. (1), contributes an error $\sim 1/kR$. But the quantity of interest appears in the exponent of Eq. (3). So the condition of its applicability becomes

$$\frac{1}{\hbar v} UR \cdot \frac{1}{kR} \ll 1, \quad \text{i.e.} \quad |U(r)| \ll E.$$

Problem 13.51

Prove that the total scattering cross-section of fast particles, $kR \gg 1$, in a potential, $U(r)$, of radius R follows from the equation:

$$\sigma(E) = 4\pi \int_0^\infty \left\{ 1 - \cos\left[\frac{m}{k\hbar^2} \int_{-\infty}^{\infty} U(\sqrt{\rho^2 + z^2}) dz\right]\right\} \rho d\rho, \quad (1)$$

independently of the relation between particle energy and the characteristic strength of the potential, *i.e.*, the validity of the equation in question does not require that the condition of applicability of the eikonal approximation, $E \gg |U(r)|$, is necessarily satisfied.[354]

Use this result to calculate the scattering cross-section from the potential barrier (or well) with $U = U_0$ for $r < R$ and $U = 0$ for $r > R$.

Solution

Eq. (1) can be derived by combining the optical theorem and the eikonal approximation, Eq. (XIII.18). The condition for its applicability for fast particles is the inequality

---

[354] This condition is needed for the applicability of the eikonal approximation for the differential scattering cross-section.

$|U(r)| \ll E$, which is valid for a large fraction of scattering angles, $\theta \lesssim 1/kR$ (see the previous problem). In the case of a "strong" potential, for which $|U(r)| \gtrsim E$ when $r \sim R$, scattering into the angles $\theta \sim 1/kR$ is not described by the eikonal expression. But for forward scattering, corresponding to the angle $\theta = 0$, the amplitude,

$$f(E, \theta = 0) = \frac{1}{2ik} \sum_l (2l+1)(e^{2i\delta_l} - 1), \qquad (2)$$

still has the eikonal form (the approximation actually remains valid for the region of small scattering angles $\theta \ll 1/kR$, for as long as we can put $P_l \approx 1$ in the amplitude expansion over partial waves, Eq. (XIII.9) and until the oscillations in the Legendre polynomials with $l \lesssim kR$ begin to play out). To see this, recall that the eikonal equation for the scattering amplitude follows from the quasi-classical expressions for the phase shifts, Eq. (XIII.14). In the case of a "strong" potential, we should use the more general expression, (XIII.13). But this fact is not reflected in the value of the forward scattering amplitude, (2). For the values of l in the sum, Eq. (2), for which $|U(r_0)| \geq E$ (r_0 is the quasi-classical turning point in Eq. (XIII.13)), the phase shifts have the following properties: they are large, $|\delta_l| \gg 1$, and change rapidly with increase in l, $|\delta_{l+1} - \delta_l| \geq 1$. These properties follow both from Eq. (XIII.13) and from[355] Eq. (XIII.14). This leads to the fact that in the corresponding part of sum (2), the contribution of the term with $\exp(2i\delta_l)$ is negligible due to fast oscillations. This happens independently of which expression is used, correct (XIII.13) or "incorrect" Eq. (XIII.14). For values of l such that $|U(r_0)| \ll E$, (XIII.14) is still valid. Therefore the amplitude of forward scattering for fast particles is still described by the eikonal expression, even if the condition $|U(r_0)| \ll E$ breaks down. The statement of the problem follows from the optical theorem. The unitary properties of the scattering amplitude in the eikonal approximation are discussed in Problem 13.76.

In the case of a potential barrier (or a well), we find:

$$\sigma(E) = 4\pi \int_0^R \left[1 - \cos\left(\xi\sqrt{1 - \frac{\rho^2}{R^2}}\right) \right] \rho d\rho =$$
$$2\pi R^2 \left[1 - \frac{2}{\xi^2}(\xi \sin\xi + \cos\xi - 1) \right], \qquad (3)$$

$$\xi = \frac{2mU_0 R}{k\hbar^2}.$$

This integral is calculated using the substitution $x = \sqrt{1 - \rho^2/R^2}$. For the values $\xi \ll 1$ from Eq. (2), the result of the Born approximation follows

---

[355] For example, according to Eq. (XIII.14) we have:

$$|\delta_{l+1} - \delta_l| \sim \left|\frac{\partial}{\partial l}\delta_l\right| \sim \frac{m}{\hbar^2 k}|U(r_0)|\frac{l}{k^2 R} \sim \frac{m|U(r_0)|}{\hbar^2 k^2}.$$

$$\sigma(E) \approx \pi R^2 \left(\frac{mU_0^2 R^2}{\hbar^2 E}\right) \ll \pi R^2$$

for fast particles (see Problem 13.1), while for $\xi \gg 1$ we have $\sigma \approx 2\pi R^2$, the known result for the scattering cross-section of fast particles from an impenetrable sphere (see Problem 13.57).

Problem 13.52

Find the total scattering cross-section in the potential $U(r) = \alpha/r^\nu$ with $\nu > 2$ and $\alpha > 0$, for energy $E \to \infty$. Compare to Problem 13.2.

Solution

Let us use the quasi-classical equation for the scattering cross-section from the previous problem. We will also use the integral:

$$\int_{-\infty}^{\infty} U(\sqrt{\rho^2 + z^2})dz = 2\alpha\rho^{-\nu+1} \int_0^\infty \frac{du}{(1+u^2)^{\nu/2}} = \frac{\sqrt{\pi}\alpha}{\rho^{\nu-1}} \frac{\Gamma((\nu-1)/2)}{\Gamma(\nu/2)}.$$

(By the substitution $1 + u^2 = 1/t$, this integral could be transformed to the Euler integral, giving the β-function, $B(x,y)$, with $x = 1/2$ and $y = (\nu-1)/2$.) We can perform the integration[356] over ρ and obtain

$$\sigma(E) = 2\pi\Gamma(\lambda) \sin\frac{\pi\lambda}{2} \left[\frac{\sqrt{\pi}\alpha}{\hbar v} \frac{\Gamma((\nu-1)/2)}{\Gamma(\nu/2)}\right]^\mu \propto E^{-1/(\nu-1)}, \quad (1)$$

where $\lambda = (\nu-3)/(\nu-1)$ and $\mu = 2/(\nu-1)$.

As is seen from Eq. (1), the decrease of the scattering cross-section for $E \to \infty$ is slower than is required for the applicability of the Born approximation, where $\sigma \propto 1/E$. This is because in the scattering of fast particles in the potential $U \propto r^{-\nu}$ with $\nu > 2$, small distances play the dominant role (small impact parameter), and the potential cannot be considered as a perturbation. For large values of energy $E \to \infty$, Eq. (1) is valid for a fairly arbitrary potential, which has the form considered only at small distances. See that for the values $\nu \to 2$, the energy dependence of the cross-section is determined by Eq. (1) and "matches" the result obtained in the Born approximation.

---

[356] First we make the substitution $x = \rho^2$ and perform integration by parts, which leads to the integral of the form $\int \frac{1}{x^s} \sin\left(\frac{a}{x^s}\right) dx$ with $s = (\nu-1)/2$. Expressing the sine as a sum of two complex exponentials and making the substitution $v = \pm ia/x^s$, we obtain the integrals that give the Γ-function.

Problem 13.53

Consider a "potential" of the form $U = g(E)e^{-r/R}$, with the interaction constant,[357] which scales as a power-law with energy, $g(E) = g_0(E/E_0)^n$. Prove the inequality

$$\sigma(E) \leq \sigma_0 \ln^2\left(\frac{E}{E_0}\right)$$

for the energy-dependence of the scattering cross-section in the large-energy limit, $E \to \infty$.

Solution

To calculate the scattering cross-section we use the result of Problem 13.51. The quasi-classical phase shift for large values of the impact parameter $\rho \gg R$ is equal to

$$\delta(\rho) = \frac{g(E)}{2\hbar v} \int_{-\infty}^{\infty} \exp\left\{-\frac{1}{R}\sqrt{\rho^2 + z^2}\right\} dz \approx$$

$$\frac{g_0}{2\hbar v} \sqrt{2\pi\rho R} \left(\frac{E}{E_0}\right)^n e^{-\rho/R}. \qquad (1)$$

To calculate this integral, use the expansion: $\sqrt{\rho^2 + z^2} \approx \rho + z^2/2\rho$.

Let us denote by ρ_0 the value of ρ for which the following condition for the phase shift is satisfied: $\delta(\rho_0) = 1$. In the case $n > 1/2$ and at large energies, we have $\rho_0 \gg R$ (for $E \to \infty$, $\rho_0 \to \infty$ as well). Now we note that the phase shift is a sharp, rapidly decreasing function of ρ. For calculation of the cross-section using the equation from Problem 13.51, we can completely neglect both the contribution of integration domain $\rho > \rho_0$ (since there, $\delta \approx 0$). On the other hand, in the other domain $\rho < \rho_0$, we can neglect the term corresponding to the fast-oscillating factor, $\cos 2\delta(\rho)$ (because, $\delta \gg 1$). As a result, we obtain $\sigma(E) = 2\pi \rho_0^2(E)$. Thus, using Eq. (1), we find that

$$\sigma(E) \approx \sigma_0 \ln^2\left(\frac{E}{E_0}\right), \quad \text{where} \quad \sigma_0 = 2\pi\left(n - \frac{1}{2}\right)^2 R^2, \qquad (2)$$

for $n > 1/2$. To calculate approximately $\rho_0(E)$ from the relation $\ln \delta(\rho_0) = 0$, we neglected the term $\ln(\rho_0/R)$ in comparison to ρ_0/R. Finally, we mention that for the values $n < 1/2$, the scattering cross-section decreases with the increase of E.

Problem 13.54

In the eikonal approximation, find the amplitude and differential scattering cross-section for the Coulomb potential, $U = \alpha/r$, in the limiting case opposite to that

---

[357] Let us note that the spin-orbital interaction, $\hat{U} = \hat{\mathbf{s}} \cdot \hat{\mathbf{l}} f(r)$, increases $\propto \sqrt{E}$ with E. Compare to Problem 13.59. The restriction on the increase of cross-section is known as the Froissart theorem in the theory of strong interactions in elementary particle physics.

required for the validity of the Born approximation: *i.e.*, consider $|\alpha|/\hbar v \gg 1$. Compare to Problem 13.1.

Hint: To calculate the scattering amplitude, introduce a cut-off for the Coulomb potential at some large, but finite, distance R (*i.e.*, put $U = 0$ for $r > R$).

Solution

According to Eq. (XIII.19), for the values $\rho \ll R$, we have

$$\delta(\rho) = -\frac{\alpha}{2\hbar v} \int_{-R}^{R} \frac{dz}{\sqrt{\rho^2 + z^2}} \approx -\frac{\alpha}{\hbar v} \ln \frac{2R}{\rho}, \tag{1}$$

and for the $q \neq 0$ scattering amplitude from Eq. (XIII.19), we obtain the relation:

$$f(E, q) = \frac{k}{2\pi i} \int_0^\infty \int_0^{2\pi} e^{iS(\rho,\varphi)} \rho \, d\rho \, d\varphi,$$

where

$$S = \frac{2\alpha}{\hbar v} \ln \frac{\rho}{2R} - q\rho \cos \varphi. \tag{2}$$

In the case $|\alpha| \gg \hbar v$, the phase in the exponential is large, and changes rapidly with a change in ρ and φ. Hence, the integral is determined mainly by the integration domains in the vicinity of the saddle points, where the phase reaches extrema as a function of the variables ρ, φ. From this saddle point condition, we find:

$$2\alpha = \hbar v \rho_0 q \cos \varphi_0, \quad q\rho_0 \sin \varphi_0 = 0.$$

Thus $\rho_0 = 2|\alpha|/\hbar vq$, and $\varphi_0 = 0$ for $\alpha > 0$ or $\varphi_0 = \pi$ for $\alpha < 0$ (here $|\alpha|/\rho_0 \ll \hbar kv$, *i.e.*, $|U| \ll E$, which allows us to use the eikonal approximation). Expanding $S(\rho, \varphi)$ in the vicinity of the extremum (ρ_0, φ_0) with quadratic accuracy, and using the value of the Poisson integral, we obtain the scattering amplitude:

$$f(E, q) \approx -i \frac{2m|\alpha|}{\hbar^2 q^2} \exp\{iS(\rho_0, \varphi_0)\}. \tag{3}$$

Therefore, the differential scattering cross-section is

$$\frac{d\sigma}{d\Omega} = \left(\frac{2m\alpha}{\hbar^2 q^2}\right)^2 \approx \frac{\alpha^2}{E^2 \theta^4},$$

which coincides with the Rutherford equation (for small scattering angles, $\theta \ll 1$).

Let us note in conclusion that the need to cut off the potential and the absence of a well-defined limit for $R \to \infty$ of the scattering amplitude (but not for the differential cross-section) are connected to the slow decrease of the Coulomb potential. At large distances, the wavefunction of the Coulomb problem is not the wavefunction of a free particle. The phase of the wavefunction gets "distorted" by the factor,

$$\Delta(r) = -\frac{\alpha}{\hbar v} \ln 2kr,$$

due to the cutting off the potential at the distance $r = R$. This distortion is then "inherited" by the scattering amplitude in the form of the factor $\exp\{2i\Delta(R)\}$. Factoring it out from Eq. (3), we obtain the scattering amplitude:

$$\tilde{f}(k,q) = f e^{-2i\Delta(R)} = -\frac{2m\alpha}{\hbar^2 q^2} e^{i\varphi(k,q)},$$

$$\varphi(k,q) = -\frac{2\alpha}{\hbar v} \ln \frac{q}{2k} + \frac{2\alpha}{\hbar v} \ln \frac{|\alpha|}{\hbar v} - \frac{2\alpha}{\hbar v} + \frac{\pi}{2} \frac{\alpha}{|\alpha|}. \tag{4}$$

It is now independent of the cutoff radius, and coincides, as expected, with the amplitude of Coulomb scattering:[358]

$$f_{Coul} = -\frac{\alpha}{2mv^2 \sin^2(\theta/2)} \frac{\Gamma(1+i\alpha/\hbar v)}{\Gamma(1-i\alpha/\hbar v)} \exp\left\{-\frac{2i\alpha}{\hbar v} \ln \sin \frac{\theta}{2}\right\}.$$

In the quasi-classical case, $|\alpha|/\hbar v \gg 1$ for all scattering angles, while the eikonal approximation is applicable only for angles $\theta \ll \hbar v/|\alpha|$.

Problem 13.55

In the eikonal approximation, express the scattering amplitude in the field of two identical potentials located at a distance **a** from one another, *i.e.*, in the potential $U(\mathbf{r}) = U_0(|\mathbf{r} - \mathbf{a}/2|) + U_0(|\mathbf{r} + \mathbf{a}/2|)$, in terms of the amplitude, f_0, of scattering from a single spherically-symmetric potential, $U_0(r)$. What is the connection between the total scattering cross-section in $U(\mathbf{r})$ and that in a single potential, σ_0?

Use the result obtained to calculate the total scattering cross-section for a weakly bound system of two centers (like a deuteron), when its characteristic size is much larger than the interaction radius of a single scattering center.

Solution
In the eikonal approximation, the scattering amplitude is described by the expression:

$$f(k, \mathbf{q}_\perp) = \frac{ik}{2\pi} \iint \left\{1 - \exp\left\{-\frac{im}{\hbar^2 k} \int_{-\infty}^{\infty} U(\boldsymbol{\rho}, z) dz\right\}\right\} e^{-i\mathbf{q}_\perp \cdot \boldsymbol{\rho}} d^2\rho. \tag{1}$$

---

[358] The ratio of Γ-functions is determined only by their phase factors, which are easily found from the known asymptotes for $\ln \Gamma(z)$.

Using the Fourier transformation, we obtain:

$$\exp\left\{-\frac{im}{\hbar^2 k}\int_{-\infty}^{\infty} U(\boldsymbol{\rho}, z)dz\right\} = 1 + \frac{i}{2\pi k}\iint f(k, \boldsymbol{\kappa}_\perp)e^{i\boldsymbol{\kappa}_\perp \cdot \boldsymbol{\rho}}d^2\kappa_\perp. \quad (2)$$

Now substituting into Eq. (1) the potential

$$U(\mathbf{r}) = U_1\left(\mathbf{r} - \frac{\mathbf{a}}{2}\right) + U_2\left(\mathbf{r} + \frac{\mathbf{a}}{2}\right),$$

and transforming potentials $U_{1,2}(\mathbf{r})$ according to Eq. (2), we obtain the scattering amplitude from two centers in the eikonal approximation:

$$f(k, \mathbf{q}_\perp) = f_1(k, \mathbf{q}_\perp)\exp\left\{-\frac{i}{2}\mathbf{q}_\perp \cdot \mathbf{a}_\perp\right\} + f_2(k, \mathbf{q}_\perp)\exp\left\{\frac{i}{2}\mathbf{q}_\perp \cdot \mathbf{a}_\perp\right\} +$$

$$\frac{i}{2\pi k}\int f_1\left(k, \boldsymbol{\kappa}_\perp + \frac{\mathbf{q}_\perp}{2}\right) f_2\left(k, -\boldsymbol{\kappa}_\perp + \frac{\mathbf{q}_\perp}{2}\right)\exp\{-i\boldsymbol{\kappa}_\perp \cdot \mathbf{a}_\perp\}d^2\kappa_\perp. \quad (3)$$

Here, $\mathbf{a}_\perp$, $\boldsymbol{\kappa}_\perp$ are vectors are perpendicular to the direction of the momentum of the incident particle, $\mathbf{n}_0$; for example, $\mathbf{a} = a_\parallel \mathbf{n}_0 + \mathbf{a}_\perp$.

Now, we use Eq. (3) to calculate the elastic scattering amplitude from a bound system of two centers (*i.e.*, on some composite particle), and average the wavefunction $\psi_0(\mathbf{a})$ of the composite system, *i.e.*, $\int d^3a|\psi_0(\mathbf{a})|^2$ (here it is assumed that the velocity of the target particle is small with respect to the velocity of the incident particle). We also emphasize that $\psi_0(\mathbf{a})$ is the wavefunction of the composite system in its center-of-mass frame (and the target as a whole remains at rest). The radius vectors of the target particles are chosen to be equal to $\pm\mathbf{a}/2$, and it is assumed that they have identical masses (as is the case for the deuteron). The integrals that appear after the averaging are expressed in terms of the form factor of the composite system, equal to

$$F(\mathbf{q}) = \int |\psi_0(\mathbf{r})|^2 \exp\left\{-\frac{i}{2}\mathbf{q} \cdot \mathbf{r}\right\} d^3r.$$

As a result of averaging, Eq. (3) takes the form:

$$\langle f(k, \mathbf{q}_\perp)\rangle = f_1(k, \mathbf{q}_\perp)F(\mathbf{q}_\perp) + f_2(k, \mathbf{q}_\perp)F(-\mathbf{q}_\perp) +$$

$$\frac{i}{2\pi k}\int f_1\left(k, \boldsymbol{\kappa}_\perp + \frac{\mathbf{q}_\perp}{2}\right) f_2\left(k, -\boldsymbol{\kappa}_\perp + \frac{\mathbf{q}_\perp}{2}\right) F(2\boldsymbol{\kappa}_\perp)d^2\kappa_\perp. \quad (4)$$

Using the optical theorem (XIII.11), the total scattering cross-section can be expressed in terms of the forward scattering amplitude for angle $\theta = 0$ (*i.e.*, $\mathbf{q}_\perp = \mathbf{0}$), so that

$$\sigma_{tot} = \sigma_1 + \sigma_2 + \frac{2}{k^2}\mathrm{Re}\int F(2\boldsymbol{\kappa}_\perp)f_1(k, \boldsymbol{\kappa}_\perp)f_2(k, -\boldsymbol{\kappa}_\perp)d^2\kappa_\perp \quad (5)$$

where $\sigma_{1,2}$ are the scattering cross-sections of free particles hitting the target. (We should emphasize that the total scattering cross-section may also include processes, where the initial composite system breaks up).

Notice that the properties of scattering from a weakly-bound system are determined by the fact that the typical momentum, where the form factor begins to noticeably decrease, is of the order of

$$q \sim \frac{1}{R} \sim \left(\frac{\mu \varepsilon_{bind}}{\hbar^2}\right)^{1/2}, \tag{6}$$

where R is the size of a system, ε_{bind} and μ are the binding energy and reduced mass of target particles, and R is assumed to be much larger than the interaction radius. Since the characteristic values of q for scattering are of the order of the inverse interaction radius, i.e., are large in comparison to (6), then in Eq. (5) the amplitudes, $f_{1,2}$, can be factored outside of the integral at the point $\boldsymbol{\kappa}_\perp = \mathbf{0}$. Next, we perform the following transformation:

$$\int F(2\boldsymbol{\kappa}_\perp) d^2\kappa_\perp = \iiint \exp\{-i\boldsymbol{\kappa}_\perp \cdot \boldsymbol{\rho}\}|\psi_0(\boldsymbol{\rho},z)|^2 d^2\rho \, dz \, d^2\kappa_\perp =$$

$$= (2\pi)^2 \int_{-\infty}^{\infty} |\psi_0(\mathbf{0},z)|^2 dz = 2\pi \int_0^\infty \frac{1}{r^2}|\psi_0(r)|^2 4\pi r^2 \, dr \equiv 2\pi \left\langle \frac{1}{R^2} \right\rangle.$$

We have assumed that the angular momentum of the composite system is equal to zero, so that the wavefunctiom $\psi_0(r)$ is spherically symmetric. $\langle R^{-2} \rangle$ is the mean value of the inverse distance between the target particles squared. We obtain:

$$\sigma_{tot} = \sigma_1 + \sigma_2 + \Delta, \quad \Delta = \frac{4\pi}{k^2}\left\langle \frac{1}{R^2}\right\rangle \text{Re}(f_1(k,0)f_2(k,0)). \tag{7}$$

In particular, if $f_{1,2}(k,0)$ are purely imaginary quantities, then, using the optical theorem for the single center amplitudes, we find:

$$\Delta = -\frac{1}{4\pi}\sigma_1\sigma_2\left\langle \frac{1}{R^2}\right\rangle. \tag{8}$$

Let us note that such a case of imaginary amplitudes corresponds to scattering from impenetrable (or "black") spheres (compare to Problems 13.57 and 13.90). Also, note that the fact that the scattering cross-section from two centers is a little smaller (since $\Delta < 0$) than $\sigma_1 + \sigma_2$, can be thought of as an effect of mutual shading one sphere by the other.

Problem 13.56

Extend the eikonal approximation to the case of an exchange interaction, $\hat{U}_{ex}\psi(\mathbf{r}) \equiv U(r)\psi(-\mathbf{r})$. What is the connection between the differential and total scattering cross-sections for the exchange potential and the corresponding ordinary potential? Compare to scattering in the Born approximation, as considered in Problem 13.3.

Solution

We will generalize the derivation of the scattering amplitude in the eikonal approximation based on the quasi-classical expression, Eq. (XIII.14), for the phase shift. If we replace the summation over partial waves by integration over the impact parameters and note that in approximation (XIII.14) for the exchange potential, the phase shift is

$$\delta_{l,ex} = (-1)^l \delta_{l,com}, \qquad (1)$$

then it is easy to obtain the following results.

In the region of small scattering angles, $\theta \ll 1$, the scattering amplitude for the exchange interaction is described by an equation analogous to (XIII.18), but with $\cos 2\delta_l$ instead of $e^{2i\delta_l}$; i.e.,

$$f_{ex}(k,\theta) \underset{\theta \ll 1}{\approx} \frac{k}{2\pi i} \iint [\cos 2\delta(\rho) - 1] \exp\{-i q \rho\} d^2\rho =$$

$$-ik \int_0^\infty [\cos 2\delta(\rho) - 1] J_0(k\rho\theta) \rho \, d\rho, \qquad (2)$$

with the same function, $\delta(\rho)$.

Note that the appearance of the factor $(-1)^l$ in Eq. (1) does not change the even-in-δ_l part of the scattering amplitude. For odd functions, $\propto \sin 2\delta_{l,ex}$, a strong mutual compensation of the neighboring terms in the sum over l appears, and this part of the amplitude is negligibly small.

Let us note that amplitude (2) is imaginary and coincides with the imaginary part of the scattering amplitude for the ordinary potential $U(r)$. If we take into account the optical theorem, then their total scattering cross-sections also coincide (the same for partial scattering cross-sections σ_l).

In the case of the exchange potential, the scattering amplitude has a sharp maximum in the region of angles close to π (for scattering backwards, compare to Problem 13.3). Taking into account the relation for the Legendre polynomials $P_l(z) = (-1)^l P_l(-z)$, we find:

$$f_{ex}(k,\theta) \underset{\pi - \theta \ll 1}{\approx} \frac{k}{2\pi} \iint \sin 2\delta(\rho) \exp\{-i\boldsymbol{\Delta} \cdot \boldsymbol{\rho}\} d^2\rho =$$

$$k \int_0^\infty \sin 2\delta(\rho) J_0(k\rho(\pi - \theta)) \rho \, d\rho, \qquad (3)$$

where $\boldsymbol{\Delta} = \mathbf{k} + \mathbf{k}_0$. In this region of angles, the amplitude is a real function, and coincides with the real part of the scattering amplitude from the corresponding regular potential, with the scattering angle $\theta' = \pi - \theta \ll 1$. From Eqs. (2) and (3), we recover the same fact that the total scattering cross-sections are the same for the exchange an the corresponding ordinary potential.

Problem 13.57

Calculate the scattering amplitude and the differential and total cross-sections for small-angle scattering ($kR\theta \lesssim 1$) of fast particles ($kR \gg 1$) on a hard sphere with radius, R. Use the quasi-classical expression (XIII.13) for the phase shift.

Solution

To calculate the scattering amplitude of fast particles, we use its expansion over partial waves, Eq. (XIII.9), and the quasi-classical expression (XIII.13) for phase shifts. Since $U = 0$ for $r > R$, then integration by parts in Eq. (XIII.13) gives:

$$\delta_l = \left(l + \frac{1}{2}\right)\arccos\frac{l+1/2}{kR} - \sqrt{k^2 R^2 - \left(l+\frac{1}{2}\right)^2}, \quad l \leq L, \tag{1}$$

where $L = kR - 1/2$ (here $r_0 = R$), while for the values $l > L$ we obtain (here $r_0 = (1+1/2)/k$):

$$\delta_l = 0, \quad l > L. \tag{2}$$

For $l > L$, the exact phase shift is exponentially small. This is beyond quasi-classical approximation, which yields $\delta_l = 0$. Let us also mention that, strictly speaking, we should put an additional term, $-\pi/4$, in Eq. (1). It has the same origin as the modification of the quantization rule considered in Problem 9.2 (for $l < L$ in our problem, the quasi-classical approximation is reliable up to the turning point). Let us also mention that the results obtained for δ_l need correction for values of l that are close to L, since in the vicinity of the point $r = R$, the quasi-classical approximation is not applicable, and the matching conditions require a modification. However, these conditions are not important for further calculations.

Using relations (1) and (2), it is convenient to write the scattering amplitude in the form:

$$f = f_{dif} + f_{cl},$$

where

$$f_{dif} = \frac{i}{2k}\sum_{l=0}^{L}(2l+1)P_l(\cos\theta), \quad f_{cl} = \frac{1}{2ik}\sum_{l=0}^{L}(2l+1)e^{2i\delta_l}P_l(\cos\theta).$$

This partitions the amplitude into "diffractive" and "classical" parts. We will note a few important points about this partition, since it will be important for this and the following few problems. For small scattering angles (i.e., for $\theta \ll (kR)^{-1/3} \ll 1$), we have $|f_{dif}| \gg |f_{cl}|$. The contribution of this region into the total cross-section is πR^2. Such small-angle scattering is often referred to as diffractive, because it is similar to the diffraction of a plane-parallel light beam falling onto a non-transparent (reflecting or absorptive) screen, *Fraunhofer diffraction* (see also Problem 13.90). For scattering angles $\theta \gg (kR)^{-1/3}$, the diffractive amplitude f_{dif} is negligibly small. f_{cl} describes the isotropic distribution of scattered particles, $d\sigma/d\Omega \approx |f_{cl}|^2 \approx R^2/4$, so that the scattering cross-section into such angles is also πR^2 (as in classical mechanics). In the region of angles $\theta \sim (kR)^{-1/3}$, the amplitudes f_{dif} and f_{cl} are of the same order. However, the contribution of these angles to scattering is negligibly small in comparison to πR^2. Therefore, the total scattering cross-section is equal to $2\pi R^2$.

For sufficiently small scattering angles, the diffraction part of the scattering amplitude becomes dominant, and $f \approx f_{dif}$. Indeed, since the phase shifts, (1), are large, $|\delta_l| \gg 1$, and change rapidly, then in the sum for f_{cl}, a mutual compensation appears due to the oscillations of $e^{2i\delta_l}$ in different terms. Using the relation $P_l(\cos\theta) \approx J_0((l+1/2)\theta)$ for $l \gg 1$, $\theta \ll 1$, and replacing the summation in f_{dif} by an integration over l, we obtain ($L \approx kR$):

$$f_{dif} \approx \frac{i}{2k} \int_0^L (2l+1) J_0\left(\left(l+\frac{1}{2}\right)\theta\right) dl \approx iR \frac{J_1(kR\theta)}{\theta}. \tag{4}$$

We have used the equation $\int x J_0(x) dx = x J_1(x)$.

Note that the diffractive part of the scattering amplitude is purely imaginary. Using the optical theorem and the relation $J_1 \approx x/2$ for $x \to 0$, we again find that the total scattering cross-section is $\sigma = 2\pi R^2$, which is two times larger than the classical cross-section (compare to Problem 13.51).

In the region of small scattering angles, the differential cross-section, $d\sigma/d\Omega = |f_{dif}|^2$, is an oscillating function of θ, and the distance between the neighboring maxima is of the order of $\Delta\theta \sim 1/kR$. Using known asymptotes of the Bessel functions, according to Eq. (4) we find:

$$\left.\frac{d\sigma_{dif}}{d\Omega}\right|_{\theta \ll 1/kR} \approx \frac{1}{4}(kR)^2 R^2;$$

$$\left.\frac{d\sigma_{dif}}{d\Omega}\right|_{1/kR \ll \theta \ll 1} \approx \frac{2R \sin^2(kR\theta - \pi/4)}{\pi k \theta^3}.$$

The differential scattering cross-section has a maximum in the region $\theta \lesssim 1/kR$, and decreases fast with the increase in θ.

The total cross-section of diffraction scattering is

$$\sigma_{dif} = \int |f_{dif}|^2 d\Omega \approx 2\pi R^2 \int_0^\infty \frac{J_1^2(kR\theta)}{\theta} d\theta = \pi R^2,$$

i.e., half of the total scattering cross-section.

Problem 13.58

The same as in the previous problem, but for not necessarily small scattering angles $\theta \gg (kR)^{-1/3}$. Compare to the result of classical mechanics.

Solution

Let us consider the classical part of the scattering amplitude,

$$f_{cl} = \frac{1}{2ik} \sum_{l=0}^{L} (2l+1) e^{2i\delta_l} P_l(\cos\theta), \quad L = kR, \tag{1}$$

where δ_l is given by Eq. (1) of the previous problem.

We use a general method of calculating sums such as in Eq. (1) (which involves replacing the sum over l with an integral and using the saddle-point approximation as discussed in great detail in, e.g., the relevant chapter of the Landau and Lifshtiz theoretical physics course) and first write down the equation

$$2 \arccos \frac{l+1/2}{kR} \pm \theta = 0$$

that determines the extremum point l_0 as follows

$$l_0 + \frac{1}{2} = kR \cos\frac{\theta}{2}, \quad \delta_{l_0} = \frac{1}{2}\left(l_0 + \frac{1}{2}\right)\theta - kR \sin\frac{\theta}{2}.$$

The final expression[359] for f_{cl} takes the form:

$$f_{cl} = -\frac{i}{2} R \exp\left\{-2ikR \sin\frac{\theta}{2}\right\}. \tag{2}$$

Therefore,

$$\frac{d\sigma_{cl}}{d\Omega} = |f_{cl}|^2 = \frac{R^2}{4}, \quad \sigma_{cl} = \pi R^2, \tag{3}$$

which coincides with the result of classical mechanics.

In conclusion, let us refer to Fig. 13.6 for an illustration of a qualitative dependence of the scattering cross-section, $d\sigma/d\Omega$, on the angle θ for scattering of fast particles, $kR \gg 1$ (or, more precisely, for $(kR)^{1/3} \gg 1$), from an impenetrable sphere according to the results of this and the previous problems.

---

[359] This corresponds to a repulsive potential. The phase factor $e^{-i\pi/4}$, which appears at the transition from the summation over l to an integration, is omitted here (which, however, is not reflected in the value of the differential scattering cross-section.)

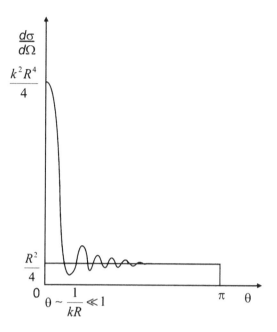

Fig. 13.6

13.5 Scattering of particles with spin

Problem 13.59

The operator describing interaction of a spin-1/2 particle with an external field has the form:[360]

$$\hat{U} = U_0(r) + U_1(r)\hat{\boldsymbol{\sigma}} \cdot \hat{\mathbf{l}}.$$

What is the energy dependence of the total scattering cross-section for fast particles in the Born approximation? Compare to the scattering of spinless particles.

Find the spin-dependent part of the electron scattering amplitude in the Coulomb field of a nucleus, $U_0 = -Ze^2/r$, taking into account the spin-orbital interaction, $U_1 = (\hbar^2/4m^2c^2r)\partial U_0/\partial r$.

Solution

Eqs. (XIII.1–XIII.5) are easily generalized to describe scattering of "spinfull" particles. A substitution of the interaction operator $\hat{U}$ (instead of the potential U) and

---

[360] A classical analog of the spin-orbital interaction $U(r)\hat{\boldsymbol{\sigma}} \cdot \hat{\mathbf{l}}$ is considered in Problem 13.60. If for a particle with spin $s = 1/2$, we write the magnetic moment in the form $\mu_0 = e\hbar/2mc + \mu'$, where e, m are the charge and mass of a particle and μ' is the *anomalous magnetic moment*, then the correct quantum-mechanical generalization of the classical result from Problem 13.60 is obtained by substituting the operator $\mu'_0\hat{\sigma}$ with $\mu'_0 = e\hbar/4mc + \mu'$ instead of μ. See Problem 15.32.

unperturbed wavefunction $\psi_{\mathbf{k}_0}^+ = e^{i\mathbf{k}_0 \cdot \mathbf{r}}\chi_l$ (χ_l is the spin function of the particle before collision) into Eq. (XIII.5) determines a spinor amplitude of the scattered wave, $F = \hat{f}\chi$, in the Born approximation. Therefore,

$$\hat{f}(\mathbf{k}, \mathbf{k}_0) = -\frac{m}{2\pi\hbar^2}\int e^{-i\mathbf{k}\cdot\mathbf{r}}[U_0(r) + U_1(r)\hat{\mathbf{l}}\cdot\hat{\boldsymbol{\sigma}}]e^{i\mathbf{k}_0\cdot\mathbf{r}}dV =$$
$$= -\frac{m}{2\pi\hbar^2}\left\{\tilde{U}_0(q) - i([\mathbf{k}_0 \times \mathbf{k}]\cdot\hat{\boldsymbol{\sigma}})\cdot\frac{1}{q}\frac{\partial}{\partial q}\tilde{U}_1(q)\right\}, \quad (1)$$

where $\tilde{U}_{0,1}(q) = \int U_{0,1}(r)\exp\{-i\mathbf{q}\cdot\mathbf{r}\}dV$.

According to Eqs. (1) and (XIII.23), the total scattering cross-section in the Born approximation is described by the expression (compare to Problem 13.1)

$$\sigma = \frac{m^2}{4\pi\hbar^4 k^2}\int_0^{4k^2}\left\{|\tilde{U}_0(q)|^2 + \left(k^2 - \frac{q^2}{4}\right)\left|\frac{\partial\tilde{U}_1(q)}{\partial q}\right|^2\right\}dq^2. \quad (2)$$

Let us note that in general, the term in the differential cross-section $\propto \boldsymbol{\nu}\cdot\mathbf{P}_0$ which describes azimuthal asymmetry in scattering (see Eq. (XII.23)) disappears from the calculation of the total scattering cross-section; the cross-section, averaging over the over spin projections, does not depend on the polarization, $\mathbf{P}_0$, of the initial state. So, for fast particles we obtain:[361]

$$\sigma(E) \approx \frac{C_0}{E} + C_1, \quad (3)$$

where

$$C_0 = \frac{m}{8\pi\hbar^2}\int_0^\infty |\tilde{U}_0(q)|^2 dq^2, \quad C_1 = \frac{m^2}{4\pi\hbar^4}\int_0^\infty \left|\frac{\partial\tilde{U}_1(q)}{\partial q}\right|^2 dq^2.$$

The terms here correspond to different interactions. For a large but finite energy, they could be of the same order. But in practice, the relativistic origin of the spin-orbit interaction for electrons, yields that the second (non-decreasing with energy) term is less important in the non-relativistic case.

For electron scattering in the Coulomb field, we obtain:

$$\hat{f} = \frac{2mZe^2}{\hbar^2 q^2}\left\{1 - i\frac{\hbar^2 k^2}{4m^2 c^2}\sin\theta\,(\boldsymbol{\nu}\cdot\hat{\boldsymbol{\sigma}})\right\}, \quad (4)$$

where $\boldsymbol{\nu} = [\mathbf{k}_0 \times \mathbf{k}]/|[\mathbf{k}_0 \times \mathbf{k}]|$. Note that the spin-dependent part of the scattering amplitude is small, $\sim \theta v^2/c^2$.

---

[361] Note that the applicability conditions of the Born approximation for the spin-orbital interaction involve a restriction on the power law decay of $U_1(r)$ at large distances, which ensures the finiteness of the total scattering cross-section. Let us emphasize that the independence of scattering cross-section from energy for $E \to \infty$ reflects its effective increase with the increase in E, since $l_{eff} \sim kR \propto \sqrt{E}$.

We now calculate the spin-dependent amplitude in the presence of spin-orbit interaction, $U_1(r) = \frac{\gamma}{r}\frac{\partial}{\partial r}U_0(r)$. Taking into account Eq. (1), let us perform the following transformations

$$\int e^{-i\mathbf{k}\cdot\mathbf{r}}U_1\hat{\mathbf{l}}e^{i\mathbf{k}_0\cdot\mathbf{r}}dV = \int U_1[\mathbf{r}\times\mathbf{k}_0]e^{-i\mathbf{q}\cdot\mathbf{r}}dV = -\gamma\int e^{-i\mathbf{q}\cdot\mathbf{r}}[\mathbf{k}_0\times\boldsymbol{U}_0(r)]\,dV =$$

$$= -i\gamma[\mathbf{k}_0\times\mathbf{q}]\tilde{U}_0(q).$$

So, for the considered spin-orbital interaction, the scattering amplitude in the Born approximation is described by the expression:

$$\hat{f} = -\frac{m}{2\pi\hbar^2}\tilde{U}_0(q)(1 - i\gamma[\mathbf{k}_0\times\mathbf{k}]\cdot\hat{\boldsymbol{\sigma}}). \tag{5}$$

Thereafter, using the expression $\tilde{U}_0 = -4\pi Ze^2/q^2$ for the Fourier component of the Coulomb potential and the parameter $\gamma = \hbar^2/4m^2c^2$, we obtain Eq. (4).

Problem 13.60

Find the Born scattering amplitude and differential cross-section of fast neutrons from the Coulomb field.

Solution

Let us first find the form of the interaction between a moving magnetic dipole and an electric field in classical electrodynamics. In the initial coordinate system, there is only an electric field $\boldsymbol{\mathcal{E}} = -\boldsymbol{\nabla}\varphi$ with $\varphi = Ze/r$. To find its interaction with a neutron, moving with the velocity $\mathbf{v} = \mathbf{p}/M$, we pass to the frame of reference moving with the neutron. In this system, a magnetic field $\boldsymbol{\mathcal{H}} \approx [\boldsymbol{\mathcal{E}}\times\mathbf{v}]/c$ appears, and the interaction energy becomes becomes equal to (with $\boldsymbol{\mu}$ being the neutron magnetic moment):

$$U = -\boldsymbol{\mu}\cdot\boldsymbol{\mathcal{H}} = \frac{1}{Mcr}\frac{\partial\varphi}{\partial r}\boldsymbol{\mu}\cdot\mathbf{L}, \quad \mathbf{L} = [\mathbf{r}\times\mathbf{p}]. \tag{1}$$

A quantum mechanical generalization of this equation is obtained by substitutions: of $\boldsymbol{\mu}$ by spin magnetic moment operator[362] $\hat{\boldsymbol{\mu}} = \mu_0\hat{\boldsymbol{\sigma}}$, where $\mu_0 = \beta e\hbar/2Mc$ (for the neutron, the experimental value is $\beta = -1.91$), and $\mathbf{L}$ by $\hbar\hat{\mathbf{l}}$. This generalization determines the interaction operator:

$$\hat{U} = \frac{\gamma}{r}\frac{\partial\varphi}{\partial r}\hat{\boldsymbol{\sigma}}\cdot\hat{\mathbf{l}}, \quad \gamma = \frac{\beta e\hbar^2}{2M^2c^2}. \tag{2}$$

Using the equations for scattering amplitude with the spin-orbit interaction in the Born approximation, obtained in the previous problem, we find:

$$\hat{f} = i\frac{2Ze^2M\gamma}{\hbar^2 q^2}\hat{\boldsymbol{\sigma}}\cdot[\mathbf{k}_0\times\mathbf{q}] = i\frac{\beta Ze^2}{2Mc^2}\cot\frac{\theta}{2}(\hat{\boldsymbol{\sigma}}\cdot\boldsymbol{\nu}), \tag{3}$$

---

[362] See the footnote of Problem 13.59.

where $\boldsymbol{\nu} = [\mathbf{k}_0 \times \mathbf{k}]/|[\mathbf{k}_0 \times \mathbf{k}]|$ is the unit vector normal to the scattering plane. The differential cross-section, summed over the neutron's spin states, is

$$\frac{d\sigma}{d\Omega} = F^*F = \chi_i^* \hat{f}^+ \hat{f} \chi_i = \left(\frac{\beta Z e^2}{2Mc^2}\right)^2 \cot^2 \frac{\theta}{2}. \tag{4}$$

Here, $F = \hat{f}\chi_i$ is the spinor amplitude of the scattered wave. For a small-angle scattering, $\theta \to 0$, we have $d\sigma/d\Omega \propto \theta^{-2}$, so that the total scattering cross-section becomes infinite (this divergence of the cross-section disappears if we take into account screening of the nuclear charge).

Problem 13.61

What constraints does the hermiticity property of the Hamiltonian impose on the spin-orbit interaction, $\hat{U} = U_0(r) + U_1(r)\hat{\mathbf{l}} \cdot \hat{\boldsymbol{\sigma}}$, of a spin-1/2 particle with an external field?

(1) Find in the first Born approximation polarization of particles scattered by such a potential, assuming that they were unpolarized initially. (2) Show also that if the particles were, on the contrary, polarized before the collision, the latter can result only in a rotation of the polarization vector.

Solution

The hermiticity of $\hat{U} = U_0(r) + U_1(r)\hat{\mathbf{l}} \cdot \hat{\boldsymbol{\sigma}}$ demands that the functions $U_{0,1}(r)$ are real. Their Fourier components, $\tilde{U}_{0,1}(q)$ are also real, and so are the invariant functions A and B in the expression below for the Born scattering amplitude

$$\hat{f} = A(k, \theta) + iB(k, \theta)\boldsymbol{\nu} \cdot \hat{\boldsymbol{\sigma}} \tag{1}$$

(see Eqs. (1) and (5) from Problem 13.59). So, according to the general equation (XIII.24), we find that the scattered particles remain unpolarized, $\mathbf{P} = \mathbf{0}$, if they were unpolarized initially, $\mathbf{P}_0 = \mathbf{0}$.

For the scattering of polarized particles, the final polarization vector is equal to[363]

$$\mathbf{P} = \frac{(|A|^2 - |B|^2)\mathbf{P}_0 + 2|B|^2\boldsymbol{\nu}(\boldsymbol{\nu} \cdot \mathbf{P}_0) - 2\mathrm{Re}\, AB^*[\boldsymbol{\nu} \times \mathbf{P}_0] + 2\mathrm{Im}\, AB^* \cdot \boldsymbol{\nu}}{|A|^2 + |B|^2 + 2\mathrm{Im}\, AB^* \boldsymbol{\nu} \mathbf{P}_0}.$$

Since A and B are real functions, then

$$\mathbf{P}_B = \frac{1}{A^2 + B^2}\{(A^2 - B^2)\mathbf{P}_0 + 2B^2\boldsymbol{\nu}(\boldsymbol{\nu} \cdot \mathbf{P}_0) - 2AB[\boldsymbol{\nu} \times \mathbf{P}_0]\},$$

and $\mathbf{P}_B^2 = \mathbf{P}_0^2$. This corresponds to a rotation of the polarization vector around the normal to the scattering plane; here, $\boldsymbol{\nu} \cdot \mathbf{P}_B = \boldsymbol{\nu} \cdot \mathbf{P}_0$.

---

[363] Note that in Eq. (1), we have adopted a convention, which includes the imaginary constant, i, in front of the coefficient, B (different conventions can be found in the literature).

Problem 13.62

Find the polarization that appears as a result of scattering of fast ($Ze^2/\hbar v \ll 1$) (initially unpolarized) electrons in the Coulomb field of a nucleus. Find also the induced polarization for the Coulomb scattering of initially unpolarized positrons.

Hint: Spin-orbital interaction was discussed in Problem 13.59. To calculate the amplitude in second-order perturbation theory, first consider a screened Coulomb potential, $U_0(r) = -(Ze^2/r)e^{-r/R}$, and take the $R \to \infty$ in the end of the calculation (compare to Problem 13.54).

Solution

To the first order in interaction

$$\hat{U} = U_0(r) + \frac{\hbar^2}{4m^2c^2 r}\frac{\partial U_0(r)}{\partial r} \hat{\boldsymbol{\sigma}} \cdot \hat{\mathbf{l}}, \quad \text{where} \quad U_0(r) = -\frac{Ze^2}{r}e^{-r/R},$$

the scattering amplitude has the form:

$$f^{(1)}(\mathbf{k}, \mathbf{k}_0) = \frac{2mZe^2}{\hbar^2[(\mathbf{k}-\mathbf{k}_0)^2 + R^{-2}]}\left\{1 - i\frac{\hbar^2}{4m^2c^2}[\mathbf{k}_0 \times \mathbf{k}]\cdot\hat{\boldsymbol{\sigma}}\right\} \equiv$$
$$A^{(1)} + iB^{(1)}\boldsymbol{\nu}\cdot\hat{\boldsymbol{\sigma}}. \tag{1}$$

See Eqs. (4) and (5) from Problem 13.59. Functions $A^{(1)}$ and $B^{(1)}$ are real, and so in the first order of perturbation theory there is no polarization from scattering (see Eq. (XIII.24)).

The general expression for the second-order amplitude, as in the case of spinless particles (see Problem 13.10), is

$$\hat{f}^{(2)}(\mathbf{k}, \mathbf{k}_0) = \frac{1}{2\pi^2}\int \hat{f}^{(1)}(\mathbf{k}, \boldsymbol{\kappa})\hat{f}^{(1)}(\boldsymbol{\kappa}, \mathbf{k}_0)\frac{d^3\kappa}{\kappa^2 - k^2 - i\varepsilon}. \tag{2}$$

To calculate a spin-dependent part of the scattering amplitude, $A^{(2)}$, it suffices to retain only the first-order terms in Eq. (2) (including $A^{(2)}$ into both amplitudes $f^{(1)}$ in (2) would result in the additional small relativistic factor $\sim (v/c)^2$, which is beyond the accuracy of the approximation used). Thus $A^{(2)}$ coincides with the scattering amplitude in second-order perturbation theory for the potential, $U_0(r)$. For a Yukawa potential (corresponding to the screened Coulomb interaction), it was calculated in Problem 13.12. Using Eq. (5) of this problem, we obtain (with substitution $\alpha \to -Ze^2$):

$$\text{Im } A^{(2)} = \frac{4(Ze^2 m)^2}{\hbar^4 k q^2}\ln qR, \quad R \gg \frac{1}{k}. \tag{3}$$

For the spin-dependent part of the amplitude, $f^{(2)}$, according to Eqs. (2) and (1) we find:

$$\hat{\boldsymbol{\sigma}} \cdot \boldsymbol{\nu} \mathrm{Im}\, B^{(2)} = \frac{1}{4\pi} \left(\frac{Ze^2}{\hbar c}\right)^2 \mathbf{k} \cdot \hat{\boldsymbol{\sigma}} \int \frac{[\mathbf{q} \times \boldsymbol{\kappa}] d\Omega_\kappa}{[(\mathbf{k}_0 - \boldsymbol{\kappa})^2 + R^{-2}][(\mathbf{k} - \boldsymbol{\kappa})^2 + R^{-2}]}. \quad (4)$$

Here, we should put $|\boldsymbol{\kappa}| = |\mathbf{k}| = |\mathbf{k}_0|$ for the calculation of the imaginary part (see, for example, Problem 13.11). The corresponding integral has the form:

$$\int \frac{\boldsymbol{\kappa}\, d\Omega_\kappa}{[(\mathbf{k}_0 - \boldsymbol{\kappa})^2 + R^{-2}][((\mathbf{k} - \boldsymbol{\kappa})^2 + R^{-2}]} = C_1(\mathbf{k} + \mathbf{k}_0) + C_2 \mathbf{q}. \quad (5)$$

After multiplying it by $(\mathbf{k} + \mathbf{k}_0)$, we obtain

$$(\mathbf{k}_0 + \mathbf{k})^2 C_1 = \frac{1}{2} \int d\Omega_\kappa \left\{ -\frac{1}{(\mathbf{k}_0 - \boldsymbol{\kappa})^2 + R^{-2}} - \frac{1}{(\mathbf{k} - \boldsymbol{\kappa})^2 + R^{-2}} + \frac{4k^2 + 2R^{-2}}{[(\mathbf{k}_0 - \boldsymbol{\kappa})^2 + R^{-2}][(\mathbf{k} - \boldsymbol{\kappa})^2 + R^{-2}]} \right\}.$$

The first two integrals are easily calculated, and the third one could be expressed in terms of the imaginary part of the scattering amplitude for the Yukawa potential (see Problem 13.12). As a result, we obtain for $kR \gg 1$ (remember that $|\boldsymbol{\kappa}| = k$):

$$(\mathbf{k}_0 + \mathbf{k})^2 C_1 = 2\pi \left[-\frac{1}{k^2} \ln 2kR + \frac{4}{q^2} \ln qR \right].$$

Taking into account the fact that $C_2 q$ in Eq. (5) does not contribute to integral (4), and that $(\mathbf{k}_0 + \mathbf{k})^2 = 4k^2 \cos^2(\theta/2)$, we find:

$$\mathrm{Im}\, B^{(2)} = \frac{1}{4} \left(\frac{Ze^2}{\hbar c}\right)^2 \frac{\sin\theta}{\cos^2(\theta/2)} \left[\frac{\ln 2kR}{k} - \frac{4k \ln qR}{q^2} \right]. \quad (6)$$

Now, using Eqs. (1), (3), and (6) along with Eq. (XIII.24), we obtain the electron polarization for scattering in the Coulomb field of the nucleus:

$$\mathbf{P} \approx \frac{2}{(A^{(1)})^2} \{ \mathrm{Im}\, A^{(2)} B^{(1)} - A^{(1)} \mathrm{Im}\, B^{(2)} \} \boldsymbol{\nu} =$$

$$2 \frac{Ze^2}{\hbar c} \frac{v}{c} \frac{\sin^3(\theta/2)}{\cos(\theta/2)} \ln\left(\sin\frac{\theta}{2}\right) \boldsymbol{\nu}; \quad \boldsymbol{\nu} = \frac{[\mathbf{k}_0 \times \mathbf{k}]}{|[\mathbf{k}_0 \times \mathbf{k}]|}. \quad (7)$$

We emphasize that the cut-off, introduced for regularization purposes, disappears from the final result.

For the scattering of positrons, the polarization vector has the opposite direction (the amplitude $f^{(1)}$ changes sign, while $f^{(2)}$ remains unchanged).

Problem 13.63

The interaction of particles with spin $s = 1/2$ in an external field has the form $\hat{U} = U_0(r) + U_1(r)\hat{\mathbf{l}} \cdot \hat{\boldsymbol{\sigma}}$. Find the phase shifts $\delta_l^{\pm}$:

a) in the Born approximation;
b) in the quasi-classical approximation.

Calculate also the scattering amplitude in the eikonal approximation, from the partial-wave expansion, Eq. (XIII.25).

Solution

The solution of the Schrödinger equation, corresponding to a definite value of the angular momentum squared, $l(l+1)$, the total angular momentum, $j = l \pm 1/2$, and its projection, j_z, has the form:

$$\psi_{kjl j_z} = \tilde{\psi}_{jl j_z} R_{kjl}(r), \quad E = \frac{\hbar^2 k^2}{2m}.$$

Here, $\tilde{\psi}_{jl j_z}$ is the spin and angular part of the wavefunction (see Problems 5.24 and 5.25, and note that in this problem its explicit form is not important). Since

$$\hat{\boldsymbol{\sigma}} \cdot \hat{\mathbf{l}} \tilde{\psi}_{jl j_z} = \left[j(j+1) - l(l+1) - \frac{3}{4} \right] \tilde{\psi}_{jl j_z},$$

then we note that the radial Schrödinger equation for R_{kjl} has the same form as in the case of a spinless particle with the angular momentum l in the potential

$$U_l^{\pm}(r) = U_0(r) \pm \left(l + \frac{1}{2} \mp \frac{1}{2} \right) U_1(r).$$

The upper and lower signs correspond to the values $j = l \pm 1/2$. Replacing the potential $U(r)$ by $U_l^{\pm}$ in Eqs. (XIII.12–XIII.14) determines the phase shifts, $\delta_l^{\pm}$, in expansions (XIII.25). Therefore the sought-after generalization of Eq. (XIII.14) has the form:

$$\delta_l^{\pm} \approx -\frac{m}{2\hbar v} \int_{-\infty}^{\infty} \left\{ U_0(\sqrt{\rho^2 + z^2}) \pm k\rho U_1(\sqrt{\rho^2 + z^2}) \right\} dz, \quad l = \rho k \gg 1.$$

Using this relation and Eq. (XIII.25), just as in the case of spinless particles, we can obtain expressions for the invariant amplitudes A and B in the eikonal approximation. It is important to remember that

$$\sin\theta P_l'(\cos\theta) \approx -\frac{\partial}{\partial\theta} J_0(l\theta) = l J_1(l\theta); \quad l \gg 1, \quad \theta \ll 1.$$

The scattering amplitude operator (*i.e.*, a matrix function in the spin space) in the eikonal approximation has the following simple form:

$$\hat{f} = \frac{ik}{2\pi} \iint \left\{1 - \exp\{2i\hat{\delta}(\mathbf{k}_0, \boldsymbol{\rho})\}\right\} e^{-i\mathbf{q}\cdot\boldsymbol{\rho}} d^2\rho, \tag{1}$$

which becomes physically more transparent if we introduce an operator for the quasi-classical phase-shift (compare to Eq. (XIII.19)):

$$\hat{\delta}(\mathbf{k}_0, \boldsymbol{\rho}) = -\frac{m}{2\hbar v} \int_{-\infty}^{\infty} \{U_0(\boldsymbol{\rho}, z) + U_1(\boldsymbol{\rho}, z)[\boldsymbol{\rho} \times \mathbf{k}_0] \cdot \hat{\boldsymbol{\sigma}}\} dz. \tag{2}$$

Using the identity

$$\exp\{i\alpha\hat{\boldsymbol{\sigma}} \cdot \boldsymbol{\nu}\} = \cos\alpha + i\hat{\boldsymbol{\sigma}} \cdot \boldsymbol{\nu} \sin\alpha, \quad \text{where} \quad \nu^2 = 1,$$

it is easy to obtain the eikonal expressions for amplitudes A and B in Eq. (XIII.22).

Problem 13.64

In the case of a spin-1/2 particle colliding with a spinless particle, find a relation between the scattering amplitude in the helicity representation (see Problem 5.20) and the *invariant functions* A and B in Eq. (XIII.22).

Solution

Let the plane, where the scattering occurs, be the plane (x, z), where the z axis is directed along momentum $\mathbf{p}_0$ of the particle with spin $s = 1/2$ in the center-of-inertia system before the collision. We have $\mathbf{p}_0 = (0, 0, p)$, $\mathbf{p} = (p\sin\theta, 0, p\cos\theta)$, and $\boldsymbol{\nu} = (0, 1, 0)$, so that $\hat{\boldsymbol{\sigma}} \cdot \boldsymbol{\nu} = \hat{\sigma}_y$.

Taking into account the fact that the helical states, φ_λ before the collision and χ_μ after the collision, are described by the spinors (see Problem 5.20):

$$\varphi_{1/2} = \begin{pmatrix} 1 \\ 0 \end{pmatrix}, \quad \varphi_{-1/2} = \begin{pmatrix} 0 \\ 1 \end{pmatrix}; \quad \chi_{1/2} = \begin{pmatrix} \cos(\theta/2) \\ \sin(\theta/2) \end{pmatrix},$$

$$\chi_{-1/2} = \begin{pmatrix} -\sin(\theta/2) \\ \cos(\theta/2) \end{pmatrix},$$

we find the helical amplitudes $f_{\lambda\mu} = \chi_\mu^* \hat{f} \varphi_\lambda$:

$$f_{1/2,1/2} = f_{-1/2,-1/2} = \cos(\theta/2) \cdot A - \sin(\theta/2) \cdot B,$$
$$f_{1/2,-1/2} = f_{-1/2,1/2} = -\sin(\theta/2) \cdot A - \cos(\theta/2) \cdot B.$$

Problem 13.65

Upon the collision of two spinless particles, a reaction occurs and two particles are created, one of which has spin $s = 1$, and the other $s = 0$. The intrinsic parities of all particles are positive.

Using the vector representation (see Problem 5.26) to describe the spin states of the particle with $s = 1$, prove that the spin structure of the amplitude for the reaction considered is described by the expression:

$$\langle f|\hat{f}|i\rangle = \mathbf{a}^* \cdot [\mathbf{p}_0 \times \mathbf{p}_1] f(E, \theta),$$

where $\mathbf{a}$ is the spin function, and $\mathbf{p}_0$ and $\mathbf{p}_1$ are momenta before and after the collision.

Present a partial-wave expansion of $f(E, \theta)$.

Also determine the spin structure of the amplitude in the case, where the particle with spin $s = 1$ has negative intrinsic parity.

Solution

The wavefunction describing the relative motion of the colliding particles at large distances has the form:

$$e^{ikz} \approx \sum_l i^l (2l+1) P_l(\cos\theta) \frac{1}{2kr} \{e^{-i(kr-\pi l/2)} - e^{i(kr-\pi l/2)}\}.$$

The term $i^l(2l+1)P_l(\cos\theta)e^{-i(kr-\pi l/2)}/2kr$ of this sum describes the state of the colliding particles (before interaction) with angular momentum $l = 1$, parity $I_l = (-1)^l$, and zero projection of the angular momentum onto the z axis (directed along the momentum $\mathbf{p}_0 = \hbar\mathbf{k}$), i.e., $j_z = l_z = 0$. The particles appearing in this reaction channel, are described (when the separation between them, r_1, is large) by the following outgoing wave:

$$\eta_l(E) \Phi_{j=l,I_l,j_z=0}(\mathbf{n}_1) \frac{1}{r_1} e^{ik_1 r_1}, \quad \mathbf{n}_1 = \frac{\mathbf{r}_1}{r_1}. \tag{1}$$

$\Phi_{jIj_z}(\mathbf{n})$ describes the spin-angular dependence of the escaping particles. The value of the parameter η_l is determined by the intensity of the interaction. To determine the explicit form of $\Phi(\mathbf{n}_1)$, we note that the angular momentum of the final state coincides with the initial one, l. Indeed, for total angular momentum $j = l$ and spin $s = 1$, the angular momentum could only take the values $l' = l, l \pm 1$. With the parity of the states accounted for, it follows that $l' = l$ (and $l \neq 0$). Thus,

$$\Phi_{lI_l 0}(\mathbf{n}_1) = \sum_{m=0,\pm 1} C^{l0}_{lm,1,-m} Y_{lm}(\mathbf{n}_1) \chi_{-m}. \tag{2}$$

Using the explicit form of the spherical functions, Y_{lm}, the components of the vector χ_m (see Problem 3.41), and the Clebsch–Gordan coefficients,

$$Y_{l,\pm 1} = \mp i^l \left[\frac{2l+1}{4\pi l(l+1)}\right]^{1/2} \sin\theta P_l'(\cos\theta)e^{\pm i\varphi};$$

$$\chi_{\pm 1} = \mp \frac{i}{\sqrt{2}}(1, \pm i, 0); \quad C_{l,\pm 1,1,\mp 1}^{l0} = \mp(-1)^l \frac{1}{\sqrt{2}}, \quad C_{l0,10}^{l0} = 0,$$

we transform Eq. (2) into the form:

$$\mathbf{\Phi}_{lI_l 0} = \gamma_l \sin\theta_1 P_l'(\cos\theta_1)(-\sin\varphi_1, \cos\varphi_1, 0),$$

where

$$\gamma_l = (-i)^l \left[\frac{2l+1}{4\pi l(l+1)}\right]^{1/2}.$$

Recognizing that the vector with components $\sin\theta_1(-\sin\varphi_1, \cos\varphi_1, 0)$ is equal to $[\mathbf{p}_0 \times \mathbf{p}_1]/p_0 p_1$, and performing a summation over l, we obtain an equation for the vector amplitude of the "scattered" wave, the coefficient in front of $e^{ik_1 r_1}/r_1$ in the asymptote of the wavefunction for $r_1 \to \infty$ in the reaction channel considered:

$$\mathbf{\Phi}(\mathbf{n}_1) = f(E, \theta)[\mathbf{p}_0 \times \mathbf{p}_1], \quad f(E, \theta) = \sum_l \tilde{\eta}_l(E) P_l'(\cos\theta), \tag{3}$$

with $\tilde{\eta}_l(E) = \eta_l(E)\gamma_l/p_0 p_1$. The differential cross-section for this reaction is summed over the spin states, $d\sigma/d\Omega = (v_1/v_B)|\mathbf{\Phi}|^2$, where $v_{0,1}$ are the velocities of the particles' relative motion in the initial and final states. The total reaction cross-section is

$$\sigma_r = \sum_l \sigma_{r,l}; \quad \sigma_{r,l} = \frac{v_1}{v_0}|\gamma_l|^{-2} p_0^2 p_1^2 |\tilde{\eta}_l|^2 = \frac{v_1}{v_B}|\eta_l(E)|^2.$$

From the unitarity condition of the S-matrix, a constraint on the partial reaction cross-section follows: $\sigma_{r,l} \leq (2l+1)\pi/k_0^2$.

The reaction amplitude for the creation of the particle with $s = 1$ in a specific spin state, described by the polarization vector $\mathbf{a}$, is determined by the expression:

$$\langle f|\hat{f}|i\rangle = (\mathbf{a}^* \cdot \mathbf{\Phi}) = f(E, \theta)(\mathbf{a}^* \cdot [\mathbf{p}_0 \times \mathbf{p}_1]). \tag{4}$$

This was evident before (and does not really require the preceding calculation) from the fact that the reaction amplitude is a scalar quantity. Indeed, it is the only possible scalar combination which could be formed from the vectors $\mathbf{p}_0$, $\mathbf{p}_1$, and $\mathbf{a}$ (due to the superposition principle, the polarization vector must enter linearly). We should take into account that $\mathbf{a}$ is an axial vector (or pseudo-vector), since under inversion, $\hat{I}\mathbf{a} = +\mathbf{a}$, due to the positive intrinsic parity of the particle with $s = 1$. Let us also note that according to Eq. (4), the particle with $s = 1$, that is created in the reaction considered, becomes linearly polarized in the direction perpendicular to the reaction plane (the spin projection on this direction has a definite value, equal to zero).

In the case of a reaction when the particle with spin $s = 1$ has negative intrinsic parity, its polarization vector, $\mathbf{v}$, is a polar vector (since under inversion $\hat{I}\mathbf{v} = -\mathbf{v}$). So now, from the condition that the transition amplitude is a scalar, it follows that its spin structure has the form:[364]

$$\langle f|\hat{f}|i\rangle = \mathbf{v}^* \cdot \{f_1(E, \theta)\mathbf{p}_0 + f_2(E, \theta)\mathbf{p}_1\}. \tag{5}$$

The appearance of the two invariant amplitudes, $f_{1,2}$, is connected with the fact that for a given angular momentum, l, of the colliding particles, the angular momentum of the particles in the final state could take two values: $l' = l \pm 1$.

Problem 13.66

A spinless particle is scattered from a system of identical spin-1/2 scattering centers, distributed about the space. Interaction with a single center is described by the expression $\hat{U} = U_0(r) + U_1(r)\hat{\mathbf{l}} \cdot \hat{\boldsymbol{\sigma}}$. Analyze the amplitude and differential scattering cross-section in the Born approximation in the case of non-polarized centers ($\mathbf{P}_n = 0$). Compare to the scattering from the same system of spinless particles.

Solution

In the Born approximation, the scattering amplitude is described by the following expression (compare to Problems 13.7 and 13.8)

$$\hat{f} = \sum_n \{A_n(q) + iB_0(q)\hat{\boldsymbol{\sigma}}_n \cdot \boldsymbol{\nu}\} \exp\{-i\mathbf{q} \cdot \mathbf{a}_n\}. \tag{1}$$

Here, $\mathbf{a}_n$ is the radius vector of the nth center, and the real functions $A_0(q)$ and $B_0(q)$ were defined in Problem 13.59.

The differential scattering cross-section, averaged over the initial spin states of the scattering centers and summed over their final spin states, has the form:

$$\left\langle \frac{d\sigma}{d\Omega} \right\rangle = A_0^2(q) \left| \sum_n e^{-i\mathbf{q} \cdot \mathbf{a}_n} \right|^2 + 2A_0(q)B_0(q) \sum_{k \neq n} \mathbf{P}_n \cdot \boldsymbol{\nu} \sin[\mathbf{q} \cdot (\mathbf{a}_n - \mathbf{a}_k)] +$$

$$NB_0^2(q) + B_0^2(q) \sum_{k \neq n} \overline{(\boldsymbol{\sigma}_n \cdot \boldsymbol{\nu})(\boldsymbol{\sigma}_k \cdot \boldsymbol{\nu})} \cos[\mathbf{q} \cdot (\mathbf{a}_n - \mathbf{a}_k)], \tag{2}$$

where the line means averaging over the initial spin states of the centers. If there is no correlation between the states of single centers, then

$$\overline{(\boldsymbol{\sigma}_n \cdot \boldsymbol{\nu})(\boldsymbol{\sigma}_k \cdot \boldsymbol{\nu})} = (\mathbf{P}_n \cdot \boldsymbol{\nu})(\mathbf{P}_k \cdot \boldsymbol{\nu}), \quad n \neq k.$$

---

[364] Here, as well as in the previous problem, it is important that all other spinless particles that take part in the reaction have positive intrinsic parity (or, more precisely, their product must be positive; in the opposite case, then Eqs. (4) and (5) must be exchanged with one another).

In the case of unpolarized centers, $\mathbf{P}_n = 0$, only the first and third terms in Eq. (2) are different from zero. The first of them is determined by the spin-independent part of the interaction, $U_0(r)$, and has the same form as for the scattering from spinless centers (compare to Problem 13.8). The term $NB_0^2(q)$ is determined by the spin-orbital interaction. Its notable property, is its proportionality to the number of scattering centers, which is due to the lack of coherence between them. This term corresponds to scattering involving a spin-flip (so that we can indicate which particular scattering center was involved in the scattering; therefore, there is no interference).

Problem 13.67

How to generalize the optical theorem to describe scattering of quantum particles with non-zero spin?

Solution

From the unitarity of the S-matrix, the optical theorem is

$$\operatorname{Im} \langle \mathbf{p}, \alpha | \hat{f} | \mathbf{p}, \alpha \rangle = \frac{k}{4\pi} \sigma_{tot}(\mathbf{p}, \alpha),$$

where $\mathbf{p} = \hbar \mathbf{k}$ is the relative momentum of the colliding particles, and α characterizes their spin state. On the left-hand side of the equation is the imaginary part of the elastic forward scattering amplitude, $\theta = 0$, without change of the particle spin state. On the right-hand side is the total scattering cross-section (including inelastic collisions) for the same spin state.

13.6 Analytic properties of the scattering amplitude

Problem 13.68

Analyze the analytical properties and dispersion relation for the scattering amplitude on a zero-range potential (see Problem 13.20). Consider separately the cases with and without a bound state in the potential. Compare to Eq. (XIII.27).

Solution

The scattering amplitude for a zero-range potential is (see Problem 13.20)

$$f(E) = \frac{1}{-1/a_0 - (i/\hbar)\sqrt{2mE}}. \tag{1}$$

As an analytical function of the complex variable E, it has branch points at $E = 0$ and $E = \infty$ and a pole at the point E_0, for which $\sqrt{E_0} = i\hbar a_0/\sqrt{2m}$. Making the branch cut along the real semiaxis, as usual (see Fig. 13.7), and choosing the phase on the upper side of the cut to be equal to $\varphi = 0$, we see that for $a_0 > 0$, pole E_0 is located on the physical sheet and corresponds to a bound state in the zero-range potential (compare to Problem 2.30). When $a_0 < 0$, the pole is located on the unphysical Riemann sheet and corresponds to a virtual level.

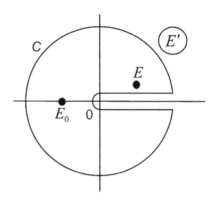

Fig. 13.7

Let us consider the integral over the contour C in Fig. 13.7:

$$\frac{1}{2\pi i} \int_C \frac{f(E')dE'}{(E'-E)}. \tag{2}$$

Using Cauchy's theorem, and sending the contour circle radius to infinity, $R_E \to \infty$, in the case of $a_0 > 0$ we obtain:

$$f(E) = -\frac{\hbar^2}{ma_0(E-E_0)} + \frac{1}{\pi}\int_0^\infty \frac{\mathrm{Im}\, f(E')dE'}{E'-E}. \tag{3}$$

We have used the fact that the value of the amplitude jump on the cut (for $E' > 0$) coincides with $2i\mathrm{Im}\, f(E')$ (on the lower side of the cut, $\sqrt{E} = -\sqrt{|E|}$), and that the value of the integral is determined by the contribution of two poles: E and E_0.

In the case $a_0 < 0$, pole E_0 is on the unphysical sheet. It does not contribute to the value of integral (2), and the dispersion relation has a form analogous to (3), but without the pole term.

The dispersion relation in the case of a zero-range potential can be found by direct calculation. Substituting the imaginary part of the scattering amplitude (1), equal to

$$\mathrm{Im}\, f(E) = \sqrt{\frac{\hbar^2 E}{2m}}\frac{1}{|E_0|+E}, \quad E > 0,$$

into the integral in Eq. (3) and integrating, we obtain:

$$\tilde{f}(E) = \frac{1}{\pi}\sqrt{\frac{\hbar^2}{2m}}\int_0^\infty \frac{\sqrt{E'}dE'}{(E'-E)(|E_0|+E')} = \frac{\hbar}{\sqrt{2m}}\frac{1}{\sqrt{|E_0|}-i\sqrt{E}}.$$

In the case $a_0 < 0$ (without a bound state), this coincides with the scattering amplitude, Eq. (1). In the case $a_0 > 0$, we obtain the scattering amplitude after adding, according to Eq. (3), the pole term.

The dispersion relations found differ from Eq. (XIII.27) by the absence of the term $f_B \propto \int U dV$. This has a simple explanation. A zero-range potential can be obtained from a potential of finite radius, R, by passage to the limit $R \to 0$ for which $U_0 R^2 = $ const, so that we would have $f_B \propto U_0 R^3 \to 0$. In a finite-range potential, the scattering amplitude in the limit $E \to \infty$ coincides with the Born one. In order to have a condition as in Eq. (2), we subtract f_B from f on a circle of the finite radius (for a zero-range potential, the scattering amplitude goes to zero for $E \to \infty$).

In conclusion, we note that the residue at the point $E = E_0$ in Eq. (3) is equal to $-\hbar^2/ma_0 = -\hbar^2 A^2/2m$. The coefficient $A = \sqrt{2/a_0} = \sqrt{2\kappa_0}$ coincides with the normalization coefficient in the wavefunction,

$$\psi_0 = A \frac{\exp\{-\kappa_0 r\}}{\sqrt{4\pi} r},$$

of the bound state in the zero-range potential (see Problem 4.10), in accordance with Eqs. (XIII.27) and (XIII.28).

Problem 13.69

Using the dispersion relation, prove that there is the following restriction on the energy-dependence of the particle-scattering cross-section in a repulsive potential, $U(r) \geq 0$:

$$\int_0^\infty \frac{\sigma(E)}{\sqrt{E}} dE < 4\pi^2 \frac{\sqrt{2m}}{\hbar} \int_0^\infty r^2 U(r) dr.$$

Solution

Since according to the optical theorem, Im $f(E, 0) = k\sigma(E)/4\pi$, then from the dispersion relation, Eq. (XIII.27), for energy $E = 0$, it follows that

$$f(E = 0) = -\frac{m}{2\pi\hbar^2} \int U(r) dV + \frac{\sqrt{2m}}{4\pi^2 \hbar} \int_0^\infty \frac{\sigma(E) dE}{\sqrt{E}}. \quad (1)$$

Here we have also used the fact that in the repulsive potential $U(r) \geq 0$, there are no bound states. Due to the fact that in such a potential we have $f(E = 0) < 0$ (see Problems 13.16 and 13.31), we immediately obtain the inequality given in the problem condition.

Let us note that in the case of a "weak" potential, $U_0 \ll \hbar^2/mR^2$, the inequality considered becomes obvious (since $f \propto U_0$, while $\sigma \propto U_0^2$) and is well-fulfilled. But for a "strong" repulsive potential,[365] $U_0 \gg \hbar^2/mR^2$, $|f_B| \gg |f(E=0)|$, and the terms of the inequality are approximately the same:

$$\int_0^\infty \frac{\sigma(E)}{\sqrt{E}} dE \approx 4\pi^2 \frac{\sqrt{2m}}{\hbar} \int_0^\infty U(r) r^2 dr. \qquad (2)$$

In conclusion, we note that from relation (1), obtained in Problem 13.16 by another method, it follows that in a repulsive potential the Born approximation for $E = 0$ greatly overestimates the scattering cross-section.

Problem 13.70

Prove the relation

$$\int_0^\infty \frac{\sigma(E)}{\sqrt{E}} dE < \pi \sqrt{\frac{2\pi \hbar^2}{m} \sigma(0)}$$

for the scattering in an attractive potential, $U(r) \leq 0$, where there are no bound states (the well is not deep enough). Here $\sigma(0) = 4\pi a_0^2$ is the scattering cross-section for $E = 0$. In what case both sides of the inequality are close?

Solution

Because there are no bound states in the attractive potential, relation (1) from the previous problem is still valid. Now, however, all three terms in it are positive. So in particular, the result from Problem 13.16 follows for the attractive potential. Then the inequality from the problem condition in the case of a "weak" potential must be conservatively valid (compare to the previous problem). When this potential approaches the "critical" one, where a bound state appears, we obtain $f(E=0) \to \infty$, so that $f(E=0) \gg f_B$. The values of both inequality parts are close to one another, so that

$$\int_0^\infty \frac{\sigma(E)}{\sqrt{E}} dE \approx \pi \sqrt{\frac{2\pi \hbar^2}{m} \sigma(0)}. \qquad (1)$$

However, this relation in the case of a "shallow" (real or virtual) s-level in the potential is evident. According to Eq. (XIII.16) in the region of small energies, the scattering cross-section is anomalously large.

$$\sigma(E) \approx \frac{2\pi \hbar^2}{m(E+\varepsilon_0)}, \quad \varepsilon_0 \ll \frac{\hbar^2}{mR^2}.$$

---

[365] As an example, for a square potential barrier of radius R, for $U_0 \to \infty$, we also have $|f_B| \to \infty$, but $f(E=0) = -R = \text{const}$.

Particle collisions

This region has the dominant contribution to the value of the integral; by calculating it, we prove relation (1).

Problem 13.71

Using only the condition of unitary and the dispersion relation (for $q^2 \neq 0$), it is possible to reconstruct[366] the scattering potential in a series over powers of the interaction potential, by the known expression for scattering amplitude in the Born approximation.

As an example, prove that the calculation of the second-order perturbative amplitude for $q^2 = 0$ reproduces the result of perturbation theory over the potential, based on the Schrödinger equation (see Problem 13.10).

Solution

The iteration procedure for the calculation of scattering amplitude $f = \sum_n f^{(n)}$ is the following. First, by the known amplitude of first approximation, $f^{(1)} = f_B$, we can use the unitarity condition, Eq. (XIII.26), to find the imaginary part of the second, order amplitude,[367] Im $f^{(2)}(E, q^2)$. Then, using the dispersion relation for $q^2 \neq 0$ (analogous to Eq. (XIII.27)), we can find the entire amplitude, $f^{(2)}$. Higher-order approximation terms are calculated in the same way.

In particular, for the value $q^2 = 0$ we easily find

$$f^{(2)}(E, q^2 = 0) = \frac{1}{4\pi} \int_0^\infty \frac{dE'}{\sqrt{E'}(E' - E)} \int_0^{4E'} f_B^2(x) dx. \tag{1}$$

The imaginary part of the scattering amplitude with the help of the optical theorem is expressed in terms of the scattering cross-section in the Born approximation. For simplicity, we use the system of units $\hbar = 2m = 1$. With this, $E = k^2$. The Born amplitude, f_B, is written in the form $f_B(q^2)$.

On the other hand, in the second order of perturbation theory according to the result from Problem 13.10, we have:

$$f^{(2)}(E, q^2 = 0) = \frac{1}{2\pi^2} \int \frac{f_B^2((k_0 - \kappa)^2)}{\kappa^2 - E} d^3\kappa =$$

$$\frac{1}{2\pi} \int_0^\infty \frac{\kappa d\kappa}{\sqrt{E}(\kappa^2 - E)} \int_{(\kappa-\sqrt{E})^2}^{(\kappa+\sqrt{E})^2} f_B^2(x) dx. \tag{2}$$

---

[366] Let us emphasize that the Schrödinger equation is not used here.
[367] Here, in principle, the relation obtained gives the imaginary part of the amplitude also for the unphysical values of energy $0 < E < \hbar^2 q^2/8m$.

To prove that Eqs. (1) and (2) are actually the same, we make the following transformations (note that their equivalence becomes obvious in the $E \to 0$ limit, if we integrate Eq. (1) by parts): First, we divide the "internal" integral in Eq. (2) into two, with the point $x = 0$ as one of the integration limits. Then, in the first of two terms that appear as a result, use the substitution $2\kappa' = \kappa + \sqrt{E}$, while in the second, use $2\kappa' = \kappa + \sqrt{E}$. In the resulting integrals over κ', again divide the integration domains into two: from 0 to ∞ and from 0 to $\pm\sqrt{E}/2$. The contributions of the latter integration domains cancel each other out, while the sum of the contributions of the former reproduces (after the substitution, $E' = (\kappa')^2$) Eq. (1).

Problem 13.72

Consider the interaction for partial waves with $l \geq L_0 \equiv kR \gg 1$ to be negligibly small, and obtain the upper bound for the scattering amplitude of spinless particles at high energies for different scattering angles.[368]

Solution

In the scattering amplitude expansion over partial waves, Eq. (XIII.9), we have the inequality $|\varphi_l| \leq 1/k$, so that in accordance with the problem condition we obtain:

$$|f(k, \theta)| \leq \frac{1}{k} \sum_{l=0}^{L_0} (2l+1) |P_l(\cos\theta)|. \tag{1}$$

Replacing the summation by integration, for angles $\theta = 0$ and $\theta = \pi$ we find:

$$|f| \leq \frac{L_0^2}{k} = kR^2, \quad \theta = 0; \, \pi. \tag{2}$$

For scattering angles, $\theta \neq 0, \pi$, we have:

$$|P_l(\cos\theta)| \leq \left(\frac{2}{\pi l \sin\theta}\right)^{1/2},$$

and so we obtain:

$$|f(\theta)| \leq \frac{4}{3} \left(\frac{2}{\pi \sin\theta}\right)^{1/2} R(kR)^{1/2}. \tag{3}$$

Note that due to the Legendre polynomials oscillation for $\theta \neq 0$, such a restriction for arbitrary scattering angle is too weak and conservatively valid. Indeed, the restriction

---

[368] In Problems 13.72–13.75, general constraints on the interaction amplitudes and cross-sections for large energies are discussed. These restrictions are connected, basically, with the possibility of neglecting interaction at distances that are larger than the radius of the potential, R. This situation occurs frequently in elementary particle physics. Here, the effective interaction radius increases with energy, but not faster than $\propto \ln(E/E_0)$ (see Problem 13.53). The necessity of using relativistic kinematics is practically not reflected in the results.

on the value of total elastic scattering cross-section which follows from inequality (3), $\sigma_{el} = \int |f|^2 d\Omega \leq CR^2\, kR$, is uninteresting, since $\sigma_{el} \leq \sigma_{tot} \leq 4\pi R^2$ and $kR \gg 1$.

Problem 13.73

Under the conditions of the previous problem, obtain the lower bound on the value of the elastic scattering cross-section, σ_{el}, of fast particles for a given value, σ_{tot}, of the total collision cross-section.

Solution

From the expressions for partial cross-sections (total and elastic cross-sections),

$$\sigma_{tot}^{(l)} = 2\pi(2l+1)\frac{1}{k^2}(1 - \text{Re } S_l),$$

$$\sigma_{el}^{(l)} = \pi(2l+1)\frac{1}{k^2}|1 - S_l|^2 = \pi(2l+1)\frac{1}{k^2}(1 - 2\,\text{Re } S_l + |S_l|^2),$$

it follows that for a given value of $\sigma_{tot}^{(l)}$, the quantity $\sigma_{el}^{(l)}$ is minimal for $\text{Im } S_l = 0$. Thus,

$$\sigma_{el} = \sum_l \sigma_{el}^{(l)} \geq \tilde{\sigma}_{el} = \frac{\pi}{k^2} \sum_{l=0}^{L_0} (2l+1)(1 - \alpha_l)^2,$$

where $\alpha = \text{Re } S_l$. To find the minimum value of $\tilde{\sigma}_{el}$ as a function of variables α_l for the given $\sigma_{tot} = \sum_l \sigma_{tot}^{(l)}$, we use Lagrange's method of undetermined multipliers and introduce $A(\alpha_l) = \tilde{\sigma}_{el} - \lambda \sigma_{tot}$. From the extremality conditions (now all variables α_l could be varied independently) for $A(\alpha_l)$, we find that $\alpha_l = \text{const} = \alpha$ (no dependence on l). Replacing summation over l by integration and excluding α from the expressions for σ_{el} and σ_{tot}, we obtain the inequality

$$\sigma_{el} \geq \min \tilde{\sigma}_{el} = \frac{1}{4\pi R^2} \sigma_{tot}^2. \tag{1}$$

As mentioned in Problem 13.53, in the theory of strong elementary particle interactions there is a restriction on the possible growth of the interaction radius R with increase in energy:

$$R \leq R_0 \ln \frac{E}{E_0} \quad \text{for } E \to \infty. \tag{2}$$

From this, inequality (1) takes the form:

$$\sigma_{el}(E) \geq C \frac{\sigma_{tot}^2(E)}{\ln^2(E/E_0)}. \tag{3}$$

Problem 13.74

For large energies, find the upper bound for the real part of the amplitude for elastic forward scattering, $\theta = 0$, assuming that the total collision cross-section is known and that particle interaction is negligibly small for distances above R. What is the restriction for $|f(E, \theta = 0)|$?

Solution

Denoting $S_l = |S_l|e^{2i\delta_l}$ for the partial amplitudes in Eq. (XIII.9), we have the following expressions:

$$\text{Im } \varphi_l = \frac{1}{2k}(1 - |S_l|\cos 2\delta_l), \quad \text{Re}\varphi_l = \frac{1}{2k}|S_l|\sin 2\delta_l.$$

For this value of Im φ_l, the value of $|\text{Re}\varphi_l|$ is maximum when $|S_l| = 1$. In this case, inelastic scattering is absent and $\sigma_{tot} = \sigma_{el}$. So, writing Im $\varphi_l = (1 - \alpha_l)/2k$, we obtain

$$|\text{Re } f(E,0)| \leq \sum_l (2l+1)|\text{Re } \varphi_l| \leq \frac{1}{2k}\sum_{l=0}^{L_0}(2l+1)\sqrt{1-\alpha_l^2}.$$

Using the method from the previous problem, we obtain the following restriction on the real part of the elastic forward scattering amplitude for the total scattering cross-section:

$$|\text{Re } f(E,0)| \leq \frac{1}{2\sqrt{\pi}}\sqrt{\sigma_{tot}}kR\left(1 - \frac{\sigma_{tot}}{4\pi R^2}\right)^{1/2} \leq \frac{1}{2\sqrt{\pi}}\sqrt{\sigma_{tot}}kR. \quad (1)$$

So, taking into account the fact that $\sigma_{tot} \leq 4\pi R^2$ and the restriction on interaction radius growth with increase in energy (see the previous problem), we obtain

$$|f(E,0)| \leq Ck\sqrt{\sigma_{tot}(E)}\ln\frac{E}{E_0}. \quad (2)$$

Here, we used the optical theorem.

As mentioned before, restriction (1) suggests the absence of inelastic processes. Analogously (using Lagrange's method of undetermined multipliers), we can obtain a less strict restriction on the scattering amplitude for given values of both total, σ_{tot}, and inelastic, σ_{inel}, collision cross-section.

Problem 13.75

Prove that for large energies, the derivative with respect to θ at $\theta = 0$ of the imaginary part of the elastic scattering amplitude satisfies the following inequalities:

$$\frac{\sigma_{tot}}{32\pi} \leq -\frac{d}{dq^2}\ln \text{Im } f(E,q)\Big|_{q=0} \leq \frac{R^2}{4}, \quad q = 2k\sin\left(\frac{\theta}{2}\right),$$

where σ_{tot} is the total collision cross-section and R is the interaction radius. (At distances larger than R, the interaction is negligibly small.)

Verify explicitly that these inequalities are indeed satisfied by the diffraction scattering amplitude from Problem 13.57.

Solution

Since for the Legendre polynomial, $P'_l(1) = l(l+1)/2$, then using the scattering amplitude expansion over partial waves, Eq. (XIII.9), for high energies we obtain

$$\frac{d}{d\cos\theta}\text{Im } f(E,\theta)\bigg|_{\theta=0} \approx \sum_l l^3 \text{Im } \varphi_l. \tag{1}$$

Using the fact that Im $\varphi_l \leq 1/k$, we see that for a given total scattering cross-section, and so the imaginary part of the scattering amplitude, Im $f(E,0)$ (due to the optical theorem), the sum (1) takes the minimum value when

$$\text{Im } \varphi_l = \begin{cases} \frac{1}{k}, & l \leq L_1, \\ 0, & l > L_1. \end{cases} \tag{2}$$

The value of L_1 is determined by the total cross-section:

$$\sigma_{tot} = \frac{4\pi}{k}\sum_l (2l+1)\text{Im } \varphi_l \approx \frac{8\pi}{k^2}\sum_{l=0}^{L_1} l.$$

Replacing summation over l by integration, we obtain $L_1^2 = k^2\sigma_{tot}/4\pi$ and the following restriction:

$$-\frac{d}{dq^2}\text{Im } f(E,q^2)\bigg|_{q^2=0} \geq \frac{1}{128\pi^2}k\sigma_{tot}^2(E). \tag{3}$$

We emphasize that relations (2) correspond to the "saturation" of the total scattering cross-section by the lower partial waves. Here $\sigma_{tot} = \sigma_{el}$ (there are no inelastic processes), and moreover, for $l \leq L_1$ all of the phases are $\delta_l = \pi/2$, so that restriction (3) must be fulfilled conservatively.

In the same way, we note that the sum in (1) takes the maximum value in the case when total scattering cross-section is saturated by the higher partial waves and

$$\text{Im } \varphi_l = \begin{cases} 0, & l < L_2. \\ \frac{1}{k}, & L_2 \leq l \leq L_0 = kR. \end{cases}$$

We obtain the following restriction (now from above):

$$-\frac{d}{dq^2}\text{Im } f(E,q^2)\bigg|_{q^2=0} \leq \frac{kR^2\sigma_{tot}}{16\pi}\left(1 - \frac{\sigma_{tot}}{8\pi R^2}\right) \leq \frac{kR^2\sigma_{tot}}{16\pi}. \tag{4}$$

As an example, consider an impenetrable sphere of the radius R (see Problems 13.57 and 13.90). The diffraction scattering amplitude is

$$f_{difr} = i\frac{kR}{q}J_1(qR).$$

The restrictions of the problem condition, which are the direct result of relations (3) and (4), take the form of the following inequality (after cancelling out the factor $R^2/4$): $1/3 < 1/2 < 1$.

In conclusion, we should emphasize that, as mentioned above, restrictions on (3) and (4) assume the absence of inelastic processes. Using the method of Lagrange multipliers, just as in Problem 13.73, we can obtain weaker restrictions when the values of both total and inelastic collision cross-sections are given independently.

Problem 13.76

Prove unitarity of the scattering amplitude in the eikonal approximation.

Solution

The eikonal approximation assumes that only small scattering angles are important. For these angles, the right-hand part of relation (XIII.26) with the eikonal expression for the amplitude, Eq. (XIII.18), taken into account could be transformed to the form:

$$\frac{ik}{2\pi}\iint [S^*(\boldsymbol{\rho}) - 1][S(\boldsymbol{\rho}) - 1]\exp\{-i\mathbf{q}_\perp \cdot \boldsymbol{\rho}\}d^2\rho. \tag{1}$$

To perform the integration over angles which leads to Eq. (1), we use the relations:

$$\mathbf{k} - \mathbf{k}_0 \approx \mathbf{q}_\perp, \quad \mathbf{k}' - \mathbf{k}_0 \approx \mathbf{q}'_\perp, \quad \mathbf{k}' - \mathbf{k} \approx \mathbf{q}'_\perp - \mathbf{q}_\perp, \quad d\Omega' \approx \frac{1}{k^2}d^2q'_\perp.$$

Although $q_\perp, q'_\perp \ll k$, due to the fast decrease of the integrand, we can integrate over $q_\perp$ in infinite limits.

Using Eq. (XIII.19) for $S(\rho)$, we find the relation:

$$[S^*(\boldsymbol{\rho}) - 1][S(\boldsymbol{\rho}) - 1] = [1 - S(\boldsymbol{\rho})] - [S^*(\boldsymbol{\rho}) - 1].$$

So now Eq. (1) takes the form $f(\mathbf{k}, \mathbf{k}_0) - f^*(\mathbf{k}_0, \mathbf{k})$, which proves the unitarity of the scattering amplitude in the eikonal approximation. In particular, according to the optical theorem, the expression for the scattering cross-section of fast particles, discussed in Problem 13.51, follows.

13.7 Scattering of composite quantum particles; Inelastic collisions

Problem 13.77

Prove that the amplitude of elastic scattering of an electron with an atom (the composite system) in the Born approximation coincides with the amplitude of electron scattering from a static local potential $U(\mathbf{r})$, if we neglect exchange effects.[369] Find its physical meaning. Compare to Problems 13.4–6.

Solution

The interaction between the incident electron and the atom has the form ($\mathbf{r}_a$ are radius vectors of the atomic electrons):

$$U(\mathbf{r}, \{\mathbf{r}_a\}) = -\frac{Ze^2}{r} + \sum_a \frac{e^2}{|\mathbf{r} - \mathbf{r}_a|}.$$

The electron's elastic scattering amplitude on the atom in the Born approximation, if we neglect exchange effects (which play a role of higher-order corrections) is described by an expression similar to Eq. (XIII.6):

$$f_B = -\frac{m}{2\pi\hbar^2} \int \psi_0^*(\xi_a) e^{-i\mathbf{k}\cdot\mathbf{r}} U(\mathbf{r}, \{\mathbf{r}_a\}) e^{i\mathbf{k}_0\cdot\mathbf{r}} \psi_0(\xi_a) dV d\tau_\xi. \quad (1)$$

Integration over τ_ξ also includes a summation over the spin variables of the atomic electrons. Performing the integration over ξ_a and also examining the "potential",

$$\varphi_{at} = \int \left(\frac{Ze}{r} - \sum_a \frac{e}{|\mathbf{r} - \mathbf{r}_a|} \right) |\psi_0(\xi_a)|^2 d\tau, \quad (2)$$

which gives the mean value of the electrostatic potential maintained by the atom, we see that expression (1) has the form of Eq. (XIII.6) for the scattering amplitude in the Born approximation, with the local potential $U(\mathbf{r}) = -e\varphi_{at}(\mathbf{r})$. Applications are considered in Problems 13.4–13.6.

Problem 13.78

A polarized electron with $s_z = +1/2$ collides with a hydrogen atom that is in the ground state, where the electron has the opposite value of spin projection, $s_z = -1/2$. In the Born approximation, find the amplitude and cross-section of a spin-flip collision (i.e., the case when $s_z = -1/2$ for a scattered electron, while $s_z = +1/2$ for the atomic

---

[369] In the Born approximation (for fast particles), the neglect of exchange effects is valid. It is important that the spin state of the free electron, as well as the atomic one, is not changed during collision. Compare to Problem 13.78.

electron), while the atom remains in the ground state. Compare to the case of elastic scattering without spin-flip (see Problem 13.77).

Solution

The system Hamiltonian (with an infinitely heavy proton nucleus) is described by the expression:

$$\hat{H} = \frac{1}{2m}(\hat{\mathbf{p}}_1^2 + \hat{\mathbf{p}}_2^2) - \frac{e^2}{r_1} - \frac{e^2}{r_2} + \frac{e^2}{|\mathbf{r}_1 - \mathbf{r}_2|}.$$

Spin projections, s_z, for each electron are conserved, and the electrons with $s_z = +1/2$ and $s_z = -1/2$ could be considered as distinguishable particles (anti-symmetrization of the wavefunction is not reflected in the results). So, we denote the electron with $s_z = +1/2$ by e_1 and the electron with $s_z = -1/2$ by e_2. Now we see that the process considered is

$$e_1 + (e_2 p) \to e_2 + (e_1 p), \tag{1}$$

where, the symbol $(e_a p)$ corresponds to the hydrogen atom with electron e_a. We see that this process involves a *redistribution* of particles, where the initial and final channels of the reaction are different.

Amplitudes of such processes are expressed in terms of matrix elements of the corresponding T-operator, which could be written in two different forms:

$$\hat{T}_1 = V_\alpha + V_\beta \frac{1}{E - \hat{H} + i0} V_\alpha, \tag{2a}$$

$$\hat{T}_2 = V_\beta + V_\beta \frac{1}{E - \hat{H} + i0} V_\alpha. \tag{2b}$$

Here α and β enumerate reaction channels, $V_{\alpha,\beta}$ describes the interaction in the corresponding channels (α initial, β final channels). Note that although $\hat{T}_1 \neq \hat{T}_2$, reaction amplitudes $\langle \beta | \hat{T}_{1,2} | \alpha \rangle$ coincide.

Further, for the plane waves, $\psi = e^{i\mathbf{p}\mathbf{r}/\hbar}$, describing relative motion in each of the two-particle channels[370] we use a probability density that is normalized to unity. The differential cross-section of a process is connected to its element of the T-matrix, $\langle \beta | \hat{T} | \alpha \rangle$, by the expression:

$$\frac{d\sigma}{d\Omega} = \frac{\mu_1 \mu_2}{(2\pi)^2 \hbar^4} \frac{p_2}{p_1} |\langle \beta | \hat{T} | \alpha \rangle|^2,$$

where $p_{1,2}$ and $\mu_{1,2}$ are the momenta, and the reduced masses for colliding particles in channels α, β; $d\Omega$ is an element of the scattering solid angle in the center-of-mass system. The amplitude of elastic scattering commonly used ($\mu_1 = \mu_2$, $p_1 = p_2$)

---

[370] These waves (but not the incoming and outgoing waves), in a product involving the wavefunctions of bound states corresponding to composite particles in the channels (compare to Eq. (3)), appear in the matrix elements, $\langle \beta | \hat{T} | \alpha \rangle$, that determine reaction amplitudes.

is connected to the T-matrix by the relation:
$$f = -\frac{\mu}{2\pi\hbar^2}\langle\alpha|\hat{T}|\alpha\rangle.$$

The same relation is valid in the case of inelastic collisions, if $\mu_1 \approx \mu_2$, $p_1 \approx p_2$.

Using Eq. (2a), we can calculate the amplitude of process (1) to the first order of approximation, *i.e.*, restricting ourselves to the term V_α in $\hat{T}_1$. In this case, $V_\alpha = -e^2/r_1 + e^2/|\mathbf{r}_1 - \mathbf{r}_2|$, and the amplitude of the process considered with electron's spin-flip takes the form:

$$f_{\downarrow\uparrow,\uparrow\downarrow} = -\frac{m}{2\pi\hbar^2}\int \psi_0^*(r_1)e^{-i\mathbf{p}_2\cdot\mathbf{r}_2/\hbar}\left[\frac{e^2}{|\mathbf{r}_1-\mathbf{r}_2|} - \frac{e^2}{r_1}\right]e^{i\mathbf{p}_1\cdot\mathbf{r}_1/\hbar}\psi_0(r_2)dV_1 dV_2, \quad (3)$$

where $\mathbf{p}_{1,2}$ are momenta of the electron becoming bound and of the scattered electron, and $\psi_0(r)$ is the wavefunction of the hydrogen atom in the ground state. In the momentum representation (below we use atomic units $e = \hbar = m = 1$),

$$U(r) = \frac{1}{r} = \int e^{i\boldsymbol{\kappa}\cdot\mathbf{r}}\tilde{U}(\kappa)d^3\kappa, \quad \psi_0(r) = \frac{1}{\sqrt{\pi}}e^{-r} = \frac{1}{(2\pi)^{3/2}}\int e^{i\boldsymbol{\kappa}\cdot\mathbf{r}}\varphi_0(\kappa)d^3\kappa,$$

$$\tilde{U}(\kappa) = \frac{1}{2\pi^2\kappa^2}, \quad \varphi_0(\kappa) = \frac{\sqrt{8}}{\pi(1+\kappa^2)^2},$$

Eq. (3) could be transformed to be

$$f_{\downarrow\uparrow,\uparrow\downarrow} = -(2\pi)^2\left\{\int \varphi_0^*(\mathbf{p}_1+\boldsymbol{\kappa})\varphi_0(\mathbf{p}_2+\boldsymbol{\kappa})\tilde{U}_{12}(\kappa)d^3\kappa + \right.$$
$$\left.\varphi_0(p_2)\int \varphi_0^*(\mathbf{p}_1+\boldsymbol{\kappa})\tilde{U}_m(\kappa)d^3\kappa\right\}. \quad (4)$$

We have $p_{1,2} \gg 1$ (the necessary condition for the Born approximation), and we can see that $\varphi_0(p)$ for $p \to \infty$ decreases faster than $\tilde{U}(p)$. Therefore, the integration domains where the argument of one of the wavefunctions, $\varphi_0(\tilde{\kappa})$, is of the order of 1 play the dominant role. In both integrals $\kappa \approx |\mathbf{p}_{1,2}| \equiv p$, and we can factor $\tilde{U}(p)$ outside of the integrals, after which they are easily calculated (compare to Problem 4.17):

$$\int \varphi_0^*(\mathbf{p}_1+\boldsymbol{\kappa})\varphi_0(\mathbf{p}_2+\boldsymbol{\kappa})d^3\kappa = \int \varphi_0^*(\boldsymbol{\kappa}')\varphi_0(\mathbf{q}+\boldsymbol{\kappa}')d^3\kappa' =$$
$$= \int \varphi_0^*(\boldsymbol{\kappa}')\hat{T}_q\varphi_0(\boldsymbol{\kappa}')d^3\kappa' = \int e^{-i\mathbf{q}\cdot\mathbf{r}}|\psi_0(r)|^2 dV =$$
$$= \frac{16}{(4+q^2)^2}, \quad \mathbf{q} = \mathbf{p}_2 - \mathbf{p}_1$$

(the integral has been reduced to the form factor of the hydrogen atom), and

$$\int \varphi_0^*(\mathbf{p}_1+\boldsymbol{\kappa})d^3\kappa = (2\pi)^{3/2}\psi_0^*(0) = \sqrt{8}\pi.$$

As a result, we obtain the amplitude of the process considered:

$$f_{\downarrow\uparrow,\uparrow\downarrow}(\mathbf{p}_2, \mathbf{p}_1) = -\frac{32}{p^2}\left\{\frac{1}{(4+q^2)^2} - \frac{1}{2(1+p^2)^2}\right\}, \quad \mathbf{q} = \mathbf{p}_2 - \mathbf{p}_1. \tag{5}$$

We can see that the region of values $q \lesssim 1$, i.e., of scattering angles $\theta \lesssim 1/p$, plays the dominant role in scattering. The second term in Eq. (5) is negligibly small, so the differential, $d\sigma/d\Omega = |f|^2$, and total scattering cross-sections become equal to

$$\frac{d\sigma}{d\Omega} = \frac{4}{p^4(1+p^2\theta^2/4)^4}, \quad \theta \lesssim \frac{1}{p}; \quad \sigma = \frac{16\pi}{3p^6} = \frac{16\pi a_B^2}{3(ka_B)^6}. \tag{6}$$

In conclusion here, note the following:

1) The electron scattering cross-section for a hydrogen atom with spin-flip, which is practically inelastic, for large energies is much lower than the elastic scattering cross-section, $\sigma_{el} = 7\pi/3p^2$. See Problem 13.4.

2) The appearance of the small factor, $\sim 1/(pa_B)^2$, in the amplitude (and $\sim 1/(pa_B)^4$ in the cross-section) compared to the case of elastic scattering (no spin flip) has a simple explanation. To "switch" places, the electrons must be scattered into one another under the angle $\approx 180°$ in the center-of-inertia system. The dependence of the Rutherford scattering amplitude on momentum transferred, $f \sim \tilde{U}(q) \sim 1/(pa_B)^2$, gives rise to this smallness (here $q \approx p$; in the case of elastic scattering $qa_B \sim 1$, the value $q_{min} \sim 1/a_B$ here is due to screening of the Coulomb potential at the distances $\sim a_B$).

3) It is essential that in the process considered, the two particles (two electrons) receive a large momentum change, and that it could be provided by a single interaction. In other reactions with a particle redistribution of the form $a + (bc) \to b + (ac)$, in the case $m_a \neq m_b$, we have $p_a \neq p_b$ and the large change of the momenta of a and b particles requires an additional large momentum transfer to particle c. This leads to the appearance of an additional small factor in the process amplitude[371] $\sim \tilde{U}(q)/q^2$. Such a situation takes place in process (1) for scattering angles that are not too small, when $q \gg 1$, as is seen from Eq. (5). The additional smallness is connected to the fact that the amplitude of the process includes the hydrogen atom wavefunction (or that of the composite system, in the more general case), $\varphi(p)$, that for large momenta ($p \sim q \gg 1$) has an asymptote of the form $\varphi(p) \sim \tilde{U}(p)/p^2 \sim 1/p^4$ (for s-states, compare to Problems 4.17 and 4.18). The factor mentioned has the same order of magnitude as the second-order terms in the T-operator. In this case, when the change of momentum for all particles is large, calculations of the amplitude and cross-section, using first order of perturbation theory, have only qualitative character. A systematic and reliable calculation of the asymptote needs the inclusion of terms of higher order in the T-operator. For processes with large momentum change for all particles in a three-particle system, as, for example, in the conditions of Problem 13.79, we should take into account terms of at least second order in the interaction, $V_{\alpha,\beta}$.

---

[371] Compare to Problem 13.79.

Problem 13.79

Evaluate the charge-exchange cross-section for the collision of a fast positron with a hydrogen atom in the ground state (*i.e.*, find the cross-section of *positronium* formation – a hydrogen-like system with an electron and a positron).

Use the *Oppenheimer–Brinkman–Kramers approximation* (OBK) for the charge-exchange processes, which neglects the mutual interaction of the nuclei involved (in this case, positron with proton); compare with the previous problem.

Solution

The Oppenheimer–Brinkman–Kramers approximation for the charge-exchange process $e_+ + (e_-p) \to (e_+e_-) + p$ is based on the following expression for the T-operator[372]

$$\hat{T} \approx V_\alpha \approx U_{e^+e^-} = -\frac{1}{|\mathbf{r}_1 - \mathbf{r}_2|},$$

where $\mathbf{r}_{1,2}$ are the radius vectors of the positron (electron). We will denote the momentum of the incident positron as $\mathbf{p}_1$ (in the rest system of the hydrogen atom, $m_p = \infty$), and the momentum of the positronium as $\mathbf{p}_2$ (with mass $m(e^+e^-) = 2$). From the energy conservation law, it follows that $p_2 = \sqrt{2}p_1$ ($E = p_1^2/2 = p_2^2/4$, and we can neglect the binding energy in the hydrogen atom and in the positronium, since $p_{1,2} \gg 1$).

The wavefunctions of the initial (channel α) and final (channel β) states have the form:

$$\psi_\alpha = e^{i\mathbf{p}_1 \cdot \mathbf{r}_1}\psi_{1s}(r_2), \quad \psi_\beta = \exp\left\{\frac{i}{2}\mathbf{p}_2 \cdot (\mathbf{r}_1 + \mathbf{r}_2)\right\} \cdot \frac{1}{\sqrt{8}}\psi_{nlm}\left(\frac{\mathbf{r}_1 - \mathbf{r}_2}{2}\right).$$

Here, $\psi_{nlm}(\mathbf{r})$ is the wavefunction of the hydrogen atom (with mass $m = 1$). Coefficient $1/\sqrt{8}$ and factor $1/2$ in the argument of the wavefunction correspond to the fact that the Bohr radius for positronium is twice that of the hydrogen atom.

The amplitude of the process in the OBK-approximation is described by the expression (after the substitution $\boldsymbol{\rho} = (\mathbf{r}_1 - \mathbf{r}_2)/2$):

$$T_{OBK}(1s \to nlm) = \langle\beta|U_{e^+e^-}|\alpha\rangle = -\sqrt{2}\int \psi_{1s}(r)e^{i(\mathbf{p}_1 - \mathbf{p}_2)\cdot \mathbf{r}}dV \times$$

$$\int \frac{1}{\rho}\psi^*_{nlm}(\boldsymbol{\rho})e^{-i(\mathbf{p}_2 - 2\mathbf{p}_1)\cdot \boldsymbol{\rho}}d^3\rho. \tag{1}$$

The first integral in this expression is equal to $(2\pi)^{3/2}\varphi_{1s}(q)$ (with $\mathbf{q} = \mathbf{q}_1 - \mathbf{q}_2$), while the second, with the Schrödinger equation taken into account, is

---

[372] The same results follow also from the choice $T \approx V_\beta \approx U_{e-p}$. See the previous problem, where we made general statements about processes with a particle redistribution. Let us note that for the charge-exchange processes, the results of both exact amplitude calculations in the first order of perturbation theory according to $T = V_{\alpha,\beta}$ and in the OBK-approximation have only qualitative character (a reliable calculation necessitates the inclusion of second-order second-order terms).

$$(2\pi)^{3/2}\left(\frac{1}{2}q_1^2 - E_n\right)\phi^*_{nlm}(\mathbf{q}_1), \quad \text{where} \quad \mathbf{q}_1 = 2\mathbf{p}_1 - \mathbf{p}_2.$$

Since $q, q_1 \gg 1$, we can use asymptotes for wavefunctions $\varphi(\mathbf{p})$ in the Coulomb potential, $U = -1/r$ (so $\tilde{U}(p) = -1/2\pi^2 p^2$), for $p \to \infty$. According to Problem 4.18:

$$\varphi_{1s}(p) = \frac{\sqrt{8}}{\pi p^4}, \quad \varphi_{nlm}(\mathbf{p}) = \sqrt{\frac{2}{\pi}}\frac{2(l!)}{p^{l+4}}(-2i)^l \tilde{R}_{nl}(0) Y_{lm}(\mathbf{n}). \tag{2}$$

To obtain this we used a relation for the coordinate Coulomb wavefunctions:

$$\psi_{nlm}(\mathbf{r}) = Y_{lm}(\mathbf{n}) r^l \tilde{R}_{nl}(r), \quad \tilde{R}_{nl}(0) = \frac{2^{l+1}}{(2l+1)! n^{l+2}}\sqrt{\frac{(n+l)!}{(n-l-1)!}}.$$

Using these asymptotes, we obtain

$$T_{OBK}(1s \to nlm) = -\frac{16\sqrt{2\pi^3}(\sqrt{2}i)^l \tilde{R}_{nl}(0) l! Y_{lm}(\mathbf{n}_0)}{(3 - 2\sqrt{2}\cos\theta)^{3+l/2} p_1^{6+l}}. \tag{3}$$

We have used $\mathbf{n}_0 = (-\mathbf{p}_2 + 2\mathbf{p}_1)/(|\mathbf{p}_2 - 2\mathbf{p}_1|)$, and also that $q_1^2 = 2q^2 = 2(3 - 2\sqrt{2}\cos\theta) p_1^2$, where θ is the angle between vectors $\mathbf{p}_2$ and $\mathbf{p}_1$.

This angular distribution of the emitted positronium,[373] $d\sigma/d\Omega = |T|^2/\sqrt{2\pi^2}$, has several interesting properties. It does not depend on the value of the incident positron's momentum. It is sharply anisotropic (the values of $3 - 2\sqrt{2}\cos\theta$ for angles $\theta = 0$ and π differ by a factor of 35). The polarization state of the positronium created is characterized by a definite projection of its angular momentum onto the direction of vector $\mathbf{n}_0$, $m_n = 0$, since $Y_{lm}(\theta = 0) = \sqrt{(2l+1)/4\pi}\delta_{m,0}$. For the total charge-exchange cross-section, after summing over the values of the projection of the angular momentum, m, of the positronium created, we obtain

$$\sigma_{OBK}(1s \to nl) = 64\pi a_B^2 \frac{2^l (2l+1)(l!)^2 \tilde{R}_{nl}^2(0)}{(l+5)(3-2\sqrt{2})^{5+l}} \frac{1}{V^{12+2l}}, \tag{4}$$

where $V = p_1 a_B/\hbar$ is the relative velocity of the colliding positron and hydrogen atom in atomic units. In particular, we can write the charge-exchange cross-section for the ns-states of the positronium as

$$\sigma_{OBK}(1s \to ns) = \frac{\pi a_B^2}{n^3}\left(\frac{2.89}{V}\right)^{12}. \tag{5}$$

Pay attention to the pronounced energy-dependence of the charge-exchange cross-section, $\sigma \propto p^{-12-2l}$ (compare with the result of the previous problem and with the case of elastic scattering; see Problem 13.4). It is explained by the fact that all particles involved experience a large change of momentum. In connection with this, we should note that the charge-exchange cross-section for collisions of fast heavy particles $m \gg m_e$ with a hydrogen atom have a similar energy-dependence. For $m \gg m_e$, the

---

[373] See Problem 13.78 for amplitude normalization. In this problem, $\mu_1 = 1$, $\mu_2 = 1$, $p_2 \approx \sqrt{2}p_1$.

charge-exchange cross-section does not depend on the mass of the incident particle (for example, proton, muon, *etc.*). In particular, the charge-exchange cross-section in the ns-state is

$$\sigma_{OBK}(1s \to ns) = \pi a_B^2 \frac{2^{18} Z^5}{5n^3 V^{12}} \approx \frac{\pi a_B^2}{n^3} Z^5 \left(\frac{2.47}{V}\right)^{12}, \tag{6}$$

where Z is the particle charge. From comparison of Eqs. (5) and (6), we see that the charge-exchange cross-section for the collision of the positron and hydrogen atom is ≈ 7 times larger than for the collision with a proton (for the same velocities).

Problem 13.80

In the Born approximation, express the amplitude of the process $A_i + B_i \to A_f + B_f$ for the collision of fast composite particles A and B, interacting electrostatically, through *electric form-factors*[374]

$$eF_{if}^{A(B)}(\mathbf{q}) = \langle \psi_{A(B)f} | \sum_a e_a \exp\{-i\mathbf{q} \cdot \mathbf{r}_a\} | \psi_{A(B)i} \rangle$$

for transitions $i \to f$.

Consider the behavior of the form-factor for $q \to 0$, depending on the quantum numbers of the initial and final states.

Calculate the form-factors for transitions $1s \to 1s$, $1s \to 2s$, $1s \to 2pm$ in the hydrogen atom. Consider their behavior as $q \to \infty$.

Find the collision cross-section for the following processes:

1) $H(1s) + H(1s) \to H(1s) + H(1s)$, elastic scattering of hydrogen atoms from one another;
2) the collision of a charged, structureless particle (electron, muon, proton, *etc.*, but not an ion) with a hydrogen atom in the ground state, accompanied by the excitation of a) $2s$ state in the atom; b) $2p$ states in the atom.

Solution

Let us denote the radius vectors of the $A(B)$ particles' centers of mass by $\mathbf{R}_{A(B)}$, their masses by $m_{A(B)}$, and the momenta of their relative motion by $\mathbf{p}_{1,2}$ (1) before and (2) after the collision. The wavefunctions of the initial and final states have the form:[375]

---

[374] Let us emphasize that $\mathbf{r}_a$ is the radius vector of the ath charged particle in the composite system $A(B)$ with respect to the system's center-of-mass.
[375] Wavefunctions $\psi_{A(B)}$ of the composite particles describe the states of the particles in the composite system, with respect to its center-of-mass.

$$\psi_i = \exp\left\{\frac{i}{\hbar}\mathbf{p}_1 \cdot \mathbf{R}\right\} \psi_{Ai}\{x'_a\}\psi_{Bi}\{r'_B\}, \quad \mathbf{R} = \mathbf{R}_A - \mathbf{R}_B,$$

$$\psi_f = \exp\left\{\frac{i}{\hbar}\mathbf{p}_2 \cdot \mathbf{R}\right\} \psi_{Af}\{x'_a\}\psi_{Bf}\{r'_B\}, \quad \hbar\mathbf{q} = \mathbf{p}_2 - \mathbf{p}_1, \tag{1}$$

The amplitude of the process considered in the Born approximation is described by the expression (compare to Problems 13.77 and 13.78):

$$f_{if} = -\frac{\mu}{2\pi\hbar^2}T_{if} = -\frac{\mu}{2\pi\hbar^2}\langle\psi_f|\sum_{a,b}\frac{e_a e_b}{|\mathbf{x}_a - \mathbf{r}_b|}|\psi_i\rangle, \quad \frac{d\sigma}{d\Omega} = |f|^2. \tag{2}$$

We use $\mu = m_A m_B/(m_A + m_B)$, the reduced mass of particles A and B. The integration in matrix element (2) is performed both over independent "inner" coordinates $\mathbf{x}'_a$, $\mathbf{r}'_b$ (in the case of atoms, these are the coordinates of all electrons), and over the radius vector, $\mathbf{R}$, of the relative motion. Writing here $\mathbf{x}_a = \mathbf{R}_A + \mathbf{x}'_a$ and $\mathbf{r}_b = \mathbf{R}_B + \mathbf{r}'_b$, using the form of the wavefunctions (1), and using the Fourier transform of the Coulomb potential

$$\frac{1}{|\mathbf{x}_a - \mathbf{r}_b|} = \frac{1}{2\pi^2}\int \exp\{i\boldsymbol{\kappa}\cdot(\mathbf{x}_a - \mathbf{r}_b)\}\frac{d^3\kappa}{\kappa^2},$$

we obtain

$$T_{if} = \frac{4\pi e^2}{q^2}F^A_{if}(-\mathbf{q})F^B_{if}(\mathbf{q}). \tag{3}$$

Here

$$eF^{A(B)}_{if}(\mathbf{q}) = \langle\psi_{A(B)f}|\sum_a e_a \exp\{-i\mathbf{q}\cdot\mathbf{r}'_a\}|\psi_{A(B)i}\rangle \tag{4}$$

are the electric form-factors for the corresponding transitions in the composite particles $A(B)$. We can associate the *Feynman diagram* in Fig. 13.8 with the amplitude, Eq. (3).

It illustrates an important property of the amplitude: it is *factorized* in terms of the particles, A and B, that take part in the process. In this figure, the wavy line between the vertices corresponds to $4\pi/q^2$, while the form-factors $eF^{A(B)}_{if}(\mp\mathbf{q})$ correspond to the vertices (note that the changes in the momenta of particles A and B differ by a sign).

Let us note some properties of the form-factors:

1) For point (structureless) particles, $F(q) = Z = $ const, where Ze is particle charge.
2) For $q \to 0$, expanding $\exp\{-i\mathbf{q}\cdot\mathbf{r}\}$ in Eq. (4) for $F(\mathbf{q})$ into a series reveals that for a system with charge Ze different from zero, only the *elastic* (without a change of composite particle's state) form-factor is different from zero, and is equal to $F_{nn}(0) = Z$. In all other cases $F_{if}(0) = 0$ (for a charged particle, due to wavefunction orthogonality). The character of how exactly the form-factor vanishes in the limit $q \to 0$ depends on the quantum numbers, angular momenta, and parity

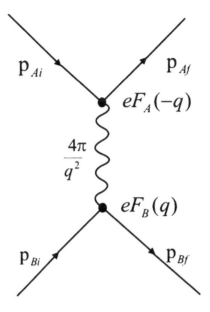

Fig. 13.8

of the initial and final states. The slowest decrease in the form factor is for the dipole transitions: $eF_{if} \approx -i\langle f|\hat{\mathbf{d}}|i\rangle q \propto \mathbf{q}$. For transitions with unchanging values of the angular momentum and parity (for example, for S-states), $F_{if} \propto q^2$. With an increase in the difference between the angular momenta of the initial and final states, the form-factor goes to zero for $q \to 0$ more abruptly.

3) For $q \to \infty$ the form-factor of any composite system vanishes.[376] The power law of this decrease depends strongly on the number of particles in the system and on the specific interaction. From physical considerations, it is evident that the faster $\overline{U}(q)$ decreases and the larger the particle number in the compound system, the faster the form-factor decreases. Compare to Problem 13.84.

For transitions of the hydrogen atom, the form-factors

$$F_{1s \to nlm}(q) = \int \psi^*_{nlm}(\mathbf{r})(1 - e^{-i\mathbf{q}\cdot\mathbf{r}})\psi_{1s}(r)dV$$

are easy to calculate. Taking into account the form of the wavefunctions (see Eq. (IV.4)), we can obtain

$$F_{1s \to 1s} = 1 - F_{at}(q) = \frac{8q^2 + q^4}{(4 + q^2)^2}, \quad F_{1s \to 2s} = -\frac{4\sqrt{2}q^2}{(9/4 + q^2)^3}. \tag{5}$$

---

[376] For an atom in the approximation of an infinite nuclear mass, for $q \to \infty$ the elastic form-factor is $F_{el} = Z$ (it is determined by the contribution from a point-like nucleus).

For the 2p-states, it is convenient to write the angular part of the wavefunction in the form $\sqrt{3/4\pi}\varepsilon(m)\cdot\mathbf{n}$, where $|\varepsilon(m)|^2 = 1$ (compare Problems 3.41 and 3.42), after which we obtain

$$F_{1s\to 2pm} = -\frac{\varepsilon^*(m)}{4\sqrt{2\pi}}\frac{i\partial}{\partial\mathbf{p}}\int\exp\left\{-\frac{3r}{2}-i\mathbf{q}\cdot\mathbf{r}\right\}dV =$$

$$= 6\sqrt{2}i\frac{\varepsilon^*(m)\mathbf{q}}{(9/4+q^2)^3}. \tag{6}$$

This form-factor is different from zero only for the states with projection of the angular momentum on the direction of the vector $\mathbf{q}$ equal to zero.

For calculating the scattering cross-sections, we first note that

$$q^2 = p_1^2 + p_2^2 - 2p_1 p_2 \cos\theta, \quad p_2 = \sqrt{p_1^2 - 2\mu(\varepsilon_1-\varepsilon_2)} \approx$$

$$\approx p_1 - \frac{\mu(\varepsilon_1-\varepsilon_2)}{p_1},$$

where $\varepsilon_{1,2} > 0$ are the binding energies of the system before and after the collision. So $d\Omega$ could be replaced by $\pi dq^2/p_1^2$, with the integration over q^2 within limits 0 and ∞ (due to the fast convergence in the upper limit, $q_{max}^2 \approx 4p_1^2$). We cannot always replace the lower integration limit

$$q_{min}^2 = (\mathbf{p}_1-\mathbf{p}_2)^2 \approx \left(\frac{\mu(\varepsilon_1-\varepsilon_2)}{p_1}\right)^2$$

by 0, because of a divergence that may appear. This complication however only arises for the inelastic collisions where one of the colliding particles has charge different from zero and its state during the collision process does not change, so that for it, $F(0) = Z \neq 0$, and for the other particle, the transition is of dipole type. Thus, $eF_{if} \approx -i\mathbf{d}_{if}\mathbf{q}$ for $q \to 0$.

Elementary integration gives the following results:

$$1)\ \sigma(1s1s \to 1s1s) = \frac{4\pi}{V^2}\int_0^\infty F_{1s\to 1s}^4(q)dq^2 = \frac{33\pi}{35V^2}, \tag{7}$$

$$2)\ \sigma(1s \to 2s) = \frac{4\pi Z^2}{V^2}\int_0^\infty F_{1s\to 2s}^2(q)dq^2 = \frac{2^{17}\pi Z^2}{5\cdot 3^{10}V^2} \approx 0.444\frac{\pi Z^2}{V^2}, \tag{8}$$

where $V = p_1/\mu$ is the relative velocity of the colliding particles, and Ze is the charge of the particle colliding with the hydrogen atom.

Now we calculate the total cross-section for collisions with transitions $1s \to 2p$, by summing over the transitions $1s \to 2pm$:

$$\sigma(1s \to 2p) = 288\pi \left(\frac{Z}{V}\right)^2 \left(-\frac{1}{5!}\frac{\partial^5}{\partial a^5}\int_{q_{min}^2}^{\infty}\frac{dx}{x(a+x)}\right) = \qquad (9)$$

$$288\pi \left(\frac{Z}{V}\right)^2 \frac{1}{5!}\left[-\frac{\partial^5}{\partial a^5}\left(\frac{1}{a}\ln\frac{a}{q_{min}^2}\right)\right] = \frac{2^{17}\pi Z^2}{3^{10}V^2}\left[\ln(16V^2) - \frac{137}{60}\right],$$

where $x = q^2$, $a = 9/4$, $q_{min}^2 = (\varepsilon_{2p} - \varepsilon_{1s})^2/V^2 = 9/64V^2$. Finally,

$$\sigma(1s \to n = 2) = \frac{2^{18}\pi Z^2}{3^{10}V^2}\left(\ln 4V - \frac{25}{24}\right) \approx 4.44\frac{\pi Z^2}{V^2}\left(\ln 4V - \frac{25}{24}\right) \qquad (10)$$

is the total cross-section of exciting the states of the hydrogen atom with the principal quantum number $n = 2$ by collision with a charged particle.

Problem 13.81

Consider the collision between a fast charged particle and a two-atom molecule in its ground state with the dipole moment, d_0. The electron term of the molecule is $^1\Sigma$. Estimate the cross-sections of collisions, which involve excitation of the rotational and oscillation molecular levels. Compare to the case of the collision with an atom (see the previous problem).

Solution

As in the previous problem, the collision cross-section is connected to the electric form-factor of the molecule,

$$eF_{0 \to vKM}(\mathbf{q}) = \langle n, \Lambda = 0, vKM| \sum_a e_a e^{i\mathbf{q}\cdot\mathbf{r}_a} |n, 0\rangle. \qquad (1)$$

In the initial state, the quantum numbers are $\Lambda = v = K = M = 0$, while n characterizes the state of molecular electrons. The wavefunctions of the moleculular states with $\Lambda = 0$, as are in matrix element (1), have the form:

$$\psi_{n\,\Lambda=0vKM} = \psi_{n\Lambda=0}^{el}(\xi_a, \mathbf{R})\psi_v^{osc}(\mathbf{R} - \mathbf{R}_0)Y_{KM}(\mathbf{n}). \qquad (2)$$

Here ξ_a are the electron variables (coordinate and spin); $\mathbf{R} = R\mathbf{n} = \mathbf{R}_1 - \mathbf{R}_2$ is the radius vector from one nucleus to the other; the masses are $M_{1,2}$; and the radius vectors in the molecule's center-of-mass system: $\mathbf{R}_1 = M_2\mathbf{R}/(M_1 + M_2)$ and $\mathbf{R}_2 = -M_1\mathbf{R}/(M_1 + M_2)$. Summation in Eq. (1) is performed over both the molecular electrons ($\mathbf{r}_a$ denotes the radius vectors relative to the molecular center of mass), and the nuclei, whose contribution is $Z_1 e^{-i\mathbf{q}\cdot\mathbf{R}_1} + Z_2 e^{-i\mathbf{q}\cdot\mathbf{R}_2}$.

758 *Exploring Quantum Mechanics*

We write the cross-section for the collision between the charged particle and the molecule,

$$\frac{d\sigma_{0 \to vKM}}{d\Omega} = \left(\frac{2\mu Ze^2}{\hbar^2 q^2}\right)^2 |F_{0 \to vKM}(\mathbf{q})|^2, \tag{3}$$

where Ze is particle charge, and μ is its and the molecule's reduced mass. We note that the easiest way to obtain these cross-sections is to neglect the change in the state of the valence electron states in the molecular formation. In this approximation, the wavefunction of the electron term of the molecule is

$$\psi_{n\Lambda=0}^{elec} \approx \psi_1(\mathbf{r}_a - \mathbf{R}_1)\psi_2(\mathbf{r}_b - \mathbf{R}_2),$$

where $\psi_{1,2}(\mathbf{r}_{a,b})$ is the wavefunction for the electron bound to a single atom of the molecule, and for its form-factor, we obtain

$$F_{0 \to \nu KM} \approx \langle \nu KM | \left(e^{-i R_1 \mathbf{q} \cdot \mathbf{n}} F_1(q) + e^{i R_2 \mathbf{q} \cdot \mathbf{n}} F_2(q)\right) | 0 \rangle, \tag{4}$$

where $F_{1,2}(q)$ are the form-factors of the atoms in the molecule (which also include contributions of the corresponding nuclei). Note that for $q = 0$, the form-factor of the molecule, as of any neutral system, is equal to zero (see a discussion of form-factors in the previous problem).

Due to the smallness of amplitude of nuclear oscillations, we can replace $\mathbf{R}_{1,2}$ by their equilibrium positions. Then, due to wavefunction orthogonality, only the form-factors for transitions with $v = 0$ (i.e., without a change of the molecular oscillation state) are non-zero. For such transitions, we can calculate the differential cross-section, Eq. (3), by summing over the values of the quantum numbers K and M of the molecule's final states. Using approximation (4) and the completeness condition for the spherical functions, according to which

$$\sum_{KM} |\langle KM|\hat{A}|0\rangle|^2 = \sum_{KM} \langle 0|\hat{A}^+|KM\rangle\langle KM|\hat{A}|0\rangle = \langle 0|\hat{A}^+\hat{A}|0\rangle,$$

we calculate the differential cross-section:

$$\sum_{KM} \frac{d\sigma_{0 \to 0KM}}{d\Omega} = \left(\frac{2\mu Ze^2}{\hbar^2 q^2}\right)^2 \langle 0||F_1(q) + F_2(q)e^{i\mathbf{q} \cdot \mathbf{R}_0}|^2|0\rangle.$$

Compare to Problem 13.7.

Collision cross-sections that do not excite molecular oscillations obey the same rules as the case of a charged particle colliding with an atom (compare to the previous problem). In particular, for collisions that excite a rotational level with the angular momentum $K \neq 1$, the cross-section is

$$\sigma_{0 \to 0KM} = \pi a_B^2 Z^2 A_{KM} \left(\frac{V_{at}}{V}\right)^2, \quad K \neq 1, \tag{5}$$

where V is the relative velocity of the colliding particles, and $A_{KM} \sim 1$ for the most important transitions.

For collisions that excite a rotational level with $K = 1$, connected to the ground level by a dipole transition,[377] for $q \to 0$ we have (compare to the previous problem):

$$\frac{d\sigma(0 \to 0\ M)}{dq^2} = \frac{4\pi Z^2 e^2 d_0^2}{3\hbar^2 q^4 V^2} |\varepsilon(M)\mathbf{q}|^2. \quad (6)$$

$\varepsilon(M)$ is the polarization vector that determines a molecular rotational state with $K = 1$, and is connected to the spherical functions by the relation $Y_{1M} = \sqrt{3/4\pi}\varepsilon(M)\mathbf{n}$, where $|\varepsilon|^2 = 1$. Eq. (6) follows from expressions (1) and (3) if we take into account that: 1) for $q \to 0$, the sum in (1) becomes $-i\mathbf{q} \cdot \mathbf{d}$, where $\mathbf{d}$ is the operator of the molecular dipole moment; 2) averaging $\mathbf{d}$ over the electron states of the molecule with $\Lambda = 0$ gives $d(R)\mathbf{n}$; 3) after averaging over the oscillation states with $v = 0$, we can replace $d(R)$ by $d(R_0) \equiv d_0$ due to the smallness of the nuclear oscillation amplitudes (compare to Problem 11.25). The summation in Eq. (6) over M gives $\sum |\varepsilon(M)\mathbf{q}|^2 = q^2$, and the following integration over q^2 with the lower limit of integration (to cut off a divergence) being

$$q_{min}^2 = \left[\frac{\hbar}{\tilde{\mu}R_0^2 V}\right]^2$$

($\tilde{\mu}$ is the reduced mass of the molecular nuclei) allows us to obtain the cross-section of transitions to states with $K = 1$ to *logarithmic accuracy*:[378]

$$\sigma(0 \to v = 0, K = 1) \approx \frac{8\pi Z^2 e^2 d_0^2}{3\hbar^2 q^4 V^2} \ln \frac{\tilde{\mu}}{m_e V_{at}}. \quad (7)$$

Let us properties of the collision processes, accompanied by excitation of the molecular oscillation levels. As was mentioned before, replacement of R by is equilibrium value R_0 due to the weak nuclear oscillations is valid only for transitions that do not excite oscillations. For states with $v \neq 0$, in this approximation, the form-factor goes to zero due to wavefunction orthogonality. So for transitions with $v \neq 0$, it is necessary to perform expansion over the small parameter $\Delta R/R_0$, where $\Delta R = R - R_0$ is of the order of the nuclear oscillation amplitudes. For the linear expansion term, which corresponds to transitions with $\Delta v = 1$, in expression (1) for $v = 1$, as opposed to $v = 0$, a small factor appears, on the order of (see Problems 11.3 and 11.25)

$$\langle v = 1 | \frac{R - R_0}{R_0} | v = 0 \rangle = \frac{a_{osc}}{\sqrt{2}R_0} \sim \left(\frac{m_e}{\tilde{\mu}}\right)^{1/4}.$$

---

[377] For this transition, approximation (4) for $q \to 0$ is not valid; the dipole moment of the molecule is determined by its valence electrons.

[378] Here as usual, for the upper integration limit we choose the value $q_{max}^2 \sim a_B^{-2}$. We should note that although the argument of the logarithm in Eq. (7) is much larger than in the case of a dipole transition in an atom, due to the small rotational energy, nevertheless the accuracy of Eq. (7) does not increase because of it. Usually, the dipole moment of the molecule, d_0, is much smaller than the value ea_B.

This leads to much lower, $\sim (\tilde{\mu}/m_e)^{1/2}$, cross-sections of the transitions with $v = 1$. With an increase of the value v, it is necessary to consider higher orders in $\Delta R/R_0$, which leads to a sharper, $\sim (\mu/m_e)^{v/2}$, suppression of the cross-section for the corresponding molecular transitions.

Problem 13.82

Find the cross-section of a fast charged particle colliding with a hydrogen atom in a metastable 2s-state, accompanied by its transition into a 2p-state.

Comment: In this problem it is necessary to take into account the relativistic splitting of the s- and p-levels (see Problem 11.62).

Solution

For collisions of a charged particle and an atom accompanied by an atomic dipole transition, the region with small values of q^2 plays the dominant role. The cross-section of the collision with transition from the s- to the p-state of the atom to logarithmic accuracy is described by the expression (compare to Problem 13.80):

$$\sigma_{s\to p} = \frac{4\pi Z^2 e^2}{\hbar^2 V^2} |\langle l = 1, m = 0 | d_z | 0 \rangle|^2 \ln \frac{1}{q_{min}^2 a_B^2}, \tag{1}$$

where Ze is the charge of the incident particle and V is the relative velocity of the colliding particles. For a transition $2s \to 2p$ in the hydrogen atom, the matrix element[379] $\langle 1, 0 | d_z | 0 \rangle = 3ea_B$ was calculated in Problem 11.33, in connection with the Stark effect for the states with $n = 2$.

If we neglect relativistic corrections, the states $2s$ and $2p$ of the hydrogen atom are degenerate; here $q_{min}^2 = 0$ and the cross-section, Eq. (1), diverges. Splittings, Δ, of the levels $2s_{1/2}$, $2p_{1/2}$ and $2p_{3/2}$, which determine the values of $q_{min}^2 = (\Delta/\hbar V)^2$, were discussed in Problem 11.62. Since here, transitions into the atomic states $2p_{1/2}$ and $2p_{3/2}$ must be considered separately, it is necessary to add the additional factors, equal $1/3$ and $2/3$, for the $p_{1/2}$- and $p_{3/2}$-states, in Eq. (1). These factors, the squares of the corresponding Clebsch–Gordan coefficients (see, for example, Problem 5.18), reflect the contribution of p-states with $l_z = 0$ to the states[380] $p_{1/2}$ and $p_{3/2}$. If we take into account these facts, we find

$$\sigma(2s \to 2p) = 72\pi a_B^2 \left(\frac{Ze^2}{\hbar V}\right)^2 \ln \frac{\hbar V}{a_B \Delta_0^{1/3} \Delta_1^{2/3}}, \tag{2}$$

where $\Delta_0 = E(2s_{1/2}) - E(2p_{1/2})$ and $\Delta_1 = E(2p_{3/2}) - E(2s_{1/2})$, with the numerical value $\Delta_0^{1/3} \Delta_1^{1/3} \approx 8 \cdot 10^{-7}$ (in atomic units).

---

[379] We emphasize that Eq. (1) describes the cross-section, summed over the projections of the angular momentum of the p-state. Although in Eq. (1) there is a state with $l_z = 0$, we should remember that the choice of a quantization axis z, along the vector $\mathbf{q}$, depends on the scattering angle.

[380] If we take into account relativistic corrections, l_z is not a constant of the motion.

As is seen from Eq. (2), the cross-section $\sigma \gg (V_{at}/V)^2 \sigma_0$, where $\sigma_0 = \pi \overline{r^2} = 40\pi a_B^2$, characterizes the transverse size of the hydrogen atom in states[381] with $n = 2$. This means that in this problem, large impact parameters ($\gg a_B$) are essential. Effective interaction at such distances is $Ze \mathbf{d} \cdot \mathbf{r}/r^3 \sim Zed/r^2$. The applicability condition of perturbation theory (see Eq. (XIII.7)) for such a potential is fulfilled for the distances $r \gg Zed/\hbar V$. Even for values $V \approx 1$, this condition holds true, which proves the applicability of Eq. (2) for such collision velocities.

In conclusion, note that the large value of the transition cross-section, (2), implies that the lifetime of the metastable 2s-state in a gas could decrease sharply due to collisions (for an isolated hydrogen atom, $\tau(2s) = 1/8$ s; while in the presence of transitions in the 2p-state, the atom "flashes up" within $\sim 10^{-9}$ s).

Problem 13.83

Calculate (with *logarithmic accuracy*) the cross-section of fast deuteron fission in the Coulomb field of a nucleus with charge Ze (for simplicity, consider the nucleus to be point-like and infinitely heavy). Choose the deuteron wavefunction as in the case of a zero-range potential (see 12.1), which is reasonable because of the smallness of the proton-neutron binding energy.

Solution

The differential cross-section of deuteron fission is described by the expression:

$$d\sigma_{fis} = \frac{4Z^2 e^4 m_d^2}{\hbar^2 q^4} |\langle \mathbf{p}, -|e^{-i\mathbf{q}\cdot\mathbf{r}_p}|0\rangle|^2 d^3k\, d\Omega_f. \tag{1}$$

Here, $\hbar \mathbf{q} = \mathbf{P}_f - \mathbf{P}_i$, $\mathbf{P}_i = m_d \mathbf{V}$ is the deuteron momentum, $\mathbf{P}_f$ is the nucleon's total momentum after the collision, $d\Omega_f$ is the infinitesimal solid angle (which contains vector $\mathbf{P}_f$ direction), and m, $m_d = 2m$ are the nucleon and deuteron masses. The Matrix element

$$F_{0\to p} = \langle \mathbf{p}, -|\exp\{-i\mathbf{q}\cdot\mathbf{r}_p\}|0\rangle$$

could be considered as an inelastic form-factor for transitions into the continuum (compare to Problem 13.80). For the wavefunction of the initial state, we write (see Problems 4.10 and 12.1):

$$\psi_0(r) = \sqrt{\frac{\kappa}{2\pi}} \frac{1}{r} e^{-\kappa r}, \quad \varepsilon_0 = \frac{\hbar^2 \kappa^2}{m},$$

where ε_0 is the deuteron binding energy, and $\mathbf{r}_p = \mathbf{r}/2$. For the wavefunctions of the final states, we should choose[382] the wavefunction $\psi_\mathbf{k}^-(r)$, that has an asymptote in the form of "a plane wave + a converging wave;" they are normalized to $\delta(\mathbf{k} - \mathbf{k}')$.

---

[381] For the ns-states of the hydrogen atom, $\overline{r^2} = \frac{1}{2}n^2(5n^2 + 1)$.
[382] For an alternative convenient choice of final state wavefunctions, see Problem 11.63.

The dominant input into the fission cross-section comes from small q^2, where transitions to states of the two-nucleon system are important, connected to the deuteron's dipole transition, since for them $d\sigma \propto dq^2/q^2$ (compare to Problem 13.80). For such q, we have

$$\langle \mathbf{p}, -|\exp\{-i\mathbf{q}\cdot\mathbf{r}_p\}|0\rangle \approx \langle \mathbf{p}, -|(-i\mathbf{q}\cdot\mathbf{r}_p)|0\rangle, \qquad (2)$$

and in the last expression we can replace the wavefunction $\psi_{\mathbf{k}}^-$ by the plane wave $\psi_{\mathbf{k}} = (2\pi)^{-3/2} e^{i\mathbf{k}\cdot\mathbf{r}}$. This is connected to the fact that in a zero-range potential, wavefunctions $\psi_{\mathbf{k}}^-$ and $\psi_{\mathbf{k}}$ differ from one another in s-wave only, which does not contribute to the dipole moment, Eq. (2). A calculation of the corresponding matrix element gives

$$\frac{\sqrt{\kappa}\mathbf{q}}{8\pi^2}\frac{\partial}{\partial \mathbf{k}}\int \frac{1}{r} e^{-\kappa r - i\mathbf{k}\cdot\mathbf{r}} d^3r = \frac{\sqrt{\kappa}\mathbf{q}}{2\pi}\frac{\partial}{\partial \mathbf{k}}\frac{1}{k^2+\kappa^2} = -\frac{\sqrt{\kappa}\mathbf{q}\cdot\mathbf{k}}{\pi(k^2+\kappa^2)^2}. \qquad (3)$$

The angular dependence $\propto (\mathbf{q}\cdot\mathbf{k})$ of the vector $\mathbf{k}$ reflects the fact that for a dipole transition, the angular momentum of the nucleon pair in the final state is $l=1$.

To obtain the differential cross-section, we substitute Eq. (3) in Eq. (1) and perform the integration over the directions of the vector $\mathbf{k}$, using the relations

$$d^3k = k^2 dk\, d\Omega, \quad \int (\mathbf{q}\cdot\mathbf{k})^2\, d\Omega = \frac{4\pi}{3}q^2 k^2.$$

After this, $d\Omega_f$ in Eq. (1) could be replaced by $\pi\hbar^2 dq^2/P_i^2$. Then the value of P_f that follows from the energy conservation law,

$$\frac{1}{2m_d}P_i^2 - \varepsilon_0 = \frac{1}{2m_d}P_f^2 + \varepsilon,$$

where $\varepsilon = \hbar^2 k^2/m$ is the energy of the nucleon pair's relative motion after the collision, determines

$$\hbar^2 q_{min}^2 = (P_f - P_i)^2 \approx \frac{1}{V^2}(\varepsilon + \varepsilon_0)^2.$$

Integrating over q^2 within the limits q_{min}^2 and[383] $q_{max}^2 \approx \kappa^2$, we find

$$d\sigma_{fis} = \frac{8(Ze^2)^2(\varepsilon_0\varepsilon^3)^{1/2}}{3m(\varepsilon+\varepsilon_0)^4 V^2}\ln\frac{\hbar^2\kappa^2 V^2}{(\varepsilon+\varepsilon_0)^2} d\varepsilon. \qquad (4)$$

As is seen, the characteristic value of energy is $\varepsilon \sim \varepsilon_0$. The integration of Eq. (4) over ε gives the total cross-section of deuteron electro-fission. With logarithmic accuracy, we can neglect the dependence on ε under the logarithm and obtain (making the substitution $x = \sqrt{\varepsilon}$; see App.1.5)

---

[383] Note that for larger values of q it is necessary to take into account the finiteness of nucleon size, whose form-factor decreases fast for $q > 1/R$. Since q_{max}^2 in the final result is inside a logarithm, its detailed value is not so important.

$$\sigma_{fis}(E) \approx \frac{\pi(Ze^2)^2}{6\varepsilon_0 E} \ln \frac{E}{\varepsilon_0}, \tag{5}$$

where $E = m_d V^2/2$ is the deuteron energy.

According to Eq. (5), the value of the cross-section for $E = 200$ MeV, $\varepsilon_0 = 2.2$ MeV is $\sigma_{fis} = 1.1 \cdot 10^{-28} Z^2$ cm^2. For all nuclei except for perhaps the heaviest, this quantity is much lower than their geometric size, which shows that the dominant role in the process of deuteron fission is nuclear interaction.

We also note a quasi-classical estimation of the fast deuteron fission cross-section accompanied by the escape of one of the nucleons, a proton or a neutron, for a collision with a nucleus with radius R. Considering the potential to be a hard sphere of radius R:

$$\sigma \sim 2\pi R \Delta R, \quad \Delta R \sim R_d \sim \frac{1}{\kappa}. \tag{6}$$

This area is that of a ring with radius R and width of the order of the deuteron size.

In conclusion, we note that each escaping nucleon takes away the energy $E_N \approx E/2$ and moves in the direction of the falling beam with the angle of dispersion $\Delta\theta \sim (\varepsilon_0/E)^{1/2}$ (the transverse component of the nucleon momentum $p_\perp \sim \sqrt{m\varepsilon_0}$ is determined by the binding energy of the deuteron).

Problem 13.84

Find the asymptotic behavior of the electric form-factor of a two-particle system as $q \to \infty$. It is assumed that the Fourier component of the interaction potential, $\tilde{U}(q)$, that is responsible for the formation of the composite system, has a power-law decay at $q \to \infty$: $\tilde{U}(q) \propto q^{-n}$ with $n > 1$. Compare to Problem 4.18. Consider applications of the results obtained to the hydrogen atom.

Solution

The electric form-factor for the transition between states ψ_1 and ψ_2 of a two-particle composite system is described by the following expression (see, for example, Problem 13.80):

$$eF_{1\to 2}(\mathbf{q}) = \langle n_2 l_2 l_{2z} | e_1 e^{-i\mathbf{q}\cdot\mathbf{r}_1} + e_2 e^{-i\mathbf{q}\cdot\mathbf{r}_2} | n_1 l_1 l_{1z} \rangle, \tag{1}$$

where $e_{1,2}$ and $m_{1,2}$ are the charges and masses of the particles, $\hbar\mathbf{q}$ is the momentum transferred between systems, and $\mathbf{r}_{1,2}$ are their radius vectors in the center-of-inertia system. Here,

$$\mathbf{r}_1 = \frac{m_2}{m_1 + m_2}\mathbf{r}, \quad \mathbf{r}_2 = -\frac{m_1}{m_1 + m_2}\mathbf{r}, \quad \mathbf{r} = \mathbf{r}_1 - \mathbf{r}_2$$

Eq. (1) includes two integrals:

$$I(\mathbf{q}_{1,2}) = \int \psi^*_{n_2 l_2 l_{2z}}(\mathbf{r}) \exp\{-i\mathbf{q}_{1,2}\mathbf{r}\} \psi_{n_1 l_1 l_{1z}}(\mathbf{r}) d^3 r, \tag{2}$$

where $\mathbf{q}_{1,2} = \pm m_{2,1}\mathbf{q}/(m_1+m_2)$. The asymptotes of these integrals for $q \to \infty$ are determined by the singular terms of the radial wavefunctions in $\psi_{1,2}$ (see Problem 4.18). Let us write

$$\psi_{nll_z}(\mathbf{r}) = \varepsilon_{i\ldots n}(l,l_z)x_i\ldots x_n\{\tilde{R}_{reg}(r) + \tilde{R}_{sin}(r)\},$$

where $\tilde{R}_{reg}(r)$ and $\tilde{R}_{sin}(r)$ are the regular and singular parts of the radial function. Remember that the regular part could be expanded into a series over even, $(r^2)^s$, powers of the variable r, and $\tilde{R}_{reg}(0) \neq 0$, while $\tilde{R}_{sin}(0) = 0$. To obtain the asymptote of integral (2), take one of the radial functions at $r=0$, and keep the singular part of the other, so that

$$I(\mathbf{q}) \underset{q\to\infty}{\approx} \varepsilon^*_{i\ldots n}(2)\varepsilon_{s\ldots l}(1) i\frac{\partial}{\partial q_i}\ldots i\frac{\partial}{\partial q_n} i\frac{\partial}{\partial q_s}\ldots i\frac{\partial}{\partial q_l} \times$$

$$\times \left\{\tilde{R}_{2\,reg}(0)\int e^{-i\mathbf{q}\cdot\mathbf{r}}\tilde{R}_{1\,sin}(r)d^3r + \tilde{R}_1(0)\int e^{-i\mathbf{q}\cdot\mathbf{r}}\tilde{R}_{2\,sin}(r)d^3r\right\}. \quad (3)$$

The singular part for both radial functions is very small for $r \to 0$, and does not effect the leading asymptotic term of the expansion $I(q)$ for $q \to \infty$. The integrals here are connected to the asymptotes of wavefunctions in the momentum representation. According to Problem 4.18, we have:

$$\frac{\hbar^2 q^2}{2\mu}\hat{D}\int e^{-i\mathbf{q}\cdot\mathbf{r}}\tilde{R}_{sin}(r)d^3r \underset{q\to\infty}{\approx} -\tilde{R}(0)\hat{D}\int e^{-i\mathbf{q}\cdot\mathbf{r}}U(r)d^3r, \quad (4)$$

where the differential operator $\hat{D} = \varepsilon_{i\ldots n}(l,l_z)\partial/\partial q_i\ldots\partial/\partial q_n$, and μ is the reduced mass of the particles.

Eqs. (3) and (4) determine the asymptote of integral (2), and the asymptote of form-factor (1) for $q \to \infty$. In particular, for a transition between the states with the angular momenta $l_1 = 0$ and $l_2 = l$, we find:

$$eF_{0\to 1m} \approx -8(2i)^l \pi^{5/2}\frac{\mu}{\hbar^2}\tilde{R}_1(0)\tilde{R}_2(0)Y^*_{lm}\left(\frac{\mathbf{q}}{q}\right) \times$$

$$\times \left\{e_1\left[q_1^l\frac{\partial^l}{\partial(q_1^2)^l}\frac{\tilde{U}(q_1)}{q_1^2} + q_1^{l-2}\frac{\partial^l}{\partial(q_1^2)^l}\tilde{U}(q_1)\right] + (-1)^l e_2[q_1 \to q_2]\right\}, \quad (5)$$

where the symbol $[q_1 \to q_2]$ means the expression written in the first square bracket, but with q_1 replaced by q_2, and

$$\tilde{U}(q) = \frac{1}{(2\pi)^3}\int U(r)e^{-i\mathbf{q}\cdot\mathbf{r}}d^3r.$$

Note that the power-law decrease of the form-factors for $q \to \infty$ is similar to that mentioned in Problem 4.18 in connection with the asymptotes of the wavefunctions in the momentum representation. In particular, for potentials such that $\tilde{U} \approx -\alpha/q^n$ with $n > 1$, the form-factor is

$$F_{1\to 2} \propto q^{-(2+n+l_1+l_2)} \quad \text{for} \quad q \to \infty. \tag{6}$$

To illustrate the result obtained, we consider the application of Eq. (5) to the hydrogen atom. In the approximation of an infinitely heavy nucleus, Eq. (5) must be slightly revised by simply putting $q_2 \equiv 0$. So, the term $e_2 e^{-i\mathbf{q}\cdot\mathbf{r}_2}$ in Eq. (1), which corresponds to the proton contribution, is now reduced to just its charge e, and inelastic transitions give no contribution to the form-factor due to the wavefunctions' orthogonality. Taking into account that for the Coulomb potential, $\tilde{U} = -e^2/2\pi^2 q^2$, and using the values of the radial functions at zero for this potential (given in Problem 13.79), we find for the transitions $1s \to ns$ and $1s \to np$ with $n \geq 2$:

$$F_{1s \to ns} \approx -\frac{16}{n^{3/2}(qa_B)^4}, \quad F_{1s \to npm} \approx \frac{16\sqrt{3}i\sqrt{n^2-1}(\boldsymbol{\varepsilon}^*(m)\cdot\mathbf{q}/q)}{n^{5/2}(qa_B)^5}.$$

These results, of course, coincide with the asymptotes of the exact expression for the form-factor, calculated in Problem 13.80.

Problem 13.85

Find the differential and total cross-sections of Coulomb excitation of an atomic nucleus (initially in a state with the angular momentum, $J = 0$) by a fast, light, charged particle[384] by a) dipole ($E1-$) and b) monopole ($E0-$) nuclear transition.

Solution

Just as in the previous problems, the sought-after cross-sections could be related to nuclear electric form-factors $F_{i\to f}(\mathbf{q}) = \langle f | \sum_p e^{-i\mathbf{q}\cdot\mathbf{r}_p} | i \rangle$ for the corresponding transitions. Summation is performed over all nuclear protons; $\mathbf{r}_p$ are their radius-vectors with respect to the nucleus' center of mass.

For non-relativistic collisions between the light particle and the nucleus, there holds the inequality $qr_p \leq pR/\hbar \ll 1$ (where R is the nucleus radius, and $p = mV$ is the incident particle momentum), so that in the expression for the form-factor, we can expand the exponent and keep only the first non-zero term. The differential cross-sections of the processes considered are $d\sigma_{E1}/dq^2 \propto q^{-2}$ and $d\sigma_{E0}/dq^2 = \text{const}$, while the total cross-sections are

$$\sigma_{E1} = \frac{8\pi e^2}{\hbar^2 V^2} |\langle J=1, J_z=0|d_z|0\rangle|^2 \ln\frac{p+p'}{p-p'} \tag{1}$$

(the cross-section is summed over projections of the angular momentum of the p-state; though the state has $J_z = 0$, we should take into account that the choice of the

---

[384] Assume that the (electron, muon) *Compton* wavelength $\hbar/mc$ is larger than the nuclear size, and the particle charge is $\pm e$.

quantization axis z along the vector $\mathbf{q}$ depends on the scattering angle; compare to Problem 13.80) and

$$\sigma_{E0} \approx \frac{4\pi}{9}\left(\frac{me^2}{\hbar^2}\right)^2 \frac{p'}{p}|Q_0|^2, \quad Q_0 = \langle f| \sum_p r_p^2 |i\rangle \tag{2}$$

(parameter Q_0 also determines the probability of inner conversion for the corresponding nuclear transition; see Problems 11.68 and 11.69). In these relations, p' is the particle momentum after collision.

Problem 13.86

Find a relation between the amplitudes and differential cross-sections for elastic scattering of a neutron on a proton and of a proton on a hydrogen atom (in the ground state). Neglect the interaction of the neutron magnetic moment with the electron.

Solution

Due to the smallness of the nuclear-force radius, the time of interaction between the proton and neutron is much smaller than the characteristic atomic time. So for the electron, the result of the interaction between the neutron and proton could be considered as a sudden change resulting in the velocity boost $\mathbf{V} = \hbar\mathbf{q}/m_p$ for the proton-nucleus. Therefore, the following relation between the amplitudes of the processes considered follows:

$$f_{nH}(E,q) \approx f_{np}(E,q)a(q), \tag{1}$$

where

$$a(q) = \int |\psi_0(r)|^2 \exp\left\{-\frac{i}{\hbar} m_e \mathbf{V} \cdot \mathbf{q}\right\} d^3r = \frac{1}{[1+(qm_e a_B/2m_p)^2]^2} \tag{2}$$

is the probability amplitude for the atom to remain in the ground state. (Compare to Problem 11.58; we note that $a(q)$ coincides with the atomic form-factor; see 13.80.)

Since $a(0) = 1$, then, using the optical theorem and relation (1), we conclude that the total cross-sections of the neutron scattering from a proton and from a hydrogen atom are the same.[385] As is seen from Eqs. (1) and (2), the differential cross-sections, $d\sigma/d\Omega = |f|^2$, start to differ only for $qa_B \gtrsim m_p/m_e$, which corresponds to neutron energies that are much larger than atomic (since $\hbar q \leq 2p$).

---

[385] It is important that $m_H \approx m_p$. A completely different situation arises for neutron scattering from a proton bound in a molecule. In the case of a heavy molecule, $M \gg m_p$, the scattering cross-section for small energies from the bound proton is four times larger than the scattering cross-section from a free nucleon.

Problem 13.87

Find the scattering cross-section for heavy charged particles (for example, protons or ions) from neutral atoms with zero angular momentum. The velocities of the scattered particles are assumed to be much lower than the velocities of atomic electrons, but at the same time, $V \gg \hbar/Ma_B$, where M is the particle mass. Use the quasi-classical expression for the scattering cross-section (see Problem 13.51).

Solution

The energy of the interaction between the charged particle and the atom at large distances, $r \gg a_B$, has the form (polarization potential, see Problem 11.49)

$$U(r) = \frac{Ze(\mathbf{d}\cdot\mathbf{r})}{2r^3} = -\frac{1}{2}\beta\frac{(Ze)^2}{r^4}, \tag{1}$$

where Ze is the particle charge, and β is the atomic polarizability. The calculation of the scattering cross-section by quasi-classical equation (see Problem 13.51)

$$\sigma = 4\pi\int_0^\infty \left\{1 - \cos\left[\frac{1}{\hbar v}\int_{-\infty}^\infty U(\sqrt{\rho^2+z^2})dz\right]\right\}\rho d\rho \tag{2}$$

for a power-law potential was performed in Problem 13.52. For polarization potential (1), it gives

$$\sigma = \pi\Gamma\left(\frac{1}{3}\right)\left(\frac{\pi\beta Z^2 e^2}{4\hbar v}\right)^{2/3}. \tag{3}$$

Let us make several comments about this result. As it follows from (3), distances essential in the scattering process are of the order $\rho_0 \sim \sqrt{\sigma} \sim v^{-1/3}$ (in atomic units), and we put $Z \sim 1$ and $\beta \sim a_B^3$ (such estimation follows from the condition that for $\rho \sim \rho_0$, the argument of the cosine, i.e., the quasi-classical phase, is of order 1; let us note that for $r \leq a_B$, Eq. (1) is not applicable, but in the problem conditions such distances do not play an essential role). To be able to use Eq. (1), these distances must be large, $\rho_0 \gg 1$, hence $v^{1/3} \ll 1$. (The condition, $\rho_0/v \gg \omega_{at}^{-1}$, that ensures the adiabaticity of the atomic electrons, is also fulfilled; when it is broken, dynamic excitation processes become essential for the atom, and the notion of an interaction potential loses its strict meaning.) On the other hand, the quasi-classical condition must also be fulfilled: $l \sim N\rho_0 v \gg 1$, so $v^{2/3} \gg 1/M$ (M is the mass of the particle considered). So, Eq.(3) is valid if the conditions

$$\left(\frac{m_e}{M}\right)^{3/2} v_{at} \ll v \ll v_{at} \tag{4}$$

are fulfilled. Hence, the particle must be heavy, $M \gg m_e$ (for electrons, Eq. (3) is not applicable).

Problem 13.88

Find the *charge-exchange* cross-section[386] for the collision of a slow, $v \ll v_{at}$, negative ion, A^-, with its atom, A. Consider the atom and ion to be in their S-states, and the ion's valence electron to be weakly bound (see Problem 11.28). The relative motion of atom and ion should be treated quasi-classically.

Solution

The specifics of the situation involving a resonant charge-exchange process, for the relative velocity of colliding particles $1/\sqrt{M} \ll v \ll 1$ (in atomic units, M is mass of atom or ion), is determined by following facts. 1) The charge-exchange cross-section is large, $\sigma_{non-rez} \gg \pi a_B^2$, i.e., large impact parameters are essential. 2) The atom's and ion's relative motion is quasi-classical, and we can only straight ballistic trajectories, since $Mv^2 \gg 1$. 3) States of the "inner" electrons of the atom and ion do not change in the collision process, while the outer-shell electron can be viewed as being in the field of two zero-range potentials. 4) Due to the adiabaticity of the collision process (for the electron subsystem), only transitions between levels that are close in energy matter, that represent, the even (g) and odd (u) states of the quasi-molecular ion. Recall that for $R \to \infty$, these states are degenerate in energy. With decrease of distance, to $R_c = 1/\alpha$, the odd term goes to the continuous spectrum, and ionization becomes possible (see Problem 11.28). But during the process of charge-exchange, distances which are much larger than R_c are essential.

In the conditions mentioned, the wavefunction of the outer-shell electron for large distances $\mathbf{R}(t) = \boldsymbol{\rho} + \mathbf{v}t$ between atoms has the form:

$$\psi(r, \mathbf{R}, t) \approx \frac{1}{\sqrt{2}} \left\{ \exp\left\{-i \int_{-\infty}^{t} E_g \, dt\right\} \psi_g + \exp\left\{-i \int_{-\infty}^{t} E_u \, dt\right\} \psi_u \right\}, \quad (1)$$

where $E_{g,u}(R)$ and

$$\psi_{g,u} \approx \frac{1}{\sqrt{2}} \left\{ \psi_0\left(\mathbf{r} - \frac{\mathbf{R}}{2}\right) \pm \psi_0\left(\mathbf{r} + \frac{\mathbf{R}}{2}\right) \right\}$$

are the energy and wavefunction of the even (odd) molecular terms on such distances, while $\psi_0(r)$ is the wavefunction of a bound state in an isolated zero-range potential (see Problem 4.10). Coefficients in the superposition, (1), were chosen so that for $t \to -\infty$, the wavefunction has the form $\psi \approx C(t)\psi_0(\mathbf{r} - \mathbf{R}/2)$, i.e., describes an electron, localized in the vicinity of one atom, which corresponds to the ion before collision. So for $t \to +\infty$, the coefficient in front of the wavefunction $\psi_0(\mathbf{r} + \mathbf{R}/2)$ determines the probability of charge-exchange:

---

[386] Compare to the charge-exchange cross-section in the collision between a slow proton and a hydrogen atom.

$$W_{ch-ex}(\rho) = \sin^2\left[\int_{-\infty}^{\infty} (E_u(R) - E_g(R))\frac{dz}{2v}\right] \tag{2}$$

$(R^2 = \rho^2 + z^2, z = vt)$; here the charge-exchange cross-section is

$$\sigma_{ch-ex}(v) = \int_0^{\infty} 2\pi\rho W_{ch-ex}(\rho)d\rho. \tag{3}$$

Compare Eqs. (2) and (3) to the expression for the elastic scattering cross-section from a potential, $U(r)$, in the quasi-classical approximation, as considered in Problem 13.51.

According to Problem 11.28, at large distances, $E_u - E_g \approx 2\alpha e^{-\alpha R}/R$, where $\varepsilon_{bind} = \alpha^2/2$ is the binding energy of an electron in the ion. Writing

$$e^{-\alpha R} \approx \exp\left\{-\alpha\rho\left(1 + \frac{z^2}{2\rho^2}\right)\right\},$$

we find that the integral in Eq. (2) in the case $\alpha\rho \gg 1$ is equal to

$$I(\rho) \approx \sqrt{\frac{2\pi\alpha}{\rho}}\frac{1}{v}e^{-\alpha\rho}.$$

Due to its pronounced dependence on ρ, the argument of the sine in Eq. (2) decreases rapidly with an increase in ρ. Thus, the dominant contribution to integral (3) is in the region of impact parameter $\rho \leq \rho_0$ (where $I(\rho_0) = 1$), where the rapidly oscillating factor $\sin^2 I(\rho)$ could be replaced by its mean value, equal to $1/2$. This gives

$$\sigma_{ch-ex}(v) \approx \frac{1}{2}\pi\rho_0^2 \approx \frac{\pi}{2\alpha^2}\ln^2\left[\frac{\alpha}{v}\sqrt{\frac{2\pi}{\ln\sqrt{2\pi\alpha^2/v^2}}}\right]. \tag{4}$$

Note that the value ρ_0 satisfying equation $I(\rho_0) = 1$, which is convenient to write in the form $\ln I(\rho_0) = 0$, can be obtained by successive iterations. The first iteration gives $\alpha\rho_0 = \ln(\sqrt{2\pi\alpha}/v)$.

Finally, we emphasize that the large value of the charge-exchange cross-section in Eq. (4) is due to the fact that for a weakly-bound electron, $\alpha \ll 1$.

Problem 13.89

For the collision of identical atoms, one of which is in the ground state while the other is in an excited state (these states are connected by a dipole transition), estimate the interaction cross-section and, in particular, the *excitation transfer* cross-section. Consider the atoms' angular momenta to be equal to 0 and 1. The velocity of the atoms' relative motion is assumed to be small in comparison to the characteristic atomic velocity, while the energy, on the other hand, is assumed to be much larger than atomic. Compare to the previous problem.

Solution

An important property of the process considered, that leads to a large excitation transfer cross-section (as well as to a large elastic scattering cross-section) is the near-degeneracy of the states corresponding to an excitation of one of the identical atoms if the distance between them is large. Therefore, for slow collisions, transitions between the close-in-energy states appear if the impact parameter is large enough. Interaction between the atoms, which has a dipole–dipole nature, was discussed in Problem 11.55. In comparison to the previous problem, here the difficulty arises due to an increase in the number of states: there are three g-terms and three u-terms. These are independent states corresponding to different polarizations of the excited atom (with the angular momentum $l = 1$).

One such state, which corresponds to the projection of the angular momentum $l_z = 0$ of the excited atom with the angular momentum $l = 1$, evolves independently of the other two. Here, as well as in the previous problem, the translational motion of atoms is considered quasi-classically in the approximation of a straight-line trajectory, and axis z is chosen perpendicular to the plane of motion. For this state, $U_{g,u} = \pm d^2/(3R^3)$ (see Problem 11.55), and calculation of the excitation transfer cross-section could be performed by Eqs. (2) and (3) from Problem 13.88. Calculating the integral

$$\int_{-\infty}^{\infty} [U_g(R) - U_u(R)] \frac{dz}{2v} = \frac{d^2}{3v} \int_{-\infty}^{\infty} \frac{dz}{(\rho^2 + z^2)^{3/2}} = \frac{2d^2}{3\rho^2 v},$$

just as in the previous problem, we find the excitation transfer cross-section with $l_z = 0$:

$$\sigma_0 = \int_0^{\infty} 2\pi\rho \sin^2\left(\frac{2d^2}{3\rho^2 v}\right) d\rho = \frac{2\pi d^2}{3v} \int_0^{\infty} \sin^2 x \frac{dx}{x^2} = \frac{\pi^2 d^2}{3v}. \tag{1}$$

As stated in the problem condition, it much exceeds the atomic length-scales (remember that $v \ll 1$).

For the two other polarized states of the excited atom (with $l_z = \pm 1$, or with $l_{x,y} = 0$; see Problem 3.21 and also Problem 3.41), calculation of the excitation transfer cross-section needs to rely on numerical methods. Note that in this case, transitions appear between such states, i.e., in the process of excitation transfer, the atom polarization may change. Diagonalization of the "instantaneous" Hamiltonian (performed in Problem 11.55) is based on the choice of a quantization axis along the direction, which passes through the atoms' centers. But due to their motion, the corresponding "rotating" system is non-inertial. Going to this rotating system gives rise to the following additional term in the Hamiltonian:

$$\hat{V}_{cor} = -\hat{\mathbf{\Omega}} \cdot \hat{\mathbf{l}} = -\frac{1}{I}\hat{L}_z\hat{l}_z = -\frac{v\rho}{R^2}\hat{l}_z.$$

This is the *Coriolis interaction*,[387] (compare to Problem 6.29). The Coriolis operator does not commute with the operators of the "instantaneous" Hamiltonian, and this leads to transitions between the eigenstates, except for the case of $l_z = 0$ mentioned above. As for the estimation of the excitation transfer cross-section, it is evident that, as in Eq. (1), it is of order $\sigma \sim \pi d^2/v$.

Problem 13.90

Find the total cross-section σ_{tot}, elastic cross-section σ_{el}, and inelastic cross-section σ_{in}, for the scattering of fast particles, $kR \gg 1$, from an absorbing ("black") sphere of radius R. Find also the differential cross-section of elastic scattering. Compare to Problems 13.57 and 13.58.

Hint

Use quasi-classical ideas to describe particle motion. Assume that all particles that reach the surface of the sphere are absorbed by it.

Solution

Since we are dealing with a fully-absorbing "black" sphere, all scattering phases in Eq. (XIII.9) for the amplitude of elastic scattering are the same: $\delta_l = i\infty$ ($e^{2i\delta_l} = 0$) for $l < l_0 = kR$ and $\delta_l = 0$ for $l > l_0$. Such values of δ_l correspond to the following physical picture (particle motion is quasi-classic, since $kR \gg 1$): particles with impact parameter $\rho = l/k < R$ are "absorbed" by the sphere, while those with $\rho > R$ move freely. Elastic scattering by its physical nature is analogous to Fraunhofer diffraction on an opaque screen (in this case, a fully absorbing screen). It is described by the scattering amplitude (compare to Problem 13.57)

$$f_{difr} = \frac{i}{2k} \sum_{l=0}^{l_0}(2l+1)P_l(\cos\theta) \approx \frac{iR}{\theta}J_1(kR\theta), \quad \theta \ll 1. \tag{1}$$

Using the optical theorem, we find total collision cross-section

$$\sigma_{tot} = \frac{4\pi}{k}\text{Im}\, f_{difr}(\theta=0) = 2\pi R^2. \tag{2}$$

The inelastic scattering cross-section (the absorption cross-section) is equal to

$$\sigma_{inel} = \frac{\pi}{k^2}\sum_{l=0}^{\infty}(2l+1)(1-|S_l|^2) = \frac{\pi}{k^2}\sum_{l=0}^{l_0}(2l+1) \approx \pi R^2, \tag{3}$$

while the elastic scattering cross-section is

$$\sigma_{el} = \sigma_{tot} - \sigma_{inel} \approx \pi R^2. \tag{4}$$

---

[387] Here $\hat{\mathbf{L}}$ is the angular momentum of the relative motion of the atoms, and $I = \mu R^2$ is the moment of inertia with respect to the center of mass. In the quasi-classical approximation, the operator $\hat{\mathbf{L}}$ could be replaced by the only non-zero component $L_z = \mu\rho v$.

This result is obtained by direct calculation of

$$\sigma_{el} = \int |f_{difr}|^2 \, d\Omega = \pi R^2,$$

just as in Problem 13.57.

Problem 13.91

Interaction between an electron and a positron may result in their mutual *annihilation*, i.e., their conversion into a pair into photons. Because of this, *positronium* levels have a width, connected to the finite lifetime of the states (see Problem 11.61).

Find a relation between the level width, Γ_{ns}, for the s-states of positronium and the annihilation cross-section of a pair, $\sigma_{an}(v)$.

Hint

Taking into account the fact that the annihilation interaction radius is small, $R_{an} \sim \hbar/m_e c \ll a_B$, use perturbation theory in terms of the scattering length for the level shift and for the phase shift. See Problems 13.36 and 13.37.

Solution

Note that in the non-relativistic case, $kR_{an} \ll 1$. Thus annihilation processes are most important for the s-states. The effect of short-range interaction on an s-state is described by only one parameter: $a_0^{(S)}$, the scattering length (for the angular momentum, $l = 0$) of an isolated center (see Problems 13.36 and 13.37). For inelastic processes, the (annihilation) scattering length has a non-zero imaginary part. So, the level shift (see Problem 13.36) has an imaginary part, which determines the level width:

$$\Gamma_{ns} = -2\text{Im } \Delta E_{ns} = -\frac{8\pi\hbar^2}{m_e}|\psi_{ns}^{(0)}(0)|^2 \text{Im } a_0^{(S)}. \tag{1}$$

Here, we have taken into account that for positronium, the reduced mass is $m = m_e/2$, and $\psi_{ns}^{(0)}(0)$ is the wavefunction at zero for the unperturbed state.

On the other hand, the change of the s-wave phase shift for the Coulomb potential, $U = -e^2/r$, under the influence of a short-range interaction, according to Problem 13.37 is equal to

$$\Delta\delta_0^{(S)}(k) = -\tilde{Q}_{k0}^2 k a_0^{(S)}, \tag{2}$$

where

$$\tilde{Q}_{k0}^2 = \frac{\pi}{ka_B(1 - \exp\{-\pi/ka_B\})}, \quad |\psi_{ns}^{(0)}(0)|^2 = \frac{1}{8\pi a_B^3 n^3}, \tag{3}$$

and $a_B = \hbar^2/m_e e^2$. (In (3), the fact that the Bohr radius for positronium is twice as large as the atomic Bohr radius has been used.) From the phase shift of the elastic scattering amplitude (its imaginary part), we find the annihilation cross-section:

$$\sigma_{an} = \frac{\pi}{k^2}(1 - |S_0|^2) = \frac{\pi}{k^2}(1 - \exp\{-4\mathrm{Im}\,\Delta\delta_0^{(S)}\}) \approx \frac{4\pi}{k^2}\mathrm{Im}\,\Delta\delta_0^{(S)} =$$
$$= -\frac{4\pi}{k^2}\tilde{Q}_{k0}^2 \mathrm{Im}\, a_0^{(S)}. \tag{4}$$

According to Eqs. (1) and (4), we obtain:

$$\Gamma_{ns} = \hbar|\psi_{ns}^{(0)}(0)|^2 \frac{1}{\tilde{Q}_{k0}^2}(v\sigma_{an}(v)), \tag{5}$$

where $v = \hbar k/m = 2\hbar k/m_e$ is the relative velocity of the electron–positron pair. In the case $\hbar v/e^2 \gg 1$, when the Coulomb potential could be considered as a perturbation, relation (5) becomes simpler, since here $\tilde{Q}_{k0}^2 \approx 1$.

In conclusion, we emphasize that both the annihilation cross-section and positronium level widths (and hence their lifetime $\tau = \hbar/\Gamma$) depend strongly on the value of the total electron–positron pair spin (see Problem 11.61), so that the ortho- and para-states of the positronium should be considered separately. Let us also mention that these results could be extended to describe hadron atoms. However, the condition of their applicability assumes that there is no s-level with small binding energy in the strong short-range potential. See Problem 11.4, and also Problem 11.74.

Problem 13.92

Using the *principle of detailed balance*, relate the cross-section of neutron *radiation capture* by a proton, $n + p \to d + \gamma$, and the cross-section of *photodisintegration* of a deuteron, $d + \gamma \to n + p$.

Hint
The relation between cross-sections of mutually inverse two-particle processes, which appear in the principle of detailed balance, is valid in the relativistic region as well. Keep in mind that although the photon spin is equal to 1, it has only two independent polarizations.

Solution
According to the principle of detailed balance for inverse reactions $A \to B$ and $B \to A$, the relation

$$\frac{\overline{\sigma}_{A \to B}}{\overline{\sigma}_{B \to A}} = \frac{g_B p_B^2}{g_A p_A^2} \tag{1}$$

must be satisfied. Here $\overline{\sigma}$ are total cross-sections of the corresponding reactions $A \to B$ and $B \to A$, averaged over the particle spins in the initial state, and summed over the particle spins in the final state; $g_{A,B}$ are the spin statistical weights, and $p_{A,B}$ are the momenta of relative motion in two-particle systems A and B, taken at the same energy in the center-of-mass system (analogous relations hold not only for the total but also for the differential reaction cross-sections).

In the case of the reactions, $n + p \rightleftharpoons d + \gamma$ we have $\sigma_{A \to B} \equiv \sigma_{cap}$ (neutron radiation capture cross-section) and $\sigma_{B \to A} \equiv \sigma_{ph}$ (deuteron photodisintegration cross-section). Since the spin (polarization) statistical weight for a particle with spin s is equal to $g_s = (2s + 1)$ (except for the photon, for which $g_\gamma = 2$ due to its transverse polarization), then

$$g_A = (2s_p + 1)(2s_n + 1) = 4, \quad g_B = g_\gamma(2s_d + 1) = 6. \qquad (2)$$

Momenta $p_{A,B}$ in Eq. (1) are equal to the momenta of the particles in the rest frames of systems A, B. For the reactions considered, $p_A \equiv p_p = p_n$ and $p_0 \equiv p_\gamma = p_d$. Considering all particles to be non-relativistic (except for the photon; for it we have $E_\gamma \ll Mc^2$, where M is the mass of a nucleon), then according to the energy conservation law we obtain

$$E_A = \frac{1}{M}p_A^2 = E_B = E_\gamma + E_d - \varepsilon_0 \approx \hbar\omega - \varepsilon_0.$$

We have defined by ε_0 the deuteron binding energy and by ω the photon frequency. We have neglected the value E_d in comparison to E_γ (for a non-relativistic deuteron, $E_d \ll E_\gamma$ for $p_d = p_\gamma$).

Taking into account the fact that $p_\gamma = \hbar\omega/c$, according to Eqs. (1)–(3) we find:

$$\frac{\sigma_{cap}}{\sigma_{ph}} = \frac{3}{2}\frac{\hbar\omega}{Mc^2}\frac{\hbar\omega}{\hbar\omega - \varepsilon_0}. \qquad (4)$$

Thus, it follows that in the non-relativistic case, $\hbar\omega \ll Mc^2$, we have $\sigma_{ph} \gg \sigma_{cap}$. The only exception is a narrow region of values $\hbar\omega$ in the vicinity of the reaction $d + \gamma \to n + p$ threshold (here $\hbar\omega \approx \varepsilon_0$), where, conversely, $\sigma_{ph} \ll \sigma_{cap}$.

In conclusion, we emphasize that Eq. (4), as well as Eq. (1), does not rely on any information or assumptions about the mechanism(s) driving the reaction, but is based solely on the symmetry of quantum mechanical equations with respect to time-reversal.

Problem 13.93

Find the relation between the cross-section of photoelectric effect (a.k.a. "photoeffect") from the ground state of a hydrogen atom and the *radiation recombination* cross-section of an electron with a proton (the process inverse to photoeffect) into the ground state of a hydrogen atom.

Solution

This problem is solved similarly to the previous one (moreover, the processes considered in these two problems are physically related; see also Problems 14.38 and 14.19). Averaged over the spins, the cross-sections of the photoeffect reaction, $\gamma + H \to e + p$, and the radiation recombination reaction, $e + p \to H + \gamma$, are connected by the relation:

$$\frac{\sigma_{ph}(\omega)}{\sigma_{rec}(\varepsilon_e)} = \frac{p_e^2}{2p_\gamma^2} = \frac{\hbar\omega - |E_0|}{\hbar\omega} \frac{m_e c^2}{\hbar\omega}. \tag{1}$$

To prove Eq. (1), we should take into account that first, the center-of-mass frames for these reactions are the rest frames of the hydrogen atom in the ground state[388] or the proton; second, the statistical spin weight is equal to 4 for both the hydrogen atom and the electron-proton system. Finally, the electron and photon energies are connected by the energy conservation law as follows:

$$\hbar\omega + E_0 = \varepsilon_e = \frac{1}{2m_e} p_e^2,$$

where E_0 is the energy of the hydrogen atom in the ground state.

---

[388] Here we neglect the hyperfine structure of the hydrogen atom (see Problem 11.2). Let us also note that for processes involving the hydrogen atom in a state with the angular momentum, l, the "spin" statistical weight is equal to $4(2l+1)$.

14
Quantum radiation theory

1) A consistent theory of photon emission and absorption relies on the occupation number representation for the photon subsystem. We describe the *radiation field* (*i.e.*, the *free* electromagnetic field) with the vector potential operator:[389]

$$\hat{\mathbf{A}}_{\text{rad}}(\mathbf{r}) = \sum_{\mathbf{k}\sigma} \left(\frac{2\pi\hbar c^2}{\omega_{\mathbf{k}} V}\right)^{1/2} (\mathbf{e}_{\mathbf{k}\sigma} \hat{a}_{\mathbf{k}\sigma} e^{i\mathbf{k}\cdot\mathbf{r}} + \mathbf{e}^*_{\mathbf{k}\sigma} \hat{a}^\dagger_{\mathbf{k}\sigma} e^{-i\mathbf{k}\cdot\mathbf{r}}). \tag{XIV.1}$$

Here, $\hat{a}^\dagger_{\mathbf{k},\sigma}$ and $\hat{a}_{\mathbf{k},\sigma}$ are the creation and annihilation operators of a photon with the wavevector $\mathbf{k}$, frequency $\omega_{\mathbf{k}} = c|\mathbf{k}|$, and polarization $\mathbf{e}_{\mathbf{k}\sigma}$. The latter, the unit vector of photon polarization, satisfies the *transversal condition* $\mathbf{k}\cdot\mathbf{e}_{\mathbf{k}\sigma} = 0$. A state of the emission field is described by the wavefunction, $\phi(n_{\mathbf{k}\sigma}, t)$, where $n_{\mathbf{k}\sigma}$ are the occupation numbers of the photon states.

The Hamiltonian operator of the free emission field is $\hat{H}^{(0)}_{\text{rad}} = \sum_{\mathbf{k}\sigma} \hbar\omega_{\mathbf{k}} \hat{a}^\dagger_{\mathbf{k}\sigma} \hat{a}_{\mathbf{k}\sigma}$. Interaction between non-relativistic particles and this field is described[390] by:

$$\hat{V}_{\text{int}} = -\frac{e_a}{mc} \hat{\mathbf{A}}_{\text{rad}}(\mathbf{r}) \cdot \hat{\mathbf{p}} + \frac{e_a^2}{2mc^2} \hat{\mathbf{A}}^2_{\text{rad}}(\mathbf{r}) - \frac{\mu}{s} \hat{\mathbf{s}} \cdot \hat{\boldsymbol{\mathcal{H}}}_{\text{rad}}(\mathbf{r}). \tag{XIV.2}$$

where e_a, m, s, μ are the charge, mass, spin and magnetic moment of a particle, and $\hat{\boldsymbol{\mathcal{H}}}_{\text{rad}}(\mathbf{r}) = \boldsymbol{\nabla} \times \hat{\mathbf{A}}_{\text{rad}}(\mathbf{r})$.

Interaction between a particle (or a system of particles) and the radiation field is characterized by the small parameter $\alpha \equiv e^2/\hbar c \approx 1/137$ (for $e_a \sim e$). For our purposes, usually only transitions accompanied by emission or absorption of the minimum possible number of photons are essential. The probability of various processes can be calculated using the methods of perturbation theory. In particular, the transition probability (per unit of time) between states in the discrete spectrum, that are connected by the emission of a single photon, is

---

[389] Compare to the $\hat{\psi}$ operators from Chapter 10. To describe the field, we used the *Coulomb gauge* $\boldsymbol{\nabla} \cdot \hat{\mathbf{A}}_{\text{rad}} = 0, \hat{\varphi}_{rad} \equiv 0$, with operators $\hat{\mathbf{A}}_{\text{rad}}(\mathbf{r})$ given in the Schrödinger representation. We assume that the system in enclosed by a volume V, which in the limit $V \to \infty$ does not enter physical observables.

[390] External fields that are described classically, as well as the Coulomb interaction, are included in the Hamiltonian in the usual manner. If the system is in an external *magnetic field*, so that $\mathbf{A} \neq 0$, then in Eq. (XIV. 2) we should replace $\hat{\mathbf{p}}$ by $\hat{\mathbf{p}} - (e_a/c)\mathbf{A}$.

$$dw_{n\sigma} = \frac{2\pi}{\hbar} \left|\langle f|\hat{V}_{\text{int}}|i\rangle\right|^2 d\rho_f. \tag{XIV.3}$$

Wavefunctions of the initial and final states of a "particle+photon" system have the form:

$$|i\rangle = \psi_i(\xi)|0\rangle_\gamma, \quad E_i = \varepsilon_i,$$
$$|f\rangle = \psi_f(\xi)|1_{\mathbf{k}\sigma}, 0, \ldots\rangle_\gamma, \quad E_f = \varepsilon_f + \hbar\omega_\mathbf{k},$$

where ξ is the set of independent particle coordinates, $\psi_{i,f}(\xi)$ and $\varepsilon_{i,f}$ are the wavefunctions and energy of the corresponding states, $|0\rangle_\gamma$ and $|1_{\mathbf{k}\sigma}, 0, \ldots\rangle_\gamma$ are state vectors of the photon subsystem, corresponding to vacuum and one-photon states. After taking Eqs. (XIV.1) and (XIV.2) (and properties of the operators $\hat{a}$, $\hat{a}^\dagger$) into account, the matrix element in Eq. (XIV.3) becomes

$$\langle f|\hat{V}_{\text{int}}|i\rangle = -\left(\frac{2\pi\hbar c^2}{\omega_\mathbf{k} V}\right)^{1/2} \langle \psi_f| \sum_a \left\{ \frac{e_a}{m_a c} \mathbf{e}^*_{\mathbf{k}\sigma} \cdot \hat{\mathbf{p}}_a + \right.$$
$$\left. i\left(\frac{\mu_a}{s_a}\right) [\mathbf{e}^*_{\mathbf{k}\sigma} \times \mathbf{k}] \cdot \hat{\mathbf{s}}_a \right\} e^{-i\mathbf{k}\cdot\mathbf{r}_a} |\psi_i\rangle. \tag{XIV.4}$$

2) In the *dipole approximation* we can replace $e^{-i\mathbf{k}\cdot\mathbf{r}}$ by 1, neglect the term with the magnetic moments of the particles, and transform[391] the perturbation matrix element to the form:

$$\langle f|\hat{V}_{\text{int}}|i\rangle = i\sqrt{\frac{2\pi\hbar\omega}{V}} \mathbf{e}^*_{\mathbf{k}\sigma} \cdot \mathbf{d}_{fi}, \tag{XIV.5}$$

where $\mathbf{d}_{fi} = \langle\psi_f|\sum_a e_a \mathbf{r}_a|\psi_i\rangle$ is the matrix element of the system's dipole moment. For each of the two independent photon polarizations, the density of final states is

$$d\rho_f = \int \delta(\varepsilon_i - \varepsilon_f - \hbar\omega) \frac{V d^3k}{(2\pi)^3} = \frac{V\omega^2}{(2\pi)^3 \hbar c^3} d\Omega_n \tag{XIV.6}$$

where $\hbar\omega = \varepsilon_i - \varepsilon_f$, $d\Omega_n$ is an element of solid angle in the direction of photon emission, and $\mathbf{k} = k\mathbf{n}$. Eq. (XIV.3) takes the form:

$$dw_{n\sigma} = \frac{\omega^3}{2\pi\hbar c^3} |\mathbf{e}^*_{\mathbf{k}\sigma} \cdot \mathbf{d}_{fi}|^2 d\Omega_n. \tag{XIV.7}$$

---

[391] Here, the following relation is used

$$\langle\psi_f|\frac{\hat{\mathbf{p}}_a}{m_a}|\psi_i\rangle = -i\omega\langle\psi_f|\mathbf{r}_a|\psi_i\rangle, \quad \hbar\omega = \varepsilon_i - \varepsilon_f.$$

It is also valid for a system that is in a magnetic field, though $\hat{\mathbf{p}}_a$ must be replaced by $\hat{\mathbf{p}}_a - (e_a/c)\mathbf{A}(\mathbf{r}_a)$.

Summation over photon polarizations, performed using the equation[392]

$$\sum_{\sigma=1,2} (\mathbf{e}_{\mathbf{k}\sigma}^*)_i \cdot (\mathbf{e}_{\mathbf{k}\sigma})_j = \delta_{ij} - \frac{k_i k_j}{k^2}, \quad \text{(XIV.8)}$$

gives the angular dependence of the emitted photon:

$$dw_n = \sum_\sigma dw_{n\sigma} = \frac{\omega^3}{2\pi\hbar c^3} |[\mathbf{n} \times \mathbf{d}_{fi}]|^2 \, d\Omega_n. \quad \text{(XIV.9)}$$

Following this with an integration over the directions of particle escape gives the probability of photon emission for the corresponding transition $i \to f$ in the dipole approximation:

$$w \equiv w_{E1} = \int dw_n = \frac{4\omega^3}{3\hbar c^3} |\mathbf{d}_{fi}|^2. \quad \text{(XIV.10)}$$

Note that under the dipole approximation, interaction (XIV.2) becomes

$$\hat{V}_{\text{int}} \approx -\frac{e}{mc}\hat{\mathbf{A}}_{\text{rad}}(\mathbf{0}) \cdot \hat{\mathbf{p}} + \frac{e^2}{2mc^2}\hat{\mathbf{A}}_{\text{rad}}^2(\mathbf{0}). \quad \text{(XIV.11)}$$

This is equivalent to the interaction[393]

$$\hat{V}_{\text{int}}' = -\hat{\mathbf{d}} \cdot \hat{\mathbf{E}}_{\text{rad}}(\mathbf{0}), \quad \text{(XIV.12)}$$

where $\hat{\mathbf{d}} = e\hat{\mathbf{r}}$ is the dipole moment, and

$$\hat{\mathbf{E}}_{\text{rad}}(\mathbf{0}) = -\frac{1}{c}\dot{\hat{\mathbf{A}}}_{\text{rad}}(\mathbf{0}) = \frac{i}{c\hbar}[\hat{H}_{\text{rad}}^{(0)}, \hat{\mathbf{A}}_{\text{rad}}(\mathbf{0})] =$$

$$i\sum_{\mathbf{k}\sigma} \sqrt{\frac{2\pi\hbar\omega_\mathbf{k}}{V}} \{\hat{a}_{\mathbf{k}\sigma}\mathbf{e}_{\mathbf{k}\sigma} - \hat{a}_{\mathbf{k}\sigma}^\dagger \mathbf{e}_{\mathbf{k}\sigma}^*\}. \quad \text{(XIV.13)}$$

In the following problems, the free-particle wave functions[394] are normalized to unity in the volume V: $\psi_\mathbf{p} = V^{-1/2} e^{i\mathbf{p}\cdot\mathbf{r}/\hbar}$. The transition probability (per unit of time),

---

[392] Here
$$\sum_\sigma (\mathbf{e}_{\mathbf{k}\sigma}^* \cdot \mathbf{a})(\mathbf{e}_{\mathbf{k}\sigma} \cdot \mathbf{b}) = \mathbf{a}\cdot\mathbf{b} - (\mathbf{n}\cdot\mathbf{a})(\mathbf{n}\cdot\mathbf{b}) = [\mathbf{a}\times\mathbf{n}][\mathbf{b}\times\mathbf{n}].$$

[393] We emphasize that Eq. (XIV.12) is equivalent to the total expression (XIV.11), not just a part linear in the operators $\hat{a}$, $\hat{a}^\dagger$ on the right-hand side.

[394] The possible values of momentum, which in a finite system form a discrete set, are determined by the orthogonality condition: $\langle \mathbf{p}|\mathbf{p}'\rangle = \delta_{\mathbf{pp}'}$. The number of independent quantum states in a momentum-volume element d^3p is $V d^3p/(2\pi\hbar)^3$.

dw, between states is related to the differential cross-section of the corresponding process by

$$d\sigma = \frac{dw}{j} = V\frac{dw}{v}. \quad \text{(XIV.14)}$$

where $j = \rho\mathbf{v}$ is the current density, $\rho = 1/V$ is the particle volume density, and $\mathbf{v}$ is the colliding relative velocity of the colliding particle. For collisions between photons and a particle (or atom), $v = c$.

14.1 Photon emission

Problem 14.1

Find the *lifetime* and *level width* for the excited 2p-states in hydrogen.

Apply the result obtained to the μ-mesoatom, and compare it to the free muon lifetime, $\tau_\mu = 2.2 \cdot 10^{-6}$ s.

Solution

Wavefunctions of the initial and final states of the hydrogen atom have the form (see Eq. (IV.4)):

$$\psi_i = \psi_{2lm} = \sqrt{\frac{3}{4\pi}}\,(\varepsilon(m) \cdot \mathbf{n})\,\frac{r}{\sqrt{24a_B^5}}\exp\left\{-\frac{r}{2a_B}\right\}, \quad (1)$$

$$\psi_f = \psi_{100} = \frac{1}{\sqrt{\pi a_B^3}}\exp\left\{-\frac{r}{a_B}\right\}, \quad \hbar\omega = \varepsilon_i - \varepsilon_f = \frac{3e^2}{8a_B}.$$

Here $\varepsilon(m)$ is the unit vector that describes polarization of a hydrogen atom with the angular momentum $l = 1$. The matrix element of dipole moment for this transition is

$$\mathbf{d}_{fi} = \int \psi_f^*(-e\mathbf{r})\psi_i d^3r = -\frac{128\sqrt{2}}{243}ea_B\varepsilon(m) \quad (2)$$

Compare to Problems 3.41 and 3.42. According to Eq. (XIV.10), the probability of the $2p \to 1s$ transition for the hydrogen atom from photon emission is

$$w = \left(\frac{2}{3}\right)^8\left(\frac{e^2}{\hbar c}\right)^4 \frac{c}{a_B} = \left(\frac{2}{3}\right)^8\left(\frac{e^2}{\hbar c}\right)^3 \frac{m_e e^4}{\hbar^3}. \quad (3)$$

The lifetime, $\tau = 1/w$, and level width, $\Gamma = \hbar/\tau = \hbar w$, for the 2p-state of the hydrogen atom are thus:

$$w \approx 0.63 \cdot 10^9 \text{ s}^{-1}, \quad \tau \approx 1.60 \cdot 10^{-9} \text{ s}, \quad \Gamma \approx 0.41 \cdot 10^{-6} \text{ eV}.$$

We calculate similarly for the μ-mesoatom. The lifetime of the 2p-level of the μ-mesoatom ($m_\mu \approx 207 m_e$) is $\approx 10^{-11}$ s. This value is much smaller than the lifetime of a free muon, and illustrates the fact that a muon, upon being captured in an atomic orbit, falls to the ground level via cascade transitions before finally decaying. This is also so for the pion atoms, where $\tau_{\pi\pm} \approx 2.6 \cdot 10^{-8}$ s.

Problem 14.2

Find the lifetime of a charged spherical oscillator in the first excited level.

Solution

The wavefunctions of the spherical oscillator were discussed in Problems 4.4 and 4.5. For the initial and final states in this case, they have the form (compare to the previous problem):

$$\psi_i = \psi_{N=1} = \left(\frac{2}{\sqrt{\pi^3 a^5}}\right)^{1/2} (\varepsilon(m) \cdot \mathbf{r}) e^{-r^2/2a^2}, \quad |\varepsilon|^2 = 1,$$

$$\psi_f = \psi_{N=0} = \left(\frac{1}{\sqrt{\pi^3 a^3}}\right)^{1/2} e^{-r^2/2a^2}, \quad a = \sqrt{\frac{\hbar}{m\omega}}.$$
(1)

Since the oscillator's first excited level has angular momentum $l = 1$, photon emission has a dipole nature. The matrix element for the dipole moment of transition, $\mathbf{d} = e\mathbf{r}$, is

$$\mathbf{d}_{fi} = e\sqrt{\frac{2}{\pi^3 a^8}} \int (\varepsilon \cdot \mathbf{r}) \mathbf{r} e^{-r^2/a^2} d^3 r = \frac{1}{\sqrt{2}} e a \varepsilon,$$
(2)

and the probability of photon emission (per unit of time) is equal to

$$w = \frac{2}{3} \frac{e^2 a^2 \omega^3}{\hbar c^3} = \frac{2}{3}\left(\frac{e^2}{\hbar c}\right)\left(\frac{\hbar \omega}{mc^2}\right)\omega.$$
(3)

As is seen, $w \ll \omega$.

Problem 14.3

Find the probability of electromagnetic transition (per unit of time[395]) for a spherical rotor on the first excited level, if the rotor has moment of inertia I and dipole moment $\mathbf{d}$ directed along its axis.

---
[395] Hereafter, this statement is omitted for brevity.

Solution

Wavefunctions of the initial ($l = 1$) and final ($l = 0$) states of the rotor have the form (compare to the two previous problems and to Problem 3.3):

$$\psi_i = Y_{lm} = \sqrt{\frac{3}{4\pi}} \varepsilon(m) \cdot \mathbf{n}, \quad \psi_f = Y_{00} = \sqrt{\frac{1}{4\pi}}.$$

The matrix element of the dipole moment, $\mathbf{d} = d_0 \mathbf{n}$, is

$$\mathbf{d}_{fi} = \int \psi_f^* d_0 \mathbf{n} \psi_i d\Omega = \frac{\sqrt{3} d_0}{4\pi} \int \mathbf{n}(\varepsilon \cdot \mathbf{n}) d\Omega = \frac{1}{\sqrt{3}} d_0 \varepsilon.$$

The probability of photon emission, according to Eq. (XIV.10), is equal to

$$w = \frac{4 d_0^2 \omega^3}{9 \hbar c^3}, \quad \hbar \omega = \varepsilon_i - \varepsilon_f = \frac{\hbar^2}{I}. \tag{1}$$

Problem 14.4

Find the probability of electromagnetic transition between two-atom molecular rotational levels (without a change in electronic and oscillatory states) that has dipole moment d_0. The molecule is $^1\Sigma$. Consider only the first excited rotational term.

Estimate transition probabilities and compare them to the probability of atomic dipole transition.

Solution

The properties of two-atom molecular rotational states with $\Lambda = 0$ are analogous to the properties of a spherical rotor with the moment of inertia $I = \mu R^2$, where μ is the reduced mass of the molecular nuclei and R is the equilibrium distance between them. This analogy, mentioned previously in Problem 11.40, translates directly to photon emission in the molecular system. The probability of the process considered is described by Eq. (1) of the previous problem.

The estimate of this molecular photon emission probability differs sharply from the typical value for atomic dipole emission, $w_{at} \sim 10^9$ s^{-1} (compare to Problem 14.1). This is due to the smallness, $\sim m_e/\mu$, of the molecular rotational energy in comparison to the electron energy (see Problem 11.25) and consequently of the frequency of the emitted photon. Since $w \sim \omega^3$, then for the emission probability of a rotational transition we obtain the estimation $w_{mol} \sim 10^{-2}$ s^{-1} (for concreteness we have used $\mu = 3 m_p$).

In conclusion, we note that emission from higher rotational levels in the dipole approximation appears only for transitions to a neighboring level, i.e., the molecular rotational level changes by 1: $K_f = K_i - 1$. The frequency of photon emission is $\omega = \hbar K_i / I$, while the emission probability, after summation over the independent final states of the rotor (different values of moment projection), is described by the expression:

$$w(K \to K-1) = \frac{4d_0^2 \omega_K^3}{3\hbar c^3} \frac{K}{2K+1}.$$

Problem 14.5

Prove that dipole transitions are *forbidden* between:

a) levels of an atom with different multiplicity (for example, between the states of ortho- and parahelium),
b) different hyperfine components of the same atomic term (*i.e.*, between different sublevels of the same multiplet with given values of L and S).

Solution

a) The operator of dipole moment, $\mathbf{d} = \sum_a e_a \mathbf{r}_a$, depends only on the spatial coordinates and so commutes with any spin operator. In particular, $\hat{\mathbf{d}}\hat{S}^2 - \hat{S}^2\hat{\mathbf{d}} = 0$, where S is the total spin. The matrix element of this relation, which corresponds to transition between states $\psi_{i,f}$ with definite values of $S_{i,f}$ gives

$$[S_i(S_i+1) - S_f(S_f+1)]\langle\psi_f|\hat{\mathbf{d}}|\psi_i\rangle = 0.$$

It follows that $\mathbf{d}_{fi} = \mathbf{0}$ for $S_i \neq S_f$, so that dipole emission for such transitions is absent.

b) All states of hyperfine structure with the same atomic terms have the same parity, since electrons in them have the same one-particle orbital states. Taking into account the anticommutativity of the inversion operators $\hat{I}$ and dipole moment, $\hat{I}\hat{\mathbf{d}} + \hat{\mathbf{d}}\hat{I} = \mathbf{0}$, and calculating the matrix element of this equality between states $\psi_{i,f}$ with definite parities, we obtain $(I_i + I_f)\mathbf{d}_{fi} = \mathbf{0}$. It follows that dipole emission for transitions between the states considered is indeed forbidden.

Problem 14.6

For the $2s_{1/2}$-state of the hydrogen atom, find the probability of electromagnetic transition to the $2p_{1/2}$-state. Compare this to the probability of transition $2s_{1/2} \to 1s_{1/2}$ through a two-photon emission, $w_{2\gamma} \approx 8$ s^{-1}, and to the result of Problem 14.8. Remember that the difference of energies between $2s_{1/2}$- and $2p_{1/2}$-levels (the *Lamb shift*) is $\Delta E_{LS} \approx 1058$ MHz $\approx 4.4 \cdot 10^{-6}$ eV.

Solution

The wavefunctions in the initial and final states in the hydrogen atom have the form (see Eq. (IV.4))

$$\psi_i = \psi_{200}\chi_i = \frac{1}{\sqrt{8\pi a_B^3}}\left(1 - \frac{r}{2a_B}\right)e^{-r/2a_B}\chi_i,$$

$$\psi_f = R_{21}(r)\frac{1}{\sqrt{4\pi}}(\hat{\boldsymbol{\sigma}} \cdot \mathbf{n})\chi_f = \frac{1}{\sqrt{96\pi a_B^5}}re^{-r/2a_B}(\hat{\boldsymbol{\sigma}} \cdot \mathbf{n})\chi_f. \quad (1)$$

The spinor χ_i characterizes the spin state of an electron in the initial $2s_{1/2}$-state of atom, while the spinor $(\hat{\boldsymbol{\sigma}} \cdot \mathbf{n})\chi_f$ determines the spin-angular part of the $2p_{1/2}$-state (see 5.21). Spinors $\chi_{i,f}$ are normalized by the condition $\langle\chi|\chi\rangle = 1$.

The matrix element of the dipole moment for the transition is

$$\mathbf{d}_{fi} = -\frac{e}{16\sqrt{3}\pi a_B^4}\chi_f^* \left\{\int r\left(1 - \frac{r}{2a_B}\right)\mathbf{r}(\hat{\boldsymbol{\sigma}} \cdot \mathbf{n})e^{-r/a_B}dV\right\}\chi_i =$$

$$\sqrt{3}ea_B\chi_f^*\hat{\boldsymbol{\sigma}}\chi_i. \quad (2)$$

The probability of photon emission according to Eq. (XIV.10) depends on the projections of the angular momentum in the initial and final atomic states. The summation of the transition probability over the two independent values, $j_z = \pm 1/2$, in the final state is performed with the help of the relation:

$$\sum_{m=\pm 1/2}|\langle\chi_f(m)|\hat{\boldsymbol{\sigma}}|\chi_i\rangle|^2 = \sum_m \langle\chi_i|\hat{\boldsymbol{\sigma}}|\chi_f(m)\rangle\langle\chi_f(m)|\hat{\boldsymbol{\sigma}}|\chi_i\rangle =$$

$$= \langle\chi_i|\hat{\boldsymbol{\sigma}} \cdot \hat{\boldsymbol{\sigma}}|\chi_i\rangle = 3\langle\chi_i|\chi_i\rangle = 3.$$

Therefore, the total probability of the transition is

$$w(2s_{1/2} \to 2p_{1/2}) = 12\frac{e^2 a_B^2 \omega^3}{\hbar c^3}. \quad (3)$$

The numerical value is ($\hbar\omega = \Delta E_{LS} \approx 4.4 \cdot 10^{-6}$ eV)

$$w \approx 0.81 \cdot 10^{-9} \text{ s}, \quad \tau = \frac{1}{w} \approx 39 \text{ years} \quad (4)$$

(1 year is $3.15 \cdot 10^7$ s). The small probability of dipole transition and large lifetime (compare to Problems 14.1 and 14.4) is due to a small frequency of the emitted photon. See also Problem 14.8.

Problem 14.7

A neutral spin-1/2 particle with the magnetic moment μ (so that $\hat{\boldsymbol{\mu}} = \mu\hat{\boldsymbol{\sigma}}$), is in an homogeneous magnetic field, $\mathcal{H}_0$, and exists in a state with a definite value of the spin projection onto the direction of the field. Find the probability of photon emission per unit of time as a result of *spin-flip*.

Solution

Interaction between the particle and electromagnetic field is described by

$$\hat{U} = -\hat{\boldsymbol{\mu}} \cdot [\mathcal{H}_0 + \mathcal{H}_{\text{rad}}(\hat{\mathbf{r}})]. \qquad (1)$$

Compare this to Eq. (XIV.2). We use

$$\mathcal{H}_{\text{rad}}(\hat{\mathbf{r}}) = \boldsymbol{\nabla} \times \mathbf{A}_{\text{rad}}(\hat{\mathbf{r}}) = \sum_{k,\sigma} \left(\frac{2\pi\hbar c^2}{\omega_k V}\right)^{1/2} i[\mathbf{k} \times (\mathbf{e}_{k\sigma} \hat{a}_{k\sigma} e^{i\mathbf{k}\cdot\mathbf{r}} - \mathbf{e}^*_{k\sigma} \hat{a}^\dagger_{k\sigma} e^{-i\mathbf{k}\cdot\mathbf{r}})].$$

Neglecting the influence of orbital motion on emission,[396] we consider only the spin degree of freedom, and put $\mathbf{r} = 0$. The term $\hat{H}_0 \equiv -\hat{\boldsymbol{\mu}} \cdot \mathcal{H}_0 = -\mu \mathcal{H}_0 \hat{\sigma}_z$ (the z axis is directed along the external magnetic field) is the unperturbed spin Hamiltonian. Its eigenfunctions and eigenvalues are described by

$$\psi_1 = \begin{pmatrix} 1 \\ 0 \end{pmatrix}, \quad E_1^{(0)} = -\mu\mathcal{H}_0; \quad \psi_2 = \begin{pmatrix} 0 \\ 1 \end{pmatrix}, \quad E_2^{(0)} = \mu\mathcal{H}_0.$$

Considering for concreteness $\mu > 0$, we find that $E_2^{(0)} > E_1^{(0)}$. Under the action of the perturbation $\hat{V} = -\hat{\boldsymbol{\mu}} \cdot \mathcal{H}_{\text{rad}}(\mathbf{0})$, a transition from the state ψ_2 to ψ_1 is possible. This happens from photon emission with energy $\hbar\omega = E_2^{(0)} - E_1^{(0)} = 2\mu\mathcal{H}_0$ (here state ψ_1 is stable with respect to emission: for $\mu < 0$, the roles of states $\psi_{1,2}$ are interchanged).

Transformations, analogous to Eqs. (XIV.3)–(XIV.7), lead to the following expression for the differential probability of the transition considered:

$$dw_{n\sigma} = \frac{2\pi}{\hbar}|V_{fi}|^2 d\rho = \frac{\mu^2 \omega}{2\pi\hbar c}|[\mathbf{k} \times \mathbf{e}^*_{k\sigma}] \cdot \psi_1^* \hat{\boldsymbol{\sigma}} \psi_2|^2 d\Omega_n. \qquad (2)$$

Denoting

$$\boldsymbol{\mu}_{1,2} = \mu\psi_1^* \hat{\boldsymbol{\sigma}} \psi_2 \equiv \mu\boldsymbol{\sigma}_{1,2},$$

which is the matrix element of the magnetic moment, and introducing vector $\mathbf{a}_{1,2} = [\mathbf{k} \times \boldsymbol{\sigma}_{12}]$, we rewrite Eq. (2) in the form:

$$dw_{n\sigma} = \frac{\omega}{2\pi\hbar c}|\mathbf{e}^*_{k\sigma} \cdot \mathbf{a}_{12}|^2 d\Omega_n. \qquad (3)$$

Compare to Eq. (XIV.7). Summation over photon polarizations, performed as in Eqs. (XIV.8) and (XIV.9), gives (using $\mathbf{k} \cdot \mathbf{a}_{12} = 0$):

$$dw_n = \sum_\sigma dw_{n\sigma} = \frac{\omega}{2\pi\hbar c}\{k^2|\boldsymbol{\mu}_{12}|^2 - (\mathbf{k}\cdot\boldsymbol{\mu}_{12})(\mathbf{k}\cdot\boldsymbol{\mu}_{12})^*\}d\Omega_n, \qquad (4)$$

---

[396] In particular, neglecting *Doppler broadening* of spectral lines.

and after integration over directions of photon escape, we obtain

$$w = \int dw_n = \frac{4\omega^3}{3\hbar c^3}|\boldsymbol{\mu}_{12}|^2. \tag{5}$$

To calculate it, we used the value of integral

$$\int (\mathbf{a}\cdot\mathbf{k})(\mathbf{b}\cdot\mathbf{k})d\Omega = a_i b_k \int k_i k_k d\Omega = a_i b_k \frac{4\pi}{3}k^2\delta_{ik} = \frac{4\pi}{3}k^2(\mathbf{a}\cdot\mathbf{b}).$$

Let us note that Eqs. (3)–(5) are general equations of magnetic-dipole emission theory.

Taking into account the explicit form of function $\psi_{1,2}$ and the Pauli matrices, we find matrix elements

$$(\sigma_x)_{12} = \begin{pmatrix}1 & 0\end{pmatrix}\begin{pmatrix}0 & 1\\1 & 0\end{pmatrix}\begin{pmatrix}0\\1\end{pmatrix} = 1, \quad (\sigma_y)_{12} = -i, \quad (\sigma_z)_{12} = 0$$

and obtain the final expression for the total probability of photon emission for spin-flip in the magnetic field:

$$w = \frac{64}{3}\frac{\mu^5\mathcal{H}_0^3}{\hbar^4 c^3}. \tag{6}$$

Problem 14.8

Estimate the probability of a single-photon transition of a hydrogen atom from the excited $2s_{1/2}$-state into the ground $1s_{1/2}$-state. Compare this value with the result of Problem 14.6. What is the transition multipolarity?

Solution

Interaction $\hat{V}_{\text{int}}$ between an electron and a radiation field is described by Eq. (XIV.2), where we put $\mu_a = -e\hbar/2mc$ ($-e$ is the electron charge). The matrix element of such a perturbation for a single-photon transition is given by Eq. (XIV.4). For the wavefunctions of initial and final states, $\psi_{i,f} = \psi_{ns}(r)\chi_{i,f}$, we should take the wavefunctions of the corresponding ns-states of the hydrogen atom with the spin state of the electron taken into account. In this problem, the first term in the matrix element $\langle f|\hat{V}_{\text{int}}|i\rangle$, which includes $\hat{\mathbf{p}}$, is zero. Indeed, due to spherical symmetry, we see that the matrix element

$$\langle\psi_{1s}|e^{-i\mathbf{k}\cdot\mathbf{r}}\hat{\mathbf{p}}|\psi_{2s}\rangle \propto \mathbf{k}, \quad \text{but } \mathbf{k}\cdot\mathbf{e}_{\mathbf{k}\sigma} = 0.$$

For the transitions considered,

$$\langle f|\hat{V}_{\text{int}}|i\rangle = -\frac{ie\hbar}{2m}\sqrt{\frac{2\pi\hbar}{\omega_k V}}\langle 1s|e^{-i\mathbf{k}\cdot\mathbf{r}}|2s\rangle \mathbf{e}_{\mathbf{k}\sigma}^* \cdot \langle\chi_1|[\hat{\boldsymbol{\sigma}}\times\mathbf{k}]|\chi_2\rangle. \tag{1}$$

Due to the fact that $ka_B \ll 1$, we can expand the exponent in the matrix element (however, the integral could be calculated exactly; see the value of form-factor $F_{1s\to 2s}(q)$

in Problem 13.80). The first two expansion terms, $e^{-i\mathbf{k}\cdot\mathbf{r}} \approx 1 - i\mathbf{k}\cdot\mathbf{r}$, give zero, the first vanishes due to wavefunction orthogonality and the second does so due to the function under the integral being odd. So we have:

$$\langle 1s|e^{-i\mathbf{k}\cdot\mathbf{r}}|2s\rangle \approx -\frac{1}{2}\int \psi_{1s}^*(\mathbf{r})(\mathbf{k}\cdot\mathbf{r})^2\psi_{2s}(\mathbf{r})d^3r.$$

This integral gives $2^9\sqrt{2}k^2a_B^2/3^6$. The differential probability of photon emission is

$$dw_{n\sigma} = \frac{2^{14}e^2\hbar a_B^2\omega^5}{3^{12}\pi m^2 c^7}|(\mathbf{e}_{\mathbf{k}\sigma}^* \cdot [\boldsymbol{\sigma}_{12} \times \mathbf{k}])|^2\, d\Omega_n. \qquad (2)$$

Compare to the transition from Eq. (XIV.5) to Eq. (XIV.7), and also to Eq. (3) from the previous problem. In Eq. (2),

$$\hbar\omega = \hbar k c = E_{2s} - E_{1s} = \frac{3e^2}{8a_B},$$

and we have introduced $\boldsymbol{\sigma}_{12} = \langle\chi_1|\hat{\boldsymbol{\sigma}}|\chi_2\rangle$. Now we sum Eq. (2) over the photon polarizations and then integrate over the direction of photon escape, as in the previous problem. This gives the transition probability:

$$w_{12} = \frac{2^{17}e^2\hbar a_B^4\omega^2}{3^{13}m^2 c^9}|\boldsymbol{\sigma}_{12}|^2 = \frac{1}{2^4\cdot 3^6}\left(\frac{e^2}{\hbar c}\right)^9|\boldsymbol{\sigma}_{12}|^2\frac{me^4}{\hbar^3}. \qquad (3)$$

It depends on the electron spin states. To find the total emission probability for the transition $2s_{1/2} \to 1s_{1/2}$ of the hydrogen atom, Eq. (3) must be summed over the two independent spin states of the $1s$-electron. As in Problem 14.6, we obtain

$$w(2s_{1/2} \to 1s_{1/2}) = \frac{1}{2^4\cdot 3^5}\left(\frac{e^2}{\hbar c}\right)^9\frac{me^4}{\hbar^3} \approx 0.62\cdot 10^{-6}\text{ s}^{-1}. \qquad (4)$$

This gives the lifetime of the $2s_{1/2}$-state with respect to the transition considered. $\tau = 1/w \approx 18$ days.

Comparison between the results of this problem and the results of Problem 14.6 with the probability of two-photon transition $w_{2\gamma} \approx 8\text{ s}^{-1}$ shows that the one-photon emission from the $2s_{1/2}$-state has a much lower probability (by several orders of magnitude), than the two-photon transition, *i.e.*, the former is strongly suppressed. In the conditions of Problem 14.6, such suppression has an obvious reason – the smallness of emitted photon frequency. The probability of emission is $w_{E1} \propto \omega^3$.

The suppression of the one-photon transition $2s_{1/2} \to 1s_{1/2}$, which has dipole magnetic character, is explained by the fact that neglecting retardation effects, $e^{-i\mathbf{k}\cdot\mathbf{r}} \approx 1$, and the transition is forbidden due to orthogonality of wavefunctions' coordinate parts, as was mentioned before. We should mention that the smallness of the matrix element (which has an order of magnitude $k^2 a_B^2 \sim \alpha^2 = (1/137)^2$ and, correspondingly, $\alpha^4 \sim 10^{-9}$ in the expression for emission probability) that appears in the exponential expansion has the same order of magnitude as relativistic corrections to the wavefunction. If we take into account the latter, we obtain a nine-fold increase of probability (4). Therefore, the analysis presented here is qualitative at best.

Problem 14.9

Find the probability of electromagnetic transition between the components of hyperfine structure of the hydrogen atom[397] (see Problem 11.2).

Solution

In the hydrogen atom, the triplet level, $S = 1$, of hyperfine structure is above the singlet one, $S = 0$, by $\Delta E_{HFS} = 1420$ MHz$\approx 5.9 \cdot 10^{-6}$ eV (see Problem 11.2). Wavefunctions of these states have the form (for concreteness we restrict ourselves to the case of the triplet state with $S_z = 0$):

$$\psi_i = \psi_{1s}\chi_{10} = \frac{1}{\sqrt{2}}\left\{\begin{pmatrix}1\\0\end{pmatrix}_e\begin{pmatrix}0\\1\end{pmatrix}_p + \begin{pmatrix}0\\1\end{pmatrix}_e\begin{pmatrix}1\\0\end{pmatrix}_p\right\}\frac{e^{-r/a}}{\sqrt{\pi a^3}}, \qquad (1)$$

$$\psi_f = \psi_{1s}\chi_{00} = \frac{1}{\sqrt{2}}\left\{\begin{pmatrix}1\\0\end{pmatrix}_e\begin{pmatrix}0\\1\end{pmatrix}_p - \begin{pmatrix}0\\1\end{pmatrix}_e\begin{pmatrix}1\\0\end{pmatrix}_p\right\}\frac{e^{-r/a}}{\sqrt{\pi a^3}}.$$

χ_{SS_z} are spin functions of the electron–proton system.

The matrix element of interaction (XIV.2) for one-photon transition between states (1) has the form:

$$\langle f|\hat{V}_{\text{int}}|i\rangle = -\frac{ie\hbar}{2m}\sqrt{\frac{2\pi\hbar}{\omega_k V}}\langle 1s|e^{-i\mathbf{k}\cdot\mathbf{r}}|1s\rangle \mathbf{e}_{\mathbf{k}\sigma}^* \cdot \langle\chi_{00}|[\hat{\boldsymbol{\sigma}}_e \times \mathbf{k}]|\chi_{10}\rangle.$$

Compare to the derivation of Eq. (1) in the previous problem. (We neglected the interaction of the proton magnetic moment with the radiation field, since it is about $m_p/m_e \approx 2 \cdot 10^3$ times weaker than that of the electron.) Replacing the exponent by unity and calculating the vector components

$$\boldsymbol{\sigma}_{12} = \langle\chi_{00}|\hat{\boldsymbol{\sigma}}_e|\chi_{10}\rangle = (0,0,1),$$

we obtain

$$\langle f|\hat{V}_{\text{int}}|i\rangle = -\frac{ie\hbar}{2m}\sqrt{\frac{2\pi\hbar}{\omega_k V}}\mathbf{e}_{\mathbf{k}\sigma}^* \cdot [\boldsymbol{\sigma}_{12} \times \mathbf{k}]. \qquad (2)$$

---

[397] Note that the emission processes considered here (corresponding to the radio range, a wavelength of 21 cm) play an important role in astrophysics research; as an example, the *redshift* of these spectral lines determines the distances to (receding) galaxies.

Now, using standard methods (compare, for example, to the solutions of the two previous problems), we find that the total probability of the transition considered is

$$w = \frac{e^2 \hbar k^2 \omega |\sigma_{12}|^2}{3m^2 c^3} = \frac{1}{3}\left(\frac{e^2}{\hbar c}\right)\frac{(\Delta E_{HFS})^3}{\hbar m^2 c^4} \approx 3 \cdot 10^{-15} \text{ s}^{-1}, \tag{3}$$

which corresponds to triplet-level lifetime $\tau = 1/w \approx 10^7$ years (such a large value is connected to the small frequency of the emitted photon, $\tau \propto \omega^{-3}$).

Problem 14.10

What is the multipolarity of emission for the dominant electromagnetic transitions between fine structure components with the same atomic terms? Estimate the probability of the corresponding transitions per unit of time.

Solution

Probabilities of one-photon electromagnetic transitions (per unit of time) of different multipolarity can be estimated as follows:

$$\text{a)} \quad w_{E1} \sim \frac{d_{12}^2 \omega^3}{\hbar c^3} \sim \alpha^3 \left(\frac{\omega}{\omega_{at}}\right)^3 \omega_{at} \tag{1}$$

for a dipole electric or $E1$-transition;

$$\text{b)} \quad w_{M1} \sim \frac{\mu_{12}^2 \omega^3}{\hbar c^3} \sim \alpha^5 \left(\frac{\omega}{\omega_{at}}\right)^3 \omega_{at} \tag{2}$$

for a dipole magnetic, or $M1$-transition;

$$\text{c)} \quad w_{E2} \sim \frac{Q_{12}^2 \omega^5}{\hbar c^5} \sim \alpha^5 \left(\frac{\omega}{\omega_{at}}\right)^5 \omega_{at} \tag{3}$$

for a quadrupole electric, or $E2$-transition.

In the above estimates, we put $d_{12} \sim ea_B$, $\mu_{12} \sim e\hbar/mc$, $Q_{12} \sim ea_B^2$ as characteristic values of the matrix elements for dipole, magnetic, and quadrupole moments, $\alpha = e^2/\hbar c = 1/137$ is fine structure constant, and $\omega_{at} = me^4/\hbar^3 = 4.13 \cdot 10^{16}$ s^{-1}. For transitions between different terms, $\omega \sim \omega_{at}$. So probabilities of $M1$ and $E2$ transitions are of the same order of magnitude (if they are not forbidden by selection rules), and are $\alpha^{-2} \sim 10^4$ times lower than $E1$-transitions.

The specifics of emission for transitions between fine structure components with the same atomic terms are determined by the two following facts. First, corresponding atomic states have the same parity, and $E1$-transitions between them are forbidden (see Problem 14.5). Second, the transition energy of photons emitted is of the order of the fine structure interval, i.e., their frequencies are small: $\omega \sim \alpha^2 \omega_{at}$. A comparison of Eqs. (2) and (3) shows that the probability of an $E2$-transition is $\alpha^{-2} \sim 10^4$ times lower than the probability of a $M1$-transition. Quadrupole emission is strongly suppressed, and magnetic dipole transitions are dominant. For such transitions, a selection rule states that $|\Delta J| = 0$ or 1. When the energy of fine structure levels changes monotonically with increase in J, $M1$-transitions appear between the neighboring components of the fine structure. Estimation of the emission probability according to Eq. (2) with $\omega \sim \alpha^2 \omega_{at}$ gives

$$w_{M1} \sim \alpha^{11} \omega_{at} = \alpha^{12} \frac{c}{a_B} \approx 10^{-7}\ \text{s}^{-1}.$$

Problem 14.11

For a particle in the field $U(r)$, prove the validity of the following relations (the so-called "sum rules," compare to Problem 6.13):

a) $\sum_m |\langle m|x|n\rangle|^2 = \langle n|x^2|n\rangle;$

b) $\sum_m \omega_{mn}|\langle m|x|n\rangle|^2 = \frac{\hbar}{2\mu};$

c) $\sum_m \omega_{mn}^2|\langle m|x|n\rangle|^2 = \frac{1}{\mu^2}\langle n|\hat{p}_x^2|n\rangle;$

d) $\sum_m \omega_{mn}^3|\langle m|x|n\rangle|^2 = \frac{\hbar}{2\mu^2}\langle n|\frac{\partial^2 U}{\partial x^2}|n\rangle.$

We use μ for particle mass, summation is performed over all stationary states, and $|n\rangle$ is a stationary state of the discrete spectrum, $\langle n|n\rangle = 1$.

Solution

To prove the "sum rules", we should use the completeness condition, $\sum_m |m\rangle\langle m| = 1$, the equality $\omega_{mn} = -\omega_{nm}$, the property $x_{mn} = (x_{nm})^*$, and also relations

$$\omega_{mn} x_{mn} = \frac{1}{\hbar}\langle m|[\hat{H}, \hat{x}]|n\rangle = -\frac{1}{\mu}\langle m|\hat{p}_x|n\rangle,$$

$$\omega_{mn}^2 x_{mn} = -i\frac{\omega_{mn}}{\mu}(p_x)_{mn} = -\frac{1}{\mu\hbar}\langle m|[\hat{H}, \hat{p}_x]|n\rangle = \frac{1}{\mu}\langle m|\frac{\partial U}{\partial x}|n\rangle.$$

Next, we make the following transformations:

a) $\sum_m |\langle m|x|n\rangle|^2 = \sum_m \langle n|x|m\rangle\langle m|x|n\rangle = \langle n|x^2|n\rangle;$

b) $\sum_m \omega_{mn}|\langle m|x|n\rangle|^2 = \frac{i}{2\mu}\sum_m \{\langle n|\hat{p}_x|m\rangle\langle m|x|n\rangle - \langle n|x|m\rangle\langle m|\hat{p}_x|n\rangle\} =$

$$= \frac{i}{2\mu}\langle n|[\hat{p}_x,\hat{x}]|n\rangle = \frac{\hbar}{2\mu};$$

c) $\sum_m \omega_{mn}^2 |\langle m|x|n\rangle|^2 = \frac{1}{\mu^2}\sum_m \langle n|\hat{p}_x|m\rangle\langle m|\hat{p}_x|n\rangle = \frac{1}{\mu^2}\langle n|\hat{p}_x^2|n\rangle;$

d) $\sum_m \omega_{mn}^3 |\langle m|x|n\rangle|^2 = -\frac{i}{2\mu^2}\sum_m \left\{\langle n|\frac{\partial U}{\partial x}|m\rangle\langle m|\hat{p}_x|n\rangle \right.$

$\left. - \langle n|\hat{p}_x|m\rangle\langle m|\frac{\partial U}{\partial x}|n\rangle\right\} = \frac{i}{2\mu^2}\langle n|\left[\hat{p}_x,\frac{\partial U}{\partial x}\right]|n\rangle = \frac{\hbar}{2\mu^2}\langle n|\frac{\partial^2 U}{\partial x^2}|n\rangle.$

Similar relations are valid for the y- and z-components. If on the left-hand sides of relations, we replace matrix elements $|\langle m|x|n\rangle|^2$ by $|\langle m|\mathbf{r}|n\rangle|^2$, then the right-hand sides become equal to

a) $\langle n|\mathbf{r}^2|n\rangle$; b) $\frac{3\hbar}{2\mu}$; c) $\frac{1}{\mu^2}\langle n|\hat{\mathbf{p}}^2|n\rangle$; d) $\frac{\hbar}{2\mu^2}\langle n|\Delta U|n\rangle.$

Note that for a particle in the Coulomb potential, $U = -\alpha/r$, we have $\Delta U = 4\pi\alpha\delta(\mathbf{r})$, so that in the case d, the sum becomes equal to

$$\sum_m \omega_{mn}^3 |\langle m|\mathbf{r}|n\rangle|^2 = \frac{2\pi\alpha\hbar}{\mu^2}|\psi_n(\mathbf{0})|^2,$$

and is zero for states $|n\rangle$ with an angular momentum different from zero.

14.2 Photon scattering; Photon emission in collisions

Problem 14.12

Find the differential and total cross-section of a photon elastic scattering off of a free charged particle. Compare to the result of classical electrodynamics.

Solution

We will calculate the probability of transition per unit of time for a "particle+photon" system from the initial state, described by the wavefunction

$$\psi_i = |1_{\mathbf{k}_1\sigma_1},0,\ldots\rangle\frac{1}{\sqrt{V}}\exp\left\{\frac{i}{\hbar}\mathbf{p}_1\cdot\mathbf{r}\right\};\quad \mathbf{p}_1 = 0,\ E_1 = \hbar\omega_1, \quad (1)$$

into a state with the wavefunction

$$\psi_f = |1_{\mathbf{k}_2\sigma_2},0,\ldots\rangle\frac{1}{\sqrt{V}}\exp\left\{\frac{i}{\hbar}\mathbf{p}_2\cdot\mathbf{r}\right\};\quad E_f = \frac{p_2^2}{2m}+\hbar\omega_2, \quad (2)$$

under the action of the perturbation (compare to Eq. (XIV.2))

$$\hat{V} = -\frac{e}{mc}\hat{\mathbf{A}}_{\rm rad}(\mathbf{r})\cdot\hat{\mathbf{p}} + \frac{e^2}{2mc^2}\hat{\mathbf{A}}_{\rm rad}^2(\mathbf{r}). \qquad (3)$$

It is determined by the general equation of the second order of perturbation theory

$$dw_{fi} = \frac{2\pi}{\hbar}\left|V_{fi} + \sum_{\nu}{}'\frac{V_{f\nu}V_{\nu i}}{E_i - E_\nu}\right|^2 d\rho_f. \qquad (4)$$

The specifics of perturbation (3) is in that it contains terms of different order (both linear and quadratic) in the expansion parameter - the fine-structure constant, $\alpha = e^2/(\hbar c)$. Therefore, first-order perturbation theory in $\propto e^2\mathbf{A}^2$ should be accompanied by second-order perturbation theory in the other linear-in-the-field term in Eq. (3). Note that the matrix element V_{fi} is determined entirely by the former (quadratic-in-the-field) term, while the intermediate matrix elements, $V_{i\nu}$ and $V_{\nu f}$, in the second-order sum are on the contrary determined by the latter. Since in our case, $\hat{\mathbf{p}}\psi_i = \mathbf{p}_1\psi_i = 0$, the sum in Eq. (4) vanishes, and we get

$$\langle f|\hat{V}|i\rangle = \frac{e^2}{2mc^2}\langle f|\hat{\mathbf{A}}_{\rm rad}(\mathbf{r})\cdot\hat{\mathbf{A}}_{\rm rad}(\mathbf{r})|i\rangle =$$

$$\frac{2\pi e^2\hbar}{m\sqrt{\omega_1\omega_2}V^2}(\mathbf{e}_2^*\cdot\mathbf{e}_1)\int\exp\left\{i\left(\mathbf{k}_1 - \mathbf{k}_2 - \frac{\mathbf{p}_2}{\hbar}\right)\cdot\mathbf{r}\right\}d^3r =$$

$$\frac{2\pi e^2\hbar}{m\sqrt{\omega_1\omega_2}V^2}(\mathbf{e}_2^*\cdot\mathbf{e}_1)\delta_{\mathbf{k}_1,\mathbf{k}_2+\mathbf{p}_2/\hbar}. \qquad (5)$$

(Compare to the derivation of Eq. (XIV.4); we used $\mathbf{e}_1$ instead of $\mathbf{e}_{\mathbf{k}_1\sigma_1}$, etc.).

The factor $\delta_{\mathbf{k}_1,\mathbf{k}_2+\mathbf{p}_2/\hbar}$ in Eq. (5) enforces momentum conservation in the process of photon scattering, and means that the final state is determined by quantum numbers $\mathbf{k}_2,\sigma_2$ of the photon. Here, we can neglect recoil energy, since

$$p_2 \sim \hbar k_{1,2} \ll mc, \quad E_2 = \frac{1}{2m}p_2^2 \sim \frac{1}{m}\hbar^2 k_{1,2}^2 \equiv \hbar kc\frac{\hbar k}{mc} \ll \hbar\omega_{1,2}.$$

Using Eq. (XIV.6) for final density of states $d\rho_f$ with $\omega = \omega_2 \approx \omega_1$, according to Eq. (4) and (5), we obtain:

$$d\sigma = \frac{V}{c}dw = \frac{e^4}{m^2c^4}|(\mathbf{e}_2^*\mathbf{e}_1)|^2 d\Omega_2. \qquad (6)$$

For the relation between differential cross-section and probability, see Problem (XIV.14).

Let us point out that for a vector particle (spin $s_v = 1$) with non-zero mass, the dependence of the elastic scattering amplitude on polarization vectors, in the form $f = A(\mathbf{e}_2^*\cdot\mathbf{e}_1)$, means that its polarization is conserved in the scattering. However, in the case of photon scattering we have a different situation because of its transverse polarization. This leads to the appearance of polarization even for scattering of initially non-polarized photons. For example, if the scattering angle is $\theta = \pi/2$, the photon becomes totally linear polarized in the direction perpendicular to the scattering plane.

Now we will calculate the scattering differential cross-section for non-polarized photons. Writing $|(\mathbf{e}_2^* \cdot \mathbf{e}_1)|^2 = e_{2i}^* e_{1i} e_{2k} e_{1k}^*$ and applying relation (XIV.8) to Eq. (6), we can the average over polarizations of the incident photons and sum over polarizations of scattered ones, to obtain a differential scattering cross-section for the non-polarized photons by a free charge (θ is the scattering angle, $\mathbf{k}_1 \cdot \mathbf{k}_2 = k^2 \cos\theta$):

$$d\sigma = \frac{1}{2} r_0^2 (1 + \cos^2\theta) d\Omega, \quad r_0 = \frac{e^2}{mc^2}. \tag{7}$$

Here r_0 is the *classical radius* of the charged particle. Integration over the angles gives the total scattering cross-section

$$\sigma = \frac{8\pi}{3} r_0^2.$$

See that this does not depend on the polarization of the incident beam.

Eq. (7) and (8) do not include the Planck constant, and coincide with corresponding results, the *Thomson formula*, of classical electrodynamics (quantum effects appear in the relativistic domain, where $\hbar\omega \gtrsim mc^2$).

Problem 14.13

Find the differential and total cross-sections of photon elastic scattering from a spherical rotor with the moment of inertia I and electric dipole moment $\mathbf{d}$ (directed along the rotor axis), and which is in the ground state (see also the following problem).

Solution
The "rotor+photon" system transitions from the initial state, described by the wavefunction,

$$\psi_i = |1_{\mathbf{k}_1\sigma_1}, 0, \ldots\rangle Y_{00}, \quad E_i = \hbar\omega_1,$$

into the final state

$$\psi_f = |1_{\mathbf{k}_2\sigma_2}, 0, \ldots\rangle Y_{00}, \quad E_f = \hbar\omega_2,$$

under the influence of the interaction between the rotor[398] and the radiation field $\hat{V} = -d_0 \mathbf{n} \cdot \boldsymbol{\mathcal{E}}_{\text{rad}}(0)$ (see Eq. (XIV.12)) and these transitions appear only in the second order of perturbation theory. The transition probability (per unit of time) is defined by the general equation, given in the previous problem. Here, we have $V_{fi} = 0$,

---

[398] We consider only intrinsic degrees of freedom of the rotor, neglecting recoil effects, and assume that it is localized at the point $\mathbf{r} = 0$ as a whole.

and only the following intermediate states contribute to the sum:

$$\psi_{\nu 1} = Y_{1m}(\mathbf{n})|1_{k_1\sigma_1}, 1_{k_2\sigma_2}, 0, \ldots\rangle, \quad E_{\nu 1} = \frac{\hbar^2}{I} + \hbar\omega_1 + \hbar\omega_2;$$

$$\psi_{\nu 2} = Y_{1m}(\mathbf{n})|0\rangle_\gamma, \quad E_{\nu 2} = \frac{\hbar^2}{I}.$$

The rotor is in the first excited level with $l = 1$ (for other states, the matrix element of the dipole moment in the perturbation operator is equal to zero). Taking into account the fact that $\omega_1 = \omega_2 = \omega$, the frequencies of incident and scattered photons are the same, and Eq. (XIV.6) for the final density of states $d\rho_f$, we find:

$$dw = \frac{d_0^4 \omega^4}{c^3 V} \left| \sum_m \langle Y_{00}|n_i|Y_{1m}\rangle \langle Y_{1m}|n_k|Y_{00}\rangle \left[\frac{e_{2i}^* e_{1k}}{\hbar\omega - \hbar^2/I} - \frac{e_{2k}^* e_{1i}}{\hbar\omega + \hbar^2/I} \right] \right|^2 d\Omega_2. \quad (1)$$

The sum over m here can be calculated if we note that summation over m for $l = 1$ could be extended to all possible values of l, m, since $\langle Y_{00}|n|Y_{lm}\rangle \neq 0$ for only $l = 1$. Using the completeness condition for the spherical functions, $\sum_{lm}|Y_{lm}\rangle\langle Y_{lm}| = 1$, and taking into account the value of the integral

$$\langle Y_{00}|n_i n_k|Y_{00}\rangle = \frac{1}{4\pi} \int n_i n_k d\Omega = \frac{1}{3}\delta_{ik},$$

we obtain the differential cross-section of photon elastic scattering by a spherical rotor in the ground state:

$$d\sigma = \frac{V}{c} dw = \frac{1}{4\pi} \sigma_0 |(\mathbf{e}_2^* \cdot \mathbf{e}_1)|^2 \frac{\omega^4}{(\omega^2 - \hbar^2/I^2)^2} d\Omega_2, \quad (2)$$

$$\sigma_0 = \frac{16\pi}{9} \frac{d_0^4}{I^2 c^4}. \quad (3)$$

For the relation between the scattering cross-section and transition probability, see Eq. (XIV.14).

Let us note that the polarization effects in photon elastic scattering from the rotor are the same as for photon-scattering from a free charge. Compare Eq. (2) to Eq. (6) of the previous problem. Following the previous problem (averaging and summing over photon polarizations), we find that the differential scattering cross-section of the non-polarized photon beam is

$$d\sigma = \frac{1}{8\pi} \sigma_0 \frac{\omega^4}{(\omega^2 - \hbar^2/I^2)^2} (1 + \cos^2\theta) d\Omega. \quad (4)$$

Integration over the angles gives the total cross-section from the unexcited rotor:

$$\sigma(\omega) = \frac{2}{3}\sigma_0 \frac{\omega^4}{(\omega^2 - \hbar^2/I^2)^2}. \tag{5}$$

In the limiting cases, we have

$$\sigma(\omega) \approx \begin{cases} \frac{2}{3}\sigma_0 \left(\frac{I\omega}{\hbar}\right)^4 & \text{for } \hbar\omega \ll \frac{\hbar^2}{I}, \\ \frac{2}{3}\sigma_0 & \text{for } \hbar\omega \gg \frac{\hbar^2}{I}. \end{cases} \tag{6}$$

For the frequency $\omega \to \hbar/I$, the scattering cross-section increases indefinitely, which reflects its resonant character (*resonant fluorescence*) and corresponds to the possibility of exciting the rotor via photon absorption (for such frequencies, Eqs. (2), (4), and (5) do not apply directly).

In connection to this problem, see also Problem 14.14, where we consider photon inelastic scattering by a rotor.

Problem 14.14

In the conditions of the previous problem, find the differential and total photon inelastic scattering cross-section by the rotor. What states of the rotor get excited as a result of this process?

Solution

The solution goes along the lines of the previous problem with minor modifications. Since in the case $|l - l'| \neq 1$, the matrix elements $\langle Y_{lm}|\mathbf{n}|Y_{l'm'}\rangle = 0$, we see that inelastic photon scattering occurs in second-order perturbation theory and these processes only excite the rotor into states with angular momentum, $l = 2$. The possible final states are described by the wavefunction,

$$\psi_f = Y_{2m}(\mathbf{n})|1_{\mathbf{k}_2\sigma_2}, 0, \ldots\rangle \quad \text{and} \quad E_f = \frac{3\hbar^2}{I} + \hbar\omega_2 = \hbar\omega_1.$$

We follow the previous problem and obtain the differential cross-section of photon inelastic scattering by the rotor as follows

$$d\sigma_{2m} = \frac{d_0^4 \omega_1 \omega_2^3}{c^4} |\langle 2m|n_i n_k|0\rangle e_{2i}^* e_{1k}|^2 \frac{I^2}{(\hbar + I\omega_2)^2(\hbar - I\omega_1)^2} d\Omega_2. \tag{1}$$

Now we sum over projections of rotor's angular momentum, m. To do this, we first write

$$|\langle 2m|n_i n_k|0\rangle e_{2i}^* e_{1k}|^2 = e_{2i}^* e_{1k} e_{2l} e_{1m}^* \langle 0|n_l n_m|2m\rangle\langle 2m|n_i n_k|0\rangle$$

and use the relation

$$\sum_m \langle 0|n_s n_l|2m\rangle\langle 2m|n_i n_k|0\rangle =$$

$$= \sum_{lm} \langle 0|n_s n_l|lm\rangle\langle lm|n_i n_k|0\rangle - \langle 0|n_s n_l|0\rangle\langle 0|n_i n_k|0\rangle =$$

$$= \langle 0|n_s n_l n_i n_k|0\rangle - \langle 0|n_s n_l|0\rangle\langle 0|n_i n_k|0\rangle = \frac{1}{45}\{3\delta_{is}\delta_{kl} + 3\delta_{il}\delta_{ks} - 2\delta_{ik}\delta_{sl}\}.$$

In the above calculation, we took into account that the terms of the sum over l, m are different from zero only for $l = 0$ and 2, the completeness condition for the spherical functions, and the values of integrals

$$\langle 0|n_i n_k|0\rangle = \frac{1}{4\pi}\int n_i n_k \, d\Omega = \frac{1}{3}\delta_{ik},$$

$$\langle 0|n_i n_k n_s n_l|0\rangle = \frac{1}{15}(\delta_{ik}\delta_{sl} + \delta_{is}\delta_{kl} + \delta_{sl}\delta_{ks}).$$

As a result of these transformations, we obtain:

$$d\sigma = \sum_m d\sigma_{2m} = \frac{d_0^4 \omega_1 \omega_2^3}{45 c^4} \frac{I^2}{(\hbar + I\omega_2)^2(\hbar - I\omega_1)^2}(3 + |\mathbf{e}_2^* \cdot \mathbf{e}_1|^2) d\Omega_2. \qquad (2)$$

After averaging and summing over photon polarizations (before and after scattering; compare to Problem 14.12), the differential cross-section of unpolarized photon inelastic scattering takes the form:

$$\frac{d\sigma_{\text{inel}}}{d\Omega} = \frac{3}{160\pi}(13 + \cos^2\theta)\sigma_{\text{inel}}(\omega_1). \qquad (3)$$

The total cross-section of inelastic scattering is

$$\sigma_{\text{inel}}(\omega_1) = \frac{16\pi}{27}\frac{d_0^4 \omega_1 \omega_2^3}{c^4}\frac{I^2}{(\hbar + I\omega_2)^2(\hbar - I\omega_1)^2}. \qquad (4)$$

In the vicinity of the rotor's excitation threshold, i.e., for $\hbar\omega_1 \to 3\hbar^2/I$, we obtain

$$\sigma_{\text{inel}} \approx \frac{4\pi}{9}\frac{Id_0^4}{\hbar^3 c^4}\left(\omega_1 - \frac{3\hbar}{I}\right) \propto \omega_2^3, \qquad (5)$$

and for large photon frequencies, it becomes

$$\sigma_{\text{inel}} \approx \frac{16\pi}{27}\frac{d_0^4}{I^2 c^4}, \quad \omega_1 \approx \omega_2 \gg \frac{\hbar}{I}. \qquad (6)$$

The latter expression, as well as the photon elastic scattering cross-section, σ_{elast}, determined in the previous problem, does not contain the Planck constant.

The total scattering cross-section

$$\sigma_{tot} = \sigma_{elast} + \sigma_{inel} \approx \frac{16\pi}{9}\frac{d_0^4}{I^2 c^4}, \quad \omega \gg \frac{\hbar}{I} \qquad (7)$$

coincides with the result of classical electrodynamics for the scattering cross-section of an electromagnetic wave by a spherical rotor.

Problem 14.15

Find the differential and total photon-scattering cross-section from a charged spherical oscillator in the ground state.

Solution

Calculation of the scattering cross-section follows Problem 14.13. It is convenient to use Eq. (XIV.12) for particle interaction with a radiation field. Taking into account the form of the wavefunction for a spherical oscillator (see Problems 4.4 and 4.5), and the fact that for a (linear) oscillator, matrix elements of the dipole moment are different from zero for transitions between the neighboring levels only (see Problem 11.3), we see that the only difference between the calculation of the photon cross-section by the oscillator in comparison to scattering by a rotor in Problem 14.13 is to replace the spherical functions Y_{00} and Y_{lm} by the wavefunctions of the oscillator $\psi_{n_r lm}$:

$$\psi_{000} = \frac{2}{(\pi a^6)^{1/4}} e^{-r^2/2a^2} Y_{00}, \quad \psi_{01m} = \frac{2\sqrt{6}}{(\pi a^{10})^{1/4}} r e^{-r^2/2a^2} Y_{1m}.$$

$a^2 = \hbar/m\omega_0$ and the rotor energy E_l is replaced by $E_N = \hbar\omega_0(N + 3/2)$. So the matrix element of the dipole moment for the rotor, $d_0 \langle Y_{1m}|\mathbf{n}|Y_{00}\rangle$, is replaced by

$$\langle 01m|e\mathbf{r}|000\rangle = \sqrt{\frac{3}{2}} ea \langle Y_{1m}|\mathbf{n}|Y_{00}\rangle.$$

As in the case of the photon scattering by a rotor in the second order of perturbation theory, only the first excited oscillator level with the angular momentum $l = 1$ contributes to the sum over intermediate states. Again, extending the summation over the $l = 1$ projections to all relevant oscillator Hamiltonian eigenfunctions is possible and useful. As a result, we obtain the differential photon elastic scattering cross-section from an oscillator (instead of Eq. (2) from Problem 14.13 in case of a rotor):

$$d\sigma = \frac{e^4}{m^2 c^4} |(\mathbf{e}_2^* \cdot \mathbf{e}_1)|^2 \frac{\omega^4}{(\omega^2 - \omega_0^2)^2} d\Omega. \qquad (1)$$

For the scattering of an unpolarized photon after averaging (summing) over photon polarizations (compare to Problem 14.12), we have

$$\frac{d\sigma}{d\Omega} = \frac{3}{16\pi}(1 + \cos^2\theta)\sigma(\omega). \qquad (2)$$

The total scattering cross-section of a photon by the oscillator is

$$\sigma(\omega) = \frac{8\pi}{3} \left(\frac{e^2}{mc^2}\right)^2 \frac{\omega^4}{(\omega^2 - \omega_0^2)^2}. \tag{3}$$

Again, the results do not contain the Planck constant, and coincide with classical results for electromagnetic waves scattered by an oscillator. Compare this to the previous problem. In the context of this comparison, we should mention that for photon-scattering by the oscillator, no inelastic scattering occurs in second-order perturbation theory, *i.e.*, the oscillator is not excited, in contrast to the scattering by a rotor.

Problem 14.16

Find the differential and total scattering cross-sections of photons from a neutral particle with spin $s = 1/2$ and magnetic moment μ. Consider the following cases:

a) No spin-flip occurs as a result of scattering and the particle remains in the initial spin state with $s_z = +1/2$ (with the z-axis along the momentum of incident photons).
b) The collision process does involve a spin-flip, *i.e.*, in the final state we have $s_z = -1/2$.
c) A spin state after collision is not detected.

Generalize the result obtained to a particle with arbitrary spin.

Solution

Focusing only on the spin degree of freedom (*i.e.*, neglecting *recoil* effects for the scattering), we consider for concreteness that the particle is localized at the point $\mathbf{r} = 0$. We now calculate the transition probability from the initial state

$$\psi_i = \chi_1 |1_{\mathbf{k}_1 \sigma_1}, 0, \ldots\rangle, \quad E_i = \hbar\omega_1$$

into the final state (here $\chi_{1,2}$ are the corresponding spin functions)

$$\psi_f = \chi_2 |1_{\mathbf{k}_2 \sigma_2}, 0, \ldots\rangle, \quad E_f = \hbar\omega_2$$

under the influence of perturbation $\hat{V} = -\hat{\boldsymbol{\mu}} \cdot \hat{\mathbf{H}}_{\text{rad}}(\mathbf{0})$. Compare to Problem 14.7.

A transition appears in the second order of perturbation theory, and its probability is given by the equation:

$$dw = \frac{2\pi}{\hbar} \left|\sum_\nu{}' \frac{V_{f\nu}V_{\nu i}}{E_i - E_\nu}\right|^2 d\rho_f. \tag{1}$$

In this problem, the sum over intermediate states, $|\nu\rangle$, contains four terms, corresponding to the following wavefunctions (with $s_z = \pm 1/2$)

$$\psi_{\nu 1} = \chi_{s_z}|0,0,\ldots\rangle, \quad E_{\nu 1} = 0;$$

$$\psi_{\nu 2} = \chi_{s_z}|1_{\mathbf{k}_1\sigma_1}, 1_{\mathbf{k}_2\sigma_2}, 0, \ldots\rangle, \quad E_{\nu 2} = 2\hbar\omega.$$

As usual, $\chi_{s_z=+1/2} = \begin{pmatrix} 1 \\ 0 \end{pmatrix}$, $\chi_{s_z=-1/2} = \begin{pmatrix} 0 \\ 1 \end{pmatrix}$ and we have taken that: $\omega_1 = \omega_2 = \omega$.

The sum in Eq. (1) takes the form:

$$\frac{2\pi\mu^2 c^2}{\omega^2 V} \sum_{s_z} \{\langle\chi_2|\hat{\boldsymbol{\sigma}}\cdot\mathbf{a}_2^*|\chi_{s_z}\rangle\langle\chi_{s_z}|\hat{\boldsymbol{\sigma}}\cdot\mathbf{a}_1|\chi_1\rangle - \langle\chi_2|\hat{\boldsymbol{\sigma}}\cdot\mathbf{a}_1|\chi_{s_z}\rangle\langle\chi_{s_z}|\hat{\boldsymbol{\sigma}}\cdot\mathbf{a}_2^*|\chi_1\rangle\}.$$

We have introduced $\mathbf{a}_{1(2)} = [\mathbf{e}_{1(2)} \times \mathbf{k}_{1(2)}]$. If we take into account the completeness condition, $\sum_{s_z}|\chi_{s_z}\rangle\langle\chi_{s_z}| = 1$, we obtain:

$$\sum_\nu{}' \frac{V_{f\nu}V_{\nu i}}{E_i - E_f} = \frac{2\pi\mu^2 c^2}{\omega^2 V} a_{2i}^* a_{1k} \langle\chi_2|\hat{\sigma}_i\hat{\sigma}_k - \hat{\sigma}_k\hat{\sigma}_i|\chi_1\rangle =$$

$$= 4\pi i \frac{\mu^2 c^2}{\omega^2 V} a_{2i}^* a_{1k} \varepsilon_{ikl} \langle\chi_2|\hat{\sigma}_l|\chi_1\rangle.$$

We have used the commutation relations for the Pauli matrices, $\hat{\sigma}_i\hat{\sigma}_k - \hat{\sigma}_k\hat{\sigma}_i = 2i\varepsilon_{ikl}\hat{\sigma}_l$. Taking into account Eqs. (XIV.6) and (XIV.14), we find the differential cross-section for a photon in a magnetic field:

$$d\sigma_{21} = \frac{4\mu^2}{\hbar^2\omega^2}|\varepsilon_{ikl}a_{2i}^* a_{1k}\chi_2^*\hat{\sigma}_l\chi_1|^2 d\Omega_2. \tag{2}$$

We perform averaging (summation) over photon polarizations. We obtain

$$|\varepsilon_{ikl}a_{2i}^* a_{1k}(\sigma_l)_{21}|^2 = (\sigma_l)_{21}(\sigma_t)_{21}^*\varepsilon_{ikl}\varepsilon_{mnt}a_{2i}^* a_{1k}a_{2m}a_{1n}^*.$$

Since $a_{1k} = \varepsilon_{ksp}e_{1s}k_{1p}$, then using Eq. (XIV.8) we obtain

$$\overline{a_{1k}a_{1n}^*} = \frac{1}{2}\sum_{\sigma_1}\varepsilon_{ksp}e_{1s}k_{1p}\varepsilon_{nuw}e_{1u}^*k_{1w} = \frac{1}{2}k_{1p}k_{1w}\varepsilon_{ksp}\varepsilon_{nuw}\left(\delta_{su} - \frac{1}{k^2}k_{1s}k_{1u}\right) =$$

$$\frac{1}{2}\varepsilon_{kps}\varepsilon_{nws}k_{1p}k_{1w}.$$

Analogously, we have

$$\sum_{\sigma_2} a_{2i}^* a_{2m} = \varepsilon_{ips}\varepsilon_{mws}k_{2p}k_{2w}.$$

Now we integrate over the angles of photon scattering:

$$\int \sum_{\sigma_2} a_{2i}^* a_{2m} d\Omega_2 = \varepsilon_{ips}\varepsilon_{mws} \cdot \frac{4\pi}{3} k^2 \delta_{pw} = \frac{8\pi}{3} k^2 \delta_{im}.$$

As a result of the transformations above, we obtain the scattering cross-section for non-polarized photons in the form:

$$\sigma_{21} = \frac{16\pi}{3} \frac{\mu^4}{\hbar^2 c^2} (\sigma_l)_{21} (\sigma_l)_{21}^* \varepsilon_{ikl}\varepsilon_{int}\varepsilon_{kps}\varepsilon_{nws} k_{1p} k_{1w},$$

or, using relation $\varepsilon_{ikl}\varepsilon_{int} = \delta_{kn}\delta_{lt} - \delta_{kt}\delta_{ln}$,

$$\sigma_{21} = \frac{16\pi}{3} \frac{\mu^4}{\hbar^2 c^2} (\sigma_l)_{21} (\sigma_t)_{21}^* \{\delta_{lt} k^2 + k_{1t} k_{1l}\} =$$

$$\frac{16\pi}{3} \frac{k^2 \mu^4}{\hbar^2 c^2} \left\{ |\boldsymbol{\sigma}_{21}|^2 + \left|\boldsymbol{\sigma}_{21} \cdot \frac{\mathbf{k}_1}{k}\right|^2 \right\}. \tag{3}$$

The cross-section depends on particle spin state before and after the collision. Let us note the following cases:

a) The cross-section from a *pure* spin state with polarization vector $\mathbf{P}$, $|\mathbf{P}| = 1$, without change of the spin state, is

$$\sigma_{\uparrow\uparrow} = \frac{16\pi}{3} \frac{\mu^4 \omega^2}{\hbar^2 c^4} \left\{ 1 + \left(\mathbf{P} \cdot \frac{\mathbf{k}_1}{k}\right)^2 \right\} = \frac{16\pi}{3} \frac{\mu^4 \omega^2}{\hbar^2 c^4} (1 + \cos^2 \alpha), \tag{4}$$

where α is the angle between the vectors $\mathbf{P}$ and $\mathbf{k}_1$. As can be seen, this cross-section is maximum if the spin is directed along the momentum of incident photons ($\alpha = 0$ or π).

b) The same as in the previous case, but with a spin-flip, *i.e.*, $\mathbf{P}_f = -\mathbf{P}$:

$$\sigma_{\uparrow\downarrow} = \frac{16\pi}{3} \frac{\mu^4 \omega^2}{\hbar^2 c^4} (2 + \sin^2 \alpha). \tag{5}$$

c) The scattering cross-section in the case, where the particle spin state after collision is not detected:

$$\sigma = \sigma_{\uparrow\uparrow} + \sigma_{\uparrow\downarrow} = \frac{64\pi}{3} \frac{\mu^4 \omega^2}{\hbar^2 c^4}. \tag{6}$$

This cross-section, unlike (4) and (5), does not depend on the initial polarization.

For a particle with arbitrary spin, s, we have $\boldsymbol{\mu} = \mu \mathbf{s}/s$. It can be shown that the total photon scattering cross-section by unpolarized particles, summed over the final spin states of the photon and particle, is equal to

$$\sigma = \frac{16\pi}{9} \frac{\mu^4 \omega^2}{\hbar^2 c^4} \frac{s+1}{s^3}, \tag{7}$$

which for $s \gg 1$ coincides with the result of classical electrodynamics for the cross-section of an electromagnetic wave scattered by a magnetic moment, averaged over different orientations of the classical moment

$$\sigma_{cl} = \frac{16\pi}{9}\left(\frac{\kappa\mu\omega}{c^2}\right)^2. \tag{8}$$

Here κ, the gyromagnetic ratio, is determined by the relation $\boldsymbol{\mu} = \kappa \mathbf{M}$, where $\mathbf{M}$ is mechanical angular momentum.

Problem 14.17

Express the photon scattering cross-section in the small-frequency limit, $\hbar\omega \to 0$, from an atom in a stationary state with zero angular momentum, in terms of atomic polarizability β_0 (which determines a level shift in a uniform electric field, $\Delta E = -\beta_0 \mathcal{E}^2/2$).

Solution

We are interested in the transition probability between the states

$$\psi_{i,f} = \psi_0 |1_{\mathbf{k}_{i,f}\sigma_{i,f}}, 0, \ldots\rangle, \quad E_{i,f} = E_0 + \hbar\omega$$

(ψ_0 is the atomic wavefunction) under the influence of interaction between the electrons and the radiation field. In the dipole approximation, it is described by Eq. (XIV.12) with $\mathbf{d} = -e\sum \mathbf{r}_a$ (summation is performed over all atomic electrons), and is calculated according to the known equation of second-order perturbation theory:

$$dw = \frac{2\pi}{\hbar}\left|\sum_\nu{}' \frac{V_{f\nu}V_{\nu i}}{E_i - E_\nu}\right|^2 d\rho_f. \tag{1}$$

In this case, $V_{fi} = 0$. Contributions to the sum come from intermediate states of two types, whose wavefunctions and energies are as follows

$$\psi_{\nu 1} = \psi_n|0,0,\ldots\rangle, \quad E_{\nu 1} = E_n;$$
$$\psi_{\nu 2} = \psi_n|1_{\mathbf{k}_1\sigma_1}, 1_{\mathbf{k}_2\sigma_2}, 0,\ldots\rangle, \quad E_{\nu 2} = E_n + 2\hbar\omega.$$

ψ_n, E_n are wavefunctions and energies of atomic stationary states. If we take into account relations (XIV.12) and (XIV.13), the sum takes the form:

$$\frac{2\pi\omega}{V}\sum_n{}' \langle 0|d_i|n\rangle\langle n|d_k|0\rangle \left\{\frac{e_{2i}^* e_{1k}}{\omega_{0n} + \omega} + \frac{e_{2k}^* e_{1i}}{\omega_{0n} - \omega}\right\}.$$

These two sums have the same tensor structure (due to spherical symmetry):

$$\sum_n{}' \frac{\langle 0|d_i|n\rangle\langle n|d_k|0\rangle}{\omega_{0n} \pm \omega} = B(\pm\omega)\delta_{ik}.$$

Performing a convolution over the indices i and k, we obtain

$$B(\pm\omega) = \frac{1}{3}{\sum_n}' \frac{|\langle n|\mathbf{d}|0\rangle|^2}{\omega_{0n} \pm \omega}. \qquad (2)$$

In the limit, $\omega \to 0$, we find

$${\sum_\nu}' \frac{V_{f\nu}V_{\nu i}}{E_i - E_\nu} \approx \frac{4\pi\omega}{V} B(0)(\mathbf{e}_2^* \cdot \mathbf{e}_1).$$

Knowing that $B(0) = -\frac{1}{2}\hbar\beta_0$ and using relations (XIV.6) and (XIV.14), we obtain the small-frequency limit of the differential cross-section of photon-scattering from an atom:

$$d\sigma = \frac{\beta_0^2 \omega^4}{c^4} |\mathbf{e}_2^* \cdot \mathbf{e}_1|^2 d\Omega. \qquad (3)$$

After averaging and summing over photon polarizations (compare to Problem 14.12), we obtain the differential cross-section and total scattering cross-section of non-polarized photon scattering:

$$\frac{d\sigma}{d\Omega} = \frac{\beta_0^2 \omega^4}{2c^4}(1 + \cos^2\theta), \quad \sigma(\omega) = \frac{8\pi}{3}\frac{\beta_0^2 \omega^4}{c^4}. \qquad (4)$$

As an illustration of this relation, see Problems 14.13 and 14.15 that discuss photon scattering on a rotor and an oscillator, respectively; polarizabilities for these systems were found in Problems 8.10 and 8.2.

In conclusion, note that Eq. (4) does not contain the Planck constant (unlike the polarizabilities of quantum systems), and coincides with the analogous results from classical electrodynamics for an electromagnetic wave scattered by a polarizable system.

Problem 14.18

Find the cross-section of the photoelectric effect for a hydrogen-like atom in the ground state. Assume that the photon frequency satisfies the condition $\hbar\omega \gg I$, where I is the atomic ionization potential.

Solution

In the "hydrogen-like atom+photon" system – that is, in the state described by the wavefunction,

$$\psi_i = \psi_0(r)|1_{\mathbf{k}\sigma}, 0, \dots\rangle; \quad \psi_0 = \frac{1}{\sqrt{\pi a^3}} e^{-r/a}, \quad E_i = \hbar\omega - \frac{m(Ze^2)^2}{2\hbar^2}$$

($a = \hbar^2/Zme^2$) – ionization of the atom could occur as a result of photon absorption by the electron, under the action of the perturbation $\hat{V} = \frac{e}{mc}\hat{\mathbf{A}}_{\text{rad}}(\mathbf{r}) \cdot \hat{\mathbf{p}}$ (compare to

Eq. (XIV.2)). Since in our case, energy is equal to $E_f \approx \hbar\omega \gg I = m(Ze^2)^2/2\hbar^2$, i.e., the escaping electron is fast, then in the final state we can neglect the effect of the nucleus on the electron and choose the corresponding wavefunctions in the form:

$$\psi_f = \frac{1}{\sqrt{V}} \exp\left\{\frac{i}{\hbar}\mathbf{p}\cdot\mathbf{r}\right\}|0\rangle_\gamma, \quad E_f = \frac{p^2}{2m}.$$

The transition probability is given the equation $dw = \frac{2\pi}{\hbar}|V_{fi}|^2 d\rho_f$, where the matrix element of perturbation is described by (see Eq. (XIV.4))

$$V_{fi} = \frac{e}{mV}\sqrt{\frac{2\hbar}{\omega a^3}} \mathbf{e}_{\mathbf{k}\sigma} \cdot \int e^{-i\mathbf{p}\cdot\mathbf{r}/\hbar + i\mathbf{k}\cdot\mathbf{r}} \hat{\mathbf{p}} e^{-r/a} d^3r. \tag{1}$$

The integral is easy to calculate, if we first act by the operator $\hat{\mathbf{p}}$ on the exponent on the left. We obtain

$$\frac{8\pi\hbar a^3}{(1+a^2\kappa^2)^2}\boldsymbol{\kappa} \approx \frac{8\pi\hbar^4}{p^4 a}\mathbf{p},$$

where $\boldsymbol{\kappa} = \mathbf{p}/\hbar - \mathbf{k}$. We have taken into account that $p \gg \hbar k$, as follows from the relations

$$\frac{p^2}{2m} \approx \hbar\omega = \hbar c k \ll mc^2.$$

Neglecting $\mathbf{k}$ corresponds to replacing $e^{i\mathbf{k}\cdot\mathbf{r}} \to 1$ in the dipole approximation. Also, $pa/\hbar \gg 1$ due to the condition $\hbar\omega \gg I$. Finally, taking into account the expression for the density of states,

$$d\rho_f = \int \delta\left(\hbar\omega - I - \frac{p^2}{2m}\right)\frac{Vp\, dp^2 d\Omega}{2(2\pi\hbar)^3} = \frac{mpV}{(2\pi\hbar)^3}d\Omega,$$

and relation (XIV.14) between the cross-section and the transition probability, we find the photo-effect differential cross-section:

$$d\sigma = 32\frac{e^2\hbar}{mc}\frac{I^{5/2}}{(\hbar\omega)^{7/2}}\left|\mathbf{e}_{\mathbf{k}\sigma}\cdot\frac{\mathbf{p}}{p}\right|^2 d\Omega. \tag{2}$$

If we average over photon polarizations using relations (XIV.8), which gives $\overline{|\mathbf{e}_{\mathbf{k}\sigma}\cdot\mathbf{p}|^2} = \frac{1}{2}p^2\sin^2\theta$ (θ is the angle between the vectors $\mathbf{p}$ and $\mathbf{k}$), we obtain the photo-effect differential cross-section for unpolarized photons:

$$d\sigma = 32Z^5\left(\frac{e^2}{\hbar c}\right)\left(\frac{\hbar^2}{me^2}\right)^2\left(\frac{I_0}{\hbar\omega}\right)^{7/2}\sin^2\theta\, d\Omega. \tag{3}$$

Note that electrons tend to escape in a direction perpendicular to photon momentum, which reflects the dominance of photon's "wave-like" properties in the non-relativistic case. The electromagnetic wave exerts the Lorentz force on the electron in this direction. In the relativistic case, for "harder" photons, their corpuscular properties begin to play out, which leads to electrons escaping primarily in the direction of photon momentum.

Integration of Eq. (3) over the angles gives the total photo-effect cross-section,

$$\sigma(\omega) = \frac{256\pi}{3} Z^5 \left(\frac{e^2}{\hbar c}\right) \left(\frac{\hbar^2}{me^2}\right)^2 \left(\frac{I_0}{\hbar \omega}\right)^{7/2}, \tag{4}$$

where $I_0 = I/Z^2 = 13.6$ eV is the ionization potential of the hydrogen atom. Cross-section (4) for the values $Z = 1$ and $\hbar\omega = 5$ keV is $\approx 6 \cdot 10^{-26}$ cm^2.

Eq. (4), multiplied by 2 (for the K-electron), could be also used for an (approximate) calculation of the photo-effect cross-section for atoms that are not hydrogen-like. The contribution into the cross-section from other atomic electrons in the excited states is lower than that of K-electrons; they are more weakly bound to the nucleus, and in the limit of free electrons, photon absorption does not take place. Estimation of the photo-effect cross-section on such atomic electrons could be performed in the way analogous to that used in the following problem for radiative electron recombination.

Problem 14.19

Find the cross-section of fast electron *radiative recombination* with a proton at rest (the inverse process of photo-effect), assuming that the final state's hydrogen atom is in the ground state.

Solution

The solution follows the previous problem with minor changes and simple substitutions. (Recall that the general relation between cross-sections of mutually inverse processes, such as the photoelectric effect and electron radiation recombination, follows directly from the principle of detailed balance; see Problem 13.93). Permutation of the initial and final states does not change the value of $|V_{fi}|^2$, since the operator, $\hat{V}$, is Hermitian. Below, we list the changes that are required compared to the solution of Problem 14.18:

1) In the expression for the final density of states, now $d\rho_f$ is described by Eq. (XIV.6) with $\hbar\omega = \varepsilon_e + I \approx \varepsilon_e$.
2) Expression (XIV.14), which relates the probability and cross-section, here we have $d\sigma = V dw/v_e$.
3) Replace averaging over photon polarizations by a summation over them, which results in an additional factor of 2 in the cross-section.

As a result, we obtain the following expressions for the differential and total cross-sections of fast electron radiative recombination into the ground state of a hydrogen-like atom (ion):

$$d\sigma_{rec,1s} = 16Z^5 \left(\frac{e^2}{\hbar c}\right)^3 \left(\frac{\hbar^2}{me^2}\right)^2 \left(\frac{I_0}{\varepsilon_e}\right)^{5/2} \sin^2\theta \, d\Omega, \tag{1}$$

$$\sigma_{rec,1s} = \frac{128\pi}{3} Z^5 \left(\frac{e^2}{\hbar c}\right)^3 \left(\frac{\hbar^2}{me^2}\right)^2 \left(\frac{I_0}{\varepsilon_e}\right)^{5/2}.$$

We should note that the recombination of fast electrons into an excited level of the atom has a much lower cross-section. Indeed, from (1) of the previous problem it follows that $V_{if} \propto (\mathbf{e}\cdot\mathbf{p})\phi_i(\mathbf{p})$, where $\phi_i(\mathbf{p})$ is the wavefunction of the appropriate electron state in the momentum representation. For ns-states, the asymptote of this wavefunction for large momenta has the form $\phi_{ns} \approx C/\sqrt{n^3}p^4$, so that[399] $\sigma_{rec,ns} \propto 1/n^3$ (for states with angular momentum $l \neq 0$, both the wavefunction $\phi_{nl}(\mathbf{p})$, and the recombination cross-section decrease in the $p \to \infty$ limit faster than those in the case of s-states; see Problem 4.18). Accounting for the recombination to excited levels is reduced to a multiplication of the cross-sections (1) by $\sum_n n^{-3} = \zeta(3) = 1.202$ ($\zeta(s)$ the Riemann ζ function), *i.e.*, these processes increase the recombination cross-section by only 20%.

Problem 14.20

Find the differential and total cross-sections of deuteron photodisintegration, *i.e.*, the process $\gamma + d \to p + n$.

Hint

Use the wavefunction in a zero-range potential to model the deuteron and approximate both the proton and the neutron as free in their final states.

Solution

The deuteron photo-disintegration process is physically reminiscent of the photo-effect, and hence the calculation here follows closely the solution of Problem 14.18. Below we mention modifications that are to be made in the current problem compared to Problem 14.18.

We approximate the deuteron wavefunction by

$$\psi_0(r) = C_{\kappa 0}\sqrt{\frac{\kappa}{2\pi}}\frac{e^{-\kappa r}}{r}, \quad \varepsilon_0 = \frac{\hbar^2\kappa^2}{M}, \tag{1}$$

where ε_0 is the deuteron binding energy, M is the nucleon mass, $\mu = M/2$ is the reduced mass of the pn-system, and $E_i = \hbar\omega - \varepsilon_0$. Note that we have included the asymptotic coefficient, $C_{\kappa 0}$, which takes into account finite-size effects (see Problem 12.1 and also Problem 11.36).

---

[399] Note that the photo effect cross-section from the excited ns-state of a hydrogen-like atom has the same dependence, $\sigma_{ns} \propto n^{-3}$.

In the expression for the interaction (now of the proton) with the radiation field, we should substitute $e \to -e$, $m \to M$, $\mathbf{r} \to \mathbf{r}_p = \mathbf{r}/2$. The form of the final wavefunction, ψ_f, does not change, but here $E_f = p^2/M$.

So the value of the integral in the perturbation matrix element becomes (see Eq. (1) from Problem 14.18)

$$\int e^{-i\mathbf{p}\cdot\mathbf{r}/\hbar} \hat{\mathbf{p}} \frac{e^{-\kappa r}}{r} d^3r = \frac{4\pi\hbar^2}{(p^2 + \hbar^2\kappa^2)} \mathbf{p}. \qquad (2)$$

The term $i\mathbf{k}\cdot\mathbf{r}/2$ in the exponent is omitted as in Problem 14.18.

The density of final states is

$$d\rho_f = \int \delta\left(\hbar\omega - \varepsilon_0 - \frac{p^2}{M}\right) \frac{V d^3p}{(2\pi\hbar)^3} = \frac{MpV\, d\Omega}{2(2\pi\hbar)^3},$$

$$p = \sqrt{M(\hbar\omega - \varepsilon_0)}.$$

Summarizing, we obtain the following expression for the differential cross-section of deuteron photo-disintegration:

$$d\sigma = 2\frac{e^2}{\hbar c} C_{\kappa 0}^2 \frac{p\sqrt{\varepsilon_0}}{\hbar M^{5/2} \omega^3} |\mathbf{e}\cdot\mathbf{p}|^2 d\Omega. \qquad (3)$$

After averaging over photons polarizations (compare to Problem 14.18), we obtain:

$$d\sigma = \frac{e^2}{\hbar c} C_{\kappa 0}^2 \frac{p^3 \sqrt{\varepsilon_0}}{\hbar M^{5/2} \omega^3} \sin^2\theta\, d\Omega. \qquad (4)$$

Integration gives the total cross-section of deuteron photo-disintegration:

$$\sigma = \frac{8\pi}{3} \frac{e^2}{\hbar c} C_{\kappa 0}^2 \frac{\sqrt{\varepsilon_0}(\hbar\omega - \varepsilon_0)^{3/2}}{M\hbar\omega^3}. \qquad (5)$$

In conclusion, we briefly comment on the applicability of these results. The above derivation was based on the zero-range approximation for the pn-interaction potential. This approximation assumes that only the distances much larger than those involved in the nuclear forces are important. As is seen from Eq. (2), this is valid for $p \sim \hbar\kappa$, i.e., for photon frequencies with $\hbar\omega \sim \varepsilon_0$. However, this condition breaks down if the momenta of escaping nucleons, $p \gtrsim \hbar/r_0$, where r_0 is the nuclear interaction radius. For large momenta, due to the exponent in integral (2), fast oscillations and small distances are important, so the details of the deuteron wavefunction are important.

Problem 14.21

Find the differential cross-section of *braking radiation (bremsstrahlung)* for electrons in the Coulomb field of a nucleus. Analyze the angle and spectral distribution of the photons emitted. Electron–nucleus interaction should be considered as a perturbation.

Solution

1) The *Bremsstrahlung* process involves a fast electron in the initial state,

$$\psi_i = \frac{1}{\sqrt{V}} \exp\left\{\frac{i}{\hbar}\mathbf{p}_1 \cdot \mathbf{r}\right\} |0\rangle_\gamma, \quad E_i = \frac{1}{2m}p_1^2.$$

The perturbation below leads to electron scattering with a single photon emission

$$\hat{V} = -\frac{Ze^2}{r} + \frac{e}{mc}\hat{\mathbf{A}}_{\text{rad}}(\mathbf{r}) \cdot \hat{\mathbf{p}}. \tag{1}$$

Compare to Eq. (XIV.2). The term $\propto \hat{\mathbf{A}}_{\text{rad}}^2$ is omitted, since it induces transitions involving an even number of photons.

The final wavefunction is

$$\psi_f = \frac{1}{\sqrt{V}} \exp\left\{\frac{i}{\hbar}\mathbf{p}_2 \cdot \mathbf{r}\right\} |1_{\mathbf{k}\sigma}, 0, \ldots\rangle, \quad E_f = \frac{1}{2m}p_2^2 + \hbar\omega.$$

The *Bremsstrahlung* transition probability in second-order perturbation theory reads

$$dw = \frac{2\pi}{\hbar}\left|\sum_\nu{}' \frac{V_{f\nu}V_{\nu i}}{E_i - E_\nu}\right|^2 d\rho_f. \tag{2}$$

Note that perturbation matrix element V_{fi} is already in first-order perturbation theory due to the second term in Eq. (1). But it contains the factor $\delta_{\mathbf{p}_1, \mathbf{p}_2 + \hbar\mathbf{k}}$, which expresses momentum conservation for the photon emission by a free electron, which when combined with energy conservation, shows that the emission of a photon by a free electron is impossible. So, interaction with the external field, which leads to momentum transfer to the nucleus, is the essential element of the process considered.

Intermediate states, $|\nu\rangle$, in the sum (2) that provide a non-zero contribution are described by the wavefunctions of two types:

$$\psi_{\nu 1} = \frac{1}{\sqrt{V}}e^{i\boldsymbol{\kappa}\cdot\mathbf{r}}|1_{\mathbf{k}\sigma}, 0, \ldots\rangle, \quad E_{\nu 1} = \frac{\hbar^2\kappa^2}{2m} + \hbar\omega;$$

$$\psi_{\nu 2} = \frac{1}{\sqrt{V}}e^{i\boldsymbol{\kappa}\cdot\mathbf{r}}|0\rangle_\gamma, \quad E_{\nu 2} = \frac{\hbar^2\kappa^2}{2m}. \tag{3}$$

Summation over $\nu_{1,2}$ is reduced to the sum over all possible values of electron wave vector κ in the intermediate state.

2) We use the explicit form of the wavefunctions and operator $\hat{\mathbf{A}}_{\text{rad}}(\mathbf{r})$ (see Eq. (XIV.1)) to find the perturbation matrix elements in Eq. (2):

$$V_{\nu 1 i} = \frac{e}{mc}(\hat{\mathbf{A}}_{\text{rad}} \cdot \hat{\mathbf{p}})_{\nu 1 i} = \frac{e}{mV}\sqrt{\frac{2\pi\hbar}{\omega V}}\mathbf{e}^*_{\mathbf{k}\sigma} \cdot \int e^{-i(\boldsymbol{\kappa}+\mathbf{k})\cdot\mathbf{r}}\hat{\mathbf{p}}e^{i\mathbf{k}_1\cdot\mathbf{r}}dV$$

$$= \frac{e\hbar}{m}\sqrt{\frac{2\pi\hbar}{\omega V}}(\mathbf{e}^*_{\mathbf{k}\sigma} \cdot \mathbf{k}_1)\delta_{\mathbf{k}_1,\boldsymbol{\kappa}+\mathbf{k}};$$

$$V_{f\nu 2} = \frac{e}{mc}(\hat{\mathbf{A}}_{\text{rad}}\hat{p})_{f\nu 2} = \frac{e\hbar}{m}\sqrt{\frac{2\pi\hbar}{\omega V}}(\mathbf{e}^*_{\mathbf{k}\sigma} \cdot \boldsymbol{\kappa})\delta_{\boldsymbol{\kappa},\mathbf{k}_2+\mathbf{k}};$$

$$V_{\nu 2 i} = -Ze^2\left(\frac{1}{r}\right)_{\nu 2 i} = -\frac{Ze^2}{V}\int \frac{1}{r}e^{i(\mathbf{k}_1-\boldsymbol{\kappa})\cdot\mathbf{r}}dV = -\frac{4\pi Ze^2}{(\mathbf{k}_1-\boldsymbol{\kappa})^2 V};$$

$$V_{f\nu 1} = -Ze^2\left(\frac{1}{r}\right)_{f\nu 1} = -\frac{4\pi Ze^2}{(\boldsymbol{\kappa}-\mathbf{k}_2)^2 V}, \quad \mathbf{p}_{1,2} = \hbar\mathbf{k}_{1,2}.$$

The factor $\delta_{\mathbf{k}_1,\boldsymbol{\kappa}+\mathbf{k}}$ in these expressions allows us to perform summation over ν in Eq. (2) (in the sum over states $\nu 1$, only the term with $\boldsymbol{\kappa} = \mathbf{k}_1 - \mathbf{k}$ is different from zero, and for states $\nu 2$, only that with $\boldsymbol{\kappa} = \mathbf{k}_2 + \mathbf{k}$) and obtain

$$\sum_{\nu}{'} \frac{V_{f\nu}V_{\nu i}}{E_i - E_\nu} = \frac{4\pi Ze^2}{\hbar\omega q^2}\frac{e\hbar}{mV}\sqrt{\frac{2\pi\hbar}{\omega V}}\mathbf{e}^*_{\mathbf{k}\sigma} \cdot (\mathbf{k}_1 - \mathbf{k}_2), \tag{4}$$

$\hbar\mathbf{q} = \hbar(\mathbf{k}_1 - \mathbf{k}_2 - \mathbf{k}) \approx \hbar(\mathbf{k}_1 - \mathbf{k}_2)$ is the momentum transferred to the nucleus. We have neglected the photon momentum, $\hbar\mathbf{k}$, in comparison to electron momenta, $\hbar\mathbf{k}_{1,2}$, (This is justified for non-relativistic electrons, and corresponds in practice to the use of the dipole approximation for photon radiation; compare to Problem 14.18), so the energy denominators become equal to $-\hbar\omega$ and $\hbar\omega$, for the states $\nu 1$ and $\nu 2$, respectively.

Finally, writing the density of final states as

$$d\rho_f = \int \delta(E_f - E_i)\frac{V\,d^3k_2}{(2\pi)^3}\frac{V\,d^3k}{(2\pi)^3} = \int \delta\left(\frac{\hbar^2 k_2^2}{2m} + \hbar\omega - \frac{\hbar^2 k_1^2}{2m}\right)\frac{V\omega^2 k_2 m}{(2\pi)^6\hbar^2 c^3} \times$$

$$\times d\Omega_\gamma d\Omega_2 d\omega\, d\frac{\hbar^2 k_2^2}{2m} = \frac{mV^2 k_2 \omega^2}{(2\pi)^6\hbar^2 c^3}d\Omega_\gamma d\Omega_2 d\omega, \quad k_2 = \sqrt{k_1^2 - \frac{2m\omega}{\hbar}},$$

and using relation (XIV.14) between the cross-section and transition probability, we find the differential cross-section of braking radiation (*Bremsstrahlung*):

$$d\sigma = \frac{e^2}{\hbar c}\frac{(Ze^2)^2 k_2}{\pi^2\hbar^2 c^2 k_1 \omega q^4}|\mathbf{e}^*_{\mathbf{k}\sigma} \cdot \mathbf{q}|^2 d\Omega_\gamma d\Omega_2 d\omega. \tag{5}$$

This equation gives the most complete information about the *Bremsstrahlung* process. To determine the spectral composition of the braking radiation (*Bremsstrahlung*), we now perform the following transformations:

3) First, we sum over two independent photon polarizations in Eq. (5), using relations (XIV.8), which gives

$$d\sigma = \frac{e^2}{\hbar c} \frac{(Ze^2)^2 k_2}{\pi^2 \hbar^2 c^2 k_1 \omega q^2} \left[1 - \frac{(\mathbf{q} \cdot \mathbf{k})^2}{q^2 k^2} \right] d\Omega_\gamma d\Omega_2 d\omega. \tag{6}$$

Now we integrate over the directions of emitted photon (which is straightforward if we choose the polar axis along the vector $\mathbf{q}$):

$$d\sigma = \frac{8}{3} \frac{e^2}{\hbar c} \frac{(Ze^2)^2}{\pi \hbar^2 c^2 k_1} \frac{k_2}{\omega q^2} d\Omega_2 d\omega. \tag{7}$$

Finally, we integrate over the angles of the scattered electron. We find the differential cross-section of braking radiation as a function of the emitted photon frequency (*i.e.*, the photon spectral distribution):

$$d\sigma_\omega = \frac{8}{3} \frac{e^2}{\hbar c} \frac{(Ze^2)^2}{mc^2 E} \frac{1}{\omega} \ln \frac{(\sqrt{E} + \sqrt{E - \hbar \omega})^2}{\hbar \omega} d\omega, \tag{8}$$

where E is the initial electron energy.

Since $d\sigma/d\omega \propto 1/\omega$ for $\omega \to 0$, then the total cross-section of *Bremsstrahlung* is infinite (the so-called *infrared catastrophe*). But this divergence is not important for the calculation of electron energy loss to radiation, characterized by "effective braking", or "effective radiation", $\kappa = \int \hbar \omega \, d\sigma_\omega$. Using Eq. (8) it is easy to obtain:

$$\kappa = \hbar \int_0^E \omega d\sigma_\omega = \frac{16}{3} Z^2 \frac{e^2}{\hbar c} mc^2 r_e^2, \tag{9}$$

where $r_e = e^2/mc^2$ is the electron classical radius.

In conclusion, we emphasize that since the nucleus field action has been considered as a perturbation, the applicability of these results, Eqs. (5)–(8), needs fulfillment of the conditions $Ze^2/\hbar v_{1,2} \ll 1$ (the electron must be fast both in the initial and final states). So, the results are not applicable if almost all the energy of the incident electron is given to the emitted photon. But Eq. (9) is valid even when only the condition $Ze^2/\hbar v_1 \ll 1$ is fulfilled, since the contribution of the upper limit of the integration domain, where Eq. (8) is not applicable, does not play an essential role in the value of integral (9).

Finally, we give a generalization of expression (8) to the case of braking radiation for a two-particle collision, with the charges e_1, e_2 and masses m_1, m_2, and in the case of purely electrostatic interaction. As can be seen, it is obtained from Eq. (8) by substitutions $-Ze^2 \to e_1 e_2$, $e/m \to (e_1/m_1) - (e_2/m_2)$, and has the form:[400]

$$d\sigma_\omega = \frac{8}{3} e_1^2 e_2^2 \left(\frac{e_1}{m_1} - \frac{e_2}{m_2}\right)^2 \frac{\mu}{\hbar c^3 E} \frac{1}{\omega} \ln \frac{(\sqrt{E} + \sqrt{E - \hbar\omega})^2}{\hbar\omega} d\omega, \qquad (10)$$

where μ is the reduced mass of the particles, v is the relative velocity of colliding particles, and $E = \mu v^2/2$.

---

[400] The fact that $d\sigma_\omega$ vanishes if $e_1/m_1 = e_2/m_2$ has an analogue in classical electrodynamics, which forbids dipole radiation by a system of classical particles with the same ratio, e/m.

15
Relativistic wave equations

The peculiar property of quantum mechanics in the relativistic domain rises from the ability of particles to transform into one another through creation and annihilation. Hence the single-particle description in terms of the probability amplitude becomes inadequate.[401] A relativistic quantum theory necessitates the use of wavefunctions with definite transformation properties under the *Lorentz transformation*. These transformation properties, as well as the form of the corresponding wave equation, depend on the value of the particle spin.

1) In the case of a spinless particle, its single-component wavefunction $\psi(\mathbf{r},t)$ is a *four-dimensional scalar*.[402] The relativistic wave equation for such a free particle – the *Klein–Gordon equation* – has the form

$$(\hat{p}_i^2 + m^2 c^2)\psi = 0, \quad or \quad \left(\Delta - \frac{1}{c^2}\frac{\partial^2}{\partial t^2}\right)\psi = \left(\frac{mc}{\hbar}\right)^2 \psi. \tag{XV.1}$$

The wave equation for a spinless charged particle with a charge, e, in an external electromagnetic field described by the potentials $\mathbf{A}$ and φ, is obtained from Eq. (XV.1) by the following substitutions: $\hat{\mathbf{p}} \to \hat{\mathbf{p}} - e\mathbf{A}/c$ and $i\hbar(\partial/\partial t) \to i\hbar(\partial/\partial t) - e\varphi$. This equation takes the form:

$$\frac{1}{c^2}\left(i\hbar\frac{\partial}{\partial t} - e\varphi\right)^2 \psi = \left[\left(\hat{\mathbf{p}} - \frac{e}{c}\mathbf{A}\right)^2 + m^2 c^2\right]\psi. \tag{XV.2}$$

From here, the continuity equation can be derived:

$$\frac{\partial \rho}{\partial t} + \boldsymbol{\nabla}\cdot \mathbf{j} = 0,$$

---

[401] Indeed, from the uncertainty relation, $\Delta p \Delta x \geq \hbar$, it follows that a particle localized in a small spatial region, $\Delta x \leq \hbar/mc$, implies a large energy transfer to the particle, *e.g.*, through the influence of strong external fields. In these conditions, particle creation becomes possible, and the single-particle description loses its meaning.

[402] With respect to spatial transformations, including inversion, the wavefunction could be either *scalar* or *pseudo-scalar*. These two possibilities correspond to particles with different (opposite) inner parities; see Problem 15.5.

where

$$\rho = \frac{i\hbar}{2mc^2}\left(\psi^*\frac{\partial\psi}{\partial t} - \frac{\partial\psi^*}{\partial t}\psi + \frac{2ie}{\hbar}\varphi\psi^*\psi\right), \quad (\text{XV.3})$$

$$\mathbf{j} = -\frac{i\hbar}{2m}\left(\psi^*\boldsymbol{\nabla}\psi - \psi\boldsymbol{\nabla}\psi^* - \frac{2ie}{\hbar}\mathbf{A}\psi^*\psi\right),$$

and the corresponding charge $Q = \int \rho(\mathbf{r},t)dV$ is conserved. Note that while these relations look similar to those of non-relativistic quantum mechanics, the quantity, ρ, here is not positive definite, and therefore cannot be interpreted as a probability density. However, since creation and annihilation processes are possible in relativistic quantum mechanics, a probability density is not a necessary ingredient of relativistic quantum theory. Some questions dealing with the interpretation of solutions to the Klein–Gordon equation and the properties of spinless particles in external fields are considered in the problems of sec. 1 of this chapter.

2) For a free particle with spin $s = 1/2$, the relativistic wave equation – the *Dirac equation* – has the form:

$$i\hbar\frac{\partial}{\partial t}\psi = \hat{H}\psi \equiv (c\boldsymbol{\alpha}\cdot\hat{\mathbf{p}} + mc^2\beta)\psi, \quad \psi = \begin{pmatrix}\varphi\\\chi\end{pmatrix} = \begin{pmatrix}\psi_1\\\psi_2\\\psi_3\\\psi_4\end{pmatrix}. \quad (\text{XV.4})$$

Here the *spinor* wavefunction, ψ, has four components,[403] and the Dirac matrices are defined as

$$\boldsymbol{\alpha} = \begin{pmatrix}0 & \boldsymbol{\sigma}\\\boldsymbol{\sigma} & 0\end{pmatrix}, \quad \beta = \gamma_4 = \begin{pmatrix}1 & 0\\0 & -1\end{pmatrix}, \quad \boldsymbol{\Sigma} = \begin{pmatrix}\boldsymbol{\sigma} & 0\\0 & \boldsymbol{\sigma}\end{pmatrix},$$

$$\boldsymbol{\gamma} = -i\beta\boldsymbol{\alpha} = i\begin{pmatrix}0 & -\boldsymbol{\sigma}\\\boldsymbol{\sigma} & 0\end{pmatrix}, \quad \gamma_5 = \gamma_1\gamma_2\gamma_3\gamma_4 = -\begin{pmatrix}0 & 1\\1 & 0\end{pmatrix}, \quad (\text{XV.5})$$

where $\boldsymbol{\sigma}$, 1, 0 are the respective two-dimensional Pauli matrices, unit, and zero matrices.

For an electron in an external electromagnetic field described by the potentials, $\mathbf{A}$ and $\varphi \equiv A_0$, the Dirac equation is obtained from Eq. (XV.4) using the same substitutions given above (with the electron charge, $-e < 0$):

$$i\hbar\frac{\partial}{\partial t}\psi = \left(c\boldsymbol{\alpha}\cdot\left(\hat{\mathbf{p}} + \frac{e}{c}\mathbf{A}\right) + mc^2\beta - eA_0\right)\psi. \quad (\text{XV.6})$$

---

[403] The increase in the number of components of the wavefunction, relative to the non-relativistic case, reflects the general fact that the relativistic wave equations describe both particles and their *antiparticles*. In the case of spinless particles, these "additional" anti-particle solutions are related to the fact that the Klein–Gordon equation, unlike the Dirac equation, contains a second-order time-derivative.

Hence it follows that the electron has a spin magnetic moment of $\mu_e = -e\hbar/2mc$, so that its gyromagnetic ratio[404] is equal to $-e/mc$, in accordance with experimental value.

From Eqs. (XV.4) and (XV.6), the continuity equations can be derived:

$$\frac{\partial \rho}{\partial t} + \boldsymbol{\nabla} \cdot \mathbf{j} = 0, \quad \rho = \psi^*\psi, \quad \mathbf{j} = c\psi^*\boldsymbol{\alpha}\psi. \qquad (XV.7)$$

Eq. (XV.6) can be cast into a covariant form:

$$\left[ic\left(\hat{p} + \frac{e}{c}\hat{A}\right) + mc^2\right]\psi(\mathbf{r}, t) = 0, \qquad (XV.8)$$

where $\hat{p} \equiv \hat{p}_\mu \gamma_\mu = \hat{\mathbf{p}} \cdot \boldsymbol{\gamma} + p_4\gamma_4 = \hat{\mathbf{p}} \cdot \boldsymbol{\gamma} - \frac{\hbar}{c}\gamma_4 \frac{\partial}{\partial t}$ and $\hat{A} = A_\mu \gamma_\mu = \mathbf{A} \cdot \boldsymbol{\gamma} + iA_0\gamma_4$.

15.1 The Klein–Gordon equation

Problem 15.1

Prove that if a wave-packet, $\psi^\pm(\mathbf{r}, t)$, consists of partial solutions to the free Klein–Gordon equation that all have the same sign of the energy (i.e., either $\varepsilon \geq mc^2$, or $\varepsilon \leq -mc^2$), then the conserved quantity.

$$Q^\pm = \int \rho^\pm(\mathbf{r}, t)dV = \frac{i\hbar}{2mc^2}\int \left\{\psi^{(\pm)*}\frac{\partial \psi^\pm}{\partial t} - \frac{\partial \psi^{(\pm)*}}{\partial t}\psi^\pm\right\}dV$$

also has a definite *sign*.

Solution

A general solution of the Klein–Gordon equation (XV.1) can be written in the form of a superposition,

$$\psi(\mathbf{r}, t) = \psi^+(\mathbf{r}, t) + \psi^-(\mathbf{r}, t), \quad \psi^\pm = \int a^\pm(k)\psi_\mathbf{k}^\pm(\mathbf{r}, t)d^3k, \qquad (1)$$

of the partial solutions

$$\psi_\mathbf{k}^\pm = e^{\pm i(\mathbf{k}\cdot\mathbf{r} - \omega(k)t)}, \quad \omega(k) = \sqrt{k^2c^2 + \left(\frac{mc^2}{\hbar}\right)^2} > 0, \qquad (2)$$

forming a complete basis.

The function $\psi_\mathbf{k}^+$ describes a particle with momentum $\mathbf{p} = \hbar\mathbf{k}$ and energy $\varepsilon = \hbar\omega \geq mc^2$, while the function $\psi_\mathbf{k}^-$ formally corresponds to a particle with energy $\varepsilon' = -\hbar\omega \leq -mc^2$ and momentum $-\hbar\mathbf{k}$. Under charge conjugation, the negative-energy solutions describe an *antiparticle* with the energy $\varepsilon = \hbar\omega \geq mc^2$ and momentum $\hbar\mathbf{k}$; see Problem 15.2. We emphasize that the general solution (1) of the

---

[404] This result, as well as Eq. (XV.6), is valid for spin-1/2 particles that have no *strong* interaction.

Klein–Gordon equation does not describe any particular single-particle state. Single-particle states are described only by functions $\psi^+(\mathbf{r},t)$ and $\psi^-(\mathbf{r},t)$, separately.

Considering superpositions of ψ^+ and ψ^- separately, we substitute them into the expression for Q, given in the problem. An elementary integration, with the use of the equation

$$\int e^{\pm i(\mathbf{k}-\mathbf{k}')\cdot\mathbf{r}} d^3r = (2\pi)^3 \delta(\mathbf{k}-\mathbf{k}'),$$

give the following relation:

$$Q^\pm = \pm(2\pi)^3 \frac{\hbar}{mc^2} \int \omega(\mathbf{k})|a^\pm(\mathbf{k})|^2 d^3k. \tag{3}$$

Therefore, the quantities $Q^\pm$ indeed have a definite sign that corresponds to the solution, $\psi^\pm$. However, in the coordinate representation the expression for $\rho^\pm(\mathbf{r},t)$, under the integral of $Q^\pm$, does not have a fixed sign. Therefore, ρ^+ and ρ^- cannot be interpreted as a probability density.

If we consider charged particles we can relate the quantity $\rho^\pm$ to a charge density. For a particle with charge e, the expression $e\rho^+(\mathbf{r},t)$ (with normalization $Q^+ = 1$) corresponds to the charge density of a one-particle state. Similarly, the quantity $e\rho^-$ describes the charge density of the corresponding anti-particle state. The normalization $Q^- = -1$ automatically provides the opposite signs of the charges for the particles and their anti-particles.

For neutral particles, interpretation of local quantities $\rho^\pm(\mathbf{r},t)$ is, generally speaking, impossible. But this fact should not be considered as a deficiency of the theory, since the local spatial characteristics have no deep physical meaning in relativistic theories. Consequently, in such relativistic theories the interpretation of a wavefunction as a probability amplitude remains only in the momentum (but not coordinate) representation; see also Problem 15.7.

Problem 15.2

Prove that the Klein–Gordon equation for a free particle is invariant under the transformation

$$\psi \to \psi_c(\mathbf{r},t) = \hat{C}\psi(\mathbf{r},t) \equiv \psi^*(\mathbf{r},t),$$

which describes *charge conjugation*. It relates the solutions, $\psi^-(\mathbf{r},t)$, of the Klein–Gordon equation that have no physical meaning individually (ψ^- is a superposition of solutions that formally have negative energy; see Problem 15.1 above) to the function, $\psi_c^+ = \hat{C}\psi^-$, that describes positive-energy states. The wavefunction, $\hat{C}\psi(\mathbf{r},t) \equiv \psi^*(\mathbf{r},t)$, can be interpreted as the wavefunction of the *antiparticle*.

Verify that if the function ψ is an eigenfunction of any of the following operators $\hat{\varepsilon} = i\hbar\frac{\partial}{\partial t}$, $\hat{\mathbf{p}}$, $\hat{l}_z$, $\hat{\mathbf{l}}^2$, then the corresponding *charge-conjugated* function, ψ_c, is also an eigenfunction. What is the relation between the corresponding charge-conjugated eigenvalues?

Solution

The invariance of the free Klein–Gordon equation under charge conjugation

$$(-\hbar^2 c^2 \Delta + m^2 c^4)\psi(\mathbf{r}, t) = -\hbar^2 \frac{\partial^2}{\partial t^2}\psi(\mathbf{r}, t) \tag{1}$$

implies that if $\psi(\mathbf{r}, t)$ is a solution of Eq. (1), then so is the function, $\psi_c = \hat{C}\psi$. It is straightforward to verify that indeed, Eq. (1) is invariant under complex conjugation.

Since the operator $\hat{C}$ satisfies the relation $\hat{C}^2 = 1$, then the general solution of the Klein–Gordon equation can be written in the form:

$$\psi = \psi^+(\mathbf{r}, t) + \psi^-(\mathbf{r}, t) = \psi^+(\mathbf{r}, t) + \hat{C}\psi_c^+(\mathbf{r}, t). \tag{2}$$

Analogously, the solution of the charge-conjugated equation can be written as

$$\psi_c = \psi_c^+ + \psi_c^- = \psi_c^+ + \hat{C}\psi^+. \tag{3}$$

General solutions of the Klein–Gordon equation include both the particle wavefunction, ψ^+, and the corresponding antiparticle wavefunction, ψ_c^+, and have only a formal mathematical meaning. (Since the transformation $\hat{C}$ is *anti-linear*, relations (2) and (3) can not be used in the framework of the quantum-mechanical superposition principle.) A physically sensible set of particle and antiparticle; states can be described as a superposition of specific solutions with frequencies (energies) of a certain sign.

Let us note that the physical meaning of the transformation, $\hat{C}$, as charge conjugation becomes clear in the presence of a coupling to an external electromagnetic field, which acts differently on the particles and antiparticles; see Problem 15.3.

Since $\psi_c = \hat{C}\psi = \psi^*$, the charge-conjugated form of the eigenvalue problem, $\hat{f}\psi_f = f\psi_f$, for a generic operator, $\hat{f}$, becomes $\hat{f}^*(\psi_f)_c = f(\psi_f)_c$. Hence, if $\hat{f}^* = \hat{f}$, then the charge-conjugated eigenfunction $(\psi_f)_c = \psi_f^*$ is also an eigenfunction of the operator, $\hat{f}$, corresponding to the same eigenvalue of f. But if $\hat{f}^* = -\hat{f}$, then $(\psi_f)_c$ is also an eigenfunction but with the eigenvalue $-f$. Hence, the eigenfunction of the operators $\hat{\mathbf{p}} = -i\hbar\nabla$, $\hat{l}_z = -i\frac{\partial}{\partial\varphi}$, and $\hat{\varepsilon} = i\hbar\frac{\partial}{\partial t}$ under charge conjugation "change" eigenvalues to those of the opposite sign, while the eigenfunctions of the operator, $\hat{\mathbf{l}}^2$, "conserve" their eigenvalues.

Problem 15.3

a) What is the form of the Klein–Gordon equation for a charged spinless particle in an external electromagnetic field under the transformation

$$\psi \to \psi_c(\mathbf{r}, t) = \hat{C}\psi(\mathbf{r}, t) \equiv \psi^*(\mathbf{r}, t)?$$

b) What transformation of the electromagnetic field should be performed simultaneously with the transformation of the wavefunction, $\psi(\mathbf{r}, t)$ in order to keep the Klein–Gordon equation invariant?

c) Using the results of parts (a) and (b), interpret the transformation $\hat{C}$ as a charge conjugation that transforms a particle into an antiparticle (compare to Problem 5.2).

Solution

a) Since $\psi_c = \hat{C}\psi = \psi^*$, the Klein–Gordon equation for a particle with the charge e in a field is

$$\left\{c^2\left(-i\hbar\boldsymbol{\nabla} - \frac{e}{c}\mathbf{A}\right)^2 + m^2c^4\right\}\psi = \left(i\hbar\frac{\partial}{\partial t} - e\varphi\right)^2\psi. \tag{1}$$

After the complex conjugation, it takes the form

$$\left\{c^2\left(-i\hbar\boldsymbol{\nabla} + \frac{e}{c}\mathbf{A}\right)^2 + m^2c^4\right\}\psi_c = \left(i\hbar\frac{\partial}{\partial t} + e\varphi\right)^2\psi_c, \tag{2}$$

which also has the form of the Klein–Gordon equation in the same electromagnetic field as Eq. (1), but for a particle with the charge $-e$.

b) Transforming the potentials $\mathbf{A} \to \mathbf{A}_c = -\mathbf{A}$ and $\varphi \to \varphi_c = -\varphi$ simultaneously with the charge conjugation of the wavefunction in Eq. (1), we obtain

$$\left\{c^2\left(-i\hbar\boldsymbol{\nabla} - \frac{e}{c}\mathbf{A}_c\right)^2 + m^2c^4\right\}\psi_c = \left(i\hbar\frac{\partial}{\partial t} - e\varphi_c\right)^2\psi_c, \tag{3}$$

which is identical to the original equation (1).

c) To interpret the transformations given above, we consider a constant electromagnetic field (*i.e.*, the potentials, $\mathbf{A}$ and φ, do not depend on time). In this case, Eq. (1) has "stationary" solutions as follows: $\psi_\varepsilon = e^{-i\varepsilon t/\hbar}\psi_c(\mathbf{r})$. All such solutions can be divided into two groups, ψ_ε^+ and ψ_ε^-, which are adiabatically connected to the free particle states, with $\varepsilon \geq mc^2$ or $\varepsilon \leq -mc^2$, upon adiabatic "removal" of the external field.[405] The solutions, ψ_ε^+, adiabatically connected to the states from the upper continuum, have the meaning of the wavefunction describing a particle with the energy, ε, in the electromagnetic field. The solutions, ψ_ε^-, are associated with the states of the antiparticle. Here, as well as in the absence of any external fields, the antiparticle wavefunction is

$$\psi_c^+ = \hat{C}\psi_\varepsilon^- = (\psi_\varepsilon^-)^*, \tag{4}$$

while its energy is equal to $-\varepsilon$.

---

[405] Let us note that such classification of the solution is worthwhile only in the case of weak external fields, when the energy spectrum of the states ψ_ε^+ lies above the upper edge of the states ψ_ε^-. In strong fields, these boundaries may "merge" together, and the single-particle language loses its meaning, since spontaneous creation of particle–antiparticle pairs becomes possible; see Problems 15.12 and 15.13 for a related discussion.

Hence, the general solution of the Klein–Gordon equation (1) for a particle in an electromagnetic field is (just as in the free-particle case, discussed in Problem 15.2)

$$\psi = \psi^+ + \psi^- = \psi^+ + \hat{C}\psi_c^+.$$

Here the antiparticle wavefunction, ψ_c^+, has the "correct" time-dependence, and corresponds to the Klein–Gordon equation (2) for a particle with charge $-e$.

Let us emphasize that both equations, (1) and (2), carry the same physical information, since solutions of either equation include a description of both particle and the corresponding antiparticle states (note that this is valid also in the case of external fields of any nature). But the description of the particle and anti-particle states in each of these equations is "non-symmetrical", since both are described by their wavefunctions, while the other state is obtained through the charge conjugation of the negative energy solution.

The interpretation given above for the transformation $\hat{C}$ as a charge conjugation operator, which converts "non-physical" particle states to "physical" states of the corresponding antiparticle, is based on the result of paragraph a) of the present solution. In this context, the invariance of the Klein–Gordon equation found in paragraph b) reflects the *charge symmetry* of the laws described by it: that is, for any physical state of a particle described by a wavefunction, $\psi^+(\mathbf{r},t)$, we have the same state of an antiparticle with the wavefunction $\psi_c^+(\mathbf{r},t) \equiv \psi^+(\mathbf{r},t)$. Here it is important that the transformation to the anti-particle involve a change of the sign of the external electromagnetic field.

Problem 15.4

Prove that an external *scalar* (with respect to the Lorentz group) field has the same effect on a spinless particle and its antiparticle. Compare this to the case of a particle in an external electromagnetic field (Problem 15.3).

Hint

The equation that describes a spinless particle in the scalar potential, $U(\mathbf{r},t)$, reads

$$\{c^2\hat{p}^2 + m^2c^4 + 2mc^2 U\}\psi = -\hbar^2 \frac{\partial^2}{\partial t^2}\psi.$$

One should not confuse a scalar field with the electrostatic field (which represents a time-component of a 4-vector). In the non-relativistic limit, $U(\mathbf{r},t)$ has the meaning of a potential energy.

Solution

Since the charge-conjugation operator, $\hat{C}$, for spinless particles is given by $\psi_c = \hat{C}\psi \equiv \psi^*$ (see Problems 15.2 and 15.3), and $U(\mathbf{r},t)$ is a real function (this constraint is analogous to the condition that a Hamiltonian must be Hermitian in the non-relativistic case), then, applying the complex conjugation operator to equation,

$$(-\hbar^2 c^2 \Delta + m^2 c^4 + 2mc^2 U)\psi = -\hbar^2 \frac{\partial^2}{\partial t^2}\psi, \qquad (1)$$

we obtain the same equation for the charge-conjugation function,

$$(-\hbar^2 c^2 \Delta + m^2 c^4 + 2mc^2 U)\psi_c = -\hbar^2 \frac{\partial^2}{\partial t^2}\psi_c, \qquad (2)$$

which proves the invariance of the Klein–Gordon equation for a particle in a scalar field with respect to charge conjugation.

Let us make a few comments. Eqs. (1) and (2) have the same form, but only the first of them (or, more accurately the positive-frequency part of its solutions) describes the particle, while the second describes the corresponding antiparticle (see the previous problem). Hence, if the wavefunction $\psi^+(\mathbf{r},t)$ is the solution of Eq. (1) and describes a physically realizable state of a particle in the field U, then the same state with wavefunction $\psi_c^+ = \psi^+$ is physically possible for the antiparticle in the same field. This shows that the action of the scalar field on the particle and its antiparticle[406] reflects the charge symmetry of Eqs. (1) and (2) (which is different from the case of a charged particle in an electromagnetic field, where to restore the charge symmetry it is necessary to change the signs of the potentials as well, as discussed in Problem 15.3).

Problem 15.5

Prove that the *intrinsic* parities of a spinless particle and its antiparticle are the same.

Solution

The parity of a spinless particle is determined by how its wavefunction transforms under spatial inversion: $\hat{P}\psi(\mathbf{r},t) = \pm\psi(-\mathbf{r},t)$. The parity is either $+1$ or -1 for *scalar* or *pseudo-scalar* functions respectively.

As mentioned in Problems 15.2 and 15.3, the wavefunctions of a particle and the corresponding anti-particle are related to special solutions of the Klein–Gordon equation, $\psi(\mathbf{r},t)$, which can be written in the form

$$\psi = \psi^+ + \psi^- = \psi^+ + \hat{C}\psi_c^+, \qquad (1)$$

where $\psi_c^+ = \hat{C}\psi^- = (\psi^-)^*$. The functions ψ^+ and ψ_c^+ are the wavefunctions of particle and antiparticle states, respectively.

Since the functions ψ^+ and ψ^- transform equivalently under space inversion (*i.e.*, they both are either scalar or pseudo-scalar functions), and it does not change under complex conjugation; hence it follows that they indeed also have the same intrinsic parities.

Note here that for particles with an arbitrary spin, the intrinsic parities of a particle and antiparticle are the same for bosons and have opposite signs for fermions.

---

[406] Compare this conclusion to a similar action of a gravitational field on the particle and antiparticle.

Problem 15.6

Using the fact that the quantity Q (defined in Problem 15.1) is conserved, discuss the orthogonality and normalization of the solutions to the Klein–Gordon equation, $\psi_{\mathbf{p},\varepsilon}(\mathbf{r}, t)$, corresponding to definite values of energy (of both signs) and momentum.

Solution

In non-relativistic quantum mechanics, the orthogonality of the eigenfunctions of an Hermitian operator, $\hat{f}$, is given by the relation:

$$\int \psi_{f'}^*(\mathbf{r})\psi_f(\mathbf{r})dV = \delta(f - f') \quad \text{(or } \delta_{f'f} \text{ for d.s.).} \tag{1}$$

Its form is closely connected to the normalization condition, $\int |\psi|^2 dV = \text{const} = 1$, that follows directly from the Schrödinger equation.

In the case of the Klein–Gordon equation (XV.1), the following quantity is conserved in time:

$$Q = \frac{i\hbar}{2mc^2} \int \left\{ \psi^* \frac{\partial}{\partial t}\psi - \left(\frac{\partial}{\partial t}\psi\right)^* \psi \right\} dV, \tag{2}$$

and it determines both a normalization constraint and eigenfunction orthogonality conditions (a generalization of Eq. (1) for the relativistic case).

Let us write the wavefunction $\psi_{\mathbf{p},\varepsilon}$ in the form (see Problem 15.1),

$$\psi_{\mathbf{p}}^{\pm} = C^{\pm}(\mathbf{p}) \exp\left\{ \pm \frac{i}{\hbar}(\mathbf{p}\cdot\mathbf{r} - \varepsilon t)\right\}, \quad \varepsilon(\mathbf{p}) = \sqrt{p^2 c^2 + m^2 c^4} \geq mc^2. \tag{3}$$

Here, $\psi_{\mathbf{p}}^+$ describes the state with momentum $\mathbf{p}$, and energy ε, while the plane-wave $\psi_{\mathbf{p}}^-$, formally corresponds to a state with momentum $(-\mathbf{p})$ and energy $(-\varepsilon)$. Physically, the latter corresponds to the antiparticle with momentum $\mathbf{p}$ and energy ε, as discussed in Problem 15.2. Substituting the respective wavefunctions, $\psi_{\mathbf{p}}^{\pm}$ and $\psi_{\mathbf{p}'}^{\pm*}$, for ψ and ψ^* into the integral in Eq. (2), we see that the integral vanishes for wavefunctions with different signs of frequency, and is proportional to $\delta(\mathbf{p} - \mathbf{p}')$ otherwise. By choosing the values of Eq. (3),

$$C^{\pm}(\mathbf{p}) = \sqrt{\frac{mc^2}{(2\pi\hbar)^3 \varepsilon(\mathbf{p})}},$$

we normalize the system functions and obtain the following orthonormality relations:

$$\frac{i\hbar}{2mc^2} \int \left\{ \psi_{\mathbf{p}'}^{\pm*} \frac{\partial}{\partial t} \psi_{\mathbf{p}}^{\pm} - \left(\frac{\partial}{\partial t} \psi_{\mathbf{p}'}^{\pm*}\right) \psi_{\mathbf{p}}^{\pm} \right\} dV = \pm \delta(\mathbf{p} - \mathbf{p}') \tag{4}$$

and

$$\frac{i\hbar}{2mc^2} \int \left\{ \psi_{\mathbf{p}'}^{\pm*} \frac{\partial}{\partial t} \psi_{\mathbf{p}}^{\mp} - \left(\frac{\partial}{\partial t} \psi_{\mathbf{p}'}^{\pm*}\right) \psi_{\mathbf{p}}^{\mp} \right\} dV = 0.$$

These relations are a generalization of Eq. (1) to the relativistic case.

Problem 15.7

Prove that for a spinless particle in the relativistic case we can keep the interpretation of a wavefunction in the momentum representation as the probability amplitude of different momenta (in sharp contrast to the coordinate representation; see Problem 15.1).

What is a connection between the particle and antiparticle wavefunctions in the momentum representation and the solutions, $\psi^{\pm}(\mathbf{r}, t)$, of the Klein–Gordon equation? Discuss the eigenfunctions of the coordinate operator. Compare this to the non-relativistic case.

Solution

1) Writing a positive-frequency solution, $\psi^{+}(\mathbf{r}, t)$, of the Klein–Gordon equation (XV.1), which describes a physical particle state, in the form of a superposition of plane waves, $\psi_{\mathbf{p}}^{+}(\mathbf{r}, t)$ (see Eq. (3) of the previous problem) with the value of the coefficient $C^{+}(p)$ chosen to be:

$$\psi^{+}(\mathbf{r}, t) = \int a^{+}(\mathbf{p}) \psi^{+}(\mathbf{r}, t) d^{3}p = \int \sqrt{\frac{mc^{2}}{(2\pi\hbar)^{3}\varepsilon(p)}} a^{+}(\mathbf{p}) e^{i(\mathbf{p} \cdot \mathbf{r} - \varepsilon t)/\hbar} d^{3}p, \quad (1)$$

gives

$$Q^{+} = \frac{i\hbar}{2mc^{2}} \int \left\{ \psi^{+*} \frac{\partial}{\partial t} \psi^{-} - \left(\frac{\partial}{\partial t} \psi^{+} \right)^{*} \psi^{+} \right\} dV = \int |a^{+}(\mathbf{p})|^{2} d^{3}p \quad (2)$$

for the quantity Q^{+} (see Problem 15.1). Hence, by analogy to the non-relativistic case, the function $a^{+}(\mathbf{p})$ (or more accurately, $a^{+}(\mathbf{p}, t) = a^{+}(\mathbf{p}) e^{-i\varepsilon t/\hbar}$) may be interpreted as a wavefunction of a particle state in the momentum representation in the usual quantum-mechanical sense, and one must use the value $Q^{+} = 1$ for its normalization.

In a similar way we can introduce a wavefunction of an antiparticle in the momentum representation, using the expansion of the negative-frequency solutions, $\psi^{-}(\mathbf{r}, t)$, over the plane-waves $\psi_{\mathbf{p}}^{-} = (\psi_{\mathbf{p}}^{+})^{*}$,

$$\psi^{-}(\mathbf{r}, t) = \int a^{-}(\mathbf{p}) \psi_{\mathbf{p}}^{-} d^{3}p = \int \sqrt{\frac{mc^{2}}{(2\pi\hbar)^{3}\varepsilon(p)}} a^{-}(\mathbf{p}) e^{-\frac{i(\mathbf{p} \cdot \mathbf{r} - \varepsilon t)}{\hbar}} d^{3}p,$$

and the standard relation between the antiparticle wavefunction $\psi_{c}^{+} = \hat{C}\psi^{-} \equiv (\psi^{-})^{*}$ and the solutions ψ^{-}; see Problem 15.2. Here the antiparticle wavefunction in the momentum representation has the form $a_{c}^{+}(\mathbf{p}, t) = a^{-*}(\mathbf{p}) e^{-i\varepsilon t/\hbar}$, while the normalization condition, $\int |a_{c}^{+}(\mathbf{p}, t)|^{2} d^{3}p = 1$, is equivalent to $Q^{-} = -1$.

The fact that the particle wavefunction in the momentum representation has the familiar meaning of a probability amplitude allows us to obtain a generalization of the corresponding quantum-mechanical equations for the coordinate representation directly from the momentum representation, as discussed in Problems 15.8–15.10.

Let us emphasize that according to Eq. (1) the transition from the momentum representation to the coordinate representation differs from the non-relativistic case by an additional factor of $\sqrt{mc^2/\varepsilon(p)}$ in the expansion of the wavefunction into the plane-wave basis. The inability to define a quantity, $\rho \geq 0$, that would have been otherwise interpreted as probability distribution in the coordinate space, is a direct consequence of this fact.

2) Despite the fact that localized particle states are poorly defined in the relativistic case, it is educative to discuss the eigenfunctions of the coordinate operator using the simplest example of a spinless particle. Consider the form of coordinate operator, $\hat{\mathbf{r}} = i\hbar \frac{\partial}{\partial \mathbf{p}}$, in the momentum representation. In this representation, the desired eigenfunctions read

$$a_{\mathbf{r}_0}^+(\mathbf{p}) = \frac{1}{(2\pi)^{3/2}} \exp\{-i\mathbf{p}\cdot\mathbf{r}_0\},$$

as in the non-relativistic case (here and below, we set $\hbar = c = 1$). In the coordinate representation we obtain from Eq. (1):

$$\psi_{\mathbf{r}_0}^+(\mathbf{r}) = \int \frac{\sqrt{m}}{(2\pi)^{3/2}(p^2+m^2)^{1/4}} a_{\mathbf{r}_0}^+(\mathbf{p}) e^{i\mathbf{p}\cdot\mathbf{r}} d^3p =$$

$$= -\frac{1}{2\pi^2} \frac{m}{\tilde{r}} \frac{\partial}{\partial \tilde{r}} \int_0^\infty \frac{\cos(mp\tilde{r})}{(p^2+1)^{1/4}} dp = \frac{m^3}{(2\pi^2)^{3/4}\Gamma(1/4)} \frac{K_{5/4}(m\tilde{r})}{(m\tilde{r})^{5/4}}, \quad (3)$$

where $\tilde{r} = |\mathbf{r} - \mathbf{r}_0|$. To perform the integration over momenta in the first integral, we used spherical coordinates with the polar axis directed along the vector $(\mathbf{r} - \mathbf{r}_0)$, and for the second (single) integral, we used its expression in terms of the MacDonald function.

Let us discuss the properties of the eigenfunctions, $\psi_{\mathbf{r}_0}^+$. First, consider the limiting cases (here we restore dimensional constants, $\hbar$ and c):

$$\psi_{\mathbf{r}_0}^+(\mathbf{r}) \propto \begin{cases} \frac{1}{|\mathbf{r}-\mathbf{r}_0|^{5/2}}, & |\mathbf{r}-\mathbf{r}_0| \ll \frac{\hbar}{mc}, \\ \frac{1}{|\mathbf{r}-\mathbf{r}_0|^{7/4}} \exp\left\{-\frac{mc}{\hbar}|\mathbf{r}-\mathbf{r}_0|\right\}, & |\mathbf{r}-\mathbf{r}_0| \gg \frac{\hbar}{mc}. \end{cases} \quad (4)$$

Both these wavefunctions reduce to $\delta(\mathbf{r} - \mathbf{r}_0)$, much as in the non-relativistic case, but are localized on the distances of the order of particle Compton wave-length, $\hbar/mc$. In the non-relativistic limit (i.e., for $c \to \infty$), the domain of localization of the function $\psi_{\mathbf{r}_0}(\mathbf{r})$ shrinks to a point, and (since $\int \psi_{\mathbf{r}_0}(\mathbf{r}) dV = 1$ to calculate the integral here, it is convenient to substitute into Eq. (3) a wavefunction expressed in terms of the plane-waves and perform integration over the variable $\mathbf{r}$ first, which gives, $\delta(\mathbf{p})$), then the eigenfunction $\psi_{\mathbf{r}_0}(\mathbf{r})$ reduces to $\delta(\mathbf{r} - \mathbf{r}_0)$ in this limit, as expected.

Problem 15.8

Obtain an expression for the mean value of the energy of a free spinless particle in an arbitrary state, described by the solution $\psi^+(\mathbf{r}, t)$ of the Klein–Gordon equation.

Solution

We use the fact that the free-particle wavefunction in momentum representation,

$$a^+(\mathbf{p}, t) = a^+(\mathbf{p}) \exp\left\{-\frac{i}{\hbar}\varepsilon(p)t\right\}, \tag{1}$$

has, as in non-relativistic quantum mechanics, the meaning of the momentum probability amplitude (see the previous problem). Hence, this wavefunction is normalized via the condition:

$$\int |a^+(\mathbf{p}, t)|^2 d^3p = 1, \tag{2}$$

and the mean value of particle energy is given by

$$\bar{\varepsilon} = \int \varepsilon(p)|a^+(\mathbf{p}, t)|^2 d^3p = \int a^{+*}(\mathbf{p}, t)\sqrt{p^2c^2 + m^2c^4}\, a^+(\mathbf{p}, t) d^3p. \tag{3}$$

Now, note that in the coordinate representation, the wavefunction, $\psi^+(\mathbf{r}, t)$ of an arbitrary free-particle state is described by a superposition of positive-frequency solutions to the Klein–Gordon equation (XV.1), and is connected to the momentum-space wavefunction by the relation:

$$\psi^+(\mathbf{r}, t) = \int \sqrt{\frac{mc^2}{\varepsilon(p)}} a^+(\mathbf{p}, t) e^{i\mathbf{p}\cdot\mathbf{r}/\hbar} \frac{d^3p}{(2\pi\hbar)^{3/2}} \tag{4}$$

(see also Problems 15.1 and 15.7). From this Eq. (4), it follows that

$$\sqrt{\frac{mc^2}{\varepsilon(p)}} a^+(\mathbf{p}, t) = \int \psi^+(\mathbf{r}, t) e^{-i\mathbf{p}\cdot\mathbf{r}/\hbar} \frac{d^3p}{(2\pi\hbar)^{3/2}}. \tag{5}$$

Using this relation, we transform Eq. (3) in the following way:

$$\bar{\varepsilon} = \frac{1}{(2\pi\hbar)^3 mc^2} \int (p^2c^2 + m^2c^4)\psi^{+*}(\mathbf{r}, t) e^{i\mathbf{p}\cdot(\mathbf{r}-\mathbf{r}')/\hbar}\psi^+(\mathbf{r}', t) d^3p\, d^3r'\, d^3r$$

$$= \frac{1}{(2\pi\hbar)^3 mc^2} \int \psi^{+*}(\mathbf{r}, t)(-\hbar^2 c^2 \Delta + m^2 c^4) \exp\left\{\frac{i}{\hbar}\mathbf{p}\cdot(\mathbf{r}-\mathbf{r}')\right\}$$

$$\times \psi^+(\mathbf{r}', t) d^3p\, d^3r'\, d^3r. \tag{6}$$

After performing the integration over momenta in Eq. (6), and using the equation,

$$\frac{1}{(2\pi\hbar)^3} \int \exp\left\{\frac{i}{\hbar}\mathbf{p}\cdot(\mathbf{r}-\mathbf{r}')\right\} d^3p = \delta(\mathbf{r}-\mathbf{r}'),$$

the integration over $\mathbf{r}'$ can be readily performed, and we obtain the mean value of the particle energy:

$$\bar{\varepsilon} = \frac{1}{mc^2} \int \psi^{+*}(\mathbf{r},t)(-\hbar^2 c^2 \Delta + m^2 c^4)\psi^+(\mathbf{r},t) d^3 r. \tag{7}$$

Here, the normalization condition (2) in the coordinate representation takes the form (Eq. (2) of the previous problem):

$$\int |a^+(\mathbf{p},t)|^2 d^3 p = \frac{i\hbar}{2mc^2} \int \left\{ \psi^{+*} \frac{\partial \psi^+}{\partial t} - \frac{\partial \psi^{+*}}{\partial t} \psi^+ \right\} d^3 r = 1. \tag{8}$$

It is possible to obtain a slightly different expression, yet still equivalent to Eq. (7), for the mean energy of the particle, if we use Eqs. (1) and (5), as follows:

$$-i\frac{\sqrt{\varepsilon}}{\hbar} a^+(\mathbf{p},t) = \frac{\partial}{\partial t} \frac{a^+(\mathbf{p},t)}{\sqrt{\varepsilon}} = \frac{1}{\sqrt{mc^2}} \int e^{-i\mathbf{p}\cdot\mathbf{r}/\hbar} \frac{\partial}{\partial t} \psi^+ \frac{d^3 r}{(2\pi\hbar)^{3/2}}. \tag{9}$$

From Eq. (9) we rewrite Eq. (3) in the coordinate representation as follows:

$$\bar{\varepsilon} = \frac{\hbar^2}{mc^2} \int \left(\frac{\partial}{\partial t} \psi^+(\mathbf{r},t) \right)^* \frac{\partial}{\partial t} \psi^+(\mathbf{r},t) d^3 r. \tag{10}$$

According to Eqs. (7) and (10), the mean energy of a spinless particle has the form:

$$\bar{\varepsilon} = \frac{\hbar^2}{mc^2} \int \left\{ \frac{\partial \psi^{+*}}{c\partial t} \frac{\partial \psi^+}{c\partial t} + (\boldsymbol{\nabla}\psi^+)^*(\boldsymbol{\nabla}\psi^+) + \frac{m^2 c^2}{\hbar^2} |\psi^+|^2 \right\} d^3 r, \tag{11}$$

which is analogous (modulo a normalization factor) to the energy of a classical scalar (or a pseudo-scalar) complex field that corresponds to the wave equation

$$\left(-\Delta + \frac{\partial^2}{c^2 \partial t^2} + \kappa^2 \right) \psi(\mathbf{r},t) = 0, \quad \kappa = \frac{mc}{\hbar}.$$

The energy of such a field, $E = \int T_{00} d^3 r$, is expressed in terms of the component T_{00} (or T_{44}) of the energy-momentum tensor, and has the form:

$$T_{00} \propto \left[\frac{\partial \psi^*}{c\partial t} \frac{\partial \psi}{c\partial t} + (\boldsymbol{\nabla}\psi^*) \cdot (\boldsymbol{\nabla}\psi) + \kappa^2 \psi^* \psi \right].$$

(See also the following problem on the relation between a particle's mean momentum and the angular momentum of a classical field:)

In conclusion, we see that all considerations above can be directly generalized to the antiparticle, using the charge-conjugated wavefunction, $\psi_c^+(\mathbf{r},t)$, instead; see Problems 15.2 and 15.3.

Problem 15.9

Consider the same conditions as in the previous problem, but find the mean value of momentum instead.

Solution

As in the previous problem, we start from the momentum representation, where the particle wavefunction has the meaning of a momentum probability amplitude. Hence, the mean value of the particle momentum is determined by the expression:

$$\bar{\mathbf{p}} = \int \mathbf{p}|a^+(\mathbf{p},t)|^2 d^3p = \int a^+(\mathbf{p},t)\mathbf{p}a(\mathbf{p},t)d^3p. \tag{1}$$

Using the relations (see Eqs. (5) and (9) of the previous problem)

$$\sqrt{\frac{mc^2}{\varepsilon(p)}}\, a^+(\mathbf{p},t) = \int \psi^+(\mathbf{r},t) e^{-i\mathbf{p}\cdot\mathbf{r}/\hbar} \frac{d^3r}{(2\pi\hbar)^{3/2}}$$

and

$$-\frac{i}{\hbar}\sqrt{mc^2\varepsilon}\, a^+(\mathbf{p},t) = \int e^{-i\mathbf{p}\cdot\mathbf{r}/\hbar} \frac{\partial}{\partial t} \psi^+(\mathbf{r},t) \frac{d^3r}{(2\pi\hbar)^{3/2}},$$

Eq. (1) here can be written in the form:

$$\bar{\mathbf{p}} = \frac{i\hbar}{(2\pi\hbar)^3 mc^2} \int \psi^{+*}(\mathbf{r},t)\mathbf{p} e^{i\mathbf{p}\cdot(\mathbf{r}-\mathbf{r}')/\hbar} \frac{\partial}{\partial t}\psi^+(\mathbf{r}',t)\, d^3p\, d^3r'\, d^3r =$$

$$\frac{\hbar^2}{(2\pi\hbar)^3 mc^2} \int \psi^{+*}(\mathbf{r},t) \frac{\partial}{\partial \mathbf{r}} e^{i\mathbf{p}\cdot(\mathbf{r}-\mathbf{r}')/\hbar} \frac{\partial}{\partial t}\psi^+(\mathbf{r}',t)\, d^3p\, d^3r'\, d^3r. \tag{2}$$

Here the integration over $\mathbf{p}$ is easily performed (and it gives rise to the factor, $\delta(\mathbf{r}-\mathbf{r}')$), and then the integration over $\mathbf{r}'$ becomes straightforward as well; so, we obtain the mean particle momentum:

$$\bar{\mathbf{p}} = \frac{\hbar^2}{mc^2} \int \psi^{+*}(\mathbf{r},t) \frac{\partial}{\partial \mathbf{r}} \frac{\partial}{\partial t} \psi^+(\mathbf{r},t)\, d^3r. \tag{3}$$

This Eq. (3) may be written in a more symmetric form:

$$\bar{\mathbf{p}} = -\frac{\hbar^2}{2mc^2} \int \left\{ \frac{\partial \psi^{+*}}{\partial \mathbf{r}} \frac{\partial \psi^+}{\partial t} + \frac{\partial \psi^{+*}}{\partial t} \frac{\partial \psi^+}{\partial \mathbf{r}} \right\} d^3r. \tag{4}$$

This expression for the mean momentum of a spinless particle has the same form, up to a normalization factor, as the equation for the momentum of a classical scalar (or a pseudo-scalar) complex field. The components of the field momentum are determined by the expression, $P_i = \int T_{i0} d^3r$, where T_{i0} (or T_{i4}) is the density of field momentum, corresponding to the component of the energy-momentum tensor, where

$$T_{i0} \propto -\left(\frac{\partial \psi^*}{\partial x_i}\frac{\partial \psi}{\partial t} + \frac{\partial \psi^*}{\partial t}\frac{\partial \psi}{\partial x_i}\right),$$

and $i = 1, 2,$ and 3.

Problem 15.10

Consider the same condition as in the two previous problems, but find the mean value of the angular momentum.

Solution

The mean value of the particle's angular momentum in the momentum representation reads:

$$\bar{\mathbf{l}} = \int a^{+*}\hat{\mathbf{l}}a^+ d^3p = -i\int a^{+*}(\mathbf{p},t)[\mathbf{p}\times\boldsymbol{\nabla}_{\mathbf{p}}]a^+(\mathbf{p},t)d^3p.$$

Using this equation and the transformations described in the previous two problems, we obtain the relation:

$$\bar{\mathbf{l}} = \frac{\hbar}{mc^2}\int \psi^{+*}(\mathbf{r},t)\left[\mathbf{r}\times\frac{\partial}{\partial \mathbf{r}}\right]\frac{\partial}{\partial t}\psi^+(\mathbf{r},t)d^3r,$$

or in more a symmetric form:

$$\bar{\mathbf{l}} = -\frac{\hbar}{2mc^2}\int\left[\mathbf{r}\times\left\{\frac{\partial\psi^{+*}}{\partial \mathbf{r}}\frac{\partial\psi^+}{\partial t} + \frac{\partial\psi^{+*}}{\partial t}\frac{\partial\psi^+}{\partial \mathbf{r}}\right\}\right]d^3r. \tag{1}$$

This form coincides with the equation for the angular momentum, **L**, of a classical scalar field. Here, the expression for density $\boldsymbol{\lambda}$ of the angular momentum of the field has a clear physical meaning, since we can write it in the form:

$$\boldsymbol{\lambda}(\mathbf{r},t) = \frac{1}{\hbar}[\mathbf{r}\times\boldsymbol{\pi}],$$

where $\boldsymbol{\pi}(\mathbf{r},t)$ is the field's momentum density, as discussed in the previous problem.

Problem 15.11

Find the energy spectrum of a relativistic charged spinless particle in a uniform magnetic field.

Solution

The energy spectrum and the corresponding wavefunctions of the stationary states are determined from the stationary solutions of the Klein–Gordon equation for a charged particle in a magnetic field. This equation has the form:

$$\left\{ c^2 \left(\hat{\mathbf{p}} - \frac{e}{c}\mathbf{A} \right)^2 + m^2 c^4 \right\} \psi = \varepsilon^2 \psi, \tag{1}$$

where e is the particle charge, and $\mathcal{H} = \nabla \times \mathbf{A}$. Eq. (1) differs from the non-relativistic Schrödinger equation

$$\frac{1}{2m} \left(\hat{\mathbf{p}} - \frac{e}{c}\mathbf{A} \right)^2 \psi = E\psi \tag{2}$$

only by the substitution $E \to (\varepsilon^2 - m^2 c^4)/2mc^2$. Hence, using the known solutions of the last equation for a particle in a uniform field (*e.g.*, given in Problem 7.1), we find:

$$\varepsilon^2_{p_z n} = m^2 c^4 + p_z^2 c^2 + 2mc^2 \hbar \omega \left(n + \frac{1}{2} \right), \quad n = 0, 1, \ldots,$$

$$\omega = \frac{|e|\mathcal{H}}{mc} > 0.$$

Hence, it follows that

$$\varepsilon_{p_z n} = \pm \sqrt{m^2 c^4 + p_z^2 c^2 + 2mc^2 \hbar \omega \left(n + \frac{1}{2} \right)} \tag{3}$$

(to be compared with the expression, $\varepsilon(p) = \pm\sqrt{m^2 c^4 + p^2 c^2}$, for a free particle).

The interpretation of the two values of $\varepsilon_{p_z n}$, which differ in sign, is the same as in the free-particle case (see Problems 15.2 and 15.3). The value $\varepsilon_{p_z n} > mc^2$ describes the energy spectrum of a particle with the charge e. The negative values, $\varepsilon_{p_z n} < -mc^2$, are associated with the antiparticle, with the charge $(-e)$. Here the energy of the antiparticle is $-\varepsilon > mc^2$. Hence, the energy spectra of the particle and antiparticle in a magnetic field are the same. (This conclusion is not unexpected, since the energy spectrum does not depend on the sign of the particle charge.)

To summarize this problem, we found the energy spectrum of a relativistic spinless particle in a magnetic field. As in the non-relativistic case, it has a continuous dependence on p_z, associated with the free longitudinal (along the magnetic field) particle motion, and it also includes a discrete dependence on the quantum number, n, connected to the particle's transverse motion. Here, the transverse motion of the particle is reflected in the kinematics of the free longitudinal motion (unlike the non-relativistic case), and according to Eq. (3) this relativistic effect may be viewed as a "change",

$$m \to m_n = m\sqrt{1 + (2n+1)\frac{\hbar\omega}{mc^2}},$$

in the particle mass.

Problem 15.12

Find the energy spectrum of the s-states of a spinless particle in the external *scalar* field (see Problem 15.4) of the form:

$$U(r) = \begin{cases} -U_0, & r \leq a \\ 0, & r > a. \end{cases}$$

What is the antiparticle spectrum in this field?

Discuss the conceptual difficulties in interpreting the spectrum in the case of a deep well.

Solution

The particle energy spectrum in a scalar field is determined by the equation

$$[-\hbar^2 c^2 \Delta + 2mc^2 U(r)]\psi = (\varepsilon^2 - m^2 c^4)\psi. \tag{1}$$

It has the form of the non-relativistic Schrödinger equation for a particle in a potential $U(r)$, where energy E is replaced by $(\varepsilon^2 - m^2 c^4)/2mc^2$. Considering only the particle s-states (so that the wavefunction is spherically symmetric), and making the substitution, $R(r) = r\psi(r)$, we transform Eq. (1) to the form:

$$-\frac{d^2}{dr^2} R + \frac{2m}{\hbar^2} U(r) R = \frac{\varepsilon^2 - m^2 c^4}{\hbar^2 c^2} R. \tag{2}$$

For the potential well considered, a solution to Eq. (2) satisfying the boundary condition $R(0) = 0$ (for $(\varepsilon^2 - m^2 c^4) < 0$) is given by the following expressions:

$$R(r) = \begin{cases} A \sin\sqrt{\frac{2mU_0}{\hbar^2} - \kappa^2}\, r, & r < a \\ B e^{-\kappa r}, & r > a, \end{cases}$$

where

$$\kappa = \frac{1}{\hbar c}\sqrt{m^2 c^4 - \varepsilon^2} > 0. \tag{3}$$

(Since $U = 0$ for $r > a$, then in the region of $\varepsilon^2 > m^2 c^4$, the energy spectrum is continuous; scattering off a scalar potential is considered in Problem 15.19.) The continuity constraints on the wavefunction and its derivative at the point $r = a$ lead to the transcendental equation,

$$\tan\sqrt{\frac{2mU_0 a^2}{\hbar^2} - \kappa_n^2 a^2} = -\frac{1}{\kappa_n a}\sqrt{\frac{2mU_0 a^2}{\hbar^2} - \kappa_n^2 a^2}, \tag{4}$$

which determines the energy spectrum of the bound s-states.

Let us discuss the main peculiarities of the energy spectrum, which are relatively easy to understand, using the analogy of our problem to that of a non-relativistic particle's discrete levels in a spherical potential well:

1) If the well is sufficiently "shallow", no bound states exist. The bound states, just as in the non-relativistic case, appear only if the strength of the potential satisfies the condition, $U_0 > \pi^2 \hbar^2 / 8ma^2$.

2) Upon deepening the well, (i.e., increasing the parameter, $U_0 a^2$), new discrete levels would appear. For the existing levels, the quantity $(m^2 c^4 - \varepsilon_n^2)$ increases, which corresponds to an increase of $|E_n|$ in the non-relativistic case; that is, ε_n^2 decreases as the well depth increases.

3) A situation, unique to the relativistic case, arises when the ground level reaches the value $\varepsilon_0^2 = 0$. Any further increase of U_0 leads to an imaginary ε_0, which indicates an instability.

To clarify the physics behind this instability, we note the following: The solution of the problem allows us to find the quantity, ε_n^2, so that $\varepsilon_n = \pm\sqrt{\varepsilon_n^2}$. The two values of the energy of the opposite sign should be interpreted in the same way as in the free-particle case: the positive energy state, $\varepsilon_n > 0$, corresponds to particles, while the other one, $\varepsilon_n < 0$, corresponds to the antiparticles with the energy $(-\varepsilon_n) > 0$. Indeed, as the well depth decreases, all the levels with $\varepsilon_n > 0$ move towards a higher continuum, $\varepsilon > mc^2$, while the levels with $\varepsilon_n < 0$ "merge" with the lower continuum, $\varepsilon < -mc^2$. Hence, the energy spectra for the particle and its antiparticle in an external scalar field are the same; i.e., the field effects them similarly (unlike, for example, the electrostatic field, as discussed in Problems 15.3 and 15.4).

For *critical* values of the well parameter (a critical combination of its depth and width), the ground-state energy of both the particle and antiparticle energies vanish, $\varepsilon_0 = 0$. This corresponds to the possibility of spontaneous creation of a "particle-antiparticle pair" (a spontaneous appearance of single particles is possible, if they are charge neutral). This is the physical explanation of the instability of the solution of the one-particle problem in a strong external field.[407] In strong fields, another interesting effect – vacuum reconstruction – also appears.

Let us discuss the dependence of the critical well depth $U_{0,cr}$ on its width a. Putting $\varepsilon_0 = 0$ in Eqs. (3) and (4), we obtain the equation:

$$\tan\left\{\frac{mca}{\hbar}\sqrt{\frac{2U_{o,cr}}{mc^2} - 1}\right\} = -\sqrt{\frac{2U_{o,cr}}{mc^2} - 1}. \tag{5}$$

For the limiting cases of "wide", $a \gg \hbar/mc$, and "narrow" wells, $a \ll \hbar/mc$, we have:

---

[407] Let us note that the single-particle problem also loses its physical meaning if the fields are not too strong but change rapidly in time. Hence, only the Fourier components of the "potential" $U(\omega)$ that are essentially different from zero correspond to frequencies $\omega \geq mc^2/\hbar$. The breakdown of the one-particle description in this case is connected to the inability to separate solutions to the wave equations into independent positive- and negative-frequency parts (due to unavoidable transitions between them). This is a fundamentally important element in the interpretation of relativistic wave equations.

a) $U_{0,cr} \approx \dfrac{mc^2}{2} + \dfrac{\pi^2 \hbar^2}{2ma^2} \left(\approx \dfrac{mc^2}{2}\right)$, $\quad a \gg \dfrac{\hbar}{mc}$;

b) $U_{0,cr} \approx \dfrac{\pi^2 \hbar^2}{8ma^2} + \dfrac{mc^2}{2} (\gg mc^2)$, $\quad a \ll \dfrac{\hbar}{mc}$. $\qquad$ (6)

(Note that independently of the well width, $U_{0,cr} > mc^2/2$, these expressions determine the smallest root $U_{0,cr}$ of Eq. (5); other roots of the equation correspond to ε_n^2 vanishing with $n \geq 1$). W see that a "wide" scalar well "demolishes" the rest energy for the depth $U_0 \approx mc^2/2$. With the decrease of the well width of critical well depth increases. In case b) of a "narrow" well, the value of $U_{0,cr}$ differs only slightly from the well depth that corresponds to the appearance of a bound state.

Problem 15.13

Find the discrete spectrum energy levels of a charged spinless particle (charge, $-e$) in the Coulomb field of a nucleus with the charge, Ze. The nucleus is to be considered point-like and infinitely heavy.

In the case of $Z\alpha \ll 1$ ($\alpha = e^2/\hbar c \approx 1/137$), compare the result to the corresponding expression of non-relativistic theory.

Pay attention to difficulties that appear in the interpretation of the energy spectrum if the nucleus charge is large, and explain the origin of the difficulties.

Solution

The energy levels and the corresponding wavefunctions are determined from the solution of the stationary Klein–Gordon equation (Eq. (XV.2) with $\mathbf{A} = 0$ and $\varphi = Ze/r$):

$$\{-\hbar^2 c^2 \Delta + m^2 c^4\}\psi = \left(\varepsilon + \dfrac{Ze^2}{r}\right)^2 \psi. \qquad (1)$$

Taking into account the spherical symmetry of the problem, we seek solutions in the form, $\psi(\mathbf{r}) = R_l(r) \times Y_{lm}(\theta, \varphi)$. Then, from Eq. (1), it follows that

$$\left\{-\dfrac{\hbar^2}{2m}\dfrac{1}{r}\dfrac{d^2}{dr^2} + \dfrac{\hbar^2[(l+1/2)^2 - 1/4]}{2mr^2} - \dfrac{Ze^2 \varepsilon}{mc^2 r} - \dfrac{Z^2 e^4}{2mc^2 r^2}\right\} R = \dfrac{\varepsilon^2 - m^2 c^4}{2mc^2} R. \qquad (2)$$

This equation has the form of the radial Schrödinger equation (IV.2) for a hydrogen-like atom in non-relativistic theory:

$$\left\{-\dfrac{\hbar^2}{2m}\dfrac{1}{r}\dfrac{d^2}{dr^2}r + \dfrac{\hbar^2[(l+1/2)^2 - 1/4]}{2mr^2} - \dfrac{Ze^2}{r}\right\} \tilde{R}_{n_r l} = E_{n_r l} \hat{R}_{n_r l},$$

and is obtained by using the following substitutions ($\alpha = e^2/\hbar c$):

$$Z \to \dfrac{Z\varepsilon}{mc^2}, \quad \left(l + \dfrac{1}{2}\right)^2 \to \left(l + \dfrac{1}{2}\right)^2 - Z^2 \alpha^2, \quad E_{n_r l} \to \dfrac{\varepsilon^2 - m^2 c^4}{2mc^2}. \qquad (3)$$

Now, using the known expression for the energy spectrum of a non-relativistic hydrogen-like atom (or ion),

$$E_{n_r l} \equiv E_n = -\frac{m(Ze^2)^2}{2\hbar^2 n^2} = -\frac{m(Ze^2)^2}{2\hbar^2 (n_r + 1/2 + l + 1/2)^2}, \quad (4)$$

and making the substitutions as in Eq. (3), we find

$$(\varepsilon^2 - m^2 c^4)\left[n_r + \frac{1}{2} + \sqrt{\left(l + \frac{1}{2}\right)^2 - Z^2\alpha^2}\right]^2 = -Z^2\alpha^2 \varepsilon^2.$$

Hence, the energy spectrum reads:

$$\varepsilon_{n_r l} = mc^2 \left\{1 - \frac{Z^2\alpha^2}{Z^2\alpha^2 + [n_r + 1/2 + \sqrt{(l+1/2)^2 - Z^2\alpha^2}]^2}\right\}^{1/2}. \quad (5)$$

(Formally, we should have put two signs, ± on the right-hand side, but the choice of the sign "−" corresponds to "extra" levels that are not in the energy spectrum. Such levels would be associated with antiparticle bound states, but there are no such levels in the conditions of this problem, *i.e.*, for a point-like nucleus; see Problem 15.16).

Let us make several statements about the result obtained in Eq. (5). If we take into account relativistic effects, the accidental degeneracy of the Coulomb potential present in non-relativistic theory is lifted. In relativistic theory, the energy levels depend on the angular momentum of the particle. In the case $Z\alpha \ll 1$, Eq. (5) yields:

$$E_{n_r l} = \varepsilon_{n_r l} - mc^2 \approx -\frac{m(Ze^2)^2}{2\hbar^2 n^2} - Z\alpha \frac{m(Ze^2)^2}{\hbar^2 n^3}\left(\frac{1}{2l+1} - \frac{3}{8n}\right). \quad (6)$$

The second term here is a relativistic correction to the non-relativistic quantum-mechanical result (see Problem 11.1).

For $Z\alpha > 1/2$, Eq. (5) leads to complex values of energy (first for s-states, and then for larger values of the angular momentum), which again signals an instability in the problem. Its origin is easy to understand if we note that the term $-Z^2 e^4/2mc^2 r^2$ in Eq. (2), which is singular for $r \to 0$, could be considered as an attractive potential. For $Z\alpha > 1/2$, this attraction is so strong that the quantum analog of the classical "falling into the center" appears (see Problem 9.14). If we take into account a finite size of the nucleus, the potential is bounded and, hence, the instability is gone. But even in the case of a finite nuclear radius, R, a further increase in the nuclear charge leads for some new threshold value of Z_{cr} (that depends on the radius, R), where another instability in the spectrum shows up. The physical reason for the latter is the same as in the previous problem: in a sufficiently strong electromagnetic field (in this case, nuclear field), spontaneous creation of particle+antiparticle pairs commences, and the single-particle description becomes fundamentally incomplete. We note that after the threshold value of the nuclear charge is reached, $Z_{cr} \approx 170$, the instability of vacuum with respect to creation of electron–positron pairs becomes important.

Problem 15.14

Prove that the free Klein–Gordon equation can be written in the form of the Schrödinger equation, $i\hbar\partial\psi/\partial t = \hat{H}_{rel}\psi$. Find the corresponding Hamiltonian and discuss its non-relativistic limit.

What is the relation between the Schrödinger wavefunction, ψ, and the solution, ψ^+ (see Problems 15.1 and 15.7) of the Klein–Gordon equation?

Solution

The Klein–Gordon equation for a free particle (XV.1) can be written in the form:

$$\left(i\hbar\frac{\partial}{\partial t} + \sqrt{c^2\hat{\mathbf{p}}^2 + m^2c^4}\right)\left(i\hbar\frac{\partial}{\partial t} - \sqrt{c^2\hat{\mathbf{p}}^2 + m^2c^4}\right)\psi_{KG} = 0. \tag{1}$$

The solutions ψ_{KG}^+, which describe particle states with positive energies (see Problem 15.1) and correspond to the equation

$$\left(i\hbar\frac{\partial}{\partial t} - \sqrt{c^2\hat{\mathbf{p}}^2 + m^2c^4}\right)\psi_{KG}^+ = 0. \tag{2}$$

This Eq. (2) has the form of the Schrödinger equation $i\hbar\frac{\partial}{\partial t}\psi = \hat{H}\psi$ with the Hamiltonian

$$\hat{H} \equiv \hat{H}_{rel} = \sqrt{c^2\hat{\mathbf{p}}^2 + m^2c^4}. \tag{3}$$

(For the negative-frequency solutions of the equation, we have $i\hbar\frac{\partial}{\partial t}\psi^- = -\hat{H}\psi^-$; after charge conjugation, $\psi_c^+ = \hat{C}\psi^-$, this equation takes the form (2), but for the wavefunction, ψ_C^+, describing an antiparticle; see Problem 15.2.)

To obtain a non-relativistic description, we make the substitution:

$$\psi_{KG}^+ = \exp\left\{-\frac{i}{\hbar}mc^2 t\right\}\psi \tag{4}$$

(the application of exponential factor here corresponds to writing the particle energy in the form $\varepsilon = mc^2 + E$, which singles out the rest energy, mc^2) and perform the expansion of the square root in Eq. (2) in powers of $\hat{\mathbf{p}}^2/m^2c^2$. As a result, we obtain the equation:

$$i\hbar\frac{\partial}{\partial t}\psi = \left(\frac{1}{2m}\hat{\mathbf{p}}^2 - \frac{1}{8m^3c^2}\hat{\mathbf{p}}^4 + \frac{1}{16m^5c^4}\hat{\mathbf{p}}^6 + \ldots\right)\psi, \tag{5}$$

where the second- and higher-order terms in the right-hand side are relativistic corrections to the Hamiltonian, $\hat{H}_0 = \hat{\mathbf{p}}^2/2m$, of a free non-relativistic particle.

In conclusion here, we draw attention to the following fact. The Klein–Gordon equation implies that the quantity $Q^+ = \int\rho^+ dV$ is conserved (in time), where ρ^+ is determined by Eq. (XV.3) with $\varphi \equiv 0$; see also Problem 15.1. At the same time, according to the Schrödinger equation (5), the value $Q = \int\rho dV$ is conserved, where

$\rho = |\psi|^2$. Let us compare Q^+ and Q. For Q^+, with Eq. (2) taken into account, we have:

$$Q^+ = \frac{1}{mc^2} \int \psi_{KG}^{+*} \sqrt{c^2 \hat{\mathbf{p}}^2 + m^2 c^4} \psi_{KG}^+ dV. \tag{6}$$

For the relation $Q^+ = Q$ ($=1$ for normalized wave-fuctions) to be valid in the transition from ψ_{KG}^+ to the Schrödinger wavefunction, ψ, we should, in addition to Eq. (3), apply the non-unitary transform

$$\psi_{KG}^+ = \hat{S}\psi, \quad \hat{S} = \left(1 + \frac{\hat{\mathbf{p}}^2}{m^2 c^2}\right)^{-1/4} \tag{7}$$

to preserve normalization (for an arbitrary unitary transformations, both the value and the form – $\int |\psi|^2 dV = $ const – of normalization integral will remain the same). However, in the case of a free particle the transformation $\hat{S}$ commutes with the Hamiltonian, and hence Eq. (2) has the same form as the equation for the Schrödinger wavefunction, $\psi = \hat{S}^{-1}\psi_{KG}^+$, (see the case of a particle in an external field, considered in Problem 15.15).

Problem 15.15

Using the stationary Klein–Gordon equation for a charged spinless particle in a constant electromagnetic field:

a) obtain the Schrödinger equation in the non-relativistic limit;
b) find the first two ($\sim 1/c^2$ and $\sim 1/c^4$) relativistic corrections to the single-particle Hamiltonian.

Prove that the correction $\sim 1/c^4$ includes terms that differ from the Hamiltonian expansion:

$$\hat{H}_{rel} = \sqrt{c^2 \left(\hat{\mathbf{p}} - \frac{e\mathbf{A}}{c}\right)^2 + m^2 c^4} + e\varphi - mc^2.$$

Solution

The stationary Klein–Gordon equation for a charged particle in an external electromagnetic field is given by

$$\left\{c^2 \left(\hat{\mathbf{p}} - \frac{e\mathbf{A}}{c}\right)^2 + m^2 c^4\right\} \psi_{KG}^+ = (\varepsilon - e\varphi)^2 \psi_{KG}^+.$$

In the case $|e\varphi| \ll mc^2$ and $|E| \ll mc^2$, where $\varepsilon = mc^2 + E$, it is convenient to write it in the form (we omit the index $(+)$ for the wavefunction, ψ_{KG}^+; see Problems 15.1 and 15.3):

$$\left\{\frac{1}{2m}\left(\hat{\mathbf{p}} - \frac{e\mathbf{A}}{c}\right)^2 + e\varphi - E\right\}\psi_{KG} = \frac{(E - e\varphi)^2}{2mc^2}\psi_{KG}. \quad (1)$$

Here, the right-hand side of the equation is much smaller than each of the left-hand-side terms, and, neglecting it in leading approximation, we obtain the Schrödinger equation of standard non-relativistic theory with the Hamiltonian, $\hat{H}_0 = \hat{\pi}^2/2m + e\varphi$, where $\hat{\pi} = \hat{\mathbf{p}} - e\mathbf{A}/c$.

The calculation of the relativistic corrections for a Hamiltonian involves successive iterations and an expansion in powers of a small parameter, $\propto 1/c^2$. It is based on the possibility of the transformation of Eq. (1) to the form of the Schrödinger equation (with the accuracy up to $1/c^2$). This situation differs from the case of a free particle, where we could easily write a closed form of the relativistic Hamiltonian; see Eq. (3) of the previous problem.

Let us begin with the calculation of the first-order correction, $\propto 1/c^2$. Taking into account that the right-hand side of Eq. (1) contains the factor $1/c^2$, we can replace $(E - e\varphi)^2\psi_{KG}$ by its zeroth-order approximation. Since in this approximation,

$$(E - e\varphi)\psi_{KG} \approx (\hat{H}_0 - e\varphi)\psi_{KG} = \frac{1}{2m}\hat{\pi}^2\psi_{KG}, \quad (2)$$

we can perform the following transformations to the right-hand side of Eq. (1):

$$(E - e\varphi)^2\psi_{KG} \approx (E - e\varphi)\frac{\hat{\pi}^2}{2m}\psi_{KG} =$$

$$= \left\{-\frac{1}{2m}[e\varphi, \hat{\pi}^2] + \frac{1}{2m}\hat{\pi}^2(E - e\varphi)\right\}\psi_{KG} \approx$$

$$\approx \left\{-\frac{1}{2m}[e\varphi, \hat{\pi}^2] + \frac{1}{4m^2}\hat{\pi}^4\right\}\psi_{KG}.$$

As a result, this equation takes the form:

$$\left\{\frac{\hat{\pi}^2}{2m} + e\varphi - \frac{\hat{\pi}^4}{8m^3c^2} + \frac{1}{4m^2c^2}[e\varphi, \hat{\pi}^2]\right\}\psi_{KG} = E\psi_{KG}, \quad (3)$$

with the accuracy of $\sim 1/c^2$. Even though it looks similar to the Schrödinger equation, it is not the case, because the operator in the curly brackets, which is supposed to represent a Hamiltonian, is not Hermitian. For the transition from Eq. (3) to the Schrödinger equation, we should transform the wavefunction according to

$$\psi_{KG} = \hat{S}\psi = \left[\frac{\varepsilon - e\varphi}{mc^2}\right]^{-1/2}\psi = \left[1 + \frac{E - e\varphi}{mc^2}\right]^{-1/2}\psi =$$

$$= \left[1 - \frac{1}{2mc^2}(E - e\varphi) + \frac{3}{8m^2c^4}(E - e\varphi)^2 + \ldots\right]\psi. \quad (4)$$

Such a non-unitary transformation preserves the wavefunction normalization

$$\int \psi_{KG}^* \frac{1}{mc^2}(\varepsilon - e\varphi)\psi_{KG} dV = \int |\psi|^2 dV,$$

as discussed in the previous problem.

Substituting Eq. (4)[408] into Eq. (3), and keeping terms of order $1/c^2$, we obtain the Schrödinger equation with the first relativistic correction:

$$\left\{\frac{1}{2m}\hat{\pi}^2 - \frac{1}{8m^3c^2}\hat{\pi}^4 + e\varphi\right\}\psi = E\psi. \tag{5}$$

This correction to the Hamiltonian, equal to $-\hat{\pi}^4/8m^3c^2$, is the same as in the free-particle case, and represents a natural quantum-mechanical generalization of the corresponding relativistic correction in the classical theory. Note, however, that this "natural" classical-to-quantum correspondence disappears in the next order of perturbation theory, $\sim 1/c^4$.

To calculate these corrections of order $\sim 1/c^4$ and higher-order terms, it is convenient to move the term $(E - e\varphi)\psi_{KG}$ in Eq. (1) to the right-hand side. Within the given accuracy, we find

$$\frac{\hat{\pi}^2}{2m}\left(1 - \frac{E - e\varphi}{2mc^2} + \frac{3(E - e\varphi)^2}{8m^3c^4}\right)\psi =$$
$$= (E - e\varphi)\left(1 + \frac{(E - e\varphi)^2}{8m^2c^4}\right)\psi. \tag{6}$$

For all the terms containing the factor $1/c^4$, we can use the leading-order approximation for $(E - e\varphi)\psi$ and simply replace it by $(\hat{\pi}^2/2m)\psi$. On the other hand, in terms of order $1/c^2$, we use the following substitution instead:

$$(E - e\varphi)\psi \quad \text{by} \quad \left(\frac{1}{2m}\hat{\pi}^2 - \frac{1}{8m^3c^2}\hat{\pi}^4\right)\psi.$$

After some algebra, Eq. (6) takes the form of the Schrödinger equation with the Hamiltonian

$$\hat{H}' = \hat{H} + \frac{1}{16m^3c^4}\left[\left(\frac{1}{2m}\hat{\pi}^2 + e\varphi\right), [\hat{\pi}^2, e\varphi]\right], \tag{7}$$

where

$$\hat{H} = \frac{\hat{\pi}^2}{2m} + e\varphi - \frac{\hat{\pi}^4}{8m^3c^2} + \frac{\hat{\pi}^6}{32m^5c^4} + \frac{1}{32m^4c^4}[\hat{\pi}^2, [\hat{\pi}^2, e\varphi]]. \tag{8}$$

This technically solves the problem. However, this Hamiltonian could be simplified further. Indeed, within the accuracy considered ($\sim 1/c^4$), in the second term in the

---

[408] Note that in the leading approximation, $\psi_{KG} = \psi$.

left-hand side of Eq. (7), we can replace $\hat{\boldsymbol{\pi}}^2/2m + e\varphi$ by $\hat{H}$. This yields

$$\hat{H}' \approx \hat{H} + \frac{1}{16m^3c^4}\left[\hat{H},[\hat{\boldsymbol{\pi}}^2,e\varphi]\right] \approx$$

$$\exp\left\{-\frac{1}{16m^3c^4}[\hat{\boldsymbol{\pi}}^2,e\varphi]\right\}\hat{H}\exp\left\{\frac{1}{16m^3c^4}[\hat{\boldsymbol{\pi}}^2,e\varphi]\right\}. \quad (9)$$

Note that the operator, $\hat{F} = i[\hat{\boldsymbol{\pi}}^2, e\varphi]$, is Hermitian, and $\hat{U} = \exp\{i\hat{F}\}$ is unitary. Therefore, Eq. (9) implies that the two operators, $\hat{H}$ and $\hat{H}'$, are connected to one another by a unitary transformation and represent physically equivalent particle Hamiltonians. Since the expression for $\hat{H}$ is simpler than that for $\hat{H}'$, it is more convenient to use the former. We see that relativistic corrections following from the Klein–Gordon equation, differ from the naïve expansion of the operator $\hat{H}_{rel} = \sqrt{\hat{\boldsymbol{\pi}}^2c^2 + m^2c^4} + e\varphi$ already in the second subleading order, $\sim 1/c^4$.

Problem 15.16

Prove that in a sufficiently strong electrostatic field a charged spinless particle experiences an attraction (in the quantum-mechanical sense) independently of the sign of its charge.[409]

Solution

Let us consider only the case where the particle energy is close to the rest energy, and write $\varepsilon = mc^2 + E$, where $|E| \ll mc^2$. The stationary Klein–Gordon equation for a particle in an electrostatic field,

$$\{-\hbar^2c^2\Delta + m^2c^4\}\psi = (\varepsilon - e\varphi)^2\psi,$$

could be written in the form:

$$\left\{-\frac{\hbar^2}{2m}\Delta + e\varphi - \frac{(e\varphi)^2}{2mc^2} + \frac{E}{mc^2}e\varphi - \frac{E^2}{2mc^2}\right\}\psi = E\psi,$$

which is analogous to the Schrödinger equation with the effective potential:

$$U_{\text{eff}} = e\varphi + \frac{E}{mc^2}e\varphi - \frac{(e\varphi)^2}{2mc^2} - \frac{E^2}{2mc^2} \approx e\varphi - \frac{(e\varphi)^2}{2mc^2}.$$

For a region of space, where $|e\varphi| > 2mc^2$, $U_{\text{eff}} < 0$, so that the interaction between the particle and the field is indeed attractive independently of the sign of the charge. Hence, we note that in the relativistic case, bound states can exist both for a spinless particle and its antiparticle if the field is sufficiently strong.

---

[409] This statement is valid for particles with non-zero spin as well.

Problem 15.17

Using the Born approximation, find the amplitude and differential scattering cross-section of a charged (charge e_1) relativistic spinless particle in the Coulomb field of a nucleus with the charge, Ze (consider the nucleus to be infinitely heavy).

Compare your result to the case of a non-relativistic particle.

Find the range of applicability of the result.

Solution

The stationary Klein–Gordon equation, corresponding to time-dependent Eq. (XV.2) with $e\varphi = Zee_1/r$ and $A = 0$, has the form

$$\left\{-\frac{\hbar^2}{2m}\Delta + \frac{Zee_1\varepsilon}{mc^2 r} - \frac{(Zee_1)^2}{2mc^2 r^2}\right\}\psi_{p_0}^+ = \frac{p_0^2}{2m}\psi_{p_0}^+, \tag{1}$$

where $\varepsilon = \sqrt{p_0^2 c^2 + m^2 c^4}$, identical to the non-relativistic Schrödinger equation with an effective potential energy (that depends on the total particle energy, ε):

$$U_{\text{eff}} = \frac{Zee_1\varepsilon}{mc^2 r} - \frac{(Zee_1)^2}{2mc^2 r^2}. \tag{2}$$

Since the free-particle wavefunction expansion into the plane-waves is the same in both relativistic and non-relativistic theories, the general approach to non-relativistic scattering problem (which is based on asymptotic solutions of the stationary wave equation in the form of a plane wave and an outgoing wave, as discussed in the introduction to Chapter 13) holds directly (or is easily generalizable) in the relativistic regime:

$$\psi_{p_0}^+(\mathbf{r}) \underset{r\to\infty}{\approx} e^{\mathbf{p}_0\cdot\mathbf{r}/\hbar} + \frac{f}{r}e^{ikr}.$$

In particular, the scattering amplitude in the Born approximation is described by Eq. (XIII.6):

$$f_B = -\frac{m}{2\pi\hbar^2}\int U_{\text{eff}} e^{-i\mathbf{q}\cdot\mathbf{r}} d^3 r, \quad \hbar\mathbf{q} = \mathbf{p} - \mathbf{p}_0. \tag{3}$$

We should consider the range of applicability of the Born approximation more accurately. The analogy between Eq. (1) and the Schrödinger equation, mentioned above, assumes the use of momentum (but not velocity or energy) to describe the free-particle states (at large distances). Hence, the well-known condition (XIII.7) of the Born approximation applicability takes the form in the relativistic case:

$$|U_{\text{eff}}| \ll \frac{\hbar p}{ma}, \quad |U_{\text{eff}}| \ll \frac{\hbar^2}{ma^2}. \tag{4}$$

For the first term in Eq. (2), the first of the conditions in Eq. (4) requires the following inequality to hold:

$$\left|\frac{Zee_1\varepsilon}{mc^2r}\right| \ll \frac{\hbar p}{mr}, \quad \text{or} \quad \frac{Ze^2}{\hbar c} \ll \frac{v}{c} < 1, \quad (v\varepsilon = pc^2, \, |e_1| \sim e). \tag{5}$$

(As well as in non-relativistic case; the necessary condition for the validity of perturbation theory is the restriction, $Z \ll 137$.) The applicability of perturbation theory for the second term in the effective potential (2) is restricted by the second condition in Eq. (4), which requires that

$$\frac{(Zee_1)^2}{2mc^2r^2} \ll \frac{\hbar^2}{mr^2}, \quad \text{or} \quad \left(\frac{Ze^2}{\hbar c}\right)^2 \ll 1. \tag{6}$$

This condition is weaker than that in Eq. (5).

Let us now note that when calculating the scattering amplitude with Eqs. (2) and (3), the second term in Eq. (2) should be omitted, because it is second-order in the small parameter $Z\alpha$. With this, and using the value of the integral below,

$$\int \frac{1}{r} e^{-i\mathbf{q}\cdot\mathbf{r}} d^3r = \frac{4\pi}{q^2},$$

we find the amplitude and differential scattering cross-section for a spinless particle in the Coulomb field:

$$f_B = -\frac{2Zee_1\varepsilon}{\hbar^2c^2q^2}, \quad \frac{d\sigma}{d\Omega} = |f|^2 \approx \left(\frac{Zee_1}{2v_0p_0}\right)^2 \frac{1}{\sin^4(\theta/2)}, \tag{7}$$

to be compared with the Rutherford equation for non-relativistic theory.

Problem 15.18

In the Born approximation, find the energy-dependence of the scattering cross-section, $\sigma(\varepsilon)$, for a charged spinless particle in an external electrostatic field, $\varphi(r)$, for $\varepsilon \to \infty$.

Find the conditions of applicability for the result obtained, and compare it to the result of non-relativistic theory.

Solution

The Born scattering amplitude for a charged spinless particle in an electrostatic field with the potential, $\varphi(r)$, is described by the expression

$$f_B = -\frac{m}{2\pi\hbar^2} \int U_{\text{eff}} e^{-i\mathbf{q}\cdot\mathbf{r}} d^3r, \tag{1}$$

where the effective potential energy is

$$U_{\text{eff}} = \frac{e\varepsilon}{mc^2}\varphi(r) - \frac{1}{2mc^2}(e\varphi(r))^2. \tag{2}$$

(For a general discussion of Eqs. (1) and (2) and the applicability of the Born approximation in the relativistic case, see the previous problem.)

In the ultra-relativistic case, when $\varepsilon \approx pc \to \infty$, we can neglect the second term in Eq. (2) and the scattering amplitude becomes

$$f_B \approx -\frac{ep}{2\pi\hbar^2 c}\int \varphi(r)e^{-i\mathbf{q}\cdot\mathbf{r}}d^3r \equiv -\frac{ep}{2\pi\hbar^2 c}\tilde{\varphi}(q).$$

Hence, the scattering cross-section is described by

$$\sigma = \int |f|^2 d\Omega = \frac{e^2}{4\pi\hbar^2 c^2}\int_0^{4p^2/\hbar^2} |\tilde{\varphi}(q)|^2 dq^2. \tag{3}$$

(We recall that $d\Omega = (\pi\hbar^2/p^2)dq^2$.)

For $p \to \infty$, the upper integration limit in Eq. (3) can be set to infinity, so that the scattering cross-section, $\sigma(\varepsilon)$, for $\varepsilon \to \infty$ is constant (in the non-relativistic case, the scattering cross-section decreases as $\sigma \propto E^{-1} \to 0$, with $E \to \infty$; see Problem 13.2). This is connected to the fact that according to Eq. (2), the interaction between a particle and an electrostatic field increases with increasing energy.

The applicability of the Born approximation in the problem considered is determined by the first of the expressions in Eq. (4) of Problem 16.17, and requires that the inequality $|e\varphi_0| \ll \hbar c/a$ is satisfied, where φ_0 and a are the characteristic strength of the potential and its characteristic radius, correspondingly. In a "strong" electrostatic field the Born approximation is not applicable. But the conclusion that the scattering cross-section is constant in the $\varepsilon \to \infty$ limit is still true. Here, the scattering cross-section could be calculated by the quasi-classical equation

$$\sigma_{\varepsilon \to \infty} = 4\pi \int_0^\infty \left\{1 - \cos\left[\frac{e}{\hbar c}\int_{-\infty}^\infty \varphi(\sqrt{\rho^2 + z^2})dz\right]\right\}\rho\, d\rho, \tag{4}$$

which is a generalization of the result of Problem 13.51 to the relativistic case. This generalization is obtained by replacing $U(r)$ by U_{eff} (see also the previous problem) in the corresponding equations, and substituting $p \approx \varepsilon/c$ for $\hbar k$.

In conclusion, we note that for the validity of the results obtained, it is necessary that the potential decreases faster than $\propto 1/r^2$. In the opposite case, the scattering cross-section becomes infinite as in the non-relativistic theory, due to the divergence of the integral in Eq. (3) at the lower limit (which corresponds to small scattering angles).

Problem 15.19

In the Born approximation, find the energy dependence of the scattering cross-section, $\sigma(\varepsilon)$, for a spinless particle in an external *scalar* field, $U(r)$ (see the note in Problem 15.4) for $\varepsilon \to \infty$.

Fine the range of applicability of the results, and compare them to the non-relativistic theory.

Solution

The stationary wave equation for a relativistic spinless particle in an external constant scalar field can be written in the form:

$$\left\{-\frac{\hbar^2}{2m}\Delta + U(r)\right\}\psi = \frac{p_0^2}{2m}\psi, \quad (c^2 p_0^2 = \varepsilon^2 - m^2 c^4),$$

which is identical in form to the non-relativistic Schrödinger equation. Due to this analogy for the scattering amplitude, we can use the known results of the non-relativistic theory (compare to Problem 15.17). In the Born approximation,

$$f_B(q) = -\frac{m}{2\pi\hbar^2}\int U(r)e^{-i\mathbf{q}\cdot\mathbf{r}}dV \equiv -\frac{m}{2\pi\hbar^2}\tilde{U}(q).$$

Hence, the scattering cross-section,

$$\sigma(\varepsilon) = \frac{m^2}{4\pi\hbar^2 p_0^2}\int_0^{4p_0^2/\hbar^2}|\tilde{U}(q)|^2 dq^2,$$

in the ultra-relativistic limit ($p_0 \approx \varepsilon/c$) is determined by the expression

$$\sigma(\varepsilon) = \frac{m^2 c^2}{4\pi\hbar^2 \varepsilon^2}\int_0^\infty |\tilde{U}(q)|^2 dq^2 \propto \frac{1}{\varepsilon^2}. \qquad (1)$$

The applicability condition for the Born approximation is $U_0 \ll \hbar p_0/ma$, where U_0 and a are the characteristic potential strength and its radius, respectively.

15.2 The Dirac equation

Problem 15.20

Determine which of the operators below commute with the Hamiltonian of a free relativistic particle with spin $s = 1/2$ (and hence are integrals of motion):

1) $\hat{\mathbf{p}} = -i\hbar\boldsymbol{\nabla}$; 2) $\hat{\mathbf{l}} = \frac{1}{\hbar}[\mathbf{r}\times\hat{\mathbf{p}}] = -i[\mathbf{r}\times\boldsymbol{\nabla}]$; 3) $\hat{\mathbf{l}}^2$;
4) $\hat{\mathbf{s}} = \frac{1}{2}\boldsymbol{\Sigma}$; 5) $\hat{\mathbf{s}}^2$; 6) $\hat{\mathbf{j}} = \hat{\mathbf{l}} + \hat{\mathbf{s}}$; 7) $\hat{\mathbf{j}}^2$;
8) $\hat{\Lambda} = \hat{\mathbf{p}}\cdot\boldsymbol{\Sigma}$; 9) $\hat{I}$[where, $\hat{I}\psi(\mathbf{r}) \equiv \psi(-\mathbf{r})$]; 10) $\hat{P} \equiv \beta\hat{I}$; 11) γ_5.

Compare to the case of a free non-relativistic particle.

Solution

The Hamiltonian of a particle is $\hat{H} = c\boldsymbol{\alpha}\cdot\hat{\mathbf{p}} + mc^2\beta$; see Eqs. (XV.4 and XV.5). For the calculation of the commutators it is convenient to use the result of Problem 1.4, and recall that an operator acting on spatial variables ($\hat{\mathbf{p}}$, $\hat{\mathbf{l}}$, etc.) commutes with an operator acting only on the spin variables ($\boldsymbol{\alpha}$, $\boldsymbol{\Sigma}$, etc.):

1) $[\hat{\mathbf{p}}, \hat{H}] = 0$,
2) $[\hat{l}_i, \hat{H}] = [\hat{l}_i, c\boldsymbol{\alpha} \cdot \hat{\mathbf{p}}] = c\alpha_k[\hat{l}_i, \hat{p}_k] = ic\varepsilon_{ikl}\alpha_k \hat{p}_l \neq 0$;
3) $[\hat{\mathbf{l}}^2, \hat{H}] = [\hat{l}_i\hat{l}_i, \hat{H}] = \hat{l}_i[\hat{l}_i, \hat{H}] + [\hat{l}_i, \hat{H}]\hat{l}_i = ic\varepsilon_{ikl}\alpha_k(\hat{l}_i\hat{p}_l + \hat{p}_l\hat{l}_i) \neq 0$;
4) Since

$$[\Sigma_i, \alpha_k] = \begin{pmatrix} \sigma_i & 0 \\ 0 & \sigma_i \end{pmatrix} \begin{pmatrix} 0 & \sigma_k \\ \sigma_k & 0 \end{pmatrix} - \begin{pmatrix} 0 & \sigma_k \\ \sigma_k & 0 \end{pmatrix} \begin{pmatrix} \sigma_i & 0 \\ 0 & \sigma_i \end{pmatrix} =$$

$$= \begin{pmatrix} 0 & \sigma_i\sigma_k - \sigma_k\sigma_i \\ \sigma_i\sigma_k - \sigma_k\sigma_i & 0 \end{pmatrix} = \begin{pmatrix} 0 & 2i\varepsilon_{ikl}\sigma_l \\ 2i\varepsilon_{ikl}\sigma_l & 0 \end{pmatrix} = 2i\varepsilon_{ikl}\alpha_l$$

and $[\Sigma_i, \beta] = 0$, then $[\hat{s}_i, \hat{H}] = \frac{c}{2}[\Sigma_i, \alpha_k]\hat{p}_k = ic\varepsilon_{ikl}\alpha_l\hat{p}_k \equiv -ic\varepsilon_{ikl}\alpha_l\hat{p}_l$;

5) $\boldsymbol{\Sigma}^2 = \begin{pmatrix} \boldsymbol{\sigma} & 0 \\ 0 & \boldsymbol{\sigma} \end{pmatrix} \cdot \begin{pmatrix} \boldsymbol{\sigma} & 0 \\ 0 & \boldsymbol{\sigma} \end{pmatrix} = \begin{pmatrix} \boldsymbol{\sigma}^2 & 0 \\ 0 & \boldsymbol{\sigma}^2 \end{pmatrix} = \begin{pmatrix} 3 & 0 \\ 0 & 3 \end{pmatrix} = 3$,

hence the operator $\hat{\mathbf{s}}^2 = (1/4)\boldsymbol{\Sigma}^2 = 3/4$ and commutes with $\hat{H}$.

Using the values of the commutators 1), 2), and 4), we find that

6) $[\hat{j}_i, \hat{H}] = 0$;
7) $[\hat{\mathbf{j}}^2, \hat{H}] = 0$;
8) $[\boldsymbol{\Sigma} \cdot \hat{\mathbf{p}}, \hat{H}] = [\Sigma_i, \hat{H}]\hat{p}_i + \Sigma_i[\hat{p}_i, \hat{H}] = 2i\varepsilon_{ikl}\alpha_l\hat{p}_k\hat{p}_i = 0$;
9) Since $\hat{I}\hat{\mathbf{p}} = -\hat{\mathbf{p}}\hat{I}$, then $[\hat{I}, \hat{H}] = [\hat{I}, c\boldsymbol{\alpha} \cdot \hat{\mathbf{p}}] = -2c(\boldsymbol{\alpha} \cdot \hat{\mathbf{p}})\hat{I}$;
10) $[\hat{P}, mc^2\beta] = 0$, $[\hat{P}, \hat{H}] = [\beta\hat{I}, c\boldsymbol{\alpha} \cdot \hat{\mathbf{p}}] = c\beta\hat{I}\boldsymbol{\alpha} \cdot \hat{\mathbf{p}} - c\boldsymbol{\alpha} \cdot \hat{\mathbf{p}}\beta\hat{I} = 0$;
11) $[\gamma_5, \hat{H}] = 2mc^2 \begin{pmatrix} 0 & 1 \\ -1 & 0 \end{pmatrix}$ (and for a particle with zero mass, $m = 0$, this commutator is equal to zero).

In the non-relativistic case, the first nine operators commute with the free-particle Hamiltonian, $\hat{H}_0 = \hat{\mathbf{p}}^2/2m$. In the relativistic case, however, we have a different situation. The commutativity between the momentum operator, $\hat{\mathbf{p}}$, and the free-particle Hamiltonian reflects translational symmetry of the problem; on the other hand, the commutativity between the total particle angular moment operator $\hat{\mathbf{j}} = \hat{\mathbf{l}} + \hat{\mathbf{s}}$ and $\hat{H}$ is due to the isotropy of free space. But the operators $\hat{\mathbf{l}}$ and $\hat{\mathbf{s}}$ do not independently commute with the Hamiltonian. This means that in the relativistic case there is some kinematic correlation between possible spin states of a particle and its angular momentum (compare to Problem 15.23). It is also seen in the non-commutativity of the square of the angular momentum operator $\hat{\mathbf{l}}^2$ with $\hat{H}$. Nevertheless, according to 1) and 8), the helicity is still a "good" quantum number. The commutativity of the inversion operator $\hat{P} = \beta\hat{I}$ with the Hamiltonian of a free particle reflects the mirror symmetry of free space (*i.e.*, the indistinguishability of "right" and "left").

Problem 15.21

Find the solutions of the Dirac equation describing a free particle with definite momentum and energy. To describe the particle spin states use the fact that the operator $\hat{\Lambda} = \mathbf{\Sigma} \cdot \hat{\mathbf{p}}$ commutes with the operators $\hat{\mathbf{p}}$ and $\hat{H}$ (see Problem 15.26).

Solution

The solutions of the Dirac equation for a free particle, corresponding to definite values of the energy, ε, and momentum, $\mathbf{p}$, have the form:

$$\psi_{\mathbf{p},\varepsilon}(\mathbf{r}, t) = u(\mathbf{p}, \varepsilon) \exp\left\{\frac{i}{\hbar}(\mathbf{p} \cdot \mathbf{r} - \varepsilon t)\right\}, \tag{1}$$

where the bi-spinor, $u(\mathbf{p}, \varepsilon)$, satisfies the stationary Dirac equation (see, Eq. (XV.4)):

$$(c\boldsymbol{\alpha} \cdot \mathbf{p} + mc^2 \beta) u(\mathbf{p}, \varepsilon) = \varepsilon u(\mathbf{p}, \varepsilon), \tag{2}$$

or, using the "two-component spinor language",

$$c\boldsymbol{\sigma} \cdot \mathbf{p}\chi + mc^2 \varphi = \varepsilon \varphi, \quad c\boldsymbol{\sigma} \cdot \mathbf{p}\varphi - mc^2 \chi = \varepsilon \chi, \tag{3}$$

$$u(\mathbf{p}, \varepsilon) = \begin{pmatrix} \varphi \\ \chi \end{pmatrix}.$$

The second equation in (3) gives

$$\chi = \frac{c}{\varepsilon + mc^2} \boldsymbol{\sigma} \cdot \mathbf{p}\varphi, \tag{4}$$

and after the substitution of this equation into the first equation in (3), we obtain (using the relation $(\boldsymbol{\sigma} \cdot \mathbf{p})^2 = p^2$):

$$\frac{c^2 p^2}{\varepsilon + mc^2} \varphi = (\varepsilon - mc^2) \varphi.$$

Hence, it follows that $\varepsilon = \pm\sqrt{p^2 c^2 + m^2 c^4}$. Here, the spinor remains undefined, and could be arbitrarily chosen in two independent ways, corresponding to the positive and negative signs of ε.

Hence, for a fixed momentum $\mathbf{p}$ there are four independent solutions to the Dirac equation of the form (1), for which the bi-spinors $u(\mathbf{p}, \varepsilon)$ are equal to

$$u(\mathbf{p}, \varepsilon = E) = \begin{pmatrix} \varphi_1 \\ \frac{c\boldsymbol{\sigma} \cdot \mathbf{p}}{E + mc^2} \varphi_1 \end{pmatrix}, \quad u(\mathbf{p}, \varepsilon = -E) = \begin{pmatrix} \varphi_2 \\ \frac{c\boldsymbol{\sigma} \cdot \mathbf{p}}{-E + mc^2} \varphi_2 \end{pmatrix}, \tag{5}$$

where $E = +\sqrt{p^2 c^2 + m^2 c^4} \geq mc^2$.

The existence of the solutions to the Dirac equation that formally correspond to a particle with negative energy is associated with the antiparticle state (see Problem 15.27).

To explicitly specify the spinor form, φ, and also the wavefunctions (1) and (5), we use the commutativity of the Hermitian operator $\hat{\Lambda} = \boldsymbol{\Sigma} \cdot \hat{\mathbf{p}}$ with the operators, $\hat{\mathbf{p}}$ and $\hat{H}$, and introduce the complete system of eigenfunctions: $\psi_{p\varepsilon\Lambda}$. Here, from the equation, $\hat{\Lambda}\psi_{p\varepsilon\Lambda} = \Lambda\psi_{p\varepsilon\Lambda}$, where

$$\psi_{p\varepsilon\Lambda} = u(\mathbf{p}, \varepsilon, \Lambda) \exp\left\{\frac{i}{\hbar}(\mathbf{p}\cdot\mathbf{r} - \varepsilon t)\right\}, \quad u(\mathbf{p}, \varepsilon, \Lambda) = \begin{pmatrix} \varphi_\Lambda \\ \frac{c\boldsymbol{\sigma}\cdot\mathbf{p}}{\varepsilon + mc^2}\varphi_\Lambda \end{pmatrix},$$

it follows that

$$(\boldsymbol{\sigma}\cdot\mathbf{p})\varphi_\Lambda = \Lambda\varphi_\Lambda, \quad or \quad \left(\frac{\boldsymbol{\sigma}}{2}\mathbf{n}\right)\varphi_\Lambda = \lambda\varphi_\Lambda, \quad \mathbf{n} = \frac{\mathbf{p}}{|\mathbf{p}|}, \quad \Lambda = 2\lambda|\mathbf{p}|. \tag{6}$$

Solutions to Eq. (6) were obtained before in the non-relativistic spin theory (see Problems 5.3 and 5.20). We remind here that particle states with a definite value of λ (with eigenvalues of λ equal to $\pm 1/2$) are called "helical states". Also note here that the helicity, λ, does not change under charge conjugation – see, Problem 15.27 – unlike $\mathbf{p}$ and ε, which do change sign.

Problem 15.22

Find 4-vector components for the current density of a free Dirac particle in a state with a definite momentum. Compare your answer to the corresponding expression from non-relativistic theory.

Solution
Substituting into the known expressions (XV.7),

$$\mathbf{j} = c\psi^*\boldsymbol{\alpha}\psi \equiv ic\overline{\psi}\boldsymbol{\gamma}\psi, \quad \rho = \psi^*\psi \equiv \overline{\psi}\beta\psi, \tag{1}$$

the wavefunction of a Dirac particle in the state with momentum $\mathbf{p}$ and energy $\varepsilon = \sqrt{p^2c^2 + m^2c^4} \geq mc^2$:

$$\psi_{p\varepsilon} = u(\mathbf{p}, \varepsilon)\exp\left\{\frac{i}{\hbar}(\mathbf{p}\cdot\mathbf{r} - \varepsilon t)\right\}, \tag{2}$$

where the bi-spinor u is equal to (see the previous problem)

$$u = N\begin{pmatrix}\varphi \\ \chi\end{pmatrix} = N\begin{pmatrix}\varphi \\ \frac{c\boldsymbol{\sigma}\cdot\mathbf{p}}{\varepsilon + mc^2}\varphi\end{pmatrix}, \quad N = \sqrt{\frac{\varepsilon + mc^2}{2\varepsilon}} \tag{3}$$

(the value of N is chosen so that the normalization of the bi-spinor, u, coincides with that of φ; i.e., $u^*u = \varphi^*\varphi$); and using the relations

$$u^* = N(\varphi^*, \chi^*) = N\left(\varphi^*, \varphi^*\frac{c\boldsymbol{\sigma}\cdot\mathbf{p}}{\varepsilon + mc^2}\right), \quad (\boldsymbol{\sigma}\cdot\mathbf{p})^2 = p^2,$$

we obtain

$$\rho = \psi^*\psi = u^*u = \varphi^*\varphi, \tag{4}$$

$$\mathbf{j} = c\psi^*\boldsymbol{\alpha}\psi = cu^*\boldsymbol{\alpha}u = cN^2\left(\varphi^*, \varphi^*\frac{c\boldsymbol{\sigma}\cdot\mathbf{p}}{\varepsilon + mc^2}\right)\begin{pmatrix} 0 & \boldsymbol{\sigma} \\ \boldsymbol{\sigma} & 0 \end{pmatrix}\begin{pmatrix} \varphi \\ \frac{c\boldsymbol{\sigma}\cdot\mathbf{p}}{\varepsilon+mc^2}\varphi \end{pmatrix} =$$

$$= cN^2\left(\varphi^*, \varphi^*\frac{c\boldsymbol{\sigma}\cdot\mathbf{p}}{\varepsilon + mc^2}\right)\begin{pmatrix} \frac{c\sigma(\boldsymbol{\sigma}\cdot\mathbf{p})}{\varepsilon+mc^2}\varphi \\ \sigma\varphi \end{pmatrix} = \frac{c^2N^2}{\varepsilon + mc^2}\varphi^*[\boldsymbol{\sigma}(\boldsymbol{\sigma}\cdot\mathbf{p}) + (\boldsymbol{\sigma}\cdot\mathbf{p})\boldsymbol{\sigma}]\phi\varphi. \tag{5}$$

Hence, using the relation

$$\sigma_i(\boldsymbol{\sigma}\cdot\mathbf{p}) + (\boldsymbol{\sigma}\cdot\mathbf{p})\sigma_i = (\sigma_i\sigma_k + \sigma_k\sigma_i)p_k = 2\delta_{ik}p_k = 2p_i,$$

it follows that

$$\mathbf{j} = c\psi^*\boldsymbol{\alpha}\psi = \frac{2c^2N^2}{\varepsilon + mc^2}\mathbf{p}\varphi^*\varphi = \frac{c^2\mathbf{p}}{\varepsilon}\varphi^*\varphi = \mathbf{v}\varphi^*\varphi = \rho\mathbf{v}, \tag{6}$$

where $\mathbf{v}$ is the velocity of a relativistic particle with momentum $\mathbf{p}$.

In conclusion, we consider the velocity operator, $\hat{\mathbf{v}}$, of a Dirac particle. The formal calculation of the commutator $[\hat{H}, \mathbf{r}]$ gives (see VI.4)

$$\hat{\mathbf{v}} = \dot{\hat{\mathbf{r}}} = \frac{i}{\hbar}[\hat{H}, \hat{\mathbf{r}}] = c\boldsymbol{\alpha}. \tag{7}$$

Remember that the appearance of the operator, $\dot{\hat{f}} = \frac{i}{\hbar}[\hat{H}, r]$ as the time derivative of a generic physical quantity, f (for which $\partial\hat{f}/\partial t = 0$) is connected in non-relativistic theory with the following relation:

$$\frac{d}{dt}\langle f \rangle \equiv \langle \frac{d\hat{f}}{dt} \rangle = \frac{i}{\hbar}\int \psi^*[\hat{H}, \hat{f}]\psi d\tau, \tag{8}$$

from which it directly follows that there are no essential restrictions on the wave-function, ψ. In the case of a relativistic particle that satisfies the Dirac equation, these restrictions do occur. They arise from the fact that only a superposition of positive-frequency solutions may have an immediate physical meaning of a particle state (compare to Problem 15.1 for a spinless particle). Such a superposition, ψ^+, can be constructed out of an arbitrary solution to the Dirac equation using the projection operator, $\hat{P}_+$:

$$\psi^+ = \hat{P}_+\psi, \quad \hat{P}_+ = \frac{c\boldsymbol{\alpha}\cdot\hat{\mathbf{p}} + mc^2\beta + \hat{\varepsilon}}{2\hat{\varepsilon}}, \quad \hat{\varepsilon} = \varepsilon(\hat{\mathbf{p}}) = \sqrt{\hat{\mathbf{p}}^2c^2 + m^2c^4}, \tag{9}$$

where $\hat{P}_+^2 = \hat{P}_+$. Using the properties of the matrices $\boldsymbol{\alpha}$ and β:

$$\alpha_i\alpha_k + \alpha_k\alpha_i = 2\delta_{ik}, \quad \beta\alpha_i + \alpha_i\beta = 0,$$

and

$$\alpha_i \alpha_k \alpha_l = \begin{pmatrix} 0 & \sigma_i \sigma_k \sigma_l \\ \sigma_i \sigma_k \sigma_l & 0 \end{pmatrix} = -i\varepsilon_{ikl}\gamma_5 + \delta_{ik}\alpha_l - \delta_{il}\alpha_k + \delta_{kl}\alpha_i,$$

we obtain

$$\hat{P}_+ \alpha \hat{P}_+ = \frac{c\hat{\mathbf{p}}}{\hat{\varepsilon}} \hat{P}_+. \tag{10}$$

Now, making the following substitutions in Eq. (8):

$$\hat{f} \to \mathbf{r}, \quad \psi \to \psi^+ \equiv \hat{P}_+ \psi^+, \quad \hat{H} = c\boldsymbol{\alpha} \cdot \hat{\mathbf{p}} + mc^2 \beta,$$

and performing the following transformation with the help of relation (10),

$$\langle \hat{\mathbf{v}} \rangle = \langle \psi^+ | \hat{P}_+ c\boldsymbol{\alpha} \hat{P}_+ | \psi^+ \rangle = \langle \psi^+ | \frac{c^2 \hat{\mathbf{p}}}{\hat{\varepsilon}} \hat{P}_+ | \psi^+ \rangle = \langle \psi^+ | \frac{c^2 \hat{\mathbf{p}}}{\hat{\varepsilon}} | \psi^+ \rangle, \tag{11}$$

we obtain the relation $\hat{\mathbf{v}} = c^2 \hat{\mathbf{p}}/\hat{\varepsilon}$ between the operators of velocity, momentum, and energy of a free relativistic particle (in particular, in the momentum representation, we obtain, $\hat{\mathbf{v}} = c^2 \mathbf{p}/\varepsilon(p)$).

Problem 15.23

Find the mean value of the spin vector for a Dirac particle with a definite momentum (and arbitrary spin state). For simplicity, assume that the momentum is directed along the z axis. Compare to the corresponding result of non-relativistic theory.

Solution
The wavefunction of the state considered has the form:

$$\psi = u(\mathbf{p}) \exp\left\{ \frac{i}{\hbar}(\mathbf{p} \cdot \mathbf{r} - \varepsilon t) \right\},$$

where the bi-spinor, $u(\mathbf{p})$, is

$$u(\mathbf{p}) = N \begin{pmatrix} \varphi \\ \chi \end{pmatrix} = N \begin{pmatrix} \varphi \\ \frac{c\boldsymbol{\sigma} \cdot \mathbf{p}}{\varepsilon + mc^2} \varphi \end{pmatrix}, \quad N = \sqrt{\frac{\varepsilon + mc^2}{2\varepsilon}},$$

(see Problem 15.21), and we used the normalization condition $u^* u = \varphi^* \varphi = 1$.

The mean value of the spin vector is calculated by the equation:

$$\bar{s} = \frac{\frac{1}{2}u^*\Sigma u}{u^* u} = \frac{N^2}{2}\left(\varphi^*, \varphi^* \frac{c(\boldsymbol{\sigma}\cdot\mathbf{p})}{\varepsilon+mc^2}\right)\begin{pmatrix}\boldsymbol{\sigma} & 0 \\ 0 & \boldsymbol{\sigma}\end{pmatrix}\begin{pmatrix}\varphi \\ \frac{c(\boldsymbol{\sigma}\cdot\mathbf{p})}{\varepsilon+mc^2}\varphi\end{pmatrix}$$

$$= \frac{N^2}{2}\varphi^*\left\{\boldsymbol{\sigma}+\frac{c^2}{(\varepsilon+mc^2)^2}(\boldsymbol{\sigma}\cdot\mathbf{p})\boldsymbol{\sigma}(\boldsymbol{\sigma}\cdot\mathbf{p})\right\}\varphi$$

$$= \frac{N^2}{2}\varphi^*\left\{\boldsymbol{\sigma}+\frac{p^2 c^2}{(\varepsilon+mc^2)^2}\sigma_z\boldsymbol{\sigma}\sigma_z\right\}\varphi, \quad \mathbf{p}=(0,0,p). \tag{1}$$

Hence, using the properties (V.3) of the Pauli matrices, we obtain

$$\bar{s}_x = \frac{N^2}{2}\varphi^*\left\{\sigma_x - \frac{p^2 c^2}{(\varepsilon+mc^2)^2}\sigma_x\right\}\varphi = \frac{mc^2}{2\varepsilon}\varphi^*\sigma_x\varphi, \tag{2}$$

$$\bar{s}_y = \frac{mc^2}{2\varepsilon}\varphi^*\sigma_y\varphi, \quad \bar{s}_z = \frac{1}{2}\varphi^*\sigma_z\varphi.$$

Note that the vector $\mathbf{s}_0 = \frac{1}{2}\varphi^*\boldsymbol{\sigma}\varphi$ has the meaning of a mean spin vector, but is expressed specifically in the coordinate system, where the particle is at rest; see Problem 15.25. Hence, the results obtained in Eq. (2) can be characterized as a sort of reduction of the transversal components in Lorentz transformation. In the ultra-relativistic limit, $\varepsilon \gg mc^2$, the vector $\bar{\mathbf{s}}$ becomes directed along the particle's momentum (compare this to the case of a particle with rest mass $m = 0$, discussed in Problem 15.24).

Problem 15.24

Consider the unitary transformation of bi-spinors given by the unitary operator (matrix)

$$\hat{U} = \frac{1}{\sqrt{2}}\begin{pmatrix}1 & 1 \\ 1 & -1\end{pmatrix}.$$

What is the form of particle spin operator and Dirac equation in the new representation for the two-component spinors

$$\psi' = \hat{U}\psi \equiv \begin{pmatrix}\xi \\ \eta\end{pmatrix}?$$

Discuss the case of a massless particle, $m = 0$.

Solution

For the unitary transformation considered,

$$\hat{U} = \hat{U}^+ = \frac{1}{\sqrt{2}}\begin{pmatrix}1 & 1 \\ 1 & -1\end{pmatrix},$$

we find

$$\beta' = \hat{U}\beta\hat{U}^+ = \frac{1}{2}\begin{pmatrix} 1 & 1 \\ 1 & -1 \end{pmatrix}\begin{pmatrix} 1 & 0 \\ 0 & -1 \end{pmatrix}\begin{pmatrix} 1 & 1 \\ 1 & -1 \end{pmatrix} = \begin{pmatrix} 0 & 1 \\ 1 & 0 \end{pmatrix},$$

$$\Sigma' = \hat{U}\Sigma\hat{U}^+ = \Sigma,$$

$$\alpha' = \hat{U}\alpha\hat{U}^+ = \frac{1}{\sqrt{2}}\begin{pmatrix} 1 & 1 \\ 1 & -1 \end{pmatrix}\begin{pmatrix} 0 & \sigma \\ \sigma & 0 \end{pmatrix}\begin{pmatrix} 1 & 1 \\ 1 & -1 \end{pmatrix} = \begin{pmatrix} \sigma & 0 \\ 0 & -\sigma \end{pmatrix}.$$

Hence, the spin vector operator in the new representation conserves its form $\frac{1}{2}\Sigma$, while the Dirac equation

$$(c\alpha' \cdot \hat{\mathbf{p}} + mc^2\beta')\psi' = i\hbar\frac{\partial}{\partial t}\psi', \quad \psi' = \hat{U}\psi = \frac{1}{\sqrt{2}}\begin{pmatrix} \varphi + \chi \\ \varphi - \chi \end{pmatrix} = \begin{pmatrix} \xi \\ \eta \end{pmatrix},$$

written in terms of the two-component spinors, ξ and η, takes the form

$$i\hbar\frac{\partial}{\partial t}\xi = c(\boldsymbol{\sigma} \cdot \hat{\mathbf{p}})\xi + mc^2\eta, \quad i\hbar\frac{\partial}{\partial t}\eta = -c(\boldsymbol{\sigma} \cdot \hat{\mathbf{p}})\eta + mc^2\xi. \qquad (1)$$

Let us discuss now the case of a particle with zero rest mass $m = 0$ (neutrino). From Eq. (1), we see that in this case the new representation is especially convenient, since the spinors, ξ and η, satisfy the equations independently. Moreover, these spinors are transformed independently from one another under the Lorentz transformations (compare to Problem 15.25). Hence, for $m = 0$ each of the equations in (1) is relativistically invariant – the *Weyl equation*. However, these equations are not invariant with respect to spatial inversion, unlike the Dirac equation. This is connected with the fact that under inversion, the spinors, ξ and η, are "permuted". This follows from the transformation $\tilde{\psi}' = \hat{P}\psi' = \hat{I}\beta'\psi'$.

Problem 15.25

Assuming that the spin state of a particle at rest is known, find the bi-spinor, $u(\mathbf{p})$, in an arbitrary coordinate system, where the particle momentum is $\mathbf{p}$.

Using the result obtained, find the connection between the mean values of the particle spin vector in the coordinate system at rest and in motion.

Solution

In the coordinate system K, where a particle is at rest and has the energy $\varepsilon = mc^2$, its spin state is described by the bi-spinor $u(0) = \begin{pmatrix} \varphi_0 \\ 0 \end{pmatrix}$, where φ_0 is some two-component spinor.

In the system K', which moves with respect to K with velocity $-\mathbf{v}$, the particle has velocity $\mathbf{v}$ and momentum $\mathbf{p} = m\mathbf{v}/\sqrt{1 - (v/c)^2}$, while its spin state is described

by a bi-spinor $u(p)$, which is expressed in terms of $u(0)$ as follows:
$$u(\mathbf{p}) \equiv u' = \hat{S}u(0),$$
with
$$\hat{S} = \exp\left\{-\frac{1}{2}\boldsymbol{\alpha}\cdot\mathbf{n}\theta\right\} = \cosh\frac{\theta}{2} - (\boldsymbol{\alpha}\cdot\mathbf{n})\sinh\frac{\theta}{2} \quad \left(\tanh\theta = \frac{v}{c}\right), \tag{2}$$

where $\mathbf{n} = -\mathbf{v}/v$ is the unit vector of the velocity of K' with respect to K.

Using the relations
$$\tanh\frac{\theta}{2} = \frac{\tanh\theta}{1+\sqrt{1-\tanh^2\theta}} = \frac{v/c}{1+\sqrt{1-(v/c)^2}} = \frac{pc}{\varepsilon(p)+mc^2},$$

$$\cosh\frac{\theta}{2} = \sqrt{\frac{1+\cosh\theta}{2}} = \sqrt{\frac{1}{2}\left(1+\frac{1}{\sqrt{1-\tanh^2\theta}}\right)} = \sqrt{\frac{\varepsilon(p)+mc^2}{2mc^2}},$$

and Eqs. (1) and (2), we obtain
$$u(\mathbf{p}) = \begin{pmatrix} \left(\cosh\frac{\theta}{2}\right)\varphi_0 \\ \left(\sinh\frac{\theta}{2}\right)\frac{\sigma\mathbf{v}}{v}\varphi_0 \end{pmatrix} = \sqrt{\frac{\varepsilon+mc^2}{2mc^2}}\begin{pmatrix} \varphi_0 \\ \frac{c\boldsymbol{\sigma}\cdot\mathbf{p}}{\varepsilon+mc^2}\varphi_0 \end{pmatrix}. \tag{3}$$

This expression for the bi-spinor, $u(\mathbf{p})$, coincides with the form of the general solution to the Dirac equation discussed in Problem 15.21. In this sense, an essential element of the solution is the fact that the spinor φ in the bi-spinor $u(\mathbf{p}) = \begin{pmatrix}\varphi \\ \chi\end{pmatrix}$ is the same in all Lorentz systems and is equal to φ_0.

Using the normalization, $\varphi_0^*\varphi_0 = 1$, we find the mean value of the particle spin vector in the system K', where it has momentum $\mathbf{p}$, as follows:
$$\bar{\mathbf{s}}_{\mathrm{p}} = \frac{\frac{1}{2}u^*\boldsymbol{\Sigma}u}{u^*u} = \frac{\varepsilon+mc^2}{4\varepsilon}\varphi_0^*\left\{\boldsymbol{\sigma} + \frac{c^2}{(\varepsilon+mc^2)^2}(\boldsymbol{\sigma}\cdot\mathbf{p})\boldsymbol{\sigma}(\boldsymbol{\sigma}\cdot\mathbf{p})\right\}\varphi_0. \tag{4}$$

It is now straightforward to calculate the mean value of the spin vector in rest frame $\bar{\mathbf{s}}_0 = \frac{1}{2}\varphi_0^*\boldsymbol{\sigma}\varphi_0$, by choosing the vector p to be directed along the z axis and using the properties of the Pauli matrices (V.3):
$$\bar{s}_{\mathrm{p},x} = \frac{mc^2}{\varepsilon}\bar{s}_{0,x}; \quad \bar{s}_{\mathrm{p},y} = \frac{mc^2}{\varepsilon}\bar{s}_{0,y}; \quad \bar{s}_{\mathrm{p},z} = \bar{s}_{0,z}, \tag{5}$$

(see Problem 15.23).

Problem 15.26

As is known (see, e.g., Problem 15.21), the wavefunction of a spin $s = 1/2$ particle with momentum $\mathbf{p}$, and energy, $\varepsilon = \sqrt{p^2c^2 + m^2c^4}$, has the form:

$$\psi_{\mathbf{p}}^{+} = u(\mathbf{p})e^{i(\mathbf{p}\cdot\mathbf{r}-\varepsilon t)/\hbar}; \quad u(\mathbf{p}) = \begin{pmatrix} \varphi \\ \frac{c\boldsymbol{\sigma}\cdot\mathbf{p}}{\varepsilon+mc^2}\varphi \end{pmatrix}.$$

This state is two-fold degenerate (there are two independent ways to choose the spinor, φ), due to the spin degree of freedom. Let us consider two such independent states, corresponding to the choice of the spinor $\varphi = \varphi_\lambda$, where

$$(\boldsymbol{\sigma}\cdot\tilde{\mathbf{n}})\varphi_\lambda = \lambda\varphi_\lambda,$$

$\tilde{\mathbf{n}}$ is an arbitrary unit vector, and $\lambda = \pm 1$; see, Problem 5.12.

Prove the orthogonality of the spin states that correspond to different values of λ for the relativistic particle.

Using the result of the previous problem, find the physical meaning of the vector $\tilde{\mathbf{n}}$ and the corresponding eigenvalues, λ.

What is the meaning of the vector $\frac{1}{2}\varphi^*\boldsymbol{\sigma}\varphi$, with the normalization condition being $\varphi^*\varphi = 1$?

Solution

Taking into account that

$$\psi_{\mathbf{p},\lambda} = u_\lambda(\mathbf{p})\exp\left\{\frac{i}{\hbar}(\mathbf{p}\cdot\mathbf{r}-\varepsilon t)\right\}, \quad u_\lambda(\mathbf{p}) = \begin{pmatrix} \varphi_\lambda \\ \frac{c\boldsymbol{\sigma}\cdot\mathbf{p}}{\varepsilon+mc^2}\varphi_\lambda \end{pmatrix}$$

$$u_\lambda^*(\mathbf{p}) = \left(\varphi_\lambda^*, \varphi_\lambda^*\frac{c\boldsymbol{\sigma}\cdot\mathbf{p}}{\varepsilon+mc^2}\right),$$

we find

$$\psi_{\mathbf{p}\lambda'}^*\psi_{\mathbf{p}\lambda} = u_{\lambda'}^*u_\lambda = \varphi_{\lambda'}^*\left[1+\frac{c^2(\boldsymbol{\sigma}\cdot\mathbf{p})^2}{(\varepsilon+mc^2)^2}\right]\varphi_\lambda = \frac{2\varepsilon}{\varepsilon+mc^2}\varphi_{\lambda'}^*\varphi_\lambda \propto \delta_{\lambda'\lambda},$$

which proves the orthogonality of the relativistic spin states corresponding to different values of λ. We used orthogonality $\varphi_{\lambda'}^*\varphi_\lambda = \delta_{\lambda'\lambda}$ of the two-component spinors as eigenfunctions of the Hermitian operator, $\boldsymbol{\sigma}\cdot\tilde{\mathbf{n}}$.

The answers to the remaining questions become evident if we use the result of the previous problem. According to the latter, the spinor, φ, in the bi-spinor, $u(\mathbf{p}) = \begin{pmatrix} \varphi \\ \chi \end{pmatrix}$, which describes a particle with a definite momentum, but in a different inertial coordinate system, is the same (modulo an overall normalization factor). In the system where the particle is at rest, the bi-spinor has the form $u(0) = \begin{pmatrix} \varphi \\ 0 \end{pmatrix}$ and $u_\lambda(0) = \begin{pmatrix} \varphi_\lambda \\ 0 \end{pmatrix}$. In this coordinate system, the equation $(\boldsymbol{\sigma}\cdot\tilde{\mathbf{n}})\varphi_\lambda = \lambda\varphi_\lambda$ is equivalent to

$$(\boldsymbol{\Sigma}\tilde{\mathbf{n}})u_\lambda(0) = \lambda u_\lambda(0). \tag{1}$$

That is, it is an equation for the eigenfunction of the operator $(\boldsymbol{\Sigma}\tilde{\mathbf{n}}) = 2(\hat{\mathbf{s}}\cdot\tilde{\mathbf{n}})$ – twice the spin projection along $\tilde{\mathbf{n}}$. Hence, the vector $\tilde{\mathbf{n}}$ has a direct physical meaning not in the initial coordinate system, where the particle momentum is equal to **p**, but in the system where it is at rest, and this vector determines the direction along which the spin projection has a definite value, equal to $\lambda/2$. Then, the vector, $\frac{1}{2}\varphi^* \boldsymbol{\sigma} \varphi$, gives the mean value of the spin vector in rest frame of the particle. In order to avoid mistakes we should reiterate that here only particle states with a definite value of momentum are considered, and hence we can speak about a rest system of the particle.

In conclusion, we mention that we can give the problem a covariant form by classifying the particle states by both a definite momentum and spin, using the quantum number, λ. We introduce the operator

$$\hat{\lambda} = i\gamma_5 \hat{\nu} \equiv i\gamma_5(\boldsymbol{\gamma} \cdot \boldsymbol{\nu} + \gamma_4 \nu_4) = \begin{pmatrix} \boldsymbol{\sigma} \cdot \boldsymbol{\nu} & -\nu_0 \\ \nu_0 & -\boldsymbol{\sigma} \cdot \boldsymbol{\nu} \end{pmatrix}, \tag{2}$$

where $\nu_i = (\boldsymbol{\nu}, \nu_4)$ is some unit 4-vector, so that $\nu_i^2 = \boldsymbol{\nu}^2 + \nu_4^2 = 1$ is orthogonal to the 4-momentum of the particle $p_i = (\mathbf{p}, i\varepsilon/c)$, i.e. $\nu_i p_i = \boldsymbol{\nu} \cdot \mathbf{p} - \nu_0 \varepsilon/c = 0$, while $\nu_4 = i\nu_0$. The equation

$$\hat{\lambda} u_\lambda(p) = \lambda u_\lambda(p) \tag{3}$$

is equivalent to the equation $(\boldsymbol{\sigma} \cdot \tilde{\mathbf{n}})\varphi_\lambda = \lambda \varphi_\lambda$, where the connection between the three-dimensional vector $\tilde{\mathbf{n}}$ and the 4-vector ν_i is determined by the condition that ν_i in the rest frame has the form $\tilde{\nu}_i = (\tilde{\mathbf{n}}, 0)$. Indeed, taking into account the expression for the bi-spinor $u_\lambda(\mathbf{p})$ and the operator $\hat{\lambda}$, we find

$$\hat{\lambda} u_\lambda \equiv \begin{pmatrix} \boldsymbol{\sigma} \cdot \left(\boldsymbol{\nu} - \frac{c\nu_0 \mathbf{p}}{\varepsilon + mc^2}\right) \varphi_\lambda \\ \left(\nu_0 - \frac{c(\boldsymbol{\sigma}\cdot\boldsymbol{\nu})(\boldsymbol{\sigma}\cdot\mathbf{p})}{\varepsilon + mc^2}\right) \varphi_\lambda \end{pmatrix}. \tag{4}$$

Then, expressing $\boldsymbol{\nu}$ and ν_0 in terms of the components, $(\tilde{\mathbf{n}}, 0)$, and using the Lorentz transformations for the 4-vector:

$$\boldsymbol{\nu} = \boldsymbol{\nu}_\perp + \boldsymbol{\nu}_\parallel = \tilde{\mathbf{n}}_\perp + \frac{\varepsilon}{mc^2}\tilde{\mathbf{n}}_\parallel, \quad \nu_0 = \frac{c}{\varepsilon}(\boldsymbol{\nu} \cdot \mathbf{p}),$$

where the labels $\perp$, $\parallel$ correspond to the perpendicular and parallel components of vectors with respect to the vector $\mathbf{p}/|\mathbf{p}|$, we obtain

$$\boldsymbol{\nu} - \frac{c\nu_0 \mathbf{p}}{\varepsilon + mc^2} = \boldsymbol{\nu} - \frac{\boldsymbol{\nu}(\varepsilon - mc^2)\mathbf{p}}{cp^2} = \boldsymbol{\nu}_\perp + \frac{mc^2}{\varepsilon}\boldsymbol{\nu}_\parallel = \tilde{\mathbf{n}}.$$

Here, as is seen, the upper spinor in the bi-spinor (4) coincides with $(\boldsymbol{\sigma} \cdot \tilde{\mathbf{n}})\varphi_\lambda$. Similarly, we can prove the relation

$$\left(\nu_0 - \frac{c(\boldsymbol{\sigma}\cdot\boldsymbol{\nu})(\boldsymbol{\sigma}\cdot\mathbf{p})}{\varepsilon + mc^2}\right)\varphi_\lambda = \frac{c(\boldsymbol{\sigma}\cdot\mathbf{p})}{\varepsilon + mc^2}(\boldsymbol{\sigma} \cdot \tilde{\mathbf{n}})\varphi_\lambda$$

for the lower spinor in (4). From these equalities, the equivalence of Eq. (3) with $(\boldsymbol{\sigma} \cdot \tilde{\mathbf{n}})\varphi_\lambda = \lambda \varphi_\lambda$ follows. Let us finally note that the equivalence of these equations is seen also from the following considerations. The operator $\hat{\lambda}$ introduced is a pseudo-scalar operator with respect to the Lorentz group. Hence, from the covariance of Eq. (3) it follows that if it is satisfied in one reference system, it must be automatically satisfied in any Lorentz system. Since in the rest system of the particle, Eq. (3) has the form $(\boldsymbol{\sigma} \cdot \tilde{\mathbf{n}})\varphi_\lambda = \lambda \varphi_\lambda$, this relation holds in all other Lorentz systems.

Problem 15.27

By applying the charge conjugation transformation, find explicitly the wavefunction, ψ_c^+, of the antiparticle state corresponding to the solution of the Dirac equation, ψ^-, with the definite momentum, equal to $-\mathbf{p}$, and the negative energy $E = -\varepsilon = -\sqrt{p^2c^2 + m^2c^4}$. Compare this to the wavefunction of the *physical* particle state with the energy, $\varepsilon \geq mc^2$, and momentum, $\mathbf{p}$ (see Problems 15.21 and 15.26).

How does the *helicity* change under charge conjugation (see Problem 15.2)?

Solution

The solutions of the Dirac equation corresponding with definite values of momentum and energy have the form:[410]

$$\psi_{\mathbf{p}\varepsilon}^\pm = \begin{pmatrix} \varphi_\mathbf{p} \\ \frac{c\boldsymbol{\sigma}\cdot\mathbf{p}}{\varepsilon \pm mc^2}\varphi_\mathbf{p} \end{pmatrix} e^{\pm i(\mathbf{p}\cdot\mathbf{r}-\varepsilon t)/\hbar} = \begin{pmatrix} \frac{c\boldsymbol{\sigma}\cdot\mathbf{p}}{\varepsilon \mp mc^2}\chi_\mathbf{p} \\ \chi_\mathbf{p} \end{pmatrix} e^{\pm i(\mathbf{p}\cdot\mathbf{r}-\varepsilon t)/\hbar}, \quad (1)$$

where $\varepsilon = \sqrt{p^2c^2 + m^2c^4} \geq mc^2$. Here the solution, $\psi_{\mathbf{p}\varepsilon}^+$, has the meaning of the wavefunction of a particle with momentum $\mathbf{p}$ and energy ε.

The solution $\psi_{\mathbf{p}\varepsilon}^-$ that corresponds to a formally negative energy and momentum, $(-\mathbf{p})$, has no direct meaning as a physical wavefunction. Such a solution is associated with an antiparticle, whose wavefunction, $\psi_c^+ = \hat{C}\psi^-$, is obtained via charge conjugation of the function $\psi_{\mathbf{p}\varepsilon}^-$. This transformation for the Dirac matrices is given in Eq. (XV.5), and can be presented as follows:

$$\psi_c^+ = \hat{C}\psi^- \equiv \gamma_2\gamma_4\overline{\psi^-} = \gamma_2\gamma_4(\psi^{-*}\beta), \quad (2)$$

or, using the bi-spinor indexes,

$$(\psi_c^+)_\alpha = (\gamma_2\gamma_4)_{\alpha\delta}(\psi^{-*}\beta)_\delta = (\gamma_2\gamma_4)_{\alpha\delta}(\psi^{-*})_\mu \beta_{\mu\delta} =$$
$$= (\gamma_2\gamma_4)_{\alpha\delta}\beta_{\delta\mu}(\psi^{-*})_\mu = (\gamma_2\gamma_4\beta)_{\alpha\mu}(\psi^{-*})_\mu = (\gamma_2)_{\alpha\mu}(\psi^{-*})_\mu \quad (3)$$

(here we used $\beta = \gamma_4$, $\beta^2 = 1$, and $\beta_{\mu\delta} = \beta_{\delta\mu}$).

---

[410] Here we use the same notations as in Problem 15.1, which discussed a spinless particle.

Using the relations

$$\psi_{p\varepsilon}^{-*} = \begin{pmatrix} \frac{c\sigma^* \cdot p}{\varepsilon + mc^2} \chi_p^* \\ \chi_p^* \end{pmatrix} e^{i(p\cdot r - \varepsilon t)/\hbar}, \quad \gamma_2 = \begin{pmatrix} 0 & -i\sigma_2 \\ i\sigma_2 & 0 \end{pmatrix},$$

$$\sigma^* = (\sigma_1, -\sigma_2, \sigma_3), \quad \sigma_2 \sigma^* = -\sigma \sigma_2,$$

and according to Eq. (3), we find the wavefunction of the antiparticle state, corresponding to the "non-physical" solution, $\psi_{p\varepsilon}^-$, of the Dirac equation:

$$\psi_c^+ = \begin{pmatrix} 0 & -i\sigma_2 \\ i\sigma_2 & 0 \end{pmatrix} \begin{pmatrix} \frac{c\sigma^* \cdot p}{\varepsilon + mc^2} \chi_p^* \\ \chi_p^* \end{pmatrix} e^{i(p\cdot r - \varepsilon t)/\hbar} =$$

$$\begin{pmatrix} -i\sigma_2 \chi_p^* \\ -\frac{c\sigma p}{\varepsilon + mc^2} \sigma_2 \chi_p^* \end{pmatrix} e^{i(p\cdot r - \varepsilon t)/\hbar}. \tag{4}$$

In this state, the antiparticle has momentum $\mathbf{p}$, and energy $\varepsilon = \sqrt{p^2 c^2 + m^2 c^4} \geq mc^2$. Denoting $\varphi_{c,p} \equiv -i\sigma_2 \chi_p^*$, we rewrite Eq. (4) in the form:

$$\psi_{c,p\varepsilon}^+ = \begin{pmatrix} \varphi_{c,p} \\ \frac{c\sigma \cdot p}{\varepsilon + mc^2} \varphi_{c,p} \end{pmatrix} e^{i(p\cdot r - \varepsilon t)/\hbar}, \tag{5}$$

which has the same form as the wavefunction of the analogous particle state with momentum $\mathbf{p}$ and energy ε.

The wavefunction of the antiparticle state with a known helicity, $\psi_{c,p\varepsilon\lambda}^+$, follows from the equation

$$\frac{1}{2}(\Sigma \cdot \mathbf{n}) \psi_{c,p\varepsilon\lambda}^+ = \lambda \psi_{c,p\varepsilon\lambda}^+, \quad \mathbf{n} = \frac{\mathbf{p}}{|p|},$$

from which it follows that

$$\frac{1}{2}(\sigma \cdot \mathbf{n}) \varphi_{c,p\lambda} = \lambda \varphi_{c,p\lambda}. \tag{6}$$

Taking into account the relation between the spinor $\varphi_{c,p}$, its antiparticle wavefunction, and the spinor, χ_p^-, in the solution, ψ_{pc}^-, of the Dirac equation ($\varphi_{c,p} = -i\sigma_2 \chi_p^{-*}$), we see that Eq. (6) is equivalent to

$$-\frac{1}{2}(\sigma \cdot \mathbf{n}) \chi_{p\lambda}^- = \lambda \chi_{p\lambda}^-.$$

To avoid any misunderstanding, we emphasize that the spinor χ_p^- corresponds to the solution of the Dirac equation with the momentum, $(-\mathbf{p})$ since the helicity operator here is $-\frac{1}{2}(\sigma \cdot \mathbf{n})$. It means that under charge conjugation, $\psi_c^+ = \hat{C} \psi^-$, the helicity remains invariant, while the momentum and energy change sign. This helicity property appears in "hole theory", where an antiparticle is interpreted as a hole among filled particle states with negative energy.

Problem 15.28

Prove that for a Dirac particle with mass $m = 0$, the operator (matrix) γ_5 commutes with the Hamiltonian of a free particle.

Find the eigenvalues of this operator, and determine their physical meaning.

Solution

The commutativity of the Hermitian operator $\gamma_5 = -\begin{pmatrix} 0 & 1 \\ 1 & 0 \end{pmatrix}$ with the Hamiltonian $\hat{H} = c\boldsymbol{\alpha} \cdot \hat{\mathbf{p}} = c \begin{pmatrix} 0 & \boldsymbol{\sigma} \\ \boldsymbol{\sigma} & 0 \end{pmatrix} \cdot \hat{\mathbf{p}}$ of a spinless Dirac particle is straightforward to verify: $[\gamma_5, \hat{H}] = 0$.

To find the physical meaning of the eigenvalue, μ, of the operator, γ_5, we find the general eigenfunctions $\psi_{\mathbf{p}\varepsilon\mu}$ of the commuting Hermitian operators $\hat{H}$, $\hat{\mathbf{p}}$, γ_5. These functions have the form:

$$\psi_{\mathbf{p}\varepsilon\mu} = \begin{pmatrix} \varphi_{\mathbf{p}\mu} \\ \frac{c}{\varepsilon}\boldsymbol{\sigma} \cdot \mathbf{p}\varphi_{\mathbf{p}\mu} \end{pmatrix} e^{i(\mathbf{p}\cdot\mathbf{r}-\varepsilon t)/\hbar}, \quad \varepsilon = \pm pc \tag{1}$$

(compare to Problem 15.21), and from the equation, $\gamma_5 \psi_{\mathbf{p}\varepsilon\mu} = \mu \psi_{\mathbf{p}\varepsilon\mu}$, it follows that

$$-\frac{c}{\varepsilon}\boldsymbol{\sigma} \cdot \mathbf{p}\varphi_{\mathbf{p}\mu} = \mu \varphi_{\mathbf{p}\mu}, \quad -\varphi_{\mathbf{p}\mu} = \mu \frac{c}{\varepsilon}\boldsymbol{\sigma} \cdot \mathbf{p}\varphi_{\mathbf{p}\mu}. \tag{2}$$

Hence $\mu^2 = 1$, i.e., the eigenvalues are equal to $\mu = \pm 1$ (which is evident *a priori*, since $\gamma_5^2 = 1$). Taking into account the relation (2) and the equality $(\boldsymbol{\sigma} \cdot \mathbf{p})^2 = p^2 = \varepsilon^2/c^2$, we see that the equations

$$\gamma_5 \psi_{\mathbf{p}\varepsilon\mu} = \mu \psi_{\mathbf{p}\varepsilon\mu} \quad \text{and} \quad (\boldsymbol{\Sigma} \cdot \mathbf{n})\psi_{\mathbf{p}\varepsilon\mu} = -\mu \frac{\varepsilon}{|\varepsilon|}\psi_{\mathbf{p}\varepsilon\mu}, \tag{3}$$

where $\mathbf{n} = \mathbf{p}/p$, are equivalent to each other. Hence, the physical meaning of the quantity $-\mu\varepsilon/|\varepsilon|$ is that it is twice the helicity of the particle, 2λ. For the positive-energy solutions of the Dirac equation, $\varepsilon = pc > 0$, we have $\mu = -2\lambda$, while for the negative-energy solutions (corresponding to an antiparticle), we have $\mu = 2\lambda$; see also Problems 15.27 and 15.29.

Problem 15.29

Prove that the matrix operators $\hat{P}_\pm = \frac{1}{2}(1 \pm \gamma_5)$, are projection operators.

For a Dirac particle with mass $m = 0$ these operators commute with the Hamiltonian. Onto what state do the operators $\hat{P}_\pm$ project?

Solution

Since $\hat{P}_\pm^2 = \hat{P}_\pm$, then the Hermitian operators $\hat{P}_\pm = \frac{1}{2}(1 + \gamma_5)$ are indeed projection operators (see Problem 1.31). Taking into account the relation between the eigenvalues of the operator γ_5 and helicity λ found in the previous problem, we see that the operator $\hat{P}_+$ acting on solutions of the Dirac equation with positive energy projects onto the states with helicity $\lambda_1 = -1/2$. Similarly, a negative energy particle (*i.e.* the anti-particle states) projects onto $\lambda_2 = +1/2$. Therefore, the operator $\hat{P}_-$ projects onto states with opposite values of helicity.

Note that the opposite signs of the particle and antiparticle helicity, corresponding to the eigenvalues μ of matrix γ_5, is due to the fact that under charge conjugation the equation $\gamma_5 \psi = \mu \psi$ becomes: $\gamma_5 \psi_c = -\mu \psi_c$, since $\hat{C}\gamma_5 = -\gamma_5 \hat{C}$.

Problem 15.30

The photon can be described quantum-mechanically through two vectors, $\boldsymbol{\mathcal{E}}(\mathbf{r}, t)$ and $\boldsymbol{\mathcal{H}}(\mathbf{r}, t)$, that satisfy[411] equations corresponding to the Maxwell equations in classical electrodynamics for a free electromagnetic field $\boldsymbol{\mathcal{E}}(\mathbf{r}, t)$, $\boldsymbol{\mathcal{H}}(\mathbf{r}, t)$ (*i.e.*, for electromagnetic waves in vacuum).

Prove that these equations can be introduced in a form analogous to the Dirac equation for two-component spinors (it is necessary to use the properties that the photon mass is zero, $m = 0$, and its spin is $s = 1$).

Solution

The Dirac equation for a two-component spinor of a massless particle, $m = 0$, has the form (see Problem 15.21, $\boldsymbol{\sigma} = 2\hat{\mathbf{s}} \equiv \hat{\mathbf{s}}/s$):

$$i\hbar \frac{\partial \varphi}{\partial t} = c\boldsymbol{\sigma} \cdot \hat{\mathbf{p}} \chi \equiv \frac{c}{s}\hat{\mathbf{s}}\hat{\mathbf{p}}\chi, \quad i\hbar \frac{\partial \chi}{\partial t} = c\boldsymbol{\sigma} \cdot \hat{\mathbf{p}} \varphi \equiv \frac{c}{s}\hat{\mathbf{s}}\hat{\mathbf{p}}\varphi. \tag{1}$$

The two spinors, φ and χ, describe the spin states of a $s = 1/2$ particle with respect to a pure spatial rotation of the coordinate system, and transform independently under such a transformation (but not under a general Lorentz transformation).

The natural generalization of Eq. (1) to the case of a particle with an arbitrary spin s (and mass $m = 0$) is to identify the equations for φ and χ with two $(2s + 1)$-component spin functions, with spin s, and take $\hat{\mathbf{s}}$ as the spin operator of the quantity s.

In the case $s = 1$, it is convenient to use a vector representation where the spin components are Cartesian coordinates of a vector (see the problems in Chapter 3, sec. 4), while the operators of the spin components are determined by the relations $\hat{s}_i a_k \equiv -i\varepsilon_{ikl} a_l$. Here,

$$\hat{\mathbf{s}}\hat{\mathbf{p}} a_k = \hat{s}_i \hat{p}_i a_k = -\hbar \varepsilon_{ikl} \frac{\partial}{\partial x_i} a_l \equiv \hbar (\boldsymbol{\nabla} \times \mathbf{a})_k,$$

---

[411] Let us emphasize that $\boldsymbol{\mathcal{E}}$ and $\boldsymbol{\mathcal{H}}$, as well as the vector potential $\mathbf{A}$, are complex quantities unlike the corresponding real functions used for the description of a classical electromagnetic field.

i.e., $(\hat{\mathbf{s}} \cdot \hat{\mathbf{p}})\mathbf{a} = \hbar \nabla \times \mathbf{a}$, and identifying in (1) φ and χ with the vectors $\boldsymbol{\mathcal{E}}$ and $i\boldsymbol{\mathcal{H}}$, we obtain the equation

$$\frac{1}{c}\frac{\partial}{\partial t}\boldsymbol{\mathcal{E}} = \nabla \times \boldsymbol{\mathcal{H}}, \quad -\frac{1}{c}\frac{\partial}{\partial t}\boldsymbol{\mathcal{H}} = \nabla \times \boldsymbol{\mathcal{E}},$$

which corresponds to the Maxwell equations in free space. The other two equations, $\nabla \cdot \boldsymbol{\mathcal{E}} = 0$ and $\nabla \cdot \boldsymbol{\mathcal{H}} = 0$, are additional conditions imposed on the vectors $\boldsymbol{\mathcal{E}}$ and $\boldsymbol{\mathcal{H}}$. In classical electrodynamics these lead to transverse electromagnetic waves, while in the quantum-mechanical case they correspond to the exclusion of photon states with zero helicity.

Problem 15.31

Find the non-relativistic limit (including terms up to order "1/c") for the charge and current densities of a Dirac particle in an external electromagnetic field.

Solution

The current and charge densities of a Dirac particle are given by Eq. (XV.7) (e is the charge of the particle):

$$\mathbf{j} = ec\psi^*\boldsymbol{\alpha}\psi \equiv iec\overline{\psi}\boldsymbol{\gamma}\psi, \quad \rho = e\psi^*\psi \equiv e\overline{\psi}\gamma_4\psi. \tag{1}$$

Note that these expressions are valid both for a free particle and a particle in an external electromagnetic field.

In the non-relativistic limit ($\varepsilon \approx mc^2$), for the spinor wavefunction $\psi = \begin{pmatrix} \varphi \\ \chi \end{pmatrix}$, the lower spinor χ corresponding to the equation

$$i\hbar\frac{\partial \chi}{\partial t} = c\boldsymbol{\sigma} \cdot \left(\hat{\mathbf{p}} - \frac{e}{c}\mathbf{A}\right)\varphi - mc^2\chi + eA_0\chi$$

is approximately equal to (since, $i\hbar\frac{\partial \chi}{\partial t} \approx mc^2\chi$)

$$\chi \approx \frac{1}{2mc}\boldsymbol{\sigma} \cdot \left(\hat{\mathbf{p}} - \frac{e}{c}\mathbf{A}\right)\varphi, \quad \text{i.e. } |\chi| \ll |\varphi|.$$

Hence, the particle wavefunction is described by the expression

$$\psi \approx \begin{pmatrix} \varphi \\ \frac{1}{2mc}\boldsymbol{\sigma} \cdot \left(\hat{\mathbf{p}} - \frac{e}{c}\mathbf{A}\right)\varphi \end{pmatrix}, \tag{2}$$

while the complex conjugate wavefunction is

$$\psi^* = \left(\varphi^*, \left(\frac{1}{2mc}\sigma\cdot\left(\hat{\mathbf{p}} - \frac{e}{c}\mathbf{A}\right)\varphi\right)^*\right)$$

$$= \left(\varphi^*, \frac{1}{2mc}\left(\hat{\mathbf{p}} - \frac{e}{c}\mathbf{A}\right)\varphi^*\sigma\right). \tag{3}$$

Substituting Eqs. (2) and (3) into Eq. (1), we find

$$\rho = e\psi^*\psi \approx e\varphi^*\varphi, \tag{4}$$

$$\mathbf{j} = ec\psi^* \begin{pmatrix} 0 & \sigma \\ \sigma & 0 \end{pmatrix} \psi = ec\left(\varphi^*, \frac{1}{2mc}\left(-\hat{\mathbf{p}} - \frac{e}{c}\mathbf{A}\right)\varphi^*\sigma\right)$$

$$\times \begin{pmatrix} \frac{\sigma}{2mc}\left(\sigma\cdot\left(\hat{\mathbf{p}} - \frac{e}{c}\mathbf{A}\right)\right)\varphi \\ \sigma\varphi \end{pmatrix}$$

$$= \frac{e}{2m}\left\{\varphi^*\sigma\left(\sigma\cdot\left(\hat{\mathbf{p}} - \frac{e}{c}\mathbf{A}\right)\right)\varphi - \left(\left(\hat{\mathbf{p}} + \frac{e}{c}\mathbf{A}\right)\cdot\varphi^*\sigma\right)\sigma\varphi\right\}, \tag{5}$$

up to terms of the order of $(1/c)^2$. Using the relation $\sigma_i\sigma_k = \delta_{ik} + i\varepsilon_{ikl}\sigma_l$ for the Pauli matrices, Eq. (5) can be simplified as follows:

$$j_i = \frac{e}{2m}\left\{\varphi^*\sigma_i\sigma_k\left(\hat{p}_k - \frac{e}{c}A_k\right)\varphi - \left(\left(\hat{p}_k + \frac{e}{c}A_k\right)\varphi^*\right)\sigma_k\sigma_i\varphi\right\}$$

$$= \frac{e}{2m}\left\{\varphi^*\hat{p}_i\varphi - (\hat{p}_i\varphi^*)\varphi - \frac{2e}{c}A_i\varphi^*\varphi + i\varepsilon_{ikl}[\varphi^*\sigma_l\hat{p}_k\varphi + (\hat{p}_k\varphi^*)\sigma_l\varphi]\right\},$$

and since

$$\varepsilon_{ikl}\left[\varphi^*\sigma_l\frac{\partial}{\partial x_k}\varphi + \left(\frac{\partial}{\partial x_k}\varphi^*\right)\sigma_l\varphi\right] = \varepsilon_{ikl}\frac{\partial}{\partial x_k}\varphi^*\sigma_l\varphi \equiv \{\nabla\times(\varphi^*\sigma\varphi)\}_i,$$

we obtain

$$\mathbf{j} = -\frac{ie\hbar}{2m}\{\varphi^*\nabla\varphi - (\nabla\varphi^*)\varphi\} - \frac{e^2}{mc}\mathbf{A}\varphi^*\varphi + \frac{e\hbar}{2m}\nabla\times(\varphi^*\sigma\varphi), \tag{6}$$

which coincides with Eqs. (VII.4–VII.6) of the non-relativistic theory for the current density of a particle with spin $s = 1/2$, charge e, and magnetic moment $\mu = e\hbar/2mc$.

Problem 15.32

The Hamiltonian of a particle with spin $s = 1/2$ in an external electromagnetic field has the form:

$$\hat{H} = c\boldsymbol{\alpha}\cdot\hat{\mathbf{p}} + mc^2\beta + \frac{i\kappa}{2}F_{\mu\nu}\gamma_\mu\gamma_\nu,$$

where κ is some parameter that characterizes the particle, and $F_{\mu\nu}$ is the electromagnetic field strength tensor.

Considering the non-relativistic limit (that is, keeping only terms of order "$1/c$") of the wave equation,[412] $i\hbar\frac{\partial}{\partial t}\psi = \hat{H}\psi$, determine the physical meaning of the parameter, κ. That is find its connection to the electromagnetic properties of the particle. Consider specifically the case of a charged Dirac particle – electron or muon, whose Hamiltonians have the form:

$$\hat{H} = c\boldsymbol{\alpha} \cdot \left(\hat{\mathbf{p}} - \frac{e}{c}\mathbf{A}\right) + mc^2\beta + eA_0.$$

Solution

Taking into account the form of Dirac matrices (XV.5) and the field strength tensor of the electromagnetic field

$$F_{\mu\nu} = \begin{pmatrix} 0 & \mathcal{H}_z & -\mathcal{H}_y & -i\mathcal{E}_x \\ -\mathcal{H}_z & 0 & \mathcal{H}_x & -i\mathcal{E}_y \\ \mathcal{H}_y & -\mathcal{H}_x & 0 & -i\mathcal{E}_z \\ i\mathcal{E}_x & i\mathcal{E}_y & i\mathcal{E}_z & 0 \end{pmatrix}, \quad \begin{array}{l} F_{4i} = -F_{i4} = i\mathcal{E}_i, \\ F_{ik} = \varepsilon_{ikl}\mathcal{H}_l, \\ i,l,k = 1,2,3, \end{array}$$

the Hamiltonian considered could be transformed to the form:

$$\hat{H} = c\boldsymbol{\alpha} \cdot \hat{\mathbf{p}} - \kappa\beta\sum\mathcal{H} + i\kappa\beta\alpha\mathcal{E} + mc^2\beta. \tag{1}$$

Here, the wave equation $i\hbar\frac{\partial}{\partial t}\psi = \hat{H}\psi$ leads to the following equations for the two-component spinors, φ and χ, of the wavefunction $\psi = \begin{pmatrix} \varphi \\ \chi \end{pmatrix}$:

$$\begin{aligned} i\hbar\frac{\partial}{\partial t}\varphi &= c\boldsymbol{\sigma}\cdot\hat{\mathbf{p}}\chi + mc^2\varphi + i\kappa\boldsymbol{\sigma}\cdot\boldsymbol{\mathcal{E}}\chi - \kappa\boldsymbol{\sigma}\cdot\boldsymbol{\mathcal{H}}\varphi, \\ i\hbar\frac{\partial}{\partial t}\chi &= c\boldsymbol{\sigma}\cdot\hat{\mathbf{p}}\varphi - mc^2\chi - i\kappa\boldsymbol{\sigma}\cdot\boldsymbol{\mathcal{E}}\varphi + \kappa\boldsymbol{\sigma}\cdot\boldsymbol{\mathcal{H}}\chi. \end{aligned} \tag{2}$$

For the transition to the non-relativistic limit, where the particle's energy is $\varepsilon \approx mc^2$, we should extract the factor $e^{-imc^2t/\hbar}$ from the wavefunction, i.e. write it in the form $\psi = e^{-imc^2t/\hbar}\begin{pmatrix}\tilde{\varphi}\\\tilde{\chi}\end{pmatrix}$ and use the inequality

$$\left|i\hbar\frac{\partial}{\partial t}\tilde{\psi}\right| \sim |E\tilde{\psi}| \ll mc^2|\tilde{\psi}|, \quad \tilde{\psi} = \begin{pmatrix}\tilde{\varphi}\\\tilde{\chi}\end{pmatrix} \tag{3}$$

in the expression

$$i\hbar\frac{\partial}{\partial t}\psi = e^{-imc^2t/\hbar}\left[mc^2\tilde{\psi} + i\hbar\frac{\partial}{\partial t}\tilde{\psi}\right],$$

---

[412] This equation can be written in a manifestly relativistically-invariant form:

$$\left(ic\hat{p} + \frac{i\kappa}{2}F_{\mu\nu}\gamma_\mu\gamma_\nu + mc^2\right)\psi = 0 \quad \left(\hat{p} \equiv \hat{p}_\mu\hat{\gamma}_\mu = \hat{\mathbf{p}}\cdot\boldsymbol{\gamma} - \frac{\hbar}{c}\gamma_4\frac{\partial}{\partial t}\right).$$

where E is the non-relativistic energy. Here the second expression of Eq. (2) takes the form:

$$2mc^2\tilde{\chi} - \kappa\boldsymbol{\sigma}\cdot\boldsymbol{\mathcal{H}}\tilde{\chi} + i\hbar\frac{\partial}{\partial t}\tilde{\chi} = (c\boldsymbol{\sigma}\cdot\hat{\mathbf{p}} - i\kappa\boldsymbol{\sigma}\cdot\boldsymbol{\mathcal{E}})\tilde{\varphi}. \tag{4}$$

From the relations given, with inequality $|\kappa\mathcal{H}| \ll mc^2$ taken into account, it follows that

$$\tilde{\chi} \approx \frac{1}{2mc}\left(\boldsymbol{\sigma}\cdot\hat{\mathbf{p}} - \frac{i\kappa}{c}\boldsymbol{\sigma}\cdot\boldsymbol{\mathcal{E}}\right)\tilde{\varphi}. \tag{5}$$

Let us note that $|\tilde{\chi}| \ll |\tilde{\varphi}|$, as in the case of a free, non-relativistic particle.

Then, substituting Eq. (5) into the first of equations (2) (having previously extracted the factor $e^{-imc^2 t/\hbar}$ from the spinors φ and χ), we obtain

$$i\hbar\frac{\partial}{\partial t}\tilde{\varphi} = \frac{1}{2m}\left(\boldsymbol{\sigma}\cdot\hat{\mathbf{p}} + \frac{i\kappa}{c}\boldsymbol{\sigma}\cdot\boldsymbol{\mathcal{E}}\right)\left(\boldsymbol{\sigma}\cdot\hat{\mathbf{p}} - \frac{i\kappa}{c\boldsymbol{\sigma}\cdot\boldsymbol{\mathcal{E}}}\right)\tilde{\varphi} - \kappa\boldsymbol{\sigma}\cdot\boldsymbol{\mathcal{H}}\tilde{\varphi}. \tag{6}$$

Writing $\boldsymbol{\sigma}\cdot\hat{\mathbf{p}} = \sigma_i\hat{p}_i$, $\boldsymbol{\sigma}\cdot\boldsymbol{\mathcal{E}} = \sigma_k\mathcal{E}_k$ and using the relation $\sigma_i\sigma_k = \delta_{ik} + i\varepsilon_{ikl}\sigma_l$ for the Pauli matrices, we find

$$(\boldsymbol{\sigma}\cdot\hat{\mathbf{p}})(\boldsymbol{\sigma}\cdot\boldsymbol{\mathcal{E}}) - (\boldsymbol{\sigma}\cdot\boldsymbol{\mathcal{E}})(\boldsymbol{\sigma}\cdot\hat{\mathbf{p}}) = (\hat{\mathbf{p}}\cdot\boldsymbol{\mathcal{E}}) - (\boldsymbol{\mathcal{E}}\cdot\hat{\mathbf{p}}) + i[\hat{\mathbf{p}}\times\boldsymbol{\mathcal{E}}]\cdot\boldsymbol{\sigma} - i[\boldsymbol{\mathcal{E}}\times\hat{\mathbf{p}}]\boldsymbol{\sigma},$$

where

$$(\hat{\mathbf{p}}\cdot\boldsymbol{\mathcal{E}}) - (\boldsymbol{\mathcal{E}}\cdot\hat{\mathbf{p}}) = -i\hbar\boldsymbol{\nabla}\cdot\boldsymbol{\mathcal{E}}, \quad [\hat{\mathbf{p}}\times\boldsymbol{\mathcal{E}}] - [\boldsymbol{\mathcal{E}}\times\hat{\mathbf{p}}] = -i\hbar\boldsymbol{\nabla}\times\boldsymbol{\mathcal{E}} - 2[\boldsymbol{\mathcal{E}}\times\hat{\mathbf{p}}] \approx -2[\boldsymbol{\mathcal{E}}\times\hat{\mathbf{p}}].$$

Note that since $\boldsymbol{\nabla}\times\boldsymbol{\mathcal{E}} = -\partial\boldsymbol{\mathcal{H}}/c\partial t$, then the term $\boldsymbol{\nabla}\times\boldsymbol{\mathcal{E}}$ can be omitted to order $1/c$ (in the stationary case it vanishes). Using these relations, the equality $(\boldsymbol{\sigma}\cdot\hat{\mathbf{p}})^2 = \hat{\mathbf{p}}^2$, and neglecting the term $\propto (\boldsymbol{\mathcal{E}}/c)^2$, we transform Eq. (6) to the form:

$$i\hbar\frac{\partial}{\partial t}\tilde{\varphi} = \frac{\hat{\mathbf{p}}^2}{2m}\tilde{\varphi} - \kappa\boldsymbol{\sigma}\cdot\boldsymbol{\mathcal{H}}\tilde{\varphi} + \frac{\kappa}{mc}\left(-\frac{\hbar}{2}\boldsymbol{\nabla}\cdot\boldsymbol{\mathcal{E}} - [\boldsymbol{\mathcal{E}}\times\hat{\mathbf{p}}]\cdot\boldsymbol{\sigma}\right)\tilde{\varphi}. \tag{7}$$

Neglecting the "small" spinor, $\tilde{\chi}$, in the wavefunction, $\tilde{\psi}$, this equation reduces to the Schrödinger equation with the Hamiltonian:

$$\hat{H} = \frac{\hat{\mathbf{p}}^2}{2m} - \kappa\boldsymbol{\sigma}\cdot\boldsymbol{\mathcal{H}} + \frac{\kappa}{mc}\left(-\frac{\hbar}{2}\boldsymbol{\nabla}\cdot\boldsymbol{\mathcal{E}} - [\boldsymbol{\mathcal{E}}\times\hat{\mathbf{p}}]\cdot\boldsymbol{\sigma}\right), \tag{8}$$

Compare to the Pauli Hamiltonian (VII.1). The absence of the term eA_0 in $\hat{H}$ means that the particle described by it is charge-neutral (A_0 is the scalar potential of an external electrostatic field), while the term $-\kappa\boldsymbol{\sigma}\cdot\boldsymbol{\mathcal{H}}$ shows that the particle has a magnetic moment equal to $\mu \equiv \kappa$.

The third term in Eq. (8) describes spin-orbital interaction. Here, the term[413] $-\frac{\kappa}{mc}[\boldsymbol{\mathcal{E}}\times\hat{\mathbf{p}}]\cdot\boldsymbol{\sigma}$ is a natural quantum-mechanical generalization of the energy

---

[413] The term $-\frac{\kappa\hbar}{2mc}\boldsymbol{\nabla}\cdot\boldsymbol{\mathcal{E}}$, which is non-zero at points of space with finite charge density, produces an external electric field, $\boldsymbol{\nabla}\cdot\boldsymbol{\mathcal{E}} = 4\pi\rho$, that has no direct classical interpretation.

describing the interaction between a moving classical magnetic dipole and an electrostatic field, as discussed in Problem 13.60.

The Hamiltonian (1) in the non-relativistic limit (8) is used to describe a neutron in an electromagnetic field. The expression $\frac{1}{2}\mu'\beta\gamma_\tau\gamma_\nu F_{\tau\nu}$ is also used for the description of the interaction between an electromagnetic field and the *anomalous* magnetic moment μ' of a charged particle with spin $s = 1/2$. Here, the interaction between the particle of charge e and the field includes a *normal* part of the magnetic field, equal to $e\hbar/2mc$, and that described by the expression $(-e\boldsymbol{\alpha} \cdot \mathbf{A} + eA_0)$. Hence, the relativistic wave equation for such a particle in the electromagnetic field has the form:

$$i\hbar\frac{\partial}{\partial t}\psi = \left\{c\boldsymbol{\alpha} \cdot \left(\hat{\mathbf{p}} - \frac{e}{c}\mathbf{A}\right) + mc^2\beta + eA_0 + \frac{i}{2}\mu'\beta\gamma_\tau\gamma_\nu F_{\tau\nu}\right\}\psi.$$

Note here that for a particle with spin $s = 1/2$, charge e, and magnetic moment μ, the "separation" of the magnetic moment into the normal and anomalous parts is determined by the relations: $\mu = \mu_{norm} + \mu_{anom}$, $\mu_{norm} = e\hbar/2mc$, and $\mu_{anom} = \mu - e\hbar/2mc$.

Problem 15.33

Find the energy spectrum of a charged Dirac particle in a uniform magnetic field.

Solution

The energy spectrum and corresponding wavefunction are determined from the Dirac equation in the magnetic field (e is the particle charge):

$$\left(c\boldsymbol{\alpha}\left(\hat{\mathbf{p}} - \frac{e}{c}\mathbf{A}\right) + mc^2\beta\right)u = \varepsilon u, \quad \psi_\varepsilon(\mathbf{r},t) = ue^{-\frac{i\varepsilon t}{\hbar}} = \begin{pmatrix}\varphi \\ \chi\end{pmatrix}e^{-\frac{i\varepsilon t}{\hbar}},$$

or, for the two-component spinors,

$$(\varepsilon - mc^2)\varphi = c\boldsymbol{\sigma}\left(\hat{\mathbf{p}} - \frac{e}{c}\mathbf{A}\right)\chi, \quad (\varepsilon + mc^2)\chi = c\boldsymbol{\sigma}\left(\hat{\mathbf{p}} - \frac{e}{c}\mathbf{A}\right)\varepsilon. \tag{1}$$

Eliminating the spinor, χ, from this system of equations, we obtain

$$(\varepsilon^2 - m^2c^4)\varphi = c^2\left(\boldsymbol{\sigma} \cdot \left(\hat{\mathbf{p}} - \frac{e}{c}\mathbf{A}\right)\right)^2\varphi. \tag{2}$$

Using the relation

$$\left(\boldsymbol{\sigma} \cdot \left(\hat{\mathbf{p}} - \frac{e}{c}\mathbf{A}\right)\right)^2 = \left(\hat{\mathbf{p}} - \frac{e}{c}\mathbf{A}\right)^2 - \frac{e\hbar}{c}\boldsymbol{\sigma} \cdot \mathcal{H}, \quad \mathcal{H} = \nabla \times \mathbf{A}$$

(see, for example, Problem 7.10), Eq. (2) can be cast into the following form:

$$\left\{\frac{1}{2m}\left(\hat{\mathbf{p}} - \frac{e}{c}\mathbf{A}\right)^2 - \frac{e\hbar}{2mc}\boldsymbol{\sigma} \cdot \mathcal{H}\right\}\varphi = \frac{\varepsilon^2 - m^2c^4}{2mc^2}\varphi. \tag{3}$$

The substitution, $(\varepsilon^2 - m^2c^4)/2mc^2 \to E$, makes Eq. (3) identical to the Pauli equation for a particle with spin $s = 1/2$, charge e, and magnetic moment $\mu = e\hbar/2mc$. Note that the Pauli equation in a uniform magnetic field $\mathcal{H}_0$ was solved in Problem 7.9. Choosing the vector potential as $\mathbf{A} = (0, \mathcal{H}_0 x, 0)$, the solution to Eq. (3) has the form:

$$\varepsilon^2_{np_z\sigma_z} - M^2c^4 = 2mc^2\left[\hbar\omega_0\left(n + \frac{1}{2}\right) - \frac{e\hbar\mathcal{H}_0}{2mc}\sigma_z + \frac{p_z^2}{2m}\right], \quad n = 0, 1, \ldots,$$

$$\varphi_{np_y p_z \sigma_z} = Ce^{i\{p_y y + p_z z\}/\hbar}\exp\left\{-\frac{1}{2a^2}\left(x - \frac{cp_y}{e\mathcal{H}_0}\right)^2\right\}\mathcal{H}_n \times$$

$$\left(\frac{1}{a}\left(x - \frac{cp_y}{e\mathcal{H}_0}\right)\right)\varphi_{\sigma_z}, \quad (4)$$

$$\omega_0 = \frac{|e|\mathcal{H}_0}{mc}, \quad a = \sqrt{\frac{\hbar c}{|e|\mathcal{H}_0}},$$

where the constant spinor, φ_{σ_z}, is an eigenfunction of the operator, σ_z, corresponding to the eigenvalues $\sigma_z = \pm 1$ (the z-axis is directed along the magnetic field).

The second expression in Eq. (1) determines spinor $\chi_{np_y p_z \sigma_z}$, and hence the wavefunction $\psi_{np_y p_z \sigma_z}$.

Recall that in non-relativistic theory the quantum number $s_z = \sigma_z/2$ determines the particle spin projection onto the z axis. In the relativistic case, σ_z no longer has this direct interpretation, because the spinor, $\chi_{np_y p_z \sigma_z}$, is no longer an eigenfunction of the operator σ_z, and neither the wavefunction $\psi_{np_y p_z \sigma_z}$ is an eigenfunction of the operator, $\hat{s}_z = \frac{1}{2}\hat{\Sigma}_z$. Nevertheless, Eq. (4) shows that the level degeneracy of the transverse motion present in non-relativistic theory (see Problem 7.9), is preserved in the relativistic case as well.

Eq. (4) gives two values of energy of the opposite sign: $\varepsilon \geq mc^2$ corresponds the particle energy spectrum, while the other, $\varepsilon \leq -mc^2$, corresponds to antiparticles with a positive energy (see Problem 15.27). Here, the energy spectra for a particle and antiparticle are the same (compare to the case of a spinless particle in a magnetic field, considered in Problem 15.11).

Problem 15.34

In first-order perturbation theory, find the differential scattering cross-section of a Dirac particle in a Coulomb field of a nucleus with charge Ze. Consider the limiting case of an infinitely heavy nucleus.

Solution

The Hamiltonian of a Dirac particle in an external electrostatic field has the form (below, e_1 is the particle charge):

$$\hat{H} = c\boldsymbol{\alpha} \cdot \hat{\mathbf{p}} + mc^2\beta + e_1 A_0(r) \equiv \hat{H}_0 + \hat{V}, \quad \hat{V} = e_1 A_0 = \frac{Zee_1}{r}.$$

Let us calculate the differential cross-section according to the pertubative equation for transitions between continuous free-particle states.[414]

The initial state with a definite momentum, $\mathbf{p}_1$, is given by (see also Problem 15.22)

$$\psi_i = \sqrt{\frac{\varepsilon + mc^2}{2\varepsilon}} \begin{pmatrix} \varphi_i \\ \frac{c\boldsymbol{\sigma}\cdot\mathbf{p}_1}{\varepsilon+mc^2}\varphi_i \end{pmatrix} \sqrt{\frac{1}{v}} e^{i(\mathbf{p}_1 \mathbf{r} - \varepsilon t)/\hbar}, \tag{1}$$

$$\mathbf{j} = c\psi_i^* \boldsymbol{\alpha} \psi_i = \frac{\mathbf{v}_1}{v}, \quad j = 1,$$

where ε and v are the particle energy and velocity. Similarly, the final wavefunction with momentum $\mathbf{p}_2$ reads:

$$\psi_f = \sqrt{\frac{\varepsilon + mc^2}{2\varepsilon}} \begin{pmatrix} \varphi_f \\ \frac{c\boldsymbol{\sigma}\cdot\mathbf{p}_2}{\varepsilon+mc^2}\varphi_f \end{pmatrix} \sqrt{\frac{1}{v}} e^{i(\mathbf{p}_2 \cdot \mathbf{r} - \varepsilon t)/\hbar}, \quad \rho = \psi_f^* \psi_f = 1. \tag{2}$$

Note that the spinors, $\varphi_{i,f}$, are normalized to unity: $|\varphi_{i,f}|^2 = 1$ and the particle energies in the initial and finite states are the same. The differential transition probability per unit time can be found from the perturbation theory formula:

$$d\sigma = dw = \frac{2\pi}{\hbar} |V_{if}|^2 d\rho_f. \tag{3}$$

The density of final states is given by:

$$d\rho_f = \int \delta(\varepsilon_f - \varepsilon_i) \frac{d^3 p_f}{(2\pi\hbar)^3} \equiv \int \delta(\varepsilon - \varepsilon_2) \frac{p_2 \varepsilon_2 d\varepsilon_2 d\Omega}{(2\pi\hbar)^3 c^2} = \frac{p\varepsilon \, d\Omega}{(2\pi\hbar)^3 c^2}. \tag{4}$$

The perturbation matrix element has the form:

$$V_{if} = \int \psi_f^* \hat{V} \psi_i d^3 r = \frac{Zee_1(\varepsilon + mc^2)}{2\varepsilon\sqrt{v}} \times$$

$$\int \frac{e^{-i\mathbf{q}\cdot\mathbf{r}}}{r} d^3 r \varphi_f^* \left[1 + \frac{c^2 (\boldsymbol{\sigma} \cdot \mathbf{p}_2)(\boldsymbol{\sigma} \cdot \mathbf{p}_1)}{(\varepsilon + mc^2)^2}\right] \varphi_i, \tag{5}$$

where, as usual, $\hbar\mathbf{q} = \mathbf{p}_2 - \mathbf{p}_1$. Introducing a scattering angle, θ, we find

$$(\boldsymbol{\sigma} \cdot \mathbf{p}_2)(\boldsymbol{\sigma} \cdot \mathbf{p}_1) = p_{2i} p_{3k} \sigma_i \sigma_k = p_{2i} p_{1k}(\delta_{ik} + i\varepsilon_{ikl}\sigma_l) =$$

$$\mathbf{p}_1 \cdot \mathbf{p}_2 + i[\mathbf{p}_1 \times \mathbf{p}_2] \cdot \boldsymbol{\sigma} = p^2 \cos\theta - ip^2 \sin\theta \sigma\nu;$$

$$\nu = \frac{[\mathbf{p}_1 \times \mathbf{p}_2]}{|[\mathbf{p}_1 \times \mathbf{p}_2]|}, \quad \nu^2 = 1, \quad \int e^{-i\mathbf{q}\cdot\mathbf{r}} \frac{d^3 r}{r} = \frac{4\pi}{q^2} = \frac{\pi\hbar^2}{p^2 \sin^2(\theta/2)},$$

---

[414] For a more consistent calculation of the differential cross-section based on the Born approximation, see Problem 15.37.

and transform Eq. (5) to the more convenient form:

$$V_{if} = \frac{\pi Z e e_1 \hbar^2}{2\varepsilon p^2 \sqrt{v}(\varepsilon + mc^2)\sin^2(\theta/2)} \varphi_f^* \{(\varepsilon + mc^2)^2 + p^2 c^2 \cos\theta - $$
$$- ip^2 c^2 \sin\theta \boldsymbol{\sigma} \cdot \boldsymbol{\nu}\} \varphi_i. \tag{6}$$

Eqs. (3), (4), and (6) determine the differential scattering cross-section as a function of the particle energy, scattering angle, and the spin states before and after scattering, described by the spinors, $\varphi_{i,f}$. Averaging the differential cross-section over the particle spin states in the incident beam (assuming it to be non-polarized), and summing over the independent spin states of the scattered particle, and using the following relation from non-relativistic theory,

$$\overline{\left|\varphi_f^*(f_1 + if_2 \boldsymbol{\sigma} \cdot \boldsymbol{\nu})\varphi_i\right|^2} = |f_1|^2 + |f_2|^2,$$

we find

$$\overline{|\varphi_f^*\{(\varepsilon + mc^2)^2 + p^2 c^2 \cos\theta - ip^2 c^2 \sin\theta \boldsymbol{\sigma} \cdot \boldsymbol{\nu}\}\varphi_i|^2} = 4\varepsilon^2 (\varepsilon + mc^2)^2 \times $$
$$\left(1 - \frac{v^2}{c^2}\sin^2\frac{\theta}{2}\right).$$

The final expression for differential scattering cross-section then reads:

$$d\bar{\sigma} = \frac{(Zee_1)^2}{4p^2 v^2 \sin^4(\theta/2)}\left(1 - \frac{v^2}{c^2}\sin^2\frac{\theta}{2}\right)d\Omega \tag{7}$$

(compare to the scattering of a spinless particle, considered in Problem 15.17). In the non-relativistic limit, $v/c \ll 1$, $p \approx mv$, and our result reproduces the Rutherford equation.

Finally, we mention that the regime of validity of this result is given by $|Zee_1| \ll \hbar v$.

Problem 15.35

To first order in perturbation theory, find the energy-dependence of the scattering cross-section $\sigma(\varepsilon)$ of a charged Dirac particle in an external electric field, $A_0(r)$, in the limit, $\varepsilon \to \infty$. Compare your result with Problem 15.18.

Solution

The expression for the differential scattering cross-section of a non-polarized particle,

$$d\bar{\sigma} = \frac{e^2 p^2 \tilde{A}_0^2(q)}{4\pi^2 \hbar^4 v^2}\left(1 - \frac{v^2}{c^2}\sin^2\frac{\theta}{2}\right)d\Omega, \tag{1}$$

can be obtained directly from Eq. (7) of the previous problem, if we make the substitution,

$$\frac{\pi Z e e_1 \hbar^2}{p^2 \sin^2(\theta/2)} = \int \frac{Z e e_1}{r} e^{-i\mathbf{q}\cdot\mathbf{r}} dV \to e \int A_0(r) e^{-i\mathbf{q}\cdot\mathbf{r}} dV \equiv e\tilde{A}_0(q).$$

Using the relation $d\Omega = (\pi\hbar^2/p^2)dq^2$, we find the total scattering cross-section as follows:

$$\bar{\sigma}(\varepsilon) = \frac{e^2}{4\pi\hbar^2 v^2} \int_0^{4p^2/\hbar^2} \tilde{A}_0^2(q) \left(1 - \frac{\hbar^2 v^2}{4p^2 c^2} q^2\right) dq^2. \tag{2}$$

In the ultra-relativistic limit, we have $p \approx \varepsilon/c$ and $v \approx c$. Noting that the integrand in (2) is dominated by the region $q^2 \leq R^{-2}$, where R is the characteristic radius of the potential $A_0(r)$, we observe that in the limit, $\varepsilon \to \infty$, the scattering cross-section approaches a constant value:

$$\bar{\sigma}(\varepsilon) \underset{\varepsilon \to \infty}{\to} \sigma_0 = \frac{e^2}{4\pi\hbar^2 c^2} \int_0^\infty \tilde{A}_0^2(q) dq^2. \tag{3}$$

This result coincides with the scattering cross-section of a charged spinless particle considered in Problem 15.18. Let us note that the integral in Eq. (3) diverges at the lower limit, $(q^2 \to 0)$, if the potential decreases as $|A_0(r)| < B/r^2$, and the total scattering cross-section diverges as well. This result is analogous to the non-relativistic case.

Problem 15.36

Find the Green functions, $\hat{G}^{\pm}_{\varepsilon,\alpha\beta}(\mathbf{r},\mathbf{r}')$, of the stationary Dirac equation for a free particle with energy, $\varepsilon \geq mc^2$. The Green functions satisfy the following equation:

$$(\hat{H} - \varepsilon)\hat{G}_\varepsilon \equiv (-i\hbar c\boldsymbol{\alpha}\cdot\boldsymbol{\nabla} + mc^2\beta - \varepsilon)\hat{G}_\varepsilon = \delta(\mathbf{r}-\mathbf{r}'),$$

and have the following asymptotic behavior, as $r \to \infty$:

$$\hat{G}^{\pm}_\varepsilon \propto \frac{1}{r} e^{\pm ikr}, \quad k = \sqrt{\frac{\varepsilon^2 - m^2 c^4}{\hbar^2 c^2}}.$$

Find also the Green functions, $\hat{f}^{\pm}_\varepsilon$, of the Dirac equation written in the symmetric form:

$$(ic\hat{p} + mc^2)\psi_\varepsilon = 0, \quad \hat{p} \equiv -i\hbar\boldsymbol{\gamma}\cdot\boldsymbol{\nabla} + \frac{i\varepsilon}{c}\gamma_4.$$

Solution

The desired Green functions can be expressed in terms of the corresponding Green functions of a free non-relativistic particle, $g^{\pm}(r, r')$. The latter satisfy the following equation:

$$(-\Delta - k^2)g^{\pm}(\mathbf{r}, \mathbf{r}') = \delta(\mathbf{r} - \mathbf{r}'), \quad g^{\pm}(\mathbf{r}, \mathbf{r}') = \frac{\exp\{\pm ik|\mathbf{r} - \mathbf{r}'|\}}{4\pi|\mathbf{r} - \mathbf{r}'|}.$$

Using the relation,

$$-\hbar^2 c^2 \Delta - \varepsilon^2 + m^2 c^4 \equiv (c\boldsymbol{\alpha} \cdot \hat{\mathbf{p}} + mc^2\beta - \varepsilon)(c\boldsymbol{\alpha} \cdot \hat{\mathbf{p}} + mc^2\beta + \varepsilon),$$

we have (note that $\varepsilon^2 = \hbar^2 k^2 c^2 + m^2 c^4$)

$$(-\hbar^2 c^2 \Delta - \hbar^2 c^2 k^2)\frac{\exp\{\pm ik|\mathbf{r} - \mathbf{r}'|\}}{4\pi\hbar^2 c^2 |\mathbf{r} - \mathbf{r}'|} \equiv$$

$$(c\boldsymbol{\alpha} \cdot \hat{\mathbf{p}} + mc^2\beta - \varepsilon)(c\boldsymbol{\alpha} \cdot \hat{\mathbf{p}} + mc^2\beta + \varepsilon)\frac{\exp\{\pm ik|\mathbf{r} - \mathbf{r}'|\}}{4\pi\hbar^2 c^2 |\mathbf{r} - \mathbf{r}'|} = \delta(\mathbf{r} - \mathbf{r}').$$

Hence, the Green function of a free Dirac particle follows:

$$\hat{G}^{\pm}_{\varepsilon}(\mathbf{r}, \mathbf{r}') = \frac{1}{4\pi\hbar^2 c^2}(c\boldsymbol{\alpha} \cdot \hat{\mathbf{p}} + mc^2\beta + \varepsilon)\frac{\exp\{\pm ik|\mathbf{r} - \mathbf{r}'|\}}{|\mathbf{r} - \mathbf{r}'|}. \tag{1}$$

In bi-spinor notations, it takes the form:

$$\hat{G}^{\pm}_{\varepsilon\alpha\beta} = \frac{1}{4\pi\hbar^2 c^2}(-i\hbar c\boldsymbol{\alpha}\nabla + mc^2\beta + \varepsilon)_{\alpha\beta}\frac{\exp\{\pm ik|\mathbf{r} - \mathbf{r}'|\}}{|\mathbf{r} - \mathbf{r}'|}.$$

Analogously, using the relation

$$-\hbar^2 c^2 \Delta - \varepsilon^2 + m^2 c^4 \equiv (ic\hat{p} + mc^2)(-ic\hat{p} + mc^2),$$

we find the "symmetric form" of the Green function

$$f^{\pm}_{\varepsilon}(\mathbf{r}, \mathbf{r}') = \frac{-ic\hat{p} + mc^2}{4\pi\hbar^2 c^2}\frac{\exp\{\pm ik|\mathbf{r} - \mathbf{r}'|\}}{|\mathbf{r} - \mathbf{r}'|} =$$

$$\frac{-\hbar c\gamma\nabla + \varepsilon\gamma_4 + mc^2}{4\pi\hbar^2 c^2}\frac{\exp\{\pm ik|\mathbf{r} - \mathbf{r}'|\}}{|\mathbf{r} - \mathbf{r}'|},$$

that corresponds to the equation

$$(ic\hat{p} + mc^2)\hat{f}^{\pm}_{\varepsilon} = \delta(\mathbf{r} - \mathbf{r}').$$

Problem 15.37

In the Born approximation, find the scattering amplitude of a Dirac particle in a constant external electromagnetic field.

Apply your results to the case of the electrostatic field, $A_0 = Ze/r$, and compare them with Problem 15.34.

Solution

Using the Green function, $\hat{G}_\varepsilon^+(\mathbf{r}, \mathbf{r}')$, from the previous problem, we can present the Dirac equation for a particle in an external electromagnetic field as

$$(c\boldsymbol{\alpha} \cdot \hat{\mathbf{p}} + mc^2\beta - \varepsilon)\psi(\mathbf{r}) = e[\boldsymbol{\alpha} \cdot \mathbf{A}(\mathbf{r}) - A_0(\mathbf{r})]\psi(\mathbf{r}). \tag{1}$$

The integral equation, corresponding to the scattering problem involving a particle with momentum $\mathbf{p}_1 = \hbar\mathbf{k}_1$, reads:

$$\psi_{\mathbf{p}_1}^{(+)}(r) = u_1(\mathbf{p}_1)e^{i\mathbf{k}_2 \cdot \mathbf{r}} + \frac{e}{4\pi\hbar^2 c^2}(c\boldsymbol{\alpha} \cdot \hat{\mathbf{p}} + mc^2\beta + \varepsilon) \times$$

$$\int \frac{e^{ik|\mathbf{r}-\mathbf{r}'|}}{|\mathbf{r}-\mathbf{r}'|}(\boldsymbol{\alpha} \cdot \mathbf{A}(\mathbf{r}') - A_0(\mathbf{r}'))\psi_{\mathbf{p}_1}^{(+)}(\mathbf{r}')dV'. \tag{2}$$

At large distances, $r \to \infty$, the second term in the right-hand side of this equation becomes

$$\frac{e}{4\pi\hbar^2 c^2}(c\boldsymbol{\alpha} \cdot \hat{\mathbf{p}} + mc^2\beta + \varepsilon)\frac{e^{ikr}}{r}\int e^{-ik\mathbf{n}\cdot\mathbf{r}'}(\boldsymbol{\alpha} \cdot \mathbf{A}(\mathbf{r}') - A_0(\mathbf{r}'))\psi_{\mathbf{p}_1}^{(+)}(\mathbf{r}')dV',$$

where $\mathbf{n} = \mathbf{r}/r$, $\mathbf{p}_2 = \hbar\mathbf{k}_2 = \hbar k\mathbf{n}$ is the momentum of the scattered particle (see Eqs. (XIII.1)–(XIII.6)). The asymptotic behavior of the wavefunction (2) at large distances is

$$\psi_{\mathbf{p}_1}^{(+)}(r) \approx u_1(\mathbf{p}_1)e^{i\mathbf{k}_1 \cdot \mathbf{r}} + F\frac{e^{ikr}}{r}, \quad r \to \infty, \tag{3}$$

where the bi-spinor,

$$F = \frac{e}{4\pi\hbar^2 c^2}(c\boldsymbol{\alpha} \cdot \mathbf{p}_2 + mc^2\beta + \varepsilon)\int e^{-i\mathbf{k}_2 \cdot \mathbf{r}'}(\boldsymbol{\alpha} \cdot \mathbf{A}(\mathbf{r}') - A_0(\mathbf{r}'))\psi_{\mathbf{p}_1}^{(+)}(\mathbf{r}')dV', \tag{4}$$

is the amplitude of the scattered wave.

In the Born approximation, instead of the asymptotically exact function, $\psi_{\mathbf{p}_1}^{(+)}(\mathbf{r})$ in (4), we should use the unperturbed value in an external field $u_1(\mathbf{p}_1)e^{i\mathbf{k}_1 \cdot \mathbf{r}}$, where

$$F \approx F_B \equiv \hat{F}_B u_1(\mathbf{p}_1), \tag{5}$$

$$\hat{F}_B = \frac{e}{4\pi\hbar^2 c^2}(c\boldsymbol{\alpha} \cdot \mathbf{p}_2 + mc^2\beta + \varepsilon)\int e^{-i\mathbf{q}\cdot\mathbf{r}'}(\boldsymbol{\alpha} \cdot \mathbf{A}(\mathbf{r}') - A_0(\mathbf{r}'))dV', \tag{6}$$

and $\mathbf{q} = \mathbf{k}_2 - \mathbf{k}_1$, $\hbar\mathbf{q} = \mathbf{p}_2 - \mathbf{p}_1$.

The operator (matrix) $\hat{F}_B$ is the scattering matrix in the Born approximation. Its matrix elements $F_{12} \equiv u_2^*(\mathbf{p}_2)\hat{F}_B u_1(\mathbf{p}_1)$, constrained by the normalization condition $u_{1,2}^* u_{1,2} = 1$, give the corresponding differential cross-section as follows:

$$d\sigma_{12} = |u_2^*(\mathbf{p}_2)\hat{F}_B u_1(\mathbf{p}_1)|^2 d\Omega_2. \tag{7}$$

Let us emphasize that this relation depends on the spin states of the scattering and scattered particles, described by the corresponding bi-spinors. If the spin of the scattered particle is not fixed, the differential scattering cross-section is determined by the expression $d\sigma = F^* F d\Omega_2$.

Eq. (7), with $\hat{F}_B$ determined by Eq. (6), can be simplified further if we use the following relations:

$$(c\boldsymbol{\alpha} \cdot \mathbf{p}_2 + mc^2 \beta) u_2(\mathbf{p}_2) = \varepsilon u_2(\mathbf{p}_2), \quad u_2^*(\mathbf{p}_2)(c\boldsymbol{\alpha} \cdot \mathbf{p}_2 + mc^2 \beta) = \varepsilon u_2^*(\mathbf{p}_2).$$

Hence,

$$u_2^*(p_2)\hat{F}_B u_1(p_1) \equiv u_2^*(p_2)\hat{G}_B u_1(p_1), \tag{8}$$

where

$$\hat{G}_B = \frac{e\varepsilon}{2\pi\hbar^2 c^2} \int e^{-i\mathbf{q}\cdot\mathbf{r}} (\boldsymbol{\alpha} \cdot \mathbf{A}(\mathbf{r}) - A_0(\mathbf{r})) dV. \tag{9}$$

The differential scattering cross-section,

$$d\sigma_{12} = |u_2^*(\mathbf{p}_2)\hat{G}_B u_1(\mathbf{p}_1)|^2 d\Omega_2, \tag{10}$$

(note that this expression is identical to Eq. (7)) in the case of a purely electrostatic field, takes the form:

$$d\sigma_{12} = \left| \frac{e\varepsilon}{2\pi\hbar^2 c^2} \int e^{-i\mathbf{q}\cdot\mathbf{r}} A_0(\mathbf{r}) dV u_2^*(\mathbf{p}_2) u_1(\mathbf{p}_1) \right|^2 d\Omega_2. \tag{11}$$

The spin-dependence of this expression, determined by the bi-spinors $u_{1,2}$, takes a more natural form when expressed in terms of the "upper" spinor components, $\varphi_{1,2}$, according to

$$u_2^*(\mathbf{p}_2) u_1(\mathbf{p}_1) = \frac{1}{2\varepsilon(\varepsilon + mc^2)} \varphi_2^*(\varepsilon + mc^2)^2 + (pc)^2 \cos\theta - i(pc)^2 \sin\theta \boldsymbol{\sigma} \cdot \boldsymbol{\nu} \varphi_1,$$

$$\boldsymbol{\nu} = \frac{[\mathbf{p}_1 \times \mathbf{p}_2]}{|[\mathbf{p}_1 \times \mathbf{p}_2]|}, \quad (\nu^2 = 1) \tag{12}$$

(where θ is the scattering angle). As a result of the summation over the particle spin states after scattering, and averaging over its initial spin state,[415] we obtain

---

[415] This operation is denoted by an overbar, and the initial spin state is assumed to be non-polarized.

$$\overline{|\varphi_2^*(\varepsilon+mc^2)^2+(pc)^2\cos\theta-i(pc)^2\sin\theta\boldsymbol{\sigma}\cdot\boldsymbol{\nu}\varphi_1|^2}=$$
$$4\varepsilon^2(\varepsilon+mc^2)^2\left[1-\frac{v^2}{c^2}\sin^2\left(\frac{\theta}{2}\right)\right] \tag{13}$$

(v is the particle velocity).

The differential scattering cross-section, described by Eqs. (11), (12) and (13), coincides with Eq. (1) of the previous problem (obtained in another way). For the case of the Coulomb field, $A_0 = Ze/r$, we have Eq. (7) from Problem 15.34.

16
Appendix

16.1 App.1. Integrals and integral relations

1.

$$\delta(x) = \frac{1}{2\pi} \int_{-\infty}^{\infty} e^{ikx} dk, \quad \delta(\mathbf{r}) = \frac{1}{(2\pi)^3} \int e^{i\mathbf{k}\cdot\mathbf{r}} d^3k \qquad \text{(App.1.1)}$$

An analogous relation holds in higher dimensions as well.

2.

$$\int_a^b \frac{F(x)dx}{x - x_0 \mp i\varepsilon} = \mathcal{P} \int_a^b \frac{F(x)dx}{x - x_0} \pm i\pi F(x_0) \qquad \text{(App.1.2)}$$

Here $a < x_0 < b$, $\varepsilon > 0$ and is infinitely small, and the integral is understood as its principle value. For a calculation of its imaginary part, see Problem 13.11.

3.

$$\int_{-\infty}^{\infty} \frac{e^{ikx}dk}{k^2 - \kappa^2 \mp i\varepsilon} = \pm i\frac{\pi}{\kappa} e^{\pm i\kappa|x|}, \quad \int_{-\infty}^{\infty} \frac{e^{ikx}dk}{k^2 + \kappa^2} = \frac{\pi}{\kappa} e^{-\kappa|x|}, \qquad \text{(App.1.3)}$$

where x and κ are real, $\kappa > 0$, and $\varepsilon > 0$ is infinitely small. Integrals are calculated using the residue theorem, by closing the integration contour in either the upper ($x > 0$) or lower ($x < 0$) complex half-plane, k.

4.

$$\int \frac{1}{r} e^{i\mathbf{k}\cdot\mathbf{r} - \kappa r} d^3r = \frac{4\pi}{k^2 + \kappa^2}, \quad \text{Re } \kappa > 0 \qquad \text{(App.1.4)}$$

The integral (which for $\kappa = 0$ gives Fourier transform of the Coulomb potential) is calculated in spherical coordinates with the choice of the polar axis along the vector $\mathbf{k}$.

5.
$$\int_{-\infty}^{\infty} \frac{dx}{(x^2+a^2)^{n+1}} = \frac{(-1)^n}{n!} \frac{\partial^n}{\partial a^{2n}} \int_{-\infty}^{\infty} \frac{dx}{x^2+a^2} =$$

$$= \frac{\pi(2n-1)!!}{2^n n! a^{2n+1}}, \quad a > 0 \qquad \text{(App.1.5)}$$

6.
$$\int_{-\infty}^{\infty} x^{2n} e^{-a^2 x^2} dx = (-1)^n \frac{\partial^n}{\partial a^{2n}} \int_{-\infty}^{\infty} e^{-a^2 x^2} dx =$$

$$= \frac{\sqrt{\pi}(2n-1)!!}{2^n a^{2n+1}}, \quad a > 0 \qquad \text{(App.1.6)}$$

7.
$$\int_a^b \frac{1}{x} \sqrt{(x-a)(b-x)}\, dx = \frac{\pi}{2}(a+b-2\sqrt{ab}), \quad 0 < a < b \qquad \text{(App.1.7)}$$

8.
$$\int_0^\infty \frac{\sin x}{x} = \frac{\pi}{2} \qquad \text{(App.1.8)}$$

16.2 App.2. Cylinder functions

The cylinder functions $Z_\nu(z)$ are solutions of the following differential equation:

$$Z_\nu''(z) + \frac{1}{z} Z_\nu'(z) + \left(1 - \frac{\nu^2}{z^2}\right) Z_\nu(z) = 0 \qquad \text{(App.2.1)}$$

The Bessel functions

$$J_\nu(z) = \sum_{k=0}^{\infty} \frac{(-1)^k}{k!\, \Gamma(k+\nu+1)} \left(\frac{z}{2}\right)^{\nu+2k} \qquad \text{(App.2.2)}$$

are a special case of the cylinder functions. If the index, ν, labeling a Bessel function is not an integer, then one can use it to construct a general solution to Eq. (App.2.1) as follows:

$$Z_\nu(z) = C_1 J_\nu(z) + C_2 J_{-\nu}(z), \quad \nu \neq 0, \pm 1, \pm 2, \ldots.$$

The asymptotic behavior of the Bessel functions near $z \to 0$ follows from their definition in terms of a series (App.2.2), while the asymptote at large arguments, $|z| \to \infty$, reads

$$J_\nu(z) \approx \left(\frac{2}{\pi z}\right)^{1/2} \cos\left(z - \frac{\pi \nu}{2} - \frac{\pi}{4}\right). \qquad \text{(App.2.3)}$$

The Neumann function is as follows:

$$N_\nu(z) \equiv Y_\nu(z) = \frac{1}{\sin \pi \nu}[\cos(\pi \nu) J_\nu(z) - J_{-\nu}(z)]. \qquad \text{(App.2.4)}$$

For the integer values of the index $\nu = n$, the Neumann function can be defined as the following limit, $N_n(z) = \lim_{\nu \to n} N_\nu(z)$, and it represents the second linearly independent solution to Eq. (App.2.1) (apart from the Bessel function, $J_\nu(z)$). Note the following asymptotic behavior of the Neumann functions,

$$N_0(z)\Big|_{z \to 0} \approx \frac{2}{\pi} \ln \frac{Cz}{2}$$

and

$$N_\nu(z)\Big|_{z \to 0} \approx -\frac{\Gamma(\nu)}{\pi}\left(\frac{2}{z}\right)^\nu \quad \text{for} \quad \nu > 0, \qquad \text{(App.2.5)}$$

here $\gamma = \ln C = 0.5772\ldots$ is the Euler constant.

The asymptote of the Neumann function for $z \to \infty$ has the form:

$$N_\nu(z) \approx \left(\frac{2}{\pi z}\right)^{1/2} \sin\left(z - \frac{\pi \nu}{2} - \frac{\pi}{4}\right). \qquad \text{(App.2.6)}$$

Closely connected to the Bessel and Neumann functions are the Hankel functions,

$$H_\nu^{(1)}(z) = J_\nu(z) + iN_\nu(z), \quad H_\nu^{(2)}(z) = J_\nu(z) - iN_\nu(z), \qquad \text{(App.2.7)}$$

and the modified Bessel functions, $I_\nu(z)$ and $K_\nu(z)$ (the latter is called the MacDonald function), determined by relations

$$I_\nu(z) = i^{-\nu} J_\nu(iz) = \sum_{k=0}^{\infty} \frac{1}{k!\Gamma(\nu+k+1)} \left(\frac{z}{2}\right)^{\nu+2k}, \qquad \text{(App.2.8)}$$

$$K_\nu(z) = \frac{\pi}{2 \sin \pi \nu}[I_{-\nu}(z) - I_\nu(z)], \quad \nu \neq 0, \pm 1, \pm 2, \ldots.$$

For all integer indices, $K_n(z) = \lim K_\nu(z)$ for $\nu \to n = 0, \pm 1, \pm 2, \ldots$, which leads to

$$K_0(z)\Big|_{z \to 0} \approx \ln \frac{2}{Cz}; \quad K_n(z) \approx_{z \to 0} \frac{(n-1)!}{2}\left(\frac{2}{z}\right)^n,$$

$$n = 1, 2, \ldots, \qquad \text{(App.2.9)}$$

See Eq. (App.2.5). A linear combination of the modified Bessel functions,

$$u_\nu(z) \equiv Z_\nu(iz) = C_1 I_\nu(z) + C_2 K_\nu(z),$$

which is a cylinder function of an imaginary argument, represents a general solution for the equation

$$u_\nu'' + \frac{1}{z} u_\nu' - \left(1 + \frac{\nu^2}{z^2}\right) u_\nu = 0. \qquad \text{(App.2.10)}$$

We also mention here that solutions to the following differential equations have important quantum-mechanical applications:

$$u'' + \alpha z^\nu u = 0, \quad u = \sqrt{z} Z_{1/(\nu+2)} \left(\frac{2\sqrt{\alpha}}{\nu+2} z^{1+\nu/2}\right); \qquad \text{(App.2.11)}$$

$$u'' + \left[\alpha z^\nu - \frac{l(l+1)}{z^2}\right] u = 0,$$

$$u = \sqrt{z} Z_{(2l+1)/(\nu+2)} \left(\frac{2\sqrt{\alpha}}{\nu+2} z^{1+\nu/2}\right); \qquad \text{(App.2.12)}$$

$$u''(z) + (\gamma^2 e^{2x} - \nu^2) u(z) = 0, \quad u = Z_\nu(\gamma e^z), \qquad \text{(App.2.13)}$$

and are related to the cylinder functions.

An effort has been made in several rounds of proofs to correct typos in both equations and the text (including reworking all remaining text obtained using automatic translation software). However, due to the large volume and dense technical presentation in the book, some undetected typos may remain. It would be much appreciated if any such instances, along with corrections, suggestions, and ideas for problems to be included in future editions of the book were emailed to the following address: GKKGbook@gmail.com

Victor Galitski,
Melbourne, January 2013

Index*

Aharonov–Bohm effect 672
adiabatic approximation, perturbation
    theory 358–73
Airy functions 38, 254
    one-dimensional motion 63, 65
    perturbation theory 367
    quasi-classical approximation 411
    time-dependent quantum mechanics 254
angular momentum
    definite mean values 91
    mean value 824
    operator 839
    quantization condition 315–16
    two particles with equal angular moments 103–4
    see also orbital angular momentum
annihilation operators 461–2
antiparticle energy spectrum 858
antiparticle state, known helicity, wavefunctions 850
antiparticle wavefunction 816
Appendix 1, Integrals and integral relations 866
Appendix 2, Cylinder functions 867
asymptotic coefficient (at infinity) 486
atom+photon system, photon scattering 800–1
atom dipole interaction operator diagonalization 563
atomic ionization energy 497–9
    ionization probability 579
    obtained at nucleus decay 566
    quasi-classical approximation 521
atomic nucleus 598–636
    Aufbau building-up principle 615
    Coulomb interaction of protons 611
    isotopic invariance 627–37
    nuclear forces 598–611
    Pauli matrices for spin 598
    shell model 599, 612–26
    see also helium; hydrogen
atomic systems
    external magnetic field 551–8
    interactions 536–65
        at large distances 558–65
    non-stationary phenomena 565–97
    statistical atomic model, many-electron atoms 509–15
    two-atom-molecule theory 525–39
atoms and molecules 485–598

external fields, interaction of atomic systems 538–65
many-electron atoms, statistical atomic model 509–25
non-stationary phenomena in atomic systems 565–98
perturbation theory 488
principles of two-atom molecule theory 525–37
stationary states of one-electron and two-electron atoms 486–509
Aufbau building-up principle 615
Auger effect, mesoatom 586
Auger S-transitions 587
autoionizational states (AIS) 588

Berry phase, coherent-state spin part-integral 205
beryllium ion 497
    atomic ionization energy 497
    polarizability values 539
Bessel function 125, 126, 134, 143–5, 154, 156, 680
    quasi-classical approximation 389, 393, 396, 425–6
    time-dependent quantum mechanics 263
    zeroth order 641
bi-spinors
    general solution to the Dirac equation 846
    particle momentum 845
    scattered wave 863
    unitary transformation 844
Bloch functions
    perturbation theory 333
    time-dependent quantum mechanics 261
Bohr radius 492
    motion in central field 116
    muon 492
Bohr–Sommerfeld quantization rule 359, 375, 378, 383, 397–9, 401–5, 432
Boltzmann constant, coherent-state spin part-integral 202
Born approximation 494
    differential scattering cross-section of charged relativistic spinless particle 835
    energy-dependence of the scattering cross-section 836, 837
    scattering amplitude 494, 835
    pseudo-potential 495
Born scattering amplitude, spinless particle 836
boron ion, polarizability values 539

---

* Index compiled by appointed indexer on behalf of Oxford University Press.

Bose gas
  energy 478–9
  non-interacting particles 472–3
  two identical Bose particles 454–6, 462–4
bosons 446, 450, 453–5, 462
  bosonic annihilation operator 456–9
  intrinsic parities 817
bound states 44, 54–5
  separable potential 51
braking radiation (Bremsstrahlung), Coulomb field 805

carbon ion 514
  atomic ionization energy 497, 499
  polarizability values 539
charge-conjugation function 817
  particle and antiparticle helicity 852
charge-conjugation operator, spinless particles 816
charge-conjugation transformation, helicity 849
commutation relation, Pauli matrices 798
commutators 838–9
Compton wave-length 820
constants of motion 227–9
Coriolis interaction, particle collisions 771
Coulomb barrier
  penetrability 536, 596
  quasi-classical approximation 435–6
Coulomb excitation of atomic nucleus 765–6
Coulomb field 488
  braking radiation 805–9
  differential scattering cross-section 864
  motion in a magnetic field 292–3
Coulomb gauge 776
Coulomb interaction 491, 625
  between electrons 505
  ns-levels 490
  protons 611
Coulomb level degeneracy 519
Coulomb potential 380, 387–8, 441–3, 491, 597
  accidental degeneracy 829
  hadron atoms 493
  long range 694–5
  motion in a central field 116, 119, 125, 135, 144
  particle collisions 700, 718
  perturbation theory 303, 309–10, 318–19, 369, 371–2
Coulomb wavefunction 493, 502
creation and annihilation operators 461–2
crystal with defect (vacancy), one-dimensional motion 80
current density, transmission and reflection of particles 64

degenerate perturbation theory 490
deuterium, rotational states 530
deuteron 598
  binding energy 598
  magnetic moment 603
  and proton 602
  proton–neutron interactions 600, 604
  quadrupole moment 604
  Schrödinger equation 606
  wavefunction 603, 605
deuteron photodisintegration, differential and total cross-sections 804
dipole approximation
dipole moments
  molecular interactions 559–61
  transition 780
    forbidden 781–2
dipole–dipole interactions 561
Dirac delta function 36
Dirac equation 812
  current density of a free Dirac particle 841
  four independent solutions 840
  free particle with definite momentum and energy 840
  Green function 861–2
  model of a hydrogen-like atom with a point nucleus 488
  model of hydrogen-like atoms 489
  motion in a magnetic field 290
  two-component spinors 852
Dirac matrices 812, 855
Dirac particle
  charge and current densities, non-relativistic limit 853
  in constant external electromagnetic field, scattering amplitude, Born approximation 862
  in Coulomb field, first-order perturbation theory 858
  energy spectrum in a uniform magnetic field 857
  first-order perturbation theory 858
  Hamiltonian 854, 858
  mean value of spin vector 843
  projection operators 851
  velocity operator 842
Dirac string 206
dynamic polarizability
  particle in a zero-range potential 579
  quasi-energy 580

effective radius 486
eigenfunctions
  and eigenvalues
    operator, spin ($s = 1/2$) 169–70
    spin-flip 783
  eigenvalues of isospin 607
  eigenvalues and mean values 8–19
  Hermitian operators 851
  orthogonality 818, 846
eigenvalues, operator, free-particle Hamiltonian 851
Eikonal approximation, scattering of fast particles 640, 713–26
elastic scattering, fast electrons 647–8, 742–46

electromagnetic transition 780
  between two-atom molecular rotational levels 781
  E1 and M1 788
electromagnetic wave, scattered by magnetic moment 799
electron Bohr radius 492
electron mutual interaction energy 520
electrostatic barrier, penetrability 582
electrostatic form-factors 753
energy levels, motion in central field 132–3
energy spectrum
  degeneracy 81–3
  identical particles 76–8
  motion in central field 128–9, 155–6
exchange integral 509
exchange potential
  large distances 562
  negative ion 562

Fermi gas 485
  ground state 475–7
  neutral particles $(s = 1/2)$ 481–4
  spinful particles $(s = 1/2)$ 479–81
fermions 446, 453–5, 462, 608
Feynman diagram 754
fine-structure constant/interval 488
Fourier components
  perturbation theory 330, 338
  quasi-classical approximation 414–17
Fourier integral 583
Fourier series, time-dependent quantum mechanics 261
Fourier transform 23
  one-dimensional motion 47
  particle collisions 645, 658
Fraunhofer diffraction, particle collisions 724
free particle in half-space, one-dimensional motion 53, 56

Galilean transformation, time-dependent quantum mechanics 239–40
Galileo's relativity principle 567
Gaussian integral, time-dependent quantum mechanics 216
Gauss's law, orbital angular momentum 90
germanium ion 514
Green function 69, 71–4, 137–8, 140–1, 151, 156–9
  charged particle 250–2
  free particle 137–8, 156–7, 247–50
  general solution 50–1
  plane rotator 157–8
  radial Schrödinger equation 140
  spherical rotator 158–9
  spin 200
  stationary Dirac equation 861
  symmetric form 862
  time-dependent quantum mechanics 213, 246–52

ground-state energy 428–30
  dissociation energy 528
  expectation value 502
  muonium 591
  neutral atom 517
  oscillation frequency 528
  rotational constant 529
  two-electron atom/ion 497, 499
  variational method 523
  variational value 537
ground-states
  mirror nuclei 622
  moments (spins) and parities 614
  ortho- and para-positronium 570–2
  wavefunction 52–3
gyromagnetic factors 619

hadrons 598
  Coulomb attractive potential 493
Hamiltonian 14
  charged particle 270–1
  Dirac particle 854, 858

  motion in central field 119–20, 148, 159
  non-relativistic limit, neutron in electromagnetic field 857
  one-dimensional motion 35, 75–6
  orbital angular momentum 86–7
  Pauli matrices 855
  relativistic corrections 832
  spin-1 system 210–12
  spin-orbit-coupled particles 198–202
Perturbation Hamiltonian, homogeneous magnetic field 556
Hankel function, quasi-classical approximation 425–6
Hartree—Fock equations, self-consistent-field method 485
Heisenberg operators, time-dependent quantum mechanics 233–7, 245–51
helical states/helicity 841, 849
  charge-conjugation transformation 849
helium atom
  atomic ionization energy 497, 499
  ground state, diamagnetic susceptibility 554
  magnetic susceptibility 554
  nuclear ground-states 616
  quantum numbers – spin, parity and isospin 616
helium-like atom 503–8
  energy and ionization energy 503
    variational method 504
  energy of state where both electrons are excited 507
  Hamiltonian in an electric field 538
  hyperfine splitting 506
  polarizability 539, 546
  stationary states, stability with respect to decay (auto-ionization) 507
Hermitian conjugate, perturbation theory 341

Hermitian operators 175–6, 487, 488
  general eigenfunctions 851
  linear operators 1–4, 9–22, 25, 27–9
  orthogonality of eigenfunctions 818, 846
  projection operators 851
  quasi-classical approximation 407
  time-dependent quantum mechanics 226–8, 242
Hilbert space 1, 6–7, 16, 266
  one-dimensional motion 56
Hulthen potential
  motion in central field 123
  perturbation theory 310
  quasi-classical approximation 387–8
Hund's law 509, 511, 513, 516
hydrogen atom
  electromagnetic transition between the components of hyperfine structure 787
  ground state 121–3
    diamagnetic shift 554
  metastable 2s-states 572
    change in lifetime 572
  one-photon transition 785
  two interacting at a large distance 559–61
hydrogen molecule
  dissociation energy 528
  oscillation frequency 528
  rotational constant 529
  Stark effect, principal quantum number $n = 2$ 539–40
hydrogen-like atom, ground state 578
hyperfine splitting 492
  ns-level 490

inversion operator 839
isospin 598, 607
  nucleon 598
isotopic invariance 627–36
  pion+deuteron 635
isotopic symmetry 598
isotropic oscillator energy levels 74

jj-coupling scheme 620

K-electrons 803
  ejection 582
Auger effect in a mesoatom 586
K-mesons 192–3
  strong interaction 488
Clebsch–Gordan coefficients 177, 188, 196, 573, 631
  addition of angular momenta 107–8
  orbital angular momentum 102, 107–8
  particle collisions 734
Klein–Gordon equation 488
  Born approximation 835
  charged particle in an external electromagnetic field 831
  energy of a spinless particle 821
  for a free particle 830
  general solution 813

  invariance 813
  orthogonality and normalization of solutions 818
  particle and antiparticle wavefunctions 819
  scalar field 816
  stationary solutions 825, 835
    particle in an electrostatic field 834
Kramer's matching conditions, quasi-classical approximation 378–80

Lagrange factor 501–2
Laguerre polynomial 116
Lamb shift 489, 492, 572, 783
Landau levels
  motion in a magnetic field 272–3
  shifts, perturbation theory 367–8, 370
  spectrum 465
Laplace operator, orbital angular momentum 84
Laplacian on sphere of unit radius 308
Larmor orbit radius, motion in a magnetic field 274–5
Legendre polynomials
  orbital angular momentum 85, 104
  oscillations 715
  perturbation theory 305
Lie group 213
linear operators 2–7
Lippmann–Schwinger equation 662
lithium atom 509
lithium ion
  atomic ionization energy 497, 499
  polarizability 546
  polarizability values 539
Lorentz force
  electromagnetic wave 803
  operator 225
  particle collisions 671
Lorentz group, pseudoscalar operator 849
Lorentz transformation 810
  spinors 645
Loschmidt number, helium gas 554
LS-coupling scheme 619, 621
Langer correction, quasi-classical approximation 378–80, 391–5, 434
Langer quantization rule 393
Langer transformation 376

MacDonald function 156, 545, 820
  quasi-classical approximation 408, 411
magic numbers, M(N) 612
magnetic field, see also motion in a magnetic field
magnetic susceptibility 482, 554–5
magnetic traps 280
many-electron atoms, statistical model 509–15
Maxwell equations, classical electrodynamics 852
mirror nuclei 622
mirror reflections 636
mirror-image charges, one-dimensional motion 54
molecular interactions, second order of perturbation theory 561

Index   875

molecule states, symmetric/anti-symmetric with respect to nuclei 532
momentum operator 839
momentum probability distribution, wavefunctions 823
momentum representation, one-dimensional motion 46–8, 73–4
monopole or $E0$ transitions 585
monotonic attractive potential 152–3
motion in a central field 116–59
   Bessel functions 125–6, 134, 143–5, 154, 156
   Bohr radius 116
   bound state of particle 129–32
   Cartesian coordinates 119
   central potential parameters 145–8
   change of energy levels 118
   Coulomb potential 116, 119, 125, 135, 144
   discrete spectrum of particle 153–5
   discrete spectrum states in central fields 117–41
   energy levels 132–3
   energy spectrum of particle 128–9, 155–6
   Fourier component 135, 137
   Gauss law 156
   Green function 137–8, 140–1, 151, 156–9
   Hamiltonian 119–20, 148, 159
   Hulthen potentials 123
   hydrogen atom ground state 121–3
   Laguerre polynomial 116
   levels with an arbitrary angular motion 125–6
   Macdonald function 156
   monotonic attractive potential 152–3
   Neumann functions 125, 134
   new bound states 144–5
   particle in attractive potential 138–9
  energy level shifts 148–51
   s-state wavefunction 134–7
   particles in a one-dimensional well 117–18
   Poisson equation 122
   s-levels 123–5
   Schrödinger equation 116–18, 119–20, 127–32, 134, 136, 142, 144–5, 153, 155
   low-energy states, combined potentials 141–53
   systems with axial symmetry 153–60
   Van der Waals forces 123
   Yamaguchi potential 129
   zero-range potential 126–8
motion in a magnetic field 270–96
   charged spherical oscillator 278–80
   Coulomb field 292–3
   Dirac equation 290
   Hamiltonian of charged particle 270–1
   Landau levels 272–3
   Larmor orbit radius 274–5
   magnetic traps 280
   neutral particle with spin $s = 1/2$ 283
   orbital currents and spin magnetic moment 291–6
   Pauli Hamiltonian 285–6

   spin $s = 1/2$ and magnetic moment 287–90, 294–5
   stationary states and energy levels of charged spinless particle 277–8
   stationary states in a magnetic field 271–86
   time-dependent Green function 290–1
   time-dependent magnetic field 286–91
muon
   Bohr radius 492
   orbital moment 586
   polarization vector 593–4
   wavefunction 568
muonium, ground-state energy 591

negative ion, exchange potential 562
Neumann functions
   motion in central field 125, 134
   particle collisions 679
neutral particle with spin $s = 1/2$ 283
neutrino
   particle with negative helicity 645
   particle with zero rest mass $m = 0$ 845
neutrons 598
   in an electromagnetic field, Hamiltonian in non-relativistic limit 857
   elastic scattering 766–7
   fast neutrons from a Coulomb field 728
nitrogen ion 514
non-relativistic theory
   charge and current densities of Dirac particle 853
   Green function 861
   neutron in an electromagnetic field 857
   quantum numbers 858
   Rutherford equation 836
non-stationary perturbation theory 336–53
normalization integral 500
np-electrons 514
ns-level, hyperfine splitting 490
nuclear magnetic moment 489
nuclear potential 491
nuclear quadrupole moment 603, 622–5
nuclear synthesis reaction rate 594
nucleons
   charge distribution for the filled shells 625
   eigenfunctions and eigenvalues of isospin 607
   energy levels 599
   many-nucleon system, wavefunctions 608
   maximum velocity 627
   nucleon density inside a nucleus 626
   radial density inside a nucleus 625
nucleon–nucleon interactions 636

one-dimensional motion 32–83
   Airy functions 63, 65
   arbitrary representation 51
   bound states 44, 54–5
   branching points 58, 69
   change of variables 39

one-dimensional motion (*cont.*)
  continuous spectrum reflection from and
    transmission through potential
    barriers 56–74
  crystal with a defect (vacancy) 80
  current density, transmission and reflection of
    particles 64
  Dirac delta function 36
  discrete spectrum 39, 52
  energy spectrum and degeneracy 81–3
  Fourier transform 47
  free particle in half-space 53, 56
  Green function 69, 71–4
    general solution 50–1
  ground state wavefunction 52–3
  Hamiltonian 35, 75–6
  Hermite polynomials 32
  Hilbert space 56
  identical particles, energy spectrum 76–8
  incident particles 70
  infinite potential well 55
  isotropic oscillator energy levels 74
  low-lying levels 41–2
  mirror-image charges 54
  momentum representation 46–8, 73–4
  particle fields 45
  periodic potential particle 77–8
  physical sheet and Riemann surface 58–9
  potential properties 56
  properties of the spectrum 42–3
  quasi momentum band structure 78
  rectangular potential well 55
  repulsive deltapotential 71–3
  Riemann surface 58
  Schrödinger equation 33–40, 43–5, 50–7, 60–1,
    63, 66–8, 74, 77–8, 80–2
  separable potential 51
  shallow potential well 53
  slow particles 66, 67–8
  square potential well 67
  stationary states with definite parity 41
  stationary states in discrete spectrum 33–46
  surface or Tamm states 81
    particle in a periodic potential 74–83
  transmission and reflection of particles 56–74
  uniform field 68
  zero-energy solution 43–4, 67
one-particle problem, strong external field 827
operators in quantum mechanics 1–31
  eigenfunctions, eigenvalues and mean
    values 8–19
  Fourier transform 23
  Hamiltonian 14
  Hermitian linear operators 1–4, 9–22, 25, 27–9
  Hilbert vector space 1, 6–7, 16
  linear operators 2–7
  projection operators 20–1
  Schrödinger equation 25
  Taylor expansion 4–6, 12

  wavefunctions unitary operators 22–31
Oppenheimer—Brinkman—Cramers
  approximation (OBC) 751–3
optical theorem, particle collisions 736–7
orbital angular momentum 84–115
  addition of angular momenta 99–110
    commutators, finding of 100
    Clebsch—Gordan coefficients 107–8
    subsystems 102–3
    three weakly interacting subsystems 106
    two particles 104–5
    weakly interacting systems 99–101, 108–9
    projection of 102
  angular momentum 94–9
  commutation relations 84
  Clebsch—Gordan coefficients 102, 107–8
  Laplace operator 84
  Legendre polynomials 85, 104
  plane rotator, stationary wavefunctions and
    energy levels 86–7
  probability overlap 93
  projection operators 92–3
  raising and lowering operators 85
  spherical rotator, stationary wavefunctions and
    energy levels 87
  tensor formalism 110–15
  vector mean values 89–90
  see also angular momentum
orbital currents, and spin magnetic moment 291–6
oxygen ion 514
  atomic ionization energy 497, 499

p-electron, wavefunction 512
particle(s)
  acted on by uniform constant force 232
  with arbitrary spin, intrinsic parities 817
  momentum representation 46–8, 73–4
  one-dimensional motion, transmission and
    reflection 58
  one-particle problem, strong external field 827
  wavefunctions 819
  with zero rest mass $m = 0$ (neutrino) 845
  zero-range potential, dynamic polarizability 579
particle collisions 637–776
  Aharonov—Bohm effect 672
  analytic properties of scattering
    amplitude 637–64, 737–47
  annihilation and positronium levels 772–3
  Born approximation 642–64
  collision between fast charged particle and
    two-atom molecule 757
  Coriolis interaction 771
  Coulomb excitation of atomic nucleus 765–6
  Coulomb potential 700, 718
  differential scattering cross-section 638–46
  effective interaction radius 702–3
  effective radius expansion 639
  elastic scattering of fast electrons 647–8, 742–6
  electrostatic form-factors 753

excitation transfer cross section 769–70
fast charged particle colliding with hydrogen
    atom 759–60
fast neutrons from a Coulomb field 728
fast particles scattering 723
Feynman diagram 754
Fourier transform 645, 658
Fraunhofer diffraction 724
invariant functions 733
Clebsch—Gordan coefficients 734
Lippmann—Schwinger equation 662
long range Coulomb potential 694–5
Lorentz force 671
low-energy scattering, scattering resonance
    phenomena 674–713
neutron elastic scattering 766–7
Oppenheimer—Brinkman—Cramers
    approximation (OBC) 750–4
optical theorem 736–7
perturbation theory in terms of scattering
    length 689–93
phase scattering theory 664–74
phase shifts 664–9, 703, 731–2
    slow particles 677–8
polarization for positron scattering 730
principle of detailed balance, neutron radiation
    capture and photodisintegration 773
proton—neutron system, value of triplet s-wave
    scattering length 698–9
Ramsauer—Townsend effect 678
resonance scattering 639
Rutherford equation for scattering angle 719
s-wave scattering 709–11
scattering amplitude of spinless particles 741
scattering of composite quantum particles,
    inelastic collisions 747–76
scattering of fast particles, Eikonal
    approximation 640, 713–26
scattering lengths in potentials 678–88
scattering of particles with spin 726–37
slow particle scattering 705
spinless particles 732–4
Thomas—Fermi model 648–9
transport cross-section 646
Yukawa potential 655–7, 730–1
identical particles, second quantization 446–84
    arbitrary one-particle state 461
    Bose gas, energy 478–9
    bosons 446, 450, 453–5, 462
    commutation relations 464
    correlation function 474, 477
    creation and annihilation operators 461–2
    electron density charge 484
    energy spectrum of two-dimensional (planar)
        oscillator 466–7
    Fermi gas
        ground state 475–7
        neutral particles ($s = 1/2$) 481–4
        spinful particles ($s = 1/2$) 479–81

    fermions 446, 453–5, 462
    identical particles 446–8
    Landau level spectrum 465
    magnetic susceptibility 482
    neutral particles ($s = 1/2$) 481–2
    occupation-number representation 456–72
    Pauli principle 481
    pseudo-scalar particle 453
    quantum statistics, symmetry of
        wavefunction 447–56
    scattering length 479
    simplest systems with large number of
        particles 472–85
    spontaneous symmetry breaking 469
    super-partners 469
    supersymmetric oscillator 467–8
    transverse motion of charged spinless
        particle 465–6
    two identical Bose particles 454–6, 462–4
particle localization domain 546
particle—antiparticle pair 827
    energy spectra 858
Pauli equation, uniform magnetic field 858
Pauli exclusion principle 198
Pauli Hamiltonian 285–6
Pauli matrices 165–9, 198, 785, 856
    commutation relation 798
    time-dependent quantum mechanics 264, 267
periodic potential particle, one-dimensional
    motion 77–8
perturbation matrix element 542–3, 547, 556–7,
    806, 859
perturbation theory
    adiabatic approximation 358–73
    Airy function 367
    Bloch functions 332
    Bohr—Sommerfeld quantization rule 359
    Born approximaton 370
    Coulomb potential 303, 309–10, 318–19, 369,
        371–2
    degenerate 490
    energy spectrum 331–6
    first-order
        Dirac particle in Coulomb field 858
        Dirac particle energy-dependence of the
            scattering cross-section 860
        ground-state energy of a two-electron
            atom/ion 497, 499
    Fourier component 330, 338
    Hermitian conjugate 341
    Hulthen potential 310
    Landau level shifts 367–8, 370
    Laplacian on a sphere of unit radius 308
    Legendre polynomials 305
    non-stationary, transitions in continuous
        spectrum 336–53
        charged linear oscillator 336–7
        homogenous electric field applied to plane
            rotator 337–9

perturbation theory (*cont.*)
 ionization probability 351-2
  periodic resonant perturbation 348-9
  periodic-in-time perturbation 343-8
 two-channel system 349-51
  wavefunction and amplitude 339
 Poisson distribution 358
 quasi-energy states (QES) 343-4, 362
 scattering length 565
 second-order 340, 546
  general equation 791
  molecular interactions 561-2
  potential of tensor forces 605
  transitions 797, 806
 stationary perturbation theory (continuous spectrum) 325-36
  amplitude of reflected wave 325-31
  energy spectrum 331-5
  reflection coefficient 329-30
 stationary perturbation theory (discrete spectrum) 298-316
  charged linear oscillator 299-300
  Coulomb potential energy levels 309-11
  energy level shifts 302
  first-order correction to energy levels 298
  particle inside an impenetrable ellipsoid of rotation 312-14
  plane isotropic oscillator 300-1
  plane rotator 303-4, 307-8
  spherical rotator 308-9
 sudden action 353-8
 variational method 316-25
  energy of ground state and of first exited state 319-20
  sudden and adiabatic action 296-374
  two particles of same mass 320
photo-effect cross-section
 hydrogen-like atom in ground state 802
 photoeffect differential cross-section for unpolarized photons 803
photo-effect, inverse, radiation recombination 803
photon, Maxwell equations in classical electrodynamics 852
photon emission 779-89
 differential photon elastic scattering cross-section 796
 magnetic-dipole emission theory 785
 one-photon electromagnetic transitions 788
 probability 781-3
 scattering cross-section for non-polarized photons 798
 scattering differential cross-section 792
 sum rules 789
 suppression of one-photon transitions 785
 total probability of photon emission for spin-flip 785
 transition probability 786
photon scattering
 atom+photon system 800

 differential and total cross-sections 792
 polarization effects 791-4
photon scattering cross-section for small frequency 800
pi-mesons, strong interaction 488
pion+deuteron system 635
pion quantum numbers 192
pions
 scattering by nucleons 635
 scattering length 493
pion–nucleon interaction 636
plane rotator, stationary wavefunctions and energy levels 86-7
Poisson distribution, perturbation theory 358
Poisson equation 485
 motion in central field 122
polarizability 543-5
 zero-range potential 543
polarization effects, photon-scattering 791-4
polarization vector, muon 593-4
positronium, ground-states (ortho- and para-) 570
positronium cross-section formation 751
projection operators 20-1, 851
proton, and neutron 598
proton magnetic moment 489
proton–neutron system, value of triplet s-wave scattering length 698-9
proton–neutron interactions 600, 604
pseudoscalar operator, Lorentz group 849

quadrupole electric one-photon electromagnetic transitions 788
quadrupole moment 603, 622-5
 deuteron 604
quantum numbers, non-relativistic theory 858
quantum radiation theory 776-810
 photon emission 779-89
quasi-classical approximation 374-445
 $1/N$-expansion 436-45
  discrete spectrum of particle 437-9
  energy spectrum for bound s-states 440-4
  short-range attractive particle 444-5
 Airy function 411
 attractive potential with a dip at origin 405-8
 Bessel function 389, 393, 396, 425-6
 Bohr–Sommerfeld quantization rule 375, 378, 383, 398-9, 401-5, 431
 classical probability 376
 Coulomb potential 380, 387-8, 441-3
 energy level shifts 398-405
 ground-state energy level 428-30
 Hankel function 425-7
 Hulthen potential 387-8
 Kramer's matching conditions 378-80
 Langer correction 378-80, 391-5, 434
 Langer quantization rule 393
 Langer transformation 376
 MacDonald function 408, 411
 matching at large energies 379-80

mean value 412–13
new bound states 395–8
particle in classically forbidden region 411–12
particle lifetime 433–4
particle reflection 427–8
penetration through potential barriers 421–37
quasi-classical energy quantization 381–409
quasi-classical wavefunctions, probabilities and mean values 382–3, 409–21
radial wavefunctions 388–91
Tietz potential 398
transmission coefficient for potential barrier 424–7
Wentzel—Kramers—Brillouin (WKB) method 374–446
Yukawa potential 445
quasi-energy states (QES), perturbation theory 343–4, 362

radiation field (free electromagnetic field) 776
radiation theory 776–809
  braking radiation 804–9
  radiation recombination 803
Ramsauer—Townsend effect, particle collisions 678
Rashba coupling 202
redshift, spectral lines, (receding) galaxies 787
relativistic wave equations 810–66
  Dirac equation 838–66
  Klein—Gordon equation 813–38
  single-particle problem 827
resonant fluorescence 794
resonant scattering, particle collisions 639
Rutherford equation, non-relativistic theory 836
Rydberg correction 503–4, 547

s-states 826
s-wave scattering 709–11
  partial amplitude 709–11
scalar field
  effect on a spinless particle and its antiparticle 816
  particle energy spectrum 826
scalar functions 817
scattering amplitude, Born approximation 494, 835
scattering cross-section, quasi-classical equation 837
scattering length 486
  perturbation theory 565
Schrödinger equation 198–9, 612
  central potential 116–17
  deuteron 606
  first relativistic correction 833
  in momentum space, Green function 46–56
  one-dimensional motion 33–40, 43–5, 50–7, 60–1, 63, 66–8, 74, 77–8, 80–2
  operators 25
  radial 828
  time independent 32
    arbitrary potential 3

asymptotic form of particles 32, 61–2
energy levels 32
linear oscillators 32
transition 832
shell model, atomic nucleus 599, 612–26
  mirror nuclei 622
  oscillator potential 612
  unperturbed oscillator 615
silicon ion 514
single-particle problem 827
slow particles 66, 67–8
spherical oscillator 278–80
spherical oscillator, charged 780
  differential and total photon-scattering cross-sections 796
spherical rotator, electromagnetic transition 780
spin 165–213
  bound states of spin-orbit-coupled particles 198–202
  density matrix, angular distribution in decays 191–8
  Dresselhaus to Rashba interaction strength ratio 199–200
  generators of Lie group 213
  Green function 200
  polarization density matrix of particle 195–6
  resolution of identity, coherent-state spin path-integral 203–9
  resting particle 193–4
  s-system, definition of coherent states 209
  spinless particles 196–7
  unstable particles 194–5
  wavefunction of particle 165–6
spin ($s = 1$) 186–9
spin ($s = 1/2$) 166–81
  angular momenta 177–8
  arbitrary rank 2 square matrix 168–9
  characterization 177
  eigenvalues and eigenfunctions of operator 169–70
  higher spins 180–91
  matrix element of form 171–2
  normalized wavefunction 168
  spin z-projection, projection operators 172–3
  system consisting of N spins 176–7
  system of three particles 178–9
  system of two particles 174–6
  system of two spins 173
  transformation law 170–1
spin ($s = 1/2$), higher spins 180–91
  helical states 180
  and magnetic moment 287–90, 294–5
  spin-angular dependence 180–2
  spin-angular wavefunction 182–4
  spinor representation 186–9
spin ($s = 3/2$) 189–91
spin-flip
  eigenfunctions and eigenvalues 783
  total probability of photon emission 785

spinless particle
   attraction independently of sign of its charge 834
   Born scattering amplitude 836
   Coulomb field 836
   energy spectrum in a uniform magnetic field 824
   external scalar field 826
   in a magnetic field 825
   mean momentum 823
   scalar potential 816
   stationary wave equation 838
   wavefunction transformation under spatial inversion 817
spinors 783
   inversion 845
   Lorentz transformation 645
   two-component spinor language 840
   two-component spinors 852
   two-fold degeneracy 846
   unitary transformation of bi-spinors 844
   wavefunction 812, 853
spin–isospin wavefunctions 616
Stark effect 539
   main quantum number $n = 2$ 539–40
   splitting, rotational components for a two-atom molecule 549
stationary states, one-electron and two-electron atoms 486–509
stationary wave equation, relativistic spinless particle 838
statistical atomic model, many-electron atoms 509–15
   coherent-state spin part-integral 207
sum rules, photon emission 789

Tietz potential, quasi-classical approximation 398
Tamm states 81
Taylor expansion 4–6, 12
tensor formalism 511
Thomas–Fermi equations, systems with large number of electrons 485
Thomas–Fermi model, particle collisions 648–9
Thomas–Fermi electrons 519
   polarizability 548
Thomas–Fermi method/model 485–6, 517, 519
Thomson formula 792
three-electron atom (or ion) state 508–9
   system ionization energy 508–9
time-dependent quantum mechanics 213–69
   asymptotic behaviour of wavefunction 218
   Bloch functions 261
   external electromagnetic field potentials 240–1
   Galilean transformation 239–40
   gauge transformation 240
   Green function 213, 246–52
   Heisenberg operators 233–7, 245–51
   Hermite polynomials 249
   interaction representation 245–6
   particle, acted on by uniform constant force 232

quasi-discrete energy levels of particles s-states 255–7
quasi-stationary and quasi-energy states Berry phase 252–69
Schrödinger equation 213–14, 231, 233–5, 239–42
Schrödinger representation 213
   motion of wave packets 214–25
spectrum of quasi-energy 260–3
spin ($s = 1/2$) particle
   arbitrary magnetic field 266–9
   interactions 230–1
sum rule 227
time-dependent Green function 247–52, 290
time-dependent unitary transformations, Heisenberg motion 232–47
two channel systems 258–60
two-level system 223–5
unitary transformation 240–3
unperturbed Hamiltonian 245–7
velocity and acceleration operators 225
virial theorem 226–8
wave-packet reflection 219–20
transmission and reflection of particles
   branching points 58, 69
   continuous spectrum potential barriers 56–74
   current density 64
   one-dimensional motion 56–74
   physical sheet and Riemann surface 58–9
   slow particles 66, 67–8
   square potential well 67
triplet terms, energy 510
tritium atom
   ground-state, beta-decay 566–8
   nuclear ground-state 616
   quantum numbers – spin, parity and isospin 616
two-atom molecule theory 525–38
   rotational levels 781
   Zeeman splitting 556
two-electron atom/ion
   ground-state energy 497, 499
   polarizability 546
two-nucleon systems 608
two-nucleon wavefunctions 616–17

unitary operators 22–31
   unitary transformation of bi-spinors 844

vacuum reconstruction 827
van der Waals force 561–2
   motion in central field 123
   two atoms 561
velocity operator, Dirac particle 842
virial theorem 505, 518, 520, 522

wavefunctions
   transition amplitude 838
   antiparticle wavefunction 816
   asymptotic behaviour 218

of continuous spectrum 543
deuteron 603, 605
free-particle in momentum representation 821
initial and final states in the hydrogen atom 783
many-nucleon system 608
momentum probability distribution 823
muons 568
quasi-stationary states 576–7
single p-electron 512
spherical oscillator 119–21
spinors 812, 853
spin–isospin 616
square well 601
two-nucleon 616–17
unitary operators 22–31
well
  infinite potential well 55
  one-dimensional well 117–18
  rectangular potential well 55
  shallow potential well 53
  square potential well 67
Wentzel—Kramers—Brillouin (WKB) method, quasi-classical approximation 374–446
Weyl equation 845

Yamaguchi potential, motion in central field 129
Yukawa potential
  particle collisions 655–7, 730–1
  quasi-classical approximation 445

Zeeman effect 550–3
Zeeman splitting, two-atom molecule rotational levels 556–7
Zeldovich effect 496
zero-energy solution, one-dimensional motion 43–4, 67
zero-range potential, motion in central field 126–8